EXPLORATIONS

EXPLORATIONS

An Introduction to Astronomy

Sixth Edition

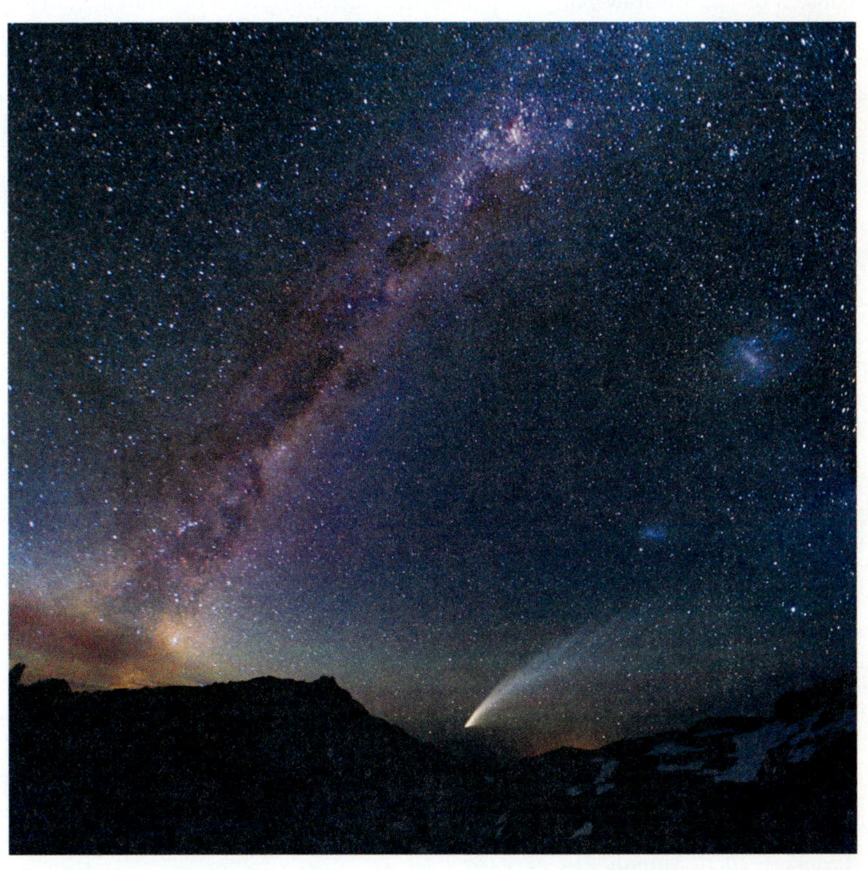

Thomas T. Arny

Professor Emeritus
Department of Astronomy
University of Massachusetts, Amherst

Stephen E. Schneider

Professor of Astronomy
University of Massachusetts, Amherst

Higher Education

Boston Burr Ridge, IL Dubuque, IA New York San Francisco St. Louis
Bangkok Bogotá Caracas Kuala Lumpur Lisbon London Madrid Mexico City
Milan Montreal New Delhi Santiago Seoul Singapore Sydney Taipei Toronto

EXPLORATIONS: AN INTRODUCTION TO ASTRONOMY, SIXTH EDITION

Published by McGraw-Hill, a business unit of The McGraw-Hill Companies, Inc., 1221 Avenue of the Americas, New York, NY 10020.

This book is printed on acid-free paper.

2 3 4 5 6 7 8 9 0 DOW/DOW 10 9 8 7 6 5 4 3 2 1 0

ISBN 978-0-07-351217-4
MHID 0-07-351217-6

Publisher: *Ryan M. Blankenship*
Sponsoring Editor: *Debra B. Hash*
Director of Development: *Kristine Tibbetts*
Developmental Editor: *Jodi Rhomberg*
Senior Marketing Manager: *Lisa Nicks*
Senior Project Manager: *Gloria G. Schiesl*
Senior Production Supervisor: *Sherry L. Kane*
Lead Media Project Manager: *Judi David*
Designer: *Tara McDermott*
Cover/Interior Designer: *Greg Nettles/Squarecrow Design*
(USE) Cover Image: *Miloslav Drückmuller, Brno University of Technology, Czech Republic*
Lead Photo Research Coordinator: *Carrie K. Burger*
Photo Research: *Mary Reeg*
Compositor: *Macmillan Publishing Solutions*
Typeface: *10/12 Minion*
Printer: *RR Donnelley Willard, OH*

Library of Congress Cataloging-in-Publication Data

Arny, Thomas.
Explorations : an introduction to astronomy / Thomas T. Arny. —6th ed. / Stephen E. Schneider.
p. cm.
Includes index.
ISBN 978-0-07-351217-4 — ISBN 0-07-351217-6 (hard copy : alk. paper)
1. Astronomy—Textbooks. I. Schneider, Stephen E. (Stephen Ewing), 1957- II. Title.
QB45.2.A76 2010
520—dc22

2009019309

www.mhhe.com

BRIEF CONTENTS

CONTENTS

CHAPTER 11

Meteors, Asteroids, and Comets 287

CHAPTER 12

The Sun, Our Star 311

CHAPTER 13

Measuring the Properties of Stars 337

CHAPTER 14

Stellar Evolution 371

CHAPTER 15

Stellar Remnants: White Dwarfs, Neutron Stars, and Black Holes 401

CHAPTER 16

The Milky Way Galaxy 423

PREFACE

Our motivations for writing *Explorations: An Introduction to Astronomy* are many, both personal and pedagogic. Perhaps foremost among these is a desire to share with students our own sense of wonder about the Universe.

That sense of wonder grows deeper when we begin to understand why things happen. Many astronomy books today seem to simply say, "This is how it is." We want instead to offer explanations that draw as much as possible on simple, everyday effects that students can see around them in the world. For example, why do some stars pulsate? A simple analogy of steam building up pressure under the lid of a pan offers a model of this phenomenon that is easy to understand and reasonably accurate. We have also tried to link complex physical processes to simple everyday experiences. An example of this is that you can see the effects of diffraction by looking at a bright light through a lock of your hair pulled over your eyes or through glasses that you have fogged with your breath. When we can thus link physical principles to everyday observations, many of the more abstract and remote ideas become more familiar. Throughout the book we have made heavy use of analogies, along with illustrations to make those analogies more concrete.

Knowing the facts about astronomical objects is important, but it is equally important to understand how astronomers deduce those facts. Thus, an additional aim throughout this text is to explain *how* astronomers have come to their understanding of our Universe. As the Science at Work boxes illustrate, observations can force astronomers to revise their ideas of how a given process occurs. As part of showing how scientists arrive at their ideas, we have set many of the modern discoveries in their historical context to illustrate that science is a dynamic process and subject to controversy—many ideas are not immediately accepted, even if they ultimately prove to be "correct." We hope that by seeing the arguments for and against various ideas, students will have a better understanding of how science works.

If we had attempted to make this textbook completely comprehensive, it would have been very long and overwhelming in detail. It is challenging to keep *Explorations* to a reasonable size because reviewers tend to suggest things that we must include but rarely suggest things to omit. To solve this problem, we cover some topics, such as timekeeping and astrobiology, in essays that the instructor might choose to skip. We also cover some background topics in later chapters, in the astronomical context where they are most often encountered. This makes it possible to jump directly to some of the later chapters without having to work through the details of all the earlier chapters.

Some astronomy textbooks maintain brevity by omitting most of the mathematics, but we feel that math is essential for understanding many of the methods used by astronomers. We have therefore included the essential mathematics in a number of places. However, because math is so intimidating to so many students, we begin these discussions by introducing the essence of the calculation in everyday language so that the basic idea can be understood without understanding the mathematics. For example, Wien's law relates the temperature of a hot object to its color by means of a mathematical law, but illustrations of the law can be seen in everyday life, as when we estimate how hot an electric stove burner is by the color of its glow. Where we do present the mathematics, we work through it step by step, explaining where terms must be cross-multiplied and so forth.

NEW TO THE SIXTH EDITION

Content Updates and Additions

In this sixth edition of *Explorations,* we have updated the areas where there have been substantial changes in our knowledge and where different ideas are emerging. Reviewers and users also had many helpful suggestions for improving the organization, expanding the coverage of particular topics, and enhancing figures. Following are some of the highlights of these changes:

Organizational Changes

- Chapter 1 had grown very large with several additions over the years, so at the suggestion of several reviewers we divided it into two chapters, one covering basic motions of the Earth, Sun, and Moon, and the other the development of the science of astronomy. As a result, former chapters 2–17 have become chapters 3–18.
- The essay on relativity has been moved to follow chapter 4 as essay 2, and it now includes some discussion of general relativity to aid in discussions of topics from exoplanet lensing to the expansion of the Universe. Former essays 2 and 3 are now essays 3 and 4.
- Some material on eclipses formerly in the Moon chapter have joined the rest of the eclipse coverage in chapter 1. Discussion of the Moon illusion also has been moved to chapter 2.
- The Looking Up figures have been moved to the front of the book, where they are easier to find and do not interrupt the flow of the chapters.

New Questions, Problems, and Projects

- We have added dozens of new end-of-chapter review questions, thought questions, worked problems, test-yourself questions, and projects.
- The end-of-chapter questions and problems are now keyed to the relevant section number to help students and instructors make connections between readings and problem solving.

Updated and Revised Topics

- Expanded discussion of global warming and the ozone layer (chapter 6)
- Expanded coverage of exoplanets (chapter 8)
- Information from recent planetary probes, including Mercury *Messenger* and the Mars *Phoenix* lander (chapter 9)
- Additional discussion of dwarf planets (chapter 10)
- More complete discussion of meteorites (chapter 11)
- Discussion of hypernova collapse and possible connection to gamma-ray bursts (chapter 15)
- Extensive reorganization of the description of galaxies and the formation of their structure in terms of current ideas (chapter 17)
- Deeper discussion of cosmological ideas throughout, including the evidence for accelerating expansion (chapter 18)
- New foldout star chart with Moon and planet finder

New and Updated Images

- Ancient observatory in Chankillo, Peru (chapter 1)
- New Moon and eclipse pictures and figures (chapters 1 and 2)
- New large telescopes (chapter 5)
- Illustrations of shifting continents and precession (chapter 6)
- Maps of Moon, Venus, and Mars, showing important features (chapters 7 and 9)
- Images and figures of exoplanetary systems (chapter 8)
- Mercury *Messenger* images, reanalyzed *Venera* lander images, new elevation maps of Venus and Mars, as well as new Mars images (chapter 9)
- *Cassini* images of Saturn and its rings and moons (chapter 10)
- More meteorite images, pictures from the *Deep Impact* mission, and illustrations of the Chicxulub Crater (chapter 11)
- Images and figures of star formation and stellar evolution (chapter 14)
- Many new galaxy images, and figures illustrating the red and blue sequences and the black hole/bulge size correlation (chapter 17)
- Diagrams of light element fusion in early Universe and expansion diagrams for various expansion scenarios (chapter 18)

Special Features

Looking Up Features The nine Looking Up figures—each one a full-page art piece—are designed to show students how some of the astronomical objects discussed in the text connect with the real sky that they can see overhead at night. The figures cover nine particularly interesting regions ranging from the North Pole to the South Pole. In particular, they show where a variety of the frequently mentioned and important astronomical objects can be seen, many with a small telescope. Each Looking Up feature presents a photograph of one or more constellations in which nebulas, star

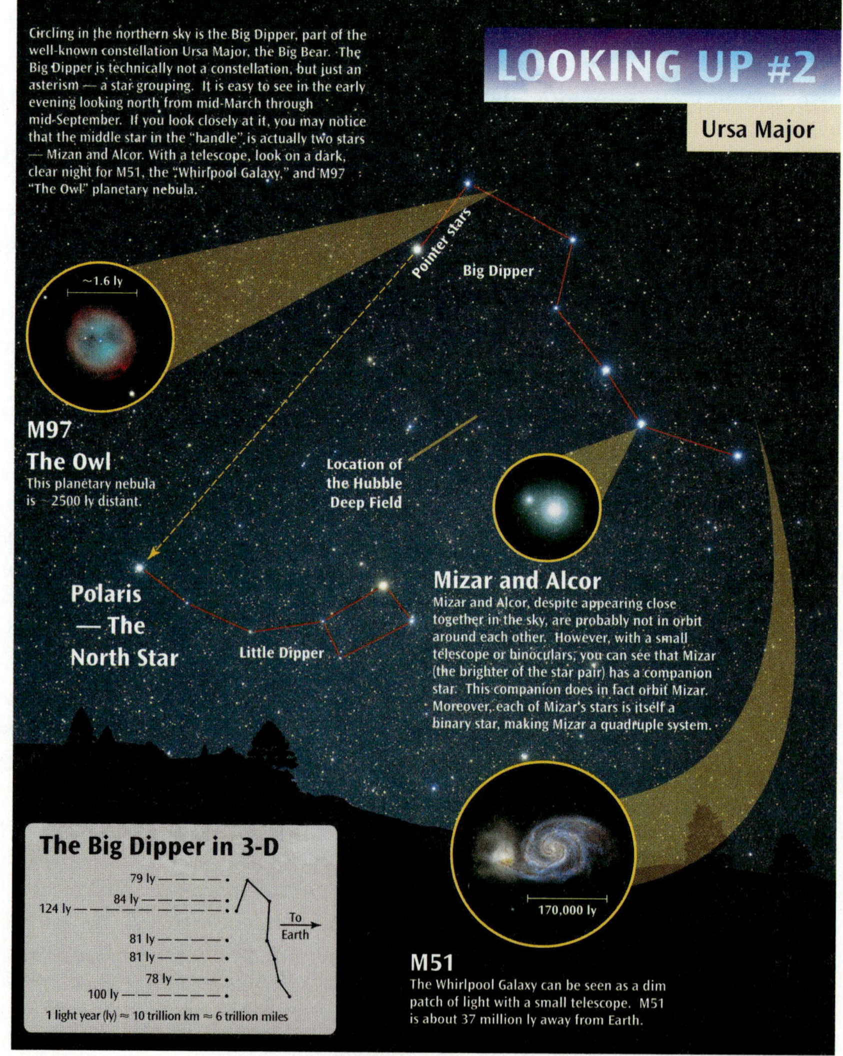
LOOKING UP #2
Ursa Major
Circling in the northern sky is the Big Dipper, part of the well-known constellation Ursa Major, the Big Bear. The Big Dipper is technically not a constellation, but just an asterism — a star grouping. It is easy to see in the early evening looking north from mid-March through mid-September. If you look closely at it, you may notice that the middle star in the "handle" is actually two stars — Mizan and Alcor. With a telescope, look on a dark, clear night for M51, the "Whirlpool Galaxy," and M97 "The Owl" planetary nebula.
~1.6 ly
Pointer stars
Big Dipper
M97
The Owl
This planetary nebula is ~2500 ly distant.
Location of the Hubble Deep Field
Polaris — The North Star
Little Dipper
Mizar and Alcor
Mizar and Alcor, despite appearing close together in the sky, are probably not in orbit around each other. However, with a small telescope or binoculars, you can see that Mizar (the brighter of the star pair) has a companion star. This companion does in fact orbit Mizar. Moreover, each of Mizar's stars is itself a binary star, making Mizar a quadruple system.
The Big Dipper in 3-D
79 ly
84 ly
124 ly
81 ly
81 ly
78 ly
100 ly
To Earth
1 light year (ly) ≈ 10 trillion km ≈ 6 trillion miles
170,000 ly
M51
The Whirlpool Galaxy can be seen as a dim patch of light with a small telescope. M51 is about 37 million ly away from Earth.

clusters, and other interesting objects are identified and illustrated. These latter illustrations include scale factors to help students visualize how even immense objects many light-years across can appear as mere dots in the sky. Along with the illustrated objects, most of the Looking Up features include a small insert to show how one of the constellation's stars are arranged in space.

In constructing the Looking Up features, we have had to draw upon many different sources for the distances and sizes of the objects presented. Some distances are poorly known. Thus, none of the numbers appearing in these Looking Up features should be assumed to be precise.

"What Is This?" Photographs and Figure Questions At the beginning of each chapter, students are presented with a photo of an astronomical object and asked to guess what it is. After reading the chapter, they should have some idea of what is shown in the photo. In addition, there are questions in red boxes about a number of other figures and images. The answers to these questions are provided at the end of each chapter under the heading "Figure Question Answers."

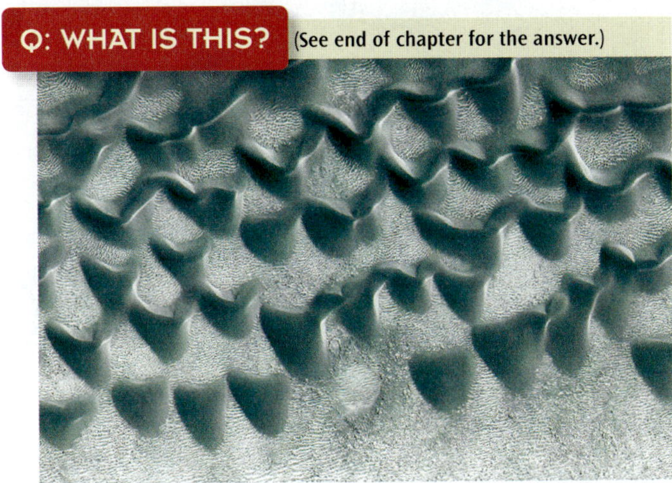

Q: WHAT IS THIS?
(See end of chapter for the answer.)

Foldout Star Chart The foldout star chart in the back of the book helps students study the sky and figure out where the Moon and planets are located in any month. The chart is useful for projects such as plotting the changing location of the Moon and planets, or the paths of meteors.

Electronic Media Integration To help students have a better grasp of key concepts, these icons have been placed near figures and selections where students can gain additional understanding through the interactives and animations on the *Explorations* website.

CUSTOM OPTIONS

Primis Online

Customize this text in print or in electronic format to meet your exact course needs. Through McGraw-Hill's Primis Online, simply select your desired chapters and preferred sequence, and supplement them with the many science items on our database. Visit www.mhhe.com/primis/online to begin today.

Volumes

Now one of the best-selling astronomy textbooks is available in smaller versions. Great for one-semester courses, these textbooks (volume 1: *Solar System,* and volume 2: *Stars and Galaxies*) contain the same great writing, the same great pedagogical tools, and the same great technology package.

SUPPLEMENTS

McGraw-Hill offers various tools and technology products to support *Explorations: An Introduction to Astronomy,* Sixth Edition. Instructors can obtain teaching aids by calling the Customer Service Department at 800-338-3987 or contacting your local McGraw-Hill sales representative.

New Connect Astronomy

McGraw-Hill's Connect Astronomy is a Web-based assignment and assessment platform that gives students the means to better connect with their coursework, with their instructors, and with the important concepts that they will need to know for success now and in the future.

With Connect Astronomy, instructors can deliver assignments, quizzes, and tests online. Questions from the text are presented in an auto-gradable format. Instructors can edit existing questions and author entirely new problems, and track individual student performance—by question, by assignment, or in relation to the class overall—with detailed grade reports. They can also integrate grade reports easily with Learning Management Systems (LMS) such as WebCT and Blackboard. And much more.

By choosing Connect Astronomy, instructors are providing their students with a powerful tool for improving academic performance and truly mastering course material. Connect Astronomy allows students to practice important skills at their own pace and on their own schedule. Importantly, students' assessment results and instructors' feedback are all saved online—so students can continually review their progress and plot their course to success.

Presentation Center

Build instructional materials wherever, whenever, and however you want! The Presentation Center is an online digital library containing assets such as photos, artworks, animations, PowerPoint presentations, and other media types that can be used to create customized lectures, visually enhanced tests and quizzes, compelling course websites, or attractive printed support materials.

Access to your book, access to all books! Illustrations from *Explorations* are available as digital files. In addition, the Presentation Center Library includes thousands of assets from many McGraw-Hill titles. This ever-growing resource gives instructors the power to utilize assets specific to an adopted textbook as well as content from all other books in the library.

Nothing could be easier! Accessed from the instructor side of your textbook's Connect website, Presentation Center's dynamic search engine allows you to explore by discipline, course, textbook chapter, asset type, or key word. Simply browse, select, and download the files you need to build engaging course materials. All assets are copyrighted by McGraw-Hill Higher Education but can be used by instructors for classroom purposes.

Instructor's Manual

The Instructor's Manual containing answers to all questions and problems within the text is found on the Instructor's side of the *Explorations* website.

Classroom Performance System and Questions

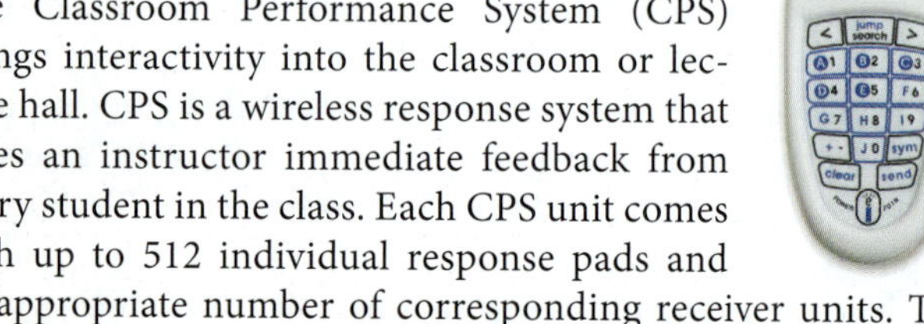

The Classroom Performance System (CPS) brings interactivity into the classroom or lecture hall. CPS is a wireless response system that gives an instructor immediate feedback from every student in the class. Each CPS unit comes with up to 512 individual response pads and an appropriate number of corresponding receiver units. The wireless response pads are essentially remotes that are easy to use and engage students. The CPS system allows instructors to create their own questions or use the astronomy questions provided by McGraw-Hill.

Interactives

McGraw-Hill is proud to bring you an assortment of 23 outstanding Interactives like no other. Each Interactive is programmed in Flash for a stronger visual appeal. These Interactives offer a fresh and dynamic method for teaching the astronomy basics. Each Interactive allows users to manipulate parameters and gain a better understanding of topics such as Blackbody Radiation, The Bohr Model, a Solar System Builder, Retrograde Motion, Cosmology, and the H-R Diagram by watching the effect of these manipulations. Each Interactive includes an analysis tool (interactive model), a tutorial describing its function, content describing its principal themes, related exercises, and solutions to the exercises. Users can jump between these exercises and analysis tools with just the click of the mouse.

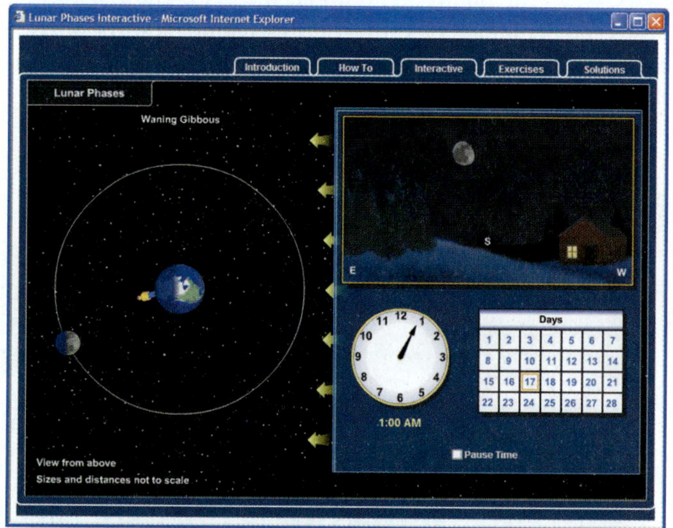

How to Study with This Book

Learning anything requires a certain amount of work. You certainly don't expect to be able to pick up a guitar and play it without practice, nor do you expect to be able to jog 5 miles without working out regularly. Learning astronomy also requires some work. The steps below may help you learn the material better and more easily.

In reading any assignment, begin by looking at the *pictures*. Turn the pages of the chapter and familiarize yourself with what the objects you will be reading about look like. Then read the introduction. Next, jump to the summary. Finally, start again and read the assigned material through. As you read, make notes of things you don't understand and ask your instructor or teaching assistant for clarification. For example, if you are puzzled about why eclipses don't happen every month, make a note. Making a few short notes is much more effective than highlighting whole paragraphs.

Look carefully at the pictures and diagrams. If the figure has a question with it, try to answer it. Make your own sketch of diagrams to be sure you understand what they represent.

In a first reading of a chapter, if you are troubled by math, you might simply skip over it for the time being. Be sure, however, to read the material leading into the math so you at least understand the basic idea. When you encounter a mathematical expression of a physical law, express the law in words. For example, the law of gravity relates the force of gravity to the mass of the objects and their distance from each other.

If you encounter words or terms as you read that you don't know, look them up in the glossary, the index, or a dictionary or encyclopedia. You are just wasting your time if you read a description of some object and you don't know what it is.

When you finish the assigned reading, try to answer the "Review Questions" for the sections you covered. They are short and are designed to help you see if you have assimilated the basic factual material of the assignment. Try to do this without looking back into the chapter, but if you can't remember, look it up rather than skip over the question. You might find it helpful to write out short answers to the questions.

Having read the material once, go back and try to work through the math parts. Then try some of the mathematical "Problems" to see if you can work through the material on your own. "Thought Questions" challenge you to think more deeply about the readings, and you can use the multiple-choice "Test Yourself" questions to check your understanding.

If you get stuck at any point, see your teaching assistant or professor for help. Don't be shy about asking questions. Learning is a thousand times easier if you ask questions when you get stuck.

Seeing a clear night sky spangled with stars is a wondrous experience. And yet the beauty and sense of wonder can be enriched even more by an appreciation of the complex processes that make the Universe work. We hope this book will similarly increase your appreciation of our Universe's wonders.

If you find mistakes or have suggestions about how to make this book better, please contact one of us. Write T. Arny at P.O. Box 545 Patagonia, AZ 85624, or by e-mail at tarny@theriver.com; or S. Schneider at Department of Astronomy, 536 Lederle Tower, Amherst, MA 01003, or by e-mail at schneider@astro.umass.edu.

ACKNOWLEDGMENTS

Many people have played an important role in bringing this book into being. For this sixth edition of *Explorations* we want to thank particularly Michael Stage, who developed most of the new end-of-chapter questions and worked with us to relate these materials more strongly to the text. We are also very grateful to many people at McGraw-Hill, but especially Debra Hash, Jodi Rhomberg, Gloria Schiesl, Carrie Burger, Tara McDermott, Lisa Nicks, Sherry Kane, Dan Wallace, and Judi David, and to Wendy Nelson and Mary Reeg for their help and patience.

We also want to thank a number of special reviewers and users who have sent comments. Linda Arny read carefully and commented thoughtfully on the entire manuscript. Bradley E. Schaefer and Roger E. Thayer also made valuable and helpful comments.

The author and McGraw-Hill would also like to extend a special thank-you to those who contributed to the supplements of this edition. They include:

Patrick Koehn, *Eastern Michigan University*
Michael Stage, *University of Massachusetts*

REVIEWERS OF THIS EDITION

Special thanks and appreciation go out to reviewers of the fifth edition. Their contributions, constructive suggestions, new ideas, and invaluable advice played an important role in the development of this sixth edition and its supplements. These reviewers include:

Jami Barnes *Owens Community College*
Michael Broyles *Collins College*
Brian Campbell *Southwestern Oklahoma State University*
Ryan Droste *Trident Technical College*
Jim W Duke *Owensboro Community College*
John Feldmeier *Youngstown State University*
Terrence Flower *College of St. Catherine*
Alina Gabryszewska-Kukawa *Delta State University*
Terry Goforth *Southwestern Oklahoma State University*
Joe Guenter *University of Arkansas at Monticello*
Jim Guinn *Georgia Perimeter College*
Kevin Lee *University of Nebraska–Lincoln*
Craig Lincoln *St. Louis Community College*
James Mackey *Harding University*
Cynthia Petersen *University of Connecticut*
Edward Sion *Villanova University*
Dale Thompson *Hillsborough Community College*

REVIEWERS OF PREVIOUS EDITIONS

This revision of *Explorations* has also been made possible by the many users and reviewers of its previous editions. The author and publisher are grateful to the following reviewers of previous editions for their critical reviews, comments, and suggestions.

Parviz Ansari *Seton Hall University*
C. Armendariz-Picon *Syracuse University*
Bruce Balick *University of Washington–Seattle*

Tom Balonek *Colgate University*
Wendy Hagen Bauer *Wellesley College*
Peter A. Becker *George Mason University*
Timothy Beers *Michigan State University*
Ralph Benbow *Northern Illinois University*
Daniel C. Boice *University of Texas at San Antonio*
Luca Bombelli *University of Mississippi*
Bernard Bopp *University of Toledo*
Michael J. Bozack *Auburn University*
Brother Guy Consolmagno, SJ *Vatican Observatory*
Jeffrey A. Brown *Washington State University*
Eugene Capriotti *Michigan State University*
Tai L. Chow *California State University–Stanislaus*
Thomas Christensen *University of Colorado at Colorado Springs*
John J. Cowan *University of Oklahoma*
David S. Curtis *University of West Georgia*
Benjamin de Mayo *West Georgia College*
Alexander Dickson *Seminole Community College*
Jess Dowdy *Abilene Christian University*
Pat Durrell *Youngstown State University*
Mark Edwards *Hofstra University*
Robert A. Egler *North Carolina State University*
Heinrich Eichhorn *University of Florida*
Don P. Engelberg *Queensborough Community College*
Eric Feigelson *The Pennsylvania State University*
Donald Foster *Wichita State University*
Robert B. Friedfeld *Idaho State University*
Aaron Galonsky *Michigan State University*
Harold Geller *George Mason University*
Robert Geller *University of California*
David J. Griffiths *Oregon State University*
Bruce Gronich *University of Texas–El Paso*
Bruno Gruber *Southern Illinois University*
Alexander Gurshtein *Mesa State College*
Martin Hackworth *Idaho State University*
Heidi Hammel *Massachusetts Institute of Technology*
Clint D. Harper *Moorpark College*
Ronald A. Hartman *Mount San Antonio College*
Andrew Harris *University of Maryland*
Harold M. Hastings *Hofstra University*
John F. Hawley *University of Virginia*
David Hedin *Northern Illinois University*
Eric R. Hedin *Ball State University*
Philip L. Hegenderfer *University of Akron*
Paul Hintzen *California State University–Long Beach*
Thomas Hockey *University of Northern Iowa*
John Greg Hoessel *University of Wisconsin*
Harry L. F. Houpis *Sierra College*
James Christopher Hunt *Prince George's Community College*
Douglas R. Ingram *Texas Christian University*
Scott B. Johnson *Idaho State University*
Terry Jay Jones *University of Minnesota*
Ronald H. Kaitchuk *Ball State University*
William Keel *University of Alabama*
Douglas M. Kelly *University of Arizona*
Marvin D. Kemple *Indiana University/Purdue University/Indianapolis*
Yong H. Kim *Saddleback College*
Patrick Koehn *Eastern Michigan University*
Robbie F. Kouri *Our Lady of the Lake University*
Joan P. S. Kowalski *George Mason University*
Robert Kren *University of Michigan*
David Kriegler *University of Nebraska at Omaha*
Jeffrey Kuhn *Michigan State University*
Claud Lacy *University of Arkansas*
Kristine Larsen *Central Connecticut State University*
John K. Lawrence *California State University–Northridge*
N. A. Levenson *University of Kentucky*
W. R. Luebke *Modesto Junior College*
Jonathan I. Lunine *University of Arizona*
Loris Magnani *University of Georgia*
Carolyn Mallory *Moorpark College*
Norman Markworth *Stephen F. Austin State University*
Nancy D. McDonald *Palm Beach Community College*
Rahul Mehta *University of Central Arkansas*
Chris Mihos *Case Western Reserve University*
Milan Mijic *California State University–Los Angeles*
Jordi Miralda-Escude *Ohio State University*
Gerald Newsom *Ohio State University*
Paul Nienaber *Marquette University*
Andrew P. Odell *Northern Arizona University*
John Oliver *University of Florida*
Melvyn Oremland *Pace University*
Peter O'Shull Jr. *Oklahoma State University*
Robert Page *University of Maine at Augusta*
Brian M. Patten *Harvard-Smithsonian Center for Astrophysics*
Jon Pedicino *College of the Redwoods*
James N. Pierce *Minnesota State University at Mankato*
Mike Reed *Southwest Missouri State University*
Richard Rees *Westfield State College*
David D. Reid *Eastern Michigan University*
Carl Rosenzweig *Syracuse University*
Dwight Russell *Baylor University*
John L. Safko Sr. *University of South Carolina*
Larry Sessions *Metropolitan State College of Denver*
Edward Schaub *Baylor University*
James Schombert *University of Oregon*
William Seeley *Massachusetts College of Liberal Arts*
Larry Sessions *Metropolitan State College of Denver*
J. Scott Shaw *University of Georgia*
Alex G. Smith *University of Florida*
George F. Spagna Jr. *Randolph-Macon College*
Norman Sperling *Chabot Observatory and Science Center*

Roger Stanley Jr. *San Antonio College*
Michael Stewart *San Antonio College*
Jack W. Sulentic *University of Alabama*
Donald Terndrup *Ohio State University*
Nilakshi Veerabathina *University of Texas at Arlington*
James Webb *Florida International University*
Walter Wesley *Moorehead State University*
Dan Wilkins *University of Nebraska*
Richard Williamon *Emory University*
Suzanne Willis *Northern Illinois University*
George Wolf *Southwest Missouri State University*
David A. Wood, Jr. *San Antonio College*
W. C. Woods *Glassboro State College*
Jon K. Wooley *Eastern Michigan University*
John Wayne Wooten *Pensacola Junior College*

LOOKING UP #1

Northern Circumpolar Constellations

For observers over most of the northern hemisphere, there are 5 constellations that are circumpolar, remaining visible all night long: Ursa Major (the Big Bear), Ursa Minor (the Little Bear), Cepheus (the King), Cassiopeia (the Queen), and Draco (the Dragon). The brightest stars in Ursa Major and Ursa Minor form two well-known asterisms: the Big and Little Dippers.

Delta Cephei
A pulsating variable star at a distance of ~300 ly.

Cassiopeia

Cepheus

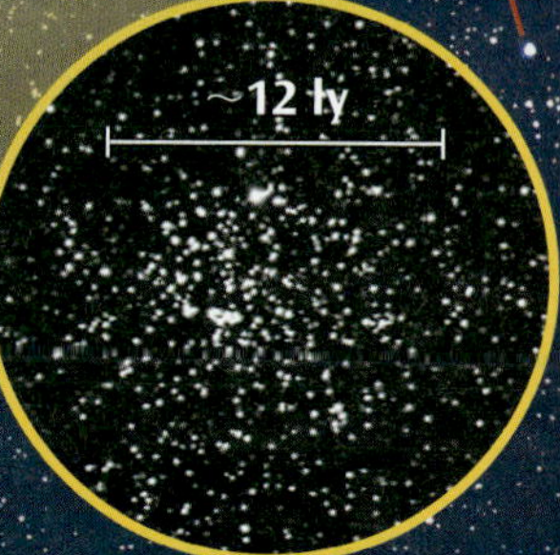

M52
This is an open star cluster. Its distance is uncertain — perhaps 3000–5000 ly.

Draco

Polaris (The North Star)

Little Dipper

M101
This spiral galaxy is ~27 million light years from us.

Thuban
This was the pole star when the pyramids were built in ancient Egypt.

M81 and M82
Gravitational interactions between M81 and M82 have triggered star formation in both galaxies.

Big Dipper

North

Cassiopeia in 3-D

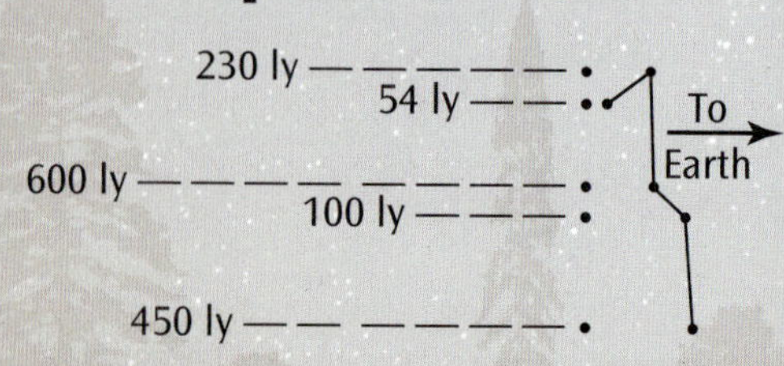

1 light year (ly) ≈ 10 trillion km ≈ 6 trillion miles

LOOKING UP #2

Ursa Major

ircling in the northern sky is the Big Dipper, part of the vell-known constellation Ursa Major, the Big Bear. The Big Dipper is technically not a constellation, but just an sterism — a star grouping. It is easy to see in the early vening looking north from mid-March through nid-September. If you look closely at it, you may notice hat the middle star in the "handle" is actually two stars — Mizan and Alcor. With a telescope, look on a dark, lear night for M51, the "Whirlpool Galaxy," and M97 The Owl" planetary nebula.

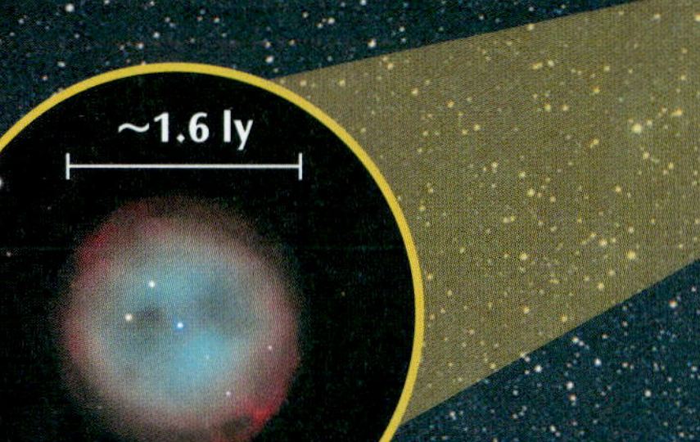

Pointer stars

Big Dipper

M97 The Owl

his planetary nebula ~2500 ly distant.

Location of the Hubble Deep Field

Polaris — The North Star

Little Dipper

Mizar and Alcor

Mizar and Alcor, despite appearing close together in the sky, are probably not in orbit around each other. However, with a small telescope or binoculars, you can see that Mizar (the brighter of the star pair) has a companion star. This companion does in fact orbit Mizar. Moreover, each of Mizar's stars is itself a binary star, making Mizar a quadruple system.

The Big Dipper in 3-D

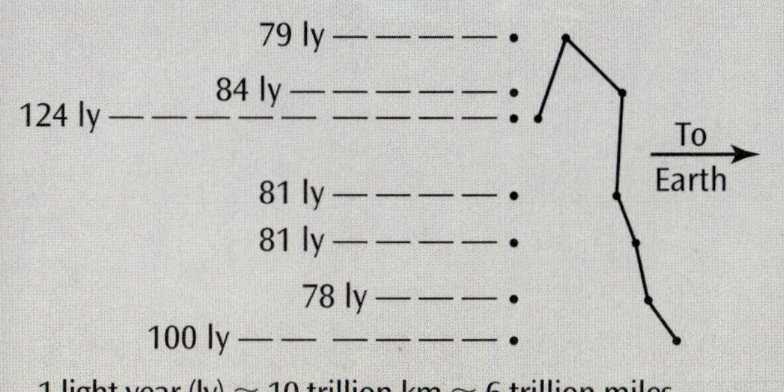

1 light year (ly) ≈ 10 trillion km ≈ 6 trillion miles

170,000 ly

M51

The Whirlpool Galaxy can be seen as a dim patch of light with a small telescope. M51 is about 37 million ly away from Earth.

LOOKING UP #3

M31 & Perseus

The galaxy M31 lies in the constellation Andromeda, near the constellations Perseus and Cassiopeia. Northern hemisphere viewers can see M31 dimly with the naked eye, but easily with binoculars. It is about 2.5 million ly from us. The best times to see it are August through December.

Andromeda

The Double Cluster

If you scan with binoculars from M31 toward the space between Perseus and Cassiopeia, you will see the Double Cluster — two groups of massive, luminous but very distant stars. The Double Cluster is best seen with binoculars. The two clusters are about 7000 ly away and a few hundred light years apart.

M31
Andromeda Galaxy

Perseus
Perseus is easy to identify: it looks a little like the Eiffel tower.

California Nebula
An emission nebula.

M45 Pleiades

Capella
The brightest star in the constellation Auriga, the Charioteer. A binary star.

Auriga

Perseus in 3-D

1300 ly
250 ly
230 ly
34 ly
110 ly
500 ly
100 ly
500 ly
560 ly
700 ly
To Earth

1 light year (ly) ≈ 10 trillion km ≈ 6 trillion miles

LOOKING UP #4

Summer Triangle

The Summer Triangle consists of the three bright stars Deneb, Vega, and Altair. These stars, the brightest to our eyes in the constellations Cygnus, Lyra, and Aquila, respectively, rise in the east shortly after sunset in late June and are visible throughout the northern summer and into late October (when they set in the west in the early evening). Deneb is intrinsically the brightest of the three, although Vega looks the brightest to us. Deneb looks dim only because it is so much farther from us than Vega and Altair.

Vega

Lyra

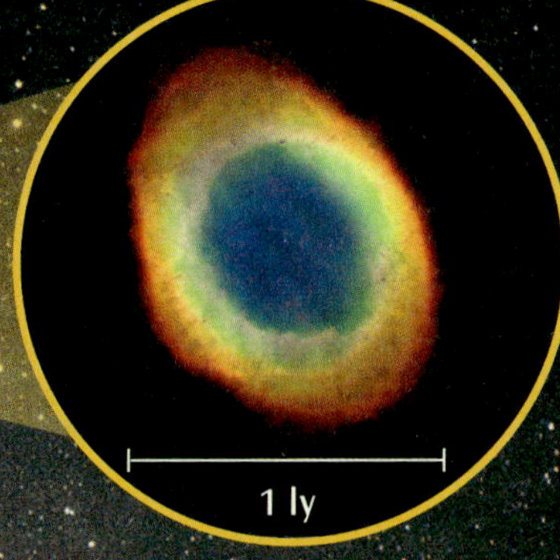

M57
Ring Nebula

This planetary nebula is about 2000 ly away from us. It is 7000 years old and has a white dwarf at its center.

Epsilon Lyra

A double, double star

Cygnus

Deneb

Deneb is a Blue Supergiant, one of the most luminous stars we can see, Deneb emits about 250,000 times more light than the Sun.

Altair

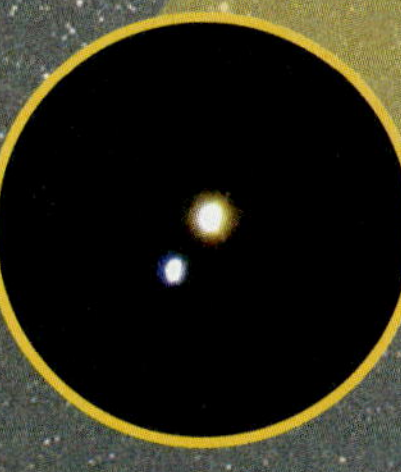

Albireo

This star pair (easily seen in a small telescope) shows a strong color contrast (orange and blue). Astronomers disagree about whether they orbit each other or just happen to lie in the same direction in the sky.

M27
Dumbbell Nebula

Another planetary nebula, The Dumbbell is about 900 ly distant and is about 2.5 ly in diameter.

The Summer Triangle in 3-D

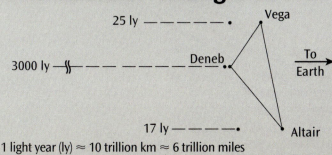

25 ly — Vega
3000 ly — Deneb
To Earth
17 ly — Altair

1 light year (ly) ≈ 10 trillion km ≈ 6 trillion miles

LOOKING UP #5

Taurus

Taurus — The Bull
One of the constellations of the zodiac and one of the creatures hunted by Orion. Taurus is visible in the evening sky from November through March. The brightest sta in Taurus is Aldebaran, the eye of the bull.

~7 ly

Zeta Tauri

M1 The Crab Nebula

The Crab Nebula is the remnant of a star that blew up in the year AD 1054 as a supernova. It is about 5000 ly away from us.

Aldebaran

Aldebaran is a red giant star. It is about 65 ly from us and its diameter is about 45 times the diameter of the Sun. It lies between us and the Hyades.

~8 ly

M45 Pleiades

This open star cluster is easy to see with the naked eye and loo like a tiny dipper. It is about 4(ly from Earth.

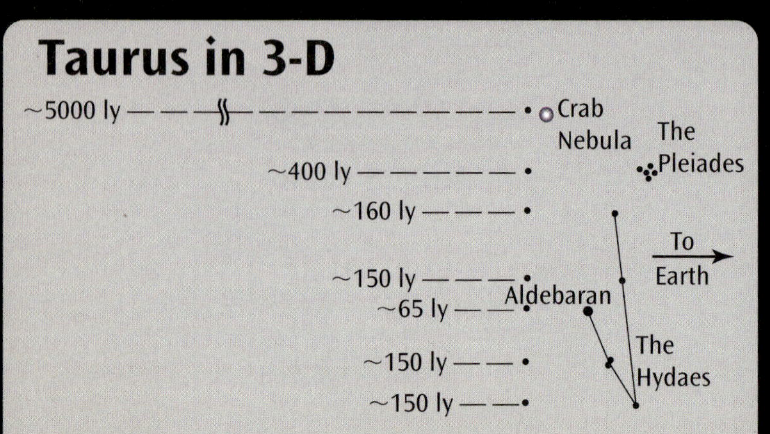

Hyades

The "V" in Taurus is a nearby star cluster, about 137 ly away. It is easy to see its many stars with binoculars.

LOOKING UP #6

Orion

Orion is easy to identify because of the three bright stars of his "belt." You can see Orion in the evening sky from November to April, and before dawn from August through September.

Betelgeuse

Betelgeuse is a Red Supergiant star that has swelled to a size that is larger than the orbit of Mars. Its red color indicates that it is relatively cool for a star, about 3000 Kelvin.

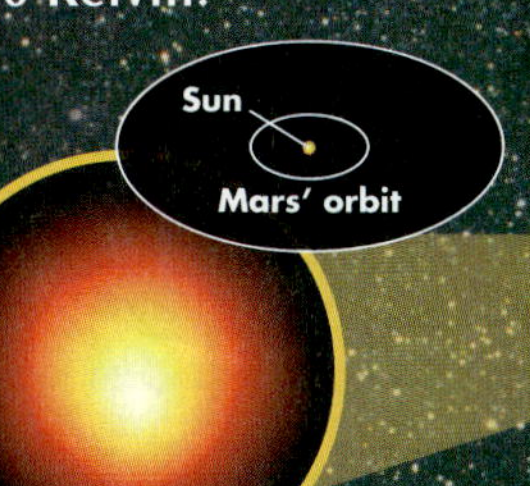

Horsehead Nebula

The horsehead shape is caused by dust in an interstellar cloud blocking background light.

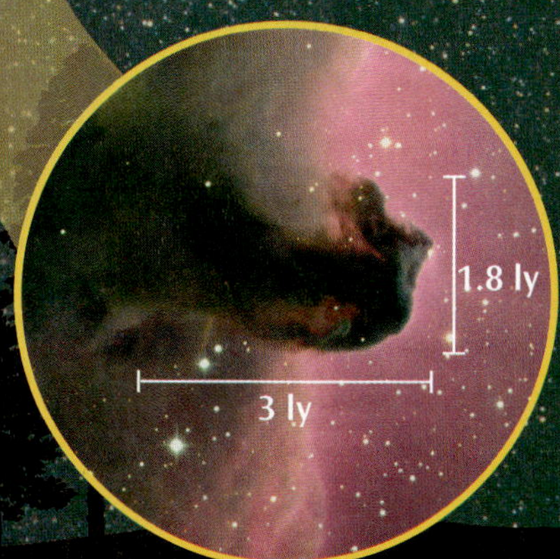

M42 Orion Nebula

The Orion Nebula is an active star-forming region rich with dust and gas.

Rigel

Rigel is a Blue Supergiant star. Its blue color indicates a surface temperature of about 10,000 Kelvin

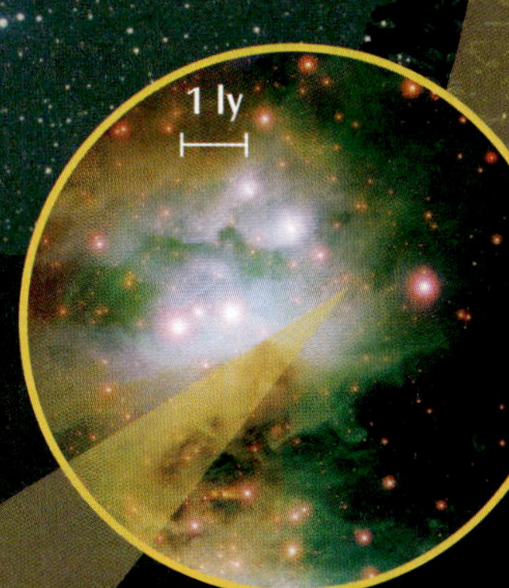

Orion in 3-D

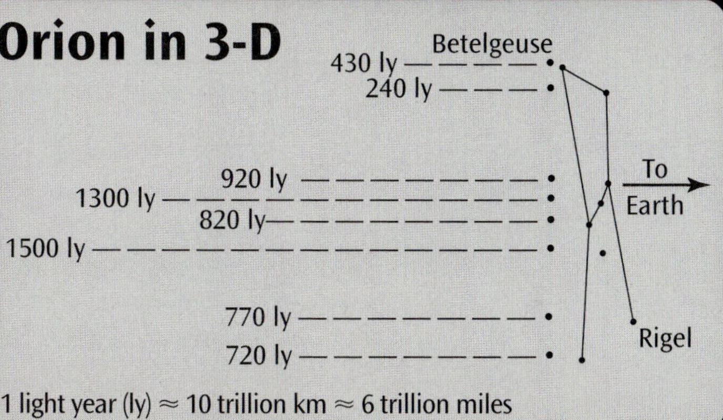

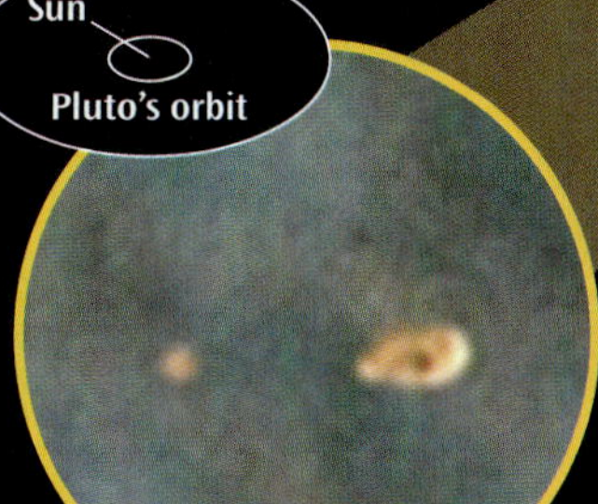

A protoplanetary disk

This is the beginning of a star; our early Solar System may have looked like this!

LOOKING UP #7

Sagittarius

Sagittarius marks the direction to the center of the Milky Way. It is best located by its "teapot" shape, with the Milky Way seeming to rise like steam from the spout. For northern latitude viewers, the constellation is best seen in the evening, July to September. For such viewers, it is low on the southern horizon. Many star-forming nebulae are visible in this region.

M16
Eagle Nebula

This young star cluster and the hot gas around it lie about 7000 ly from Earth.

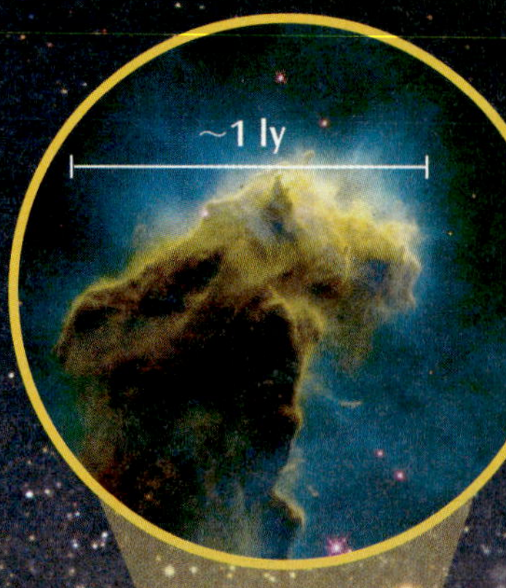

M17
Swan Nebula

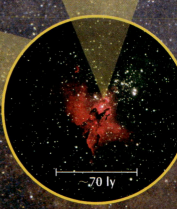

M20
Trifid Nebula

The distance to the Trifid Nebula (so named because of the black streaks that divide it into thirds) is very uncertain. It lies between 2000 and 8000 ly away. This makes its size very uncertain, too.

M8
Lagoon Nebul

Center of the Milky Way

M22

M22 is one of the many globular clusters that are concentrated toward the core of our galaxy. Easy to see with binoculars, it is just barely visible to the naked eye. It is about 11,000 ly away from us.

The "teapot" of Sagittarius

Sagittarius in 3-D

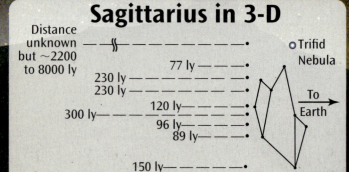

Distance unknown but ~2200 to 8000 ly

Trifid Nebula

77 ly
230 ly
230 ly
120 ly
300 ly
96 ly
89 ly
150 ly

To Earth

1 light year (ly) ≈ 10 trillion km ≈ 6 trillion miles

LOOKING UP #8

Centaurus and Crux The Southern Cross

hese constellations are best observed rom the southern hemisphere. Northern hemisphere viewers can ee Centaurus low in the southern ky in May – July. Crux may be een just above the southern horizon in May and June rom the extreme outhern US (Key West nd South Texas)

Centaurus A

This unusual galaxy, ~11 million ly distant, is one of the brightest radio sources in the sky

~50,000 ly

Alpha Centauri

Proxima Centauri

This dim star is the nearest star to the Sun, 4.22 ly distant

The Jewel Box

NGC 4755, an open star cluster ~500 ly from us.

Crux
The Southern Cross

The Coal Sack

An interstellar dust cloud

Eta Carinae

A very high-mass star doomed to die young ~8000 ly distant

Omega Centauri

The largest globular cluster in the Milky Way ~17,000 ly distant and containing millions of stars

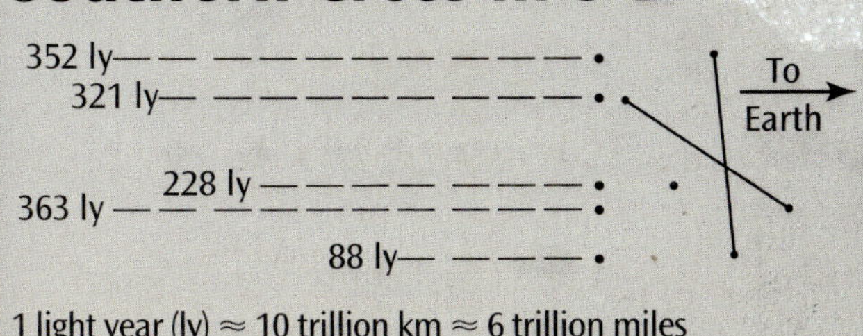

LOOKING UP #9

Southern Circumpolar Constellations

Crux
The Southern Cross

Musca

Most of the constellations in this part of the sky are dim, but observers in much of the southern hemisphere can see the Magellanic Clouds circling the south celestial pole throughout the night.

Hourglass Nebula
A planetary nebula
~8000 ly distant

Apus

Octans
The constellation closest to the south celestial pole is named after a navigational instrument, the octant.

~0.5 ly

Chamaeleon

Thumbprint Nebula
A Bok globule
about 600 ly distant

The South Celestial Pole
No bright stars lie near the south celestial pole, but the southern cross points toward it.

Volans

Mensa

Hydrus

Small Magellanic Cloud
A dwarf galaxy orbiting the Milky Way at a distance of about 200,000 ly

Large Magellanic Cloud
A small galaxy orbiting the Milky Way at a distance of about 160,000 ly

Tarantula Nebula
A star-formation region in the Large Magellanic Cloud larger than any known in the Milky Way.

PREVIEW

The Cosmic Landscape

Astronomy is the study of the heavens, the realm extending from beyond the Earth's atmosphere to the most distant reaches of the Universe. Within this vast space we discover an amazing diversity of planets, stars, and galaxies. That creatures as tiny as ourselves can not only contemplate but also understand such diversity and immensity is amazing. But even more amazing are the objects themselves: planets with dead volcanos whose summits dwarf Mount Everest, stars a hundred times the diameter of the Sun, and galaxies—slowly whirling clouds of stars—so vast that they make the Earth seem like a grain of sand in comparison. All this is the cosmic landscape in which we live, a landscape we will explore briefly now to familiarize ourselves with its features and to gain an appreciation for its vast scale.

THE EARTH, OUR HOME

We begin with the Earth, our home **planet** (fig. P.1). This spinning sphere of rock and iron circling the Sun is huge by human standards, but it is one of the smaller bodies in the cosmic landscape. Nevertheless, it is an appropriate place to start because, as the base from which we view the Universe, it influences what we can see. We cannot travel from object to object in our quest to understand the Universe. Instead, we are like children who know their neighborhood well but for whom the larger world is still a mystery, known only from books and television.

But just as children use knowledge of their neighborhood to build their image of the world, so astronomers use their

FIGURE P.1
The planet Earth, our home, with blue oceans, white clouds, and multihued continents.

A

B

FIGURE P.2
The Moon as seen (A) with the unaided eye and (B) through a small telescope.

knowledge of Earth as a guide to more exotic worlds. For example, we can deduce from the glowing lava of an erupting volcano and the boiling water shooting from a geyser that the interior of our planet is hot. That heat creates motion inside the Earth, much the way heat makes soup in a pot bubble and churn. Although the motions inside Earth are far slower than those we see in bubbling soup, over millions of years they buckle the seemingly firm rock of our planet's crust to heave up mountains and volcanoes. Deeper inside Earth, similar motions generate magnetic forces that extend through the surface and into space. On Earth's surface these forces tug on the needle of a compass so that it points approximately north–south. High in our atmosphere, these same magnetic forces shape the northern lights.

Looking outward to our planetary neighbors, we find landscapes on Venus and Mars that bear evidence of many of the same processes that sculpt our planet and create its diversity. Likewise, when we look at the atmospheres of other planets, we see many of the same features that occur in our atmosphere. For example, winds in the thin envelope of gas that shelters us swirl around our planet much as similar winds sweep the alien landscapes of Venus and Mars.

THE MOON

The Moon is our nearest neighbor in space, a **satellite** that orbits the Earth some quarter million miles (384,000 km) away. Held in tow by the Earth's gravity, the Moon is much smaller than Earth—only about one-quarter our planet's diameter.

With the naked eye, and certainly with a pair of binoculars or small telescope, we can clearly see that the Moon's surface is totally unlike Earth's. Instead of white whirling clouds, green-covered hills, and blue oceans, we see an airless, pitted ball of rock that shows us the same face night after night (fig. P.2).

Why are the Earth and Moon so different? Their differences arise in large part from the great disparity in their masses. The Moon's mass is only about 1/80th the Earth's, and it was therefore unable to retain an atmosphere. Without wind and rain, there has been relatively little erosion of the Moon's surface. Because of the Moon's smaller bulk it was also less able to retain heat. Without that strong internal heat, the crustal motions that are so important in shaping Earth are absent on the Moon. In fact, the Moon has changed so little for billions of years that its surface provides important clues to what Earth was like when it was young. In addition to this scientific importance, the Moon has symbolic significance for us—it is the farthest place from Earth that humans have traveled.

THE PLANETS

Beyond the Moon, circling the Sun as the Earth does, are seven other planets, sister bodies of Earth. In the order of their average distance from the Sun, working outward, the eight planets are Mercury, Venus, Earth, Mars, Jupiter, Saturn, Uranus, and Neptune. These worlds have dramatically different sizes and landscapes. For example:

- Ancient craters blasted out by asteroid impacts scar the airless surface of Mercury.
- Dense clouds of sulfuric acid droplets completely shroud Venus.

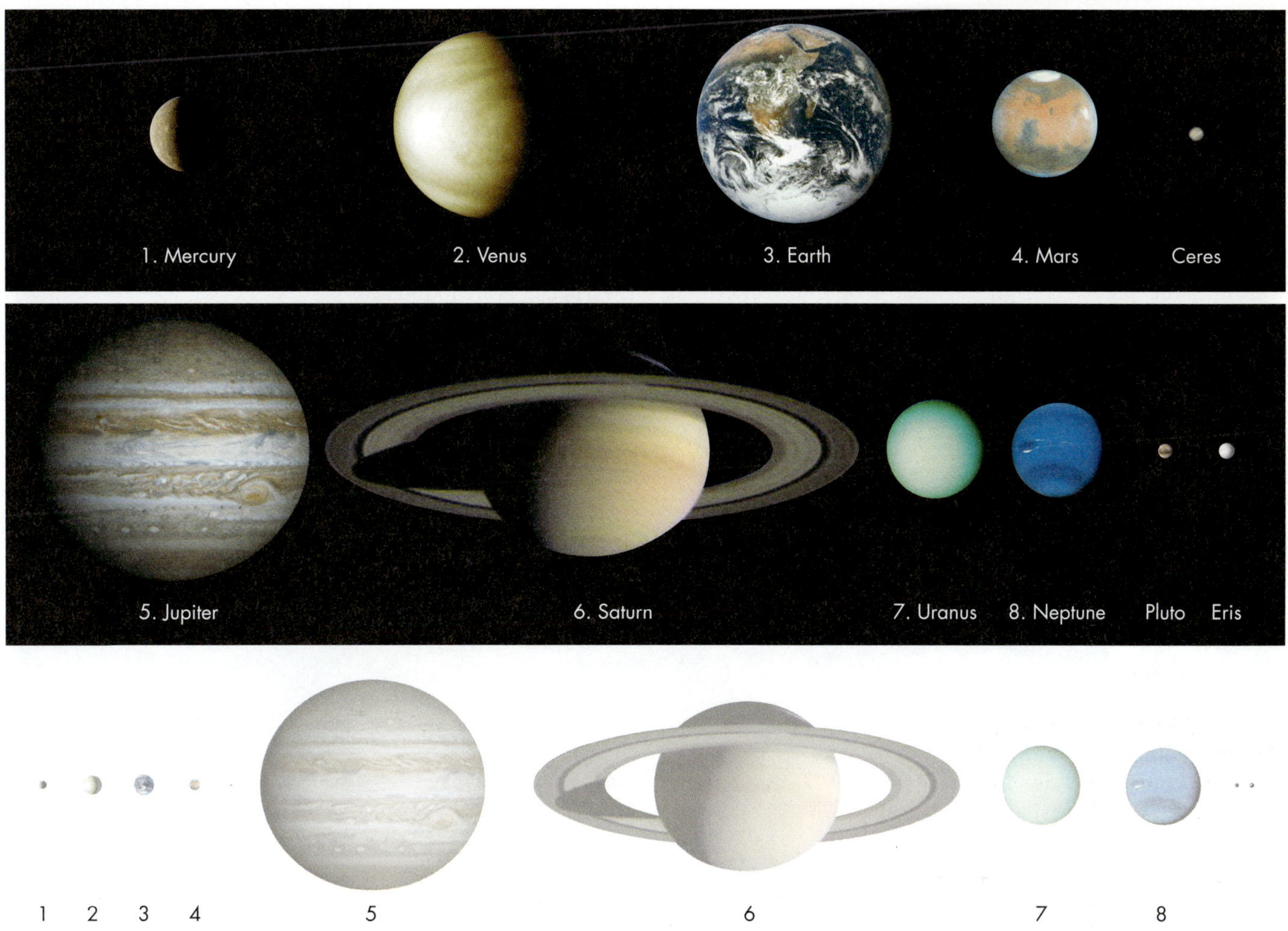

FIGURE P.3
Portraits of the eight planets along with silhouettes show their correct relative size. Also shown are three dwarf planets.

- White clouds, blue oceans, green jungles, and red deserts tint Earth.
- Huge canyons and deserts spread across the ruddy face of Mars, where possibly, long ago, lakes and even oceans may have reflected a pink sky.
- Immense atmospheric storms with lightning sweep across Jupiter.
- Trillions of icy fragments orbit Saturn, forming its bright rings.
- Dark rings girdle Uranus, its spin tipped by some cosmic catastrophe.
- Choking methane clouds whirl in the deep blue atmosphere of Neptune.

Mercury, Venus, Mars, Jupiter, and Saturn are visible to the naked eye as bright points of light, much like stars. But whereas stars do not noticeably change their positions relative to one another, the planets, because of their orbital motion around the Sun, drift slowly and regularly against the pattern of the background stars. This regular motion gave the planets a special significance to people in ancient times—a significance that has been carried forward to today in the names of many of the days of the week. For example, in English, Saturday gets its name from Saturn, while in Spanish, miércoles gets its name from Mercury.

The planets are significant to us today because with modern telescopes and spacecraft we can see that they are truly other worlds. Figure P.3 shows pictures of these eight distinctive bodies and reveals something of their relative size and appearance. Some are far smaller and others vastly larger than Earth, but all are dwarfed by the Sun, whose immense gravity holds them in orbit.

THE SUN

The Sun is a **star,** a huge ball of gas more than 100 times the diameter of the Earth and more than 300,000 times more massive: if the Sun were a volleyball, the Earth would be about the size of a pinhead, and Jupiter roughly the size of a nickel (fig. P.4). The Sun contains about 1000 times more matter than all of the Solar System's planets combined. The Sun, of course, differs from the planets in more than just size: it generates energy in its core by nuclear reactions that convert hydrogen into helium. From the core, the energy flows to the Sun's surface, and from there it pours into space to illuminate and warm the planets.

The Sun's energy output cannot last forever. It has been warming the planets for more than 4 billion years—long enough for life to arise on Earth and for intelligent creatures to evolve who can marvel at such wonders. Studies of other stars teach us that the Sun will run out of fuel in another 5 or 6 billion years, then finally fade away like a cooling ember. Thus astronomy helps us not only to examine unusual objects at huge distances, but to look deep into the past and far into the future.

THE SOLAR SYSTEM

The Sun and the eight planets orbiting it form the **Solar System.** But other objects orbit the Sun as well. These include the recently defined dwarf planets,* such as Pluto and Eris, as well as even smaller objects such as the asteroids and comets. Most asteroids orbit between Mars and Jupiter in the so-called asteroid belt, home also to the dwarf planet Ceres. Comets orbit mainly in the outermost fringes of the Solar System, where several other dwarf planets are found too.

*In mid-2006, the International Astronomical Union voted to define a planet in such a way as to make Pluto no longer a full-fledged planet. By the new definition (described in more detail in chapter 7), Pluto and several other objects in our Solar System are now officially known as "dwarf planets."

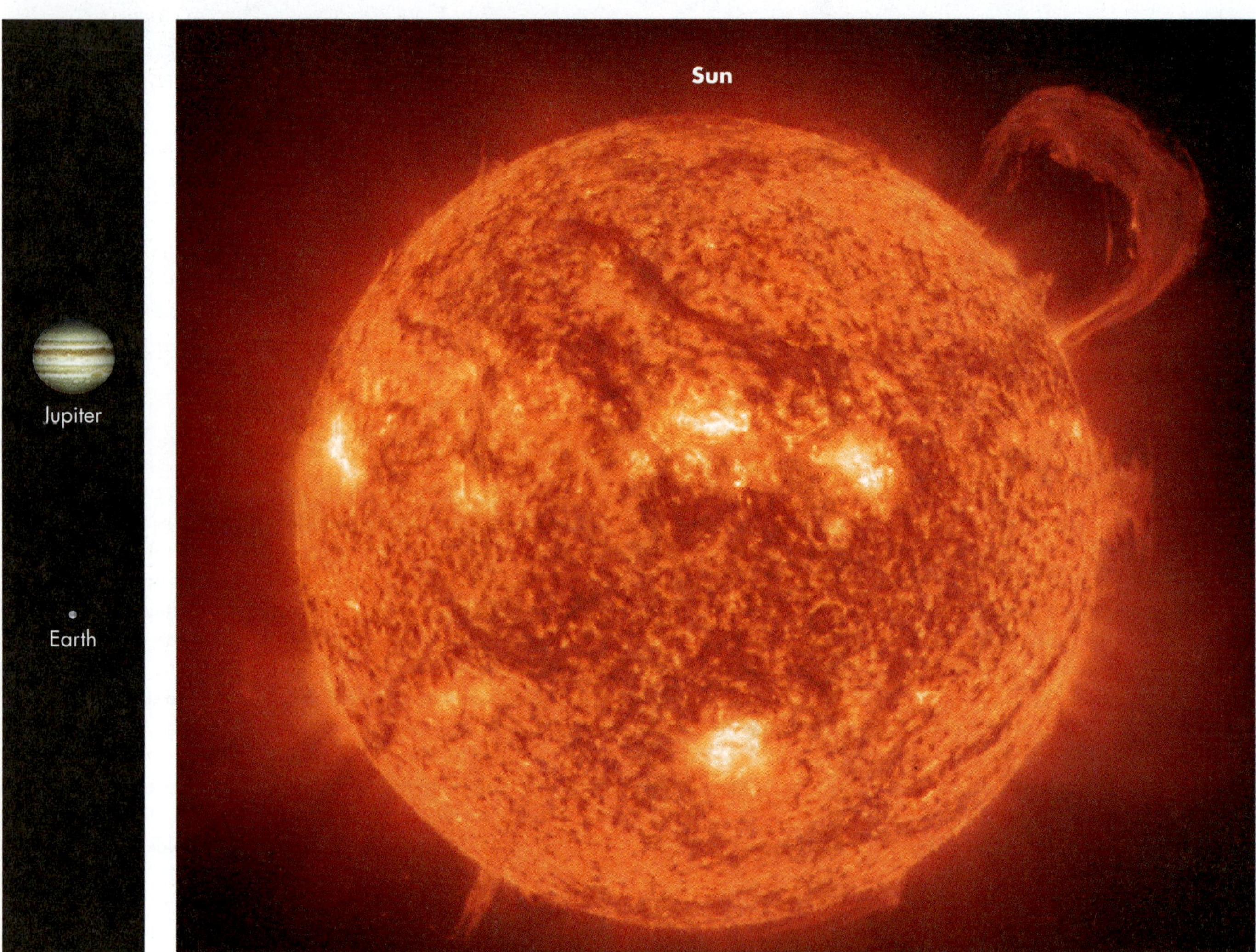

FIGURE P.4
The Sun as viewed through a filter that allows its hot outer gases to be seen. The Earth and Jupiter are shown to scale beside it for comparison. (The filter allows astronomers to see extremely hot helium gas.)

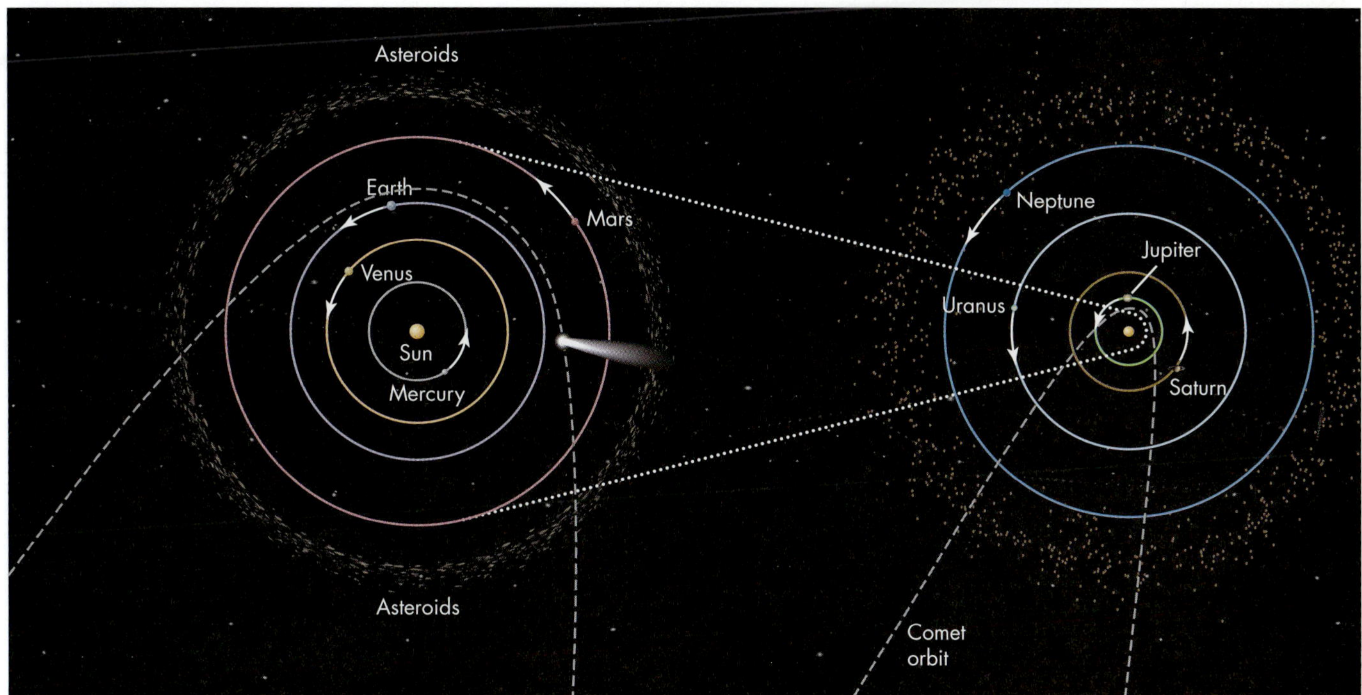

FIGURE P.5
Sketch of the positions and orbits of the planets and a variety of smaller bodies in our Solar System on March 20, 2011. The orbits of three of the largest "dwarf planets," Halley's comet, and another typical comet are also shown. The approximate location of small bodies in the asteroid belt and Kuiper belt are indicated. To show the orbits to scale, the inner and outer Solar System are shown separately.

If the paths that the planets follow around the Sun were visible, we would see that the Solar System is like a huge set of nested, nearly circular rings, centered approximately on the Sun and extending about 4 billion miles outward to the dwarf planet Pluto's orbit (fig. P.5). It is hard to imagine such immense distances measured in miles. In fact, it is as foolish to use miles to measure the size of the Solar System as it is to use inches to measure the distance between New York and Tokyo. Whenever possible, we try to use units appropriate to the scale of what we seek to measure. For example, in earlier times people used units that were quite literally at hand, such as finger widths or the spread of a hand to measure a piece of cloth and paces to measure the size of a field. In the same tradition, although on a different scale, astronomers use distance scales related to familiar objects, such as the Earth. As we shall see in later chapters, the Earth's radius and mass are convenient units for measuring the other planets. Likewise, the Earth's distance from the Sun is a good unit for measuring the scale of the Solar System.

THE ASTRONOMICAL UNIT

The **astronomical unit,** abbreviated as *AU,* is defined by the distance from the Earth to the Sun.* This translates into about 93 million miles (150 million kilometers). If we use the AU to measure the scale of the Solar System, Mercury turns out to be 0.4 AU from the Sun, while the Neptune is about 30 AU (fig. P.5). Figure P.6 shows a picture of the Solar System made by the spacecraft *Voyager I.* Notice how *empty* space is.

The Solar System extends far beyond the planets. For example, small icy objects, such as comets, drift along immense orbits that stretch up to about 100,000 AU away from the Sun. The *Voyager* spacecraft has only penetrated a small way into this remote region, and its signal is becoming extremely weak as it travels so far from the Sun. The Solar System still remains the limit to our direct exploration of the Universe, but telescopes extend our view far beyond the Solar System to reveal that just as the Earth is but one of many planets orbiting the Sun, so too the Sun is but one of a vast swarm of stars orbiting the center of our galaxy, the Milky Way.

THE MILKY WAY GALAXY

The **Milky Way Galaxy** is a cloud of several hundred billion stars with a flattened shape like the Solar System (fig. P.7). The Sun and other stars orbit the Milky Way at some 140 miles per second (220 kilometers per second), but so vast is our galaxy that it still takes the Sun about 240 million years to complete one trip around this immense disk. The Milky Way's myriad stars come in many varieties, some hundreds of times larger than the Sun, others hundreds of times smaller. Some stars are much

*Because the Earth's orbit is actually an ellipse, the AU is technically defined slightly differently, a point we will discuss further when we consider planetary orbits.

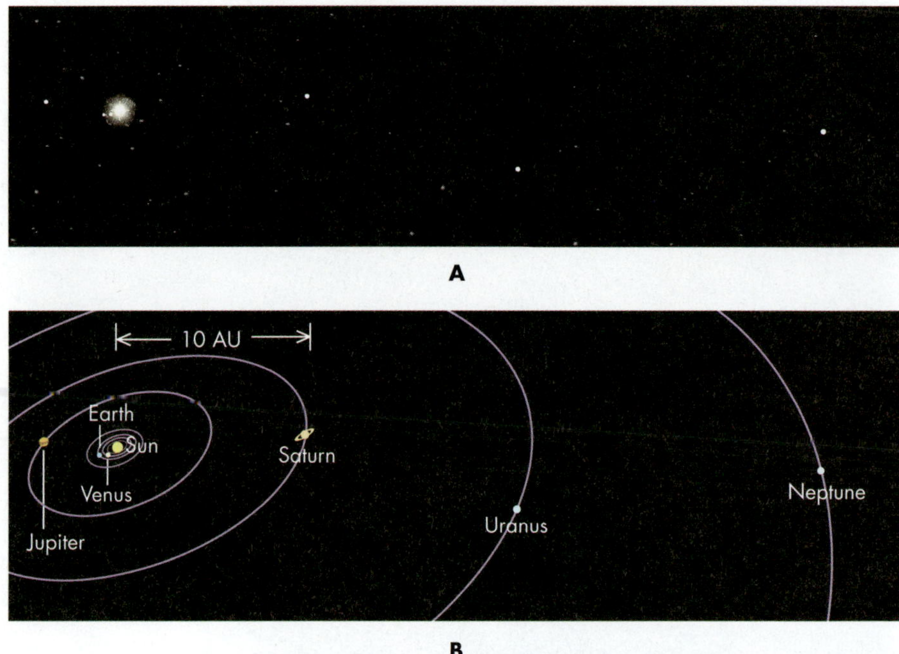

FIGURE P.6
(A) This view of the Solar System is based on a series of real images made by the *Voyager I* spacecraft. The craft was about 40 AU from the Sun and about 20 AU above Neptune's orbit. The images of the planets (mere dots because of their immense distance) and the Sun have been made bigger and brighter in this view to allow you to see them more clearly. Mercury is lost in the Sun's glare and Mars happened to lie nearly in front of the Sun at the time the image was made, so it too is invisible. (B) A sketch of the orbits of the planets, showing where each was located at the time the image was made in February 1990.

hotter than the Sun and shine a dazzling blue-white, while others are cooler and glow a deep red.

In the Milky Way, as in many other galaxies, stars intermingle with immense clouds of gas and dust. These clouds, enormously larger than the Solar System, are the sites of stellar birth and death. Deep within their cold, dark gas, gravity draws their matter into dense clumps that eventually turn into new stars, lighting the gas and dust around them. Stars eventually burn themselves out and explode, spraying matter outward to mix with the surrounding clouds. This matter from exploded stars is ultimately recycled into new stars (fig. P.8).

In this huge swarm of stars and clouds, the Solar System is all but lost—like a single grain of sand on a vast beach—forcing us again to grapple with the problem of scale. Stars are almost unimaginably remote: the nearest one to the Sun is over 25 trillion miles away. Such distances are so immense that analogy is often the only way to grasp them. For example, if we think of the Sun as a pinhead, the nearest star would be another pinhead about 35 miles away and the space between them would be nearly empty. In fact, distances between stars are so immense that even astronomical units are inappropriately small, and so we again choose a new unit of length—the light-year.

THE LIGHT-YEAR

Measuring a distance in terms of a time may at first sound peculiar, but we do it often. We may say, for example, that our town is a 2-hour drive from the city, or our dorm is a 5-minute walk from the library, but expressing a distance in this fashion implies that we have a standard speed.

Astronomers are fortunate to have a superb speed standard: the speed of light in empty space, which is a constant of nature and equal to 299,792,458 meters per second (about 186,000 miles

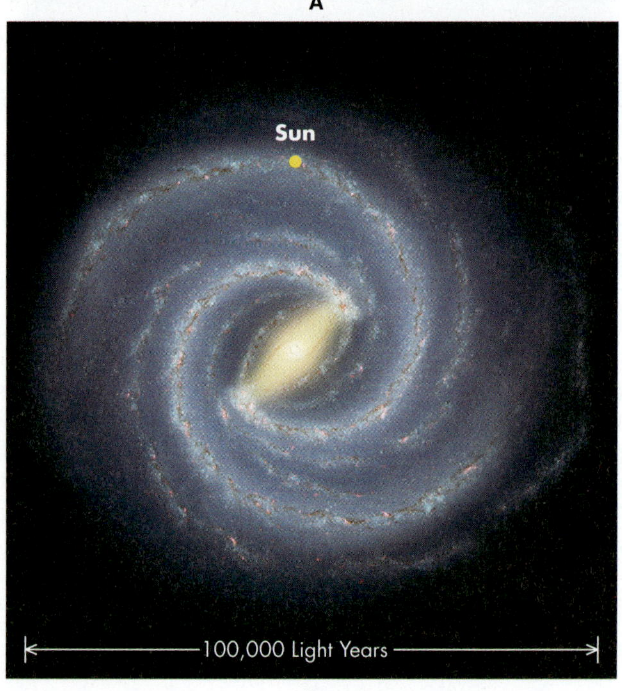

FIGURE P.7
The Milky Way Galaxy. (A) A side view made by plotting stars in the 2MASS star catalog. (B) The approximate structure of the Milky Way if it were seen from above, as mapped out by the Spitzer Space Telescope.

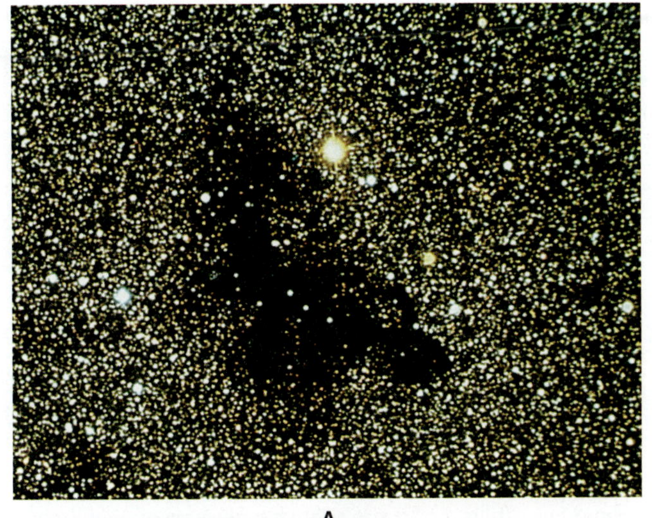
A

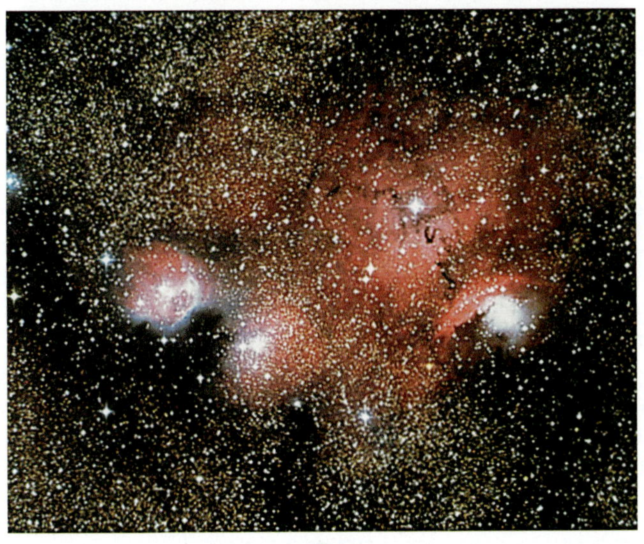
B

FIGURE P.8
Interstellar clouds in the Milky Way. On the scale of these pictures, the Solar System out to Pluto is about 1000 times smaller than the period ending this sentence. (A) A cold, dark cloud seen in silhouette. Dust in the cloud blocks our view of the stars behind it. (B) The hydrogen in clouds heated by young stars glows red.

per second). Moving at this constant and universal speed, light in 1 year travels a distance defined to be 1 **light-year,** abbreviated as ly. As we will show below, this works out to be about 6 trillion miles (10 trillion kilometers). To demonstrate that, however, is cumbersome because it involves multiplying a series of large numbers. We therefore will use a more concise way to write them called scientific notation.

In **scientific notation** (also sometimes called powers-of-ten notation), we write numbers using ten to an exponent, or power. Thus we write $100 = 10 \times 10 = 10^2$ and 1 million (1,000,000) as $10 \times 10 \times 10 \times 10 \times 10 \times 10 = 10^6$. Instead of writing out all the zeros, therefore, we use the exponent to tell us the number of zeros. A number like the speed of light (186,000 miles per second) may also be written in scientific notation, becoming 1.86×10^5 miles per second. Likewise, the astronomical unit (150 million kilometers) can be written as 1.5×10^8 km.

One reason to use scientific notation is that multiplying and dividing becomes enormously easier. For example, to multiply two powers of ten we just add the exponents, and to divide we subtract them. Thus $10^2 \times 10^5 = 10^7$, and $10^8/10^3 = 10^5$. More details on using scientific notation are given in the appendix, but as an illustration of its usefulness, let us now calculate the number of miles in a light-year.

To find how far light travels in a year, we multiply its speed by the travel time. One year is approximately 31,600,000 (or 3.16×10^7) seconds.* Multiplying this by the speed of light we get

$$3.16 \times 10^7 \text{ seconds} \times 1.86 \times 10^5 \text{ miles/second}$$

$$= 3.16 \times 1.86 \times 10^{12} \text{ seconds} \times \text{miles/second}$$

$$= 5.88 \times 10^{12} \text{ miles,}$$

or about 6 trillion miles (about 10^{13} kilometers), as we previously claimed. In these units, the star nearest the Sun is 4.2 light-years away. Although we achieve a major convenience in adopting such a huge distance for our scale unit, we should not lose sight of how truly immense such distances are. For example, if we were to count off the miles in a light-year, one every second, it would take us about 186,000 years!

We can now use the light-year for setting the scale of the Milky Way Galaxy. In light-years, our galaxy is about 100,000 light-years across, with the Sun orbiting roughly 30,000 light-years from the center. Within the Milky Way's disk, stars are separated by a few light-years. The gas clouds, such as those in figure P.8, which we mentioned earlier, range in size from a fraction of a light-year to hundreds of light-years across.

GALAXY CLUSTERS AND THE UNIVERSE

Having gained some sense of scale for the Solar System and the Milky Way, we now resume our exploration of the cosmic landscape, pushing out to the realm of other galaxies. Here we find that just as stars assemble into galaxies, so galaxies themselves assemble into **galaxy clusters.**

The cluster of galaxies to which the Milky Way belongs is called the **Local Group.** It is "local," of course, because it is the one we inhabit. It is termed a "group" because it is small as galaxy clusters go, containing just several dozens of galaxies as members, but it is still several million light-years in diameter. Yet despite such vast dimensions, the Local Group is itself part of a still larger assemblage of galaxies known as the **Virgo Supercluster.** Figure P.9 puts this in perspective for us.

*Numbers here and elsewhere are rounded off.

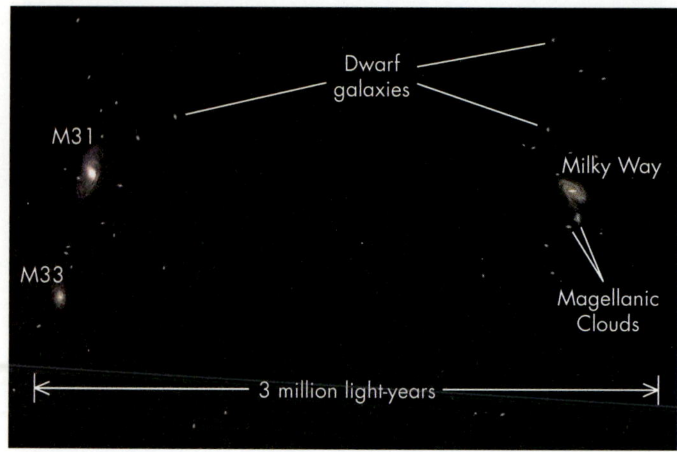

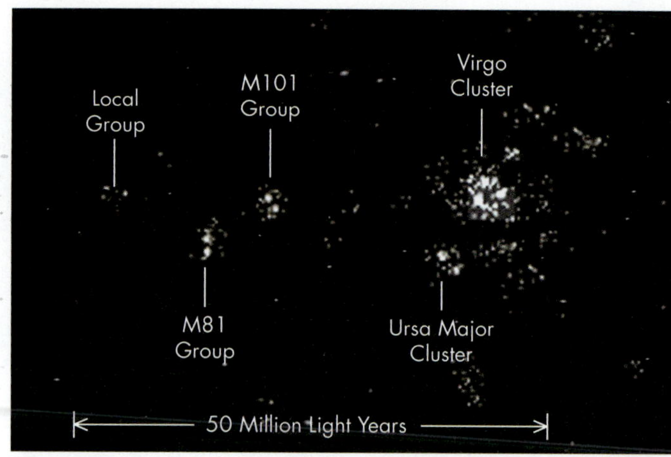

FIGURE P.9
(A) A sketch of the central region of the Local Group.* Galaxies are shown at about twice their correct size relative to their separation. Only a few of the numerous dwarf members are shown. (B) A sketch of the Virgo Supercluster. Only the larger groups of galaxies are labeled. Clusters are shown at approximately correct size relative to their separation.

*The names of the galaxies M31 and M33 in the Local Group stand for Messier 31 and 33. They get their names from a list of galaxies and other astronomical objects that was compiled in the late 1700s by French astronomer Charles Messier ("Mess-yay"). We will encounter many other Messier objects later in our exploration of the Universe.

Our supercluster consists of hundreds of galaxy groups and clusters, spread over some 100 million light-years, but it is perhaps itself part of an even larger structure known as the Great Attractor region, a cluster of superclusters, probably more than 300 million light-years across. Structures of such vast size are about the largest objects we can see before we take the final jump in scale to the **Universe** itself.

The visible Universe is the largest astronomical structure of which we have any knowledge. From the observations presently available to them, astronomers have deduced that the Universe is about 14 billion years old. This limits the distance we can see, even in principle, to about 14 billion light-years, a number often taken as the approximate radius of the *visible* Universe. However, that does not mean the Universe ends at such a distance. Rather, it means we cannot see what lies beyond. But regardless of our uncertainty about the known Universe's size, we can observe that its structure is similar throughout the visible Universe. Small objects are clustered into larger systems, which are themselves clustered: planets around stars, stars in galaxies, galaxies in clusters, clusters in superclusters, and superclusters into even larger groups. Although astronomers do not yet understand completely how this orderly structure originated, they do know that gravity plays a crucial role.

GRAVITY

Gravity gives the Universe structure because it creates a force of attraction between *all* objects. You experience gravity's attraction in everyday life. For example, if you drop a book, the Earth's gravitational force makes the book fall. Moreover, that same force spans the vast distance between the Earth and the Moon to hold our satellite in its orbit. Similarly, gravity holds our planet in its orbit around the Sun and the Sun in its orbit around the Milky Way.

Although gravity dominates the large-scale structure of the Universe, other forces dominate on smaller scales. To understand these forces, we need to look briefly at the small-scale structure of matter.

ATOMS AND OTHER FORCES

Matter is composed of submicroscopic particles called **atoms.** Atoms are incredibly small. For example, a hydrogen atom is about one ten-billionth of a meter (10^{-10} m) in diameter. Ten million hydrogen atoms could be put in a line across the diameter of the period at the end of this sentence. But despite this tiny size, atoms themselves have structure. Every atom has a central core called the **nucleus** that is orbited by smaller particles called **electrons** (fig. P.10). The nucleus is in turn composed of two other kinds of particles called **protons** and **neutrons.**

Although the particles in an atom exert a gravitational attraction on one another, atoms are not held together by gravity. Instead, an electric force gives them their structure. That force arises because protons and electrons have a property called **electric charge.** A proton has a positive electric charge, and an electron has a negative electric charge. A neutron, as its name suggests, has no charge.

The **electric force** can either attract or repel, depending on the charges. Opposite charges attract, and like charges repel. Thus, two electrons (both negative) repel each other, while an

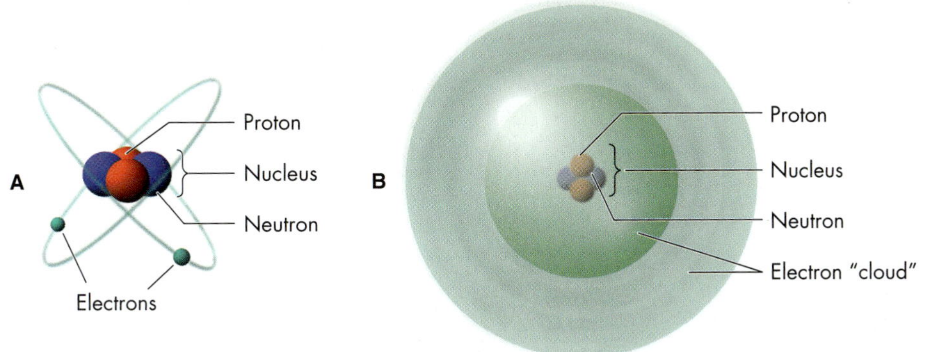

FIGURE P.10
An atom consists of a nucleus around which electrons orbit. The nucleus is itself composed of particles called protons and neutrons. (A) A classical view of an atom. (B) A more modern view with the electrons orbiting as "clouds" called orbitals.

electron and a proton (negative and positive, respectively) attract each other. That attraction is what holds the electrons in their orbits around the nucleus of an atom.

You can see the electric force at work in many ways. For example, the static electric charges generated when a clothes drier tumbles your laundry creates an attraction that may make clothes cling together. The crackling sound you hear as you pull fuzzy socks away from a shirt is the electric charges jumping and making tiny sparks.

The electric force is closely linked with the magnetic force that makes a compass work or holds the little magnets to the door of your refrigerator. In fact, the theory of relativity demonstrates that electric and magnetic forces are fundamentally the same, and scientists generally refer to them jointly as the electromagnetic force.

Yet another force plays a critical role in atoms. This force, called the **strong force,** binds the protons and neutrons together to form the atom's nucleus. Although the effects of the strong force cannot be seen directly in everyday life, without it the nuclei of atoms, and with them our familiar world, would disintegrate.

In addition, a fourth force, known as the **weak force***, operates on the atomic scale and plays a role in radioactive decay. The weak force is so weak that interactions involving it are extremely rare. Their rareness is important in determining how long stars live. Stars would burn themselves out much more quickly, or would not shine at all, if the weak force were much stronger or weaker. In fact, astronomers are beginning to suspect that the weak force may play an even bigger role in shaping the Universe than previously thought.

THE STILL-UNKNOWN UNIVERSE

Our quick trip from Earth outward has shown us a Universe of planets, stars, and galaxies. But astronomers today have evidence that, although these objects are the ones we notice, the bulk of the Universe must consist of something completely different. That evidence comes from many sources, the most convincing of which are the findings that (1) stars within galaxies, and (2) galaxies within clusters of galaxies, experience a far stronger gravitational force than can be explained by the directly observable matter. That is, both galaxies and galaxy clusters appear to contain huge amounts of what astronomers call **dark matter.**

Dark matter is so named because it emits no as-yet-observed radiation. But from its gravitational effects, astronomers deduce that it outweighs luminous matter by a factor of about ten to one. What is the dark matter? Astronomers do not know, but it may be made up of particles that interact only through the weak and gravitational forces. For example, there are billions of weakly interacting particles called **neutrinos** passing through your body each second. These were generated by the Universe in its early stages, by nuclear reactions in the Sun, and by other cosmic events. You do not sense these particles because normal matter is more transparent to them than a glass window is to light. Astronomers suspect that there may be particles much more massive than neutrinos that fill space, generating a much stronger gravitational pull than all of the stars in all of the galaxies that we can see.

On the largest scales, galaxies throughout in the Universe are moving away from each other in a great cosmic expansion. This expansion began almost 14 billion years ago in an unimaginable explosion called the **Big Bang** that created time and space and sent hot matter flying apart everywhere. During the last decade, astronomers studying the expansion have discovered a great mystery—the rate of expansion is speeding up. Something is overcoming the gravitational attraction between galaxies, causing them to accelerate away from each other.

It is as if empty space contains a sort of energy that drives the expansion to grow faster. Because its nature is still unknown, astronomers have named it **dark energy.** If we compare the mass of the dark matter and the dark energy with the mass that we directly detect (such as stars, galaxies, and gas clouds), those luminous objects amount to a mere 1% of the Universe's total mass. What we see of the Universe is therefore much like the footprints of an invisible man: a being who leaves tracks, but whose build and nature we do not yet know.

*The weak force is linked to the electromagnetic force, and their combination is known technically as the electroweak force.

THE SCIENTIFIC METHOD

Our scientific understanding of the Universe has not come easily. It has grown out of the work of thousands of men and women over thousands of years. Their work is part of the broad field that we call science.

By "science" we mean the systematic study of things and the search for the underlying principles that govern them, be they living things, matter, or, in our case, the astronomical universe. An essential part of that study is the rigorous testing of ideas. We call the process of such testing the **scientific method.** In using the scientific method, a scientist typically proposes an idea—a hypothesis—about some property of the Universe and then tests that hypothesis by experiment. In fact, whether an idea is "scientific" depends to some extent on whether it can be verified by either a real or an imagined experiment. Ideally the experiment either confirms the hypothesis or refutes it. If refuted, the hypothesis is rejected. On the other hand, if the experiment confirms the hypothesis, the scientist may then go on to develop related hypotheses or perhaps to make predictions about some as-yet-undiscovered aspect of the subject.

Once a set of ideas has been thoroughly tested and verified, the ideas may be incorporated into a theory or law. When we use the term *theory* here, we do *not* mean that the ideas are unproven or tentative. Rather, we mean that they have achieved wide acceptance by successful testing. For example, scientists have subjected the quantum theory of atomic structure and the theory of relativity to numerous tests, and these theories have passed all such tests with high precision.

Astronomers face a special difficulty in applying the scientific method because usually they cannot experiment with their subject matter directly: in virtually all cases, they can only passively observe. Nevertheless, they try—like all scientists—to use the scientific method. You will find some specific examples of this method in later chapters, where *Science at Work* boxes show how this process has led to new ideas and revision of old ones.

Application of the scientific method is no guarantee that its results will be believed. For example, we will see in chapter 1 that even before 300 B.C., the Greek philosopher Aristotle taught that the Earth is a sphere. Yet despite the proofs he offered to support that hypothesis, many people continued to believe the Earth to be flat. Today, too, some scientific hypotheses might be rejected despite their experimental verification, and others might be accepted though untrue. For example, one astronomer might find evidence supporting some hypothesis, but another astronomer might claim that the experiment was done incorrectly or the data were analyzed improperly. Therefore, throughout this book, whenever we discuss our knowledge of a given topic, we must keep in mind the fact that such knowledge is not always proved or even universally accepted. This is especially true of topics at the frontiers of our understanding, such as the origin and structure of the Universe or the properties of black holes. Therefore, keep in mind that some of what we discuss in this book will be proved wrong in the future. That is not a failing of science, however. It is its strength.

SUMMARY

The Earth is one of eight planets orbiting the Sun, and the Sun is one of about a hundred billion stars that make up the Milky Way Galaxy. The Milky Way, two other similar-size galaxies, and dozens of smaller galaxies compose the Local Group, which in turn is part of the Local Supercluster of galaxies. Superclusters seem to be grouped into even larger systems that fill the visible Universe. We can speak with some certainty about the size and properties of objects in our immediate neighborhood, but the farther we move from Earth, the less certain we become.

Astronomers use the astronomical unit (AU) and light-year (ly) to measure the immense sizes and distances of astronomical systems. The AU is defined by the average distance between the Earth and the Sun, and the light-year is defined as the distance light travels in a year, which is about 6 trillion miles. Using these units, we can see the immense scale of the Universe in figure P.11 and table P.1. The former shows a series of drawings to help you visualize how enormous the Universe is.

Matter is made up of atoms in which charged particles called electrons orbit a nucleus. The nucleus is itself composed of smaller particles called protons and neutrons. Four forces give the Universe its structure: the electromagnetic, strong, and weak forces on the scale of atoms, and gravity on the cosmic scale. Astronomers have uncovered evidence that most of the Universe is made of types of matter and energy that we have not yet been able to detect.

QUESTIONS FOR REVIEW

1. About how much bigger in radius is the Sun than the Earth?
2. How big is an astronomical unit?
3. Roughly how big across is the Milky Way Galaxy?
4. How is a light-year defined?
5. What force holds the different astronomical systems described in this section together? What other forces exist in nature?
6. What was the Big Bang? What are dark matter and dark energy?
7. What particles make up an atom?
8. What force holds the electrons to an atom's nucleus?
9. What is meant by the scientific method?
10. What is the difference between a hypothesis and a theory?

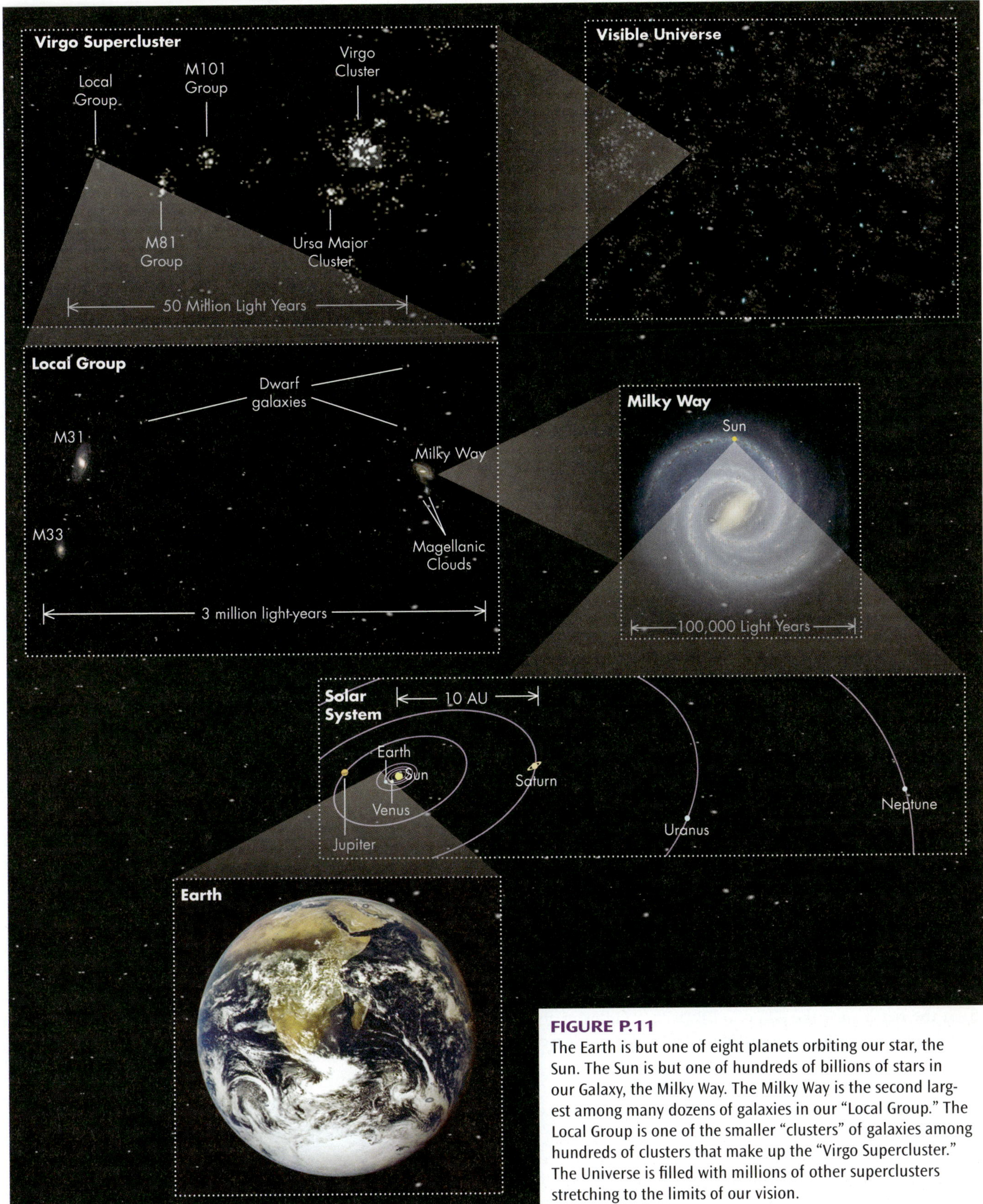

FIGURE P.11
The Earth is but one of eight planets orbiting our star, the Sun. The Sun is but one of hundreds of billions of stars in our Galaxy, the Milky Way. The Milky Way is the second largest among many dozens of galaxies in our "Local Group." The Local Group is one of the smaller "clusters" of galaxies among hundreds of clusters that make up the "Virgo Supercluster." The Universe is filled with millions of other superclusters stretching to the limits of our vision.

TABLE P.1 THE SCALE OF THE UNIVERSE

Object	Approximate Radius
Earth	6400 km (about 4000 miles)
Sun	700,000 km (about 100 times radius of the Earth)
Earth's orbit	150 million km (about 200 times radius of Sun) = 1 AU
Solar System out to Neptune	40 AU (about 8600 times radius of the Sun)
Milky Way galaxy	50,000 ly (about 10^8 times radius of the Solar System)
Local Group	2.5 million ly (about 50 times radius of the Milky Way)
Local Supercluster	50 million ly (about 20 times radius of the Local Group)
Visible Universe	14 billion ly (about 300 times radius of the Local Supercluster)

THOUGHT QUESTIONS

1. To what systems, in increasing order of size, does the Earth belong?
2. Propose a hypothesis about something you can experiment with in everyday life and try to verify or disprove the hypothesis. For example, what kind of surfaces will the little magnetic note holders people use on refrigerators stick to? Any smooth surface? Any metal surface?

PROBLEMS

1. The radius of the Sun is 7×10^5 kilometers, and that of the Earth is about 6.4×10^3 kilometers. Use scientific notation to show that the Sun's radius is about 100 times the Earth's radius.
2. Given that an astronomical unit is 1.5×10^8 kilometers and a light-year is about 10^{13} kilometers, how many AU are in a light-year?
3. Imagine building a model of the Solar System on your campus. Work out the diameter and spacings of the planets in millimeters and meters, respectively.
4. Calculate approximately how long it takes light to travel from the Sun to the dwarf planet Pluto.
5. If the Earth were the size of a BB (with a radius of about 0.2 centimeter), how big would the Sun be? How big would the Milky Way be?
6. If the Milky Way were the size of a nickel (about 2 cm), how big would the Local Group be? How big would the Local Supercluster be? How big would the visible Universe be? The data in table P.1 may help you here.
7. Suppose two galaxies move away from each other at 6000 km/sec and are 300 million (3×10^8) light-years apart. If their speed has remained constant, how long has it taken them to move that far apart? Express your answer in years.
8. A typical bacterium has a diameter of about 10^{-6} meters. A hydrogen atom has a diameter of about 10^{-10} meters. How many times smaller than a bacterium is a hydrogen atom?
9. Using scientific notation, numerically evaluate $(4\times10^8)^3$ / $(5\times10^{-6})^2$.
10. Use scientific notation to numerically evaluate $(3\times10^4)^2$ / $(4\times10^{-6})^{1/2}$.

TEST YOURSELF

1. Judging from the lower part of figure P.3, about how much larger is Jupiter's diameter than the Earth's?
   (a) 2 times (b) 5 times (c) 10 times (d) 25 times (e) 100 times
2. Ancients believed the planets to be special compared to stars because
   (a) the surface of each planet is very different from Earth.
   (b) planets repeat the same paths on the sky each week.
   (c) over time the planets appear to move against background stars.
   (d) they could see Jupiter's moons and Saturn's rings.
3. The light-year is a unit of
   (a) time. (b) distance. (c) speed. (d) age. (e) weight.
4. You write your home address in the order of street, town, state, and so on. Suppose you were writing your cosmic address in a similar manner. Which of the following is the correct order?
   (a) Earth, Milky Way, Solar System, Local Group
   (b) Earth, Solar System, Local Group, Milky Way
   (c) Earth, Solar System, Milky Way, Local Group
   (d) Solar System, Earth, Local Group, Milky Way
   (e) Solar System, Local Group, Milky Way, Earth

5. Which of the following astronomical system is/are held together by gravity?
   (a) The Sun
   (b) The Solar System
   (c) The Milky Way
   (d) The Local Group
   (e) All of them are.

6. Which of the following statements can be tested for correctness using the scientific method? (There may be more than one correct answer).
   (a) An astronaut cannot survive on the Moon without life-support systems.
   (b) The Moon is an uglier place than the Earth.
   (c) Electrons are charged particles.
   (d) The Sun's diameter is about 100 times larger than the Earth's diameter.
   (e) The sky is sometimes blue.

KEY TERMS

astronomical unit (AU), 5
atom, 8
Big Bang, 9
dark energy, 9
dark matter, 9
electric charge, 8
electric force, 8
electron, 8
galaxy cluster, 7
gravity, 8
light-year (ly), 7
Local Group, 7
Milky Way Galaxy, 5
neutrino, 9
neutron, 8
nucleus, 8
planet, 1
proton, 8
satellite, 2
scientific method, 10
scientific notation, 7
Solar System, 4
star, 3
strong force, 9
Universe, 8
Virgo Supercluster, 7
weak force, 9

FURTHER EXPLORATIONS

Boorstin, Daniel J. *The Discoverers.* New York: Random House, 1983.

Bronowski, Jacob. *A Sense of the Future.* Cambridge, Mass.: MIT Press, 1977.

Morrison, Philip, and Phylis Morrison. *Powers of Ten.* New York: W. H. Freeman, 1982.

Pine, Ronald C. *Science and the Human Prospect.* Belmont, Calif.: Wadsworth Publishing, 1989.

Sagan, Carl. *Cosmos.* New York: Random House, 1980.

Video

Powers of Ten. Pyramid Film and Video, Santa Monica, Calif., 1989. (Available at http://www.powersof10.com.)

Website

Visit the *Explorations* website at **http://www.mhhe.com/arny** for additional online resources on these topics.

PROJECTS

1. **Proportions of the Earth, Sun, and Moon:** Get a long roll of paper and draw a scale model of the Earth, Sun, and Moon on it. If you make the Earth a blue dot about 1 millimeter (the size of a poppy seed, or about 1/32 inch) in diameter, the Moon should be about $1\frac{3}{16}$ inches away from it. The Sun should be a circle about 4 inches in diameter and about 38 feet away from it. Shelf paper will work fine, and art stores have very long rolls of paper for use on children's easels. A roll of paper towel would work too.
2. **Scale Models:** Make scale models of the Solar System and the Local Group. Use the information given in table P.1 and/or the appendix tables to determine the sizes and separations of the objects in each. To make the model, you will need to establish a scale. For example, a 5-inch diameter grapefruit is about 10 billion (10^{10}) times smaller than the Sun. If you use this to represent the Sun, you will need to find other objects to represent each planet that are smaller by the same factor, and then separate them by distances 10 billion times smaller as well. This might require an athletic field. For the Local Group you could use frisbees and coins to represent the galaxies. You might make a video of the project to share with students in a local elementary school or invite them to watch you set up the model.

Stonehenge is a stone monument built by the ancient Britons on Salisbury Plain, England over 4000 years ago. Its orientation marks the seasonal rising and setting points of the Sun. The photo shows sunrise on the first day of summer.

KEY CONCEPTS

- We see the Sun, Moon, and stars rise along the eastern horizon, move across the sky, and set along the western horizon.
- The stars form patterns (constellations) whose shapes do not change.
- We see different constellations at different times of the year—Orion in January, Scorpius in August, for example.
- Cycles of the Moon, Sun, and stars are the basis for keeping time and our calendar.
  - The day is based on the Earth's spin (rotation about its own axis).
  - The month is based on the Moon's orbital motion around the Earth.
  - The year is based on the Earth's orbital motion around the Sun.
- The shape of the lit portion of the Moon seen from Earth changes during the month.
- Seasons are caused by the tilt of the Earth's axis relative to its orbit, which produces different amounts of heating in different regions during the year.
- Eclipses occur when the Earth, Sun, and Moon all align.

1 The Cycles of the Sky

CONCEPTS AND SKILLS TO REVIEW

- The properties of Earth and Moon (Preview, pp. 1–2)
- The orbit of the Earth (Preview, pp. 4–5)

We do not know when people of antiquity first began studying the heavens, but it was certainly many thousands of years ago. Astronomical observations are part of virtually every culture and include events that anyone who watches the sky can see, such as the rising of the Sun in the eastern sky and its setting toward the west, the changing appearance of the Moon throughout the month, and the beautiful and awe-inspiring occurrences of eclipses.

For many prehistoric people, observations of the heavens had more than just curiosity value. Because so many astronomical phenomena are cyclic—that is, they repeat day after day and year after year—they can serve as timekeepers. For example, when is it safe to set out on a sea voyage? When is it time to harvest crops? When will an eclipse occur? Moreover, the cyclic behavior of the heavens implies that many events seen in the sky are predictable. The desire to foretell these changes in the sky and on Earth probably motivated early cultures to study the heavens, and it may have led them to build monumental stone structures such as Stonehenge (facing page).

Sadly, many of the astronomical phenomena well known to ancient people are not nearly so familiar to people living today because the smog and bright lights of cities make it hard to see the sky and its rhythms. Perhaps more important, we no longer rely upon direct astronomical observations to tell us what season it is, when to plant, and so on. Therefore, if we are to appreciate the growth of astronomical ideas, we need to first understand what our distant ancestors knew and what we ourselves can learn by watching the sky over the course of a year.

In the following discussion, you might imagine yourself as a shepherd in the Middle East, a hunter-gatherer on the African plains, a trader sailing along the coast of the Mediterranean, or even a flight navigator in the early twentieth century. Whichever role you choose to assume, try to get out and actually look at the sky.

Q: WHAT IS THIS? See end of chapter for the answer

1.1 The Celestial Sphere

One of nature's spectacles is the night sky seen from a clear, dark location with the stars scattered across the vault of the heavens. Many of the patterns and motions of the stars have been all but forgotten in our hectic modern world, so our first goal is to familiarize ourselves with some general aspects of the sky at night.

The thousands of stars visible on a clear night are at vastly different distances from us. For example, the nearest star is about 4 light-years away, but others are more than a thousand times farther away. Such huge distances prevent us from getting any sense of their true three-dimensional arrangement in space. For purposes of naked-eye observations, we can therefore treat all stars as if they are at the same distance from the Earth, imagining that they lie on the inside of a gigantic dome that stretches overhead. Astronomers describe this dome as part of the **celestial sphere** and picture it as completely surrounding the Earth, with the Earth at its center. Thus, if you were suspended in space far from Earth, you would see the entire celestial sphere surrounding you, as depicted in figure 1.1.* However, if you are on the Earth and look around, you will see that the ground blocks your view of approximately half the celestial sphere. The line where the sky meets the ground, and below which your view of the celestial sphere is blocked, is called the **horizon** (fig 1.1B).

The celestial sphere has no physical reality, but it serves as a **model** of the heavens—a way of simplifying the arrangement and motions of celestial bodies so they are easier to visualize. We use the term *model* to mean a representation of some aspect of the Universe. That is, the celestial sphere represents a way of thinking about or viewing the location and motions of stars and planets. The celestial sphere is the first of many models we will encounter that humans have used to describe the known Universe. In later chapters, we will use models to enhance our understanding whenever the size or other properties of what we study fall outside the range of

*In figure 1.1 and in many other figures throughout the book, distances and sizes of astronomical bodies are exaggerated for clarity.

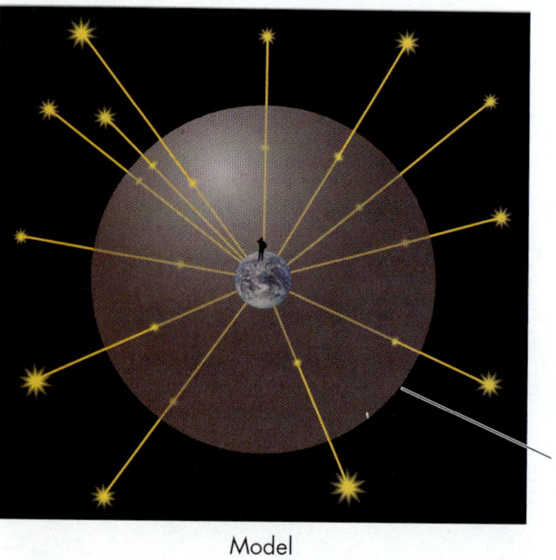
Stars are scattered throughout space in different directions and at different distances.
Stars *appear* to all lie at the *same* distance on what we call the *celestial sphere.*
Model
A

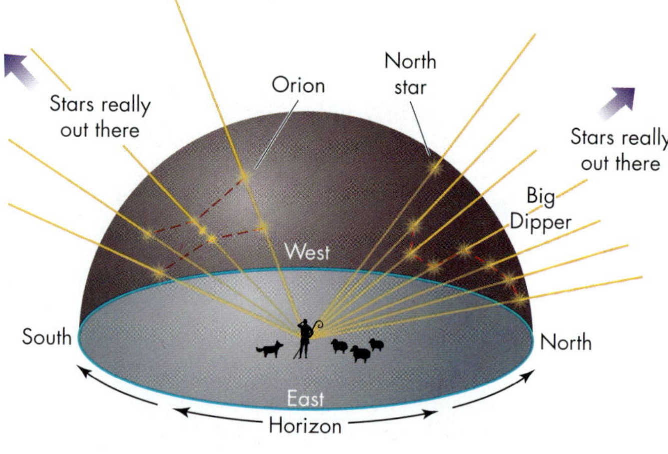
Stars really out there
Orion
North star
Stars really out there
Big Dipper
West
South
North
East
Horizon
Our Experience of the Celestial Sphere.
B

FIGURE 1.1
(A) Although in reality the stars are scattered through space at very different distances, they *appear* to lie at the same distance from us on what we call the celestial sphere. Note: Sizes and distances are drastically exaggerated. For Earth at the size shown, the *nearest* star would really be 6000 miles away. (B) The celestial sphere meets the ground at the horizon.

our everyday experience. Thus, we will speak of models of atoms, models of stars, and models of the Universe itself.

Constellations

As human beings, we seek order in what we look at. When ancient people looked at the night sky, they noticed that the stars form fixed patterns on the celestial sphere, what we today call **constellations.** Some of these constellations resemble animals if we use a little imagination. For example, the pattern of stars in Leo looks a little like a lion, whereas that of Cygnus looks like a swan in flight, as depicted in figure 1.2. However, you will discover, as you learn to identify the constellations, that many have shapes that bear little resemblance to their namesakes. Also, keep in mind that stars in a constellation generally have no physical relation to one another. They simply happen to be in more or less the same direction in the sky.

All stars move, but as seen from Earth, their positions change very slowly, taking tens of thousands of years to make any noticeable shift. Thus, we see today virtually the same pattern of stars that was seen by ancient peoples. A shepherd who lived 5000 years ago in the Middle East would have no trouble recognizing the star patterns of the night sky we see and might even call them by the same names.

We do not know how the constellation names were chosen, but most of them date back thousands of years. In fact, we don't know when the names were first given to the constellations or why, although there is some evidence they served as mnemonic devices for keeping track of the seasons and for navigating. For example, the beginning of the stormy winter months, when sailing was dangerous and ships were often wrecked, was foretold by the Sun's appearance in the constellations Pisces and Aquarius, the water constellations. Likewise, the harvest time was indicated by the Sun's appearance in Virgo, a constellation often depicted as the goddess Proserpine, holding a sheaf of grain.

Daily Motions of the Sun and Stars

Take a look at the night sky, and you will see stars rise along the eastern horizon, move across the sky, and set along the western horizon, exactly as the Sun does. You can verify this by watching the night sky for as little as 10 minutes. A star seen just above the eastern horizon will have risen noticeably higher, and stars near the western horizon

FIGURE 1.2
The two constellations Leo (A) and Cygnus (B) with figures sketched in to help you visualize the animals they represent.

will have sunk lower or disappeared. Likewise, if you look at a constellation, you see its stars rise as a fixed pattern in the eastern sky, move across the sky, and set in the western sky.

In terms of our model of the heavens based on the celestial sphere, we can visualize the rising and setting of stars as rotation of the celestial sphere around us (fig. 1.3). Ancient peoples would have found it far easier to believe in that rotation than to believe that the Earth moved. Thus, they attributed all celestial motion—that of the Sun, Moon, stars, and planets—to a vast sphere slowly turning overhead. Today, we still say the Sun *rises* and *sets*, but of course we know that it is the Earth's rotation that makes the Sun, Moon, and stars rise and move westward across the sky each day. It is not the celestial sphere that spins but the Earth.

If you look at the celestial sphere turning overhead, two points on it do not move, as you can see in figure 1.3. These points are defined as the north and south **celestial poles.** The celestial poles lie exactly above the North and South Poles of the Earth, and just as our planet turns about a line running from its North to South Poles, so the celestial sphere rotates around the celestial poles. Over the course of a night, stars appear to circle the north celestial pole in a counterclockwise direction for observers in the Earth's northern hemisphere.

Because it lies directly above the Earth's North Pole, the north celestial pole always marks the direction of true north. Near the position of the north celestial pole, there happens to be a moderately bright star, Polaris, which is therefore known as the North Star.* This is an important guide for travelers on land and sea and was widely used by early peoples for this purpose.

Another useful sky marker frequently used by astronomers is the **celestial equator.** The celestial equator lies directly above the Earth's equator, just as the celestial poles lie above the Earth's poles, as figure 1.3 shows.

Q. Stars appear to move counterclockwise around the north celestial pole. Looking from above the Earth's north pole, what direction does the Earth rotate, clockwise or counterclockwise?

*The direction of the Earth's axis gradually shifts or *precesses* over tens of thousands of years, so Polaris has not always been the North Star. This is discussed further in chapter 6.

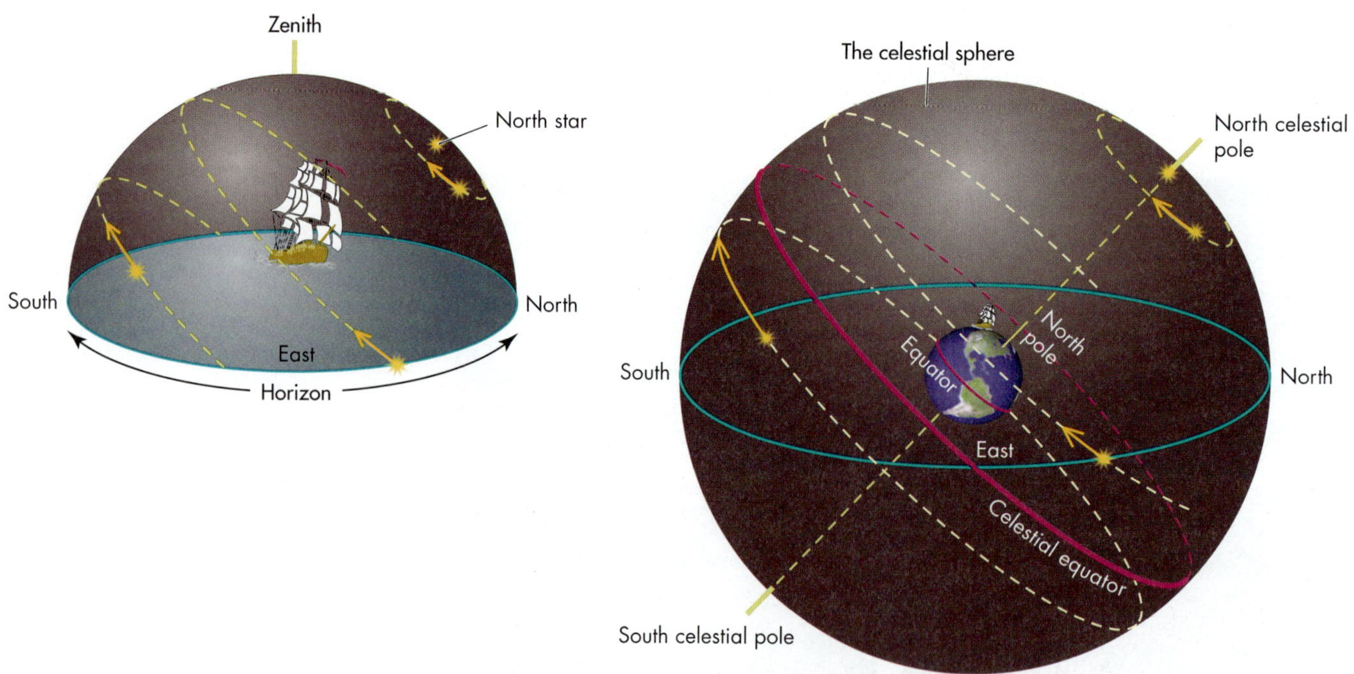

FIGURE 1.3
Stars appear to rise and set as the celestial sphere rotates overhead. Also shown are the celestial equator and poles and where they lie on the celestial sphere with respect to the Earth's equator and poles.

Annual Motion of the Sun

At the same time that the Earth's spin causes the apparent daily motion of the Sun and stars across the sky, the Earth's orbital motion around the Sun also causes changes in the sky. This motion is much slower and harder to observe without making records of star positions. Nevertheless, astronomers today think that many ancient peoples built monuments to keep track of such motions. Because these movements repeat on a yearly cycle, they are called *annual motions.*

In discussing the motion of astronomical objects, astronomers often find it helpful to use different terms to distinguish an object's spinning motion from its orbital motion. For example, we say that the Earth rotates on its axis (spins) at the same time as it revolves around the Sun (moves along its orbit). Likewise we use the words *rotation* and *revolution* to distinguish between spin motion and orbital motion.

If you compare the sky at the same time each evening for a few months, you will discover that different constellations are visible. For example, on an early July evening across most of North America, Europe, and Asia, the constellation Scorpius will be visible in the southern half of the sky. On December evenings, Scorpius is not visible. Instead, you will see the brilliant constellation Orion, the hunter. A year later, the same constellations will again be visible as they were originally.

The discovery that different stars are visible at different times of the year was extremely important to early people because it provided a way to measure the passage of time other than by carefully counting days. Moreover, the stars demonstrated that many celestial events are predictable and that they may be used to order our lives on Earth. For example, using the constellations just mentioned, if the evening sky shows Scorpius, it must be summer, and if it shows Orion, it must be winter. Knowing the exact season can be crucial for such things as planting crops. A brief warm spell might have tricked an ancient farmer into sowing seeds too early, but by studying the the sky for many years, she might have discovered that when the constellation Taurus is visible just before dawn, it is time to plant.

The change of the constellations with the seasons is caused by the Earth's motion around the Sun. As the Earth moves around the Sun, the Sun's glare blocks our view of the part of the celestial sphere that lies toward the Sun, making the stars that lie beyond the Sun invisible, as figure 1.4 shows. For example, in early June, a line from the Earth to the Sun points toward the constellation Taurus, and so its stars are lost in the Sun's glare. After sunset, however, we can see the neighboring constellation, Gemini, just above the western horizon. By early August, the Earth has moved to a new position in its orbit. At that time of year, the Sun lies in the direction of Cancer, causing this constellation to disappear in the Sun's glare. Looking to the west just after sunset, we now see Leo just above the horizon.

Month by month, the Sun covers one constellation after another. It is like sitting around a campfire and not being able to see the faces of the people on the far side. But if we get up and walk around the fire, we can see faces that were previously hidden. Similarly the Earth's motion allows us to see stars previously hidden in the Sun's glare. Our planet's motion also makes a given star rise 3 minutes and 56 seconds earlier each night. That 3 minutes and 56 seconds, when added up each night over an entire year, amounts to 24 hours. Thus a year later, when the Earth returns to the same spot in its orbit, the sky looks the same again.

The Ecliptic and the Zodiac

If we could mark on the celestial sphere the path traced by the Sun as it moves through the constellations, we would see a line that runs around the celestial sphere, as illustrated in figure 1.4. Astronomers call the line that the Sun traces across the celestial sphere the **ecliptic.** The name *ecliptic* arises because only when the new or full moon

Evening
August twilight – looking westward
Leo
Sun

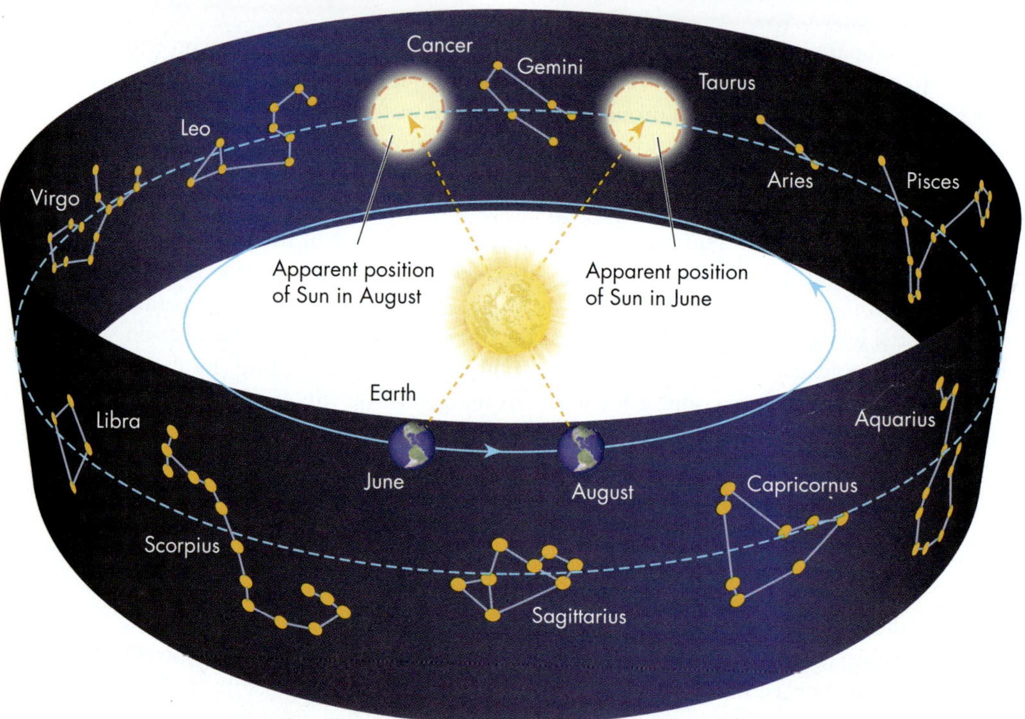

FIGURE 1.4
As the Earth orbits the Sun, the Sun appears to move around the celestial sphere through the background stars. The Sun's path is called the ecliptic. The Sun appears to lie in Taurus in June, in Cancer during August, in Virgo during October, and so forth. Therefore the constellations we see after sunset change with the seasons. Note that the ecliptic is the extension of the Earth's orbital plane out to the celestial sphere. (Sizes and distances of objects are not to scale.)

crosses this line can an eclipse occur, as you will learn later in this chapter. If you look at figure 1.4, you can see that the ecliptic is the extension of the Earth's orbit onto the celestial sphere, just as the celestial equator is the extension of the Earth's equator onto the celestial sphere.

The belt-shaped region of the sky surrounding the ecliptic passes through twelve constellations and is called the **zodiac.** The word *zodiac* is from the Greek *zoidion,* "little animal," leading to "animal sign." That is, zodiac refers to a circle of animals, which for the most part its constellations represent. The names of these constellations are Aries, Taurus, Gemini, Cancer, Leo, Virgo, Libra, Scorpius, Sagittarius, Capricornus, Aquarius, and Pisces.*

It is not easy to track the Sun's position relative to the stars because the Sun's glare makes it impossible to see the stars. However, if you look just before sunrise or just after sunset, you can identify the stars near the Sun on any given date. By watching over several weeks, you will be able to see that the Sun changes its position with respect to the stars.

*The zodiac also passes through part of the constellation Ophiuchus, the Serpent Bearer. However, this is not one of the traditional constellations of the zodiac.

Another complication we will discover when trying to track the Sun relative to the stars is that the Sun's path does not run parallel to the daily motion of the stars. Instead, the path is tipped with respect to the celestial equator. As a result, the Sun lies north of the celestial equator for half the year and south of the celestial equator for the other half of the year. Why is the Sun's path tipped in this way? To understand, it will help if we first discuss another change that happens with a yearly cycle: the seasons.

1.2 The Seasons

Many people mistakenly believe that we have seasons because the Earth's orbit is elliptical. They suppose that summer occurs when we are closest to the Sun and winter when we are farthest away. It turns out, however, that the Earth is nearest the Sun in early January, when the Northern Hemisphere is coldest. Clearly, then, seasons must have some other cause.

To see what *does* cause seasons, we need to look at how our planet is oriented in space. As the Earth orbits the Sun, our planet also spins. That spin is around an imaginary line—the **rotation axis**—that runs through the Earth from its North Pole to its South Pole. The Earth's rotation axis is *not* perpendicular to its orbit around the Sun. Rather, it is tipped by 23.5° from the vertical, as shown in figure 1.5A.

As our planet moves along its orbit, its rotation axis maintains nearly exactly the same tilt and direction, as figure 1.5B shows. That is, the Earth behaves much like a giant spinning top. This tendency of the Earth to preserve its tilt is shared by all spinning objects. For example, it is what keeps a rolling coin upright and a thrown football pointed properly (fig. 1.6). Moreover, you can easily feel this tendency of a spinning object to resist changes in its orientation by lifting a spinning bicycle wheel off the ground and trying to twist it from side to side.

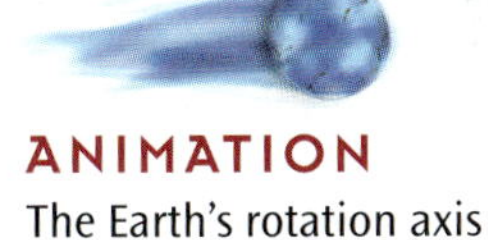

ANIMATION
The Earth's rotation axis

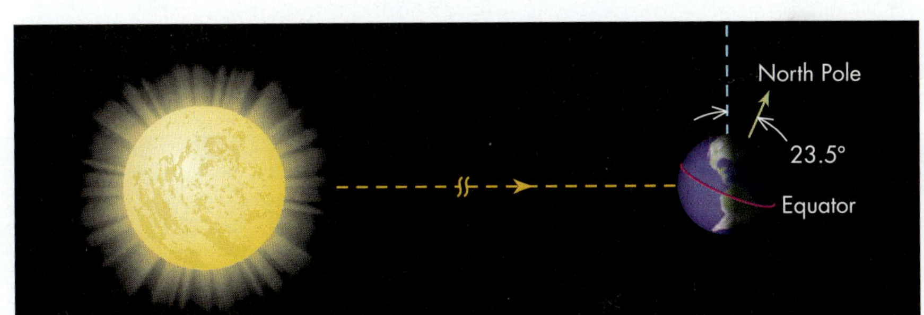

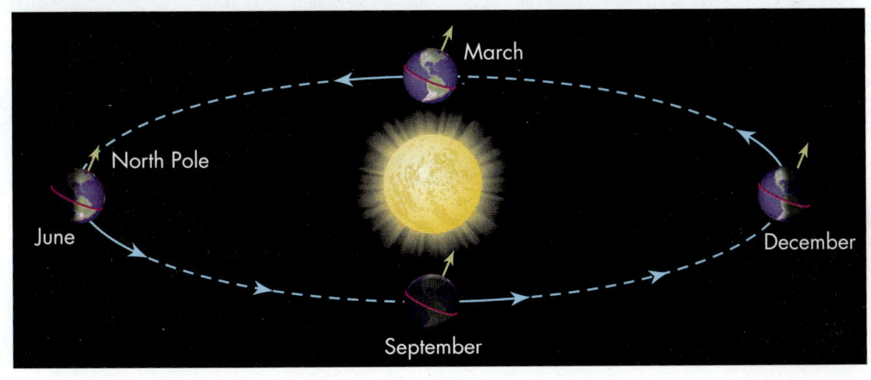

FIGURE 1.5
The Earth's rotation axis keeps nearly the same tilt and direction as it moves around the Sun. The rotation axis is tilted 23.5° to the Earth's orbit around the Sun. (Sizes and distances are not to scale.)

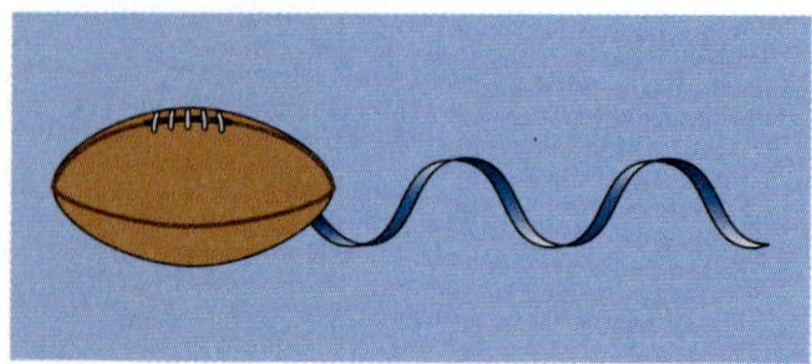

FIGURE 1.6
The tendency of a spinning object to keep its orientation is called "conservation of angular momentum," and it is the principle on which gyroscopic compasses operate and the reason a quarterback puts "spin" on a football.

Because the Earth's tilt remains nearly constant as we move around the Sun, sunlight falls more directly on the Northern Hemisphere for part of the year and more directly on the Southern Hemisphere for another part of the year, as can be seen in figure 1.7. This changes the amount of heat we receive from the Sun over the course of a year.

A flat surface area presented directly to the Sun intercepts a larger portion of the Sun's radiation, and hence its heat, than the same surface area when tilted, as figure 1.8 illustrates. You take advantage of this effect instinctively when you warm your hands at a fire by holding your palms flat toward the fire, not edgewise. Similarly, the Northern Hemisphere receives its greatest heating at the time of year when the Sun shines most directly on it, making it summer. Six months later, the Northern Hemisphere receives its sunlight least directly, and so it is colder and therefore winter. This heating difference is enhanced because the Earth's tilt leads to many more hours of daylight in the summer than in winter. As a result, not only do we receive the Sun's light more directly, we receive it for a longer time. Thus,

INTERACTIVE
Seasons

the seasons are caused by the tilt of the Earth's rotation axis.

Figure 1.8 also illustrates that this makes the seasons reversed between the Northern and Southern Hemispheres; when it is summer in one, it is winter in the other.

The Ecliptic's Tilt

The tilt of the Earth's rotation axis not only causes seasons, it also is why the Sun's path across the celestial sphere—the ecliptic—is tilted with respect to the celestial equator.

FIGURE 1.7
Because the Earth's rotation axis keeps the same tilt as we orbit the Sun, sunlight falls more directly on the Northern Hemisphere during part of the year and on the Southern Hemisphere during the other part of the year. (Sizes and distances are not to scale.)

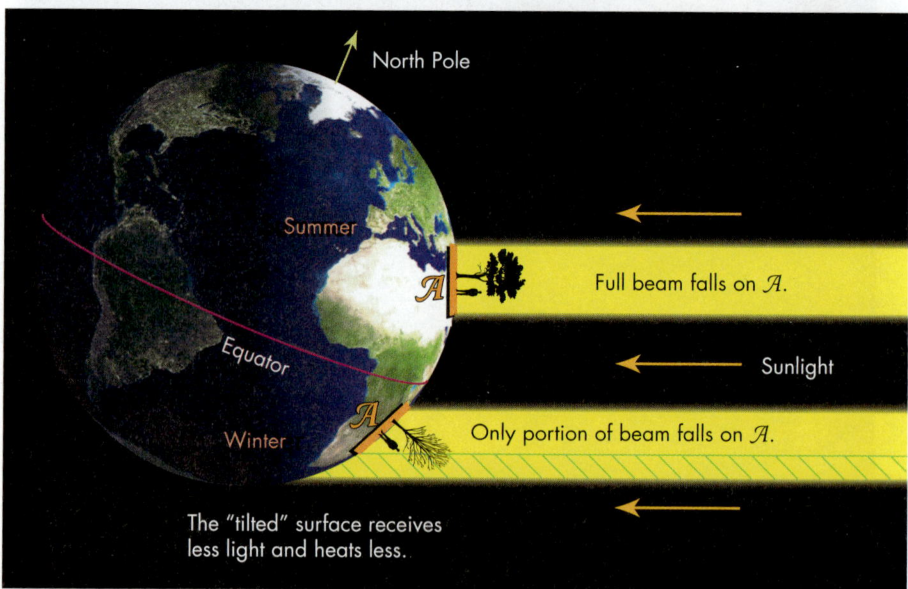

FIGURE 1.8
A portion of the Earth's surface directly facing the Sun receives more concentrated light (and thus more heat) than other parts of the Earth's surface of equal area. The same size "beam" of sunlight (carrying the same amount of energy) spreads out over a larger area where the surface is "tilted."

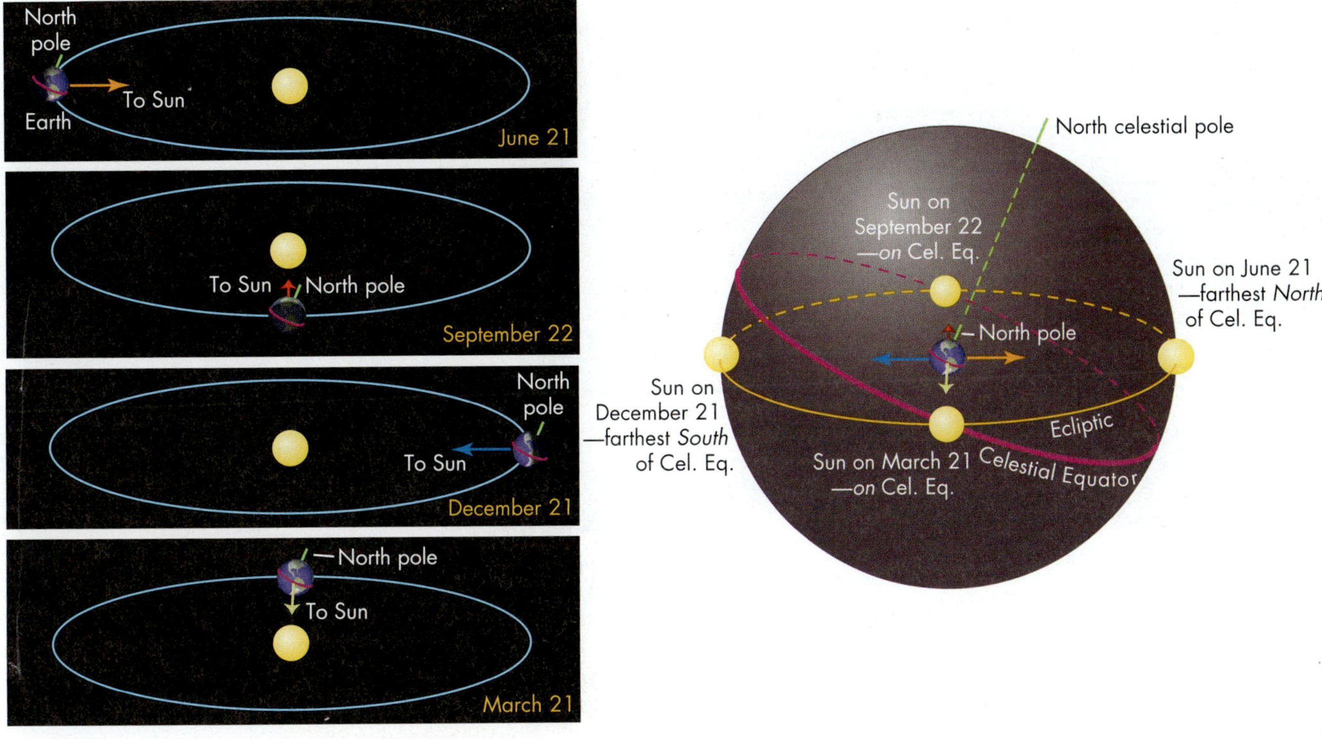

FIGURE 1.9
As the Earth orbits the Sun, the Sun's position with respect to the celestial equator changes. The Sun reaches 23.5° north of the celestial equator on June 21 but 23.5° south of the celestial equator on December 21. The Sun crosses the celestial equator on about March 20 and September 22 each year. The times when the Sun reaches its extremes are known as the solstices; the times when it crosses the celestial equator are the equinoxes. (The dates can sometimes vary because of the extra day inserted in leap years.)

Because the Earth's axis remains oriented in a fixed direction, there is a point in its orbit when the North Pole is tipped most closely toward the Sun. This occurs on about June 21, as illustrated in figure 1.9. On this date, the North Pole is tilted 23.5° toward the Sun, so the Sun lies 23.5° north of the celestial equator. Half a year later, around December 21, the Earth is on the other side of the Sun, and the Sun lies 23.5° south of the celestial equator.

As a result of this north–south motion, the Sun's path—the ecliptic—crosses the celestial equator, and therefore the ecliptic must be tilted with respect to that line, as the sequence of sketches in figure 1.9 shows. The only dates when the Sun lies on the celestial equator are around March 20 and September 22.

The motion of the Sun north and south in the sky over the course of the year is also why the Sun is so high in the summer sky at noon and so low in the winter sky at noon (fig. 1.10). For example, on June 21 at a midnorthern latitude of 40°, the noon Sun is about 73.5° above the horizon. On December 21, on the other hand, it is only about 26.5° above the horizon.

ANIMATION
The Sun's motion north and south in the sky as the seasons change

Solstices and Equinoxes

The dates when the Sun reaches its extremes north and south are used to mark the beginning of summer and of winter, while the dates when the Sun crosses the celestial equator mark the beginning of spring and of autumn. Astronomers give these dates special names. When the Sun is on the celestial equator, the days and nights are of equal length (approximately), so these dates are called the

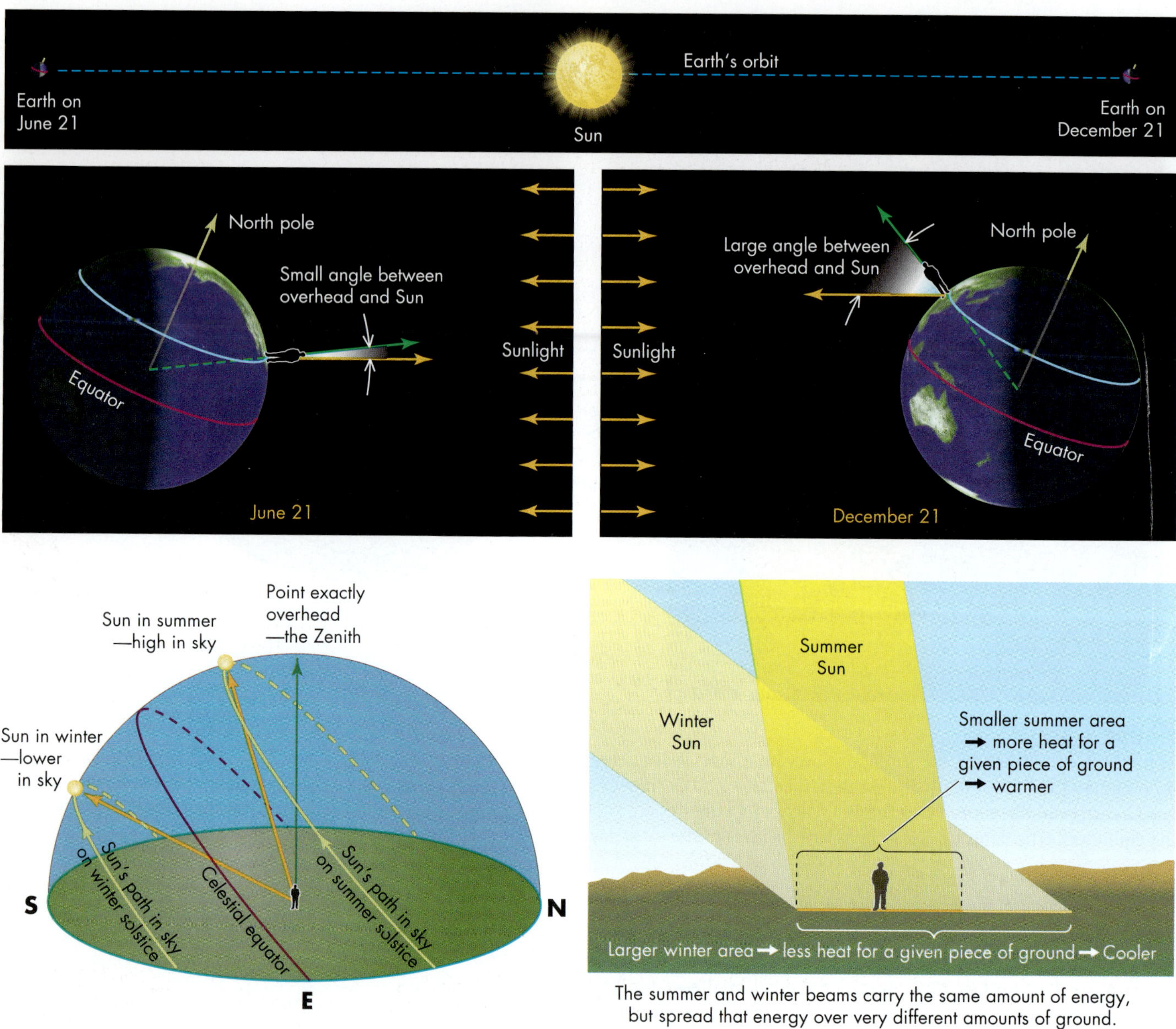

FIGURE 1.10
Why the Sun at noon is high in the sky in summer and low in the sky in winter.

equinoxes, for "equal nights." The spring or vernal equinox occurs near March 20, while the fall or autumnal equinox occurs near September 22. (These dates vary a little from year to year, mostly because the year is about one quarter day longer than a whole number of days—which is also what causes us to insert leap years.)

The beginning of summer and of winter mark the times of year when the Sun pauses in its north–south motion and changes direction. Accordingly, these times are called the **solstices,** meaning the Sun (sol) has stopped its motion north or south and is *static* and about to reverse direction. The dates of the solstices (summer and winter) also change slightly from one year to the next, but they are always close to June 21 and December 21.

Because the Sun moves north and south of the celestial equator during the year, the Sun generally does *not* rise due east or set due west. Rather, over a year, the direction to the rising and setting position of the Sun constantly changes.

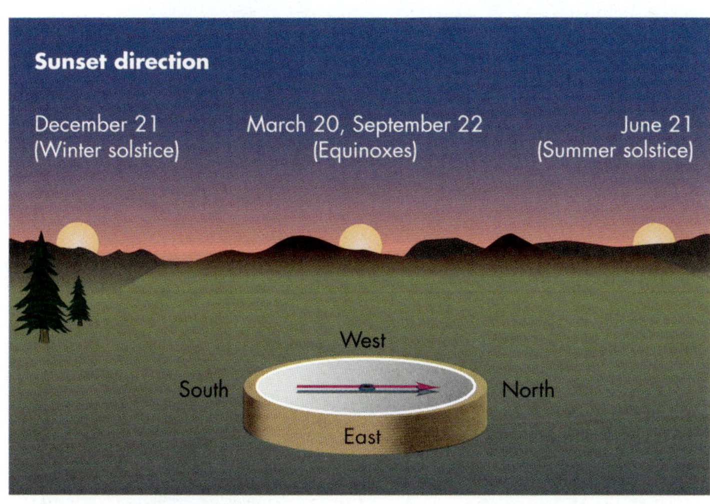

A

B

FIGURE 1.11
(A) The direction of the rising and setting Sun changes throughout the year. Only on the equinoxes are the rising and setting points due east and due west. The sunset direction shifts slowly northward from the winter solstice to the summer solstice, after which it shifts back, moving southward until the winter solstice. (B) A pair of photographs taken eight days apart from the same location. The pictures were taken just before sunset, close to the date of the autumnal equinox. The sunset position changed by more than 4° during this time—just about the size of the Sun's apparent diameter each day. The width of the outstretched thumb in the bottom picture helps to indicate a scale of about 2°.

When the Sun is on the celestial equator, on the vernal equinox, it rises due east and sets due west. From this date up to the summer solstice, the Sun's rising and setting points shift a little farther northward each day, as illustrated in figure 1.11A. From the summer solstice to the winter solstice, the position shifts southward each day, rising and setting due east and due west again on the autumnal equinox. After the winter solstice the southward drift reverses, and the Sun begins to move north again. The shift of the Sun's position is particularly obvious near the equinoxes, when the Sun's position on the horizon shifts by almost its own diameter each day (figure 1.11B).

Just as the changing position of the Sun against the constellations could be used as an indicator of the seasons, so too could the position of the rising and setting Sun. One well-known example is Stonehenge, the ancient stone circle in England. Although we do not know for certain how this ancient monument was used, it was laid out so that such seasonal changes in the Sun's position could be observed by noting through which stone arches the Sun was visible when it rose or set. For example, the illustration that opens this chapter (page 14) shows a photograph taken on the summer solstice at sunrise. On this day only, an observer standing at the center of this circle of immense standing stones would see the rising Sun framed by an arch, as illustrated in Figure 1.12A. Similarly, some ancient Egyptian temples and pyramids have astronomical alignments, such as the Temple of Amen-Ra at Karnak, whose main chamber points toward the position of sunrise at the winter solstice.

Astronomical alignment of buildings occurs in many other places as well. For example, in Chankillo, Peru, a series of towers was built on a ridge about 2300 years ago. From an ancient observatory at the base of the ridge, the towers span the shift on the horizon of the rising Sun (Figure 1.12B). The Maya, native peoples of Central America, and their neighbors built pyramids from the summits of which they could get a clear view of the sky over the surrounding rain forest. The pyramid at Chichén Itzá was

> Although the seasons begin on the solstices and equinoxes, the hottest and coldest times of year occur roughly 6 weeks after the solstices. The delay, known as the lag of the seasons, results from the oceans and land being slow to warm up in summer and slow to cool down in winter.

A

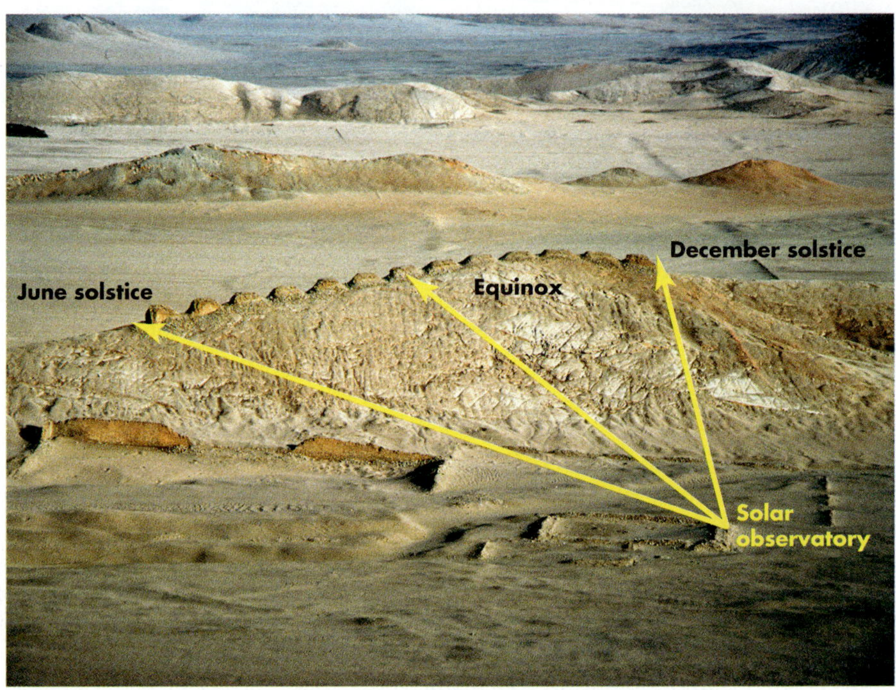

B

FIGURE 1.12
(A) Stonehenge, built more than 4000 years ago on the Salisbury plain in Britain. The enormous stones are arranged to frame various positions of the Sun on the horizon, helping to mark dates such as when the Sun reaches its point farthest north on the summer solstice. (B) The oldest known astronomical observatory in the Americas is found in Chankillo, Peru. This ancient observatory marked the shifting position of sunrise with a series of 13 towers built along a ridge about 2300 years ago.

specially designed so that sunlight would create the image of a snake slithering down the steps on the equinoxes.

Many cultures also built monuments that appear to have been used to track another important celestial body: the Moon. Like the Sun, the Moon shifts relative to the stars, and its cyclic changes formed the basis for calendar systems around the world. Some archaeo-astronomers claim that sites such as Stonehenge were used to track the moonrises and moonsets and perhaps even used to predict eclipses.

1.3 The Moon

Like all celestial objects, the Moon rises in the east and sets in the west. Also, like the Sun, the Moon shifts its position across the background stars from west to east. You can verify this motion by observing the Moon at the same time each evening and noticing its change in position with respect to nearby stars. In fact, if the Moon happens to lie close to a bright star, its motion may be seen in only a few minutes, because in 1 hour the Moon moves a distance, on the sky, that is approximately equal to its own apparent diameter.

Perhaps the most striking feature of the Moon is that, unlike the Sun, its shape seems to change throughout the month in what is called the "cycle of lunar **phases.**" That is, the Moon appears alternately as a thin crescent, a fully illuminated disk, and then as a crescent again. During a period of approximately 29.5 days, the Moon goes through a complete set of phases from invisibility (new) to fully lit (full) and back to invisibility. This is the origin of the month as a time period and also the source of the name "month," which was derived from the word *moon.*

INTERACTIVE
Lunar phases

The cycle of the phases and the Moon's changing position against the stars are caused by the Moon's orbital motion around the Earth. Like the Sun and the planets, the Moon moves through the constellations of the zodiac. In other words, its orbit is close to the orbital plane of the Earth around the Sun.

A new moon occurs when the Moon lies approximately between us and the Sun. A full moon occurs when the Moon is on the other side of the Earth from the Sun, opposite it in the sky. Many people believe these changes in shape are caused by the Earth's shadow falling on the Moon. However, that is *not* the explanation, as you can deduce from the fact that crescent phases occur when the Moon and Sun lie approximately in the same direction in the sky and the Earth's shadow must therefore point *away* from the Moon. In fact, we see the Moon's shape change because, as it moves around us, we see different amounts of its illuminated half. For example, when the Moon lies approximately opposite the Sun in the sky, the side of the Moon toward the Earth is fully lit. On the other hand, when the Moon lies approximately between us and the Sun, its fully lit side is turned nearly completely away from us, and therefore we glimpse only a sliver of its illuminated side, as illustrated in figure 1.13. This figure also shows the cycle of the lunar phases and the names used to describe the Moon's appearance.

If you watch the Moon go through a cycle of its phases, you will notice a change in the times at which it is visible. For example, shortly after the new phase, you can see the Moon low in the western sky after sunset. But a few hours later that same night it will have set and become invisible. On the other hand, when the Moon is full, it rises at about sunset and doesn't set until dawn. Thus, the full moon is visible throughout the night. At some of its phases, you can see the Moon even during the day, if you know where to look. The different times when the Moon is visible are explored further in the Extending Our Reach box: "Observing the Moon."

The Moon's motion around the Earth has other effects as well. For example, because the Moon shifts eastward through the stars, the Earth itself must rotate eastward a little extra each day to bring the Moon above the horizon. This extra rotation takes about 50 minutes each day, on average. So if the Moon rises at 8 p.m. one evening, the next evening it will rise about 8:50 p.m., the following night about 9:40 p.m., and so forth.

One further consideration is that as the Moon revolves around the Earth, the Earth is revolving around the

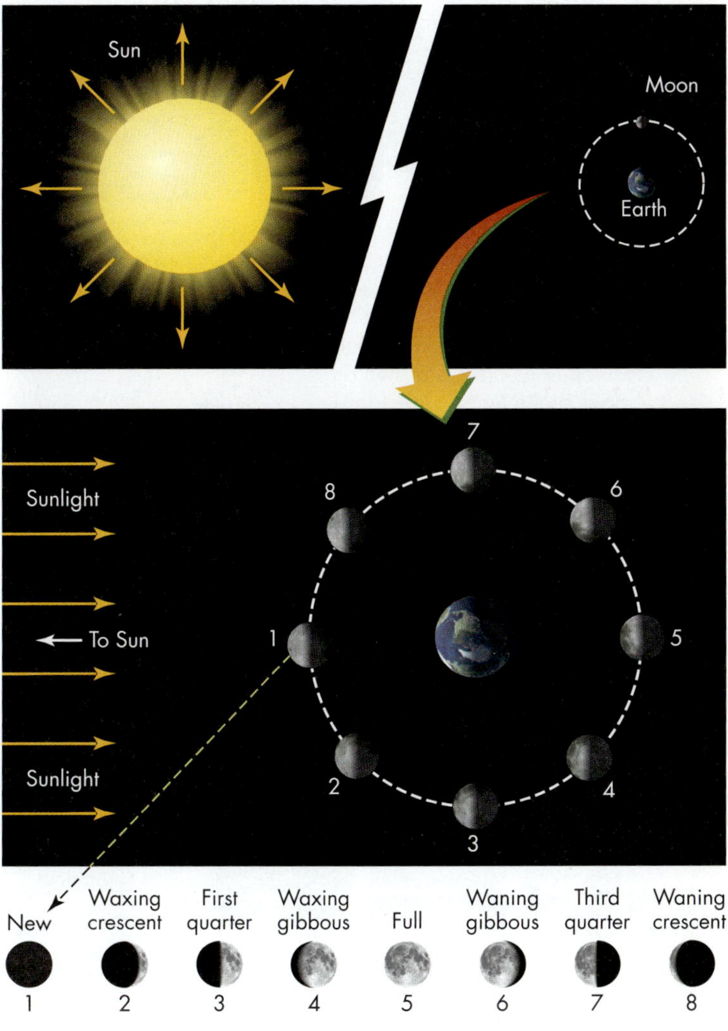

FIGURE 1.13
The cycle of the phases of the Moon, from new to full and back again. The phases are caused by our seeing different amounts of the half of the Moon's surface that is illuminated by the Sun. Sizes and distances of objects are *not* to scale.

EXTENDING *our reach*

OBSERVING THE MOON

When the Moon is full, it lies approximately opposite to where the Sun lies, but when the Moon is a thin crescent, it lies in nearly the same direction as the Sun (see the middle of Box figure 1.1). These connections between the Moon's phase and its position with respect to the Sun are the key to understanding when the Moon is visible from Earth.

Because the full Moon is approximately opposite the Sun, it *rises* above the eastern horizon at about the same time that the Sun *sets* below the western horizon. Likewise, the full Moon *sets* at about the time the Sun *rises*. Thus, the full Moon is visible all night and highest in the sky near midnight.

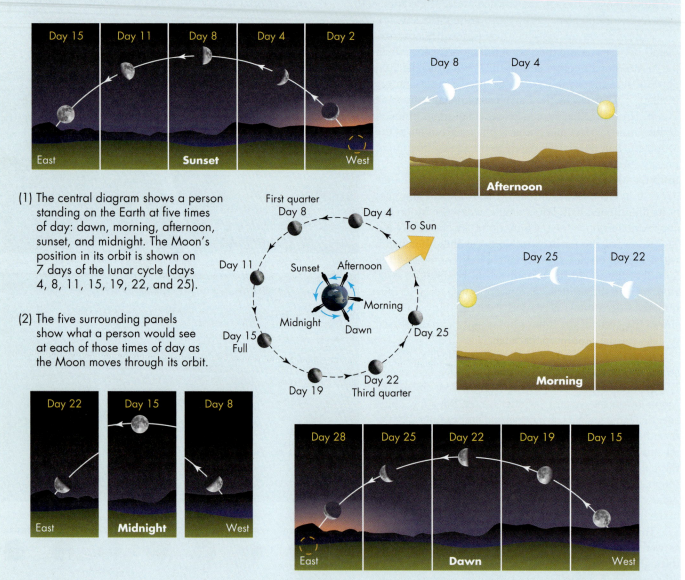

(1) The central diagram shows a person standing on the Earth at five times of day: dawn, morning, afternoon, sunset, and midnight. The Moon's position in its orbit is shown on 7 days of the lunar cycle (days 4, 8, 11, 15, 19, 22, and 25).

(2) The five surrounding panels show what a person would see at each of those times of day as the Moon moves through its orbit.

BOX FIGURE 1.1
When the Moon Is Visible. This figure shows where to look for the Moon and how it appears at different times of day as it goes through its monthly cycle of phases. The central sketch shows the Earth and Moon and where the Moon is in its orbit at different phases, indicated here by the number of days since the new Moon. The yellow arrow points toward the Sun.

On the other hand, the crescent moon is not visible during most of the night. Because it lies in nearly the same direction as the Sun, once the Sun is well below the horizon, the crescent Moon must be below the horizon too. Moreover, the crescent Moon is hard to see during the day because it is only a sliver of light, so it is lost in the brightness of the daytime sky. Therefore, when the Moon is a few days past its new phase and is a thin crescent, you can see it low in the western sky at sunset. This crescent moon will set shortly after the Sun and not be visible again until after sunrise the next day.

As a further example, let's look at the Moon when it is at first quarter (half lit). At this phase (as you can see in Box figure 1.1), it is about one-quarter of the way around the sky and eastward of the Sun by about 90° on the sky. Thus the Earth must turn about an additional 90° to bring the Moon above the horizon. How long will it take the Earth to rotate those extra 90°? We can easily figure that out by noting that it takes the Earth 24 hours to rotate once (360°). Thus, to rotate 90° degrees (one-quarter of 360°) takes one-quarter of 24 hours, or 6 hours. Thus, the first-quarter Moon rises about 6 hours after the Sun rises. Moreover, because it is about 90° from the Sun, it will be at its highest point in the sky at sunset and by midnight it will have moved to the western horizon and be setting.

With similar arguments, you can figure out when and where the Moon is easiest to see in other phases, as illustrated in Box figure 1.1. Note that at a given time, the Moon lies a little farther east in the sky each night and also that its phase changes. This shift to the east from night to night is simply the result of the Moon's orbital motion around the Earth. Also note that because the Moon orbits approximately in the plane of the ecliptic, it shifts north and south of the celestial equator, just as the Sun does. As a result, the full Moon behaves opposite to the Sun, only reaching a relatively low height in the sky in the summer and high in the sky in winter. This also affects the moonrise and moonset times, depending upon how far north or south of the Sun the Moon is on the celestial sphere.

Sun. We saw that the Moon takes about 29.5 days to go through its cycle of phases. However, the Moon requires only 27.3 days to complete its motion through the constellations of the zodiac. This is illustrated in figure 1.14, where you can see that because the Earth has shifted its position in orbit, the Sun is in a different direction after a month has passed. As a result, after the Moon comes back into alignment with distant stars, it must still travel a little farther around its orbit to come back into alignment with the Sun. Alignments with the Sun can result in some of the most beautiful and dramatic effect of the Moon's motion: eclipses.

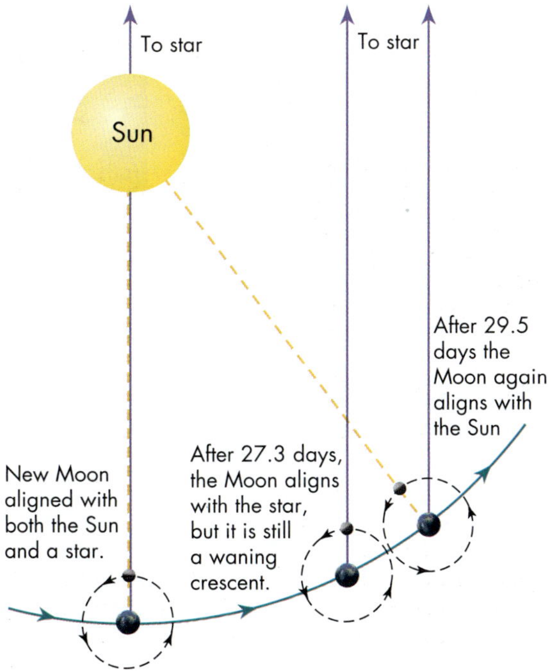

FIGURE 1.14
The sidereal month is the time the Moon takes to complete an orbit relative to the distant stars. This is about 27.3 days, less than the lunar month because as the Moon is orbiting the Earth, the Earth is orbiting the Sun. It takes about two additional days for the Moon to come back in alignment with the Sun.

1.4 Eclipses

INTERACTIVE
Lunar phases

An eclipse occurs when the Moon happens to lie exactly between the Earth and the Sun, or when the Earth lies exactly between the Sun and the Moon, so that all three bodies lie on a straight line. Thus, there are two types of eclipses: solar and lunar.

A **solar eclipse** occurs whenever the Moon passes directly between the Sun and the Earth and blocks our view of the Sun, as depicted in figure 1.15. A **lunar eclipse** occurs when the Earth passes between the Sun and the Moon and casts its shadow on the Moon, as shown in figure 1.16.

Appearance of Eclipses

Eclipses take a few hours to go through all of their stages. Sometimes an eclipse is *partial*, with only a portion of the Sun or the Moon ever being covered over. These partial eclipses sometimes pass unnoticed unless you know to look for them. However, *total* eclipses are beautiful and marvelous events.

During a total lunar eclipse, the Earth's shadow gradually spreads across the full Moon's face, cutting an ever deeper dark semicircle out of it. The shadow takes about an hour to completely cover the Moon and produce totality. At totality, the Moon generally appears a deep ruddy color, almost as if dipped in blood. Sometimes it even disappears. After totality, the Moon again becomes lit, bit by bit, reverting to its unsullied, silvery light.

A little light falls on the Moon even at totality because the Earth's atmosphere bends some sunlight into the shadow, as shown in figure 1.17. The light reaching the Moon is red because interactions with air molecules remove the blue light as it passes through our atmosphere and is bent, exactly as happens when we see the setting Sun.

Total lunar eclipses are visible if you are anywhere on the night side of the Earth when the eclipse is occurring. It is far rarer to see a total solar eclipse because the Moon's shadow on the Earth is quite small. In fact, you are unlikely to ever see a total solar eclipse in your lifetime unless you travel to see it, because on average they occur in any location only once every several centuries.

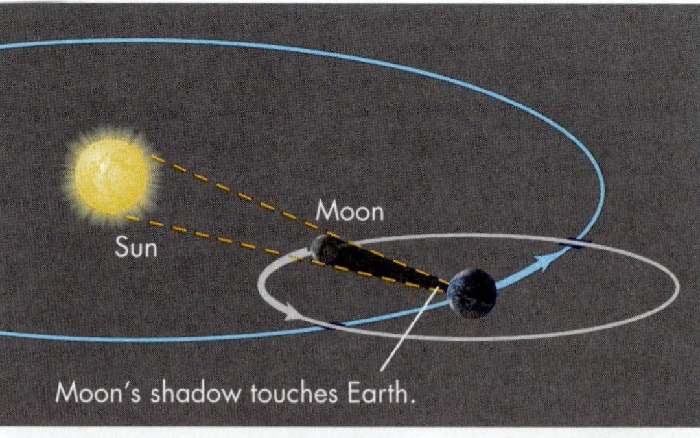

FIGURE 1.15
A solar eclipse occurs when the Moon passes between the Sun and the Earth so that the Moon's shadow touches the Earth. The photo inset shows what the eclipse looks like from Earth.

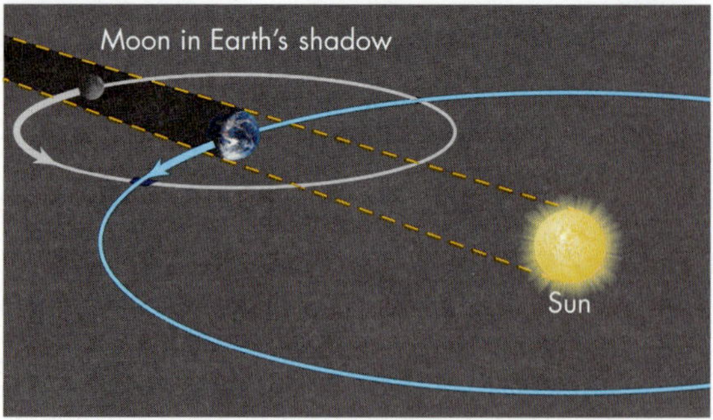

FIGURE 1.16
A lunar eclipse occurs when the Earth passes between the Sun and Moon, causing the Earth's shadow to fall on the Moon. Some sunlight leaks through the Earth's atmosphere, casting a deep reddish light on the Moon. The photo inset shows what the eclipse looks like from Earth.

FIGURE 1.17
As sunlight falls on the Earth, some passes through the Earth's atmosphere and is slightly bent so that it ends up in the Earth's shadow. In its passage through our atmosphere, most of the blue light is removed, leaving only the red. That red light then falls on the Moon, giving it its ruddy color at totality.

Q. Sometimes you see clouds after sunset that are lit red. How is this like the red color you see on the totally eclipsed Moon?

A total solar eclipse begins with a small black "bite" taken out of the Sun's edge as the Moon cuts across its disk. Over the next hour or so, the Moon gradually covers over more and more of the Sun.While the Sun is only partially covered, you must be careful when viewing it, so you don't hurt your eyes. If you are fortunate enough to be at a location where the eclipse is total, you will see one of the most amazing sights in nature.

As the time when the Moon's disk completely covers the Sun (totality) approaches, the landscape takes on an eerie light. Shadows become incredibly sharp and black: even individual hairs on your head cast crisp shadows. Sunlight filtering through leaves creates tiny bright crescents on the ground. Seconds before totality, pale ripples of light sweep across the ground, and to the west the deep purple shadow of the Moon hurtles toward you at more than 1000 miles an hour. In one heartbeat you are plunged into darkness. Overhead, the sky is black, and stars become visible. The corona of the Sun—its outer atmosphere—gleams with a steely light around the Moon's black disk. Perhaps a solar prominence—a tiny, glowing, red flamelike cloud in the Sun's atmosphere—may protrude beyond the Moon's black disk (fig. 1.18). Birds call as if it were evening. A deep chill descends, because for a few minutes the Sun's warmth is blocked by the Moon. The horizon takes on sunset colors: the deep blue of twilight with perhaps a distant cloud in our atmosphere glowing orange. As the Moon continues in its orbit, it uncovers the Sun, and instantly it is daylight again. Now the cycle continues in reverse. If you ever have the chance to see a total eclipse, do it!

Be extremely careful when watching a partial solar eclipse. Looking at the Sun through improper filters will blind you. A safer way is to *not* look directly at the Sun but to use eyepiece projection to view the Sun. Hold a piece of paper about a foot from the eyepiece of a small telescope (or even binoculars), and a large image of the Sun will be visible on it. This method also allows many people to watch the eclipse simultaneously.

FIGURE 1.18
Pictures of a total solar eclipse. (A) The landscape is eerily lit during a total solar eclipse. The dark color of the sky is the Moon's shadow. (B) The bright halo of light is the Sun's corona, its outer atmosphere. The red protrusions are prominences, clouds of glowing gas in the Sun's upper atmosphere.

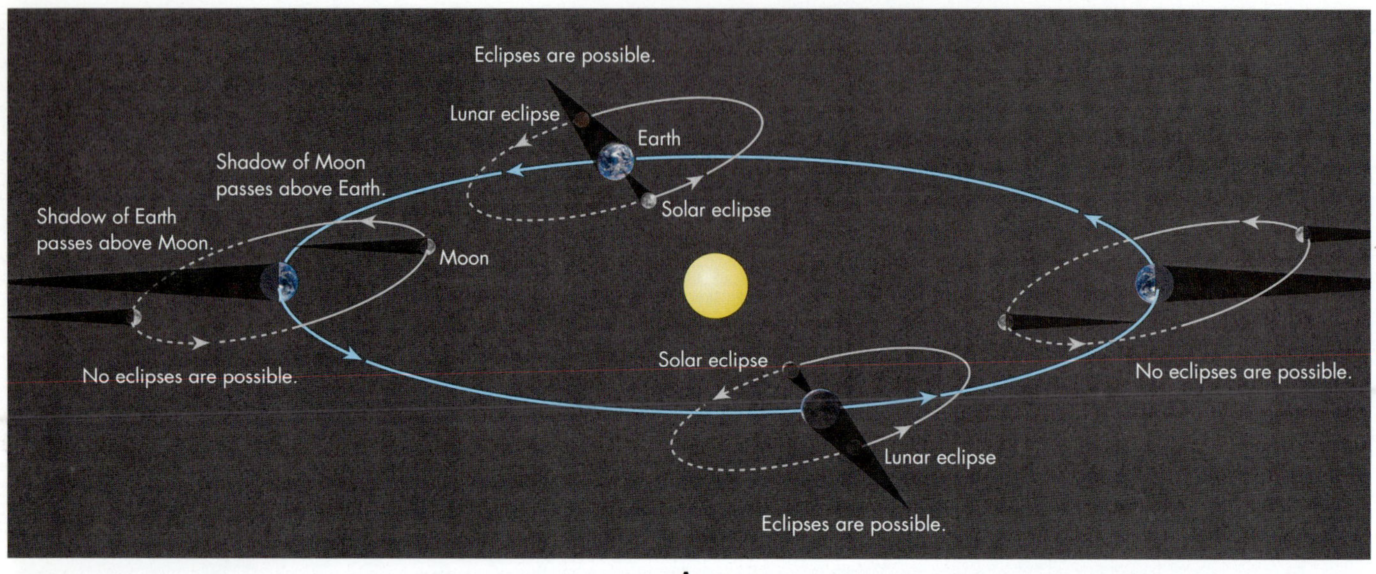

A

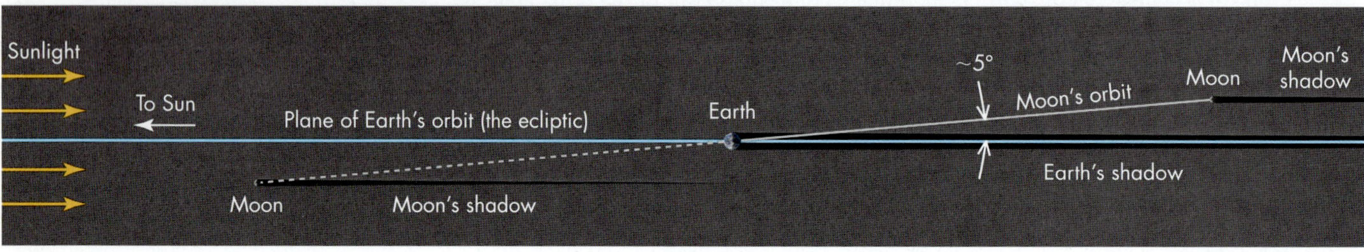

B

FIGURE 1.19
(A) The Moon's orbit keeps approximately the same orientation as the Earth orbits the Sun. Because of its orbital tilt, the Moon generally is either above or below the Earth's orbit. Thus, the Moon's shadow rarely hits the Earth, and the Earth's shadow rarely hits the Moon. Eclipse seasons are when the Earth is in either of two places in its orbit, about 6 months apart, when the Moon's orbital plane, if extended, intersects the Sun. (B) The The Earth and Moon are drawn to correct relative size and separation with their orbits seen here edge on. Note how thin their shadows are.

ANIMATION
Eclipses and the Moon's orbital inclination

Rarity of Eclipses

Given that the lunar cycle is about 29.5 days, you may wonder why we do not have eclipses every month. The answer is that the Moon's orbit is tipped with respect to the Earth's orbit (fig. 1.19). Because of this tip, even if the Moon is new, the Moon's shadow may pass above or below Earth, as you can see in figure 1.19A. As a result, no eclipse occurs. Similarly, when the Moon is full, the Earth's shadow may pass above or below the Moon so that again no eclipse occurs. Only a nearly exact alignment of the Earth, Moon, and Sun leads to eclipses, a point that is easier to appreciate if you look at figure 1.19B, which shows the Earth and Moon drawn to scale.

The great beauty of eclipses and their rarity make them eagerly awaited, and table 1.1 lists some upcoming ones. Total solar eclipses can only be seen within a narrow path where the Moon's shadow crosses the Earth. The approximate locations of these paths are shown for total eclipses from 2001 to 2025 in figure 1.20.

The tilt of the Moon remains fixed—like that of the spinning Earth—by a gyroscopic effect or, more technically, by the conservation of angular momentum. The result is that twice each year, the Moon's orbital plane (if extended) passes through the Sun, as shown in figure 1.21A. At those times—**eclipse seasons**—eclipses will happen when the Moon crosses the Earth's orbital plane, the ecliptic. Table 1.1 shows that in 2010, the eclipse seasons are within about two weeks of the start of January and July. Only at those times can eclipses happen: at other times, the shadows of the Earth and Moon always fall on empty space. You can also see from figure 1.21A that when a solar eclipse occurs

TABLE 1.1 SOME UPCOMING SOLAR AND LUNAR ECLIPSES

Solar Eclipses			Lunar Eclipses		
2010 January 15	Annular†	Africa, India, China	2009 December 31	Partial*	Europe, Africa, Asia, Australia
2010 July 11	Total	s. Pacific, Chile, Argentina	2010 June 26	Partial	e. Asia, Australia, w. Americas
2012 May 20	Annular	China, Japan, w. U.S.	2010 December 21	Total	e. Asia, Australia, Americas, Europe
2012 November 13	Total	n. Australia, s. Pacific	2011 June 15	Total	S. America, Europe, Africa, Asia, Aus.
2013 May 10	Annular	n. Australia, Pacific	2011 December 10	Total	Europe, e. Africa, Asia, Aus., N. A.
2013 November 3	Total/Annular	Atlantic, central Africa	2012 June 4	Partial	e. Asia, Aus., Americas, Europe
2014 April 29	Annular	Antarctica	2013 April 25	Partial	Europe, Africa, Asia, Australia
			2014 April 15	Total	Australia, Pacific, Americas

Data from NASA's eclipse website: http://eclipse.gsfc.nasa.gov/.

† A total solar eclipse occurs if the Moon completely covers the Sun's disk. A solar eclipse is annular if it occurs when the Moon is near the point in its orbit farthest from Earth. Under that condition, the Moon will not quite cover the Sun, even though it is exactly in line with it. The edge of the Sun, therefore, remains uncovered all around the Moon and appears as a bright ring, or annulus.

*A partial lunar eclipse occurs when the Earth's shadow falls on only part of the Moon. A total lunar eclipse occurs if the Earth's shadow completely covers the Moon.

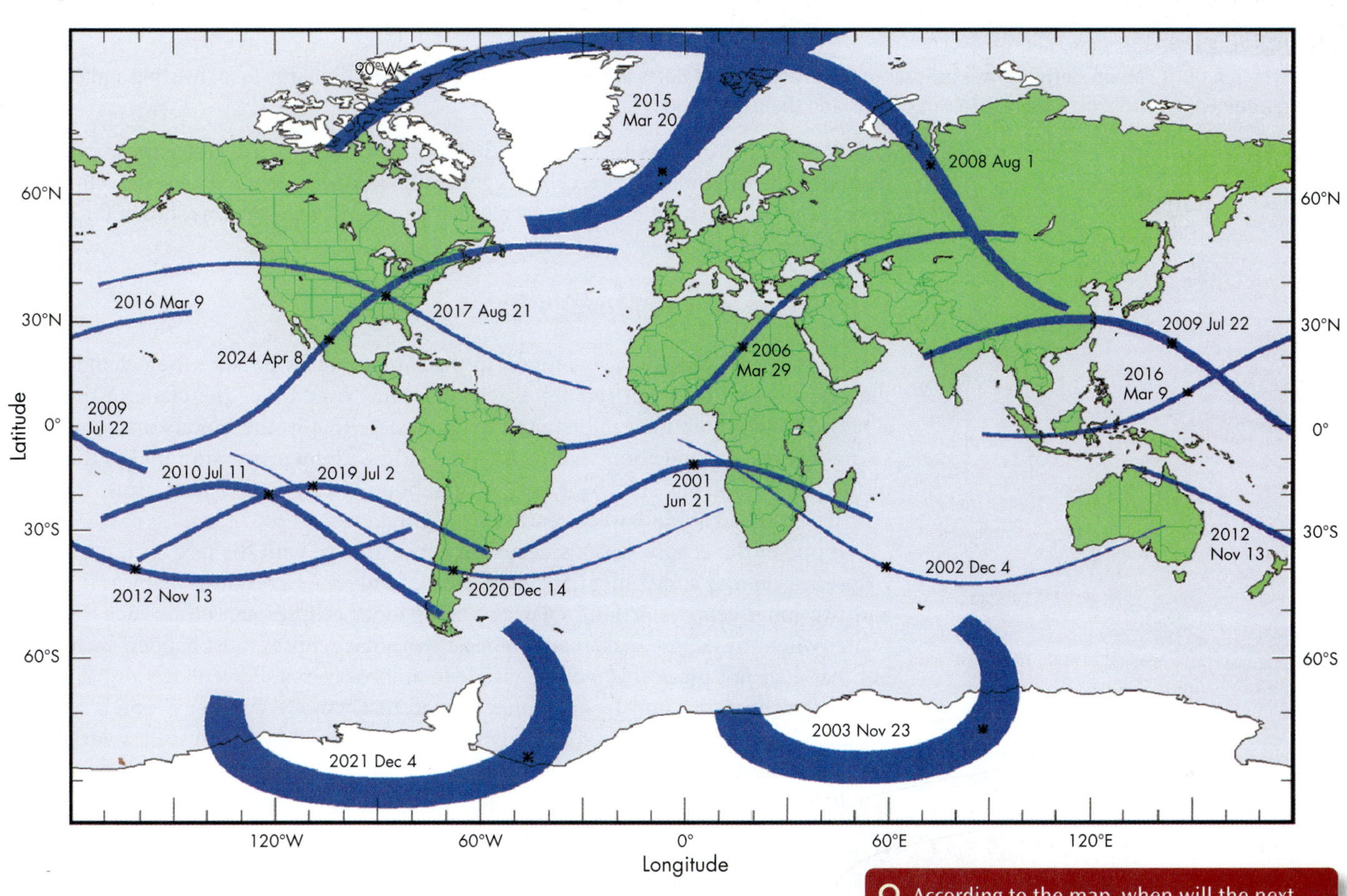

FIGURE 1.20
Location of recent and upcoming total solar eclipses. The paths show where totality can be observed. In regions outside of these paths, a partial eclipse may be visible.

Q. According to the map, when will the next total solar eclipse occur in North America? In Australia?

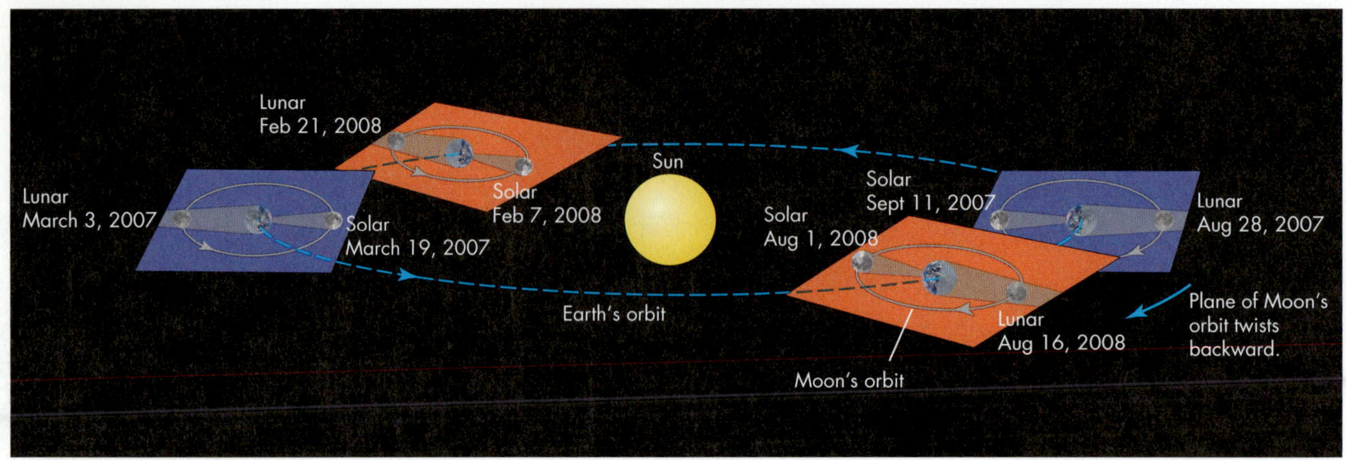

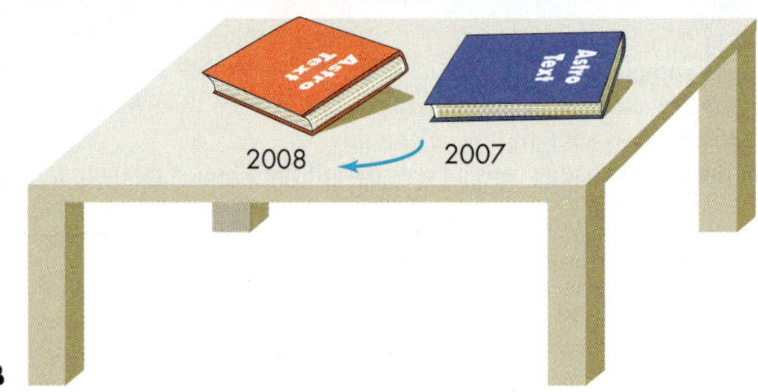

FIGURE 1.21
(A) Precession of the Moon's orbit causes eclipses to come at different dates in successive years. Notice its similarity to (B) twisting a tilted book that has one edge resting on a table. (Sizes of objects and their separations are not to scale.)

at new moon, conditions are right for a lunar eclipse to happen at either the previous or the following full moon. Thus, eclipses generally occur in pairs, with a solar eclipse followed approximately 14 days later by a lunar eclipse, or vice versa, as shown for several of the eclipses listed in table 1.1.

Precession of the Moon's Orbit

Eclipse seasons do not always remain in the same months because the orientation of the Moon's orbit does not remain exactly the same over time. The plane of the orbit slowly changes direction as illustrated in fig. 1.21A. That is, the Moon's orbit *precesses*, swinging once around about every 18.6 years. This orbital **precession** makes the dates of the eclipse seasons shift by 1/18.6 year (about 20 days) each year. Thus, in 2008, eclipses occurred about 3 weeks earlier on average than in 2007.

If one of the eclipse seasons occurs in early January with the next in June, a third eclipse season may sometimes happen in late December. As a result, as many as five solar and two lunar eclipses or four solar and three lunar eclipses can occur each year. No matter when the eclipse season falls, at least two solar eclipses must happen each year,* but that does not mean they will be visible to an observer at a given location, since the eclipse may be visible only from another part of the Earth. Because the Moon is so small compared with the Earth, its shadow is small, and therefore you can see a solar eclipse only from within a narrow band, as illustrated in figure 1.20. Lunar eclipses, however, are visible from anywhere the Moon is above the horizon at the time of the eclipse.

*At least two solar and two lunar eclipses happen each year. However, sometimes one or both of the solar eclipses is only partial, and sometimes in one or both of the lunar eclipses only the outer part of the Earth's shadow (technically, the penumbra) falls upon the Moon. In these cases the Sun or Moon is only slightly dimmed and the eclipse is not particularly noticeable.

SUMMARY

We create a mental model of the night sky overhead as a giant dome, forming part of the celestial sphere. Star patterns on the celestial sphere are called *constellations*. According to this model, stars rise in the east and set in the west as the celestial sphere rotates overhead. This apparent motion is actually caused by the Earth's spin. The points about which the celestial sphere appears to turn are called the *celestial poles*.

At any given time of year, the Sun's glare hides those stars lying near it on the celestial sphere. However, as the Earth moves around the Sun, it looks to us as if the Sun changes its position with respect to the stars. Thus, stars previously visible are lost in the Sun's glare, whereas stars previously hidden become visible, making different constellations visible at different times of year. We call the path that the Sun follows around the celestial sphere the *ecliptic*, and the 12 constellations through which it runs, the *zodiac*.

The ecliptic and zodiac are tipped at an angle of 23.5° to the *celestial equator* because the Earth's rotation axis is tipped by that amount with respect to its orbit. The solstices and equinoxes mark when the Sun reaches its maximum distance from the celestial equator and when it crosses the equator, respectively. These points define the onsets of the seasons.

The Earth's spin keeps its rotation axis pointing in nearly a fixed direction as we orbit the Sun. Because the axis is tipped, the Sun shines more directly on the Northern Hemisphere for half the year and on the Southern Hemisphere for the other half of the year. This difference in exposure to the Sun's light and warmth creates the seasons.

Ancient peoples noted the basic patterns of the night sky, and they built monuments to trace the motions of the Sun through the seasons. They also tracked the position of the Moon, which moves through a cycle of *phases* every 29.5 days. The plane of the Moon's orbit around the Earth is close to, but at a small angle to, Earth's orbital plane around the Sun (the ecliptic). As a result, when a new or full Moon is crossing the ecliptic, there can be a solar or lunar eclipse, respectively.

Because of the small size of the Moon relative to the Earth, the full Moon can be completely in the Earth's shadow during a lunar eclipse, but during a solar eclipse the Moon's shadow covers only a narrow path across the Earth. The dates when the orbital planes of the Moon and the Earth cross come about 6 months apart during what are called *eclipse seasons*, but the dates of these seasons gradually shift as the orientation of the Moon's orbit changes over time.

QUESTIONS FOR REVIEW

1. (1.1) What is the celestial sphere? What are the celestial equator and the ecliptic?
2. (1.1) What is a constellation, and what is special about the zodiac constellations?
3. (1.2) What causes the seasons?
4. (1.3) How long does it take the Moon to go through a cycle of phases?
5. (1.3) How much later does the Moon rise each day?
6. (1.4) What is the difference between lunar and solar eclipses?
7. (1.4) Why aren't there eclipses each month?
8. (1.4) What is an eclipse season?

THOUGHT QUESTIONS

1. (1.1) If you were standing on the Earth's equator, where would you look to see the north celestial pole? Could you see this pole from Australia?
2. (1.1) Draw a sketch of the Earth and a distant North Star, and show that your latitude is the angle of the north celestial pole above the northern horizon.
3. (1.1) Can you think of an astronomical reason why the zodiac may have been divided into 12 signs rather than 8 or 16?
4. (1.1/1.4) Some people still believe the Earth is flat. What "proof" would you offer them that it is spherical? Do you think you could you persuade them?
5. (1.1/1.2) When it is winter in New York, what season is it in Australia? Is it then October in Paris? If you can see Orion in the evening in New York, can you see it in the evening in Australia or Paris?
6. (1.2) If the shape of the Earth's orbit were unaltered but its rotation axis were shifted so that it had no tilt with respect to the orbit, how would seasons be affected?
7. (1.2) Why does the position of sunrise along the eastern horizon change during the year?
8. (1.2) Why do we have time zones? Sketch and label a diagram to justify your answer.
9. (1.3) Provide two or three pieces of evidence you could use to explain to someone that the moon's phases are not caused by the Earth's shadow.
10. (1.3) If the Moon orbited the Earth in the opposite direction, but everything else remained the same, how would the sidereal and solar months change (if at all)? Create a drawing like figure 1.14 representing this situation.
11. (1.2/1.4) In addition to the distortions that result because the map is a flat projection, explain why the eclipse tracks in figure 1.20 show variations in width. (Hint: there is more than one possible reason.)

PROBLEMS

1. (1.1) If the Earth turns one full rotation in approximately 24 hours, how many degrees per hour does the sky turn?
2. (1.2) From a latitude of 35°, what is the highest and lowest altitude above the horizon of the noon sun? What will be the altitude on March 20?
3. (1.3) Make a sketch to calculate what times the waxing gibbous moon will rise and set. Indicate the observer's location and lines of sight to the moon for these times.
4. (1.3/1.4) The Moon's orbital period is 27.21222 days. Its synodic period, the period of the phases, is 29.5306 days. Show that 242 orbital periods very nearly equals 223 synodic periods. How long is this in years? What does this suggest about eclipses and why? (This match of cycles is called the saros and was used by ancient astronomers to predict eclipses.)
5. (1.4) Find how many hours it takes the Moon to move in its orbit a distance equal to the Earth's diameter. (You will need to determine the speed of the Moon in its orbit. You can find values for the diameter of the Earth and the radius and period of the Moon's orbit in the appendix.) How does this relate to the time it takes for a lunar eclipse to occur?
6. (1.4) List some of the details left out of problem 5 that you would need to consider to exactly calculate the length of an eclipse. What effect would each have on the final answer?
7. (1.4) The Moon's shadow at the Earth is much smaller than the Moon's diameter—it's only a few hundred kilometers wide. Is the Moon's speed still a good estimate of how fast the shadow moves? Repeat problem 5 to estimate the duration of a solar eclipse.

TEST YOURSELF

1. (1.1) If you are standing at the Earth's North Pole, which of the following will be directly overhead?
   (a) The celestial equator
   (b) The ecliptic
   (c) The zodiac
   (d) The north celestial pole
   (e) The Sun
2. (1.1) Ancients noticed patterns in the motion of the objects in the sky, which led to the development of units of time like hours, days, weeks, months, and years. Which of the following moves the least in the sky, regardless of the time period being considered?
   (a) The Moon
   (b) The Sun
   (c) Polaris
   (d) Sirius, the dog star
3. (1.1) If you observe Polaris to be 55° above the horizon, you are at a latitude of approximately
   (a) 31.5° (b) 45° (c) 55° (d) 65° (e) 78.5°
4. (1.1/1.2) For this question, choose as many answers as are correct. If the Earth reversed its direction of spin,
   (a) the Sun would rise in the west and set in the east.
   (b) the seasons would be reversed.
   (c) the stars would circle Polaris clockwise.
   (d) the Moon would rise in the west and set in the east.
   (e) the Moon would rise in the east and set in the west.
5. (1.2) In the Northern Hemisphere, summertime is warmer than wintertime because
   (a) the Earth's orbit is an ellipse.
   (b) the Sun is visible for more hours.
   (c) sunlight is more concentrated on the ground.
   (d) both b and c.
   (e) All the answers are true.
6. (1.3) If there is a waning gibbous moon visible in Chicago, that night in Australia there will be a
   (a) waxing crescent moon.
   (b) waning gibbous moon.
   (c) waxing gibbous moon.
   (d) waning crescent moon.
7. (1.3) You observe the Moon rising at 6 p.m., around sunset. Its phase is
   (a) 1st quarter (b) new (c) full (d) 3rd quarter
8. (1.3) You observe the Moon rising at 3 a.m., a few hours before sunrise. Its phase is
   (a) between new and first quarter.
   (b) between first quarter and full.
   (c) between full and third quarter.
   (d) between third quarter and new.
9. (1.3) If you see a full Moon at midnight, about how long will it be until there is a new Moon?
   (a) 12 hours (b) 3 days (c) 2 weeks (d) 6 months
10. (1.4) Figure 1.15 shows an eclipse of the Sun. The black circle in the middle is
   (a) the Earth's shadow on the Sun.
   (b) the Sun's shadow on the Moon.
   (c) the Moon covering the Sun.
   (d) the Earth's shadow on the Moon.
   (e) a dark cloud in our atmosphere.
11. (1.4) If the Moon were twice as large in diameter as it is, we would have total solar eclipses
   (a) every month.
   (b) more often than now but less than (a).
   (c) never.
   (d) less often than now but more than (c).

KEY TERMS

celestial equator, 18
celestial poles, 18
celestial sphere, 16
constellations, 17
eclipse seasons, 32
ecliptic, 19
equinoxes, 24
horizon, 16
lunar eclipse, 30
model, 16
phases, 27
precession, 34
rotation axis, 21
solar eclipse, 30
solstices, 24
zodiac, 20

FURTHER EXPLORATIONS

Ahmad, Imad A. "The Science of Knowing God: Astronomy in the Golden Era of Islam." *Mercury* 24 (March/April 1995): 28.

Aveni, Anthony F. "Emissaries to the Stars: The Astronomers of the Ancient Maya." *Mercury* 24 (January/February 1995): 15.

Aveni, Anthony F. "Native American Astronomy." *Physics Today* 37:24, 1984.

Krupp, E. C. *In Search of Ancient Astronomies.* New York: McGraw-Hill, 1978.

Krupp, E. C. *Echoes of the Ancient Skies: The Astronomy of Lost Civilizations.* New York: Harper & Row, 1983.

Schaefer, Bradley E. "The Origin of the Greek Constellations." *Scientific American* 295 (November 2006): 96.

Website

Visit the *Explorations* website at **http://www.mhhe.com/arny** for additional online resources on these topics.

FIGURE QUESTION ANSWERS

WHAT IS THIS? (chapter opening): The figure at the start of the chapter shows the Moon's shadow on the Earth's surface. The shadow is usually a few hundred kilometers across. People within the region of the shadow would be able to see a total solar eclipse.

FIGURE 1.3: counterclockwise

FIGURE 1.17: In both cases they are lit by sunlight that has passed through our atmosphere, which has removed most of its blue light.

FIGURE 1.20: 2017 for North America; 2012 for Australia.

PROJECTS

1. **The changing location of sunset:** Find a spot, perhaps a window facing west, where you can see the western horizon in the evening. Make a sketch of the horizon, noting hills, buildings, or trees that might serve as reference marks. From your chosen viewing spot, watch the sunset and mark on your sketch where the Sun goes down. Label the date and time. Make an observation each week over a period of a month or so. Does the Sun set at the same point each night? If not, which direction along the horizon is the Sun moving? Does the Sun go down at the same time each night? Does it set earlier or later? If you enjoy photography, you might try taking a photograph or videotaping the sunset or sunrise.
2. **Motion of the stars across the sky:** Many people are startled when they are told that the stars move across the sky the same way that the Sun does. However, it is easy to show that they do.

   Get a stick that you can poke into the ground so it will stand upright. Get a second, smaller stick that you can tape or affix to the upright stick in some manner. A ruler taped to a camera tripod would be ideal, as sketched in Project figure 1.1.

   Find a bright star and sight along the smaller stick toward the star. If you now wait 5 or so minutes and again sight along the stick, you will see that it no longer points to the star. That is, the star has moved so that it now lies west of where the stick is pointing. You could do this experiment indoors if you have a window on which you can put a small mark. Set up a chair by the window so that you can watch a star through the glass. While you remain seated, have a friend place a mark on the glass with a grease pencil or piece of tape where the star appears to be. Again, wait a few minutes. It will be clearly visible that the star has moved.

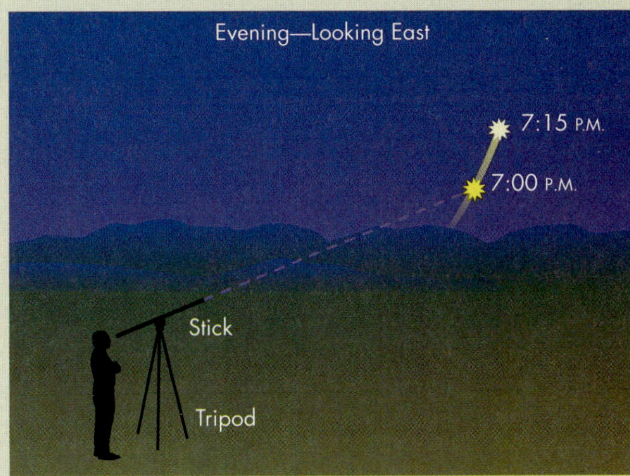

PROJECT FIGURE 1.1
Sketch illustrating how to observe the motion of the stars across the sky by sighting along a stick.

This device was built in the early 1800s to model the motions of the Earth, planets, and Moon as they move around the Sun.

KEY CONCEPTS

- In the classical period (about 500 B.C.–A.D. 1400), scientist-philosophers began to make measurements of the heavens and, with their knowledge of geometry, constructed idealized models that could account for the motion of heavenly bodies.
- Their models allowed these ancient astronomers to deduce the shape and size of the Earth, and even to estimate the distance to the Moon and the Sun.
- The motions of the planets proved to be difficult to explain in the Earth-centered models of the Universe, although complicated models were developed that made fairly accurate predictions.
- In the Renaissance period (about 1400–1650), those geometrical models were reassessed and found inadequate. Astronomers therefore devised new models that took into account a much greater body of data based on observational records accumulated over centuries. Astronomers also benefited from a technological advance that allowed them to observe even more—the telescope.
- Starting around A.D. 1500, scientists such as Copernicus, Galileo, Kepler, and Newton deduced laws to describe and explain planetary motion.
- Finally, in the modern period (1650 to the present), scientists began the search for the physical laws (such as the law of gravity) that underlie the observed movements in the heavens. Other important contributions to our understanding of the Universe came from technological advances (for example, in optics, electronics, and computers) and better mathematical techniques (such as calculus). Such factors continue to be important today.

2 The Rise of Astronomy

CONCEPTS AND SKILLS TO REVIEW

- The seasonal motion of the Sun (1.2)
- The Moon's motions and phases (1.3)
- Lunar eclipses (1.4)

Our understanding of the Universe has been assembled bit by bit from many separate discoveries—discoveries made by scientists from many parts of the world, at many times in the past, and in many disciplines. How those discoveries led to our current knowledge is the subject of this chapter.

The astronomical phenomena that we discussed in chapter 1 (the rising and setting of Sun, Moon, and stars; the constellations; annual motion of Sun; phases of the Moon and eclipses) were the basis of ancient knowledge of the heavens. With these observations, we can now describe people's early attempts to explain the heavens. We will see that some of their conclusions were incorrect, just as we today are probably in error about some aspects of modern astronomy. We study ancient ideas of the heavens not so much for what they tell us about the heavens but to learn how observation and reasoning can lead us to an understanding of the Universe.

Much of what we know about the Universe can be shown by carrying out simple observations and making a few logical deductions. For example, by observing the shape of the Earth's shadow during a lunar eclipse, it is possible to deduce the shape of the Earth and its size relative to the Moon. This was understood by ancient Greek philosophers more than 2000 years ago. It is only a myth that the Earth was widely believed to be flat until recent times.

Astronomers of classical times determined a remarkable amount about the Earth, Moon, Sun, and stars. However, they struggled to understand the motions of the planets. The puzzling motions of these objects in the sky finally forced humans to consider the possibility that they did not live at the center of the Universe. This revolution of thinking during the Renaissance led to the development of new mathematical and scientific ideas and the birth of astrophysics.

Q: WHAT IS THIS? See end of chapter for the answer.

2.1 Early Ideas of the Heavens: Classical Astronomy

As far as we know, the ancient Greek astronomers of classical times were the first people to try to explain the workings of the heavens in a careful, systematic manner, using observations and models. Given the limitations of naked-eye observation, these astronomers were extraordinarily successful, and their use of logic, mathematics, and geometry as tools of inquiry created a method for studying the world around us that we continue to use even today. This method is in many ways as important as the discoveries themselves.

The Shape of the Earth

The ancient Greeks knew that the Earth is round. As long ago as about 500 B.C., the mathematician Pythagoras (about 560–480 B.C.) was teaching that the Earth is spherical, but the reason for his belief was as much mystical as rational. He, like many of the ancient philosophers, believed that the sphere was the perfect shape and that the gods would therefore have utilized that perfect form in the creation of the Earth.

By 300 B.C., however, Aristotle (384–322 B.C.) was presenting arguments for the Earth's spherical shape that were based on simple naked-eye observations that anyone could make. Such reliance on careful, firsthand observation was the first step toward acquiring scientifically valid knowledge of the contents and workings of the Universe. For instance, Aristotle noted that if you look at an eclipse of the Moon when the Earth's shadow falls upon the Moon, the shadow can be clearly seen as curved, as figure 2.1A shows. He wrote in his treatise "On the Heavens":

> **The shapes that the Moon itself each month shows are of every kind—straight, gibbous, and concave—but in eclipses the outline is always curved: and, since it is the interposition of the Earth that makes the eclipse, the form of this line will be caused by the form of the Earth's surface, which is therefore spherical.**

Another of Aristotle's arguments that the Earth is spherical was based on the observation that a traveler who moves south will see stars that were previously hidden

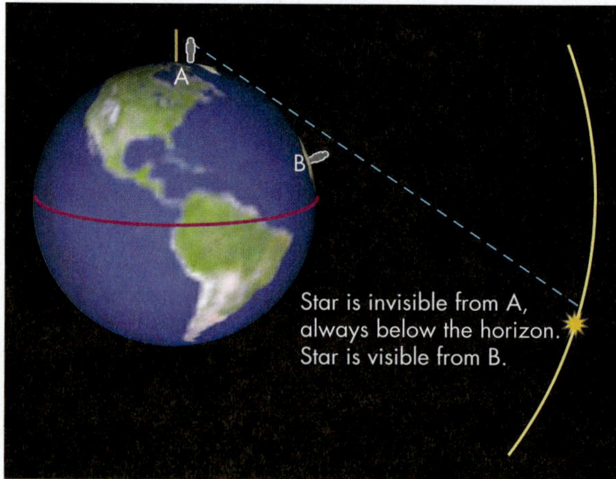
A
B
Star is invisible from A,
always below the horizon.
Star is visible from B.

FIGURE 2.1
(A) A sequence of photographs during a partial lunar eclipse. The edge of the Earth's shadow on the Moon is always a portion of a circle, showing that the Earth must be round. (B) As a traveler moves from north to south on the Earth, the stars that are visible change. Some stars that were previously hidden become visible above the southern horizon. This variation would not occur on a flat Earth.

below the southern horizon, as illustrated in figure 2.1B. For example, the bright star Canopus is easily seen in Miami but is invisible in Boston. This could not happen on a flat Earth.

The Size of the Earth

Knowing that the Earth is spherical, the Greeks wanted to know its size. Eratosthenes (276–195 B.C.), head of the famous Library at Alexandria in Egypt, made the first measurement of the Earth's size. He obtained a value for its circumference of about 25,000 miles, which is roughly the presently known value. Eratosthenes's demonstration is one of the most beautiful ever performed. Because it so superbly illustrates how science links observation and logic, the demonstration is worth describing in some detail.

You must realize that at this point in time, astronomers were very well acquainted with the yearly movement of the Sun and could predict accurately the times of the solstices and equinoxes. The summer solstice marked the day of the year in Alexandria when the Sun would reach its highest point in the noonday sky. However, despite its height, the Sun still cast a shadow at noon. Eratosthenes, a geographer as well as an astronomer, heard that lying to the south, in the Egyptian town of Syene (the present city of Aswân), the Sun would be directly overhead at noon and cast no shadow. Proof of this was the fact that at that time the Sun shone exactly down a well near there. Knowing the distance between Alexandria and Syene and appreciating the power of geometry, Eratosthenes realized he could deduce the circumference of the Earth. He analyzed the problem as follows: Because the Sun is far away from the Earth, its light travels in parallel rays toward the Earth. Thus, two rays of sunlight, one hitting Alexandria and the other shining down the well, are parallel lines, as depicted in figure 2.2.

Now imagine drawing a straight line from the center of the Earth outward so that it passes vertically through the Earth's surface in Alexandria. The angle between that line and the Sun's rays in Alexandria is the same as the angle between that line and the line from the center of the Earth up through the well in southern Egypt (see fig. 2.2). The reason is that a single line crossing two parallel lines forms the same angle to both (a geometric theorem).

The angle between sunlight and vertical directions in Alexandria can be measured with sticks and a protractor (or its ancient equivalent) and is the angle between the direction to the Sun and the vertical to the ground (see fig. 2.2). Eratosthenes found this angle to be about 1/50th of a circle. Therefore the angle formed by a line from

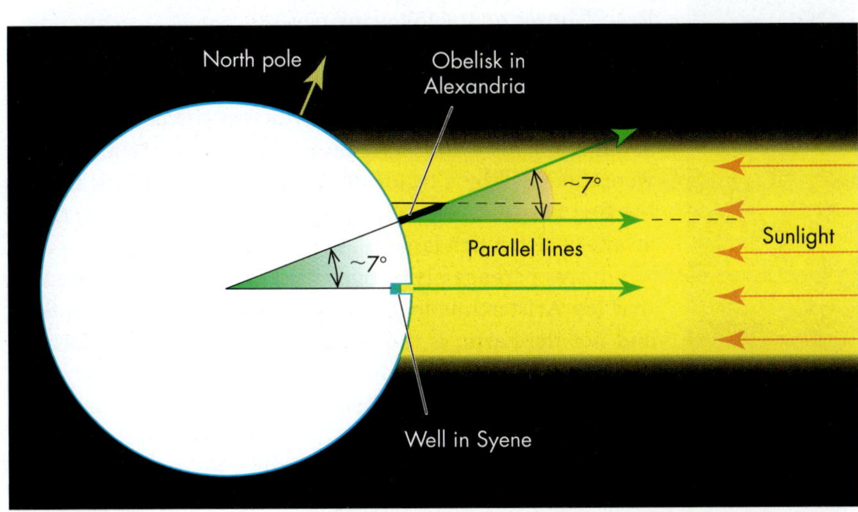

FIGURE 2.2
Eratosthenes's calculation of the circumference of the Earth. The Sun is directly overhead at local noon on the summer solstice at Syene, in southern Egypt. On that same day and time, Eratosthenes found the Sun to be 1/50 of a circle (about 7°) from the vertical in Alexandria, in northern Egypt. Eratosthenes deduced that the angle between two verticals placed in northern and southern Egypt must be 1/50 of the circumference of the Earth.

Alexandria to the Earth's center and a line from the well to the Earth's center must also be 1/50th of a circle, as deduced from the parallel line–equal angle theorem.

To find the circumference of the Earth, all that is needed is to note that the distance between Alexandria and the well thus represents 1/50th of the distance around the Earth. Because Eratosthenes knew the distance between Alexandria and the well to be 5000 stadia (where a stadium is about 0.1 mile), the distance around the entire Earth is 5000 stadia times 50, or 250,000, stadia. When expressed in miles, this is roughly 25,000 miles, which is approximately the circumference of the Earth as we know it today.

By modern standards, there is absolutely nothing wrong with this technique. You can use it yourself to measure the size of the Earth, as described in the project at the end of this chapter. Eratosthenes's success was a triumph of logic and the scientific technique, but it suffered from one weakness. The method required that he assume the Sun is so far away that its light reaches Earth along parallel lines. That assumption, however, was supported by another set of measurements made by the ancient Greeks—namely, a rough measurement of the relative diameters and distances of the Sun and Moon.

Distance and Size of the Sun and Moon

About 75 years before Eratosthenes measured the Earth's size, Aristarchus of Samos (an island in the Mediterranean) had estimated the relative size of the Earth, Moon, and Sun, and the relative distances to the Moon and Sun. His values for these numbers gave at least the correct sense of their proportionate sizes and their relative distances from Earth. For example, by comparing the size of the Earth's shadow on the Moon during a lunar eclipse to the size of the Moon's disk, illustrated in figure 2.3, Aristarchus calculated that the Moon's diameter is about one-third of the Earth's.

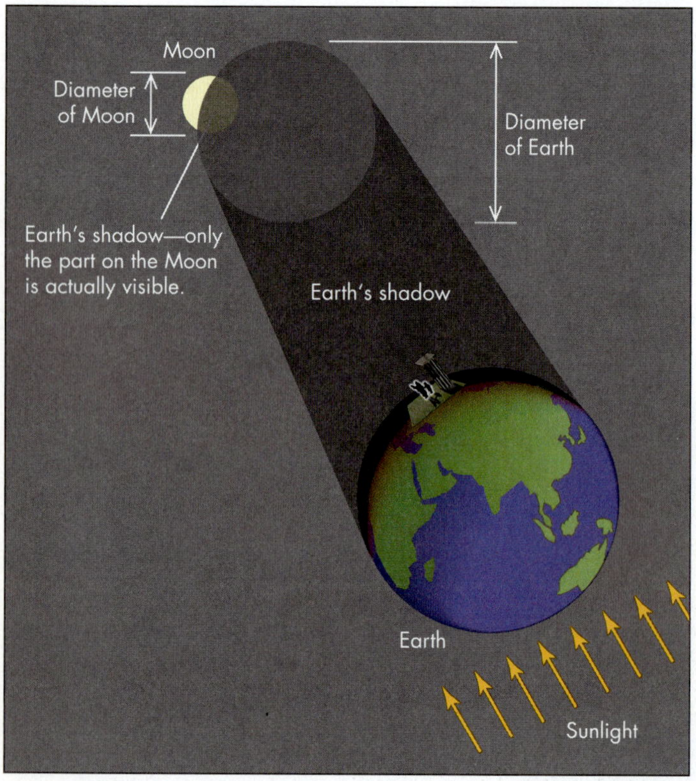

Aristarchus also calculated the Sun to be about 20 times farther away from the Earth than the Moon is. He did this calculation by measuring the angle between the Sun and the Moon when the Moon is exactly half lit, as shown in figure 2.4. From the apparent size in the sky of the Sun and Moon and their relative distances, and his calculation of the Moon's diameter relative to the Earth's, Aristarchus was able to deduce that the Sun's diameter is about 7 times that of the Earth's—far too small, as we know today (it is actually about 100 times greater), but nevertheless evidence that the Sun is bigger than the Earth. The relationship between distance, size, and the *angular size* of objects is explored further in the "Extending Our Reach" boxes on the following pages.

Although Aristarchus's determination of the Moon's relative size was quite accurate (he found the Moon's diameter to be about 0.35 that of the Earth, whereas the correct ratio is about 0.27), his distance to the Sun and consequently its size were too small by a factor of about 20. Nevertheless, he was the first person to establish that the Sun is not only very far away but also much larger than the Earth.

It was perhaps his recognition of the vast size of the Sun that led Aristarchus to the revolutionary idea that the Sun, and not the Earth, is the center of the heavens. Aristarchus was right, of course, but his idea was too revolutionary, and another 2000 years passed before scientists became convinced of its correctness.

However, this was not mere pigheadedness. There was a good reason for not believing that the Earth moves around

Moon
Diameter of Moon
Diameter of Earth
Earth's shadow—only the part on the Moon is actually visible.
Earth's shadow
Earth
Sunlight

FIGURE 2.3
Aristarchus used the size of the Earth's shadow on the Moon during a lunar eclipse to estimate the relative size of the Earth and Moon.

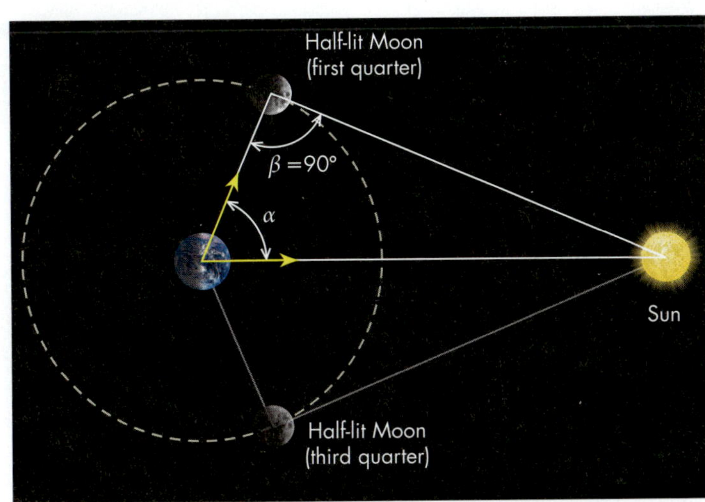

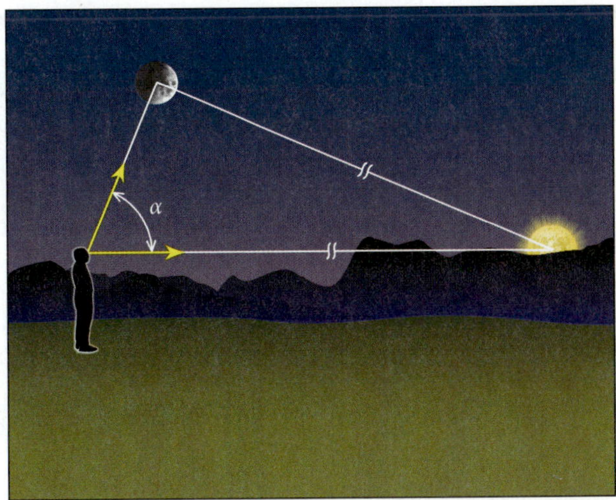

FIGURE 2.4
Aristarchus estimated the relative distance of the Sun and Moon by observing the angle α between the Sun and the Moon when the Moon is exactly half lit. Angle β must be 90° for the Moon to be half lit. Knowing the angle α, he could then set the scale of the triangle and thus the relative lengths of the sides. (Sizes and distances are not to scale.)

the Sun. If it does, the positions of stars should change during the course of the year. Looking at figure 2.5, you can see that the nearby star should appear to lie in a different position in January and in July.

This shift in a star's apparent position resulting from Earth's motion is called **parallax,** and Aristarchus's critics were absolutely right in supposing that it should occur. So, when they could observe no parallax caused by the Earth's motion, they concluded that Aristarchus's Sun-centered system must be wrong. But Aristarchus's critics failed to appreciate how tiny parallax is. In particular, they did not realize that stars are so enormously far away that parallax is much too small to be detected without a powerful telescope. As a result, in Aristarchus's time (about 2000 years before the telescope was invented) there was no hope of detecting the parallax of stars. It was not until 1838 that astronomers had telescopes of sufficient power and accuracy to measure this parallax. Aristarchus's idea was rejected for reasons that were logically correct but were based on inaccurate data.

INTERACTIVE
Stellar parallax

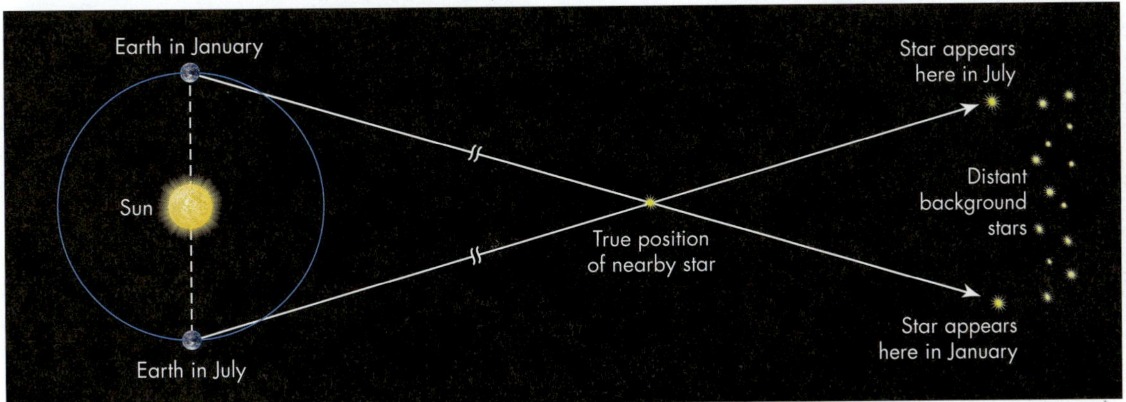

FIGURE 2.5
Motion of the Earth around the Sun causes stellar parallax—a shift in where in the sky a star appears to lie. Because the stars are so remote, this shift is too small to be seen by the naked eye. This lack of observable parallax led the ancient Greeks incorrectly to deduce that the Sun could not be the center of the Solar System. (Sizes and distances are not to scale.)

EXTENDING *our reach*

MEASURING THE DIAMETER OF ASTRONOMICAL OBJECTS

One of the most basic properties of an astronomical body is its diameter. But there is no astronomical body for which this fundamental observation can be made directly. Since we cannot stretch a tape measure across the disk of the Sun or Moon, how do we know the size of such heavenly bodies?

The basic method for measuring the size of a distant object was worked out long ago and is still used today. This method involves first measuring how big an object *looks,* a quantity called its **angular size.**

We can measure an object's angular size by drawing imaginary lines to each side of it, as shown in Box figure 2.1, and then measuring the angle between the lines using a protractor. For example, we can measure the angular diameter of the Moon with two straight sticks connected at one end with a nut and bolt. We sight along the edge of one stick to one edge of the Moon and then sight along the other stick to the Moon's opposite edge. Measuring the angle between the sticks, we will find that the Moon's angular diameter is about 1/2°.

We can find an astronomical body's true diameter from its angular diameter if we know its distance. We need the distance because a body's angular size changes with distance. For example, a building looks big when it is near us and small when it is far away, as shown in Box figure 2.2. Furthermore, it is easy to verify that the angular size of a distant object changes inversely with the object's distance. That is, if we double the distance to an object, its angular size is halved. These basic features of angular size were known in antiquity and were in fact used by Aristarchus to determine the size of the Moon and Sun compared to the Earth.

To find an object's true diameter from its angular diameter and distance, imagine we are at the center of a circle passing through the object, as illustrated in Box figure 2.3. Let ℓ be the diameter of the body and d the distance to the body, which is the *radius* of the circle. Next draw lines from the center to each end of ℓ, letting the angle between the lines be α, the object's angular diameter.

We now determine the object's true size, ℓ, by forming the following proportion: ℓ is to the circumference of the circle as α is to the total number of degrees around the circle, which we know is 360.

Thus,

$$\frac{\text{Object's diameter}}{\text{Circumference}} = \frac{\text{Angle between lines}}{360}$$

$$\frac{\ell}{\text{Circumference}} = \frac{\alpha}{360}$$

However, we know from geometry that the circle's circumference is $2\pi d$. Thus,

$$\frac{\ell}{2\pi d} = \frac{\alpha}{360}$$

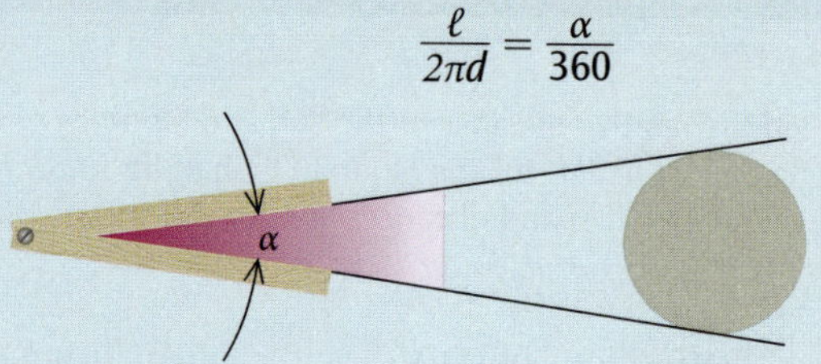

BOX FIGURE 2.1
Definition of angular size.

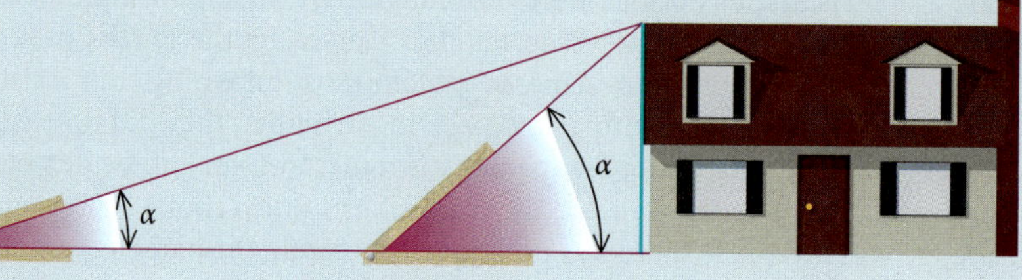

BOX FIGURE 2.2
How angular size varies with distance.

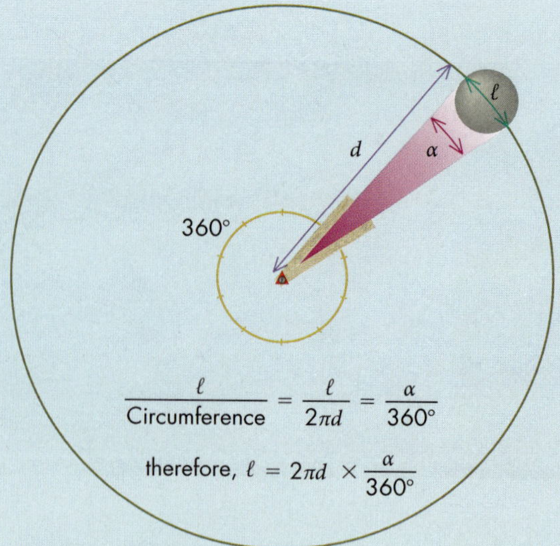

BOX FIGURE 2.3
How to determine linear size from angular size.

We can now solve for ℓ and find that

$$\ell = \frac{2\pi d\alpha}{360}$$

Thus, given a body's angular diameter and distance, we can calculate its true diameter. For example, suppose we apply this method to measure the Moon's diameter. We stated previously that the Moon's angular diameter is about 1/2°. Its distance is about 384,000 kilometers. Thus, its real diameter is

$$\ell = \frac{2\pi(384{,}000)(0.5)}{360}$$

$$= \text{about } 3350 \text{ km}$$

Suppose we try another example. The angular diameter of the Sun also turns out to be about 1/2°, an interesting coincidence. The Sun's distance is 150 million kilometers. The Sun's diameter must therefore be $2\pi(150{,}000{,}000)(0.5)/360$, or about 1,310,000 kilometers.

An easy way to get at this same answer is to recognize that because the Sun and Moon have the same angular size, their true sizes must be in direct proportion to their distances. The Sun is about (150,000,000·384,000) = 391 times farther away. Therefore, it must be about 391 times larger than the Moon.

EXTENDING *our reach*

THE MOON ILLUSION

The Moon sometimes appears to be huge when you see it rising. In fact, if you measure the Moon's angular diameter carefully, you will find it to be smaller when it is near the horizon than when it is overhead, regardless of how huge it looks. This misperception, known as the **Moon illusion,** is still not well understood but is an optical illusion caused, at least in part, by the observer's comparing the Moon with objects seen near it on the horizon, such as distant hills and buildings. You know those objects are big even though their distance makes them appear small. Therefore, you unconsciously magnify both them and the Moon, making the Moon seem larger. You can verify this sense of illusory magnification by looking at the Moon through a narrow tube that blocks out objects near it on the sky line. Seen through such a tube, the Moon appears to be its usual size.

Box figure 2.4 shows a similar effect. Because you know that the rails are really parallel, your brain ignores the apparent convergence of the railroad tracks and mentally spreads the rails apart. That is, your brain provides the same kind of enlargement to the circle near the rails' convergence point as it does to the rails, causing you to perceive the middle circle as larger than the lower one, even though they are the same size.

BOX FIGURE 2.4
Circles beside converging rails illustrate how your perception may be fooled. The bottom circle looks smaller than the circle on the horizon but is in fact the same size. Similarly, the circle high in the sky looks smaller than the circle on the horizon.

2.2 The Planets

Many ancient cultures noted that there are several bright objects in the night sky that look like stars but do not follow the same cyclic motion associated with the rest of the stars in the sky. The Greeks are credited with giving them the name *planētai,* meaning "wanderers," from which our word *planet* comes.

Planets move across the background stars because of a combination of the Earth's and their own orbital motion around the Sun. One of the more striking features of this motion is that the planets *always* remain close to the ecliptic, within the constellations of the zodiac.

The motion of the planets lies in the same narrow zone as the Sun because their orbits, like that of the Earth, all lie in nearly the same plane, as illustrated in figure 2.6. Thus, like the path of the Sun through the stars, the paths of the planets are tilted by about 23.5° to the celestial equator, moving into our northern and southern skies depending on their position in their orbits.

The motions of the planets relative to the stars are very slow, detectable only through observations over many nights. Therefore, just like the Sun, the planets rise and set each day—reflecting, of course, the rotation of the Earth.

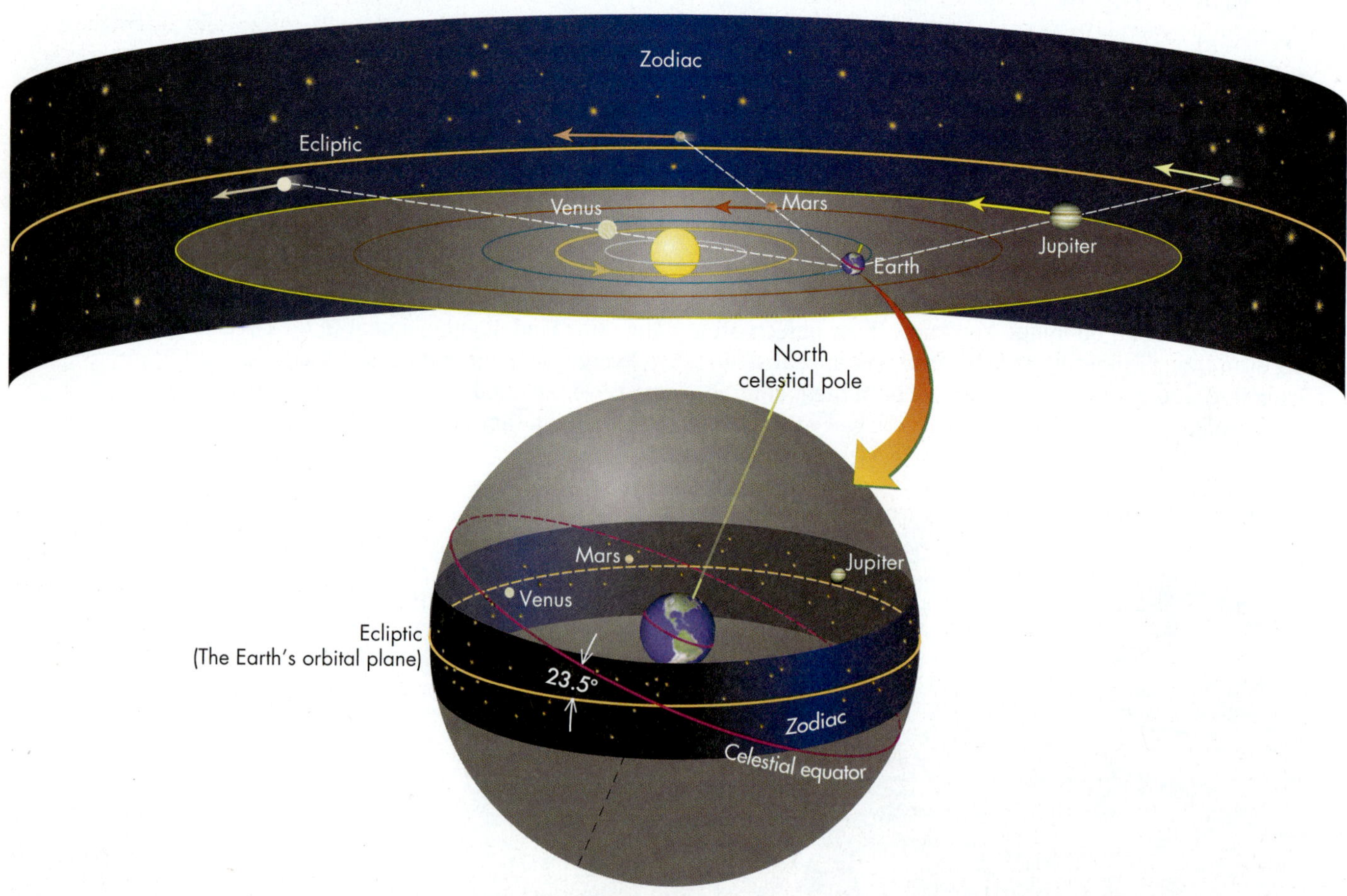

FIGURE 2.6
To the naked eye, the planets look like bright stars that "wander" through the sky. Although they move, they always remain near the ecliptic in the constellations of the zodiac, like the Sun and the Moon.

The motion of the planets through the zodiac can be seen by marking off the position of a planet on the celestial sphere over a period of a week or more. Figure 2.7 illustrates such a plot and shows that planets normally move eastward through the stars as a result of their orbital motion around the Sun.

Although the apparent motion is usually from west to east through the stars, this does *not* mean that the planet rises in the west and sets in the east. As seen from Earth, planets *always* rise in the east and set in the west because they are carried across the sky—just as the stars are—by the Earth's rotation. However, the motion of the planets is usually slower than that of the stars because their orbital motion partly offsets the rotation of the Earth that causes the apparent motion of the stars. Thus, if a star and planet rise side by side, at some later time the planet will not be as far above the horizon as the star. Therefore, *with respect to the stars,* the planet has moved to the east because of its orbital motion around the Sun.

This simple pattern of movement is sometimes interrupted. Occasionally a planet will move west with respect to the stars, a condition known as **retrograde motion** and shown in figure 2.8. The word *retrograde* means "backward," and when a planet is in retrograde motion, its path through the stars bends backward, sometimes even forming a loop, for a few months. All planets undergo retrograde motion for a portion of their paths around the sky. This motion greatly complicates the otherwise straightforward idea that the celestial sphere and its bodies rotate around the Earth. In fact, the need for a simple, plausible explanation of retrograde motion was a major reason astronomers ultimately rejected models of the Solar System with the Earth at the center.

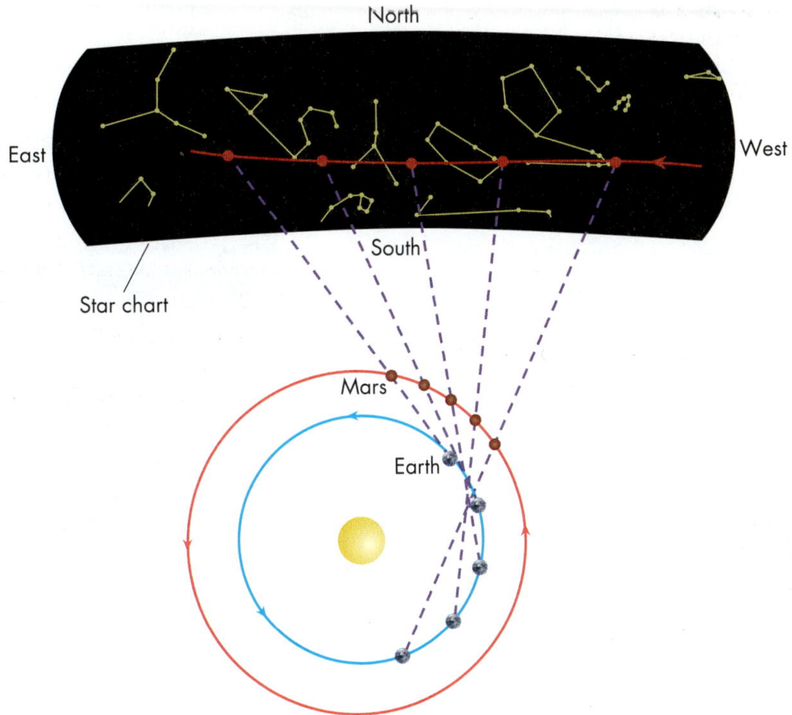

FIGURE 2.7
A planet's eastward drift against the background stars plotted on the celestial sphere. Note: Star maps usually have east on the left and west on the right, so that they depict the sky when looking south.

INTERACTIVE
Retrograde motion

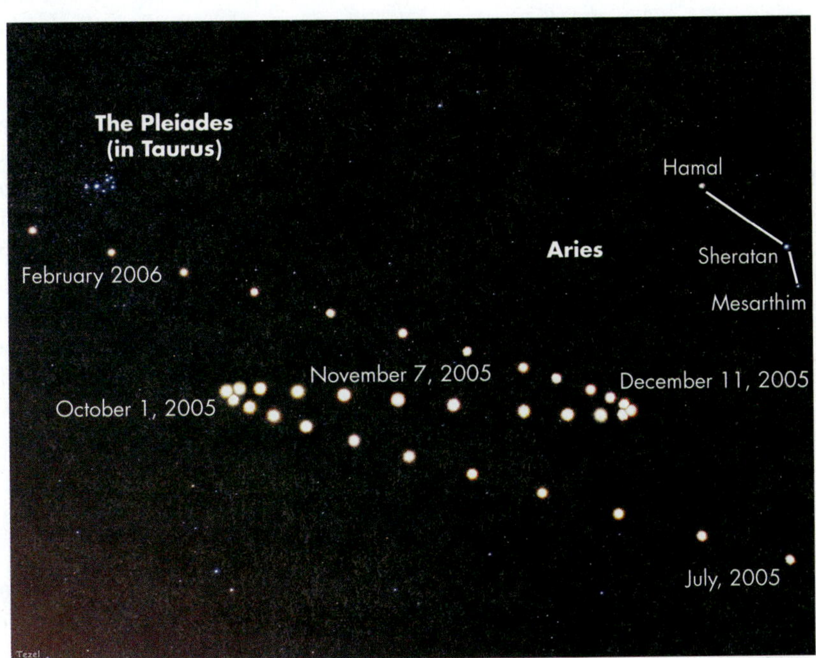

FIGURE 2.8
A sequence of images of Mars made in late 2005, showing its motion relative to the background stars. The pictures were taken roughly a week apart. Mars underwent retrograde motion in October and November of that year.

Q. Why does the brightness of Mars change in the image? (Hint: Draw a sketch of the positions of Mars and the Earth as Mars undergoes retrograde motion.)

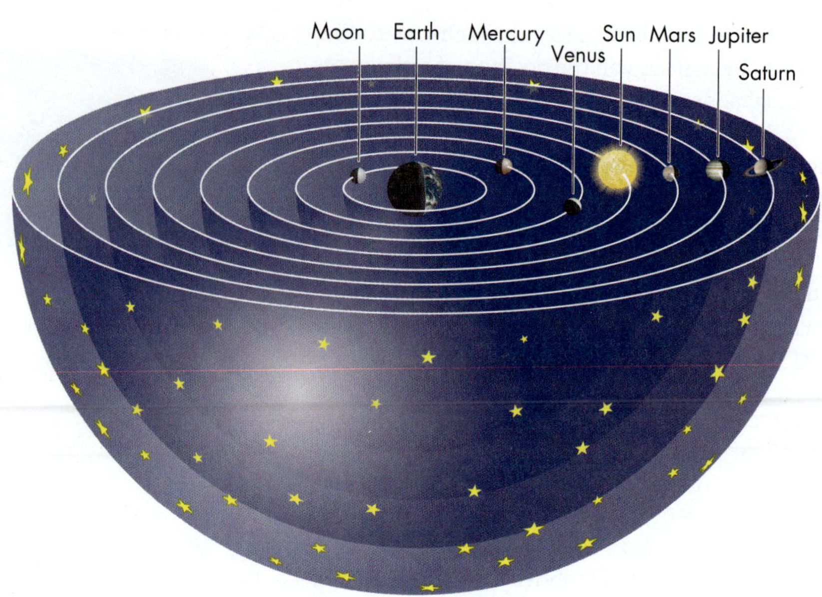

FIGURE 2.9
A cutaway view of the geocentric model of the Solar System according to Eudoxus.

Explaining the Motion of the Planets

Following the basic discoveries about the size and distance of the Sun and Moon, the main thread of astronomical research for almost the next 2000 years centered on the motion of the planets. From earliest Greek times, one simple model formed the basis for understanding those motions. It was based on the observation that as one looks at the sky, everything seems to move around the Earth from east to west. This led naturally to the idea that the Earth was at the center of the Universe and the planets and stars moved around it. Descriptions of the Universe of this type are called **geocentric models.**

Figure 2.9 shows a typical geocentric model based on the work of the Greek astronomer Eudoxus, who lived about 400–347 B.C. In this model, the Sun, Moon, and planets all revolve around the Earth. The bodies that move fastest across the sky are those that are nearest to the Earth. Thus, the Moon, whose path through the stars takes only about 27 days, is nearest to the Earth, whereas Saturn, whose path through the stars takes roughly 29 years, is located the farthest out of the planets known then. By assuming that each body was mounted on its own revolving transparent (crystalline) sphere and by tipping the spheres slightly with respect to one another, Eudoxus was able to give a quite satisfactory explanation of the motions of the heavenly bodies at any given time.

Unfortunately, such models do not work well if they are used to explain retrograde motion, unless one believes that the planets sometimes stop moving, reverse direction, pause, and then resume their original motion. This idea is clumsy and unappealing. Eudoxus was able to explain retrograde motion only by requiring that each planet moved on *two* spheres, one inside the other. By adjusting their rotation rates and sizes, he was able to get reasonable agreement with the observed positions of the planets as they shifted across the sky. However, about A.D. 150, far more elaborate models, which could be used to predict with much better accuracy, were developed by Ptolemy, the great astronomer of Greco-Roman times.

Ptolemy

Ptolemy lived in Alexandria, Egypt, which at that time was one of the intellectual centers of the world, in part because of its magnificent library. Ptolemy fashioned a model of planetary motions in which each planet moved on one small circle, which in turn moved on a larger one (fig. 2.10). The small circle, called an **epicycle**, was supposed to be carried along on the large circle like a Frisbee spinning on the rim of a bicycle wheel.*

According to Ptolemy's model, the motion of the planet from east to west across the night sky is caused by the rotation of the large circle (the bicycle wheel, in our analogy). Retrograde motion occurs when the epicycle carries the planet in a reverse direction (caused by the rotation of the Frisbee, in our model). Thus, with epicycles, it is possible to account for retrograde motion, and Ptolemy's model was able to predict planetary motions with fair precision.

Unfortunately, discrepancies remained between the predicted and true positions of the planets. This led to further modifications of the model, each of which led to slightly better agreement but at the cost of adding much greater complexity. Nevertheless, Ptolemy's theory survived until the 1500s and ultimately collapsed from both its failure to adequately predict where planets should be and also a loss of faith in its complexity. It had become far too complex to be plausible, and simplicity is an important element of scientific theory. As the medieval British philosopher William of Occam wrote in the 1300s, "Entities must not be unnecessarily multiplied," a principle known as "Occam's razor."

Ptolemy's era was one of decay and general political instability for the Greco-Roman civilization, which accounts for our uncertainty about the year of his birth or death. We know of him mainly through his great book, *Almagest,* a compendium of the astronomical knowledge of the ancient Greeks. The book includes tables of star positions and brightnesses. In fact, a great deal of what we know of Ptolemy (and of Greek and Roman civilization in general) we owe to the Islamic civilization that flourished around the southern edge of the Mediterranean from about 700 to 1200.

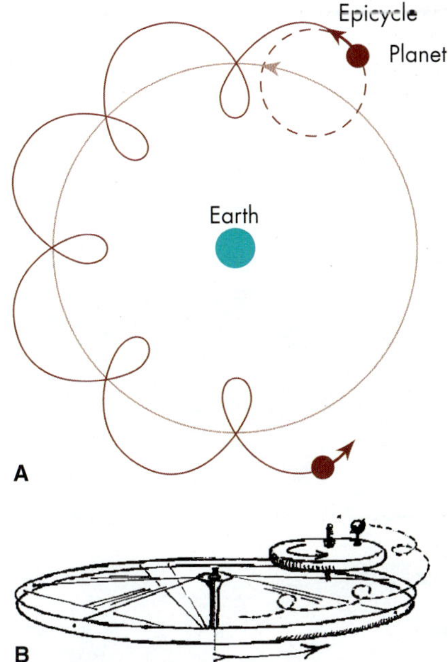

FIGURE 2.10
Epicycles are a bit like a bicycle wheel with a Frisbee bolted onto its rim.

Islamic Contributions

Much of what we know today about ancient Greek astronomy comes to us from Islamic scholars who studied and expanded upon the ancient texts while most of Europe struggled through the Middle Ages. Islamic civilization, like so many others, relied on celestial phenomena to set its religious calendar, and Islamic astronomers made many detailed studies of the sky and the motions of Sun, Moon, and planets. Islam's influence is very evident in astronomy through Arabic words such as *zenith* and the names of nearly all the bright stars—Betelgeuse, Aldebaran, and so on. In addition, Islamic scholars revolutionized mathematical techniques through innovations such as algebra (another Arabic word) and Arabic numerals.

ANIMATION
Ptolemy's model of motion of a planet

Asian Contributions

The early people of Asia, like their contemporaries to the west, studied the heavens. They too devised constellations, but based on their own mythologies, and they too made maps of the sky. Although the ancient astronomers of the East did not devise elaborate geometric models of the heavens, their careful observations of celestial events nevertheless prove useful to astronomers even today. For example, Chinese, Japanese, and Korean astronomers kept detailed records of unusual celestial events, such as eclipses, comets, and exploding stars. Based on such records, Chinese astronomers devised ways to predict eclipses. They even noted dark spots on the Sun (sunspots) that they could occasionally see with the naked eye when the Sun was low in the sky and its glare was dimmed by dust or haze. These records have allowed astronomers to discover ancient patterns of variation in the Sun's behavior. Their records of exploding stars also allow today's astronomers to determine the dates of many of these celestial outbursts.

---

*Ptolemy probably got the idea of epicycles from the writings of Hipparchus, who lived about 150 B.C. Hipparchus is best known to astronomers for his invention of the magnitude system (see chapter 13) for measuring stellar brightness, and for his discovery of precession (see chapter 6). The latter was made possible by his meticulous observations of star positions and the care with which he compared his data to those of his predecessors.

2.3 Astronomy in the Renaissance

FIGURE 2.11
Nicolaus Copernicus

Nicolaus Copernicus

The man who began the demolition of the geocentric model and the revolution in astronomical ideas that continues to this day was a Polish physician and lawyer by the name of Nicolaus Copernicus (1473–1543). Copernicus (fig. 2.11) made many attempts to reconcile the centuries of data on planetary positions that had been collected since Ptolemy's geocentric model, but all such attempts failed. Thus, he was led to reconsider Aristarchus's ancient idea that the Earth moves around the Sun.

You will remember that **heliocentric models** in which the Sun (*hēlios,* in Greek) was the center of the system had been proposed nearly 2000 years earlier by Aristarchus but had been rejected partly because the observational tools available at that time were inadequate to detect stellar parallax. Nevertheless, such models offer an enormously simpler explanation of retrograde motion. In fact, if the planets orbit the Sun, retrograde motion becomes a simple consequence of one planet on a smaller orbit overtaking and passing another on a larger orbit, as Copernicus was able to show.

To see why retrograde motion occurs, look at figure 2.12. Here we see the Earth and Mars moving around the Sun. The Earth completes its orbit, circling the Sun in 1 year, whereas Mars takes 1.88 years to complete an orbit, with the Earth overtaking and passing Mars every 780 days.

If we draw lines from the Earth to Mars, we see that Mars will appear to change its direction of motion against the background stars each time the Earth overtakes it. A very similar phenomenon happens when you drive on a highway and pass a slower car. Both cars are, of course, moving in the same direction. However, as you pass the slower car, it *looks* as if it is shifting backward against the stationary objects behind it.

ANIMATION
The retrograde motion of Mars according to the heliocentric model

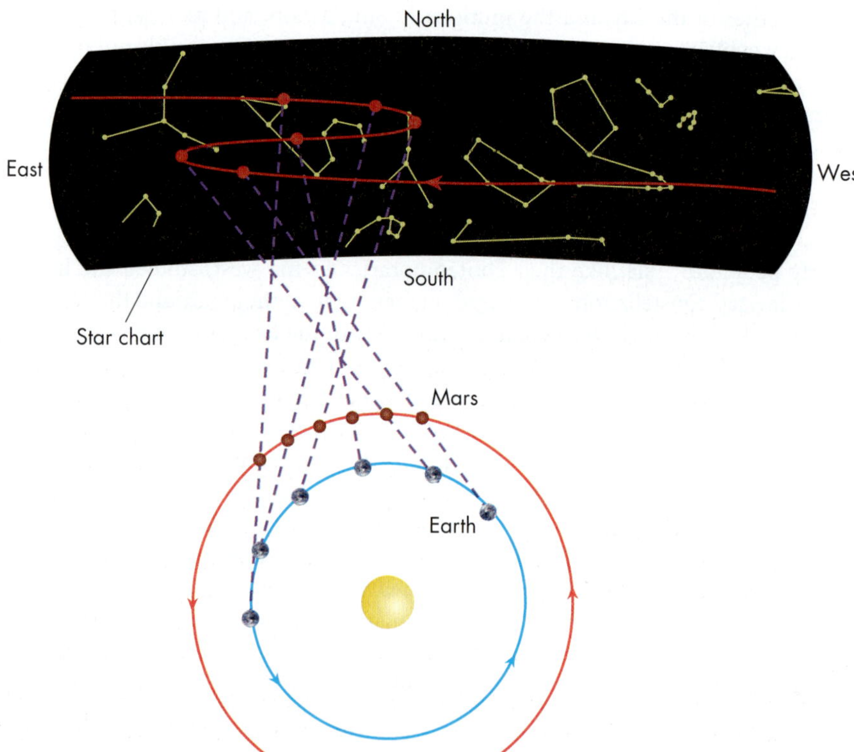

FIGURE 2.12
Why we see retrograde motion. (Object sizes, positions, and distances are exaggerated for clarity.)

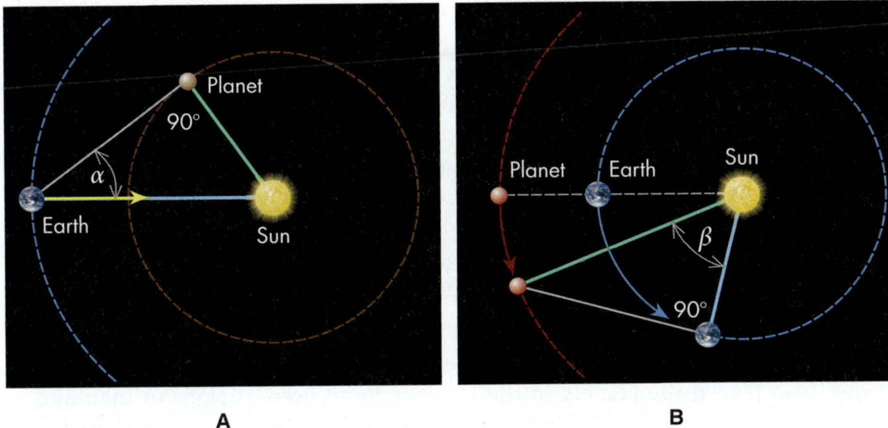

FIGURE 2.13
How Copernicus calculated the distance to the planets. (A) When an inner planet appears farthest from the Sun, the planet's angle on the sky away from the Sun, α, can be measured. You can see from the figure that at the same time the planet makes an angle of 90° with the Sun. The planet's distance from the Sun can then be calculated with geometry, if one knows the measured value of angle α and the fact that the Earth–Sun distance is 1 AU. (B) The distance to an outer planet can be found by determining interval of time from when the planet is opposite the Sun in the sky (the planet rises at sunset) to when the planet is 90° away from the Sun in the sky. From that time interval we can determine the fraction of their orbits that the Earth and planet moved in that time. Multiplying those fractions by 360° gives the angles for those movements. The difference between those angles gives angle β. Finally, using geometry and the value of angle β, the planet's distance from the Sun can be calculated.

With his heliocentric model, Copernicus not only could give a simple explanation of retrograde motion but also could determine each planet's distance from the Sun by a geometric construction, as shown in figure 2.13. The distances found in this manner must be expressed in terms of the Earth's distance from the Sun (whose value was not known until several hundred years later), but table 2.1 illustrates that they agree well with modern values. These distances became an important starting place for those who came later.

Copernicus described his model of a Sun-centered Universe in one of the most influential scientific books of all time, *De revolutionibus orbium coelestium (On the Revolutions of the Celestial Orbs)*. Because his ideas were counter to the teaching of the Catholic Church, they were met with hostility and skepticism. The book itself was not published until shortly before Copernicus's death (which was perhaps just as well), and he saw the first copy while on his deathbed.

Ironically, some of the criticism of Copernicus's work was justified. Although his model was basically correct, it did not account for the observed positions of the planets any more accurately than did Ptolemy's more complicated but incorrect model. This lack of complete agreement between model and observation arose at least in part because Copernicus insisted that the planetary orbits were circles. Furthermore, his model again raised the question of why no stellar parallax could be seen. Finally, his views of planetary motion ran counter to the teachings of Aristotle, views supported both by "common sense" and by the Catholic Church at that time. After all, when we observe the sky, it *looks* as if it moves around us. Moreover, we do not detect any sensations caused by the Earth's motion—it feels at rest. This mixture of rational and irrational objections made even scientists slow to accept the Copernican view.

However, by this time there was a growing recognition of the immensity of the Universe. Astronomers such as the Englishman Thomas Digges and the Italian Giordano Bruno went so far as to claim that the stars were other suns, perhaps with other worlds around them. This new scientific open-mindedness, coupled with the aesthetic appeal of the simpler system, led to a growing belief in the Copernican system.

TABLE 2.1 PLANETARY DISTANCES ACCORDING TO COPERNICUS

Planet	Copernicus's Distance	Actual Distance
Mercury	0.38 AU	0.39 AU
Venus	0.72 AU	0.72 AU
Earth	1.00 AU	1.00 AU
Mars	1.52 AU	1.52 AU
Jupiter	5.22 AU	5.20 AU
Saturn	9.17 AU	9.54 AU

FIGURE 2.14
Tycho Brahe

Tycho Brahe

Copernicus's model, although not the only stimulus, marked the opening of a new era in the history of astronomy. Conditions were favorable for new ideas: the cultural renaissance in Europe was at its height; the Protestant Reformation had just begun; the New World was being settled. In such an intellectually stimulating environment, new ideas flourished and, at least among scientists, found a more receptive climate than in earlier times.

One scientist whose ideas flourished in this more intellectually open environment was the sixteenth-century Danish astronomer Tycho Brahe (1546–1601). Born into the Danish nobility, Tycho (fig. 2.14) utilized his position and wealth to indulge his passion for study of the heavens, a passion based in part on his professed belief that God placed the planets in the heavens to be used as signs to mankind of events on Earth. Driven by this interest in the skies, Tycho designed and had built instruments of far greater accuracy than any yet devised in Europe. Tycho then used these devices to make precise measurements of planetary positions. His meticulous observations turned out to be crucial not only for distinguishing the superiority of the heliocentric over the geocentric system but also for revealing the true shape of planetary orbits.

Tycho was more than just a recorder of planetary positions; he recognized opportunity when he saw it. In 1572, when an exploding star (what we would now call a supernova) became visible, Tycho demonstrated from its lack of motion with respect to the other stars that it was far beyond the supposed spheres on which the planets move. Likewise, when a bright comet appeared in 1577, he showed that it lay far beyond the Moon, not within the Earth's atmosphere, as taught by the ancients. These observations suggested that the heavens were both changeable and more complex than was previously believed.

Although Tycho could see the virtues of the simplicity offered by the Copernican model, he was also unconvinced of its validity because he could find no evidence for stellar parallax. Therefore, he offered a compromise model in which all of the planets except the Earth went around the Sun, while the Sun, as in earlier models, circled the Earth. Tycho was the last of the great astronomers to hold that the Earth was at the center of the Universe.

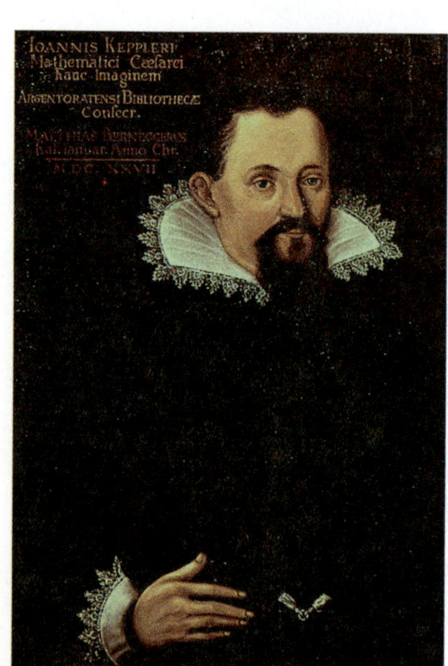

FIGURE 2.15
Johannes Kepler

Johannes Kepler

Upon Tycho Brahe's death, his observational data were passed to his young assistant, Johannes Kepler. Kepler (1571–1630) was able to derive from this huge set of precise information a detailed picture of the path of the planet Mars. Whereas all previous investigators had struggled to fit the planetary paths to circles, Kepler (fig. 2.15), by using Tycho's superb data, was able to show that the path of Mars was not circular but elliptical.

The shape of an **ellipse** is described by its long and short dimensions, called its major and minor axes, respectively. An ellipse can be drawn with a pencil inserted in a loop of string that is hooked around two thumbtacks. If you move the pencil while keeping it tight against the string, as shown in figure 2.16A, you will draw an ellipse. Each point marked by a tack is called a **focus** of the ellipse.

Having calculated from the observations that the orbit of Mars was elliptical, Kepler noticed that the Sun was located at a spot that was *not* the center of the ellipse but was off center at a focus. With the elliptical shape of the orbit now established, he was able to obtain excellent agreement between the calculated and the observed position not only of Mars but also of the other planets. Thus, Kepler's seemingly simple proposal that planetary orbits are ellipses and not circles was a critical step in understanding planetary motion.

TABLE 2.2 TABLE ILLUSTRATING KEPLER'S THIRD LAW FOR THE PLANETS KNOWN AT HIS TIME

Planet	Distance from Sun (a) (in Astronomical Units)	Orbital Period (P) (in Years)	a^3	P^2
Mercury	0.387	0.241	0.058	0.058
Venus	0.723	0.615	0.378	0.378
Earth	1.0	1.0	1.0	1.0
Mars	1.524	1.881	3.54	3.54
Jupiter	5.20	11.86	141.0	141.0
Saturn	9.54	29.46	868.0	868.0

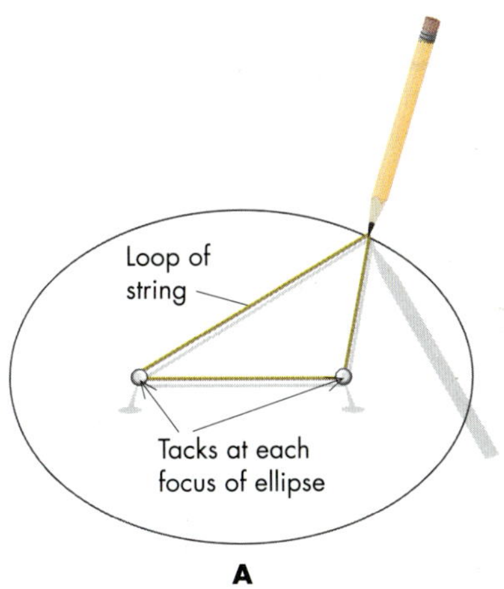

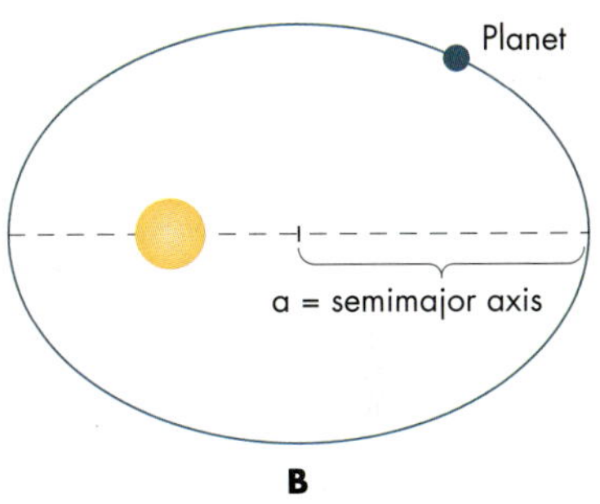

FIGURE 2.16
(A) Drawing an ellipse. (B) The Sun lies at one focus of the ellipse.

Along with discovering the shape of planetary orbits, Kepler also measured the relative sizes of the orbits. Because the orbits are elliptical, he could not describe them by a single radius. He therefore used the orbit's **semimajor axis**—half the major axis, as shown in figure 2.16B—as a measure of its size. When he compared these orbital sizes with how long it takes the planets to orbit the Sun, their periods, Kepler noticed the relation illustrated in table 2.2: the square of the orbital period was proportional to the cube of the orbital size, as measured by the semimajor axis.

Kepler's discoveries of the nature of planetary motions are expressed in what are known today as **Kepler's three laws:**

I. Planets move in *elliptical* orbits with the Sun at one focus of the ellipse (see fig. 2.17A).
II. The orbital speed of a planet varies so that a line joining the Sun and the planet will sweep over equal areas in equal time intervals (see fig. 2.17B).
III. The amount of time a planet takes to orbit the Sun is related to its orbit's size, such that the period, P, squared is proportional to the semimajor axis, a, cubed (fig. 2.17C). Mathematically,

$$P^2 = a^3$$

where P is measured in years and a is measured in astronomical units.

ANIMATION
Kepler's laws

These three laws describe the essential features of planetary motion around our Sun. They describe not only the shape of a planet's path but also its speed and distance from the Sun. For example, the second law—in its statement that a line from the planet to the Sun sweeps out equal areas in equal times—implies that when a planet is near the Sun, it moves more rapidly than when it is farther away. We can see this by considering the shaded areas in figure 2.17B. For the areas to be equal, the distance traveled along the orbit in a given time must be larger when the planet is near the Sun. Thus, according to Kepler's second law, as a planet moves along its elliptical orbit, its speed changes, increasing as it nears the Sun and decreasing as it moves away from the Sun.

INTERACTIVE
Kepler's second law

The third law also has implications for planetary speeds, but it deals with the relative speeds of planets whose orbits are at different distances from the Sun, not the speed of a given planet. Because the third law states that $P^2 = a^3$, a planet far from the Sun (larger a) has a longer orbital period (P) than one near the Sun (see table 2.2). For example, the Earth takes 1 year to complete its orbit, but Jupiter, whose distance from the Sun is slightly more than 5 times Earth's distance, takes about 12 years. Thus, a planet orbiting near the Sun overtakes and passes a planet orbiting farther out, leading to the phenomenon of retrograde motion, as discussed earlier.

INTERACTIVE
Kepler's third law

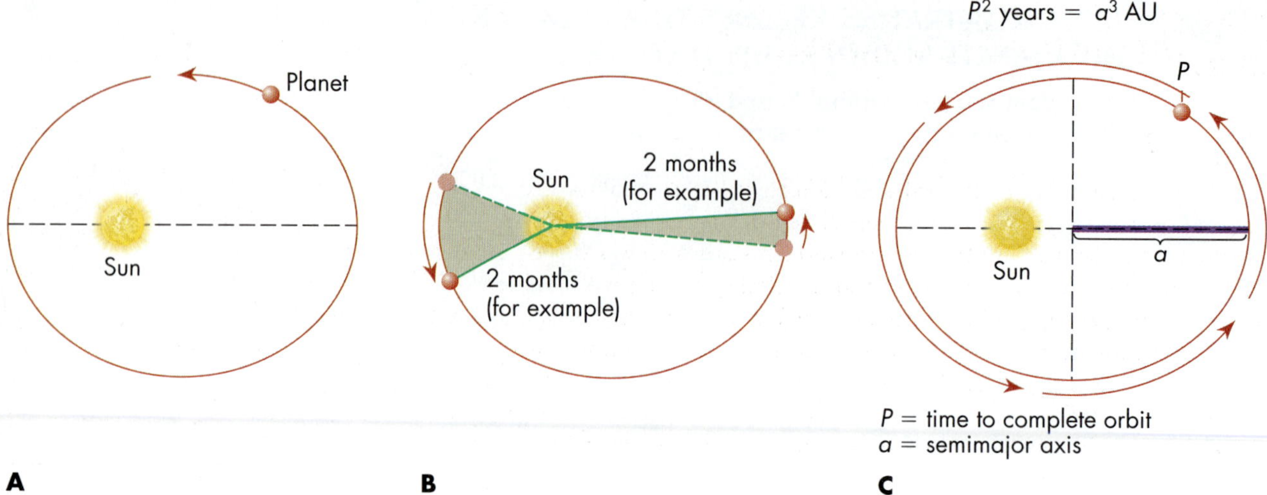

FIGURE 2.17
Kepler's three laws. (A) A planet moves in an elliptical orbit with the Sun at one focus. (B) A planet moves so that a line from it to the Sun sweeps out equal areas in equal times. Thus, the planet moves fastest when nearest the Sun. For purposes of the drawings a 2-month interval is chosen. (C) The square of a planet's orbital period (in years) equals the cube of the semimajor axis of its orbit (in AU), the planet's distance from the Sun if the orbit is a circle.

Kepler's third law has other implications. For example, as we shall see in section 2.4, the law gives information about the nature of the force holding the planets in orbit. Also, it implies that a planet close to the Sun moves along its orbit faster than a planet far from the Sun. Finally, the third law allows us to calculate the distance from the Sun of any body orbiting it if we measure the body's orbital period. The distance we obtain will be relative only to the Earth's distance, but the law thereby gives us at least the relative scale of the Solar System.

For example, suppose we wish to determine how far Pluto is from the Sun compared with the Earth's distance from the Sun. We measure Pluto's motion against the background stars to determine how long Pluto takes to circle the Sun, finding from such observations that its orbital period, P, is 248 years. Kepler's third law tells us that $P^2 = a^3$. If $P = 248$ yrs, then $(248)^2 = a^3$. Squaring 248 and then taking the cube root gives us $a = 39.5$ astronomical units. That is, Pluto is about 40 times farther from the Sun than the Earth is.

Apart from such astronomical applications, Kepler's laws have an additional significance. Kepler's laws are the first mathematical formulas to describe the heavens correctly, and as such they revolutionized our way of thinking about the Universe. Without such mathematical formulations of physical laws, much of our technological society would be impossible. These laws are therefore a major breakthrough in our quest to understand the world around us.

It is perhaps ironic that such mathematical laws should come from Kepler, because so much of his work is tinged with mysticism. For example, as a young man he sought to explain the spacing of the planets as described in Copernicus's work in terms of nested geometrical figures, the sphere, the cube, and so on. In fact, it was Tycho's notice of this work that led to his association with Kepler. Moreover, Kepler's third law evolved from his attempts to link planetary motion to music, using the mathematical relations known to exist between different notes of the musical scale. Kepler even attempted to compose "music of the spheres" based upon such a supposed link. Nevertheless, despite such excursions into these nonastronomical matters, Kepler's discoveries remain the foundation for our understanding of how planets move. The work of Tycho Brahe and Johannes Kepler was the pinnacle of pre-telescopic astronomy. Even as Kepler was developing his geometric and mathematical laws describing the motion of the planets, the nature of astronomy was about to change dramatically.

2.4 The Birth of Astrophysics

Galileo Galilei

At about the same time that Tycho Brahe and Johannes Kepler were striving to understand the motion of heavenly bodies, the Italian scientist Galileo Galilei (1564–1642) was likewise trying to understand the heavens. However, his approach was entirely different.

Galileo (fig. 2.18) was interested not just in celestial motion but in all aspects of motion. He studied falling bodies, swinging weights hung on strings, and so on. In addition, he used the newly invented telescope to study astronomical objects. Galileo did not invent the telescope himself. That invention seems to have been the work of the Dutch spectacle-maker Johannes (Hans) Lippershey. However, Galileo was the first person we know of who used the telescope to study the heavens and interpret his findings. What he found was astonishing.*

FIGURE 2.18
Portrait of Galileo

In looking at the Moon, Galileo saw that its surface had mountains and was in that sense similar to the surface of the Earth. Therefore, he concluded that the Moon was not some mysterious ethereal body but a ball of rock. He looked (without taking adequate precaution) at the Sun and saw dark spots (now known as sunspots) on its surface. He noticed that the position of the spots changed from day to day, showing not only that the Sun had blemishes and was not a perfect celestial orb but that it also changed. Both these observations were in disagreement with previously held conceptions of the heavens as perfect and unchangeable. In fact, by observing the changing position of the spots from day to day, Galileo deduced that the Sun rotated.

Galileo looked at Jupiter and saw four smaller objects orbiting it, which he concluded were moons of the planet. We owe credit to Kepler for the word *satellite*. When he saw the moons of Jupiter with a small telescope, their motion around the planet made him think of attendants or bodyguards—*satellēs*, in Latin. These bodies are known today as the Galilean satellites in honor of Galileo's discovery. They proved unambiguously that there were at least *some* bodies in the heavens that did not orbit the Earth. But perhaps even more important, the motion of these bodies raised the crucial problem of what held them in orbit.

When Galileo looked at Saturn, he discovered that it did not appear as a perfectly round disk but that it had blobs off the edge. However, his telescope was too small and too crudely made (inferior to inexpensive modern binoculars) to show these as rings. That discovery that had to wait until 1656, when they were first seen by the Dutch scientist Christiaan Huygens as features that were detached from the planet.

Galileo also examined the Milky Way and saw that it was populated with an uncountable number of stars. This single observation, by demonstrating that there were far more stars than previously thought, shook the complacency of those who believed in the simple Earth-centered Universe.

Galileo observed that Venus went through a cycle of phases, like the Moon, as illustrated in figure 2.19. The relation between the phase of the planet and its position with respect to the Sun left absolutely no doubt that Venus must be in orbit around the Sun exactly as the Moon is in orbit around the Earth. Perhaps more than any other observation, this one dealt the death blow to the old geocentric model of planetary motion.

Galileo's contributions to science would be honored even had he not made all these important observational discoveries, for he is often credited with originating the

*Thomas Harriott (1560–1621), an English mathematician–scientist, appears to have used a telescope to study the heavens a little before Galileo. He too saw sunspots and the moons of Jupiter, but he failed to publish his discoveries at the time.

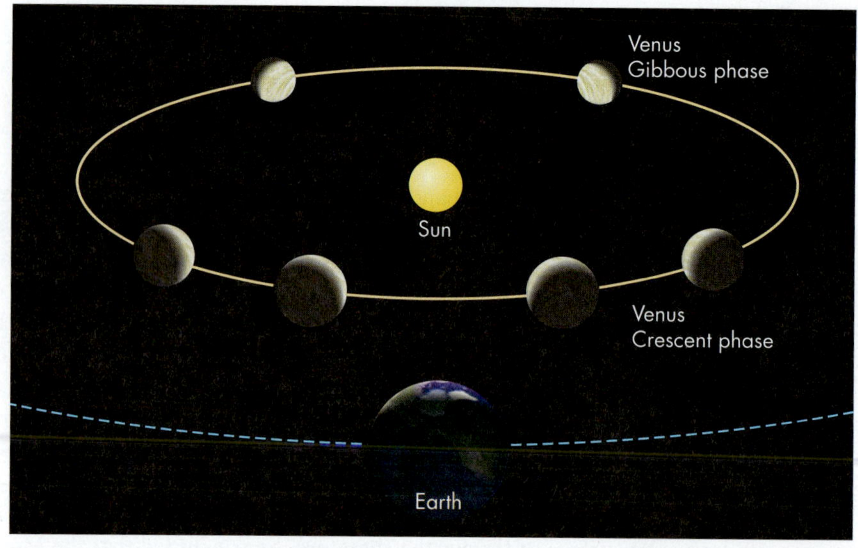

A

B

FIGURE 2.19
As Venus orbits the Sun, it goes through a cycle of phases (A). The phase and its position with respect to the Sun show conclusively that Venus cannot be orbiting the Earth. The gibbous phases Galileo observed occur for the heliocentric model but cannot happen in the Earth-centered Ptolemaic model (B), where Venus is shown on its epicycle.

experimental method for studying scientific problems. From his experiments on the manner in which bodies move and fall, Galileo deduced the first correct "laws of motion," laws that ultimately led Newton to his explanation of why the planets obey the laws of planetary motion that Kepler discovered.

Galileo's probings into the laws of nature led him into trouble with religious "law." He was a vocal supporter of the Copernican view of a Sun-centered Universe and wrote and circulated his views widely and somewhat tactlessly. His exposition followed the style of Plato, presenting his arguments as a dialog between a wise teacher (patterned after himself) and an unbeliever in the Copernican system whom he called Simplicio and who, according to his detractors, was patterned after the pope. Although the pope was actually a friend of Galileo, more conservative churchmen urged that Galileo be brought before the Inquisition because his views that the Earth moved were counter to the teachings of the Catholic Church. Considering that his trial took place at a time when the papacy was attempting to stamp out heresy, Galileo escaped lightly. He was made to recant his "heresy" and was put under house arrest for the remainder of his life. Only in 1992 did the Catholic Church admit it had erred in condemning Galileo for his ideas.

ANIMATION
The phases of Venus according to the Ptolemaic and Copernican systems

Isaac Newton

FIGURE 2.20
Isaac Newton

Isaac Newton (1642–1727), who was born the year Galileo died, is arguably the greatest scientist of all time. Newton's contributions span mathematics, physics, and astronomy. Moreover, Newton (fig. 2.20) pioneered the modern studies of motion, optics, and gravity. In his attempts to understand the motion of the Moon, Newton not only deduced the law of gravity but also discovered that he needed mathematical methods for calculating the gravitational force of a spherical body and that no such methods were then available. This realization led him to invent what we now know as calculus.

What is especially remarkable about Newton's work is that the discoveries he made in the seventeenth century still form the core for most of our understanding of gravity and the motion of bodies, discoveries we will discuss in more detail in chapter 3. In chapter 4, we will discuss some of Newton's ideas and discoveries about light, ideas that also are still in use.

Newton was a fascinating individual. He came from very modest origins and rose to high positions not only in academia but also in the government. He was Warden of the Mint and is alleged to have invented milling, the process whereby grooves are cut in the edge of coins to detect metal being pared off them, which would debase their value. He was also a deeply religious man and wrote prolifically on theological matters as well as science.

Newton's laws of motion, when combined with his law of gravity, were successfully applied for the next 200 years to essentially all problems of the motion of astronomical bodies. They still form the foundation for space flight today. These laws are so fundamental that chapter 3 is devoted to understanding them.

Astronomy and Astrology

Astrology is an ancient belief that the positions and patterns of celestial bodies in the sky exert an influence on the course of human events. Astronomy and astrology were not considered separate subjects before the seventeenth century. Actually, part of the motivation behind the astronomical discoveries of the Renaissance was the hope of better understanding the motions of celestial bodies in order to cast more accurate *horoscopes*. The idea of a horoscope is that the positions of celestial bodies at the time of a person's birth (particularly the position of the Sun in the zodiac) along with their current positions could provide predictive power over human events. Kepler and Galileo both cast horoscopes, and both pondered whether their new discoveries, such as the existence of satellites around the planets, might provide new insights into astrology.

Newton's discoveries of the laws of motion and gravity removed the mystery of the motions of the Sun, Moon, and planets. He and subsequent astronomers gave little or no credence to astrology, because carefully conducted studies show that it has no predictive power. This is not to say that astrologers never offer useful advice or even cast horoscopes that seem to be accurate, but they are no more accurate or predictive than guesses in which the birth dates and planetary positions are unknown.* This makes astrology a belief system rather than a science.

New Discoveries

Newton's enormous contributions tend to overshadow other advances in astronomy during the eighteenth and nineteenth centuries. That period began with

*You might be surprised to discover that your horoscope "sign" does not even reflect where the Sun is located among the stars anymore. Because of the precession of the Earth's axis (see Chapter 6), the Sun's position is shifted into the constellation where it used to be about a month earlier when astrology was devised over 2000 years ago.

observational discoveries that increased astronomers' confidence in using physical laws to understand the structure and workings of astronomical bodies. However, by the end of the period, newly found physical laws gave astronomers totally new tools for studying the heavens. In fact, the increasing use of the word *astrophysics* describes that shift well.

The shift of stars due to parallax as the Earth orbits the Sun was not detected until 1838, but proof of the Earth's motion was discovered in 1729. The motion of the Earth actually causes the observed positions of all stars to shift throughout the year because as the Earth moves through space, the angle of the light entering a telescope changes. This is the same effect that causes droplets to hit the front of your body more than your back as you run through falling rain. Your motion causes the rain to appear to fall at an angle toward your front side, and likewise the Earth's motion makes the light appear to come in at an angle toward the direction of the Earth's motion as it orbits the Sun.

Unexpected discoveries play a major role today in expanding our knowledge of the heavens, no less so than in the time immediately after Newton's death. For example, in 1781 the English astronomer Sir William Herschel discovered the planet Uranus. He also discovered that some stars have companion stars in orbit around them. The motion of such double stars offered additional tests of Newton's laws, but the most striking triumph of these laws of motion was their explanation of irregularities in the orbital motion of Uranus. Such irregularities hinted that another body was exerting a gravitational force on Uranus, and from Newton's laws, astronomers could calculate the position of the unseen body. As we will discuss further in chapter 10, a search of the sky near the calculated position revealed the planet Neptune.

New Technologies

Steady improvements in telescopes played an important role during this period. For example, improvements in optics allowed astronomers to build bigger telescopes and thereby observe much fainter objects. Among these objects, astronomers found dim, fuzzy patches of light—the so-called *nebulae*. Some of these were gas clouds within the Milky Way; others turned out to be external star systems similar to the Milky Way.

Another important technological advance was the application of photography to astronomy, starting in the middle of the nineteenth century. Photographic film gave astronomers permanent records of what they saw, and because film could store light during long exposures, astronomers were now able to detect objects much fainter than the eye could see in a single moment.

The scientific and technical advances described above have a direct bearing on astronomy, but scientific discoveries often influence totally unconnected areas. For example, during the eighteenth and nineteenth centuries, many scientists were studying the nature of matter and heat. The study of heat was prompted, at least in part, by a desire to improve the newly invented steam engine. Understanding the generation of heat and energy in turn gave new insights into how stars work, but it also presented a mystery—stars were generating far more power than could be explained by any known source of energy. This conundrum was finally resolved with the discovery of nuclear energy in the twentieth century.

It was also not until the twentieth century that the discovery of a tiny discrepancy in the motion of Mercury, as calculated using Newton's work, showed scientists that Newton's laws were not the last word on planetary motion. His descriptions of motion require modification if we are to correctly describe motion at speeds near that of light or where gravitational fields are very intense. These modifications are incorporated in Einstein's theories of relativity described in essay 2.

SUMMARY

Ancient peoples noted the basic patterns of the night sky, but the Greeks appear to have been the first to give explanations of planetary motion based on a combination of observational and geometric analysis. The Greeks pictured the planets, Sun, and Moon all orbiting the Earth on crystalline spheres.

Through the work of Aristotle and Eratosthenes, respectively, the Greeks knew the shape and size of the Earth. Aristarchus measured the relative size and distance of the Moon and Sun and proposed about 300 B.C. that the Earth orbited the Sun. However, his model was rejected because the expected shift in star positions (parallax) was unobservable at that time.

Planets also move with respect to the constellations but always within the narrow band of the zodiac. The zodiac follows the Sun's path around the sky, but it is slightly tilted with respect to the celestial equator, an imaginary line on the celestial sphere that lies directly above the Earth's equator. The usual direction of planetary motion is from west to east with respect to the stars. However, during part of the year, each planet's observed position shifts in the other direction, undergoing apparent retrograde motion.

Ptolemy (about A.D. 150) developed a complex model of planetary motion with the Earth at the center (geocentric) and with retrograde motion explained by planets moving on epicycles.

The geocentric model began to crumble in the 1500s with Copernicus's revival of the heliocentric model. Better observations by Tycho Brahe and detailed mathematical models by Kepler based on those observations placed the heliocentric model on a firmer basis. Galileo's observations with the recently invented telescope helped prove the heliocentric model.

Newton's discovery in the 1600s of the law of gravity and the laws of motion allowed him to explain why Kepler's laws worked, thereby completing the understanding of planetary motions.

QUESTIONS FOR REVIEW

1. (2.1) What is meant by the phrase *angular diameter*?
2. (2.1) If you triple your distance from an object, what happens to its angular size?
3. (2.1) What is parallax and how is it measured?
4. (2.2) If you see a bright "star" in the sky, how could you tell whether it is a star or, instead, for example, Venus?
5. (2.2) Where on the celestial sphere would you look for the planets?
6. (2.2) Sketch the path on the sky that a planet makes when undergoing retrograde motion.
7. (2.2) Will a planet in retrograde motion rise in the east or west?
8. (2.2/2.3) What is the difference between a heliocentric and a geocentric model?
9. (2.3) What are the three laws of planetary motion?
10. (2.4) How does astrology differ from astronomy?
11. (2.1-2.4) Describe the major astronomical contribution(s) of the following in a sentence or two for each: Aristotle, Aristarchus, Eratosthenes, Ptolemy, Copernicus, Tycho, Kepler, Galileo, and Newton.

THOUGHT QUESTIONS

1. (2.1) Suppose the stars were very much closer than they really are. How might that have made it easier for Aristarchus to persuade people that the Earth moves around the Sun?
2. (2.1-4) Make a table listing the astronomers named in review question 11 above, and then add the approximate dates of their births and deaths. Then add a few historic events of each period, as well as names of famous artists, writers, musicians, or politicians who lived at about the same time.
3. (2.1/2.2) Can you see Venus both before sunrise in the morning and just after sunset on the same day? Why?
4. (2.2/2.3) Tycho argued that the Sun orbits the Earth but that the other planets orbit the Sun. Could Tycho's model explain the phases of Venus as observed by Galileo? Why?
5. (2.3) We know from Kepler's laws that the periods of the outer planets are very long. Jupiter, for example, has a period of almost 12 years. How then is it that, over a matter of months, you could observe Jupiter's position on the sky move from one side of the Sun, then closer to the Sun, then past it to the other side? (Drawing a sketch might be helpful).
6. (2.3) You may have noticed that although every ten years or so there is a comet visible in the night sky, the same comet is only seen once or twice during a human lifetime. Use this fact and Kepler's third law to deduce how the semimajor axis and shape of a comet's orbit must compare to the Earth's orbit.
7. (2.4) Describe how modern astrophysics differs from ancient astronomy, with examples based on the work of specific astronomers or astrophysicsts.

PROBLEMS

1. (2.1) Suppose you are an alien living on the fictitious warlike planet Myrmidon and you want to measure its size. The Sun is shining directly down a missile silo 1000 miles to your south, while at your location, the Sun is 36° from

straight overhead. What is the circumference of Myrmidon? What is its radius?

2. (2.1) The great galaxy in Andromeda has an angular diameter along its long axis of about 5°. Its distance is about 2.2 million light-years. What is its linear diameter?
3. (2.1) Suppose that the shell of gas blown out of a star has an angular diameter of 0.1° and a linear diameter of 1 light-year. How far away is it?
4. (2.1) A small probe is exploring a spherical asteroid. As the probe creeps over the surface, it drills holes to take soil samples. Scientists on Earth notice that the Sun shines straight down into one of the holes. At the same time, 10 kilometers due "north," the shadow of the vertical antenna on the main landing craft allows the scientists to deduce that the Sun is 15° from directly overhead. What is the radius of the asteroid? How many times smaller or bigger than Earth's is its radius?
5. (2.3) Suppose a planet is found with an orbital period of 64 years. How might you estimate its distance from the Sun? If its orbit is circular, what is its radius?
6. (2.3) Suppose you receive a message from aliens living on a planet orbiting a star identical to our Sun. They say they live 4 times farther from their star than the Earth is from the Sun. What is the length of their year compared to ours?
7. (2.3) A planet is discovered orbiting a nearby star once every 125 years. If the star is identical to the Sun, how could you find the planet's distance from its star? If the planet's orbit is a perfect circle, how far from the star is the planet in AU?
8. (2.3) Suppose that future observations with a new telescope reveal a planet about 16 AU from a star whose mass is the same as our Sun's. How long does it take the planet to orbit the star?

TEST YOURSELF

1. (2.1) A total solar eclipse demonstrates that the Moon and Sun are very nearly the same angular size. If the Sun is 400 times farther from us than the Moon, then the radius of the Moon must be __________ the radius of the Sun.
   (a) 1600 times as (b) 400 times as (c) the same as (d) 1/400th of (e) 1/1600th of
2. (2.2) A planet in retrograde motion
   (a) rises in the west and sets in the east.
   (b) shifts westward with respect to the stars.
   (c) shifts eastward with respect to the stars.
   (d) will be at the north celestial pole.
   (e) will be exactly overhead no matter where you are on Earth.
3. (2.2) "Occam's razor" refers to
   (a) a device used by the ancient Greeks to measure the angle between the Sun and planets.
   (b) a metaphor for the process of discriminating between models based on their simplicity.
   (c) another term to describe the heliocentric model.
   (d) a method used to execute heretics.
   (e) a description of retrograde motion of planets.
4. (2.3) If an asteroid has an average distance from the Sun of 4 AU, what is its orbital period?
   (a) 1 year (b) 2 years (c) 4 years (d) 8 years (e) 16 years
5. (2.3) Kepler's third law
   (a) relates a planet's orbital period to the size of its orbit around the Sun.
   (b) relates a body's mass to its gravitational attraction.
   (c) allowed him to predict when eclipses would occur.
   (d) allowed him to measure the distance to nearby stars.
   (e) showed that the Sun is much farther away than the Moon.
6. (2.4) Galileo used his observations of the changing phases of Venus to demonstrate that
   (a) the Sun moves around the Earth.
   (b) the Universe is infinite in size.
   (c) the Earth is a sphere.
   (d) the Moon orbits the Earth.
   (e) Venus follows an orbit around the Sun rather than around the Earth.
7. (2.4) A major objection to the heliocentric model not resolved until the development of high-quality telescopes was that
   (a) the speed of light had been thought to be infinite.
   (b) the Moon was believed to shine by its own light, not reflected light from the Sun.
   (c) the stars did not exhibit parallax.
   (d) Jupiter did not show a crescent phase.
   (e) Earth's gravitational pull was originally estimated to be stronger than the Sun's.

KEY TERMS

angular size, 44
ellipse, 52
epicycle, 49
focus, 52
geocentric model, 48
heliocentric model, 50
Kepler's three laws, 53
Moon illusion, 45
parallax, 43
retrograde motion, 47
semimajor axis, 53

FURTHER EXPLORATIONS

Ahmad, Imad A. "The Science of Knowing God: Astronomy in the Golden Era of Islam." *Mercury* 24 (March/April 1995): 28.

Farrell, Charlotte. "The Ninth-Century Renaissance in Astronomy." *Physics Teacher* 34 (May 1996): 268.

Gingerich, Owen. "Astronomy in the Age of Columbus." *Scientific American* 267 (November 1992): 100.

Gingerich, Owen. "Islamic Astronomy." *Scientific American* 254 (April 1986): 74.

Gingerich, Owen. "Copernicus and Tycho." *Scientific American* 229 (December 1973): 86.

Gingerich, Owen. *The Great Copernicus Chase.* New York: Cambridge University Press, 1992.

Gould, Stephen J. "The Persistently Flat Earth." *Natural History* 103 (March 1994): 12.

Hoskin, Michael, ed. *The Cambridge Illustrated History of Astronomy.* New York: Cambridge University Press, 1997.

Website

Visit the ***Explorations*** website at **http://www.mhhe.com/arny** for additional online resources on these topics.

Q FIGURE QUESTION ANSWERS

WHAT IS THIS? (chapter-opening): Photograph of a partial lunar eclipse. The person in the photo is holding up a hoop, which has about the same angular diameter as the shadow of the Earth as seen by the photographer. Early astronomers recognized that the shape of Earth's shadow is always part of a circle, so they realized that the Earth must be spherical.

FIGURE 2.8: Mars appears to move backward when Earth is passing it at the same time as Earth is closest to Mars, so Mars appears its brightest.

PROJECT

Measure the diameter of the Earth: This project duplicates the method used by Eratosthenes; he used the length of a shadow at two different locations to determine the Earth's size. Thus, you need either to collaborate with someone relatively far away (at least several hundred miles north or south of you) or to travel such a large distance yourself. The larger the distance, the better.

Begin by setting up a vertical stick in a sunny spot. Tape down paper beside the stick and mark off on it the length of the stick's shadow over a few hours around noontime. Record the length of the stick and the smallest shadow.

On the same day as you do your measurement, have your collaborator in the other location do the same. If you are going to travel yourself—for example, during a school break—make your measurements at the two different locations as close together in time as possible.

From a road atlas, measure the straight-line, north–south distance between the two places where you made the measurements. On a piece of paper, make a careful scale drawing of the stick and shadow at each location. A sample pair of drawings is shown in Project figure 2.1A. From these drawings, measure the angles A and B at the top of the triangles.

Now look at Project figure 2.1B. This shows a beam of sunlight striking the Earth and the sticks you used in your experiment at the two different locations. Lines 1, 2, and 3 represent rays of light from the Sun, with lines 1 and 2 just cutting the top of the stick at the two locations, and line 3 passing through the center of the Earth. Notice that the difference in latitude between the two locations is just the angle AB. To see why this is so, notice that a line from the center of Earth to A makes equal angles with lines 1 and 3 because these lines are parallel. Similarly, a line from the center to B makes equal angles with lines 2 and 3. If we let l be the difference in latitude between the two locations, then $l = A - B$. You could, of course, just look up that latitude difference in an atlas, but there were no atlases when Eratosthenes lived.

Once you know the angle of the difference in latitude, l, you can find the circumference of the Earth. A little geometry will show you that $360/l =$ the Earth's circumference$/d$, where d is the distance in miles or kilometers between the two places at which you made the observations. Thus, you can solve for the Earth's circumference. With the Earth's circumference known, you can then find its radius from the formula Circumference $= 2\pi R$. How does your answer compare with the value you find in the appendix? What might explain the difference? How might you conduct the experiment more accurately?

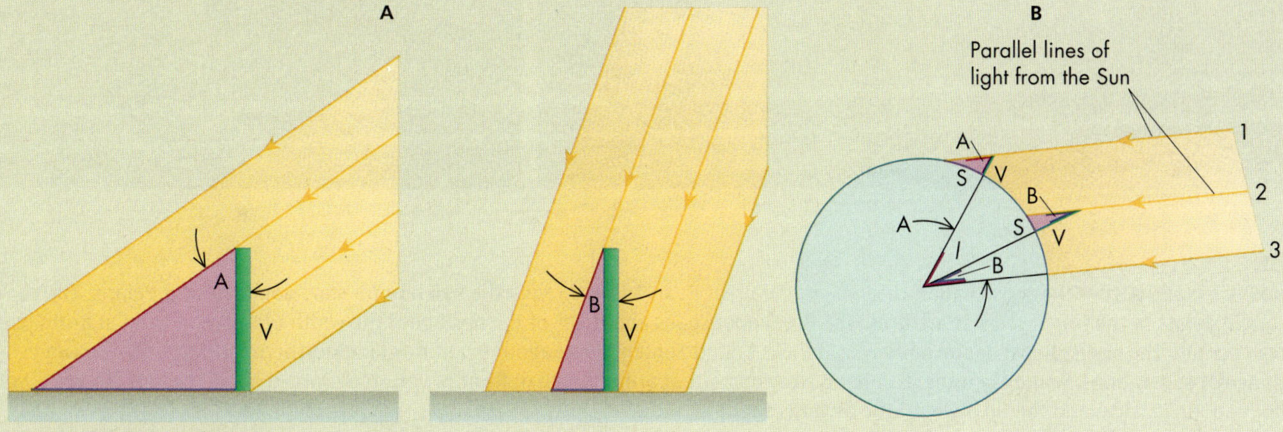

PROJECT FIGURE 2.1
Sketch illustrating a method for finding the radius of the Earth.

ESSAY 1

Backyard Astronomy

You can learn many of the same things classical astronomers did by simply watching the night sky. But there is a bonus as well. Backyard astronomy is just plain fun, as evidenced by the many thousands of amateur astronomers who in their spare time pursue activities ranging from simply stargazing to searching for new comets.

This section is intended to give you some hints on how to become an amateur astronomer, beginning with learning the constellations and some of the stories associated with them. We will then briefly discuss small telescopes and star charts, introducing some of the terms used to describe the location of the planets. We will conclude with a description of some of the physical changes in your eye when you are observing in very dim light.

LEARNING THE CONSTELLATIONS

One of the best ways to get started as a backyard astronomer is to learn the constellations. All it takes is a star chart (such as the ones provided at the back of this book), a dim flashlight, and a place that is dark and has an unobstructed view of the night sky. The star chart will tell you how to hold it so that it matches the sky for the date and time that you are observing.

Start by determining which way is north, using a compass if necessary. Then try to locate a few of the brighter stars, matching them up with the chart. This will give you some sense of how big a piece of the sky the chart corresponds to. Next, try to identify a few of the constellations. Focus at first on just a few of the brighter ones. For example, if you live at midlatitude in the Northern Hemisphere, the Big Dipper—part of the constellation Ursa Major—is a good group to start with because it is always visible in the northern part of the sky.

As you attempt to find and identify stars, your spread hand held at arm's length makes a useful scale. For most people, a fully spread hand at arm's length covers about 20° of sky, or about the length of the Big Dipper from tip of handle to bowl, as shown in figure E1.1. For smaller distances, you can use your thumb's width, which is about 2°, or your little finger's width, which is about 1°.

This scaling of sky distances with your hand makes it easy to point out stars to other people. For example, you can

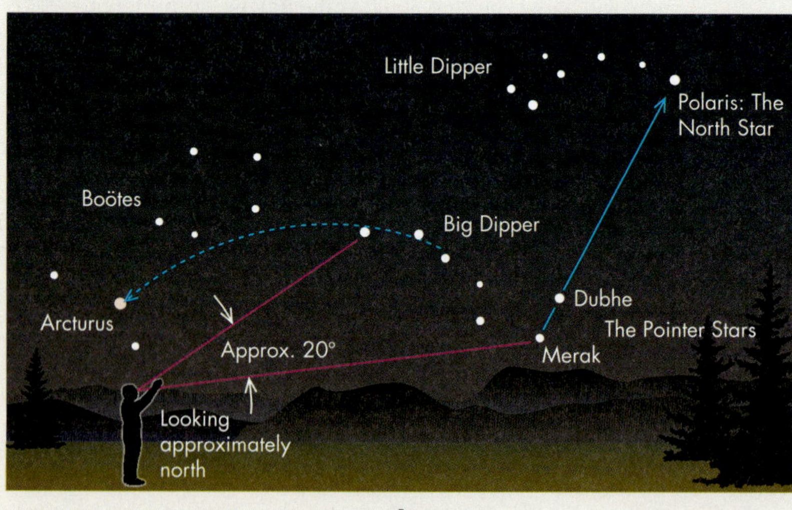

A

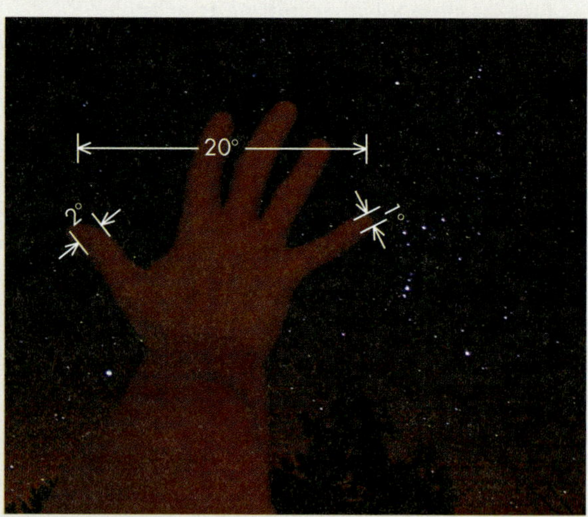

B

FIGURE E1.1
(A) The Big Dipper, part of the constellation Ursa Major, the Great Bear. A line through the two pointer stars points toward Polaris. The arc of the handle points to the bright red star Arcturus. The Big Dipper spans about 20° of the sky, about the width of a spread hand at arm's length for most people. The sky is shown approximately as it looks in mid-September at about 9 p.m.from midnorthern latitudes. (B) You can estimate angular separations on the sky using your hand stretched out at arm's length in front of you. Your spread hand is about 20° wide, your thumb is about 2° wide, and the tip of your little finger is about 1° wide.

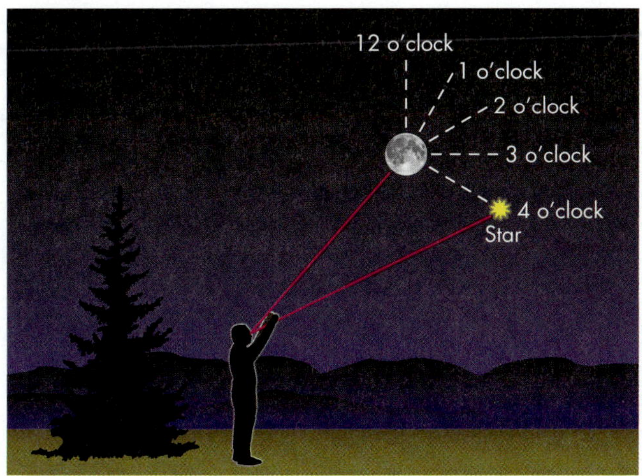

FIGURE E1.2
Describing the location of stars by clock position. The star is two hands from the Moon and at 4 o'clock position.

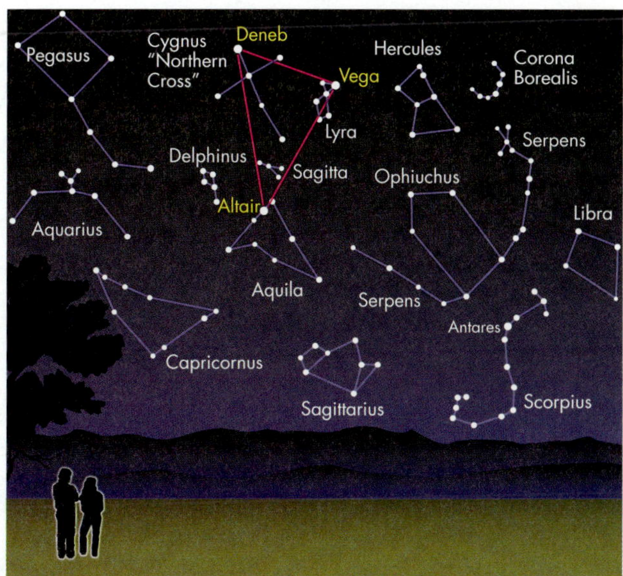

FIGURE E1.3
Dominating the night sky in July, August, and September are the three bright stars Vega, Altair, and Deneb, which form the Summer Triangle. This sketch shows how the sky looks looking south (from midnorthern latitudes) at about 9 p.m. in early September.

say that a star is two hands away from the Moon and at the 4 o'clock position, as illustrated in figure E1.2.

Once you have come to recognize a few of the constellations, you may be interested in the stories associated with them—that is, star lore.

STAR LORE

Star lore is part of virtually all cultures. The ancient Greeks, the Pawnee tribes of the American Midwest, the Australian aborigines, for example, all created stories about the star groupings they saw in the sky.

Because the star groupings do not change except on time scales of tens of thousands of years, the night sky we see is essentially the same night sky that ancient peoples saw. Star lore can therefore link us to the remote past. Furthermore, star lore can help us to remember constellations. In fact, it has been suggested that many of the stories were created as aids to memory, which were especially important in a time when familiarity with the stars could be literally a matter of life or death to a farmer or a navigator. Scientists have shown that baby birds learn to recognize star patterns and movements and use them to navigate safely—unguided by their parents—across thousands of miles of ocean to their winter homes. Perhaps we too have such instinctive faculties that help us learn the stars.

Probably the most familiar star grouping is the Big Dipper. It is not a constellation but rather is called an **asterism.** An asterism is an easily recognized grouping of stars that may be part of one constellation or may incorporate pieces of several. For example, the Big Dipper is part of the constellation Ursa Major, the Great Bear. The asterism of the Summer Triangle, on the other hand, spans three constellations. It consists of the three bright stars conspicuous in the summer evening: Deneb (in Cygnus, the Swan), Altair (in Aquila, the Eagle), and Vega (in Lyra, the Harp), shown in figure E1.3. "Looking Up #4" (in the front matter) shows a photograph of the Summer Triangle and some of the other constellations that lie near it.

The native inhabitants of North America had a story about the Big Dipper. Like the ancient Greeks, they too saw a bear in this set of stars. The dipper's "bowl" represented a huge bear, and the "handle" represented three warriors in pursuit of the bear. They had wounded it, and it was bleeding. The red color of the leaves in autumn was said to be caused by the bear's blood dripping on them when the constellation lies low in the sky during the evening hours of the autumn months.

The Big Dipper not only is easy to spot, but is also an excellent signpost to other asterisms and stars. For example, if you extend the arc of the handle of the Big Dipper, you come to the bright red star Arcturus ("follow the arc to Arcturus") as illustrated in figure E1.1A. The two stars farthest from the handle in the bowl of the Big Dipper (see fig. E1.1A) are called the "pointers" because they point, roughly, to the North Star, Polaris.

Polaris marks the end of the handle of the Little Dipper, another asterism and part of the constellation Ursa Minor, the Little Bear. The Little Dipper is not easy to see because its stars are dim, but it curves back toward the Big Dipper. Look for these two star groupings near the north celestial pole.

Polaris is an important star for navigation because it lies almost exactly above the Earth's North Pole. Because of its position there, it is the only star in the northern sky that shows, to the naked eye, no obvious motion during the night. Its relatively fixed position is illustrated by the time exposure in figure E1.4, showing the other stars rotating around it. Because Polaris

FIGURE E1.4
A time exposure showing how Polaris remains essentially fixed while the sky pivots around it.

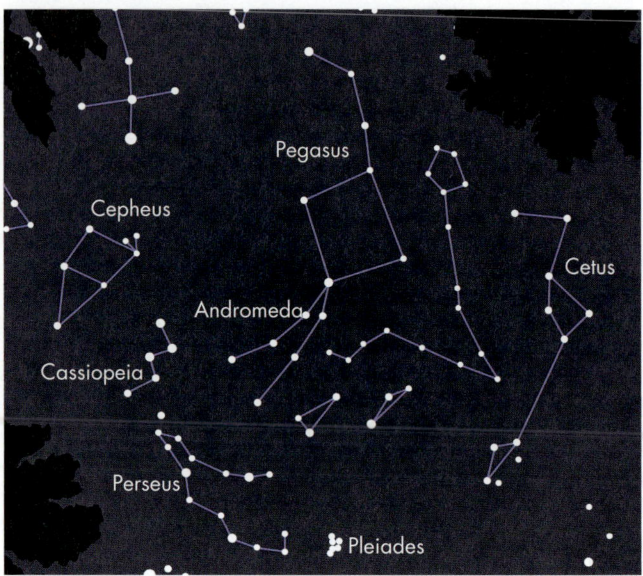

FIGURE E1.5
Perseus, Andromeda, Cassiopeia, Cepheus, Cetus, and Pegasus. The sky is drawn as it looks in mid-November at about 8 p.m., looking approximately straight overhead from midnorthern latitudes.

always lies nearly true north, it is useful in orienting yourself to compass directions.

If you follow the pointer stars in the Big Dipper past the Little Dipper and Polaris, you will come to a set of constellations tied together by an ancient Greek myth, the story of Perseus and Andromeda. The cast includes a king (Cepheus), a queen (Cassiopeia), the hero (Perseus), the princess (Andromeda), a sea monster (Cetus), and the winged horse (Pegasus). The constellations are shown in figure E1.5, and their story goes as follows:

In ancient days there lived a queen of Ethiopia, Cassiopeia, who was very beautiful but also very vain. She and king Cepheus, her husband, and their daughter, Andromeda, lived happily until one day the queen boasted that she was more beautiful than the daughters of Nereus, a sea god. In punishment for such pride, the sea-god Neptune sent a sea monster, Cetus, to ravage the kingdom. To save his people and appease the gods, Cepheus was instructed to tie his daughter, Andromeda, to a rock for the monster to devour. Meanwhile, Perseus was returning home from a quest in which he slew the snake-haired Gorgon, Medusa. On her death, her blood dripped into the sea and turned into the flying winged horse, Pegasus. Medieval versions of the story have Perseus riding Pegasus, but the classical myth has him borne on winged sandals. Regardless of his means of travel, Perseus saw the maiden's peril and landed, slaying the monster and delivering the kingdom. They all lived as happily ever after as most mythological families, and their astronomical depiction is most easily seen in the late autumn, when the brighter of its constellations are high in the sky.

Stories are also told about stars in other parts of the sky. For example, in the winter sky you can see the Hunter, Orion, and the maiden who, in the sad legend, refused to fall in love with him. The story also involves Orion's hunting dogs (Canis Major and Canis Minor), a bull (Taurus), a rabbit (Lepus), the maiden's sisters (the Pleiades—a cluster of stars in the constellation Taurus), and a scorpion (Scorpius).

The king of the island Chios had a lovely daughter, Merope. His island was filled with savage beasts, and to rid his kingdom of these dangerous animals, the king called upon Orion to kill the beasts and make his kingdom safe. When the task was done, Orion met Merope and made unwelcome advances. In punishment, he was blinded by the king, but after he did penance his sight was restored. After reaching an old age, however, Orion one day stepped on a scorpion, which stung and killed him. Upon his death, the gods placed him in the sky with his faithful dogs (one of whom chases Lepus, the rabbit), forever attacking the wild bull, Taurus. Beyond the bull, Merope and her sisters (the Pleiades) run from the hunter, who pursues them each night across the sky. The scorpion was also placed in the sky, but on the other side of the heavens so that Orion would never again be threatened by it (Orion is visible in the evening only in the winter, whereas Scorpius is visible in the evening only in the summer).

The Orion myth has several versions, but the one described here fits together many of the astronomical references. Another myth with several versions involves the late summer constellations Boötes and Virgo. They are linked by the gloomy story of Icarius, the first cultivator of grapes for wine. According to legend, in recompense for sharing this knowledge, he was killed by drunken peasants, who buried his body under a tree. His dog, Maera, led Erigone, his daughter, to the spot, where, on discovering her father's body, she killed herself out of grief. Icarius

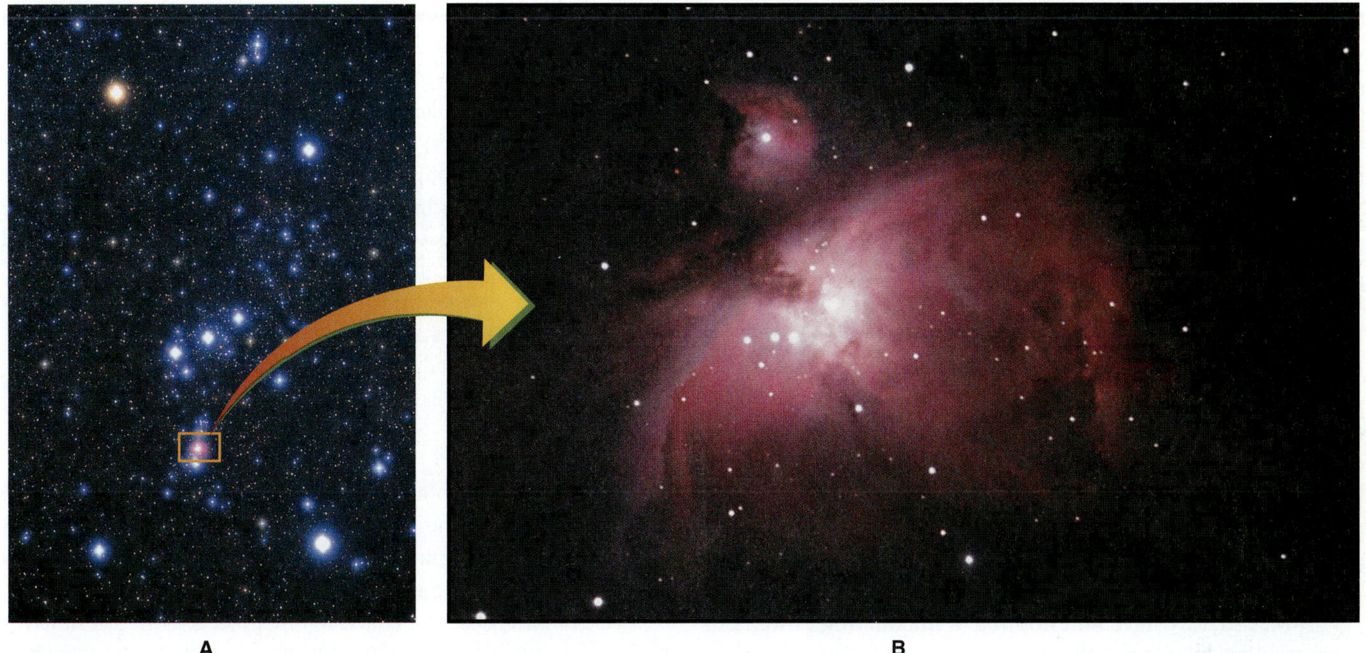

FIGURE E1.6
(A) A photograph of the constellation Orion made with an ordinary camera. (B) A picture of the Orion Nebula taken with a small backyard telescope.

was placed in the sky as Boötes, and Erigone as Virgo. The dog Maera became the star Procyon.

There are many other stories about constellations, but the above may give you some sense of the ones that have been handed down for thousands of years of written and oral history. Pass them on.

AMATEUR ASTRONOMY

Anyone with access to even very modest equipment such as binoculars or low-powered telescopes has better equipment than Galileo ever had. With such equipment and access to a dark sky, that person can become an amateur astronomer. The pleasures of the hobby can range from the aesthetic satisfaction of taking a lovely photograph (fig. E1.6) to the thrill of discovering a new comet or an exploding star.

Many amateur astronomers now use digital cameras for astrophotography. With a middle- to upper-price-range camera mounted on a tripod, you can make surprisingly good images of constellations. If you use still more expensive, special-purpose CCDs,* you can take pictures with a backyard telescope that rival those made 40 years ago with a 200-inch telescope.

To take astronomy pictures with a digital camera, you generally must change the camera's settings from automatic. For example, because you will typically need an exposure time of 10 or so seconds, the camera should be mounted on a tripod. You should also set the lens aperture as wide open as possible and the focus distance at infinity. In addition, if you can change the ISO setting, make it at least 400, preferably higher (higher ISO values allow you to image dimmer objects). Your camera's instruction guide will tell you how to make these changes. You can also find helpful suggestions at a number of websites that describe how to do astrophotography. A cable release will help you expose without shaking the camera but is not essential.

If you expose for more than about 15 seconds, the Earth's rotation will smear the star image into a streak. Deliberately allowing the smearing to occur can produce dramatic pictures of what are called "star trails" (see fig. E1.4). To make star-trail pictures, leave the shutter open for 20 minutes or so.

To take untrailed long exposures or to use a telephoto lens, you will need a way to compensate for the Earth's rotation. You can make a simple device to do this from hinged boards with a carriage bolt through them. Mount the camera on one board and set the other on a firm surface. Slowly turning the bolt will shift the camera, allowing you to take exposures of half an hour or so. For best results, however, you need a motor drive. Check your local library or the references at the end of the chapter for books or articles on drives and astrophotography. You may also find in such sources suggestions for scientifically useful projects, such as variable star observing or comet hunting. Projects like these are ideal for amateur astronomers because the results are not only valuable but also require basically only patience and care without the need for a big telescope.

*CCD = "charge-coupled device." See chapter 5 for more details on these electronic cameras.

SMALL TELESCOPES

A small telescope will greatly increase the number and interest of objects you can observe. Such telescopes come in a wide range of styles and prices, but selecting the best one for your needs can be confusing. Moreover, it is a sad truth that you generally get what you pay for.

Many amateur astronomers begin with a 3- to 4-inch reflecting telescope. Such a telescope uses a mirror to collect and focus the light; thus its name, "reflecting." The numbers refer to the diameter of the mirror, important because a larger mirror collects more light, allowing you to see fainter objects. A larger mirror also permits seeing finer details, other things being equal (see chapter 5 for how a telescope works). With such a telescope, you can easily see the moons of Jupiter, the rings of Saturn, and many lovely star clusters and galaxies. However, these latter objects will not look like the pictures in books because your eyes—unlike a photograph or electronic detector—cannot store up light. Moreover, your eyes are far less sensitive to colors in very dim light.

Notice that we have said nothing of magnifying power. For most amateur telescopes, the maximum useful power is limited by distortions to the light as it passes through our atmosphere. These distortions make a magnification of about 100 to 200 the useful limit. Beyond that, the distortions dominate and higher power gives no increase in clarity.

Whatever type of telescope you choose, be sure to get a sturdy mounting for it. Even at 100 power, tiny vibrations of the telescope caused by wind or the touch of your hand will make the image jiggle, hopelessly blurring it.

Before you actually buy a telescope, you might want to talk with your instructor or a local amateur astronomer. Such people often belong to an astronomy club, some of whose members may have secondhand telescopes they are willing to sell at reduced prices. For information about new telescopes, browse through the magazines *Astronomy* or *Sky and Telescope*—publications widely read by both amateur and professional astronomers—which contain many advertisements for small telescopes.

STAR CHARTS

One of the pleasures of using a telescope is to look at objects too faint to be seen by the naked eye, such as most galaxies, faint star clusters, and remnants of dying stars. To find many of these dim objects, you will need a good star chart.

Astronomers use star charts to find objects in the sky much as navigators use charts to find places on Earth. A typical star chart (fig. E1.7) shows the location of the constellations, the stars, and other objects. It also gives some indication of the relative brightness of the stars. Finally, many charts also have information about the season and time of night at which the stars are visible.

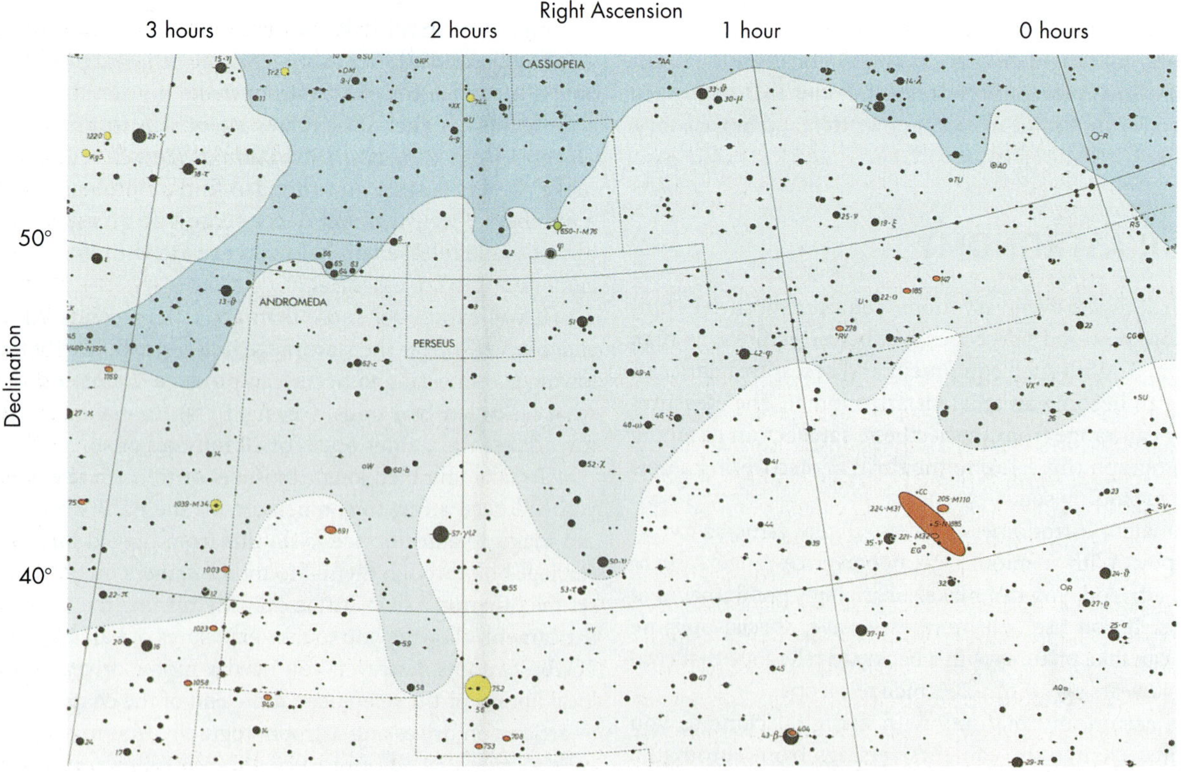

FIGURE E1.7
A star chart showing stars, galaxies, and coordinates. Black circles are stars. Their size indicates their brightness—larger are brighter. Red ellipses are galaxies; the large one at lower right is M31, the nearest large galaxy to our own. Blue shading indicates the brightness of the Milky Way.

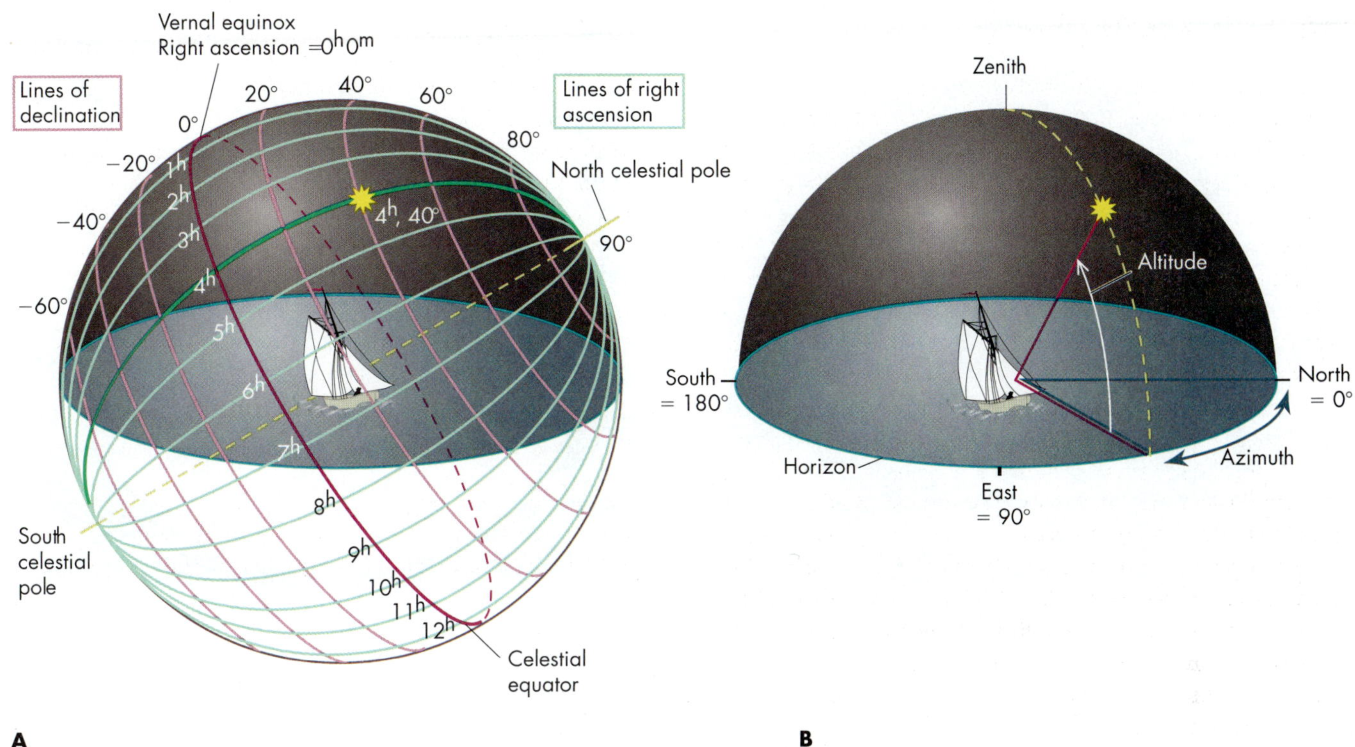

FIGURE E1.8
(A) Locating a star according to right ascension and declination. (B) Locating a star according to altitude and azimuth.

Star charts are designed much like maps of the Earth. For example, both represent on a flat surface a map of something curved—in the one case the celestial sphere; in the other the surface of the Earth. Also, both use a coordinate grid.

Nowadays, star charts are as likely to be on a computer as on paper. These offer tremendous ease of use, and many are very beautifully done with many additional features.

CELESTIAL COORDINATES

The coordinate grid used by astronomers is similar to that used by navigators. The grid consists of one set of lines running east–west on the celestial sphere, parallel to the celestial equator, and another set running north–south, connecting one celestial pole to the other. The east–west lines play the same role as latitude on the Earth, but to avoid confusion with terrestrial coordinates, they are called lines of **declination,** or "dec" for short. The north–south lines play the same role as longitude on the Earth and are called lines of **right ascension,** or "RA" for short. Declination values run from +90° to −90° (the north and south celestial poles), with 0° being the celestial equator. Right-ascension lines divide the celestial sphere into 24 equal zones that are labeled not in degrees but in units of time. Thus, the right ascension of an object is given in hours (h), minutes (m), and seconds (s). Because the 360° around the sky is divided into 24 segments, each hour of RA equals 15°; that is, 360° divided by 24 = 15°. The point $0^h\,0^m\,0^s$ of RA is arbitrarily chosen to be where the Sun's path, the ecliptic, crosses the celestial equator as the Sun moves north (fig. E1.8A).

Right ascension and declination are not the only way to locate objects on the sky. For example, a celestial object may be located by its altitude and azimuth, as illustrated in figure E1.8B. In this so-called horizon system, **altitude** is an object's angle above the horizon. **Azimuth** is generally defined as the angle measured eastward along the horizon from North to the point directly below the object, although sometimes it is measured from South. Whichever convention is used for azimuth, these coordinates are useful for pointing out objects or tracking their motion. However, they have a serious drawback compared to the otherwise seemingly cumbersome right ascension and declination: an astronomical body's altitude and azimuth constantly change as it moves across the sky, whereas, because they are defined with respect to the rotating celestial sphere, a body's right ascension and declination remain the same as a body rises and sets.

With a set of coordinate lines established, we can now locate astronomical objects on the sky the same way we can locate places on the Earth. For example, M31, a galaxy in the Local Group, is at right ascension 0 hours 42.7 minutes ($0^h\,42.7^m$) declination +41°16′, as you can see in figure E1.7.

PLANETARY CONFIGURATIONS

Because planets move across the stellar background, astronomers have invented some terms to help describe where they are located at any given time. These terms describe a planet's

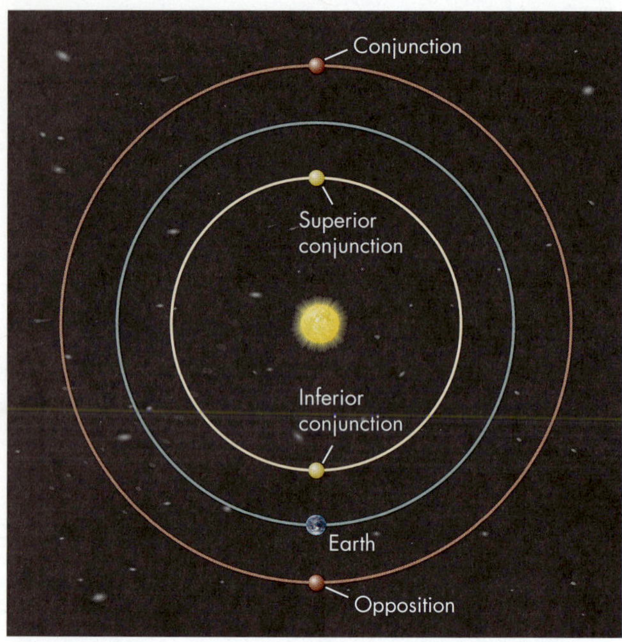

FIGURE E1.9
Planetary configurations: opposition, superior conjunction, and inferior conjunction.

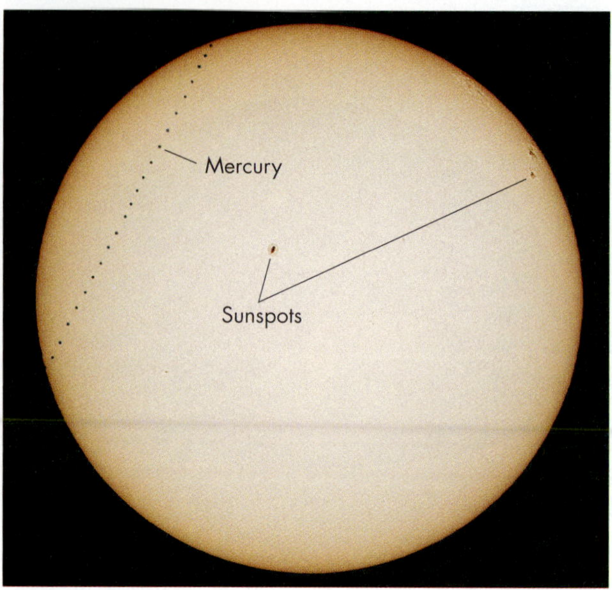

FIGURE E1.10
Transit of Mercury, May 7, 2003, photographed by Dominique Dierick of Belgium. Twenty-three separate images, taken about 15 minutes apart, were combined to form this picture. The next three transits will occur on May 9, 2016; Nov. 11, 2019; and Nov. 13, 2032.

position with respect to the Earth and the Sun—planetary configurations—and are shown in figure E1.9. Understanding these terms when they are used can help you find planets.

If a planet lies in the sky in the same direction as the Sun, it is said to be at **conjunction.** If it lies approximately between us and the Sun, it is at **inferior conjunction.** If it is on the other side of the Sun, it is at **superior conjunction.**

Planets are very hard to see at either conjunction because they are hidden in the Sun's glare. On some very rare occasions, a planet may pass directly between us and the Sun. We may then see it silhouetted against the Sun's bright disk, as shown in figure E1.10. Such an event is called a **transit.** Only Mercury and Venus can transit the Sun as seen from Earth, but we can imagine talking with an astronaut on Mars who has just witnessed the Earth transiting the Sun. This would occur when Mars is directly opposite the Sun in the sky, or at what seen from Earth is called **opposition.**

A planet's configuration strongly affects how easily it can be viewed from Earth. For example, when an outer planet is at opposition, it is at its nearest to the Earth. Other things being equal, the planet is therefore also at its brightest. Being opposite the Sun, a planet at opposition rises at sunset. Inner planets, on the other hand, are easiest to see when they lie far from the Sun in our sky, so that they are not lost in the Sun's glare. However, there is a limit to how far from the Sun in our sky an inner planet can be. As figure E1.11 shows, Mercury can never be more than 28° from the Sun and Venus never more than 47° as seen on the sky. It is for this reason that Mercury and Venus are usually visible only in the morning or evening sky when the Sun is just below the horizon.

A bright planet like Venus seen close to the Sun at dawn or dusk is sometimes called the **Morning Star** or **Evening Star.** When Venus or Mercury is at its greatest angular separation from the Sun, it is said to be at **greatest elongation**—which can be either western (morning) or eastern (evening).

The time interval between successive planetary configurations of the same type is called the **synodic period.** The synodic period differs from the planet's orbital period because both the Earth and the other planets move around the Sun. Thus, the interval between oppositions is neither an Earth year nor another planet's orbital period. For example, the Earth takes about 2 years to catch up to and overtake Mars after an opposition. The Earth overtakes the slower moving, more distant planets more quickly, and the interval between oppositions is close to a year. Thus, the Martian synodic period is about 780 days, whereas the Saturnian synodic period is 378 days.

YOUR EYES AT NIGHT

You will soon discover that the longer you stay outside in dim light, the more sensitive your eyes will become and the fainter the stars you will be able to see. This is the result of physiological changes in your eye referred to as **dark adaption.**

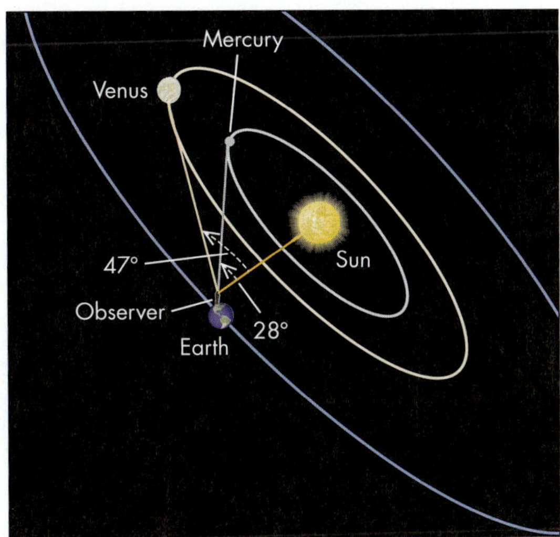

View from Space

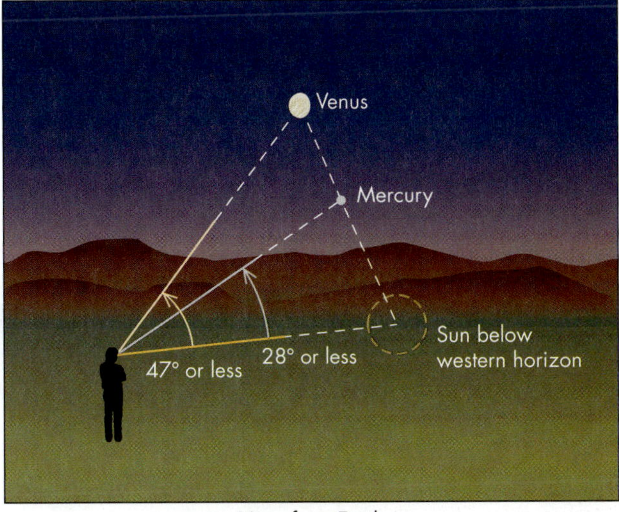

View from Earth

FIGURE E1.11
The greatest elongations of Mercury and Venus and the Evening Star phenomenon. The left-hand diagram also shows that Mercury and Venus can *never* appear more than 28° and 47°, respectively, from the Sun.

The simplest change in your eye occurring in dim light is that the pupil opens wider. This is easy to verify by looking at yourself in a mirror in a dimly lit room. In full sunlight, your pupil normally has a diameter of about 2 millimeters, but in total darkness its diameter may expand to about 7 or 8 millimeters, thereby allowing more light to enter your eye. The same principle is used in cameras when you adjust the aperture to match the available light.

Your eyes undergo another change in the dark. Chemical changes make the dark-adapted retina about 1 million times more sensitive to light than under full daylight conditions. The process takes about 20 minutes to get well established but is undone by even a few seconds' exposure to bright light. Thus, once you are dark-adapted, you should stay away from bright lights for as long as you intend to observe.

In addition to becoming more sensitive to light, your eye also changes its sensitivity to color slightly, a phenomenon known as the "Purkinje effect." In full daylight, the eye responds best to greenish colors. At low light levels, it responds best to slightly bluer colors. This is probably the result of natural selection, because starlight is much bluer than sunlight and eyes responsive to blue will therefore aid survival. It is certainly the case that night-flying insects see blue light better than yellow. That is the reason bug-zappers use a blue light to attract insects and why yellow lightbulbs used for outdoor night lighting are less attractive to insects.

You may also notice that it is easier to see very faint objects if you don't look directly at them but instead look a little to one side. The greater sensitivity you enjoy from this so-called **averted vision** arises because the center of your field of view is densely packed with receptors designed to allow you to see fine details. These receptors are better at showing fine structure than at seeing faint light. Thus, looking slightly to one side of a faint object makes it easier to see.

SUMMARY

Looking at the night sky is not only fun, it will help you understand some of the phenomena described in chapters 1 and 2. Star maps will help you identify constellations and bright stars, and by learning the mythology of the stars, you will be able to find your way around the night sky more easily. In addition, you will forge a link to distant and ancient cultures. For many people, backyard astronomy—even with simple equipment—is an enjoyable and exciting hobby. Perhaps you will discover a comet and have it named for you!

QUESTIONS FOR REVIEW

1. Do we see the same constellations today as ancient cultures saw?
2. What are right ascension and declination?
3. What is the main advantage of the celestial coordinate system over altitude-azimuth coordinates?
4. What is azimuth?
5. Approximately where would you look for Mercury in the sky at about sunset?
6. What is meant by "Morning Star"?
7. What is meant by "dark adaption"?
8. Why does the pupil of your eye grow wider in dim light?
9. What is averted vision?

THOUGHT QUESTIONS

1. If a planet is at opposition and you see it high in the sky, about what time of night must it be?
2. Can you see Mercury in the western sky at dawn?
3. Pirates are often depicted wearing an eye patch. Can you think of any reason a pirate with two good eyes might wear an eye patch while on a ship?

TEST YOURSELF

1. As a star rises and moves across the sky, which of the following change? (more than one may be correct)
   (a) Its right ascension
   (b) Its declination
   (c) Its azimuth
   (d) Its altitude
   (e) None of the above
2. A planet is at inferior conjunction. It therefore rises at approximately
   (a) sunset.
   (b) sunrise.
   (c) midnight.
   (d) 2 hours before the Sun.
   (e) You can't tell from the available information.
3. If Mercury is at greatest elongation to the west of the Sun, it will be easiest to see
   (a) just before dawn.
   (b) just after sunset.
   (c) about midnight.
   (d) just before sunset.
   (e) None of the above
4. Which of the following planets can be at inferior conjunction?
   (a) Jupiter
   (b) Mars
   (c) Uranus
   (d) Venus
   (e) All of them
5. When your eye is dark-adapted, (select all that apply)
   (a) your pupils are smallest.
   (b) your pupils are biggest.
   (c) your color vision is at its most sensitive.
   (d) your eyes are most sensitive to light.

KEY TERMS

altitude, 67
asterism, 63
averted vision, 69
azimuth, 67
conjunction, 68
dark adaption, 68
declination, 67
Evening Star, 68
greatest elongation, 68
inferior conjunction, 68
Morning Star, 68
opposition, 68
right ascension, 67
superior conjunction, 68
synodic period, 68
transit, 68

FURTHER EXPLORATIONS

Allen, Richard Hinckley. *Star Names: Their Lore and Meaning.* New York: Dover, 1963.

Arnold H. J. P. *Astrophotography: An Introduction to Film and Digital Imaging.* Willowdale, Ontario: Firefly, 2002.

Aveni, Anthony. *Conversing with the Planets,* rev. ed. Boulder: University Press of Colorado, 2002.

Berman, Bob. *Secrets of the Night Sky: The Most Amazing Things in the Universe You Can See with the Naked Eye.* New York: W. Morrow, 1995.

Covington, Michael. *Astrophotography for the Amateur,* 2nd ed. New York: Cambridge University Press.

Hamilton, Edith. *Mythology.* Boston: Back Bay Books, 1998.

Mechler, Gary. *Galaxies and Other Deep-Sky Objects.* New York: Knopf, 1995. (One in a series of National Audubon Society Pocket Guides.)

Muirden, James, ed. *Sky Watcher's Handbook: The Expert Reference Source for the Amateur Astronomer.* New York: W. H. Freeman, 1993.

Nagler, A. "Choosing Your Telescope's Magnification." *Sky and Telescope* 81 (May 1991): 553.

Newton, Jack. *Guide to Amateur Astronomy.* Cambridge, UK: Cambridge University Press, 1995.

Observer's Handbook. Toronto: Royal Astronomical Society of Canada.

Pasachoff, Jay. *Field Guide to the Stars and Planets,* 4th ed. Boston: Houghton Mifflin, 2000.

Ridpath, Ian. "The Origin of Our Constellations." *Mercury* 19 (November/December 1990): 163.

Schaaf, Fred. *Wonders of the Sky: Observing Rainbows, Comets, Eclipses, the Stars, and Other Phenomena.* New York: Dover, 1983.

Schaefer, Bradley E. "The Origin of the Greek Constellations." *Scientific American* 295 (November 2006): 96.

Sesti, Giuseppe M. *The Glorious Constellations: History and Mythology.* New York: Harry N. Abrams, 1991. (This beautiful book contains extensive descriptions of the constellations and their mythology from many ancient cultures.)

Staal, Julius D. W. *The New Patterns in the Sky: Myths and Legends of the Stars.* Blacksburg, Va.: McDonald & Woodward, 1988.

Strom, Karen D. "Photographing the Stars." *Physics Teacher* 34 (September 1996): 340.

Tirion, Wil. *Uranometria 2000.0: Deep Sky Atlas,* 2nd ed. Richmond, Va.: Willmann-Bell, 2001.

Two excellent magazines for the amateur astronomer are the following:

Astronomy, published by Kalmbach Publishing Co., 21027 Crossroads Circle, P.O. Box 1612, Waukesha, WI 53187.

Sky and Telescope, published by Sky Publishing Corporation, P.O. Box 9111, Belmont, MA 02178-9111.

Websites

"Backyard Astronomy: Tips on Observing the Universe," by *Sky and Telescope:* http://www.skyandtelescope.com/howto/howtoequipment/3304526.html. (An excellent site with many useful links and good advice about choosing telescopes.)

Visit the *Explorations* website at **http://www.mhhe.com/arny** for additional online resources on these topics.

PROJECT

Motion of the Moon and Planets: You can use the fold-out star chart at the back of the book to track the positions of the Moon and planets as they move in their orbits. Determine from the table at the bottom of the chart when the Moon is a couple of days past new so that it will be visible in the early evening. Go outside shortly after sunset and look for the Moon in the west near where the Sun went down. Next check the dates along the bottom of the star map to find out what region of the sky is visible at that date and time. As the sky darkens, locate the brighter stars near the Moon and mark on the chart where the Moon is with respect to those stars. Finally, make a sketch of the Moon's shape.

Repeat this process for the next 4 or 5 nights. The Moon will set a little later each evening, so you need to adjust your observing time accordingly. After watching for a few nights, mark out the Moon's path on the star map. Ideally, you might want to follow the Moon's track for about 2 weeks, although as the Moon reaches third quarter (3 weeks after new moon), you'll have to stay up pretty late because the third-quarter moon does not rise until midnight. If you are really ambitious, get up before dawn and watch the crescent moon shrink as it approaches the new phase.

You can also use this method to study the motion of the planets. Venus is a good choice because it is bright and moves rapidly across the sky. Mercury also moves rapidly, but it usually lies too close to the Sun to see the stars around it.

To locate the planets, use the tables along the bottom of the fold-out starchart, which will tell you what constellation the planet is in for the month when you make your observations. The table also indicates whether each planet is more easily seen in the morning or evening. Because the outer planets move relatively slowly across the sky, you should space out your observations, perhaps marking positions once a week rather than every night.

As you progress in your project, you might ask whether the planets really follow the ecliptic. Can you see retrograde motion?

The space shuttle blasts off at night.

KEY CONCEPTS

- If no forces act on an object, inertia keeps the object moving in a straight line at a constant speed.
- If an object is not moving along a straight line at a constant speed, a force must be acting on it.
- Newton's laws of motion allow us to describe and predict an object's motion if we know the forces acting on it.
- Gravity controls the motion of most astronomical objects.
- The force of gravity between two objects depends on their mass and the distance between them:
  - The greater their separation, the weaker the force.
  - The greater their mass, the greater the force.
- Orbital motion allows us to determine the mass of ("weigh") astronomical objects.
- The mass and radius of an object determine the strength of the gravitational pull at its surface (surface gravity) and the speed needed to escape its gravitational pull (escape velocity).

3 Gravity and Motion

CONCEPTS AND SKILLS TO REVIEW

- Kepler's laws (2.3)
- Forces (Preview, p. 8)

Gravity gives the Universe its structure. It is a universal force that acts on all the objects in the Universe so that every particle is drawn toward every other particle by its pull. Gravity holds together astronomical bodies of all sizes, from the Earth to the Universe itself. But the role of gravity extends beyond giving structure to astronomical bodies. Gravity also controls their motions, holding the Earth in orbit around the Sun, the Sun in orbit around the Milky Way, and the Milky Way within the Local Group. Thus, gravity and motion are tightly connected in the Universe. This connection is the theme of this chapter.

Astronomers of antiquity did not make the connection between gravity and astronomical motion that we recognize today. They were puzzled about what kept the planets moving and why. They could not understand why, if the Earth moved, objects thrown into the air were not left behind.

The solutions to these mysteries began with a series of careful experiments conducted by Galileo in the 1600s. Apart from his famous—and perhaps fictitious—demonstration of weights dropped from the Leaning Tower of Pisa, Galileo experimented with projectiles and with balls rolling down planks. Such experiments led him to propose several laws of motion. More important, perhaps, these experiments demonstrated the power and importance of the experimental method for verifying scientific conjectures.

Perhaps the most remarkable discovery of these investigations is that the same laws that govern the motions of objects here on Earth also govern the motions of objects in space. If we really understand what makes objects move or change direction at home, we have the keys to understanding the motions of planets, stars, and galaxies throughout the Universe.

Q: WHAT IS THIS? See end of chapter for the answer

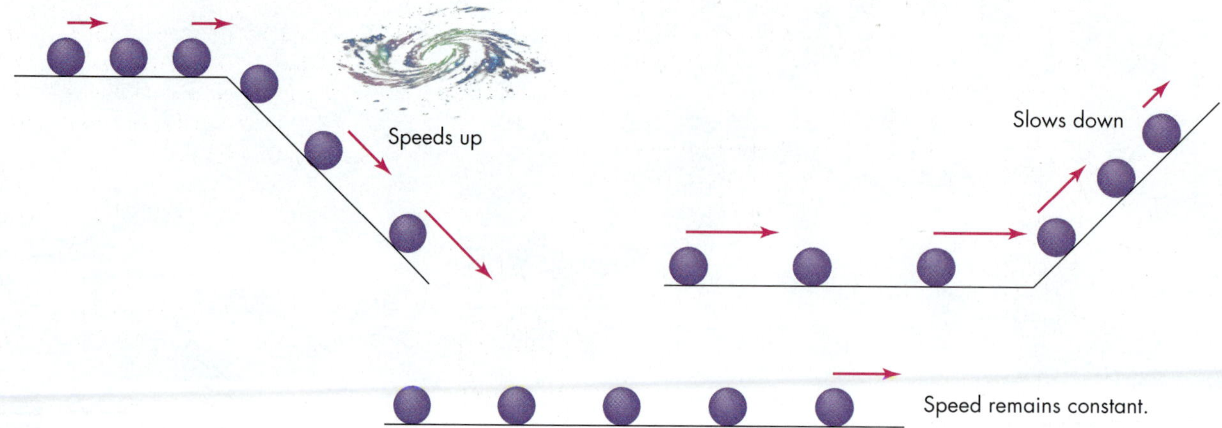

FIGURE 3.1
A ball rolling down a slope speeds up. A ball rolling up a slope slows down. A ball rolling on a flat surface rolls at a constant speed if no forces act on it.

3.1 Inertia

Central to Galileo's laws of motion is the concept of **inertia.** Inertia is the tendency of a body at rest to remain at rest or of a body in motion to keep moving in a straight line at a constant speed. Aristotle noted that bodies at rest resist being moved, but he failed to link this property to the tendency of objects to keep moving once they are set in motion. Kepler also recognized inertia's importance and in fact was first to use that term. However, Galileo not only proposed this property of matter but also demonstrated it by real experiment.

In one such experiment, Galileo rolled a ball down a sloping board over and over again and noticed that it always sped up as it rolled down the slope (fig. 3.1). He next rolled the ball up a sloping board and noticed that it always slowed down as it approached the top. He hypothesized that if a ball rolled on a flat surface and no forces—such as friction—acted on it, its speed would neither increase nor decrease but remain constant. That is, in the absence of forces, inertia keeps an object already in motion moving at a fixed speed. Inertia is familiar to us all even in everyday life. Apply the brakes of your car suddenly, and the inertia of the bag of groceries beside you keeps the bag moving forward at its previous speed until it hits the dashboard or spills onto the floor.

Newton recognized the special importance of inertia. He described it in what is now called **Newton's first law of motion** (sometimes referred to simply as the law of inertia). The law can be stated as follows:

A body continues in a state of rest or of motion in a straight line at a constant speed unless made to change that state by forces acting on it.

FIGURE 3.2
Balanced forces lead to no acceleration.

In applying Newton's first law, we should note two important points. First, we have not defined force yet but have relied on our intuitive feeling that a force is anything that pushes or pulls. Second, we need to note that when we use the term *force,* we are talking about *net* force; that is, the total of all forces acting on a body. For example, if a brick at rest is pushed equally by two opposing forces, the forces are balanced. Therefore, the brick experiences no net force and accordingly does not move (fig. 3.2).

Newton's first law may not sound impressive at first, but it carries an idea that is crucial in astronomy: that if a body is *not* moving in a straight line at a constant speed, some net force *must* be acting on it.

Actually, Newton was preceded in stating this law by the seventeenth-century Dutch scientist Christiaan Huygens. However, Newton went on to develop additional physical laws and—more important for astronomy—showed how to apply them to the

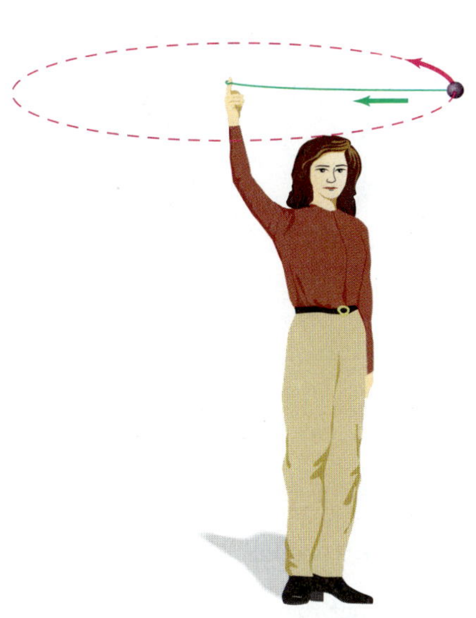
Side view

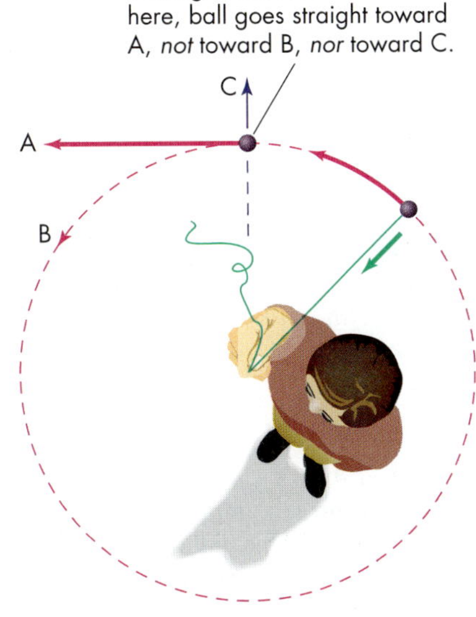
If string is released when ball is here, ball goes straight toward A, *not* toward B, *nor* toward C.
C
A
B
Top view

FIGURE 3.3
For a mass on a string to travel in a circle, a force (green arrow) must act along the string to overcome inertia. Without that force, inertia makes the mass move in a straight line.

Universe. For example, let us look at what happens if we swing a mass tied to a string in a circle. Newton's law tells us that the mass's inertia will carry it in a straight line if no forces act. What force, then, is acting on the circling mass? The force is the one exerted by the string, preventing the mass from moving in a straight line and keeping it in a circle. We can feel that force as a tug on the string, and we can see its importance if we suddenly let go of the string. With the force no longer acting on it, the mass flies off in a straight line, demonstrating the first law, as illustrated in figure 3.3.

We can translate this example to an astronomical setting and apply it to the orbit of the Moon around the Earth, or of the Earth around the Sun, or to the Sun in orbit around the Milky Way. Each of these bodies follows a curved path. Therefore, each must have a force acting on it, the origin and nature of which we will now describe.

3.2 Orbital Motion and Gravity

Newton was not the first person to attempt to discover and define the force that holds planets in orbit around the Sun. Nearly 100 years before Newton, Kepler recognized that some force must hold the planets in their orbits and proposed that something similar to magnetism might be responsible. Newton was not even the first person to suggest that gravity was responsible. That honor belongs to Robert Hooke, another Englishman, who noted gravity's role in celestial motions several years before Newton published his law of gravity in 1687. Newton's contribution is nevertheless special because he demonstrated the *properties* that gravity must have if it is to control planetary motion. Moreover, Newton went on to derive equations that describe not only gravity but also its effects on motion. The solution of these equations allowed astronomers to predict the position and motion of the planets and other astronomical bodies.

According to legend, Newton realized gravity's role when he saw an apple falling from a tree. The falling apple drawn downward to the Earth's surface presumably made him speculate whether Earth's gravity might extend to the Moon. Influenced by an apple or not, Newton correctly deduced that Earth's gravity, if weakened by distance, could explain the Moon's motion.

Most of Newton's work is highly mathematical, but as part of his discussion of orbital motion, he described a thought experiment to demonstrate how an object can move in

Q. Why are space vehicles normally launched from regions near the equator? Why are space vehicles normally launched to the east?

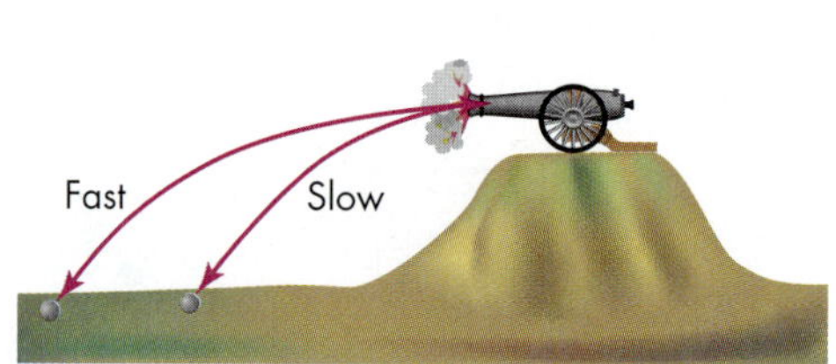

A

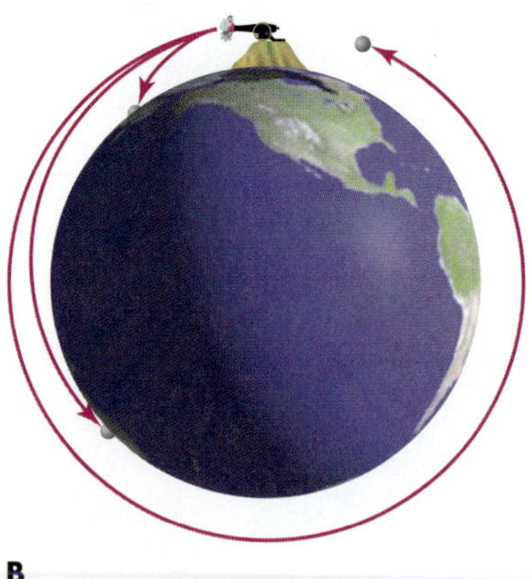

B

FIGURE 3.4
(A) A cannon on a mountain peak fires a projectile. If the projectile is fired faster, it travels farther before hitting the ground. (B) At a sufficiently high speed, the projectile travels so far that the Earth's surface curves out from under it, and the projectile is in orbit.

orbit. Thought experiments are not actually performed; rather, they serve as a way to think about problems. In Newton's thought experiment, we imagine a cannon on a mountain peak firing a projectile (fig. 3.4A). From our everyday experience, we know that whenever an object is thrown horizontally, gravity pulls it downward so that its path is an arc. Moreover, the faster we throw, the farther the object travels before striking the ground.

ANIMATION
Newton's cannon

We now imagine increasing the projectile's speed more and more, allowing the projectile to travel ever farther. However, as the distance traveled becomes very large, we see that the Earth's surface curves away below the projectile (fig. 3.4B). Therefore, if the projectile moves sufficiently fast, the Earth's surface may curve away from it in such a way that the projectile will never hit the ground. Such is the nature of orbital motion and how the Moon orbits the Earth. The balance between inertia and the force of gravity maintains the orbit.

We can analyze this thought experiment more specifically with Newton's first law of motion. According to that law, in the absence of forces, the projectile will travel in a straight line at constant speed. But because a force, gravity, is acting on the projectile, its path is not straight but curved. Moreover, the law helps us understand that the projectile does not stop, but continues moving, because its inertia carries it forward.

If we apply the same reasoning to the Moon's motion, we conclude that

- the Moon moves along its orbit because its inertia carries it forward, and
- its path is curved (not a straight line) because the force of gravity deflects it.

Notice that in the above discussion we used *no* formulas. All we needed was Newton's first law and the idea that gravity supplies the deflecting force. However, if we are to understand the particulars of orbital motion, we require additional laws. For example, to determine how rapidly the projectile must move to be in orbit, we need laws that have a mathematical formulation.

3.3 Newton's Second Law of Motion

We stated that an object's inertia causes it to move at a constant speed in a straight line in the absence of forces. However, suppose forces do act on the object. How much deviation from straight-line motion will such forces produce? To answer that question, we need to define more carefully what we mean by motion.

Motion of an object is a change in its position, which we can characterize in two ways: by the direction the object is moving and by the object's speed. For example,

Uniform motion
(Same speed (*V*), same direction)

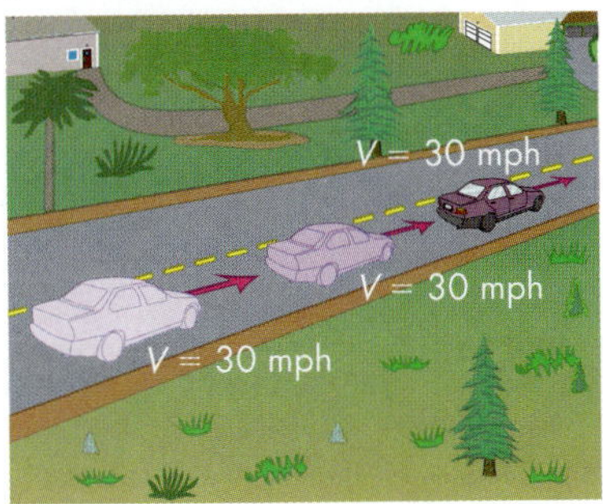

A

Acceleration
(A change in speed)

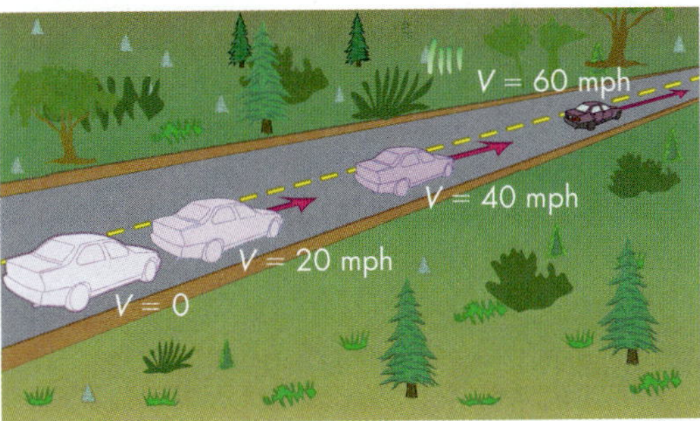

B

Acceleration
(A change in direction)

C

FIGURE 3.5
Views of a car in uniform motion and accelerating. (A) Uniform motion implies no change in speed or direction. The car moves in a straight line at a constant speed. If an object's (B) speed or (C) direction changes, the object undergoes an acceleration.

Q. Is it possible for an object to travel with constant speed and still accelerate?

suppose a car is moving east at 30 miles per hour. If the car's speed and direction remain constant, we say it is in uniform motion (fig. 3.5A). If the car changes either its speed or its direction, it is no longer moving uniformly, as depicted in figure 3.5B and C. Such nonuniform motion is defined as **acceleration.**

Acceleration

We are all familiar with acceleration as a change in speed. For example, when we step on the accelerator in a car and it speeds up from 30 to 40 miles per hour, we say the car is accelerating. Its speed has changed, and its motion is therefore nonuniform. Although in everyday usage acceleration implies an increase in speed, scientifically any change in speed is an acceleration. Thus, technically, a car "accelerates" when we apply the brakes and it slows down.

In the example above, we produced an acceleration by changing an object's speed.* We can also produce an acceleration by changing an object's direction of motion. For example, suppose we drive a car around a circular track at a steady speed of 30 miles per hour. At each moment, the car's direction of travel is changing, and therefore it is not in uniform motion. Similarly, a mass swung on a string or a planet orbiting the Sun is experiencing nonuniform motion and is therefore accelerating. In fact, a body moving in a circular orbit constantly accelerates, even though its speed is not changing.

How do we produce an acceleration? Newton realized that for an object to accelerate, a force must act on it. For example, to accelerate—change the direction of—the mass whirling on a string, we must constantly exert a pull on the string. Similarly, to accelerate a shopping cart, we must exert a force on it. In addition, experiments show that the acceleration we get is proportional to the force we apply. That is, a larger force produces a larger acceleration. For example, if we push a shopping cart gently, its acceleration is slight. If we push harder, its acceleration is greater. But experience shows us that more than just force is at work here. For a given push, the amount of acceleration also depends on how full the cart is. A lightly loaded cart may scoot away under a slight push, but a heavily loaded cart hardly budges, as illustrated in figure 3.6. Thus, the acceleration produced by a given force also depends on the amount of matter being accelerated.

---

*In our discussion, we use the word *speed* to denote the rate of motion, irrespective of its direction. Were we to be more technically correct, we would use the term *velocity,* which means the speed in a *given direction.* Thus, a body's velocity changes if *either* its speed or its direction changes. With velocity so defined, acceleration is simply the change in velocity over some interval of time.

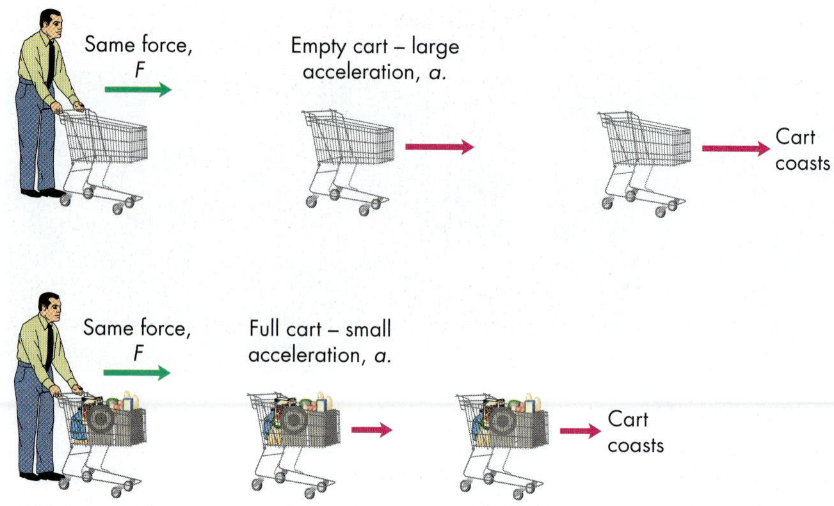

FIGURE 3.6
A loaded cart will not accelerate as easily as an empty cart.

Mass

The amount of matter an object contains is determined by a quantity that scientists call mass. Technically, **mass** measures an object's inertia. The more inertia, the more mass, and vice versa.

Scientists generally measure the mass of ordinary objects in kilograms (abbreviated kg). One kilogram equals 1000 grams or about 2.2 pounds of mass. For example, under normal conditions, a liter of water (roughly a quart) has a mass of 1 kilogram, but it is important to remember that mass is not the same as weight. Because an object's mass describes the amount of matter in it, its mass in kilograms is a fixed quantity. An object's weight, however, measures the force of gravity on it, a point we will explore more later. Thus, although an object's mass is fixed, its weight changes if the local gravity changes. For example, on Earth we have one weight but on the Moon, where gravity is less, we have a lesser weight, but no matter where we are, we have the same mass.

Mass is the final quantity needed to understand **Newton's second law of motion** in its full form. Mathematically, the law states that when a force, F, acts on an object whose mass is m, it produces an acceleration, a, according to the equation:

$$F = ma$$

F = Any force
m = Mass of the body being accelerated
a = Amount of acceleration

or, in other words:

> **The net force exerted on a body is proportional to the force but inversely proportional to the mass of the accelerated body.**

This astonishingly simple equation allows scientists to predict virtually all features of a body's motion. With $F = ma$ and with knowledge of the masses and the forces in action, engineers and scientists can, for example, drop a spacecraft safely between Saturn and its rings or use a computer to design an airplane that will fly successfully without being tested.

3.4 The Law of Gravity

Using Newton's second law, we now can determine an object's motion if we know its initial state of motion and the forces acting on it. For astronomical bodies, that force is often limited to gravity, and so to predict their motion, we need to know how to calculate gravity's force. Once again we encounter Newton's work, for it was he who first worked out the **law of gravity.** On the basis of his study of the Moon's

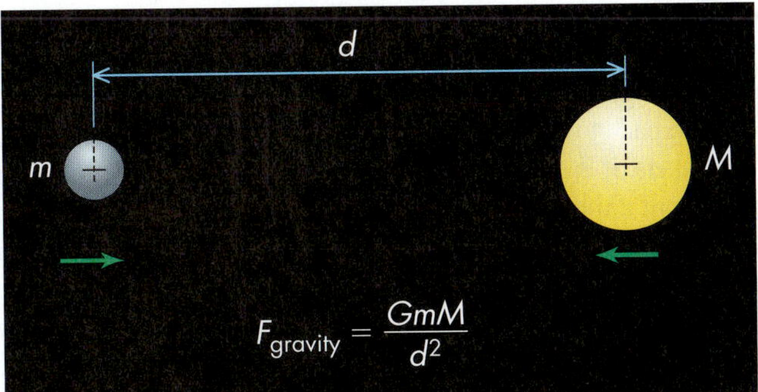

FIGURE 3.7
Gravity produces a force of attraction (green arrows) between bodies. The strength of the force depends on the product of their masses, *m* and *M*, and the square of their separation, d^2. G is the universal gravitational constant.

motion, Newton concluded the following:

> **Every mass exerts a force of attraction on every other mass. The strength of the force is directly proportional to the product of the masses divided by the square of their separation.**

We can write this extremely important result in a shorthand mathematical manner as follows:

Let *m* and *M* be the masses of the two bodies (fig. 3.7) and let the distance between their centers be *d*. Then the strength of the force between them, *F*, is

$$F = \frac{GMm}{d^2}.$$

F = Strength of the gravitational force between two bodies
M = Mass of one body
m = Mass of second body
d = Separation between the bodies' centers
G = Gravitational constant

The factor G is a constant whose value is found by measuring the force between two bodies of known mass and separation. The resulting number for G depends on the units chosen to measure *M*, *m*, *d*, and *F*, but once determined, G is the same as long as the same units are used. For example, if *M* and *m* are measured in kilograms, *d* in meters, and *F* in SI units,* then $G = 6.67 \times 10^{-11}$ meters^3/(kilogram·second^2) [$m^3 \cdot kg^{-1} \cdot s^{-2}$].

Writing the law of gravity as an equation helps us see several important points. If either *M* or *m* increases and the other factors remain the same, the force increases. If *d* (the distance between two objects) increases, the force gets weaker. Moreover, the force weakens as the square of the distance. That is, if the distance between two masses is doubled, the gravitational force between them decreases by a factor of four, not two. Finally, although one object's gravitational force on another weakens with increasing distance, the gravitational force never completely disappears. Thus, the gravitational attraction of an object reaches across the entire Universe, so the Earth's gravity not only holds you on to Earth's surface but also extends to the Moon and exerts the force that holds the Moon in orbit around the Earth.

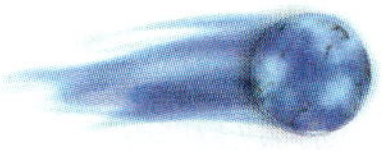

ANIMATION
Gravity produces a force of attraction between bodies.

3.5 Newton's Third Law of Motion

Newton's studies of motion and gravity led him to yet another critical law, which relates the forces that bodies exert on each other. This additional relation, **Newton's third law of motion,** is sometimes called the law of action–reaction. This law states:

> **When two objects interact, they create equal and opposite forces on each other.**

Two skateboarders side by side may serve as a simple example of the third law (fig. 3.8). If A pushes on B, *both* move. According to Newton's third law, when A exerts a force on B, B exerts a force on A, so that both are accelerated.

---

*SI (System International) units use the kilogram, meter, and second as the fundamental units.

FIGURE 3.8
Skateboarders illustrate Newton's third law of motion. When A pushes on B, an equal push is given to A by B.

The same principle allows an astronaut who accidentally floats free from a spacecraft to redirect her motion back to the spacecraft. By throwing an object in a direction away from the spacecraft, an equal and opposite force is exerted on the astronaut, who is thereby pushed back toward the ship. This is also why a spacecraft accelerates in response to firing its rockets. It is often incorrectly believed that the propellant "pushes against" the ground or air, but if that were true, a spacecraft would be unable to accelerate once it was in space. Newton's third law explains the thrust produced when firing a rocket—by pushing propellant out the rocket nozzle, the ship experiences an equal push in the opposite direction.

The gravitational force between the Earth and the Sun affords another astronomical example of Newton's third law and at the same time leads us a step closer to understanding orbital motion. The gravitational force of the Earth on the Sun is exactly equal to the gravitational force of the Sun on the Earth. We can see this perhaps surprising result from Newton's law of gravity where the gravitational force between two objects depends on the product of their masses. We thus get the same force regardless of whether we let the Earth act on the Sun or vice versa. Why, then, does the Earth orbit the Sun and not the other way around?

The second law supplies the answer. When we translate $F = ma$ into $a = F/m$, we see that the acceleration an object feels is inversely proportional to its mass; that is, the more massive it is, the more force is required to accelerate it. Thus, even though the forces acting on the Earth and Sun are precisely equal, the Sun accelerates 300,000 times *less* because it is 300,000 times *more* massive than the Earth. As a result, because the Earth's acceleration is so much larger than the Sun's, the Earth does most of the moving. In fact, however, the Sun does move a little bit as the Earth orbits it, much as you must move if you swing a child around you in play.

3.6 Measuring an Object's Mass Using Orbital Motion

Knowledge of orbital motion is important for more than just understanding the paths of astronomical objects. From the orbit's properties (such as size and orbital period), astronomers can deduce physical properties of the orbiting objects, such as their mass.

The basic method for determining an astronomical object's mass uses a modified form of Kepler's third law and was first worked out by Newton using his laws of motion and gravity.* The underlying idea is very simple:

- The masses of the orbiting objects determine the gravitational force between them.
- The gravitational force in turn sets the properties of the orbit.

Thus, from knowledge of the orbit, astronomers can work backward to find the masses of the objects. To see how this can be done, we consider a very simple case: orbital motion in a perfect circle with the orbiting object having a mass so small that it can be ignored compared with that of the central body. These restrictions are met to high precision in many astronomical systems, such as the Earth's motion around the Sun and the Sun's motion around the Milky Way. By assuming that the mass of the central body is large compared with the orbiting body, we can ignore the acceleration of the central body (as we just discussed) and assume it is at rest. These assumptions simplify the problem but are not essential for its solution.

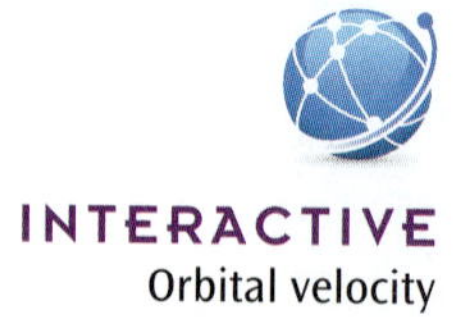

INTERACTIVE
Orbital velocity

To work out the orbital properties of an object moving around another, we use Newton's laws of motion and his law of gravity. From the first law, we know that if an object moves along a circular path, there is a net force (an unbalanced force) acting on

*In fact, Newton drew upon Kepler's third law in deriving his law of gravity. Although this is not simple to see, it turns out that the exponents in Kepler's law (the square of *P* and the cube of *a*) are set by the power (square of *d*) in the law of gravity.

it because balanced forces give straight-line motion. This force† must be applied to any object moving in a circle, whether it is a car rounding a curve, a mass swung on a string, or the Earth orbiting the Sun.

Using Newton's second law of motion and some algebra and geometry gives us an equation that shows that if a mass (m) moves with a velocity (V) around a circle at a distance (d) from the center, the force needed to hold it in a circular orbit is

$$F = \frac{mV^2}{d}$$

F = Force needed to hold a body in orbit
m = Mass of the body
d = Radius of the orbit
V = Velocity of the body

Using the above equation, we can find the orbital velocity of a planet around the Sun. Let the Sun's mass be M and the planet's mass be m, the latter of which is assumed to be much smaller than M. Assume the planet moves in a circular orbit at distance (d) at a velocity (V) so that the force required to hold the planet in orbit is mV^2/d. That force—the force that deflects the planet from its tendency to move in a straight line—is supplied by the gravitational force between the Sun and the planet. We have already defined that force as GMm/d^2.

Because mV^2/d is the force required to hold the planet in orbit, it must equal the force of gravity.

$$\frac{mV^2}{d} = \frac{GMm}{d^2}$$

We can cancel out m, and one of the d's, to obtain

$$V^2 = \frac{GM}{d}$$

Finally, we take the square root of both sides to obtain the orbital velocity.

$$V = \sqrt{\frac{GM}{d}}$$

The above equation, giving the orbital velocity in terms of M and r, can be used to determine the mass of the central object if the orbiting object's velocity and distance from it are known. The same method can be used to find the mass of any object around which another object orbits. This is illustrated for the Extending Our Reach box "Weighing the Sun." Thus, gravity becomes a tool for determining the mass of astronomical objects, and we shall use this method many times throughout our study of the Universe.

There are other ways we can use the law of gravity. For example, we can use it to find out how much we would weigh on another planet and how fast a spacecraft must move to escape from a planet's surface.

3.7 Surface Gravity

Surface gravity measures the gravitational attraction at a planet's or star's surface. It is the *acceleration* on a mass created by the local gravitational force, not the force itself. This acceleration determines how fast objects fall. To understand the importance of surface gravity, recall that mass measures the amount of material an object contains and is therefore constant. On the other hand, an object's weight depends on its mass and the acceleration of gravity. Thus, surface gravity determines what a mass weighs. In addition to determining weight on a planet, surface gravity also influences a planet's shape and whether it has an atmosphere. For example, small objects such as asteroids are not spherical because their surface gravity is too weak to crush them into round shapes. Likewise, a small surface gravity makes it hard for a planet to keep an atmosphere.

† This force is called a "centripetal force," which is the force applied to draw an object *inward* toward the center of the orbit.

EXTENDING *our reach*

WEIGHING THE SUN

To find the mass (M) of the Sun, we go back to our equation for V^2, cross-multiply by d, and divide by G to obtain

$$M = \frac{V^2 d}{G}$$

Therefore, to evaluate M, we need V, the Earth's orbital velocity (the speed with which it is moving in its orbit). We find V by dividing the circumference of its orbit, C, by the time it takes the Earth to complete one orbit, a length of time we call the orbital period, P. Thus

$$V = \frac{C}{P}$$

The formula for the circumference of a circle of radius d is $C = 2\pi d$ (note that in this case the radius is d). Thus

$$V = \frac{2\pi d}{P}$$

Putting this expression for V into the equation for M, above, gives

$$M = \left(\frac{2\pi d}{P}\right)^2 \frac{d}{G} = \frac{4\pi^2 d^3}{GP^2}$$

This is a modified form of Kepler's third law and it is important because we can use it to measure the mass of an object given the orbital period, P, and orbital distance, d, of a much smaller object moving around it. For example, suppose we want to find the Sun's mass, M.

According to the expression for M, we need P and d for the Earth's orbit. If we measure P in seconds and d in meters, then the gravitational constant $G = 6.67 \times 10^{-11}\ \text{m}^3\cdot\text{kg}^{-1}\cdot\text{s}^{-2}$.

To find P in seconds, we remember that it takes the Earth one year to orbit the Sun. Thus, P is one year. We can express P in seconds by multiplying the number of seconds in a minute (60), times the number of minutes in an hour (60), times the number of hours in a day (24), times the number of days in a year (365.25). The result of that calculation, rounded off to three significant figures, is

$$P = 3.16 \times 10^7 \text{ seconds}$$

Similarly, we need d, the Earth–Sun distance, in meters, which we can look up in the appendix and find is 1.50×10^{11} m. Putting these values and the value of π and G into the expression for M we find

$$M = \frac{4(3.14)^2 \times (1.50 \times 10^{11}\ \text{m})^3}{6.67 \times 10^{-11}\ \text{m}^3\cdot\text{kg}^{-1}\cdot\text{sec}^{-2} \times (3.16 \times 10^7\ \text{sec})^2}$$

$$= \frac{4 \times 9.86 \times 3.38 \times 10^{33}\cancel{\text{m}^3}}{6.67 \times 9.99 \times 10^{14-11}\cancel{\text{m}^3}\cdot\text{kg}^{-1}\cancel{\text{sec}^{-2}}\ \cancel{\text{sec}^2}}$$

$$= 2 \times 10^{30}\ \text{kg}$$

We determine the strength of a planet's surface gravity as follows.

The law of gravity states that a planet of mass M exerts a gravitational force F on a body of mass m at a distance d from its *center* given by

$$F = \frac{GMm}{d^2}$$

At the planet's surface, $d = R$, the planet's radius, so

$$F = \frac{GMm}{R^2}$$

Newton's second law tells us that for any force, F, the acceleration it produces on a body of mass m is given by $F = ma$. Therefore,

$$ma = \frac{GMm}{R^2}$$

Canceling out the m's then gives

$$a = \frac{GM}{R^2}$$

INTERACTIVE
Gravity variations

Thus, a planet's surface gravity depends on its mass and radius. That dependence is such that two planets with the same radius but different masses will have different surface gravities—the planet with the larger mass will have the larger surface gravity. Similarly, if two planets have the same mass but different radii, the planet with the larger radius will have a smaller surface gravity.

Surface gravity is usually denoted by the letter g (the origin of the phrase "Pulling g's," which is used by pilots). We can therefore write that

$$g = \frac{GM}{R^2}$$

g = Surface gravity
G = Gravitational constant
M = Mass of the attracting body
R = Radius of the attracting body

Because the surface gravity depends on the mass and radius of the attracting body, the strength of the surface gravity is different from body to body. For example, we can compare the surface gravity of the Moon with that of the Earth to show why you would weigh less on the Moon than on the Earth.

To make the calculation, we need to know the mass and radius of the Earth and the Moon. Those numbers are given in table 3.1.

TABLE 3.1 MASS AND RADIUS OF THE EARTH AND MOON*

	Mass	Radius
Earth	6.0×10^{24} kg	6.4×10^6 m
Moon	7.3×10^{22} kg	1.7×10^6 m
Earth/Moon	81	3.8

*Numbers have been rounded slightly.

Because the Earth has greater mass, we might guess that its surface gravity, g, is greater than the Moon's. However, the Moon's radius is smaller, so its surface gravity might be greater. Thus, to determine which body has the larger surface gravity, we need to evaluate g mathematically.

We can make this calculation two ways. The first way is simply to plug into the equation for g the values of M and R appropriate for the Earth and then repeat the process with values appropriate for the Moon. A far easier way is to work with ratios. To do that, we write out on one line the expression for g on the Earth and then write out on the line below it the value for g on the Moon. Next we draw a horizontal line between them to indicate division, as shown below. This gives

$$\frac{g_{\text{Earth}}}{g_{\text{Moon}}} = \frac{GM_{\text{Earth}}/(R_{\text{Earth}})^2}{GM_{\text{Moon}}/(R_{\text{Moon}})^2}$$

Notice that by doing this, the value of G, the gravitational constant, cancels out. If we then group the terms in M and the terms in R separately, we find

$$= \frac{M_{\text{Earth}}/M_{\text{Moon}}}{(R_{\text{Earth}}/R_{\text{Moon}})^2}$$

$$= \frac{(81)}{(3.8)^2} = 5.6 \approx 6$$

This shows that the ratio of g on the Earth to g on the Moon is about 6:1, so that you weigh about 6 times more on the Earth than you would on the Moon. That fact allowed the astronauts to make such large leaps on the Moon (fig. 3.9) despite carrying a load in their suits and backpacks that weighed about 180 pounds on Earth.

FIGURE 3.9
An astronaut can make huge leaps in the Moon's low gravity.

3.8 Escape Velocity

To overcome a planet's gravitational force and escape into space, a rocket must achieve a critical speed known as the **escape velocity.** Escape velocity is the speed an object needs to move away from a body and not be drawn back by its gravity. We can understand how such a speed might exist if we think about throwing an object into the air. The faster the object is tossed upward, the higher it goes and the longer it takes to fall back. Escape velocity is the speed an object needs so that it will *never* fall back, as depicted in figure 3.10. Thus, escape velocity is of great importance in space travel if craft are to move away from one object and not be drawn back to it.

INTERACTIVE
Escape velocity

V_{esc} = Escape velocity
G = Gravitational constant
M = Mass of the body to be escaped from
R = Radius of the body to be escaped from

The escape velocity, V_{esc}, for a spherical body, such as a planet or star, can be found from the law of gravity and Newton's laws of motion and is given by the following formula:

$$V_{esc} = \sqrt{2GM/R}$$

Here, M stands for the mass of the body from which we are attempting to escape, and R is its radius, as shown in figure 3.10.

Notice in the equation for V_{esc} that if two objects of the same radius are compared, the larger mass will have the larger escape velocity. Likewise, if two objects of the same mass are compared, the one with the smaller radius will have the greater escape velocity.

To illustrate the use of the formula, we calculate the escape velocity from the Moon. From the data in table 3.1, we find the Moon's radius and mass. We insert these values in the formula for escape velocity and find

$$V_{esc}(\text{Moon}) = \sqrt{2GM/R}$$

$$V_{esc}(\text{Moon}) = \sqrt{2 \times 6.7\times10^{-11}\ \text{m}^3\cdot\text{sec}^{-2}\cdot\text{kg}^{-1} \times 7.3 \times 10^{22}\ \text{kg}/1.7 \times 10^{6}\ \text{m}}$$

We can simplify the math by collecting the numerical values, powers of ten, and units separately to get

$$V_{esc}(\text{Moon}) = \sqrt{2 \times 6.7 \times 7.3/1.7 \times 10^{-11} \times 10^{22}/10^{6} \times \text{m}^3\cdot\text{sec}^{-2}\cdot\text{kg}^{-1}\ \text{kg/m}}$$
$$= \sqrt{58 \times 10^{5}\ \text{m}^2/\text{sec}^2}$$

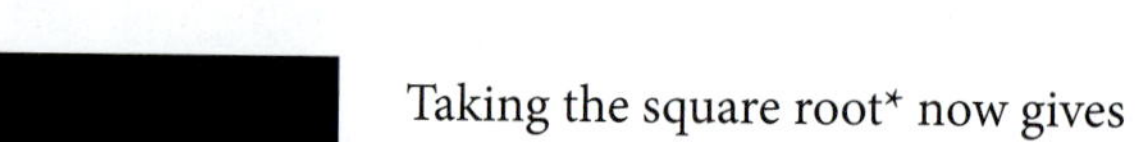

Taking the square root* now gives

$$= 2.4 \times 10^{3}\ \text{m/sec} = 2.4\ \text{km/sec}$$

A similar calculation shows that the escape velocity from the Earth is 11 kilometers per second. Thus, it is much easier to blast a rocket off the Moon than off the Earth.† In chapter 7, we will see that this low escape velocity is partly responsible for the Moon's lack of an atmosphere, as illustrated in figure 3.11.

While escape velocity is usually calculated from the surface of a body, where R is the body's radius, the escape velocity can also be found at any larger distance. For example, at Earth's distance from the Sun (1 AU), we would find that a spacecraft would need to be launched at 42 km/sec to escape the Sun's gravity. Thus a spacecraft launched just fast enough to escape the Earth would not have enough speed to leave the Solar System.

Escape velocity is particularly important in understanding the nature of black holes. In chapter 15, we will see that a black hole is an object whose escape velocity equals the speed of light. Thus, light cannot escape from it, thereby making it black. A black hole has such a huge escape velocity, not so much because it has an unusual mass, but because it has an abnormally small radius.

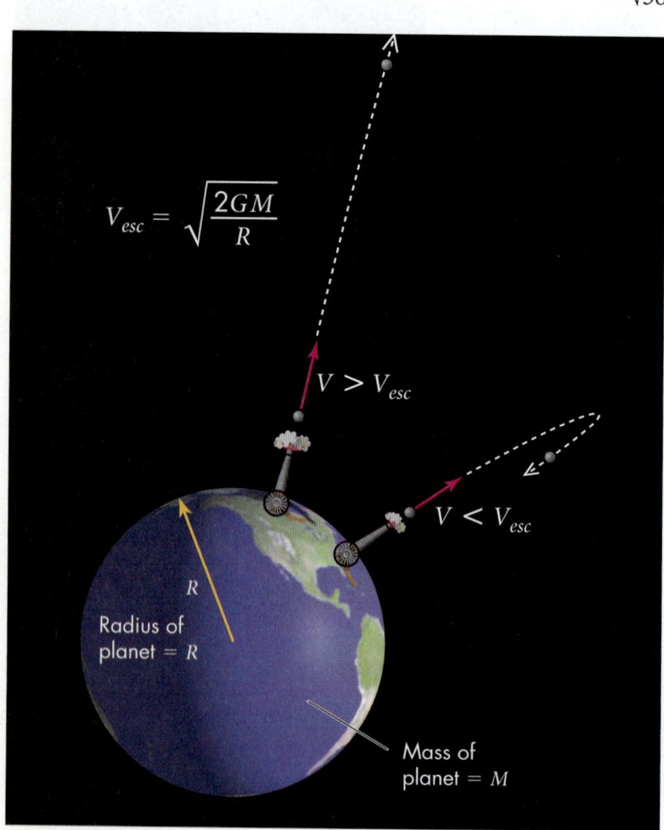

FIGURE 3.10
Escape velocity is the speed an object must have to overcome the gravitational force of a body such as a planet or star and not fall back. In general, the larger the mass of the planet or star and the smaller the distance from its center, the greater the escape velocity will be.

*Note: You must take the square root of the units too. This is easy to do. Just divide their exponents by 2. Thus: $\sqrt{\text{m}^2/\text{sec}^2} = \text{m/sec}$.

† If an object's mass and radius are given in units of the Earth's, then $V_{esc} \approx 11\sqrt{M/R}$ km/sec.

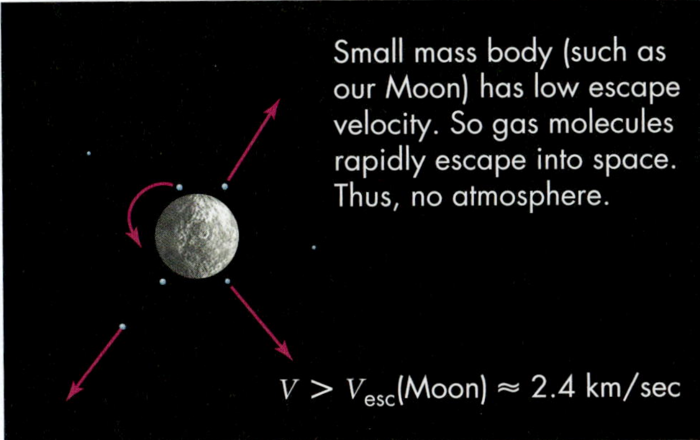

Small mass body (such as our Moon) has low escape velocity. So gas molecules rapidly escape into space. Thus, no atmosphere.
$V > V_{esc}$(Moon) ≈ 2.4 km/sec

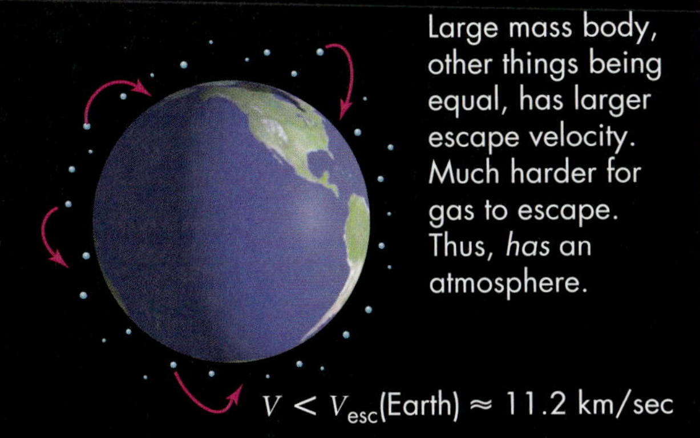

Large mass body, other things being equal, has larger escape velocity. Much harder for gas to escape. Thus, *has* an atmosphere.
$V < V_{esc}$(Earth) ≈ 11.2 km/sec

FIGURE 3.11
The Moon's escape velocity is about five times smaller than the Earth's. A low escape velocity, in general, leads to the absence of an atmosphere on a planet or satellite.

SUMMARY

A gravitational force exists between any two objects in the Universe. The strength of this force depends on the masses of the bodies and their separation. Gravitational forces hold astronomical bodies together and in orbit about one another.

An object's inertia makes it remain at rest or move in a straight line at a constant speed unless the object is acted on by a net force. Thus, for a planet to orbit the Sun, the Sun's force of gravity must act on it. The law of inertia and Newton's other laws of motion, when combined with the law of gravity, allow us to relate the size and speed of orbital motion to the mass of the central object.

The gravitational force exerted by a planet determines its surface gravity and escape velocity. The former determines your weight on a planet. The latter is the speed necessary to leave the surface and escape without falling back.

QUESTIONS FOR REVIEW

1. (3.1) What is meant by inertia?
2. (3.1) What does Newton's first law of motion tell you about the difference between motion in a straight line and motion along a curve?
3. (3.2) Explain how inertia and gravity are both involved in an orbit.
4. (3.3) How does mass differ from weight?
5. (3.3) If your mass is 70 kilograms on Earth, what is it on the Moon?
6. (3.4) If you weigh 110 pounds on the Earth, do you weigh 110 pounds on the Moon? Why?
7. (3.4) What is Newton's law of gravity?
8. (3.7) What does surface gravity measure?
9. (3.8) What is meant by escape velocity?

THOUGHT QUESTIONS

1. (3.1/3.3) Which do you think has more inertia: a small, inflated balloon or a cinder block? If each were moving toward you at 1 meter/second, which would be easier to catch?
2. (3.1/3.3) A cinder block can be weightless in space. Would you want to kick it with your bare foot? Even if it is weightless, does it have mass?
3. (3.1/3.2) In some amusement park rides, you are spun in a cylinder and are pressed against the wall as a result of the spin. People sometimes describe that effect as being due to "centrifugal force." What is really holding you against the wall of the spinning cylinder? Drawing a sketch and using Newton's first law may help you answer the question.
4. (3.2) Is there a force of gravity between the orbiting International Space Station and the Earth? If so, is it large enough that the ISS is affected by it? Why do the astronauts in the ISS float freely?
5. (3.3) Use Newton's second law of motion to explain why smaller cars tend to get better mileage than larger ones.
6. (3.4) Using $F = ma$ and $F = GMm/d^2$, deduce the units of Newton's gravitational constant G if masses are measured in kilograms, times in seconds, and distances in meters.
7. (3.4) Is there a force of gravity between you and this textbook? What is the dependence of the force of gravity on distance? If the book is much closer to you than the center of the Earth, why does it accelerate to the ground instead of toward you when you drop it?
8. (3.5) How many times greater is the Earth's gravitational force on the Moon than the Moon's gravitational force on

the Earth? Think about Newton's third law of motion before answering this.

9. (3.5) When you walk, does the ground push on you, or do you push on the ground? Explain clearly how you are accelerated forward. Why is it harder to walk across the beach than down a road?
10. (3.5) If one of the two children in figure 3.8 were much larger than the other, which would move farther? Why does this happen?
11. (3.7) Explain how a larger planet could have a smaller surface gravity than a smaller planet, or why it could not.

PROBLEMS

1. (3.3) If you apply a force F to a mass m, it results in an acceleration a. What acceleration would result if you applied a force of (a) $2F$ to m, (b) $2F$ to $2m$, (c) $10F$ to m, and (d) $10F$ to $3m$.
2. (3.3) You are working at the hockey rink and your resurfacing machine breaks down in the middle of the ice. Assuming you can get it moving, how much force will you need to apply to accelerate it to 2 m/sec in 25 seconds if it has a mass of 2500 kg?
3. (3.3/3.7) Calculate your weight on the Moon.
4. (3.6) Given that Jupiter is about 5 times farther from the Sun than the Earth, calculate its orbital speed. Use this speed to calculate how many years it takes Jupiter to complete an orbit around the Sun?
5. (3.6) Assuming that the mass of the Milky Way Galaxy is 10^{11} times that of the Sun and that the Sun is 2.6×10^{20} meters from its center, what is the Sun's orbital speed around the center of the Galaxy? How long does it take the Sun to orbit the Milky Way? (In this problem, we assume that the Galaxy can be treated as a single body. Strictly speaking, this isn't correct, but the more elaborate math needed to calculate the problem properly ends up giving almost the same answer.)
6. (3.6) Use the modified form of Kepler's third law to find the mass of the imaginary star 57 Fungaloid, given that a planet is in a circular orbit around it at a distance of 3×10^{11} meters with an orbital period of 3 years. Divide your answer by the Sun's mass to see how much more (less) massive the star is than our Sun.
7. (3.7) Using the method of section 3.7, compare the surface gravity of the Earth with the surface gravity of Jupiter and Pluto.
8. (3.8) Calculate the escape velocity from the Earth, given that the mass of the Earth is 6×10^{24} kilograms and its radius 6×10^{6} meters. In this problem, round off G to 7×10^{-11} meters^3/(kg·sec^2).
9. (3.8) Convert the escape velocity of Earth (problem 8) into miles per hour.
10. (3.8) Calculate the escape velocity from the Sun, given that its mass is 2×10^{30} kg and its radius is 7×10^{8} meters.
11. (3.8) Calculate the ratio of the escape velocities from the Moon and Earth.
12. (3.7/3.8) Which body has a larger escape velocity, Mars or Saturn? Solve this problem using ratios in a way similar to the comparison of the Earth's and the Moon's gravity in section 3.7. Show your work. In the appendix you can find values for Mars and Saturn's radii and masses in terms of Earth's.
13. (3.8) A good baseball pitcher can throw a ball at 100 miles/hour (about 45 meters/sec). If the pitcher were on the fictional asteroid Cochise, could the pitcher throw the ball fast enough so it would escape from Cochise? Assume Cochise is a sphere and that its mass is $M = 9.6 \times 10^{16}$ kg and its radius is 20 km = 2.0×10^{4} meters.

TEST YOURSELF

1. (3.1) Which of the following demonstrate the property of inertia? Select all that apply.
   (a) A car skidding on a slippery road
   (b) The oil tanker *Exxon Valdez* running aground
   (c) A brick sitting on a tabletop
   (d) Whipping a tablecloth out from under the dishes set on a table
2. (3.1–3.3) If an object moves along a curved path at a constant speed, you can infer that
   (a) a force is acting on it. (c) it is in uniform motion.
   (b) it is accelerating. (d) both (a) and (b) are true.
3. (3.2) Newton's work added ________ to Kepler's laws.
   (a) the concept of elliptical orbits
   (b) an equation for ellipses
   (c) a physical reason
   (d) planetary data
   (e) an experimental verification
4. (3.3) An astronaut has a mass of 60 kilograms before she takes off in her ship. When she reaches Earth orbit, her mass is _________; when she lands on the Moon, her mass is _________.
   (a) zero; the same as on Earth
   (b) much smaller than on Earth; smaller than on Earth
   (c) smaller than on Earth; larger than on Earth
   (d) zero; smaller than on Earth
   (e) the same as on Earth; the same as on Earth
5. (3.4) If the distance between two bodies is increased by a factor of 4, the gravitational force between them is________ by a factor of____.
   (a) increased; 4 (c) decreased; 8 (e) decreased; 64
   (b) decreased; 4 (d) decreased; 16
6. (3.4/3.5) The strength of the gravitational force exerted by the Sun on the Earth is the same as the strength of the gravitational force exerted by the Earth on the Sun.
   (a) True (b) False

7. (3.5) A rocket blasts propellant out of its thrusters and "lifts off," heading into space. What provided the force to lift the rocket?
   (a) The propellant pushing against air molecules in the atmosphere
   (b) The propellant heating and expanding the air beneath the rocket, and so pushing the rocket up
   (c) The action of the propellant accelerating down, giving a reaction force to the rocket
   (d) The propellant reversing direction as it strikes the ground below the rocket, then bouncing back and pushing the rocket up
8. (3.7) If the Earth were 9 times as far from the Sun as it is, the Earth's orbital velocity would be about ___ times slower.
   (a) 2 (b) 3 (c) 4 (d) 9 (e) 81
9. (3.8) Two planets have identical diameters but differ in mass by a factor of 25. The more massive planet therefore has an escape velocity ________ than the other.
   (a) 5 times larger
   (b) 625 times larger
   (c) 25 times larger
   (d) 50 times larger
   (e) 25 times smaller

KEY TERMS

acceleration, 77
escape velocity, 83
inertia, 74
law of gravity, 78
mass, 78
Newton's first law of motion, 74
Newton's second law of motion, 78
Newton's third law of motion, 79
surface gravity, 81

FURTHER EXPLORATIONS

The following books discuss the laws of motion in greater detail:

Casper, Barry M., and Richard J. Noer. *Revolutions in Physics.* New York: W. W. Norton, 1972.

Hecht, Eugene. *Physics in Perspective.* Reading, Mass.: Addison-Wesley, 1980.

Hewitt, Paul G. *Conceptual Physics.* 10th edition. San Francisco: Pearson Addison-Wesley, 2006.

Kirkpatrick, Larry D., and Gregory E. Francis. *Physics: A World View.* Belmont, CA: Thomson Brooks/Cole, 2007.

Website

Visit the *Explorations* website at **http://www.mhhe.com/arny** for additional online resources on these topics.

Q FIGURE QUESTION ANSWERS

WHAT IS THIS? (chapter opening): This is a time-lapse photo of the launch of the space shuttle Atlantis on July 12, 2001.

FIGURE 3.4: To take advantage of the Earth's spin in helping the vehicle reach orbital velocity.

FIGURE 3.5: Yes—by traveling along a curved path.

PROJECT

Measure the surface gravity of Earth: You can directly measure the acceleration of gravity at the Earth's surface by measuring the change in velocity over time of a falling object. You will need a video camera or a webcam capable of recording at least 30 frames per second. You will also need to be able to import the video into a computer and use basic video and image editing programs to capture and analyze individual frames of the movie. Finally, you will need to find a small, brightly colored object to drop, perhaps a plastic soda bottle cap. You will also need a dark backdrop about 1 m wide by 1 to 2 m tall, perhaps a flattened cardboard box. Mark a series of bright horizontal lines on the backdrop separated by 10 cm. These will enable you to convert your image measurements into real distances.

Mount the camera on a tripod or set it on a table, and start recording. Hold the object still for a few seconds near the top of the image and just in front of the backdrop—make sure you can see it in the video frame—then release the object. Repeat this a few times. Make sure you have plenty of light, which will reduce the exposure time of the camera and improve the quality of the images of the dropping object. A desk light placed to the side can help light up the object without lighting up the backdrop.

In the computer, save each frame where the object is falling. Find out the frame rate of the camera and invert it to determine the time between frames (10 frames per second would be 1/10 sec per frame). Measure how many pixels are between two of the marked lines on the backdrop and determine how many meters per pixel. You might want to check that this is the same at the top, middle, and bottom of your image. Most image editing programs will report the pixel position of the cursor. Then for each frame find the position in pixels of the falling object. Make a table listing: frame, vertical pixel position, pixel difference from previous frame, difference in meters, velocity (difference/time per frame), acceleration (difference from previous velocity/time per frame). Note that you will not be able to calculate a velocity until the second frame, or an acceleration until the third.

Light reflects inside raindrops, and the paths of different colors (different wavelengths) are bent by different amounts. This produces a spectrum (rainbow) of the Sun's light.

KEY CONCEPTS

- Light can be thought of either as a wave of energy or a stream of particles called photons.
- The color of light is determined by its wavelength.
- White light is a mix of all colors.
- Our eyes can see only a narrow range of wavelengths.
- Atoms can emit or absorb light when their electrons shift to lower- or higher-energy orbitals.
- Each kind of atom emits (or absorbs) light at a unique set of wavelengths (colors).
- For many kinds of objects, the color and intensity of their light are set by their temperature:
  - The hotter the object, the bluer the light it emits.
  - The hotter the object, the more light it emits.
  - These properties allow us to deduce an object's temperature from its color.
- The wavelengths of the radiation emitted by a moving object are seen to be shifted if the object is moving toward or away from the observer.

4 Light and Atoms

CONCEPTS AND SKILLS TO REVIEW

- Structure of atoms (Preview, p. 8)
- Forces (Preview, p. 8)

Our home planet is separated from other astronomical bodies by such vast distances that, with few exceptions, we cannot learn about them by direct measurements of their properties. For example, if we want to know how hot the Sun is, we cannot stick a thermometer into it. Similarly, we cannot directly sample the composition of the atmosphere of Saturn or a distant star. However, we can sample such remote bodies indirectly by analyzing their light. Light from a distant star or planet can tell us what the body is made of, its temperature, and many of its other properties. Light, therefore, is our key to studying the Universe. To use the key, however, we need to understand some of its properties.

In this chapter, we will discover that light is a form of energy that can be thought of either as a wave or as a stream of particles. Furthermore, we will discover that the light we see is just part of the radiation emitted by astronomical objects. We will also learn that light can be produced within an atom by changes in its electrons' energies. These changes imprint the atom's "signature" on the light. However, the light may also bear unwanted messages. For example, when light reaches our atmosphere, gases there alter its properties, blocking some rays, and bending and blurring others. These distortions limit what astronomers can learn from the ground.

The goal of this chapter is to explain the nature of light, how it is produced, and how it interacts with our atmosphere. Our first step toward this goal is to better understand what light is.

Q: WHAT IS THIS? See end of chapter for the answer

4.1 Properties of Light

Light is radiant energy; that is, it is energy that can travel through space from one point to another without the need of a direct physical link. Therefore, light is very different in its basic nature from, for example, sound. Sound can reach us only if it is carried by a medium such as air or water, whereas light can reach us even across empty space. In empty space, we can see the burst of light of an explosion, but we will hear no sound from it at all.

Light's capacity to travel through the vacuum of space is paralleled by another very special property: its high speed. In fact, the speed of light is an upper limit to all motion. In empty space, light travels at the incredible speed of 300,000* kilometers per second. An object traveling that fast could circle the Earth seven times in under a second.

The speed of light in empty space is a constant and is denoted by "c." However, in transparent materials, such as glass, water, and gases, the speed of light is reduced. Furthermore, different colors of light are slowed differently. For example, in nearly all materials, blue light travels slightly more slowly than red light. As we will see in chapter 5, lenses and prisms work because they slow the light as it travels through them.

The Nature of Light—Waves or Particles?

Observation and experimentation on light throughout the last few centuries have produced two very different models of what light is and how it works. According to one model, light is a wave that is a mix of electric and magnetic energy, changing together, as depicted in figure 4.1. Because light is a mix of electric and magnetic energy, it is often called an **electromagnetic wave** or **electromagnetic radiation.** The ability of such radiation to travel through empty space comes from the interrelatedness of electricity and magnetism.

You can see this relationship between electricity and magnetism in everyday life. For example, when you start your car, turning the ignition key sends an electric current from the battery to the starter. There, the current generates a magnetic force that turns over the engine. Similarly, when you pull the cord on a lawn mower, you spin a magnet that generates an electric current that creates the spark to start its engine.

ANIMATION
Photons stream away from a light source at the speed of light.

This interrelatedness between electricity and magnetism is what allows light to travel through empty space. A small disturbance of an electric field creates a magnetic disturbance in the adjacent space, which in turn creates a new electric disturbance in the space adjacent to it, and so on. Thus, a fluctuation of electric and magnetic field spreads out from its source carried by the fields. In this fashion, light can move through empty space "carrying itself by its own bootstraps."

*More precisely, 299,792.458 km/sec.

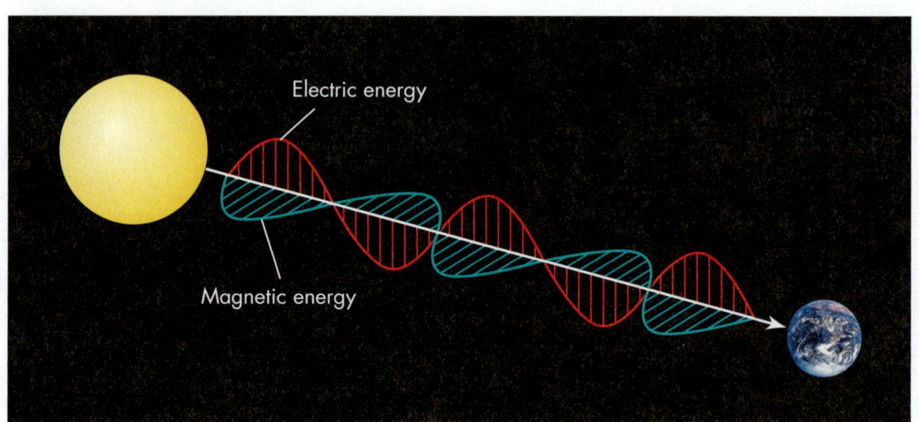

FIGURE 4.1
A wave of electromagnetic energy moves through empty space at the speed of light, 299,792.5 kilometers per second. The wave carries itself along by continually changing its electric energy into magnetic energy and vice versa. (The curving lines illustrate the changing strength of electric and magnetic energy as the light travels through space.)

As the electromagnetic wave travels through matter, it may disturb the atoms, causing them to vibrate the way a water wave makes a boat rock. It is from such disturbances in our eyes, a piece of film, or an electronic sensor that we detect the light.

The model of light as a wave works well to explain many phenomena, but it fails to explain some of light's properties. In those circumstances, it is necessary (and easier!) to use a different model.

In this model, light may be thought of as a stream of particles called **photons** (fig. 4.2). The photons are packets of energy, and when they enter your eye, they produce the sensation of light.

In empty space, photons move in a straight line at the speed of light. Although they are described as particles, photons can behave as waves, but they are not unique in this respect. According to the laws of quantum physics, subatomic particles such as electrons and protons can also behave like waves. For this reason, scientists often speak of light and subatomic particles as having a **wave-particle duality,** and they use whichever model—wave or particle—best describes a particular phenomenon. For example, reflection of light off a mirror is easily understood if you imagine photons striking the mirror and bouncing back just the way a ball rebounds when thrown at a wall. On the other hand, the focusing of light by a lens is best explained by the wave model.

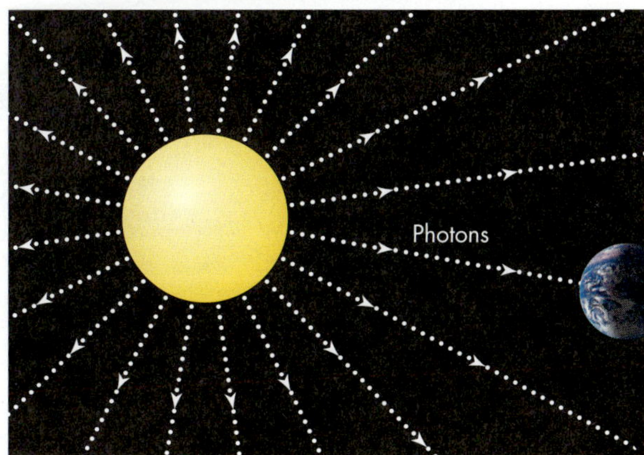

FIGURE 4.2
Photons—particles of energy—stream away from a light source at the speed of light.

Usually we will discuss light using the wave model, so that we do not have to constantly refer first to photons and then to waves. But regardless of which model we use, light has two important properties that we need to describe. The first of these properties is its brightness (or intensity). Brightness is a measure of the total amount of energy carried by the light. In the wave picture, brightness is related to the height of the wave. In the photon picture, brightness is related to the number of photons traveling in a given direction. The second important property of light is its color.

Light and Color

Human beings can see colors ranging from deep red through orange and yellow into green, blue, and violet, and we call these colors the **visible spectrum.** But what property of photons or electromagnetic waves corresponds to light's different colors?

According to the wave theory, the color of light is determined by the light's **wavelength,** which is the spacing between wave crests (fig. 4.3). That is, instead of describing a quality (color), we specify a quantity (wavelength), which we usually denote by the Greek letter lambda, λ. For example, the wavelength of deep red light is about 7×10^{-7} meters. The wavelength of violet light is about 4×10^{-7} meters. Intermediate colors have intermediate wavelengths. The wave-particle duality model allows us to make a similar connection between wavelength and color for photons. Thus, we can also characterize photons by their wavelength.

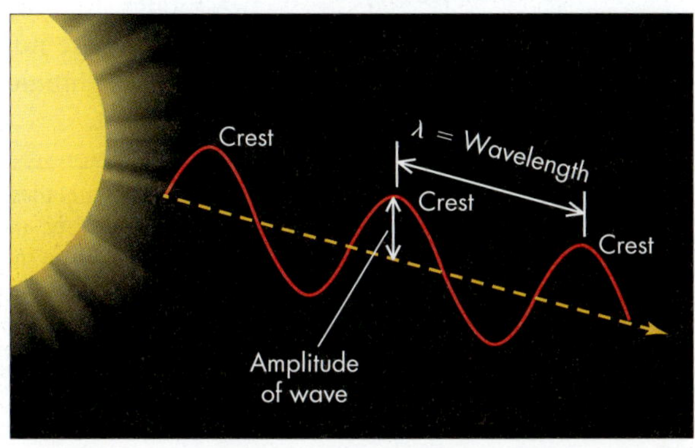

FIGURE 4.3
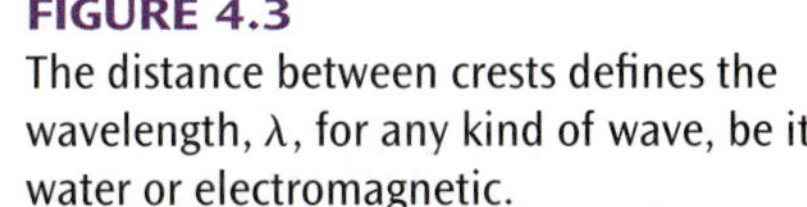
The distance between crests defines the wavelength, λ, for any kind of wave, be it water or electromagnetic.

TABLE 4.1	COLORS AND WAVELENGTHS*	
Red	700 nm	0.7 micrometers
Yellow	580	0.58
Blue	480	0.48
Violet	400	0.40

*These color equivalences are only approximate.

The wavelengths of visible light are very small (roughly the size of a bacterium). They are therefore usually measured not in meters but in billionths of a meter, a unit called the **nanometer,** abbreviated nm. Thus, the wavelength of red light is about 700 nanometers and that of violet light is about 400 nanometers. Table 4.1 lists the wavelengths of the primary colors in nm and micrometers, which are sometimes used to measure wavelengths.*

The relation between wavelength and color is important, and we will refer to it repeatedly in later chapters. In doing so, we will often mention only the red and blue colors of visible light. The other colors are not missing: they are just not referred to explicitly. Given this simplification, you should be sure to remember that *red* colors refer to *long* wavelengths of visible light and *blue* colors refer to *short* wavelengths.

Note that the wavelength of light is independent of the intensity, or *amplitude,* of the electromagnetic wave (figure 4.3). Thus, the strength of the variations of the electromagnetic radiation do not change its color. In the particle description of light, we might say that more-intense red light contains a larger number of red photons.

Characterizing Electromagnetic Waves by Their Frequency

Sometimes it is useful to describe electromagnetic waves by their frequency rather than their wavelength. **Frequency** is the number of wave crests that pass a given point in 1 second. It is measured in hertz (abbreviated as Hz) and is usually denoted by the Greek letter nu, ν. You can see an everyday example of this on a radio dial, where you tune in a station by its frequency rather than its wavelength.

For all kinds of waves, the frequency and wavelength are related to the wave speed, because in one vibration a wave must travel a distance equal to one wavelength. This implies that for light, the product of the wavelength ($\lambda\nu$) and the wave frequency (ν) equals the speed of light (c): that is, $\lambda\nu = c$.† Because all light travels at the same speed (in empty space), we can treat c as a constant. Thus, specifying λ determines ν and vice versa. We will generally use λ to characterize electromagnetic waves, but ν is just as good.

You can see how colors of light mix if you look at a color television screen close up. You will notice that the screen is covered with tiny red, green, and blue dots. In a red object, only the red spots are lit. In a blue one, only the blue spots. In a white object, all three are lit, and the brain mixes these three colors to form white. Other colors are made by appropriate blending of red, green, and blue. Notice this is very different from the way that pigments of paint mix. Red, green, and blue paint when mixed give a brownish color.

White Light

Although wavelength is an excellent way to specify most colors of light, some light seems to have no color. For example, the Sun when it is seen high in the sky and an ordinary lightbulb appear to have no dominant color. Light from such sources is called **white light.**

---

* Scientists sometimes use other units of length to measure wavelengths. For example, astronomers have traditionally used angstrom units and micrometers. One angstrom unit is 10^{-10} meters (1 ten-billionth of a meter) and is thus the same as 1/10 of a nanometer. The wavelength of red light is thus about 7000 angstroms. The micrometer (also called a micron and abbreviated μm) is used especially at infrared wavelengths (wavelengths longer than visible light that we perceive as heat). One micrometer is 10^{-6} meters. The wavelength of red light is about 0.7 micrometers.

† In using this equation, if ν is in hertz, the units of λ will be set by the units you choose for c. Thus, if c is in meters per second, λ will be in meters.

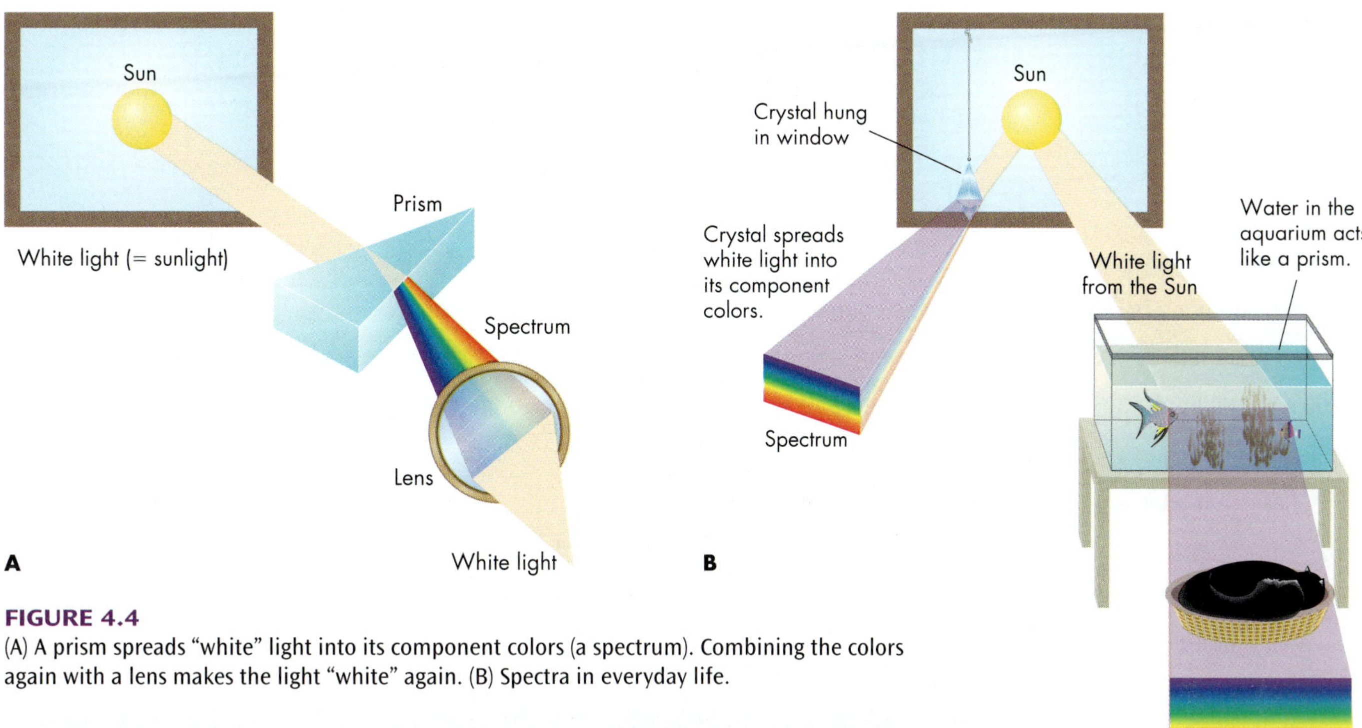

FIGURE 4.4
(A) A prism spreads "white" light into its component colors (a spectrum). Combining the colors again with a lens makes the light "white" again. (B) Spectra in everyday life.

White light is not a special color of light; rather it is a mixture of all colors. That is, the sunlight we see is made up of all the wavelengths of visible light—a blend of red, yellow, green, blue, and so on—and our eyes perceive this blend as white. Newton demonstrated this property of sunlight by a very simple but elegant experiment. He passed sunlight through a prism (fig. 4.4A) so that the light was spread out into the visible spectrum (or rainbow of colors). He then recombined the separated colors with a lens and reformed the beam of white light.

Why do we see sunlight as white? Presumably because our senses have evolved to make us aware of *changes* in our surroundings. Thus, we ignore the ambient "color" of sunlight just as we come in time to ignore a steady background sound or smell.

But there is more to light than what meets the eye. Just as red is but one part of the visible spectrum, so too the visible spectrum itself is but one part of a much wider spectrum of electromagnetic waves.

4.2 The Electromagnetic Spectrum: Beyond Visible Light

Visible light is just one of the many kinds of electromagnetic waves that exist in our Universe. For example, radio waves, X rays, and ultraviolet radiation are also electromagnetic waves. They differ from visible light only in their wavelengths, and to indicate this fundamental unity, scientists refer to them in general as the **electromagnetic spectrum.**

The electromagnetic spectrum has been studied over a huge range of wavelengths. The longest electromagnetic waves yet detected have wavelengths thousands of kilometers long.* The shortest have wavelengths of 10^{-18} meters or less. Ordinary visible light falls in a very narrow section in about the middle of the known spectral range (see table 4.2 and fig. 4.5). Although our eyes can detect only a tiny portion of the electromagnetic spectrum—namely, the part we call visible light—various instruments allow us to explore

*Such long waves have not been detected from astronomical sources, however, and cannot pass through our atmosphere.

TABLE 4.2 ELECTROMAGNETIC SPECTRUM

Wavelength	Kind of Radiation	Astronomical Sources
100–500 meters	Radio (AM broadcast)	Pulsars (remnants of exploded stars—also emit X rays)
10–100 meters	Short-wave radio	Active galaxies
1–10 meters	TV, FM radio	Solar radio outbursts, interstellar gas
10–100 centimeters	Radar, cell phones	Planets, active galaxies
1–100 millimeters	Microwaves	Interstellar clouds, cosmic background radiation
700 nm–1000 μm	Infrared (heat)	Young stars, planets, interstellar dust
400–700 nanometers	Visible light	Stars, Sun
10–400 nanometers	Ultraviolet	Stars
0.01–10 nanometers	X rays	Collapsed stars, hot gas in galaxy clusters
10^{-7}–0.01 nanometers	Gamma rays	Active galaxies and gamma-ray bursters

Note: 1 mm = 1000 μm; 1 μm = 1000 nm

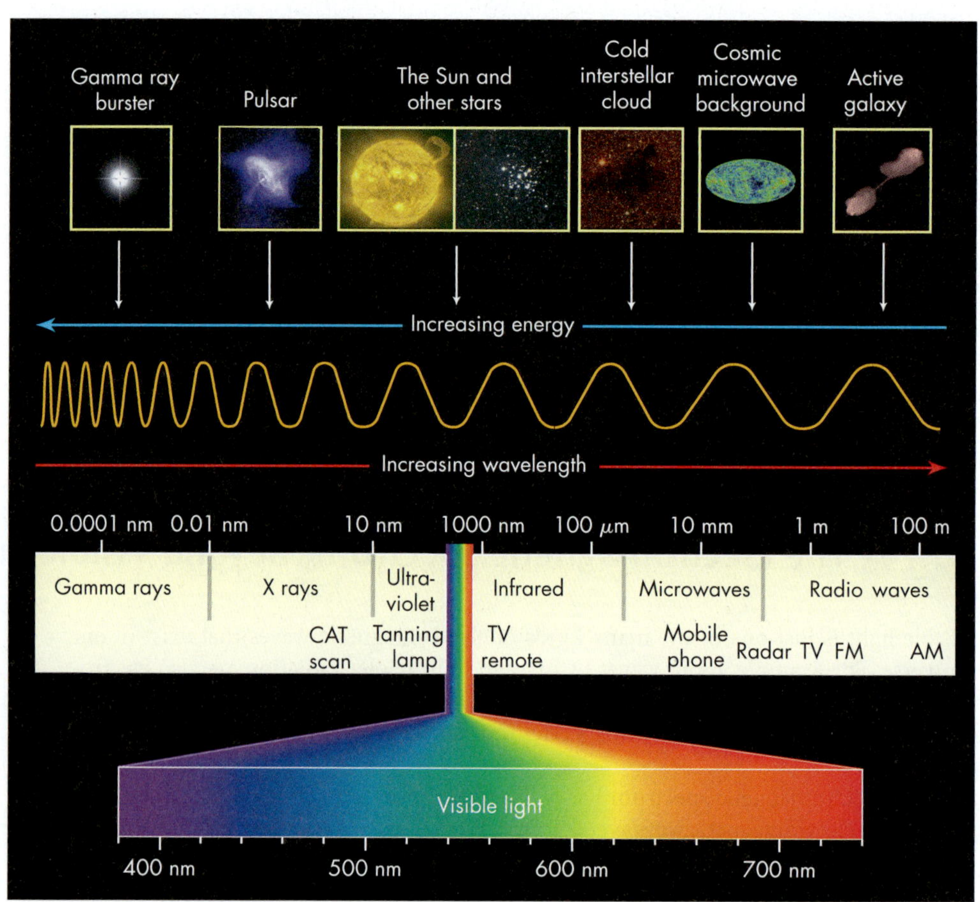

FIGURE 4.5
The electromagnetic spectrum.

most of the other wavelength regions, too. In fact, new instruments allow astronomers to "see" such astronomically important events as the formation of stars, the remnants left behind when stars die, and, indirectly, black holes. In the following sections, we discuss these different wavelength regions in the order in which they were discovered.

Infrared Radiation

The exploration of the electromagnetic spectrum began in 1800, when Sir William Herschel (discoverer of the planet Uranus) showed that heat radiation, such as you feel from the Sun or from a warm radiator, though invisible, was related to visible light.

Herschel was trying to measure heat radiated by astronomical sources. He projected a spectrum of sunlight onto a table top and placed a thermometer in each color to measure its energy. He was surprised that when he put a thermometer just off the red end of the visible spectrum, the thermometer registered an elevated temperature there just as it did in the red part of the spectrum. He concluded that some form of invisible energy perceptible as heat existed beyond the red end of the spectrum and he therefore called it **infrared.** Even though our eyes cannot see infrared light, nerves in our skin can feel it as heat.

Although humans cannot see infrared radiation, several kinds of snakes, including the rattlesnake, have special infrared sensors located just below their eyes. These allow the snake to "see" in total darkness, helping it to find warm-blooded prey such as rats.

Ultraviolet Light

Another important part of the electromagnetic spectrum, **ultraviolet** radiation, was discovered in 1801 by J. Ritter while he was experimenting with chemicals that might be sensitive to light. Ritter noted that when he shined a spectrum of sunlight on a layer of silver chloride, the chemical blackened most strongly in the region just beyond the violet end of the spectrum, implying the presence there of some invisible radiation.

Infrared and ultraviolet radiation differ in no physical way from visible light except in their wavelengths. Infrared has longer wavelengths and ultraviolet shorter wavelengths than visible light (see table 4.2). Exploration of those parts of the electromagnetic spectrum with wavelengths much longer and much shorter than visible light had to await the growth of new technology, as the development of radio astronomy demonstrates.

Radio Waves

James Clerk Maxwell, a Scottish physicist, predicted the existence of radio waves in the mid-1800s. It was some 20 years later, however, before Heinrich Hertz produced them experimentally in 1888, and another 50 years had to pass before Karl Jansky discovered naturally occurring radio waves coming from cosmic sources. Jansky's discovery in the 1930s that the center of the Milky Way was a strong source of radio emission was the birth of radio astronomy.

Radio waves range in length from about 10 centimeters to hundreds of meters, making them much longer than visible and infrared waves. Today we can generate radio waves and use them in many ways, ranging from communication to radar. Astronomers detect radio waves using radio telescopes. Radio signals, generated by natural processes, allow astronomers to obtain radio "views" of forming stars, exploding stars, active galaxies, and interstellar gas clouds. Radio wavelengths are even being searched for signals that might reveal the existence of extraterrestrial civilizations (see essay 3).

Other Wavelength Regions

Many decades also passed between the discovery of X rays by Wilhelm Roentgen in 1895 and the detection in the late 1940s of X rays coming from the Sun. X-ray wavelengths are far shorter than those of visible light, typically between 0.01 and 10 nanometers, but they too are important. Doctors and dentists use X rays to probe our bones and organs. Astronomers use X-ray telescopes to detect X rays emitted by the hot gas surrounding black holes and the tenuous gas in distant groups of galaxies.

Two parts of the electromagnetic spectrum were relatively unexplored until recently: microwaves, the region between infrared and radio waves, and gamma rays, which are the shortest wavelengths known. Both of these wavelength regions are difficult to study from the ground because, as we will see in section 4.7, they fall in wavelength bands that are strongly blocked by the Earth's atmosphere. However, orbiting telescopes have now given astronomers preliminary views of the sky in these wavelengths.

Despite the enormous variety of electromagnetic radiation, it is all the same physical phenomenon: the vibration of electric and magnetic energy traveling at the speed of light, or a stream of photons. (In fact, scientists refer to radio photons, visible photons, ultraviolet photons, etc.) The essential difference between these many kinds of electromagnetic radiation is simply their wavelength (or frequency). This difference alters not only how we perceive them but also how much energy they can carry.

Energy Carried by Electromagnetic Radiation

The warmth we feel on our face from a beam of sunlight demonstrates that light carries energy, but not all wavelengths carry the same amount of energy. It turns out that the amount of energy, E, carried by electromagnetic radiation depends on its wavelength, λ. Each photon of wavelength λ is an energy packet that carries an amount of energy, E, given by

E = Energy carried by a photon of wavelength = λ
h = Planck's constant
c = Speed of light (constant)

$$E = \frac{hc}{\lambda}$$

The speed of light c, and the constant, h, are unchanging,* so if the wavelength of the light decreases, the energy it carries increases. Thus:

> **Short-wavelength photons carry proportionally more energy than long-wavelength photons.**

Ultraviolet photons with their short wavelength therefore carry more energy than infrared photons. In fact, an ultraviolet photon of sufficiently short wavelength carries so much energy that it can break apart molecular bonds. As we will see in chapter 13, this can cause intense heating of gas near stars. Nearer to home, it is the reason ultraviolet light gives you a sunburn while an infrared heat lamp does not.

We have now described the main properties of light itself. We now need to discuss how light is produced and what it can tell us about the objects emitting it.

4.3 The Nature of Matter and Heat

Like so many of our ideas about the nature of the Universe, our ideas of matter date back to the ancient Greeks. For example, Leucippus, who lived about the fifth century B.C. in Greece, and his student Democritus taught that matter was composed of tiny indivisible particles. They called these particles *atoms,* which means "uncuttable" in Greek.

Our current model for the nature of atoms dates back to the early 1900s, with the work of the British physicist Ernest Rutherford. Rutherford showed with a series of experiments that atoms have a tiny core, the nucleus, around which yet smaller particles, called electrons, orbit. Electrons have a negative electric charge while the nucleus of an atom has a positive charge; atoms are held together by the electrical attraction between oppositely charged particles. This electrical attraction is what causes clothes to stick together in a dryer—electrons can rub off from one garment to another, building up static electricity. The attraction between the nucleus of one atom and the electrons of a neighboring atom also can link atoms together to form *molecules.*

The presence of electrical charges in atoms allows them to generate electromagnetic radiation and to interact with photons. These interactions leave an "imprint" on electromagnetic radiation that allows us to determine many properties of a material, including its temperature and the kinds of atoms and molecules out of which it is made.

---

*If E is measured in joules and λ in meters, h = 6.63×10^{-34} joule·second (known as Planck's constant) and hc = 1.99×10^{-25} joule·meters.

Before we examine these interactions we need to introduce the scale that astronomers and other scientists use to measure the temperature of materials.

The Kelvin Temperature Scale

One of the most important contributors to the understanding of heat and molecular motion was the English physicist Lord Kelvin. Kelvin studied numerous problems in physics and astronomy, ranging from the motion of fluids to the properties of gases. Much of this latter work was motivated by his attempts to improve the energy efficiency of steam engines. In the course of studying the energy content of gases, Kelvin devised a temperature scale that is used today in virtually all the physical sciences. The reason for this wide usage is that on the Kelvin scale, a body's temperature is directly related to its energy content and to the speed of its molecular motion. That is, the greater a body's Kelvin temperature, the more rapidly its atoms move and the more energy it possesses. Similarly, if the body is cooled toward a temperature of zero on the Kelvin scale, molecular motion within it slows to a virtual halt and its energy approaches zero. Partly as a result of this, the Kelvin scale has no negative temperatures.

Temperatures on the Kelvin scale are not given in degrees but are simply called "Kelvin." For example, the freezing and boiling points of water are very nearly 273 and 373 Kelvin, respectively. Room temperature is about 300 Kelvin. Relatively simple formulas allow conversion between Kelvin and the more familiar Fahrenheit and Celsius scales. Celsius temperatures are simply Kelvin temperatures minus 273. Fahrenheit temperatures, F, can be calculated by using the formula $F = 9/5\,K - 459.4$, where K is the Kelvin temperature. Figure 4.6 shows how the Kelvin, Celsius, and Fahrenheit temperature scales compare. Because of its direct relation to so many physical processes, we will use the Kelvin scale in most of the remainder of this book.

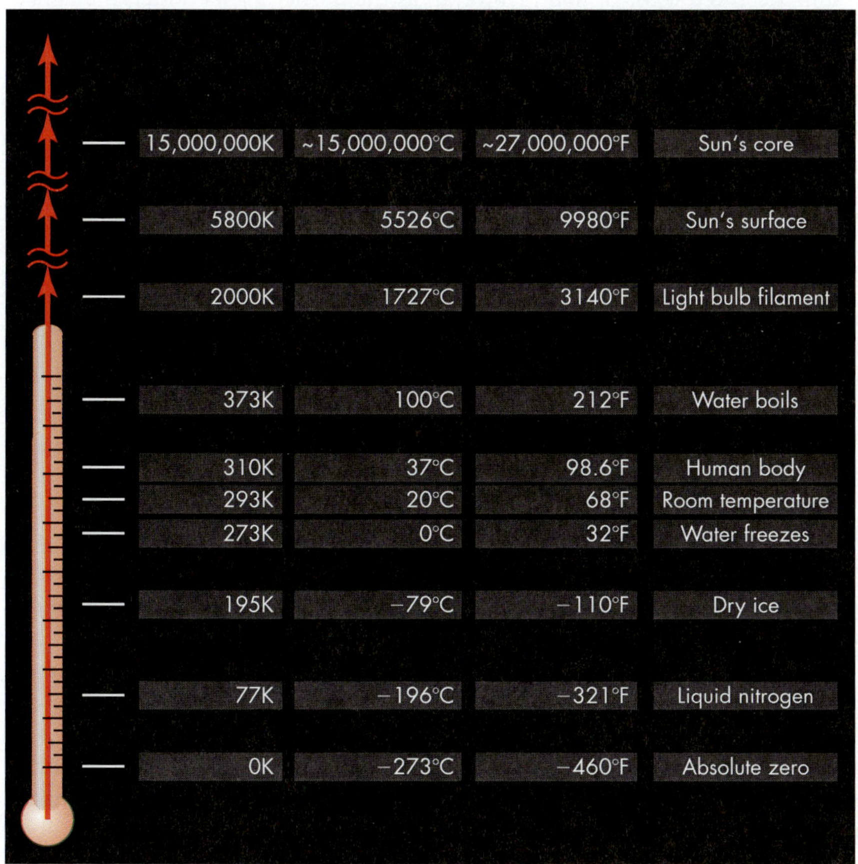

15,000,000K
~15,000,000°C
~27,000,000°F
Sun's core
5800K
5526°C
9980°F
Sun's surface
2000K
1727°C
3140°F
Light bulb filament
373K
100°C
212°F
Water boils
310K
37°C
98.6°F
Human body
293K
20°C
68°F
Room temperature
273K
0°C
32°F
Water freezes
195K
−79°C
−110°F
Dry ice
77K
−196°C
−321°F
Liquid nitrogen
0K
−273°C
−460°F
Absolute zero

FIGURE 4.6
Temperatures in Kelvin (K) and on the Celsius and Fahrenheit scales.

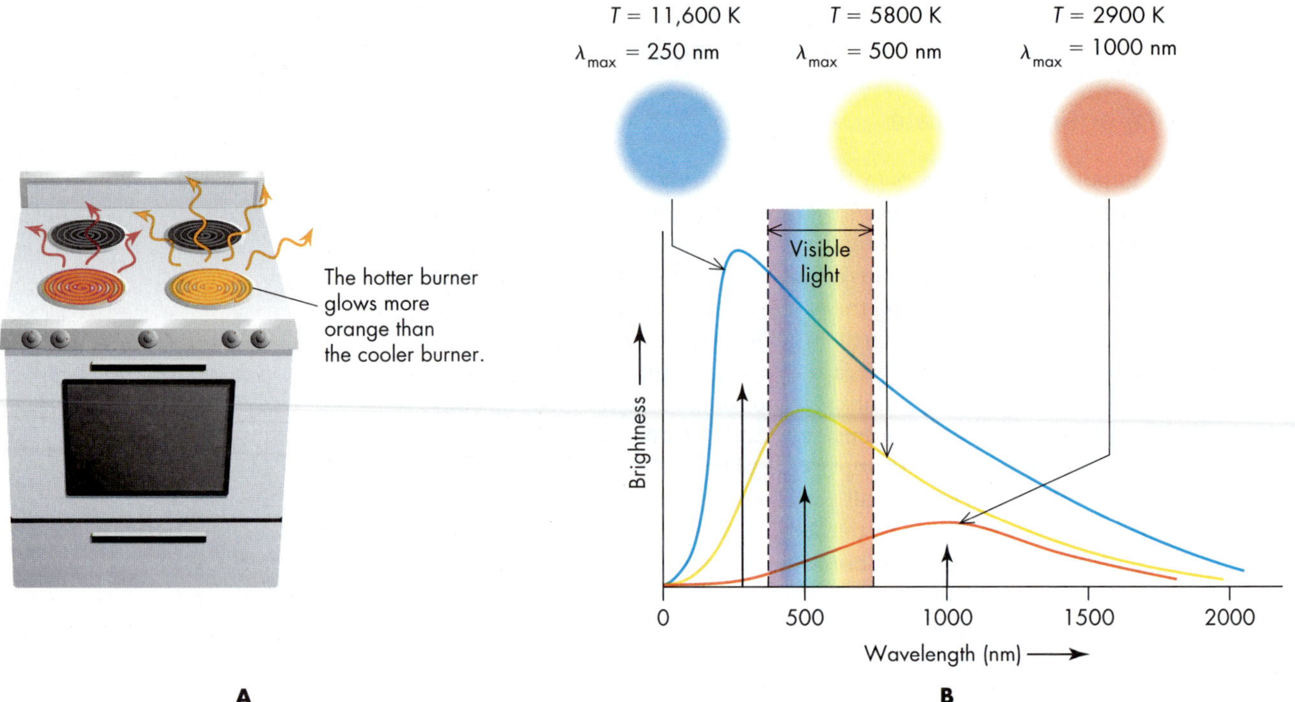

FIGURE 4.7
(A) The hotter burner glows more yellow than the cooler burner. (B) As an object is heated, the wavelength at which it radiates most strongly, λ_{max}, shifts to shorter wavelengths, a relation known as "Wien's law." Note also that as the object's temperature rises, the amount of energy radiated increases at *all* wavelengths.

Temperature and Radiation

Hot objects emit light, as you can easily demonstrate if you turn on a burner on an electric stove. As the stove's burner warms up, it begins to glow. Initially the burner emits only a dim, deep red color of light. However, as the burner grows hotter, the light it emits becomes brighter and its color changes (figure 4.7A), becoming brighter red and eventually yellow. If we could make the burner even hotter, it would glow blue.

Recalling that blue light has a shorter wavelength than red light, you can see from this simple demonstration an important relation between an object's temperature and the color of the light it emits:

> **As an object's temperature increases, the object radiates light more strongly at shorter wavelengths.**

That is, as an object heats up, the color of the light it emits shifts from red (long) wavelengths toward yellow (medium) wavelengths and, if hot enough, to blue (short) wavelengths. This connection between an object's temperature and the color of the light it emits applies to more than just electric stove burners. It is a general property of many hot objects, including stars, and it allows astronomers to measure the temperature of stars (and many other astronomical objects) from their light.

In the above discussion we have not given a value to the temperature but merely said "hot" or "hotter." However, we can find a numerical value for the temperature using a relation first worked out by the German physicist Wilhelm Wien (pronounced "veen") about 1900. **Wien's law** states that the wavelength (color) at which an object radiates most strongly is inversely proportional to the object's temperature.

In figure 4.7B we illustrate this principle by plotting the amount of energy radiated at each wavelength (color) for three objects of different temperatures. You can see from the curves for these objects that the hottest one is brightest (emits its greatest amount of energy) at 250 nanometers. That is, its curve is highest at that wavelength. On the other hand, the coolest one emits its greatest amount of energy at 1000 nanometers. Finally, the object of intermediate temperature has the peak of its curve at an

EXTENDING our reach

TAKING THE TEMPERATURE OF THE SUN

To measure the temperature of an object using Wien's law, we proceed as follows. First we measure the object's brightness at many different wavelengths to find at which particular wavelength it is brightest (that is, its wavelength of maximum emission). Then we use the law to calculate the object's temperature. To see how this is done, however, we need a mathematical expression for the law.

If we let T be the object's temperature measured in Kelvin, and λ_{max} be the wavelength in nanometers at which it radiates most strongly (fig. 4.7), Wien's law can be written in the form

$$T = \frac{2.9 \times 10^6}{\lambda_{max}}$$

The subscript "max" on λ is to remind us that it is the wavelength of maximum emission. The constant 2.9×10^6 is more accurately 2.898×10^6 K·nm. We round it off here to make calculations easier. The error this creates is small.

As an example, let's measure the Sun's temperature. The Sun turns out to radiate most strongly at a wavelength of about 500 nanometers. That is, its $\lambda_{max} = 500$ nm. Then, substituting that value for λ_{max} in our expression for T, we find

$$T = \frac{2.9 \times 10^6}{500} = \frac{2.9 \times 10^6}{5 \times 10^2}$$

$$= 0.58 \times 10^4 = 5800 \text{ Kelvin}$$

This is within a hundred degrees of the actual value.

intermediate wavelength (500 nm). You can therefore see in this figure the relation between temperature and wavelength we mentioned above—namely, that hotter objects emit more strongly at shorter wavelengths. In the Extending Our Reach box, we show how to *use* this relation to take the temperature of the Sun.

You might note that the wavelength at which the Sun radiates most strongly corresponds to a blue-green color, yet the Sun looks yellow-white to us. The reason we see it as whitish is related to how our eyes perceive color. Physiologists have found that the human eye interprets sunlight (and light from all extremely hot bodies) as whitish, with only tints of color. Keep in mind that although hot bodies emit most strongly at a wavelength indicated by Wien's law, they emit at all other wavelengths too. Thus, cool stars look white tinged with red, while very hot stars look white tinged with blue.

INTERACTIVE
Blackbody radiation and stellar luminosity

Although Wien's law works accurately for most stars and planets, it has some important exceptions. For example, the red color of an apple and the green color of a lime come from the light they *reflect* and have nothing to do with their temperature. The apple does emit some radiation, but if it is at normal room temperature, its radiation will be mostly in the infrared.

Wien's law makes good sense if you think about the relation between energy and temperature. Hotter things carry more energy (other quantities being equal) than cooler things. Also, bluer light carries more energy than red. Thus, it is reasonable to expect that hotter bodies emit bluer light.

Our discussion above has been qualified several times by terms such as *usually* and *most*. The reason for these qualifications is that Wien's law is exact only for a class of objects known as blackbodies. A **blackbody** is an object that absorbs all the radiation falling upon it. Because such a body reflects no light, it looks black to us when it is cold; hence its name. Experiments show that when blackbodies are heated, they radiate more efficiently than other kinds of objects. Thus, they are both excellent absorbers and excellent emitters. Moreover, the intensity of their radiation changes smoothly from one wavelength to the next with no gaps or narrow peaks of brightness, as illustrated by the curves in figure 4.7B. Very few objects are perfect blackbodies, but many of the objects we will study are near enough to being blackbodies that we can use Wien's law with little fear of its being in error. For example, the electric stove burner, the Sun, and the Earth all obey Wien's law quite satisfactorily.

4.4 Radiation from Individual Atoms

While solid matter and dense collections of atoms emit blackbody radiation, gases generally behave quite differently. We see these contrasting kinds of emission in ordinary lightbulbs. Incandescent lightbulbs emit light by heating a solid filament to high temperature, which emits light according to Wien's law. However fluorescent lights and neon signs are not blackbodies. They instead produce light by first pulling electrons free from the atoms in the gas, which then emit light when the electrons recombine with the atoms.

This same difference is found in nature. Interstellar clouds, for example, radiate strongly only at specific wavelengths, such as a narrow wavelength range in the red part of the visible spectrum or the millimeter wavelength part of the radio spectrum. The clouds' colors are determined by characteristics of the individual atoms in the gas more than by temperature.

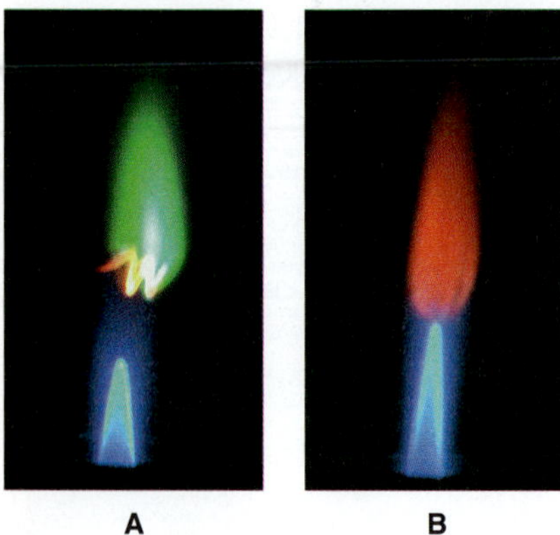

FIGURE 4.8
These photographs show the effect of (A) copper (green) and (B) strontium (red) on the burner's flame.

You can easily demonstrate the importance of composition in determining color with a gas flame on a stove or Bunsen burner. Normally the flame has a blue part and a yellow part. The yellow part is blackbody radiation from very hot specks of carbon soot. However, the blue part is caused by nonblackbody emission from carbon atoms. If you add chemicals to the flame, the flame's color may change dramatically. For example, if you hold some copper sulfate crystals in the flame with a pair of pliers, the flame will take on a greenish-blue color caused by the emission wavelengths of copper (fig. 4.8A). Likewise, the strontium in a highway emergency flare gives its light a strong red color (fig. 4.8B).

The structure of atoms determines both their chemical properties and their light-emitting and light-absorbing properties. For example, iron and hydrogen not only have very different atomic structures but also emit very different wavelengths of light. From those differences astronomers can deduce whether an astronomical body—a star or a planet—contains iron, hydrogen, or whatever chemicals happen to be present. Therefore, an understanding of the structure of atoms ultimately leads us to an understanding of the nature of stars.

The Chemical Elements

Iron and hydrogen are examples of what are called chemical **elements.** A chemical element is a substance composed only of atoms that all have the same electrical charge in their nucleus. We described in section 4.3 how an atom has a dense core called a *nucleus* around which particles called *electrons* orbit. The nucleus is in turn composed of particles called *protons* and *neutrons;* the protons have a positive charge, the neutrons have no charge. The number of protons therefore determines the kind of chemical element the atom is.

For example, hydrogen consists exclusively of atoms that contain 1 proton; helium, of atoms that contain 2 protons; carbon, 6; oxygen, 8; and so forth. Although the identity of an element is determined by the number of protons in its nucleus, the chemical properties of each element are determined by the number of electrons orbiting its nucleus. However, the number of electrons normally equals the number of protons. The protons attract an equal number of the oppositely charged electrons until the atom is electrically neutral.†

Table 4.3 lists some of the more important elements we will discuss during our exploration of the Universe and the number of protons each contains. Most elements can have various forms with different numbers of neutrons, called *isotopes.* Isotopes have the same chemical properties, but different masses. The table lists stable isotopes of each kind of atom. Other numbers of neutrons are possible, but the resulting nucleus is unstable, or *radioactive.*

TABLE 4.3 ASTRONOMICALLY IMPORTANT ELEMENTS

Element	Number of Protons	Number of Neutrons*
Hydrogen	1	0, 1
Helium	2	2, 1
Carbon	6	6, 7
Nitrogen	7	7, 8
Oxygen	8	8, 10, 9
Silicon	14	14, 15, 16
Iron	26	30, 31, 32

*The number of neutrons listed is the number found in stable forms of the element, the most abundant listed first. Different neutron numbers lead to what are called isotopes of the element. Isotopes with different numbers of neutrons than those listed are unstable.

† Under some circumstances, an atom may lose or gain one or more electrons. Such atoms are said to be ionized, as we will discuss in chapters 12 and 13.

Electron Orbitals

The orbits of electrons in an atom are generally extremely small. For example, the diameter of the smallest electron orbit in a hydrogen atom is only about 10^{-10} (1 ten-billionth) meter. This infinitesimal size leads to effects that operate at an atomic level and have no counterpart in larger systems. The most important of these effects is that the electron orbits may have only certain prescribed sizes. Although a planet may orbit the Sun at any distance, an electron may orbit an atomic nucleus at only certain distances, much as when you climb a set of stairs, you can be only at certain discrete heights. For example, in a hydrogen atom the electron must have an orbital radius of $0.053\ n^2$ nanometers, where $n = 1$, or 2, or 3, . . . etc. That is, the radius can be 0.053, 0.21, 0.48, etc. nanometers, but it cannot have intermediate values. We describe this restriction on the allowable sizes of orbits by saying that they are **quantized.** Figure 4.9 illustrates this property of electron orbits and compares it to painters being confined to only certain levels when they work on a scaffold.

The above restriction on orbital sizes results from the electron's acting not just as a particle but also as a wave. That is, just as light itself has a wave–particle duality, so too does an electron. The electron's wave nature forces the electron to move only in orbits whose circumference is a whole number of wavelengths. If it were to move in other orbits, the electron's wave nature would "cancel" it out.

The wave nature of the electron has another important effect. It "smears" the electrons. As a result, although we have described the electrons as orbiting like tiny planets around the nucleus, most scientists prefer to think of them as moving within an electron cloud, which is called an **orbital.** The pattern of the orbital describes the probability of finding the electron at different positions. Simplified depictions of several orbitals are illustrated in figure 4.9.

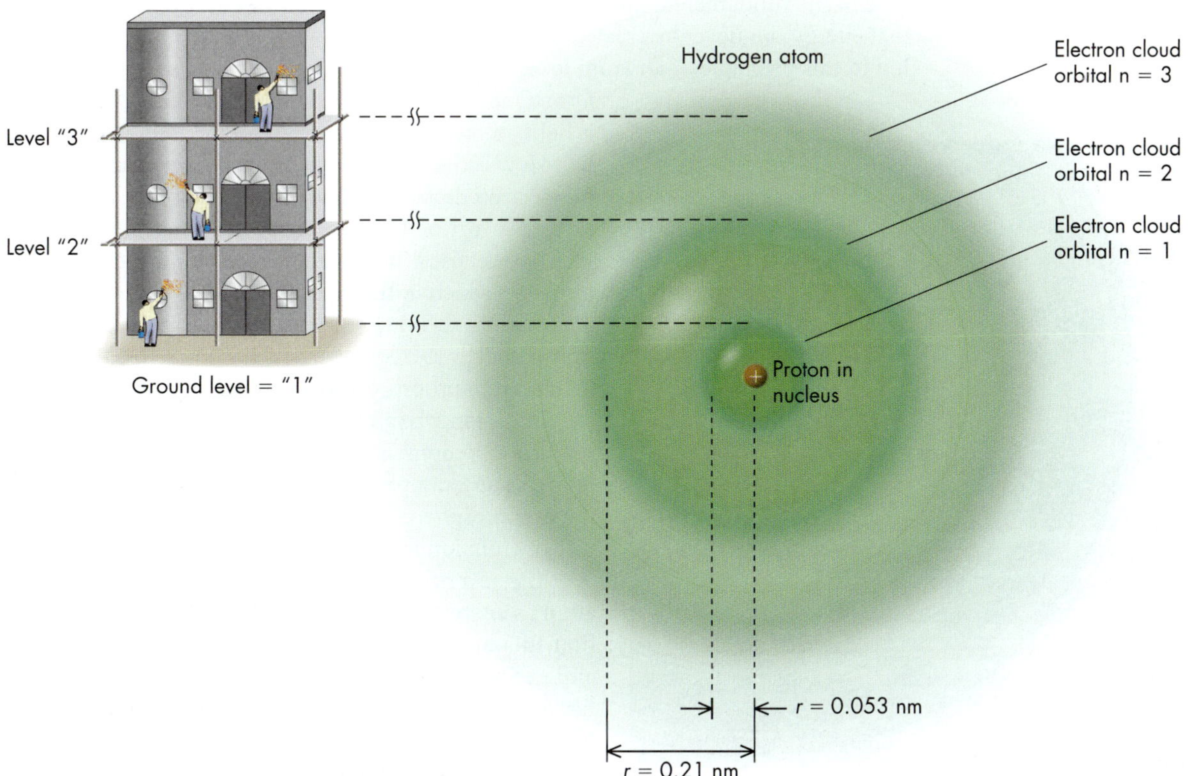

FIGURE 4.9
Just as the painters can only be at levels 1, 2, 3, . . . of the scaffold (and cannot "float in between"), so too an electron must be in orbital 1, 2, 3, . . . etc.

ANIMATION
Energy is released when an electron drops from an upper to a lower orbital.

ANIMATION
Absorption

Electrons in orbitals have another property totally unlike those of planets in orbit: they routinely shift from one orbital to another. This shifting changes their energy, as can be understood by a simple analogy. The electrical attraction between the nucleus and the electron creates a force between them like a spring. If the electron increases its distance from the nucleus, the spring must stretch. This requires giving energy to the atom. Likewise, if the electron moves closer to the nucleus, the spring contracts and the atom must give up, or emit, energy. We perceive that emitted energy as light or, more generally, electromagnetic radiation. The wavelength of that radiation is not the same for all atoms because of the different charges of the nucleus and the interactions between electrons. Therefore the electrons in hydrogen atoms behave in one way, but the electrons in iron atoms have a very different pattern of behaviors.

In summary then, atoms consist of a nucleus containing protons and neutrons surrounded by electrons in orbitals. The identity of the atom—the element—is determined by the number of protons in its nucleus. The electrons are bound to the nucleus by the electric attraction between the protons and electrons. Electrons may shift from one orbital to another accompanied by a change in the atom's energy. With this picture of the atom in mind, we can now turn to how light is generated within atoms.

The Generation of Light by Atoms

We saw above that when an electron moves from one orbital to another, the energy of the atom changes. If the atom's energy is increased, the electron moves outward from an inner orbital. Such an atom is said to be **excited.** On the other hand, if the electron moves inward toward the nucleus, the atom's energy is decreased.

Although the energy of an atom may change, the energy cannot just disappear. One of the fundamental laws of nature is the **conservation of energy.** This law states that energy can never be created or destroyed, it can only be changed in form. According to this principle, if an atom loses energy, that energy *must reappear in some other form.* One important form in which the energy reappears is light, or, more generally, electromagnetic radiation.

How is the electromagnetic radiation created? When the electron drops from one orbital to another, it alters the electric energy of the atom. As we described in section 4.1, such an electrical disturbance generates a magnetic disturbance, which in turn generates a new electrical disturbance. Thus, the energy released when an electron drops from a higher to a lower orbital becomes an electromagnetic wave, a process called **emission** (fig. 4.10).

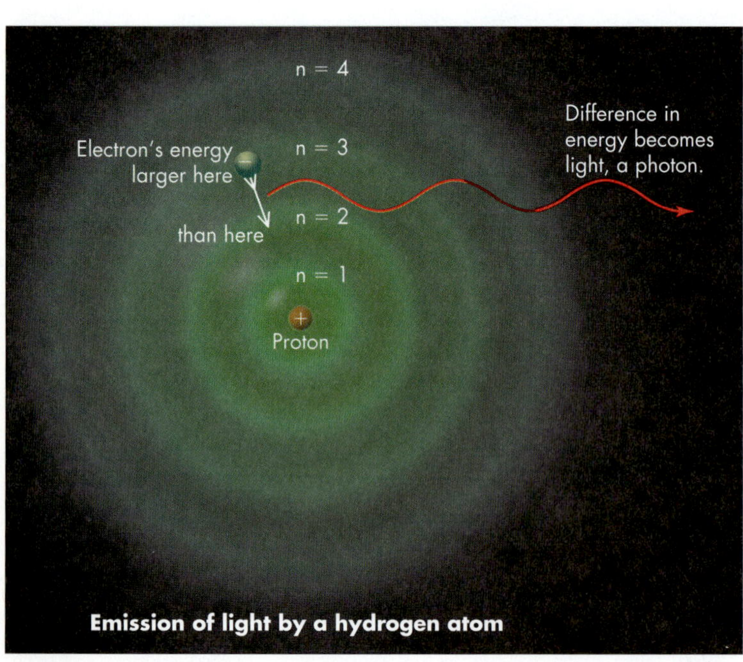

FIGURE 4.10
Energy is released when an electron drops from an upper to a lower orbital, causing the atom to emit electromagnetic radiation.

Emission plays an important role in many astronomical phenomena. The aurora borealis (northern lights) is an example of emission by atoms in the Earth's upper atmosphere, and sunlight and starlight are examples of emission in those bodies.

The reverse process, in which light is stored in an atom as energy, is called **absorption** (fig. 4.11). Absorption lifts an electron from a lower to a higher orbital and excites the atom by increasing the electron's energy. Absorption is important in understanding such diverse phenomena as the temperature of a planet and the identification of star types, as we will discover in later chapters.

Emission and absorption are particularly easy to understand if we use the photon model of light.

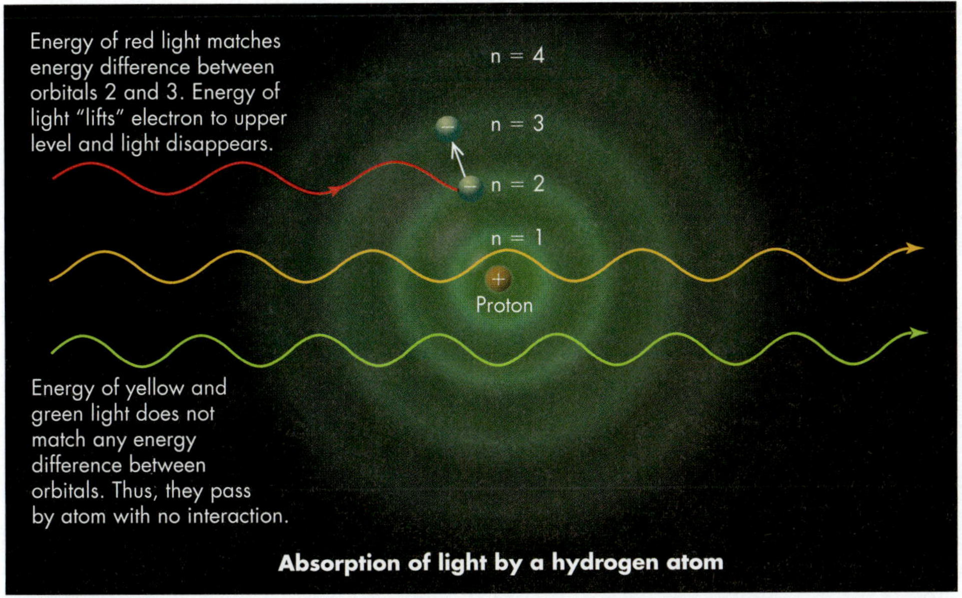

Energy of red light matches energy difference between orbitals 2 and 3. Energy of light "lifts" electron to upper level and light disappears.
n = 4
n = 3
n = 2
n = 1
Proton
Energy of yellow and green light does not match any energy difference between orbitals. Thus, they pass by atom with no interaction.
Absorption of light by a hydrogen atom

FIGURE 4.11
An atom can absorb light if its energy matches the energy difference between two orbitals.

According to this model, an atom emits a photon when one of its electrons drops from an upper to a lower orbital. Similarly, an atom absorbs light when a photon collides with it and "knocks" one of its electrons into an upper level.

You may find it helpful in understanding emission and absorption if you think of an analogy. Absorption is a bit like drawing an arrow back preparatory to shooting it from a bow. Emission is like the arrow being shot. In one case, energy of your muscles is transferred to and stored in the flexed bow. In the other, it is released as the arrow takes flight.

4.5 Formation of a Spectrum

The key to determining the composition and conditions of an astronomical body is its spectrum. The technique used to capture and analyze such a spectrum is called **spectroscopy.** In spectroscopy, the light (or more generally the electromagnetic radiation) emitted or reflected by the object being studied is collected with a telescope and spread into its component colors to form a spectrum by passing it through a prism or a grating consisting of numerous, tiny, parallel lines. Figure 4.12 shows not only how to form a spectrum but also what the spectrum looks like, in this case a band of rainbow colors. In general, we can show a spectrum as it would look to us or as a plot of the light's brightness at each color. In the case shown in figure 4.12, all colors are present and are more or less equally bright. As we will discover later, not all spectra look like this. Sometimes only a few colors are present.

Because light is emitted from atoms as electrons shift between orbitals, we might expect that the light will bear some imprint of the kind of atom that creates it. That is usually the case, and astronomers can search for the atom's "signature" by measuring how much light is present at each wavelength.

Spectroscopy is such an important tool for astronomers that we should look in greater detail at how it works. Specifically, why does an atom produce a unique spectral signature? To understand that, we need to recall how light is produced.

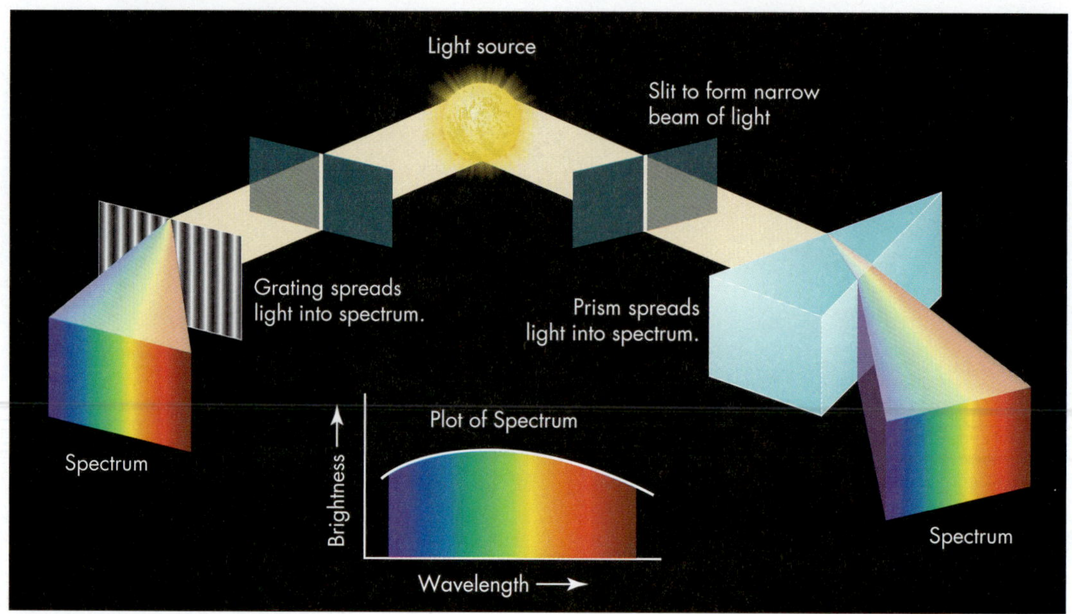

FIGURE 4.12
Sketch of a spectroscope and how it forms a spectrum. Either a prism or a grating may be used to spread the light into its component colors.

How a Spectrum Is Formed

We saw earlier that each kind of atom has a different number of electrons. This means that each kind of atom has a different set of electron orbitals. We also saw that the orbital in which an electron is located at any given moment sets the atom's energy. For this reason scientists sometimes refer to the orbitals as **energy levels.**

When an electron moves from one energy level (orbital) to another, the atom's energy changes by an amount equal to the *difference* in the energy between the two levels. As an example, suppose we look at light from heated hydrogen. Heating speeds up the atoms, causing more forceful and frequent collisions, knocking each excited atom's electron to outer orbitals. However, the electrical attraction between the nucleus and the electron draws the electron back almost at once. Suppose we look at an electron shifting from orbital 3 to orbital 2, as shown in figure 4.13. As the electron shifts downward, the atom's energy decreases, and the energy lost appears as light.

The wavelength of the emitted light can be calculated from the energy difference of the levels and the relation we mentioned earlier between energy (E) and wavelength (λ) ($E = \mathrm{hc}/\lambda$). If we evaluate the wavelength of this light, we find that it is 656 nanometers, a bright red color. An electron dropping from orbital 3 to orbital 2 in a hydrogen atom will always produce light of this wavelength.

If, instead, the electron moves between orbital 4 and orbital 2, there will be a different change in energy because orbital 4 has a different energy from that of orbital 3. That different energy will have a wavelength different from 656 nanometers. A calculation of its energy change leads in this case to a wavelength of 486 nanometers, a turquoise blue color.

Similar calculations show that when the electron jumps from orbital 5 to orbital 2 or from orbital 6 to orbital 2, other spectrum lines are emitted. However, we will see no light at most other wavelengths because hydrogen has no electron orbitals corresponding to those energies. Therefore, the hydrogen spectrum shows a set of brightly colored lines separated by wide, dark gaps. This is how an emission-line spectrum is formed.*

You can see these emission lines in figure 4.14, which shows not only what the spectrum of hydrogen looks like but also a plot of how bright the spectrum is at each wavelength.

*The model for the atom that we have used here and that so successfully explains its spectrum was developed by Danish physicist Niels Bohr. He won the 1922 Nobel Prize for Physics for his work.

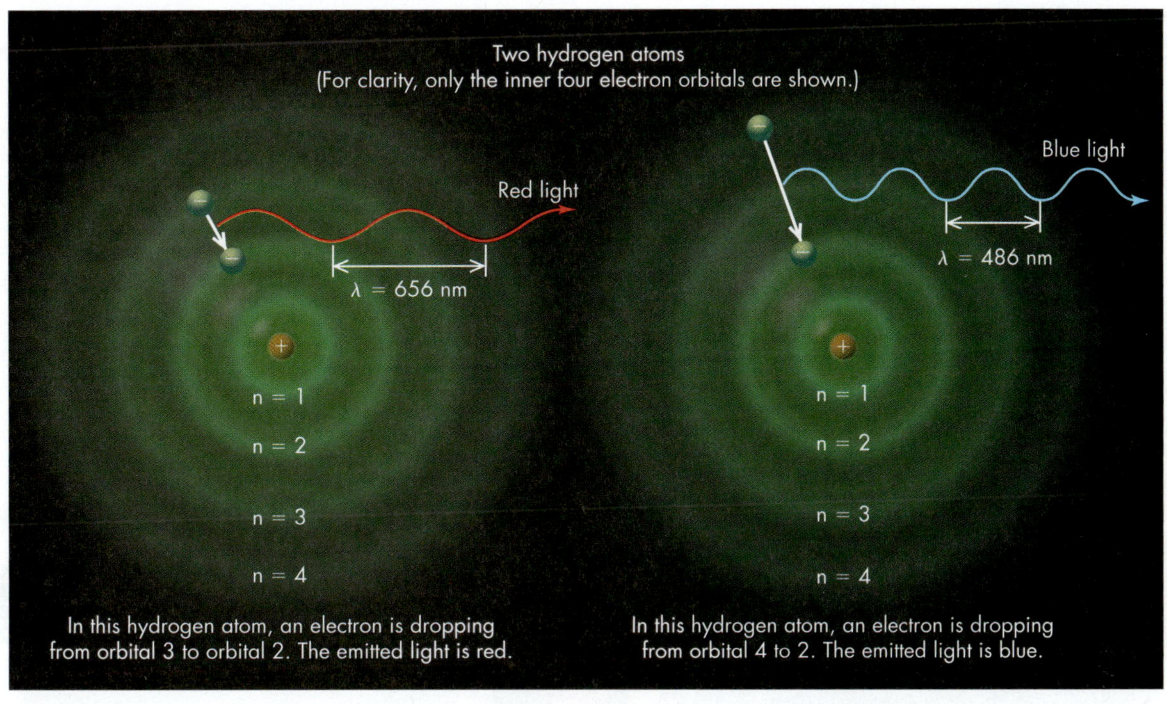

FIGURE 4.13
Emission of light from a hydrogen atom. The energy of an electron dropping from an upper to a lower orbital is converted to light. The light's color depends on the orbitals involved.

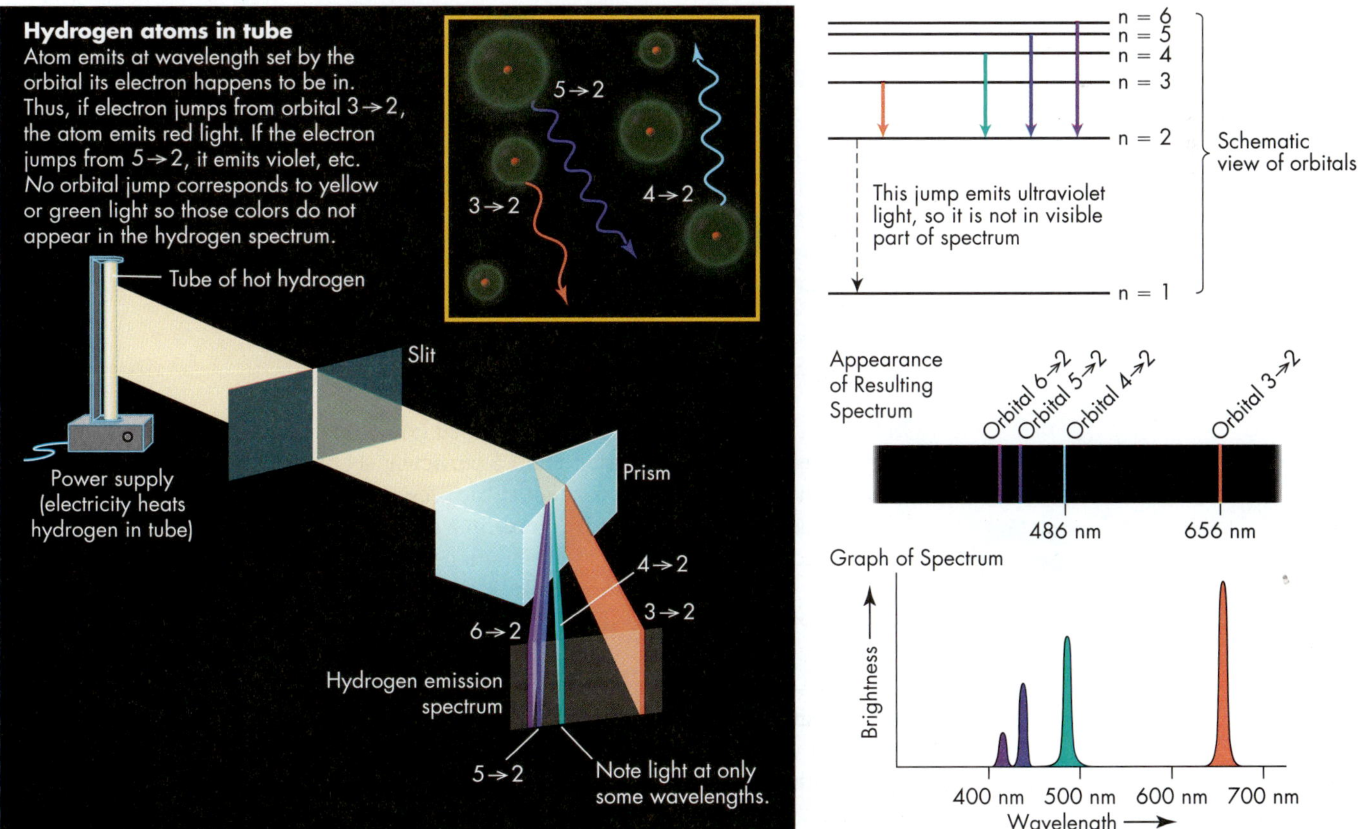

FIGURE 4.14
The spectrum of hydrogen in the visible wavelength range.

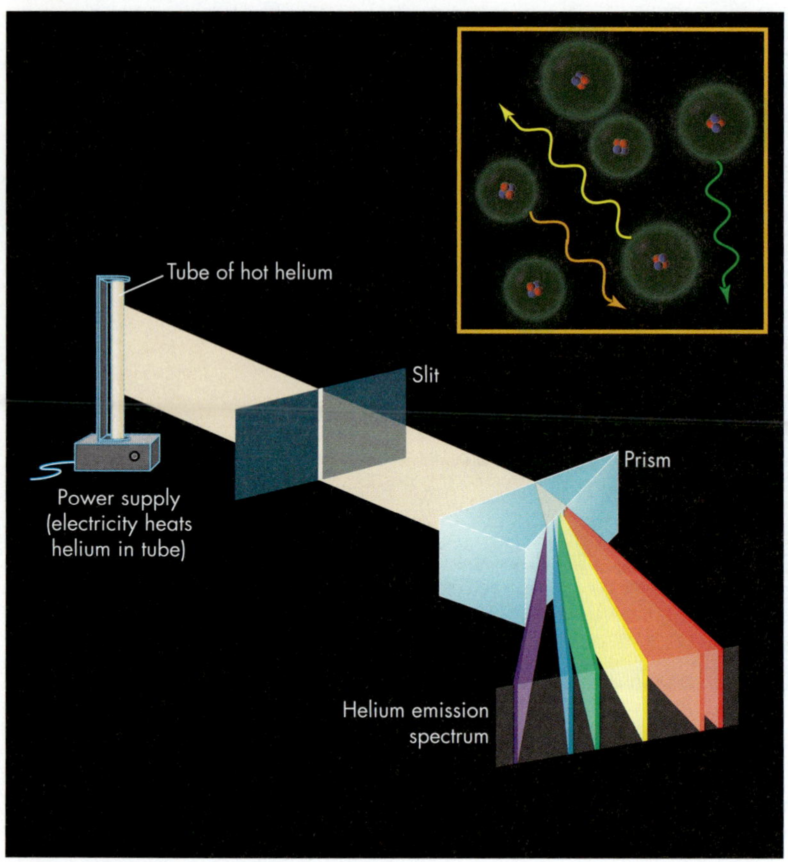

Helium atoms in tube
The electron orbitals for helium atoms are different from the orbitals in hydrogen. The light they emit therefore differs from that of hydrogen.

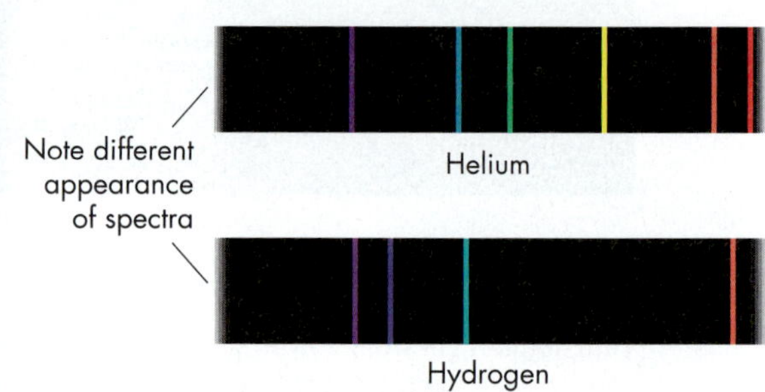

FIGURE 4.15
The spectrum of helium in the visible wavelength range.

We might now imagine making the same kind of calculations for a different chemical element, such as helium. If we did, we would discover that its wavelengths are in general different from those of hydrogen (see fig. 4.15). Thus, hydrogen's signature is its red 656-nanometer and blue 486-nanometer lines, and that signature offers astronomers a way to determine what astronomical objects are made of.

Identifying Atoms by Their Light

In the previous paragraphs we have described how atoms emit light. Moreover, we have seen that each chemical element emits a particular set of spectrum lines and that these emission lines provide a way to identify the presence of that element in a hot gas.

ANIMATION
Atomic emission and absorption

It is also possible to identify atoms in a gas from the way they absorb light. Light is absorbed if the energy of its wavelength corresponds to an energy that matches the difference between two energy levels in the atom. If the wavelength *does not* match, the light will not be absorbed, and it will simply move past the atom, leaving itself and the atom unaffected.

For example, suppose we shine a beam of light that initially contains all the colors of the visible spectrum through a box full of hydrogen atoms. If we examine the spectrum of the light after it has passed through the box, we will find that certain wavelengths of the light have been removed and are missing from the spectrum (fig. 4.16). In particular, the spectrum will contain gaps that appear as dark lines at 656 nanometers and 486 nanometers, precisely the wavelengths at which the hydrogen atoms emit. The absorption spectrum is, in effect, the opposite of the emission

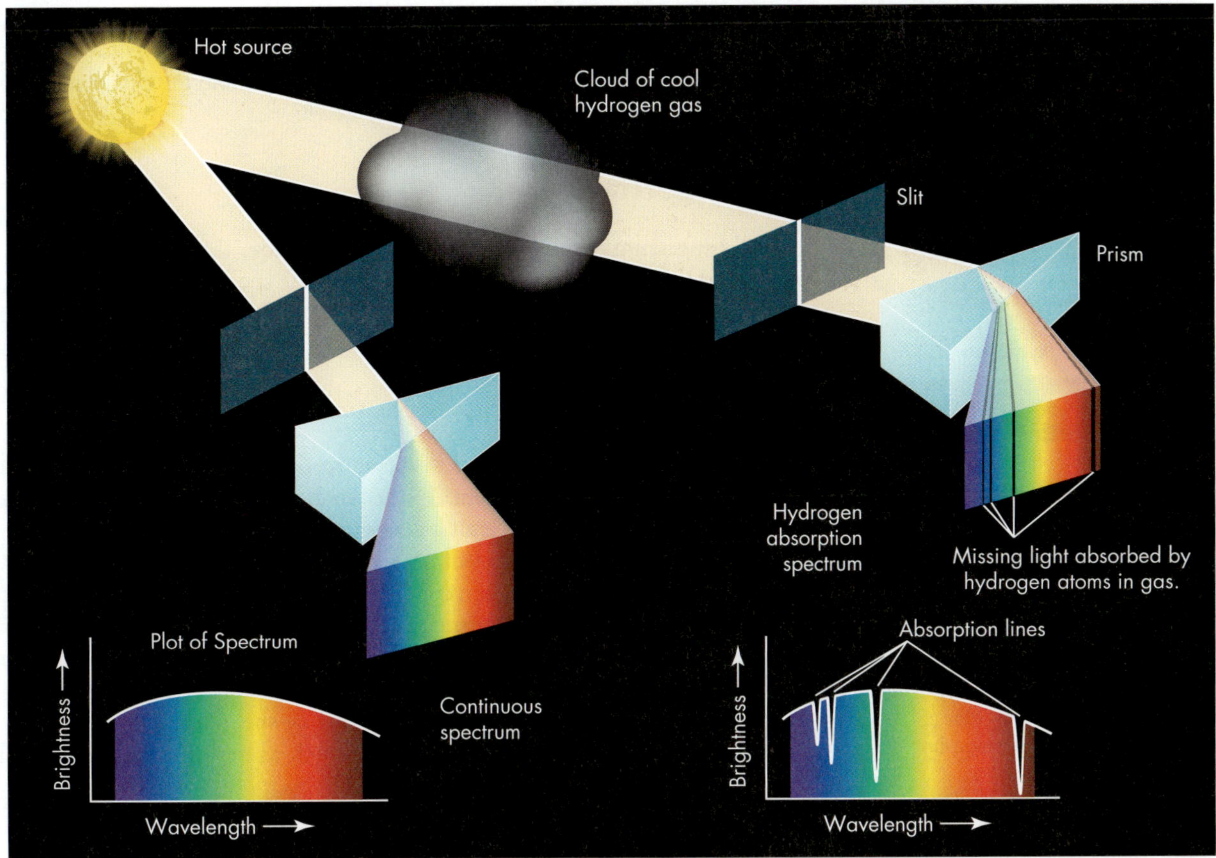

FIGURE 4.16
A hot, dense substance produces a continuous spectrum. Atoms in a gas cloud between an observer and the source of continuum emission absorb only those wavelengths whose energy equals the energy difference between their electron orbitals. The absorbed energy lifts the electrons to upper orbitals. The lost light makes the spectrum darker at the wavelengths where it is absorbed.

spectrum and an atom's absorption lines have exactly the same wavelengths as its emission lines.

These gaps are created by the light at 656 nanometers interacting with the hydrogen and lifting the electron in some atoms from orbital 2 to orbital 3 while the light at 486 nanometers lifts the electron in other hydrogen atoms from orbital 2 to orbital 4. Light at other wavelengths in this range has no effect on the atom. Thus, we can tell that hydrogen is present from either its emission or its absorption spectral lines.

In our discussion above, we have considered light emitted and absorbed by individual atoms in a gas. If the atoms are linked to one another to form molecules, such as water or carbon dioxide, the molecules too produce emission and absorption lines. In fact, even solid objects may imprint spectral lines on light that reflects off them.

In the above examples, we have considered light emitted from or absorbed by the body we wish to identify. But even reflected light generally bears some imprint of the surface from which it came. For example, when light from the Sun reflects from an asteroid, spectral features appear that were not present in the original sunlight. This gives astronomers information about the surface composition of bodies too cool to emit significant light of their own.

We conclude that in general we can identify the kind of atoms or molecules that are present by examining either the bright or the dark spectrum lines. Gaps in the spectrum at 656 nanometers and 486 nanometers imply that hydrogen is present. Similar gaps at other wavelengths would show that other elements are present. By matching the observed gaps to a directory of absorption lines, we can identify the atoms that are present. *This is the fundamental way astronomers determine the chemical composition of astronomical bodies.*

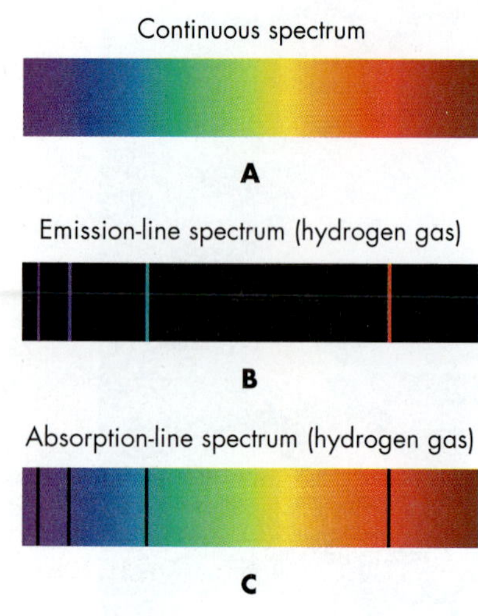

FIGURE 4.17
Types of spectra: (A) continuous, (B) emission-line, and (C) absorption-line.

Types of Spectra

Although a spectrum may bear the imprint of the matter that emitted the light, it may also have certain general properties. For example, the spectrum of a hot, tenuous gas is usually different from that of a hot, dense solid, regardless of the composition of either the gas or the solid. Therefore, it will be helpful to learn to recognize these general properties of spectra as well.

Spectra have the following three basic forms:

A. For some sources, the brightness of the emitted light changes smoothly with wavelength and all colors are present. We say such a light has a **continuous spectrum** (fig. 4.17A). For a source to emit a continuous spectrum, its atoms must in general be packed so closely that the electron orbitals of one atom are distorted by the presence of neighboring atoms. Such conditions are typical of solid or dense objects such as the heated filament of an incandescent lightbulb or a nail heated by a blowtorch.
B. Some heated objects have a spectrum in which light is emitted at only a few particular wavelengths while most of the other wavelengths remain dark (fig. 4.17B). This type of spectrum is called an **emission-line spectrum.** Emission-line spectra are usually produced by *hot, tenuous gas,* such as that in a fluorescent tube, the aurora, and many interstellar gas clouds.
C. A still different type of spectrum arises when light from a hot, dense body passes through cooler gas between it and the observer. In this case, nearly all the colors are present, but light is either missing or much dimmer at some wavelengths, as we discussed earlier. This causes the bright background to be crossed with narrow dark lines where the light of some colors is fainter or absent altogether (fig. 4.17C). The resulting spectrum is therefore called a dark-line or **absorption-line spectrum.**

Absorption lines were first detected astronomically in 1802, when the English scientist William H. Wollaston viewed sunlight through a prism and a narrow slit (figure 4.18A). He noticed dark lines between some of the colors but paid little attention to them. These dark lines in the Sun's spectrum were independently discovered a few years later by the German scientist Joseph Fraunhofer, who catalogued them and discovered similar lines in other stars. In fact, because nearly all stars have absorption-line spectra, this spectrum type is especially important in astronomy. However, if we consider the physical process of spectrum formation, such absorption-line spectra are really just a special case of continuous spectra with light missing at some wavelengths.

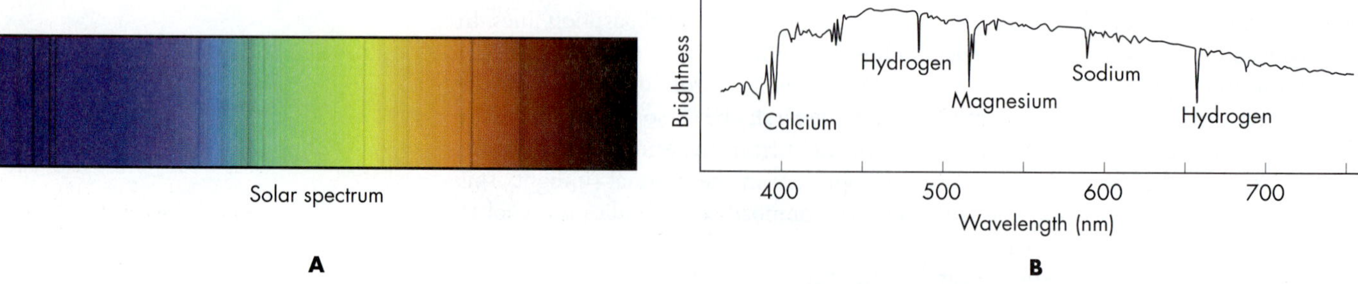

FIGURE 4.18
(A) The spectrum of the Sun. Note the narrow dark absorption lines. (B) A graphical representation of the spectrum.

Astronomical Spectra

The first step facing an astronomer who wants to analyze a spectrum is to identify the spectrum lines. This is done by measuring the wavelengths of the lines and then consulting a directory of spectrum lines. By matching the wavelength of the line of interest to a line in the table, astronomers can determine what kind of atom or molecule created the line. A look at a typical spectrum will show you that some lines are hard to see, being faint and weak. On the other hand, some lines may be very obvious and strong. The strength or weakness of a given line turns out to depend on the number of atoms or molecules absorbing (or emitting, if we are looking at an emission line) at that wavelength. Unfortunately, the number of atoms or molecules that can absorb or emit depends not just on how many of them are present but also on their temperature, as we will discuss more fully in chapter 12. Nevertheless, astronomers can deduce from the strength of emission and absorption lines the relative quantity of each atom producing a line and thereby deduce the composition of the material in the light source. Table 4.4 shows the result of such an analysis for our Sun, a typical star.

TABLE 4.4 COMPOSITION OF A TYPICAL STAR, OUR SUN*

Element	Relative Number of Atoms	Percent by Mass
Hydrogen	10^{12}	71.1%
Helium	9.64×10^{10}	27.4%
Carbon	2.88×10^{8}	0.25%
Nitrogen	7.94×10^{7}	0.08%
Oxygen	5.75×10^{8}	0.65%
Neon	8.91×10^{7}	0.13%
Silicon	4.07×10^{7}	0.06%
Iron	3.47×10^{7}	0.14%
Gold	8	0.00000011%
Uranium	0.4	0.000000007%

*The table lists eight of the most common elements along with gold and uranium to illustrate how extremely rare they are. Data on relative number of atoms drawn from Lodders (2003) *The Astrophysical Journal*, vol. 591, pp. 1220–1247.

Let us now apply what we know about spectra to astronomical bodies. We begin by using a telescope to obtain a spectrum of the object of interest. Next we measure the wavelengths and identify the lines. As an example, consider the spectrum of the Sun in figure 4.18A.

We can see from the spectral lines that the Sun contains hydrogen. In fact, when a detailed calculation is made of the strength of the lines, it turns out that about 71% of the Sun's mass is hydrogen. (This is about 90% of the atoms because hydrogen is so light.) Similar observations show that the spectrum of a comet consists mainly of emission lines from such substances as the molecules carbon dioxide and cyanogen (CN). Thus, we know that comets contain these substances. Moreover, recalling our earlier discussion of types of spectra (continuous, emission-line, or absorption-line), we can tell that the CN and carbon dioxide must be gaseous because the spectrum consists of emission lines. There may be other gases present too, but without seeing their spectral features, we cannot tell for sure.

Although the examples we have used above involve spectra of visible light, one of the most useful features of spectroscopy is that it may be used in *any* wavelength region. For example, figure 4.19 shows emission-line spectra from gases at a variety of wavelengths. Regardless of the wavelength region we use, the spectrum allows us to determine what kind of material is present. In addition, it can sometimes reveal to us in what direction and how fast that material is moving.

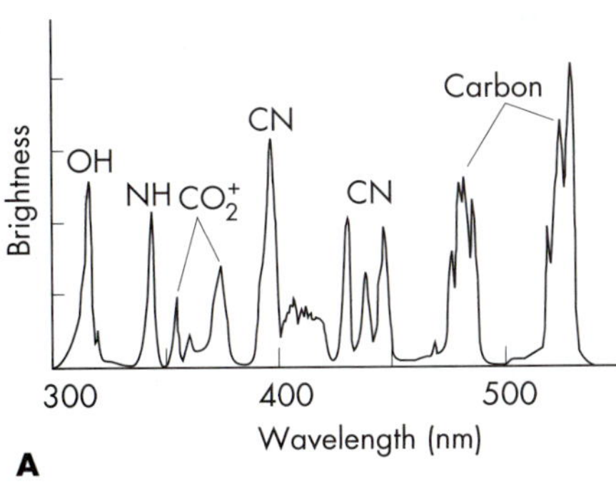

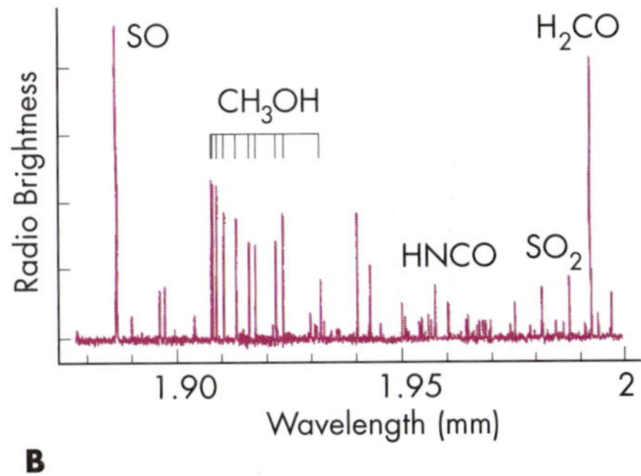

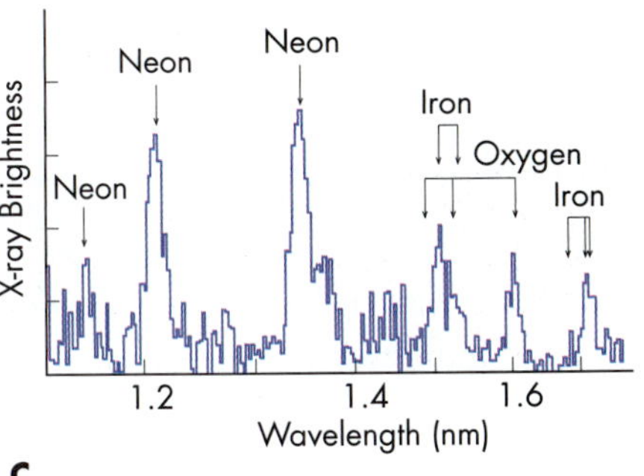

FIGURE 4.19
Emission-line spectra at a variety of different wavelengths. (A) A spectrum of a comet at visible and ultraviolet wavelengths. (B) A microwave-radio spectrum of a cold interstellar cloud. (C) An X-ray spectrum of hot gas from an exploding star.

4.6 The Doppler Shift: Detecting Motion

INTERACTIVE
Doppler shift

ANIMATION
The Doppler effect

If we observe light from a source that is moving toward or away from us, we will find that the wavelengths we receive from it are altered by the motion. If the source moves toward us, the wavelengths of its light will be shorter. If it moves away from us, the wavelengths will be longer, as illustrated in figure 4.20A. Furthermore, the faster the source moves, the greater those changes in wavelength will be. This change in wavelength caused by motion is called the **Doppler shift,** and it is a powerful tool for measuring the speed and direction of motion of astronomical objects.

The Doppler shift occurs for all kinds of waves. You have perhaps heard the Doppler shift of sound waves from the siren of an emergency vehicle as it passes you and moves away: the siren's pitch drops as the wavelength of its sound increases (fig. 4.20B). Likewise, the Doppler shift of a radar beam that bounces off your car reveals to a law enforcement officer how fast your car is moving (fig. 4.20C).

It is easy to see why the Doppler shift occurs. Imagine that the waves from the moving source are a wiggly line. If the source is moving away from you, the wiggles get stretched so the spacing of the waves increases. If the source is moving toward you, the wiggles are scrunched up so the spacing of the waves decreases (fig. 4.20D).

Mathematically, the Doppler shift arises because the wavelength we observe (λ) is the original wavelength (λ_0) plus the distance the source travels during the time a single wave is emitted. That distance depends on the speed, V, of the source along the

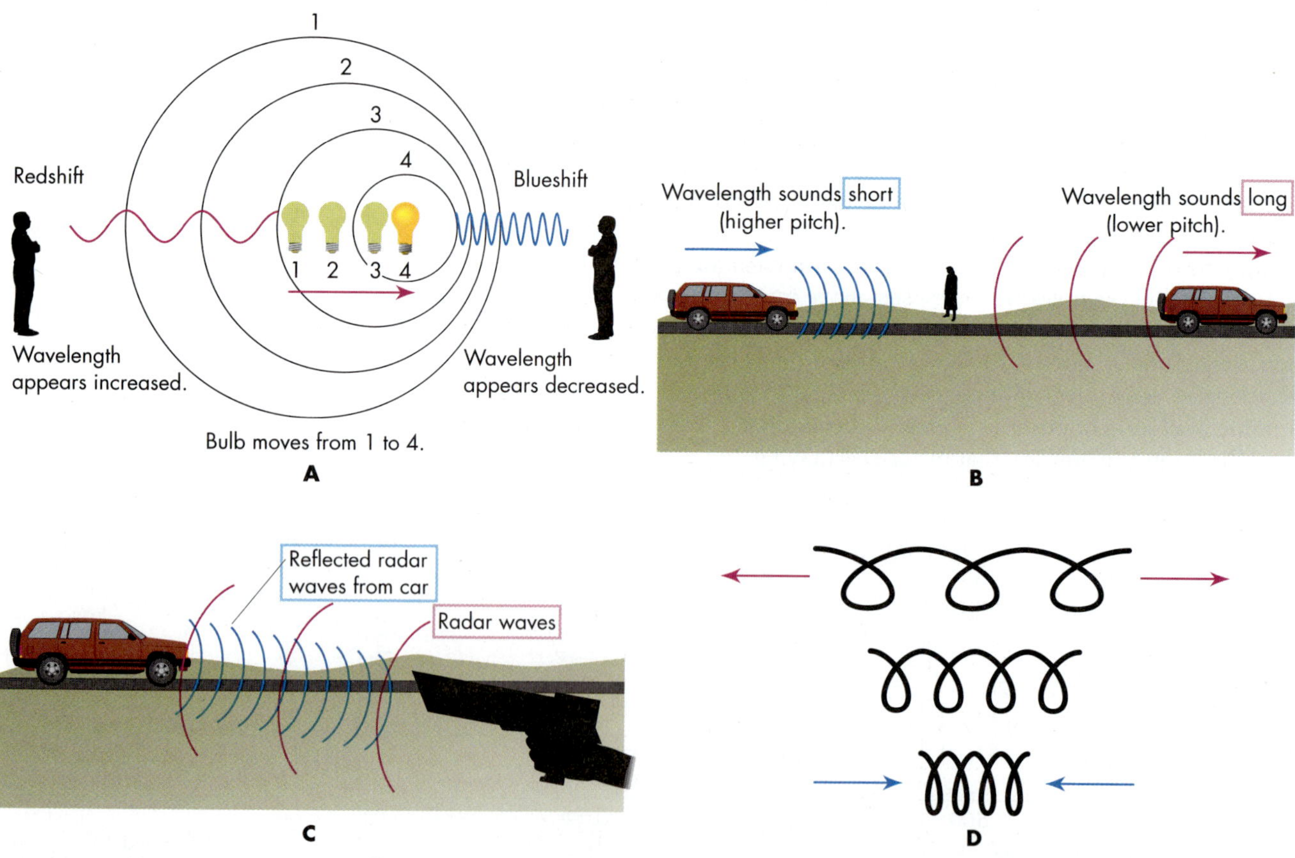

FIGURE 4.20
(A) The Doppler shift: waves appear to shorten as a source approaches and to lengthen as it recedes. (B) The Doppler shift of sound waves from a passing car. (C) The Doppler shift of radar waves in a speed trap. (D) A Slinky illustrates the shortening of the space between its coils as its ends move toward each other and a lengthening of the space as the ends move apart.

line from the source to the observer, a speed astronomers call the radial velocity. Some mathematics (omitted here) then leads to the Doppler shift formula

$$V = c(\lambda - \lambda_0)/\lambda_0 = c(\Delta\lambda/\lambda_0)$$

where c is the speed of the wave and $\Delta\lambda = (\lambda - \lambda_0)$ is simply shorthand for the change in wavelength.* We will see applications of this law in later chapters. Here, our goal is simply to indicate that the Doppler shift allows us to find out how fast a source is moving away from (positive V) or toward (negative V) us.

$\Delta\lambda$ = Wavelength shift
λ = Measured wavelength (what we observe)
λ_0 = Emitted wavelength
c = Speed of light
V = Velocity of source along the line of sight (radial velocity)

Doppler-shift measurements can be made at any wavelength of the electromagnetic spectrum. But regardless of the wavelength region observed or whether the waves are of visible light, astronomers generally refer to shifts that increase the measured wavelength as redshifts and those that decrease the measured wavelength as blueshifts. Thus, even though we may be describing radio waves, we will say that an approaching source is blueshifted and a receding one is redshifted.

The previous sections show how important a source of information electromagnetic radiation is. It provides information about the temperature, composition, and motion of astronomical objects. The Extending our reach box on the next page describes one example of how astronomers used observations at many wavelengths to understand the "Crab Nebula." One final thing we must consider is how the Earth's own atmosphere affects the radiation passing through it.

4.7 Absorption in the Atmosphere

Gases in the Earth's atmosphere absorb electromagnetic radiation, affecting the flow of heat and light through it. The amount of this absorption depends strongly on wavelength. For example, the gases affect visible light hardly at all, and so our atmosphere is nearly completely transparent at the wavelengths we see with our eyes. On the other hand, some of the gases strongly absorb infrared radiation while others strongly block ultraviolet radiation.

ANIMATION
Absorption by atmosphere

This nearly total blockage of infrared and ultraviolet radiation results from the ability of molecules such as carbon dioxide, water, and ozone to absorb at a wide range of wavelengths. For example, carbon dioxide and water molecules strongly absorb infrared wavelengths. Likewise, ozone (O_3) and ordinary oxygen (O_2) strongly absorb ultraviolet radiation, while oxygen and nitrogen absorb X rays and gamma radiation.

*The Greek letter Δ, or delta, is widely used to stand for "the change in quantity."

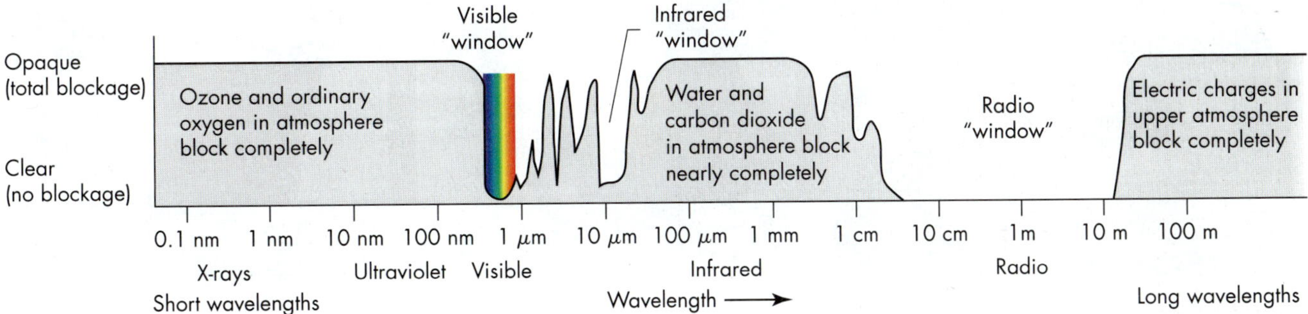

FIGURE 4.21
Atmospheric absorption. Wavelength regions where the atmosphere is essentially transparent, such as the visible spectrum, are called "atmospheric windows." Wavelengths and atmosphere are not drawn to scale.

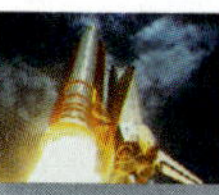

EXTENDING *our reach*

OBSERVING THE CRAB NEBULA AT MANY WAVELENGTHS

In midsummer A.D. 1054 just after sunset, astronomers in China and other East Asian countries noticed a brilliant star near the crescent moon in a part of the sky where no bright star had previously been seen. They wrote of this event: "In the last year of the period Chih-ho, . . . a guest star appeared. . . . After more than a year it became invisible."

We know today that these astronomers of long ago witnessed a supernova explosion, the violent event that marks the death of a massive star. Their record—nearly 1000 years old—begins a story that continues today as astronomers try to understand what causes a star explosion. Although the story began with naked-eye observations, it continued with observations made with telescopes on the ground and in space. Moreover, the story illustrates how astronomers have come to rely on observing radiation at many wavelengths, not just visible light.

Despite its initial brilliance, the dying star seen so long ago faded and disappeared from the sky and astronomical records. Then, in 1731, John Bevis, a British physician and amateur astronomer, noticed with his telescope a faint dim patch of light in the constellation Taurus. (You can see Taurus and where the Crab Nebula lies in "Looking Up #5 in the front matter.) Twenty-seven years later, Charles Messier, a French astronomer and comet hunter, rediscovered the glowing cloud and made it the first entry in his catalog of fuzzy patches of light that were not comets. In 1844, Lord Rosse, a British astronomer and telescope builder, noticed that the fuzzy patch contained filaments (Box fig. 4.1A) that to his eye resembled a crab. He therefore named it the Crab Nebula.

In 1921, John Duncan, an American astronomer, compared two photographs of the nebula taken 12 years apart and noticed that it had increased slightly in diameter. He therefore deduced that the nebula was expanding. At the same time, several other astronomers came across the ancient Chinese records and noticed the coincidence in position of the nebula with the report of the exploding star. Then, 7 years later, Edwin Hubble, at Mount Wilson Observatory in California, measured the increase of size more accurately and calculated from the rate of expansion that the nebula was about 900 years old—roughly the same age as the dying star seen nearly a millennium earlier by the Chinese astronomers.

Since then, astronomers have examined the Crab Nebula at virtually all wavelengths and, in doing so, have added yet more to their understanding of a star's demise. For example, in 1948, Australian astronomers discovered that the Crab Nebula is a powerful source of radio waves (Box fig. 4.1B). In 1968, further observations at radio wavelengths revealed that a faint peculiar star near the center of the nebula is spinning about 30 times per second and that it is the core of the star whose explosion created the Crab Nebula. Likewise, astronomers discovered that it is a source of X-ray radiation (Box fig. 4.1C).

What have all these observations shown? They have given astronomers their best view yet of the last moments of a star's life. From visible-wavelength observations, astronomers measure that the gas ejected when the star exploded is expanding with a speed of about 1000 km/sec. From radio-wavelength observations, they deduce that the nebula contains charged particles moving at nearly the speed of light and that the central star pulses on and off about 30 times per second. The X-ray observations confirm this picture. Thus, by observing the Crab Nebula and its stellar corpse at a variety of wavelengths, astronomers have shown that it is a far richer and more mysterious object than could be deduced from observations at one wavelength alone.

A 

B 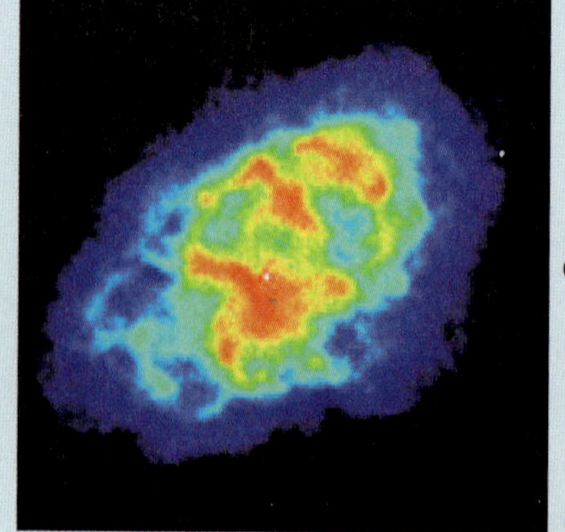

C 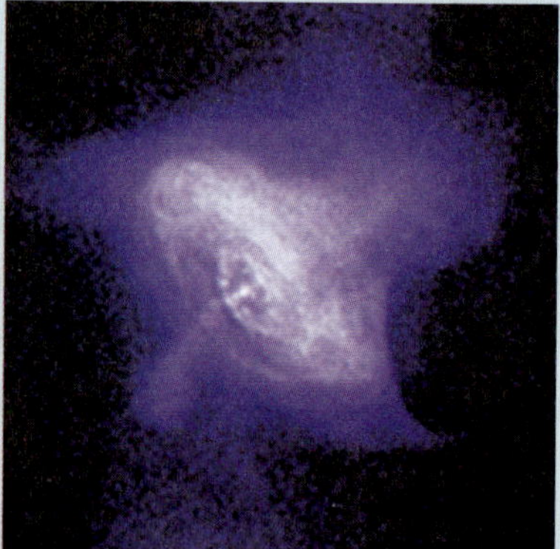

BOX FIGURE 4.1
(A) Visible-light photograph of the Crab Nebula. (B) Radio image of the Crab Nebula. (C) X-ray image in false-color of the core of the Crab Nebula.

Q. What does the flattened round shape of the glowing gas in Box figure 4.1C suggest about the gas's motion?

As a result of this absorption by molecules, virtually no infrared, ultraviolet, X-ray, or gamma-ray radiation can pass through our atmosphere.

The transparency of the atmosphere to visible light compared to its opacity (non-transparency) to infrared and ultraviolet radiation creates what is called an **atmospheric window.** An atmospheric window is a wavelength region in which energy comes through easily, compared to other wavelengths (fig. 4.21).

Without atmospheric windows, it would be impossible for us to study astronomical objects from the ground. As it is, the visible window allows us to study stars and galaxies (which radiate lots of visible energy), but the lack of ultraviolet and the rarity of infrared windows makes it very difficult to observe objects that radiate strongly in those spectral regions. This is one of several reasons why astronomers so badly need telescopes in space, where there is no absorption by our atmosphere.

Molecules in general are excellent absorbers (and emitters) because they can store energy in more ways than atoms can. As we have described earlier, atoms store energy by lifting electrons to upper orbitals. Molecules can store energy not only by lifting electrons but also by the spinning and vibrating motions of the molecule as a whole. These added ways to store energy are what make molecules such powerful blockers of radiation at many wavelengths.

SUMMARY

Light can be described in two complementary ways: as a stream of particles called photons, or as electromagnetic waves. In the wave picture, the energy increases as the wavelength decreases. The wavelength of light determines its color. Red light has a longer wavelength than blue light. In addition to the electromagnetic radiation that we see as visible light, many electromagnetic waves are invisible to the eye, such as infrared, ultraviolet, radio, X, and gamma rays. The entire assemblage of electromagnetic waves is called the electromagnetic spectrum.

Energy can be absorbed by or released from an atom when an electron moves to a higher or lower orbital, respectively. If an electron drops from an upper to a lower orbital, the energy appears as light. If light of the appropriate energy (wavelength) hits an atom, it may lift an electron in the atom from a lower to an upper orbital. The generation of light is called emission. The removal of light from a beam of radiation is called absorption.

Each kind of atom has a unique set of wavelengths at which it emits and absorbs. These create the atom's spectrum and allow it to be identified. Motion of the emitting material alters the wavelengths, creating a Doppler shift from which the material's speed and direction of motion can be deduced.

Atmospheric gases also absorb light, although there is little absorption in the wavelengths of the visible atmospheric window. Carbon dioxide and water absorb infrared radiation, and oxygen and ozone in our upper atmosphere absorb ultraviolet radiation. These gases therefore strongly hinder observations of astronomical objects at infrared and ultraviolet wavelengths from the ground.

QUESTIONS FOR REVIEW

1. (4.1) Why is light called electromagnetic radiation?
2. (4.1) What is a photon? How fast can photons travel?
3. (4.1/4.3) How are color and wavelength related? What about temperature and wavelength of a glowing body?
4. (4.2) What is meant by the electromagnetic spectrum?
5. (4.2) Name the regions of the electromagnetic spectrum from short to long wavelengths.
6. (4.3) How does the color of dense materials change with temperature? How does this relate to the idea of a blackbody?
7. (4.4) What makes elements different from each other? What is the arrangement of the parts of an atom?
8. (4.5) What is the differenc between emission and absorption in terms of what happens to an electron in an atom?
9. (4.6) Explain how the Doppler shift affects waves reflected by or emitted from a moving body.
10. (4.5/4.6) What are some of the things astronomers can learn about astronomical objects from their spectra?
11. (4.7) Which gases in the atmosphere absorb infrared radiation? Which gases absorb ultraviolet?

THOUGHT QUESTIONS

1. (4.1–3) If red stars are cooler than blue stars, and red light has less energy than blue light, why do you suppose we associate the color red with hot and the color blue with cold in everyday life?
2. (4.2) Why do night-vision cameras use infrared detectors?
3. (4.3/4.4) Through a telescope, you see a red object. Is that enough information to tell what temperature it is? Explain.
4. (4.4) If you were to look at the spectrum of the gas flame of a stove or the blue part of a Bunsen burner flame, what sort

of spectrum would you expect to see: absorption, emission, or continuous? Why?

5. (4.3–5) Why don't atoms emit a continuous spectrum?
6. (4.3–5) Given that water absorbs microwaves very strongly, can you explain why a Pop-Tart gets very hot inside while its crust stays cool if you heat it in a microwave oven?
7. (4.5) How can you tell what sort of gas is emitting light?
8. (4.5) How would a spectrum help you learn what the atmosphere of Venus is made of ?
9. (4.7) If you added more water or carbon dioxide to our atmosphere, how would it alter the loss of heat from our planet? Would you expect the Earth to get warmer or colder? Why?
10. (4.3/4.4/4.7) Can you explain why the atmospheric layer containing ozone is much warmer than the levels above and below it?

PROBLEMS

1. (4.1) Use the Sun's distance of 150 million kilometers to calculate how long light takes to travel from the Sun to the Earth.
2. (4.1) Suppose you are operating a remote-controlled spacecraft on Mars from a station here on Earth. How long will it take the craft to respond to your command if Mars is at its nearest point to Earth? Use data in the appendix for your calculations.
3. (4.2) A solar flare emits X rays and radio waves simultaneously. Which reaches the Earth first? If the X rays have a wavelength of 0.25 nanometers and the radio waves have a wavelength of 6 cm, how many times larger is the frequency of the X rays than the radio waves?
4. (4.3) Your body temperature is about 300 K. At what wavelength do you radiate most strongly? What region of the electromagnetic spectrum is this? Do you understand now how a rattlesnake can bite you in the dark?
5. (4.3) A lightbulb radiates most strongly at a wavelength of about 3000 nanometers. How hot is its filament?
6. (4.3) An electric stove burner on "high" radiates most strongly at about 2000 nanometers. What is its temperature?
7. (4.3) The Earth's temperature averaged over the year is about 300 Kelvin. At what wavelength does it radiate most strongly? In what part of the electromagnetic spectrum does this wavelength lie? Can you see it?
8. (4.4) Sketch an atom emitting light. Does the electron end up in a higher or lower orbit? Repeat for an atom absorbing light.
9. (4.6) Calculate the Doppler shift for red light (wavelength of 650 nanometers) reflected off a sports car traveling away from you at 140 km/hr. What is $\Delta\lambda$? What is the wavelength you see? Could we see the shift in color with our eyes?
10. (4.6) You are analyzing a radio spectrum of an outer part of a distant spiral galaxy. A spectral line expected to be at 21 centimeters is instead measured to be at 21.015 centimeters. Is the outer part of the galaxy rotating toward or away from you? How fast is that part of the galaxy moving?

TEST YOURSELF

1.(4.1) Which kind of light travels fastest?
 (a) ultraviolet (b) visible (c) gamma rays (d) radio waves
 (e) They all travel at the same speed.

2.(4.2) Which type of electromagnetic radiation has the longest wavelength?
 (a) Ultraviolet (b) Radio (c) Infrared (d) X ray (e) Visible

3.(4.2) Which kind of photon has the highest energy?
 (a) Ultraviolet (b) Visible (c) X ray (d) Infrared (e) Radio

4.(4.3) A star's radiation is brightest at a wavelength of 400 nanometers. Its temperature is about
 (a) 4000 K. (b) 12,000 K. (c) 1500 K. (d) 750 K. (e) 7500 K.

5.(4.3–5) Suppose we detect red photons at 656 nanometers emitted by electrons dropping from the n=3 to n=2 orbital in hydrogen. The hydrogen is in an interstellar cloud at 5000 K. If the cloud were heated to 10,000 K, what would be the wavelength of the photons emitted by the transition?
 (a) 328 (b) 656 (c) 1312 (d) 658 (e) 654

6.(4.5) An astronomer finds that the visible spectrum of a mysterious object shows bright emission lines. What can she conclude about the source?
 (a) It contains cold gas.
 (b) It is an incandescent solid body.
 (c) It is rotating very fast.
 (d) It contains hot, relatively tenuous gas.
 (e) It is moving toward Earth at high speed.

7.(4.5) Most stars have spectra showing dark lines against a continuous background of color. This observation indicates that these stars
 (a) are made almost entirely of hot, low-density gas.
 (b) have a warm interior that shines through hotter, high-density gas.
 (c) have a hot interior that shines through cooler, low-density gas.
 (d) are made almost entirely of cool, low-density gas.

8.(4.6) If an object's spectral lines are shifted to longer wavelengths, the object is
 (a) moving away from us. (b) moving toward us. (c) very hot. (d) very cold.

KEY TERMS

absorption, 102
absorption-line spectrum, 108
atmospheric window, 113
blackbody, 99
conservation of energy, 102
continuous spectrum, 108
Doppler shift, 110
electromagnetic radiation, 90
electromagnetic spectrum, 93
electromagnetic wave, 90
elements, 100
emission, 102

emission-line spectrum, 108
energy levels, 104
excited, 102
frequency, 92
infrared, 95
light, 90
nanometer, 92
orbital, 101
photons, 91
quantized, 101
spectroscopy, 103
ultraviolet, 95
visible spectrum, 91
wavelength, 91
wave–particle duality, 91
white light, 92
Wien's law, 98

FURTHER EXPLORATIONS

Kiang, Nancy Y. "The Color of Plants on Other Worlds." *Scientific American* 298 (April 2008): 48.

Holloway, Marguerite. "What Visions in the Dark of Light." *Scientific American* 297 (September 2007): 50.

Steffy, Philip C. "The Truth about Star Colors." *Sky and Telescope* 84 (September 1992): 266.

Zajonc, Arthur. *Catching the Light: The Entwined History of Light and Mind.* New York: Bantam Books, 1993.

Website

Visit the *Explorations* website at **http://www.mhhe.com/arny** for additional online resources on these topics.

Q FIGURE QUESTION ANSWERS

WHAT IS THIS? (chapter opening): The ring of light around the Sun is a halo and is caused by sunlight refracted in tiny atmospheric ice crystals. Haloes are quite common (perhaps one a week) and may be seen around both the Sun and the Moon. They are easier to see around the Sun if you cover the Sun with your hand or block its direct light with a building or tree, as shown here.

BOX FIGURE 4.1: It is spinning.

PROJECTS

1. **Make a simple spectroscope:** Plans are sketched below to build a spectroscope using aluminum foil, a cardboard tube, and a grating. Put aluminum foil over a small hole at one end of the tube. Cut a very thin slit in the foil with a razor blade or sharp knife. The slit should be straight, about an inch long, and as narrow as you can make it and still have light get through it. At the other end of the tube mount a small square of grating material. Tape it over a small hole cut in the tube's end. If you use a toilet-paper tube, just bunch the foil over one end and tape it on. Fold some thin cardboard over the other end and cut a small square hole about an inch across. Hold the grating material up to light and notice to which side it creates a rainbow. Tape the grating along its edges oriented so that the rainbow runs perpendicular to the slit. Ask your instructor about details of construction and how to get the grating. To use, hold the grating end of tube right up to your eye and look at the light source. Rotate the tube so that you see a rainbow off to the side. Sketch the spectra you see from (a) a fluorescent light, (b) a mercury vapor street light, (c) an ordinary incandescent lightbulb, (d) a white surface reflecting sunlight.*
2. **Spectroscopy:** Use your spectroscope to look at the spectrum of flames. Use a gas-burning stove and add a pinch of table salt to see sodium emission lines (yellow) or use copper sulfate crystals or a scrap of copper wire to see green copper emission lines.

*Do not look directly at the Sun.

Old oatmeal box–or even toilet paper tube. A square box will work as well. Length should be at least 4 inches.

Narrow slit–as thin as possible (approximately 1 inch long).

Grating material taped over hole. Hole should be about 1 inch square.

PROJECT FIGURE 4.1

ESSAY 2

Special and General Relativity

A favorite theme in science fiction is human space travel. At present such travel is limited to flights orbiting the Earth, although in the 1960s and 1970s astronauts traveled to the Moon, landed there, and explored some of its surface. The Moon is figuratively on our doorstep, however. Can we expect to ever be able to travel the vastly longer distances to other stars or galaxies? In science fiction stories, these immense interstellar distances are crossed by craft using faster-than-light travel (fig. E2.1).

Scientists have excellent reasons for concluding that faster-than-light travel is impossible and that travel even at near-light speed requires long spans of time. For example, it takes light more than 4 years to reach us from even the nearest star beyond the Sun, and tens to hundreds of thousands of years to reach us from nearby galaxies. This seems to mean that for astronauts to conquer interstellar distances, they would have to be prepared to live in space for decades, perhaps having the journey completed by their descendants. But it might surprise you to learn that astronauts could in principle travel *millions* of light-years to another galaxy and return to Earth in their own lifetime! The science that explains how such an immense journey could be made is one theme of this essay. Along the way, we will learn a little about two of science's most important and intriguing theories, the theories of special and general relativity. Before we deal with travel at near the speed of light, we need to look at the far simpler problem of describing motion.

REST FRAMES

Astronomical objects are in constant motion: the Earth moves through space around the Sun; the Sun moves through space within the Milky Way Galaxy; and so forth. To describe such motion we need a frame of reference, or **rest frame.** In everyday life we often use the ground as our rest frame. For example, we drive a car at 60 mph (~100 km/hr) along the freeway or walk to class across the campus at 4 mph (~6 km/hr) with respect to the ground.

Suppose, however, we are traveling in an airplane and we walk down the aisle of the plane from the back to the front. How fast are we moving? With respect to the plane, we are moving at a walking pace, say a few miles per hour. However, we could also measure our speed in the plane with respect to the ground, in which case our speed would be that of the plane plus our walking speed with respect to the plane, or hundreds of miles per hour. In other words, our measurement of an object's motion depends on the rest frame that we use for our observation.

FIGURE E2.1
Space ships that can travel interstellar distances in a short period of time are popular in science fiction. The realities of such travel are very different, but no less strange.

Describing an object's motion in one rest frame when it is viewed from another rest frame is not difficult. If we consider again a person walking in an airplane, we simply add the speed of the person to the speed of the plane. For another example, suppose a person is running at 10 mph and throws a javelin forward at 40 mph in the direction of his or her motion—the javelin will move across the ground at the speed 10 mph + 40 mph = 50 mph. Addition of speeds in this fashion (thrower plus javelin) (fig. E2.2) is an example of what is sometimes called **Galilean relativity,** in honor

FIGURE E2.2
The speed of a javelin with respect to the ground is found by adding the speed of the thrower across the ground to the speed with which the javelin is thrown. That is, the velocities add.

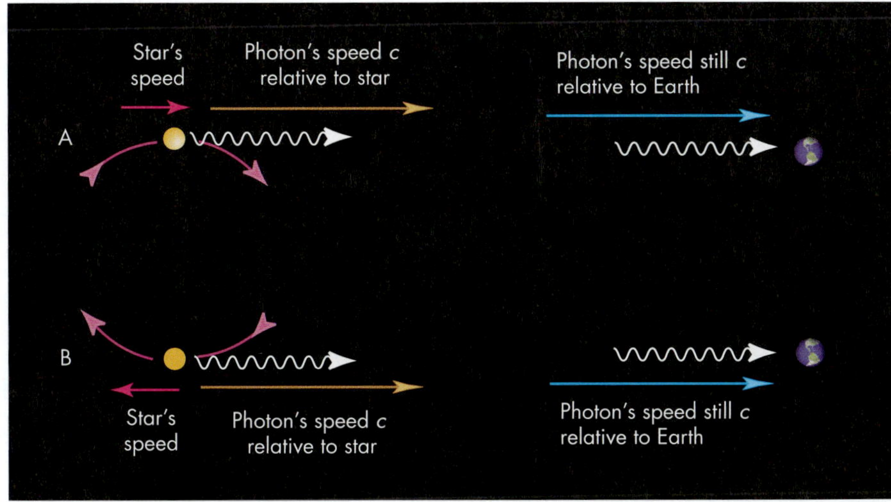

FIGURE E2.3
Light from orbiting stars reaches Earth at the same speed whether the star is moving toward or away from us.

of Galileo, one of the first scientists to recognize how motions combine with one another. But Galilean relativity fails for light.

THE SPEED OF LIGHT FROM MOVING OBJECTS

Suppose we have two stars, A and B, in orbit around each other (fig. E2.3). At some point in the orbit, one of the stars, say star A, is moving toward us while the other, say star B, is moving away from us. Light leaves the surface of each star at the speed of light, c. We might then expect that light from the approaching star A will travel toward us a little faster than c, while light from star B, the one swinging away from us, will travel toward us a little slower than c. However, when scientists tried to measure this effect, they found no change in the light's speed: the light arrived at the same speed, c, regardless of the motion of the stars. This is not so odd as it might first seem. We see a similar effect in waves generated by a boat moving across a lake. The waves travel through the water at the same speed irrespective of what speed or direction the boat moves. Such arguments led scientists in the 1800s to conclude that light moves through space like the waves on the water. But ordinary waves require some material to move through. What substance plays the role of the water for light? Scientists at that time concluded that space was filled with some transparent substance, which they call the **æther.**

Calculations showed that the æther must have very special properties. It must be very rigid for light waves to have their high observed velocity (the stiffer a substance, the faster waves travel in it). At the same time, it must freely allow astronomical objects to plow through it with no resistance. In particular, it must flow past the Earth as we orbit the Sun.

THE MICHELSON-MORLEY EXPERIMENT

In 1887 the American scientists Albert Michelson and Edward Morley designed an apparatus to search for the æther. They predicted that light traveling along the direction of the Earth's motion should move at a different speed from that of light traveling perpendicular to the Earth's motion. But when Michelson and Morley conducted their experiment, they detected no difference in the speed of light along the perpendicular paths. At first they thought this might just be bad luck: that during the experiment, the combination of the Earth's motion around the Sun and the Sun's motion through space might just have happened to make the Earth stationary relative to the æther at that point in its orbit. So they repeated their experiment many times over the course of a year. At some point they should have easily detected the motion of Earth relative to the æther, but they found no sign of the Earth's motion at all.

The Michelson-Morley experiment has been called the most famous "failed" experiment in history because it led to a revolution in physics. The results implied that there was no æther regulating the speed of light. However, if light was not moving relative to an æther, then physicists had no explanation for the constancy of the speed of light reaching us from sources moving at different speeds toward or away from us.

All experiments made then and now show that no matter how fast the source generating the light is moving, and no matter how fast the observer measuring the light is moving, the speed of light through space is always measured to be c = 299,792,458 meters per second. How can this be?

One explanation offered in the late 1800s was that motion through space somehow caused matter to contract* in the direction of motion. If matter contracts when it is moving, this would change our perception of length so that we might be tricked into thinking that the speed of light had not changed. For example, Michelson and Morley were searching for a difference in the speed of light in two perpendicular directions. If the apparatus were compressed in the direction it was moving through space but not in the perpendicular direction, the path the light traveled would be shorter. Such a contraction could potentially cancel out the effect that Michelson and Morley were searching

*The idea of a contraction caused by motion was first proposed by Irish physicist George Fitzgerald. The Dutch physicist Hendrik Lorentz developed a model for explaining how the contraction might arise.

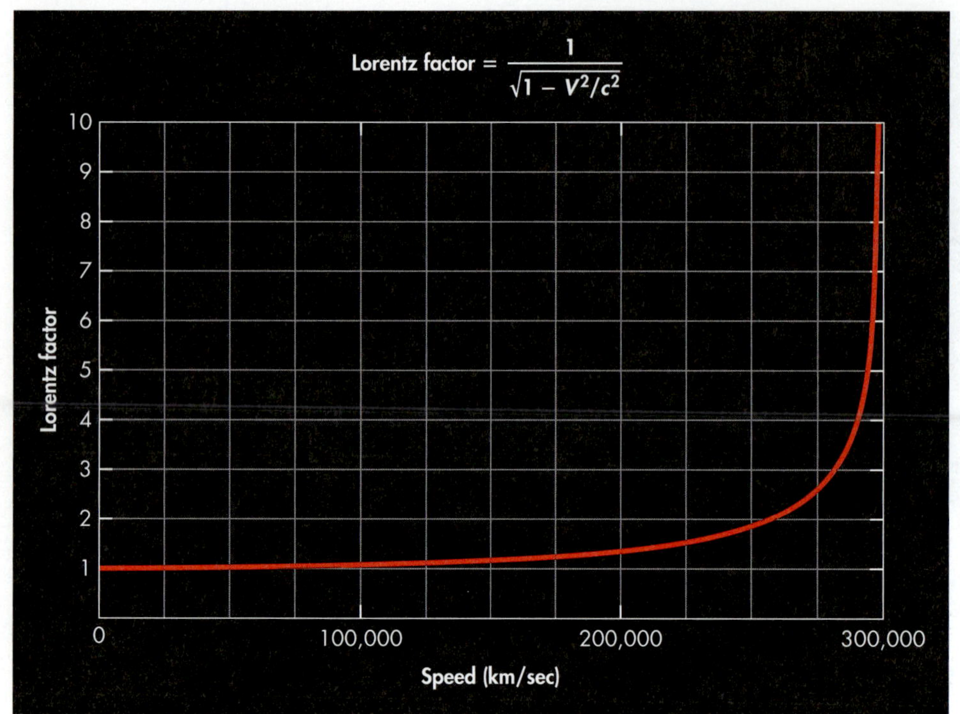

FIGURE E2.4
The Lorentz factor expresses the amount by which objects appear to compress in the direction of their motion. It is also the factor by which a moving clock appears to slow down, and the factor by which a moving object's mass appears to increase.

FIGURE E2.5
Albert Einstein (1879–1955).

for. The factor by which the apparatus would need to contract is known today as the **Lorentz factor,** usually denoted by the Greek letter gamma (γ). Lorentz hypothesized that an object's length shortens in the direction it is moving by a factor equal to

$$\gamma = \frac{1}{\sqrt{1 - V^2/c^2}}$$

where V is the speed of the object and c is the speed of light. The value of γ for different speeds V is plotted in figure E2.4.

The Lorentz factor is close to 1 at small speeds. For example, at the speed at which the Earth is orbiting the Sun, 30 kilometers per second, the Earth would contract by only a few centimeters in its direction of motion. At much higher speeds the contraction factor becomes very large, growing to infinity if the speed V were to reach the speed of light. The Lorentz contraction factor can explain the Michelson-Morley experiment, but it cannot explain a number of other conflicting results that were found. Still, it contains an important idea that grew into a whole new concept about the nature of motion.

EINSTEIN'S THEORY OF SPECIAL RELATIVITY

In 1905 a 26-year-old graduate student named Albert Einstein (fig. E2.5) took on the problem of the seemingly inexplicable measurements of the speed of light. He was completing his physics degree while working in the Swiss patent office and supporting a family. Yet, in that one year alone, he completed his doctorate degree and wrote four papers in several areas of physics. Physicists widely agree that three of these papers were each worthy of a Nobel Prize! He was little known at the time and had few colleagues with whom to discuss his ideas; but nonetheless in one of these papers he came up with a brilliant new approach to the question of the motion of light.

Einstein began by concentrating on the finding that light travels at the same speed no matter what the speed of its source or of the observer measuring the light. Even though many experiments had come to this conclusion, most physicists had assumed it was an impossibility and so were seeking other explanations—such as errors in the experiments or the Lorentz contraction. Einstein, instead of thinking of a constant speed of light for all observers as an impossibility, accepted this notion as correct, and proceeded to work out its consequences. He found that this led inevitably not only to a Lorentz-like contraction of space, but also to a stretching of time, or **time dilation,** by the same factor. Even more important, he found that this contraction was not relative to some imagined æther filling space, but that these effects depended only on the relative motions of any two objects.

These alterations of our perception of space and time affect everything that we see that has any speed relative to us, forming the foundation of what is known as the theory of **special relativity.** According to this theory, if a star or a rocket

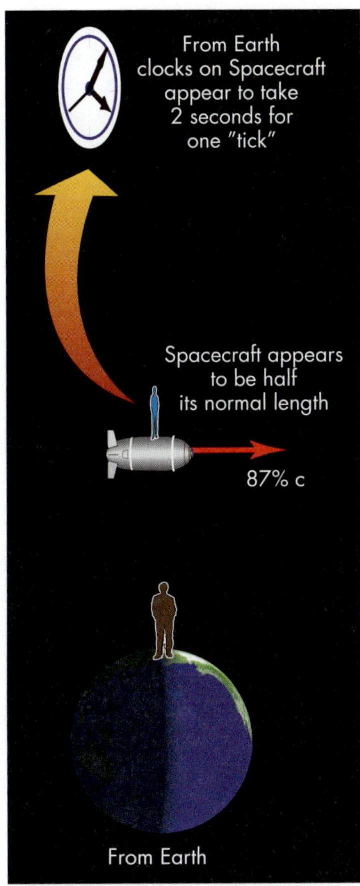

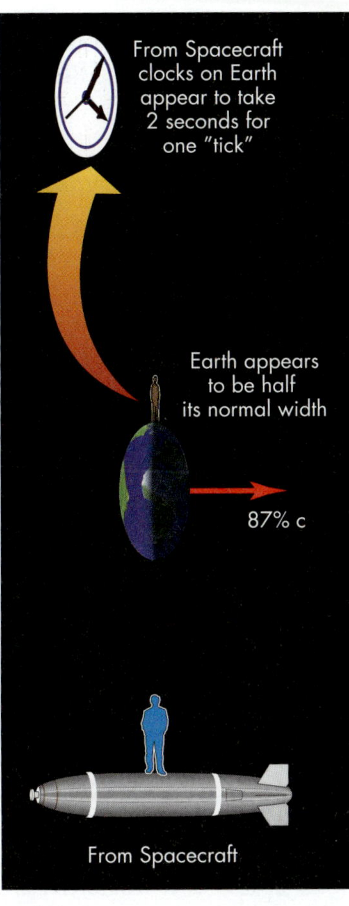

FIGURE E2.6
When a spacecraft travels by us, its length is contracted by the Lorentz factor, and clocks on board run more slowly by the same factor. From on board the ship it appears that the Earth is compressed and time is running slowly on Earth.

moves by us at high speed, we will see it squashed in its direction of motion by the Lorentz factor (fig. E2.6). If we could watch a clock tick or measure the speed of an astronaut's heart in the spaceship, we would discover that all these processes occur more slowly by the Lorentz factor.

What is remarkable is that the mathematics Einstein worked out showed that the situation is exactly symmetrical for the astronaut moving by us at high speed. That astronaut will sense herself as being the one who is stationary and will see us as moving by her at high speed. She will measure us and the Earth as being contracted in the direction of "our" motion, and she will measure our clocks and our hearts as running slowly (fig. E2.6). So what two observers in relative motion see is parallel—each would find that the other was the one undergoing the distortions of space and time. Thus, there is no preferred rest frame.

An especially important feature of Einstein's work is the behavior of light. Suppose we or the passing astronaut shine a light beam at each other. We will each measure that the light is moving past us at the same speed, c. However, as we watch each other making these measurements, we will each think that the other is measuring a shorter distance and using a clock that runs too slow.

This theory of special relativity is far-reaching in its implications. The theory is the basis for Einstein's discovery of the relationship between energy and mass: $E = mc^2$. It also predicts that other fundamental quantities change for moving objects. For instance, a moving object grows more massive by the same Lorentz factor that describes how lengths grow shorter and time slows.

These effects mean that nothing can reach the speed of light in our rest frame, let alone exceed it. The rocket ship traveling by us can fire its rockets and accelerate forever, but even though the ship goes ever faster, its mass grows ever greater, making it harder and harder to accelerate. The ship may approach the speed of light, but because the Lorentz factor goes to infinity, it would require infinite energy to reach the speed of light.

SPECIAL RELATIVITY AND SPACE TRAVEL

The theory of special relativity may seem strange, but it has been tested by a century of high-precision experiments. There is not a single verified contradiction to it, and its predictions about such things as the slowing of time have been verified directly.

Special relativity is also about more than just perceived differences in space, time, and mass. For example, when atomic clocks (the most accurate clocks, used for establishing time worldwide) are flown between locations, it is found that their travel in airplanes leaves them a little bit slow relative to the network of fixed clocks maintained around the world. The moving clocks must be readjusted after each trip. The Lorentz factor for traveling at airplane speeds of about 1000 km/hour (600 mph) is just $\gamma = 1.0000000000004$, so every tick of the clock on an airplane takes about 4 ten-trillionths of a second longer than a tick of the clock in the ground-based network; but after several hours of flying the effect is measurable.

More intriguing is what happens at such high speed that the Lorentz factor is large. Experiments have demonstrated, for example, that a subatomic particle called a *muon* normally has a lifetime of about 2 millionths of a second before it decays. However, when muons are traveling at 99% of the speed of light (and therefore have a Lorentz factor of about 7), they live about 14 millionths of a second. This longer lifetime allows them to travel distances that would be impossible for them within their normal lifetimes.

When speaking of such brief times, this change in the rate at which time passes seems minor; but the same slowing factor would apply for a human traveling at 99% of the speed of light. If a spaceship could be built that traveled that fast,

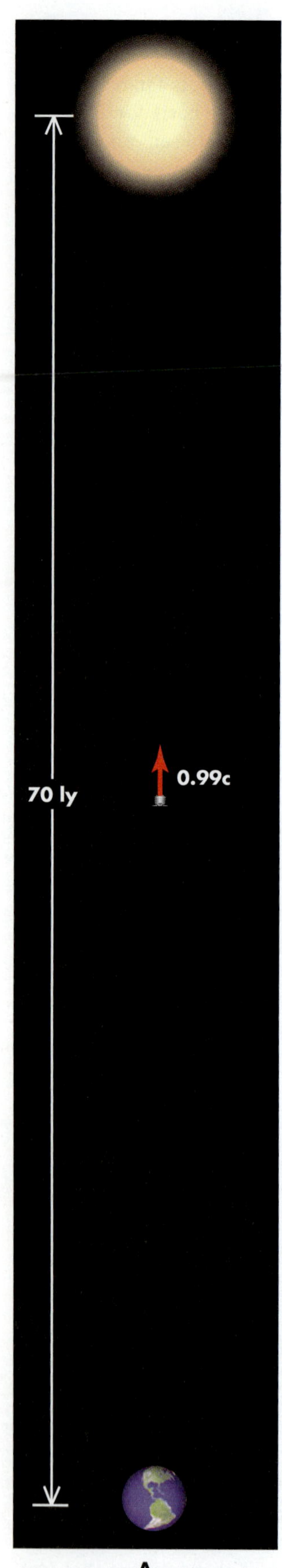

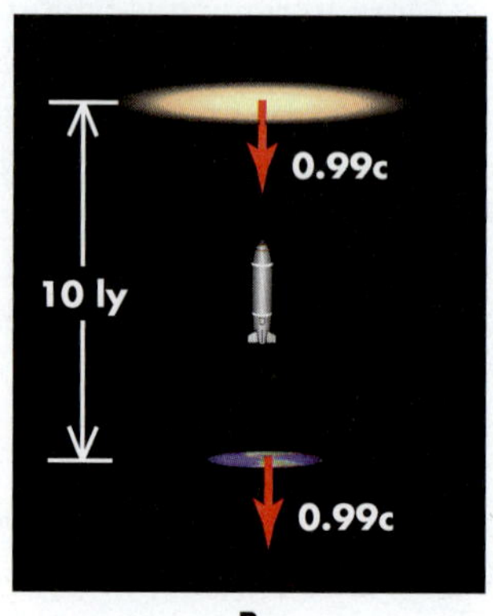

FIGURE E2.7
(A) From the rest frame of Earth and a star 70 light-years away, it appears that a spaceship traveling at 99% of c is shortened to one-seventh of its original length. (B) On the spaceship, once it is up to speed, it appears that the Earth and star are both moving at 99% of c. They and the distance between them are shortened to one-seventh of their original length.

TABLE E2.1 THE LORENTZ FACTOR AT HIGH SPEEDS

Speed	Lorentz Factor
0.87 c	2.0
0.97 c	4.1
0.99 c	7.1
0.999 c	22.4
0.9999 c	70.7
0.99999 c	223.6

time would effectively run seven times more slowly on board compared with time here on Earth. If astronauts had food and air supplies for a 10-year trip, the Lorentz factor of $\gamma \approx 7$ (corresponding to $V = 99\%$ the speed of light) would mean they could travel for $\gamma \times 10$ years $\approx$ 70 years in Earth time. At their speed relative to Earth of 0.99 c, they would be able to reach a distance of almost 70 light-years (fig. E2.7A) before they ran out of supplies. At speeds even closer to c the Lorentz factor becomes even larger (see table E2.1) and the potential distances greater.

From the perspective of astronauts on a craft traveling at 0.99 c, it would not seem that time was passing any more slowly than normal. Nor would they feel that they or their ship was foreshortened or more massive. From their perspective, the Earth and the star they are visiting and the distance between the two are *contracted* by the Lorentz factor (fig. E2.7B). The distance would look like 70 light-years when they were stationary (in the rest frame of Earth); but at their high speed, the distance would look only one-seventh as large!

These marvelous "tricks" of relativity open up possibilities for traveling distances far greater than we might once have imagined. Theoretically, one could travel a million light-years within a human lifetime—although it is far beyond current technologies. It is also important to realize that such travel would have major challenges beyond simply reaching such high speeds. From the perspective of the spaceship traveling among the stars at near the speed of light, every atom and every dust particle in space along the ship's path has a mass increased by the Lorentz factor, and it is heading toward the ship at nearly the speed of light! This can give a pebble the impacting force of a ship-destroying asteroid.

Also, as intriguing as these possibilities are, they offer no time savings for the rest of us back on Earth. Consider again the astronauts traveling at 99% of the speed of light for 10 years to visit a distant star. If they then turned around and came home in another 10 years (by their reckoning), they would find that 140 years had passed on Earth. Everyone they knew would have grown old and died. They might even be younger than their great-great-grandchildren!

THE TWIN PARADOX

Something may seem wrong with the description of space travel in the previous section. When we first introduced relativity, we noted that the time stretching appears symmetrical. That is, from the rest frame of the Earth it appears that the astronauts' time is running slowly, while from the rest frame of the astronauts it appears that time on Earth is running slowly. This is sometimes presented as the **twin paradox:** If one of the astronauts left a twin back on Earth, how can we say that one ages more than the other?

The explanation of this seeming paradox is that the situations are *not* in fact symmetric. The astronauts experience accelerations as their ship speeds up and then slows down at their destination. Furthermore, when they turn their ship around to return to Earth, they experience yet more accelerations as the ship again speeds up and later slows down as it reaches Earth again. But people remaining back on Earth experience none of these accelerations because they always remain in the same rest frame. Thus, their progression of time remains constant. By contrast, the astronauts' rest frame keeps changing, and in the end they return to the rest frame of the Earth.

Imagine, for example, that one of the astronaut twins sends messages once each day to her twin back on Earth, and meanwhile the twin sends messages once each day to his astronaut sister on the ship. As they part, they each receive the other's messages much more slowly—both because of the Lorentz factor *and* because the separation between the ship and Earth is growing larger and the messages take longer to reach each other. When the astronaut twin reaches the distant star, she sends her 10th-year message. Until this time, the situations are symmetrical. Both have received only a small fraction of each other's messages because of the Lorentz factor and growing separation.

The astronaut twin turns around and starts back, receiving the messages from her brother that were sent years earlier and have been on their way to her across the 70-light-year gap. They come more quickly now because she is approaching Earth, cutting the distance each message has to travel. She reads about her brother getting older and older, now seemingly very rapidly. In the meantime, her brother back on Earth is still reading her messages from the outgoing trip. By the time he dies, he would not even have received the 10th-year message announcing that the ship had reached the other star. At the speed of light in Earth's rest frame, that 10th-year message would take 70 years to reach Earth. In fact, the spaceship, traveling at 99% of the speed of light, returns to Earth just shortly after the message arrives at Earth saying that the astronauts reached the star.

In the movies, space travel is fast and everyone ages at the same rate. In reality, traveling at high speeds means that the travelers must leave behind not only their homes but their own times and the people in them.*

*With everyone's time running at different speeds, can you imagine a story line that would make a good movie if it portrayed space travel accurately?

RETHINKING GRAVITY

Despite the success of the theory of special relativity, Einstein saw that it was not yet complete. The ideas of motion and different rest frames did not really account for the effects of gravity. For example, the International Space Station orbits the Earth under the pull of Earth's gravity. Astronauts inside the station feel weightless, as though no forces are acting on them, even though they are constantly changing direction. If there were no windows, the astronauts could easily believe that they were traveling in a straight line, far from any massive objects.

Gravity has the unique ability to change the direction or speed of objects without them feeling a force being applied. Imagine that you had the bad fortune to be inside an elevator whose cable broke. As it was falling, you would find yourself floating inside the elevator, like an astronaut in a space capsule far from any massive objects (fig. E2.8). On the other hand, if you were on a stationary elevator in Earth's gravitational field, the acceleration you feel is no different from what you would experience in a space ship that was accelerating at 9.8 meters per second per second, or 1 *g* (fig. E2.9).

Einstein had the remarkable insight that gravity actually alters the nature of space. When you are near a massive object, it causes space to, in effect, flow past you. Thus, when you let go of an apple, it moves downward as if you were moving upward through space. What you observe is the same as letting go of an apple in an accelerating space ship (fig. E2.9). In the spaceship we would say that the apple remains stationary as the floor of the ship accelerates upward to meet it . This idea that gravity is the same as being in an accelerating frame of reference is known as the **principle of equivalence.**

Newton described gravity as a force that causes the path of objects to accelerate or their paths to curve. By contrast, Einstein described space itself as being curved so that an object would follow a curved path in much the way a golf ball follows

FIGURE E2.8
If you are inside a free-falling elevator, you feel weightless, just as you would feel in a space ship that is drifting in deep space.

FIGURE E2.9
The principle of equivalence. If you are stationary on the surface of the Earth, objects fall in a way that is equivalent to being aboard a spaceship accelerating at 1*g*.

a curved path or accelerates as it rolls along the hills and slopes of a putting green. At first this idea may sound like a semantic difference, but it proves to have fundamental consequences.

GENERAL RELATIVITY

In 1915, a decade after publishing the theory of special relativity, Einstein arrived at his theory of **general relativity.** General relativity is a theory of gravity in that it replaces Newton's law with a mathematical description of how space is curved by mass. This curvature affects the trajectories of objects as they move through space. General relativity is broader than Newton's theory in that it also describes how energy as well as mass can affect space and how the presence of mass and energy also both alter the flow of time.

While it is mathematically complex to carry out the calculations of how mass and energy "curve" space and time, a simple analogy illustrates many of the important features of general relativity. Imagine a large artificial lake, in the middle of which there is a drain that draws out water at a steady rate. If you sit in a boat on the lake, the outflow of water draws you toward the drain, with a stronger pull the closer you get to the drain. This is analogous to how gravity exerts a stronger pull the closer you get to an object. If you were to travel across the lake in a motorboat, holding your steering wheel in a fixed direction, you would discover that your path was curved by the flow of water toward the drain. This is similar again to the trajectory of a spacecraft passing by a planet.

The water in this analogy represents space, and it illustrates an important idea of general relativity: space itself can have motion. While the notion of "nothing" having motion seems strange, it is demonstrated in a number of surprising ways. For example, clocks on the surface of the Earth run slower than clocks in deep space. In fact, standing still on the surface of the

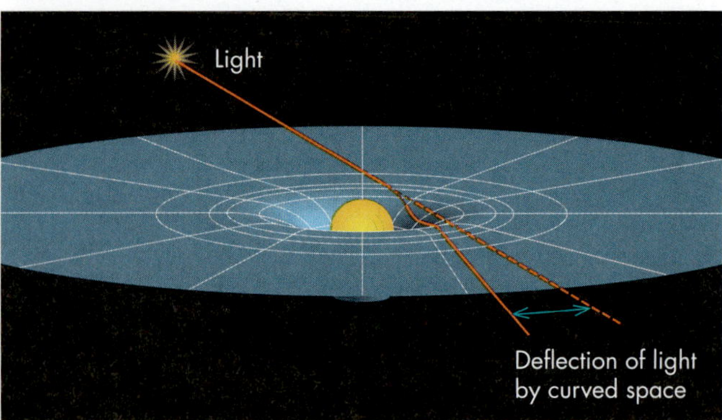

FIGURE E2.10
The "flow" of space toward a massive object like the Sun deflects the path of starlight passing near it. This prediction of general relativity was confirmed in 1919.

Earth, we experience a gravitational time dilation the same as the Lorentz factor for an object moving at the escape velocity (11 km/sec) from the Earth's surface.*

The "flow" of space also bends the paths of photons. For example, as starlight passes close to the Sun, where its gravity is strongest, it is deflected like the boat in our analogy, even though photons have no mass (fig. E2.10). This was confirmed in 1919 during a total eclipse of the Sun, where the positions of stars showed a shift that agreed exactly with Einstein's prediction. Einstein further showed that the changes in space and time close to the Sun explained some peculiarities of Mercury's orbit that had long puzzled astronomers when using the Newtonian understanding of gravity.

General relativity has become a cornerstone of modern physics. It is more than a theoretical curiosity. For instance, if we did not correct for the effects of general relativity between our position on the Earth's surface and satellites orbiting in space, global positioning system (GPS) units would not work. Furthermore, the flow of space and time in general relativity are central to our understanding of black holes and the expansion of the entire Universe, as we shall discover in later chapters.

*The gravitational time dilation factor at any point is equal to $1/\sqrt{1 - V_{esc}^2/c^2}$ where V_{esc} is the escape velocity at that position.

SUMMARY

Experiment shows that the speed of light is a constant and is not affected by the motion of its source or the observer. This puzzling result became the basis for Einstein's theory of special relativity. A consequence of that theory is that an observer moving past an object sees its length shrunk (the so-called Lorentz contraction) and the rate of passage of its time slowed (time dilation).

These changes in space and time cause someone in a non-accelerating rest frame to observe anyone moving relative to them as having clocks that run slower than their own. On the other hand, anyone undergoing acceleration finds that their clocks have in fact run slower, even if the acceleration occurred as a result of being in a gravitational field. Einstein developed a new theory of gravity, general relativity, that showed how mass and energy "curve" space and time, changing the flow of both and producing the effects of gravity described by Newton's laws.

QUESTIONS FOR REVIEW

1. What is Galilean relativity? Give an example of how it is used.
2. Describe what the Michelson-Morley experiment was trying to detect.
3. What is the Lorentz factor?
4. How are length, time, and mass altered according to special relativity?
5. What is the twin paradox, and how is it resolved?
6. What is general relativity?
7. How does gravity affect space and time?

THOUGHT QUESTIONS

1. Would you be willing to travel to a nearby galaxy if it meant you would return to Earth one million years in the future?
2. Given the Lorentz factor, does time pass for a photon? What about for a place where gravity is so strong that the escape velocity equals the speed of light?

PROBLEMS

1. To travel 100,000 light-years in 10 years of your own time, at what fraction of the speed of light would you have to travel?
2. Mercury orbits the Sun at speeds ranging from 59 km/sec to 39 km/sec when it is nearest and farthest from the Sun, respectively. What are the Lorentz factors for these two speeds?
3. The escape velocity from the Sun is 76 km/sec at Mercury's closest distance to the Sun and 62 km/sec at Mercury's farthest distance. Using the gravitational time dilation formula $1/\sqrt{1 - V_{esc}^2/c^2}$, find how slowly time runs in these two places. Compare your results to the previous question, and discuss how time running at different speeds in the orbit might affect Mercury's orbit.

TEST YOURSELF

1. When a spaceship is traveling at 99% of the speed of light (Lorentz factor = 7), an astronaut on board the ship will find that

   (a) everything in the ship weighs 7 times more.
   (b) the ship is very cramped—only 1/7th its original length.
   (c) everyone onboard talks 7 times more slowly than normal.
   (d) All of the above
   (e) None of the above. Everything seems normal to the astronaut on board.

2. Suppose Tom and Molly are both flying in spaceships toward each other at half the speed of light (0.5 c). If Tom shines a light toward Molly, what speed will Molly measure for the light coming toward her?

   (a) 0.25 c
   (b) 0.5 c
   (c) 1.0 c
   (d) 1.5 c
   (e) 2.0 c

3. If Bob travels at close to the speed of light to another star and then returns, he will find that his twin sister Alice who remained on Earth is

   (a) younger than him.
   (b) older than him.
   (c) the same age as him.
   (d) He cannot return to Earth because it would violate special relativity.

4. Sort the following in order of where time runs slowest to where it runs fastest.

   (a) on the Earth's surface
   (b) on the Moon's surface
   (c) in deep space.
   (d) on the Sun's surface.

KEY TERMS

æther, 117
Galilean relativity, 116
general relativity, 122
Lorentz factor, 118
principle of equivalence, 121
rest frame, 116
special relativity, 118
time dilation, 118
twin paradox, 121

FURTHER EXPLORATIONS

Gamow, George. *Mr. Tompkins in Paperback.* Cambridge: Cambridge University Press, 1993. (This book, a reprint of Gamow's superb Mr. Tompkins books, explores the ideas of special relativity and quantum physics at an easily accessible level. Originally published in the 1940s, this reprint contains *Mr. Tompkins in Wonderland* and *Mr. Tompkins Explores the Atom.*)

Website

Visit the *Explorations* website at **http://www.mhhe.com/arny** for additional online resources on these topics.

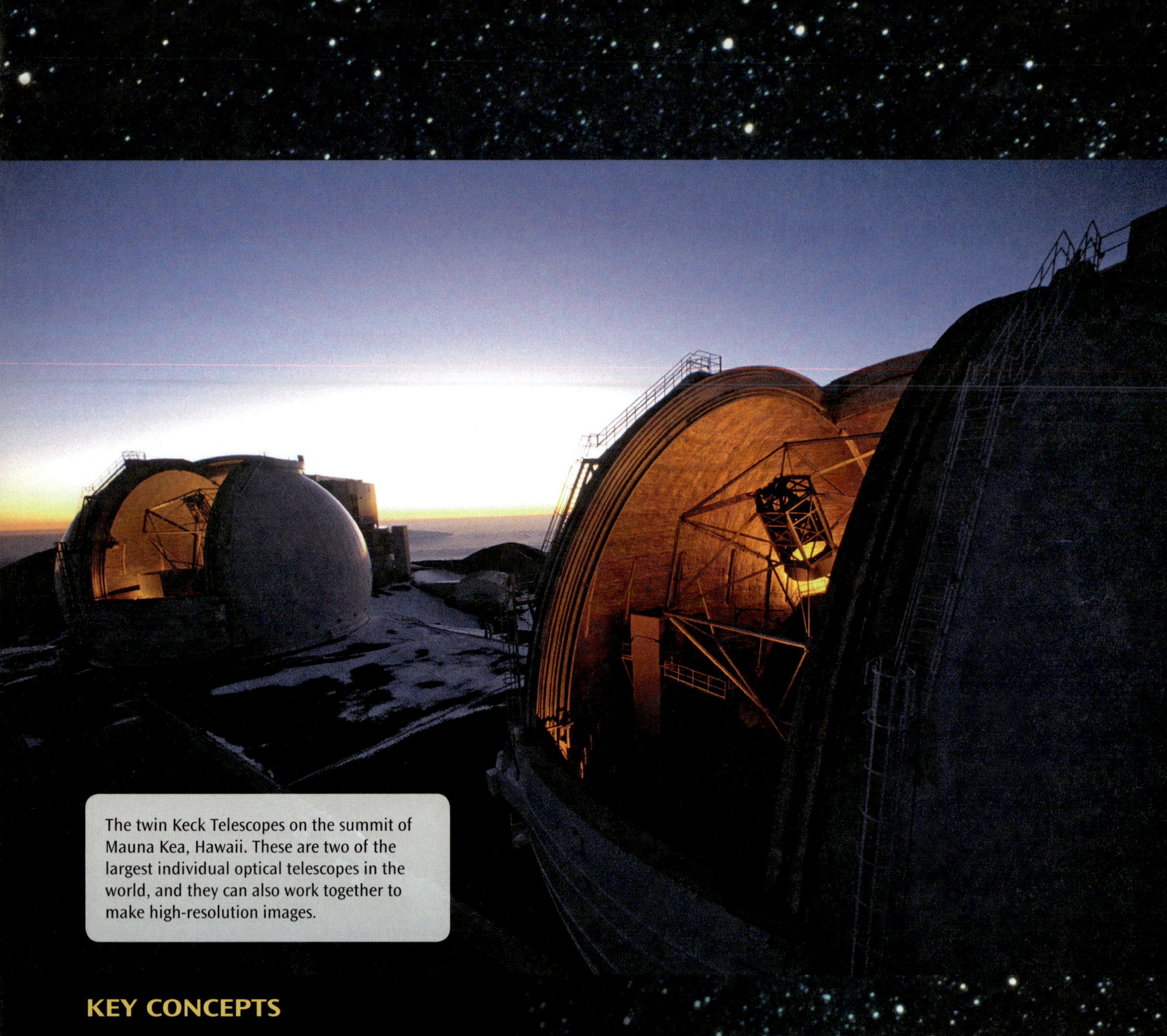

The twin Keck Telescopes on the summit of Mauna Kea, Hawaii. These are two of the largest individual optical telescopes in the world, and they can also work together to make high-resolution images.

KEY CONCEPTS

- Astronomers use telescopes to gather light and thereby make dim objects observable.
- Telescopes use lenses and mirrors to redirect a large amount of light to a focused spot.
- The larger the diameter of the main lens or mirror used in a telescope:
  - The greater its ability to gather light.
  - The finer the detail that the instrument will show.
- The Earth's atmosphere interferes seriously with observing.
  - It absorbs and totally blocks many wavelengths.
  - It blurs details.
- Astronomers use special-purpose telescopes to observe non-visible wavelengths. Many of these are in orbit, so they are not affected by blurring or atmospheric absorption.
- Astronomers are developing new ground-based telescopes using special techniques to produce high-resolution images.

5 Telescopes

CONCEPTS AND SKILLS TO REVIEW

- The nature of light (wave or photons) (4.1)
- Red light has a longer wavelength than blue (4.1)
- The parts of the electromagnetic spectrum (4.2)
- Absorption in Earth's atmosphere (4.7)

Astronomers, like all scientists, rely heavily on observations to guide them in theorizing and in testing theories already developed. Unlike most scientists, however, astronomers cannot directly probe the objects they study. Rather, they must perform their observations from vast distances and can only passively collect radiation emitted by the bodies they seek to study. Collecting enough radiation to be useful in studying astronomical objects is difficult because most objects are so remote that their radiation is extremely faint by the time it reaches Earth. Moreover, extracting the desired information from the radiation requires special instruments—instruments that can measure the brightness, the spectrum, and the position of objects to high precision. For example, to collect enough light to detect remote galaxies, astronomers use telescopes with mirrors the size of a small swimming pool. To avoid the blurring and blocking effects of our atmosphere, they use orbiting observatories. To analyze and display the observations, they use a computer. This chapter describes some of the more important devices and how they work. We will see that modern telescopes bear little resemblance to the long tubes depicted in cartoons. Moreover, modern astronomers rarely sit at the eyepiece of a telescope. They are more likely to be sitting at a computer terminal operating a telescope remotely, examining the data collected, or solving equations that describe such things as the paths stars take when two galaxies collide.

Q: WHAT IS THIS? See end of chapter for the answer

5.1 Telescopes

A telescope enables the astronomer to observe things not visible to the naked eye. Although our eyes are superb detectors, they cannot see extremely faint objects or fine details on distant sources. For example, we are unable to read standard newspaper print on the opposite wall in a dimly lit room. A telescope overcomes these difficulties, first, by collecting more light than the eye can collect and, second, by increasing the detail discernible. The first of these properties is called "light-gathering power"; the second is called "resolving power."

Light-Gathering Power

For our eyes to see an object, photons (light) from it must strike the retina. How bright that object appears to us depends on the number of its photons that enter our eye per second, a number limited by the size of our eyes. Astronomers overcome that limit by "collecting" photons with a telescope, which then "funnels" the photons to our eye. The bigger the telescope's collecting area, be it a lens or mirror, the more photons it collects, as shown in figure 5.1. Thus, a larger-diameter mirror or lens gives a telescope a greater **light-gathering power.** The result is a brighter image, which allows us to see dim objects that are invisible in telescopes with smaller gathering areas. Because the gathering area of a circular lens or mirror of radius R is πR^2, a small increase in the radius of the gathering area gives a large increase in the number of photons caught. For example, doubling the radius of a lens or mirror increases its light-gathering area by a factor of 4. Because the light-gathering area is so important to a telescope's performance, astronomers often describe a telescope by the diameter of its lens or mirror. Thus, the 10-meter Keck Telescopes in Hawaii have mirrors spanning 10 meters (roughly 30 feet) in diameter.

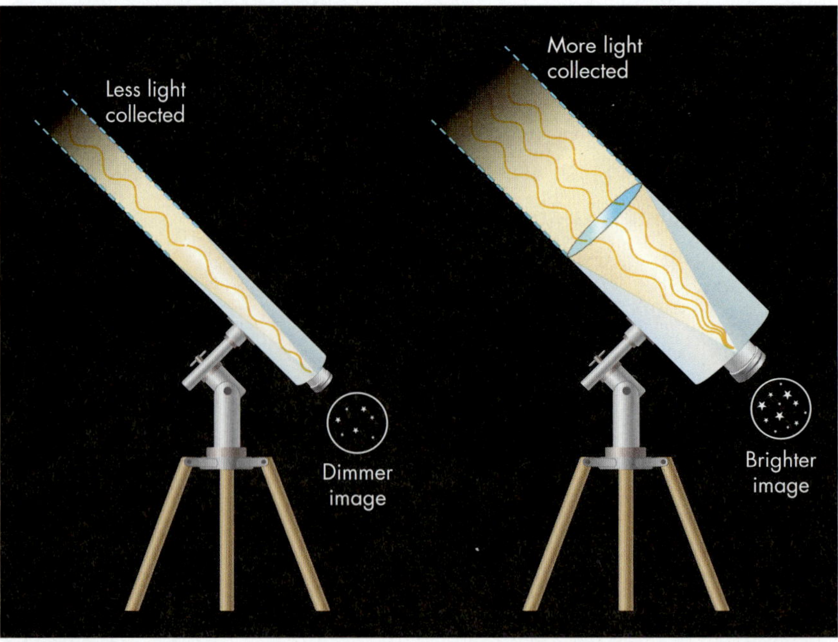

FIGURE 5.1
A large lens collects more light (photons) than a small one, leading to a brighter image. We therefore say that the larger lens has a greater "light-gathering power."

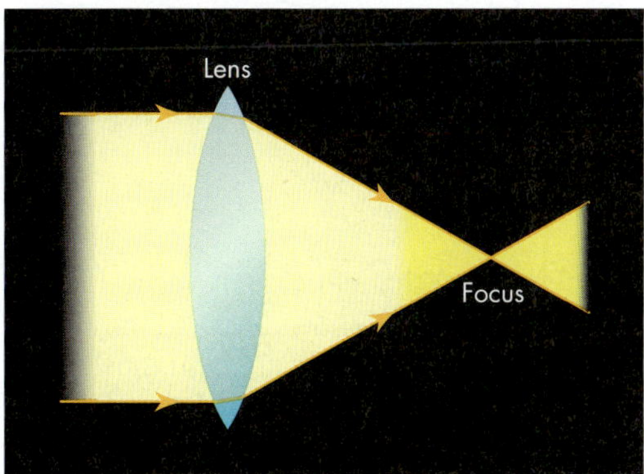

FIGURE 5.2
How a lens focuses light.

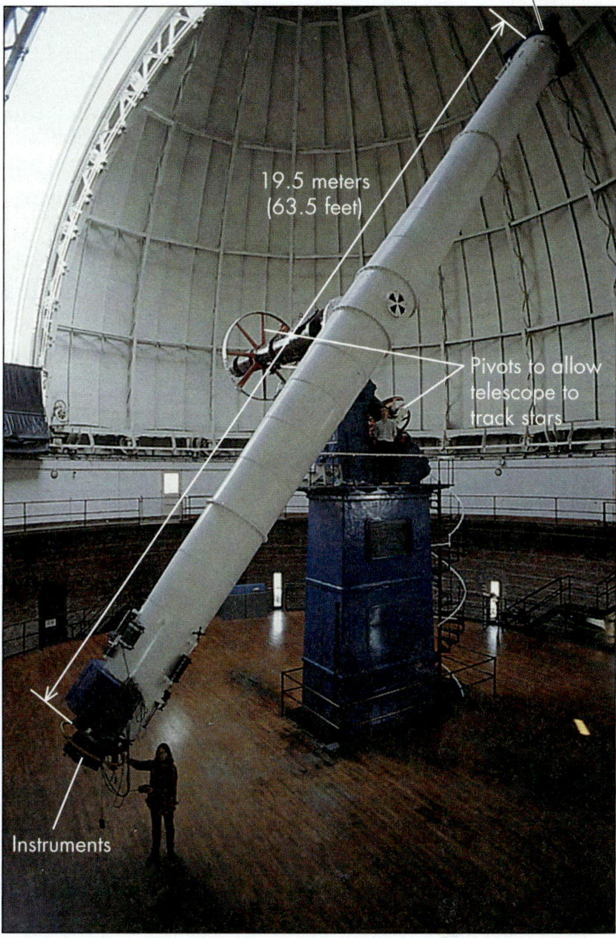

FIGURE 5.3
A refracting telescope. Completed in 1897 for the University of Chicago's Yerkes Observatory in Williams Bay, Wisconsin, this refractor has a lens approximately 1 meter (40 inches) in diameter, making it the world's largest refracting telescope.

Focusing the Light

Once light has been gathered, it must be focused to form an image or to concentrate it on a detector. Telescopes in which light is gathered and focused by a lens are called "refracting telescopes," or **refractors** for short. The lens of a refractor focuses the light by bending the rays, as shown in figure 5.2. This bending is called **refraction,** and it happens when light moves from one substance (such as air) into a different substance (such as glass), as discussed in the Extending Our Reach box on refraction. It is this refraction that gives such telescopes their name. Figure 5.3 shows a photograph of the world's largest refractor, the 1-meter (40-inch) diameter Yerkes Telescope of the University of Chicago.

Lenses have many serious disadvantages in large telescopes, however. First, large-diameter lenses are extremely expensive to fabricate. Moreover, a lens must be supported at its edges so as not to block light passing through it. This makes the lens "sag" in the middle (though by only tiny amounts), distorting its images. A third difficulty with lenses is that most transparent materials bring light of different colors to a focus at slightly different distances from the lens. This creates images fringed with color, a flaw called "chromatic aberration." Finally, many lens materials completely absorb short-wavelength light, making them, for some purposes, as useless in a telescope as a chunk of concrete.

To avoid such difficulties, most modern telescopes use mirrors rather than lenses to gather and focus light, and they are therefore called **reflectors.** The mirrors are made of glass that has been shaped to a smooth curve, polished, and then coated with a thin layer of aluminum or some other highly reflective material. As figure 5.4 shows, such a curved mirror can focus light rays reflected from it, creating an image just as well as a lens can. Moreover, because the light does not pass through the mirror, it focuses all colors equally well and does not absorb short-wavelength light. Furthermore, because the light does not have to pass through the mirror, the mirror can be supported from behind, thereby reducing the sagging problem that affects large lenses. For these and other reasons, astronomers

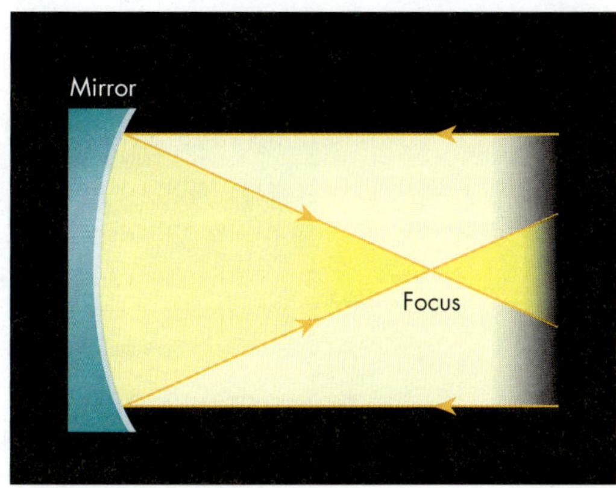

FIGURE 5.4
How a curved mirror focuses light.

EXTENDING our reach

REFRACTION

When light moves at an angle from one material into another (for example, from air into water), its direction of travel generally bends. This phenomenon is called refraction. Refraction is the principle by which our eyes and eyeglasses focus light. You can easily see its effects by sticking a pencil in a glass of water and noticing that the pencil appears bent, as seen in Box figure 5.1. The pencil in water also illustrates an important property of refraction. If you look along a pencil placed partly in the water and change the pencil's tilt, you will see that the amount of bending (refraction) changes. Exactly vertical rays are not bent at all, nearly vertical rays are bent only a little, and rays entering at a grazing angle are bent most.

BOX FIGURE 5.1
Refraction of light in a glass of water. Note how the pencil appears to be bent.

Refraction occurs because the speed of light changes as it enters matter, generally becoming slower in denser material. This decrease in the speed of light arises from its interaction with the atoms through which it moves. To understand how this reduction in light's speed makes it bend, imagine a light wave approaching a slab of material. The part of the wave that enters the material first is slowed while the part remaining outside is unaffected, as depicted in Box figure 5.2A. To see why slowing part of the wave makes it bend, imagine what would happen if the wheels on the right-hand side of your car turned more slowly than those on the left. Your car would swerve to the right, a result that lies behind the reason why cars have a differential. By allowing one wheel to turn faster than the other, the differential "swings" your car smoothly around corners. A similar effect occurs if you walk hand-in-hand with a friend, and your friend walks more slowly than you do. You will soon find yourself traveling in a curve (see Box fig. 5.2B). So, too, if one portion of a light wave moves more slowly than another, the light's path will bend.

Refraction not only bends light but also generally spreads the light into its component colors, breaking white light into a spectrum, or rainbow. This spreading occurs because different colors of light travel at different speeds in most materials and are therefore bent by

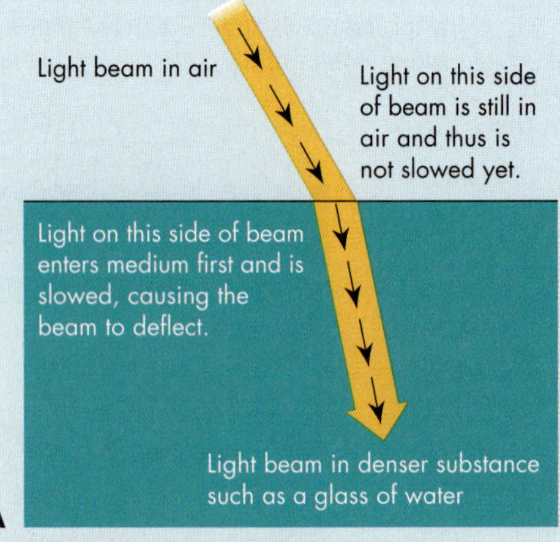

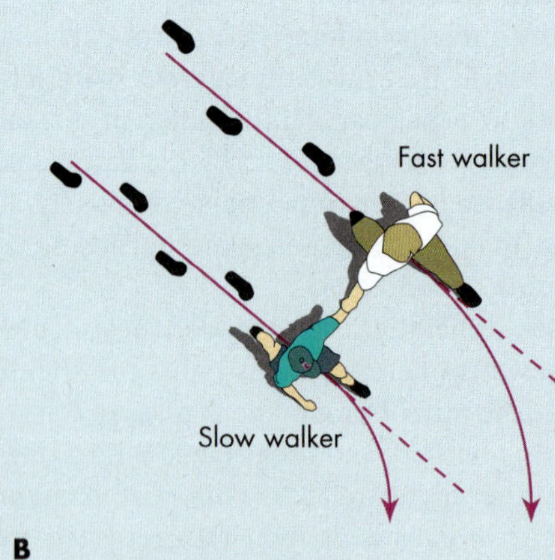

BOX FIGURE 5.2
Cause of refraction. (A) Light entering the denser medium is slowed, while the portion still in the less dense medium proceeds at its original speed. (B) A similar effect occurs when you walk hand-in-hand with someone who walks slower than you do.

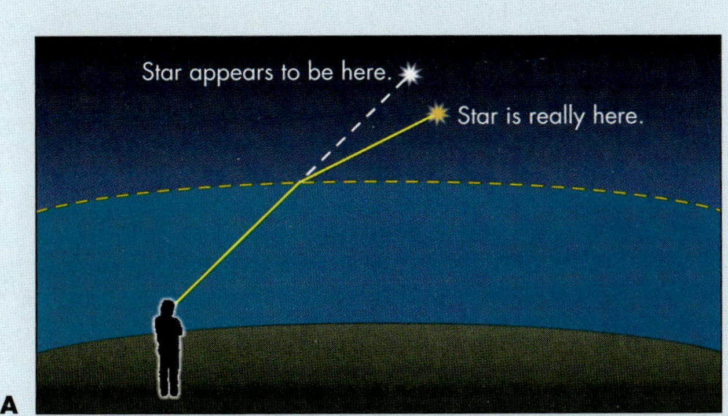

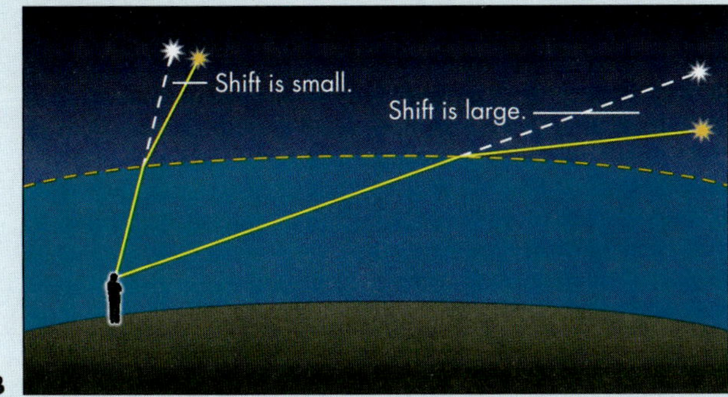

BOX FIGURE 5.3
(A) Atmospheric refraction makes the Sun or a star look higher in the sky than it really is. (B) Refraction is stronger for objects nearer the horizon. (C) The Sun is flattened because refraction "lifts" its lower edge more than its upper edge. The Sun's reflection in a band below it on the water is called the glitter path.

different amounts in a process called **dispersion.** Thus, if light consisting of a mix of colors enters a block of glass, each color is slowed to a different speed and is therefore deflected differently. The result is that colors initially traveling together separate into different beams. This is how a prism creates a spectrum.

Atmospheric Refraction
Refraction is easy to see in the Earth's atmosphere where it distorts the shape of the Sun when it rises and sets and makes the stars twinkle at night.

Distortion of the Sun's Shape
Refraction distorts the shape of the rising or setting Sun because when sunlight enters the Earth's atmosphere from space, it is refracted and bent slightly toward the ground. Thus, if you are on the ground and look back along the light ray, the light seems to come from slightly higher than it really does (Box fig. 5.3A). That is, refraction makes astronomical objects look higher in the sky than they actually are. This effect is greatest when objects are near the horizon, as shown in Box figure 5.3B. The result is that light from the lower edge of the Sun is refracted more than light from its upper edge, which "lifts" the lower edge more than the upper and makes the Sun look flattened (Box fig. 5.3C).

Refraction also slightly alters the time at which the Sun seems to rise or set. When it is at the horizon, the Sun is "lifted" by about the height of its diameter. Thus, at the moment when we see the Sun touch the horizon, it has actually already set. By "lifting" the Sun's image above the horizon, even though it has set, refraction slightly affects the length of the daylight hours. As a result, the day of the year the Sun is above the horizon for exactly 12 hours is not the equinox, but rather a few days before the spring and a few days after the autumnal equinox. It turns out that near latitude 40° N, St. Patrick's Day (March 17) is the day with almost exactly 12 hours between sunrise and sunset in the spring.

FIGURE 5.5
The largest telescope mirrors yet built are 8.4 meters (about 27 feet) in diameter. Two of these huge mirrors are used together on the Large Binocular Telescope on Mount Graham, Arizona.

now use reflecting telescopes almost exclusively. Figure 5.5 shows a photograph of the largest telescope mirrors yet built—a pair of 8.4-meter diameter reflectors in Arizona. The mirrors will eventually work together to give the same collecting area as a single 11.8-meter diameter mirror.

Light striking a mirror is focused in front of the mirror. Thus, to see the image, the observer would ordinarily have to stand in front of the mirror, thereby blocking some of the light. To overcome this difficulty, a secondary mirror is often used to deflect the light either off to the side or back toward the mirror and out through a hole in its center (fig. 5.6).

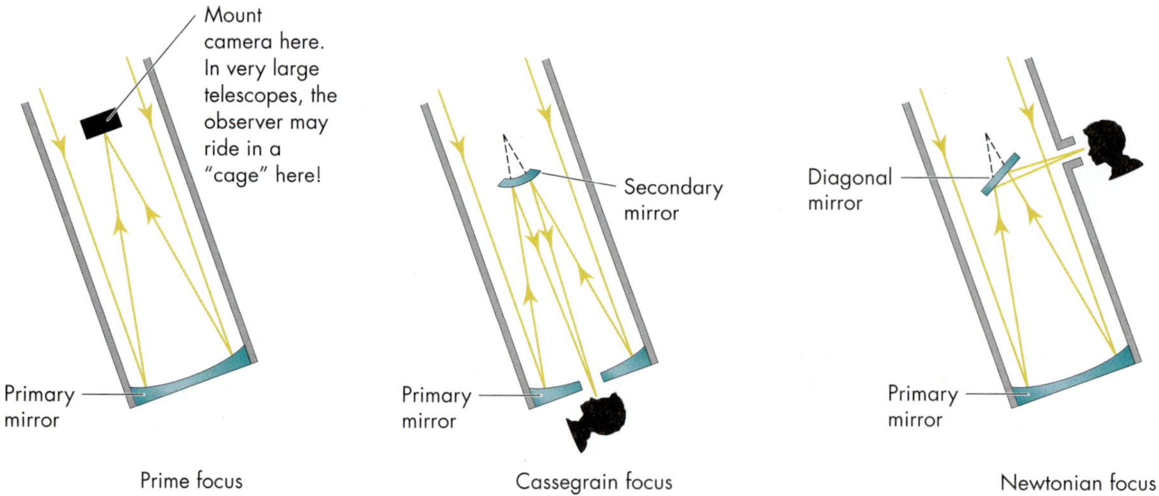

FIGURE 5.6
Sketches of different focus arrangements for reflectors.

Most telescopes are mounted on huge pivots that allow them to follow astronomical objects as they move across the sky. Swinging the many tons of metal and glass smoothly and with precision requires great care in construction and design. Moreover, as the telescope moves, its lenses or mirrors must keep their same precise shapes and relative positions if the images are to be sharp. This is one of the most technically demanding parts of building a large telescope, because the large pieces of glass used in lenses and mirrors bend slightly when their positions are shifted.

In the past, astronomers made mirrors thick to make them stiff and reduce sagging of the glass. Large pieces of glass, however, weigh more than smaller pieces and thus sag even worse, just as a smaller gob of whipped cream will keep its shape on a tilted plate whereas a large gob will slump under its own weight. As a result, a 5-meter diameter telescope completed in 1949 on Mount Palomar, California, remained the largest telescope for decades. The structural limits of glass required astronomers to develop a different approach. They discovered that a thin piece of glass, if properly supported, could keep its shape better than a thick piece. Thus, astronomers have sought ways to make thin mirrors that are then kept precisely shaped by alignment systems on the back side of the mirror.

While thin mirrors have allowed astronomers to build larger telescopes than in the past, another approach shows promise for much larger telescopes. Instead of using a single mirror, telescopes are being designed with many smaller mirrors aligned to collect and focus the light as if they were a single mirror. Telescopes designed this way are called multi mirror instruments. Currently, the largest multi mirror instrument is a 10.4-meter telescope in the Canary Islands (fig. 5.7). The mirror consists of 36 separate mirrors that are kept aligned by lasers that measure precisely the tilt and position of each mirror. If any misalignment is detected, tiny motors shift the offending separate mirror segment to keep the image sharply in focus. Astronomers think that this method should permit the building of telescopes perhaps 30, 50, or possibly even 100 meters in diameter in the future.

Up to now, we have concentrated on the light-gathering power of telescopes: their ability to make dim objects bright enough to see. Telescopes, however, serve another important function—they increase our eyes' ability to see details.

A

B

FIGURE 5.7
The Gran Telescopio Canarias, currently the largest visual-wavelength telescope in the world. (A) "Fisheye" view inside the telescope dome shows the 10.4-meter diameter telescope pointing up. (B) A view of the mirror under construction shows the individual mirrors that are precisely aligned to make the overall large mirror size.

5.2 Resolving Power

If you mark two tiny black dots close together on a piece of paper and look at them from the other side of the room, your eye may see them as a single dark mark, not as separate spots. Likewise, stars that lie very close together or markings on planets may not be clearly distinguishable. A telescope's ability to discern such detail depends on its **resolving power.**

Resolving power is limited by the wave nature of light. For example, suppose two stars are separated by a very tiny angle. For them to be discernible as separate images, their light waves must not get mixed up. Such mixing, however, always occurs when waves pass through an opening, because as each wave passes the opening, smaller, secondary waves are produced in a phenomenon called **diffraction.** Figure 5.8A shows how water waves are diffracted as they pass through a narrow opening. Light waves are similarly diffracted as they enter a telescope. The result of diffraction is that point sources of light become surrounded by rings of light. You can observe diffraction by looking at a tiny, bright light source, such as a study lamp, through a piece of cloth, such as a T-shirt. The light will be surrounded by colored diffraction rings produced as the light waves pass through the tiny openings in the weave of the fabric. Similarly, if you look at a bright light source through a few strands of hair pulled over your eye, you may see diffraction fringes with rainbow-like colors around the hairs. An even better way to see the effects of diffraction is to fog a piece of clear glass with your breath and hold it close to your eye while you look at a small, bright light source. If you wear glasses, just breathe on them, put them back on, and look at a bright light. You will see colored diffraction rings around the light, as figure 5.8B illustrates.

Diffraction seriously limits the detail visible through a telescope. In fact, diffraction theory shows that if two points of light that are separated by an angle α (measured

A

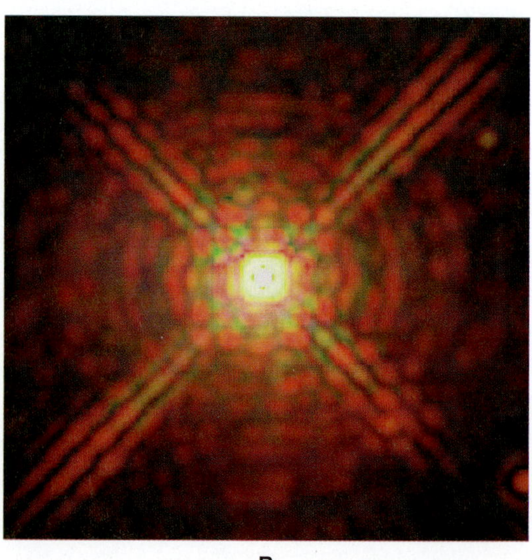

B

FIGURE 5.8
(A) Water ripples diffracted as they pass through a narrow opening. (B) A highly magnified image of a star made with the Hubble Space Telescope. The light from a single point is spread out by diffraction at the edges of the mirror and structures within the telescope. (These diffraction features are very faint relative to the star, but are amplified in this image.)

in arc seconds*) are observed at a wavelength λ through a telescope whose diameter is D, the points cannot be seen as separate sources unless D is greater than (>)

$$2.5 \times 10^5 \lambda/\alpha$$

If D is expressed in centimeters, λ in nanometers, and α in seconds of arc, then

$$D > 0.02 \lambda/\alpha$$

For example, to resolve two stars separated by 0.1 seconds of arc when observing with visible light (λ = 500 nanometers), you need a telescope whose lens has a diameter greater than 100 centimeters (about 40 inches). Notice that the diameter needed to resolve two sources increases as the sources get closer together.†

Interferometers

Diffraction effects can never be totally eliminated, but they can be reduced by enlarging the opening through which the light passes, so that its waves do not mix as severely. Astronomers sometimes accomplish this with a device called an **interferometer.** With an interferometer, observations are made simultaneously through two or more widely spaced telescopes (fig. 5.9) that direct the light to a common detector that combines the separate light beams.

---

*An arc second is a unit of angle and is equal to 1/3600 of a degree.

† As we will discover later in this chapter, our atmosphere seriously blurs fine details in astronomical objects, degrading the resolving power of large ground-based telescopes to far below their diffraction limits.

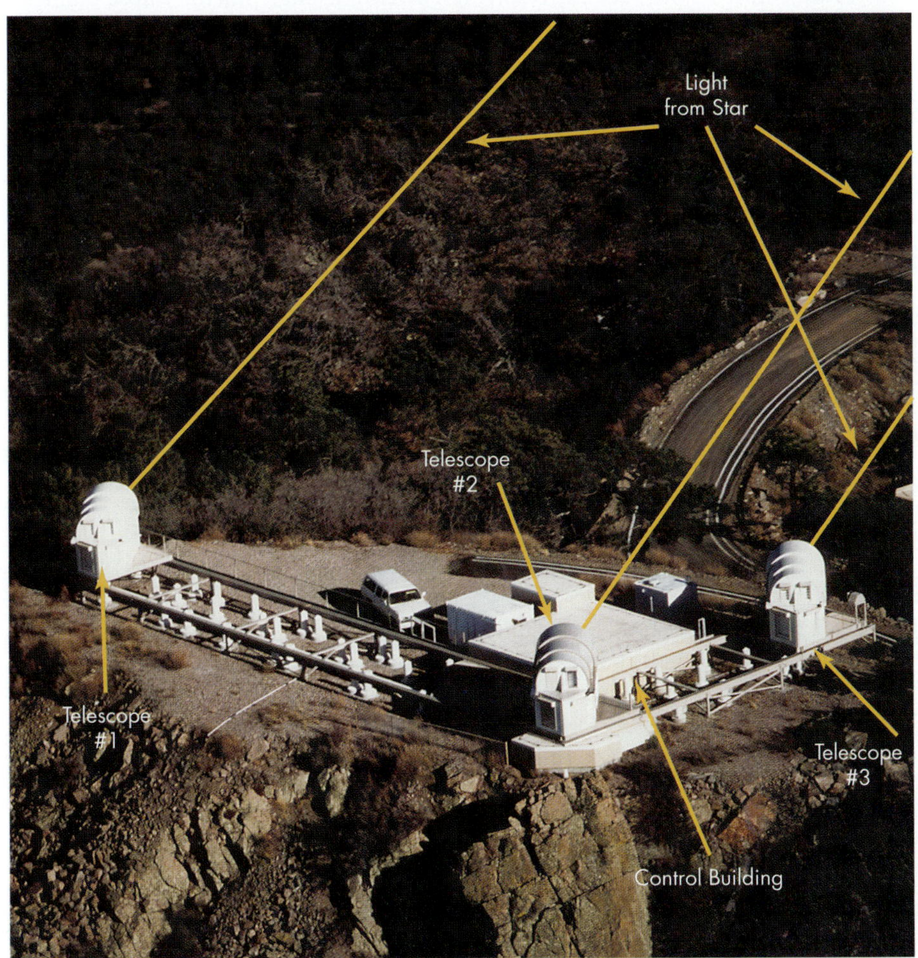

FIGURE 5.9
Photograph of an infrared and optical wavelength interferometer (IOTA). Light from the object of interest is collected by the three telescopes and sent to a control room. Computers there combine the light and reconstruct an image of the object.

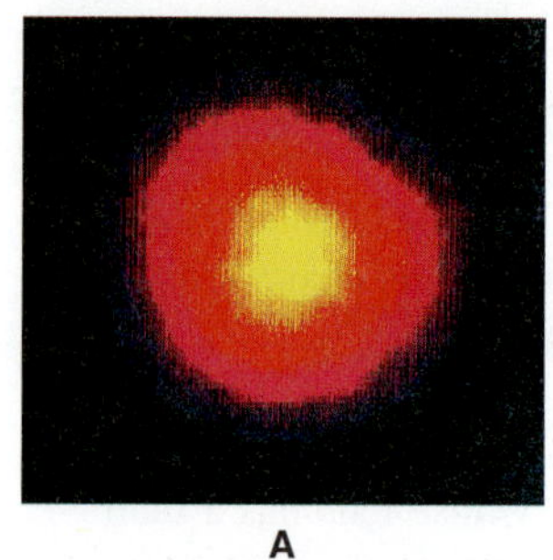

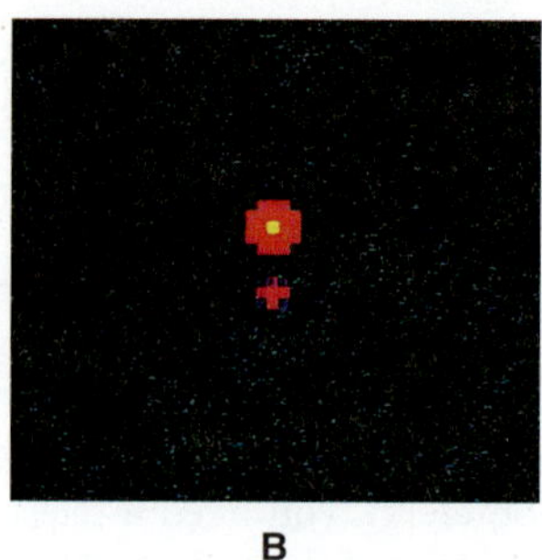

FIGURE 5.10
(A) A young star observed with an ordinary telescope. (B) The same star observed with an interferometer. The higher resolution of the interferometer reveals that the "star" is actually two stars in orbit around one another.

The interferometer is so named because when it mixes the separate beams, the light waves of one "interfere" with the waves from the other. Where the crests of two waves arrive together, they create a bright region. Where the crest of one wave arrives simultaneously with the trough of another, they cancel and create a dark patch. The result is a complex pattern of alternating light and dark regions, which can be analyzed by a computer to create an image of the object observed.

The result of this complicated process is an image in which the resolution is set not by the size of the individual mirrors but rather by their separation. If the mirrors are 100 meters apart, for instance, the interferometer has the same resolving power as a telescope 100 meters in diameter. The resulting high resolution is far beyond what can be obtained in other ways. For example, figure 5.10A shows a view of two closely spaced stars as observed with a small telescope. Their images are severely blended as a result of diffraction. Figure 5.10B shows the same stars observed with an interferometer and after the image has been processed by a computer. The two stars can now be easily distinguished: the two separate mirrors produce the resolving power of a single mirror whose diameter equals the spacing between them.

5.3 Detecting Light

Visible Light

Once light has been collected, it must be detected and recorded. In olden days, the detector was the eye of an astronomer who sat at the telescope eyepiece and wrote down data or made sketches of the object being observed. The human eye, marvel that it is, has difficulty seeing very faint light. Many astronomical bodies are too distant or too dim for their few photons to create a sensible effect on the eye. For example, if you were to look at any but the nearest galaxies through even the Gran Telescopio Canarias, the galaxies would appear merely as dim smudges. Only by storing up their light, sometimes for hours, can a quality picture of them be made. Thus, to see very faint objects, astronomers use detectors that can store light in some manner. Such storage can be done chemically with film or electronically with detectors similar to those used in video camcorders.

From the late 1800s until the 1980s, astronomers generally used photographic film to record the light from the bodies they were studying. Film forms an image by absorbing photons that cause a chemical change, making the film dark where light has fallen on it. This process, however, is very inefficient: fewer than 4% of the photons striking the film produce a useful image. The result of such low efficiency is that many hours are needed to accumulate enough light to create an image of faint objects. Moreover, the film must be developed, thereby delaying the observing process even further.

Astronomers today use many kinds of electronic detectors. One of the best types is the **CCD** (charge-coupled device). Modern CCDs can make pictures virtually indistinguishable from photographs in their detail and with a sensitivity to faint light approximately 20 times greater. In these devices, the incoming light strikes a semiconductor surface, allowing electrons to move within the material. The surface is divided into millions of little squares called pixels, in which the electrons are temporarily stored. The number of electrons in each pixel is proportional to the number of photons hitting it (that is, proportional to the intensity of the light). An electronic device coupled to a computer then scans the detector, counting the number of electrons in each pixel and generating a picture, in much the same way as a TV screen or printed photo creates a picture made up of many separate tiny dots.*

Such electronic devices are extremely efficient, recording 75% (or more) of the photons striking them, allowing astronomers to record images much faster than with film. Electronic detectors have other advantages as well. For example, they record the signal digitally, essentially counting every photon that falls on each part of the detector. Such digital images can be processed by computers to sharpen them, remove extraneous light, and enhance contrast.

Observing at Nonvisible Wavelengths

Visible light, which we can see because its wavelengths are detectable by our eyes, is just one of many wave bands of electromagnetic radiation. Many astronomical objects, however, radiate at wavelengths that our eyes cannot see, and so astronomers have devised ways to observe such objects. For example, cold clouds of gas in interstellar space emit little visible light but large amounts of radio energy. To observe them, astronomers use radio telescopes. Radio-wavelength detectors are similar to radio receivers used for man-made broadcasts but are much more sensitive. They are also made highly directional by building huge radio "mirrors" (figure 5.11), just as we build large mirrors to make higher-resolution observations with visible light. Radio telescopes can also be joined together to form interferometers for even higher resolutions. Some radio interferometers use telescopes in different continents to form a telescope nearly as large as the whole Earth.

*This is easy to see if you look closely at a color picture printed in a newspaper.

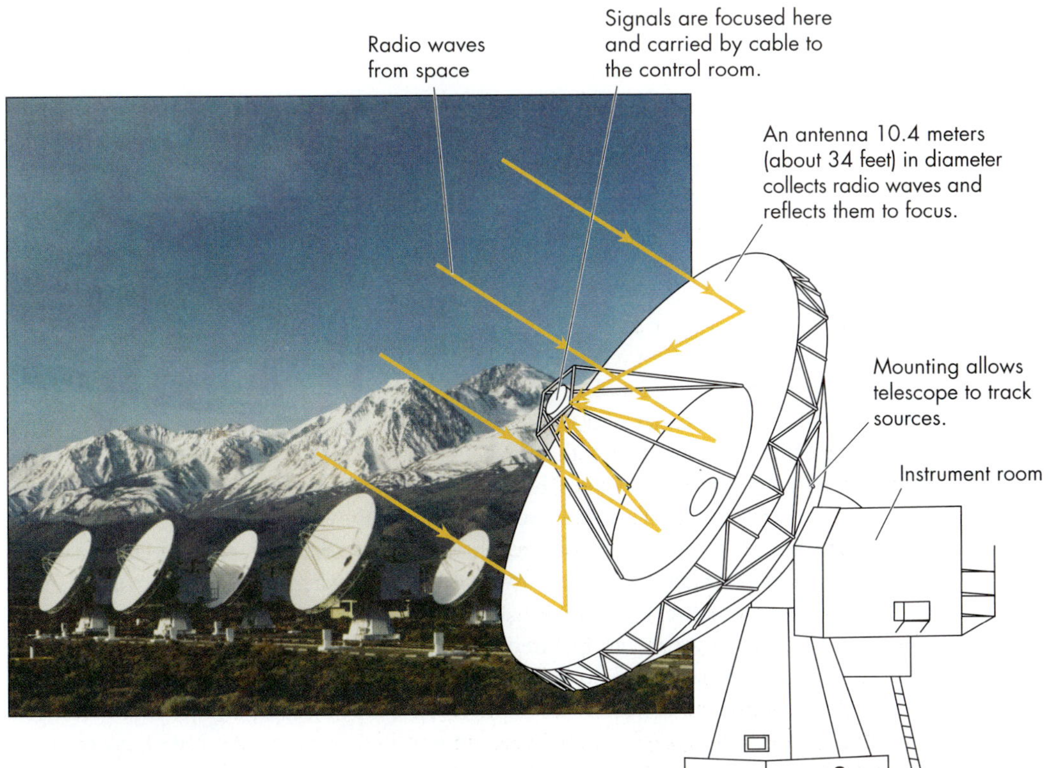
Radio waves from space
Signals are focused here and carried by cable to the control room.
An antenna 10.4 meters (about 34 feet) in diameter collects radio waves and reflects them to focus.
Mounting allows telescope to track sources.
Instrument room

FIGURE 5.11
Photograph of the radio telescope at the Owens Valley Radio Observatory, operated by the California Institute of Technology. In the background you can see the Sierras. The "telescope" is an array of six separate dishes that collect the radio waves. The captured radiation is then combined by computer to increase the resolution of the instrument.

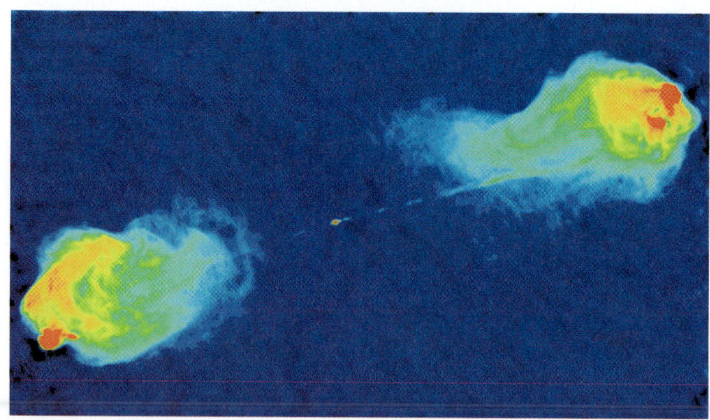

A

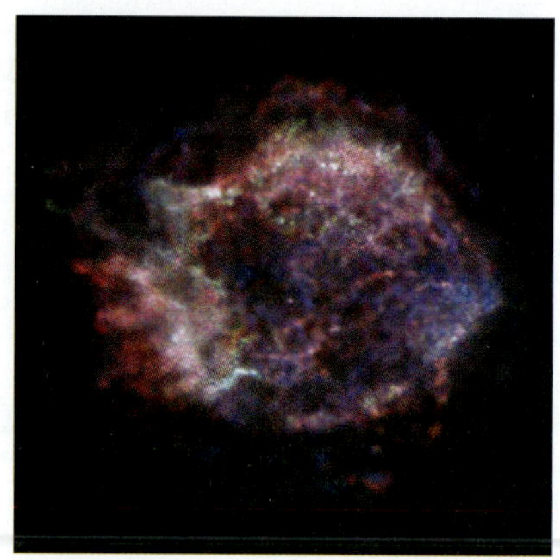

B

FIGURE 5.12
(A) A false-color picture of a radio galaxy. We can't see radio waves, so colors are used to represent their brightness—red brightest, blue dimmest, (B) A false-color X-ray picture of Cas A, an exploding star. In this case colors represent different wavelengths of X-ray photons (bluer colors corresponding to more energetic photons).

Different challenges face astronomers when building telescopes in different wave bands. For example, dust clouds in space are too cold to emit visible light, but they do radiate infrared energy, which astronomers observe with infrared telescopes. One of the challenges for infrared telescopes is that the telescope itself may emit infrared radiation that can mask the objects being observed. These telescopes must be carefully designed to prevent that local radiation from entering the detectors, and parts that cannot be shielded are kept at extremely low temperatures.

Designing a telescope for observing X rays presents different challenges. X rays entering a normal telescope would strike the mirror surface and be absorbed, making the telescope no more effective for observing than a slab of concrete. Astronomers have found, however, that X rays can be reflected if they strike a smooth surface at a very shallow angle, somewhat as a rock can skip over the surface of water if thrown nearly horizontally. X-ray telescopes are like curved funnels, gradually redirecting the X-ray photons toward the detector. As for visual photons, CCDs are again used as detectors, but because X-ray photons carry so much energy, a single photon frees up many electrons when it strikes a CCD pixel. This allows astronomers to measure the energy of each X-ray photon by reading out the CCD quickly enough to avoid multiple-photon hits.

Because our eyes cannot see these other wavelengths, astronomers must devise ways to depict what such instruments record. The most common way to illustrate the radiation is with false-color pictures, as shown in figure 5.12. In a false-color picture, the colors represent different properties of the radiation. For example, in figure 5.12A (a radio "picture" of a radio galaxy and the jet of hot gas spurting from its core), astronomers color the regions emitting the most intense radio emission red; they color areas emitting somewhat weaker emission yellow and the faintest areas blue. Thus, if we could "see" radio waves, the red areas would look brightest and the blue areas dimmest. Another approach is to "translate" the energies of the photons into colors. For example, figure 5.12B shows a false-color X-ray "photograph" of the gas shell ejected by an exploding star. In this image the highest-energy photons are colored blue, intermediate ones yellow, and the lowest-energy photons red. In this case, if our eyes were sensitive to the X-ray band instead of the visible band, this is what X rays might look like to us. Astronomers sometimes use false-color images to bring out particular features or even to depict calculated quantities, such as magnetic field strength or pressure—quantities that we could never directly see with our eyes.

Telescopes operating at infrared, ultraviolet, and X-ray wavelengths face an additional obstacle: most of the radiation they seek to measure cannot penetrate the Earth's atmosphere. If astronomers want to view an object in a blocked wavelength, they must use a telescope in space, operated remotely from the ground or, more rarely, by a scientist-astronaut in space.

5.4 Telescopes on the Ground and in Space

Visible light is special not only because our eyes can detect it, but also because it is one of the few wavelength regions, called **atmospheric windows,** in which it is possible to peer out into space from the ground (see section 4.7). Gases in our atmosphere such as ozone, carbon dioxide, and water strongly absorb infrared, ultraviolet, and shorter wavelengths, as shown in figure 5.13. For example, infrared radiation with a wavelength of 50 micrometers is strongly absorbed by water and carbon dioxide in our atmosphere. Astronomers can make some observations from high-flying airplanes or balloons, but other wavelength ranges are so strongly absorbed that it is necessary to launch telescopes into space. Some of the most exciting discoveries

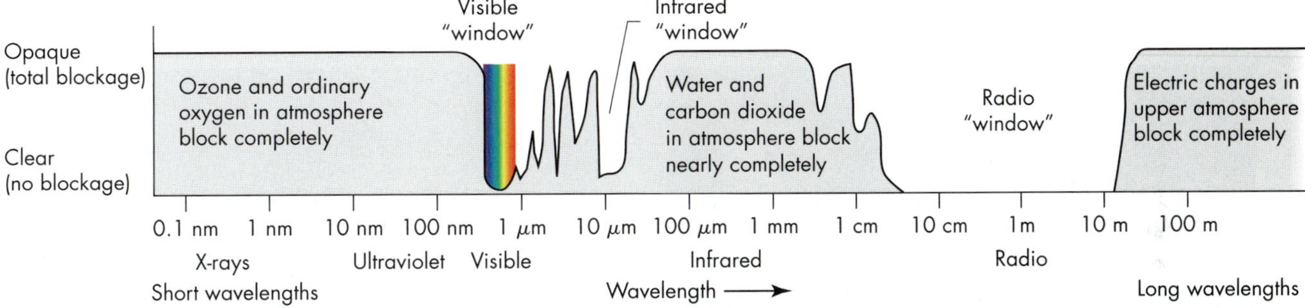

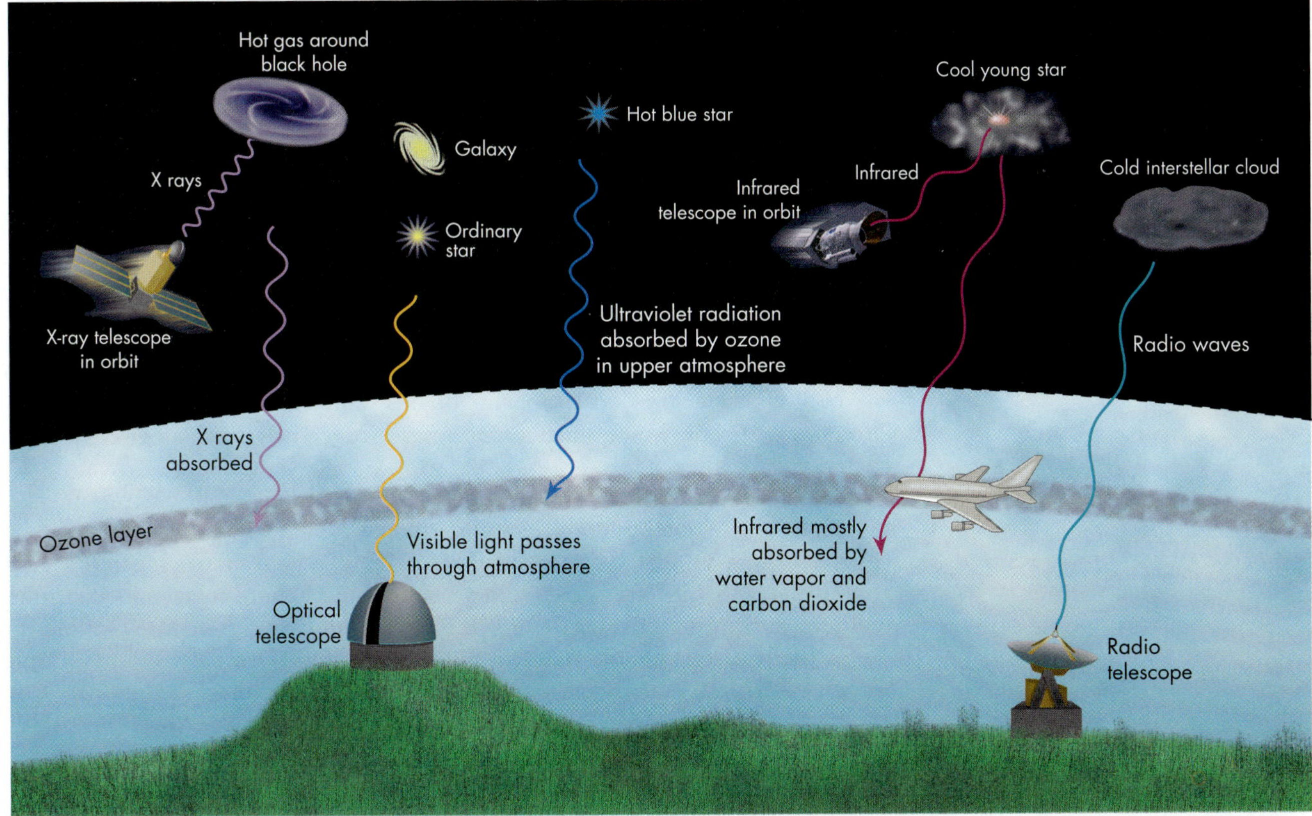

FIGURE 5.13
Diagram showing the transmission of light through the Earth's atmosphere.

Q. Would 100 µm be a good wavelength at which to try to observe a cool star with a ground-based telescope? Would 10 µm be better?

Hubble Space Telescope (13.6 m long) – HST

Extreme Ultraviolet Explorer – EUVE

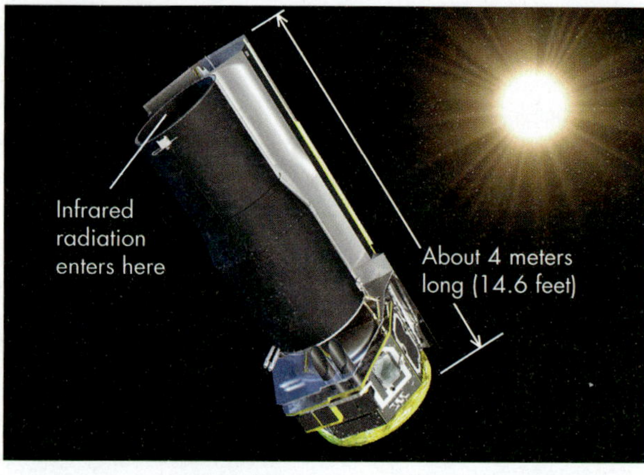

Spitzer Infrared Space Telescope

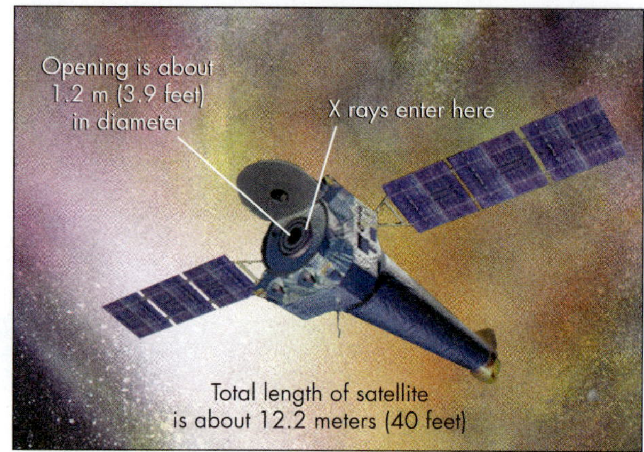

Chandra X-ray Telescope Satellite

FIGURE 5.14
Photograph of the Hubble Space Telescope and drawings of some other orbiting observatories: EUVE, Spitzer, and Chandra. "Chandra" is short for Chandrasekhar, the Indian astrophysicist who won the Nobel Prize in Physics in 1983 for work he did in the 1930s on collapsed stars (discussed further in chapter 15).

have been made as astronomers explore wavelength regions, where they sometimes find objects never previously imagined. For example, in the gamma-ray range, astronomers encountered a phenomenon never previously seen, as discussed in the Extending Our Reach box.

Figure 5.14 shows several of the many orbiting telescopes astronomers use. Some, like the Hubble Space Telescope, which was launched in 1990, have operated for many years. Others, such as the Spitzer Space Telescope, have a much shorter lifetime. This NASA infrared observatory needs liquid helium to keep its instruments cold for maximum sensitivity. Once the helium on board the craft was used up in 2009, the telescope could no longer operate with full sensitivity. In the meantime, other wavelength regions not observable from the ground are being probed, such as the X-ray and gamma-ray parts of the electromagnetic spectrum. Probably the best known of all space telescopes, however, operates in one of the atmospheric windows—it was launched because of another problem that our atmosphere causes.

Atmospheric Blurring

Of the many orbiting telescopes used by astronomers, the Hubble Space Telescope (HST) is the most ambitious to date. The HST can observe at visible wavelengths and into portions of the infrared and ultraviolet bands. Its mirror, 2.4 meters (about 94 inches) in

EXTENDING *our reach*

EXPLORING NEW WAVELENGTHS: GAMMA RAYS

Astronomers have made many of their most exciting discoveries when new telescopes allowed them to observe the sky at wavelengths not previously detectable. Gamma-ray wavelengths are among the last of the wavelength regions to be explored, and astronomers are still trying to interpret what they see.

Gamma-ray astronomy began in 1965 when a small and (by modern standards) primitive satellite detected cosmic gamma rays. A few years later, a slightly more advanced satellite detected gamma rays coming from the center of our Galaxy, the Milky Way. By the 1970s, astronomers had discovered that many familiar sources, such as the Crab Nebula and the remnants of other exploded stars, emit gamma rays. Ironically, perhaps the most interesting gamma-ray sources were discovered accidentally, earlier.

In 1967 the United States placed several military surveillance satellites in orbit to watch for the gamma rays produced when a nuclear bomb explodes. The satellites were designed to monitor the United States–Soviet Union ban on nuclear bomb tests in the atmosphere. Curiously, on a number of occasions the satellites detected gamma-ray bursts coming not from the Earth but from space. Unfortunately for astronomers, the discovery of the bursts was top secret at the time and was not made public until 1973.

Astronomers' thirst for more information about these high-energy sources was unsatisfied for many years because our atmosphere absorbs gamma rays, and ordinary telescopes cannot focus gamma rays. Nevertheless, with ever more complex instruments in satellites, astronomers discovered that gamma-ray sources—apart from the bursts—coincided with known astronomical objects such as dying stars and some peculiar galaxies. The gamma-ray burst, on the other hand, would appear suddenly in otherwise blank areas of the sky, flare in intensity for a few seconds, and then fade to invisibility. It has taken nearly 30 years of study to answer even the simple question "Are they near or far?" The breakthrough came in December 1997, when astronomers detected a gamma-ray burst that coincided with a distant galaxy. This solved the mystery of the bursters' distance but leaves unanswered what they are. Theories to explain the bursts abound. According to a favored hypothesis, the bursts are "hypernovas," stellar explosions caused when massive stars run out of fuel and collapse to form black holes. This latter proposal gained support when, in 2002, astronomers obtained spectra of a burst that showed emission lines suggestive of the explosion of a massive star. Today, the source of some gamma-ray bursts is still mysterious, but that is what makes doing science exciting.

diameter, directs light to a set of instruments. The strikingly sharp images from the HST lead many people to imagine that it must be the largest telescope in the world, but it is actually quite modest in size compared to most ground-based research telescopes. The quality of its images has to do with its ability to avoid the blurring effects of our atmosphere.

Anyone who has ever watched the stars flicker and "twinkle" on a clear night has seen the blurring that our atmosphere creates. Twinkling, more properly called **scintillation,** is caused by atmospheric irregularities refracting the star's light. These irregularities result from slight variations in the air's density caused by small temperature differences. As light moves through these irregularities, its speed changes by a tiny amount, and the light is slightly refracted. If there were no irregularities, a star's light would reach your eye along a single path and be steady. However, the many irregularities create many paths by which the light reaches you. Moreover, these paths change direction rapidly and erratically as the irregularities move, carried by the wind. As a result, the starlight you see at any instant is a blend of light from many slightly different directions, a blend that smears the star's image and makes it dance (fig. 5.15). You can see a similar effect looking at something through water. If the surface of the water has even slight disturbances, a pebble or coin on the bottom seems to dance around. Atmospheric irregularities also slightly disperse the light, making the

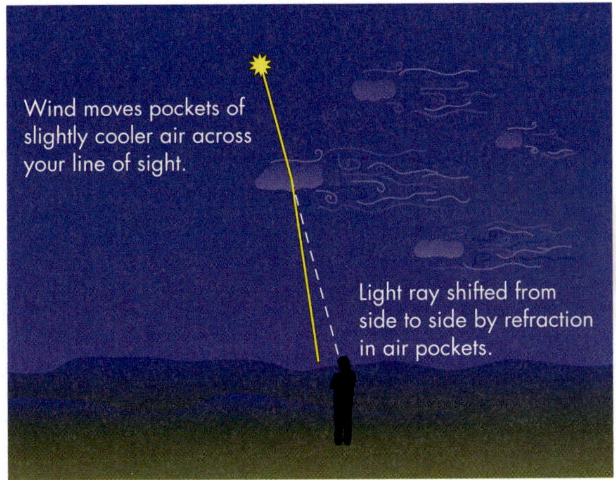

FIGURE 5.15
Twinkling of stars (seeing) is caused by moving atmospheric irregularities that refract light in random directions.

Q. Why do stars twinkle more when they are low in the sky and close to the horizon than when they are nearly overhead?

Sombrero Galaxy

Hourglass Nebula
A dying star

FIGURE 5.16
Some images obtained with the Hubble Space Telescope.

star's color dance, too. Such refractive twinkling, though very pretty to watch, seriously limits the ability of Earthbound observers to see fine details in astronomical objects. The dancing image of a star or planet distorts its picture when recorded by a camera or other device. Outside the atmosphere (in space) this distortion, technically called **seeing,** disappears. The elimination of seeing problems caused by the turbulence of our atmosphere is one reason why astronomers find space observatories so useful.

Although the HST initially had a number of problems, astronauts repaired the major defects, and now the clarity of its images is remarkable (fig. 5.16). These images reveal details never before seen by telescopes on the ground because such telescopes must peer through the blurring effects of our atmosphere. The HST has several different instruments, including cameras for wide-field views and for detailed images, as well as spectrographs for analyzing light from stars and galaxies. The HST has remained an exceptional instrument for so long because it was designed to be serviced by astronauts aboard the space shuttle. They have replaced and repaired instruments, allowing the HST to remain a state-of-the-art instrument. A final servicing mission in 2009 should keep the HST running for about five more years until a larger, new space telescope can take its place.

Space Telescopes Versus Ground-Based Telescopes

Until recently, ground-based astronomers had to simply accept the distortions of seeing, but now they can partially compensate for such seeing in several ways. One technique involves observing a known "reference" star simultaneously with the object of interest. By measuring carefully how the atmosphere distorts the known star's image, corrections can be made in the pictures of other objects. Unfortunately, it is rare that there is a bright enough star close enough to an object of interest for this technique to work. Astronomers have therefore developed a technique using a powerful laser beam to create an artificial star where they need it.

The laser beam is projected on the atmosphere, as shown in figure 5.17. The distortions of the artificial star image are recorded by a computer that then triggers tiny actuators on a correcting mirror placed in the telescope's light beam. The actuators create tiny adjustments in the correcting mirror that cancel out those created by the atmosphere. This technique, called **adaptive optics,** has already given astronomers dramatically improved views through the turbulence of our atmosphere, as illustrated by the inset images in figure 5.17.

Even though visible-wavelength space observatories can produce sharper images, techniques like adaptive optics, along with the great expense of launching an object into space, guarantee that much astronomical work, and the largest telescopes, will

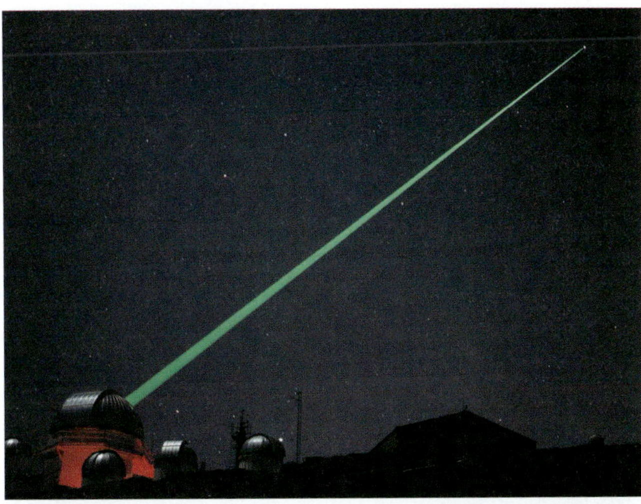

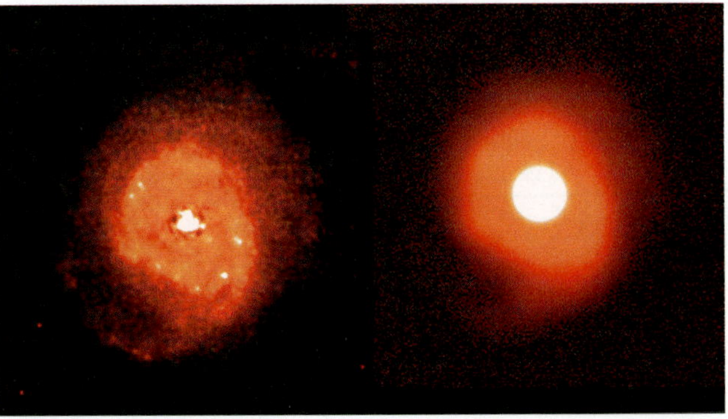

FIGURE 5.17
A laser beam creates an artificial star whose image serves as a reference to eliminate the atmosphere's distortion of real stars (Starfire Optical Range of the Phillips Laboratory at Kirtland Air Force Base in New Mexico). The images at right illustrate the difference in image quality for a galaxy with adaptive optics turned on and off (images made at the Canada-France-Hawaii Telescope on Mauna Kea).

be ground-based for the foreseeable future. Moreover, with ground-based equipment, problems can be corrected easily without the expense, delay, danger, and complexity of a space-shuttle launch.

Because huge telescopes in space or even on the Moon will remain dreams for years to come, astronomers choose with care the location of ground-based telescopes. Sites are picked to minimize clouds and the inevitable distortions and absorption of even clear air. Nearly all observatories are built in dry, relatively cloud-free regions of the world such as the American Southwest, the Chilean desert, Australia, and a few islands such as Hawaii and the Canaries. Moreover, astronomers try to locate observatories on mountain peaks to get them above the haze that often develops close to the ground in such dry locales.

Recently, astronomers have had to contend with another factor that affects the location of observatories: light pollution. Most inhabited areas are peppered with nighttime lighting such as street lights, advertising displays, and automobile headlights (fig. 5.18A and B). Although some such lighting may increase safety, much of it is wasted energy, illuminating unessential areas and spilling light upward into the sky, where it serves no purpose at all. Figure 5.18C shows a satellite view of North America at night and illustrates the waste of energy involved in light pollution. Such stray light can seriously interfere with astronomical observations. In fact, some observatories have been essentially shut down because of light pollution. In some places, astronomers have persuaded regional planning bodies to develop lighting codes to minimize light pollution. Light pollution, however, not only wastes energy and interferes with astronomy, it also destroys a part of our heritage—the ability to see stars at night. The night sky is a beautiful sight, and it is shameful to deprive people of it.

5.5 Observatories

The immense telescopes and the associated equipment astronomers use are extremely expensive. Therefore, the largest telescopes are often national or international facilities, such as the National Optical Astronomical Observatory of the United States and the Anglo-Australian

FIGURE 5.18
Photographs illustrating light pollution. (A) Los Angeles basin viewed from Mount Wilson Observatory in 1908. (B) Los Angeles at night in 1988. (C) Notice the pattern of the interstate highway system visible in the satellite picture of North America at night.

Telescope. Despite the expense of such facilities, many colleges and universities have their own large research telescopes (in addition to smaller ones near campus for instructional purposes). In addition, some large private groups such as the Carnegie Institution operate observatories. Altogether, several thousand observatories exist around the world, on every continent. There are even telescopes at the South Pole in Antarctica to take advantage of the extreme dryness of the bitterly cold Antarctic air.

The largest optical telescopes in the United States at this time are the twin 10-meter Keck Telescopes pictured in the chapter-opening figure. These telescopes pioneered the use of segmented mirrors, and the slightly larger Gran Telescopio Canarias is based on their design. The two Keck Telescopes can be operated individually or as an interferometer with double the collecting area. The Large Binocular Telescope (fig. 5.5) is the next-largest telescope in the United States, both as individual mirrors or when combined as an interferometer.

The optical telescope with the largest collecting area in the world is the VLT (for "very large telescope"), a group of four 8.2-meter telescopes that work individually or as an interferometer. The VLT is operated by a consortium of European countries and Chile and is located in the extremely dry northern part of Chile. Several other large

FIGURE 5.19
Designs for a 42-meter multimirror telescope are currently under study at the European Southern Observatory.

telescopes have begun operation recently, such as the 8.3-meter Subaru* Telescope operated by Japan and located in Hawaii. Others are the Gemini telescopes—two identical 8.1-meter instruments. One is located in Hawaii; the other is in Chile. These twin telescopes are run by a consortium consisting of the United States, the United Kingdom, Canada, Chile, Australia, Brazil, and Argentina.

To observe the faint trickle of light from forming galaxies billions of light-years away or small dimly lit bodies in the outer Solar System, astronomers seek to build even bigger telescopes, such as the 42-meter European Extremely Large Telescope (EELT, for short) (fig. 5.19). Its proposed collecting area would be nearly half the size of a football field. However, a number technical problems need to be overcome before this colossus can be built, and no starting date for it has been set.

Observatories operating at other wavelengths, from radio waves to gamma rays are also designed and operated by international consortiums. For instance, an X-ray space telescope such as Chandra contained detectors designed by teams in Germany, the United Kingdom, and the United States. New ground-based telescopes increasingly rely on international collaborations to share the expense of building the best possible instruments. An example of this is the Atacama Large Millimeter Array (ALMA), a large interferometer being built in the high desert of Chile, jointly funded by agencies in Canada, Chile, Europe, Japan, Taiwan, and the United States. Among other objects, ALMA will be able to study galaxies forming near the beginning of the Universe. By pooling their resources, astronomers can build a far more powerful instrument than would be possible if each country built a separate instrument.

Going Observing

When astronomers want to observe at a telescope, they do not just run off to the observatory. Generally, they must submit an observing proposal that describes what objects they wish to look at and why such observations are important. They must also show

*Subaru is the Japanese name for the Pleiades star cluster. Look carefully at the logo on a Subaru car and you will see the stars in the design.

that all necessary equipment will be available and that it have the sensitivity necessary to complete the proposed observations. At all major observatories and space facilities, proposals are invited from astronomers from around the world to encourage the best science possible. Proposals are screened by a committee that then allocates telescope time according to the scientific merits of the proposals.

For ground-based telescopes, observing sessions are typically assigned in blocks of several nights so that the astronomer's risk of clouds or equipment failure is minimized. If the astronomer is using an optical telescope and must therefore observe at night, he or she must also become accustomed to switching schedules to be awake all night and sleep during the day. Most observatories have small dorm rooms with special shades to make the room dark and quiet during daylight hours for astronomers trying to catch up on sleep. Many also have small cafeterias where food (and coffee) is available at odd hours. Sometimes observing runs go smoothly, and the astronomers can return to their home institutions laden down with data. At other times weather or equipment will not cooperate, and the runs are a total loss.

For space-based telescopes, astronomers have developed techniques to allow the telescope to be run remotely by computer programs. These techniques are increasingly being applied to ground-based telescopes so that astronomers are not required to travel to the observatory, but instead may operate the telescope from their desk or prepare computer files that allow the telescope to run on its own. While this style of observing lacks much of the adventure of traveling to remote mountain peaks, it is generally more efficient.

Computers

In the last few decades, the computer has become one of the astronomer's most important tools. In fact, for many astronomers today, operating a computer and being able to program is more important than knowing how to use a telescope. Astronomers use computers not only to solve equations but also to move the telescope, feed the information to detectors, convert the data obtained to a useful form, and communicate with other astronomers.

The developments in computers and the Internet have also opened up new "observing" possibilities. A number of projects have been undertaken with the idea of systematically observing the sky at one or more wavelengths, then storing the data in large archives that can be accessed over the Internet. Instead of gathering data on a few particular objects, such maps record information about *every* astronomical object they can detect. For example, astronomers working on 2MASS (the Two Micron All Sky Survey) mapped the entire sky at short infrared wavelengths that get through our atmosphere, compiling a database of several *million* images. You can see these images of gas clouds, galaxies, star clusters, and so on at the relevant websites, where they can be downloaded, then carry out analyses of the data. Similarly, astronomers working on the Sloan Digital Sky Survey are mapping portions of the sky at visible wavelengths and carrying out spectroscopic observations of more than a million objects. The use of archives does not make direct observing obsolete, but it can give a first look at a problem so that when the astronomer later goes to a telescope, he or she can use the time there more efficiently.

Modern computers have opened up a new age in modeling astronomical objects. When astronomers attempt to interpret observational data, they now use computers to examine the physical processes that they suspect are taking place. This may involve simulating the gravitational interactions between two galaxies that appear to be colliding, calculating what kinds of nuclear interactions must be occurring in the core of a star, or perhaps tracing how photons are expected to move through an interstellar cloud of a particular molecular composition. These computer simulations allow astronomers to test and refine their hypotheses, and make predictions for what they expect to see when making future observations.

SUMMARY

Astronomers use telescopes to collect radiation from astronomical sources. Telescopes generally have large-diameter mirrors or lenses to collect as much radiation as possible and allow faint objects to be seen. This gives them large light-gathering power. A large mirror or lens also increases the telescope's ability to resolve detail, giving sharper images, but such gains are seriously limited for ground-based visible-light telescopes by the blurring effects of our atmosphere. Interferometers give the resolving power of a single large area by combining radiation detected with two or more small but widely separated collectors.

Although many astronomical objects radiate visible light, some of the most interesting bodies radiate in radio, infrared, X-ray, or gamma-ray regions of the electromagnetic spectrum. Astronomers use special telescopes to observe in these other wavelength regions. Because many of these wavelengths do not penetrate our atmosphere, telescopes have been put in space to observe them.

QUESTIONS FOR REVIEW

1. (5.1) What is light-gathering power? How does it affect the ability to see faint objects?
2. (5.1) What is the difference between a reflecting and a refracting telescope? What are some advantages of a reflecting telescope?
3. (5.2) What is resolution of a telescope? What physical process limits it?
4. (5.2) How is resolution affected by the size of a telescope's mirror or lens?
5. (5.2) What is the purpose of an interferometer?
6. (5.3) What is a CCD and how is it better than film?
7. (5.4) Why do astronomers put X-ray observatories in space rather than just on a high mountain?
8. (5.4) What kinds of astronomy can be done from the ground?
9. (5.4) What is meant by adaptive optics?
10. (5.5) How do astronomers get to use observatories?

THOUGHT QUESTIONS

1. (5.1) Apart from magnification, how do binoculars help you see better? All else being equal, what difference will you see with 50-millimeter lenses versus 25-millimeter lenses?
2. (5.1) Put a pencil straight down into a glass of water. Notice whether the pencil looks bent. Now tilt the pencil and note how its apparent bending changes. How does this illustrate why light from objects near the horizon is refracted more strongly?
3. (5.1) Why isn't there a hole in the image from a reflecting telescope because the secondary mirror blocks some light? Is your answer true even if there is a hole in the center of the primary mirror, like in the Cassegrain focus shown in figure 5.6?
4. (5.2) Is it better to have a telescope with a high resolving power or a high magnification? Explain why.
5. (5.2/5.4) The Very Large Array is a radio interferometer observatory in New Mexico with twenty-seven 25-meter telescopes. In its widest arrangement, it has the resolving power of a telescope 36 kilometers in diameter. What is better about this arrangement than a single 36-kilometer diameter telescope? What is lacking compared to a single 36-kilometer dish?
6. (5.2/5.4) Why does the useful resolving power of a ground-based telescope with a 2-meter diameter mirror not match its theoretical value?
7. (5.4) It is difficult to observe 1-nanometer, 1-millimeter, and 100-meter radiation with ground-based telescopes. What are the reasons, for each?
8. (5.4) If you look with binoculars down a beach on a hot day, you will see that distant objects appear shimmery. How is this related to astronomical "seeing"?

PROBLEMS

1. (5.1) Compare the light-gathering power of a telescope with a 10-centimeter (about 4-inch) diameter mirror to that of the human eye. (Take the diameter of the pupil of the eye to be about 5 millimeters.)
2. (5.2) Estimate your eye's resolving power by drawing two lines 1 millimeter apart on a piece of paper. Put the paper on the wall and then step back until the two lines appear as one, measuring that distance. From the distance and the separation of the lines (1 millimeter), estimate their angular separation. How does your result for the eye's resolving power compare with that calculated from the resolving-power formula, using a pupil diameter of 5 millimeters and a wavelength of 500 nanometers?
3. (5.2) Can the unaided human eye resolve a crater on the Moon whose angular diameter is 1 minute of arc (=60 seconds of arc)? (Take the diameter of the pupil of the eye to be about 5 millimeters and the wavelength of the light to be 500 nanometers.)

4. (5.2) Determine the resolving power of a 25 meter radio telescope observing 10 centimeter radio waves. Compare this to its resolving power for 1 meter radio waves. (Remember to convert units for the equation in Section 5.2).
5. (5.2) Using ratios or proportionalities, determine how large a diameter "eye" a person would need to see as well in the (1) infrared at wavelength of 12 micrometers, and (2) in the radio at a wavelength of 10 centimeters, as we can in the visible at 500 nanometers with a 5 millimeter pupil.
6. (5.1/5.3) Compute the collecting area of the 27 telescopes in the Very Large Array radio interferometer if each has a diameter of 25 meters. If this was the collecting area of a single dish, what would be its diameter?
7. (5.1/5.3) If a CCD could record 75% of the photons striking it, and a photograph about 5%, how many times larger diameter telescope would you need to take a photograph of equal quality to a CCD image in the same amount of time?
8. (5.4) The altitude of Hubble's orbit is about 569 km above the surface of the Earth. Calculate the circumference of the orbit, the orbital velocity, and the period of the orbit (see Section 3.6). How does this period affect observations?

TEST YOURSELF

1. (5.1) Telescope A's mirror has three times the diameter of telescope B's. How much greater is A's light-gathering power?
   (a) 3 times (b) 6 times (c) 8 times (d) 9 times (e) 27 times
2. (5.2) A telescope's resolving power measures its ability to see
   (a) fainter sources.
   (b) more distant sources.
   (c) finer details in sources.
   (d) larger sources.
   (e) more rapidly moving sources.
3. (5.2) One way to increase the resolving power of a telescope is to
   (a) make its mirror bigger.
   (b) make its mirror smaller.
   (c) replace its mirror with a lens of the same diameter.
   (d) use a mirror made of gold.
   (e) observe objects using longer wavelengths.
4. (5.2) Astronomers use interferometers to
   (a) observe extremely dim sources.
   (b) measure the speed of remote objects.
   (c) detect radiation that otherwise cannot pass through our atmosphere.
   (d) enhance the resolving power (see fine details) in sources.
   (e) measure accurately the composition of sources.
5. (5.3) Which of the following are advantages of a CCD over photographic film? (more than one may be correct)
   (a) CCDs can collect light for a long time.
   (b) CCDs do not need to be changed out for each exposure.
   (c) CCDs are not affected by blurring of the Earth's atmosphere.
   (d) CCDs do not need to be corrected for instrumental effects.
   (e) CCDs record a greater percentage of the photons striking them.
6. (5.1/5.3) Suppose you were examining a pulsing radio signal from a stellar remnant in a distant part of the Milky Way. Knowing that ionized gas in interstellar space causes dispersion of radio waves, what effect would you expect this to have on the signal?
   (a) The time when the pulses arrived would be different for different wavelengths.
   (b) The path would be bent so the signal would come from a different direction than it started from.
   (c) The wavelengths would all grow longer as they ran out of energy.
   (d) The signal would be slowed down—stretched out to fill a much longer time.
7. (5.4) A ground-based telescope to observe X rays would
   (a) be a powerful tool for studying abnormally cold stars or distant planets.
   (b) give astronomers the chance to study the insides of stars and planets.
   (c) be worthless because no astronomical objects emit X rays.
   (d) be worthless because X rays cannot get through the Earth's atmosphere.
   (e) be worthless because astronomers have not yet devised detectors for X rays.
8. (5.4) To use ground based optical telescopes to their theoretical specifications, astronomers must use
   (a) much larger mirrors than we have today.
   (b) adjustable mirrors that can adapt to correct for the atmosphere.
   (c) far more sensitive detectors.
   (d) space satellite surveys to plan observations.

KEY TERMS

adaptive optics, 140
atmospheric window, 137
CCD, 134
diffraction, 132
dispersion, 129
interferometer, 133
light-gathering power, 126
reflectors, 127
refraction, 127
refractors, 127
resolving power, 132
scintillation, 139
seeing, 140

FURTHER EXPLORATIONS

Berger, David H., Jason P. Aufdenberg, and Nils H. Turner. "Resolving the Faces of Stars." *Sky and Telescope* 113 (February 2007): 40.

Chen, P. K. "Visions of Today's Giant Eyes." *Sky and Telescope* 100 (August 2000): 34.

Fugate, Robert Q., and Walter J. Wild. "Untwinkling the Stars—Part I." *Sky and Telescope* 87 (May 1994): 24.

Jayawardhana, Ray. "The Age of Behemoths." *Sky and Telescope* 103 (February 2002): 30.

Kellerman, Kenneth I. "Radio Astronomy in the 21st Century." *Sky and Telescope* 93 (February 1997): 26.

Lowe, Jonathan. "Mirror, Mirror." *Sky and Telescope* 114 (December 2007): 22.

Lowe, Jonathan. "Next Light: Tomorrow's Monster Telescopes." *Sky and Telescope* 115 (April 2008): 20.

Naeye, Robert. "NASA's New Gamma-Ray Trailblazer." *Sky and Telescope* 115 (June 2008): 18.

Schilling, Govert. "Giant Eyes of the Future." *Sky and Telescope* 100 (August 2000): 52.

Shubinski, Raymond. "Who Really Invented the Telescope?" *Astronomy* 36 (August 2008): 84.

Sinnott, Roger W., and David Tytell. "A Galaxy of Telescopes: Coming of Age in the 20th Century." *Sky and Telescope* 100 (August 2000): 42.

Tytell, David. "The Heat Is On." *Sky and Telescope* 107 (April 2004): 30.

Waller, William H. "NASA's Space Infrared Telescope Facility: Seeking Warmth in the Cosmos." *Sky and Telescope* 105 (February 2003): 43.

Wanjek, Christopher. "Gamma-Ray Astronomy Achieves Stardom." *Sky and Telescope* 106 (August 2003): 38.

Website

Visit the *Explorations* website at **http://www.mhhe.com/arny** for additional online resources on these topics.

FIGURE QUESTION ANSWERS

WHAT IS THIS? (chapter opening): This is alleged to be Galileo's telescope. It is in the Florence Museum of Science in Italy.

FIGURE 5.14: 100 μm would not be good because the transmission of our atmosphere is very low at that wavelength. 10 μm would be much better because there is a narrow "window" of transmission there.

FIGURE 5.16: Their light passes through much more atmosphere and thus is more likely to encounter atmospheric irregularities. (Box figure 5.3B shows the greater path length through the atmosphere for a star that is low in the sky.)

PROJECTS

1. **Research an observatory:** Most of the major ground- and space-based observatories have substantial public websites. Pick one of the observatories listed in this chapter and look up the type of the telescope, the path light takes, the size of the mirror, the kinds of instruments and cameras available, what specific wavelengths of light can be studied with the telescope, and a couple of discoveries made by astronomers using the observatory.
2. **Telescope orbits:** Space telescopes have a variety of orbits. Using their public websites, look up and compare the shape, period, and the type of orbit of Hubble, Chandra, and Spitzer. Why are they different? What are the advantages of each?

The Earth seen from Apollo 11 as it traveled to the Moon. East Africa is visible near the center of the crescent.....

KEY CONCEPTS

- Understanding the Earth helps us better understand other planets.
- Scientists can determine properties of the Earth's interior by analyzing earthquake (seismic) waves.
- The Earth's structure consists of:
  - a solid crust,
  - a mantle of rocky material, and
  - a core of mainly iron, part of which is liquid.
- The Earth's interior is very hot. The heat is partly left over from the time of the Earth's formation.
- Radioactive material in the Earth contributes heat to its interior.
- Analysis of the amount of radioactivity gives the Earth's age—about 4.6 billion years.
- Heat inside the Earth causes motions in the interior, which cause the surface to creep slowly.
- Such surface motions build mountains and alter our planet's surface.
- Earth's atmosphere contains mostly nitrogen and oxygen.
- The atmosphere formed from gases liberated mainly during Earth's formation and possibly from comets. Oxygen, however, has been added by green plants.

6 The Earth

CONCEPTS AND SKILLS TO REVIEW

- Determining the shape and size of the Earth (2.1)
- Newton's law of gravity (3.5)
- The law of inertia (3.1)
- Absorption in the Earth's atmosphere (4.7)

Earth is a beautiful planet. Even from space we can see its beauty—blue seas, green jungles, red deserts, white clouds. Much of our appreciation of the Earth comes from knowing that it is home for us and the billions of other living things that share this special and precious corner of the Universe. But why study Earth in an astronomy course? The reason is that we know Earth better than any other astronomical body, and from it we can learn about many of the properties that shape other worlds.

In the simplest terms, Earth is a huge, rocky sphere spinning in space and hurtling around the Sun. In the time it took you to read that sentence, the Earth carried you about 100 miles through the black hostile space around the Sun. But you were protected by a blanket of air, a screen of magnetism, and filters of molecules that blocked most of the hazards of interplanetary space. Other planets share many of these properties but not in the right mix to make it possible for life as we know it to live on them.

Earth's special characteristics result in large measure from its dynamic nature. The Earth is *not* a dead ball of rock; both its surface and its atmosphere have changed greatly during its vast lifetime. Even today, the ground below our feet sometimes trembles and wrenches in response to dynamic forces, crumpling our planet's crust into mountains, stretching it, and tearing it open to form new ocean basins.

These slow but violent motions within the Earth arise as heat generated deep within flows toward the surface. That heat also drives volcanic eruptions, which vent gases and molten rock. Over billions of years, such gases accumulated, in part creating our atmosphere—an atmosphere that has itself been changed by the presence of abundant water and life.

Q: WHAT IS THIS? See end of chapter for the answer

6.1 The Earth as a Planet

Shape and Size of the Earth

Astronomers in ancient Greece knew that the Earth is a sphere with a radius of about 6400 kilometers or 4000 miles (chapter 2). Aristotle even argued that the Earth formed by material being pulled together in such a way that the jostling of the pieces led to a spherical shape. He didn't refer to gravity, specifically, but we know today that it is gravity, in fact, that makes the Earth round.

Gravity is the great leveler. Over millions of years, the force of gravity crushes and deforms rock, pulling high points down and rounding large bodies off. However, this shaping process is effective only if the body exceeds a critical size that depends on the object's composition. For bodies made of rock, the critical radius is about 350 kilometers. An object with a radius larger than this has strong enough gravity that it can pull itself into a sphere even if initially it was irregular. Smaller objects retain their irregular shape, as you can see by comparing the Earth with the asteroid Gaspra (fig. 6.1), which has a length of about 19 kilometers (about 12 miles).

Although the Earth is approximately a sphere, it is not a perfect one. It bulges out at the equator, as illustrated in figure 6.2A. The existence of this bulge was first demonstrated in the eighteenth century, when detailed mappings of the Earth showed that its equatorial radius is about 21 kilometers (about 13 miles) greater than its polar radius. This discovery was not a complete surprise, because astronomers such as Newton and Hooke had already suggested that the Earth's spinning motion might make its equator bulge into a shape technically known as an oblate spheroid. Moreover, telescopic views of other planets showed that they bulged (fig. 6.2B), making it likely the Earth did too.

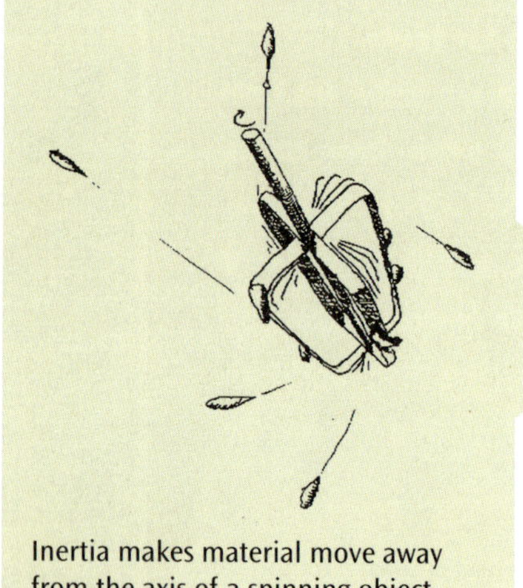

Inertia makes material move away from the axis of a spinning object.

To understand why the Earth's equator bulges, think of what happens when you lift a spinning electric beater from a bowl of cake batter. Particles on such a spinning object fly outward as a result of their inertia, the tendency of all moving objects to keep moving in a straight line, as described by Newton's laws of motion. This same tendency moves matter away from the Earth's rotation axis and is strongest at the equator because the Earth rotates fastest there.

All points on the Earth take the same time (one day) to rotate once around its axis, but because points near the equator travel farther than points near the pole, they must travel faster. At the equator, a point on the Earth's surface moves at about 1000 miles per hour, while at middle latitudes, such a point moves at about 700 miles per hour. The greater speed of the equator is harder for the Earth's gravity to overcome, so the equator bulges outward.

A 12,800 kilometers (about 8000 miles)

FIGURE 6.1
Photographs show that the Earth is round but the asteroid Gaspra is not. Gaspra is too small for its gravity to make it spherical.

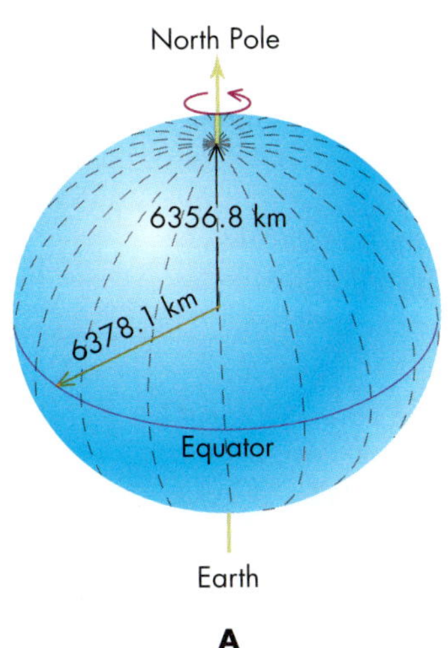

A

B

You can demonstrate rotationally caused bulges with a water balloon. If you toss a water balloon (gingerly!) into the air, it will take on an almost spherical shape. If you set it spinning as you toss it up, it will become noticeably bulged. This is a great demonstration, but choose an appropriate place.

FIGURE 6.2
(A) Rotation makes the Earth's equator bulge. (B) Jupiter's rapid rotation creates an equatorial bulge visible in this photograph.

Composition of the Earth

Although we may call the Earth a ball of rock, the statement is not very informative because so many different kinds of rock exist. Rocks are composed of minerals, and minerals in turn are composed of chemical elements. Analysis of the surface rocks of the Earth shows that the most common elements in them are oxygen, silicon, aluminum, magnesium, and iron. Furthermore, silicon and oxygen usually occur together as **silicates.** For example, ordinary sand (particles of the silicate mineral quartz) is nearly pure silicon dioxide (SiO_2). Table 6.1 lists a few of the most abundant elements in the Earth.

Other kinds of minerals are more complicated, with atoms of calcium, magnesium, or iron included with the silicates. For example, much of the Earth's interior is composed of the mineral olivine, which is an iron-magnesium silicate. It gets its name from its color: in some solidified lavas, it forms tiny green crystals that look like little pieces of chopped olive (fig. 6.3).

At this point you might ask how we can tell what the interior of the Earth is made of. We can infer what lies deep inside our planet in several ways. One is by studying earthquake waves, a point we will take up in more detail in the next section. Another way is by analyzing the Earth's density.

TABLE 6.1 COMPOSITION OF EARTH'S CRUST

Chemical Element (Symbol)	% of Element in Crust by Mass	Density (g/cm^3)
Oxygen (O)	46%	*
Silicon (Si)	28%	2.42
Aluminum (Al)	8%	2.70
Iron (Fe)	6%	7.9
Calcium (Ca)	4%	1.55
Magnesium (Mg)	2%	1.74
Sodium (Na)	2%	0.97
Potassium (K)	2%	0.87
Titanium (Ti)	0.6%	4.5
Hydrogen (H)	0.1%	*
Others	1%	

*Oxygen and hydrogen in crustal rock are not gases, but are part of various minerals.

FIGURE 6.3
Olivine (the greenish crystals) in a sample of solidified lava.

Density of the Earth

Density is a measure of how much material is packed into a given volume. It is defined as an object's mass divided by its volume and is usually measured in terms of the mass in grams of 1 cubic centimeter of the substance. In these units, for example, water has a density of 1 gram per cubic centimeter.* Ordinary rocks, on the other hand, have a density of about 3 grams per cubic centimeter, and iron has a density of almost 8 grams per cubic centimeter. In other words, a volume of iron has about 8 times as much mass as a similar volume of water; that is, iron is much denser than water.

We see from this that density gives some clue to an object's composition. For instance, it would be easy to tell if a closed box contained a block of iron or a block of wood because an equal mass of a dense substance takes up a smaller volume. This is the basis of the famous story of the ancient scientist Archimedes leaping from his bath and shouting "Eureka!" Seeing how his body displaced the water, he realized that was how he could test the density of the king's crown to determine if it was pure gold. Likewise, we can use the density of the Earth to estimate *its* composition.

We find the Earth's density by dividing its mass by its volume. Its mass is 6.0×10^{27} grams (see box below), and its radius is about 6400 kilometers, or 6.4×10^{8} centimeters. To actually make the calculation, we divide the mass of the Earth by its volume, $\left(\frac{4}{3}\right)\pi R^3$, assuming it is a sphere. Thus, the density is

M = Mass of the Earth
R = Radius of the Earth
$\frac{4}{3}\pi R^3$ = Volume of a sphere

$$\frac{M}{\left(\frac{4}{3}\right)\pi R^3} = \frac{6.0 \times 10^{27}\text{gm}}{\left(\frac{4}{3}\right)\pi(6.4 \times 10^{8}\text{ cm})^3} = \frac{6.0}{\left(\frac{4}{3}\right)\pi \times 6.4^3} \times \frac{10^{27}\text{ gm}}{10^{24}\text{ cm}^3} = 5.5\text{ gm/cm}^3$$

That is about twice the density of ordinary rock.

The density as defined here is really an *average density* over the whole planet. Because we can measure directly that the average density of surface rocks is much less than 5.5 grams per cubic centimeter, we can therefore infer that other parts of the Earth must have a density much greater than 5.5 grams per cubic centimeter. That by itself does not tell us what lies inside the Earth, but if we ask ourselves what substances are both dense and abundant in nature, we find that iron is a likely choice, as table 6.1 shows. We therefore deduce that the Earth has an iron core. But we can do better than merely deduce. We can test that hypothesis by taking advantage of one of nature's most violent phenomena: earthquakes.

*Note that the number of grams per cubic centimeter is the same as the number of kilograms per liter or the number of metric tons per cubic meter. For example, water's density can also be written as 1 kg/L.

EXTENDING *our reach*

MEASURING THE EARTH'S MASS

If we know the value of Newton's gravitational constant, G, we can measure the Earth's mass using Newton's laws (chapter 3). But how can we measure G in the first place?

The first accurate measurement was made by Henry Cavendish in 1797. He placed large lead masses next to two carefully balanced weights (Box fig. 6.1). By observing how much the weights twisted toward the lead masses, he was able to calculate G, which in turn allows us to measure the mass of the Earth, Sun, or any other body whose gravity causes an acceleration that we can measure.

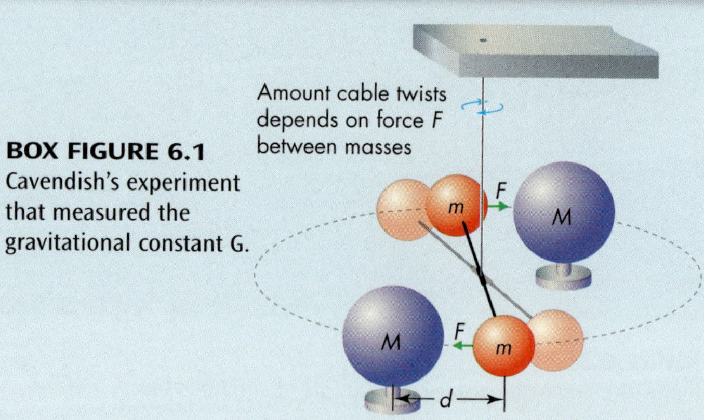

BOX FIGURE 6.1 Cavendish's experiment that measured the gravitational constant G.

6.2 The Earth's Interior

If we ask how the Earth's interior can be studied, your first reaction might be to say, "Why not drill a very deep hole and take a look or pull out samples?" Unfortunately, the deepest hole yet drilled in the Earth penetrates only 12 kilometers, a mere scratch when compared to the Earth's 6400-kilometer radius. If the Earth were an apple, the deepest holes yet drilled would not have broken the apple's skin. Thus, to study the Earth's interior, we rely on indirect means such as earthquake waves.

Probing the Interior with Earthquake Waves

ANIMATION
S and P waves generated by earthquakes

When earthquakes shake and shatter rock within the Earth, they generate **seismic waves** that travel outward from the location of the quake through the body of the Earth (fig. 6.4). Seismic waves slightly compress rock or cause it to vibrate side to side. The speed of the waves depends on the properties of the material through which they move. A wave's speed can be determined by carefully timing its arrival at remote points of the world. From that speed, scientists can deduce a picture of the Earth's interior along the path of that wave. Thus, seismic waves allow us to "see" inside the Earth much as doctors use sound waves to "see" inside our bodies.

To make a picture of your internal organs, sound waves are sent through your body and are then picked up with a sensitive microphone. Because the sound travels at different speeds in bone, tissue, cartilage, and so forth, a medical technician can analyze the signals with a computer to make a picture of your anatomy or of an unborn child (fig. 6.5). You can use a similar, though obviously much cruder, technique to locate wall studs by thumping areas of a plaster wall with your knuckle, and petrogeologists use small explosions in a similar way to hunt for underground oil deposits.

Seismic waves in the Earth are of two main types: S waves and P waves.* P waves form as matter in one place pushes against adjacent matter—whether solid or liquid—compressing it. They travel easily through both solids and liquids. By contrast, S waves form as matter "jerks" adjacent material up and down or from side to side, like a wriggle in a shaken rope. In a liquid, material easily slips past adjacent matter, preventing S waves from spreading. Thus, S waves can travel only through solids. Therefore, if a laboratory on the far side of the Earth from an earthquake

*P and S stand for "primary" and "secondary" and refer to the waves' arrival time at a distant site. That is, the primary waves arrive first.

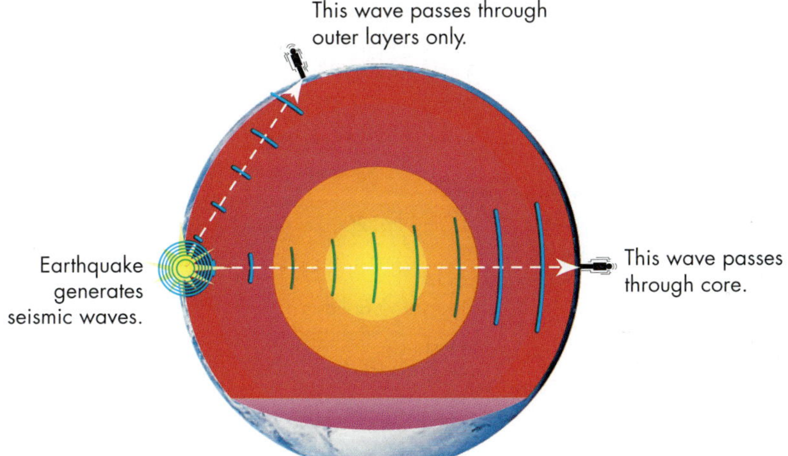
This wave passes through outer layers only.
Earthquake generates seismic waves.
This wave passes through core.

FIGURE 6.4
Seismic waves spread out through the Earth from an earthquake.

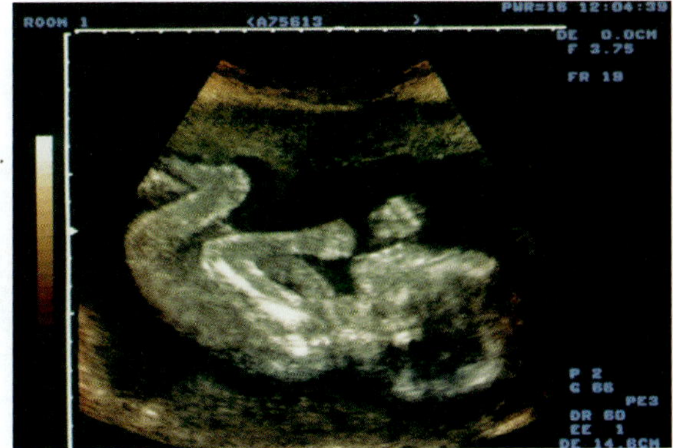

FIGURE 6.5
A sonogram allows a doctor to "see" inside a patient. Here, a developing fetus in the womb is visible.

Q. P-type seismic waves have velocity of about 10 km-s^{-1}. About how long does it take such a wave to travel straight through the Earth from surface to core to the opposite surface?

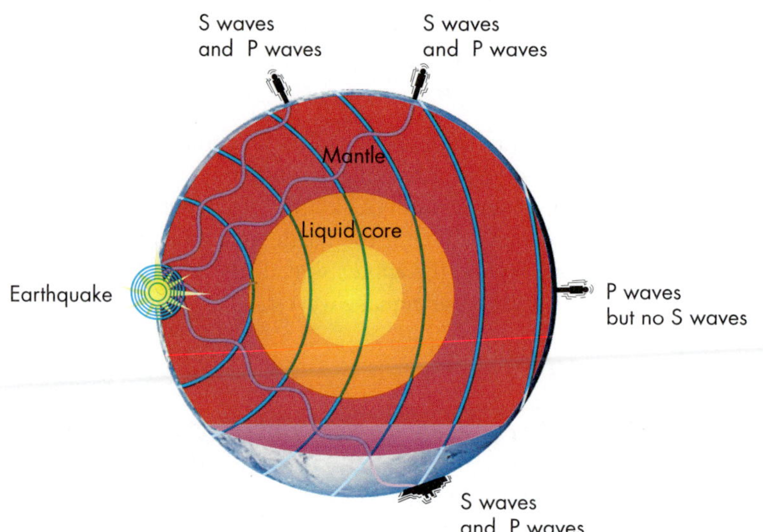

FIGURE 6.6
P and S waves move through the Earth, but the S waves cannot travel through the liquid core.

detects P waves but no S waves, the seismic waves must have encountered a region of liquid on their way from the earthquake to the detecting station (fig. 6.6), an indication that the Earth has a liquid interior. Observations show precisely this effect, from which we infer that the Earth has a liquid core. More complicated analyses can then reveal the density of the material and give clues to its composition.

Seismic studies show that the Earth's interior has four distinct regions. The surface layer is a solid, low-density **crust** about 20 to 70 kilometers (12 to 43 miles) thick and composed of rocks that are mainly silicates. Beneath the crust is a region of hot, essentially solid rock called the **mantle.** This region is also composed of silicates, the most common of which is the mineral olivine. The mantle extends roughly halfway to the Earth's center and, despite being basically solid, is capable of slow flow when stressed, much the way a wax candle can be bent by a steady pressure.

Beneath the mantle is a region of dense liquid material, probably a mixture of iron, nickel, and perhaps sulfur, called the **liquid** (or **outer**) **core.** At the very center is a **solid** (or **inner**) **core,** probably also composed of iron and nickel. Figure 6.7 illustrates these different layers and their relative sizes.

You can see from this discussion that the Earth's interior structure is layered so that the dense heavier materials (the iron and nickel) are at the center and the lower-density lighter materials (silicates) are near the surface in the crust and mantle. Scientists call this separation of materials by density **differentiation.**

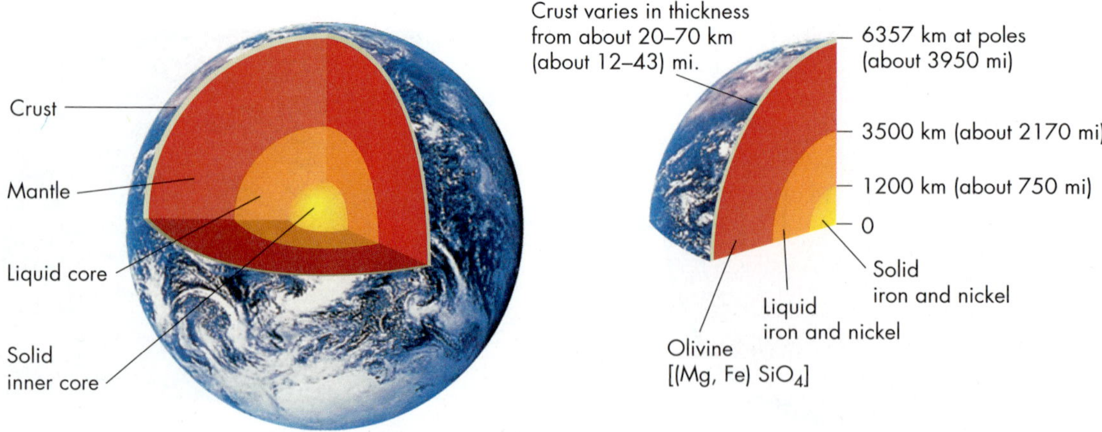

FIGURE 6.7
An artist's view of the Earth's interior.

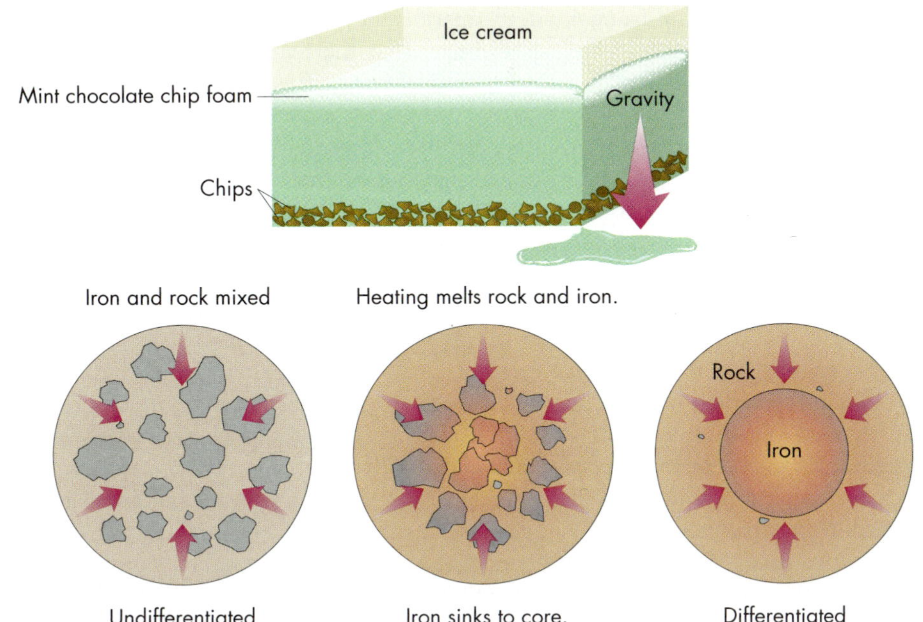

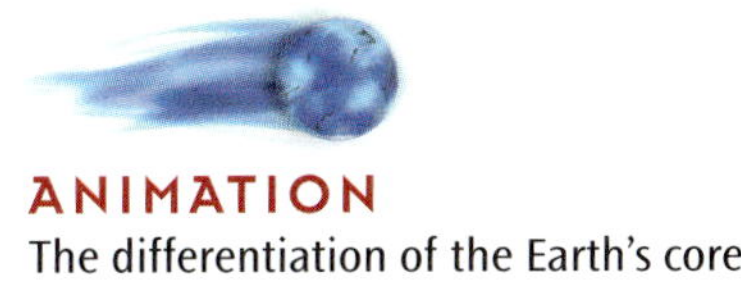

ANIMATION
The differentiation of the Earth's core

FIGURE 6.8
Melting ice cream "differentiates" as the dense chocolate chips sink to the bottom of the carton. So too, melting has made much of the Earth's iron sink to its core.

Differentiation can occur if a mixture of high- and low-density material melts, allowing the dense substances to sink and the lighter ones to rise. You have seen differentiation at work if you have ever had the misfortune to have a half gallon of mint chocolate chip ice cream melt. When you open the box, you find all the chips have sunk to the bottom and the air in the ice cream has risen to the top as foam. Because the Earth is differentiated, we can infer that it must have been almost entirely melted at some time in the past (fig. 6.8).

You may be puzzled as to why there is a solid core inside a liquid core at the Earth's center. If it is hot enough to melt part of the interior of the Earth, why is the very center not liquid as well? The solid core is not cooler, but rather it has a higher melting point because it is under greater pressure. At very high pressures, a previously melted material may resolidify. You can understand why this happens in the following way. For a solid to form, the atoms composing it must be able to link up to their neighbors to form rigid bonds. Heating makes the atoms move faster and breaks the bonds between neighboring atoms. With the bonds broken, the solid has nothing to hold it together, and so it becomes liquid. However, if the material is highly compressed, the atoms may be forced so close together that, despite the high temperature, bonds to neighbors may hold and keep the substance solid.

The compression needed to solidify the Earth's inner core comes from the weight of the overlying material. The thousands of miles of rock above the Earth's deep interior generate an enormous pressure there. To help visualize that pressure, imagine what it would feel like to have a pile of cinderblocks a mile high put on your stomach. In the Earth's core, the huge pressure squeezes what would otherwise be molten iron into a solid.

Heating of the Earth's Core

The Earth's interior is much hotter than its surface (a fact that figures in folklore and theology!). Anyone who has ever seen a volcano erupt can hardly doubt this. In fact, the rise in temperature as you move deeper into the Earth can be measured easily in deep mines where air conditioning must be used to create a tolerable working environment. Near the surface, the temperature rises about 2 K every 100 meters you descend. If this temperature increase continued at this same rate all the way to the center, the Earth's core would be a torrid 120,000 K. However, by measuring the

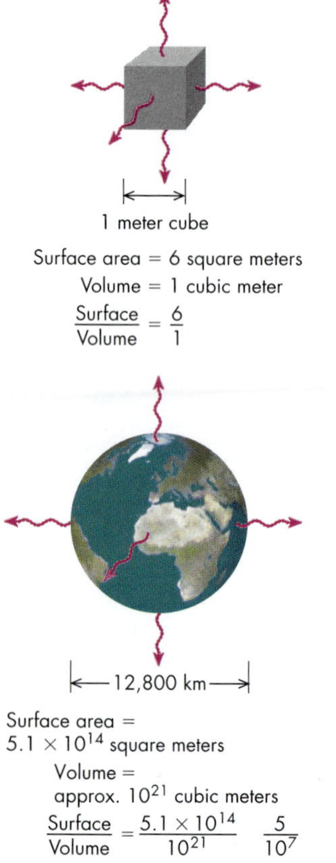

FIGURE 6.9
Heat readily escapes from small rocks but is retained in larger bodies.

amount of heat escaping from the deep interior and from laboratory studies of the properties of heated rock, geologists estimate that the temperature in the core of the Earth is "only" about 6500 K (hotter than the Sun's surface). What makes our planet's core so hot?

Scientists think that the Earth was born hot. According to this theory, the Earth formed from many smaller bodies drawn together by their mutual gravity. As each body hit the accumulating young Earth, the impact generated heat.* When the bombardment stopped, the Earth's surface cooled, but its interior has remained hot. That is, the Earth has behaved much like a baked potato taken from the oven, cooling on the outside but remaining hot inside because heat leaks only slowly from its interior to its surface.

According to another theory, even if the Earth were cold at its birth, it has heated up since its formation, with the heat being supplied by small amounts of natural radioactivity in the Earth's interior. That is, the Earth generates heat much as a nuclear reactor does. All rock contains trace amounts of naturally occurring **radioactive elements** such as uranium. A radioactive element is one that breaks down into another element by ejecting a subatomic particle from its nucleus, a process called **radioactive decay.** Radioactive decay releases energy, generating heat. If that heat is created in a small piece of rock at the Earth's surface, it simply escapes into the surroundings. Thus, the rock's temperature does not increase. However, in the Earth's deep interior, the heat is trapped by the outer rocky layers, slowing its escape. The amount of heat lost depends on the surface area, but the amount of heat contained depends on the volume. Because a smaller body has proportionately a much larger surface area compared to its volume than a larger body, the smaller body cools quickly, but the larger one remains hot (fig. 6.9). Thus, with the heat trapped, the temperature of the Earth's interior gradually rises and the rock eventually melts.

Scientists are unsure which of the above heating mechanisms is more important.† In fact, many scientists now think that the most important effect of radioactive heating is that the extra heat it supplies simply slows our planet's cooling. But regardless of its exact role in heating the Earth, radioactivity in rocks is important because it gives scientists a powerful tool for measuring our planet's age.

6.3 The Age of the Earth

One can find the age of a rock sample by measuring the amount of radioactive material it contains. As the rock ages, its radioactive atoms decay into so-called **daughter atoms.** For example, uranium decays into lead, and radioactive potassium decays into calcium and the gas argon. The more daughter atoms a rock contains relative to the original number of radioactive atoms, the older the rock is. For instance, suppose we have a rock sample containing both potassium and argon trapped within the rock crystals. To apply the method, we assume that when the rock formed from molten material, any argon—being a gas—in it escaped.

Suppose a rock sample happened to contain 100,000 atoms of radioactive potassium and no argon when it solidified, as illustrated in figure 6.10. Laboratory studies of radioactive potassium show that it decays into calcium and argon at a constant rate such that half the potassium present at a particular time will decay within the next

*You can demonstrate that impact generates heat by hitting a small piece of metal repeatedly with a hammer and then feeling the metal. It will be warmer than it was before you began hitting it. For the forming Earth, the impacting objects act as the hammer, and gravity is the force that drives them onto the young Earth. Technically, therefore, impact heating is release of gravitational potential energy.

† Some recent proposals suggest that chemical changes during the formation of the Earth's core may also have been important heat sources.

1.28 billion years.* About 90% of the decaying potassium atoms turn into calcium, the other 10% turn into argon. Neither the argon nor the calcium is radioactive. Although argon is the rarer of the daughter atoms, it is easier to identify as a decay product, so we will ignore the calcium from now on in this discussion.

The argon is trapped within the rock unless the rock melts, and so at the end of 1.28 billion years, the sample will contain 50,000 potassium atoms. The other 50,000 potassium atoms have decayed into 45,000 calcium atoms plus 5,000 argon atoms (10% of the 50,000). The argon/potassium ratio at this point will be 5,000/50,000 = 0.1. In another 1.28 billion years, half the surviving potassium will decay, leaving 25,000 potassium atoms and creating another 2,500 argon atoms (again, 10% of the decays). The argon/potassium ratio will now be (2,500 + 5,000)/25,000 = 0.3. The ratio of argon to potassium in the rock therefore changes with time from 0, when it solidified, to 0.1, after 1.28 billion years, and so on. This ratio gives us the age of the rock sample, as figure 6.10 shows.

From such studies, scientists have found that the oldest rocks on the Earth have an age of over 4 billion years. These ancient rocks are found in such diverse places as northern Canada, southern Africa, and Australia. Thus, the Earth must be at least 4 billion years old.

However, scientists think that the Earth is even older, because rock samples from other Solar System bodies, such as the Moon and asteroids (fragments of which fall to the Earth as meteorites), have ages of a little more than 4.5 billion years. Moreover, some small mineral crystals within old Earth rocks can be individually dated to about 4.4 billion years old. In addition, based on other lines of evidence the Sun also appears to be about this old, as we will see in chapter 14. All of this evidence suggests a common age for the Solar System of nearly 4.6 billion years, an age that we will also take to be the Earth's. Why, then, are there no rocks this old on Earth? They were probably destroyed by processes we will discuss in the next section.

An age of 4.6 billion years is immense. To illustrate, if those billions of years were compressed into a single year, all of human existence would be a mere 3 hours, all human recorded history would have happened during the last minute of the year, and a human life span would be less than 1 second.

The brevity of human life compared to the vast age of the Earth prevents us from seeing how dynamic our planet is. Mountains and seas appear to us permanent and unchanging, but even they change over the vast epochs Earth has existed. Such changes have their ultimate cause in the heat slowly flowing from Earth's interior, a flow that creates motions in the Earth's interior and crust.

---

*Technically, the time it takes for half of a radioactive element's atoms to decay is called its half-life. Thus, 1.28 billion years is the half-life of this form of potassium.

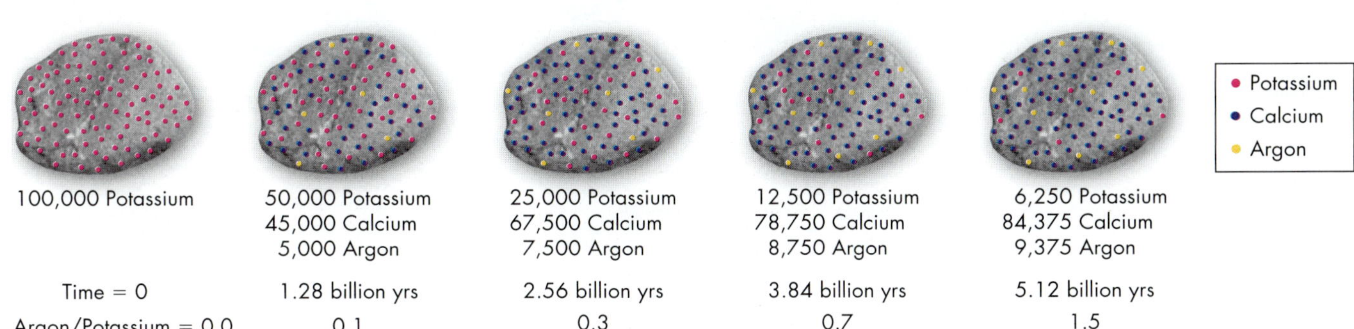

FIGURE 6.10
The amount of argon compared to potassium in a sample of rock gives information about the rock's age.

6.4 Motions in the Earth's Interior

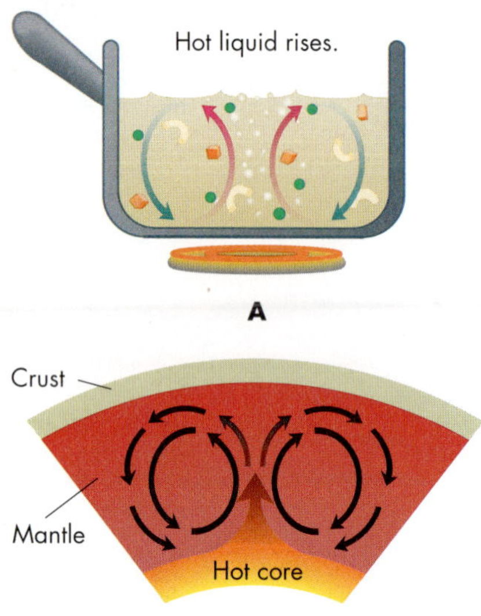

FIGURE 6.11
Examples of convection. (A) When the soup in a pan is heated on a stove, the heated liquid drops in density, rises, cools at the surface, then sinks. (B) Convection occurs in the Earth's mantle, but over vastly longer times.

Heat in the Earth's interior, whether left over from our planet's birth or generated by radioactive decay, creates movement of the material inside the Earth. In fact, heating often causes motion, as you can see by watching a pan of soup on a hot stove. If you look into the pan as it heats, you will see some of the soup, usually right over the burner, slowly rising from the bottom to the top, while some will be sinking again (fig. 6.11A). Such circulating movement of a heated liquid or gas is called **convection.**

Convection in the Earth's Interior

Convection occurs because heated matter expands and becomes slightly less dense than the cooler material around it. Being less dense, it rises, the basis on which a hot-air balloon operates. As the hotter material flows upward, it carries heat along with it. Thus, convection not only causes motion but also carries heat.

Convective motions in a pot of soup are easy to see—here a lima bean rises, there a noodle sinks. Such motions are less obvious in the Earth. Our planet's crust and mantle are not bubbling and heaving like the soup; rather, they are solid rock. Nevertheless, when rock is heated, it too may develop convective motions, though they are very, very slow and therefore difficult to observe. Despite its slowness, the results of convective motion are evident around us. They create such diverse phenomena as earthquakes, volcanoes, the Earth's magnetic field, and perhaps even the atmosphere itself.

Plate Tectonics

Deep within the Earth, hot molten material rises in great, slow plumes. When such a plume nears the crust, it spreads out and flows parallel to the surface below the crust (fig. 6.11B). There, the hot material drags the surface layers, shifting and stretching the crust in a process called **plate tectonics.** In some places, the stretching breaks the crust apart in a process called **rifting** (fig. 6.12A). Molten rock rises into these rifts, producing long ridges along ocean floors, almost resembling the seams on a baseball

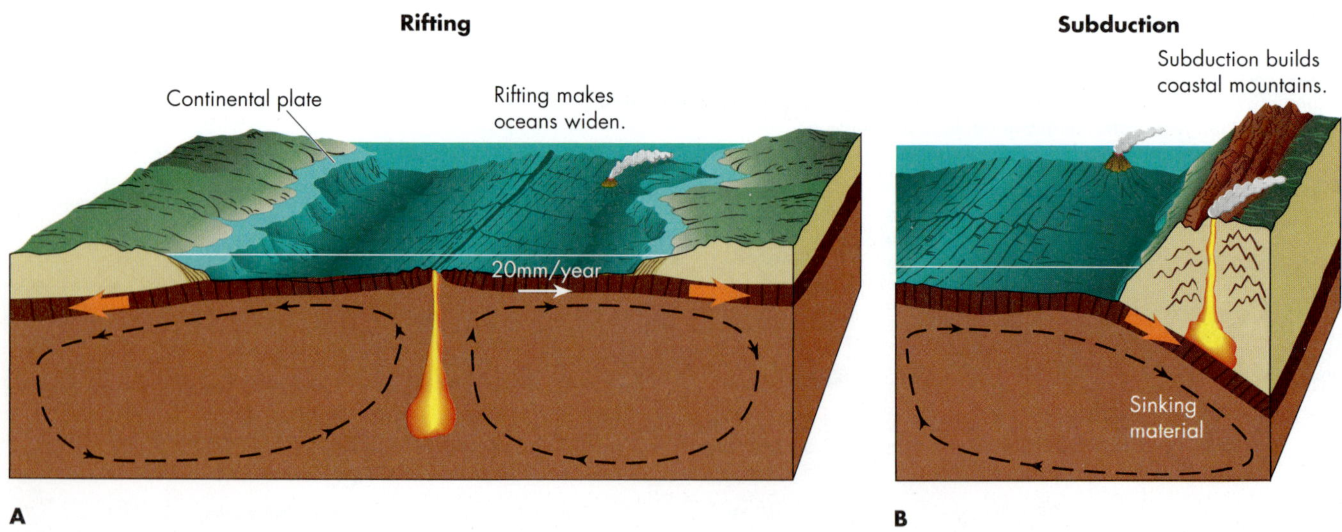

FIGURE 6.12
(A) Rifting may occur where rising material in the mantle nears a planet's surface. (B) Subduction builds mountains where material sinks back toward the interior of the Earth.

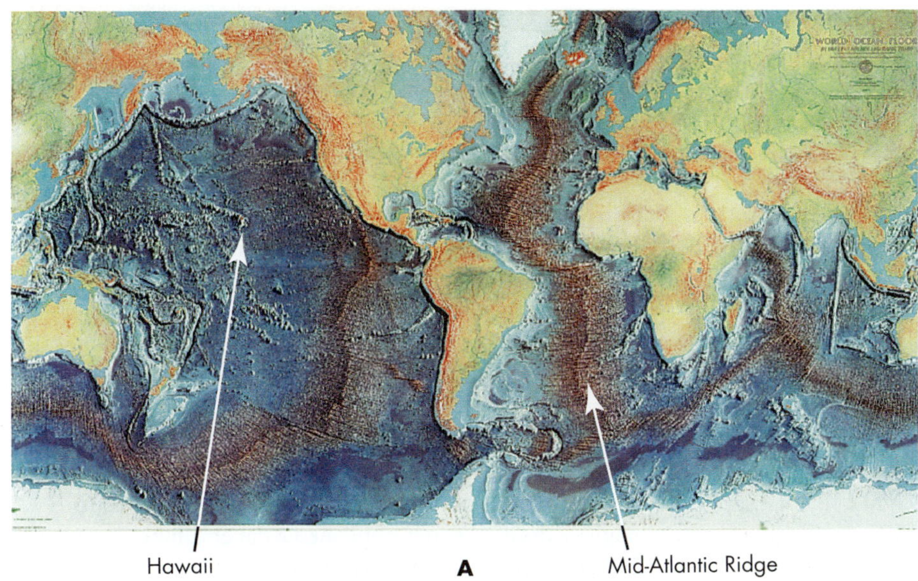

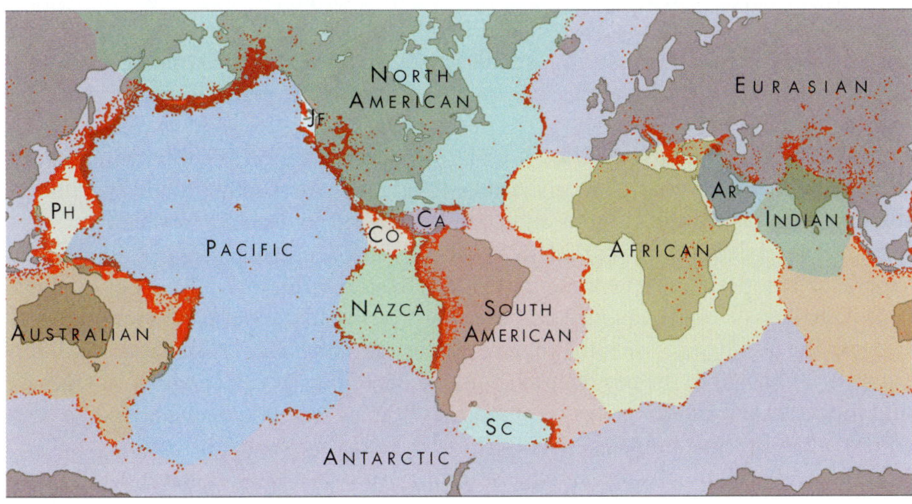

FIGURE 6.13
(A) A map of the Earth's ocean floors. Rifting occurs along long ridges on the ocean floor, where plates are moving apart, such as the Mid-Atlantic Ridge. Deep trenches occur where one plate is forced down under another, which can be seen around much of the rim of the Pacific Ocean.
(B) Locations of earthquakes (red dots) identify most plate boundaries. Smaller plate names are abbreviated: Cocos (Co), Caribbean (Ca), Juan de Fuca (Jf), Arabia (Ar), Philippines (Ph), and Scotia (Sc).

Source (A): Copyright by Marie Tharp 1977/2003. Reproduced by permission of Fiona Schiano-Yacopino of Marie Tharp Maps, LLC, 8 Edward Street, Sparkill, NY 10976.

(figure 6.13A). Gradually, the hot material loses its heat to the outside and then cools and sinks downward again to be rewarmed and rise once more.

Where cool material sinks, it may drag pieces of continental crust together, buckling them upward where they collide to form mountain ranges such as the Rockies and Andes along the western coasts of North and South America. Sometimes one piece of crust may slip beneath another and be pushed under it in a process called **subduction** (fig. 6.12B). Rifting and subduction are the dominant forces that sculpt the landscape of our planet and contribute to Earth's uniqueness in the Solar System. These motions also trigger earthquakes and cause volcanoes. For example, friction may temporarily stop two plates from sliding past each other, causing them to stick. Pressure may then build until the rock breaks, freeing the stuck plates and generating a sudden lurch in the crust, an earthquake. A map of the Earth showing the locations of such quakes (fig. 6.13B) helps us see the shapes and locations of the plates.

The shifting of large blocks of the Earth's surface used to be called *continental drift*, but contemporary geologists prefer the term *plate tectonics*. The term *plate* is used because the pieces of the Earth's crust that move are very thin (perhaps only 50 kilometers deep) but many thousands of kilometers across. Plate motion is a little like the movement of the crust on bubbling oatmeal. The crust breaks apart where hot oatmeal bubbles up and collects where the slightly cooler oatmeal sinks. Some of the more important plates are those that contain the continents, such as the North American and African plates, as seen in figure 6.13B. The ocean floors also are composed of plates, such as the Pacific and Caribbean plates.

The development of plate tectonic theory is an interesting example of how science works. As early as 1596, Abraham Ortelius, a geographer from Antwerp, in what is now Belgium, noticed that maps depicting the new discoveries of the period showed that the coastlines of South America and Africa approximately matched like two pieces of a giant jigsaw puzzle. In 1858 a French scientist, Antonio Snider-Pellegrini, also remarked how similar the coastlines were and noted that fossils found at matching locales on both sides of the Atlantic were also very alike. He conjectured that the continents had broken apart, creating the Atlantic Ocean in the opening rift between, but apart from the similarities in fossils, he offered little in the way of supporting evidence for his idea. Similarly, in 1910 the American geologist F. B. Taylor published a paper proposing that South America and Africa had once been joined, but he too offered only slight evidence supporting his hypothesis. The first significant evidence came in 1912, when the German meteorologist Alfred Wegener published a paper called "The Origin of the Continents," in which he developed the modern theory.

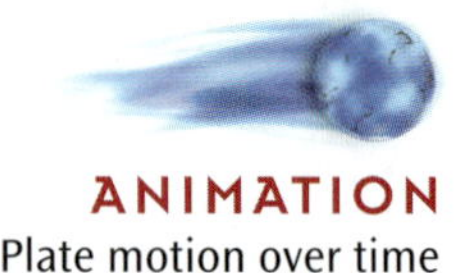

ANIMATION
Plate motion over time

Wegener proposed that all the continents were originally assembled in a single supercontinent, which he called Pangea (literally "all-Earth"). For reasons that are still obscure, Pangea began to split into smaller plates that became the familiar continents of today (fig. 6.14), taking about 250 million years for the plates to move into their present locations.

Although he had amassed fossil and geological evidence to support his theory, Wegener's ideas were not well received at first. In fact, there are still problems in understanding what drives the motions and causes a new rift to develop. Nevertheless, over the last several decades, the cumulative effect of many separate pieces of evidence that support the model of plate tectonics have made the theory almost universally accepted. In fact, geologists now recognize that there was a long history of plate motion before Pangea. The low-density rock that makes up the continents has broken apart in different ways far back into the remote past. Some of the highly eroded mountain ranges we find in scattered locations today were produced by the collision of ancient plates with completely different configurations of the material that makes up the continents.

Today we can directly measure the shifting positions of the continents using global positioning system (GPS) satellites. The continents can be seen to move up to about 10 centimeters per year relative to each other. Our world is literally changing beneath our feet, growing new crust at mid-oceanic ridges and devouring it at subduction zones. But this devoured rock is not lost. As it is carried downward, it is heated, and eventually its lower density causes it to rise again toward the surface. These motions within the mantle gradually change the face of the Earth, but the circulation within the core gives rise to another phenomenon—a magnetic field.

6.5 The Earth's Magnetic Field

Many astronomical bodies have magnetic properties, and the Earth is no exception. The English natural philosopher William Gilbert (1540–1603) was the first to appreciate that the Earth acts like a magnet, though the ancient Chinese had used the Earth's magnetism in their invention of the compass many hundreds of years earlier.

Magnetic forces are communicated by what is called a **magnetic field.** Although some forces are transmitted directly from one body to another (as when two billiard balls collide), other forces, such as gravity or magnetism, need no such direct physical link.

Magnetic fields are often depicted by a diagram showing **magnetic lines of force.** Each line represents the direction in which a tiny compass would point in response to

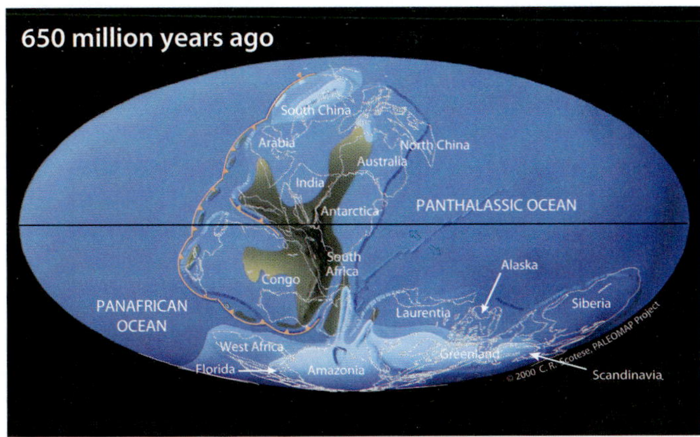

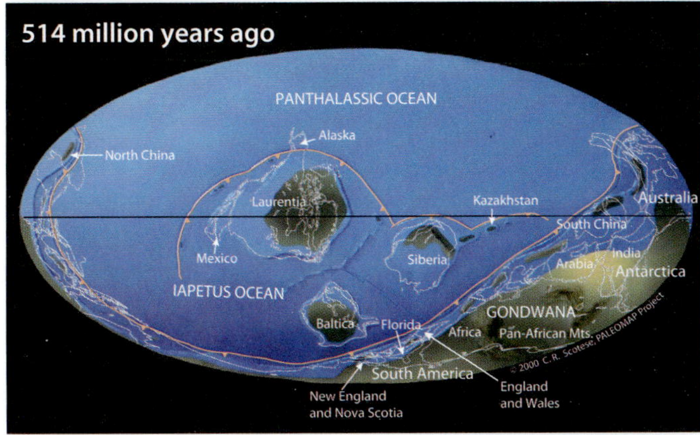

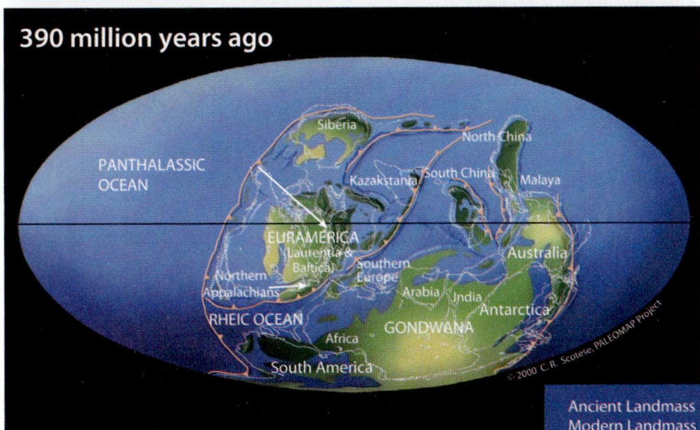

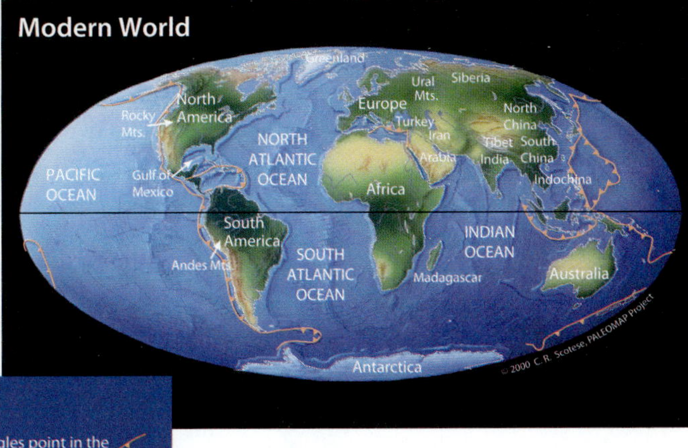

Ancient Landmass
Modern Landmass
Subduction Zone (triangles point in the direction of subduction)
Sea Floor Spreading Ridge

FIGURE 6.14
Continental masses have shifted throughout the Earth's history because of plate tectonics. The changes that geologists can trace over the last 650 million years (just 14% of Earth's history) are illustrated above. Most of the continental material was joined in a single large continent called Pangea several hundred million years ago, but this began to break apart to form the modern continental features about 250 million years ago.

Q. 94 Million years ago, large portions of North America were under water. What else back then was different from today that could explain this?

the field. The concentration of lines indicates the field's strength, with more lines implying a stronger field. For example, the field lines of an ordinary toy magnet emanate from one end of the magnet, loop out into the space around it, and return to the other end. The Earth's magnetic field has a similar shape, as represented in figure 6.15.

Magnetic fields have an important property called **polarity.** Polarity gives field lines a direction, and so they always start at a north magnetic pole and end at a south magnetic

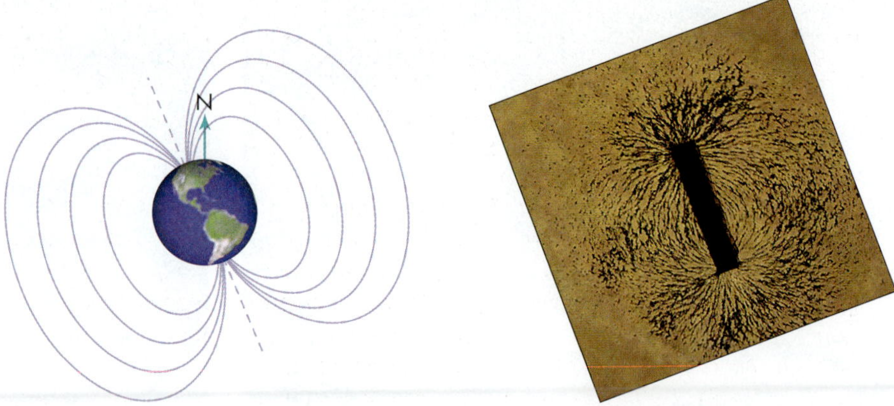

FIGURE 6.15
Schematic view of Earth's magnetic field lines and photograph of iron filings sprinkled on a toy magnet, revealing its magnetic field lines. "N" indicates the Earth's rotational North Pole. The magnetic north pole lies fairly close to the rotation axis, but its orientation changes over time.

pole. Thus, all magnets must have both a north pole and a south pole. The existence of north and south poles allows magnets to either attract or repel. Two north poles or two south poles repel each other, but a north and a south pole attract. A compass works on this principle. Its needle is a magnet, and its north pole is attracted to the Earth's magnetic south pole, and its south pole is attracted to the Earth's magnetic north pole.

An important question about any field is, How is it generated? Gravitational fields are generated by masses. Magnetic fields are generated by electric currents, either large-scale currents or currents on the scale of atoms. You can easily demonstrate the former by wrapping a few coils of insulated wire around an iron nail and attaching the wire ends to a battery. The nail will now act like a magnet and be able to pick up small pieces of iron and will deflect a compass needle. Microscopic electric currents on the atomic scale create the magnetism of toy magnets.

Origin of the Earth's Magnetic Field

The magnetic field of the Earth is generated by electric currents flowing in its molten iron core. Scientists are still unsure how these currents originate but hypothesize that they develop from a combination of rotational motion and convection.* Studies of the magnetic fields of other Solar System bodies support this view. For example, bodies with weak or no magnetic fields, such as the Moon and Venus, are either too small to have a large convecting core or rotate very slowly. On the other hand, bodies with large magnetic fields, such as Jupiter and Saturn, rotate very rapidly and have very hot cores.

The Earth's magnetic poles do not coincide with the poles determined by its rotation axis (true north and south). Therefore, a compass needle does not in general point true north but, instead, several degrees away to what is called "magnetic north." Both the position and the strength of the Earth's magnetic poles change slightly but measurably, from year to year, even reversing their polarity about every 250,000 years on average. Thus, at some time in the future, a compass that now points north will point south. These changes in our planet's magnetic field probably come from irregular motions of the molten iron swirling in the Earth's core, and they are recorded in the rocks on the ocean floor (see the Extending Our Reach box on the Earth's magnetic fields).

INTERACTIVE
Planetary variations

Magnetic Effects on Cosmic Particles

The Earth's magnetic field does more than make a compass work. It partially screens us from electrically charged particles emitted by the Sun and even more energetic particles called *cosmic rays* produced during violent cosmic events, such as the explosion of a dying star. Many of these particles are energetic enough to damage living cells and are therefore potentially harmful to us. The Earth's magnetic field protects us because when charged

*Astronomers call this process a magnetic dynamo.

EXTENDING our reach

MEASURING REVERSALS OF THE EARTH'S MAGNETIC FIELD

When molten rock cools, magnetic minerals within the rock align to the direction of the Earth's magnetic field much as a compass needle does. This leaves rocks slightly magnetic after they solidify, and the orientation of Earth's magnetic field at the time the lava solidified is frozen into them.

In the early 1900s, geologists studying ancient lava flows discovered that Earth's magnetic field was sometimes oriented in the opposite direction—its polarity was reversed—so a compass would point in the opposite direction from what we currently expect.

In the 1960s, geologists discovered that the ocean floor showed stripes of opposite magnetic polarity running parallel to the mid-ocean ridges (see Box figure 6.2). The theory of plate tectonics explains how this occurs. New crust is created at the ridge and spreads away from it like paper peeling off a roll. The molten rock cooling along the ridge becomes magnetically aligned with the Earth's magnetic field at the time it emerges. This preserves a recording of Earth's magnetic field going back millions of years.

By dating the rock on the ocean floor, geologists can determine how fast the crust has moved and how often the magnetic field has reversed. One can calculate the speed of the spreading motion by dividing a rock's distance from the ridge by its age. For example, if rocks 50 kilometers from the ridge are 5 million years old, the plates must have shifted at an average speed of 50 kilometers per 5 million years, or 1 centimeter per year, a fairly typical plate speed.

The magnetic reversals occur erratically. Sometimes the orientation remains the same for tens of millions of years, but at other times just thousands of years. On average, it reverses once every 250,000 years. The current orientation has persisted for 780,000 years. It is suspected that during a reversal, we would lose much of our shielding from cosmic rays, but our ancestors have survived this many times before. It may, however, be a bigger problem for our satellites and technologies.

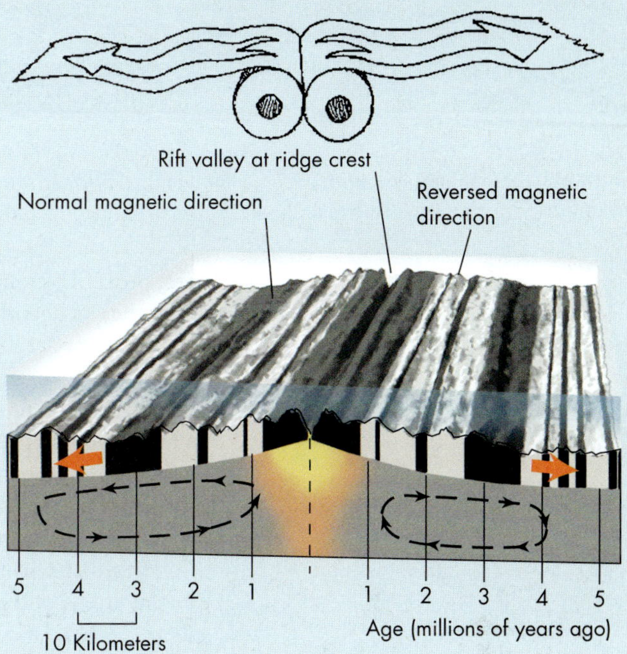

BOX FIGURE 6.2
Magnetic reversals recorded in the ocean floor. As plates spread apart, the upwelling lava cools and records the direction of the magnetic field at that time. Portions of the ocean floor having the same magnetic orientation as we have currently are shown in black, while those with reversed orientation are in white.

particles encounter it, they are deflected into a spiraling motion around the field lines (fig. 6.16). This diverts the particles streaming from the Sun and space, causing many of them to flow toward the polar regions.

The magnetic field traps some of these charged particles in two doughnut-shaped regions called the **Van Allen radiation belts** (fig. 6.16). The particles trapped in the Van Allen belts are energetic enough to penetrate spacecraft and could be a hazard to space travelers, damaging their genetic material or other tissue as well as sensitive electronic equipment. Astronauts therefore try to either avoid passing through the belts or go through them as quickly as possible.

As the charged particles flow toward the magnetic poles, they generate electric currents in the upper atmosphere. These currents, circulating around the magnetic poles, drive electrons along the magnetic field lines. The moving electrons spiral around the field lines, colliding with molecules of nitrogen and oxygen. Such collisions excite atmospheric gases, lifting their electrons to higher energy orbitals. As the electrons drop back to lower orbitals, they emit the lovely light we see as the **aurora** (fig. 6.17). The

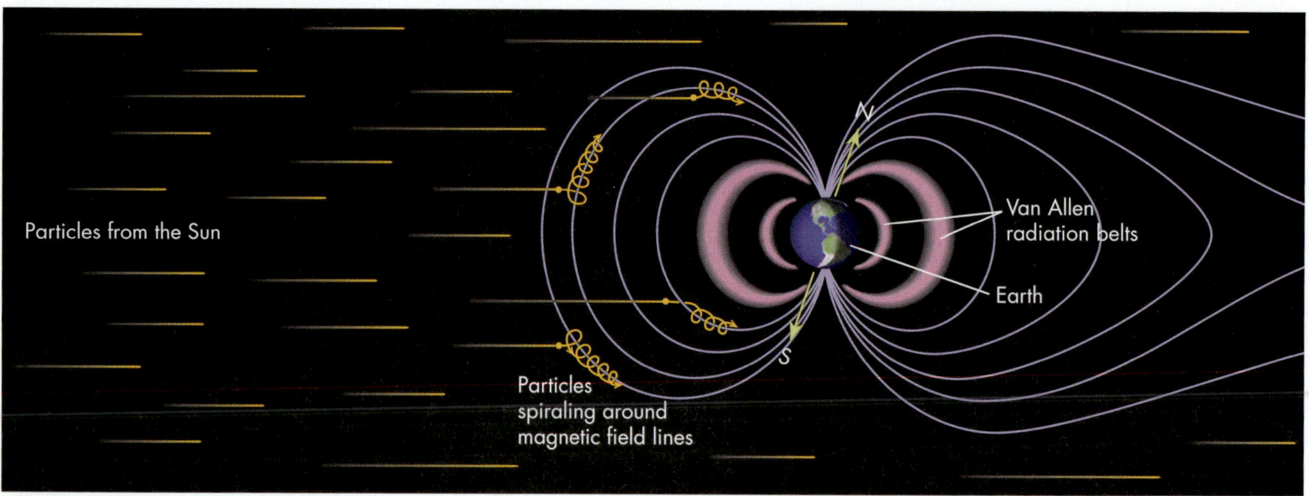

FIGURE 6.16
Electrically charged particles from the Sun spiral in the Earth's magnetic field. Some of these particles become trapped in two regions ranging out to many times the Earth's radius known as the Van Allen radiation belts. "N" and "S" indicate the Earth's North and South poles.

exact process by which the aurora forms is still not completely certain, but there is no doubt that its beautiful streamers are shaped by the Earth's magnetic field.

Once a spaceship is far outside of the Earth's magnetic field, astronauts on board are exposed to a steady stream of energetic particles. This is one of the major concerns for future astronauts traveling to Mars. Astronauts on board a deep-space vehicle will be bombarded by hundreds of times more of these particles than we are exposed to on Earth's surface, greatly increasing their cancer risk over a long mission. Providing shielding that could protect the astronauts may require a spacecraft so massive as to make a Mars mission much more difficult than previously thought.

The most powerful cosmic rays can penetrate the Earth's magnetic field, but fortunately those of us living on Earth's surface have a second line of defense, the atmosphere. Most of a cosmic ray's energy is spent as it slams into molecules in the upper atmosphere. The atmosphere also blocks certain types of electromagnetic radiation, and this plays a critical role in making the Earth a hospitable planet for life.

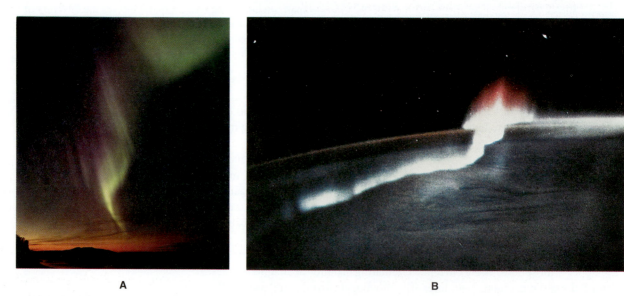

FIGURE 6.17
Photographs of an aurora (A) from the ground and (B) from space.

6.6 The Earth's Atmosphere

Structure of the Atmosphere

Surrounding the solid body of the Earth is a veil of gases that constitutes our atmosphere. Most planets in the Solar System have an atmosphere, but the Earth's has many unique features. The Earth's atmosphere is of interest not just for its unique properties but also for what it can tell us about atmospheres in general.

Our atmosphere extends from the ground to an elevation of hundreds of kilometers, but at the highest altitudes, the air is extremely tenuous. In fact, the density of the atmospheric gases decreases steadily with height, as illustrated in figure 6.18. Gases near the ground are compressed by the weight of gases above them. Thus, the atmosphere is a little like a tremendous pile of pillows. The pillow at the bottom is squashed by the weight of all those above it. Likewise, a block of air near sea level is more compressed and therefore has a greater density than a block of air near the top of the atmosphere. That is why it is so difficult to breathe at 30,000 feet (~10,000 m), where the air is far less dense.*

*One cubic centimeter (a volume roughly the size of the end of your little finger) of the air around you contains about 10^{19} molecules. When you take even a tiny sniff, you inhale a number of molecules roughly comparable to the number of grains of sand in a pile the size of the Astrodome.

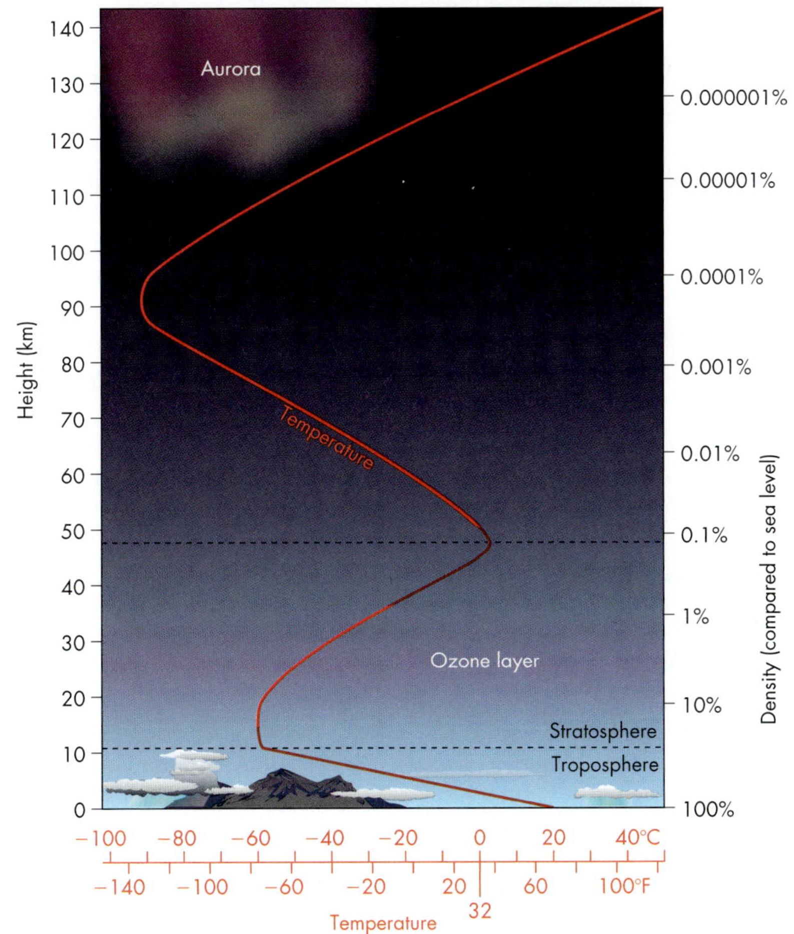

FIGURE 6.18
The Earth's atmosphere becomes steadily less dense with height, but it has a number of layers and varying temperatures, as illustrated in this diagram. The weather we experience is located in the lowest layer, known as the troposphere. Above this is the stratosphere, where most of the ozone is located. The charged particles associated with the aurora occur far above even the stratosphere where the gas is less than one-millionth as dense as at sea level, while the International Space Station orbits about 300 kilometers above the surface.

TABLE 6.2 ATMOSPHERIC GASES

Gas	% of Molecules by Number (excluding water vapor)
Nitrogen (N_2)	78.08
Oxygen (O_2)	20.95
Argon (Ar)	0.93
Water (H_2O)	Variable—typically between about 0.1 and 3.0
Carbon dioxide (CO_2)	0.03
Trace gases (less than 100 parts per million) Neon, helium, ozone, krypton, hydrogen, methane, carbon monoxide, and many pollutants both natural and human-made.	

The lowest layer of the atmosphere is known as the **troposphere.** which extends up to about 12 kilometers. This is the part of the atmosphere most familiar to us, where clouds and airplanes are generally found. Because of compression by the overlying layers, roughly three quarters of the mass of the atmosphere is within the troposphere. Above the troposphere is the **stratosphere,** a region particularly important to us because of the protective ozone layer located there. Above this are other, even less dense layers gradually merging with the near vacuum of interplanetary space.

The very rarified gas located far above Earth's surface has a cumulative effect on spacecraft in low orbits, slowing them down and gradually causing them to spiral into the denser lower regions where they burn up. Even the International Space Station, orbiting about 330 kilometers (about 180 miles) above the surface requires periodic boosts from docked space shuttles to keep from spiralling back down to the surface.

Composition of the Atmosphere

One of the most striking differences between the atmosphere of the Earth and that of other planets is its composition. For example, the atmospheres of Mars and Venus are nearly completely carbon dioxide, while the atmospheres of Jupiter and Saturn are mostly hydrogen and helium. On the other hand, the atmosphere of the Earth is primarily a mixture of nitrogen and oxygen. Nitrogen molecules make up about 78% of our atmosphere's gas and oxygen about 21%. The remaining 1% includes carbon dioxide, ozone, and water, gases crucial for protecting us and making life possible. Table 6.2 lists the main gases in our atmosphere.

It may surprise you to see water listed as a gas. But even at low temperatures liquid water evaporates into individual water molecules that mingle with the other gases in our atmosphere. We call such free water molecules *water vapor.* Water vapor is almost completely concentrated in the troposphere. Being a lightweight molecule, air rich with water vapor—humid air—tends to rise until its temperature gets so low that the water condenses, forming clouds of tiny ice crystals and water droplets, and rain that falls back to the surface (fig. 6.19). The circulation of water from liquid to gaseous and back to liquid form drives much of our weather. It has also played an important role in removing carbon dioxide from our atmosphere. Because carbon dioxide dissolves in water, rain scrubs the air of carbon dioxide. In fact, there is about 50 times more carbon dioxide dissolved in Earth's oceans than is present in the atmosphere.

FIGURE 6.19
In our atmosphere, puffy cumulus clouds form when the Sun heats the ground and warms the air, making it rise. When the air reaches a low enough temperature, water vapor condenses into small droplets, forming a visible cloud.

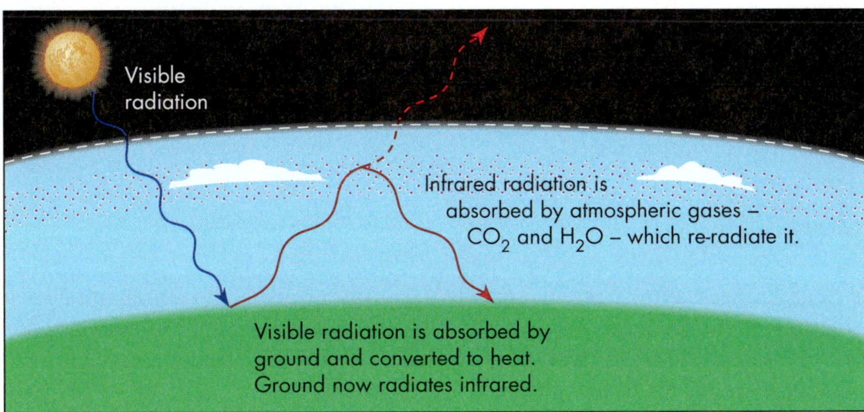

FIGURE 6.20
The greenhouse effect. Radiation at visible wavelengths passes freely through the atmosphere and is absorbed at the ground. The ground heats up and emits infrared radiation. Atmospheric gases absorb the infrared radiation and warm the atmosphere, which in turn warms the ground.

Q. The red light waves in this figure are drawn with longer "wiggles" than the blue light waves. Why might this have been done?

The circulation of our atmosphere is similar to the circulation of rock in the Earth's mantle, in that both are convection processes driven by heating from below. However, in the atmosphere, the heating originates not from the interior of the Earth, but from sunlight.

The Greenhouse Effect

The transparency of the Earth's atmosphere to visible radiation allows sunlight to enter the atmosphere and reach the Earth's surface. The energy of the photons is absorbed by molecules in the surface and converted to heat. The warmed surface radiates infrared energy, but the atmosphere is not very transparent at infrared wavelengths. This reduces the heat loss and makes the surface warmer than it would be if the infrared energy could escape freely, a phenomenon illustrated in figure 6.20 and known as the **greenhouse effect.**

You can get some idea of how effectively water vapor traps heat by noticing how the temperature drops dramatically at night in desert regions or on clear nights. All gardeners know that it is clear nights (with no clouds and little water vapor) that are most likely to have frost. On humid or cloudy nights, heat is retained.

It is important to recognize that the greenhouse effect does not generate heat; rather, it limits the heat loss to space. The greenhouse effect therefore warms the Earth the same way a blanket warms you. The blanket doesn't make you generate more heat; it simply slows down the loss of heat already there. Likewise, the water and carbon dioxide do not create heat of their own; they simply slow down the loss of heat from the ground by absorbing the infrared energy. Eventually they re-emit it, but much of it is re-emitted back down toward the ground so it is not lost into space. That extra infrared energy reradiated to the ground helps keep the surface warm at night.

We can see how important our atmosphere is as a heat blanket by comparing the temperature of the Earth and the Moon. Although they are the same distance from the Sun, the average temperature of the Earth is much higher than the average temperature of the Moon. When averaged over seasons and latitude, the Earth's average temperature is 59° F (15° C), whereas the Moon's average temperature is a frigid −4° F (−20° C). A certain amount of greenhouse warming is critical for making Earth a habitable planet, but many scientists are concerned that humans are adding so much carbon dioxide to the atmosphere that we might trap so much heat that Earth's temperature will climb—a process called **global warming.**

Ironically, an atmosphere's trapping heat by blocking infrared energy is not what makes an actual greenhouse warm. Air inside a greenhouse is warmer than air outside because the greenhouse confines the air within it, preventing it from rising and cooling the way outside air can. Accordingly, some scientists prefer the term *atmosphere effect* rather than greenhouse effect.

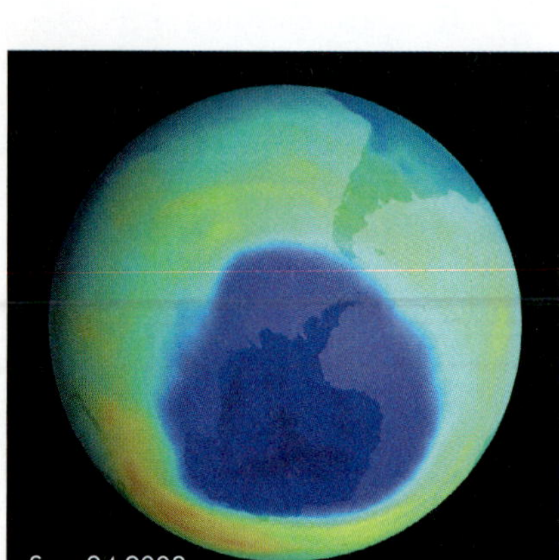

FIGURE 6.21
The Antarctic ozone hole in September 2003. The colors in this image indicate the amount of ozone above the Earth's surface. The violet colors over Antarctica indicate only about one-third the normal amount of ozone.

The Ozone Layer

The oxygen in our atmosphere is important to us not only for breathing but also as a vitally protective blanket that shields us from harsh solar ultraviolet radiation. Some of that shielding is provided by O_2 (the normal form of oxygen), but much of it comes from another molecular form of oxygen, O_3, or **ozone.**

Most of the ozone in our atmosphere is located in the ozone layer* at an altitude of about 25 kilometers (80,000 feet) in the stratosphere. Ozone is formed because at these upper levels solar ultraviolet radiation splits O_2 into individual oxygen atoms.† The splitting occurs because the ultraviolet radiation makes the molecule vibrate so energetically that it flies apart. These individual oxygen atoms then combine with other O_2 molecules to form O_3.

Ozone is important because it is a strong absorber of ultraviolet radiation; without the ozone layer, solar ultraviolet radiation would pour into the lower atmosphere. The short wavelength (and therefore the high energy) of the radiation would damage living organisms. Without the protective ozone layer, you would get a severe sunburn on exposed skin simply by stepping outside. In fact, it is doubtful that life could exist on the Earth's surface without the ozone layer to shield us.

Concerns about the ozone layer were triggered by studies of the atmosphere over Antarctica. During the long polar winter, atmospheric pollutants build up over Antarctica. Some pollutants destroy ozone, creating an "ozone hole" (fig. 6.21). Ozone levels in the Antarctic spring have been seen to decline by more than half over the last few decades, providing us with a warning of what might happen to the rest of the ozone layer if the amount of ozone-destroying pollutants continued to increase. For this reason, governments around the world are cooperating to limit the use of chemicals that can rise into the stratosphere where they can chemically combine with ozone and destroy it.

Oxygen and ozone are similar to water vapor and carbon dioxide in the sense that all are gases that can absorb radiation that might otherwise pass through our atmosphere. This may explain some of the common confusion between the problems of global warming and ozone depletion. Both problems are related to how humans are changing the atmosphere, but they are caused by different gases in different layers of the atmosphere. It is important to keep in mind that ozone blocks ultraviolet light from the Sun as it is entering the Earth's atmosphere. By contrast, greenhouse gases block infrared radiation from the Earth on its way out of the atmosphere.

Origin of the Atmosphere

ANIMATION
The origin of Earth's atmosphere by volcanoes, comet impacts, and planetesimal collision

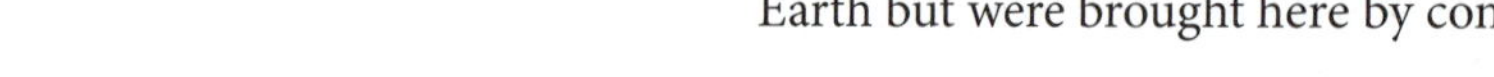

Astronomers have proposed several theories to explain how our atmosphere formed. According to one theory, the gases of our atmosphere were originally trapped inside the solid material that eventually became the Earth. When that material was heated—either by volcanic activity (fig. 6.22A) or by the violent impact of asteroids hitting the surface of the young Earth (fig. 6.22B)—the gases escaped and formed our atmosphere.

Recently, some astronomers have proposed a very different explanation to account for our atmosphere. According to this theory, the gases were not originally part of the Earth but were brought here by comets.

*Ozone is, in fact, found throughout our atmosphere, but its strongest concentration is at 25 km.
† Recall that ultraviolet radiation, having shorter wavelengths, is more energetic than visible light.

A

B

C

FIGURE 6.22
Sources of our atmosphere. (A) Volcanic gas vent today. Gas from ancient eruptions built some of our atmosphere. (B) Planetesimals collide with young Earth and release gas—another source of our atmosphere. (C) Comets striking young Earth and vaporizing. The released gases also contributed to our atmosphere.

As we will see in chapter 11, comets are made mostly of a mixture of frozen water and gases. When a comet strikes the Earth, the impact melts the ices and vaporizes the frozen gas. Given a large enough number of impacts, comets could have delivered enough gas to form the atmosphere (fig. 6.22C). We know from the collision of Comet Shoemaker-Levy 9 with Jupiter in July 1994 that comets collide with planets even today. But such collisions were almost certainly far more common billions of years ago, when the Earth was young, because at that remote time the Solar System was full of smaller objects—the pieces from which the planets themselves grew.

In both of these theories, the early atmosphere had a very different composition than the air we breathe today. For example, our planet's ancient atmosphere probably contained far more methane (CH_4) and ammonia (NH_3) than it does now. Although these gases are still abundant in the giant planets such as Jupiter and Saturn, they have all but disappeared from our atmosphere, which is fortunate because both methane and ammonia are poisonous.

What has rid Earth of these noxious gases? Astronomers think that sunlight is responsible. Solar ultraviolet radiation is intense enough at Earth's distance from the Sun to break the hydrogen atoms out of both methane and ammonia, leaving carbon and nitrogen atoms, respectively. The nitrogen and carbon remain behind, supplying at least some of the nitrogen in our atmosphere. The hydrogen, however, gradually escapes into space because Earth's gravity is too weak to hold it. Only huge planets such as Jupiter and Saturn have strong enough gravities to retain their hydrogen and thus preserve the large amounts of methane and ammonia we see there today.

Which of these theories of the origin of our atmosphere—delivered by comet or liberated from Earth's own material—is correct? Scientists have tried hard to test these very different theories. For example, they have studied whether volcanoes erupt enough gas to have supplied our atmosphere and whether the composition of these gases can explain the mix of molecules in the air around us.

It requires a certain nonchalance to walk up to the lip of a bubbling volcano and hold a collecting tube over the edge to sample the foul-smelling exhalation, but when geologists make such dangerous tests, they find that nitrogen, water, and carbon dioxide are added to the atmosphere even now by volcanic eruptions. Thus, the theory can account for several of the gases in our atmosphere. The theory requires, however, a second important test: Can it account for the amount of these gases we see today? Again the answer is yes, if we assume that eruptions have been about as frequent in the past as they are now and that they ejected comparable amounts of gas. Thus, these tests confirm that the gases of our atmosphere might have been released by heating the material from which our planet formed.

Testing the comet delivery theory is difficult because of our lack of precise knowledge about what frozen gases and ices comets contain. The evidence currently available, however, confirms that comet impacts could account for at least part of our planet's atmospheric gases. Thus, scientists remain divided about which theory is correct. Perhaps, as with so many differences of opinion, each side is partly right and some gases came from comets while others came from volcanoes and the heating of rocks during our planet's birth.

Neither of the above hypotheses for the origin of our atmosphere—volcanic exhalations or comet impacts—can account for the large amount of oxygen in our atmosphere. Where, therefore, did that vital ingredient originate? Chemical analysis of ancient rocks, particularly those rich in iron compounds that react with oxygen, shows that our atmosphere once contained much less oxygen than it does today. In fact, over the past 3 billion years, the amount of oxygen in our atmosphere has steadily increased, a rise paralleled by the spread of plant life across our planet. Most scientists therefore agree that the bulk of the free oxygen, which we breathe, was created from H_2O and CO_2 by photosynthesis of ancient plants. This intimate connection between life and the environment of our planet is a fact that we ignore at our peril.

Plants have created most of our oxygen by photosynthesis, but not all. Some has come from water molecules split by solar ultraviolet radiation into hydrogen and oxygen. The lighter hydrogen slowly drifts to the top of the atmosphere and escapes, leaving oxygen behind. This mechanism for adding oxygen to the atmosphere was probably the dominant source of that gas in the early history of the Earth.

6.7 Motions of the Earth

Considering what happens to many of us on an amusement park ride, it is just as well that we are unaware of the Earth's many motions. Our planet spins on its axis, orbits the Sun, is dragged along by the Sun around the Galaxy, and moves through the Universe with the Milky Way (fig. 6.23). We have already discussed how the Earth's rotational and orbital motions define the day and year, and cause seasons. But our planet's motions have other effects. For example, its spin strongly influences winds and ocean currents, and over thousands of years a slow "wobble" of its rotation axis plays a role in the onset of ice ages.

Air and Ocean Circulation: The Coriolis Effect

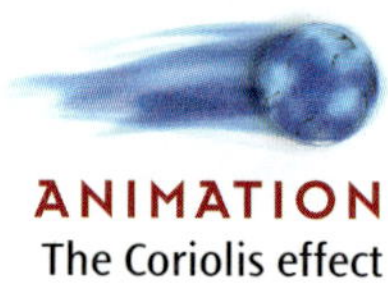

ANIMATION
The Coriolis effect

If you sit with a friend on a rotating schoolyard merry-go-round and toss a ball back and forth, you will discover that the ball does not travel in the direction in which you aim but instead curves off to the side. Similarly, ocean and air currents sweeping across a spinning planet like Earth are deflected from their original direction of motion. This phenomenon is called the **Coriolis effect.**

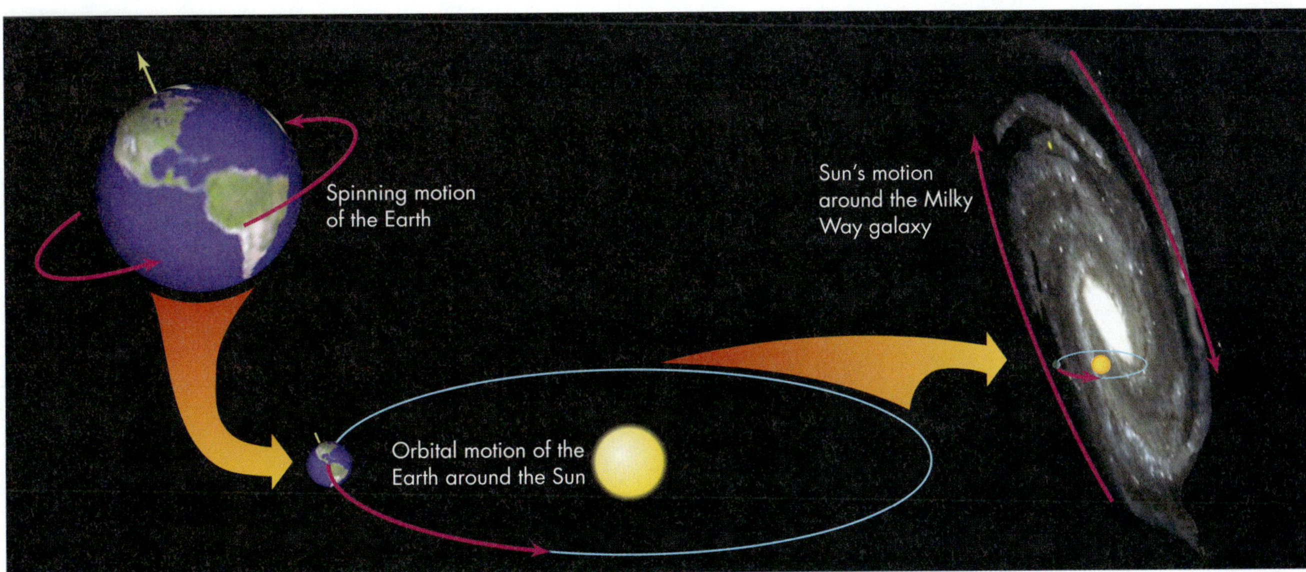

FIGURE 6.23
The Earth's many motions in space.

The Coriolis effect, named for the French engineer who first studied it, alters the path of objects moving over a rotating body, such as the Earth, other planets, or stars. To understand why the Coriolis effect occurs, imagine standing at the North Pole and throwing a rock as far as you can toward the equator (fig. 6.24). As the rock arcs through the air, the Earth rotates under it. Thus, if you were aiming at a particular point on the equator, you will miss because the surface has turned beneath the rock's path, making the rock appear to have been pushed to the right.* Air, water, rockets, and anything else moving across the rotating Earth in any direction is affected similarly.

You can clearly see the results of the Coriolis effect when you see weather satellite pictures of a large storm system. The Coriolis effect spins the air flowing into

*The Coriolis effect deflects objects to the *left* in the Southern Hemisphere. There is no truth to the story that water spirals down the drain clockwise in the Northern Hemisphere and counterclockwise in the Southern Hemisphere. The Coriolis effect is totally overwhelmed by the persistence of the motion from filling a basin unless you allow the water to come to a complete rest, a state that can take weeks to achieve.

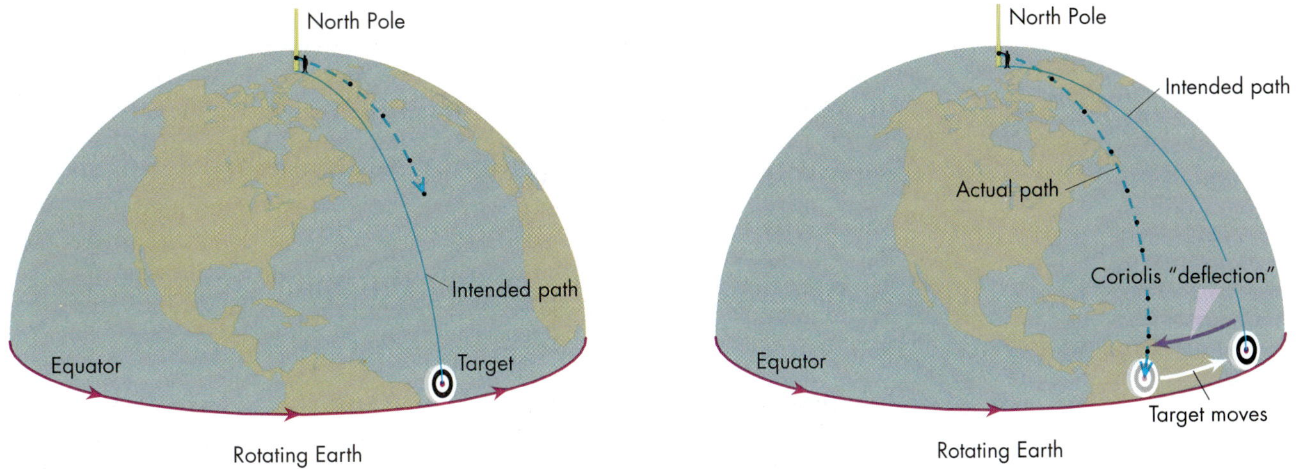

FIGURE 6.24
Coriolis effect on a rock thrown toward the equator from the North Pole.

FIGURE 6.25
Weather satellite pictures show clearly the spiral pattern of spinning air around a storm that results from the Coriolis effect.

FIGURE 6.26
Cloud bands on Jupiter created in part by the Coriolis effect.

the storm, creating the spiral pattern seen in figure 6.25. Likewise, the Coriolis effect deflects ocean currents, creating flows such as the Gulf Stream. It also creates atmospheric currents such as the trade winds.

The trade winds blow relatively steadily around our planet from east to west in two broad bands, one north and the other south of the equator. They form in response to the Sun's heating of equatorial air, which rises and expands, flowing away from the equator toward the poles. However, before this air can reach the poles, it cools and sinks toward the surface, where it flows back toward the equator. As that air approaches the equator, the Coriolis force deflects it into the pattern we observe in the trade winds.

The Coriolis force also establishes the direction of the **jet streams,** which are narrow bands of rapid, high-altitude winds. Jet streams are an important feature of the Earth's weather and are found on other planets as well. On rapidly rotating planets like Jupiter, Saturn, Uranus, and Neptune, the Coriolis effect is much stronger than on the Earth, creating extremely fast jet streams. The striking cloud bands we see on Jupiter, for example, are partly caused by this effect (fig. 6.26).

At one time, the Coriolis effect was of special interest because it is an indirect proof that the Earth rotates. Another such indirect proof is the Foucault pendulum, the huge swinging ball on a long wire that you may have seen at a science museum. (fig. 6.27). If you watch the pendulum for half an hour or so, you will notice that its swing changes direction. This can be made more evident by setting pegs in a circle around the pendulum; the pendulum will knock the pegs over, one by one. However, if you could watch the pendulum from space, its swing would not change direction. This is because the ball's inertia keeps its path fixed in space. It is the Earth (and therefore you and the building) that is turning beneath the swinging ball. Today, such indirect proofs are unnecessary: we can *see* the spin of our planet from pictures taken in space.

Precession

As the Earth moves around the Sun over long periods of time, the direction in which its rotation axis points changes very slowly. This motion, similar to the wobble that occurs when a spinning coin or toy top begins to slow down, is called **precession.** If the Earth

ANIMATION
The Foucault pendulum

FIGURE 6.27
A Foucault pendulum.

were perfectly spherical, precession would not occur. But the Earth's spin makes its equator bulge slightly, so the Sun and Moon exert an unbalanced gravitational attraction on our planet, twisting it slightly. That twisting makes the Earth's rotation axis slowly change direction, completing one swing in about 26,000 years (fig. 6.28). Currently, the North Pole points almost at the star Polaris. In about A.D. 14,000, the North Pole will point instead nearly at the bright star Vega. Thirteen thousand years later, the North Pole will again point nearly at Polaris. Looking Up #1 (in the front matter) shows part of this region of the sky, including the star Thuban in the constellation Draco. Thuban

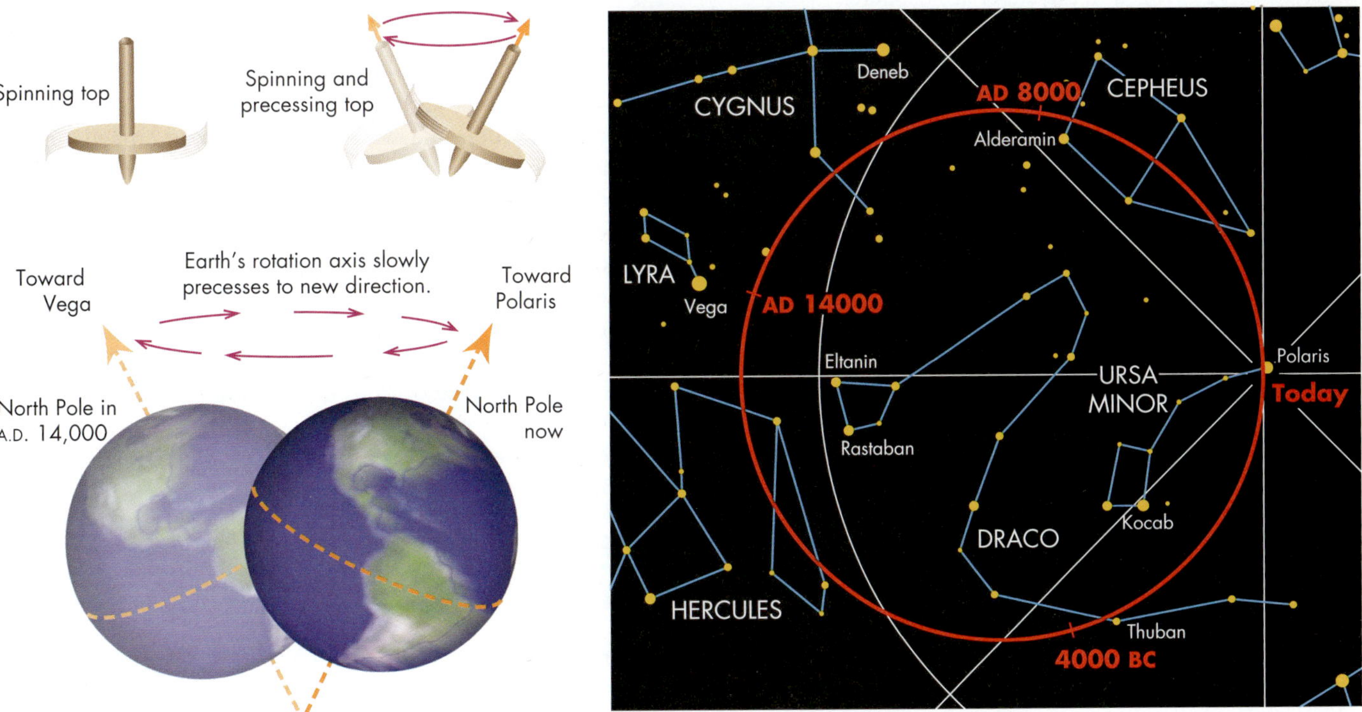

FIGURE 6.28
Precession makes the Earth's rotation axis swing slowly in a circle, similar to the "wobble" of a spinning top.

is of interest because it was the pole star at the time the pyramids were built in ancient Egypt, and the passage into the Great Pyramid pointed at that star.

Precession is of minor importance in day-to-day life, but over long periods of time it may cause climate changes. As noted earlier, we are currently nearest the Sun when it is winter in the Northern Hemisphere. In 13,000 years, the Earth will be farthest from the Sun in Northern Hemisphere winter. That will make winters slightly colder in the Northern Hemisphere and perhaps trigger a new ice age.

SUMMARY

The Earth is roughly spherical, and its radius is about 6400 kilometers (4000 miles). Trace amounts of radioactivity in rocks here and elsewhere in the Solar System reveal that the Earth formed about 4.6 billion years ago. Radioactive material also adds heat to its interior. The flow of heat to the Earth's surface stirs slow convective motions, which shift the Earth's crust (plate tectonics), creating mountains, volcanoes, ocean basins, and earthquakes.

The waves generated by earthquakes (seismic waves) allow us to study the Earth's interior. They show it is stratified into four distinct regions: a very thin crust of ordinary rock, a mantle of hot but essentially solid silicates, an outer core of liquid iron and nickel, and an inner core of solid iron and nickel.

Currents created by motions in the Earth's core generate the Earth's magnetic field. That field in turn affects the motion of charged particles in the upper atmosphere. Such particles may create auroral displays when they collide with oxygen and nitrogen in the upper atmosphere.

The nitrogen, carbon dioxide, and water of our atmosphere may have come from volcanic gases vented over the Earth's history. Alternatively, these atmospheric gases may be the evaporated remains of comets that hit the Earth in its infancy. Plant life has created the atmosphere's oxygen by photosynthesis.

Some atmospheric gases absorb radiation. High-altitude ozone absorbs ultraviolet radiation, thereby protecting us from its biologically harmful effects. Carbon dioxide and water vapor absorb infrared radiation, trapping heat radiated from the Earth's surface. By slowing heat loss from our planet into space, these gases create the greenhouse effect and make Earth slightly warmer than it would be if the infrared radiation could escape freely.

The Earth's spinning motion creates a Coriolis effect that deflects objects moving over its surface. The Coriolis effect makes large storm systems rotate and is essential for driving the circulation of the atmosphere and the oceans.

QUESTIONS FOR REVIEW

1. (6.1) Why is the Earth not perfectly round?
2. (6.1) What are some of the most common elements composing the Earth?
3. (6.2) How do we know that the Earth has a liquid core? Why is the inner core solid even though it is hotter than the outer liquid core?
4. (6.2) What are two explanations that scientists offer for why the interior of the Earth is hot? How hot is it?
5. (6.3) How can scientists determine the age of the Earth?
6. (6.4) What is convection? What are some other examples of convection besides hot soup?
7. (6.4) What is the relation between rising and sinking material in the Earth's interior and subduction and rifting?
8. (6.4) On what plate of the crust are you located? Which way is it taking you?
9. (6.4) What is happening where one tectonic plate is smashing into another?
10. (6.5) What factors are thought to be responsible for the Earth's magnetic field?
11. (6.5) How is the aurora related to the Earth's magnetic field?
12. (6.5) How does the fact that the Earth has a magnetic field help provide evidence for the theory of plate tectonics?
13. (6.6) What were the main components of the atmosphere when the Earth formed, and what are the main components today? How and why did they change?
14. (6.6) Explain how the greenhouse effect works and how it relates to global warming.
15. (6.6) What is ozone? Why is it important?
16. (6.7) What is the Coriolis effect? How does it affect life on Earth?
17. (6.7) What is precession? What are some of its possible consequences?

THOUGHT QUESTIONS

1. (6.1) When you choose fruit at a supermarket, you might heft the fruit in your hand to test its weight. How does this tell you whether the fruit is dried out inside? How is that similar to using mean density as an indicator of the composition of the Earth's interior?

2. (6.1) Submarines contain "ballast tanks" that can take on or expel seawater. Explain how these tanks allow a submarine, which is largely constructed of steel with a density much higher than that of water, to rise and submerge at will.
3. (6.1) According to the *Guinness Book of Mountains and Mountaineering,* the summit of the volcano Chimborazo in Ecuador is the point on the Earth's surface farthest from the center. However, the book also states that the summit of Mount Everest is the highest point above sea level. Are these claims inconsistent? Why?
4. (6.2) Flicking your finger against your cheek makes a different sound from flicking it against your forehead. How is that similar to studying the interior of the Earth with seismic waves?
5. (6.4) How does the eventual acceptance of the plate tectonic theory illustrate some aspects of the scientific method?
6. (6.5) If the Earth rotated more slowly, would you expect it to have as strong a magnetic field?
7. (6.6) Astronomers are still uncertain about how the Earth's atmosphere formed. How does this illustrate the workings of the scientific method?
8. (6.7) Think about a stone thrown from the pole toward the equator. If the Earth rotated faster, would the Coriolis effect be larger?
9. (6.7) As seen from above the North Pole, the Earth rotates counterclockwise. Using the thrown-stone argument of question 8, explain why the Coriolis effect deflects objects to the right of their motion in the Northern Hemisphere.

PROBLEMS

1. (6.1) Suppose the Earth's radius were only half of its real value. If the Earth's mass remained the same, what would be the average density? What if the Earth's radius were twice its real value?
2. (6.1) Using the periodic table in the back of the book, determine the astrophysical source(s) of the elements in the Earth's crust listed in table 6.1. Compare this source to the source of "precious" metals like gold, silver, and platinum.
3. (6.2) Seismic waves are partly reflected when they cross a boundary such as that between the mantle and the liquid core. Suppose that a P wave has a constant velocity of 8.0 km/sec. Suppose further that 700 seconds (about 12 minutes) after an earthquake near the surface, a seismometer detects a reflected P wave. How far below the surface is the liquid core–mantle boundary? Compare your answer to the distance in figure 6.7. Hint: Remember this is an echo.
4. (6.3) The half-life of carbon 14, which is commonly used to date organic materials, is 5700 years. What is the minimum age of a sample in which less than 1% of the original carbon 14 is left?
5. (6.4) Studies of the South American and African plates indicate that for tens of millions of years they have spread apart at a roughly constant rate of approximately 4 centimeters per year. How many kilometers farther apart are the two continents now than they were 80 million years ago? How does this compare to 6000 kilometers, approximately the distance between some matching parts of the South American and African coastlines?
6. (6.6) The total mass of the Earth's atmosphere is about 5.1×10^{18} kg. If you assume it is entirely made of nitrogen (N_2) and oxygen (O_2) gas molecules, what is the mass of oxygen gas in the atmosphere? The mass of one molecule of gas is equal to the sum of the masses of each atom (its atomic mass times 1.66×10^{-27} kg) in the molecule. Consult table 6.1 and the periodic table in the back of the book.
7. (6.7) Use the principle of the Coriolis effect to predict what will happen to the air that flows away from a "high-pressure system." Draw a diagram to predict the circulation of air around the system.

TEST YOURSELF

1. (6.1) Scientists think the Earth's core is composed mainly of
   (a) silicate rocks.
   (b) uranium.
   (c) lead.
   (d) sulfur.
   (e) iron.
2. (6.2) What evidence indicates that part of the Earth's interior is liquid?
   (a) With sensitive microphones, sloshing sounds can be heard.
   (b) We know the core is lead, and we know the core's temperature is far above lead's melting point.
   (c) Deep bore holes have brought up liquid from a depth of about 4000 kilometers.
   (d) No S-type seismic waves are detectable at some locations after an earthquake.
   (e) S-type waves are especially pronounced at all locations around the Earth after an earthquake.
3. (6.3) Scientists use radioactivity in rock samples to measure
   (a) the temperature in the Earth's core.
   (b) the depth of the oceans.
   (c) the Earth's age.
   (d) the composition of the mantle.
   (e) the composition of the inner core.
4. (6.4) The slow shifts of our planet's crust are believed to arise from
   (a) the gravitational force of the Moon pulling on the crust.
   (b) the gravitational force of the Sun pulling on our planet's crust.
   (c) the Earth's magnetic field drawing iron in crustal rocks toward the poles.

(d) heat from the interior causing convective motion, which pushes on the crust.
(e) the great weight of mountain ranges forcing the crust down and outward from their bases.

5. (6.4) Plate motion at subduction zones can cause (more than one answer may be correct)
(a) earthquakes.
(b) convection currents in the Earth's mantle.
(c) plates to grow larger.
(d) volcanic activity.
(e) the creation of mountains.
6. (6.5) The presence of a strong magnetic field around a planet like the Earth is evidence for
(a) rotational and convective motion in a liquid core.
(b) the presence of an atmosphere.
(c) a slow rotational period.
(d) intense heat in the core.
7. (6.6) Why is carbon dioxide called a "greenhouse gas"?
(a) it is generated when plants are burned.
(b) it is needed by plants to grow.
(c) it absorbs infrared light.
(d) it appears greenish when concentrated.
(e) All of the above.
8. (6.7) The Coriolis effect is why the Earth rotates counterclockwise as seen from above.
(a) True.
(b) False.
9. (6.7) At what location would a pendulum's direction appear to change the most over a day?
(a) On the ice at the North Pole.
(b) On a high mountain at midlatitude.
(c) On an island at the equator.
(d) On the ice at the north magnetic pole.
(e) It would change the same amount at all locations.

KEY TERMS

FURTHER EXPLORATIONS

Akasofu, Syun-Ichi. "The Dynamic Aurora." *Scientific American* 260 (May 1989): 90.

Allègre, Claude J., and Stephen H. Schneider. "The Evolution of the Earth." *Scientific American* 271 (October 1994): 66.

Davis, Neil. *The Aurora Watcher's Handbook.* Fairbanks: University of Alaska Press, 1992.

Fernie, J. D. "The Shape of the Earth," part 1. *American Scientist* 79 (March/April 1991): 108; part 2, 79 (September/October 1991): 393; part 3, 80 (March/April 1992): 125.

Murphy, J. Brenden, and R. Damian Nance. "Mountain Belts and the Supercontinent Cycle." *Scientific American* 266 (April 1992): 84.

Video

Earth Revealed (Intelecom, 1992) and *Planet Earth,* produced by WQED in association with the National Academy of Sciences and funded by the Annenberg/CPB Project, 1986. Both are available for streaming at Annenberg Media, http://www.learner.org/resources/series49.html.

Website

Visit the *Explorations* website at **http://www.mhhe.com/arny** for additional online resources on these topics.

FIGURE QUESTION ANSWERS

WHAT IS THIS? (chapter opening): This is a satellite picture of the volcanic island Santorini in the Mediterranean Sea. The central bay in the island was created when the volcano exploded violently about 3600 years ago, possibly giving rise to the legend of Atlantis. A new volcanic cone can be seen growing in the middle of the bay.

FIGURE 6.6: The time, t, for something moving at a speed v to travel a distance D is given by $t = D/v$. The Earth's diameter is about 12,000 km. Therefore t = 12,000 km/(10 km/s) = 1200 seconds, or about 20 minutes.

FIGURE 6.14: There were no polar ice caps, so sea level was much higher.

FIGURE 6.20: Red light has a longer wavelength than blue light.

PROJECTS

1. **Plate tectonics:** Make a photocopy or tracing of a map of the world (fig. 6.13 will do). Cut out North and South America and place them against the western edge of Europe and Africa. Do they fit? Does this help you to see what led early scientists to the idea that the continents shift?

2. **Radioactive decay:** You can simulate the process of radioactive decay by tossing coins. You'll need at least 25 coins, but a larger number (50 or 100) will work better—you might even try this with your whole class. Set the coins down heads up to start. Each represents one atom of uranium, which has a half-life of almost 4.5 billion years. Flip each coin once; each atom that comes up heads is still uranium, but each that comes up tails has decayed into daughter elements. Record the number of heads remaining. Now, repeat the process—only flipping each of the coins that is still heads (daughter elements—tails—are no longer Uranium, although in fact the daughter elements of Uranium are themselves radioactive). Again record the number of remaining heads. Repeat the process until there are no heads remaining. Make a graph of the number of heads remaining at each round of tosses. How much time is represented in between each round? What does this suggest to you about the problem of storing radioactive waste?

ESSAY 3

Keeping Time

From before recorded history, people have used events in the heavens to mark the passage of time. The day, the month, and the year were all originally defined in terms of obvious astronomical phenomena. The day was the time interval from sunrise to sunrise. The month was the interval from new moon to new moon. The year was the time it takes for the Sun to complete one circle of the zodiac. Astronomical events are not perfect time markers, however: even the day and year need to be defined with care if we are to have reliable clocks and calendars. Otherwise, we may end up with snow in "summer" and heat waves in "winter."

HOURS OF DAYLIGHT

Although each day lasts 24 hours, the number of hours of daylight, or the amount of time the Sun is above the horizon, changes greatly throughout the year unless you are close to the equator. FoDat a latitude of 40° (approximately that of New York, Rome, or Beijing), summer has about 15 hours of daylight and only 9 hours of night. In the winter, the reverse is true.

This variation in the number of daylight hours is caused by the Earth's tilted rotation axis. Remember that as the Earth moves around the Sun, its rotation axis points in roughly a fixed direction. Thus, the Sun shines more directly on the Northern Hemisphere during its summer and less directly during its winter. The result (as you can see in fig. E3.1) is that only a small part of the Northern Hemisphere is unlit in the summer, but a large part is unlit in the winter. Thus, as the Earth's rotation carries us around, only a relatively few hours of a summer day are unlit, but a relatively large number of winter hours are dark. Figure E3.1 also shows that on the first day of spring and of autumn (the equinoxes), the hemispheres are equally lit, so that day and night are of equal length everywhere on Earth.

If we change our perspective and look out from the Earth, we see that during the summer, the Sun's path is high in the sky, so that the Sun spends a larger portion of the day above the horizon. This gives us not only more heat (because the sunlight falls more directly on the ground) but also more hours of daylight. On the other hand, in winter the Sun's path across the sky is much shorter, giving us less heat (because the sunlight falls less directly on the ground) and fewer hours of light.

THE DAY

The length of the day is set by the Earth's rotation speed on its axis. One day is defined as one rotation. However, we must be careful how we measure our planet's rotation. For example,

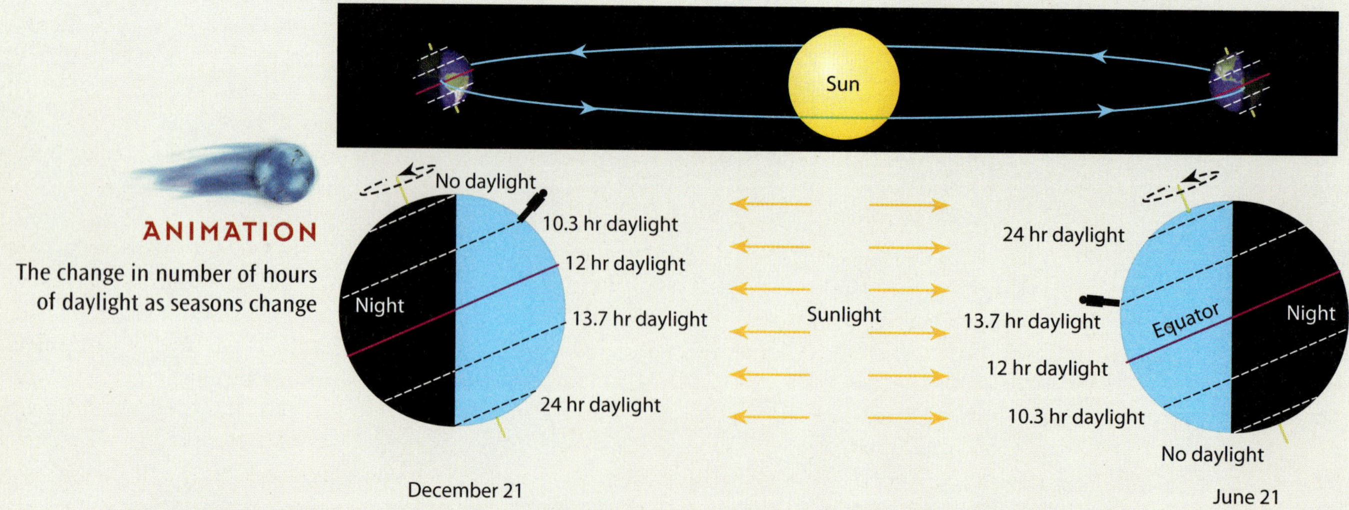

FIGURE E3.1
The tilt of the Earth affects the number of daylight hours. Locations near the equator always receive about 12 hours of daylight, but locations toward the poles have more hours of dark in winter than in summer. In fact, above latitudes 66.5°, the Sun never sets for part of the year and never rises for another part of the year (the midnight sun phenomena). At the equinoxes, all parts of the Earth receive the same number of hours of light and dark. (Sizes and separation of the Earth and Sun are *not* to scale.)

we might use the time from one sunrise to the next to define a day. That, after all, is what sets the day–night cycle around which we structure our activities. We would soon discover, though, that the time from sunrise to sunrise changes steadily throughout the year as a result of the seasonal change in the number of daylight hours. A better time marker might be the time it takes the Sun to move from its highest point in the sky on one day (what we technically call apparent noon) to its highest point in the sky on the next day—a time interval that we call the **solar day.**

If we measure the length of the solar day, however, we will discover that it is, in general, *not* exactly 24 hours. Moreover, its length changes by several minutes over the course of the year. As we will discover later, this variation arises from the Earth's motion around the Sun. Thus, although the Sun's motion across the sky determines the day–night cycle, the Sun's motion is *not* a good reference for the actual time it takes our planet to complete one spin.

We can avoid most of this variation in the day's length if, instead of using the Sun, we use a star as our reference. For example, if we pick a star that lies exactly overhead at a given moment and measure the time it takes for that same star to return to exactly overhead, we will find that the time interval is not 24 hours but an essentially unchanging 23 hours and 56 minutes. This day length, measured with respect to the stars, we call a **sidereal day.**

Why do the solar and sidereal day differ in length? We can see the reason by looking at figure E3.2, where we measure the interval between successive apparent noons—a solar day. Let us imagine that while we are watching the Sun, we can also watch a star, and that we measure the time interval between the star's passages overhead, a sidereal day.

As we wait for the Sun and star to move back overhead, the Earth moves along its orbit. The distance the Earth moves in one day is so small compared with the star's distance that we see the star in essentially the same direction as on the previous day. However, we see the Sun in a measurably different direction, as figure E3.2 shows. The Earth must therefore rotate a bit more before the Sun is again overhead. That extra rotation, needed to compensate for the Earth's orbital motion, makes the solar day slightly longer than the sidereal day.

It is easy to figure out how much longer, on average, the solar day must be. Because it takes us $365\frac{1}{4}$ days to orbit the Sun and because a circle has by definition 360°, the Earth moves approximately 1° per day in its orbit around the Sun. That means that for the Sun to reach its noon position, the Earth must rotate approximately 1° past its position at the previous noon.* In 24 hours $= 24 \times 60 = 1440$ minutes, the Earth rotates 360°. Therefore, to rotate 1° takes 1440/360 minutes, or about 4 minutes. The solar day is therefore about 4 minutes (3 minutes 56 seconds, to be precise) longer than the sidereal day.

*Another way of thinking about this is that the Sun is slowly moving eastward across the sky through the stars at the same time the Earth is rotating. Thus, in a given "day," the Earth must rotate a bit more to keep pace with the Sun than it would to keep pace with the stars.

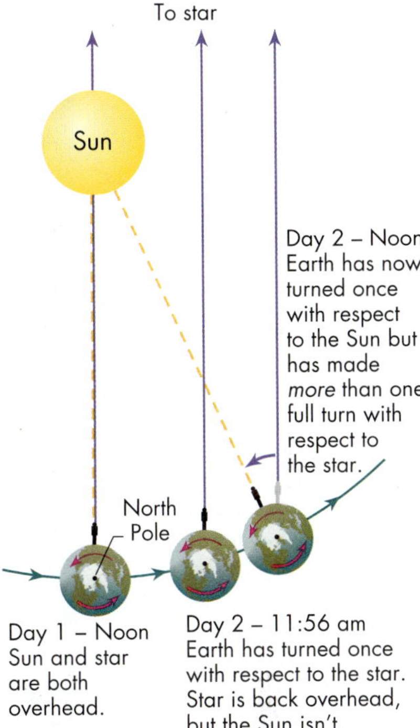

FIGURE E3.2
The length of the day measured with respect to the stars is not the same as the length measured with respect to the Sun. The Earth's orbital motion around the Sun makes it necessary for the Earth to rotate a tiny bit more before the Sun will be back overhead. (Motion is exaggerated for clarity.)

As mentioned earlier, the motion of the Earth around the Sun alters the length of the solar day. If you measure carefully the time interval from one apparent noon to the next, in general it is not 24 hours but may differ slightly either way. This variation arises because the Earth's orbit is not circular and therefore our orbital velocity changes according to Kepler's second law.

The Earth moves along its orbit faster when it is near the Sun and slower when it is farther away. This means that it takes a little longer for the Earth to swing you around into the morning Sun (slightly lengthening the interval between successive noons) when the Earth is moving rapidly in its orbit than when it is moving slowly (fig. E3.3). Hence, the solar day is longer when we are near the Sun and shorter when we are farther away. The amount by which the length of the solar day varies is small, but it must be accounted for if our clocks are to always read about noon when the Sun is highest in the sky.

We could design clocks so that the hour is of different lengths at different times of the year. That could be done to ensure that our clocks advance to conform with the changing length of the solar day. However, it is much easier to define the length of the day differently, using not the true interval from one apparent noon to the next, but the *average* value of that interval over the year. That average daylength is called the **mean solar day,** and it has, by definition, 24 hours of clock time. We therefore use mean solar time in our daily timekeeping.

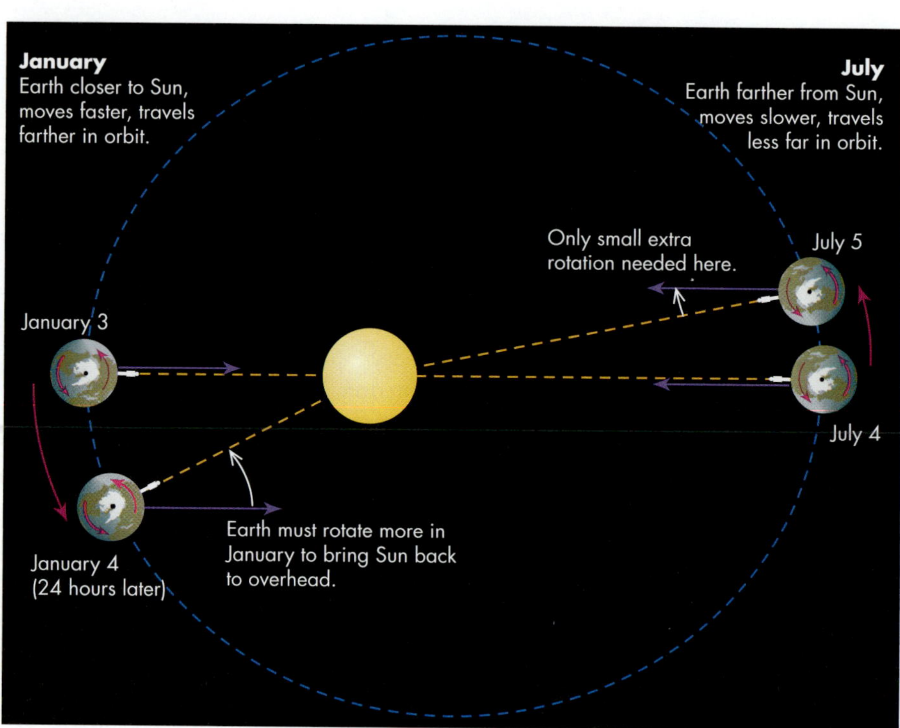

FIGURE E3.3
As the Earth moves around the Sun, its orbital speed changes as a result of Kepler's second law of motion. For example, the Earth moves faster in January when it is near the Sun than in July when it is far from the Sun. Thus, in 24 hours the Earth moves farther along its orbit in January than in July. As a result, the Earth must turn slightly more in January to bring the Sun back to overhead. This makes the interval between successive noons longer in January than in July and means they are not exactly 24 hours. For that reason, time is kept using a "mean Sun" that moves across the sky at the real Sun's average rate. (Note that the ellipticity of the Earth's orbit has been exaggerated to make the differences clearer in this figure.)

Over the course of the year, the difference in length between the mean solar day and the true solar day accumulates and leads to a difference of several minutes between clock time and time based on the position of the Sun. This difference is called the "equation of time" and is shown graphically in figure E3.4. The equation of time gives the correction needed on a sundial if it is to give the same time as your watch.

Although we use solar time in regulating our daily activities, astronomers find sidereal time extremely useful. One reason is that, at a given location, a given star always rises at the same sidereal time. To avoid the nuisance of A.M. and P.M., sidereal time is measured on a 24-hour basis. For example, the bright star Betelgeuse in the constellation Orion rises at about 10 P.M. in November but at about 8 P.M. in December. However, on a clock keeping sidereal time, it always rises at the same time at a given location, about 23 hours 50 minutes.

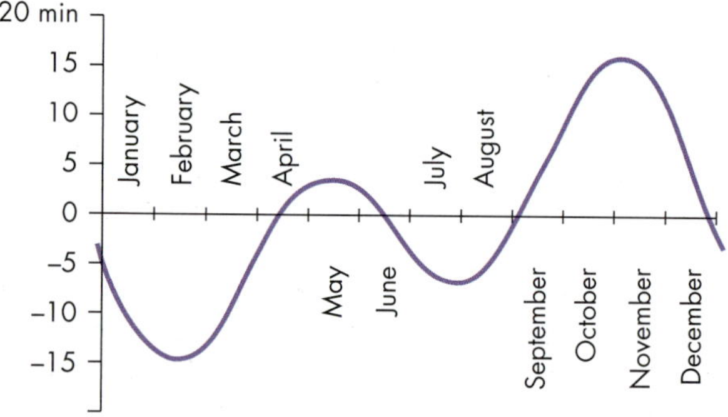

FIGURE E3.4
The equation of time is the correction that must be applied to the true Sun to determine mean solar time. It can be shown as a graph (as here) or as a figure-8 shape called an "analemma," often seen on globes of the world.

TIME ZONES

Because the Sun is our basic timekeeping reference, most people like to measure time so that the Sun is highest in the sky at about noon. This is unnecessary now that we have good electronic clocks that can keep time independent of the Sun. Nevertheless, it is a tradition that is hard to break, and as a result, clocks in different parts of the world are set to read different times. Because the Earth is round, the Sun can't be "overhead" everywhere at the same time, so it can't be noon everywhere at the same time.

The Earth is therefore divided into 24 major **time zones** in which the time differs by one hour from one zone to the next. There are a few exceptions, but we'll ignore them here.

Across the contiguous 48 United States, the time zones are, from east to west, Eastern, Central, Mountain, and Pacific (see fig. E3.5). The time within each zone is the same and is called **standard time.** Thus, in the eastern zone, the time is denoted Eastern Standard Time (EST), whereas in the central zone, it is denoted Central Standard Time (CST). As you travel across the country, it is therefore necessary to reset your watch if you cross from one time zone to another, adding 1 hour for each time zone as you move from west to east and subtracting 1 hour when you move from east to west.

If you travel through a very large number of time zones, you may need to make such a large time correction that you shift your watch past midnight. For example, if you could

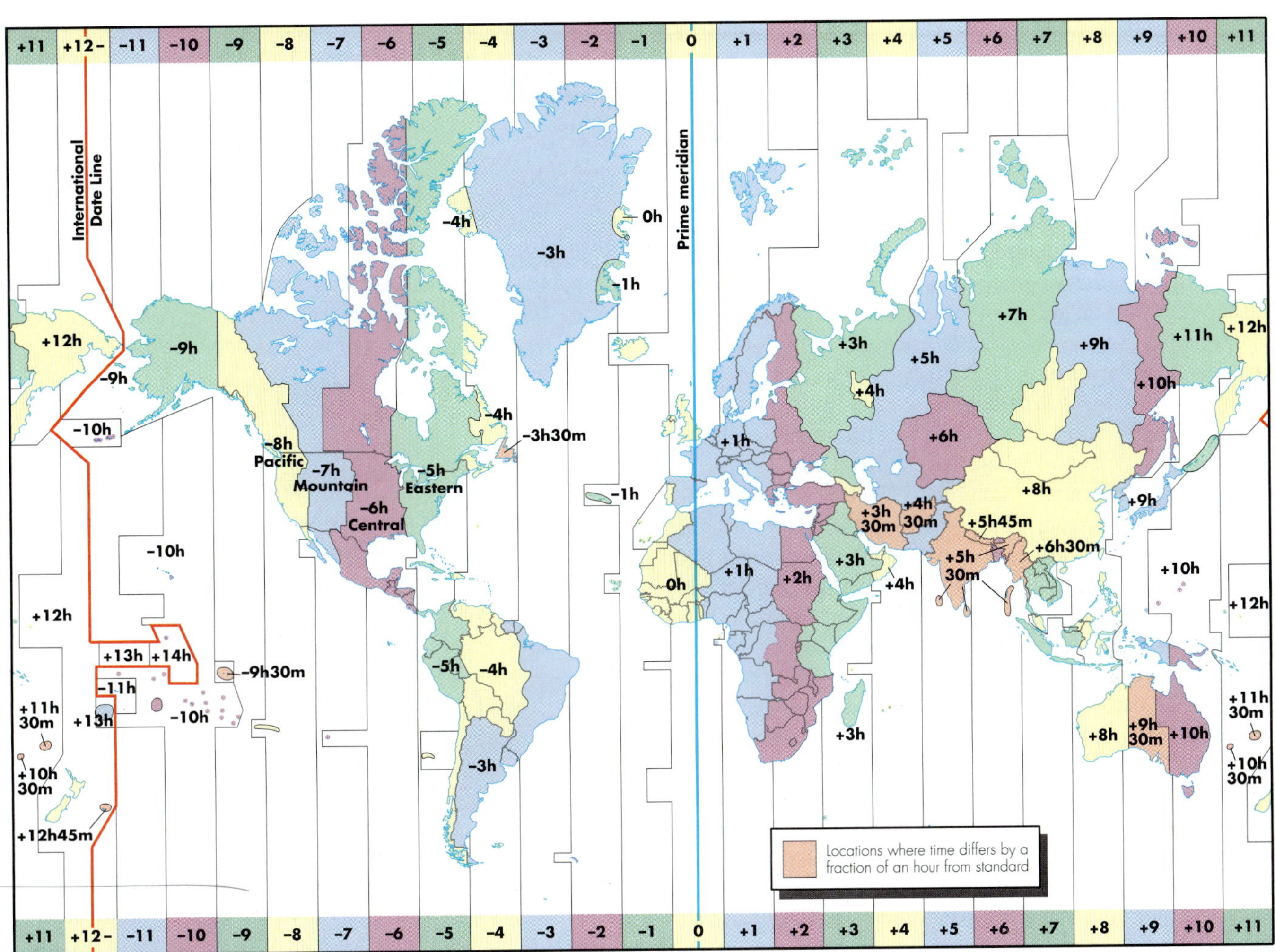

FIGURE E3.5
Time zones of the world and the international date line. Local time = Universal time + numbers on top or bottom of chart.

travel fast enough and far enough, setting your watch back each time you cross a time zone, you could end up at your starting point with your watch set to a time 24 hours before you left!

A high-speed traveler cannot actually "gain" a day, however, because when you cross longitude 180° (roughly down the middle of the Pacific Ocean), you add a day to the calendar if you are traveling west and subtract a day if you are traveling east. The precise location where the day shifts is called the **international date line** (fig. E3.5). It generally follows 180° longitude but bends around extreme eastern Siberia and some island groups to ensure they keep the same calendar time as their neighbors.

UNIVERSAL TIME

The nuisance of having different times at different locations can be avoided by using **Universal time,** abbreviated as UT. Universal time is the time kept in the time zone containing the longitude zero, which passes through Greenwich, England. By using UT, which is based on a 24-hour system to avoid confusion between A.M. and P.M., two people at remote locations can decide to do something at the same time without worrying about what time zone they are in.

DAYLIGHT SAVING TIME

In many parts of the world, people set clocks ahead of standard time during the summer months and then back again to standard time during the winter months. This has the effect of shifting sunrise and sunset to later hours during the day, thereby creating more hours of daylight during the time most people are awake. Time kept in this fashion is therefore called **daylight saving time** in the United States. In other parts of the world, it is called "Summer Time."

Daylight saving time was originally established during World War I as a way to save energy. With clocks set ahead, less artificial light was needed during work hours late in the day. Nowadays, it allows us more daylight hours for recreation after work during the summer. As of March 2007, daylight saving time in the United States runs from the second Sunday in March to the first Sunday in November.

THE WEEK

That there are 7 days in the week is probably a result of there being 7 visible objects that move across the sky with respect to the stars: the Sun, the Moon, Mercury, Venus, Mars, Jupiter, and Saturn. We can see the names of some of these bodies in our English day names (*Sunday, Monday,* and *Saturday*). The influence is even clearer in the romance languages, such as Spanish (*lunes, martes, miércoles, jueves, viernes*).

Some English day names come to us through the names of Germanic gods (many of whom have a direct parallel with the Greco-Roman gods after whom the planets are named). For example, Tuesday is from *Tīw,* god of war, like Mars (matching Spanish *martes*). Wednesday is named for *Wōden,* the chief god of Germanic peoples and identified with the Roman Mercury (matching Spanish *miércoles*). Thursday is named for *Thōr* the thunder god (matching Spanish *jueves,* "Jove's day"). Friday is named for *Freya,* a love goddess, like Venus (matching *viernes*).

While a 7-day week was independently adopted by many cultures around the world, other time periods have also been used. For example, the ancient Romans had an 8-day week between market days, and after the French revolution, people experimented with a 10-day week. Other collections of days, such as sets of 20 days used by ancient Mayans are also known, but almost all cultures have marked the passage of time using the Moon's cycles.

THE MONTH

As you know from looking at a calendar, the month is usually about 30 days long. This time interval, and its name, derives from the Moon's cycle of phases. The time interval between full moons is about 29.5 days, which for use on the calendar is rounded off to 30. Because the year has about 365 days in it, there are about 12 lunar cycles per year. That is the reason that we have 12 months in the year.

You will notice, however, that 12 lunar cycles of 30 days ends up 5 days short of a full year. For that reason, some of the months are made 31 days long. In fact, if every other month, starting with January, were a 31-day month, the year would total 366 days. To make the days add up to 365, February was trimmed 1 day, to 29.

But, you protest, that is not the way the calendar looks. Although January, March, May, and July have 31 days, the sequence is broken in the later months. You see at work there the politics of ancient Rome.

THE CALENDAR

Our calendar is based on one developed about 200 B.C. by the Romans. In fact, the word *calendar* is itself of Roman origin.

TABLE E3.1 ORIGIN OF THE NAMES OF THE MONTHS

Month	Origin
January	Janus (gate), the two-faced god looking to the past and future; hence, beginnings.
February	Februa (expiatory offerings).
March	The god Mars.
April	Etruscan *apru* (April), probably shortened from the Greek Aphrodite, goddess of love and earlier of the underworld.
May	Maia's month; Maia ("she who is great"), the eldest of the Pleiades and the mother of Hermes by Zeus.
June	Junius, an old Roman noble family (from Juno, wife and sister of Jupiter, equal to Greek Hera).
July	Julius Caesar (Julius, "descended from Jupiter"; The *Ju* of June and July are the same: Jupiter, "Sky-father").
August	Augustus Caesar (*augustus* means sacred, grand).
September–December	Seventh- to tenth-month. The *-ember* may come from the same root as *month.*

There is some controversy about how the original Roman calendar was organized. It may have had only 10 months, and it probably began on the first day of spring (the vernal equinox) rather than in January.

The names of our months date from that calendar and its modifications. For example, if the year began in March, then September, October, November, and December were the 7th (Sept.), 8th (Oct.), 9th (Nov.), and 10th (Dec.) months, respectively. The 5th and 6th months (*Quintilis* and *Sextilis* in ancient Rome) were renamed later in honor of Roman emperors. Possible origins of the names of other months are listed in table E3.1.

Because it did not contain the right number of months and days to match the astronomical phenomena, this original calendar became a form of political patronage. The priests who regulated the calendar would add days and even months to please one group, and take days off to punish another. Such confusion resulted from these abuses that in 46 B.C., Julius Caesar asked the astronomer Sosigenes to design a calendar that would fit the astronomical events better and give less room for the priests and politicians to tinker with it. The resulting calendar, known as the **Julian calendar,** consisted of 12 months, which with the exception of February, alternated between 31 and 30 days in length.

The Julian calendar barely survived Caesar before the politicians were at it again. First, the name of the seventh month was changed to *Julio* to honor Julius Caesar—hence our *July.* Next, on the death of Julius Caesar's successor, Augustus Caesar, a very able and highly respected leader, it was decided to name the eighth month in *his* honor—hence, the name *August*. However, because it would have been impolitic to have his month a day shorter than Julius's, August became a 31-day month, and all the following months had the number of their days changed to maintain the alternation. Unfortunately, this led to using up one more day than there were days in the year. Thus, poor February, already one day short, was trimmed a second day, leaving it with only 28 days. With only minor modifications, this is the calendar we use today. However, those modifications are important, as we will see.

LEAP YEAR

The ancient Egyptians knew that the year is not exactly 365 days long. It turns out that it takes about 365 and $\frac{1}{4}$ days for the Earth to complete an orbit around the Sun, which is how we measure a year. Because we can't have fractions of a day in the calendar, a calendar based on a year of 365 days will come up 1 day short every 4 years.

Your first reaction might be, So what? However, the seasons are set by the orientation of the Earth's rotation axis with respect to the Sun, not by how many days have elapsed. We therefore want to make sure that we start each year with the Earth having the proper orientation. Otherwise, the seasons get out of step with the calendar. For example, because in 4 years you will lose 1 day, in 120 years you will lose a month, and in 360 years, you will lose an entire season. With a 365-day year, in a little over three centuries April would be coming in what is now January.

This problem is corrected by leap year, a device implemented by the Julian calendar to keep the calendar in step with the seasons. Leap year corrects by adding a day to the calendar every fourth year. The extra day is traditionally added to February because it is the shortest month.

Unfortunately, the year is actually a little bit shorter than $365\frac{1}{4}$ days. Thus, having leap year every four years corrects a tiny bit too much. For nearly 1600 years after the adoption of the Julian calendar, the small errors accumulated, adding up to about 10 days, making it obvious that the calendar was out of synch with the seasons. To prevent further accumulation of errors, three leap years needed to be dropped every four centuries. Therefore it was decided that centuries not divisible by 400 would be eliminated as leap years. Thus, 1900 was not a leap year, but 2000 was. This modification of omitting leap year for all century years not divisible by 400 was added in 1582 at the direction of Pope Gregory XIII. The calendar we use today is thus known as the **Gregorian calendar.**

The inauguration of the Gregorian calendar in 1582 was not a peaceful affair. To bring the calendar back into synchrony with the seasons, Pope Gregory simply eliminated 10 days from the year 1582 so that the day after October 4 became October 15.

Although the changeover went smoothly in most places, non-Catholic countries such as Protestant England refused to abide by the Pope's edict. The calendar in England and in a few other northern European countries was not altered. This made commerce between Catholic and non-Catholic countries very difficult because the day and sometimes even the month and year were different from one country to the next.

Eventually, the Gregorian calendar was adopted essentially worldwide, but the change was not made in England until 1752.* In Russia, the change was not made until the revolution in the early part of the twentieth century. Other countries (Greece and Turkey, for example) changed in the 1920s.

RELIGIOUS CALENDARS

Although the Gregorian calendar is used nearly worldwide, many religions use traditional calendars for setting their feast and holy days. Two of the best-known examples are the Jewish and Islamic calendars, which are particularly interesting to astronomers because of the role played in them by the lunar cycle.

The Islamic calendar is in fact a purely lunar one consisting of 12 months of either 29 or 30 days. The resulting year comes out about 11 days shorter than a year in the Gregorian calendar. This means that the Muslim calendar is totally out of synchrony with the seasons. As a result, the holy month of Ramadan can fall at any time during the year, irrespective of the season. This may seem odd to people living in climates with strong seasonal variations, but for people living in the Middle East where seasons are not so extreme, it makes little difference.

Likewise, the Jewish calendar is based on the lunar cycle. To correct for the missing days, from time to time an extra month is added in the middle of the year to keep it in step with the seasons. The Jewish calendar is especially interesting astronomically because it begins near the autumnal equinox and the extra month is added near the vernal equinox. Also, the holy days of Yom Kippur and Passover are located near the equinoxes, a feature shared by the Gregorian calendar, wherein Easter is near the vernal equinox (actually the first Sunday after the first full moon after the equinox) and Christmas is near the winter solstice.

OTHER CALENDARS

Another calendar of note is the Chinese one, which, like the Jewish calendar, combines lunar and seasonal aspects. The Chinese calendar year contains 12 or 13 lunar months, and it generally begins on the second new moon after the winter solstice. The years are grouped into 60-year cycles composed of 5 cycles repeating every 12 years. The years in each 12-year cycle are given names such as the Year of the Rat, the Year of the Dog, and so forth. Months are added so that it averages out to match the cycle of the seasons.

MOON LORE

The Moon figures prominently in folklore around the world. Most stories concerning its powers are false. For example, people often claim that the full moon triggers antisocial

*This elimination of some 10 days from the calendar supposedly led to riots by people fearing they would be charged a full month's rent for only 20 days.

TABLE E3.2 NAMES USED FOR FULL MOONS*

January	old moon	July	thunder or hay
February	hunger	August	grain or green corn
March	sap or crow	September	harvest
April	egg or grass	October	hunter's
May	planting	November	frost or beaver
June	rose or flower	December	long night

*Most of these names derive from Native American usage.

behavior, hence the term *lunatic*. All studies to look for such effects have found nothing. Automobile accidents, murders, admissions to clinics, and so forth show no increase when the Moon is full.

On the other hand, "once in a blue moon," indicating a rare event, is a phrase with a basis in fact, because on rare occasions the Moon may look blue. This odd coloration comes from particles in the Earth's atmosphere. Normally our atmosphere filters the blue colors from light better than the red ones. For example, light from the rising or setting Sun passes through so much atmosphere that little blue light remains by the time it reaches us. Therefore, the Sun looks red when it is low in the sky. However, if the atmosphere contains particles whose size falls within a very narrow range, the reverse may occur. Dust from volcanic eruptions or smoke from forest fires may have just the right size to filter out the red light, allowing mainly the blue colors to pass through. Under these unusual circumstances, we may therefore see a "blue Moon."

A different meaning for "blue moon" has appeared more recently, referring to months with two full moons. Because the cycle of phases is 29.5 days, it is rare for a second full moon to occur in the same month, and some calendars have printed the second full moon in blue.

Another well-known phrase is "harvest moon," the full moon nearest the time of the autumn equinox. As it rises in the east at sunset, the light from the harvest moon helps farmers see to get in the crops. Full moons in other months also have popular names, but only the harvest and hunter's moon are widely known. Other names occasionally used in American folklore are listed in table E3.2.

THE ABBREVIATIONS A.M., P.M., B.C., A.D., B.C.E., AND C.E.

Four abbreviations are used frequently in the measure of time and calendars. They are the familiar letters A.M., P.M., B.C., and A.D. The first two have specific astronomical meaning. The last two have cultural meaning.

A.M. and P.M. stand for "ante meridian" and "post meridian," respectively. The meridian is the line passing from due north to due south and passing directly through the point exactly overhead (also called the zenith).

As the Sun moves across the sky, it crosses the meridian at the time called apparent noon. Before noon, it lies before (ante) the meridian. After noon, it lies past (post) the meridian. Hence, A.M. and P.M.

B.C. stands for "before Christ," referring to the year of his birth. Oddly, by convention 1 B.C. refers to the year before his birth, to avoid having a year "0." Most historians believe this chronology was inaccurate and that Jesus would have been born about 5 B.C.

A.D. stands, not for "after death," but for *anno Domini*, meaning "in the year of the Lord." The term A.D. was introduced by the sixth-century monk Dionysius Exiguus, about A.D. 528, in his attempts to trace the chronology of the Bible.

Recently, two different abbreviations have begun to replace A.D. and B.C. They are B.C.E. and C.E., which stand, respectively, for "before the common era" and "common era." "Common era" refers to our present calendar, which is used nearly worldwide for most business purposes and thereby avoids reference to a particular religion. Yet another abbreviation—B.P.—is used, especially in anthropological and geological works. B.P. stands for "before present (era)" and is used for dates determined by analyzing the radioactive carbon in the object of interest. It takes 1950 C.E. as its base year.

SUMMARY

Our system for keeping time is based on the motion of the Earth, Moon, and Sun. The day is determined by the Earth's spin, the month by the Moon's orbital motion around the Earth, and the year by the Earth's orbital motion around the Sun.

The solar day is based on the time interval between one apparent noon and the next. The sidereal day, or the interval between the time of star-rise for a given star and the time of its next rising, is about 4 minutes shorter than the solar day. This difference arises because as the Earth moves along its orbit, the direction to the Sun shifts slightly. We must therefore wait a little longer to allow the Earth's rotation to carry us into the same position with respect to the Sun.

Time zones divide the Earth into regions such that the time differs by 1 hour (in general) from zone to zone. The resulting time difference allows the Sun to be approximately at its highest point above the horizon at noon in each zone.

The Earth makes approximately 365.25 rotations in the time it takes it to complete one orbit around the Sun. Thus, every 4 years, an extra day accumulates, which in leap years we add to the calendar as February 29.

QUESTIONS FOR REVIEW

1. How does the number of hours of daylight vary with location and time of year?
2. How is the solar day defined? How is the sidereal day defined?
3. Why do the sidereal and solar days differ in length?
4. Why isn't the solar day always exactly 24 hours long?
5. Why do we need a leap year?
6. What do A.M., P.M., B.C., A.D., B.C.E.,and C.E. stand for?

THOUGHT QUESTIONS

1. Why do you suppose that ancient mathematicians choose to divide a circle into 360°?
2. One might logically conclude that adding a leap day would make the most sense at the end of the calendar year. Thinking about the history of the calendar, reconcile this idea with the fact February has 28 days.
3. Speculate on the role of the Moon's importance versus the seasons in defining a calendar for cultures at different latitudes.
4. Suppose you were asked to revise the calendar. What changes would you make?

PROBLEMS

1. Compare the 7-day week to the Roman 8-day week or the French Revolution experiment with a 10-day week. In trying to match lunar and solar cycles, with a whole number of "weeks," which works best? Is there another number of days that works better?
2. If there were no leap days, after how many years would the seasons align correctly with the calendar again?

TEST YOURSELF

1. Suppose the Earth's rotation axis were not tilted with respect to its orbit. How would the number of daylight hours change throughout the year?

   (a) The number would be no different.
   (b) Days would be longer and nights shorter at all year.
   (c) Days and nights would be of equal length all year.
   (d) Days would be shorter and nights longer all year.
   (e) None of the above.

2. If on a given date there are 24 hours of night at the North Pole, how many hours of night are there at the South Pole?

   (a) 12 hours
   (b) 24 hours
   (c) 36 hours
   (d) 48 hours
   (e) There is no night then.

3. On what day(s) of the year are nights longest at the equator?

   (a) They are the same length throughout the year there.
   (b) The solstices
   (c) The equinoxes
   (d) Approximately June 21
   (e) Approximately December 21

4. Why is February the shortest month?

   (a) The Earth is moving most slowly in its orbit then.
   (b) The Earth is moving fastest in its orbit then.
   (c) The Earth spins faster in February than at other times of the year.
   (d) The Earth spins slower in February than at other times of the year.
   (e) When the calendar was revised, days were taken from February to make other months longer.

5. Suppose that the length of the year were 365.2 days instead of 365.25 days. How often would we have leap year? Every

   (a) 2 years.
   (b) 5 years.
   (c) 10 years.
   (d) 20 years.
   (e) 50 years.

KEY TERMS

daylight saving time, 181
Gregorian calendar, 183
international date line, 181
Julian calendar, 182
mean solar day, 179
sidereal day, 179
solar day, 179
standard time, 180
time zone, 180
Universal time, 181

FURTHER EXPLORATIONS

Bartky, Ian R., and Elizabeth Harrison. "Standard and Daylight-Saving Time." *Scientific American* 240 (May 1979): 46.

Cleere, G. S. "Eleven Lost Days." *Natural History* 100 (September 1991): 78.

Duncan, David E. *Calendar: Humanity's Epic Struggle to Determine a True and Accurate Year.* New York: Avon Press, 1998.

Jespersen, James, and Jane Fitz-Randolph. *From Sundials to Atomic Clocks: Understanding Time and Frequency.* National Bureau of Standards Monograph 155. Washington, D.C.: U.S. Government Printing Office, 1977.

Monson, B. "A Simple Method of Measuring the Length of the Sidereal Day." *Physics Teacher* 30 (December 1992): 558.

Moyer, Gordon. "The Gregorian Calendar." *Scientific American* 246 (May 1982): 144.

Website

Visit the *Explorations* website at **http://www.mhhe.com/arny** for additional online resources on these topics.

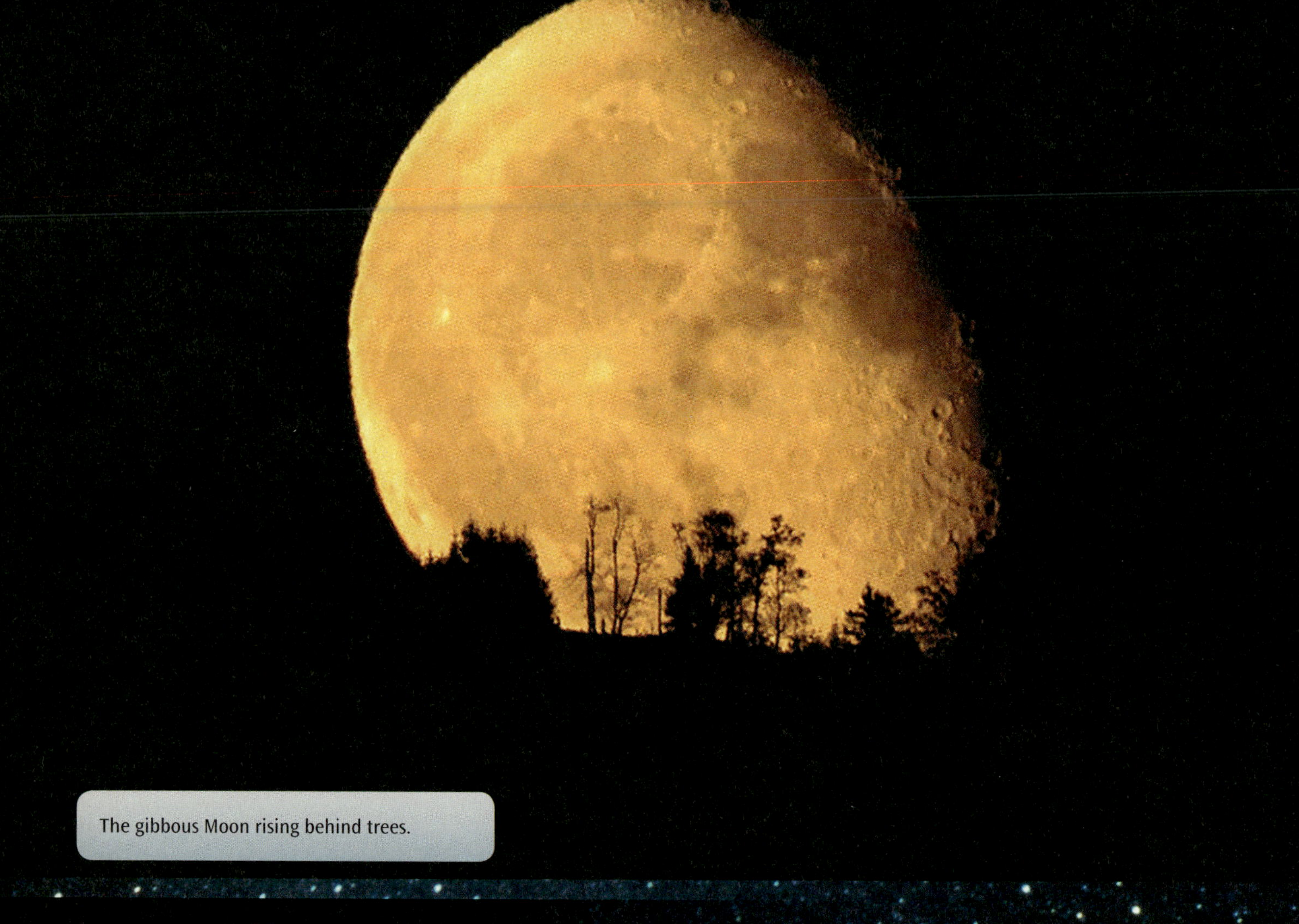

The gibbous Moon rising behind trees.

KEY CONCEPTS

- The Moon is much smaller than the Earth.
- The Moon's small size gives it a weak gravity, so that today it lacks an atmosphere.
- Because of the Moon's small size It has cooled much more since the time of its formation than the Earth has.
- Because its interior is so relatively cool, the Moon is not active and lacks crustal motions and active volcanoes.
- The Moon's surface is covered with craters blasted into its surface by asteroid impacts.
- The Moon was probably formed from debris blasted out of the Earth by the impact of a planet-size object shortly after the Earth's formation.
- The Moon's gravity creates tides in the Earth's oceans.
- Because tides depend on differences in the gravitational force, high tides occur on the sides of Earth nearest and fathest from the Moon.
- Tidal braking is slowing the spin of the Earth and probably caused the Moon to slow its spin until one side always faced the Earth.

7 The Moon

CONCEPTS AND SKILLS TO REVIEW

- Lunar phases (1.3)
- Density (6.1)
- Law of gravity (3.4)
- Escape velocity (3.8)

The Moon is our nearest neighbor in space, a natural satellite orbiting the Earth. It is a barren ball of rock, with about one-fourth the diameter of Earth, possessing no air, water, or life. In the words of lunar astronaut Buzz Aldrin, the Moon is a place of "magnificent desolation."

The Moon is the frontier of direct human exploration, an outpost that we reached more than a quarter century ago but from which we have since drawn back. But despite our retreat from its surface, the Moon remains of great interest to astronomers. Although originally it was molten, its small mass and radius made the Moon unable to generate or retain any appreciable internal heat. It is therefore a dead world, with neither plate tectonic nor volcanic activity. That inactivity, coupled with the Moon's lack of atmosphere, means that its surface features are essentially unaltered since its youth.

The Moon has not always been inactive. Shortly after its formation, it was pelted with a hail of rocky fragments up to 200 kilometers (about 100 miles) in diameter. The small fragments made craters, and the big fragments made huge basins. The basins subsequently flooded with lava (long since congealed) to create several dark, nearly circular plains easily visible to the naked eye. The Earth probably once bore such features, but erosion and plate motions have erased them. On the Moon's windless, rainless, airless surface, they remain as a record of events in the early Solar System, giving clues to the birth of not only the Moon but also the Solar System.

In this chapter, we will describe the Moon's surface and why astronomers conclude that so many of its features were carved by impact. We will see that lunar rocks differ significantly from terrestrial rocks and how they point to the Moon's having been born in a cataclysmic event early in the Earth's history. We will also discuss how the Moon affects Earth today through tides and eclipses. But we will begin with a short physical description of the Moon to help us visualize this nearest world.

Q: WHAT IS THIS? See end of chapter for the answer

7.1 The Surface of the Moon

Surface Features

To the naked eye, the Moon is a world of grays. Some patches are darker than others, creating a vague impression of what some see as a face ("the man in the moon") or maybe a rabbit. Closer examination reveals that the dark patches are in fact quite different from their lighter surroundings. Through a small telescope or even a pair of binoculars, you can see that the dark areas are smooth while the bright areas are covered with numerous large circular pits called **craters,** as illustrated in figure 7.1. Craters usually have a raised rim and range in size from tiny holes less than a centimeter across to gaping scars in the Moon's crust such as Clavius,* about 240 kilometers (150 miles) across. Some of the larger craters have mountain peaks at their center.

The large, smooth, dark areas are called **maria** (pronounced MAR-ee-a), from the Latin word for "seas." However, these regions, like the rest of the Moon, are totally devoid of water. This usage comes from early observers who believed the maria looked like oceans and who gave them poetic names such as Mare (pronounced MAR-ay), Serenitatis (Sea of Serenity), and Mare Tranquillitatis (Sea of Tranquility), the site where astronauts first landed on the Moon.

The bright areas that surround the maria are called **highlands.** The highlands and maria differ in brightness because they are composed of different rock types. The maria are basalt, a dark, congealed lava rich in iron, magnesium, and titanium silicates. The highlands, on the other hand, are mainly anorthosite, a rock type rich in calcium and aluminum silicates. This difference has been verified from rock samples obtained

*Most lunar craters are named for famous scientists. For example, Cristoph Clavius (1537–1612) was a German astronomer and mathematician.

FIGURE 7.1
Photograph showing the different appearance of the lunar highlands and maria. The highlands are heavily cratered and rough. The maria are smooth and dark and have few craters. The long, narrow, white streaks radiating away from some of the craters bottom are lunar rays.

Q. Some rays cross maria. What does this imply about the relative age of the rays and the maria?

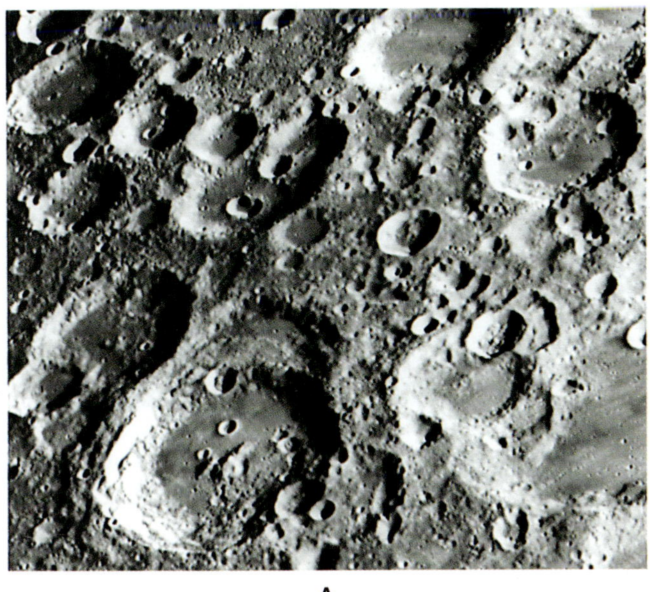

FIGURE 7.2
(A) Overlapping craters in the Moon's highlands. (B) Isolated craters in the smooth mare.

Q. In (A) a small crater lies at the edge of a larger one. Which formed more recently: the small one or the large one?

by astronauts. Moreover, the samples also show that the highland material is generally less dense than mare rock and considerably older.

The highlands are not only brighter and their rocks less dense than the maria, they are also more rugged, being pitted with craters. In fact, highland craters are so abundant that they often overlap, as shown in figure 7.2A. Contrast this picture with the mare region shown in figure 7.2B, in which only a few, small craters are visible.

From many craters, long, light streaks of pulverized rock called **rays** radiate outward, as can be seen in figure 7.1. A particularly bright set spreads out from the crater Tycho near the Moon's south pole and can be seen easily with a pair of binoculars when the Moon is full.

A small telescope reveals still other surface features. Lunar canyons known as **rilles,** perhaps carved by ancient lava flows, wind away from some craters, as shown in figure 7.3. Elsewhere, straight rilles gouge the surface, probably the result of crustal cracking. Drying mud and chocolate pudding left too long in the refrigerator show similar cracks.

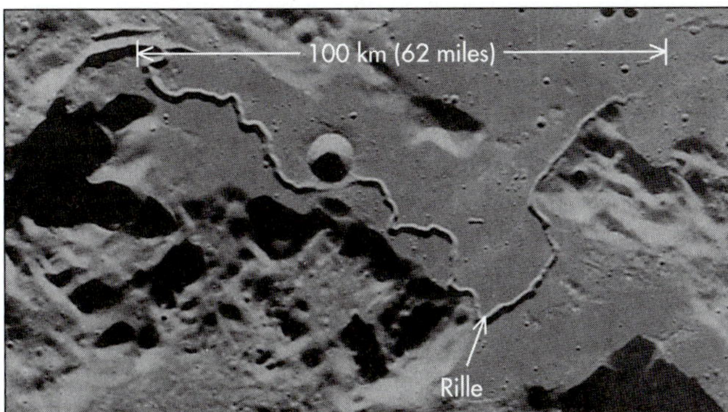

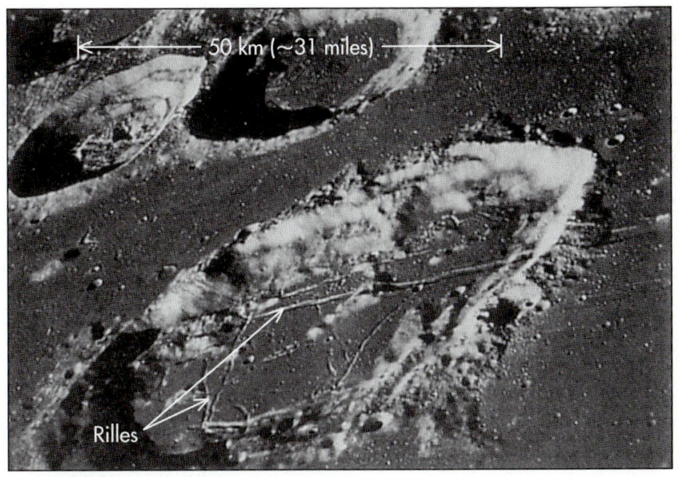

FIGURE 7.3
Photographs of some lunar rilles.

A

B

FIGURE 7.4
(A) Central peak in a crater and slumped inner walls. Apollo astronauts took this photograph of the crater Eratosthenes on the last manned flight to the Moon. This crater is 58 kilometers (approximately 36 miles) in diameter. (B) A drop of rain falling into water produces an effect similar to the one that creates central peaks in lunar craters. The drop falls into the water but is then pushed up again as the water rebounds, as shown in this high-speed photo.

Origin of Lunar Surface Features

Nearly all the surface features we see on the Moon—craters, maria, and lunar rays—were made by the impact of solid bodies on its surface. When such an object hits a solid surface at high speed, it disintegrates in a cloud of vaporized rock and fragments. The resulting explosion blasts a hole whose diameter depends on the mass and velocity of the impacting object. The hole's shape is circular, however, unless the impact is grazing.

As the vaporized rock expands from the point of impact, it forces surrounding rock outward, piling it into a raised rim. Pulverized rock spatters in all directions, forming rays. Sometimes the impact compresses the rock below the crater sufficiently that it rebounds upward, creating a central peak, as shown in figure 7.4A. Figure 7.4B shows a similar process happening in a high-speed picture of a raindrop falling into water. In other cases, material blasted out into the crater's rim may slump back into the crater. As it slides in toward the crater's center, the material pushes up a peak.

Astronomers think that the maria are also impact features, but to understand their formation we must briefly describe the early history of the Moon. From the great age of the highland rocks (in some cases as old as 4.5 billion years), astronomers deduce that these rugged uplands formed shortly after the Moon's birth. At that time, the Moon was probably molten, allowing dense, iron-rich material to sink to its interior while less-dense material floated to the lunar surface. On reaching the surface, the less-dense rock cooled and congealed, forming the Moon's crust. A similar process probably formed the Earth's continents. The highlands were then heavily bombarded by solid bodies from space, forming the numerous craters we see there.

Before the Moon solidified completely, a small number of exceptionally large bodies (with diameters of more than about 100 kilometers or 60 miles) struck the surface, blasting huge craters and pushing up mountain chains along their edges, as you can see in figure 7.5. Mare Orientale (fig. 7.5A) exhibits the multiple rings from a major impact, the shock waves frozen in place around the impact site. Subsequently, molten material from within the Moon flooded the vast crater and congealed to form the smooth, dark lava plains that we see now, as illustrated in figure 7.6. Because the denser material sunk to the Moon's interior during its molten stage, the erupted lava from those depths was denser than the crustal rock into which it flooded. Moreover, because it was molten more recently, the mare material is therefore younger than the highlands. By the time the maria formed, most of the impacting bodies were gone—collected into the Earth and Moon in earlier collisions. Therefore, too few bodies remained to crater the maria, which therefore remain relatively smooth, even to this day.

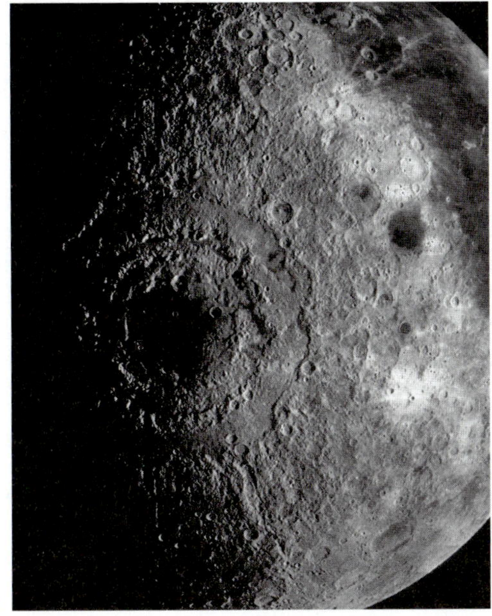
A

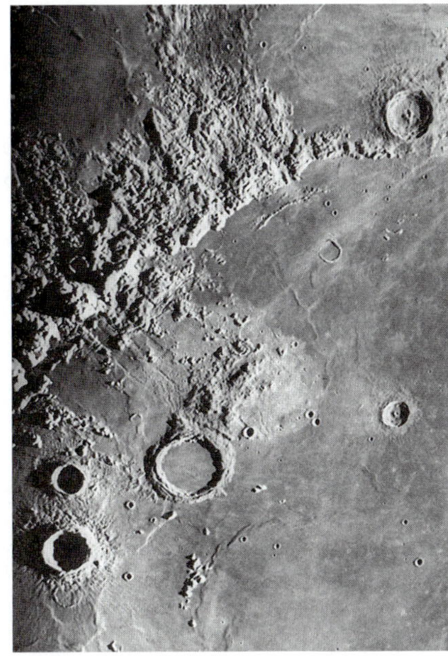
B

FIGURE 7.5
(A) Mare Orientale shows the multiple ring structure from a major impact. The central area of the impact crater was flooded by lava after the impact. (B) Mountains along the edge of a mare were probably thrown up by the impact that created the mare.

Our home planet Earth furnishes additional evidence that most lunar craters formed early in the Moon's history. Like the Moon, Earth too was presumably battered by impacts in its youth. Although the vast multitude of these craters have been obliterated by erosion and plate tectonics, a few remain in ancient rock whose measured age is typically hundreds of millions of years. From the scarcity of such craters, astronomers can deduce that the main bombardment must have ended billions of years earlier.

Craters and maria so dominate the lunar landscape that we might not notice the absence of folded mountain ranges and the great rarity of volcanic peaks, land forms common on Earth. Why have such features not formed on the Moon?

Asteroid

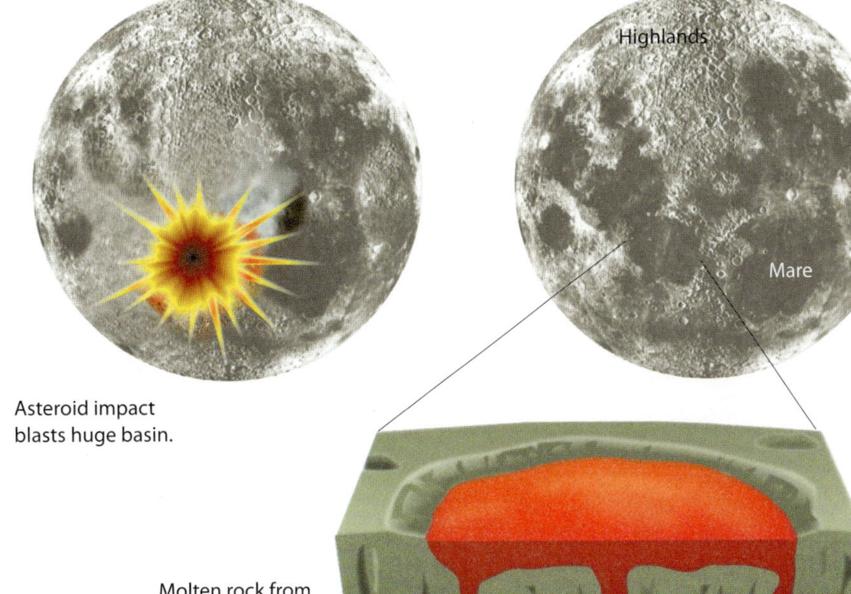
Highlands
Mare
Asteroid impact blasts huge basin.
Molten rock from Moon' s interior rises into basin and floods it.

FIGURE 7.6
A rain of debris creates craters on the young Moon. Large impacts late in the process form the maria basins. Lava floods the basins to make the maria.

7.2 Structure of the Moon

The Moon's small size relative to the Earth explains the differences between the two bodies. Because its volume compared to its surface area is small relative to the Earth's, heat escapes far more easily from the Moon. Thus, the Moon has cooled far more than the Earth. (Think of how a small baked potato cools much faster than a big one.) Moreover, because its mass is much smaller than the Earth's, the Moon contains much less radioactive material and so cannot generate as much heat. Thus, having a much cooler interior, the Moon lacks the convection currents that drive plate tectonic activity on the Earth. Confirmation of this comes from studies of the Moon's interior.

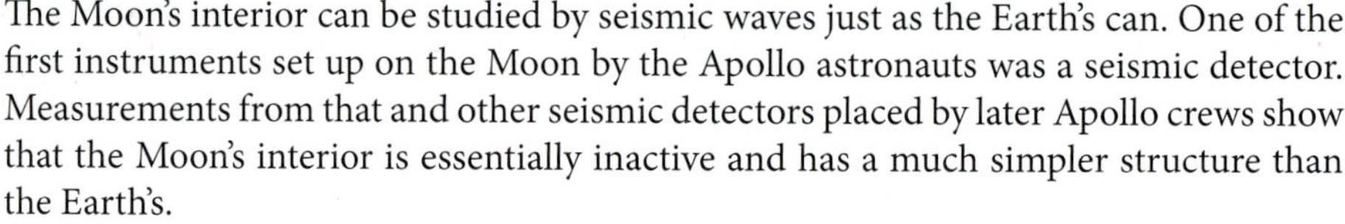

Crust and Interior

FIGURE 7.7
Footprint of an astronaut on the Moon.

The Moon's interior can be studied by seismic waves just as the Earth's can. One of the first instruments set up on the Moon by the Apollo astronauts was a seismic detector. Measurements from that and other seismic detectors placed by later Apollo crews show that the Moon's interior is essentially inactive and has a much simpler structure than the Earth's.

The Moon's surface layer is shattered rock that forms a **regolith**—meaning "blanket of rock"—tens of meters deep. The regolith consists of both rock chunks and fine powder, the result of successive impacts breaking rock into smaller and smaller pieces. This powdery nature is easily seen in the crispness of the astronauts' footprints (fig. 7.7). Samples of the regolith picked up by astronauts show that these surface rocks are typically the same type as the underlying rock. That is, the regolith on maria is generally broken-up basalt, whereas that on the highlands is broken-up highland material. In places the regolith may be several hundred meters thick. Analysis of the regolith shows that over time, its rocks have been broken up by high-velocity impacts, supporting the interpretation that the surface has been bombarded by meteoritic bodies.

Below this surface layer of rocky rubble is the Moon's crust, about 100 kilometers (60 miles) thick, on the average. The crust is much thinner (about 65 kilometers) on the side of the Moon that faces the Earth than on the far side (about 150 kilometers), as shown in figure 7.8. The reason for this difference is not clear, but it may have resulted from the Earth's gravity shifting the Moon's core slightly toward Earth billions of years ago, when the Moon's interior was hotter. The crust on the near side—being slightly closer to the Moon's core because of that shift—might therefore have become hotter and as a result thinner than that on the far side. Subsequently, the Moon cooled, leaving the crust thinner on one side than on the other.

The thinner crust on the near side made it much easier for maria to form, as you can see in figure 7.9, which compares the two sides of the Moon. Note that the largest impact

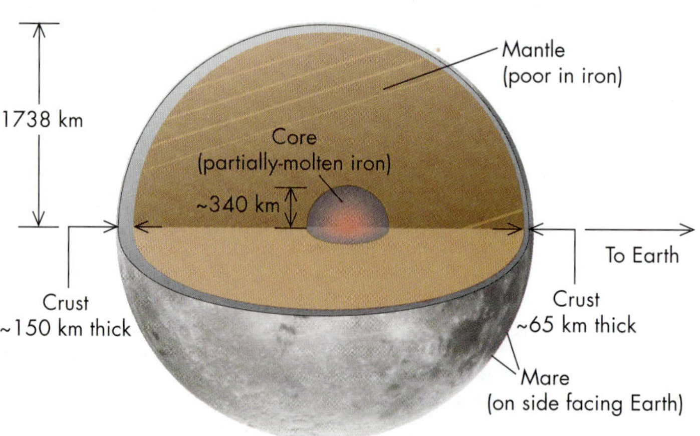

FIGURE 7.8
An artist's impression of the Moon's interior. Notice the thinner near-side crust and the displacement (exaggerated for clarity) of the core toward the Earth.

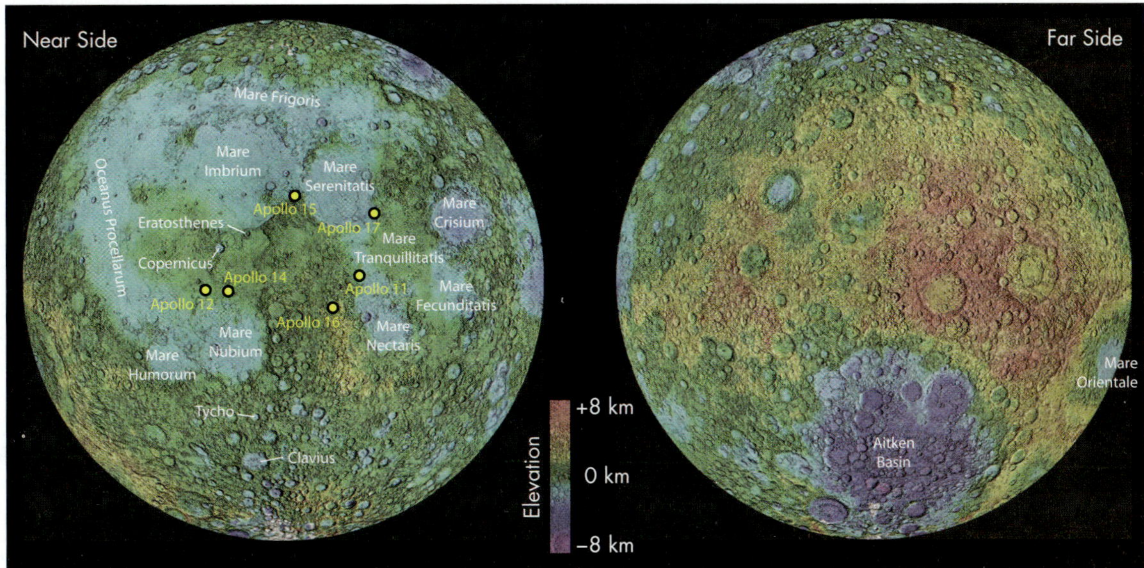

FIGURE 7.9
Topographic map showing the near and far sides of the Moon. The elevations were mapped by the Clementine satellite and are shown in different colors. The maria are generally at lower elevations, but the largest impact feature, the Aitken Basin, is at an even lower elevation. Several major features and the locations of the six Apollo landing sites are labeled.

Q. Why can't we see the far side of the Moon from Earth?

feature, Aitken Basin, is on the far side of the Moon. It reaches more than 10 kilometers beneath the height of the surrounding terrain, but it was not flooded with basalt.

The Moon's crust, like the Earth's, is composed of silicate rocks relatively rich in aluminum and poor in iron. Beneath the crust is a thick mantle of solid rock, extending down a little more than 1000 kilometers (600 miles). The Moon's mantle is probably rich in olivine, the same type of dense, greenish rock that composes most of the Earth's mantle. Unlike the Earth's mantle, however, it appears too cold and rigid to be stirred by the Moon's feeble heat.

The Moon's low density (3.3 grams per cubic centimeter) tells us its interior contains little iron. Recall that in chapter 6 we saw that the Earth's high density (about 5.5 grams per cubic centimeter) is an indication that it has a large iron core. Some molten material may lie below the mantle, as illustrated in figure 7.8, but the Moon's core is smaller and contains far less iron and nickel than the Earth's. These factors, plus the fact that the Moon rotates very slowly (as we will see in section 7.3), lead astronomers to think that the Moon's core is unable to generate a magnetic field as the Earth's does. Measurements made by the Apollo astronauts confirm that the Moon has essentially no magnetic field. Thus, a compass would be of no use to an astronaut lost on the Moon.

The Absence of a Lunar Atmosphere

The Moon's surface is never hidden by lunar clouds or haze, nor does the spectrum of sunlight reflected from it show obvious signs of gases. With no atmosphere* to absorb

*Lunar scientists have found tiny quantities of gas, mostly helium, above the Moon's surface. This gas is so tenuous (less than one-quadrillionth the density of our atmosphere) that we will ignore it. Perhaps more interesting, however, is the detection of traces of hydrogen near the Moon's poles by a spacecraft orbiting the Moon. Some scientists think the hydrogen comes from the breakdown of frozen water mixed with rock in several craters near the lunar poles. The water itself may have come from comets striking the Moon and vaporizing. The water vapor then condensed in the coldest places on the Moon (the polar craters into which sunlight never shines). You may have seen this tendency for frost to form in cold spots if you have taken something out of a freezer and left it for a while on a table.

and trap heat, temperatures on the Moon soar during the day and plummet at night. Likewise, no wind blows. The lunar surface lies dead and silent under a black sky.

The Moon has no atmosphere for two reasons. First, its interior is too cool to cause volcanic activity, which as we saw in chapter 6 was probably the source of much of the Earth's early atmosphere. Second, and more important, even if volcanos had created an atmosphere in its youth, the Moon's small mass creates too weak a gravitational force for it to retain the erupted gas. In chapter 3, we learned that the Moon's escape velocity is only about a fourth that of the Earth's (2.4 kilometers per second versus 11 kilometers per second), and so atoms in the Moon's atmosphere would have found it easy to escape its gravity. With no atmosphere and no plate tectonics, the Moon has been essentially unchanged for billions of years. The footprints left on the Moon by the astronauts in 1969 will probably still be there a million years from now.

7.3 Orbit and Motions of the Moon

By watching the Moon for a few successive nights, you can see it move against the background stars as it follows its orbit around the Earth. The Moon's orbit is elliptical, with an average distance from Earth of 380,000 kilometers (about 250,000 miles) and a period of 27.3 days.* Its distance can be measured by triangulation, radar, or laser beams (fig. 7.10). To triangulate its distance, astronomers observe the Moon from two different spots on the Earth. The distance between the locations, the angles to the Moon, and a little trigonometry give the Moon's distance.

A more accurate method is to bounce either a radar pulse or a laser beam off special reflectors that were placed on the Moon by the Apollo astronauts. Half the time interval between the transmission and the return of the reflected radiation multiplied by the speed of light gives the Moon's distance to an accuracy of centimeters.

The Moon's Rotation

As it orbits, the Moon keeps the same side facing the Earth, as you can see by watching it through a cycle of its phases. You might think from this that the Moon does not rotate. Figure 7.11 shows, however, that the mountain on the side facing the Earth points

*This is the time to complete an orbit around the Earth and is shorter than the cycle of the phases, as discussed in chapter 1.

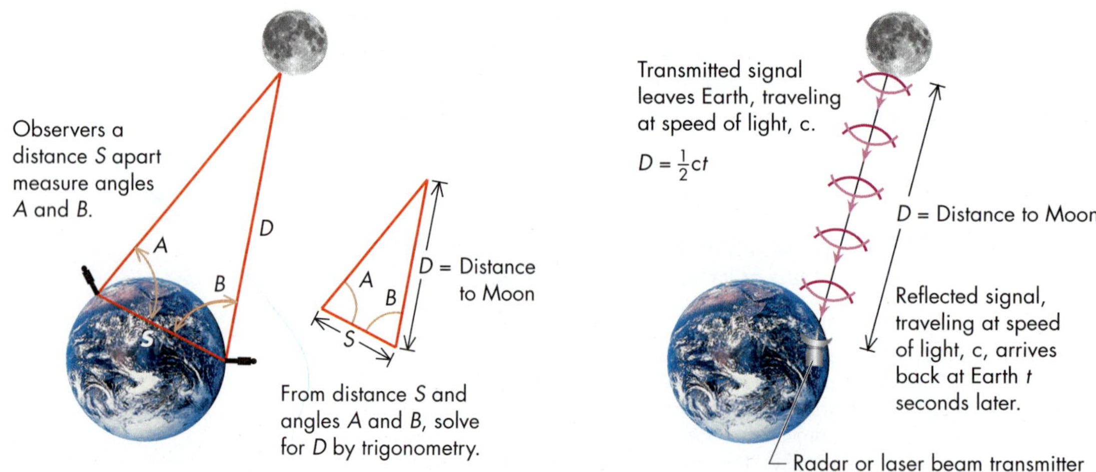

FIGURE 7.10
Finding the distance from the Earth to the Moon by triangulation, radar, and laser ranging.

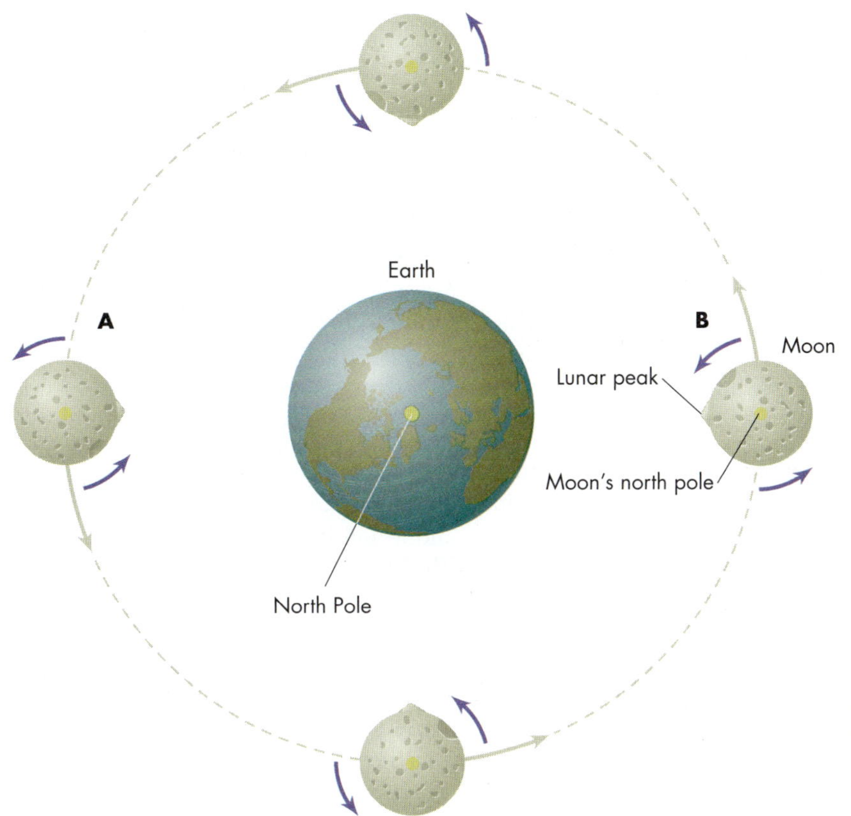

ANIMATION
The rotation of the Moon

FIGURE 7.11
The Moon rotates once each time it orbits the Earth, as can be seen from the changing position of the exaggerated lunar mountain. Notice that at (A) the lunar peak is to the right, while at (B) it is to the left. Thus, from the Earth, we always see the same side of the Moon even though it turns on its axis.

Q. To help see that the Moon rotates even though it keeps the same face toward the Earth, put a coin on the figure of the Moon and move it around the Earth so that the same edge of the coin always faces the Earth.

to the right when the Moon is at A and to the left when the Moon is at B. Thus, the Moon *does* turn on its axis but with a rotation period exactly equal to its orbital period, a condition known as **synchronous rotation.** The Earth's gravity causes this locking of the Moon's spin to its orbital motion, as we will discuss in section 7.6.

Oddities of the Moon's Orbit

The Moon's orbit is tilted by about 5° with respect to the Earth's orbit around the Sun, as illustrated in figure 7.12. It is also tilted with respect to the Earth's equator and is thus unlike most of the moons of Jupiter, Saturn, and Uranus, which lie nearly exactly in their planet's equatorial plane. Also, most moons are tiny compared to their planets. Even the largest of the moons of Jupiter and Saturn have masses less than 1/1000 that of their planet. But our Moon's mass is 1/81 that of the Earth. These oddities suggest that our Moon formed differently from the moons of other planets, but how?

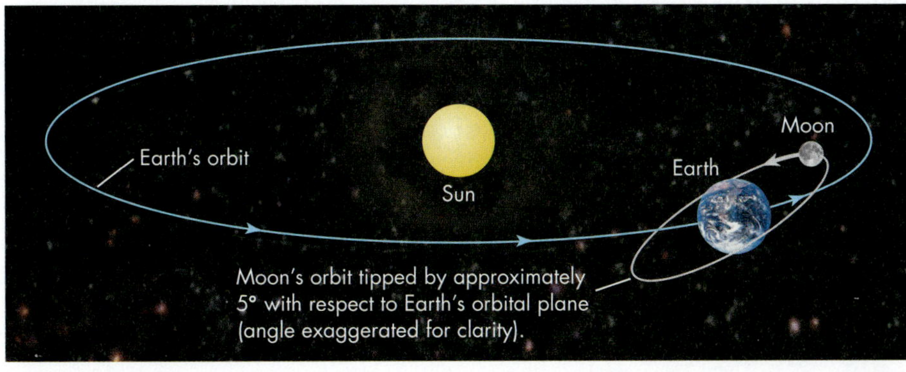

FIGURE 7.12
The Moon's orbit is tipped 5° with respect to the Earth's. The angle is exaggerated in this picture for clarity. Sizes and separations of objects are not to scale.

7.4 Origin and History of the Moon

Lunar rocks brought back to Earth by the Apollo astronauts caused astronomers to radically revise their ideas of how the Moon formed. Before the Apollo program, lunar scientists had three hypotheses regarding the Moon's origin:

- The Moon was originally a small planet orbiting the Sun; it approached the Earth and was captured by Earth's gravity (capture theory).
- The Moon and Earth were "twins," forming side by side from a common cloud of dust and gas (twin formation theory).
- The Earth initially spun enormously faster than now and formed a bulge that ripped away from the Earth to become the Moon (fission theory).

Each of these hypotheses led to different predictions about the composition of the Moon. For example, had the Moon been a captured planet, its composition might be very unlike the Earth's. If the Earth and Moon had formed as twins, their overall composition should be similar. Finally, if the Moon was once part of the Earth, its composition should be nearly identical to the Earth's crust. When the rock samples were analyzed, astronomers were surprised that for some elements the composition was the same, but for others it was very different. For example, the Moon has a relatively high abundance of high-melting-point materials such as titanium and an almost complete lack of low-melting-point materials such as water. It also has much less iron than the Earth, as we pointed out when discussing its interior and low density.

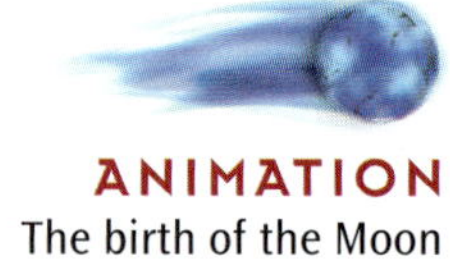

ANIMATION
The birth of the Moon

The failure of evidence based on lunar surface samples to confirm any of the three hypotheses led astronomers to consider alternatives, and now a completely different picture of the Moon's origin has emerged. According to the new hypothesis, the Moon formed from debris blasted out of the Earth by the impact of a Mars-sized body, as shown in figure 7.13A. The great age of lunar rocks and the absence of any impact feature on the Earth indicate that this event must have occurred during the Earth's own formation, at least 4.5 billion years ago. The colliding body melted and vaporized millions of cubic kilometers of the Earth's surface rock and hurled it into space in an incandescent plume. As the debris cooled, its gravity gradually drew it together into what we now see as the Moon.

This violent-birth hypothesis explains many of the oddities of the Moon. The impact would vaporize low-melting-point materials and disperse them, leaving, for example, little water to be incorporated into the lunar body. Computer models (fig. 7.13B) of such an event also show that only surface rock would be blasted out of the Earth, leaving our planet's iron core intact, thereby also explaining the low iron content of lunar rocks. The splashed-out rock would condense in an orbit whose shape and orientation were determined by the collision rather than by the orientation of the Earth's equator. Furthermore, we would expect both similarities and differences in composition between the Earth and Moon because the Moon was made partly from Earth rock and partly from rock of the impacting object.

After the Moon's birth, stray fragments of the ejected rock pelted its surface, creating the craters that blanket the highlands, as described earlier. A few huge fragments plummeting onto the Moon later in its formation process blasted enormous holes that later flooded with molten interior rock to become the maria. That rock was probably melted in the Moon's interior by radioactive decay, as happened in the Earth. During the time it took the rock to melt, about half a billion years, most of the debris remaining in the Moon's vicinity fell onto its surface. Thus, by the time the maria flooded, little material was left to fall on them, and so they are only lightly cratered. Since that time, the Moon has experienced no major changes. It has been a virtually dead world for all but the earliest times in its history, but its gravity continues to influence the Earth through the action of tides.

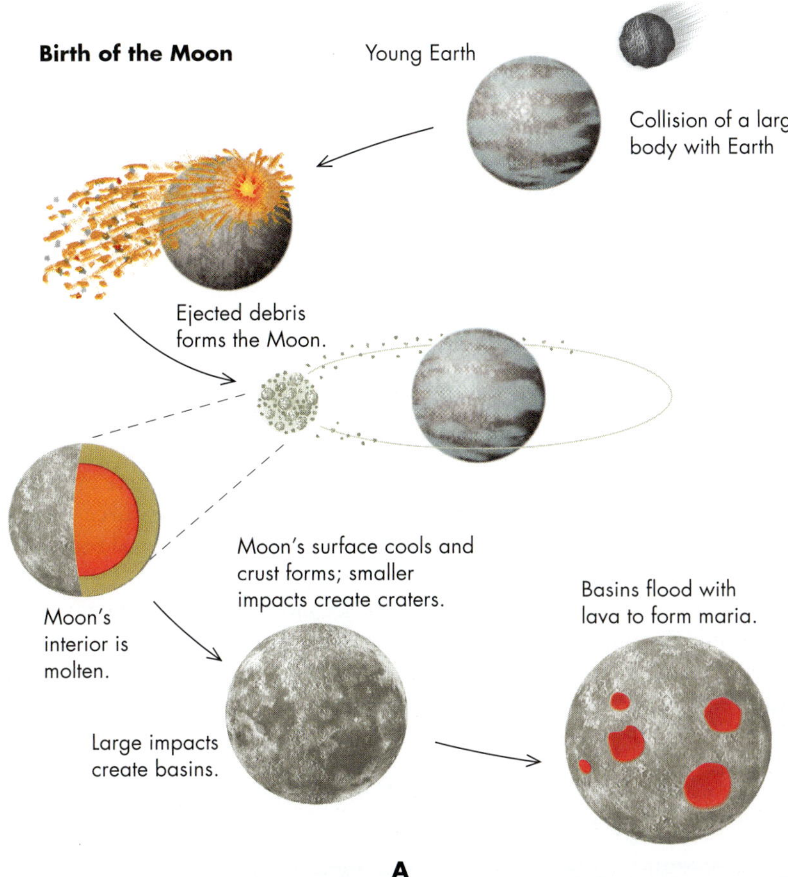
Birth of the Moon
Young Earth
Collision of a large body with Earth
Ejected debris forms the Moon.
Moon's interior is molten.
Moon's surface cools and crust forms; smaller impacts create craters.
Basins flood with lava to form maria.
Large impacts create basins.
A

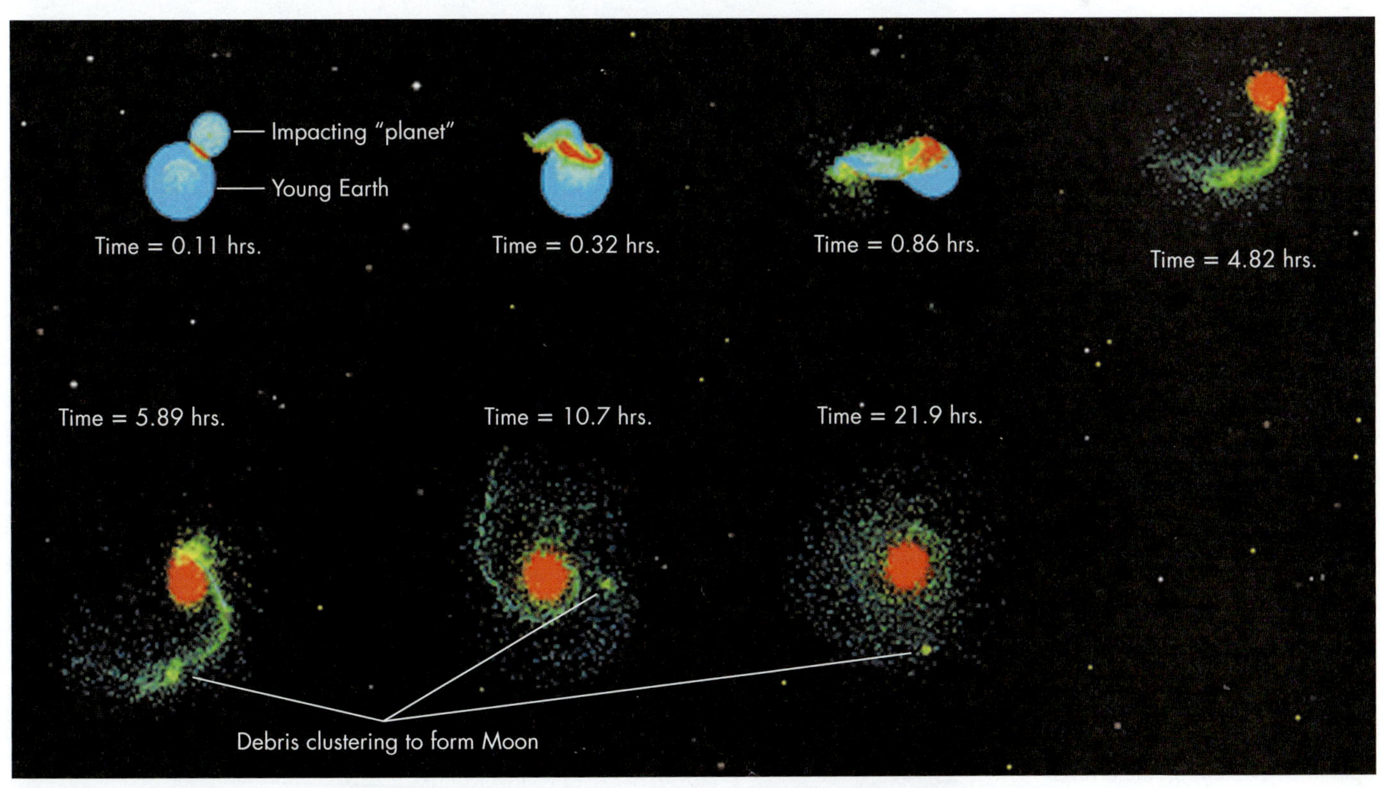
Impacting "planet"
Young Earth
Time = 0.11 hrs.
Time = 0.32 hrs.
Time = 0.86 hrs.
Time = 4.82 hrs.
Time = 5.89 hrs.
Time = 10.7 hrs.
Time = 21.9 hrs.
Debris clustering to form Moon
B

FIGURE 7.13
Origin of the Moon. (A) Artist's sketch illustrating main stages in the birth of the Moon. (B) This computer simulation shows how the Moon might have formed when a Mars-size object hit the young Earth and splashed out debris that later assembled into the Moon.

7.5 Tides

Anyone who has spent even a few hours by the sea knows that the ocean's level rises and falls during the day. A blanket set on the sand 10 feet from the water's edge may be inundated an hour later, or a boat pulled ashore may be left high and dry. This regular change in the height of the ocean is called the **tides** and is caused mainly by the Moon.

Cause of Tides

Just as the Earth exerts a gravitational pull on the Moon, so too the Moon exerts a gravitational attraction on the Earth and its oceans and draws material toward it. The attraction is stronger on the side of the Earth near the Moon and weaker on the far side (see fig. 7.14) because the force of gravity weakens with distance (recall Newton's law of gravity, section 3.4). The difference between the strong force on one side and the weaker force on the other is called a **differential gravitational force.**

The differential gravity draws water in the oceans into a **tidal bulge** on the side of the Earth facing the Moon, as shown in figure 7.14.* But curiously, it creates an identical tidal bulge on the Earth's far side. This second tidal bulge can be viewed as a result of the Moon's gravity pulling the Earth "out from under" the water on the far side. A better approach, however, is to examine the Moon's gravitational forces on the Earth and its oceans as seen by a person on the Earth, as shown in figure 7.15.

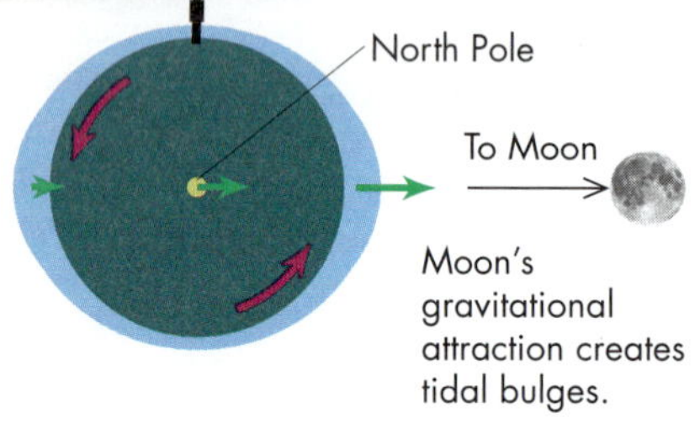

FIGURE 7.14
Tides are caused by the Moon's gravity creating tidal bulges.

*The tidal bulges do not exactly align with the Moon, for reasons we will discuss in the section on tidal braking.

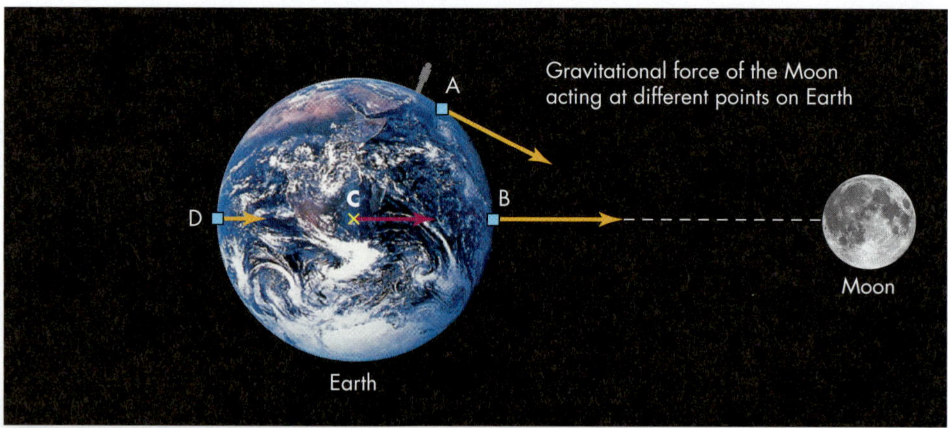

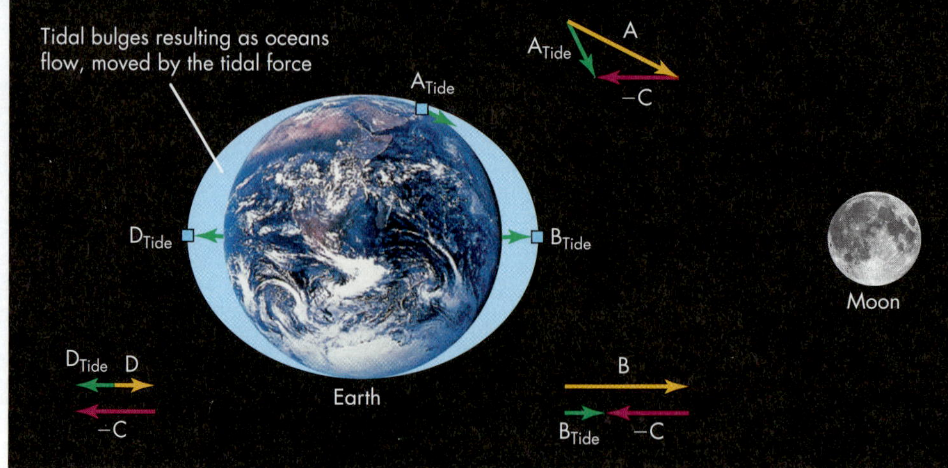

FIGURE 7.15
(Top) Arrows schematically show the Moon's gravitational force at different points on the Earth. (Bottom) Tidal forces from the point of view of an observer on the Earth. These arrows represent the difference between the Moon's gravitational force at a given point and its force at the Earth's center (C). Graphically, you can find the tidal force by "adding" the arrows. The figure shows schematically how to do this, but details are omitted.

The arrows in figure 7.15 (top) represent the Moon's gravitational pull at several points on the Earth. Points on the side of the Earth near the Moon (B) undergo a stronger pull toward the Moon than points on the far side (D), and so the arrow from point B is longer than the arrow from point D. Likewise, because point C is closer to the Moon than point D, the arrow from point C (which is the Moon's pull on the center of the Earth) is longer than the arrow from point D.

To see how the tidal bulges form, we need to look at the *difference* between the gravitational force at a given point and at the center of the Earth. For example, at point B, the force is larger than at point C, and so matter at point B will be pulled *away* from point C. This creates one tidal bulge. But matter at point C is in turn pulled away from point D, which creates a second tidal bulge. If we now draw a second set of arrows to represent the *difference* between the force at C and at every other point (the differential gravitational force), we find the forces illustrated in figure 7.15 (bottom). These drive the oceans into the bulges* that we see.

In the previous discussion, we have ignored the Earth's rotation. The tidal bulges are aligned approximately with the Moon, but the Earth spins. Its rotation therefore carries us first into one bulge and then the next. As we enter the bulge, the water level rises, and as we leave it, the level falls. Because there are two bulges, we are carried into high water twice a day, creating two high tides. Between the times of high water, as we move out of the bulge, the water level drops, making two low tides each day (fig. 7.16A).

This simple picture must be altered to account for the inability of the ocean to flow over land areas. Thus, water tends to pile up at coastlines when the tidal bulge reaches shore. In most locations, the tidal bulge has a depth of about 2 meters (6 feet), but it may reach 10 meters (30 feet) or more in some long narrow bays (as you can see in the photographs of high and low tides along the Maine coast in fig. 7.16B) and may even rush upriver as a tidal bore—a cresting wave that flows upstream. On some rivers, surfers ride the bore upstream on the rising tide.

The motion of the Moon in its orbit makes the tidal bulge shift slightly from day to day. Thus, high tides come about 50 minutes later each day, the same delay as in moonrise, discussed in chapter 1.

Solar Tides

The Sun also creates tides on the Earth, but although the Sun is much more massive than the Moon, it is also much farther away. The result is that the Sun's tidal force on the Earth is only about one-half the Moon's. Nevertheless, it is easy to see the effect of their tidal cooperation in spring tides, which are abnormally large tides that occur at new and full moon. At those times, the lunar and solar tidal forces work together, adding their separate tidal bulges, as illustrated in figure 7.17A. Notice that spring tides have nothing to do with the seasons; rather, they refer to the "springing up" of the water at new and full moon.

*Tides also occur in the atmosphere and solid ground, but tides in the ground are very small because the ground is rigid and cannot move as easily as water or air.

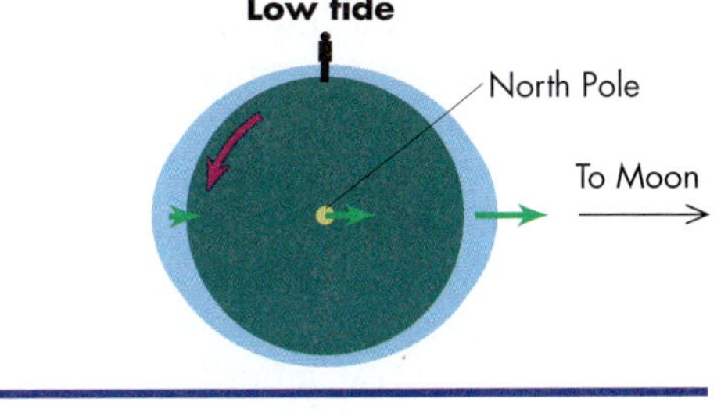

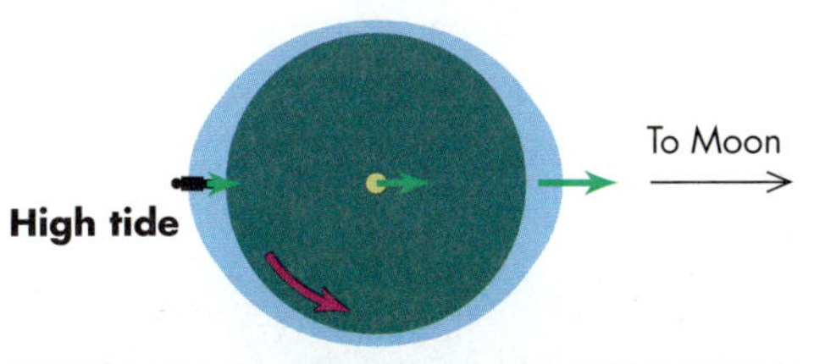

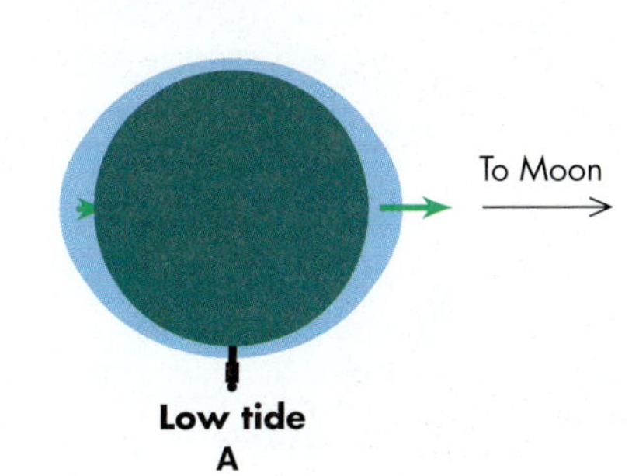

High tide

Low tide

B

FIGURE 7.16
As the Earth rotates, it carries points along the coast through the tidal bulges. Because there are two bulges where the water is high and two regions where the water is low, we get two high tides and two low tides each day at most coastal locations.

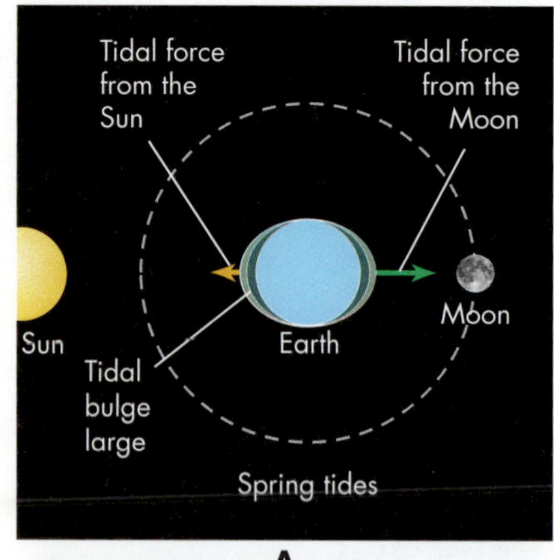

A

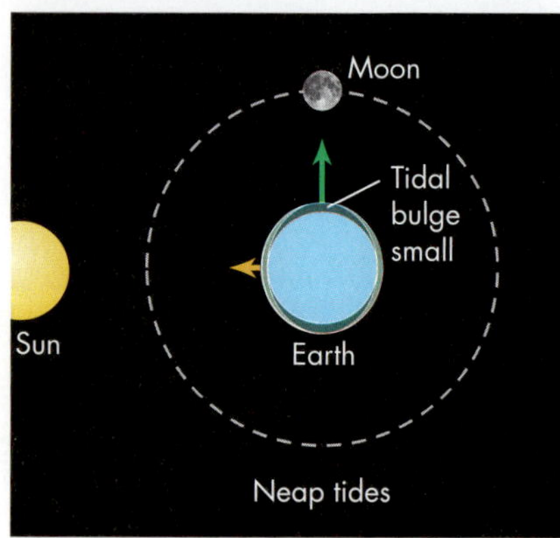

B

FIGURE 7.17
The Sun's gravity creates tides too, though its effect is only about one-half that of the Moon. (A) The Sun and Moon both create tidal bulges on the Earth. When the Sun and Moon are in line, their tidal forces add together to make larger-than-normal tides. (B) When the Sun and Moon are at 90° as seen from Earth, their tidal bulges are at right angles and partially nullify each other, creating smaller-than-normal tidal changes.

It may seem odd that spring tides occur at both new and full moon, because the Moon and Sun pull together when the Moon is new but in opposite directions when it is full. However, the Sun and the Moon both create two tidal bulges, and the bulges add together regardless of whether the Sun and Moon are on the same or opposite sides of the Earth. On the other hand, at first and third quarters, the Sun and Moon's tidal forces work at cross-purposes, creating tidal bulges at right angles to one another, as shown in figure 7.17B. The so-called neap tides that result are therefore not as extreme as normal high and low tides.

Tidal Braking

Tides create forces on the Earth and Moon that slow their rotation, a phenomenon known as **tidal braking.** Figure 7.18 shows how the Moon tidally brakes the Earth. As the Earth spins, friction between the ocean and the solid Earth below drags the tidal bulge ahead of the imaginary line joining the Earth and Moon, as depicted in figure 7.18. The Moon's gravity pulls on the bulge, as shown by the long green arrow in the figure, and holds it back. The resulting drag is transmitted through the ocean to the Earth, slowing its rotation the way a brake shoe on a car or your hand placed on a spinning bicycle wheel slows the wheel.

As the Earth's rotation slows, the Moon accelerates in its orbit, moving farther from the Earth, as required by the need to conserve angular momentum. The Moon accelerates because the tidal bulge it raises on the Earth exerts a gravitational force back on the Moon (as predicted by Newton's third law of motion), which pulls the Moon ahead in its orbit, as shown by the short green arrow in figure 7.18. That acceleration makes the Moon move away from the Earth at about 4 centimeters (roughly $1\frac{1}{2}$ inches) per year, a tiny increase in the Earth–Moon distance, but nevertheless detectable with laser range finders. Thus, the Moon was once much closer to the Earth and the Earth spun much faster, perhaps as rapidly as once every 5 hours several billion years ago. Over that immense period of time, the Moon has receded to its present distance, and the Earth's rotation has slowed to 24 hours. These processes occur even now: tidal braking lengthens the day by about 0.002 seconds each century.

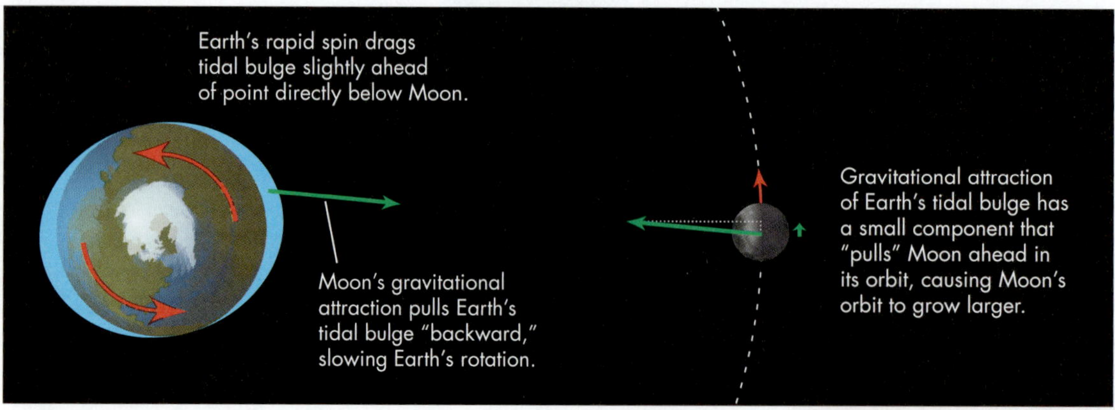

FIGURE 7.18
Tidal braking slows the Earth's rotation and speeds up the Moon's motion in its orbit. Friction between the oceans and Earth's solid crust "drags" the bulges of water "ahead" of the Earth–Moon line.

Q. Why doesn't the pull of the bulge on the far side of the Earth cancel the effects of the bulge closer to the Moon?

Tidal braking is also the reason the Moon always keeps the same face to the Earth. Just as the Moon raises tides, which slow the Earth, the Earth raises tides on the Moon, which slow it. These lunar tides distort the Moon's crust and have braked the Moon severely, locking it into synchronous rotation. The Moon's braking of the Earth will eventually make the Earth rotate synchronously with the Moon's orbital motion. Billions of years from now, the Earth and Moon will orbit so that each constantly presents the same face to the other: the Moon will then be visible only from one side of the Earth! Similar tidal effects have locked some of the moons of other planets into synchronous rotation, but the planets themselves have not been noticeably slowed.* On the other hand, tidal braking by the Sun probably slowed the rotation of Mercury and Venus.

The Moon's gravitational pull on the Earth may also stabilize our climate. Astronomers have recently discovered with computer simulations that the tilt of a planet's rotation axis may change erratically by many tens of degrees if the planet has no moon. Because the tilt causes seasons, changes in the tilt will alter the severity of the seasons. Our Moon is large enough that its gravitational attraction on Earth's equatorial bulge helps hold the Earth's tilt relatively fixed, sparing us catastrophically large climate changes.

*The dwarf planet Pluto is an exception. Its moon Charon has locked it into synchronous rotation.

SUMMARY

The Moon is the Earth's satellite. It is much smaller than the Earth: it has about one-fourth the Earth's radius and about 1/81 its mass. Its small size has allowed its internal heat to escape, keeping its core cool, thereby preventing plate tectonic motions. The Moon has no atmosphere because it is too cool to create one by volcanic outgassing and too small for its low gravity to retain gases that may have been present in the past.

With neither atmosphere nor tectonic activity, the Moon's surface is unaltered except by impact features: craters, rays, and the maria. Maria are enormous lava flows that have flooded into basins made by large impacting bodies late in the Moon's formation.

The Moon is asymmetrical, internally and externally, and keeps the hemisphere containing its major maria always facing the Earth. The crust is thicker on the far side, where no maria are present even though the largest impact basin is located there.

The Moon may have formed when a Mars-size body collided with the Earth and splashed material from the Earth into orbit. That debris, drawn together by its own gravity, would then have reassembled into the Moon.

The Moon's gravity creates tides, and as the Earth rotates beneath the tidal bulge of the ocean, our planet's rotation is slowed. Similar tidal braking exerted by the Earth on the Moon probably slowed the Moon's spin long ago, making its spin synchronous with its orbital motion around the Earth.

QUESTIONS FOR REVIEW

1. (7.1) Describe a crater and how it is formed. Why do some contain maria?
2. (7.1) How do the maria differ from the highlands?
3. (7.1) What are lunar rilles? What are rays?
4. (7.1) What formed the maria? Why are they smooth?
5. (7.2) List the structure and composition of the Moon from surface to core. How is it different from Earth?
6. (7.2) Why does the Moon lack an atmosphere?
7. (7.3) What are two ways to determine the distance to the Moon?
8. (7.4) How do astronomers think the Moon formed? What supports this theory? How does the theory explain why the Earth and Moon have such different densities?
9. (7.4) Why is the Moon's surface cratered but the Earth's is not?
10. (7.5) How are tides formed on the Earth?
11. (7.5) Why does the Moon form two tidal bulges on the Earth?
12. (7.3/7.5) How does the Moon rotate? Why does it spin in this manner?

THOUGHT QUESTIONS

1. (7.1) Highway surfaces develop "potholes" over time. How can you use the number of potholes as an indication of the "age" of the paving? How is this like using craters to estimate the age of the Moon's surface?

2. (7.2) Bergmann's rule states that individuals of a given species—for example, bears—will be larger in cold climates than in warmer climates. How is an explanation of this rule similar to an explanation of the temperature difference between the Earth's interior and the Moon's interior?
3. (7.2) How is the apparent lack of water on the lunar surface an argument against the idea that comets were a significant source of Earth's water and atmosphere? What is a good counterargument to your answer?
4. (7.2) Why will an astronaut's footprint on the Moon last so long?
5. (7.3/7.5) If the Moon were not in synchronous rotation with the Earth, would its phases be affected? What if both the Earth and the Moon were in synchronous rotation?
6. (7.4) Would the Moon be a good source of mineral ores? Mining in space would be very expensive—considering the lunar composition compared to the Earth's, what would make lunar mining attractive?
7. (7.5) If the day were 12 hours long, what would be the approximate time interval between high and low tide?
8. (7.5) As the Moon recedes from the Earth, are the tides getting taller or shorter? If the Moon is someday twice as far from the Earth, how many high tides will there be each day?

PROBLEMS

1. (7.1) Use data from the appendix to calculate the ratio of the Moon's mass to the Earth's, and the ratio the Moon's radius to the Earth's radius.
2. (7.1) If a lunar crater has an angular diameter of 1 minute of arc, what is its diameter in kilometers? (You may want to review section 2.1)
3. (7.1) The crater Eratosthenes is 60 kilometers wide. What is its angular diameter from Earth? Could you see a crater this size with the naked eye?
4. (7.2) Calculate the Moon's density (see the end of section 6.1 in the Earth chapter for how to calculate density). The Moon's mass and radius can be found in the appendix. On the basis of your value for the density, what can you say about the amount of iron in the Moon? (See table 6.1 for iron's density.)
5. (7.2) The density of Swiss cheese is about 1.1 g/cm^3. If the Moon were in fact made of (incompressible) cheese, what would be its mass?
6. (7.2) A spacecraft orbits the Moon 100 kilometers above its surface and with an orbital period of 114 minutes. What is the Moon's mass? (You may want to review section 3.6).
7. (7.3) A laser pulse takes 2.56 seconds to travel from Earth to the Moon and return. Use this to calculate how far away the Moon is. How might this time delay affect conversations between an astronaut on the Moon and someone back on Earth?
8. (7.2/7.4) Since the Earth and Moon are both rocky spheres, we can make a crude estimate of how much faster the Moon cooled than the Earth. Compute the ratio of the surface area to the volume of the Moon, and compare it to the same ratio for the Earth (formulas for surface area and volume, and values of the radii, can be found in the appendix; also review fig. 6.9).
9. (7.5) If the Earth constantly slowed down at a rate of 0.002 seconds / century, how many years ago would the Earth's day have been only 5 hours long?

TEST YOURSELF

1. (7.1) The large number of craters on the lunar highlands compared to those on the maria is evidence that
   (a) the maria have a liquid surface, so craters disappear there.
   (b) the highlands are composed of soft, easily cratered material.
   (c) the bodies that struck the Moon and made the craters were clumped, and missed hitting the maria.
   (d) the maria are much younger than the highlands.
   (e) the maria are much older than the highlands.
2. (7.2) What evidence indicates that the Moon lacks a large iron core?
   (a) The Moon has a very strong magnetic field.
   (b) The Moon always keeps the same side facing the Earth.
   (c) The Moon has no atmosphere.
   (d) The Moon's average density is about 3.3 grams per cubic centimeter, similar to that of rock.
   (e) The Moon has so many volcanos that all the iron in its core has been erupted onto its surface.
3. (7.3) If the Moon did not rotate on its own axis, we would observe
   (a) both sides of the Moon.
   (b) the Moon remaining stationary against the stars.
   (c) a lack of tides on Earth.
   (d) the Moon from only one hemisphere of Earth.
   (e) everything the same as now—it doesn't rotate.
4. (7.4) The Moon does not undergo plate tectonics because
   (a) it has no areas of thin crust (like the Earth's ocean floors) where spreading ridges can form.
   (b) it does not have a substantial magnetic field.
   (c) its mantle is cold and rigid.
   (d) it has no active volcanoes.
   (e) its mantle is made of iron.
5. (7.5) The photographs in figure 7.16 were taken at high tide and the next low tide. About how much time elapsed between the pictures?
   (a) 3 hours (c) 12 hours (e) 1 month
   (b) 24 hours (d) 6 hours
6. (7.5) As a result of the Moon's gravitational pull, when would you weigh the least?
   (a) when it is high tide locally
   (b) when it is low tide locally
   (c) when the Moon is overhead

(d) when you are near one of the Earth's poles
(e) your weight is the same at all of these locations and times

KEY TERMS

craters, 188
differential gravitational force, 198
highlands, 188
maria, 188
rays, 189
regolith, 192
rilles, 189
synchronous rotation, 195
tidal braking, 200
tidal bulge, 198
tides, 198

FURTHER EXPLORATIONS

Brueton, Diana. *Many Moons: The Myth and Magic, Fact and Fantasy of Our Nearest Heavenly Body.* Englewood Cliffs, N.J.: Prentice-Hall, 1991.

Dingell, Charles, Johns, William A., and White, Julie Kramer. "To the Moon and Beyond." *Scientific American* 297 (October 2007) 62.

Foust, Jeffrey A. "The Moon Rediscovered." *Sky and Telescope* 96 (December 1998): 32.

Goldreich, Peter. "Tides and the Earth-Moon System." *Scientific American* 226 (April 1972): 42.

Hiscock, Philip. "Once in a Blue Moon." *Sky and Telescope* 97 (March 1999): 52.

Nicholson, T. D. "The Moon Illusion." *Natural History* (August 1991): 66.

Spudis, Paul D. "The New Moon." *Scientific American* 289 (December 2003): 86.

Website

Visit the *Explorations* website at **http://www.mhhe.com/arny** for additional online resources on these topics.

Q FIGURE QUESTION ANSWERS

WHAT IS THIS? (chapter opening): This picture was taken by Apollo 15 astronauts orbiting the Moon shortly before their return to the Earth. The bright crescent is the Earth seen above the Moon's surface. Note that when the Moon is near its full phase, the Earth appears to be near its new phase from the perspective of the Moon.

FIGURE 7.1: The rays are younger than the maria.

FIGURE 7.2: The smaller one.

FIGURE 7.9: The Moon rotates on its axis so that it always keeps the same face toward the Earth. That is, it makes exactly one turn for each orbit around the Earth.

FIGURE 7.18: The bulge on the far side of Earth is farther away, so its gravitational pull is weaker.

PROJECTS

1. **Crater sizes:** Look at the Moon on an evening when it is nearly full. Make a sketch of the light and dark markings you see on its surface with the naked eye. Then observe the Moon with binoculars and make an enlarged sketch that shows more detail. Mark a few of the craters you can see. Estimate the diameter of these craters from your knowledge that the Moon's radius is about 1000 miles (1700 km). How big is the largest crater you can see, compared to the size of your town? Can you see any lunar rays? If so, sketch them on your drawing. How long are the rays?
2. **Moon hoax:** An assortment of frivolous claims have been made saying that the Apollo Moon landings were a hoax. The evidence offered for many of these claims is based on aspects of photos taken by the Apollo astronauts, and it is possible to test the validity of these claims by simulating the situation. For example, one of the arguments used by conspiracy theorists is that photos taken by the astronauts do not show stars in the backgrounds. On Earth, the brightness of the atmosphere prevents us from seeing stars during the day, but since there is no atmosphere on the Moon, this argument goes, you should be able to see stars.

   Even an inexpensive modern digital camera likely has a better response to low-light sources than the equipment used on the Moon landings, so it should be even easier to capture stars with such a camera, but it is not as easy as you might expect. The Moon landings occurred during the "day" on the Moon, so pictures by astronauts had to be taken using exposure settings appropriate to daytime photography.

   Use a camera that shows you the "f-stop" and exposure duration. The f-stop gets bigger as the diameter of your camera's aperture gets smaller, so the area of your aperture decreases as the inverse square of your f-stop. For example, a picture taken at f/8 gets 4 times less light than one taken at f/4. Record and make a table of your exposure settings of pictures of the night sky alone, of objects lit by lamplight and of a gray surface lit by sunlight (simulating Moon rock).

   If the camera is compensating for the light level, how many times less light is present in each of the different situations compared to the sunlit Moon rock? Try taking a photo outside at night with stars and some brightly illuminated object in the foreground. Can you make a picture that shows both? Can you explain why stars are not seen in the astronaut's photos?

Artist's depiction of a solar system in its early stages of formation.

KEY CONCEPTS

- The Solar System consists of the Sun, the eight planets, their moons, and smaller objects (such as dwarf planets, asteroids, and comets).
- The planets orbit the Sun within a flat disk-shaped region, held there by the Sun's gravity.
- The planets form two families:
  - The four inner planets (Mercury, Venus, Earth, and Mars) are rocky with iron cores and are similar to the Earth in size.
  - The four outer planets (Jupiter, Saturn, Uranus, and Neptune) are much larger than the Earth and are balls of gas and liquid, rich in hydrogen and its compounds.
- Pluto, which fits into neither family, is now considered a dwarf planet. It and a number of similar small icy objects orbit beyond Neptune.
- The Solar System formed about 4.6 billion years ago from an interstellar gas cloud that collapsed and shrank under the force of its own gravity.
- Rotation of the cloud made it flatten into a disk.
- Within the disk, gravity drew dust and gas into clumps that became the planets.
- Hundreds of planets around other stars have been detected.
- The patterns of these other planetary systems look quite different from the Solar System, but the techniques used to find *exoplanets* are not yet sensitive enough to detect systems like the Solar System.
- Some giant exoplanets orbit so close to their stars that they must have "migrated" to these positions—a process now thought to have been important in the young Solar System.

8 Survey of Solar Systems

CONCEPTS AND SKILLS TO REVIEW

- Law of gravity (3.4)
- Density (6.1)
- Modified form of Kepler's third law (3.6)

The Solar System consists of the Sun and the bodies in its gravitational domain: the eight planets, dozens of dwarf planets, and swarms of moons, asteroids and comets. Although earthlings have not walked on any objects except the Earth and Moon, we have detailed pictures sent to us from spacecraft of most of the planets and their satellites. Some are naked spheres of rock; others are mostly ice. Some have thin, frigid atmospheres so cold that ordinary gases crystallize as snow on their cratered surfaces; others have thick atmospheres the consistency of wet cement and no solid surface at all. Despite such diversity, the Solar System possesses an underlying order, an order from which astronomers attempt to read the story of how our Solar System came to be.

The Solar System formed in the extremely remote past, about 4.6 billion years ago. Astronomers hypothesize that the Sun and planets formed from the collapse of a huge, slowly spinning cloud of gas and dust. Most of the cloud's material fell inward and ended up in the Sun, but in response to rotation, some settled into a swirling disk around it. Then, within that disk, dust particles coagulated—perhaps aided by electrostatic effects such as those that make lint cling to your clothes—to form pebble-size chunks of material, which in turn collided and sometimes stuck together, growing ever larger to become the planets we see today. The objects that formed in the disk retained the motion of the original gas and dust, and so we see them today, moving in a flattened system, all orbiting the Sun in the same direction.

Seeing planets around other stars is much more challenging—something like trying to see a mosquito flying around a light bulb hundreds of miles away. However, astronomers have developed an array of techniques that have revealed hundreds of planets around other stars. Astronomers can even observe other "solar systems" in their first stages of formation. The other systems detected so far look very different from our own, challenging astronomers' understanding of how solar systems form.

In this chapter, we will survey the general properties of our Solar System and others. In later chapters we will explore the components of our Solar System in much more detail.

Q: WHAT IS THIS? See end of chapter for the answer

8.1 Components of the Solar System

The Sun

The Sun is a star, a ball of incandescent gas (fig. 8.1) whose light and heat are generated by nuclear reactions in its core. It is by far the largest body in the Solar System—more than 700 times the mass of all the other bodies put together—and its gravitational force holds the planets and other bodies in the system in their orbital patterns about it. This gravitational domination of the planets by the Sun justifies our calling the Sun's family the **Solar System.**

The Sun is mostly hydrogen (about 71%) and helium (about 27%), but it also contains very small proportions of nearly all the other chemical elements (carbon, iron, uranium, and so forth) in vaporized form, as we can tell from the spectrum of the light it emits.

The Orbits and Spins of the Planets

The planets are much smaller than the Sun and orbit about it. They emit no visible light of their own but shine by reflected sunlight. In order of increasing distance from the Sun, they are Mercury, Venus, Earth, Mars, Jupiter, Saturn, Uranus, and Neptune. The planets move around the Sun in approximately circular* orbits all lying in nearly the same plane, as shown in the view from above in figure 8.2 and the side view in figure 8.3. Thus, the Solar System is like a spinning pancake, with the planets traveling around the Sun in the same direction: counterclockwise, as seen from above the Earth's

*The circularity of the orbits is only approximate, because we know from Kepler's laws that in reality the orbits are ellipses. However, the amount of ellipticity is, in general, very small.

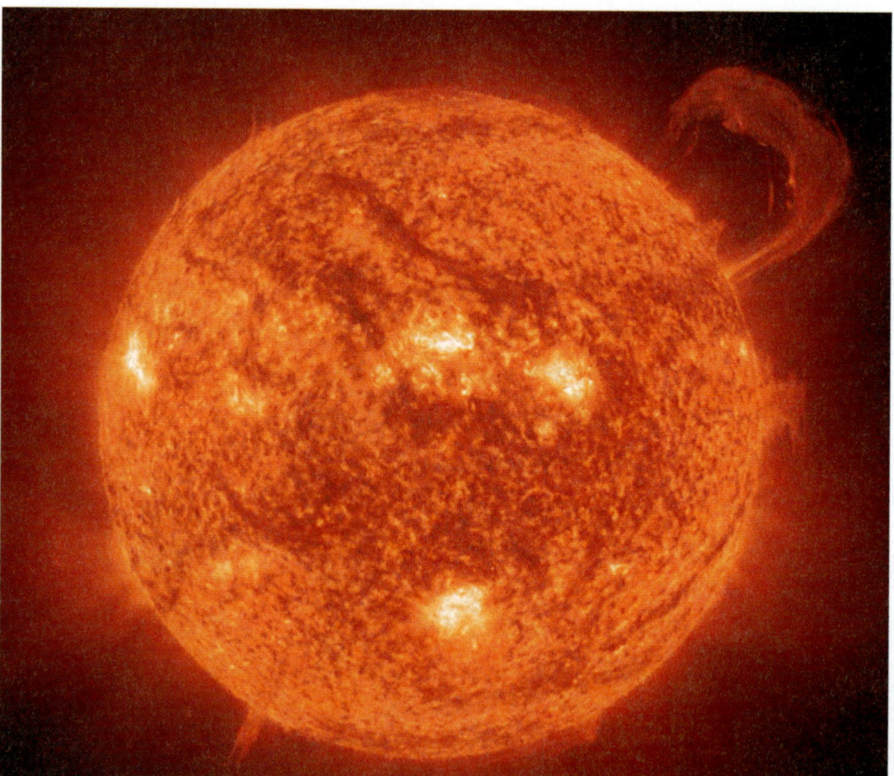

FIGURE 8.1
Image of the Sun made with an ultraviolet telescope that reveals high-temperature gases in the Sun's atmosphere.

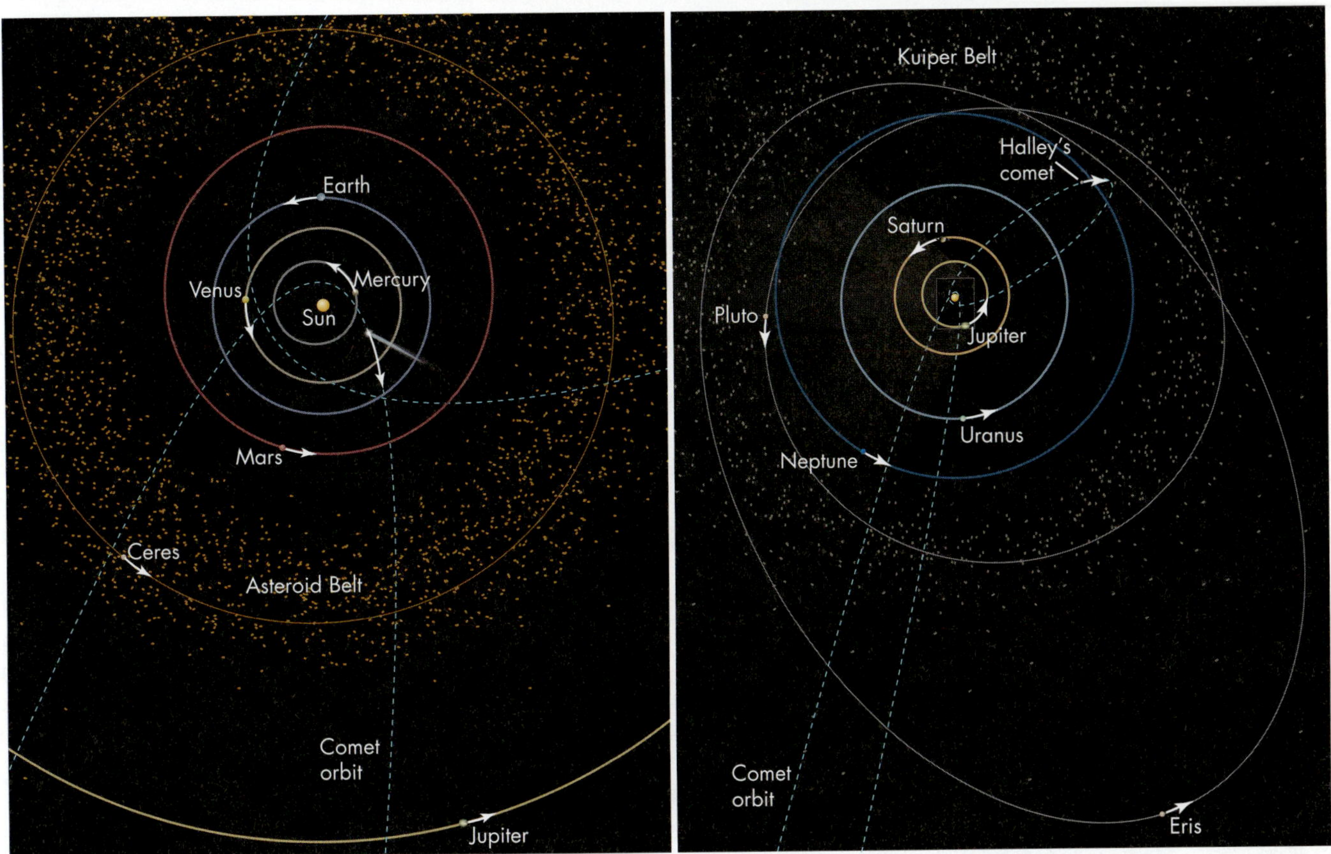

FIGURE 8.2
Diagrams of the Solar System from above. The orbits are shown in the correct relative scale in the two drawings. Because of the great difference in scale, the inner and outer Solar System are displayed separately.

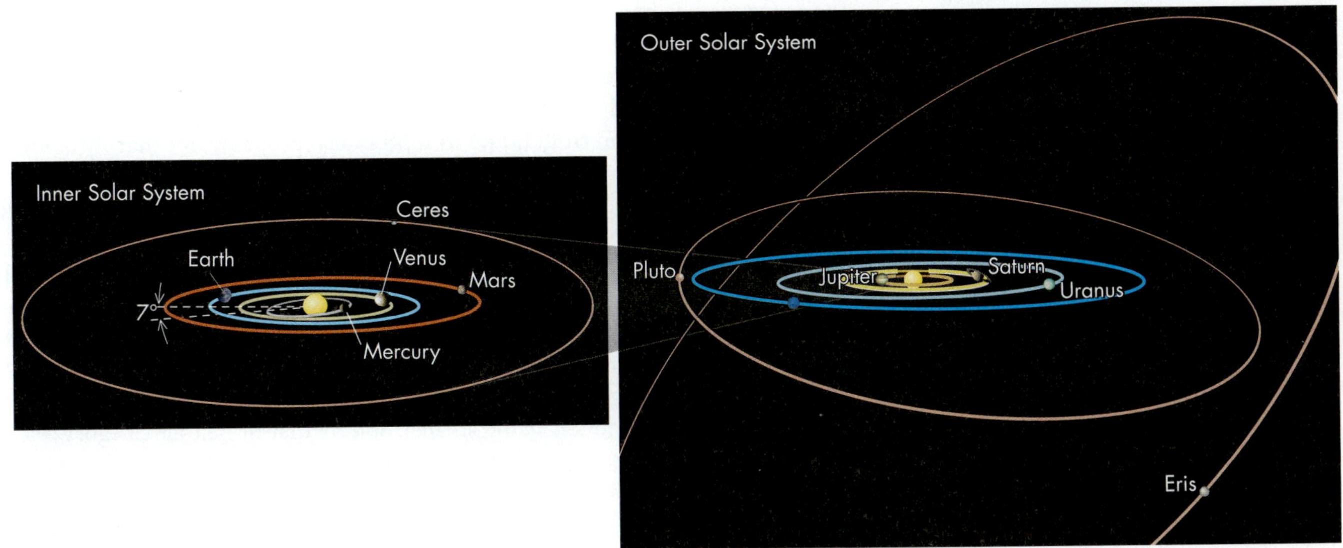

FIGURE 8.3
The planets' orbits from the side. The dwarf planets Ceres, Pluto, and Eris are also shown, illustrating their highly inclined orbits.

The Solar System out to Neptune has about the same relative thickness as 3 CDs stacked together.

North Pole. We can see this flatness from Earth: it makes the planets all lie in an approximate line in the sky (fig. 8.4).

As the planets orbit the Sun, each also spins on its rotation axis. The spin is generally in the same direction as the planets' orbital motion around the Sun (again, counterclockwise, as seen from above the Earth's North Pole), and the tilt of the rotation axes relative to the plane of planetary orbits is generally not far from the perpendicular. However, there are two exceptions: Venus and Uranus. Uranus has an extremely large tilt to its rotation axis, which lies nearly in its orbital plane (fig. 8.5). Venus's rotation axis has such a large tilt that it spins backward, a motion technically called "retrograde rotation." However, despite this backward spin, Venus orbits the Sun in the same direction as the rest of the planets.

Its flattened structure, and the orderly orbital and spin properties of its planets, are two of the most fundamental features of the Solar System, and any theory of the Solar System must explain them. But a third and equally important feature is that the planets fall into two families called inner and outer planets based on their size, composition, and location in the Solar System.

Two Types of Planets

The **inner planets**—Mercury, Venus, Earth, and Mars—are small rocky bodies with relatively thin or no atmospheres. The **outer planets**—Jupiter, Saturn, Uranus, and Neptune—are gaseous and liquid. They are much larger than the inner planets and have deep, hydrogen-rich atmospheres. For example, Jupiter is more than 10 times larger in diameter than the Earth and has 318 times its mass. These differences can be seen in figure 8.6, which also shows a small part of the edge of the Sun to illustrate how the Sun dwarfs even the large planets.

In describing the planets, we have used the terms *rock* and *ice*. By rock, we mean material composed of silicates, which are composed of silicon (Si) and oxygen (O) with an admixture of other heavy elements such as aluminum (Al), magnesium (Mg), sulfur (S), and iron (Fe). By ice, we mean frozen liquids and gases such as ordinary water ice (H_2O), frozen carbon dioxide (CO_2), frozen ammonia (NH_3), frozen methane (CH_4), and so on. If we consider the Solar System as a whole, rock is rare, because the silicon atoms that compose it are outnumbered more than 25,000 to 1 by hydrogen. However, in the warmth of the inner Solar System, rock dominates because intrinsically more abundant materials such as hydrogen, water, methane, and ammonia cannot condense to mingle with it. Thus, the inner planets are composed mainly of rock.

The outer planets have no true "surface"; rather, their atmospheres thicken with depth and eventually compress to liquid form despite high temperatures. They have no distinct boundary between "atmosphere" and "crust" as we have on the Earth. In the deep interior, the liquid may be compressed into a solid, as happens in the Earth's inner core, but the transition from liquid to solid is also probably not sharply defined. Thus, we can never "land" on Jupiter because we would simply sink ever deeper into its interior. By contrast, the inner planets have at most a thin layer of gas over their solid surface, and the least massive have too little gravity to retain any atmosphere at all.

Instead of "inner" and "outer" planets, astronomers sometimes use "terrestrial" and "Jovian" to describe the two types of planets. The **terrestrial planets** (Mercury to Mars) are so named because of their resemblance to the Earth. The **Jovian planets*** (Jupiter to Neptune) are named for their resemblance to Jupiter.

Although the two categories of planets neatly describe the larger objects that orbit the Sun, astronomers have found many smaller objects that fit neither category. Pluto has long failed to fit, because of its small size, composition of ice and rock, and odd orbit. (Not only is its orbit highly tilted with respect to the other planets, it also crosses Neptune's orbit). Moreover, in the last decade astronomers have discovered more than a

FIGURE 8.4
Sunset view of four planets strung along the zodiac on March 1, 1999. Their straight-line arrangement results from the flatness of the Solar System. From top to bottom, you can see Saturn, Venus, Jupiter, and Mercury (nearly lost in the twilight).

*Some astronomers go further and divide the Jovian planets into *gas giants* (Jupiter and Saturn) and *ice giants* (Uranus and Neptune).

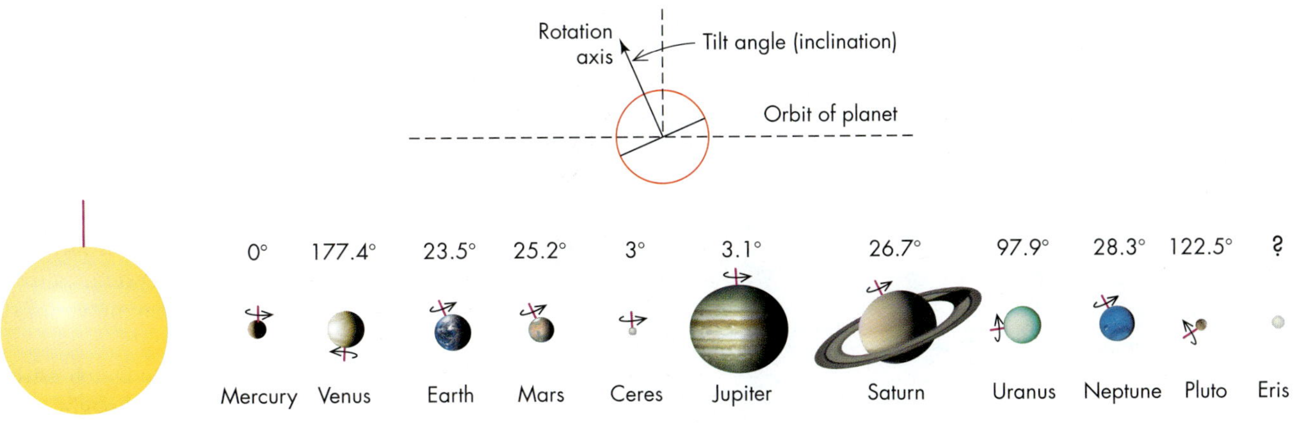

FIGURE 8.5
Sketches showing the orientation of the rotation axes of the planets and Sun. The figure illustrates that most of them spin in the same direction, counterclockwise as seen from above the Earth's North Pole. The dwarf planets Ceres, Pluto, and Eris are also shown. The bodies are not shown to the same scale.

thousand icy objects orbiting at similar distances from the Sun as Pluto. In 2005 it was discovered that one of these objects, named Eris, is an icy world slightly *larger* than Pluto that orbits about 68 AU from the Sun, roughly half again Pluto's distance from our star.

In response to the discovery of Eris and half a dozen other objects similar in size to Pluto, the International Astronomical Union introduced in 2006 a new category of Solar System objects called **dwarf planets.** Dwarf planets are objects that orbit the Sun, are massive enough that their gravity compresses them into an approximately spherical shape, but have not swept their orbital region clear of other objects that add up to a comparable mass as the planet. To recognize Pluto's important place in the history of the discovery of these objects, astronomers decided in 2008 to call dwarf planets that orbit beyond Neptune *plutoids.*

Satellites

As the planets orbit the Sun, most are themselves orbited by satellites. Jupiter, Saturn, Uranus, and Neptune have large families of 63, 60, 27, and 13 clearly identified moons, respectively.* Mars has 2, Earth has 1, while Venus and Mercury have none. Even some of the dwarf planets have moons. Pluto has 3 and Eris has 1.

The larger moons generally move along approximately circular paths that are roughly in the planet's equatorial plane, their orbits tilted like the planets themselves. Thus, each planet and its moons resemble a miniature Solar System—an important clue to the origin of these satellites. Some large moons and many of the smaller moons have much more irregular orbits, suggesting that they may have been captured.

Asteroids and Comets

Asteroids and comets are far smaller than planetary bodies. The **asteroids** are rocky or metallic bodies with diameters that range from a few meters up to about 1000 km (about one-tenth the size of the Earth). The **comets,** on the other hand, are icy bodies about 10 km (about 6 miles) or less in diameter that grow huge tails of gas and dust as they near the Sun and are partially vaporized by its heat. Thus, these minor bodies exhibit the same split into two families that we see for the planets; that is, rocky bodies and icy bodies.

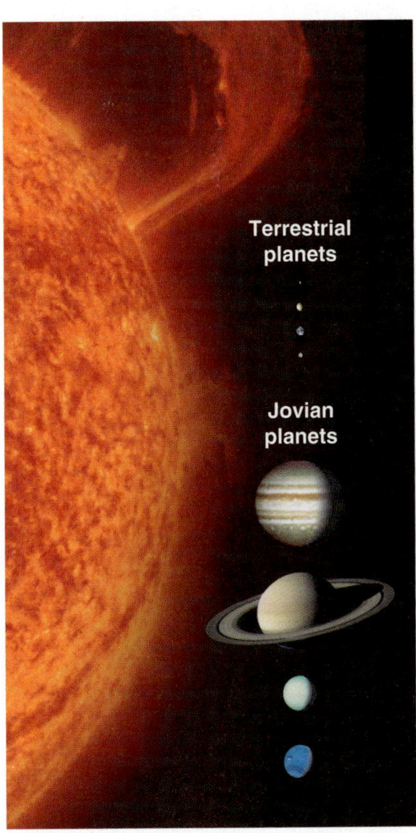

FIGURE 8.6
The planets and the Sun to scale.

*Astronomers are finding new moons around these planets so rapidly that it is difficult to keep the numbers up-to-date. Most of the new discoveries are of very small bodies, sometimes just a few kilometers in diameter.

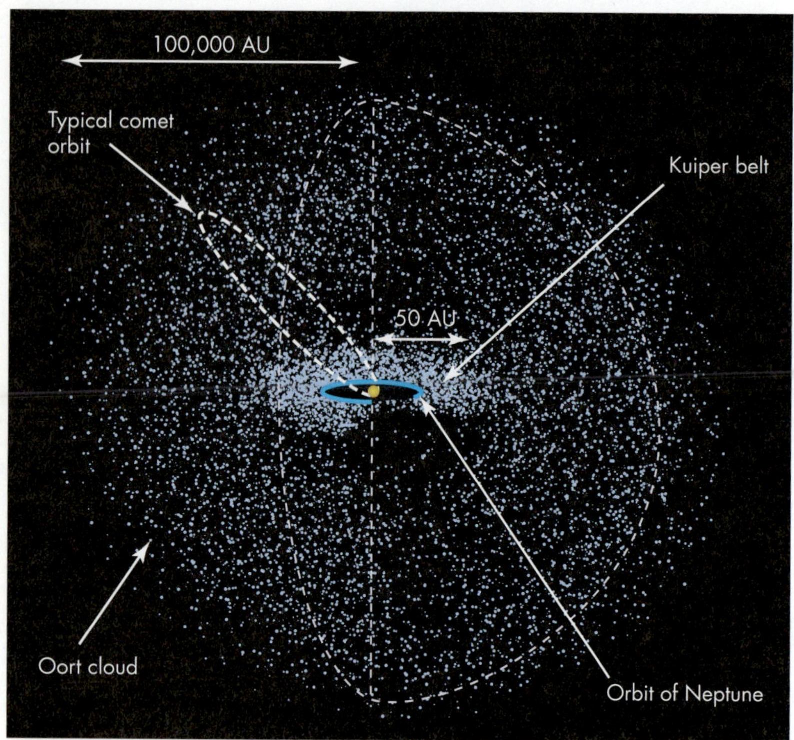

FIGURE 8.7
Sketch of the Oort cloud and the Kuiper belt. The scale shown is only approximate. Orbits and bodies are not to scale.

Asteroids and comets differ not only in their composition but also in their location within the Solar System. Most asteroids circle the Sun in the large gap between the orbits of Mars and Jupiter, a region called the **asteroid belt.** They are probably material that failed—perhaps as a result of disturbance by Jupiter's gravity—to aggregate into a planet.* Most comets, on the other hand, orbit far beyond Neptune in a region of the Solar System called the Oort cloud, and only rarely do they move into the inner Solar System.

The **Oort cloud,** named for the Dutch astronomer who proposed its existence, is thought to be a spherical region that completely surrounds the Solar System and extends from about 40,000 to 100,000 AU from the Sun (fig. 8.7). Although the majority of comets probably originate in the Oort cloud, some come from a disklike swarm of icy objects that lies just beyond the orbit of Neptune and extends to perhaps 50 AU from the Sun, a region called the **Kuiper belt.** We will discuss more details of the Oort cloud and Kuiper belt in chapter 11, but for now we simply note that together they probably contain more than 1 trillion (10^{12}) comet nuclei, thousands of larger objects, and several dozen dwarf planets, including Pluto.

ANIMATION
Oort cloud and Kuiper belt

Composition Differences Between the Inner and Outer Planets

We stated earlier that the inner planets are rocky and the outer planets are hydrogen-rich. These composition differences are so important to our understanding of the history of the Solar System that we should look at them more closely and see how we determine them.

Astronomers can deduce a planet's composition in several ways. From its spectrum, they can measure its atmospheric composition and get some information about the nature of its surface rocks. However, spectra give no clue as to what lies deep inside a planet where light cannot penetrate. To learn about the interior, astronomers must therefore use alternative methods.

We saw in chapter 6 how earthquake waves reveal what lies inside the Earth, but this method has not yet been used for other planets. Although quake detectors were landed on Mars, they did not work properly, and such detectors would require very special modification to work on the Jovian planets, which have no surface on which to land! Thus, we must try other means to study the interior of planets. One such technique uses the planet's density.

Density as a Measure of a Planet's Composition

The average density of a planet is its mass divided by its volume. Both of these quantities can be measured relatively easily. For example, we showed in chapter 3 how to determine a body's mass from its gravitational attraction on a second body orbiting around it by applying Newton's modification of Kepler's third law. Thus, from this law,

*A regular pattern of increasing spacing between planets suggested the existence of a body between Mars and Jupiter. See the Extending Our Reach box on p. 213.

Volume

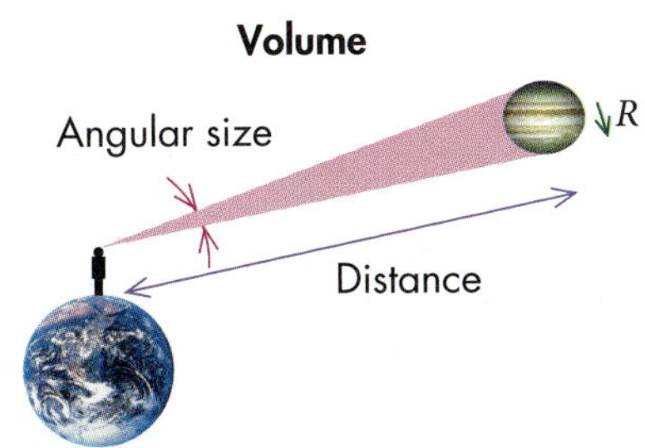

Measure angular size of planet, and use relation between angular size and distance to solve for planet's radius, R. Calculate volume, $\mathcal{V}$, of planet:

$$\mathcal{V} = \frac{4\pi R^3}{3}$$

for a spherical body of radius R.

Mass

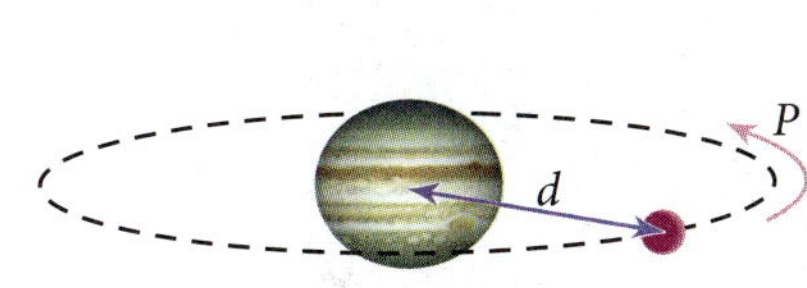

Observe motion of a satellite orbiting planet. Determine satellite's distance, d, from planet and orbital period, P. Use Newton's form of Kepler's third law:

$$M = \frac{4\pi^2 d^3}{GP^2}$$

Insert measured values of d and P, and value of constant G. Solve for M.

Average Density

Average density, ρ, equals mass, M, divided by volume, $\mathcal{V}$:

$$\rho = \frac{M}{\mathcal{V}}$$

FIGURE 8.8
Measuring a planet's mass, radius, and average density. Volume can be determined from the radius of a planet, which in turn is found from its distance and angular size (chapter 2). Mass can be determined from the orbit of a satellite (chapter 3).

Q. Suppose you are given a tiny box that has a volume of 10 cubic centimeters and a mass of 30 grams. What is its density? Is it more likely to contain solid iron or rock?

we can calculate a planet's mass by observing the orbital motion of one of its moons or a passing spacecraft. We can determine a planet's volume ($\mathcal{V}$) from the formula $\mathcal{V} = 4\pi R^3/3$, where R is the planet's radius. We can measure R in several ways—for example, from its angular size and distance, a technique we used in chapter 2 to measure the radius of the Moon. With the planet's mass, M, and volume, $\mathcal{V}$, known, we can calculate its average density straightforwardly by dividing M by $\mathcal{V}$ (fig. 8.8).

Once the planet's average density is known, we can compare it with the density of abundant, candidate materials to find a likely match. For example, we saw in chapter 6 that the average density of the Earth (5.5 grams per cubic centimeter) was intermediate between silicate rock (about 3 grams per cubic centimeter) and iron (7.9 grams per cubic centimeter). Therefore, we inferred that the Earth has an iron core beneath its rocky crust, a supposition that was verified from studies using earthquake waves.

Although density comparison is a powerful tool for studying planetary composition, it also has drawbacks. First, there may be several different substances that will produce an equally good match to the observed density. Second, the density of a given material can be affected by the planet's gravitational force. For example, a massive planet may crush rock whose normal density is 3 grams per cubic centimeter to a density of 7 or 8 grams per cubic centimeter. Thus, in making a match to determine the composition, we must take into account compression by gravity.

All the terrestrial planets have an average density similar to the Earth's (3.9 to 5.5 grams per cm^3). On the other hand, all the Jovian planets have a much smaller average density (0.7 to 1.7 grams per cm^3), similar to that of ice. After correcting for the above gravitational compression, we conclude that all the inner planets contain large amounts of rock and iron and that the iron has sunk to the core, as shown in figure 8.9. Likewise, the outer planets contain mainly light materials, as borne out by their spectra, which show them to be mostly hydrogen, helium, and hydrogen-rich material such as methane (CH_4), ammonia (NH_3), and water (H_2O).

The outer planets probably have cores of iron and rock about the size of the Earth beneath their deep atmosphere, as illustrated in figure 8.9. Astronomers deduce the existence of these cores in two ways. First, if the outer planets have the same relative amount of heavy elements as the Sun, they should contain several Earth masses of iron and silicates, and because these substances are much denser than hydrogen, they must sink to the planet's core. Secondly, detailed analyses of these planets' gravitational fields, determined from their effect on space probes, are best explained by dense cores. In the case

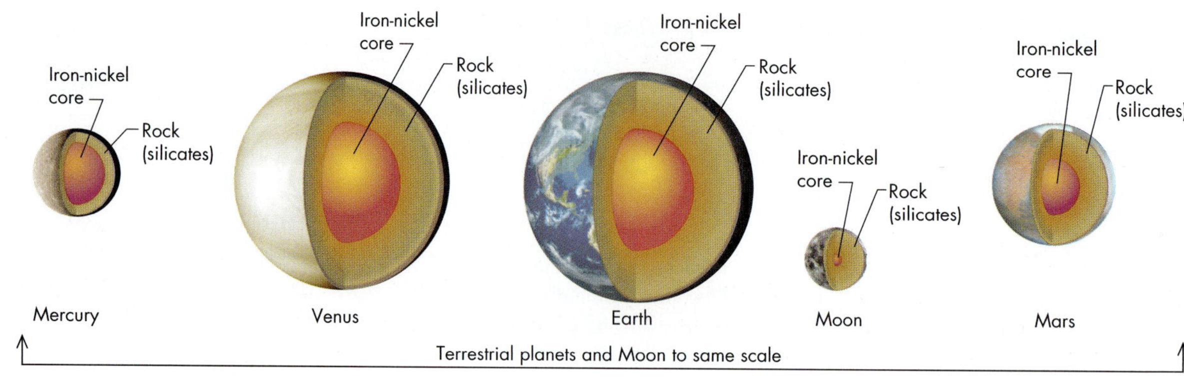

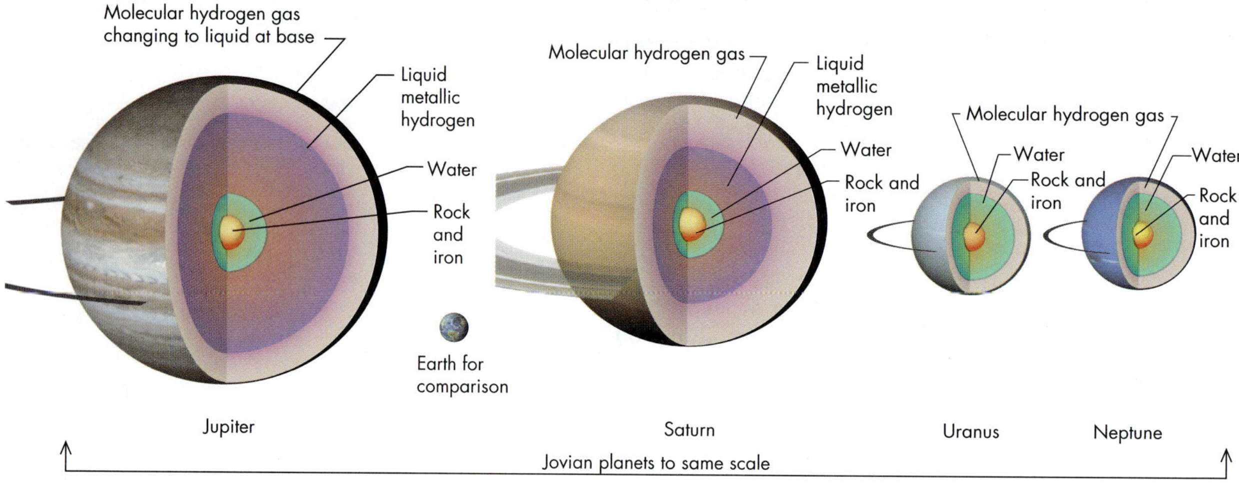

FIGURE 8.9
Sketches of the interiors of the planets. Details of sizes and composition of inner regions are uncertain for many of the planets.

of Jupiter, a core of about 7 times the Earth mass is estimated. However, there is a large uncertainty in the exact value, with some recent models estimating twice as much.

Our discussion of the composition of the planets not only underlines the differences between the two families of planets but also furnishes another clue to their origin: the planets and Sun were all made from the same material. Astronomers come to this conclusion because Jupiter and Saturn have a composition almost identical to that of the Sun, and the inner planets have a similar composition *if we were to remove the Sun's hydrogen and helium.** Thus, we can explain the compositional difference between the inner and outer planets by proposing a process that would keep the inner planets from collecting and capturing these light gases.

Age of the Solar System

An important clue to the origin of the Solar System comes from its age. Despite great differences in size, structure, and composition, the planets, asteroids, and comets all seem to have formed at nearly the same time. We can directly measure that date for the Earth, Moon, and some asteroids from the radioactivity of their rocks, and we find that none is more than about 4.6 billion years old. (Rocks from asteroids are the origin of many of the meteors that have reached Earth and are thus available for study, as we will discuss in chapter 11). Likewise, we find a similar age for the Sun, based on its current brightness and temperature and its presumed rate of nuclear fuel consumption.

*Carbon, nitrogen, neon, and other elements normally in gaseous compounds are also relatively rare in the inner planets.

EXTENDING *our reach*

BODE'S RULE: THE SEARCH FOR ORDER

A curious and as yet unexplained feature of the orbits of the planets is their regular spacing. Very roughly, each planet is about twice as far from the Sun as its inner neighbor. This progression of distance from the Sun can be expressed by a simple mathematical relation known as **Bode's rule,** which works as follows: write down 0, 3, and then successive numbers by doubling the preceding number until you have nine numbers. That is, 0, 3, 6, 12, 24, and so on. Next, add 4 to each, and divide the result by 10, as shown in Box table 8.1. The resulting numbers, with two exceptions, are very close to the distances of the planets from the Sun in astronomical units.

Bode's rule was worked out before the discovery of Uranus, Neptune, and Pluto, and when Uranus was discovered and found to fit the law, interest was focused on the "gap" at 2.8 AU. Astronomers therefore began to search for a body in the gap, and, as we will see in chapter 10, Giuseppi Piazzi, a Sicilian astronomer, soon discovered the dwarf planet Ceres, which fitted the rule splendidly.

Ironically, the next planet to be found, Neptune, did not fit the rule at all, though the dwarf planet Pluto does, at least approximately. These irregularities show that Bode's rule is not a law like the "law of gravity," which is why we prefer to call it "rule" to emphasize this difference. It is not based on any (known) physical principles, but computer simulations of planet formation sometimes produce planets at similar spacing patterns. It may tell us that systems of planets are not likely to remain in stable orbits for billions of years unless their orbits are a factor of 1.5 to 2 times larger than the next planet interior to them. Or perhaps it merely shows the human fascination with patterns and our tendency to see order where none may actually exist.

BOX TABLE 8.1 BODE'S RULE

Bode's Rule	Number	Object	True Distance
$(0 + 4)/10 =$	0.4	Mercury	0.39
$(3 + 4)/10 =$	0.7	Venus	0.72
$(6 + 4)/10 =$	1.0	Earth	1.00
$(12 + 4)/10 =$	1.6	Mars	1.52
$(24 + 4)/10 =$	2.8	Ceres	2.77
$(48 + 4)/10 =$	5.2	Jupiter	5.2
$(96 + 4)/10 =$	10.0	Saturn	9.5
$(192 + 4)/10 =$	19.6	Uranus	19.2
		Neptune	30.1
$(384 + 4)/10 =$	38.8	Pluto	39.5

8.2 Formation of Planetary Systems

How did the Solar System form? What processes gave it the features we discussed in the previous section, such as its flatness and two main families of planets? Given that we were not around 4.6 billion years ago to witness its birth, our best explanation of its origin must be a reconstruction based on observations that we make now, billions of years after the event. Those observations, discussed above, are summarized below. Each must be explained by whatever theory we devise.

1. The Solar System is flat, with all the planets orbiting in the same direction.
2. There are two types of planets, inner and outer; the rocky ones are near the Sun and the gaseous or liquid ones are farther out.
3. The composition of the outer planets is similar to the Sun's, while that of the inner planets is like the Sun's minus the gases that condense only at low temperatures.
4. All the bodies in the Solar System whose ages have so far been determined are less than about 4.6 billion years old.

We have listed only the most important observed features that our theory must explain. There are many additional clues from the structure of asteroids, the number of craters on planetary and satellite surfaces, and the detailed chemical composition of surface rocks and atmospheres.

INTERACTIVE
Solar System builder

The currently favored theory for the origin of the Solar System derives from theories proposed in the eighteenth century by Immanuel Kant, the great German philosopher, and Pierre-Simon Laplace, a French mathematician. Kant and Laplace independently proposed what is now called the **solar nebula theory** that the Solar System originated from a rotating, flattened disk of gas and dust, with the outer part of the disk becoming the planets and the center becoming the Sun. This theory offers a natural explanation for the flattened shape of the system and the common direction of motion of the planets around the Sun. Since there is nothing about these processes that appears to be unique to the Solar System, we would expect to find evidence of similar processes occurring as other stars form. Therefore we can test this idea by searching for stars at various stages of this process.

FIGURE 8.10
Photograph of an interstellar cloud (the dark region at center) that may be similar to the one from which the Solar System formed. The dark cloud is known as Barnard 86.

Interstellar Clouds

The modern form of the solar nebula theory proposes that the Solar System was born 4.6 billion years ago from an **interstellar cloud,** an enormous rotating aggregate of gas and dust like the one shown in figure 8.10. Such clouds are common between the stars in our Galaxy even today, and astronomers now think all stars have formed from them. Thus, although our main concern in this chapter is with the birth of the Solar System, we should bear in mind that our theory applies more broadly and implies that *most* stars could have planets, or at least surrounding disks of dust and gas from which planets might form.

Because interstellar clouds are the raw material of the Solar System, we need to describe them more fully. Although such clouds are found in many shapes and sizes, the one that became the Sun and planets probably was a few light years in diameter and contained about twice the present mass of the Sun. If it was like typical clouds we see today, it was made mostly of hydrogen (71%) and helium (27%) gas, with tiny traces of other chemical elements, such as gaseous carbon, oxygen, and silicon. In addition to the gases, interstellar clouds also contain tiny dust particles called **interstellar grains.**

Interstellar grains range in size from large molecules to micrometers or larger and are believed to be made of a mixture of silicates, iron compounds, carbon compounds, and water frozen into ice. Astronomers deduce the presence of these substances from their spectral lines, which are seen in starlight that has passed through dense dust clouds. Moreover, a few hardy interstellar dust grains, including tiny diamonds, have been found in ancient meteorites. This direct evidence from grains and the data from spectral lines shows that the elements occur in proportions similar to those we observe in the Sun. This is additional evidence that the Sun and its planets could have formed from an interstellar cloud.

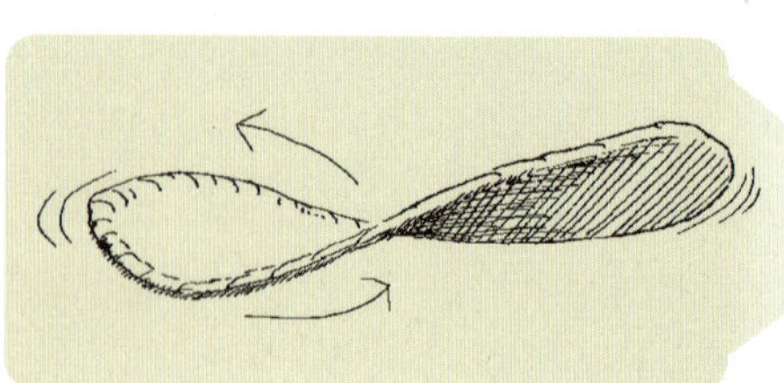

The cloud began its transformation into the Sun and planets when the gravitational attraction between the particles in the densest parts of the cloud caused it to collapse inward, as shown in figure 8.11. The collapse may have been triggered by a star exploding nearby or by a collision with another cloud. But regardless of its initial cause, the infall was not directly to the center. Instead, because the cloud was rotating, it flattened. Flattening occurred because rotation retarded the collapse perpendicular to the cloud's rotation axis. A similar effect happens in an old-fashioned pizza parlor where the chef flattens the dough by tossing it into the air with a spin.

ANIMATION
Flattening and spreading up of a collapsing interstellar cloud

Formation of the Solar Nebula

It took a few million years for the cloud to collapse and become a rotating disk with a bulge in the center. The disk is called the **solar nebula,** and it eventually condensed into the

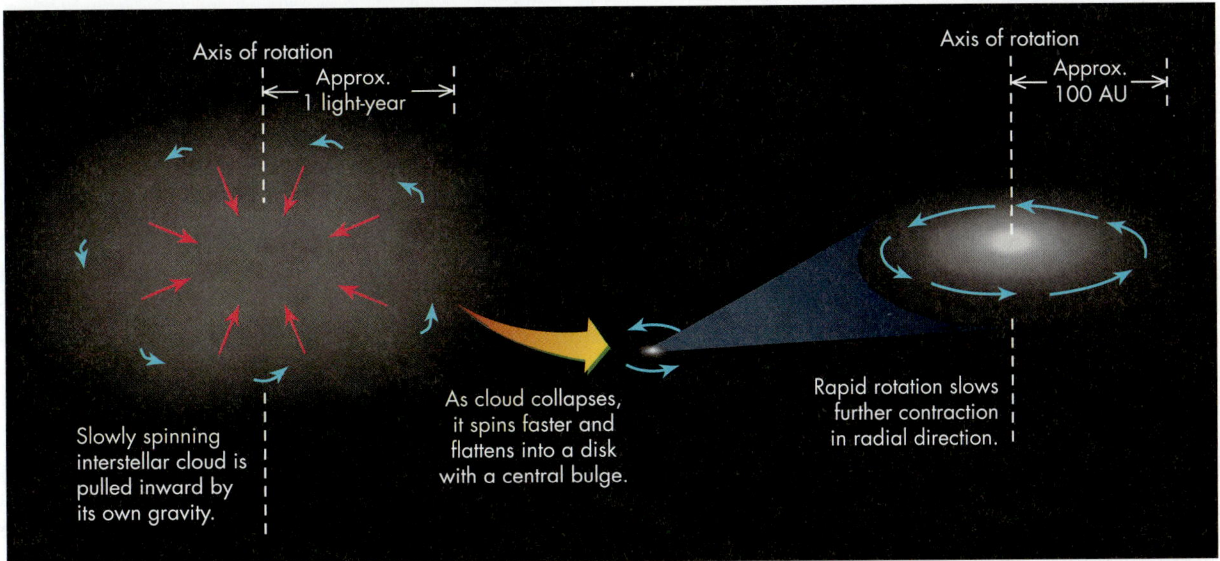

FIGURE 8.11
A sketch illustrating the collapse of an interstellar cloud to form a rapidly spinning disk. Note that the final size of the disk is not shown to scale—in actuality it would be thousands of times smaller than the cloud from which it formed.

planets while the bulge became the Sun. This explains the first obvious property of the Solar System—its disklike structure—which we mentioned at the beginning of this section.

The solar nebula was probably about 200 AU in diameter and perhaps 10 AU thick. Its inner parts were hot, heated by the young Sun and the impact of gas falling on the disk during its collapse, but the outer parts were cold, far below the freezing point of water. We are fairly certain of these dimensions and temperatures because we can observe disks around other stars and, in a few cases, can even detect other planets. For example, figure 8.12A shows a picture made with the Hubble Space Telescope of gas and dust disks near the Orion Nebula. The stars at the centers of these disks have not yet become hot enough to emit much visible light. Figure 8.12B, on the other hand, shows a disk (seen nearly edge-on) in which the star has reached full brilliance. Although the picture is grainy and in false color in order to emphasize the limited detail, you can see the disk edge-on.

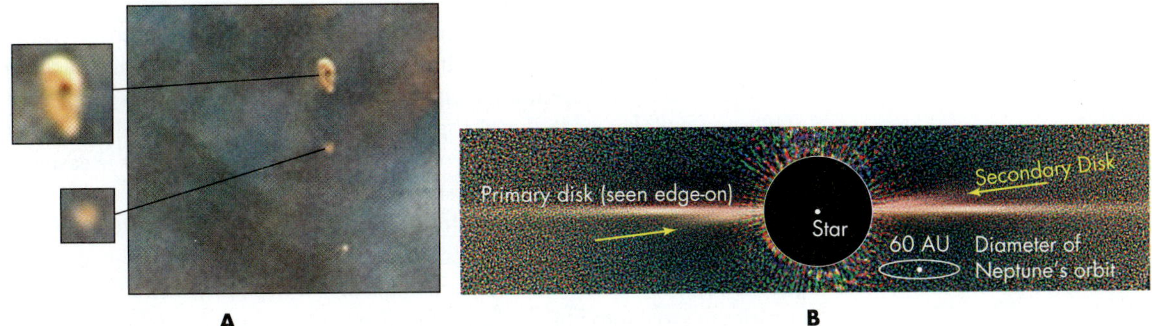

FIGURE 8.12
(A) The small blobs in this picture are stars in the process of formation (protostars) and their surrounding disk of dust and gas. These are in the Orion Nebula, a huge gas cloud about 1500 light years from Earth. (B) Picture in false color of a disk of dust around the young star β Pictoris made with the Hubble Space Telescope. A small mask in the telescope blots out the star's direct light, which would otherwise overexpose the image. Note the second faint dust disk, which is probably produced by a planet orbiting at a small angle relative to the primary disk.

Q. How does the process illustrated in figure 8.13 explain why you can see your breath on a cold morning?

FIGURE 8.13
Water vapor cools as it leaves the kettle. The cooling makes the vapor condense into tiny liquid water droplets, which we see as the "steam."

Condensation in the Solar Nebula

Condensation occurs when a gas cools and its molecules stick together to form liquid or solid particles.* For condensation to happen, the gas must cool below a critical temperature (the value of which depends on the substance condensing and the surrounding pressure). For example, suppose we start with a cloud of vaporized iron at a temperature of 2000 K. If we cool the iron vapor to about 1300 K, tiny flakes of iron will condense from it. Likewise, if we cool a gas of silicates to about 1200 K, flakes of rocky material will condense.

At lower temperatures, other substances will condense. Water, for instance, can condense at room temperature, as you can see as steam escapes from a boiling kettle (fig. 8.13). Here, water molecules in the hot steam come into contact with the cooler air of the room. As the vaporized water cools, its molecules move more slowly, so that when they collide, electrical forces can bind them together, first into pairs, then into small clumps, and eventually into the tiny droplets that make up the cloud we see at the spout.

An important feature of condensation is that when a mixture of vaporized materials cools, the materials with the highest vaporization temperatures condense first. Thus, as a mixture of gaseous iron, silicate, and water cools, it will make iron grit when its temperature reaches 1300 K, silicate grit when it reaches 1200 K, and finally water droplets when it cools to only a few hundred degrees K. It is a bit like putting a jar of chicken soup in the freezer. First the fat freezes, then the broth, and finally the bits of chicken and celery.

However, the condensation process stops if the temperature never drops sufficiently low. Thus, in the example above, if the temperature never cools below 500 K, water will not condense and the only solid material that forms from the gaseous mixture will be iron and silicates.

This kind of condensation sequence occurred in the solar nebula as it cooled after its collapse to a disk. But because the Sun heated the inner part of the disk, the temperature from the Sun to almost the orbit of Jupiter never dropped low enough for water and other substances with similar condensation temperatures to condense there. On the other hand, iron and silicate, which condense even at relatively high temperatures, could condense everywhere within the disk. Thus, the nebula became divided into two regions: an inner zone of silicate-iron particles, and an outer zone of similar particles on which ices also condensed, as illustrated schematically in figure 8.13. Water, hydrogen, and other easily vaporized substances were present as gases in the inner solar nebula, but they could not form solid particles there. However, some of these substances combined chemically with silicate grains so that the rocky material from which the inner planets formed contained within it small quantities of water and other gases.

Accretion and Planetesimals

In the next stage of planet formation, the tiny particles that condensed from the nebula must have begun to stick together into bigger pieces in a process called **accretion.**

The process of accretion is a bit like building a snowman. You begin with a handful of loose snowflakes and squeeze them together to make a snowball. Then you add more snow by rolling the ball on the ground. As the ball gets bigger, it is easier for snow to stick to it, and it rapidly grows in size.

Similarly in the solar nebula, tiny grains stuck together and formed bigger grains that grew into clumps, perhaps held together by electrical forces similar to those that make lint stick to your clothes. Subsequent collisions, if not too violent, allowed these smaller particles to grow into objects ranging in size from millimeters to kilometers. These larger objects are called **planetesimals** (that is, small, planetlike bodies) (fig. 8.14).

Because the planetesimals near the Sun formed from silicate and iron particles, while those farther out were cold enough that they could incorporate ice and frozen gases as well,

*Technically, condensation is the change from gas to liquid, and deposition is the change from gas to solid. However, we will not make that distinction here.

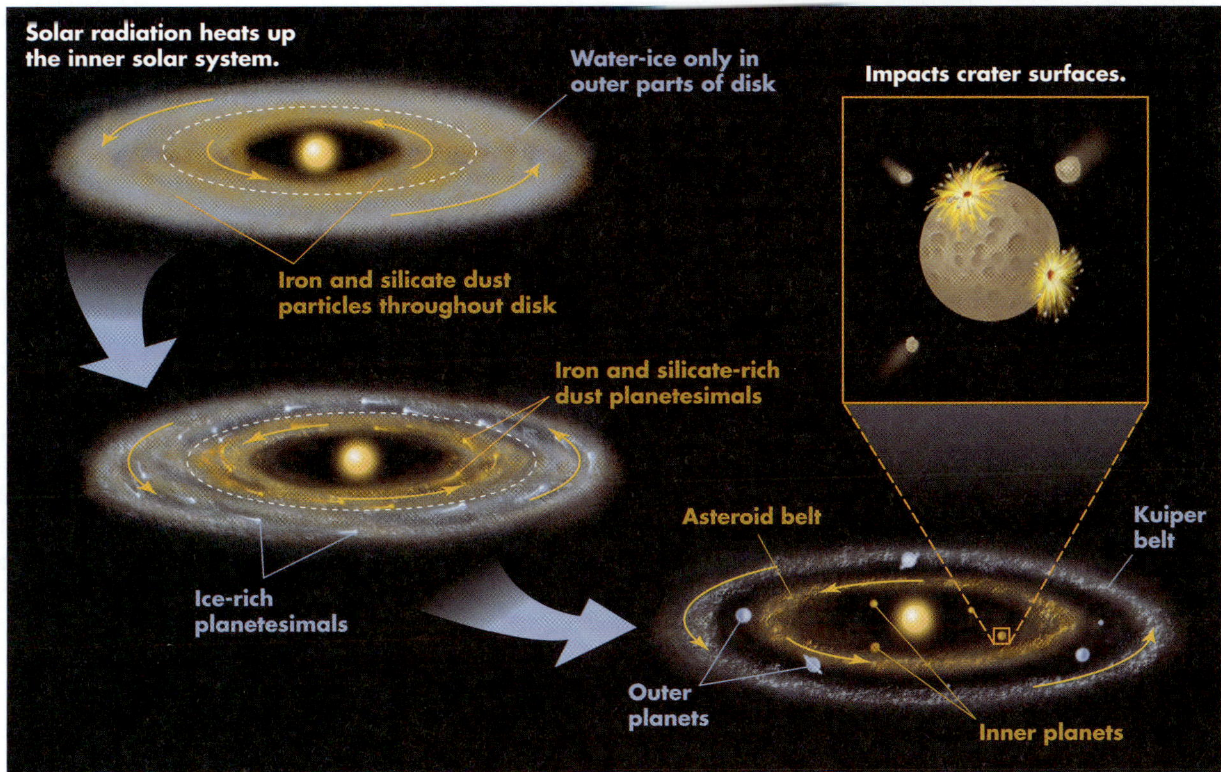

FIGURE 8.14
An artist's depiction of how the planets may have formed in the solar nebula.

there were two main types of planetesimals: rocky-iron ones near the Sun and icy-rocky-iron ones farther out. This then explains the second observation we described at the beginning of this section—that there are two types of planets—as described below.

Formation of the Planets

As planetesimals moved within the disk and collided with one another, planets formed. Computer simulations show that some collisions led to the shattering of both bodies, but gentler collisions led to merging, with the planetary orbits gradually becoming approximately circular. Moreover, in some such simulations, the distance between orbits is similar to that given by Bode's rule.

Merging of the planetesimals increased their mass and thus their gravitational attraction. That, in turn, helped them grow even more massive by drawing planetesimals into clumps or rings around the Sun. Within these clumps, growth went even faster, so that over a time lasting several million years, larger and larger objects formed.

Planetesimal growth was especially rapid beyond 4 or 5 AU from the Sun. Planetesimals there had more material from which to grow, because there ice could condense and it is about 10 times more abundant than silicate and iron compounds. Thus, planetesimals in the outer solar nebula could in principle become 10 times larger than those in the inner nebula.

Additionally, once a planet grew somewhat larger than the diameter and mass of the Earth, it was able to attract and retain gas by its own gravity. Because hydrogen was overwhelmingly the most abundant material in the solar nebula, planets large enough to tap that reservoir could grow vastly larger than those that formed only from solid material. Thus, Jupiter, Saturn, Uranus, and Neptune may have begun as Earth-size bodies of ice and rock, but their gravitational attraction resulted in their becoming surrounded by the huge envelopes of hydrogen-rich gases that we see today. The smaller and warmer

ANIMATION
Planet formation from the solar nebula

bodies of the inner Solar System could not capture hydrogen and therefore remained small and lack that gas. This explains the third observation we mentioned at the beginning of the section—that the outer planets have a composition similar to the Sun's.

As planetesimals struck the growing planets, their impact released gravitational energy that heated both the planetesimal and the planet. Gravitational energy is liberated whenever something falls. For example, when a cinder block falls onto a box of tennis balls, the impact scatters the balls in all directions, giving them kinetic energy—energy of motion. In much the same manner, planetesimals falling onto a planet's surface give energy to the atoms in the crustal layers, energy that appears as heating. You can easily demonstrate that motion can generate heat by hitting a steel nail a dozen or so times with a hammer and then touching the nail to your lip: the metal will feel distinctly hot. Imagine now the vastly greater heating created as mountain-size masses of rock plummet onto a planet. The heat so liberated, in combination with the radioactive heating described earlier, melted the planets and allowed matter with high density (such as iron) to sink to their cores, while matter with lower density (such as silicate rock) "floated" to their surfaces. We saw in chapter 6 that the Earth's iron core probably formed by this process, and astronomers believe that the other terrestrial planets formed their iron cores and rocky crusts and mantles the same way. A similar process probably occurred for the outer planets when rock and iron material sank to their cores.

Final Stages of Planet Formation

The last stage of planet formation was a rain of planetesimals that blasted out the huge craters such as those we see on the Moon and on all other bodies in the Solar System with solid surfaces. Figure 8.15 shows some of the planets and moons bearing vivid testimony to this role of planetesimals in planet building.

Occasionally an impacting body was so large that it did more than simply leave a crater. For example, we saw in chapter 7 that the Moon may have been created when the Earth was struck by a Mars-size body. Likewise, as we will discuss in chapter 9, Mercury may have suffered a massive impact that blasted away its crust. The peculiar rotation of Uranus and Venus may also have arisen from planetesimal collisions. In short, planets and satellites were brutally battered by the remaining planetesimals.

Although planet building consumed most of the planetesimals, some survived to form small moons, the asteroids, and comets. Rocky planetesimals and their fragments remained between Mars and Jupiter, where, stirred by Jupiter's gravitational force, they were unable to assemble into a planet. We see them today as the asteroid belt. Jupiter's gravity (and that of the other giant planets) also disturbed the orbits of icy planetesimals, tossing some in toward the Sun and others outward in elongated orbits to form the swarm of comet nuclei of the Oort cloud. The few that remain in the disk from Neptune's orbit out to about 50 AU form the Kuiper belt.

Formation of Satellite Systems

The large systems of satellites around the outer planets probably were formed from planetesimals orbiting the growing planets. Once a body grew massive enough that its gravitational force could draw in additional material, it became ringed with debris. Thus, moon formation was a scaled-down version of planet formation, and so the satellites of the outer planets have the same regularities as the planets around the Sun.

All four giant planets have flattened satellite systems in which the larger satellites (with few exceptions) orbit in the same direction. Many of these satellites are about as large as Mercury, and they would be considered full-fledged planets were they orbiting the Sun along an isolated orbit. A few of these bodies even have atmospheres, but they have too little mass (and thus too weak a gravitational attraction) to have accumulated

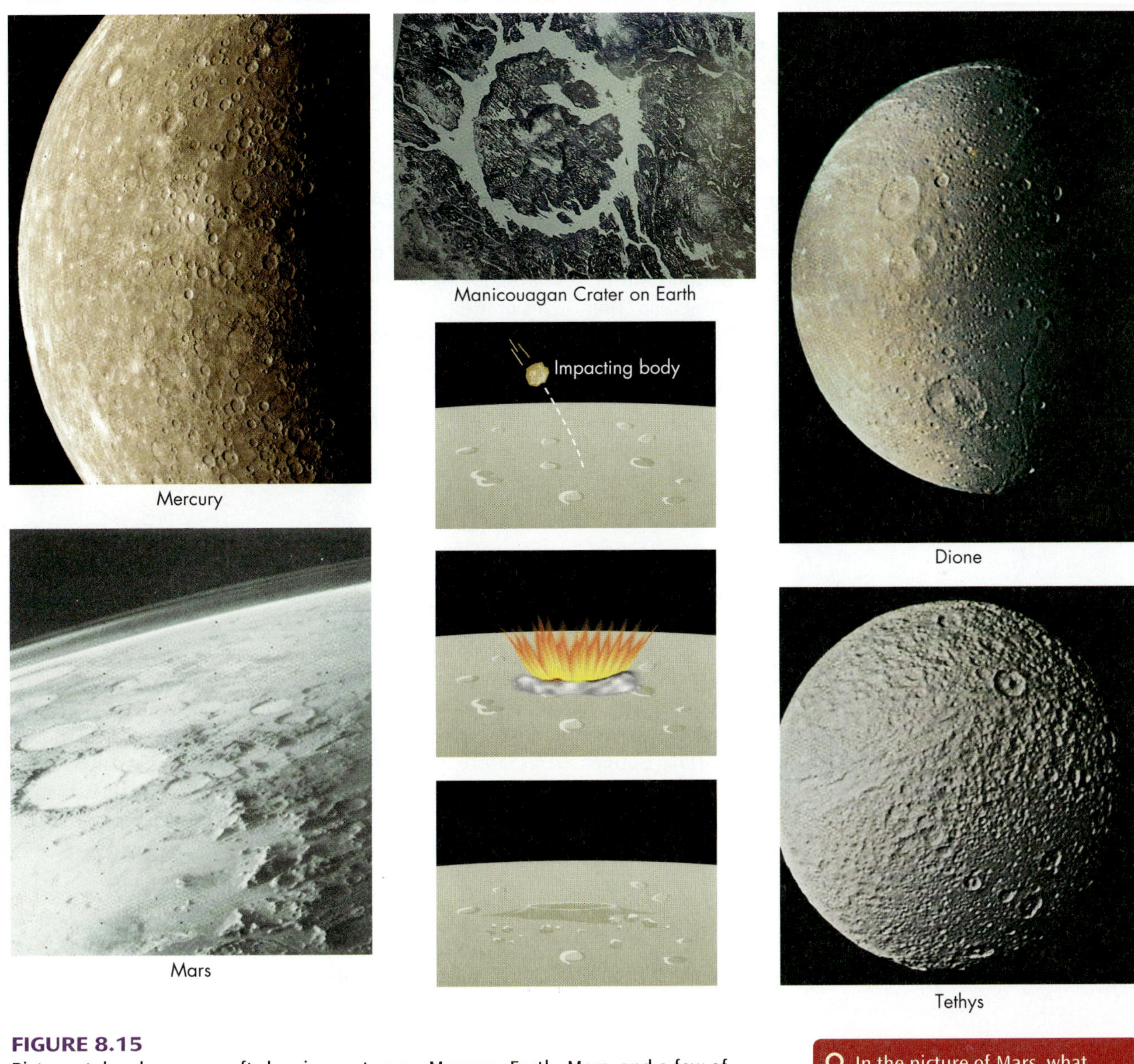

FIGURE 8.15
Pictures taken by spacecraft showing craters on Mercury, Earth, Mars, and a few of the moons of the outer planets. Included are Dione and Tethys, moons of Saturn. Note: Objects are shown to different scales.

Q. In the picture of Mars, what evidence can you see that Mars has an atmosphere?

large quantities of hydrogen and other gases from the solar nebula as their parent planets did. Thus, these moons are composed mainly of rock and ice, giving them solid surfaces—surfaces that are generally cratered and that, in a few cases, show signs of volcanic activity. These distant moons might in the future be ideal bases for studying those planets that have no surface to land on.

Formation of Atmospheres

Atmospheres were the last part of the planet-forming process. The inner and outer planets are thought to have formed atmospheres differently, a concept that explains their very different atmospheric composition. The outer planets probably captured most of their atmospheres from the solar nebula, as mentioned above; because the nebula was rich in hydrogen, so are their atmospheres.

SCIENCE *at work*

DIRECT FORMATION OF GIANT PLANETS

Because astronomers have no direct way to observe how the Solar System formed, they rely heavily on computer simulations to study that distant time. Computer simulations try to solve Newton's laws of motion for the complex mix of dust and gas that we believe made up the solar nebula. The solutions then can reveal what might have happened as the dust particles stuck together to form planetesimals and how the planetesimals then drew together under the influence of their gravity to form planets.

One of the more interesting findings of such calculations is that Jupiter may have formed directly from slightly denser regions of gas in the disk. Far from the Sun, where the gas is cold, gravity can more easily overcome the resistance of warmer gas to being squeezed into a smaller region. (Think of how a balloon resists being squeezed.) This may have allowed gravity to pull gas together to make a giant planet without the need to first form cores from planetesimals. Does this make the planetesimal theory wrong? No, just incomplete. Moreover, because this is an area of active research, astronomers still await definitive answers.

The inner planets were not massive enough and were too hot to capture gas from the solar nebula (as mentioned earlier) and are therefore deficient in hydrogen and helium. Venus, Earth, and Mars probably created their original atmospheres—by volcanic eruptions and by retaining gases from infalling comets and icy planetesimals that vaporized on impact. In fact, as a general rule, bodies too small to have captured atmospheres directly but that show clear signs of extensive volcanic activity (now or in the past) have atmospheres. More quiescent ones do not. Moreover, small bodies such as Mercury and our Moon keep essentially no atmosphere at all because their weak gravitational force means that their escape velocity is rather small, and atmospheric gases tend to escape easily from them.

Cleaning Up the Solar System

Only a few million years were needed to assemble most of the mass of the planets from the solar nebula, though the rain of infalling planetesimals lasted several hundred million years. Such a time is long in the human time frame but short in the Solar System's. All the objects within the Solar System are about the same age—the fourth property of the Solar System, mentioned at the beginning of this section.

One process still had to occur before the Solar System became what we see today: the residual gas and dust must have been removed. Just as a finished house is swept clean of the debris of construction, so too was the Solar System. In the sweeping process, the Sun was probably the cosmic broom, with its intense heat driving a flow of tenuous gas outward from its atmosphere. As that flow impinges on the remnant gas and dust around the Sun, the debris is pushed away from the Sun to the fringes of the Solar System. Such gas flows are seen in most young stars, and astronomers are confident the Sun was no exception. Even today, some gas flows out from the Sun, but in its youth, the flow was more vigorous.

The above theory for the origin of the Solar System explains many of its features, but astronomers still have many questions about how the Sun and its family of planets and moons formed. Is there any way, therefore, we can confirm the theory? For example, according to the solar nebula theory, planet formation is a normal part of star formation. So, do other planetary systems exist and do they resemble ours? Might we even be able to discover very young stars in the process of forming their planets and see the process at work?

8.3 Other Planetary Systems

Astronomers have long searched for planets orbiting stars other than the Sun. Their interest in such **exoplanets*** (as these distant worlds are called) is motivated not merely by the wish to detect other planets. Equally important is the hope that study of such systems will help us better understand the formation of the Solar System.

INTERACTIVE
Exoplanets

Planets are, of course, very small, so the light they reflect is drowned by that of their star. However, as we saw in section 8.2, young stars are often surrounded by a disk of dust and gas tens of astronomical units across. In 2008, the Hubble Space Telescope succeeded in detecting a large planet orbiting within a debris disk orbiting the star Fomalhaut (fig. 8.16). Fomalhaut is estimated to be about 200 million years old, so most of the dusty disk of material has been consumed, probably in forming planets. The ring of material that remains appears similar to our Kuiper belt, and the planet orbiting at the inner edge of the ring has probably grown by accreting material from the ring. The planet is more than a billion times fainter than Fomalhaut itself, making it extremely difficult to see in images; however, astronomers have discovered another way of detecting exoplanets.

Most present evidence for exoplanets comes from their effect on the star they orbit. As a planet orbits its star, the planet exerts a gravitational force back on the star as a result of Newton's third law—the law of action–reaction. That force makes the star's position wobble slightly, just as you wobble a little if you swing a heavy weight around you. The wobble creates a Doppler shift in the star's light that astronomers can measure (fig 8.17). From that shift and its change in time, astronomers can deduce the planet's orbital period, mass, and distance from the star. Using this Doppler method, astronomers had discovered over 300 exoplanets by early 2009, and figure 8.18 shows diagrams of many of the systems where more than one planet has been detected orbiting the star.

Looking at figure 8.18, you can see that none of these exoplanetary systems looks much like our own. First, most of the planets found so far have a mass comparable to

*A number of astronomers use the term extra-solar planets. However, this is a bit peculiar because, after all, Earth is extra-solar too, in the sense that it is orbiting outside the Sun.

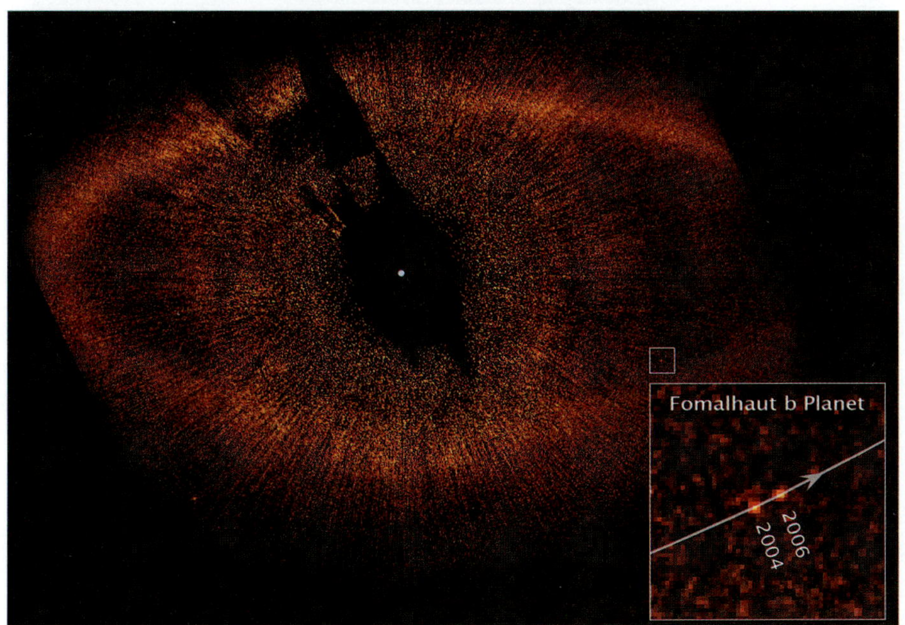

FIGURE 8.16
An icy ring surrounding the star Fomalhaut. Fomalhaut has about twice the Sun's mass and is about 200 million years old. The star itself has been blotted out by a small disk in the telescope so that its glare will not hide the faint ring of material, which is about twice the diameter of the Kuiper belt. A planet, magnified in the inset image, orbits along the inside edge of the ring. The planet's mass is estimated to be a few times Jupiter's mass.

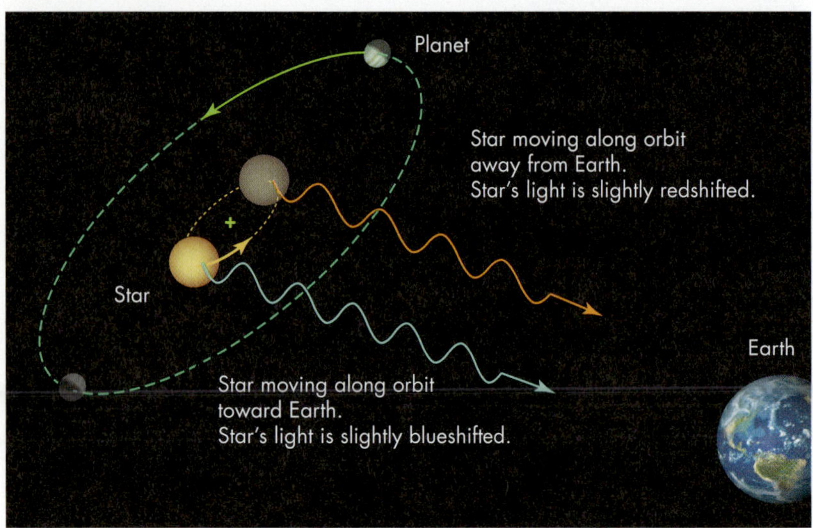

FIGURE 8.17
Detecting a planet from the motion of the star it orbits.

Jupiter's.* However, unlike Jupiter, many of these objects orbit extremely close to their star. Does this imply that our model for planet formation is wrong? Perhaps, but more likely we are seeing what is simply the outcome of the way we search. That is, the technique we use to find planets works best if the planet is massive and near its star. The reason for this selection is that only for such planets is the tug they exert on the star large enough to detect by the Doppler-shift method. Given that limitation, is there some way we can search for planets that more closely resemble our Earth? One way is to use a discovery made by Einstein in the early 1900s.

Einstein showed, as part of his general theory of relativity, that a mass bends space in its vicinity and that this bending creates the mass's gravity. As a result, if a ray of light passes near a mass, the bent space around the mass deflects the light and can bring it to a focus, as figure 8.19 schematically shows. Such bending of light by gravity may sound like science fiction, but as long ago as 1916, astronomers, following Einstein's suggestion, detected the bending of light from a distant star as the star's light traveled past our Sun (see essay 2). By analogy with the focusing ability of an ordinary lens, astronomers call such deflection of light **gravitational lensing.**

*Using the Doppler technique, astronomers have detected two planets with masses of about 4 or 5 Earth masses, but they are orbiting so close to their stars that they complete an orbit in under two weeks.

ANIMATION
The position and Doppler shift of a star orbiting its common center of mass with a planet

Star	Planet masses (left to right)
HD 169830 (1.40 $M_\odot$)	2.9, 4.0
HD 38529 (1.39 $M_\odot$)	0.78, 12.7
Ups And (1.27 $M_\odot$)	0.69, 2.0, 3.9
HD 74156 (1.24 $M_\odot$)	1.9, 0.40, 6.2
HD 82943 (1.18 $M_\odot$)	2.0, 1.8
HD 160691 (1.08 $M_\odot$)	0.044, 0.52, 1.67, 3.1
HD 12661 (1.07 $M_\odot$)	2.3, 1.6
HD 190360 (1.04 $M_\odot$)	0.057, 1.5
55 Cnc (1.03 $M_\odot$)	0.034, 0.82, 0.17, 0.14, 3.9
HD 217107 (1.02 $M_\odot$)	1.3, 2.5
HD 73526 (1.02 $M_\odot$)	2.9, 2.5
Sun (1.00 $M_\odot$)	Earth = 0.0031, Jupiter = 1.0
HD 108874 (1.00 $M_\odot$)	1.4, 1.0
HIP 14810 (0.99 $M_\odot$)	3.8, 0.76
HD 37124 (0.91 $M_\odot$)	0.61, 0.60, 0.68
HD 155358 (0.87 $M_\odot$)	0.89, 0.50
HD 69830 (0.86 $M_\odot$)	0.033, 0.038, 0.058
HD 128311 (0.80 $M_\odot$)	2.2, 3.2
Gliese 876 (0.32 $M_\odot$)	0.018, 0.56, 1.9
Gliese 581 (0.31 $M_\odot$)	0.049, 0.016, 0.024

Distance from star (AU): 1, 2, 3, 4, 5

FIGURE 8.18
Comparison of the orbital radii and relative sizes of exoplanets with the Solar System. Most of the systems with two or more known exoplanets are shown, organized according to the mass of the star that they orbit. The sizes of the dots are based on the mass of each planet, and approximately indicate their true relative size. The numbers indicate the mass of each planet in units of Jupiter's mass.

Gravitational lensing has proved to be a powerful tool for detecting low-mass planets. The method works approximately as follows. Suppose we look at some distant star and measure its brightness. Suppose further that, by chance, a star at an intermediate distance moves between us and the distant star. Rays of the distant star's light that would have traveled past us in the absence of the intermediate star are now bent so that they reach us (see fig. 8.19). Thus, we observe *more* light from the distant star when an intermediate-distance star is present. Moreover, we receive even more light (although only a very tiny amount more) if a planet is orbiting the intermediate-distance star.

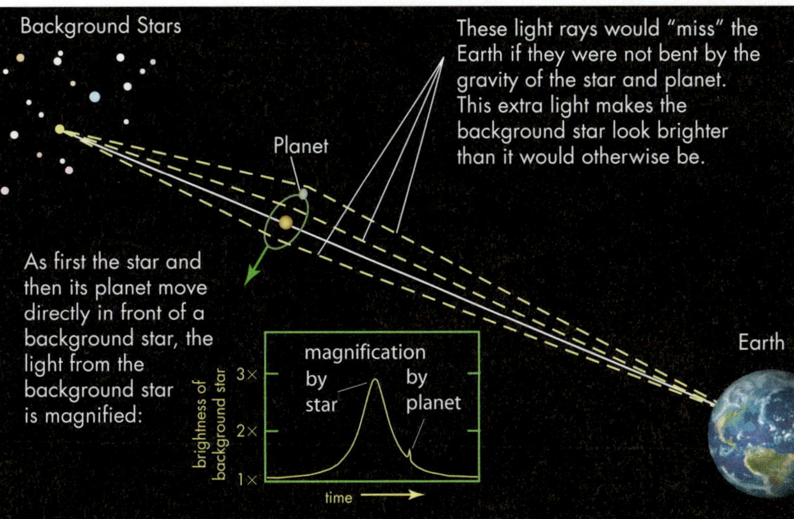

FIGURE 8.19

Detecting a planet by the slight bending of light from a background star caused by the planet's gravity.

It is very rare to find an intermediate star with an orbiting planet that is properly positioned. Thus, to search for planets by this method, astronomers monitor the brightness of millions of stars, and computers scan millions of bits of data for the tiny increase in brightness of a lensing event. For example, in 2005, astronomers detected a brightening event in OGLE-2005-BLG-390.* Calculations based on the light change show that the dim (and cool) star contains a little less than one-quarter as much matter as our Sun and that the planet is only about six times more massive than the Earth and orbits its star at distance of about 2.9 AU (roughly three times the Earth–Sun distance). The planet's small mass implies that it could not have drawn in a significant quantity of hydrogen or helium, so it cannot be a gas giant planet like Jupiter. Moreover, because the star turns out to be so cool and because the planet is farther from its star than the Earth is from the Sun, the planet must also be very cold. Thus, it is probably an icy planet, perhaps like a huge Pluto. Six more of these brightening events were detected in the following three years, including one planet that is only about 40% more massive than Earth.

Our inability, so far, to directly see the majority of exoplanets greatly limits what we can learn about them. However, astronomers can partly overcome this limitation for planets whose orbits carry them in front of their stars from our vantage point. For example, a planet of the Sun-like star HD209458 orbits so that it passes between us and the star every 3.5 days. At such times, the planet blocks a tiny amount of the star's light. From the amount of dimming, astronomers can deduce the diameter of the planet, which turns out to be about 1.3 times the diameter of Jupiter. Moreover, a tiny fraction of the star's light leaks through the planet's atmosphere so that gas in the planet's atmosphere imprints very weak absorption lines on the spectrum. The lines are from hydrogen, sodium, carbon, oxygen, and even water vapor. Analysis of the line strengths suggests that the planet is a gas giant planet similar to Jupiter. Notice, however, the extremely short orbital period of 3.5 days. Using Kepler's third law, astronomers deduce that this planet orbits a mere 0.05 AU from its star, roughly one-tenth the distance that Mercury orbits from our Sun—vastly nearer than where we expect a giant planet to orbit. In fact, from the extent of the absorption lines seen, it appears that the planet is surrounded by a cloud of evaporating gas. The planet may have lost as much as a quarter of its mass over several billion years, according to some estimates.

Astronomers have found no system of exoplanets yet that looks particularly like our own. The nearest match so far is the system of planets orbiting the star 55 Cancri. This Sun-like star has five planets orbiting within 6 AU of the star, just as in the Solar System (see fig. 8.20). However, all of these planets are massive, at least 10 times

*This name identifies the star as having been found by the *O*ptical *G*ravitational *L*ensing *E*xperiment against the bulge of stars (= BLG) in the center of our Galaxy.

SCIENCE *at work*

MIGRATING PLANETS

With the discovery of massive planets orbiting close to other stars, astronomers have been forced to think more critically about how our own Solar System formed.

Until very recently, most astronomers assumed that the Solar System's planets move along orbits that lie close to where the planets formed. But the discovery of planets with masses comparable to Jupiter's orbiting close to their star, like our terrestrial planets, has led astronomers to propose that planets may form at one distance from a star and then "migrate" to a new distance.

Can this proposal be tested? We can't watch real planets shift their orbits, but we can make computer simulations that follow a planet for millions of years under conditions similar to the early stages of our Solar System. These simulations show that interactions between the forming planets and leftover material in the disk of dust and gas can shift the planets' orbits either inward or outward, depending on the circumstances. The amount of such shifting in our Solar System is unknown, but according to some models, Neptune may have formed less than 20 AU from the Sun and moved past Uranus to its present distance of about 30 AU during the first several hundred million years of the Solar System.

Migration of planets has important consequences for planetary systems. For example, if a giant planet migrates inward toward its star, it will probably destroy smaller, Earth-size planets as it passes them. Thus, small planets, suitable for life as we know it, may form but fail to survive in such systems.

Yet another consequence of planet migration is that as a planet changes its orbital distance from its star, it encounters regions still rich with small bodies. These may be captured or flung into new orbits. This in turn may explain a surge of impacts that appears to be recorded on the Moon's surface about 600 million years after the Solar System's formation. This late bombardment may mark the time when the giant planets reached their final orbital positions.

Earth's mass, and three of the planets orbit at distances much closer than Mercury's distance from the Sun.

The presence of gas giant planets so near their stars presents a challenge to our understanding of how the Solar System formed. According to the solar nebula theory, gas giants should form only beyond several astronomical units from a star, where hydrogen-rich compounds can condense. If gas giants can form so close to a star, we need to understand what is different in these systems, or find a new mechanism to

FIGURE 8.20
The 55 Cancri system contains five known planets around a star that is very similar to the Sun. The estimated masses (compared to Jupiter) for these planets and their orbits are compared with the five innermost planets in the Solar System. The figure also shows the approximate relative sizes of the planets. The fourth planet out in the 55 Cancri system orbits its star at about the same distance as the Earth from the Sun, but its mass is more than twice that of Neptune.

explain their formation. However, perhaps these planets did not actually form so close to their star, but instead "migrated" inward from an initial location much farther out (see Science at Work: "Migrating Planets").

Not only are the giant planets "too close" to their star, many also are in very elliptical orbits (rather than the essentially circular ones in our own system). That so many extra-solar systems have a massive planet on a very elliptical orbit does not bode well for the existence of Earth-like planets in these systems. As a massive planet sweeps into the inner portion of a star system, it will, over time, disturb the orbit of smaller planets, either ejecting them from the system or causing them to fall into their star. Some evidence suggests this fate may have befallen planets in a few of these remote systems. A number of the stars with exoplanets are appreciably richer in iron than our Sun. One suggestion for why these stars are so iron-rich is that they have swallowed Earth-like planets and vaporized them. The iron from the vaporized planet's core then enriches the star, making its spectrum lines of iron stronger. This is not the only interpretation, however. Perhaps it is easier to make planets in the first place if a star has a higher-than-average concentration of iron. Which interpretation is correct? We do not yet know.

SUMMARY

The Solar System consists of a star (the Sun) and planets, asteroids, and comets, which orbit it in a broad, flat disk. All the planets circle the Sun in the same direction, and most of them spin in the same direction. Their moons also form flattened systems, generally orbiting in the same direction. The planets fall into two main categories: small, high-density bodies (the inner, or terrestrial, planets) and large, low-density bodies (the outer, or Jovian, planets). The former are rich in rock and iron; the latter are rich in hydrogen and ice.

These features of the Solar System can be explained by the solar nebula theory. In this theory, the Solar System was born from a cloud of interstellar gas that collapsed to a disk called the solar nebula. The center of the nebula became the Sun, and the disk became the planets. This explains the compositional similarities and the common age of the bodies in the system.

The flat shape of the system and the common direction of motion around the Sun arose because the planets condensed within the nebula's rotating disk. Planet growth occurred in two stages: dust condensed and clumped to form planetesimals; and then later the planetesimals aggregated to form planets and satellites. Two kinds of planets formed because lighter gases and ice could condense easily in the cold outer parts of the nebula but only rocky and metallic material could condense in the hot inner parts. Impacts of surviving planetesimals late in the formation stages cratered the surfaces and may have tilted the rotation axes of some planets. Some planetesimals (and/or their fragments) survive to this day as the asteroids and comets.

Astronomers have found many planets orbiting other stars. Study of these exoplanets helps us better understand the origin of planetary systems, although most systems found so far indicate major differences from the patterns seen in the Solar System. One reason for these differences is that current methods for detecting planets are mainly able to detect only massive planets close to their stars. Nevertheless, it is surprising to find that so many giant exoplanets do orbit very close to their star.

QUESTIONS FOR REVIEW

1. (8.1) Name the eight planets in order of increasing distance from the Sun.
2. (8.1) Make a sketch of the Solar System showing top and side views.
3. (8.1) Make a table listing separately the inner and outer planets.
4. (8.1) What properties, apart from position, distinguish the inner and outer planets?
5. (8.1) What is the Oort cloud? Where is it located, and what kind of objects come from it?
6. (8.1) How do we know the composition of Jupiter?
7. (8.1) What is Bode's rule?
8. (8.2) What is an interstellar cloud? What does it have to do with the Solar System?
9. (8.2) What is the solar nebula? What is its shape and why?
10. (8.2) Why are there two main types of planets?
11. (8.2) What is the difference between condensation and accretion?
12. (8.2) What are planetesimals?
13. (8.2) Describe the planetesimal theory of planet formation.
14. (8.2) How does the planetesimal theory of planet formation explain the asteroids?

15. (8.2) How did the craters we see on many of the planets form?
16. (8.2) Describe a theory of how planets may have formed their atmospheres.
17. (8.2) How would you describe the formation of the Solar System to a little brother or sister?
18. (8.3) What observations of other solar systems have been made that support the solar nebula hypothesis?
19. (8.3) What are two methods used to find exoplanets?
20. (8.3) How do some exoplanets differ from what we might expect? Does this prove the nebula theory is wrong?

THOUGHT QUESTIONS

1. (8.2) What kinds of physics would be important to include in a computer simulation of solar system formation?
2. (8.3) How are the kinds of exoplanets found by the Doppler method a biased sample of exoplanets? Give an example of a survey method that might give a biased result in everyday life.
3. (8.3) How do some exoplanets differ from what we might expect? Does this prove the nebula theory is wrong?

PROBLEMS

1. (8.1) Calculate the densities of Venus and Jupiter (use the masses and radii given in the appendix). How do these numbers compare with the density of rock (about 3 grams per cm^3) and water (1 gram per cm^3)? (Note: Be sure to convert kilometers to centimeters and kilograms to grams if you are expressing your answer in grams per cm^3.)
2. (8.1) Look up the orbital periods of the Earth and Jupiter in the appendix. If you started measuring when the Earth and Jupiter were at their closest to each other, how many years would it be until they returned to that position again?
3. (8.1/3.8) Look up the mass and radius of Mercury and Jupiter and calculate their escape velocities, using the expression in chapter 3. Does this help you see why the one body has an atmosphere but the other doesn't? (Note: Be sure to convert kilometers to meters or the appropriate unit.)
4. (8.1/3.8) Look up the mass and radius of Neptune and Mars and calculate their escape velocities, using the expression in chapter 3. Compare both with that of the Earth (see section 3.8). What is different about the atmospheres of these three planets? (Note: Be sure to convert kilometers to meters or the appropriate unit.)
5. (8.3/2.3) Suppose an exoplanet orbits a star of 1 solar mass. Suppose the orbital period is 5 days. What is the semimajor axis of the orbit in AU?
6. (8.3) Calculate the maximum Doppler shift that could be observed for the planet in question 5.
7. (8.3) Using the modified form of Kepler's laws given in figure 8.8, calculate the orbital period for Gliese 851d, an exoplanet with a mass about 7.1 times the mass of Earth. The star Gliese 851 is a red dwarf with a mass of 0.31 solar masses and the planet orbits with a semi-major axis of 0.22 AU. (Remember to convert distances to meters and masses to kilograms when using the equation).

TEST YOURSELF

1. (8.1) Which of the following objects are primarily rocky with iron cores?
   (a) Venus, Jupiter, and Neptune
   (b) Mercury, Venus, and Pluto
   (c) Mercury, Venus, and Earth
   (d) Jupiter, Uranus, and Neptune
   (e) Mercury, Saturn, and Eris
2. (8.2) One explanation of why the planets near the Sun are composed mainly of rock and iron is that
   (a) the Sun's magnetic field attracted all the iron in the young Solar System into the region around the Sun.
   (b) the Sun is made mostly of iron. The gas ejected from its surface is therefore iron, so that when it cooled and condensed, it formed iron-rich planets near the Sun.
   (c) the Sun's heat made it difficult for other substances such as ices and gases to condense near it.
   (d) the statement is false. The planets nearest the Sun contain large amounts of hydrogen gas and subsurface water.
   (e) the Sun's gravitational attraction pulled iron and other heavy material inward and allowed the lighter material to float outward.
3. (8.2) Which of the following features of the Solar System does the solar nebula theory explain?
   (a) All the planets orbit the Sun in the same direction.
   (b) All the planets move in orbits that lie in nearly the same plane.
   (c) The planets nearest the Sun contain only small amounts of substances that condense at low temperatures.
   (d) All the planets and the Sun, to the extent that we know, are the same age.
   (e) All of the above
4. (8.2) The numerous craters we see on the solid surfaces of so many Solar System bodies are evidence that
   (a) they were so hot in their youth that volcanos were widespread.
   (b) the Sun was so hot that it melted all these bodies and made them boil.
   (c) these bodies were originally a mix of water and rock. As the young Sun heated up, the water boiled, creating hollow pockets in the rock.

(d) they were bombarded in their youth by many solid objects.
(e) all the planets were once part of a single, very large and volcanically active mass that subsequently broke into many smaller pieces.

5. (8.3) The Doppler-shift method for detecting the presence of exoplanets is best able to detect
(a) massive planets near the star.
(b) massive planets far from the star.
(c) low-mass planets near the star.
(d) low-mass planets far from the star.

6. (8.3) The transit method for detecting exoplanets works best for
(a) very massive planets.
(b) solar systems seen face-on.
(c) planets very far from their stars
(d) solar systems seen edge-on.
(e) planets very close to their stars.

KEY TERMS

accretion, 216
asteroid, 209
asteroid belt, 210
Bode's rule, 213
comet, 209
condensation, 216
dwarf planet, 209
exoplanet, 221
gravitational lensing, 222
inner planets, 208
interstellar cloud, 214
interstellar grains, 214
Jovian planet, 208
Kuiper belt, 210
Oort cloud, 210
outer planets, 208
planetesimals, 216
solar nebula, 214
solar nebula theory, 214
Solar System, 206
terrestrial planets, 208

FURTHER EXPLORATIONS

Beatty, J. Kelly, Carolyn Collins Petersen, and Andrew Chaikin. *The New Solar System.* 4th ed. Cambridge: Sky Publishing Company, 1999.

Gingerich, Owen. "Losing It in Prague: The Inside Story of Pluto's Demotion." *Sky and Telescope* 112 (November 2006): 34.

Jayawardhana, Ray. "Are Super-Sized Earths the New Frontier?" *Astronomy* 36 (November 2008): 26.

Lin, Douglas N.C. "The Genesis of Planets." *Scientific American* 298 (May 2008): 50.

Malhotra, Renu. "Migrating Planets." *Scientific American* 281 (September 1999): 56.

Seager, Sara. "Unveiling Distant Worlds." *Sky and Telescope* 111 (February 2005): 28.

Seager, Sara. "Alien Earths from A to Z." *Sky and Telescope* 115 (January 2008): 22.

Soter, Steve. "What Is a Planet?" *Scientific American* 296 (January 2007): 34.

Weinberger, Alycia J. "Building Planets in Disks of Chaos." *Sky and Telescope* 116, (November 2008): 32.

Website

Visit the *Explorations* website at **http://www.mhhe.com/arny** for additional online resources on these topics.

Q FIGURE QUESTION ANSWERS

WHAT IS THIS? (chapter opening): This is a disk of gas and dust around a forming star. The dark disk is visible in silhouette against the glow of emission from the Orion nebula. The forming star glows red at its center.

FIGURE 8.8: The density is the mass (30 grams) divided by the volume (10 cm^3). 30 grams/10 cm^3 = 3 grams/cm^3. Iron's density is about 8 gm/cm^3, while a typical silicate rock's density is about 3 gm/cm^3. It is thus more likely to be rock.

FIGURE 8.13: When you breathe out, warm moist air from your lungs comes in contact with the cold air outside. The moisture in your breath condenses and makes a tiny "cloud."

FIGURE 8.15: Above the edge of the planet, especially on the left, you can see thin wispy clouds.

PROJECT

Extrasolar planets: Look on the Web for results of searches for exoplanets and young solar systems. Try to find an example of a star system in each of the primary phases believed to have occurred in the formation of the Solar System, starting with an interstellar cloud. For example, you can easily find images of protoplanetary disks in the Hubble Space Telescope archive. For established solar systems, actual images are rare, but you may be able to find a diagram or chart of the planets in the system. Record the mission, masses of the planet(s), and method of detection for each. Of the stars you can find that have planets, which star has a mass closest to that of the Sun? Of the planets that have currently been found, which has a mass closest to that of the Earth?

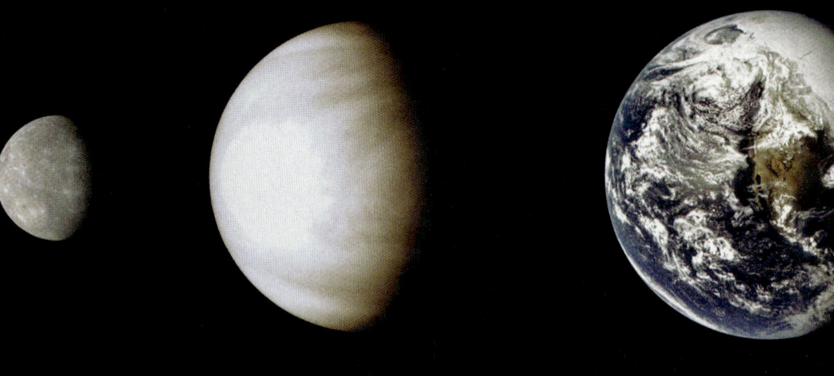

Images of Mercury, Venus, Earth, and Mars, showing representative surface features.

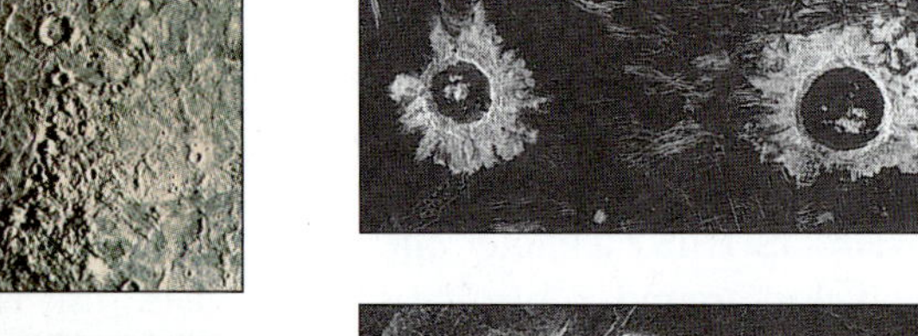

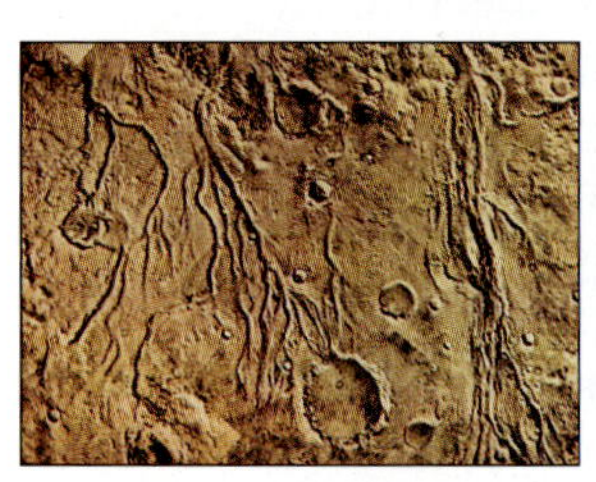

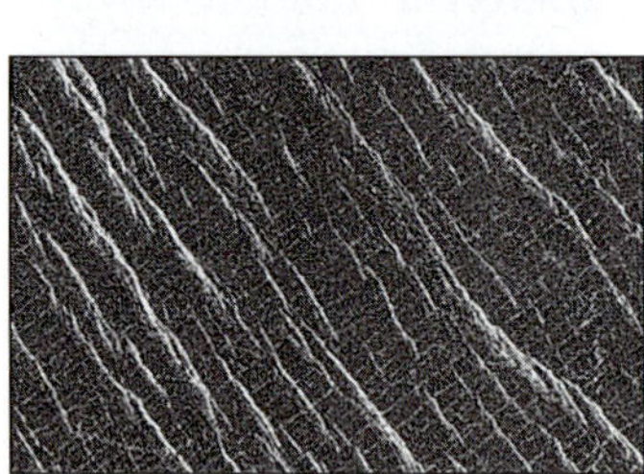

KEY CONCEPTS

- The terrestrial (or inner) planets are Mercury, Venus, Earth, and Mars.
- These planets are similar in size and composition to the Earth, with an iron core surrounded by rocky outer layers.
- As for Earth, radioactivity and heat left over from their formation make their centers hot.
- The interiors of smaller objects (Mercury and Mars) have cooled more than Earth and Venus have.
- The flow of heat drives motions inside Earth and Venus.
- These motions determine the types of surface features found on these planets.
- Their original atmospheres were probably mainly carbon dioxide and nitrogen.
- Liquid water in Earth's early atmosphere gradually removed CO_2. Green plants produced Earth's oxygen by photosynthesis.

9 The Terrestrial Planets

CONCEPTS AND SKILLS TO REVIEW

- Density (6.1)
- Absorption in Earth's atmosphere (4.7)

Terrestrial planets, as their name suggests, have a size and structure similar to Earth's. Within our Solar System, the planets Mercury, Venus, Earth, and Mars are terrestrial. Orbiting in the inner part of the Solar System, close to the Sun, these rocky worlds are too small and too warm to have captured massive hydrogen envelopes such as those that cloak the outer planets.

Despite their similarity in size and composition, the surface conditions on the terrestrial planets differ enormously. Our goal in this chapter is to better understand how these neighboring planets came to be so different from Earth. We will discover that size plays a major role. For example, because Mercury is so small, it generates little internal heat to create surface activity, and so its crust is essentially unchanged from its birth. Venus, on the other hand, is large enough to have held in the heat from its formation and from the decay of radioactive elements that are present in every planet. Its hot interior is rather like Earth's, and so it has a surface with mountains and volcanic peaks.

Size, coupled with distance from the Sun, creates the great atmospheric differences among these terrestrial worlds. For example, Mercury is too small, and its surface is too hot, to retain an atmosphere, while Mars, only slightly larger but farther from the Sun and therefore cooler, has retained one. Venus and Earth are both large enough to have sizable atmospheres, but Earth's slightly greater distance from the Sun has made it cool enough to have liquid water in its atmosphere. That simple fact has led to the profound difference between the atmospheres of Earth and Venus, because liquid water can remove carbon dioxide from the atmosphere. Moreover, liquid water has allowed life to form and flourish here, and life has not only removed additional carbon dioxide from our air but has also added oxygen.

Q: WHAT IS THIS? (See end of chapter for the answer.)

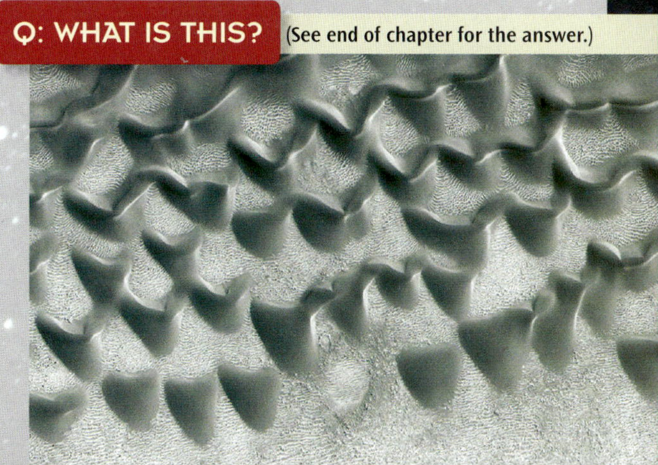

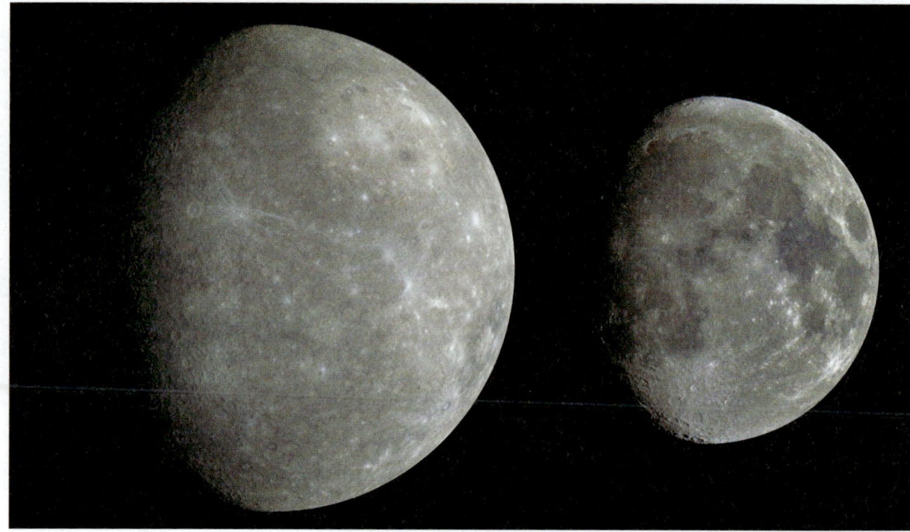

FIGURE 9.1
Mercury (on the left) and the Moon (on the right) shown in their relative sizes. Note that these airless objects are both heavily cratered, but the Moon has large maria.

9.1 Mercury

Mercury is the smallest terrestrial planet. It is named for the Roman deity who was the speedy messenger of the gods because it changes its position on the sky faster than any other planet. Mercury resembles our Moon in both size and appearance (fig. 9.1), with its radius being about one-third and its mass about one-twentieth that of the Earth. Astronomers can determine Mercury's radius from its angular size and distance, as we discussed in chapter 8 (see fig. 8.8). They calculated its mass from its gravitational attraction on the *Mariner 10* spacecraft, which flew past the planet in 1974 and 1975, taking pictures of its surface. Those images showed that circular craters like those on the Moon cover Mercury's surface, as illustrated in figure 9.2.

Mercury's surface is not totally Moonlike, however. Congealed lava flows flood many of Mercury's old craters and pave much of its surface. On our Moon, such flows are found almost exclusively within the maria. In addition, enormous scarps—cliffs formed where the crust has shifted—run for hundreds of kilometers across Mercury's surface, as seen in figure 9.2. The scarps may have formed as the planet cooled and shrank, wrinkling like a dried apple.

The largest crater on Mercury by far is the vast Caloris Basin, shown in figure 9.3. With a diameter of 1300 kilometers (about 800 miles), this

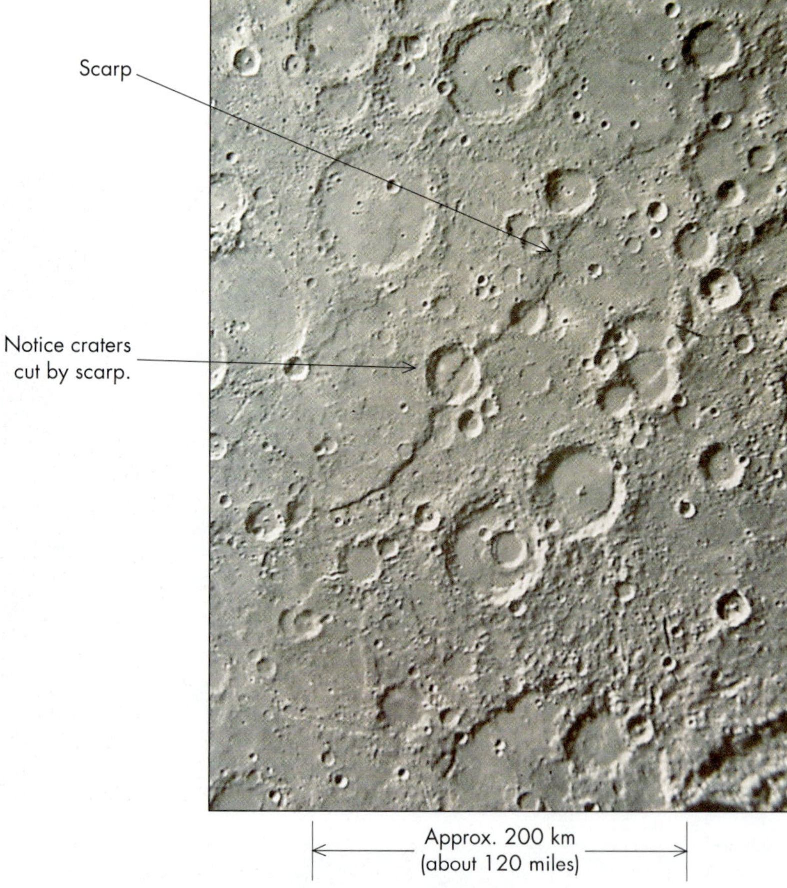

FIGURE 9.2
Photograph of craters and a scarp, a cliff, running across Mercury's surface.

Q. Which formed first, the scarp or the craters it passes through? On what do you base your answer?

A

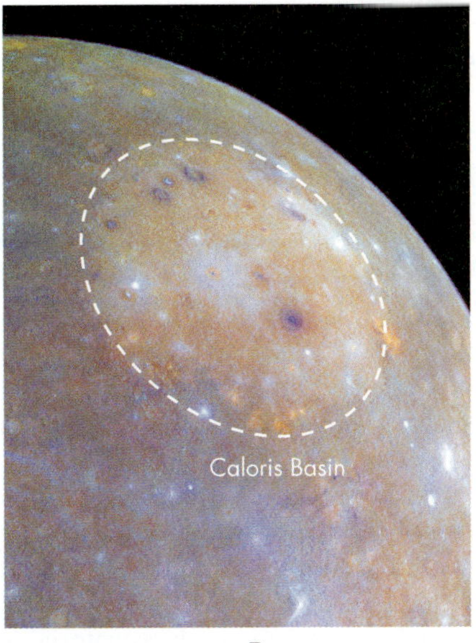

B

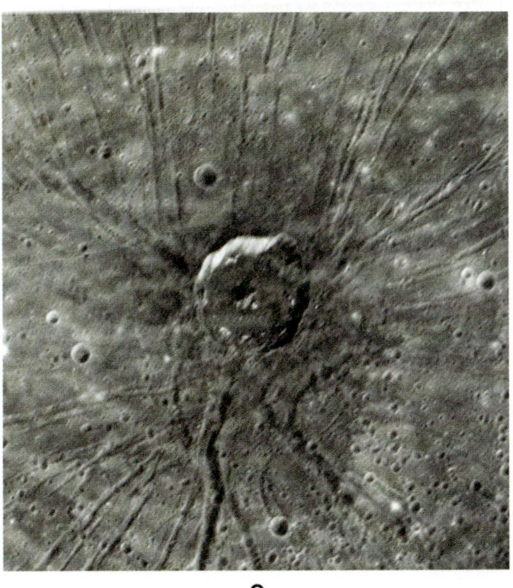

C

FIGURE 9.3
(A) The edge of the Caloris Basin is seen in the semicircular set of rings surrounding Mercury's largest impact feature. Only this edge of the basin was seen by the passing *Mariner 10* spacecraft in 1974. (B) In 2008 the Messenger spacecraft imaged the rest of Mercury. The Caloris Basin appears light orange in this false-color image, which enhances color differences to show different mineral compositions. (C) The center of the Caloris Basin has a strange spidery pattern of troughs surrounding a 40-kilometer-diameter crater near the center of the basin.

Q. Does Caloris Basin look like lunar maria? Compare figure 9.3 (above) to figure 7.5. How are they similar? How are they different?

mountain-ringed depression is reminiscent of lunar maria. Moreover, its circular shape and surrounding hills indicate that, like the maria, it was formed by impact. The impact spawned volcanoes around the edge of the basin, several of which are visible as bright orange spots in the false-color image shown in figure 9.3B.

By counting craters within the Caloris Basin and comparing the number to other regions of Mercury, it appears that the basin formed during a collision about 3.8 billion years ago. This is about the time when some of the last major collisions occurred on the Moon, producing somewhat similar-looking features such as Mare Orientale (see fig. 7.5).

Mercury also possesses the curious landscape illustrated in figure 9.4, which lies on the far side of the planet exactly opposite Caloris Basin. Astronomers call this hummocky, jumbled surface "chaotic terrain" and think it was churned up by earthquake waves generated by the impact that created Caloris. As the waves traveled around Mercury, they converged on its far side, heaving up the rock, much as dropping cream into coffee creates a tiny splash as the ripples reconverge.

FIGURE 9.4
Image of the chaotic terrain that lies on the side of Mercury opposite the Caloris Basin.

Mercury's Temperature and Atmosphere

Mercury's surface is one of the hottest places in the Solar System. At its equator, noon temperatures reach approximately 710 K (about 820°F). On the other hand, nighttime temperatures are among the coldest in the Solar System, dropping to approximately 80 K (about −320°F). These extremes result from Mercury's closeness to the Sun, its slow rotation, and its lack of atmosphere. The slow rotation allows the surface to heat up strongly during the Mercurian day and to cool down greatly at night. The absence of an atmosphere means that nothing moderates the inflow of sunlight during the day or retains heat during the night. Despite a high equatorial temperature, however, Mercury's poles are very cold. Sunlight shines so indirectly on them that they receive little heat, and with no atmosphere to distribute warmth, the poles have grown very cold—so cold, in fact, that radar telescopes have detected features that are thought to be small ice caps.

Mercury has almost no atmosphere for the same reason as the Moon. Mercury's small mass makes its gravitational attraction too small to retain much gas around it. Moreover, its proximity to the Sun makes keeping an atmosphere difficult because the resulting high temperature in the equatorial regions causes molecules to move so fast that they readily escape into space. Traces of gas have been detected spectroscopically. Some of this gas is probably captured temporarily from gas flowing away from the Sun, while some may be escaping from the interior of the planet. Enhancements of gases detected in the vicinity of Caloris Basin and the chaotic terrain opposite it may be the result of fracturing of rock by the collision that formed these features.

Mercury probably never had an extensive atmosphere—the Sun is so close to it that most of the gases would have been "baked out" of the materials from which Mercury was formed. Where, then, has the ice at its poles come from? One theory suggests the ice came from comets that struck its surface. Comets occasionally plunge into the Sun, and so some must occasionally hit Mercury. Such impacts vaporize the comet, creating a wispy and ephemeral atmosphere, most of which quickly escapes into space. Some gas, however, may drift toward the cold polar regions and freeze there, much as frost condenses on automobile windshields on a subfreezing morning. Over time, such frost deposits might have built up the caps seen on this otherwise hot planet.

Mercury's Interior

Mercury probably has an iron core beneath its silicate crust, but astronomers have little proof because no spacecraft has landed there to deploy seismic (earthquake) detectors. Their conclusion is therefore based on Mercury's density and gravitational field. A massive planet's gravity can compress its interior to high density, but Mercury is too small for this effect to be significant. Thus, its high density (5.4 grams per cm^3) indicates an iron-rich interior with only a thin rock (silicate) mantle, as depicted in figure 9.5.

Why Mercury is so relatively rich in iron but poor in silicates is unclear. One possibility is that silicates did not condense as easily as iron compounds in the hot, inner solar nebula where Mercury formed. Another possibility is that Mercury once had a thicker rocky crust, which was blasted off by the impact of an enormous planetesimal, as the computer simulation in figure 9.6 illustrates. (Recall that a similar collision may have happened to Earth and formed our Moon.)

Astronomers had not expected Mercury's core to be molten, because Mercury is not much bigger than the Moon, whose interior appears to have solidified relatively early in its history. A small radius allows heat from the interior to escape more readily. However, Mercury has at least a partially liquid (molten) core like the Earth. The existence of a liquid core was shown in 2007 by the way Mercury "wobbles." A similar effect can be seen by spinning hard-boiled and raw eggs on a tabletop. When you stop the hard-boiled egg from spinning, it remains stopped, but if you briefly stop the raw

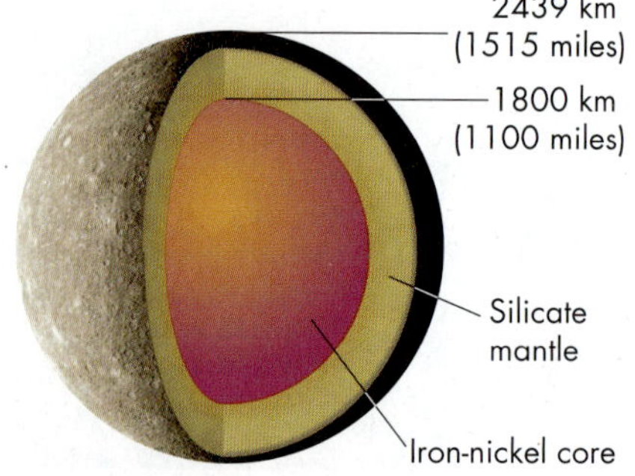

FIGURE 9.5
Artist's depiction of Mercury's interior.

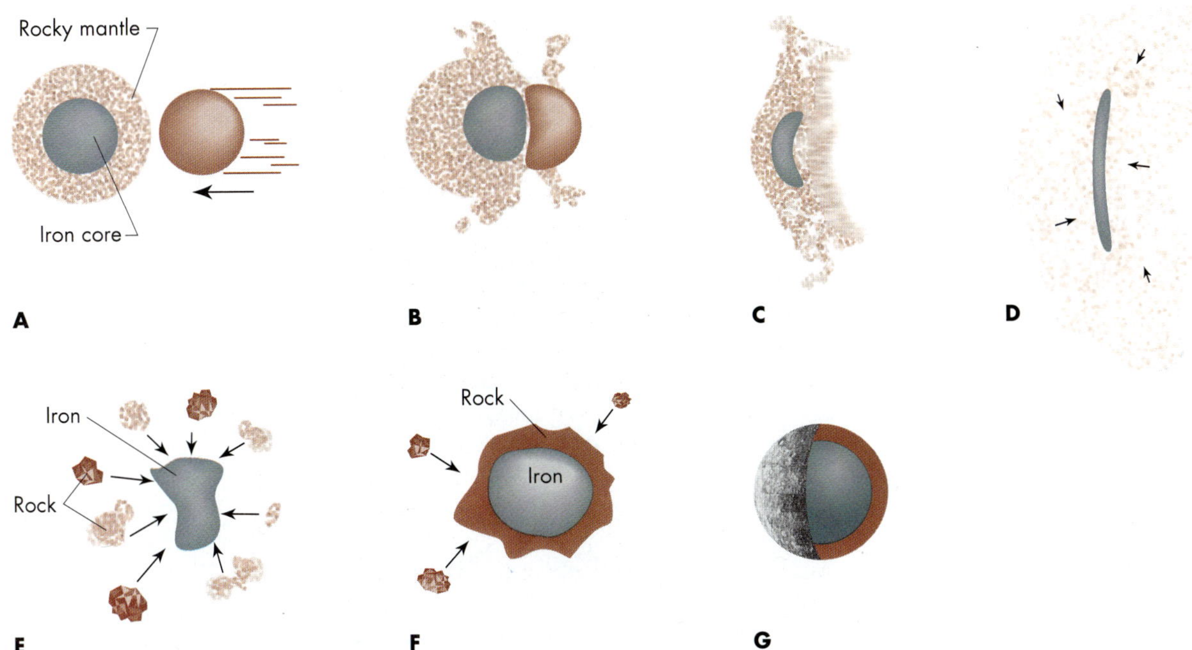

FIGURE 9.6
Computer simulation of a collision between Mercury and a large planetesimal. The impact strips away most of the outer rocky layers and leaves a highly distorted iron core surrounded by rocky debris. Gravity eventually reshapes the planet into a sphere.

egg then let it go, it will begin spinning again. This is because the liquid material in its interior continues to rotate even after the shell is stopped. Precise measurements of the way Mercury's rotation responds to gravitational tugs by the Sun show a similar effect.

Mercury's magnetic field is only about 1% as strong as Earth's. As we discussed in chapter 6, the Earth's magnetic field is probably generated by circulating motions in the Earth's molten iron core combined with our planet's spin. Given that Mercury has a molten core, the weakness of its magnetic field is probably a consequence of the slowness of the planet's rotation.

Mercury's Rotation

Mercury spins very slowly. Its rotation period is 58.646 Earth days, which is exactly two-thirds of its orbital period around the Sun of 87.969 Earth days. This means that it spins exactly three times for each two trips it makes around the Sun. How did this proportion occur, or is it just coincidence?

The Sun has affected Mercury's rotation by tidal forces, just as the Earth has affected the Moon's. That is, the Sun's gravity exerts a force on Mercury, which tends to twist the planet and make it rotate with the same period in which it orbits the Sun. Mercury's orbit, however, is very elliptical. Thus, in accordance with Kepler's second law of planetary motion, Mercury's orbital speed changes as it moves around the Sun. Because of that changing speed, the Sun cannot lock Mercury into a purely synchronous spin; the closest to synchrony it can get is three spins for each two orbits (fig. 9.7). Such an integer ratio of periods is called a *resonance.*

Resonance occurs when a force that acts repeatedly on a body causes its motion to grow ever larger. For example, pushing a child on a swing can create a resonance. If you push just as the swing starts to move forward, the child will swing higher and higher, and the pushing force will be in resonance with the motion of the swing. On the other hand, if you push before the backward motion is stopped, the swinging motion will

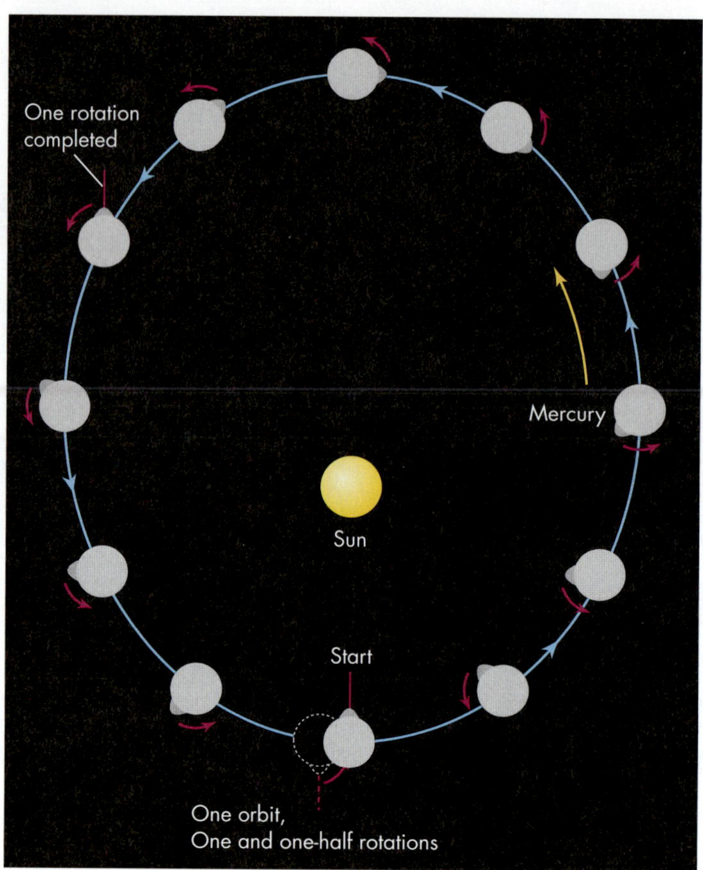

FIGURE 9.7
Mercury's odd rotation. The planet spins three times for each two orbits it makes around the Sun.

Q. What resemblance do you see between Mercury's motions and those of our Moon?

ANIMATION
The rotation of Mercury during 2 Mercury years

decrease, and no resonance occurs. Likewise, applying power resonantly to a car stuck in a ditch may "rock" it out. A similar resonance exists between the Sun's changing gravitational tug on Mercury as it moves along its elongated orbit and its rotation. The result is the 2:3 relation between its orbital and spin periods.

Mercury's odd rotation gives it an extremely long solar day (the time between successive sunrises) of 176 Earth days. During that time, the Sun sometimes changes its direction of motion across the sky. For example, if sunset occurs when Mercury is at the point of its orbit nearest the Sun, the Sun will set and then briefly rise again before setting a second time!

9.2 Venus

Of all the planets, Venus is most like the Earth in diameter and mass. It is named for the Roman goddess of beauty and love. Because Venus is so similar to the Earth in size, we might therefore expect it to be like the Earth in other ways. However, Venus and the Earth have radically different surfaces and atmospheres. Many of these differences we have discovered only recently, because dense clouds perpetually cloak Venus, as shown in figure 9.8. Nevertheless, astronomers know that its surface looks nothing like the Earth's, and its atmosphere is much hotter and denser than ours and has a very different composition.

The Venusian Atmosphere

The atmosphere of Venus is 96% carbon dioxide. Astronomers know the atmospheric composition from its spectrum and from measurements with space probes. Gases in its atmosphere absorb some of the sunlight falling on the planet and create absorption lines in addition to those of the Sun itself. These lines then reveal the composition and density of the gas. Moreover, spacecraft have descended through the atmosphere to the surface and have sampled its atmosphere. Thus, we have learned that, in addition to carbon dioxide, Venus's atmosphere contains about 3.5% nitrogen and very small amounts of water vapor and other gases.

Spectra also reveal the nature of the Venusian clouds: they are composed of sulfuric acid droplets that formed when sulfur compounds—perhaps ejected from volcanos—combined with the traces of water in the atmosphere. These clouds permanently cover the planet and are very high and thick, beginning at about 30 kilometers (19 miles) above the surface and extending upward to about 60 kilometers (37 miles). In fact, the clouds are so thick that no surface features can be seen through them with ordinary telescopes. This deep cloud layer strongly reflects sunlight falling on it, making Venus very bright as seen from Earth (fig. 9.9)—so bright that if the air is very clear, you can see Venus in broad daylight. Below the clouds, the Venusian atmosphere is relatively clear, and some sunlight penetrates to the surface. The light is tinged orange, however, because the blue wavelengths are absorbed in the deep cloud layer.

Venus's atmosphere is extremely dense. It exerts a pressure roughly 100 times that of the Earth's, equivalent to the pressure you would feel under 1000 meters (3000 feet) of water. We will discuss in section 9.4 why Venus has such a massive atmosphere, but there are other features of the planet that we need to understand first. One of the most important of those features is that its lower atmosphere is extremely hot. Observations made with radio telescopes from Earth show that the surface temperature is more than

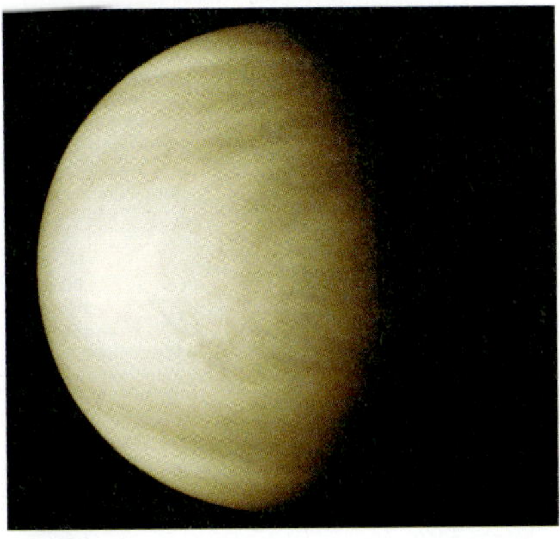

FIGURE 9.8
Photograph through an ultraviolet filter of the clouds of Venus. The picture is artificially colored and enhanced to show the clouds clearly.

Q. If this were the view of Venus from Earth, what would it imply about the position of Venus in its orbit?

FIGURE 9.9
The Moon and Venus over Ottawa Lake, Wisconsin. Venus is the bright dot below and to the right of the Moon. The left part of the Moon is dimly lit by sunlight reflected from the Earth—earthlight.

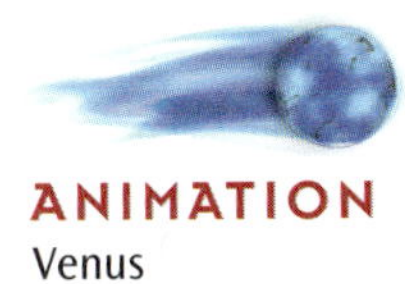

ANIMATION
Venus

750 K (about 900°F), a value confirmed by spacecraft landers. What makes the atmosphere of a planet so similar to the Earth and only slightly nearer the Sun, so very hot?

The Runaway Greenhouse Effect

Venus's carbon dioxide atmosphere creates an extremely strong **runaway greenhouse effect.** We discussed in chapter 6 how gases in a planet's atmosphere allow sunlight to enter and warm the surface but retard the heat so generated from escaping to space. That is, certain gases trap a planet's heat by hindering or blocking its infrared radiation from escaping to space, as was illustrated in figure 6.20.Carbon dioxide in the Earth's atmosphere is especially effective at trapping heat, creating a weak, but necessary, greenhouse effect here. Venus, however, has about 300,000 times more carbon dioxide than Earth, and so its greenhouse effect is correspondingly much stronger. Whatever liquid water was present on Venus in its early history was boiled away, adding further to the greenhouse effect. In fact, the heat trapping is so effective that Venus's surface is hotter than Mercury's, even though Venus is farther from the Sun.

The Surface of Venus

Despite the extremely hostile conditions on Venus's surface, several Russian *Venera* spacecraft landed there in the 1970s and 1980s and transmitted pictures back to Earth. These robotic spacecraft made a variety of measurements and sampled the rocks, showing them to be volcanic. The landers lasted at most about two hours before succumbing to the high temperatures and atmospheric pressure. The pictures show a barren surface covered with flat, broken rocks and lit by the pale orange glow of sunlight diffusing through the deep clouds (fig. 9.10). The cameras on these spacecraft scanned a narrow strip from horizon to horizon, providing a peculiar perspective. Using the views from the two cameras on the spacecraft and artistically filling in some missing areas, it is possible to produce a view of what Venus's surface might look like to someone standing on the surface (fig. 9.11).

Magellan met a deliberately engineered fiery doom in 1994. Its orbit was altered so that it plunged into Venus's atmosphere. Analysis of its final tumblings gave astronomers data on the density of Venus's upper atmosphere.

The surface of Venus is hidden beneath its thick clouds, but planetary scientists can map its ground features with radar from both Earth and spacecraft orbiting Venus. Just as radar penetrates terrestrial clouds to show an aircraft pilot a runway even through fog, so too radar penetrates the Venusian clouds, revealing the planet's surface. Figure 9.12 shows a radar map of Venus made by the *Magellan*, a U.S. spacecraft previously orbiting the planet. Such maps show that Venus is less mountainous and rugged than Earth, with most of its surface being low, gently rolling plains. Only two major highland regions, Ishtar and Aphrodite, rise above the lowlands to form land masses similar to terrestrial continents. Ishtar, named for the Babylonian goddess of love, is

FIGURE 9.10
Picture made on the surface of Venus by the Russian spacecraft *Venera* that landed there. Sunlight filtering through the thick clouds gives the landscape its orange tint. The view is highly distorted by the camera so that the horizon is seen angling across the upper left and upper right corners. The lens cap lies at the foot of the lander.

FIGURE 9.11
Images from Venera 13 (left) and Venera 14 (right), showing a more natural perspective. The two cameras on each spacecraft scanned strips that intersected at each end. Distant hills are visible as well as nearby volcanic rock.

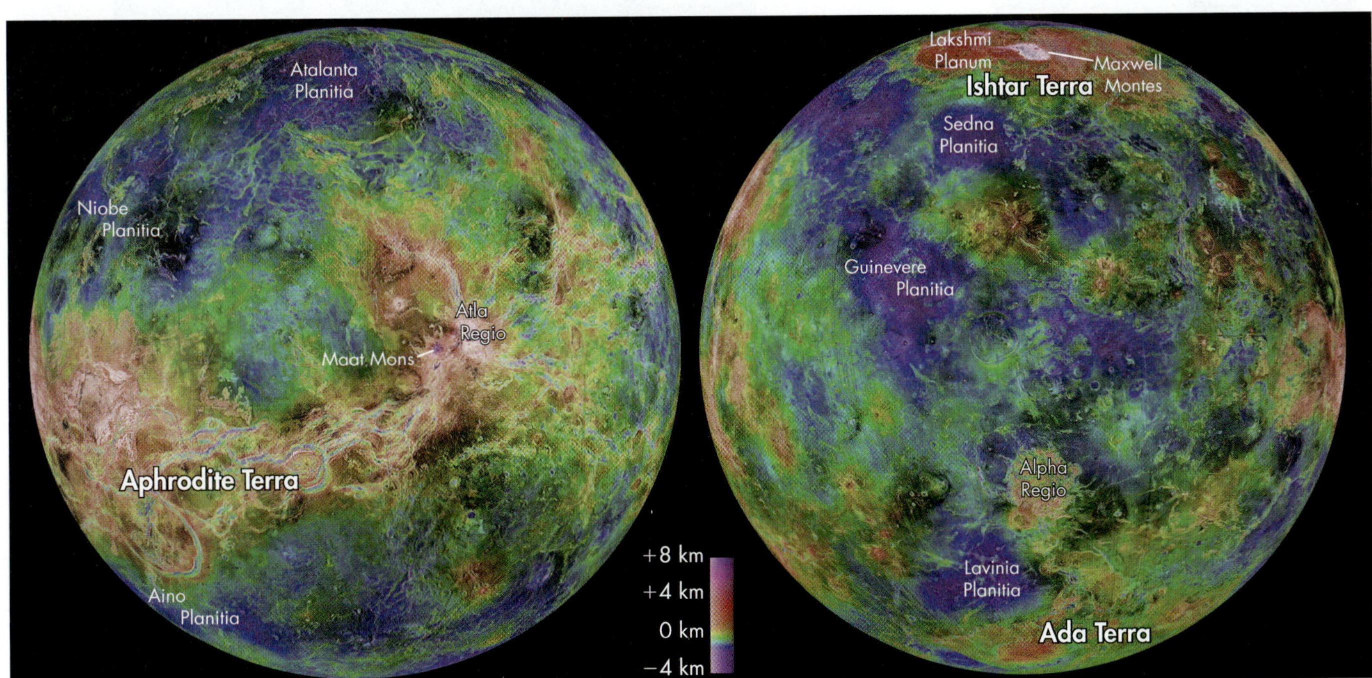

FIGURE 9.12
Global radar map of Venus made by the *Magellan* Venus-orbiting satellite. Colors indicate the relative height of surface features. Lowlands are blue; high elevations are red.

about the size of Australia and is studded with volcanic peaks, the highest of which, Maxwell Montes, rises more than 11 kilometers (about 6.8 miles) above the average level of the planet. (Notice that because no oceans exist on Venus, "sea level" has no meaning as a reference height.) The other major highland region, Aphrodite, bears the ancient Greek name for Venus and is about the size of South America. Together, Ishtar and Aphrodite compose only about 8% of Venus's surface, a far smaller fraction than for Earth, where continents and their submerged margins cover about 45% of the planet.

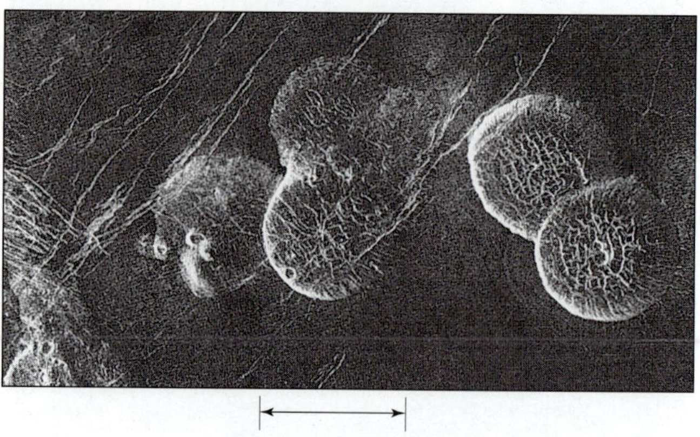

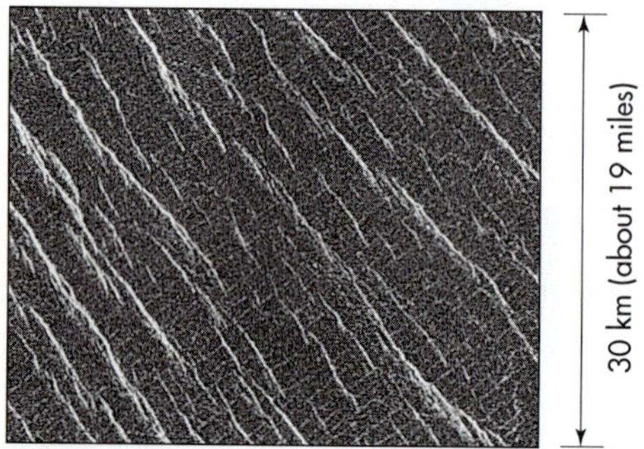

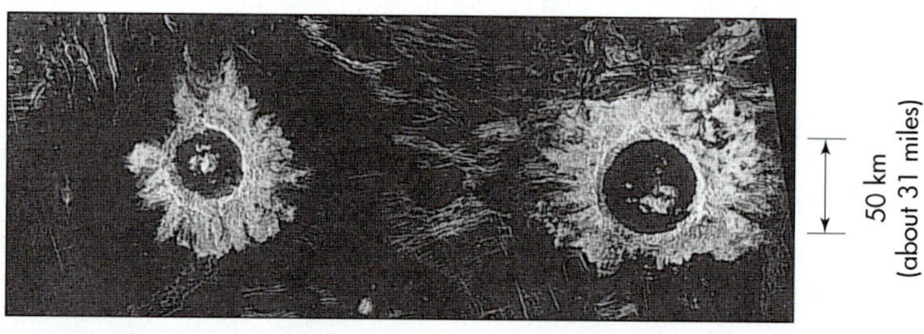

FIGURE 9.13
Gallery of *Magellan* radar pictures. The image at upper left is a perspective image of Venus's volcano Maat Mons, which, based on radar data, appears to have relatively recent lava flows surrounding it. An orange color is artificially added to match the color of the landscape observed by the Russian *Venera* lander. The other images are views looking straight down at the surface, showing lava domes, fractures, and impact craters.

The *Magellan* spacecraft was able to detect features as small as about 100 meters (about 300 feet) across. Figure 9.13 shows close-ups of some of the more intriguing images that *Magellan* transmitted. Brighter regions reflected radar more strongly, generally indicating that the surface is rougher than the areas that appear darker in these radar images. Thanks to spacecraft, we have better maps of Venus than we had of the Earth itself 50 years ago.

The many odd and unique structures seen in the radar maps have proved puzzling to planetary geologists. Venus is so similar in diameter and mass to the Earth that they expected to see landforms there similar to those on the Earth. For example, some planetary geologists predicted that there would be evidence of plate tectonics, such as continental blocks, crustal rifts, and trenches at plate boundaries. But few such features are visible. Instead, Venus has a surface almost totally unlike the Earth's. Although Venus has some craters (often weirdly distorted, however) and crumpled mountains, volcanic landforms dominate. These include peaks with immense lava flows, "pancake-shaped domes" of uplifted rock, long narrow faults (cracks), and peculiar lumpy terrain. All these features indicate a young and active surface, a deduction borne out by the scarcity of impact craters. From the small number of craters, scientists have concluded that virtually all of Venus's original surface has been destroyed by volcanic activity. The surface we see is probably at most half a billion years old, much younger than Earth's continental surface, and some regions may be less than 10 million years old. Such estimates of crustal age are difficult to make, however, because the Venusian atmosphere is so dense that all but the largest infalling bodies (bigger than a few hundred meters) are broken up in it.

Are the Venusian volcanos (such as Maat Mons in fig. 9.13) still active? Eruptions have not been seen directly, but some lava flows appear very fresh. Moreover, electrical discharges, perhaps lightning, have been detected near some of the larger peaks. On Earth, volcanic eruptions frequently generate lightning, and some astronomers think the electrical activity indicates that Venus's volcanos are still erupting. Such eruptions might also explain brief increases in sulfur content detected in the Venusian atmosphere, changes similar to those produced on Earth by eruptions here.

The numerous volcanic peaks, domes, and uplifted surface regions suggest to some scientists that heat flows less uniformly within Venus than within the Earth. Although some locations on Earth (Yellowstone Park and the Hawaiian Islands, for example) are heated anomalously by "plumes" of rising hot rock, such plumes seem to dominate on Venus. As hot rock wells upward, it bulges the crust, stretching and cracking it. We may be viewing on our sister planet what Earth looked like as its crust began to form and before smooth heat flows were established. Alternatively, Venus and Earth may differ for deeper reasons.

The Interior of Venus

The deep interior of Venus is probably like the Earth's, an iron core and rock mantle. Planetary geologists have no seismic information to confirm this conjecture, so, as with Mercury, they must rely on deductions from its gravity and density, which are similar to the Earth's.

Astronomers have proposed several hypotheses for why Venus and Earth differ so much in their surface features and subsurface activity. Although these hypotheses differ in detail, they tend to agree that the amount of water in the rocks of these two worlds plays a key role. For example, rocks that contain water trapped in their structure melt at a lower temperature than similar rocks that lack water. Moreover, when they become molten, they are "runnier," which makes it easier for the melted rock to flow. As a result, convection can be more vigorous in a planet whose rocks are rich in water compared to a planet with drier rocks. Furthermore, because the water-rich rock melts at a lower temperature, the solid crust of such a planet will be thinner. Other things being equal, we therefore expect wet Earth to have a thinner crust than dry Venus.

Why does the thickness of the crust affect a planet's surface features? A thin crust, such as we have on Earth, allows a steady loss of heat, so our planet's interior cools steadily. Moreover, its thin crust breaks easily into many small plates. This breaking allows crustal motion and activity to occur more or less continuously and at many places on the surface. For Venus, the thicker crust holds heat in, keeping the interior hot, but the convective motions that develop are unable to break up the thick crust.

Ultimately, however, the trapped heat must escape, and it is here that theories differ strongly. According to one theory, at points where the hot, rising material reaches the crust, the surface bulges upward and weakens, and volcanoes may form.* Where cooler material sinks, the thick crust crumples into a continent-like region. This creates a planet with few and small continents and whose surface is active but only in isolated spots.

According to another theory, the trapped heat gradually melts the bottom of the thick crust, thinning it and allowing it to break up. This may happen over widespread areas, flooding almost the entire planetary surface with lava in a brief time. The heat then rapidly escapes to space, the interior cools, and the crust again thickens. Surface

*A similar process occurs on Earth in some regions, such as the Hawaiian Islands.

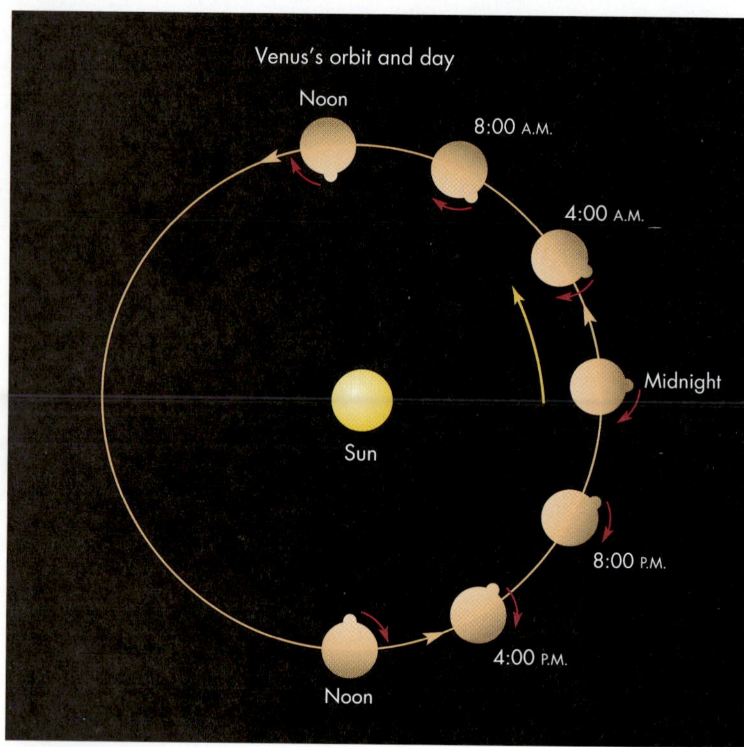

FIGURE 9.14
Venus spins slowly backward, so that a day lasts a little more than half its orbit.

activity therefore subsides, but heat is trapped once more, and the process may repeat but only at intervals of hundreds of millions of years.

Rotation of Venus

Venus spins on its axis more slowly than any other planet in the Solar System, taking 243 days to complete one rotation relative to the stars. Moreover, its spin is retrograde ("backward") compared with the direction of rotation of the other terrestrial planets. This slow and retrograde spin has led some astronomers to hypothesize that Venus was struck shortly after its birth by a huge planetesimal; the impact slowed Venus and set it spinning backward. A less dramatic explanation of the spin is that Venus has been affected by a combination of tidal forces exerted by the Sun—and perhaps the Earth—so that the tilt angle of its rotation axis may have shifted in time.

Venus rotates so slowly that it cannot generate a strong magnetic field as the Earth does. The slow rotation also makes the solar day there very long, approximately 117 Earth days (fig. 9.14). Furthermore, because the planet's spin is retrograde (backward), the Sun rises in the west and sets in the east.

9.3 Mars

FIGURE 9.15
Picture of Mars made by the Hubble Space Telescope orbiting Earth.

Q. What season is it in this picture? (North is at top of image.)

Mars is named for the Roman god of war, presumably because of its distinctly reddish color. Compared with Mercury and Venus, Mars seems positively Earthlike. Although its diameter is only about half and its mass about one-tenth the Earth's, many other features are similar. The Martian day is just 39 minutes longer than an Earth day, and the tilt of its axis is almost the same as Earth's, so it experiences a similar sequence of seasons. On a warm day, the temperature at the Martian equator reaches about 50°F (10°C), and although winds sweep dust and patchy clouds of ice crystals through its sky, the Martian atmosphere is generally clear enough for astronomers on Earth to view its surface clearly. Such views show a world of familiar features. Polar caps of sparkling white contrast with the reddish color of most of the planet and are visible from Earth, as seen in figure 9.15.

The similarities of Mars to Earth have excited interest in the planet for centuries, perhaps even as a place we might someday inhabit. A series of spacecraft we've sent to explore the planet—*Mariner, Viking I* and *II, Mars Global Surveyor, Mars Odyssey, Mars Express, Mars Reconnaissance Orbiter, Phoenix*—have revealed the true marvels of the planet.

The Surface of Mars

Mars has some of the most dramatic surface features of any of the terrestrial planets. Along the equator runs a rift—Valles Marineris—that stretches 4000 kilometers (2500 miles) long, 100 kilometers (60 miles) wide, and 7 kilometers (4 miles) deep, as shown in figure 9.16. This canyon, named for the *Mariner* spacecraft whose pictures led to its discovery, dwarfs the Grand Canyon and would span the continental

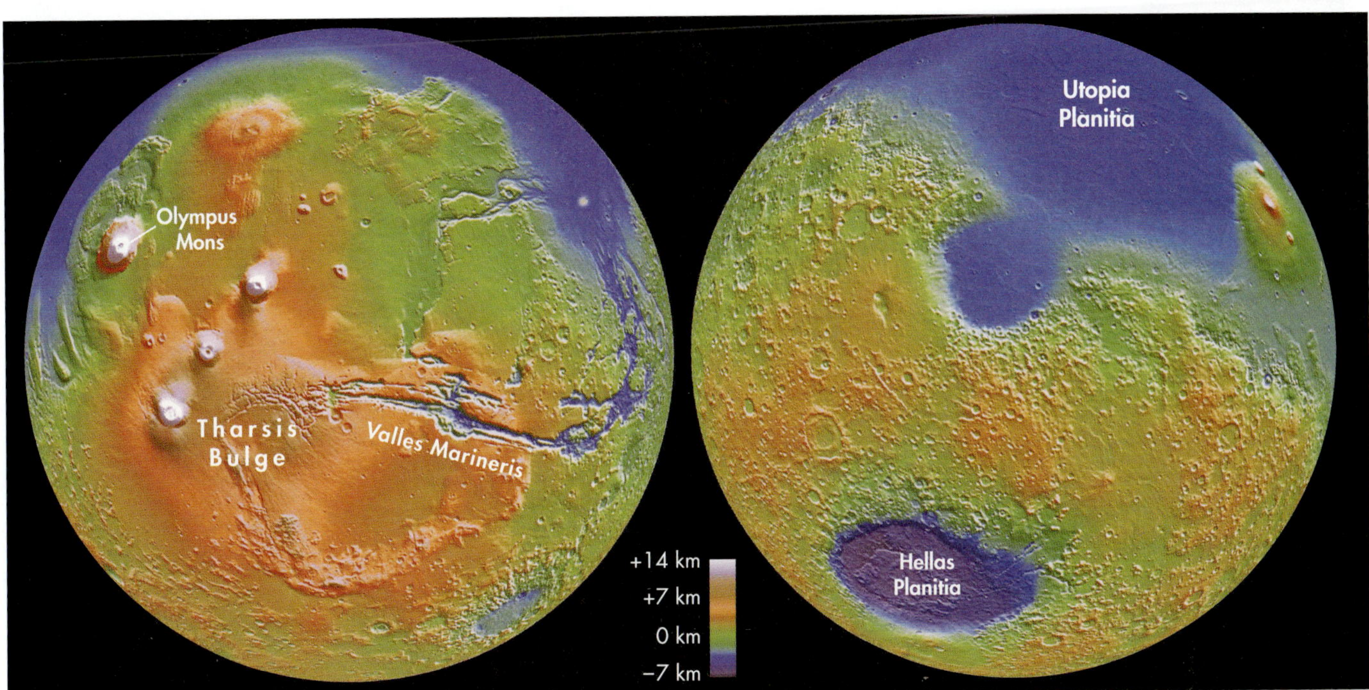

FIGURE 9.16
Topographic map of Mars showing its major features. The map is color coded according to elevations. Olympus Mons is the largest volcano in the Solar System, while Valles Marineris, the Grand Canyon of Mars, is an enormous gash in Mars's crust about 4000 kilometers (approximately 2500 miles) long. Were it on Earth, it would stretch from California to Florida.

FIGURE 9.17
A reconstructed view down Valles Marineris, the Grand Canyon of Mars. This image was constructed from hundreds of thousands of laser altimeter measurements made by the *Mars Odyssey* orbiter. This enormous gash may be a rift that began to split apart the Martian crust but failed to open farther.

United States. Planetary scientists used satellite data to reconstruct in figure 9.17 what Valles Marineris would look like from a high-flying airplane.

At midlatitudes, a huge uplands called the Tharsis bulge (see fig. 9.16) is dotted with enormous volcanic peaks, two of which can be seen in figure 9.18A. Another volcano at the edge of Tharsis, Olympus Mons, rises about 25 kilometers (about 16 miles) above its surroundings, nearly three times the height of Earth's highest peaks, and is illustrated in figure 9.18B. If ever interplanetary parks are established, Olympus Mons should lead the list.

Planetary geologists think that the Tharsis region formed as hot material rose from the deep interior of the planet and forced the surface upward as it reached the crust. The hot matter then erupted through the crust to form the volcanos, some of which appear

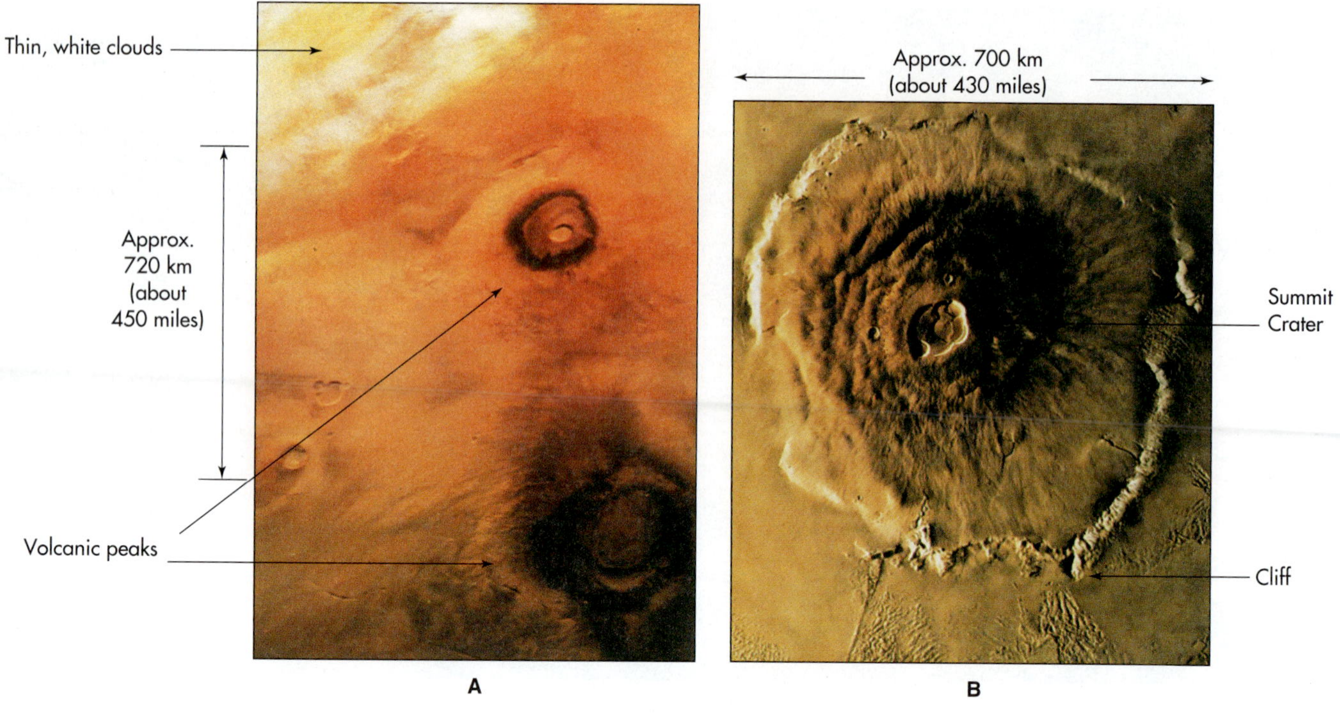

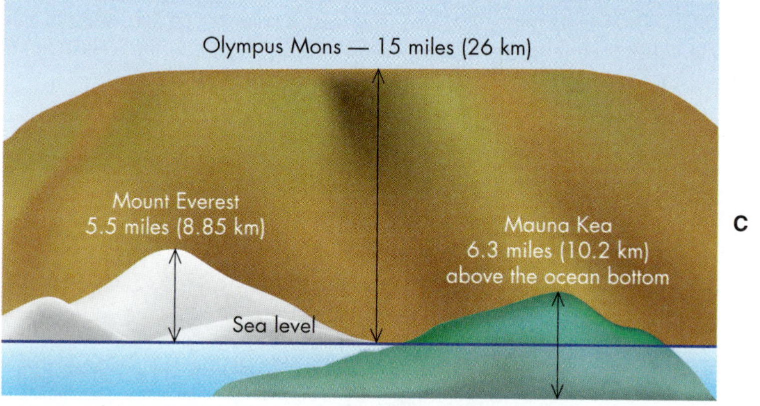

FIGURE 9.18
Pictures of (A) two volcanos on the Tharsis bulge and (B) Olympus Mons, the largest known volcano (probably inactive) in the Solar System. (C) Profile of Olympus Mons; Mount Everest, the highest mountain above sea level on Earth; and the Hawaiian volcano Mauna Kea, rising from the sea floor.

relatively young. For example, the small number of impact craters in its slopes implies that Olympus Mons is no older than 250 million years and that it may in fact have been active much more recently. Some planetary geologists think the Tharsis bulge may also have created the gigantic Valles Marineris, which lies to the southeast. According to this theory, Valles Marineris formed as the Tharsis region swelled, stretching and cracking the crust. Other planetary scientists think that this vast chasm is evidence for plate tectonic activity, like that of Earth, and that the Martian crust began to split, but the motion ceased as the planet aged and cooled.

Figure 9.19 shows the Martian polar caps. These frozen regions change in size during the cycle of the Martian seasons, a cycle resulting from the tilt of Mars's rotation axis in the same way that our cycle of seasons is caused by the tilt of Earth's rotation axis. The Martian seasons are more extreme than terrestrial ones because the Martian atmosphere is much less dense than Earth's, and therefore it does not retain heat as well. Because Mars's seasonal changes are so extreme, its polar caps vary greatly in size, shrinking during the Martian summer and growing again during the winter. Much of the visible part of the southern cap is frozen carbon dioxide—dry ice—and in winter its frost extends in a thin layer across a region some 5900 kilometers (about 3700 miles) in diameter, from the

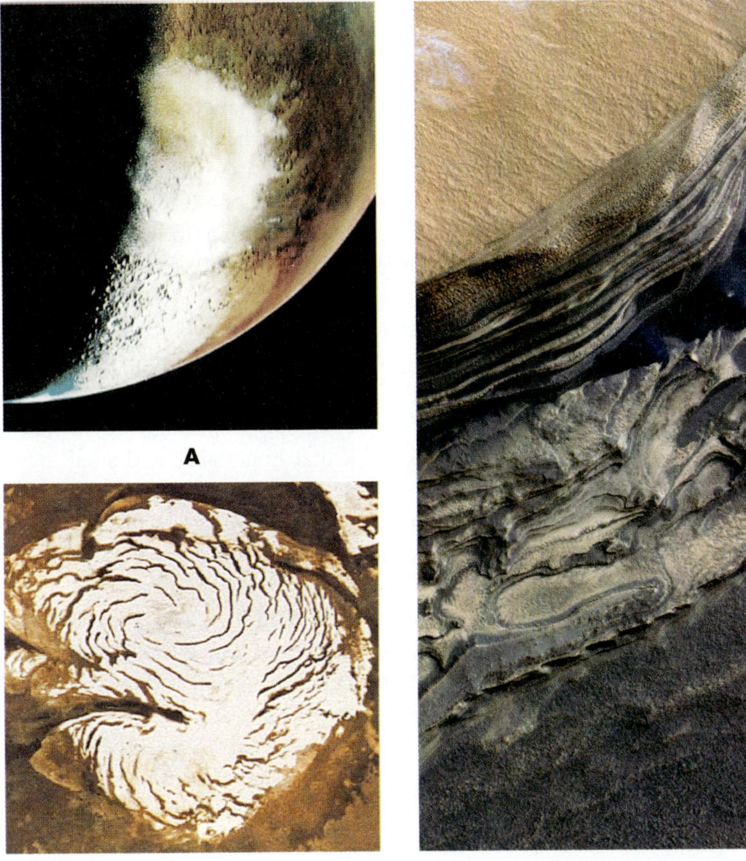

FIGURE 9.19
Pictures of (A) the south Martian polar cap and (B) the north Martian polar cap. (C) Note the layered structure of the north polar cap visible in this view from the *Mars Reconnaissance Orbiter*. This high-resolution image shows a portion of a chasm wall about 1.3 kilometers wide. Layers of dust and ice alternate in the chasm wall.

south pole to latitude 40°, much as snow cover extends to middle latitudes such as New York in our winters. But because the frost is very thin over most of this vast cap, it shrinks in the summer to a diameter of about 350 kilometers (approximately 220 miles). The northern cap shrinks to a diameter of about 1000 kilometers (approximately 600 miles). Although the caps have a surface layer of CO_2, the bulk of the frozen material is ordinary water ice, as deduced from its temperature and radar studies. From the depth of ice measured by radar, there is enough water in the caps to cover the entire surface of Mars with water to a depth of at least 10 meters (30 feet).

The northern cap consists of numerous separate layers, as can be seen in figure 9.19C. These strata indicate that the Martian climate changes cyclically. Thus, Mars may have "ice ages" similar to those on Earth.

Why do the Martian polar caps differ so? Altitude measurements made by the *Mars Global Surveyor* spacecraft show that Mars's south pole is considerably higher in elevation and is thus much colder than its north pole. This creates a strong wind pattern that carries water vapor and carbon dioxide away from the south pole toward the north pole. There, it precipitates out, leading to a larger north polar cap.

The Martian poles are bordered by immense deserts with dunes blown into parallel ridges by the Martian winds, as illustrated in figure 9.20. Huge dust storms blow the fine red dust over the entire surface of the planet, giving the planet its characteristic color. What makes it so red? The color comes from the iron minerals in its surface rocks. We know from everyday experience that a piece of iron will become rust-colored when exposed to air. Here on Earth, such rusting occurs because the iron combines with oxygen in our atmosphere to form iron oxides and other compounds. On Mars, even though there is little oxygen in its atmosphere, other chemical reactions with the iron in its surface minerals lead to the same effect.

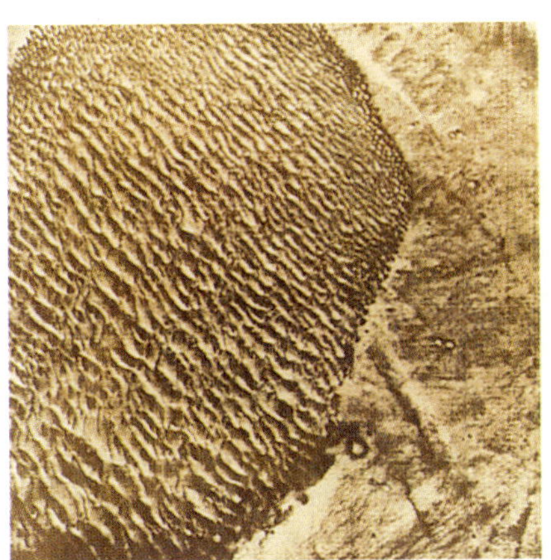

FIGURE 9.20
Picture of dune fields in the Martian desert.

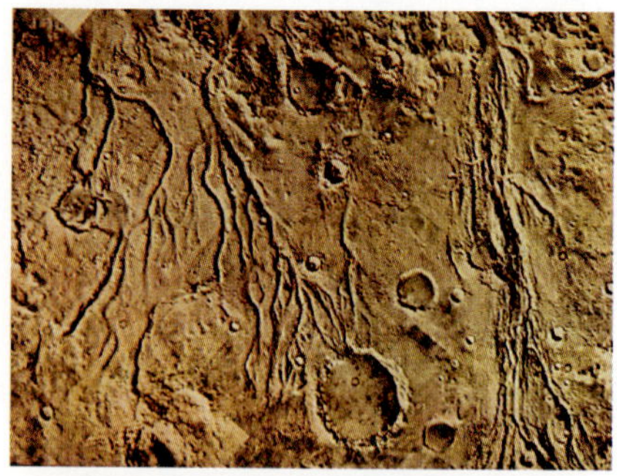

A

B

FIGURE 9.21
(A) Picture of channels probably carved by running water on Mars. (B) "Islands" formed as water flowed (from bottom toward top of figure) around the rims of craters.

Water on Mars

Over the last few decades, scientists have sent dozens of spacecraft to Mars. Six have successfully landed on the surface and sent back to Earth pictures and measurements of the Martian landscape and close-up images of rocks. These missions have many goals, but perhaps the major one is searching for evidence that liquid water was once present on Mars. Why is liquid water of such interest? The answer is simple: scientists who study the possibility of life elsewhere in the Universe think that liquid water is a critical ingredient for living organisms of almost any type. Thus the search for water is a first step in the search for life.

Perhaps the most intriguing features revealed by the two *Viking* spacecraft in the 1970s were huge channels and dry riverbeds, such as those seen in figure 9.21A. We infer from these features, which wind across the Martian surface and often contain "islands" (fig. 9.21B), that liquid water once flowed on Mars, even though no surface liquid is present now. In fact, many astronomers think that huge lakes and small oceans once existed on Mars. Further evidence for these ancient bodies of water are smooth terraces that look like old beaches around the inner edges of craters and basins, as you can see in the lower left part of figure 9.22. Narrow canyons breach this crater's rim, showing where water flowed in and drained out into lowland areas.

The *Opportunity* and *Spirit* spacecraft, sent by NASA in 2003 to explore Mars's surface, landed at sites that were chosen because pictures and spectral data taken from orbit suggested that water might have been present there long ago. *Spirit* landed in Gusev Crater, a smooth-floored feature at the end of a narrow Martian valley. *Opportunity* landed on the flat plains along the Martian equator on the opposite side of Mars from *Spirit*. Both craft then deployed rovers—small, wheeled vehicles that can move away from the landing site and explore interesting features at close range. For example, in figure 9.23 you can see the view at the *Spirit* lander site.

Both rovers have been highly successful in their searches. For example, *Opportunity* took the picture shown in figure 9.24, which shows a rock outcropping thought to be material deposited in an ancient, now dried-up small sea or ocean. Examination of the rocks shows that they contain layers that are typical of sediment that sank to the bottom of a body of water and later was transformed into rock. *Opportunity* has found more than just rock layers, however. It has also found in those layers features that are similar

FIGURE 9.22
A Martian crater thought to have once been a "crater lake." The crater is roughly 50 kilometers (about 30 miles) across. Note the inflow channel on the upper right and the outflow channel on the left. The smooth floor (apart from a few small craters) suggests that the crater bottom is covered with sediment left behind as the lake dried out.

FIGURE 9.23
A panoramic view of the Bonneville Crater, lying in the floor of the much larger Gusev Crater on Mars. This spot is near where the *Spirit* spacecraft landed.

FIGURE 9.24
A close-up image of a rock outcropping at the *Opportunity* landing site. The rocks show thin layers and contain minerals that suggest that they were formed on the bottom of a salty lake or ocean. The lines in the rock suggest that the rock formed from sediment in flowing water.

to the ripple marks you see at the beach as water washes back and forth across the sand. Moreover, minerals in the rocks at the *Opportunity* site have a chemical makeup consistent with their having been deposited in a salty lake or small ocean. Although scientists are still analyzing the images and the readings from the many instruments carried on these landers, the evidence has already convinced many scientists that the locations on Mars that *Spirit* and *Opportunity* have explored were indeed once under liquid water. Ice was dug up by the *Phoenix* lander, which reached the northern polar region of Mars in 2008. *Phoenix* carried out experiments on the soil, showing that it is quite different from Earth's—about as alkaline as baking soda with high levels of oxidizing chemicals.

Meanwhile, the *Mars Reconnaissance Orbiter* and *Mars Odyssey* orbiters (launched by NASA) and the *Mars Express* orbiter (launched by the European Space Agency) continue to take detailed pictures of the surface of Mars, such as the "mesa" shown in figure 9.25 and the fan of sediment in figure 9.26, which is thought to be the delta of an ancient river. But in addition to charting such surface features, these orbiters also carry instruments that can measure the spectrum of light reflected from Mars. From such spectra, astronomers can deduce what minerals are present at a given point on the Martian surface. Matching the composition of those minerals with data on whether

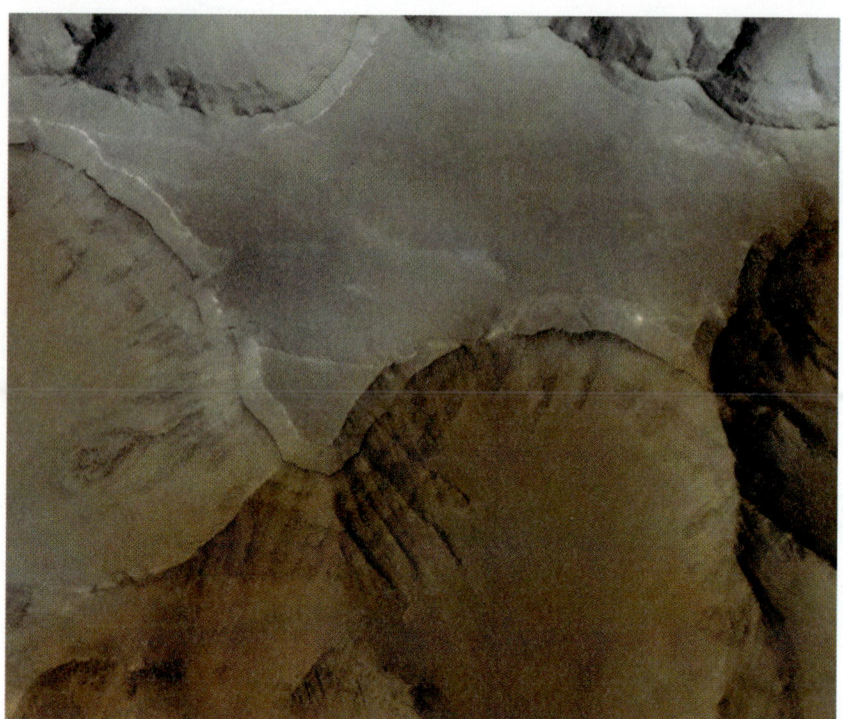

FIGURE 9.25
Image from the *Mars Global Surveyor.* Note the layering visible in several places along the edges of the flat-topped mesa. Such layers often form when sediment is deposited by liquid water.

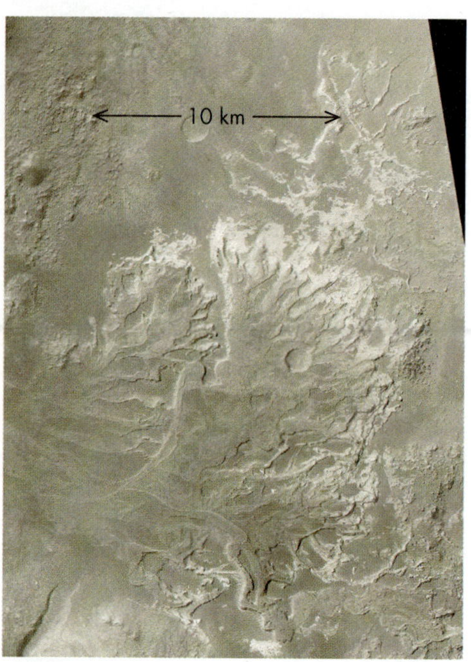

FIGURE 9.26
A view of what appears to be a dried-up river delta. The fan of sediment is about 13 kilometers (approximately 8 miles) wide.

water is needed to produce those minerals gives additional evidence that Mars was once much wetter. For example, the presence of the mineral hematite, a compound of iron that forms by minerals precipitating in water, is an indicator that a region now dry was once covered by a salty lake or sea.

Other instruments on the *Mars Odyssey* orbiter can detect hydrogen atoms locked up in water in the Martian soil. Data from the *Odyssey* orbiter has astonished astronomers by revealing what appear to be huge amounts of water, presumably ice, buried under a thin layer of Martian dust.

Astronomers are both excited and puzzled by the possibility that water was present on Mars. They are excited because of the belief that where there is water, there may be life—even if only microbes. They are puzzled because Mars does not have conditions that allow water to be present on the planet. To understand why, we need to look at the properties of the Martian atmosphere.

The Martian Atmosphere

Clouds and wind-blown dust are visible evidence that Mars has an atmosphere. Spectra and direct sampling by spacecraft landers confirm this and show that the atmosphere is mostly (95%) carbon dioxide with small amounts (3%) of nitrogen and traces of oxygen and water. From this, astronomers can determine the density of Mars's atmosphere, which turns out to be very low—only about 1% the density of Earth's.

This density is so low that, although the Martian atmosphere is mostly carbon dioxide, it creates only a very weak greenhouse effect. The consequent lack of heat trapping and Mars's greater distance from the Sun make the planet very cold. Temperatures at noon at the equator may reach a little above the freezing point of water, but at night they plummet to far below zero on the Fahrenheit scale. The resulting average

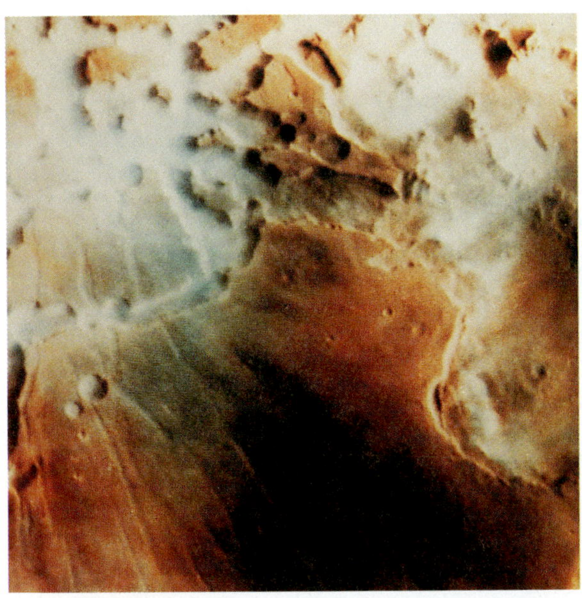

A

B

FIGURE 9.27
(A) Fog in Martian valleys and (B) frost on surface rocks near the *Viking* lander.

temperature is a frigid 218 K (−67°F). Thus, although water exists on Mars, it is frozen solid, locked up either below the surface in the form of permafrost or in the polar caps as solid water ice.

Clouds of dry ice (frozen CO_2) and water-ice crystals (H_2O) drift through the Martian atmosphere carried by the Martian winds. These winds, like the large-scale winds on Earth, arise because air that is warmed near the equator rises and moves toward the poles. This flow from equator to poles, however, is deflected by the Coriolis effect arising from the planet's rotation. The result is winds that blow around the planet approximately parallel to its equator. The Martian winds are generally gentle, but seasonally and near the poles they become gales, which sometimes pick up large amounts of dust from the surface. The resulting vast dust storms occasionally cover the planet completely and turn its sky pink.

No rain falls from the Martian sky, despite its clouds, because the atmosphere is too cold and contains too little water. In fact, there is so little water in the Martian atmosphere that even if all of it were to fall as rain, it would make a layer only about 12 micrometers deep (less than 1/2000 inch). For comparison, Earth's atmosphere holds enough water to make a layer a few centimeters (inches) deep. Despite such dryness, however, fog sometimes forms in some Martian valleys, and frost condenses on the ground on cold nights, as seen in figure 9.27. In addition, during the Martian winter, CO_2 "snow" falls on the Martian poles.

Mars has not always been so dry, as we saw from the numerous channels in its highlands and other evidence from many Mars missions. But for a planet to have liquid water, it must have a warm atmosphere with a pressure similar to that of Earth's. If the pressure on a liquid is very low, molecules can break free from its surface, evaporating easily because no external force restrains them. On the other hand, if the pressure is high, molecules in a liquid must be heated strongly to turn them into gas. For example, at normal atmospheric pressure, water boils at 100° Celsius. If the pressure is reduced, however, the boiling point drops, an effect used by food producers to make "freeze-dried" foods, such as instant coffee. Coffee is brewed normally, then frozen and placed in a chamber from which the air is pumped out. The reduced pressure makes the liquid "boil" without heating and evaporate, leaving only a powder

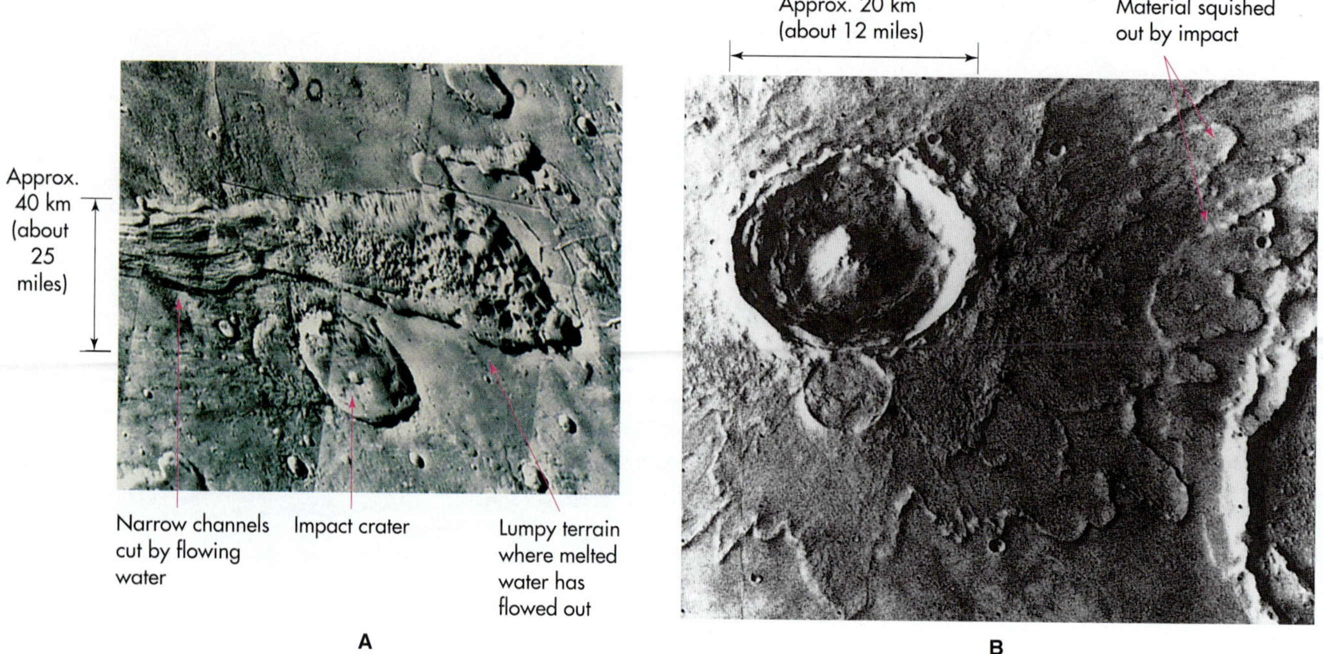

FIGURE 9.28
Pictures of (A) a channel cut by water released as subsurface ice melted. Note the lumpy terrain at the broad end of the channel where the surface collapsed as the water drained away. (B) Crater with surrounding flow patterns. Heat released by the impact that formed the crater has melted subsurface ice. The thawed material has oozed out.

residue—instant coffee. Similarly, on Mars, any liquid water on its surface today would evaporate.

The existence of channels carved by liquid water on Mars is therefore strong evidence that in its past, Mars was warmer and had a denser atmosphere. However, that milder climate must have ended billions of years ago. How do we know? The large number of craters on Mars shows that its surface has not been significantly eroded by rain or flowing water for about 3 billion years. Why did Mars dry out, and where have its water and atmosphere gone?

Some water probably lies buried below the Martian surface as ice, as indicated by measurements from the *Mars Odyssey* orbiting spacecraft. If the Martian climate was once warmer and then cooled drastically, water would condense from its atmosphere and freeze, forming sheets of surface ice. Wind might then bury this ice under protective layers of dust, as happens in polar and high mountain regions of Earth. In fact, figure 9.28A shows indirect pictorial evidence of such buried ice. Planetary geologists think that subsurface ice melted, perhaps from volcanic activity, and drained away, causing the ground to collapse and leaving the lumpy, fractured terrain visible at the broad end of the channel. The water from the melted ice then flowed downstream, carving the 20-kilometer (12 mile)-wide canyon. Similarly, figure 9.28B shows a crater from which partially melted matter "squished" out on impact. Thus, Mars's water may now mostly be subsurface ice.

If Mars had a denser atmosphere in the past, as deduced from the higher pressure needed to allow liquid water to exist, then the greenhouse effect might have made the planet significantly warmer than it is now. The loss of such an atmosphere would have weakened the greenhouse effect and plunged the planet into a permanent ice age. Such a loss could happen in at least two ways. According to one theory, repeated asteroid impacts on Mars when it was young may have blasted its original atmosphere off into space. Such impacts, although rare now, do occur, and our own planet may have been struck about 65 million years ago with results nearly as dire, as we shall learn in chapter 11.

A less dramatic explanation for how Mars lost most of its atmosphere is that Mars's low gravity allowed gas molecules to escape over the first 1 to 2 billion years of the planet's history. Regardless of which explanation is correct, the loss of its atmosphere would have cooled the planet and locked its remaining water up as permafrost. But why have the Martian volcanos not replenished its atmosphere, keeping the planet warm? Astronomers believe that the blame lies with Mars's low level of tectonic activity, a level set by conditions in its interior.

The Martian Interior

Astronomers think that the interior of Mars is differentiated like the Earth's into a crust, mantle, and iron core. However, because Mars is so small compared with the Earth, its interior is cooler. Mars's smaller mass supplies less heat, and its smaller radius allows the heat to escape more rapidly. Unfortunately, astronomers have no direct confirmation of Mars's interior structure because the seismic detectors landed there by the *Viking* spacecraft failed. Thus, as is the case for Mercury and Venus, astronomers must rely on indirect evidence from its density and gravitational field to learn about the interior of Mars. Having a mass between that of dead Mercury and lively Earth and Venus implies that Mars should be intermediate in its tectonic activity. Such seems to be the case. Using the *Mars Global Surveyor* spacecraft in orbit around the planet, astronomers have measured Mars's magnetic field and internal structure. They concluded that Mars has a metallic core whose radius is between 1200 km and 2400 km, or about 0.4 to 0.7 the overall radius of the planet. (For comparison, Earth's iron core is about 0.5 our planet's radius.) But Mars, unlike Earth, has no planetwide magnetic field, so its core is probably no longer molten.

Although it possesses numerous volcanic peaks and uplifted highlands, implying that it had an active crust, at least in the past, Mars bears no evidence of large-scale crustal motion like the Earth's. For example, it has no folded mountains. Astronomers therefore think that Mars has cooled and its crust thickened to perhaps twice the thickness of the Earth's crust. As a result, the now-weak interior heat flow can no longer drive tectonic motions. A thick Martian crust may also explain why Mars has a small number of very large volcanos, while the Earth has a large number of small ones. Mars's immense volcanos are thus mute testimony to a more active past.

Mars's current low level of tectonic activity is also demonstrated by the many craters that cover its older terrain, far more than are seen on either Earth or Venus. The number of those craters implies that Mars has been geologically quiet for billions of years. Mars is probably not dead, however, because some regions (for example, the slopes of Olympus Mons and other volcanos) are essentially free of craters. Thus, these immense peaks may still occasionally erupt. They do not erupt often enough, however, to replace the gas lost to space because of the planet's low gravity. It appears that Mars has entered a phase of planetary old age. Recently, however, using Earth-based telescopes, astronomers have detected methane in spectra of the Martian atmosphere. Since methane is rapidly destroyed in the Martian environment, this means that Mars must be producing it currently, indicating that there is still at least a low level of geological activity.

The Martian Moons

Mars has two tiny moons, Phobos and Deimos, which are named for the demigods of Fear and Panic. These bodies are only about 20 kilometers across and are probably captured asteroids. They are far too small for their gravity to have pulled them into spherical shapes. Both moons are cratered, implying bombardment by smaller objects. Phobos (see fig. 9.29) has cracks, suggesting that it may have been struck by a body large enough to split it nearly apart.

FIGURE 9.29
Picture of Phobos and Deimos, the moons of Mars. These tiny bodies are probably captured asteroids.

Q. What does the irregular shape of these bodies tell you about the strength of their surface gravity? Is it likely these moons have any atmosphere of their own?

Phobos and Deimos were discovered in 1877, but by chance they appeared in literature nearly two centuries earlier in Jonathan Swift's book *Gulliver's Travels*. Gulliver stops at the imaginary country Laputa whose inhabitants include numerous astronomers. Among the accomplishments of these people is the discovery of two tiny moons of Mars. Even earlier, Kepler guessed that Mars might have two moons because the Earth has one moon and Jupiter, at least in Kepler's time, was known to have four. Mars, lying between these two bodies should therefore (according to Kepler's mystic argument) have a number of moons lying between 1 and 4, and he chose 2 as the more likely case.

Life on Mars?

Scientists have long wondered whether living organisms developed on Mars. Much of that interest grew from a misinterpretation of observations made in 1877 by the Italian astronomer Giovanni Schiaparelli. Schiaparelli saw what he took to be straight-line features on Mars and called them *canali,* by which he meant "channels." In English-speaking countries, the Martian *canali* became canals, with the implication that Mars must be inhabited by intelligent beings who built them. The interest in these canals had become so great by 1894 that the wealthy Bostonian Percival Lowell built an observatory in northern Arizona to study Mars and search for signs of life there.

Most astronomers could see no trace of the alleged canals, but they did note seasonal changes in the shape of dark regions, changes that some interpreted as the spread of plant life in the Martian spring. By the early 1970s, scientists were excited by satellite photographs of water-carved canyons and old riverbeds because water—at least on Earth—is so important for life. Therefore, to further the search for life on Mars, the United States landed two *Viking* spacecraft on the planet in 1976. These craft carried instruments to search for signs of carbon chemistry in the soil and to look for metabolic activity in soil samples that were put in a nutrient broth carried on the lander. All tests either were negative or ambiguous.

Then, in 1996, a group of American and English scientists reported possible signs of life in rocks from Mars. These were not samples returned to Earth by a spacecraft but samples of meteorites found on Earth. They arrived here after being blasted off the surface of Mars, presumably by the impact of a small asteroid. Such impacts are not uncommon, but most fragments are scattered in space or fall back to Mars. Moreover, of those that are shot into space, only a tiny fraction have just the right combination of speed and direction to reach Earth.

How can astronomers tell if a meteorite has come from Mars? One way is to sample the gas trapped in tiny bubbles in the meteorite and see if it matches the composition of

Mars's atmosphere as measured by the *Viking* Mars landers. For the meteorite in question, the match was excellent, assuring that it came from Mars. What was the evidence suggesting life? That turns out to be far more controversial. Microscopic examination of samples from the interior of the meteorite revealed many tiny, rod-shaped structures (fig. 9.30). These look very much like ancient terrestrial bacteria but are much smaller. To some scientists, they look like "fossilized" primitive Martian life. Moreover, the meteorite contains traces of organic chemicals known as polycyclic aromatic hydrocarbons (PAHs, for short). Terrestrial bacteria make such chemicals when they die and decay, but PAHs can also form spontaneously, given the proper mix of chemicals. In fact, they have been found in a number of non-Martian meteorites and have also been detected by their spectrum lines in the radio emission from interstellar gas and dust clouds. Other structures in the meteorite can also be interpreted as having a biological origin. But over the last few years, scientists have shown that ordinary chemical weathering can form very similar structures. As a result, most scientists today are unconvinced that any meteorite yet studied shows evidence of Martian life.

It will take years of further analysis of these and other Martian meteorites or perhaps a robot Mars explorer to tell us for sure whether life exists or once existed on that remote red world.

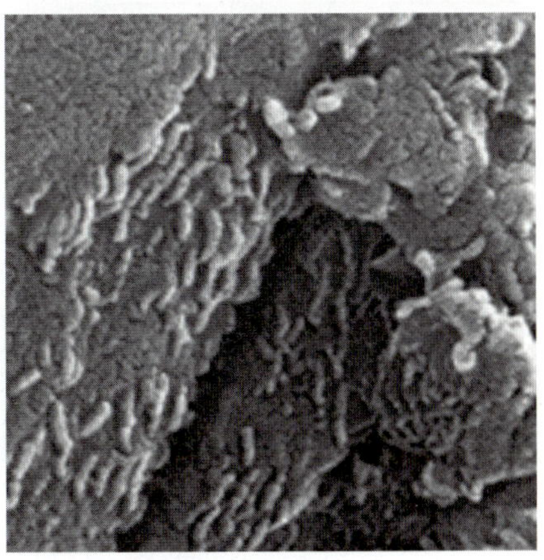

FIGURE 9.30
Fossils of ancient Martian life? The tiny rod-shaped structures look similar to primitive fossils found in ancient rocks on Earth. However, some scientists think these structures formed chemically.

9.4 Why Are the Terrestrial Planets So Different?

We have seen in sections 9.1–9.3 that the terrestrial planets have little in common with Earth, apart from being rocky spheres. They have different surfaces, atmospheres, and interiors. Astronomers think these differences arise from their different masses, radii, and distance from the Sun.

Role of Mass and Radius

As discussed earlier, a planet's mass and radius affect its interior temperature and thus its level of tectonic activity, with low-mass, small-radius planets being cooler inside than larger bodies. We see, therefore, a progression of activity from small, relatively inert Mercury, to slightly larger and once-active Mars, to the larger and far more active surfaces of Venus and Earth, as illustrated in figure 9.31. Mercury's surface still bears the craters made as it was assembled from planetesimals. Mars has some craters, from which we infer that much of its surface is very old, but being larger and more tectonically active, it also has younger surface features such as volcanos, canyons formed by surface cracking as hot material rose inside it, and erosional features such as canyons and riverbeds carved by running water. In contrast, Earth and Venus retain essentially none of their original crust; their surfaces have been enormously modified by activity in their interiors over the lifetimes of these planets.

Role of Internal Activity

Internal activity, as we have seen, also affects a planet's atmosphere. In fact, the atmospheres of the terrestrial planets, though now greatly modified, are probably mostly volcanic gases vented as the result of their internal activity. Thus small, inactive Mercury probably never had much atmosphere, and Mars, active once but now geologically quiet, likewise could create and maintain only a thin atmosphere. Moreover, these planets have so little mass and consequently such a small surface gravity relative to Venus and Earth that they have difficulty retaining what little gases they might once have had. As a result, Mercury is virtually without an atmosphere today, and Mars has only a small remnant of its original atmosphere. On the other hand, highly active

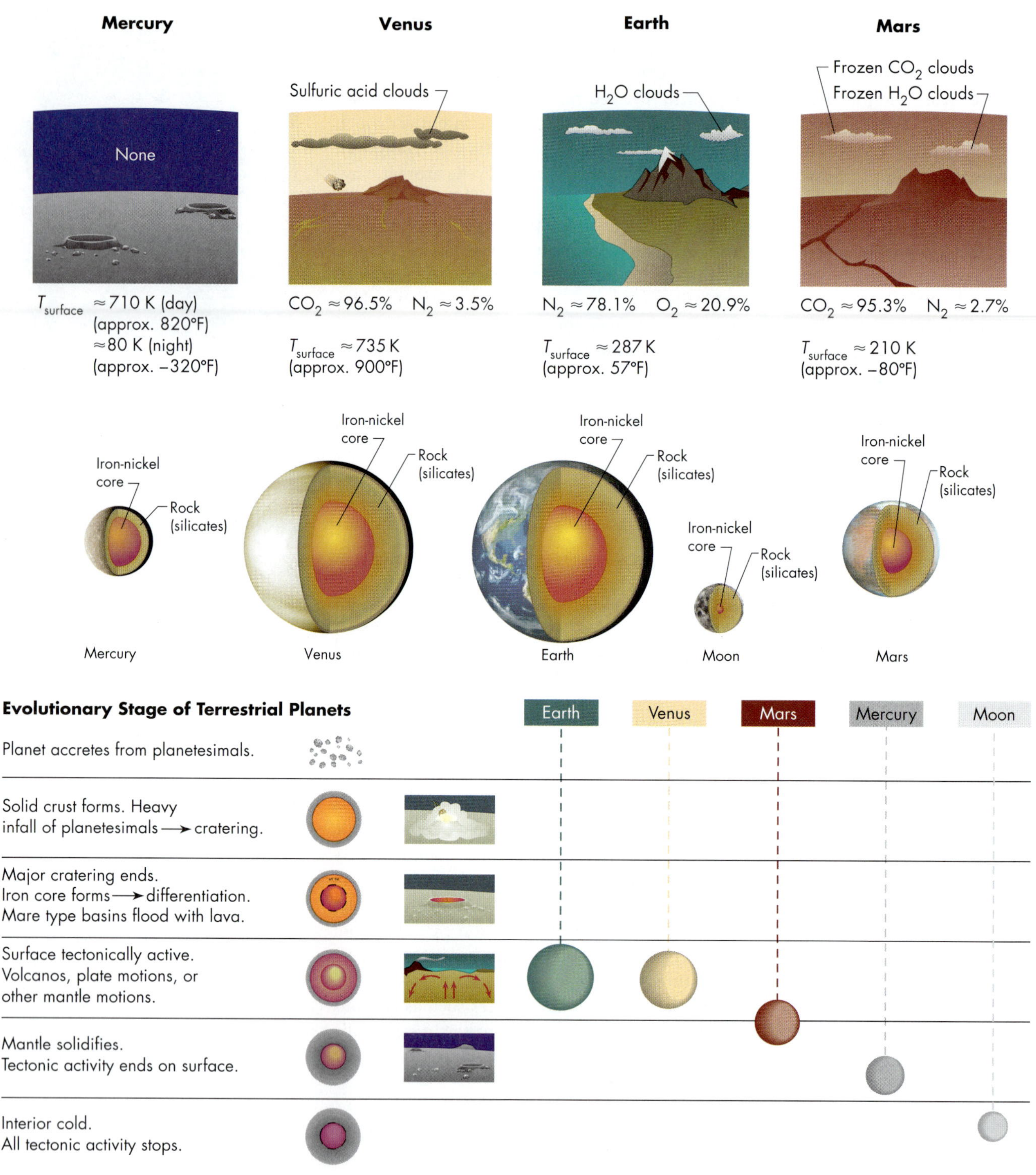

FIGURE 9.31
Gallery comparing interiors, atmospheres, and surfaces of the terrestrial planets.

Venus and Earth have extensive atmospheres. These properties are summarized in table 9.1 as well as figure 9.31.

Astronomers think that the atmospheres that remain on Venus, Earth, and Mars have changed their composition appreciably over time from chemical processes. The atmospheres of all three bodies were probably originally nearly the same composition: primarily CO_2 but with small amounts of nitrogen and water. But these original atmospheres have been modified by sunlight, tectonic activity, and, in the case of the Earth, life.

TABLE 9.1 COMPARISON OF THE TERRESTRIAL PLANETS

	Mercury	Venus	Earth	Mars	Moon
Radius (Earth units) (km)	0.383 2439	0.950 6052	1.000 6378	0.533 3398	0.273 1738
Mass (Earth units) (kg)	0.055 3.30×10^{23}	0.815 4.87×10^{24}	1.000 5.98×10^{24}	0.107 6.42×10^{23}	0.012 7.35×10^{22}
Density (grams/cm^3)	5.43	5.25	5.52	3.94	3.34
Atmospheric composition	None	CO_2 (96.5%) N_2 (3.5%)	N_2 (78.1%) O_2 (20.9%)	CO_2 (95.3%) N_2 (2.7%)	None
Pressure (bars)	0.0	about 90	1.0	about 0.007	0.0
Surface features	Craters, scarps	Mountains, volcanic peaks, congealed lava plains	Oceans, mountains, volcanic peaks	Deserts, craters, canyons, volcanic peaks	Craters, maria
Sidereal day	58.65 Earth days	243.02 Earth days	23.9345 hours	24.62 hours	27.32 Earth days
Solar day	176 Earth days	116.8 Earth days	24 hours	24.66 hours	29.53 Earth days
Satellites	None	None	Moon	Phobos and Deimos	—
Distance from Sun	0.387 AU	0.723 AU	1 AU	1.524 AU	1 AU
Orbital period	87.969 Earth days	224.70 Earth days	365.26 Earth days	686.98 Earth days	—
Axial tilt	0°	177.4°	23.45°	23.98°	5.15°

Role of Sunlight

Sunlight affects a planet's atmosphere in several ways. First, of course, it warms a planet by an amount that depends on the planet's distance from the Sun, and so we expect that Venus will be warmer than Earth and Earth will be warmer than Mars. These expected temperature differences are increased, however, by the atmospheres of these bodies. Moreover, even relatively small differences in temperature can lead to large differences in physical behavior and chemical reactions within an atmosphere. For example, Venus is just enough nearer the Sun than the Earth that even without a strong greenhouse effect, most of its atmosphere would be so warm that water would have difficulty condensing and turning to rain. Moreover, with the Venusian atmosphere being warmer throughout, water vapor can rise to great heights in the Venusian atmosphere, whereas on cooler Earth, water vapor condenses to ice at about 30,000 feet, making our upper atmosphere almost totally devoid of water. You may have noticed this extreme dryness on a long plane trip, because after such a flight your skin may feel itchy and the inside of your nose may feel cracked or stuffy. That dryness is the result of bringing upper-atmospheric air into the cabin and compressing it to make it breathable but not adding moisture.

Role of Water Content

The great difference in the water content of the upper atmospheres of Earth and Venus has led to a drastic difference between their atmospheres at lower levels. At high altitudes, ultraviolet light from the Sun is intense enough to break apart any water molecules in that region into their component oxygen and hydrogen atoms, a process called **photo dissociation.** Being very light, the hydrogen atoms so liberated escape into space, while the heavier oxygen atoms remain. Because water can rise to great heights in the Venusian atmosphere, over billions of years it has steadily been dissociated there and almost completely lost from our sister planet. In our atmosphere, however, water has survived.

Water makes possible chemical reactions that profoundly alter the composition of our atmosphere. For example, CO_2 dissolves in liquid water and creates carbonic acid, which in dilute form we drink as "soda water." In fact, the bubbles in soda water are just carbon dioxide that is coming out of solution. As rain falls through our atmosphere, it picks up CO_2, making it slightly acidic, even in unpolluted air. As the rain falls on the ground, it reacts chemically with rocks to form carbonates, locking some CO_2 into the rock. The amount of carbon dioxide that can be removed from an atmosphere by this process is impressive. For example, if here on Earth the CO_2 that is now bound into carbonates were to be freed again, our atmosphere's pressure would increase by a factor of about 60. This would make our atmosphere, apart from its oxygen, comparable in composition and mass to the existing atmosphere of Venus.

Role of Biological Processes

Biological processes also remove some CO_2 from the atmosphere. For example, plants use it to make the large organic molecules such as cellulose, of which they are composed. This CO_2 is usually stored for only short periods of time, however, before decay or burning releases it back into the atmosphere. More permanent removal occurs when rain carrying dissolved CO_2 runs off into the oceans, where sea creatures use it to make shells of calcium carbonate. As these creatures die, they sink to the bottom, where their shells form sediment that eventually is changed to rock. Thus, carbon dioxide is swept from our atmosphere and locked up both chemically and biologically in the crust of our planet. With most of the carbon dioxide removed from our atmosphere, mostly nitrogen is left. In fact, our atmosphere contains roughly the same total amount of nitrogen as the atmosphere of Venus.

Our atmosphere is also rich in oxygen, a gas found in such relative abundance nowhere else in the Solar System. Our planet's oxygen is almost certainly the product of green plants breaking down the H_2O molecule during photosynthesis, as we discussed in chapter 6. Thus, much of the cause of the great difference between the atmospheres of Earth and Venus may be that life was able to start on Earth. On our planet, liquid water removed much of the carbon dioxide from our atmosphere (as we discussed in the previous paragraph) and life swept out most of the rest of it, leaving only the tiny amount (0.03%) we have today. In removing the carbon dioxide, plants also produced the oxygen on which all animal life depends. What evidence supports this idea that liquid water and living things removed most of the early Earth's carbon dioxide? It is the carbonate rock we mentioned earlier. If the carbon dioxide locked chemically in that rock were released back into our atmosphere and the oxygen were removed, our planet's atmosphere would closely resemble Venus's.

If water is so effective at removing carbon dioxide from our atmosphere, why does any CO_2 remain? We add small amounts of CO_2 by burning wood and fossil fuels, but the major contribution is from natural processes. Atmospheric chemists hypothesize that tectonic activity gradually releases CO_2 from rock back into our atmosphere. At plate boundaries, sedimentary rock is carried downward into the mantle, where it is melted. Heating breaks down the carbonate rock so that it releases carbon dioxide, which then rises with the heated rock to the surface and reenters the atmosphere. A similar process may once have occurred on Mars to remove carbon dioxide from its atmosphere, locking it up in rock there. Mars's lower level of tectonic activity, however, prevents its CO_2 from being recycled. Thus, with so little of its original carbon dioxide left, Mars has grown progressively colder. Our Earth, because it is active, has retained enough CO_2 in its atmosphere to maintain a moderate greenhouse effect, making our planet habitable. Thus, poised between one planet that is too hot and another that is too cold, Earth has been blessed with a relatively stable atmosphere, one factor in the complex web of our environment to which we owe our existence.

SUMMARY

The terrestrial planets—Mercury, Venus, Earth, and Mars—are alike in being rocky bodies of comparable size and internal structure, but they nevertheless differ dramatically from one another. Their slight differences in mass and diameter have led to great differences in their internal heat and surface activity. In fact, when ordered by mass, the terrestrial planets are also ordered by activity. Mercury, the least massive, is least active and has the oldest surface. The many impact craters created late in its formation make Mercury look much like our Moon.

Mars, intermediate in mass and radius between Mercury and Earth, shows a surface with an intermediate level of activity. Much of it is cratered, implying a great age. Some regions, however, show volcanic uplift like that seen on Venus, implying that its interior was once much hotter than now. Moreover, its surface shows an amazing variety of landforms: polar caps of ordinary ice and frozen carbon dioxide, immense deserts with dune fields, and canyons created both by crustal cracking and erosion from running water.

Venus and Earth have young surfaces, extensively altered by volcanic activity. On Earth, the flow of heat from its core has led to plate motion. On Venus, perhaps because of its hotter surface, the flow of heat to its surface is less uniform. As a result its surface is marked by isolated regions of intense volcanic uplift. In fact, Venus is probably as active geologically as the Earth.

Mercury, Venus, and Mars have very different atmospheres. Mercury was too small to either make or retain an atmosphere and is now essentially airless. Venus and Mars both have carbon dioxide atmospheres, but Venus's is about 100 times denser than Earth's, while Mars's is about 100 times less dense. The great thickness of the Venusian atmosphere creates a strong greenhouse effect, heating it to about 900°F. Thick clouds of sulfuric acid droplets high in the Venusian atmosphere block our view of the Venusian surface. Astronomers have nevertheless been able to study the Venusian surface with radar telescopes both on Earth and in orbit around Venus.

Mars's atmosphere is too thin now for a strong greenhouse effect, but the old river channels, cut presumably by running water, imply that Mars once had a denser and warmer atmosphere, perhaps even resembling Earth's.

The great differences among the atmospheres of Venus, Earth, and Mars can probably be explained by the differences in their water content and the fact that life evolved on Earth.

QUESTIONS FOR REVIEW

1. (9.1) How do Mercury's radius and mass compare with the Earth's?
2. (9.1) What is the surface of Mercury like?
3. (9.1) Does Mercury have an atmosphere? Why or why not?
4. (9.1) What is peculiar about Mercury's rotation? What causes this oddity?
5. (9.2) How does Venus compare with the Earth in mass and diameter?
6. (9.2) What is the dominant gas in Venus's atmosphere? How do astronomers know this?
7. (9.2) What are the clouds of Venus made of?
8. (9.2) Why is Venus so hot?
9. (9.2) Can we see the surface of Venus? Why or why not? How do astronomers know what the surface of Venus is like?
10. (9.2) What sort of features are seen on Venus's surface? Is the surface young or old? Some of the youngest surfaces on Earth are from plate ridges—is the same true on Venus?
11. (9.3) How does Mars compare with the Earth in mass and diameter?
12. (9.3) Describe some of the surface features seen on Mars.
13. (9.3) What are the Martian polar caps composed of?
14. (9.3) What is the Martian atmosphere like?
15. (9.3) What is the evidence that Mars once had running water on its surface?
16. (9.3) What is the likely origin of Mars's two moons? What evidence supports this?
17. (9.3) What is the evidence that leads some scientists to believe life may have existed on Mars?
18. (9.4) What explanations have been offered for why the atmospheres of the terrestrial planets are so different?
19. (9.4) How do astronomers explain why the Earth's atmosphere ended up with so little CO_2, compared with that of Mars and Venus?

THOUGHT QUESTIONS

1. (9.1) What are some examples of resonances in everyday life?
2. (9.2) How does the surface temperature of Venus relate to concerns about global warming of the Earth?

3. (9.3) Suppose scientists discover simple life forms on Mars. Would that alter the way you look at life here on Earth?
4. (9.2–9.4) Some scientists have proposed "terraforming" Venus and Mars. That is, they propose introducing organisms that would use carbon dioxide to photosynthesize oxygen and make both planets more suitable for humans. Do you think this is a good idea? Why or why not?
5. (9.2–9.4) Thinking about the relative roles of sunlight and greenhouse effect on temperature, volcanic activity and the gases it liberates, liquid water and a liquid water cycle, the presence or removal of CO_2 in the atmosphere, and the ability to retain an atmosphere, make an argument as to whether Mars or Venus would be easier to "terraform" into a condition suitable for colonization.
6. (9.2–9.4) Reconcile the idea that Mars, Earth, and Venus all have approximately the same internal structure and composition, and the fact that the average density of Mars is nearly 1 g/cm^3 less than Earth or Venus.
7. (9.1–9.4) What evidence do we see from the terrestrial planets that supports the theory for the origin of the Solar System discussed in chapter 8?
8. (9.1–9.4) What role might planetesimal impacts have played in the history of the terrestrial planets?

PROBLEMS

1. (9.1/3.8) Calculate the escape velocity needed to launch a rocket from the surface of Mercury.
2. (9.2) Venus rotates "backward" when seen from above—spinning clockwise, while orbiting the Sun in a counterclockwise motion. Draw a diagram to show how, for a "retrograde" planet like Venus, the time from solar noon to solar noon is shorter than the time required to spin 360 degrees (the opposite is true for Earth—see essay 3 and fig. E3.2).
3. (9.3/3.7) Olympus Mons is 26 kilometers above the surrounding surface of Mars. Calculate the surface gravity on Mars at the surface and on top of Olympus Mons (neglecting any effects on gravity from the mountain itself). What percentage of the surface value is the difference in the acceleration of gravity?
4. (9.2–9.4) In chapter 13, we will learn that the brightness of light decreases in proportion to the square of the distance from the source. We know sunlight plays an important role in the conditions on the planets and also is used to power equipment like the Mars rovers. How many times brighter is sunlight at Venus's distance from the Sun than at Earth? How many times dimmer is it at Mars than at Earth? (Distances from the Sun compared to Earth's can be found in table 9.1 or in the appendix).

TEST YOURSELF

1. (9.1) Why does Mercury have so many craters and the Earth so few?
   (a) Mercury is far more volcanically active than the Earth.
   (b) Mercury is much more massive than the Earth and therefore attracts more impacting bodies.
   (c) The Sun has heated Mercury's surface to the boiling point of rock, and the resulting bubbles left craters.
   (d) Erosion and plate tectonic activity have destroyed most of the craters on the Earth.
   (e) Mercury's iron core and its resulting strong magnetic field have attracted impacting bodies.
2. (9.1) Mercury's average density is about 1.5 times greater than the Moon's, even though the two bodies have similar radii. What does this suggest about Mercury's composition?
   (a) Mercury's interior is much richer in iron than the Moon's.
   (b) Mercury contains proportionately far more rock than the Moon.
   (c) Mercury's greater mass has prevented its gravitational attraction from compressing it as much as the Moon is compressed.
   (d) Mercury must have a uranium core.
   (e) Mercury must have a liquid water core.
3. (9.2) Why is Venus's surface hotter than Mercury's?
   (a) Venus rotates more slowly, so it "bakes" more in the Sun's heat.
   (b) Clouds in Mercury's atmosphere reflect sunlight back into space and keep its surface cool.
   (c) Carbon dioxide in Venus's atmosphere traps heat radiating from its surface, thereby making it warmer.
   (d) Venus is closer to the Sun.
   (e) Venus's rapid rotation generates strong winds that heat the ground by friction as they blow.
4. (9.2) What are the Venusian clouds made of?
   (a) Mostly water droplets
   (b) Frozen carbon dioxide
   (c) Sodium chloride
   (d) Sulfuric acid droplets
   (e) Ice crystals
5. (9.3) Astronomers can measure the time since the last eruption of a volcano on Mars by
   (a) using radioactive dating.
   (b) counting the number of craters on its slopes.
   (c) measuring how much the lava has cooled
   (d) searching ancient records of Chinese astronomers.
   (e) estimating based on the time since an eruption on Earth.
6. (9.3) Why is there no liquid water on the surface of Mars today?
   (a) The CO_2 in the atmosphere has caused a runaway greenhouse effect.
   (b) The fine surface sand would immediately absorb any water.

(c) Liquid water is immediately broken down by chemical reactions with iron (rusting) on the surface.
(d) The atmospheric pressure is so low, any liquid water would immediately evaporate.

7. (9.4) Which body in the inner Solar System has the densest atmosphere?
(a) Mercury (b) Venus (c) Earth (d) Mars
8. (9.4) Which of the following features are shared by all of the terrestrial planets? (select all that apply)
(a) A silicate mantle
(b) An iron core
(c) A strong magnetic field
(d) Volcanic activity

KEY TERMS

photo dissociation, 253
resonance, 233
runaway greenhouse effect, 236

FURTHER EXPLORATIONS

Albee, Arden L. "The Unearthly Landscapes of Mars." *Scientific American* 288 (June 2003): 44.

Beatty, J. Kelly. "Reunion with Mercury." *Sky and Telescope* 115 (May 2008): 24.

Beatty, J. Kelly, Carolyn Collins Petersen, and Andrew Chaikin. *The New Solar System.* 4th ed. Cambridge: Sky, 1999.

Bell, Jim. "The Red Planet's Watery Past." *Scientific American* 295 (December 2006): 62.

Bullock, Mark A., and David H. Grinspoon. "Global Climate Change on Venus." *Scientific American* 280 (March 1999): 50.

Clifford, Stephen M. "The Iceball Next Door." *Sky and Telescope* 106 (August 2003): 30.

DiGregorio, Barry E. "Life on Mars? 27 Years of Questions." *Sky and Telescope* 107 (February 2004): 40.

Hartmann, William K. "Mysteries of Mars." *Sky and Telescope* 106 (July 2003): 36.

Kargel, Jeffrey S., and Robert G. Strom. "Global Climatic Change on Mars." *Scientific American* 275 (November 1996): 80.

Kaula, W. M. "Venus: A Contrast in Evolution to Earth." *Science* 247 (9 March 1990): 1191.

Musser, George. "The Spirit of Exploration." *Scientific American* 290 (March 2004): 52.

Naeye, Robert. "Europe's Eve on Mars." *Sky and Telescope* 110 (December 2005): 30.

Robertson, Donald. "Parched Planet." *Sky and Telescope* 114 (April 2008): 26.

Schubert, G., and C. Covey. "The Atmosphere of Venus." *Scientific American* 245 (July 1981): 66.

Squyres, Steve. *Roving Mars.* New York: Hyperion (2005). See also NASA/JPL's rover website home at http://marsrovers.jpl.nasa.gov/home/index.html.

Stofan, Ellen R. "The New Face of Venus." *Sky and Telescope* 86 (August 1993): 22.

Strom, R. G. "Mercury: The Forgotten Planet." *Sky and Telescope* 80 (September 1990): 256.

Talcott, Richard. "Mercury Reveals Its Hidden Side." *Astronomy* 36 (May 2008): 26.

Videos

Flying by the Planets: The Videos (available from the Astronomical Society of the Pacific, 390 Ashton Ave., San Francisco, CA 94122).

Website

Visit the *Explorations* website at **http://www.mhhe.com/arny** for additional online resources on these topics.

Q FIGURE QUESTION ANSWERS

WHAT IS THIS? (chapter opening): These are sand dunes on Mars imaged by the Mars Global Surveyor.

FIGURE 9.2: The scarp must be younger because it cuts the craters. If the craters had formed after the scarp, they would have destroyed the scarp within the craters.

FIGURE 9.3: Caloris Basin is round and large, but its interior is much rougher than lunar mare.

FIGURE 9.7: Mercury's rotation, like our Moon's, has been slowed by tidal interactions. Mercury tries to rotate so that its spin and orbital motions match, but because its orbit is so elliptical, and its orbital speed therefore changes, it can't make the match occur.

FIGURE 9.8: Venus does not look round in this image because we are seeing it with the Sun off to the left. Basically, we are seeing here a gibbous phase of Venus, which is visible from Earth only when Venus is on the far side of the Sun.

FIGURE 9.15: Because the south pole is illuminated, it must be spring or summer in the southern hemisphere. The polar cap has not yet shrunk to its smallest, so you are correct if you guessed spring.

FIGURE 9.29: Their irregular shape implies that they have only a weak surface gravity. If they had a stronger surface gravity, it would crush them into a rounder shape. Their weak surface gravity would make it virtually impossible for them to retain any atmosphere.

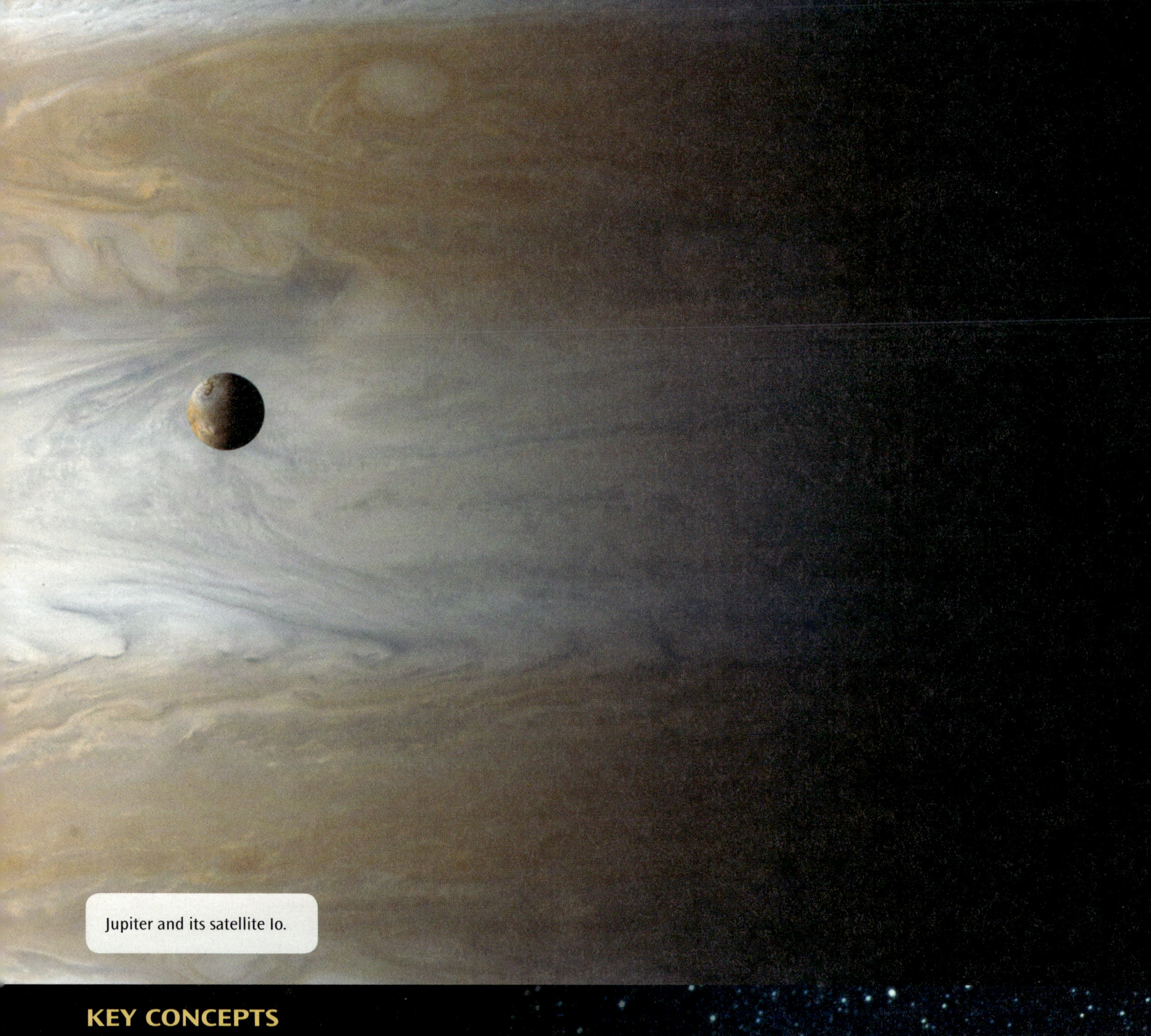

Jupiter and its satellite Io.

KEY CONCEPTS

- The outer planets are Jupiter, Saturn, Uranus, and Neptune.
- The outer planets (also called the Jovian planets) are much larger than the Earth.
  - They are rich in hydrogen gas and/or its compounds, such as water, methane, and ammonia.
  - They have deep, thick atmospheres and no solid surface.
  - They have many moons and rings.
- Pluto was formerly considered one of the outer planets. It and the slightly-larger Eris are now officially called *dwarf planets.* They are composed mainly of rock and ice and are far smaller than even Mercury.
- The Jovian planets have large satellite systems, with some moons even larger than Mercury.
- The larger of these icy satellites have unusual properties, including volcanoes, geysers, dense atmospheres, and sub-surface oceans.

10 The Outer Planets

CONCEPTS AND SKILLS TO REVIEW

- Density (6.1)
- Coriolis force (6.7)

Beyond Mars, the Solar System is a realm of ice and frozen gas. In this frigid zone, far from the Sun, where solar heat is only a vestige of what we receive on Earth, the giant planets formed. The low temperatures—about −100° Fahrenheit and colder—allowed objects condensing there, within the solar nebula, to capture hydrogen-rich materials such as methane, ammonia, and especially water. Because these substances were far more abundant in the young Solar System than the silicate- and iron-rich material from which the terrestrial planets condensed, planets that formed in this cold environment had more material available for their growth. As a result, these cold planets became vastly larger than those near the Sun, and they developed very different structure and composition. The four largest planets—Jupiter, Saturn, Uranus, and Neptune—are composed mainly of gaseous and liquid hydrogen and its compounds. Although these giant bodies may have cores of molten rocky matter, they lack solid surfaces and consequently have no surface features, such as mountains and valleys. Rather, it is their atmospheric features that give them such different appearances.

The dwarf planets Pluto and Eris are also members of this remote part of the Solar System. They are far smaller than the giant planets and are composed of ice and rock. They closely resemble the larger moons of these planets.

The moons of the outer Solar System range in size from very small to larger than Mercury. As they orbit their parent planets, they form families rather like miniature Solar Systems. Some of the larger moons have brightly colored surfaces, others have numerous craters, and a few have surface features unlike anything seen elsewhere in the Solar System. A few moons even have active volcanoes. In fact, astronomers consider these diverse bodies, virtually unknown before the space age, to be some of the most interesting members of the Sun's family.

Q: WHAT IS THIS? See end of chapter for the answer.

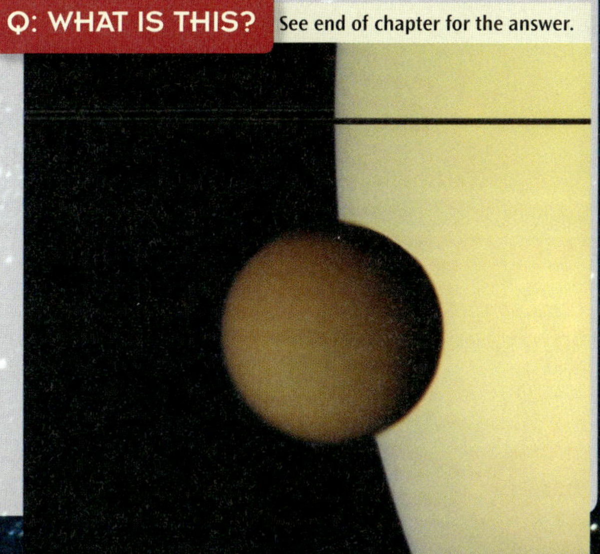

10.1 Jupiter

To the ancient Romans, Jupiter was the king of the gods. Although they did not know how immense this planet is, they nevertheless chose its name appropriately.

Jupiter's Appearance and Physical Properties

Jupiter is the largest planet in our Solar System, both in radius and in mass. In fact, its mass is larger than that of all other planets in the Solar System combined. It is slightly more than 10 times the Earth's diameter and more than 300 times its mass. Richly colored parallel bands of clouds cloak the planet, as shown in figure 10.1. Spectra of the sunlight reflected from these clouds show that Jupiter's atmosphere consists mostly of hydrogen, helium, and hydrogen-rich gases such as methane (CH_4), ammonia (NH_3), and water (H_2O). These gases were also directly detected in December 1995 when a probe dropped from the *Galileo* spacecraft parachuted into Jupiter's atmosphere. The clouds themselves are harder to analyze, but theoretical calculations of the chemistry of Jupiter's atmosphere suggest they are particles of water-ice, and ammonia compounds. Their colors may come from complex organic molecules or compounds of sulfur or phosphorus. Time-lapse pictures show that Jupiter's clouds move swiftly, sweeping around the planet in jet streams that are far faster than those of Earth. Moreover, Jupiter itself rotates once every 10 hours, spinning so fast that its equator bulges significantly.

Jupiter's Interior

Astronomers cannot see through Jupiter's cloud layers to its interior, nor can they probe its interior with seismic detectors. Instead, they must rely on theory to tell them what lies inside this giant planet. For example, despite its great mass, Jupiter is much less dense, on the average, than the Earth. To determine a planet's average density, astronomers divide its mass by its volume. The mass can be found by calculating the planet's gravitational attraction acting on one of its moons, using the method described in chapter 8 (see fig. 8.8). We also described in that figure how to find the planet's radius and, hence, its volume. When such calculations are made for Jupiter, we find that its average density is only slightly greater than that of water—1.3 grams

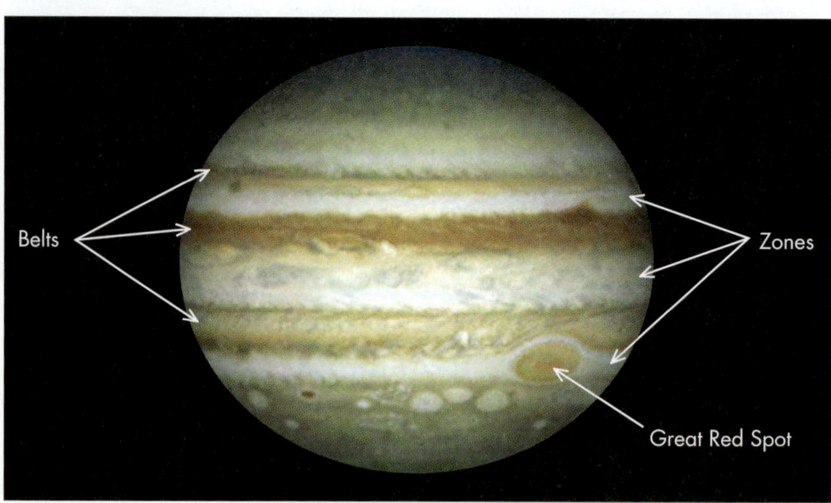

FIGURE 10.1
Jupiter and its clouds.

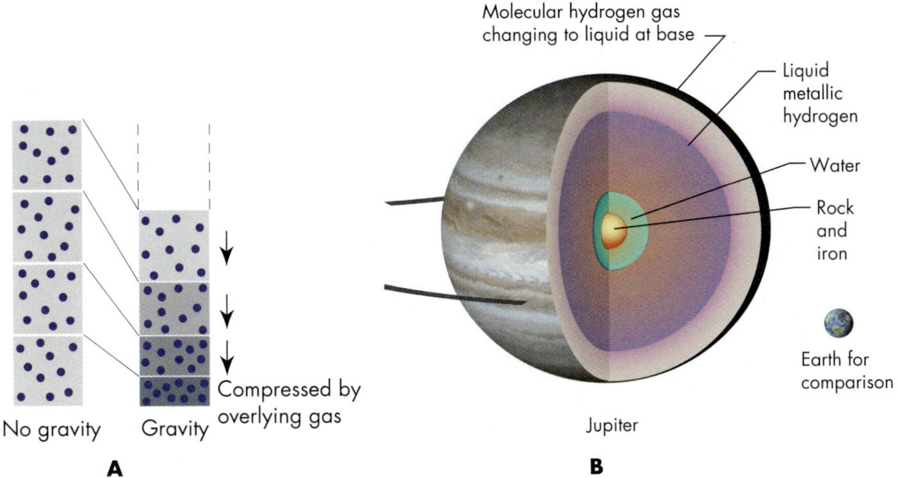

FIGURE 10.2
(A) The density of gas increases with depth because the overlying gas compresses the matter. (B) A sketch of what astronomers think Jupiter's interior is like.

Q. From the change in volume shown here in (A), how many times more dense is the bottom, compressed box than the top box?

per cubic centimeter—showing that the bulk of the planet, not just its atmosphere, must be composed mainly of very light elements, such as hydrogen. Calculations based on Jupiter's low density, its shape, and measurement of its gravitational attraction on its satellites and passing spacecraft all give astronomers information on what lies below its clouds.

Jupiter's immense mass exerts a tremendous gravitational force on its atmosphere and interior, holding the planet together and compressing its gas. Near the cloud tops, this compression is only slight, so the gas density there is low. Deep in the interior, however, the weight of thousands of kilometers of gas compresses the matter to a high density (approximately 3 times the density of iron). Thus, the density of the gas increases with depth, as illustrated in figure 10.2A. Everywhere in the planet, however, the forces of gravity and pressure are nearly in balance, just as they are in our own atmosphere.

Deep within Jupiter, the compression created by its gravity presses molecules so close together that the gas changes to liquid. Thus, about 10,000 kilometers below the cloud tops—about one-sixth of the way into the planet—Jupiter's interior is a vast sea of liquid hydrogen. Deeper still, the weight of the liquid hydrogen compresses matter below it into a state known as liquid metallic hydrogen, a form of hydrogen that scientists on Earth may have created in tiny high-pressure chambers. Astronomers theorize that if the material from which Jupiter formed has the same overall composition as the Sun's, then Jupiter must also contain heavy elements such as silicon and iron. Because of their high density, these elements have probably sunk to Jupiter's center, forming a core of iron and rocky material a few times larger than that of the Earth (see fig. 10.2B). Detailed measurements of Jupiter's gravitational field made by orbiting satellite also indicate the presence of a core with a mass perhaps as high as 18 times the mass of the Earth.

Jupiter's interior is extremely hot, perhaps 30,000 K, which is about five times hotter than the Earth's core. This heat rises slowly to the planet's surface and escapes into space as low-energy infrared radiation. Astronomers have long known from measurements of this infrared radiation that Jupiter emits more energy than it receives from the Sun, but they are still uncertain what supplies that heat. Much of that energy is probably left over from Jupiter's formation. As we saw in discussing the birth of the Earth, planet building is a hot process. The hail of gas and planetesimals onto the forming world releases a huge amount of gravitational energy. Subsequently, additional heat may be generated by slow but steady shrinkage of the planet as it adjusts to the greater gravity created by the mass added in the last stages of its formation. In fact, giant gas planets such as Jupiter may still be shrinking (and therefore heating) slightly. Yet more

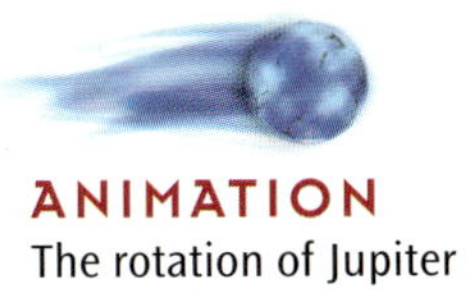

ANIMATION
The rotation of Jupiter

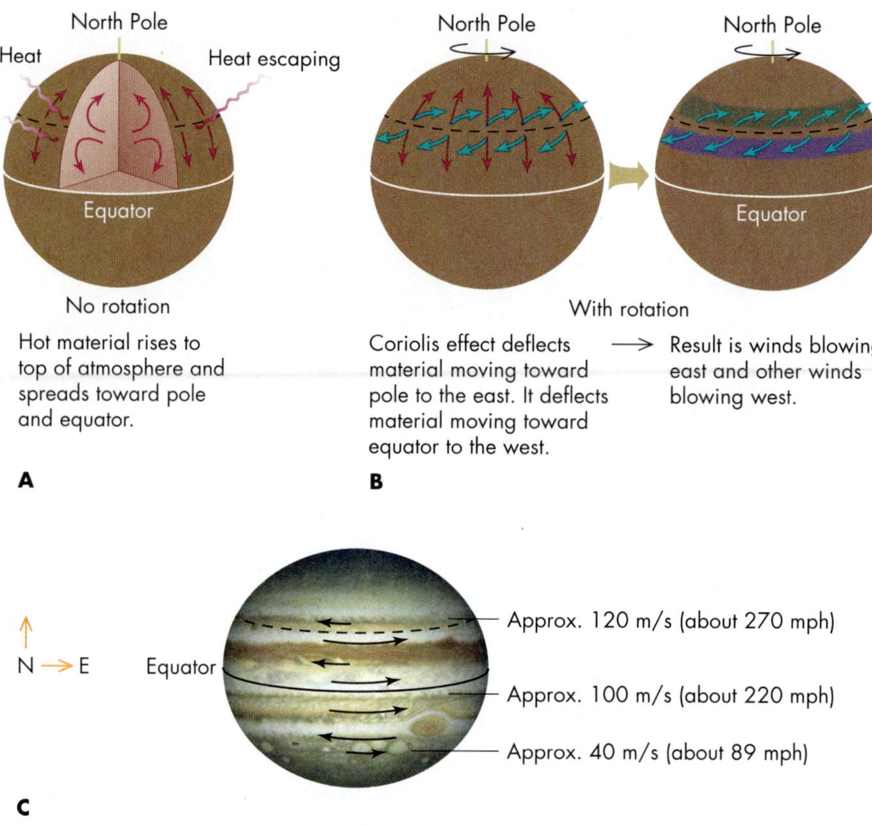
North Pole
Heat
Heat escaping
Equator
No rotation
Hot material rises to top of atmosphere and spreads toward pole and equator.
A
North Pole
North Pole
Equator
With rotation
Coriolis effect deflects material moving toward pole to the east. It deflects material moving toward equator to the west.
Result is winds blowing east and other winds blowing west.
B
N E
Equator
Approx. 120 m/s (about 270 mph)
Approx. 100 m/s (about 220 mph)
Approx. 40 m/s (about 89 mph)
C

FIGURE 10.3
(A) Rising gas from Jupiter's hot interior cools near the top of the atmosphere and sinks. (B) The Coriolis effect, arising from Jupiter's rotation, deflects the gas, creating winds that blow as narrow jet streams. (C) The wind varies widely in speed and direction from region to region.

heat is released as matter that is denser than hydrogen, such as helium, sinks toward Jupiter's core. Whatever its source, the heat generates convection currents similar to those in the Earth's interior. These currents stir both Jupiter's deep interior and its atmosphere.

Jupiter's Atmosphere

Heat within Jupiter generates convection currents in its outer layers that carry warm gas upward to the top of its atmosphere. Here, the gas radiates heat to space, becoming cooler and sinking again, as illustrated in figure 10.3A. Because of its motion, this rising and sinking gas is subject to a Coriolis effect* created by Jupiter's rapid rotation that deflects the gas into powerful winds called jet streams. We see these winds as the cloud belts, illustrated in figure 10.3B. Winds in these jet streams can have velocities of up to 300 kilometers per hour (nearly 200 miles per hour) with respect to the planet's overall rotation. Yet despite such high speeds, winds in adjacent regions may blow in opposite directions, as shown in figure 10.3C. Such reversals of wind direction (wind shear) from place to place also occur on Earth, where equatorial winds generally blow from east to west, while midlatitude winds generally blow from west to east.

These reversals occur because the Coriolis effect deflects northern hemisphere winds to the right and southern hemisphere winds to the left of the direction they would travel in the absence of the Coriolis effect. As a result, winds moving toward the equator are deflected to the west, while winds moving toward the poles are

*The Coriolis effect is a consequence of the conservation of angular momentum. It is not some new force. Recall our discussion in section 6.7.

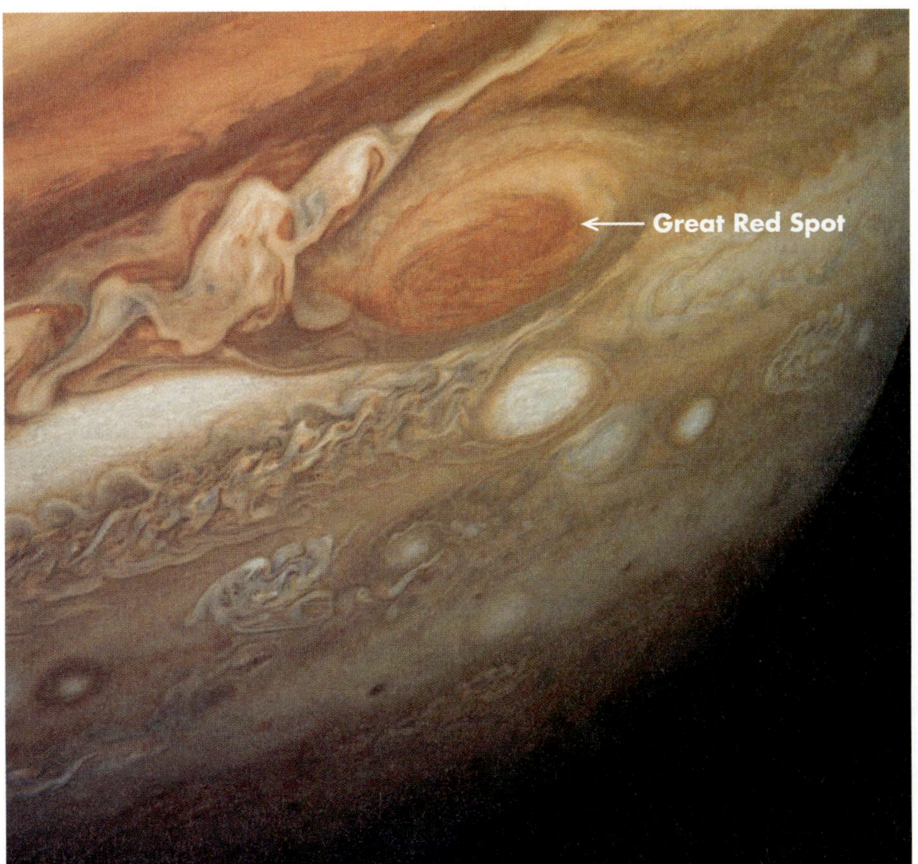

FIGURE 10.4
Vortices form between atmospheric streams of different velocities. Note the Great Red Spot and the white oval below it.

deflected to the east, as shown in figure 10.3B. The cumulative effect of these deflections is a set of jet streams blowing in opposite directions with respect to Jupiter's overall rotation.

As the various jet streams circle Jupiter, gas between them rising from below may be spun into huge, whirling atmospheric vortices, much as a pencil between your palms twirls as you rub your hands together. Some of these spinning regions are brightly colored, as figure 10.4 shows. Brown and shades of white dominate, but one exceptionally large vortex—bigger across than the Earth—is nearly brick red. Known as the Great Red Spot, this vortex was discovered in the seventeenth century. Since then, it has changed slightly in appearance and location, but both direct observations of the spot and theoretical models show that it is an essentially permanent feature of Jupiter's atmospheric circulation.

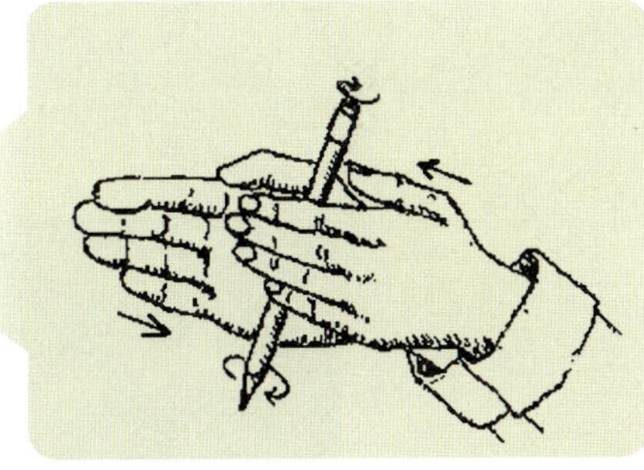

Matter also circulates in Jupiter's deep interior. There, convection in the metallic liquid hydrogen combines with the planet's rapid rotation to generate a magnetic field by a natural dynamo process similar to that which generates the Earth's magnetic field. Jupiter's dynamo process is far more powerful than Earth's, however, and creates the strongest magnetic field of any planet in the Solar System. Measurements taken by passing spacecraft show that the field near the top of Jupiter's cloud layer is about 20 times the strength of the magnetic field on the Earth's surface. When Jupiter's larger radius is taken into account, its field is some 20,000 times as powerful as Earth's, as was deduced even before it was directly measured. For example, astronomers inferred that Jupiter had a strong magnetic field from its auroral activity—just barely detectable from Earth—and from the intense radio emission from the giant planet.

The Earth's auroral activity is linked to its magnetic field, which steers incoming energetic particles from the Sun into our upper atmosphere, where the particles

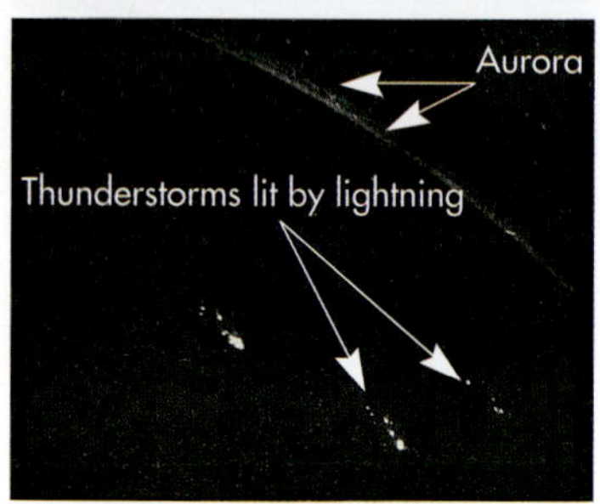

FIGURE 10.5
(A) Jupiter's auroral zone as seen by the Hubble Space Telescope. (B) Aurora and lightning on Jupiter, as observed by the *Voyager* spacecraft as it flew over Jupiter's nightside.

trigger the lovely pale glow of the northern and southern lights. Our planet's magnetic field also traps such particles in the Van Allen radiation belts. Jupiter's magnetic field does much the same: particles trapped magnetically in radiation belts far above the planet emit radio waves, making Jupiter a powerful source of radio emission. Some of those particles descend into Jupiter's upper atmosphere, where they create an aurora, as you can see in figure 10.5A, a picture made with the Hubble Space Telescope. Jupiter's aurora also shows in figure 10.5B as the pale glow along the edge of the planet in a picture taken by the passing *Voyager* spacecraft. This figure also depicts another familiar phenomenon in Jupiter's atmosphere: lightning. Cloud particles carried up and down by the rising and sinking atmospheric motions collide and generate atmospheric electricity—thunderstorms—just as such motions do on Earth. In the picture, you can see many storms lighting up the Jovian night.

Jupiter's Rings

For centuries, astronomers believed that the only Solar System planet with rings was Saturn. But in 1977, thin rings were detected around Uranus, leading astronomers to wonder if similar rings might surround Jupiter. The opportunity to look for such rings came with the *Voyager I* spacecraft, which flew by Jupiter in 1979. Pictures taken from the craft clearly show that Jupiter has rings, although very thin ones, as shown in figure 10.6. Jupiter's ring system is thought to be made of tiny particles of rock dust held in orbit by Jupiter's immense gravitational attraction. These particles are so tiny, however, that radiation from the Sun and collisions with gas trapped in Jupiter's magnetic field exert frictional forces on them, making them gradually drift down into Jupiter's atmosphere. There, they mingle with the swirling gas and are lost from sight. Thus, to maintain the ring, new dust particles must constantly be added to it. Where does this new ring material come from? According to current theories, the tiny satellites of Jupiter's system occasionally collide and fragment, creating new dust to replenish the rings.

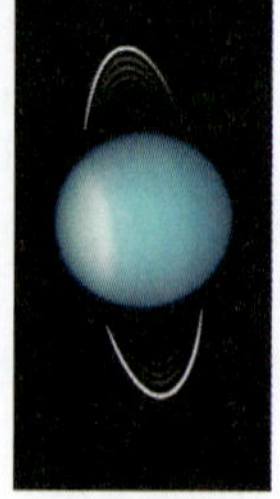

FIGURE 10.6
Rings of the outer planets shown to their correct relative sizes. The image of Saturn's rings was made by the *Cassini* spacecraft, and Uranus's by the Hubble Space Telescope. Jupiter's and Neptune's are shown in montages of *Voyager* images from behind the planet, which show the small dusty particles more clearly.

FIGURE 10.7
Photographs of the four Galilean satellites of Jupiter and our Moon. From left to right, they are: our Moon, Io, Europa, Ganymede, and Callisto.

Q. Why are these moons called the Galilean satellites?

Jupiter's Moons

When Galileo first viewed Jupiter with his telescope, he saw four moons orbiting the planet, bodies now called the "Galilean satellites." Over the past three centuries, astronomers have found additional Jovian moons, with the total now reaching 63. Most of these are too small to be readily seen from Earth and were discovered by examining pictures taken by the *Voyager* and *Galileo* spacecraft.

The Galilean satellites—Io, Europa, Ganymede, and Callisto—are very large: all but Europa are larger than our Moon, and Ganymede has a diameter bigger than Mercury's, making it the largest moon in the Solar System. They orbit along with most of the rest of Jupiter's moons, approximately in Jupiter's equatorial plane, forming a flattened disk, rather like a miniature Solar System. In fact, the Galilean moons most likely formed by a scaled-down version of the process that created the Solar System. That is, they probably aggregated from planetesimals and gas that collected around Jupiter during its formation. Some astronomers believe that Jupiter heated this orbiting debris, affecting the composition and density of the moons that formed from it. Such heating would melt ices and partially evaporate them,* so that the moons nearest Jupiter would have less ice and gas and therefore be denser, much as the Sun heated the solar nebula, thereby creating the difference between the terrestrial and outer planets. Evidence that Jupiter heated its moons comes from their density: the densest Galilean satellites are those nearest Jupiter. In order of increasing distance, the densities (in units of grams per cm^3) are 3.53 for Io, 2.99 for Europa, 1.94 for Ganymede, and 1.85 for Callisto. Further evidence of heating comes from their strange, often colorful surface features, as illustrated in figure 10.7, and in the case of Io, from its active volcanoes.

Io (pronounced† *eye-oh*) is named for a mythological maiden with whom Jupiter fell in love and whom he changed into a heifer (a young cow) so that his wife, Hera, would not suspect his infidelity. Io is the nearest to Jupiter of the Galilean moons and therefore is subject to a strong tidal force created by Jupiter's gravity. That tidal force locks Io's spin to its orbital motion the way our Moon's spin is locked to its motion around the Earth. But Io also undergoes a strong gravitational attraction from Europa, the Galilean satellite next closest to Jupiter, and Europa's gravitational tug twists Io from side to side. Moreover, Europa's pull forces Io into an orbit whose shape constantly changes. The outcome of these two effects is that Io is subject to a strong and changing

---

*Technically, we should use the word *sublimate* to describe the change of ice to vapor. For simplicity, we will continue to use *evaporate* for this change.

† American astronomers tend to say *eye-oh,* but European ones prefer *ee-oh*.

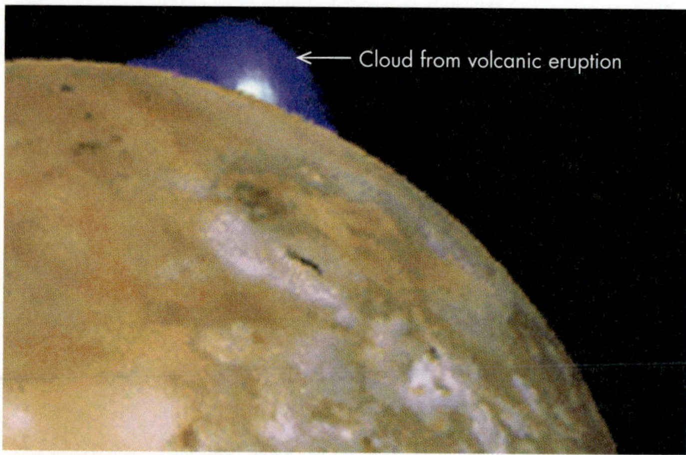

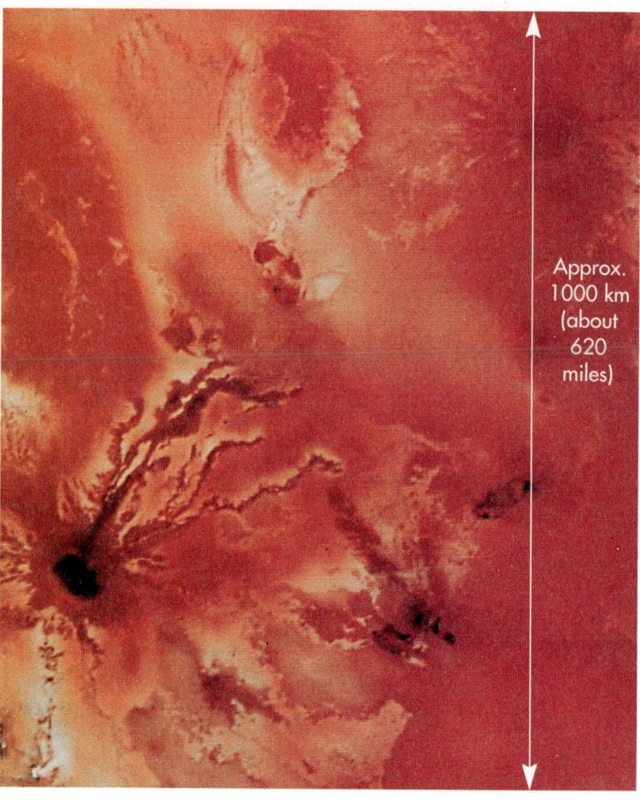

FIGURE 10.8
A volcanic eruption and lava flow on Io. The lava is believed to be mostly molten sulfur. These pictures were taken by the *Voyager* spacecraft as it flew past Jupiter.

Q. What explanation has been offered for the source of heat in Io?

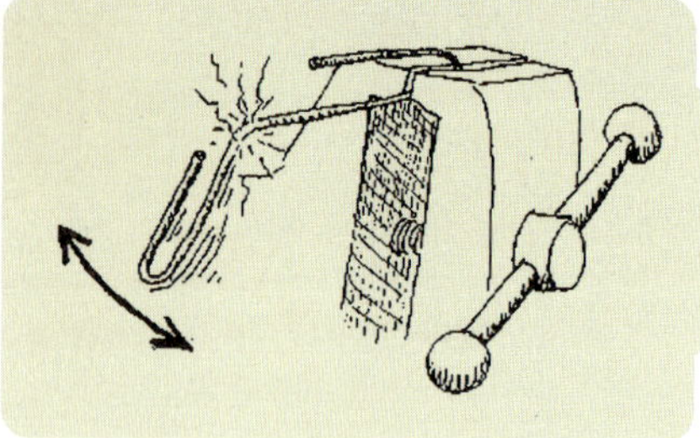

gravitational force from Jupiter that distorts its shape. This deformation heats it by internal friction, much as bending a paperclip until it breaks heats the wire (touch the freshly broken end and it will feel hot). Over billions of years, the heating has melted the rocky matter in Io's interior. As molten matter oozes to the surface, it erupts, creating volcanic plumes and lava flows, as shown in figure 10.8. Sulfur, common in terrestrial volcanoes as well, is now the major component of Io's volcanic outpourings,* and the erupted sulfur compounds give Io its rich red, yellow, and orange colors. A recent sulfur lava flow can be seen in figure 10.8 where the molten material has snaked down the slope from the dark volcanic peak. (The peak's darker color indicates that the sulfur was molten there.)

Europa, the smallest of the Galilean moons, is named for another maiden whom Jupiter pursued. According to the legend, Jupiter disguised himself as a bull and carried her on his back across the Hellespont from Asia to Europe, thereby giving the continent its name. Despite this romantic myth, Europa looks rather like a cracked egg. Long, thin lines score its surface, as shown in figure 10.9A. The white material is probably a crust of ice, while the red material is probably mineral-rich water that oozed to the surface through the cracks and then froze. The absence of large craters on its surface suggests that some process has eradicated them, for Europa's surface—like that of Ganymede and Callisto—must surely have been cratered during and following its formation. To explain the lack of craters, astronomers think that Europa may be heated by Jupiter's gravitational forces deforming it, although not as strongly as Io is. That heat, in combination with a small amount from radioactive decay of rocky material in its core, may be sufficient to keep a layer of water melted beneath Europa's crust, forming an ocean (fig. 10.9B). Moreover, the heat may soften the surface ice, allowing it to "flow," glacier-like, and obliterate craters as they form. Close-up images made by the *Galileo* spacecraft show features

*Although eruptions of sulfur-rich material dominate, astronomers have observed silicate lava (molten rock) in a few eruptions.

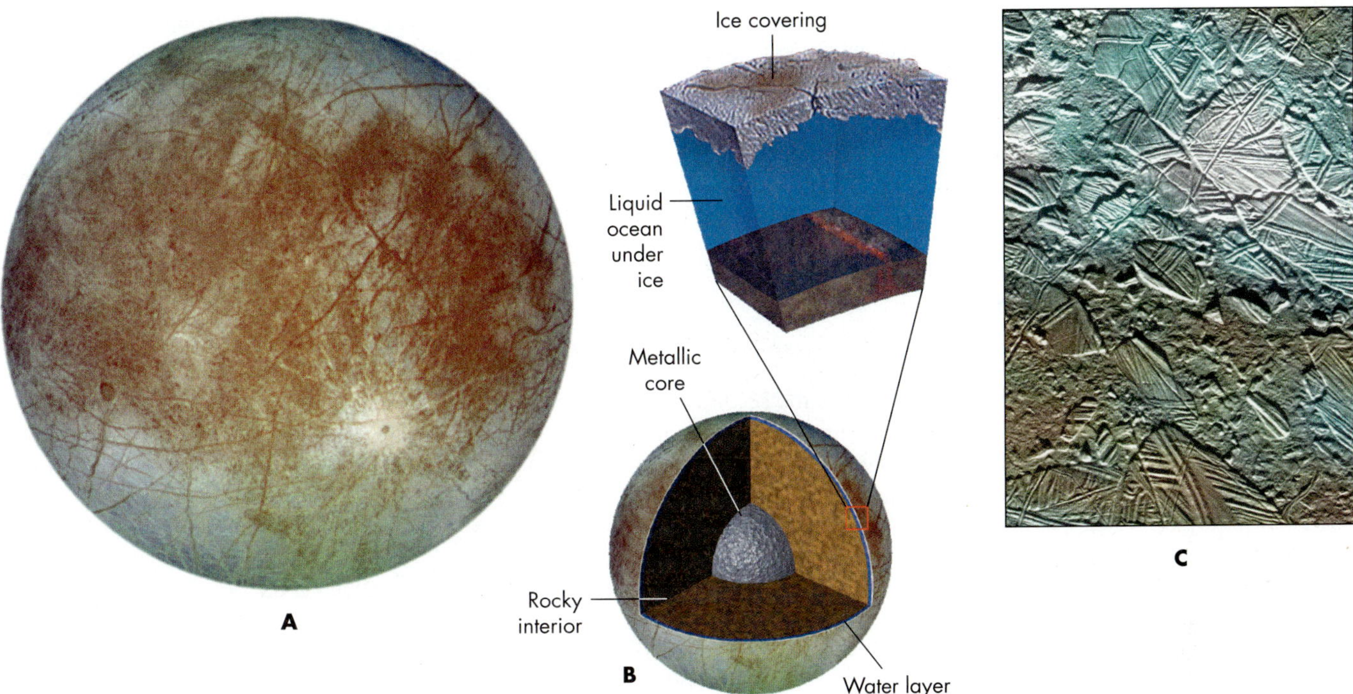

FIGURE 10.9
Pictures of Europa. (A) Europa, as imaged by the *Voyager* spacecraft. (B) A model of Europa's interior. Note possible liquid water ocean beneath ice. (C) Close-up picture of ice chunks on Europa's surface.

reminiscent of ice floes (fig. 10.9C). Some astronomers speculate that Europa's ocean might harbor life forms, and the possibility of a mission that could send a probe through the ice crust to explore the ocean beneath it is being examined.

Ganymede and Callisto, named for yet two more of Jupiter's loves, look somewhat like our own Moon, being basically grayish brown and covered with craters made during the late stages of their formation, much as our own Moon was pockmarked by infalling debris. But these similarities to our Moon are only superficial, because the surfaces of Ganymede and Callisto are probably mostly ice, the white of the craters being similar to the white you see on ice cubes when they are shattered. Recent observations suggest that Callisto, like Europa, may have a liquid water ocean below its icy crust. Ganymede's surface is less heavily cratered than Callisto's, implying that craters in the smoother areas have been destroyed, perhaps much as happened on our own Moon. That is, astronomers think that after Ganymede's surface solidified and was cratered by infalling debris, a few large bodies may have hit it. These large bodies created basins that subsequently flooded with water and then froze into a relatively smooth surface with few craters. However, the smooth regions do show curious parallel ridges, possibly created by tectonic forces as the water in them froze.

From the average density of the Galilean satellites, astronomers deduce that their interiors are composed of a mix of icy and rocky material. Heating of these moons during their youth may have allowed iron in them to sink to their centers and form cores, as happened with our own planet. Indirect evidence for just such a dense core in Ganymede came from the *Galileo* spacecraft, which orbited Jupiter.* As the craft swung

*When the *Galileo* probe ran low on the fuel with which to maneuver, NASA deliberately crashed it into Jupiter, where it burned up. This eliminated the chance of a crash into one of Jupiter's moons (especially Europa). That is important because some scientists think that life might have developed in the liquid water below Europa's icy crust, and they don't want to risk accidentally contaminating it or any of Jupiter's moons with terrestrial microorganisms.

past that moon, its trajectory was slightly altered by Ganymede's gravity—in a way that suggests that Ganymede has an ice mantle surrounding a rocky core. Moreover, sensors on *Galileo* detected a magnetic field around Ganymede. The existence of such a field is strong evidence that inside its rocky core Ganymede may have a small iron core.

Jupiter's other moons are much smaller than the Galilean satellites, and are heavily cratered. The orbits of the outermost of these moons are steeply tilted relative to the others, suggesting that these objects may be captured asteroids.

10.2 Saturn

Saturn is the second largest planet in our Solar System and lies about 10 AU from the Sun. Surrounded by its lovely rings (illustrated in fig. 10.10), Saturn bears the name of an ancient Roman harvest god. In later mythology, Saturn came to be identified with Cronus (also spelled Kronos), whom the ancient Greeks considered the father of the gods.

Saturn's Appearance and Physical Properties

Saturn's diameter, like Jupiter's, is about 10 times larger than the Earth's. Saturn's mass is about 95 times that of the Earth, and its average density is very small—only 0.7 grams per cubic centimeter, which is less than the density of water. Such a low density suggests that Saturn, like Jupiter, is composed mostly of hydrogen and hydrogen-rich compounds. Spectra of the planet bear this out, and astronomers think that Saturn is similar to Jupiter in its composition and internal structure, depicted in figure 10.11.

Saturn radiates more energy than it gains from the Sun, implying that, like Jupiter, it has an internal heat source. Based on the predicted conditions inside Saturn, astronomers think that a major source of Saturn's heat comes from deep beneath Saturn's cold clouds, where helium droplets condense in its atmosphere, much as water droplets condense in Earth's atmosphere. As the helium droplets fall toward Saturn's core, they release gravitational energy that heats the planet's interior.

If Jupiter and Saturn both have hot interiors and similar compositions, why do they look so different externally? In particular, why does Saturn show only faint cloud belts and markings, compared with the striking patterns seen on Jupiter? Saturn's greater distance from the Sun and its consequently lower temperature may provide an answer. Saturn's atmosphere is cold enough for ammonia gas to freeze into cloud particles that veil its atmosphere's deeper layers, making markings below the clouds indistinct.

FIGURE 10.10
Saturn as pictured by the *Cassini* spacecraft.

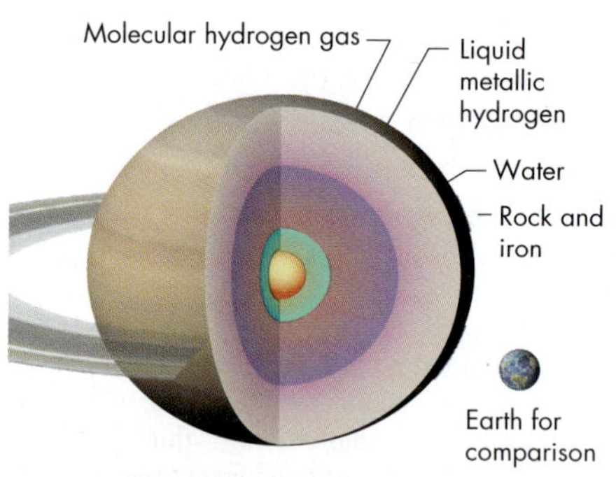

FIGURE 10.11
Internal structure of Saturn.

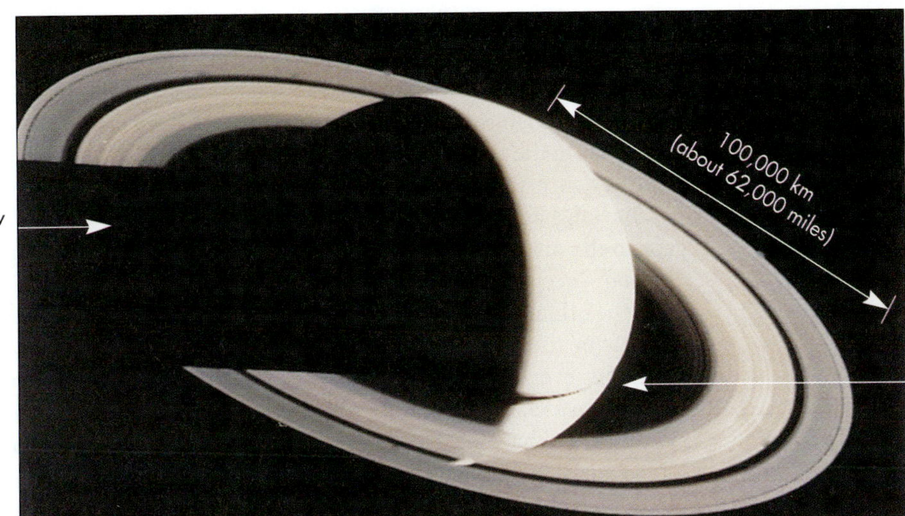

FIGURE 10.12
Rings of Saturn.

Saturn's Rings

Saturn's spectacular rings, illustrated in figure 10.12, were first seen by the astronomer Galileo. Through his small, primitive telescope, however, they looked like "handles" on each side of the planet, and it was not until 1659 that Christiaan Huygens, a Dutch scientist, observed that the rings were detached from Saturn and encircled it.

The rings are very wide but very thin. The main band extends from about 30,000 kilometers above the top of Saturn's atmosphere to a little more than twice the planet's radius (136,000 kilometers, or about 84,000 miles), as illustrated in figure 10.12. Some faint inner rings can be seen even closer to Saturn, and faint outer rings extend considerably farther from the planet. Yet despite the rings' immense breadth, they are probably less than a few hundred meters thick—thin enough to allow stars to be seen through them.

The British physicist James Clerk Maxwell, a pioneer in the study of electromagnetism, demonstrated that the rings must be a swarm of particles. He showed mathematically that no material could plausibly be strong enough to hold together in a solid sheet of such vast size. Spectra of the rings support Maxwell's theory: the inner and outer parts orbit Saturn at different velocities, obeying Kepler's third law, as shown by their Doppler shift. Thus, the rings must be a swarm of individual bodies.

Astronomers have since discovered that the ring particles are relatively small, only a few centimeters to a few meters across. Although these particles are far too small to be seen individually with telescopes, they reflect radar signals bounced off them. From the strength of the radar "echo," astronomers can estimate the particle sizes. More precise measurements were made with radio signals from the *Voyager* spacecraft. As *Voyager* transmitted data to Earth, the signals passed through the rings, scattering slightly from the particles. From the amount of scattering of the *Voyager* signal and the radar waves, astronomers can deduce not only the size of the ring particles but also their composition. Better information about the composition of the rings, however, comes from analyzing the spectrum of sunlight reflected from them. Such analyses show that the rings are composed primarily of water-ice. However, the *Voyager* and *Cassini* spacecraft pictures, such as figure 10.13, show that some

FIGURE 10.13
Image of the ringlets, showing the substructure of Saturn's rings.

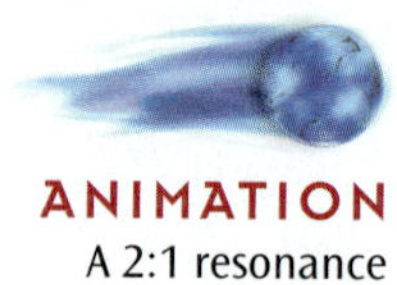

ANIMATION
A 2:1 resonance

parts of the rings are much darker than others, implying that the composition of the rings is not the same everywhere. Particles in the darker ring segments may be rich in carbon compounds similar to those found in some asteroidal material. Figure 10.13 also shows that the rings are not uniformly filled: they consist of numerous separate ringlets. Large gaps in the rings had been seen from Earth, but the many narrow gaps in the rings came as a surprise. What causes these gaps that create the multitude of ringlets?

As long ago as 1866, Daniel Kirkwood, the astronomer who also discovered gaps in the distribution of asteroids, noticed that the largest gap—known as Cassini's division—occurs where ring particles orbit Saturn in exactly one-third the time (one-third the orbital period) of its moon, Enceladus. Thus, any ring particle that attempted to orbit in the gap would undergo a strong and repeated gravitational force from Enceladus every third orbit. Kirkwood concluded that, over long periods of time, the cumulative effect of Enceladus's force would pull particles from the gap. He therefore hypothesized that Enceladus's gravitational attraction creates Cassini's division, just as Jupiter's creates the gaps in the asteroid belt. A few years later, Kirkwood revised his theory of the gap to account for the action of the four largest Saturnian moons. Today, astronomers believe that Saturn's moon Mimas, whose period is twice that of particles in Cassini's division, causes that large gap, but they think the many narrow gaps apparent in figure 10.13 have a different cause.

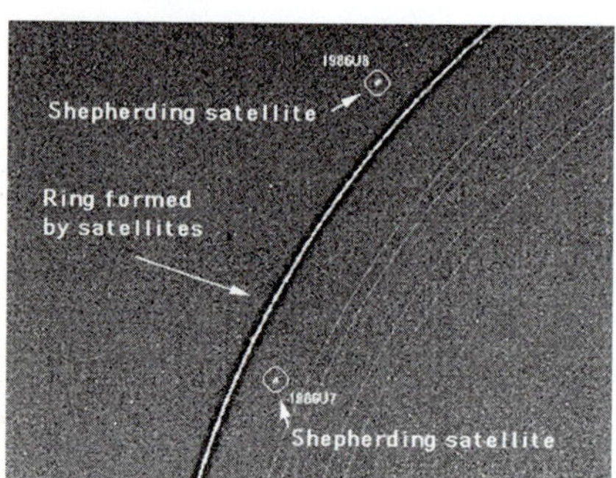

FIGURE 10.14
Two shepherding satellites (the dots inside the small circles) and a portion of the narrow ring around Uranus that they created. The picture's contrast is strongly enhanced to show these faint moons.

Narrow gaps in the rings probably arise from a complex interaction between the ring particles and the tiny moons that orbit within the rings. As these moonlets—only tens of kilometers or less in size—orbit Saturn, their gravitational attraction on the ring particles generates waves. These waves spread through the rings much like ripples in a cup of coffee that is lightly tapped. Such ripples are circular in a cup or on a still pond, but in a planetary ring system they take a different form. Because the inner part of the ring orbits faster than the outer part—a consequence of Kepler's laws—the spreading waves wrap into a tightly wound pattern called "spiral density waves." The crests of these density waves form the narrow rings.

Moonlets may generate gaps within the rings in yet another way. If two moonlets move along orbits that lie very close together, their combined gravitational force may deflect ring particles into a narrow stream between them (fig. 10.14). Such **"shepherding satellites"** occur not only in Saturn's rings but also in Uranus's. But what created planetary rings in the first place?

Origin of Planetary Rings

Earlier in the 1900s, some astronomers thought that planetary rings were material left over from a planet's formation, perhaps matter that had failed to condense into a satellite. But they now realize that rings are short-lived, because they are subject to forces in addition to gravity. For example, gas trapped in a planet's magnetic field may exert a frictional force on the ring particles, gradually causing them to spiral into the planet's atmosphere, as may happen to the material in Jupiter's ring. Thus, new material must be added to the rings from time to time, for without such replenishment they would disappear in a few million years. One source of new material is the satellites orbiting the planet. A moon in a satellite system as complex as those of the Jovian planets is subject not only to the gravitational forces of the planet but also to that of the other moons. The cumulative effect of such forces alters the satellites' orbits and may lead to collisions between them. Alternatively, a moon may drift so close to its planet that tidal forces break the moon apart, a theory first suggested more than a century ago.

The Roche Limit

In 1849, the French scientist M. E. Roche (pronounced *rohsh*), while studying the problem of a planet's gravitational effect on its moons, demonstrated mathematically that

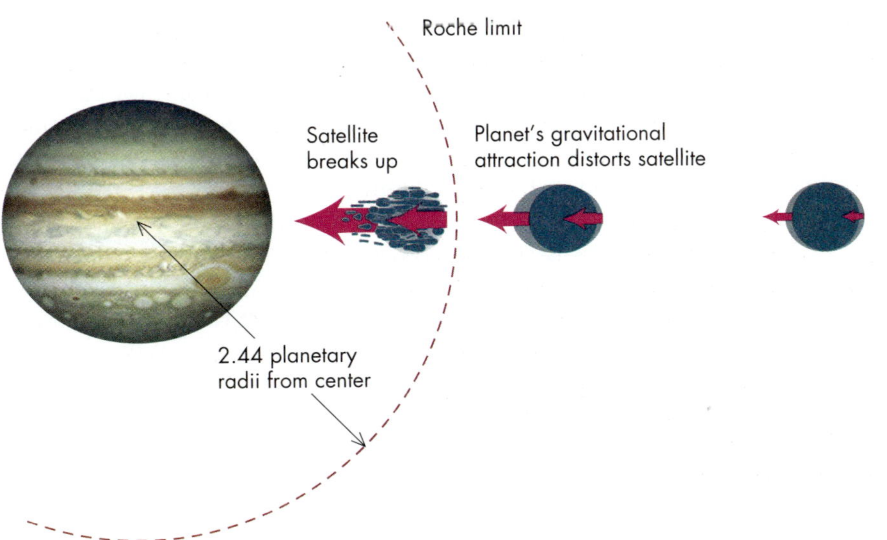

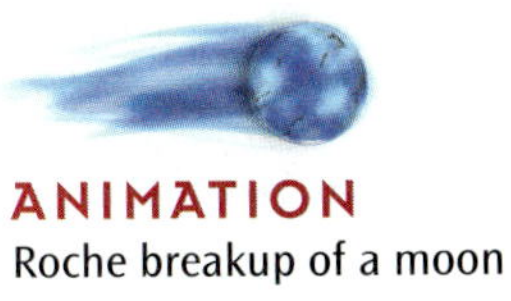

ANIMATION
Roche breakup of a moon

FIGURE 10.15
The Roche limit. A planet's gravity pulls more strongly on the near side of a satellite than on its far side, stretching it and, if strong enough, pulling it apart.

if a moon gets too close to its planet, the planet's gravity could rip the moon apart. This disruption occurs because the planet pulls harder on one side of the satellite than the other. If the difference in this pull exceeds the moon's own internal gravitational force, the moon will be pulled apart, as shown in figure 10.15. Thus, if a moon—or any body held together by gravity—approaches a planet too closely, the planet raises a tide so large it pulls the encroaching object to pieces. Roche calculated the distance at which the tide becomes fatally large and showed that for a moon and planet of the same density, breakup occurs if the moon comes nearer to its planet than 2.44 planetary radii, a distance now called the **Roche limit.** All planetary rings lie near their planet's Roche limit, suggesting that rings might be caused by satellite disruption. But moons are not the only bodies that can stray into the danger zone. Asteroids or comets may occasionally pass too close to a planet, and shattered fragments of such bodies (fig. 10.16) may help keep rings filled. The existence side-by-side of ringlets with different compositions (some rich in ice, others rich in carbon) is additional evidence that the rings formed by the breakup of many different small objects.

The Roche limit applies only to bodies held together by gravity, however. Artificial satellites or small bodies bonded together by chemical forces can pass safely through the Roche limit without effect.

Saturn's Moons

Saturn has one very large moon and 60 smaller ones that have been cataloged. The larger of these bodies, like Jupiter's moons, orbit in a flat "mini–Solar System" aligned with Saturn's equator. Saturn's moons have a smaller average density than the Galilean satellites of Jupiter, from which astronomers deduce that their interiors must be mostly ice. Moreover, all have about the same density, implying that they were not strongly heated by Saturn as they formed. (Recall that the Galilean satellites nearest Jupiter have a higher density than those farther away, presumably a result of Jupiter's heat driving away much of the icy material from the nearer bodies, leaving them with a higher proportion of rocky material.)

Titan, the largest Saturnian moon, has a diameter of about 5000 kilometers (3000 miles), making it slightly bigger in diameter than the planet Mercury and comparable in mass and radius to Jupiter's large moons Ganymede and Callisto. Because

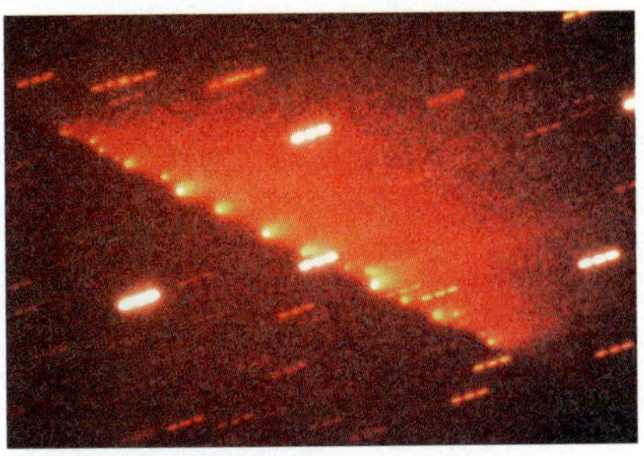

FIGURE 10.16
Comet Shoemaker-Levy 9. This image shows 20 or so of the fragments into which it was broken by Jupiter's tidal force when it passed close to Jupiter in 1992. These pieces struck Jupiter in 1994.

FIGURE 10.17
Titan, as seen from the *Cassini* spacecraft. This false-color image shows some surface features, glimpsed through the clouds of this mysterious moon.

Titan is farther from the Sun than these bodies, it is much colder. Thus, as gas molecules leak from its interior, they move relatively slowly and are unable to escape Titan's gravitational attraction. This immense moon therefore possesses its own atmosphere, which spectra show to be mostly nitrogen. Clouds in Titan's atmosphere hide its surface, but in 2005, during one of several flybys of Titan by NASA's *Cassini* probe, astronomers began mapping Titan's surface with radar and infrared cameras (fig. 10.17). At the same time, a small robot lander, carried by *Cassini* and built by the European Space Agency, parachuted into Titan's atmosphere. This probe, named *Huygens* in honor of Christiaan Huygens, descended through Titan's clouds and landed on the surface. As it drifted down to the surface, the probe took pictures of river networks (fig. 10.18A). On the surface it sent back pictures of a rock-strewn plain (fig. 10.18B), the "rocks" probably made of water ice.

Astronomers had shown on the basis of calculations that, because of Titan's extreme cold (94 K, or about −290°F), its clouds would be composed of methane (CH_4) and other hydrocarbons rather than water. Such models also predicted

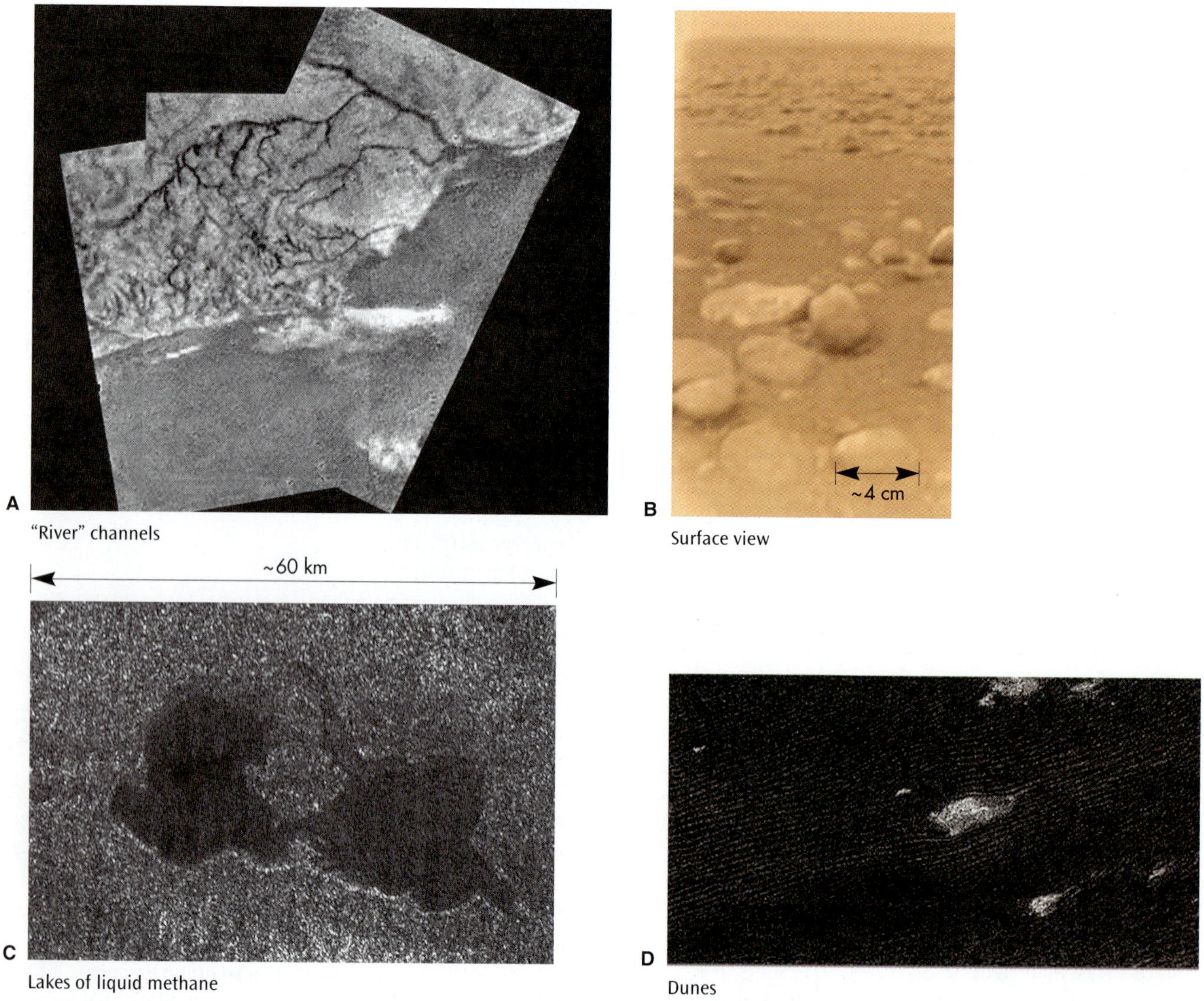

A "River" channels

B Surface view ~4 cm

~60 km

C Lakes of liquid methane

D Dunes

FIGURE 10.18
The landscape of Titan: (A) river networks carved by flowing liquid methane imaged by the *Huygens* lander from a height of 16 kilometers (10 miles); (B) "rocks" of ice on the surface; (C) radar image of lakes of liquid methane; (D) dunes.

FIGURE 10.19
Pictures of Saturn's seven largest moons shown to scale. Note the thin haze layer in Titan's upper atmosphere.

Q. Based on their surface features, which of these moons might be geologically active? Why?

that liquid methane or ethane (C_2H_6) would "rain" from Titan's clouds and might even form oceans on this frigid moon. No oceans have been found, but radar maps of Titan's polar regions (fig. 10.18C) clearly show lakes, which appear to be liquidmethane and ethane. In addition to the lakes and channels, other radar images show rows of dunes (fig. 10.18D), presumably built and aligned by winds, and at least one "ice" volcano. But these seemingly familiar surface features differ in composition from similar looking features here on Earth. On Titan, the bitter cold makes ordinary water-ice as hard as rock, so Titan's dunes may be ice crystals (not particles of sand) and its river channels may have been cut by methane rain falling and eroding water-ice (not rock).

Figure 10.19 shows pictures of Saturn's seven largest moons taken by the *Cassini* spacecraft. Notice that most of these bodies are heavily cratered, implying that they have been extensively bombarded by infalling bodies. The smoother surface seen, for example, on Enceladus is the result of water that has erupted from its interior, flooding old craters and drowning them as it freezes. The *Cassini* spacecraft has observed these eruptions, which appear to be powered by tidal flexing, the same mechanism that causes Jupiter's satellite Io to be so active.

Many of Saturn's satellites have bright and dark streaks. Iapetus is particularly extreme, with the side of the moon that faces forward as it orbits Saturn covered with an extremely dark black material (fig. 10.19). Scientists are unsure of the origin of this material, but it may have come from eruptions on some satellites or collisions that scattered fine debris in space. This material was then swept up by the moons, coating parts of their surfaces with thin layers of dark dust or fine icy particles.

10.3 Uranus

Uranus, although small compared with Jupiter and Saturn, is much larger than the Earth. Its diameter is about 4 times that of the Earth, and its mass is about 15 Earth masses. Lying approximately 19 AU from the Sun (more than twice Saturn's distance), Uranus is difficult to study from Earth, visible only as a blue but featureless disk. Even pictures of it taken by the *Voyager* spacecraft show few details (fig. 10.20), although computer processing of the images shows that it has faint cloud bands.

FIGURE 10.20
Uranus as pictured from the *Voyager* spacecraft. Because of Uranus's odd tilt, we are viewing it nearly pole-on. Note the lack of clearly defined cloud belts.

Uranus was unknown to the ancients, even though it is just visible to the naked eye. It was discovered by Sir William Herschel, a German émigré to England. Herschel, a musician, was at the time only an amateur astronomer interested in hunting comets, a task in which he collaborated with his sister Caroline. In 1781, he observed a pale blue object whose position in the sky changed from night to night. Herschel at first thought he had discovered a comet, but observations over several months showed that the body's orbit was nearly circular, and he therefore concluded that he had found a new planet. For this discovery, King George III named Herschel his personal astronomer, and to honor the king, Uranus was briefly known as "Georgium Sidus," or "George's Star." In ancient Greek mythology, Uranus was the father of Cronus and was identified as the God of the Heavens.

Uranus's Structure

Uranus, like Jupiter and Saturn, is rich in hydrogen and its compounds water and methane. We know this from spectra of its atmosphere, which show very strong absorption lines of methane. In fact, it is methane that gives the planet its deep blue color. When sunlight falls on Uranus's atmosphere, the methane gas strongly absorbs the red light. The remaining light, now blue, scatters from cloud particles in the Uranian atmosphere and is reflected into space, as depicted in figure 10.21. The cloud particles that cause the scattering are thought to be primarily crystals of frozen methane. Such crystals can form in Uranus's atmosphere because, being so far from the Sun, it is extremely cold.

Astronomers rely on indirect methods to study the interior of Uranus, using, for example, its density and shape. From its mass and radius, astronomers can calculate that Uranus has an average density of about 1.2 grams per cubic centimeter. This density is nearly twice that of Saturn and almost as large as Jupiter's. But a planet's density depends on both the

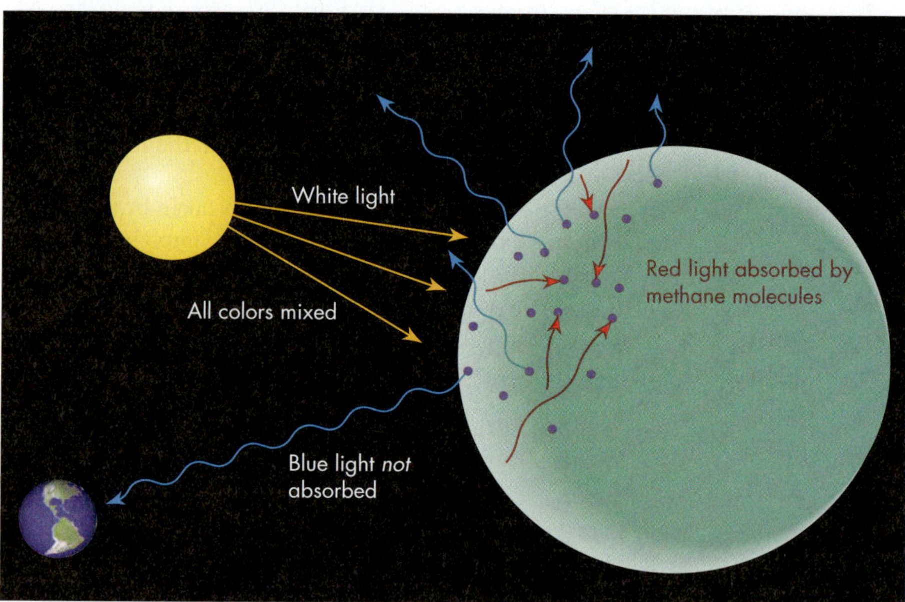

FIGURE 10.21
Sketch illustrating why Uranus is blue. Methane absorbs red light, removing the red wavelengths from the sunlight that falls on the planet. The surviving light—now missing its red colors—is therefore predominantly blue. As that light scatters off cloud particles in the Uranian atmosphere and returns to space, it gives the planet its blue color.

planet's composition and the amount by which its gravity compresses it. Given that some part of the density of Jupiter and Saturn is simply a result of their greater mass, astronomers deduce that Uranus must contain proportionally fewer light elements, such as hydrogen, than those more massive worlds. On the other hand, the density is too low for Uranus to contain much rock or iron material. Astronomers therefore believe it must be composed of material that is light and abundant, such as ordinary water mixed with methane and ammonia. This mix satisfactorily explains both the density and the spectrum of Uranus.

Confirmation of the abundance of water also comes from studies of the planet's shape. Uranus, like Jupiter and Saturn, rotates moderately fast, with its equator rotating faster than its poles. At its equator, Uranus spins once every 17 hours, bulging the planet's equator. The size of such a bulge depends in part on the planet's gravitational attraction and therefore on how the mass generating that attraction is distributed inside the planet. Thus, astronomers can deduce the density and composition deep inside Uranus from the size of its bulge. Such studies are consistent with the hypothesis that Uranus is composed of mostly water and hydrogen-rich gases and that it may have a core of rock and iron-rich material, as illustrated in figure 10.22. It is not known whether the core formed first and then attracted the lighter gases and ices that condensed around it, or whether the core formed by heavy material sinking to the center after the planet formed. In fact, some astronomers think Uranus's core may be simply highly compressed ices with little rocky material. It appears that Uranus did not attract as large an amount of hydrogen and helium gas as Jupiter and Saturn when it formed. Its composition can be explained if it formed primarily from planetesimals rich in ices of water, methane, and ammonia. Astronomers consequently often describe Uranus (and Neptune as well) as "ice giants," although it should be stressed that there is no ice inside them today.

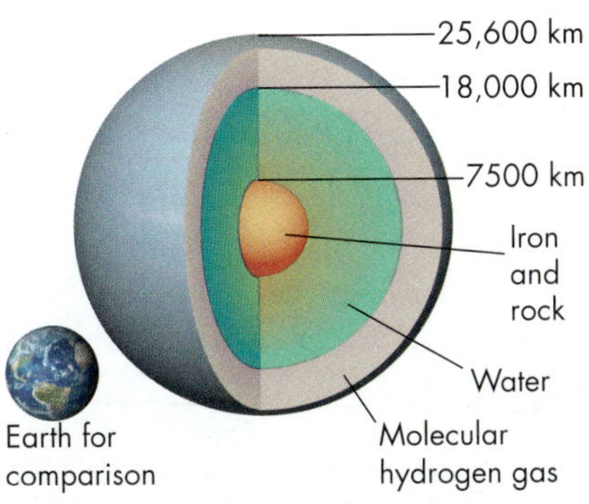

FIGURE 10.22
Artist's view of a recently suggested model for the interior of Uranus. Note how different it is from Jupiter and Saturn, which both contain large regions of liquid and metallic hydrogen.

Uranus's Odd Tilt

Uranus's rotation axis is tipped so that its equator is nearly perpendicular to its orbit. That is, it spins nearly on its side, as illustrated in figure 10.23. Moreover, the orbits of Uranus's moons are similarly tilted. They orbit Uranus in its equatorial plane, and, as a result, their orbits are also tilted at approximately 90° with respect to the planet's orbit. Some astronomers therefore hypothesize that during its formation, Uranus was struck by an enormous planetesimal whose impact tilted the planet and splashed out material to create its family of moons. Other astronomers think Uranus was tilted by gravitational tugs exerted on it by neighboring planets (especially Saturn).

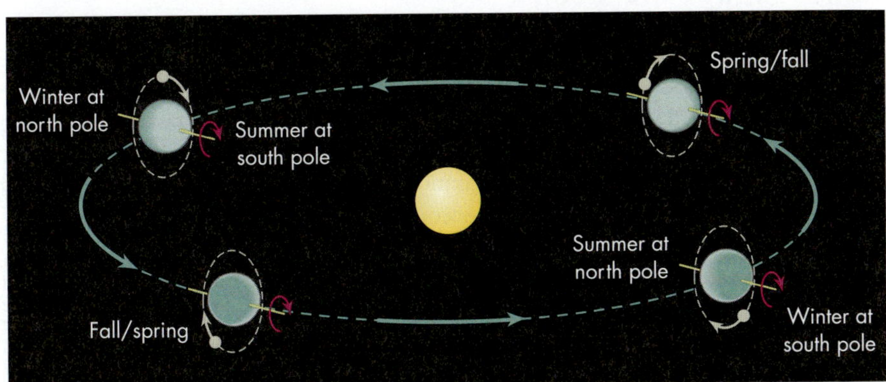

FIGURE 10.23
Sketch of the odd tilt of Uranus and its satellites, perhaps caused by a huge impact early in the planet's history. Because of this tilt, when Uranus's north pole points toward the Sun (its northern summer), the Sun will be above the horizon for many Earth years. Likewise, the other pole will be in night for a corresponding period. On the other hand, during the Uranian spring and fall, the Sun rises and sets approximately every 17 hours. Size of bodies and orbits are not to scale.

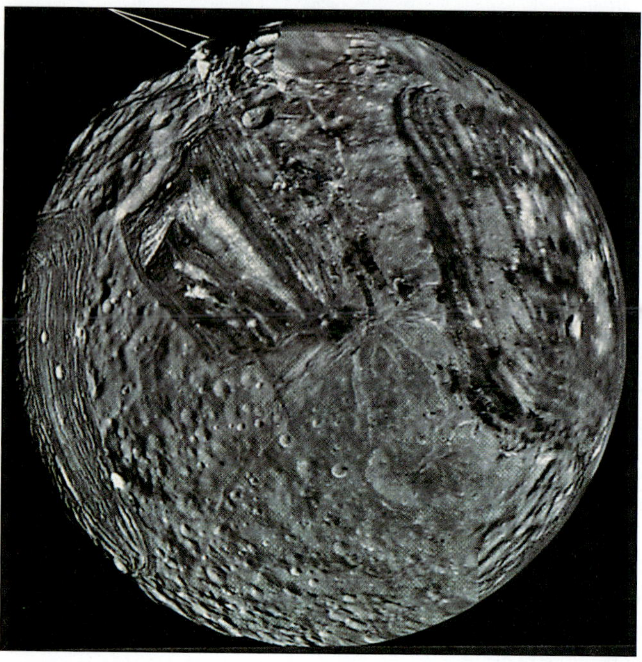

FIGURE 10.24
Miranda, an extremely puzzling moon, as observed by the *Voyager* spacecraft. Note the enormous cliffs glinting in the sunlight at the top of the picture.

Regardless of what caused it, the strong tilt of its axis gives Uranus an odd pattern of day and night. For part of its orbit, one pole is in "perpetual" day and the other pole is in "perpetual" night. Thus, sunlight heats the planet very unevenly, perhaps explaining why Uranus lacks the cloud bands seen on the other giant, gaseous planets. The lack of cloud bands may be temporary, however. Recent observations with an infrared telescope on Earth reveal an odd dark marking in Uranus's atmosphere, perhaps a feature that fades and reappears.

> The five large Uranian moons are named for characters in English literature. Titania and Oberon are the queen and king of the fairies in Shakespeare's *A Midsummer Night's Dream*. Ariel and Miranda are characters in Shakespeare's *The Tempest*. Umbriel is a sprite in Alexander Pope's poem "The Rape of the Lock," in which Ariel also appears. The smaller Uranian moons are also named for characters from Shakespeare, such as Puck and Cordelia.

ANIMATION
Miranda

Uranus's Rings and Moons

Uranus is encircled by a set of narrow rings, as illustrated in figure 10.6. The rings, like those of Saturn, are composed of a myriad of small particles, perhaps a meter or so in diameter, moving in individual orbits. The Uranian rings are very dark, however, implying that they are not made of, or coated with, ice like the bright rings of Saturn; instead, they may be rich in carbon particles or organic molecules. Uranus's rings are also extremely narrow compared with those of Saturn. According to one theory, the ring particles are held in such narrow zones by "shepherding satellites," as discussed for Saturn's rings.

Uranus has five fairly large moons and about 20 smaller ones. Like the moons of Jupiter and Saturn, they form a regular system and are probably composed mainly of ice and rock. Many of the Uranian moons are heavily cratered, but Miranda, the smallest of the five large moons, has a surface totally unlike that of any other Solar System body (fig. 10.24). The surface is broken into distinct areas that seem to bear no relation to one another. One region is wrinkled, while an adjacent region has small hills and craters, rather like our Moon. Miranda's patchwork appearance leads some astronomers to think it may have been shattered by impact with another large body. The pieces were subsequently drawn back together by their mutual gravity, giving this peculiar moon its jumbled appearance. Alternatively, the curious surface might have been caused by rising and sinking motions driven by heat in Miranda's interior. Regardless of the cause, Miranda has some extremely curious and unexplained surface features, such as a set of cliffs, visible in the top corner of figure 10.24, that are twice the height of Mount Everest.

10.4 Neptune

Neptune is the outermost of the large planets and is very similar to Uranus in size, with a diameter about 3.9 times that of the Earth and a mass about 17 times the Earth's. Through a telescope on the Earth, it looks like a chip of sapphire, a lovely blue color, but because of its great distance from the Sun—about 30 AU—it is difficult to study. Pictures taken of it by passing spacecraft show it to be a deep blue world, with markings reminiscent of Jupiter's, as shown in figure 10.25. Cloud bands encircle it, and it even has briefly shown a Great Dark Spot, a huge dark blue atmospheric vortex that looks a little like Jupiter's Great Red Spot but is blue instead of red.

Neptune, named for the Roman god of the sea, was discovered in the 1840s from predictions made independently by a young English astronomer, John Couch Adams, and a French astronomer, Urbain Leverrier. Adams and Leverrier both noticed that Uranus was not precisely following its predicted orbit, and they therefore inferred that its motion was being disturbed by the gravitational force of an as yet unknown planet. From the size of these orbital disturbances, Adams and Leverrier predicted where the unseen body must lie.

Adams completed his calculations in 1845, but when he reported his results, the astronomer royal, Sir George Airy, was unconvinced and gave a low priority to the search for the unseen planet. In 1846, however, Airy was startled to read a paper by Leverrier detailing calculations nearly identical to those made by Adams. This spurred Airy to begin a search in earnest, but by then it was too late: Leverrier had given his predicted positions to Johann Galle, a German astronomer who that same night pointed his telescope to the predicted location and saw Neptune. Assignment of credit for the discovery of the new planet led to a rancorous dispute tinged with national pride that lasted decades. The discovery is now credited equally to Adams and Leverrier. Ironically, Galileo had seen Neptune in 1613 while observing Jupiter's moons. His observation notes record a dim object whose position changed with respect to the stars, as would be expected for a planet. Galileo failed, however, to appreciate the significance of that motion, so Neptune eluded discovery for another two centuries.

Q. Approximately how big is this spot, compared to the Earth?

FIGURE 10.25
Image of Neptune taken by the *Voyager* spacecraft. Notice the dark blue oval, an atmospheric vortex whose origin is probably similar to that of Jupiter's Great Red Spot.

Neptune's Structure

Neptune's structure is probably similar to Uranus's. That is, the planet probably is composed mostly of ordinary water surrounded by a thin atmosphere rich in hydrogen and hydrogen compounds, such as methane. Our understanding of the atmosphere's composition is derived from the planet's spectrum. The composition of the interior is inferred from the planet's low average density, which is 1.67 grams per cubic centimeter. As we have seen for the other giant planets, such a low density implies that the planet must be composed mostly of light atoms. But during its formation, Neptune must also have collected heavy elements such as silicon and iron. These denser materials probably sank to the center of Neptune (just as they did in the other giant planets), where they now form a core of rock and iron a little smaller than the Earth's.

Neptune's Atmosphere

ANIMATION
The rotation of Neptune

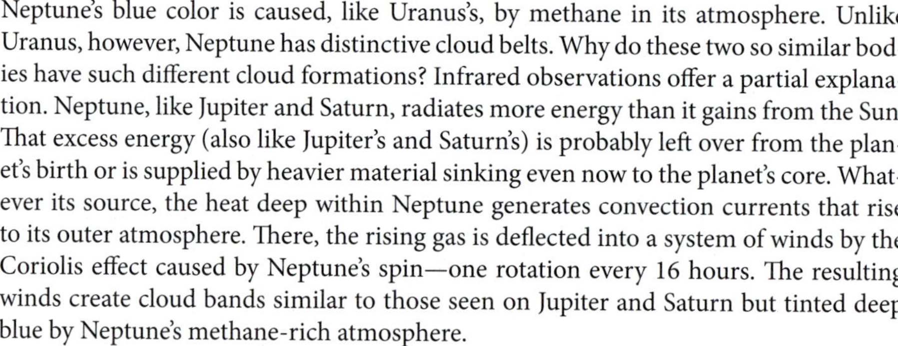

Neptune's blue color is caused, like Uranus's, by methane in its atmosphere. Unlike Uranus, however, Neptune has distinctive cloud belts. Why do these two so similar bodies have such different cloud formations? Infrared observations offer a partial explanation. Neptune, like Jupiter and Saturn, radiates more energy than it gains from the Sun. That excess energy (also like Jupiter's and Saturn's) is probably left over from the planet's birth or is supplied by heavier material sinking even now to the planet's core. Whatever its source, the heat deep within Neptune generates convection currents that rise to its outer atmosphere. There, the rising gas is deflected into a system of winds by the Coriolis effect caused by Neptune's spin—one rotation every 16 hours. The resulting winds create cloud bands similar to those seen on Jupiter and Saturn but tinted deep blue by Neptune's methane-rich atmosphere.

Neptune's winds are extremely fast. For example, near its equator, strong easterly winds blow opposite to the planet's direction of rotation, much like the trade winds on Earth. But these Neptunian winds reach speeds of nearly 2200 kilometers per hour (about 1300 miles per hour). As these gales sweep around Neptune, gas sandwiched between adjacent streams spins, as described in our discussion of Jupiter's atmosphere. Such localized spinning motion creates the Great Dark Spot, as the large, dark blue spot illustrated in figure 10.25 is called. Curiously, pictures of Neptune taken by the Hubble Space Telescope in the mid-1990s showed that the spot is gone. Perhaps such spots grow and dissipate like terrestrial storms. Support for that possibility comes from recent Hubble Space Telescope observations of Uranus, which show that its previously bland appearing atmosphere now has a small dark spot.

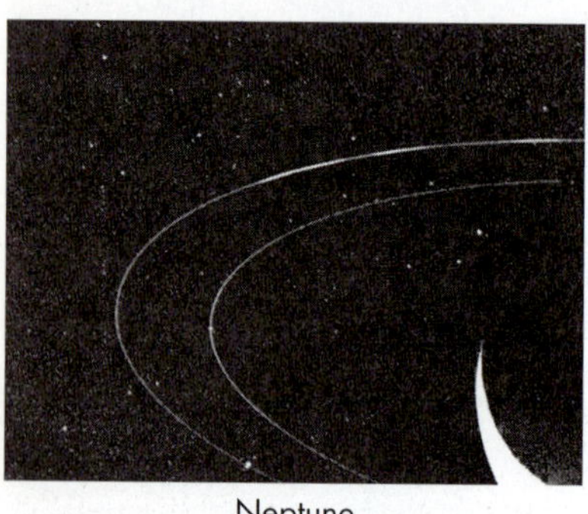

FIGURE 10.26
Ring arcs around Neptune. This *Voyager* image shows dense regions along one of Neptune's rings, probably caused by gravitational interactions with one of its moons.

Neptune's Rings and Moons

Neptune, like the other giant planets, has rings, but they are very narrow, more like those of Uranus than those of Saturn, as shown in figure 10.6. They are probably composed of debris from small satellites or comets that have collided and broken up, and they contain proportionally more dust than the rings of either Saturn or Uranus. As a result, they are difficult to see from Earth, but sunlight scattered by the small dust particles is more easily visible from behind the planet. Moreover, photographs of the rings show that in some places, the ring particles are not distributed uniformly around the ring but are gathered into arcs (fig. 10.26). This clumping may be the result of gravitational interactions with some of the satellites with eccentric orbits.

Neptune has six small moons orbiting close to the planet in roughly circular orbits and seven other moons at much greater distances. One of these, Triton, is nearly as big as Jupiter's moon Europa, but it is highly unusual among large satellites in that it orbits "backward" (counter to Neptune's rotation) in a clockwise direction. This orbital

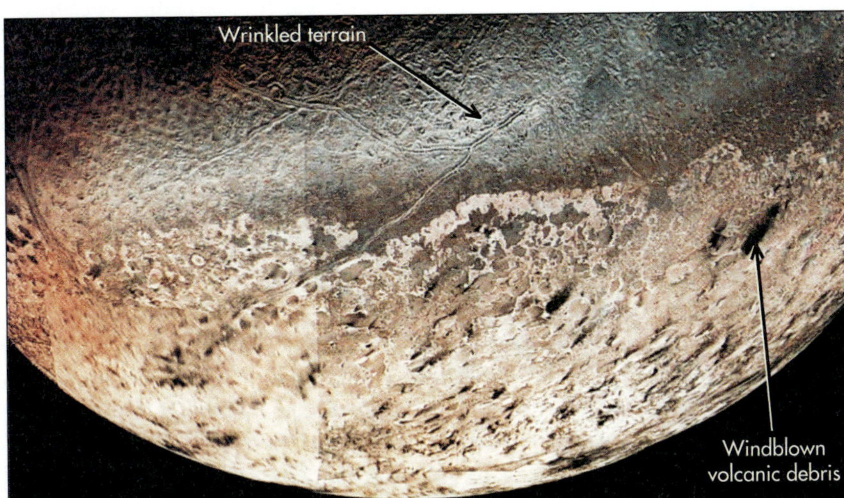

FIGURE 10.27
The surface of Triton.

peculiarity leads many astronomers to think that Triton is a surviving icy planetesimal from the inner Kuiper belt. When Neptune captured it, the encounter destroyed or expelled any midsize moons that Neptune originally possessed. In fact, Neptune's outermost moon, Nereid, follows such a highly elliptical path far beyond the other moons that it just barely missed escaping from its parent world.

Neptune's moons are named for mythological sea deities, in keeping with the planet's own name.

Triton intrigues astronomers for more than its orbital oddity. It is massive enough that its gravity, in combination with its low temperature, allows it to retain gases around it. Triton is thus one of the few moons in the Solar System with an atmosphere.

Triton's surface also has many unusual features, as shown in figure 10.27. Wrinkles give much of the surface a texture that looks like a cantaloupe. Craters pock the surface elsewhere, and dark streaks extend from some of them. At least one of these streaks originates from a geyser caught in eruption by the passing *Voyager II* spacecraft. Astronomers think that the matter ejected by the geysers is a mixture of nitrogen, ice, and carbon compounds. Sunlight, pale though it is at Triton's immense distance from the Sun, may warm such matter trapped below the surface and make it expand and burst through surface cracks. The erupted material cools and condenses in Triton's cold, thin atmosphere, where winds carry and deposit it as a black "soot."

10.5 Pluto and Beyond

Beyond Neptune no more major planets are known, but there are many smaller objects. Many of these **Trans-Neptunian Objects, or TNOs,** orbit the Sun in the Kuiper belt. Relatively little is known about them since none has been visited by spacecraft, although Neptune's moon Triton may be a captured TNO. The TNOs are so far from the Sun that they are difficult to detect from Earth. Most have been identified only in the last decade using large telescopes, though one was found about 80 years ago.

Pluto was discovered in 1930 by Clyde Tombaugh, an astronomer at Lowell Observatory in Flagstaff, Arizona. Tombaugh painstakingly examined pairs of photographs of the sky for over a year—scanning millions of star images—searching for objects whose position changed between the exposures, the telltale motion that distinguishes a planet from a star. Pluto is named for the Greek and Roman god of the underworld, and for more than 75 years it was considered a planet, although almost from the time of its discovery it was clear that it differed in many ways from the previously known planets. Pluto's great distance from us and the Sun, combined with its small size, make it very dim and thus difficult to study. Even in the largest telescopes on the ground, Pluto looks like a dim star, as illustrated in figure 10.28.

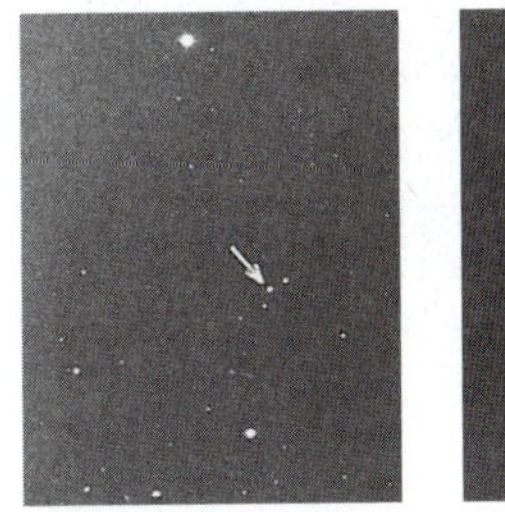

FIGURE 10.28
Pluto looks like merely another dim star in this photograph taken at Lick Observatory in California. Only its change in position from night to night shows it to be a planet.

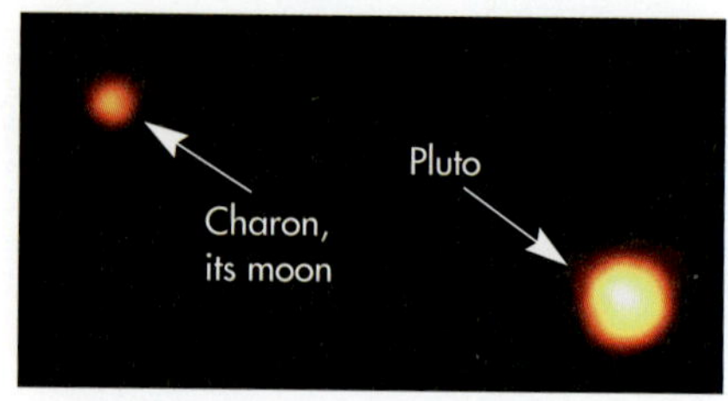

FIGURE 10.29
A picture of Pluto and its moon Charon obtained with the Hubble Space Telescope and therefore free of the blurring effects of our atmosphere.

Recently astronomers have discovered that Pluto has two other moons in addition to Charon. These moons, called Nix and Hydra, are both very small (40 to 160 km in diameter), and little is yet known of them except that they orbit about two to three times farther from Pluto than Charon orbits.

For years after its discovery, astronomers knew only that Pluto was smaller than the Earth. Then in 1978, James Christy of the U.S. Naval Observatory discovered a moon orbiting Pluto, which made it possible for astronomers to measure Pluto's radius and mass. Pluto's moon, Charon, is named for the boatman who, in mythology, ferries dead souls across the river Styx to the underworld. Circling Pluto in a tiny orbit whose average radius is 19,600 kilometers, Charon takes only 6.4 days to complete a trip around the planet. From this orbital data, Pluto's mass* can be calculated using Newton's modified form of Kepler's third law of planetary motion (see fig. 8.8). Such a calculation shows that Pluto's mass is about 0.002 times the Earth's, or less than $\frac{1}{20}$th the mass of Mercury, the smallest planet.

Observations of Charon's motion around Pluto show that its orbit is tilted steeply with respect to Pluto's orbital plane and that Pluto's rotation axis is similarly tipped, reminiscent of Uranus's odd tilt. As Charon orbits Pluto, it occasionally† eclipses the planet from our view. From these eclipses, the diameters of both Pluto and Charon can be measured. Pluto's diameter turns out to be a little less than one-fifth the Earth's, and Charon, though smaller, turns out to be surprisingly large. Satellites are generally dwarfed by their planets, but Charon's diameter is slightly more than half Pluto's. Moreover, its orbit is locked in synchrony with Pluto's rotation, suggesting a tight gravitational coupling between the bodies. Such eclipse studies are not the only way to study Pluto and Charon, however. Observations of them made with interferometers (see chapter 5) and from the Hubble Space Telescope show Pluto's and Charon's tiny disks, nicely confirming the eclipse data (fig. 10.29).

From Pluto's mass and radius, astronomers can deduce its density to be about 2.1 grams per cubic centimeter, a value suggesting that Pluto must be mainly a mix of rock, ordinary water-ice, and frozen nitrogen and methane. From spectra of its surface, they can directly determine that Pluto is covered with a mixture of water-ice and ices of methane, nitrogen, and carbon monoxide. Despite this knowledge of the dwarf planet's composition, however, astronomers still lack information about its surface features. Although astronomers have made pictures of Pluto and its moon, Charon, with the Hubble Space Telescope, the best "images" of this distant system come from computer analysis of its changing brightness during eclipses by Charon. These images (fig. 10.30) show that Pluto's south pole is much brighter than its equator, implying the presence of a polar cap there. Spectra suggest that the cap is frozen methane.

*Technically, this method gives the *combined* mass of Pluto and Charon.
† Because of Charon's orbital tilt, Pluto-Charon eclipses occur about every 6 days for a few years but then do not recur for half a Plutonian year, or about 124 Earth years!

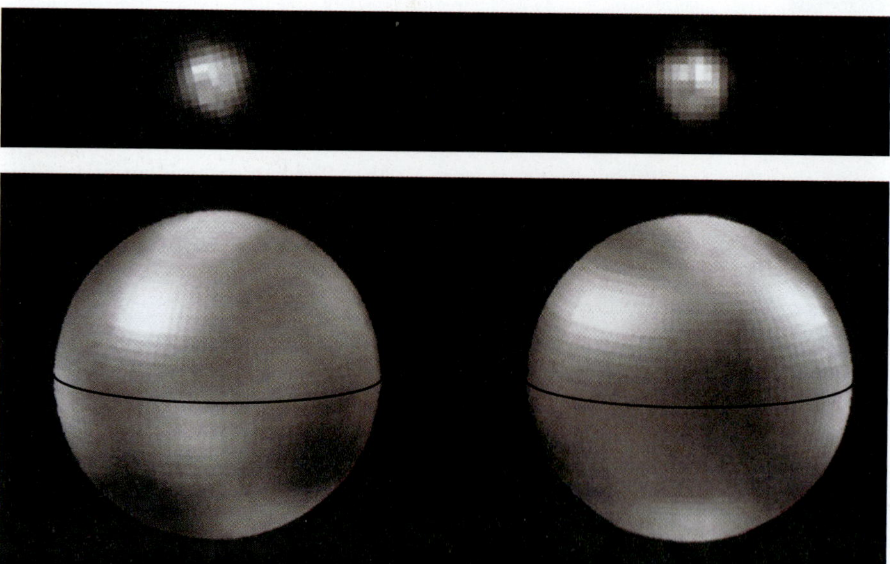

FIGURE 10.30
(Top) Pictures of Pluto made with the Hubble Space Telescope.
(Bottom) Computer-generated maps of Pluto based on such pictures.

Astronomers have also detected a very tenuous atmosphere on Pluto. From measurements of the temperature and surface composition, they deduce that its atmosphere is mostly nitrogen and carbon monoxide, with traces of methane. In the bitter cold (40 K, or about −387°F) of this remote world, molecules move so slowly that, despite its tiny mass, Pluto's gravity is strong enough to retain a thin atmosphere. In 2007, NASA launched the *New Horizons* spacecraft, which will reach Pluto in 2015 and give us our first detailed pictures of this icy world.

Pluto's small size and peculiar orbit (it crosses Neptune's orbit, as shown in fig. 10.31) once led some astronomers to hypothesize that it was originally a satellite of Neptune that escaped and now orbits the Sun independently. Today, however, astronomers think almost the reverse—that Neptune has "captured" Pluto. Pluto's orbital period is 247.7 years, very close to one and one-half times Neptune's. Thus, Pluto makes two orbits around the Sun for every three made by Neptune. This match of orbital periods has created a cumulative gravitational attraction on Pluto that "tugged" it into its current orbit.

In fact, several hundred other objects a few hundred kilometers in diameter orbit at nearly the same distance from the Sun as Pluto. For example, Orcus, discovered in 2004, is about the size of Pluto's moon Charon and also orbits the Sun twice for every three orbits of Neptune. Even larger icy objects orbit farther out—presumably surviving icy planetesimals—but are more difficult to detect because of their large distance from the Sun. Some of the larger TNOs are illustrated in figure 10.32. The illustrations shown are mostly sketches, since astronomers can only estimate sizes and colors from the dim light we see. Astronomers think there are dozens more objects at least 1000 kilometers (600 miles) in diameter. Sedna is of particular interest to astronomers because its orbit extends well beyond the standard Kuiper belt, and it has a strange red color that is not well understood. It currently happens to be relatively close to the Sun along its elliptical orbit, and it would not have been detectable through most of its orbit.

In 2005 astronomers reported the discovery of a dim object orbiting at about 97 AU from the Sun. Now officially named Eris (for the Greek goddess of discord), this remote world turns out to have a diameter of about 2400 kilometers (about 1500 miles), making it slightly larger than Pluto. Its orbit is both highly elliptical (ranging from

Q. If Pluto's orbit crosses Neptune's, why don't these planets collide?

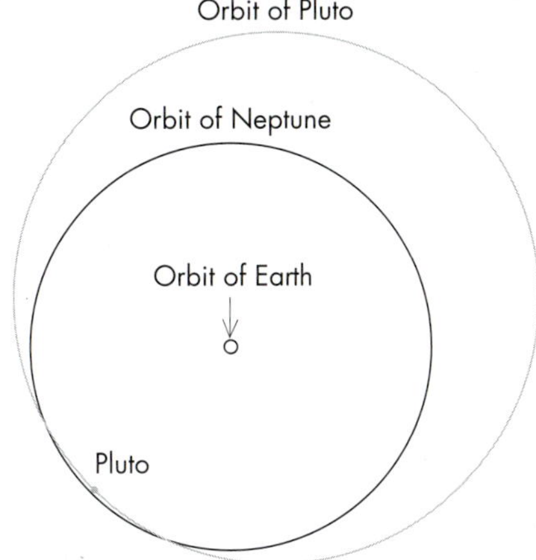

FIGURE 10.31
Pluto's odd orbit is highly tilted and highly eccentric (elongated)—so much so that it crosses Neptune's orbit.

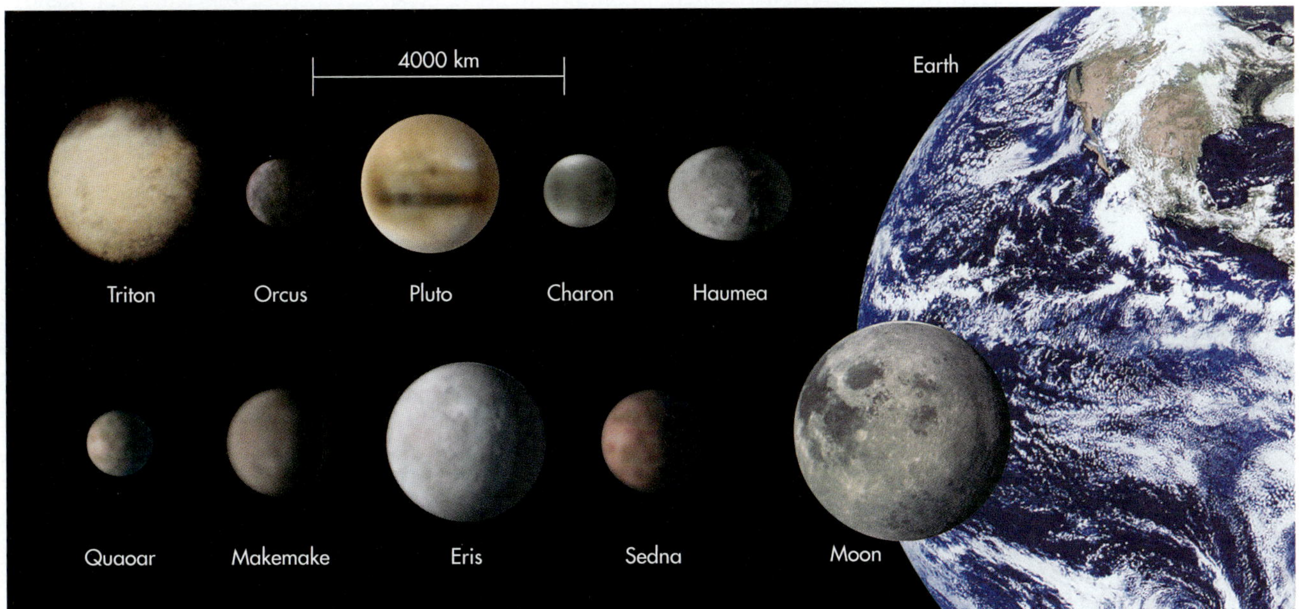

FIGURE 10.32
Some of the larger Trans-Neptunian Objects shown to the same relative scale, along with the Moon and Earth. Most of these are artist's representations, because these icy worlds are unresolved specks in even in the largest telescopes.

38 to 97 AU from the Sun) and highly inclined (about 44°) to the main disk of the Solar System. Because of its size, some astronomers initially described it as a tenth planet, despite its orbit being so different from that of the other planets. Eris is like Pluto in many ways besides its size. Its surface is covered with frozen methane, and it has a moon approximately the same size as Pluto's moon Charon.

With the discovery of an object larger and more massive than Pluto, and faced with the prospect of dozens of more objects that were not significantly different from Pluto, astronomers decided to define a new category for Pluto and these other moderately large icy bodies. In 2006 the International Astronomical Union came up with the designation *dwarf planets* to describe bodies like Pluto and Eris, which are large enough that their gravity has pulled them into a roughly spherical shape, but which do not dominate all other masses orbiting at the same distance from the Sun. They also created the designation *plutoid* for dwarf planets orbiting beyond Neptune. Two more plutoids, Haumea and Makemake (fig. 10.32), have been identified recently, but there will probably be many more. Terminology is not the only issue posed by the TNOs, though. Many of them are too large to have formed in the low-density regions of the outer Solar System where we see them now. Thus, astronomers hypothesize that the TNOs formed nearer the Sun, perhaps near Neptune, and have been ejected into their present orbits by successive gravitational tugs exerted by the massive outer planets.

SUMMARY

The outer planets formed in the solar nebula far from the Sun, where it was cold enough for them to accumulate hydrogen-rich material. Because of hydrogen's abundance relative to other elements, Jupiter, Saturn, Uranus, and Neptune grew much larger than the terrestrial planets in both diameter and mass (see table 10.1) and are composed mostly of hydrogen and its compounds, such as methane (CH_4), ammonia (NH_3), and water (H_2O). These gases form a deep atmosphere, often richly colored, that grows denser and hotter with depth and eventually is compressed into liquid form. The interiors of Jupiter and Saturn are composed mainly of hydrogen—liquid just below the atmosphere and metallic deeper down. Each is believed to have a rocky core whose diameter—within a factor of 2 or 3—is comparable to the Earth's. Uranus and Neptune have methane-rich atmospheres, but they are probably mostly water, ammonia, and methane surrounding a rocky core. Because they are smaller bodies, their weaker gravity has not allowed them to capture or retain as much hydrogen and helium, but they have retained the heavier gases.

The giant planets are probably heated by continued gravitational contraction and settling of heavier matter toward their cores. As this heat flows outward, it generates convective motions that stir them. The planets' rotation creates a Coriolis effect on the rising gas, drawing it into the cloud belts seen on Jupiter, Saturn, and Neptune.

Rotation also creates an obvious equatorial bulge on Jupiter and Saturn, which spin in about 10 hours. The spinning motion, when combined with convection in their metallic hydrogen cores, generates powerful magnetic fields that in turn contribute to the formation of auroras and energetic, charged particle belts like the Van Allen belts of our Earth.

Each of the four giant planets has a ring system composed of small orbiting particles. The rings are probably debris from small objects that have broken up as the result of either collisions or tidal forces exerted by the planet. Collisions continue to break up small particles, replacing ring material that falls into the planets' atmospheres.

Many satellites orbit each of the four giant planets, resembling miniature Solar Systems. Some of these moons are comparable in diameter to Mercury. They are thought to be made mostly of water, ice, and rock, although Io, an exceptionally active moon of Jupiter, may be mostly sulfur and rock. Many of the moons are heavily cratered, but a few are smooth, indicating surface activity (Io's volcanos of sulfur, for example) that has filled in old craters. Saturn's moon, Titan, has an atmosphere of mainly nitrogen, but because of its great distance from the Sun, it is bitterly cold. "Rain" falling from Titan's clouds is not water, but liquid ethane, which flows across Titan's surface, carving rivers and collecting in lakes.

The dwarf planets Pluto and Eris are like some of the moons of the outer planets in their size and mass, particularly Neptune's moon Triton, which may have been captured. They are probably mostly ice and rock, with thin methane atmospheres. These and many other icy worlds orbit beyond Neptune and probably are surviving planetesimals.

TABLE 10.1 COMPARISON OF THE FOUR GIANT PLANETS

	Jupiter	Saturn	Uranus	Neptune
$R_{equator}$ (Earth units) (km)	11.19 (71,492)	9.46 (60,268)	3.98 (25,559)	3.81 (24,764)
M (Earth units) (kg)	317.9 (1.9×10^{27})	95.18 (5.68×10^{26})	14.54 (8.68×10^{25})	17.13 (1.02×10^{26})
Density (gm/cm^3)	1.33	0.69	1.32	1.64
Sidereal day (hours)	9.924	10.656	17.24	16.11
Distance from Sun (AU)	5.203	9.539	19.19	30.06
Orbital period (years)	11.8622	29.4577	84.014	164.793
Axial tilt (degrees) (with respect to orbit)	3.12	26.73	97.86	29.56

QUESTIONS FOR REVIEW

1. (10.1) How do Jupiter's mass and radius compare with the Earth's? How do they compare with those of the other outer planets?
2. (10.1) What does Jupiter look like?
3. (10.1) How do astronomers know what lies inside the outer planets?
4. (10.1–10.2) What are the major gaseous substances that make up Jupiter and Saturn?
5. (10.1–10.2) What is the interior structure of Jupiter and Saturn thought to be?
6. (10.1–10.2) Do Jupiter and Saturn have solid surfaces?
7. (10.1) What sorts of atmospheric motion and activity are observed in Jupiter?
8. (10.1) What is the Great Red Spot?
9. (10.1)What sort of activity has been seen on Io? What is Io's heat source thought to be?
10. (10.2) What are the rings of Saturn made of? How do astronomers know this?
11. (10.2) What creates the gaps between the rings?
12. (10.2) How might the rings have formed?
13. (10.2) What is the Roche limit? Why does such a limit exist?
14. (10.3) What is unusual about Uranus's rotation axis? What might explain this peculiarity?
15. (10.3–10.4) How do Uranus and Neptune differ from Jupiter in their interiors?
16. (10.3–10.4) Why are Uranus and Neptune so blue?
17. (10.1–10.4) Why are the outer planets so large?
18. (10.1–10.4) What are the satellites of the outer planets thought to be composed of ? Which have atmospheres? What might be special about Europa?
19. (10.5) What evidence makes astronomers believe that Pluto is strongly influenced by Neptune?
20. (10.5) How did the discovery of a moon orbiting Pluto help astronomers better understand this object?
21. (10.5) Where did Pluto and other TNOs form? How did they get to where they are today?

THOUGHT QUESTIONS

1. (10.1) If Jupiter were moved closer to the Sun, what do you think would happen to it?
2. (10.1–10.2) Why do the relatively uncratered surfaces of Europa and Enceladus imply that these moons may have been "active" recently?
3. (10.1–10.2) Ganymede and Callisto orbiting Jupiter and Tethys and Dione orbiting Saturn appear in many ways similar to our Moon. If one of these satellites were located close to the Earth, would the similarities remain? What might happen? How would the gas giant satellites then differ from our Moon?
4. (10.2) Saturn's moon Titan exhibits an atmosphere, lakes, clouds, rocks, dunes, and at least one volcano. Explain how these features are substantially similar to or different from what we expect based on Earth's geology.
5. (10.3) Is Uranus's sky blue for the same reason our sky is blue? Compared to the yellow-white we see on Earth, would the Sun look redder or bluer from inside Uranus's atmosphere?
6. (10.3) How is the fact that Uranus's moons orbit its equator evidence that Uranus was in a collision with an asteroid or planetesimal? What if the moons had formed prior to the collision?
7. (10.3–10.4) Why might Uranus and Neptune have proportionally less hydrogen than their more massive companions, Jupiter and Saturn?

8. (10.3–10.4) Describe in detail why Neptune clearly shows bands of atmospheric circulation and Uranus does not. Explain when, if ever, Uranus could show such features.
9. (10.1–10.4) If only one new mission could be funded to visit or revisit one of the planets in this chapter, and land a rover on one of that planet's moons, make a case for which planet and moon would be most worth exploring.
10. (10.5) Why do some astronomers think that Pluto should be considered a planet? Do you agree? What happens if you apply this logic to Sedna and Eris?

PROBLEMS

1. (10.1–10.5) How long does it take sunlight to reach Jupiter? Uranus? Pluto? How might this affect satellite or manned missions to these planets?
2. (10.1) Use the modified form of Kepler's third law, illustrated in figure 8.8 (and discussed in detail in chapter 3), to calculate Jupiter's mass using the orbital data for any of Jupiter's moons given in the appendix. Be sure to convert the orbital period to seconds and the orbital radius to meters before putting those numbers into the formula.
3. (10.2) At closest approach, Saturn is about 8.5 AU from the Earth. If the rings are 270,000 kilometers in diameter, what is their angular size seen from Earth?
4. (10.2) Use the modified form of Kepler's third law and Saturn's mass to calculate the period of the material at the inner and outer edges of Saturn's rings, with values for "a" of 90,000 and 136,000 kilometers. Can you see why the rings cannot be solid?
5. (10.4) What is the "surface" gravity of Neptune (at the top of the clouds)?
6. (10.4–10.5) Show that Pluto's orbital period is very close to one and one-half times Neptune's. Use the data in table 4 of the appendix.
7. (10.5) Use the modified form of Kepler's third law to calculate the mass of Pluto and Charon from the orbital data for Charon given in the text (unlike in problem 2, the satellite's mass is about 12% of the planetoid's mass, so we will not neglect it here). Compare your result to the sum of the masses given for Pluto and Charon in the appendix. Be sure to convert the orbital period to seconds and the orbital radius to meters before putting those numbers into the formula.
8. (10.5) Calculate the density of Charon, given that its radius is approximately 593 kilometers and its mass is about 1.1×10^{24} grams. (Be sure to convert kilometers to centimeters or meters.) Is it likely that Charon has a large iron core? Why?

TEST YOURSELF

1. (10.1) The low average densities of Jupiter and Saturn compared with the Earth's suggest that
   (a) Jupiter and Saturn are hollow.
   (b) the gravitational attraction of Jupiter and Saturn has compressed their cores into a rare form of iron.
   (c) Jupiter and Saturn contain large quantities of light elements, such as hydrogen and helium.
   (d) Jupiter and Saturn are very hot.
   (e) volcanic eruptions have ejected all the iron that was originally in Jupiter's and Saturn's cores.
2. (10.1) Astronomers think that the inner core of Jupiter is composed mainly of
   (a) hydrogen.
   (b) helium.
   (c) uranium.
   (d) rock and iron.
   (e) water.
3. (10.1) Why would it be foolish to send astronauts to land on Jupiter?
   (a) It has no solid surface for them to land on.
   (b) Its clouds are so hot that any spacecraft getting near it would burn up.
   (c) Its gravitational attraction is so weak that they would float off.
   (d) All of the above
   (e) The idea is perfectly reasonable.
4. (10.2) What is the Roche limit?
   (a) The mass a planet must exceed to have satellites.
   (b) The smallest mass a planet can have and still be composed mainly of hydrogen.
   (c) The greatest distance from a planet that its satellites can orbit without falling into the Sun.
   (d) The distance at which a moon held together by gravity will be broken apart by the planet's gravitational attraction.
   (e) The distance astronomers can see into a planet's clouds.
5. (10.1–10.2) What source of energy allows Jupiter and Saturn to radiate more heat than they receive from the Sun?
   (a) A strong greenhouse effect
   (b) A high concentration of radioactive elements
   (c) A strong magnetic dynamo
   (d) Tidal stresses from their moons
   (e) Gravitational energy of sinking material
6. (10.3) What makes some astronomers think that Uranus was hit by a large body early in its history?
   (a) It goes around the Sun in a direction opposite to the other planets.
   (b) Its rotation axis has such a large tilt.
   (c) Its composition is so different from that of Neptune, Jupiter, and Saturn.
   (d) It has no moons.
   (e) All of the above.

7. (10.4) Like the other gas giants, Neptune's rotational period is closest to two-thirds of an Earth ______.
(a) day. (b) month. (c) year. (d) decade. (e) century.
8. (10.5) Which of the following are reasons why we don't have a picture of Pluto? (There may be more than one correct answer.)
(a) Pluto's surface reflects too little light to get a good image.
(b) We have not yet sent a mission to Pluto.
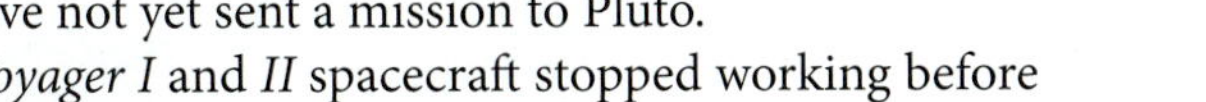
(c) The *Voyager I* and *II* spacecraft stopped working before they got to Pluto.
(d) Its angular size is too small for even the Hubble Space Telescope to get a clear image of it.

KEY TERMS

Roche limit, 271
shepherding satellites, 270
trans-Neptunian object (TNO), 279

FURTHER EXPLORATIONS

Barnes, Jason. "Titan: Earth in Deep Freeze." *Sky and Telescope* 116 (December 2008): 26.

Beatty, J. Kelly. "Celebrating Cassini." *Sky and Telescope* 114 (December 2007): 30.

______, and Stuart J. Goldman. "The Great Crash of 1994: A First Report." *Sky and Telescope* 88 (October 1994): 18.

______, Carolyn Collins Petersen, and Andrew Chaikin. *The New Solar System*. 4th ed. Cambridge: Sky, 1999. (This fine book covers the Solar System in great detail.)

Binzel, Richard. "Pluto." *Scientific American* 262 (June 1990): 50.

Burns, Joseph A., Douglas P. Hamilton, and Mark R. Showalter. "Bejeweled Worlds." *Scientific American* 286 (February 2002): 64.

Gingerich, Owen. "Losing It in Prague: The Inside Story of Pluto's Demotion." *Sky and Telescope* 112 (November 2006): 34.

Johnson, Torrence V. "The Galileo Mission to Jupiter and Its Moons." *Scientific American* 282 (February 2000): 40.

Lunine, Jonathan I. "Neptune at 150." *Sky and Telescope* 92 (September 1996): 38.

Maugeri, Joseph, and MacRobert, Alan. "Tiny Players on Jupiter's Stage." *Sky and Telescope* 116 (July 2008): 62.

Pappalardo, Robert T., James W. Head, and Ronald Greeley. "The Hidden Ocean of Europa." *Scientific American* 281 (October 1999): 54.

Schilling, Govert. "Discovery of Charon: A Bump in the Night." *Sky and Telescope* 115 (June 2008): 24.

Stern, S. Alan. "Journey to the Farthest Planet." *Scientific American* 286 (May 2002): 56.

______. "The 3rd Zone: Exploring the Kuiper Belt." *Sky and Telescope* 106 (November 2003): 30.

Tytell, David. "Titan: A Whole New World." *Sky and Telescope* 109 (April 2005): 34.

Website

Visit the *Explorations* website at **http://www.mhhe.com/arny** for additional online resources on these topics.

FIGURE QUESTION ANSWERS

WHAT IS THIS? (chapter opening): This is a picture made by the Cassini spacecraft of Titan in front of Saturn. The thin dark lines are Saturn's rings seen nearly edge-on.

FIGURE 10.2: The bottom box is about twice as dense as the upper box.

FIGURE 10.7: They are called the Galilean satellites because they were discovered by Galileo.

FIGURE 10.8: Io is heated by the gravitational (tidal) force exerted on it by Jupiter, which distorts and thereby heats it.

FIGURE 10.19: The moons completely covered with craters probably are inactive, or the craters would have been largely covered up or otherwise destroyed by eruptions or surface tectonics. Enceladus has some smooth areas, and it has been discovered to have eruptions.

FIGURE 10.25: Measure the spot and Neptune's disk, and divide the spot's size by Neptune's. The spot is not round, but is roughly 0.2 times the size of Neptune. Neptune is about 3.8 times the size of the Earth, so the spot is $0.2 \times 3.8 = 0.76$ times the size of the Earth (about three-fourths Earth's size).

FIGURE 10.31: Pluto's orbit is relatively steeply tilted with respect to Neptune's. As a result, when Pluto "crosses" Neptune's orbit, Pluto is actually well above or below Neptune's path.

Picture of Comet McNaught as seen from Australia in January 2007.

KEY CONCEPTS

- Numerous objects much smaller than the planets orbit the Sun.
  - Rocky objects—the asteroids
  - Icy objects—the comets
- Asteroid orbits are affected by Jupiter, whose gravity makes certain orbits unstable.
- Comets generally orbit beyond Neptune. If they enter the inner Solar System, their ices "boil" away.
- The lost gas is pushed away by the solar wind and sunlight, creating the comet's tail.
- Asteroids and comets are left over from the formation of the Solar System and provide information about early conditions.
- Fragments of these objects reach Earth from time to time as meteoroids, most of which are small.
- When such a fragment enters our atmosphere, it is heated by friction with the air and glows, creating a meteor or "shooting star."
- Earth has been struck in the past by very large objects, which have produced craters and caused mass extinctions.

11 Meteors, Asteroids, and Comets

CONCEPTS AND SKILLS TO REVIEW

- Density (6.1)

Orbiting the Sun and scattered throughout the Solar System are numerous bodies much smaller than the planets—the asteroids and comets. The asteroids are generally rocky objects in the inner Solar System. The comets are icy bodies and spend most of their time in the outer Solar System. These small members of the Sun's family, remnants from the formation of the Solar System, are of great interest to astronomers because they are our best source of information about how long ago and under what conditions the planets formed. In fact, some asteroids and comets may be planetesimals—the solid bodies from which the planets were assembled—that have survived nearly unchanged from the birth of the Solar System.

Apart from their scientific value, asteroids and comets merit study because they can be both beautiful and deadly. A comet in the dawn sky with its tail a shining plume is a sight not to miss. But "miss" is precisely what we hope will happen if a large comet or asteroid is on a collision course with Earth. Such a collision in the past may have exterminated much of Earth's ancient life, and such an event in the future could well have equally disastrous effects on today's living things.

In this chapter, we will see why astronomers think asteroids and comets are related to planetesimals. We will also discover why meteorites (which are simply fragments of asteroids and comets that by chance fall into our atmosphere) are such important clues to the time of birth and the structure of the ancient Solar System. Likewise, we will study how a comet changes from a 10-kilometer diameter ball of ice into a beautiful banner of light in the night sky and why an asteroid may be the reason you have hair rather than scales.

Q: WHAT IS THIS? See end of chapter for the answer

11.1 Meteors, Meteoroids, and Meteorites

FIGURE 11.1
A time-exposure photograph captures a "shooting star" (meteor) flashing overhead. The curved streaks are star trails.

Q. Why is the meteor's track straight, whereas the star images are curved?

If you have spent even an hour looking at the night sky, you have probably seen a "shooting star," a streak of light that appears in a fraction of a second and as quickly fades (fig. 11.1). Astronomers call this brief but lovely phenomenon a **meteor.**

A meteor is the glowing trail of hot gas and vaporized debris left by a solid object heated by friction as it moves through the Earth's atmosphere. Most of the heating occurs between about 100 and 50 kilometers in the outer fringes of the atmosphere. The solid body, while in space and before it reaches the atmosphere, is called a **meteoroid.**

Heating of Meteoroids

Meteoroids heat up on entering the atmosphere for the same reason a reentering spacecraft does. When an object plunges from space into the upper layers of our atmosphere, it collides with atmospheric molecules and atoms. These collisions convert some of the body's energy of motion (kinetic energy) into heat, as shown in figure 11.2. In a matter of seconds, the outer layer of the meteor reaches thousands of degrees Kelvin and glows. Given that reentry speeds are typically at least 10 kilometers per second and often 30 to 40 kilometers per second, the collisions with air molecules are extremely violent and tear atoms off the body, vaporizing the surface layers. The trail of hot evaporated matter and atmospheric gas emits light, making the glow that we see.

If the meteoroid is larger than a few centimeters, it creates a ball of incandescent gas around it and may leave a luminous or smoky trail. Such exceptional meteors, sometimes visible in daylight, are called "fireballs."

Meteoroids bombard the Earth continually: a hail of solid particles that astronomers estimate amounts to hundreds of *tons* of material each day. Slightly more strike between midnight and dawn than in the evening hours, and so it is best to watch for meteors in the early morning. This difference arises for the same reason that if you run through rain, your front will get wetter than your back. That is, the dawn side of our planet advances into the meteoritic debris near us in space, while the night side moves away from it.

Most meteors that we see last only a few seconds and are made by meteoroids the size of a raisin or smaller. These tiny objects are heated so strongly that they completely vaporize. Larger pieces, though heated and partially vaporized, are so drastically slowed by air resistance that they may survive the ordeal and reach the ground.* We call these fragments found on the Earth **meteorites.**

*People are sometimes hit by meteorites, but only rarely. In one well-documented case, a meteor crashed through a roof and hit a woman, badly bruising her, and a number of cars and houses have also been hit. See section 11.4 for other reports.

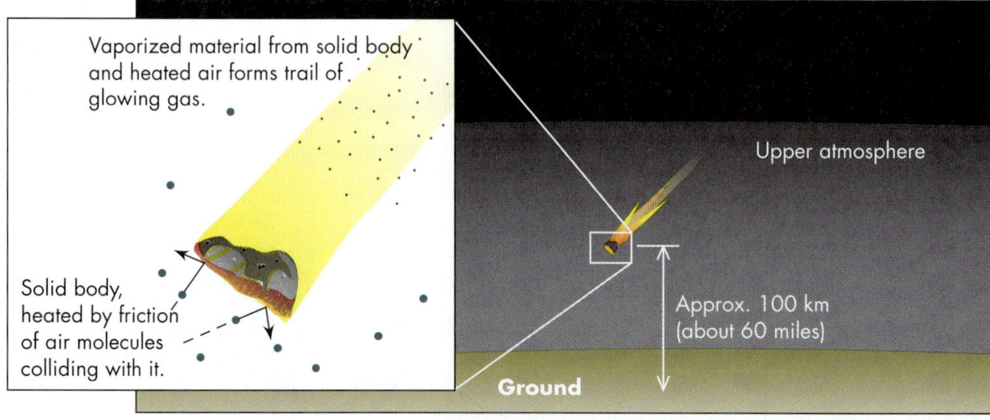
Vaporized material from solid body and heated air forms trail of glowing gas.
Solid body, heated by friction of air molecules colliding with it.
Upper atmosphere
Approx. 100 km (about 60 miles)
Ground

FIGURE 11.2
Sketch depicting how air friction heats an object entering our atmosphere, creating the glow we see as a meteor.

Types of Meteorites

Astronomers classify meteorites into three broad categories based on their composition: stony (that is, composed mainly of silicate compounds), iron, and stony-iron. Most stony meteorites are composed of smaller rounded chunks of rocky material stuck together, as shown in figure 11.3A. The grains are called **chondrules,*** and meteorites that have this lumpy structure are called **chondrites.** Chondrules appear to have been rapidly melted and cooled in the solar nebula. The cause of the heating is not known, although it may have been caused by outbursts from the Sun as it formed, collisions between planetesimals, or the explosion of a nearby star—perhaps an explosion that caused the solar nebula to begin collapsing.

Chondrules contain traces of radioactive material, which can be used to measure their age, as described in chapter 6. They are almost 4.6 billion years old, and are believed to be the first solid material that condensed within the solar nebula. Many chondrules contain even older material: dust grains that have survived from before the

* *Chondrule* is pronounced "KON-drool" and means "small grain" in Greek.

FIGURE 11.3
Sliced samples of various kinds of meteorites. (A) Chondrites are stony meteorites made up of small round bits of rock called chondrules. Slices show the chondrules inside an ordinary chondrite and a carbonaceous chondrite, which has a matrix of carbon-rich material. (B) Iron and stony-iron meteorites as well as rocky achondrites come from fragments of differentiated asteroids. After the material has melted, the iron sinks to the core and the pattern of the chondrules is lost. Note the crystalline pattern in the iron meteorite. These patterns form only when the metal cools very gradually over millions of years.

Amino acids occur in two molecular forms—right-handed and left-handed. All living things on Earth contain only left-handed amino acids. The amino acids found in meteorites contain both forms, however. Their presence indicates that they are not contamination by terrestrial organisms but were formed in space by nonbiological processes.

birth of the Solar System. Thus, chondritic meteorites offer us valuable information about the early history of the Solar System.

In some chondritic meteorites, the chondrules are embedded in a black, carbon-rich, coal-like substance and are therefore called **carbonaceous chondrites.** This carbonaceous matter contains organic compounds, including amino acids, the same complex molecules used by living things for the construction of the proteins on which they depend. Thus, the presence of amino acids in meteoritic matter indicates that the raw material of life can form in space and that it might therefore have been available right from the start within the Solar System. Irrespective of the existence of amino acids in meteorites, we can still ask how and where these bodies formed, and what brought them to Earth. Astronomers think that most of them are fragments of asteroids* and comets.

Some stony meteorites have no chondrules and are therefore called **achondrites.** They appear to have melted and been transformed like the rock found on Earth, and therefore probably come from asteroids that were large enough to heat up and melt. In such bodies, iron would sink to the core, which would then explain the existence of iron meteorites. The unusual history of the iron in these meteorites is revealed when the iron is etched with acid. This shows a crystalline pattern that can only form when metals cool over millions of years (see fig. 11.3B). Finally, the rarest meteorites, stony-irons, contain silicate rock embedded in a matrix of iron, and may come from the transition region between the core and mantle of a large asteroid.

11.2 Asteroids

Asteroids are small, generally rocky bodies that orbit the Sun. They are found throughout the Solar System, but most lie in the **asteroid belt,** a region between the orbits of Mars and Jupiter, stretching from about 2 to 4 AU from the Sun, as shown in figure 11.4.

*Although most meteorites are fragments of asteroids or comets, some are chunks of rock from the surface of the Moon and Mars, blasted into space by the collision of an asteroid with these bodies.

Q. With a ruler, estimate the distance (in AU) from the center of the Trojan swarms to the Sun and to Jupiter, and compare this with Jupiter's distance from the Sun.

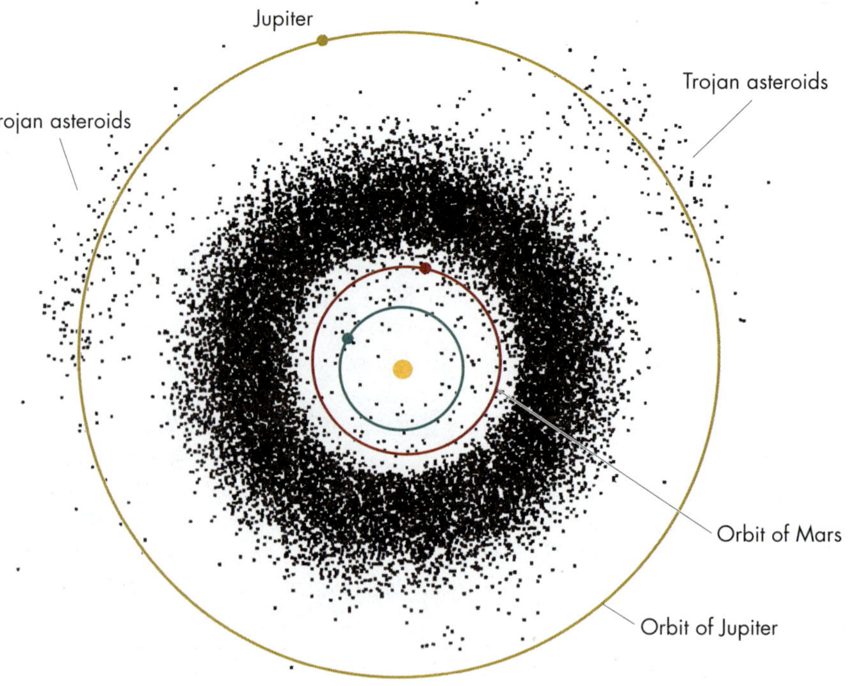

FIGURE 11.4
Diagram showing the distribution of 21,785 known asteroids. Notice that most lie between Mars and Jupiter, but a small number form two loose clumps—the Trojan asteroids—located on Jupiter's orbit. Making the plotted points large enough to see causes them to appear far more closely packed than they really are.

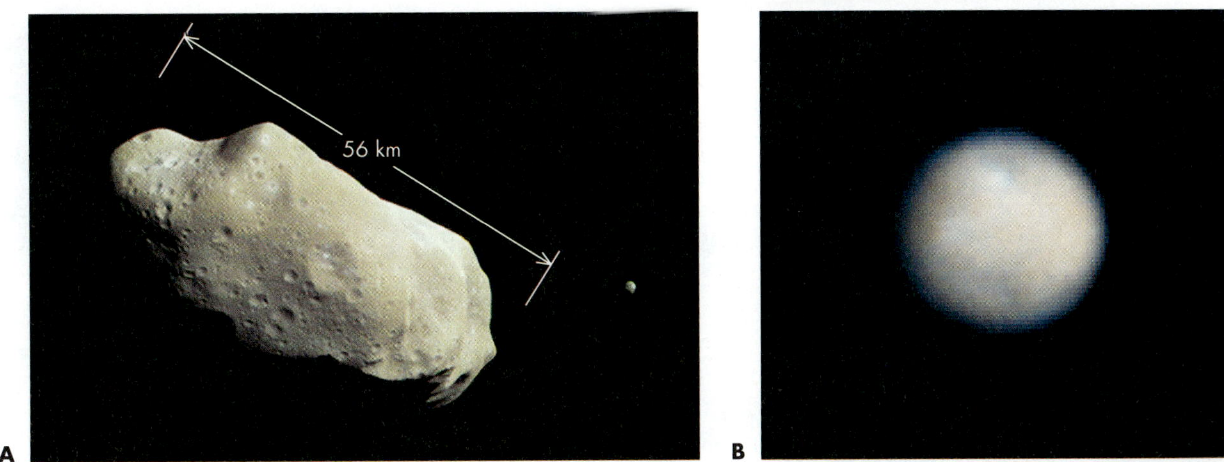

FIGURE 11.5
(A) The asteroid Ida and its "moon" Dactyl. (B) The asteroid Ceres, as viewed by the Hubble Space Telescope. Note that Ceres looks approximately spherical.

Giuseppe Piazzi discovered the first of this swarm of bodies in 1801 during his search for the "missing planet," which, according to Bode's rule (p. 213), should have been at 2.8 AU from the Sun. He named the asteroid that he found Ceres in honor of the patron goddess of Sicily, his home. More than 100,000 other asteroids have subsequently been found, but Ceres remains the largest one we know of. Despite the huge number of asteroids, their combined mass is very small, amounting to probably less than $\frac{1}{1000}$ the mass of the Earth.

Size and Shape

The diameter of an asteroid is difficult to measure because nearly all are so small that they appear in ground-based telescopes merely as points of light. Moreover, the amount of light reflected from an asteroid is not a good clue to its size because a large, poorly reflective object will look as bright as a small, highly reflective one. For this reason, the emitted infrared radiation is a better measure of diameter; bigger bodies emit more than smaller ones of the same temperature. From such measurements, astronomers have found that asteroids range tremendously in diameter, from Ceres—about 1000 kilometers (less than one-tenth the Earth's size) across—down to bodies of a kilometer or less. For example, the diameter of a tiny asteroid 1991 BA, which passed about 170,000 kilometers (less than half the distance to the Moon) from Earth in January 1991, is probably less than 9 meters (approximately 30 feet).

The smallest asteroids are far too tiny to observe unless they come extremely close to Earth or strike it. In that case, we see them as meteors. That is, meteorites are in a real sense just the very smallest (we hope!) asteroids that happen to reach our planet.

Most asteroids—such as Ida (fig. 11.5A)—are irregularly shaped. Only Ceres (fig. 11.5B) and a few other large asteroids are approximately spherical.* Since the gravitational force of Ceres is strong enough to crush its material into a sphere, it has been reclassified as a dwarf planet. Smaller asteroids with weak gravities remain irregular and are made more so by collisions blasting away pieces. Collisions leave the parent body pitted and lumpy, and the fragments become smaller asteroids in their own right, or sometimes even satellites of larger asteroids, such as Ida's moon Dactyl.

Although we have described asteroids as rocky or iron-rich, that does not mean they are all solid chunks of such material. For example, the density of the asteroid Mathilde is

*Because of its approximately spherical shape, Ceres is now officially also a dwarf planet.

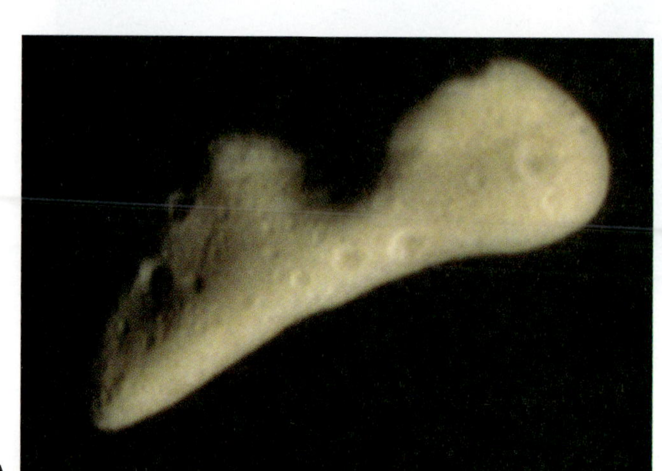

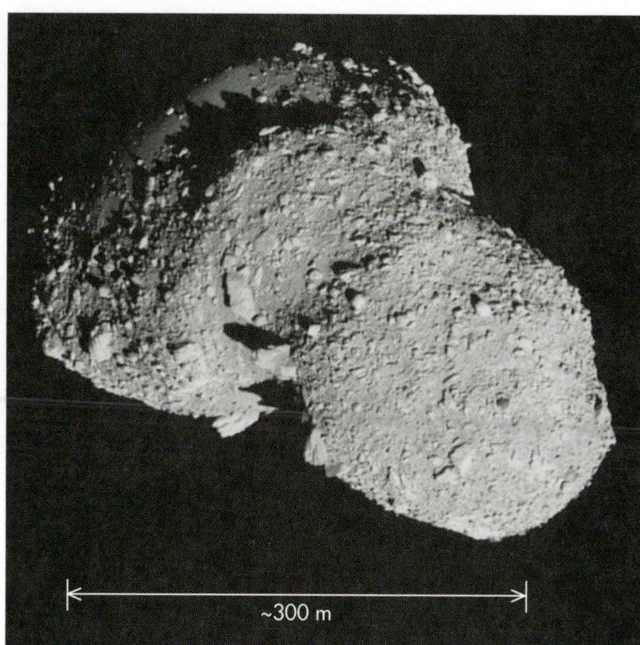

FIGURE 11.6
(A) The asteroid Eros, as observed by the *NEAR* spacecraft. It is about 33 kilometers long. (B) Asteroid Itokawa. Note its rubbly appearance.

very low (1.4 gm/cm^3), suggesting that it is a loose pile of rubble, just barely held together by its gravity. Such a loose structure might also explain the extremely irregular shapes of the asteroids shown in figure 11.6: Eros, which was visited by the *NEAR* spacecraft in 2000, and Itokawa, whose picture (fig. 11.6B) was taken by the Japanese space probe *Hayabusa*.* The *Hayabusa* probe landed on Itokawa in 2005, and from measurement of the asteroid's gravitational attraction on the probe, astronomers calculated the asteroid's mass. From the mass and with knowledge of the asteroid's dimensions, astronomers calculated its density, which turned out to be 1.9 gm/cm^3. This too is so low a density that the asteroid must be mainly loose rock lumps, only weakly held together by its gravity. Figure 11.6B clearly shows this lumpy structure.

Following its landing, the probe attempted to scoop up a sample from the asteroid and then lifted off to return to Earth. It is scheduled to arrive here in 2010, but scientists will not know until then if it succeeded in picking up a sample.

Composition

When sunlight falls on an asteroid, the minerals in its surface create absorption features in the spectrum of the reflected light, from which we can determine the asteroid's composition. Such spectra show that asteroids belong to three main compositional groups similar to those of meteorites: carbonaceous bodies, silicate bodies, and metallic iron-nickel bodies. The groups are not mixed randomly throughout the asteroid belt: inner-belt asteroids tend to be silicate-rich, and outer-belt ones tend to be carbon-rich.

Origin of Asteroids

The properties of asteroids that we have just discussed (composition, size, and their location between Mars and Jupiter) give us clues to their origin and support the solar nebula hypothesis for the origin of the Solar System. As we described in chapter 8, the asteroids are probably fragments of planetesimals, the bodies from which the planets were built.

---

* *Hayabusa* means "Falcon" in Japanese.

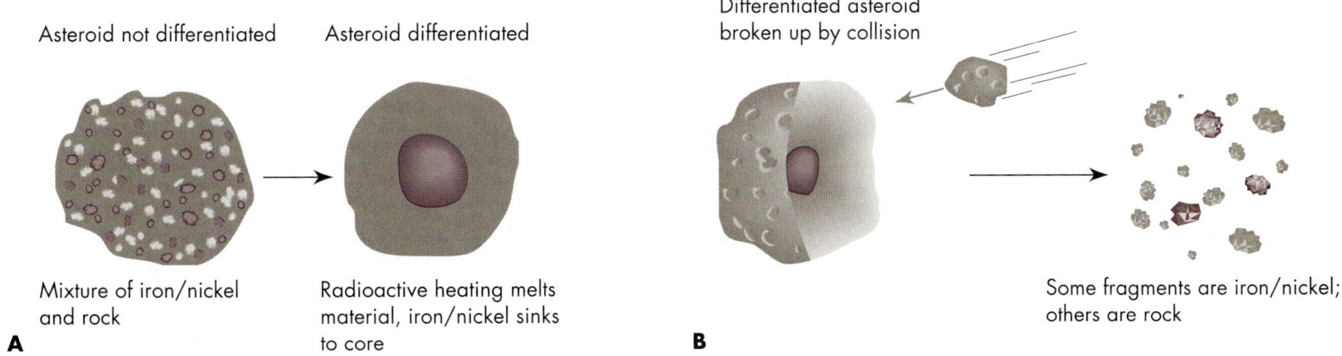

FIGURE 11.7
Sketch depicting (A) differentiation in asteroids and (B) their subsequent breakup by collision to form iron and stony bodies.

According to the solar nebula hypothesis, bodies that condensed in the inner asteroid belt have a different composition from those that condensed farther out. The inner belt, being warmer, is richer in the easy-to-condense silicate and iron materials and contains less of the hard-to-condense water and carbon-rich materials, as is indeed observed.

The existence of stony and iron asteroids might at first seem to be evidence *against* the solar nebula theory. How could a swirling mass of gas and dust have separated so as to form some bodies of rock and some bodies of iron? Such separation would be a bit like shaking a piece of cake and having it disintegrate into eggs, flour, sugar, and milk.

ANIMATION
Differentiation in asteroids and their subsequent breakup by collisions to form iron and stony bodies

We saw in chapter 6, however, that chemical elements *can* be separated by differentiation. That is, astronomers think that the Earth's rocky crust and iron core formed as a result of melting, with the iron then sinking to the core and the lighter rock floating. At least some asteroids have also differentiated (fig. 11.7A), and at least one (Vesta) has had volcanic eruptions—this activity, now long dormant, is deduced from its spectrum, which shows the presence of basalt, a volcanic rock.

After differentiation, collisions with neighboring asteroids broke up most of the large bodies (fig. 11.7B). The fragments are what we see today: pieces of crust became stony asteroids, while pieces of core became iron asteroids. But for a body to differentiate, it must be large enough to be able to melt from the heat liberated within it by radioactivity. Thus, the existence of stony and iron asteroids is strong evidence that the early Solar System contained intermediate-size bodies, that is, planetesimals.

The solar nebula hypothesis also offers an explanation for the asteroids' being concentrated between Mars and Jupiter. Any small planet there would have to compete for material with Jupiter, whose immense gravity would disturb the accretion process and prevent the planet's growth.*

Unusual Asteroids

Even today, Jupiter affects the asteroid belt. Figure 11.8 shows a partial census of the number of asteroids found at each distance from the Sun within the belt. Gaps can be seen at about 2.1, 2.5, and 2.8 AU, for example.

The seemingly empty regions in the asteroid belt are called **Kirkwood gaps,** and they are caused by the same process that creates the gaps in Saturn's rings: gravitational forces of an outlying body. Saturn's moons, particularly Mimas, create the gaps in the rings. Jupiter creates the Kirkwood gaps. As it orbits the Sun, Jupiter exerts a gravitational force on asteroids in the belt that slightly alters their orbits. If an asteroid

*This is a less dramatic view than the one that was common 50 years ago, when asteroids were believed to be fragments of a terrestrial planet destroyed by collision. This no longer seems likely, given the tiny amount of material in the asteroid belt.

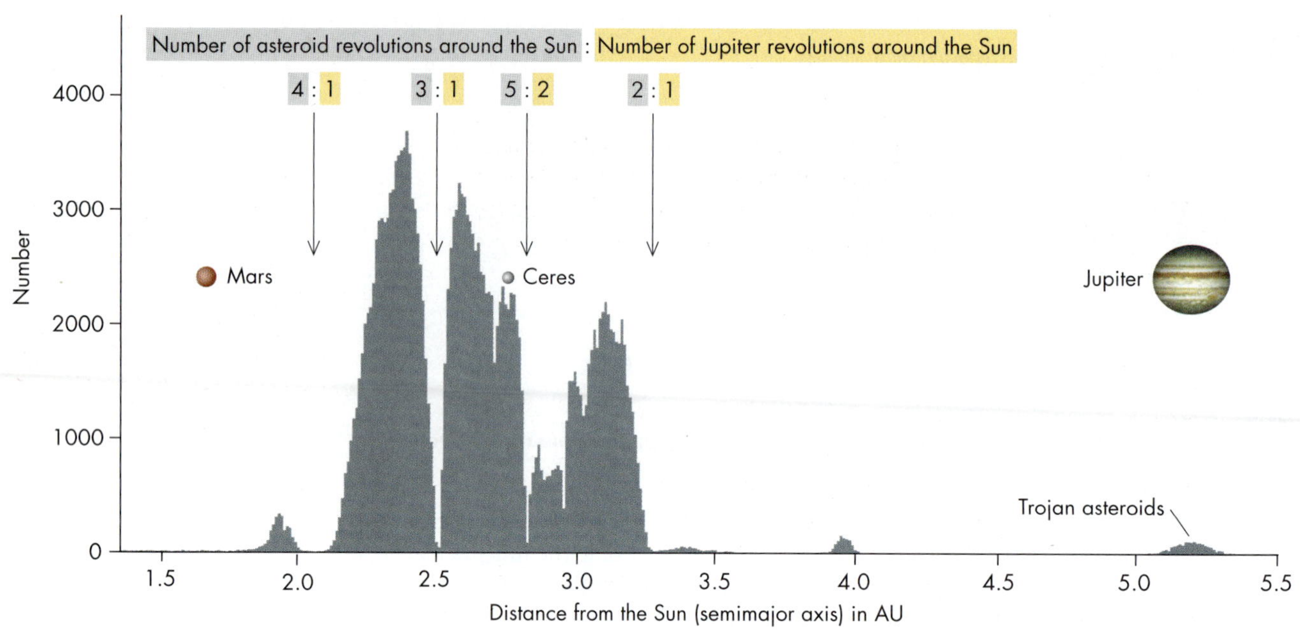

FIGURE 11.8
The number of asteroids at each distance from the Sun within the asteroid belt. Notice the conspicuous gaps where there are few, if any, objects. These empty zones are called Kirkwood gaps.

Q. What feature in Saturn's rings is similar to these gaps?

has an orbital period that when multiplied by an integer equals Jupiter's period, the asteroid may be subject to a cumulative force that makes it drift to a new orbit. The gap at 2.5 AU arises because, according to Kepler's third law, an asteroid at that distance would have an orbital period exactly one-third that of Jupiter. Every third trip around the Sun, the asteroid would undergo exactly the same tug from Jupiter. Over time, it will be shifted to a new orbit, leaving a gap.

Not all asteroids are found in the main belt. A few, the so-called Trojan asteroids, travel along Jupiter's orbit in two loose swarms, 60° ahead and 60° behind it, as illustrated in figure 11.4.

Nearer the Sun from the main belt are the **near-Earth objects,** whose orbits carry them into the inner Solar System across or close to the Earth's orbit. Fortunately for us, there are relatively few of these. The near-Earth objects are mostly small, with diameters of less than 1 kilometer, such as Itokawa (figure 11.6B). Over 5000 have been identified, mostly the larger ones, and accurate orbits have been determined. None is likely to collide with the Earth over the next century, although several have been found that will approach to within one-tenth the distance of the Moon. On average, astronomers estimate that a body 100 meters in diameter or larger hits the Earth about every 10,000 years. The near-Earth objects have probably been shifted into their peculiar orbits by gravitational interactions with Jupiter and other planets. Some may be related to comets, but "dead" ones, stripped of ice and gas by their repeated passage around the Sun. We will describe how this happens in more detail later in this chapter.

An asteroid 1 kilometer in diameter striking the Earth will release on impact about the same amount of energy as a 40,000-megaton nuclear explosion. Some astronomers have therefore urged the construction of an international monitoring system to warn of the approach of small asteroids. For example, LINEAR (the Lincoln Near Earth Asteroid Network, operated by MIT's Lincoln Laboratory and the U.S. Air Force) and the British NEO (Near Earth Object) office are now in operation. The hope is that if an asteroid on a collision course with Earth is detected early enough, it might be deflected away from Earth.

Chiron is another odd "asteroid." Its orbit stretches from just inside that of Saturn almost to Uranus, putting it far outside the main asteroid belt. In addition to this odd orbit, Chiron changes brightness oddly; sometimes it flares up and ejects gas. Such behavior is more like that of a comet than a "normal" asteroid. From its brightness, astronomers deduce that Chiron has a diameter of about 180 kilometers (approximately 110 miles), making it much larger than most comets that enter the inner Solar System. It may well be a trans-Neptunian object (chapter 10) that has moved from the Kuiper belt because of gravitational interactions with the giant planets.

11.3 Comets

A bright comet is a stunning sight, as illustrated by the chapter-opening figure. Sadly, such sights are now rare because light pollution from our cities drowns the view for most people. Comets have long been held in fear and reverence, and their sudden appearance and equally sudden disappearance after a few days—or, in some cases, weeks—have added to their mystery.

Light pollution not only spoils our view of the night sky, it also wastes money. Groups such as the International Dark-Sky Association have shown that better-designed streetlights and other outdoor lights make the ground brighter, the sky darker, and use less electricity.

Structure of Comets

Comets consist of three main parts—the tail, the coma, and the nucleus—as illustrated in figure 11.9. The largest part is the long **tail,** a narrow column of dust and gas that may stretch across the inner Solar System for as much as 100 million kilometers (nearly an AU!).

The tail emerges from a cloud of gas called the **coma,** which may be some 100,000 kilometers in diameter (10 times or so the size of the Earth). However, despite the great volume of the coma and the tail, these parts of the comet contain very little mass. The gas and dust are extremely tenuous, and so a cubic centimeter of the gas contains only a few thousand atoms and molecules. By terrestrial standards, this would be considered a superb vacuum. This extremely rarified gas is matter that the Sun's heat has boiled off the heart of the comet, its nucleus.

The comet **nucleus** is a block of ice and gases that have frozen in the extreme cold of the outer Solar System into an irregular mass whose diameter is typically about 10 kilometers. The nucleus of a comet has been described as a giant "iceberg" or "dirty snowball," and it contains most of the comet's mass.

Astronomers' first close look at a comet nucleus came from *Giotto*,* a spacecraft launched by the European Space Agency as part of an international study of Comet Halley.† *Giotto* approached to within 600 kilometers of Halley's nucleus and sent

*The name *Giotto* was given to the spacecraft in honor of the Italian artist who painted a portrait of a comet as part of a Christmas scene for a church altarpiece. Some astronomers believe his painting depicts Halley's comet as it appeared in 1301. Other astronomers believe that Giotto based it on a different comet.

† Comets are generally named for their discoverer or for the year they were first seen. Halley's gets its name because Sir Edmund Halley was the first to propose that some comets move around the Sun like planets, and he predicted that the great comet he saw in 1682 would reappear in 1759. It did, but Halley did not live to see his prediction verified. There is controversy about how to pronounce *Halley*. Most astronomers say it rhymes with *Sally*, although there is some evidence that Halley himself pronounced his name as *haw-lee*.

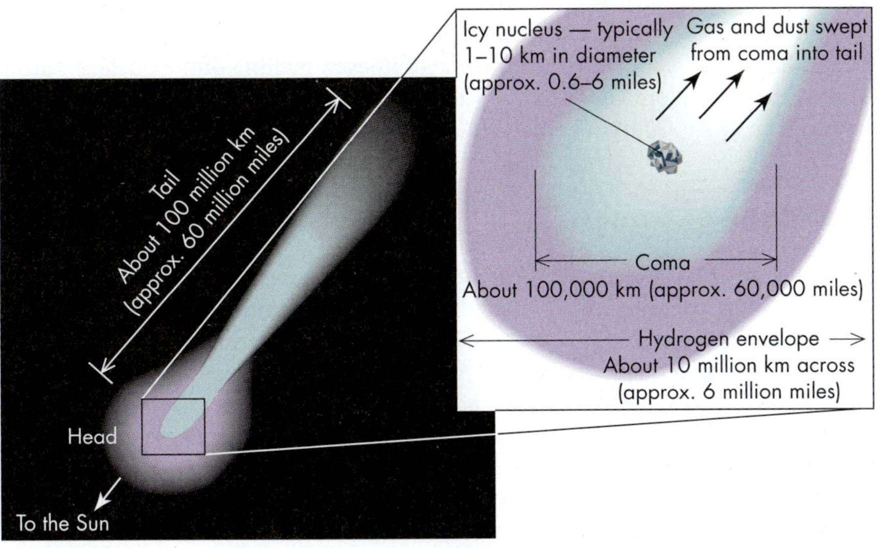

FIGURE 11.9
Artist's depiction of the structure of a comet, showing the tiny nucleus, surrounding coma, and long tail.

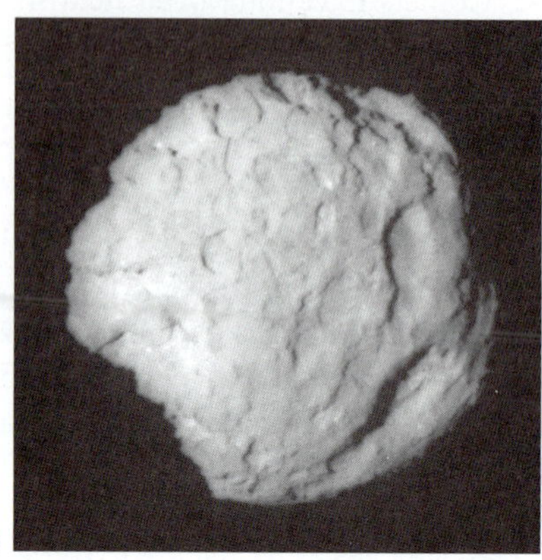

FIGURE 11.10
(A) Picture made by the *Giotto* spacecraft of the nucleus of Comet Halley. The spacecraft was approximately 1000 kilometers (about 600 miles) from the nucleus—deep inside Halley's coma of gas—during the time this picture was made. (B) The nucleus of Comet Wild 2 (*Wild* is said "vilt"), which is about 5 kilometers (about 3 miles) in diameter. Note the many craters. The image was made by NASA's *Stardust* spacecraft, which looped back past Earth in 2006 and parachuted into our atmosphere a canister containing samples of dust collected from near the comet.

pictures back to Earth of the nucleus. From these its size could be measured. From the size and estimates of the mass of the nucleus, its density can be calculated. This turns out to be about 0.6 grams per cubic centimeter, a value implying that the icy material of the nucleus is "fluffy," like snow, not hard and compacted like pure ice. Unfortunately, the mass estimates are not very accurate, and so the density we infer from them is uncertain.

Despite its icy composition, the nucleus is extremely dark, as you can see in figure 11.10A, which is one of the pictures made by the *Giotto* spacecraft. Astronomers think that the dark color comes from dust and carbon-rich material (similar to that of the carbonaceous chondritic meteorites) coating the surface of the nucleus. Other visible features of the nucleus are its irregular shape and the jets of gas erupting from the frozen surface. The jets form when sunlight heats and vaporizes the icy material. The irregular shape is probably the outcome of uneven melting of the nucleus during passage by the Sun on previous orbits. You can see similar features on Comet Wild 2 in a picture taken by the *Stardust* spacecraft (fig. 11.10B).

Composition of Comets

The escaped gas from the comet offers astronomers a way to probe the comet's composition. The dust particles reflect sunlight, and the gases emit light of their own by a process called **fluorescence.** Fluorescence is produced when light at one wavelength is converted to light at another wavelength. A familiar example is the so-called black light that you may have seen for illuminating posters.

Black light is really ultraviolet radiation that we have difficulty seeing because of its short wavelength. When such ultraviolet radiation falls on certain paints or dyes, the chemicals in the pigment absorb the ultraviolet radiation and convert it into visible light.

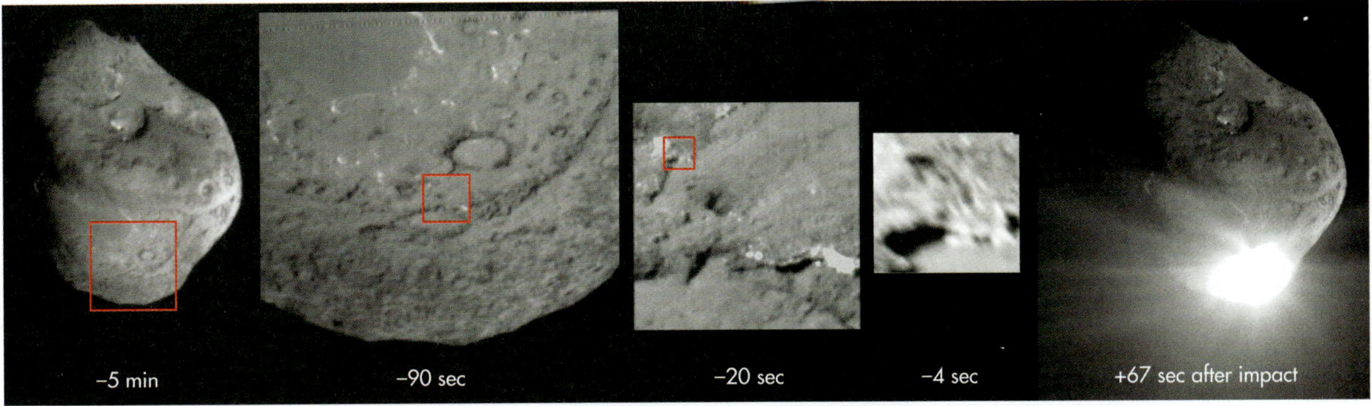

FIGURE 11.11
Sequence of images from NASA's *Deep Impact* mission. The impactor probe sent images as it approached the comet Tempel 1. The comet nucleus is about 8 × 5 kilometers in size, and the final image before impact shows features just meters in size. An image made by the main spacecraft about a minute after impact shows a spray of fine particles blasted out by the impact and brightly lit by sunlight.

A major part of a comet's light is created by fluorescence. A photon of ultraviolet—and thus energetic—radiation from the Sun lifts electrons in the atoms of the comet's gas molecules to an upper, excited level in a single leap. The electron then returns to its original level in two or more steps, emitting a photon each time it drops. The combined energy of these photons must equal that of the absorbed ultraviolet photon to conserve energy. Thus, the energy of each emitted photon must be less than that of the original ultraviolet one. That smaller energy then gives them a longer wavelength, which we can see with our eyes. Thus, fluorescence creates the soft glow of the comet's light. In addition, the spectrum of the fluorescing gas tells us of what the comet is made.

Some laundry detergents have "whiteners" in them that fluoresce. They convert ultraviolet radiation into visible light and thereby make a white shirt or blouse look brighter. You can see this effect strongly if you shine a black light on a freshly washed shirt in a darkened room.

Spectra of gas in the coma and tail show that comets are rich in water, CO_2, CO, and small amounts of other gases that condensed from the primordial solar nebula. Evaporating water is broken up by solar ultraviolet radiation to create oxygen and hydrogen gas, and most comets are surrounded by a vast cloud of hydrogen created in this way.

Astronomers have recently gotten even closer looks at comets. The craft that took pictures of Comet Wild 2 collected samples of dust from near the comet and successfully returned them to Earth by parachute in early 2006. The dust particles included small crystals of silicate rock as well as a wide range of organic compounds.

A more intrusive comet sampling was made by NASA's *Deep Impact* mission, which in 2005 smashed a 370-kilogram (~800-lb) probe into Comet Tempel 1 at a relative speed of just over 10 kilometers per second (about 23,000 mph). The impact was designed to break through the comet's outer crust and stir up and release dust and gas. The impact event is shown in figure 11.11. The first four images are from the point of view of the impact probe, which sent pictures up until a few seconds before it smashed into the surface. The final image is from the main spacecraft, showing a cloud of very fine dust blasted out by the impact, which created a crater estimated to be about 100 meters across. The spectra of the material blasted out by the impact showed the presence of water and silicates as well as clays and other water-based crystals.

If a comet passes by the Sun too often, the escape of its gas eventually erodes it away. Also, some comets literally fall into the Sun. For example, the *SOHO* satellite (which observes the Sun's outer atmosphere) has taken pictures of dozens per year falling into the Sun. Since new comets show up frequently, there must be a source to replace those devoured by the Sun, and it is to their origin that we now turn.

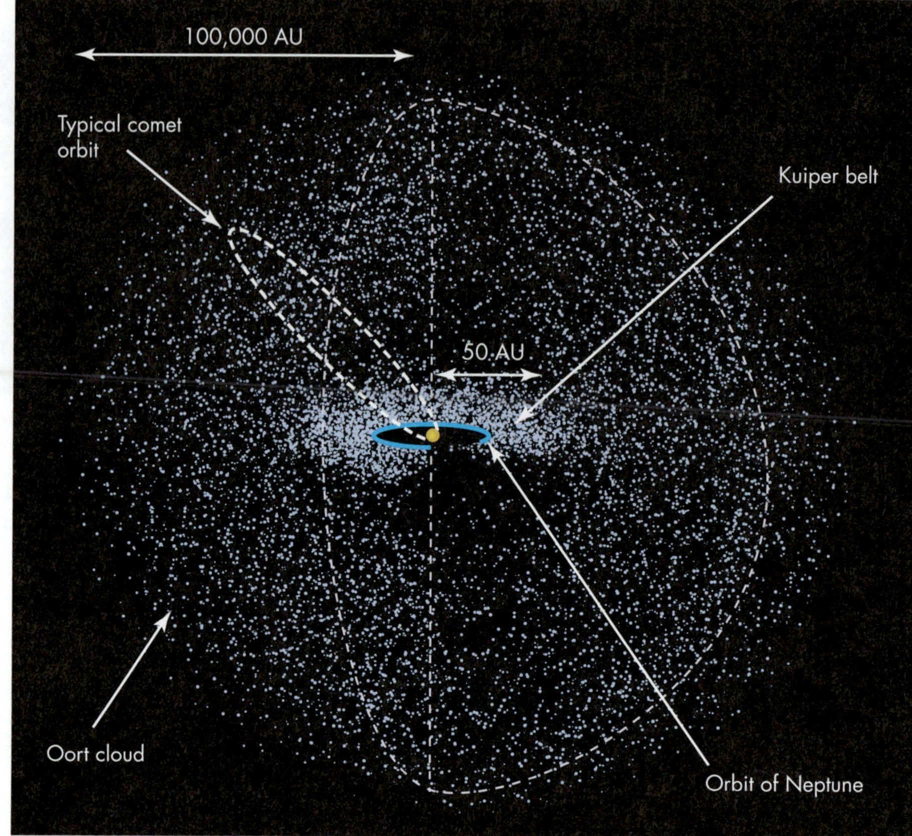

FIGURE 11.12
Schematic drawing of the Oort cloud, a swarm of icy comet nuclei orbiting the Sun out to about 100,000 AU. Also shown is the Kuiper belt, another source of comet nuclei, with a size exaggerated for clarity.

ANIMATION
Oort cloud and Kuiper belt

Origin of Comets

Astronomers think that most comets come from the **Oort cloud,** the swarm of trillions of icy bodies* believed to lie far beyond the orbit of Neptune, as we discussed in chapter 8. You may recall that astronomers think the Oort cloud formed from planetesimals that originally orbited near the giant planets and were tossed into the outer parts of the Solar System by the gravitational force of those planets. There, they form a spherical shell that completely surrounds the Solar System and extends to perhaps as much as 150,000 AU from the Sun, as shown in figure 11.12. Astronomers deduce this shape for the Oort cloud from the many comet orbits that are highly tilted with respect to the main plane of the Solar System. However, as we will discuss later, some comets also seem to come from a flatter, less remote region—the so-called **Kuiper belt,** also shown schematically in figure 11.12. The Kuiper belt begins at about the orbit of Neptune and extends to approximately 50 AU.

Each comet nucleus moves along its own path, and those in the Oort cloud take millions of years to complete an orbit. With orbits so far from the Sun, these icy bodies receive essentially no heat from the Sun, and calculations indicate that their temperature is a mere 3 K, or about −454°F. Thus, the gases and ices remain deeply frozen.

Such cold and distant objects are invisible to us on Earth; so, if we are to see a comet, its orbit must somehow be altered to carry it closer to us and the Sun. Astronomers think that such orbital changes may arise from the chance passage of a star far beyond the outskirts of the Solar System or from tidal forces exerted on the Oort cloud by the Milky Way. Such gravitational effects disturb the orbits of the comet nuclei in the Oort cloud, altering their paths and making some drop in toward the inner Solar System, as shown in figure 11.13. A single disturbance may shift enough orbits to supply comets to the inner Solar System for tens of thousands of years.

*Astronomers call these cold and inert bodies "comets," even though they do not have tails.

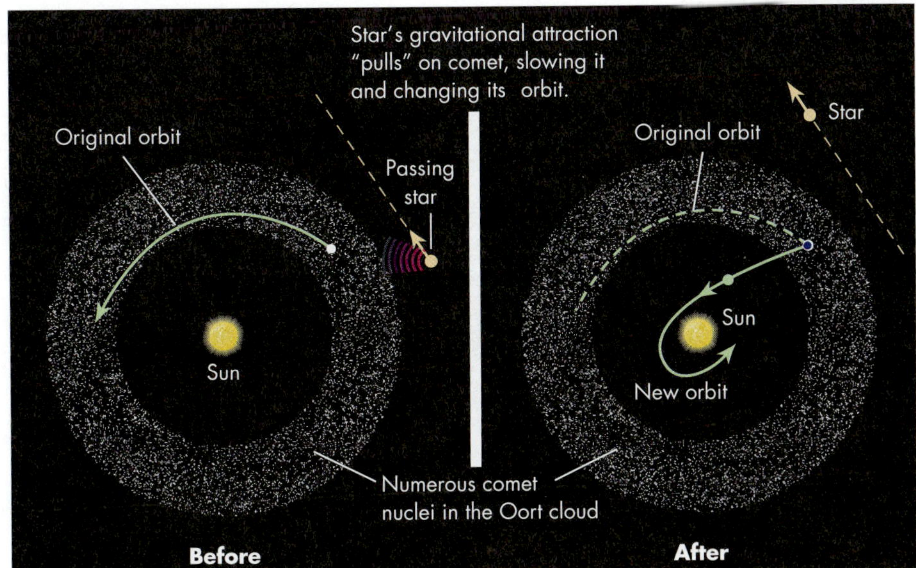

FIGURE 11.13
Sketch of how a passing star alters the orbit of a comet nucleus. On its new orbit, the comet will pass by the Sun and be visible from Earth. Although only one comet's orbit is shown, such encounters typically will affect many comets. Note: The distance and size scales are greatly exaggerated. No star presently gets anywhere near this close to the Oort cloud.

As the comet falls inward toward the inner Solar System, the Sun's radiation heats it and begins to melt the ices. At a distance of about 5 AU from the Sun (Jupiter's orbit), the heat is enough to vaporize the ices, forming gas that escapes to make a coma around the comet nucleus. The escaping gas carries with it tiny dust grains that were frozen into the nucleus. The comet then appears through a telescope as a dim, fuzzy ball. As the comet falls ever nearer the Sun, its gas boils off even faster, but now the Sun begins to exert additional forces on the cometary gas and dust.

Formation of the Comet's Tail

Sunlight striking dust grains imparts a tiny force to them, a process known as **radiation pressure.** We don't feel radiation pressure when sunlight falls on us because the force is tiny and the human body is far too massive to be shoved around by solar photons. However, the microscopic dust grains in the coma do respond to radiation pressure and are pushed away from the Sun, as shown in figure 11.14. Because all the grains move in the same direction, away from the Sun, a tail begins to form.

The tail pushed out by radiation pressure is made of dust particles, but figure 11.15 shows that comets often have a second tail. That tail is created by an outflow of gas that streams from the Sun into space, a flow called the **solar wind.**

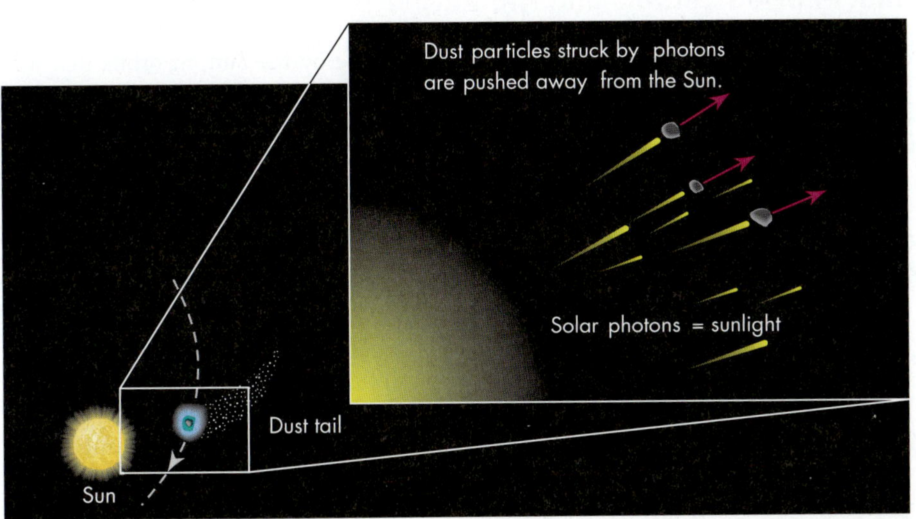

FIGURE 11.14
Sketch of how radiation pressure pushes on dust particles. Photons hit the dust, and their impact drives the dust away from the Sun, forming a dust tail. Sizes and distances are not to scale.

FIGURE 11.15
Photograph of Comet Hale-Bopp in the evening sky, clearly illustrating the two tails, one of dust, one of gas. These two tails point in slightly different directions because the dust particles feel the effect of the Sun's gravity as well as its radiation pressure. Responding to the gravity, the dust particles follow orbits around the Sun, but according to Kepler's laws, particles farther from the Sun move more slowly, so the dust tail arcs behind the comet. The gas tail, on the other hand, is effected almost exclusively by the solar wind, so it points nearly directly away from the Sun.

ANIMATION
The orientation of comet tails

The solar wind blows away from the Sun at about 400 kilometers per second. It is very tenuous, containing only a few atoms per cubic centimeter. But the material in the comet's coma is tenuous too, and the solar wind is dense enough by comparison to blow it into a long plume.

Magnetic fields carried along by the solar wind enhance its effect on the comet's tail, helping to drag matter out of the coma and channel its flow, just as magnetic fields in the Earth's atmosphere channel particles to form the aurora. Thus, two forces, radiation pressure and the solar wind, act on the comet to drive out a tail. Because those forces are directed away from the Sun, the comet's tail always points away from the Sun, and the tail even points out ahead of the comet as it moves away from the Sun (fig. 11.16).

It might help you understand this seemingly odd phenomenon if you think of a runner carrying a torch. If the air is still, the smoke from the torch will, of course, trail behind the runner. However, if a strong wind is blowing at, say, 40 miles per hour, the smoke will be carried along in the direction of the wind regardless of which way the runner moves. Likewise, the high velocity of the solar wind (400 kilometers per second versus about 40 kilometers per second for the comet) carries the tail of a comet outward from the Sun regardless of the comet's motion. Therefore the tails always point away from the Sun.

Short-Period Comets and the Kuiper Belt

Although most comets that we see from Earth swing by the Sun on orbits that will bring them back to the inner Solar System only after millions of years, a small number of comets reappear at time intervals less than 200 years. These **short-period comets** include Halley's, which has a period of 76 years.

The origin of short-period comets is still under study. At one time it was thought that they came from the Oort cloud, but as they moved through the region of the Solar System containing the giant planets, their orbits were shifted by a close encounter with one of the planets into smaller orbits with periods of centuries rather than millennia. However, astronomers now think that short-period comets come from the icy nuclei, orbiting beyond Neptune, in the Kuiper belt. Support for this origin comes from the detection of hundreds of small, presumably icy, bodies orbiting near and somewhat beyond Pluto. Astronomers estimate that the Kuiper belt contains well over 30,000 icy objects bigger than 100 kilometers in diameter, and its total mass is hundreds of times larger than the asteroid belt between

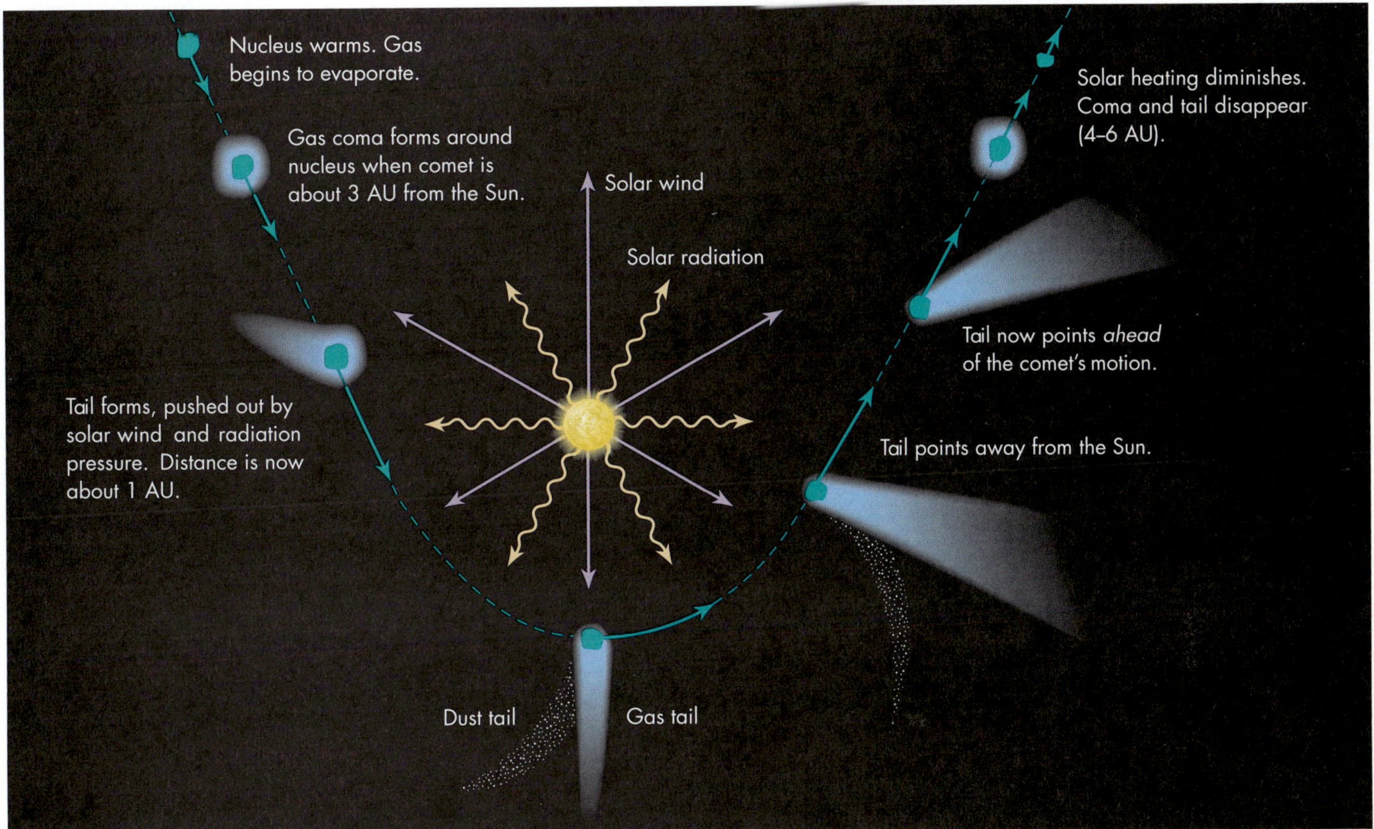

FIGURE 11.16
Sketch illustrating how radiation pressure and the solar wind make a comet's tail always point approximately away from the Sun. Sizes and distances are not to scale. Comets may orbit in any direction around the Sun.

Mars and Jupiter. These frozen objects are probably survivors of the Solar System's birth—icy planetesimals still orbiting in the disk—but they are too far apart to form additional planets.

Astronomers at one time thought the outer edge of the Kuiper belt might lie as much as 1000 AU from the Sun. However, searches for extremely dim objects have failed to find significant numbers more than about 50 AU from the Sun. On the other hand, in such searches they have discovered a small number of much larger icy objects, such as Sedna, Orcus, Quaoar, and the dwarf planet Eris, which, as we mentioned in chapter 10, is larger than Pluto. Thus, much still remains to be learned about this remote part of our Solar System.

Fate of Short-Period Comets

A short-period comet's repeated orbits past the Sun gradually whittle it away: all the ices and gases evaporate, and only the small amount of solid matter, dust and grit, remains. This fate is like that of a snowball made from snow scooped up alongside the road, where small amounts of gravel have been packed into it. If such a snowball is brought inside, it melts and evaporates, leaving behind only the grit accidentally incorporated in it. So too, the evaporated comet leaves behind in its orbit grit that continues to circle the Sun.

A comet is a bit like the "Peanuts" cartoon character Pigpen, who trails dirt wherever he goes.

As a comet orbits the Sun and its icy and gaseous material evaporates, it leaves in its path a trail of dust and small bits of solid material ejected from its nucleus. When we cross through or closely approach such a trail, our planet is blasted by this microscopic debris that rains into our atmosphere, burning up, and creating a meteor shower.

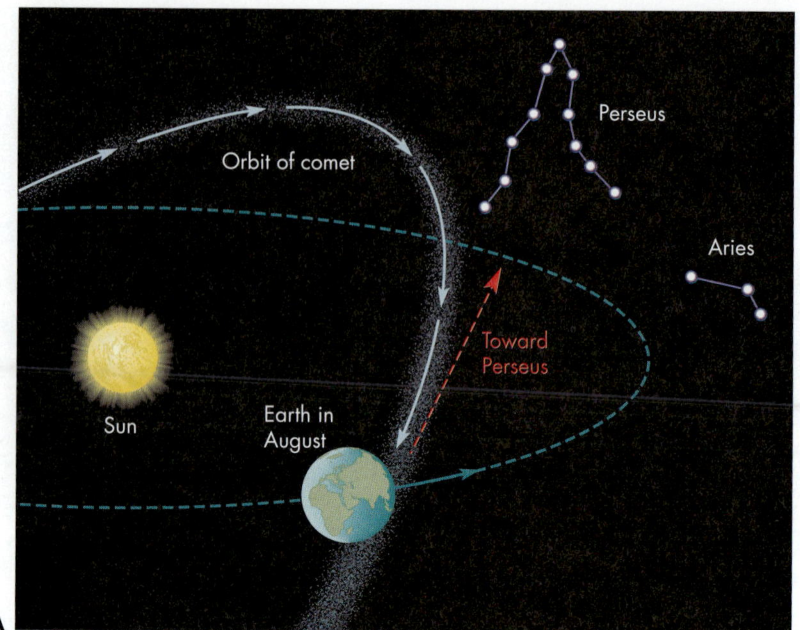

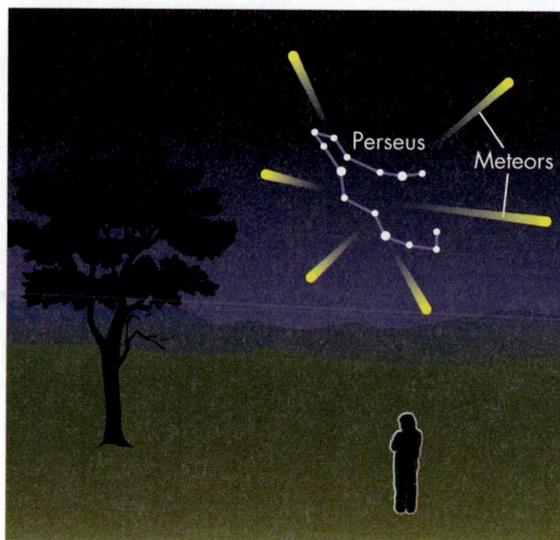

FIGURE 11.17
Sketch showing how (A) in mid-August, at the time of the Perseid meteor shower, the Earth is moving along its orbit. When the Earth crosses the debris strewn along a comet's orbit, the scattered material plunges into our atmosphere, producing (B) the diverging pattern of meteors characteristic of a meteor shower. (Bodies and orbits are not to scale.)

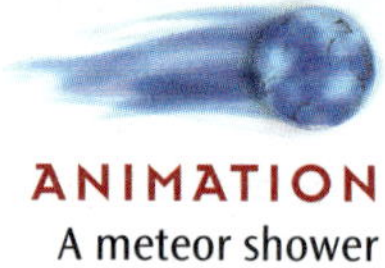

ANIMATION
A meteor shower

Meteor Showers

If you go outside on a clear night and have an unobscured view of the sky, you will see on the average one meteor every 15 or so minutes. Most of these meteors are stray fragments of asteroids that arrive at the Earth randomly.

At some times of year, instead of one meteor per quarter hour or so, you may observe one every few minutes. Furthermore, if you watch such meteors carefully, you will see that they appear to come from the same general direction in the sky. Meteors of this type are called "shower meteors," and the events that produce them are called **meteor showers.**

The most famous meteor shower occurs each year in mid-August. From August 11 to 13, meteoroids rain into our atmosphere from a direction that lies toward the constellation Perseus. The meteors themselves have no association with Perseus. Rather, they are following an orbit around the Sun that happens to lie roughly in that direction (fig. 11.17A), and the Earth happens to cross their orbits in mid-August. Thus, at that time we encounter far more meteoroids than usual.

This encounter creates an effect similar to what you observe when you drive at night through falling snow: the flakes seem to radiate from a point in front of you, the location of which depends on a combination of the direction and speed of both the wind and your car. Thus, during the time that the Earth crosses the path followed by the meteoroids, they seem to diverge from a common point (fig. 11.17B), called the **radiant.** Meteor showers are generally named for the constellation from which they appear to diverge, and appendix table 7 lists several of the brighter and more impressive showers and their dates. Each shower therefore marks when the Earth crosses the path of a comet.

On rare occasions, the Earth will pass through a particularly dense clump of material left by the comet. If that happens, thousands of meteors per hour may spangle the sky. Such a display happened in November 1966, when dawn observers on the west coast of the United States and Canada saw literally dozens of meteors per second! The sky looked as if someone had lit a celestial Fourth of July sparkler. Likewise, in November 2001 and 2002, observers were treated to similarly amazing displays.

Spectacles of this kind are one of the delights of astronomy. However, on even rarer occasions, far more sinister meteoritic events may occur.

11.4 Giant Impacts

Every few thousand years, the Earth is hit by a large meteoroid, a body tens of meters or more in size. Such bodies will produce not only a spectacular glare as they pass through the atmosphere but also an enormous blast on impact.

As we discussed earlier in this chapter, meteoroids may have a very large kinetic energy, that is, energy of motion. If the meteoroid does not burn up on passage through the atmosphere, its remaining kinetic energy is released when it hits the ground or when it breaks up in the atmosphere.

The energy so released can be huge, as we can easily show from the formula for a body's kinetic energy, *E*, given by the expression

$$E = \frac{mV^2}{2}$$

E = Kinetic energy of an object
V = Velocity of the object
m = Mass of the object

where *m* is its mass and *V* is its velocity. An iron meteoroid 3 meters (10 feet) in diameter has a mass of about 100,000 kilograms (about 100 tons). If it were traveling at 30 kilometers per second ($= 3 \times 10^4$ meters per second) relative to the Earth, the kinetic energy of impact would be $10^5 \times (3 \times 10^4)^2/2 = 4.5 \times 10^{13}$ joules, which is about the same as that released when a 10,000 tons of dynamite are exploded—comparable to the nuclear bomb that destroyed Hiroshima. Such an object hitting the Moon would make a crater roughly 100 meters (300 feet) in diameter. Fortunately, the Earth's atmosphere significantly slows down meteoroids this small. Even so, they can cause damage if they hit an object such a house or car.

One well-documented event that proved harmless only because of its remoteness was a brilliant flash detected in February 1994 over the South Pacific ocean by a spy satellite. Analysis of the flash suggests that it marked the breakup of a 7-meter (20-foot) diameter meteoroid. The resulting blast had an energy between 10 and 20 kilotons of TNT, comparable to the bomb that destroyed Hiroshima. Had the object burst over a city, the casualties might have been in the hundreds of thousands.

Our atmosphere is less able to slow meteoroids that are much larger than this, and so although rare, they can potentially cause a large amount of damage. For example, the impact of a body 100 meters in diameter, the size of a football stadium, would have the explosive power of the largest thermonuclear bombs ever built. It would make a crater about a kilometer in diameter, as illustrated in figure 11.18. Were such a body to hit a heavily populated area, the results would be catastrophic.

Fortunately, we have been spared such disasters recently, but there have been some close calls and some truly horrific impacts in the distant past.

Giant Meteor Craters

Many giant meteor craters scar our planet. One of the most famous is in northern Arizona. About 50,000 years ago, a meteoroid estimated to have been some 50 meters in diameter hit the Earth about 40 miles east of what is now Flagstaff. Its impact vaporized

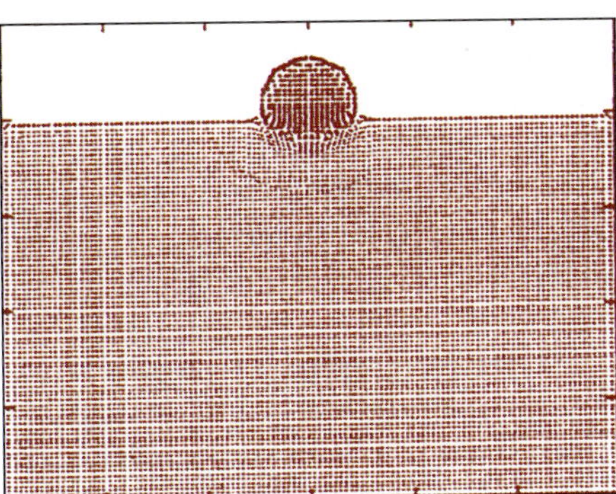
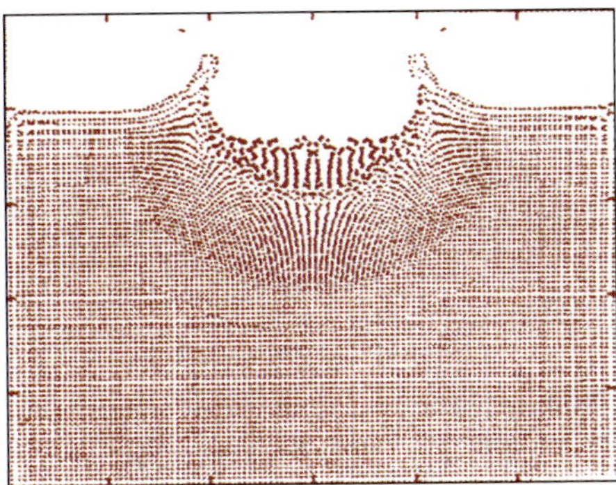
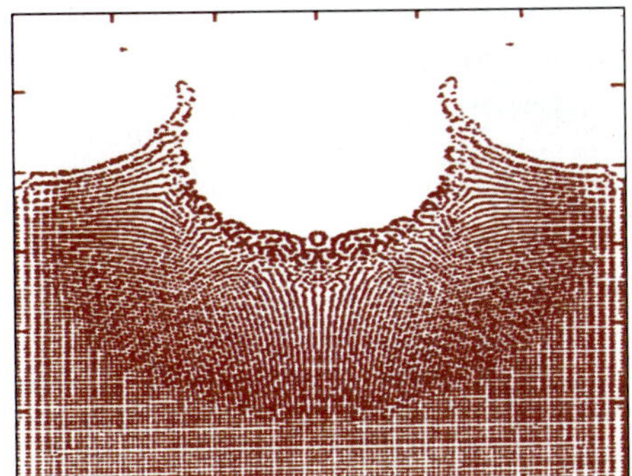

FIGURE 11.18
A computer simulation of a crater's formation.

SCIENCE *at work*

METEORITES CAN BE DEADLY

As astronomers have studied the details of the impact believed responsible for the Cretaceous extinction, they have come to realize that meteor impacts may be more common (and deadly!) than once believed. Although we have no recent, well-documented cases of people being killed by meteorites, such deaths surely do occur. Careful searches through newspapers have also turned up many very near misses. Moreover, reports of meteorites hitting buildings and even cars are surprisingly common. Searching old records from Europe and China, astronomers have found many instances of fatalities attributed to meteorites. One of the grimmer of these reports is from a fifteenth-century Chinese chronicle that describes "stones that fell like rain" that killed 10,000 people. Even allowing for exaggeration in such old records, we can conclude that meteor falls can be far from harmless. In fact, astronomers conclude that although meteoritic disasters are rare, when they occur, they have the potential to kill enormous numbers of people.

FIGURE 11.19
Photograph of the Arizona meteor crater, which is more than 1 kilometer across. The inset image shows one of the iron meteorites found at the site.

Q. What might explain why Manicouagan Crater is so much less sharply defined than the Arizona meteor crater?

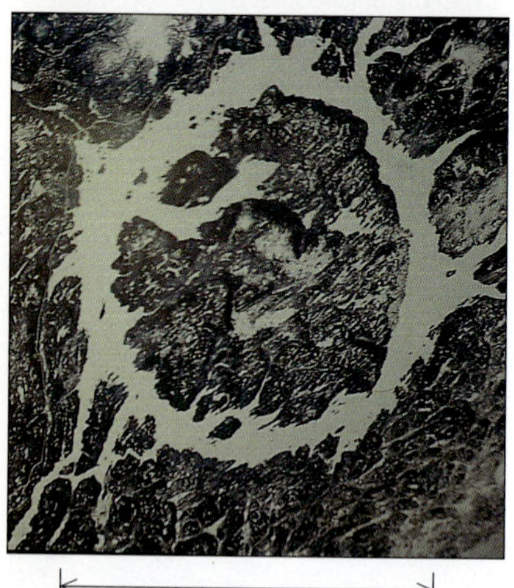

FIGURE 11.20
Picture (from Earth orbit) of the Manicouagan Crater in Quebec. This winter view shows the lake that fills the crater covered with snow.

tons of rock, which expanded and peeled back the ground, creating a crater about 1.2 kilometers across and 200 meters deep (fig. 11.19).

More recently, in 1908, an asteroid broke up in our atmosphere over a largely uninhabited part of north-central Siberia. This so-called Tunguska event, named for the region where it hit, leveled trees radially outward from the blast point to a distance of some 30 kilometers. The blast was preceded by a brilliant fireball in the sky and was followed by clouds of dust that rose to the upper atmosphere. Sunlight reflected off this dust gave an eerie glow to the night sky for several days. According to some accounts, the blast killed two people. Casualties were few because the area was so remote.

Unfortunately, scientists didn't visit the site until about two decades later, because of the political turmoil in Russia at that time. When they did reach the site, they found no crater, just the felled trees. Interestingly, trees at the center of the damaged area were left standing vertically but with their branches stripped off. Thus, the explosion must have occurred in the air.

More ancient impact scars occur in many places on our planet. The huge, ring-shaped Manicouagan Lake, about 70 kilometers (approximately 43 miles) in diameter and pictured in figure 11.20, is a meteor crater, as is Wolf Creek Crater in northwestern Australia. Astronomers have found even larger craters but are not sure they are impact features. Two such craters are the vast arc (nearly 500 kilometers across) on the east

SCIENCE *at work*

GHOST CRATERS, OR NO TELL-TALE FRAGMENTS

Hypotheses to explain the Tunguska event offer a good example of how science tests and retests ideas. The lack of a big crater and the absence of meteorites at the site at first seemed to be evidence against the idea that it was caused by a small asteroid. This led some astronomers to propose that a comet might have caused the blast. Being icy, a comet would break up more easily—perhaps even while still in the atmosphere—so there would be no crater. Moreover, any fragments that survived would rapidly disappear, explaining the absence of large rock or iron meteorites.* However, computer simulations of how an asteroid travels at high speed through our atmosphere showed what at first seemed a surprising result: stony or iron asteroids can also create devastation without leaving a crater or tell-tale fragments. A typical asteroid approaches Earth at a speed of 10 to 30 kilometers per second. When it hits our atmosphere, it compresses and heats the air ahead of it. The hot compressed air obeys Newton's law of action–reaction and exerts a tremendous force back on the asteroid that may shatter it. The resulting fragments plunge deeper into our atmosphere, where air resistance heats them further until they vaporize, creating a fireball at a height of 20 kilometers (12 miles) or so above the ground. No trace of the asteroid survives to form a crater or fragments at the ground. All that remains is the 6000 K incandescent air of the fireball that blasts downward and out, crushing and igniting whatever lies below it, exactly the way a nuclear bomb burst would.

Such calculations do not prove that an asteroid caused the Tunguska event, but they do show that the lack of meteorites does not imply that a comet was to blame.

*Stories of radioactivity at the site are false but persistent, especially in the kinds of supermarket tabloids that feature articles about 165-pound babies and "How Aliens Abducted My Wife."

edge of Hudson Bay and a basin (about 300 kilometers across) in central Europe. Still other craters may lie hidden under sediment on land or beneath the oceans.

Mass Extinction and Asteroid/Comet Impacts

At the end of the Cretaceous period, about 65 million years ago, an asteroid or comet hit the Earth. Its impact and the subsequent disruption of the atmosphere are blamed for exterminating the dinosaurs and many less conspicuous but widespread creatures and plants. In fact, the sudden disappearance of large numbers of life forms at the end of the Cretaceous period defines the end of the Mesozoic era.

The evidence that an extraterrestrial body caused this devastation comes from the relatively high abundance of the otherwise rare element iridium found in sediments from that time. Iridium, a heavy element similar to platinum, is very rare in terrestrial surface rocks because it dissolves easily in molten iron and is one of the densest elements, so most of it sank to the Earth's deep interior at the time our planet formed its core. On the other hand, samples of meteoritic material contain moderate amounts of iridium because most of these bodies have not formed iron cores. Thus, the presence of so much iridium in a layer of clay laid down 65 million years ago is suggestive of a link to meteoritic material. The amount of iridium in the earth sediments at that layer (fig. 11.21) is the quantity that would be dispersed from a piece of meteoritic material 10 kilometers in diameter. Therefore, many astronomers and paleontologists believe the Earth was hit by an asteroid of that size.

FIGURE 11.21
A layer of dark, iridium-rich clay marks the end of the Cretaceous period on Earth. The layer was probably formed when an asteroid struck the Earth. Fossils of many species are found below the layer (running diagonally across the image), but have disappeared from the more recent layers above it.

A 10-kilometer asteroid hitting the Earth would produce an explosion on impact equivalent to that of several billion nuclear weapons. The impact would not only make an immense crater, but it would also blast huge amounts of dust and molten rock into the air. The molten rock raining down would raise the surface temperature as high as that under an electric broiler and ignite global wildfires. The hot fragments and blast would also create nitrogen oxides, which would combine with water to

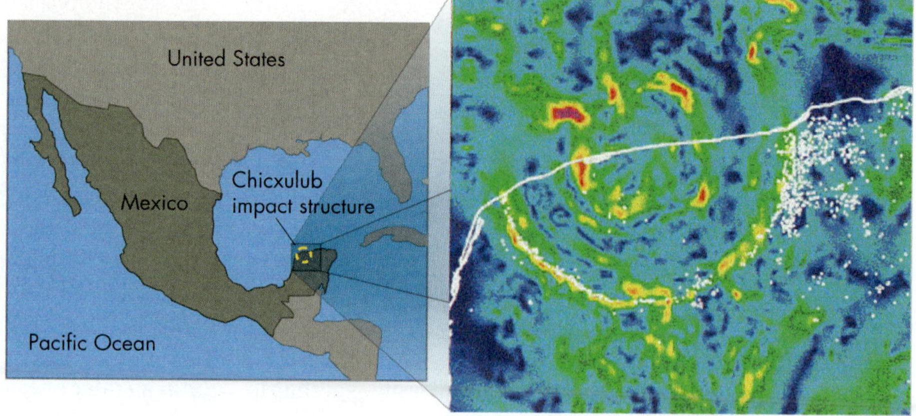

FIGURE 11.22
The Chicxulub impact structure from 65 million years ago is buried under a kilometer of sedimentary rock, so it is difficult to trace on the surface. However, geological maps of subsurface features clearly trace the edge of a 180-kilometer (110-mile) diameter impact crater.

form a rain of highly concentrated nitric acid. This devastating combination of heat, acid rain, and blast would then be followed by months of darkness and intense cold caused by the dust shroud blotting out the Sun. It seems likely that the biosphere would be devastated, leading to mass extinctions, just as the fossil record shows.

This frightening picture is supported not only by the iridium layer and the sudden disappearance of dinosaurs but also by a layer of soot, as well as a layer of tiny quartz pellets believed to have been created by the melting and blast of a violent impact. Moreover, geologists have found a large circular impact feature (figure 11.22) about 65 million years old near Chicxulub* (*cheek-shoo-loob*) in the Yucatán region of Mexico.

The Cretaceous mass extinction may have played an especially important role in our own evolution. Before that event, reptiles were the largest animals on Earth. Subsequently, mammals have assumed that niche. Small mammals may have escaped the fury of the heat and acid rain by remaining in burrows, and they may have survived the subsequent cold by virtue of their fur. You may be running your fingers through your hair, rather than your claws across your scales, because of that impact.

Other mass extinctions have occurred earlier and later than the Cretaceous event, but many scientists believe that massive volcanic eruptions or major changes in sea level might have caused them. Thus, like so many of the most interesting issues in science, no single accepted explanation has yet emerged.

*"Flea of the devil" in Mayan.

SUMMARY

Our Solar System contains numerous small bodies: asteroids, meteors, and comets. They are important astronomically because they give us information about the time of formation, composition, and physical conditions in the solar nebula.

Asteroids are rocky, metallic, or carbon-rich objects found mainly in the asteroid belt between the orbits of Jupiter and Mars. Comets are icy bodies found mainly in the Oort cloud, far beyond the orbit of Pluto.

A comet becomes visible if a passing star or some other event disturbs the comet's orbit so that it drops in toward the inner Solar System. There, solar heating thaws the frozen nucleus and evaporates gases. Radiation pressure and the solar wind then sweep the liberated gas and dust into a tail.

Some comets are caught in short-period orbits where they may be melted away to a collection of dust and grit. If the Earth crosses or nears the path of such a skeletal comet, the debris falling into the atmosphere causes a meteor shower.

The Earth is hit frequently by small pieces of asteroidal material which may reach the ground and are called meteorites. Occasionally a large asteroidal or cometary body strikes the Earth, producing craters or, in very rare instances, mass extinctions.

QUESTIONS FOR REVIEW

1. (11.1) What makes a "shooting star"?
2. (11.1) What is the difference between a meteor, a meteoroid, and a meteorite?
3. (11.1) How is a meteor heated?
4. (11.1) What kinds of meteorites are there?
5. (11.2) Where are most asteroids found?
6. (11.2) What shape are typical asteroids and how do we know? Why does Ceres not have this shape?
7. (11.2) How do we know that asteroids have a composition similar to that of some meteorites?
8. (11.2) What do asteroids tell us about the formation of the Solar System?
9. (11.2) What are near-Earth objects?
10. (11.3) What parts make up a comet? What are they made of? How do we know?
11. (11.3) Why are there two tails to some comets? What are they made of?
12. (11.3) What is the Oort cloud? What is the Kuiper belt?
13. (11.3) What is the life history of a comet from the Oort cloud that has become an object we see?
14. (11.1/11.3) What creates meteor showers? When do some occur?
15. (11.4) What evidence is there that the Earth has been hit by asteroids or comets?
16. (11.4) What was the Tunguska event?
17. (11.4) Why do some scientists believe that asteroids and comets play a role in mass extinctions?

THOUGHT QUESTIONS

1. (11.1//11.3) Explain the difference between the "tail" of a meteor and the "tail" of a comet. Are both of them hot gas and debris?
2. (11.2) The total mass of the asteroid belt is much less than the mass of any of the planets. If there were many, many more asteroids, do you think they could form a planet?
3. (11.2) The coast guard is monitoring satellite pictures of the ocean, looking for vessels traveling at night in faint moonlight. The images do not have enough detail to resolve the boats—they appear as a single dot in the image. If they see a dot, how can they estimate the size of the boat? How would this depend on the color of the boat? How would infrared images help determine the size of the boat? Compare this situation to the visual and infrared study of asteroids described in this chapter.
4. (11.2) Compare the compositions of asteroids with the compositions of Mars and of Jupiter. How does the composition of bodies across the asteroid belt support the solar nebular hypothesis?
5. (11.3) Examining the images of the nuclei of comets shown in section 11.3, do you think any of the crater-like depressions were caused by impacts like the ones that cause craters on asteroids? Explain your reasoning.
6. (11.4) If an asteroid were heading toward a possible future impact with Earth, what might be some advantages and disadvantages of setting off a nuclear explosion on it to deflect it to a different path?

PROBLEMS

1. (11.1) The speed with which a meteoroid hits the atmosphere is roughly the speed of the Earth in its orbit (you might say the Earth hits the meteoroid). Show that the speed of the Earth in its orbit is about 30 kilometers/second.
2. (11.2) Calculate the surface gravity and escape velocity for Ceres, assuming it has a radius of 950 kilometers and a mass of 9.5×10^{20} kilograms.
3. (11.3) Use Kepler's third law to find the semimajor axis of Comet Halley, given that its orbital period is 76 years.
4. (11.3) With the result of the previous problem and the fact that for a very elliptical orbit the distance farthest from the Sun is roughly twice the semimajor axis, estimate how far from the Sun Halley's comet gets. What planet is about that same distance from the Sun? How does this distance compare with the distance to the Kuiper belt?
5. (11.3) Use Kepler's third law to determine the period of a comet whose orbit extends to 50,000 AU, within the inner Oort cloud.
6. (11.3) Given that the temperature of a body decreases as the square root of its distance from the Sun increases, estimate the temperature of a comet nucleus in the Oort cloud. Take the temperature at 1 AU to be 300 Kelvin.
7. (11.1/11.4) Use the formula for kinetic energy of a moving body to estimate the energy of an SUV (mass about 2700 kilograms) traveling at 65 miles per hour (you'll need to do a few conversions on units). Compare this to the kinetic energy of a 0.010-kilogram meteoroid (about as heavy as two quarters) that collides with the Earth at 30 kilometers/second.
8. (11.4) Calculate the kinetic energy of impact of a 1000-kilogram (roughly 1-ton) object hitting the Earth at 30 kilometers per second. Express your answer in kilotons of TNT, using the conversion that 1 kiloton is about 4×10^{12} joules. Be sure to convert kilometers/second to meters/second.

TEST YOURSELF

1. (11.1) The bright streak of light we see as a meteoroid enters our atmosphere is caused by
(a) sunlight reflected from the solid body of the meteoroid.
(b) radioactive decay of material in the meteoroid.

(c) a process similar to the aurora that is triggered by the meteoroid's disturbing the Earth's magnetic field.
(d) frictional heating as the meteoroid speeds through the gases of our atmosphere.
(e) the meteoroid's disturbing the atmosphere so that sunlight is refracted in unusual directions.

2. (11.1/11.3) Meteor showers such as the Perseids in August are caused by
(a) the breakup of asteroids that hit our atmosphere at predictable times.
(b) the Earth passing through the debris left behind by a comet as it moves through the inner Solar System.
(c) passing asteroids triggering auroral displays.
(d) nuclear reactions in the upper atmosphere triggered by an abnormally large meteoritic particle entering the upper atmosphere.
(e) none of the above.
3. (11.1) Meteoroids originate from (there may be more than one correct answer):
(a) comet debris.
(b) material left over from the formation of the Solar System.
(c) asteroids.
(d) material blasted off other planets in collisions.
4. (11.2) The asteroid belt lies between the orbits of
(a) Earth and Mars.
(b) Saturn and Jupiter.
(c) Venus and Earth.
(d) Mars and Jupiter.
(e) Pluto and the Oort cloud.
5. (11.2) Asteroids in the asteroid belt are made up of
(a) iron. (b) silicates (rock). (c) organic compounds.
(d) a and b only, in amounts that vary depending on where they orbit (and were formed).
(e) a, b and c, in amounts that vary depending on where they orbit (and were formed).
6. (11.3) The tail of a comet
(a) is gas and dust pulled off the comet by the Sun's gravity.
(b) always points away from the Sun.
(c) trails behind the comet, pointing away from the Sun as the comet approaches it and toward the Sun as the comet moves out of the inner Solar System.
(d) is gas and dust expelled from the comet's nucleus by the Sun's heat and radiation pressure.
(e) Both (b) and (d)
7. (11.3) Astronomers think that most comets come from
(a) interstellar space.
(b) material ejected by volcanic eruptions on the moons of the outer planets.
(c) condensation of gas in the Sun's hot outer atmosphere.
(d) small icy bodies in the extreme outer parts of the Solar System that are disturbed into orbits that bring them closer to the Sun.
(e) luminous clouds in the Earth's upper atmosphere created when a small asteroid is captured by the Earth's gravitational force.
8. (11.3) Short period comets have a period of around ________ and mostly come from the ________.
(a) decades to a few hundred years; Kuiper belt
(b) a few hundred to a thousand years; Kuiper belt
(c) decades to a few hundred years; Oort cloud
(d) a few hundred to a thousand years; Oort cloud
(e) a thousand to a million years; Oort cloud
9. (11.4) A moving object's kinetic energy depends on its ________ and ________.
(a) size, density
(b) density, velocity
(c) velocity, mass
(d) magnetic field, velocity
(e) It depends only on the velocity
10. (11.4) Strong evidence that the dinosaurs were killed by a meteor impact is provided by (there may be more than one correct answer, select all that apply):
(a) mass extinctions 65 million years ago.
(b) a large crater in Arizona.
(c) pieces of the asteroid that have been recovered.
(d) an unusually rich layer of a rare element in the rock record at 65 million years ago.
(e) a layer of 65-million-year-old soot in the rock record.

KEY TERMS

achondrite, 290
asteroid, 290
asteroid belt, 290
carbonaceous chondrite, 290
chondrite, 289
chondrule, 289
coma, 295
fluorescence, 296
Kirkwood gaps, 293
Kuiper belt, 298
meteor, 288
meteor shower, 302
meteorite, 288
meteoroid, 288
near-Earth object, 294
nucleus, 295
Oort cloud, 298
radiant, 302
radiation pressure, 299
short-period comet, 300
solar wind, 299
tail, 295

FURTHER EXPLORATIONS

Asphaug, Erik. "The Small Planets." *Scientific American* 282 (May 2000): 46.

Becker, Luan. "Repeated Blows." *Scientific American* 286 (March 2002): 76.

Binzel, Richard P., M. Antonietta Barucci, and Marcello Fulchignoni. "The Origin of the Asteroids." *Scientific American* 265 (October 1991): 88.

Cowen, Ron. "The Day the Dinosaurs Died." *Astronomy* 24 (April 1996): 34.

Davis, Joel. "250 Million Years Ago, Did an Asteroid Nearly End Life on Earth?" *Astronomy* 36 (April 2008): 34.

Dodd, R. T. *Thunderstones and Shooting Stars*. Cambridge, Mass.: Harvard University Press, 1986.

Gallant, Roy A. "Journey to Tunguska." *Sky and Telescope* 87 (June 1994): 38.

Gasperini, Lucas, Enrico Bonatti, and Guiseppe Longo. "The Tunguska Mystery." *Scientific American* 298 (June 2008): 80.

Gehrels, Tom. "Collision with Comets and Asteroids." *Scientific American* 274 (March 1996): 54.

Grieve, Richard A. "Impact Cratering on the Earth." *Scientific American* 262 (April 1990): 60.

Lewis, John S. *Rain of Iron and Ice.* Reading, Mass.: Addison-Wesley, 1996.

Littmann, Mark. "From Chaos to the Kuiper Belt." *Sky and Telescope* 114 (September 2007): 22.

Littmann, Mark. "Dark Beasts of the Trans-Neptunian Zoo." *Sky and Telescope* 114 (November 2007): 26.

Luu, Jane X., and David C. Jewitt. "The Kuiper Belt." *Scientific American* 274 (May 1996): 46.

Schaffer, Bradley E. "Comets That Changed the World." *Sky and Telescope* 93 (May 1997): 46.

Semeniuk, Ivan. "Asteroid Impact." *Mercury* 31 (November/December 2002): 24.

Stern, Alan. "The Sun's Fab Four [Ceres, Pallas, Juno, and Vesta]." *Astronomy* 23 (June 1995): 30.

———. "The 3rd Zone: Exploring the Kuiper Belt." *Sky and Telescope* 106 (November 2003): 30.

Warner, Elizabeth M., and Greg Redfern. "Deep Impact: Our First Look Inside a Comet." *Sky and Telescope* 109 (June 2005): 40.

Weissman, Paul R. "The Oort Cloud." *Scientific American* 279 (September 1998): 84.

Wynn, Jeffrey C., and Eugene M. Shoemaker. "The Day the Sands Caught Fire." *Scientific American* 279 (November 1998): 64.

Website

Visit the *Explorations* website at **http://www.mhhe.com/arny** for additional online resources on these topics.

Q FIGURE QUESTION ANSWERS

WHAT IS THIS? (chapter opening): This image was made by combining a number of separate images of the Leonid meteor shower taken just before dawn on November 19, 2002. You can clearly see that the meteors appear to radiate from a small area on the sky. That area, it turns out, lies in the constellation Leo, hence the name of the shower. The bright spot near the center of the image is the planet Jupiter.

FIGURE 11.1: The star trails show that this was a fairly long time exposure, during which the Earth rotated, smearing the stars' images into long arcs. The meteor passed by very quickly; the Earth didn't rotate very far during its passage, and as a result the meteor's track is straight.

FIGURE 11.4: The Trojan asteroids are about equidistant from the Sun and Jupiter, forming a nearly equilateral triangle. This position is known as a Lagrange point, and is a stable location for orbits. There are similar Lagrange points relative to the Moon orbiting the Earth or the Earth orbiting the Sun.

FIGURE 11.8: The gaps in the distribution of the asteroids are very like the gaps in Saturn's rings, such as Cassini's division.

FIGURE 11.20: Manicouagan Crater is far enough north that it was buried by glaciers during the last ice age. Moreover, being in a wet climate, it erodes faster and more extensively than the Arizona Meteor Crater, which is in a dry climate.

PROJECTS

1. **Meteor Hunt:** Pick a clear night and watch the night sky with a friend for half an hour. Count the number of meteors you see. Note the direction they come from. From your knowledge of these bodies, were the ones you saw (if any!) cometary or asteroidal? Note that meteors are generally more frequent before dawn, so you should try to observe then if possible.
2. **Meteor Shower:** Watch a meteor shower and count the number you see. Estimate the paths of the meteors on a star map and see if you can find the position they seem to radiate from and mark it on a star map. If it is summer, you might try for the Perseids in mid-August (August 10–14); in autumn, the Orionids (October 18–23) or the Geminids (December 10–13); in the spring, the Lyrids (April 20–22). Other showers are listed in appendix table 7. Most are best when viewed before dawn in a dark sky without bright moonlight. What was the average time interval between meteors? Does your estimate of the radiant match the predicted position?

Magnetic loops in the Sun's lower atmosphere. Gas trapped along the loops is heated by magnetic activity and in turn heats the Sun's corona.

KEY CONCEPTS

- The Sun is an immense ball of gas.
- Gravity holds it together and compresses the core to an extremely high density and temperature.
- Gas pressure prevents it from collapsing under the force of its gravity.
- Nuclear fusion in its core supply its power.
- The fusion of 4 hydrogen nuclei into a helium nucleus results in a small reduction in mass.
- The mass lost during fusion is converted into energy according to Einstein's formula $E = mc^2$.
- Motions in the Sun's interior generate magnetic effects, such as sunspots, prominences, and flares.
- The Sun's magnetism changes over time.
- Solar magnetism affects Earth:
  - It creates the aurora.
  - It appears to affect our climate.

12 The Sun, Our Star

CONCEPTS AND SKILLS TO REVIEW

- Relation between temperature of a hot object and the color of the light it emits (Wien's law) (4.2)
- Doppler shift (4.6)
- Structure of atoms (4.4)

The Sun is a star, a dazzling, luminous ball of gas more than 100 times bigger in diameter than the Earth. The Sun is the source of light that heats the planets and maintains the comfortable temperatures that allow life to fluorish on Earth. In fact, the Sun is the source of most of the energy that we use in our everyday lives. For example, the hydropower from a dam relies on solar heating to evaporate water so it can rise into the atmosphere, where it is carried by winds (also driven by the Sun) to precipitate over high elevations and then run downhill again. Even fossil fuels represent solar energy stored up by plants in chemical bonds millions of years ago.

Although it gleams peacefully in our daytime sky, a telescope reveals the Sun's surface to be violently agitated, with rising fountains of incandescent gas and a twisted magnetic field. Even greater violence wracks its core. There, a nuclear "furnace" burns 600 million tons of hydrogen into helium every second, producing in one heartbeat the energy of 100 billion nuclear bombs.

How the Sun releases vast quantities of energy while managing to hold itself together is one of the main themes of this chapter. We will begin by describing the Sun: its radius, mass, and so on. Then we will discuss how the crushing force of its gravity balances the explosive power in its core. Finally, we will see how energy escaping from the core stirs its atmosphere into the inferno we see.

Q: WHAT IS THIS? See end of chapter for the answer.

12.1 Size and Structure

The Sun is immense, dwarfing the Earth and even Jupiter, and its immensity is what makes it shine. With a radius over 100 times that of the Earth and a mass 300,000 times the Earth's, the Sun has an enormous gravity that crushes the material in its interior. To offset that crushing force and prevent its own collapse, the Sun must be extremely hot. But hot objects always lose energy, and the Sun is no exception. We see that lost energy as sunshine and welcome it as the source of our life. But sunshine is a death warrant for the Sun because the energy it carries off must be replenished or the Sun will collapse. Fortunately for us, the Sun does replace its lost energy, but only at the cost of consuming itself—a dilemma that is not unique to the Sun but is shared by most stars.

Before we discuss how the Sun replaces its lost energy, we need to understand better some of its overall properties. How much mass does it contain? How rapidly does it lose energy? What resources are available to supply its energy needs? Table 12.1 lists some of these vital statistics, and the portrait of the Sun in figure 12.1 shows its main features. It will be helpful many times in this chapter to refer to this picture.

Astronomers can measure some of the Sun's properties, such as its diameter and surface temperature. However, they can only deduce most of its other properties, such as its internal temperature and density. Such deductions are based on computer models that use the laws of physics to calculate the Sun's properties. The correctness of the model's predictions is then judged by whether the predictions agree with observable properties. For example, if a model predicted the Sun's surface temperature to be 10,000 K, it would be rejected as incorrect. Thus, our understanding of the Sun's properties comes from a combination of theory and measurement.

Measuring the Sun's Properties

The Sun is about 150 million kilometers (1 AU, or 93 million miles) from Earth. Astronomers originally measured its distance by triangulation, but they now use radar, bounced either directly from the Sun or from other bodies whose distance is known in AU from Kepler's third law. Once the Sun's distance is known, we can find its radius from its angular size, as we showed in chapter 2. We also need to know the Sun's distance from Earth if we are to measure its mass with the help of Kepler's third law, as we showed in chapter 2. From its mass and radius, we can calculate that its surface gravity is about 30 times the Earth's.

TABLE 12.1 PROPERTIES OF THE SUN

Radius ($R_\odot$*)	7×10^8 m $= 7 \times 10^5$ km, or about 109 R_{Earth}
Mass ($M_\odot$*)	2×10^{30} kg, or about 333,000 M_{Earth}
Distance from Earth	1.5×10^8 km, or 1 AU
Temperature of surface	5780 Kelvin (approx. 9900°F)
Temperature of core	15 million Kelvin (approx. 27 million °F)
Composition	71% hydrogen, 27% helium, 2% heavier elements
Power output	4×10^{26} watts

*Astronomers use the symbol ⊙ to stand for the Sun. Thus, $R_\odot$ is the Sun's radius and $M_\odot$ is its mass. The symbol is the ancient Egyptian hieroglyph for "Sun."

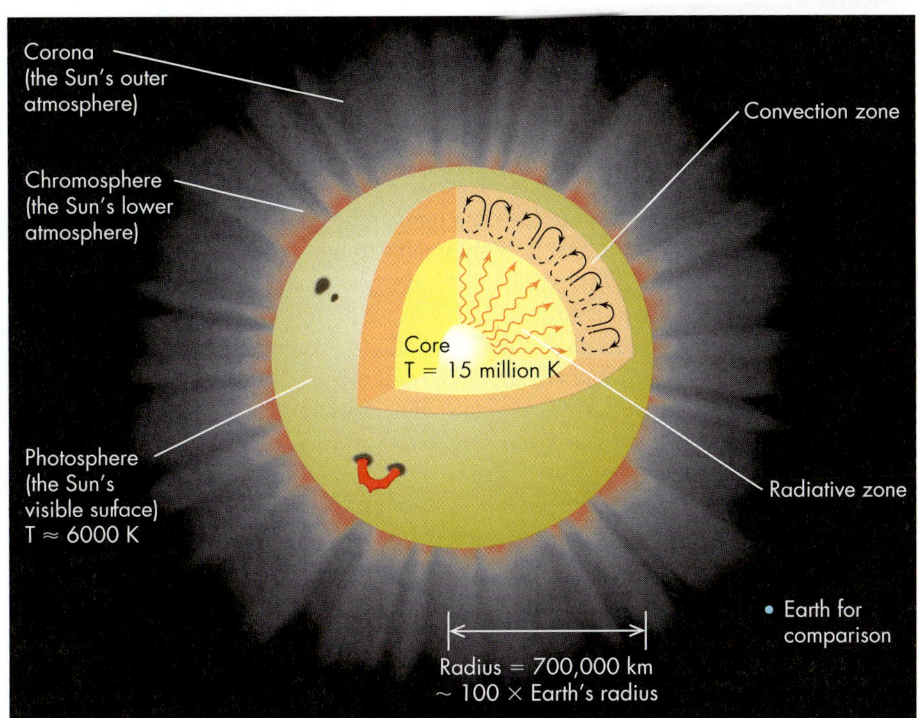

FIGURE 12.1
A cutaway sketch of the Sun.

The Sun's internal heat balances this crushing force of gravity. Its surface temperature can be found from its color and Wien's law, using the methods discussed in chapter 4. Astronomers cannot measure its interior temperature directly; rather, from calculations and indirect measurements (discussed in section 12.3), they deduce that its core temperature is about 15 million K (approximately 27 million °F).

From measurements of the amount of solar energy that reaches the Earth, astronomers calculate that the Sun radiates a total of 4×10^{26} watts of power into space from its surface, energy that must be replenished by the fusion of hydrogen in its core, as we will see in section 12.2. Fortunately, the Sun has a plentiful supply of hydrogen: its spectrum shows it is about 71% hydrogen, 27% helium, and 2% vaporized heavier elements such as carbon and iron, a composition similar to Jupiter's and Saturn's. But unlike these mostly liquid bodies, the Sun is gaseous throughout because its high temperature breaks most molecular bonds, vaporizing even iron, and allowing the atoms to move freely as a gas.

The Solar Interior

When we observe the Sun, we see through the low-density, tenuous gases of its outer atmosphere. Our vision is ultimately blocked, however, as we peer deeper into the Sun. There, the material is compressed to high density by the weight of the gas above it. In this dense material, the atoms are sufficiently close together that they strongly absorb the light from deeper layers, blocking our view of them much like frosted glass obscures what lies behind it. Above these layers, however, the absorption of light is weaker and the light emitted by hot material there escapes freely into space as the sunlight we see. These layers where the Sun's gases change from transparent to opaque form what is called the **photosphere,** the visible surface of the Sun. Although the blocking of radiation by denser gas below the photosphere limits our view into the Sun, it helps the Sun retain heat and, like a well-insulated house, thereby reduces the amount of fuel it must consume.

If we could see into the Sun, we would find that its density and temperature rise steadily inward. In the photosphere, the density is comparable to that of the air

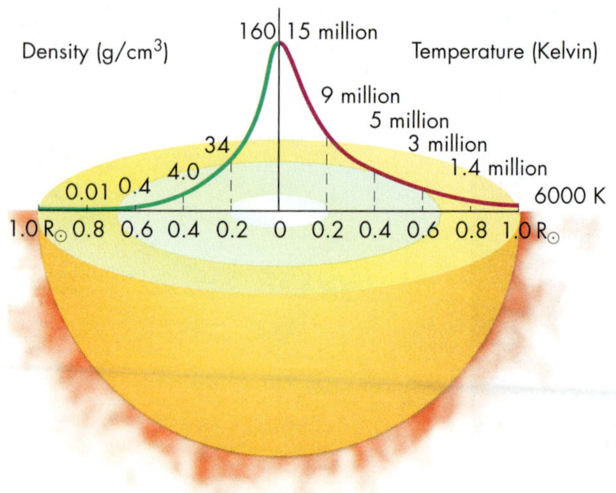

FIGURE 12.2
Plots of how density and temperature change through the Sun.

around us; deeper, the material above pushes down on that below, compressing the gas like a pile of pillows. A similar compression occurs in the atmosphere of the Earth and other planets, as we discussed in chapters 6 and 10. But the greater mass of the Sun leads to a vastly greater compression of its gas, and so, near its core, the density is more than 100 times that of water. Despite this great density, the Sun is gaseous throughout because its high temperature gives the atoms so much energy of motion that they are unable to bond with one another to form a liquid or solid substance.

The temperature also rises as we plunge into the Sun's interior. The photosphere is almost 6000 K. Below it, heat is partially trapped and the temperature soars to about 15 million K at the core. Figure 12.2, based upon theoretical calculations, illustrates how the temperature and density change through the Sun. No spacecraft has made, or is ever likely to make, the measurements directly, but we are confident that they are correct because the Sun needs such high temperatures and densities to keep it from collapsing under its own gravity.

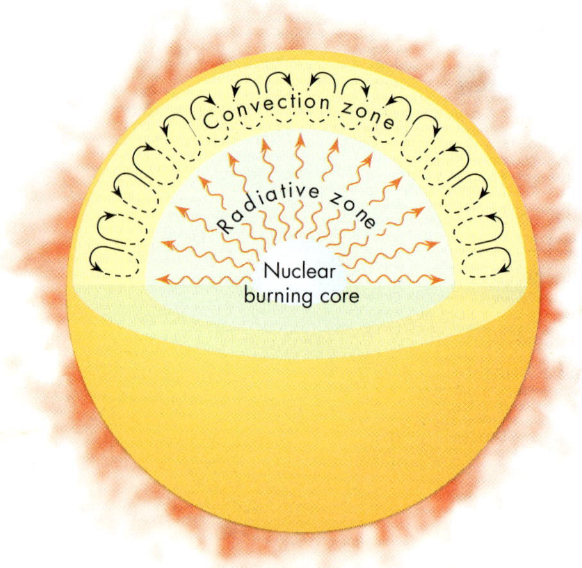

FIGURE 12.3
Sketch of how energy flows from the Sun's core to its surface. In the deep interior, radiation carries the energy. Near the surface, convection carries the energy.

Energy Flow in the Sun

Your own experience and experiments in the laboratory show that heat always flows from hot to cold. Applying this principle to the Sun, we can infer that because its core is hotter than its surface, heat will flow outward from its center, as illustrated in figure 12.3. Near the core, the energy moves by radiation carried by photons through what is called the **radiative zone.** Because the gas there is so dense, a photon travels less than an inch before it is absorbed by an atom and stopped. The photon is eventually re-emitted, but it will be almost immediately reabsorbed. The constant absorption and re-emission slows photons like cars caught in stop-and-go traffic. Even though photons travel at the speed of light between absorptions, it takes them about 16 million years to move from the core to the surface.* Thus, today's sunshine was born in the Sun's core before we existed as a species!

The flow of energy from the core toward the Sun's surface is slowed in the region just below the photosphere, where the gas is cooler and less transparent. Here, photons are even less effective in moving energy, and convection currents like those in the Earth and giant planets carry the energy to the surface. The rising and sinking gas occupies the **convection zone,** and we can infer the gas's motion there from the numerous tiny bright regions surrounded by narrow darker zones, called **granulation** (fig. 12.4). The bright areas are bubbles of hot gas many hundreds of kilometers across, rising from deep within the Sun. They are brighter because they are hotter than the gas around them. On reaching the surface, these hot bubbles radiate their heat to space and cool. The cooler matter then sinks back toward the hotter interior, where it is reheated and rises again to radiate away more heat. Astronomers can measure the speed of these up-and-down motions using the Doppler effect and find that the bubbles rise at about 1 kilometer per second.

*The question of how long it takes a photon to travel from the Sun's center to its surface turns out to be tricky. One reason is that a single photon never makes such a journey. A typical photon leaving the core is a gamma ray, which will be converted to many lower-energy photons before its energy reaches the surface. Therefore, it is perhaps better to ask how long it takes energy to travel from the Sun's core to the Sun's surface. That time turns out to be about 16 million years, according to one calculation.

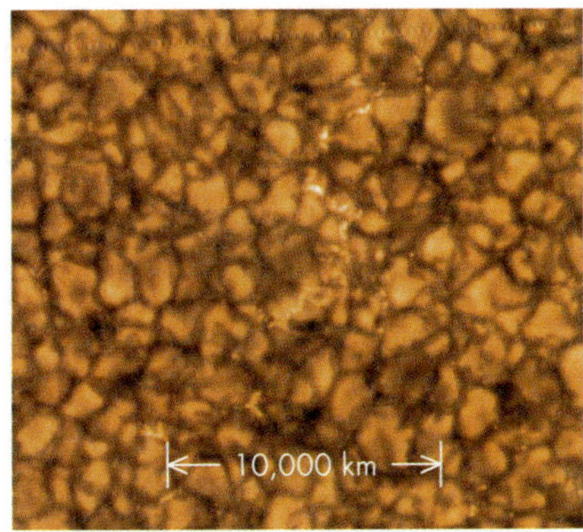

FIGURE 12.4
Granulation in the Sun's photosphere. Each brighter patch is an immense bubble of hot gas rising from deep within the Sun. The darker material outlining each bubble is cooler gas sinking back into the Sun's interior.

Q. How big across is one of the rising bubbles that forms the granulation? What has a similar size on Earth?

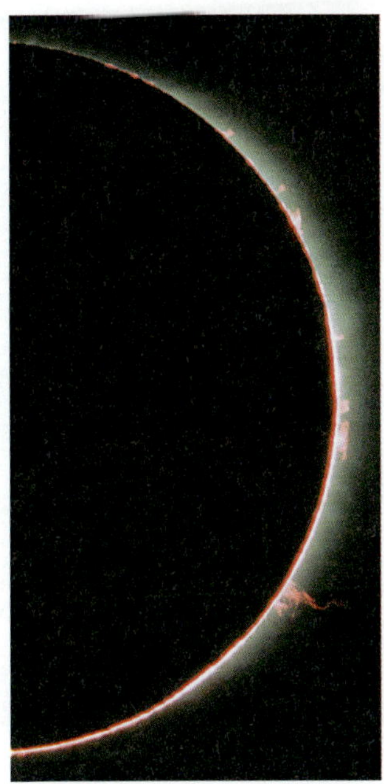

FIGURE 12.5
Photograph of a portion of the solar chromosphere at a total solar eclipse.

FIGURE 12.6
Photograph of spicules in the chromosphere. The spicules are the thin, stringy features that look like tufts of grass.

The Solar Atmosphere

Astronomers refer to the low-density gases that lie above the photosphere as the Sun's atmosphere. This region marks a gradual change from the dense gas of the photosphere to the extremely low-density gas of interplanetary space. A similar transition occurs in our own atmosphere, where the gas density decreases steadily with altitude and eventually merges with the near-vacuum of space.

Although the gas density in the Sun's atmosphere decreases above the photosphere, the gas temperature behaves very differently. Immediately above the photosphere the temperature decreases, but at higher altitudes the gas grows hotter, reaching temperatures of several million Kelvin. Why the Sun's atmosphere is so hot remains a mystery, though astronomers believe that the Sun's magnetic field plays a role in somehow heating these low-density gases, as we will discuss in section 12.4.

The Sun's atmosphere consists of two main regions. Immediately above the photosphere lies the **chromosphere,** the Sun's lower atmosphere. It is usually invisible against the glare of the photosphere but can be seen at a total eclipse of the Sun as a thin red zone around the Sun (fig. 12.5). With a telescope equipped with an appropriate filter, you can see that the chromosphere contains millions of thin columns called **spicules** (fig. 12.6), each a jet of hot gas thousands of kilometers long.

The chromosphere's color* comes from the strong red emission line of hydrogen, Hα. We saw in chapter 4 that emission lines arise in hot, low-density gas. From these lines astronomers can infer the gas's temperature. Just above the photosphere, the temperature is about 4500 K, but 2000 kilometers higher, it reaches 50,000 K. Here, the chromosphere ends and the temperature shoots up to about 1 million degrees as we enter the **corona,** the Sun's outer atmosphere.

The corona's extremely hot gas has such low density that under most conditions we look right through it. But like the chromosphere, it can be seen during a total

*The prefix *chromo-* means "colored."

FIGURE 12.7
Photograph of the corona at a total eclipse of the Sun.

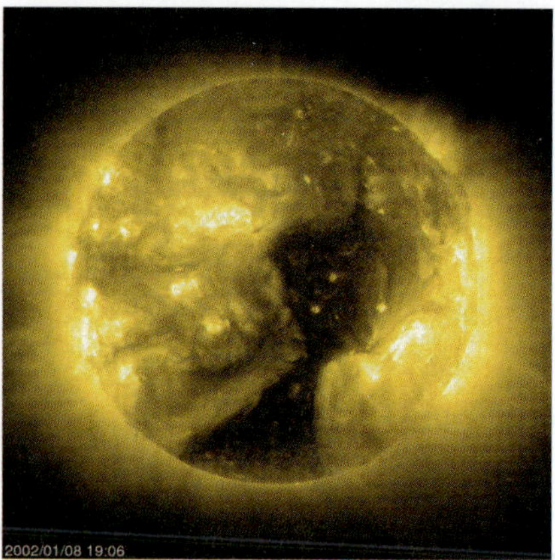

FIGURE 12.8
Image of the Sun in ultraviolet light made with the *SOHO* satellite. The dark region is a coronal hole where gas is streaming out from the Sun.

solar eclipse when the Moon covers the Sun's brilliant disk. Then the pale glow of the corona can be seen to extend beyond the Sun's edge to several solar radii (fig. 12.7). Pictures of the Sun made at ultraviolet and X-ray wavelengths show that the corona is not uniform but has extremely hot streamers along the Sun's magnetic field. As shown in figure 12.8, the corona also contains huge low-density regions called **coronal holes** through which gas escapes from the Sun into space, as we will discuss in section 12.4.

Because the corona is so tenuous, it contains very little energy despite its high temperature. It is like the sparks from a Fourth of July sparkler: despite their high temperature, you hardly feel them if they land on your hand because they are so tiny and thus carry very little total heat.

12.2 How the Sun Works

Our discussion so far has centered on the structure of the Sun. Our task now is to understand *why* it has that structure and how it works. For example, why is the Sun hot? What makes it shine? How does it generate energy?

Internal Balance (Hydrostatic Equilibrium)

There are many examples of hydrostatic equilibrium in everyday life. Our atmosphere is in hydrostatic equilibrium, its gases pulled downward by the Earth's gravitational force but supported by collisions of the air molecules, creating pressure. Likewise, pressure in an automobile tire supports the weight of the car.

The structure of the Sun depends on a balance between its internal forces. One force holds the Sun together. A second force prevents the Sun from collapsing. This balance is technically called **hydrostatic equilibrium.**

The Sun's inward force arises from its own gravity. The outward force arises from the rapid motion of its atoms, a motion that gives rise to a **pressure.** Thus, in the Sun, as in virtually all stars and planets, the balance of hydrostatic equilibrium requires that the *outward* force created by pressure exactly balance the *inward* force of the Sun's gravity (fig. 12.9). Without such a balance, the Sun would rapidly change. For example, if its pressure were too weak, the Sun's own gravity would rapidly crush it. Therefore, to understand the Sun, we need to discuss in more detail how its pressure arises.

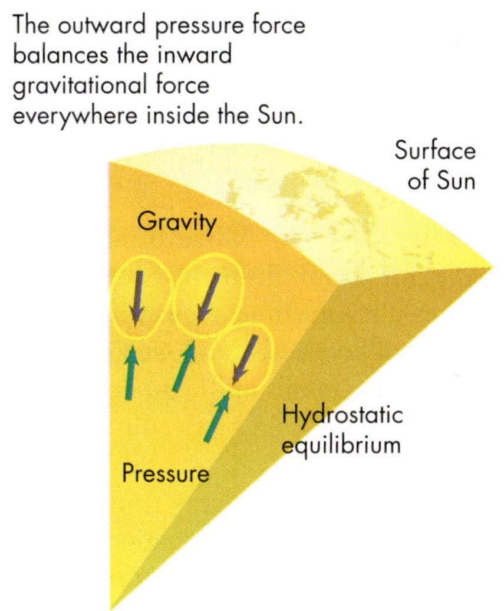

FIGURE 12.9
A sketch illustrating the condition of hydrostatic equilibrium, the balance of pressure (blue arrows) and gravitational force (purple arrows) in the Sun.

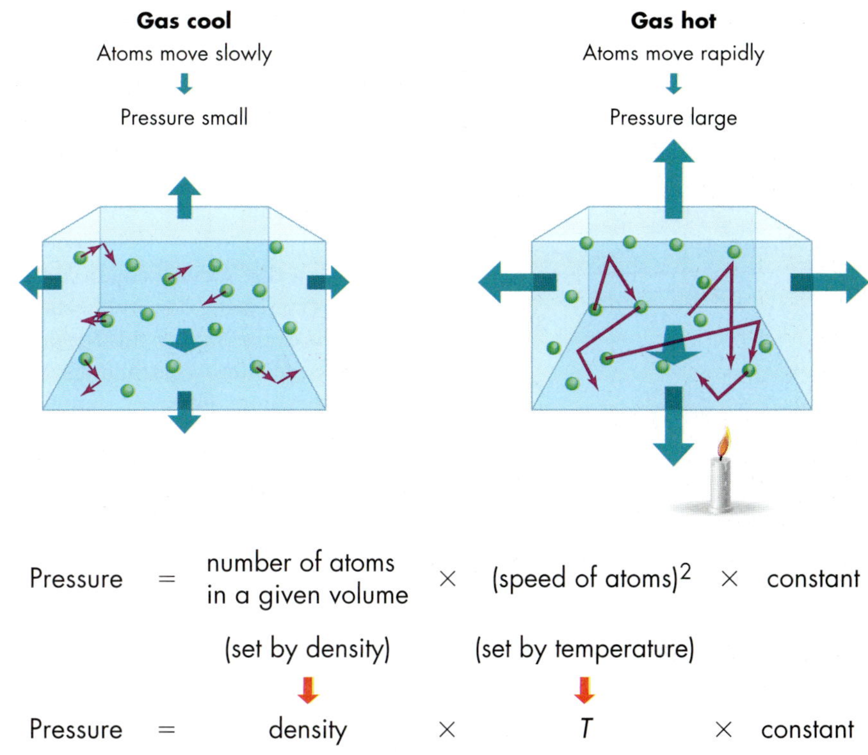

FIGURE 12.10
Sketch illustrating the perfect, or ideal, gas law. Gas atoms move faster at the higher temperature, so they collide both more forcefully and more often than atoms in a cooler gas. These effects combine to create a higher pressure. Thus, other things being equal, a hotter gas exerts a greater pressure.

Pressure in the Sun

Pressure in a gas comes from collisions among its atoms and molecules. If the gas is squeezed, atoms are pushed toward each other. As they collide, they rebound, resisting the compression, as you can feel by squeezing a balloon. The strength of the pressure depends on how often and how hard the collisions occur. Raising the density increases collisions by moving atoms closer together. Raising the temperature speeds atoms up, making them collide harder and more often (fig. 12.10). Thus, the strength of the pressure is proportional to the density times the temperature of the gas; that is:

$$\text{Pressure} = \text{Density} \times \text{Temperature} \times \text{a constant}^*$$

a result known as the **perfect,** or **ideal, gas law.**

Although the perfect gas law shows us that the pressure inside the Sun depends on temperature, it does not by itself show us how hot that temperature needs to be. We might reasonably guess that because the Sun has a huge mass (and therefore a huge gravity crushing it), it needs a huge temperature to offset that crushing force.

Indeed, the Sun's core is very hot. To determine exactly how hot, astronomers solve a computer model of the Sun, and the answer turns out to be about 15 million K for the temperature and about 160 grams per cubic centimeter for the density, as we claimed earlier and showed in figure 12.2. But because heat flows out of the hot core, the Sun needs a mechanism to keep replacing its core heat.

Powering the Sun

Energy that leaves the core eventually escapes into space as sunshine: heat and light. That heat loss must be replaced, or the Sun's internal pressure would drop and the Sun

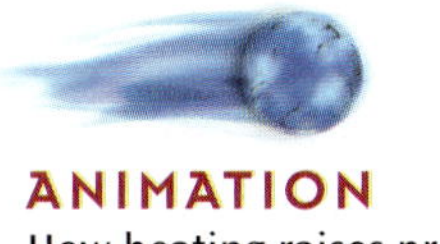

ANIMATION
How heating raises pressure

*The value of the constant depends on the units used to measure the pressure, density, and temperature.

would begin to shrink under the force of its own gravity. The Sun is therefore like an inflatable chair with tiny leaks through which the air escapes. If you sit in the chair, it will gradually collapse under your weight unless you pump air in to replace that which escapes. What acts as the energy pump for the Sun?

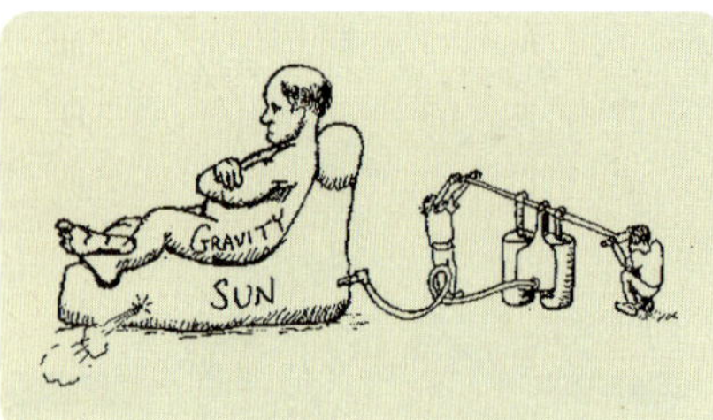

Early astronomers believed that the Sun might burn ordinary fuel such as coal. But even if the Sun were pure coal, it could burn only a few thousand years. At the end of the nineteenth century, the English physicist Lord Kelvin and the German scientist Hermann Helmholtz independently proposed that the Sun is not in hydrostatic equilibrium but that gravity slowly compresses it, making it shrink. In their theory, compression (increasing density) heats the gas and makes the Sun shine. However, gravity could power the Sun by this mechanism only for 10 million years, and we know the Sun has been shining for billions of years because we have fossils of life on Earth that old. Furthermore, if gravity were the power source, the Sun would be steadily shrinking, and it isn't. Therefore, something else must supply energy.

In 1899, T. C. Chamberlin suggested that subatomic energy—energy from the reactions of atomic nuclei—might power stars, but he could offer no explanation of how the energy was liberated. In 1905, Einstein proposed that energy might come from a body's mass. His formula,

c = Speed of light: 3×10^8 m/sec
m = An amount of mass
E = An amount of energy

$$E = mc^2$$

states that a mass, m, can become an amount of energy, E, equal to the mass multiplied by the square of the speed of light, c, where $c = 3 \times 10^8$ meters per second. The amount of energy available from reactions involving atomic nuclei is vastly larger than that from chemical processes such as burning. For example, if 1 gram of mass—less than the amount in an aspirin tablet—is converted to energy, it releases 9×10^{13} joules, the equivalent of about 20 kilotons of TNT* or a small nuclear weapon. If the Sun could convert even a tiny amount of its mass into energy, it would have an enormous source of power.

In 1919 the English astrophysicist A. S. Eddington, a pioneer in the study of the physics of stars, showed that the conversion of hydrogen into helium would release enough energy to power the Sun, but even his theory lacked the necessary details, which were supplied independently in the late 1930s by the physicists Hans A. Bethe and Carl F. von Weizsäcker. Their work showed that the Sun could generate its energy by converting hydrogen into helium by a process called **nuclear fusion,** a process that bonds two or more nuclei into a single, heavier one.

Nuclear Fusion. Fusion is possible in the Sun because its interior is so hot. Under normal conditions, hydrogen nuclei repel each other, pushed apart by the electrical charge of the protons. However, at high temperatures, nuclei move so fast that when they collide, they are driven extremely close together. At such a collision, the nuclei may get close enough to each other that the electrical repulsion between their protons is overwhelmed by the **nuclear force,** or **strong force,** the force that holds the protons together in a normal atom.† Thus, the two separate nuclei can merge, or fuse, into a single new nucleus. Because this fusing process requires such a high temperature, the only place in the Sun hot enough for fusion to occur is its core. The core therefore is where the Sun makes its energy. However, before we can understand how fusion creates energy, we need to look at the structure of hydrogen and helium.

The Structure of Hydrogen and Helium. Hydrogen consists of one proton and an orbiting electron, and helium consists of two protons, two neutrons, and two orbiting

*One kiloton is about 4.2×10^{12} joules of energy.

†Although the bonding of protons and neutrons into nuclei results from the strong force, the first step in the proton–proton chain requires the operation of the weak force to convert one of the protons into a neutron.

electrons. In the Sun's hot interior, however, atoms collide so violently that the electrons are generally removed. Atoms that are missing one or more electrons are said to be ionized. A gas in which the atoms are ionized is technically called a "plasma." Because the electrons are removed from most atoms and move independently in the gas, we can therefore ignore them in much of what follows.

Hydrogen and helium always have one and two protons respectively, but they have other forms (called "isotopes") with different numbers of neutrons (fig. 12.11). To identify the isotopes, we write their chemical symbol with a superscript that shows the total number of protons and neutrons. The usual form of hydrogen with one proton and no neutrons is ^{1}H, while the form of hydrogen containing a proton and a neutron is ^{2}H, which is also known as deuterium or "heavy hydrogen." Likewise, the normal form of helium with two protons and two neutrons is ^{4}He, while helium with two protons but only one neutron is ^{3}He. These are often called "helium-4" and helium-3," respectively. As we will see, these isotopes play a critical role in the Sun's energy supply.

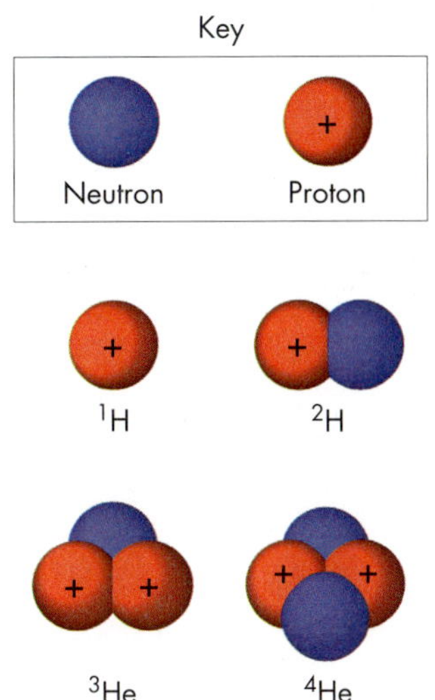

FIGURE 12.11
Schematic diagrams of the nuclei of hydrogen, its isotope deuterium, and two isotopes of helium.

The Proton–Proton Chain. Hydrogen fusion in the Sun occurs in three steps, called the **proton–proton chain.** In the first step, two ^{1}H nuclei collide and fuse to form the isotope of hydrogen ^{2}H. In the collision, one proton becomes a neutron by ejecting two particles: a positron (denoted as e^-) and a **neutrino** (denoted by ν). The neutrinos so generated play no further role in the Sun's energy generation, but we will encounter them again because they can be detected when they leave the Sun and may help astronomers learn more about conditions in the Sun's interior.

This first step in converting the mass of an atom into the Sun's energy is depicted in figure 12.12A and can be written symbolically as

$$^1\mathrm{H} + {}^1\mathrm{H} \rightarrow {}^2\mathrm{H} + e^- + \nu$$

The terms to the left of the arrow are the normal hydrogen nuclei that start the process. The terms to the right are the heavy isotope of hydrogen that results (^{2}H), plus the positron and the neutrino. Energy is released by this reaction because the mass of ^{2}H is slightly less than the mass of the two ^{1}H's that were used to make it. The missing mass is converted to energy, as described in Einstein's formula $E = mc^2$, and ultimately that energy becomes the electromagnetic radiation of sunlight.

In the second step, the ^{2}H nucleus collides with a third ^{1}H to make the isotope of helium containing a single neutron, ^{3}He. This process releases a high-energy photon (gamma ray), which we denote by γ. Figure 12.12B shows this step, which can be written as:

$$^1\mathrm{H} + {}^2\mathrm{H} \rightarrow {}^3\mathrm{He} + \gamma$$

Here again, the resulting particle, ^{3}He, has a smaller mass than the particles from which it was made, and again energy is released.

The third and final step is the collision and fusion of two ^{3}He nuclei. Here, the fusion results not in a single particle but rather in one ^{4}He and two ^{1}H nuclei. You can think about this reaction as the attempt to form a nucleus with 4 protons and 2 neutrons, except that 2 protons are ejected by their electric repulsion, as shown in figure 12.12C. Symbolically, we can write it as:

$$^3\mathrm{He} + {}^3\mathrm{He} \rightarrow {}^4\mathrm{He} + {}^1\mathrm{H} + {}^1\mathrm{H}$$

where again, the final mass is less than the initial mass.

We can find the quantity of energy released by comparing the initial and final masses and using the mass–energy relationship $E = mc^2$. Steps 1 and 2 use three ^{1}H, but the first two steps must occur twice to make the two ^{3}He's for the last step. Therefore, six ^{1}H's are used, but two are returned in step 3, and so a total of four ^{1}H's are needed to make each ^{4}He. If we now add up their masses, we find the following:

The mass of a single hydrogen nucleus is 1.673×10^{-27} kilograms. The mass of a helium nucleus is 6.645×10^{-27} kilograms. Thus,

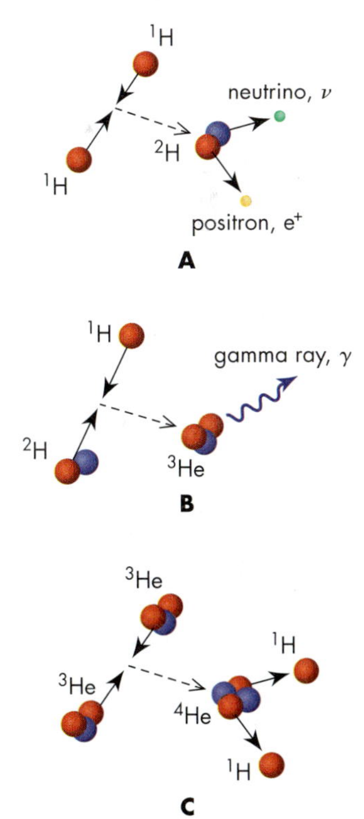

FIGURE 12.12
Diagram of the proton–proton chain. (A) Hydrogen (H) nuclei first combine to make deuterium (^{2}H). (B) Deuterium and hydrogen combine to make ^{3}He. (C) ^{3}He nuclei combine to form ^{4}He.

ANIMATION
The reactions of the proton–proton chain

$$4 \times 1 \text{ hydrogen} = 6.693 \times 10^{-27} \text{ kg}$$
$$-1 \text{ helium} = \underline{6.645 \times 10^{-27} \text{ kg}}$$
$$\text{Mass lost} = 0.048 \times 10^{-27} \text{ kg}$$

Multiplying this by c^2 gives the energy yield per helium atom made, E. That is,

$$E = 0.048 \times 10^{-27} \times (3 \times 10^8)^2 = 0.048 \times 10^{-27} \times 9 \times 10^{16} = 4.3 \times 10^{-12} \text{ joules}$$

This tiny number by itself is insignificant, but when multiplied by the vast number of hydrogen atoms undergoing fusion in the Sun, 4 million tons (4×10^9 kg) of mass is converted into energy. The total energy released is equivalent to exploding 100 billion-megaton H-bombs per second! Our sunshine has a violent birth.

It is important to keep in mind that although the Sun's power comes from the conversion of its mass to energy, the change in the Sun's mass is extremely small compared to the overall mass of the Sun. Nuclear fusion creates such a large amount of energy from even a tiny amount of mass that the decrease in the Sun's mass over its entire lifetime will be less than 1%. Far more mass.

12.3 Probing the Sun's Core

Although the immense bulk of the Sun hides its core from view, two types of experiments allow us an indirect look. Solar neutrinos escaping from the interior carry information about conditions there, and waves on the Sun's surface reveal the structure of the gas far below as waves in a stream reveal the presence of submerged rocks.

Solar Neutrinos

We saw in section 12.2 that the Sun makes neutrinos as it converts hydrogen into helium. The number of released neutrinos therefore tells us how rapidly hydrogen is being converted into helium, and from that we can deduce the temperature and density in the Sun's core.

But counting neutrinos is extremely difficult. Neutrinos have no electric charge and only a tiny mass,* which gives them phenomenal penetrating power. They escape from the Sun's core through its outer 700,000 kilometers and into space like bullets through wet Kleenex. They pass straight through the Earth and anything on the Earth, such as you, and keep going. In fact, roughly a trillion neutrinos from the Sun pass harmlessly through your body every second.

The penetrating power of neutrinos is the theme of a poem called "Cosmic Gall" by novelist and poet John Updike:

Neutrinos, they are very small.
They have no charge and have no mass
And do not interact at all.
The earth is just a silly ball
To them, through which they simply pass,
Like dustmaids down a drafty hall
Or photons through a pane of glass.
They snub the most exquisite gas,
Ignore the most substantial wall,
Cold-shoulder steel and sounding brass,
Insult the stallion in his stall,
And scorning barriers of class,
Infiltrate you and me! Like tall
And painless guillotines, they fall
Down through our heads into the grass.
At night they enter at Nepal
And pierce the lover and his lass
From underneath the bed—you call
It wonderful: I call it crass.

The elusiveness that allows neutrinos to slip so easily through the Sun makes them slip with equal ease through detectors on Earth. Nevertheless, because so many neutrinos are produced, we need trap only a tiny fraction of them to get useful information. This trapping is done with neutrino detectors.

Currently, the largest neutrino "telescopes" are the Super-Kamiokande detector located deep in a zinc mine west of Tokyo and the Sudbury Neutrino Observatory (fig. 12.13) more than a mile underground in a nickel mine in northern Ontario, Canada. The Super-Kamiokande detector contains 50,000 tons of water. The Sudbury detector contains about 1000 tons of heavy water, water consisting of molecules in which one of the ordinary hydrogen atoms (^{1}H) is replaced by "heavy hydrogen," ^{2}H.

Neutrino detectors are constructed in mines deep underground to shield them from the many other kinds of particles besides neutrinos that constantly bombard the Earth. For example, protons, electrons, and a variety of other subatomic particles constantly shower our planet. These particles, traveling at nearly the speed of light, are

*Exactly how tiny a mass is still under study.

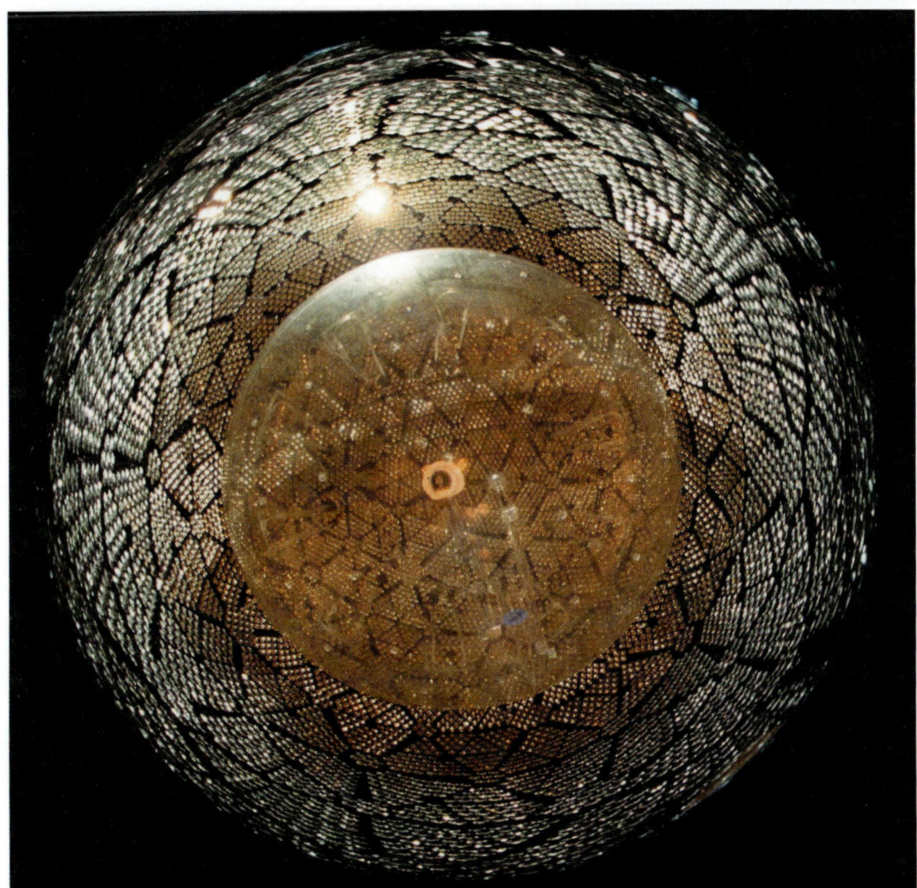

FIGURE 12.13
An inside view of the neutrino detector located in Sudbury, Canada. The sphere is 12 meters (about 40 feet) in diameter. It is filled with heavy water, a form of water in which each hydrogen atom contains a neutron as well as the usual proton. When a neutrino strikes one of the neutrons, the neutron may break down into a proton and electron. As the electron streaks off, it produces a tiny flash of light. Ten thousand detectors (which form the grid visible around the sphere) record the emitted light, thereby allowing scientists to detect the neutrino's passage.

called **cosmic rays** and are thought to be particles blasted across space when a massive star explodes.

Cosmic rays can penetrate only a short distance into the Earth; so if a detector is located deep underground, nearly all the cosmic rays are filtered out. Neutrinos, however, are unfazed by a mere mile of solid ground: on the average they could travel through a light-year of lead!

Given their virtual unstoppability, how are neutrinos detected? Scientists have several ways. In the Sudbury detector, occasionally a neutrino collides with a neutron in the heavy hydrogen, breaking the neutron into a proton and an electron. As the electron shoots off, it emits a tiny flash of light, which is recorded by photodetectors. But when astronomers first tried to detect solar neutrinos, they found only one-third the expected number. What was wrong? Did the Sun not fuse hydrogen into helium, as predicted? Did neutrinos somehow escape detection? The solution to the puzzle offers us another chance to see the scientific process at work, as explained in the Science at Work box "Solving the Neutrino Puzzle."

Solar Seismology

Solar seismologists study the Sun's interior by analyzing waves in the Sun's atmosphere. We learned in chapter 6 that scientists can study the Earth's interior by analyzing earthquake waves. Astronomers can also learn about the Sun's interior by analyzing waves in its gases.

Waves similar to those of earthquakes travel through the Sun and make its surface heave like the ocean or a bubbling pan of hot oatmeal. The rising and falling surface

FIGURE 12.14
Computer diagram of solar surface waves.

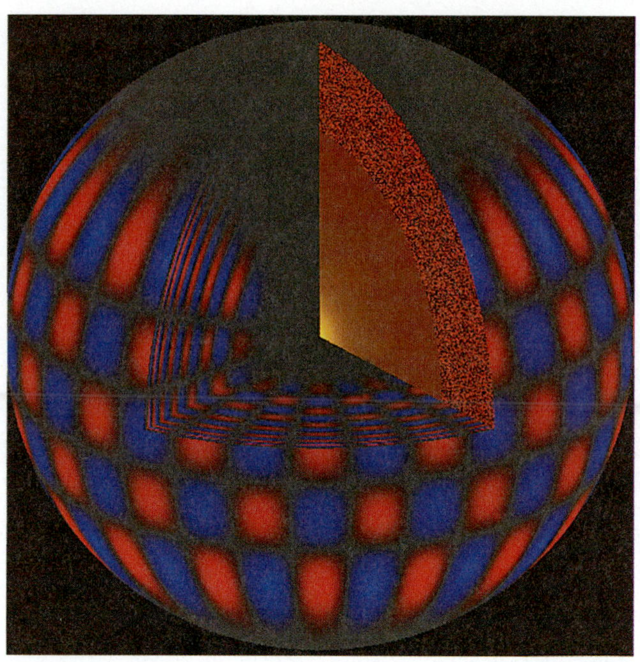

gas makes a regular pattern (fig. 12.14), which can be detected as a Doppler shift of the moving material. Astronomers next use computer models of the Sun to predict how the observed surface waves are affected by conditions in the Sun's deep interior. With this technique, astronomers can measure the density and temperature deep within the Sun from the pattern and speed of the waves in its atmosphere. The results they find agree well with those of the current models, indicating that we have a good understanding of the Sun's structure.

SCIENCE *at work*

SOLVING THE NEUTRINO PUZZLE

As a first step toward resolving the solar neutrino discrepancy, astronomers checked that their calculations for the predicted number of neutrinos were correct. All such checks led to roughly the same result, implying that there was no obvious flaw in their understanding of how the Sun works. Was it possible, then, that neutrinos had undiscovered properties that affected their detectability on Earth or their production in the Sun?

As scientists looked more closely at the properties of neutrinos, they found that one theory suggested the existence of three kinds of neutrinos, not just one. This seemed promising because the first neutrino detectors could detect only one of the three kinds, thus perhaps explaining why only one-third the expected number were observed. Moreover, the revised theory of neutrinos implied that these elusive particles were not massless and that one kind could change into another. With this new hypothesis about the nature of neutrinos, scientists built a second generation of neutrino detectors that could detect all three varieties. The results from these new detectors agreed perfectly with the predictions of the solar theory. Thus, astronomers have a new and greater confidence in their understanding of the Sun. But in addition, they now know more about neutrinos, including the fact that they are not massless.

In 2002, the Nobel Prize in Physics was shared by the American scientist Raymond Davis and the Japanese scientist Masatoshi Koshiba for their work on neutrinos from astronomical sources.* Davis led the first experiment to detect solar neutrinos and revealed the solar neutrino discrepancy. Koshiba headed the Kamiokande group whose work led to a resolution of the discrepancy.

*The American scientist Riccardo Giacconi, who made pioneering discoveries in X-ray astronomy, also shared the 2002 prize.

12.4 Solar Magnetic Activity

The surface waves described above are but one of many kinds of disturbances in the Sun's outer regions. A wide class of dramatic and lovely phenomena on the Sun are caused by its magnetic field. This magnetic activity is also of interest because it affects the Earth, where it triggers auroral displays and appears to alter the climate.

Sunspots

Sunspots are the most common type of solar magnetic activity. They are large, dark-appearing regions (fig. 12.15) ranging in size from a few hundred to many thousands of kilometers across. Spots last from a few days to over a month. They are darker than the surrounding gas because they are cooler* (4500 K as opposed to 6000 K of the normal photosphere), and they are cooler because they contain strong magnetic fields. Why do magnetic fields create cool regions? To understand, we must look at how such fields interact with the Sun's hot gas.

Solar Magnetic Fields

The magnetic field of sunspots is more than a thousand times stronger than the Earth's or the "normal" field of the Sun. In such intense fields, electrons and other charged particles spiral around the field, "frozen" to it, as shown in figure 12.16. They are thereby forced to follow the magnetic field as they spiral in it. On the Earth, the field deflects particles toward the poles, where they create auroras. In the Sun, the field slows the ascent of hot gas

*Spattering a few drops of water onto a hot electric stove burner will show a similar effect. Each drop momentarily cools the burner, making a dark spot.

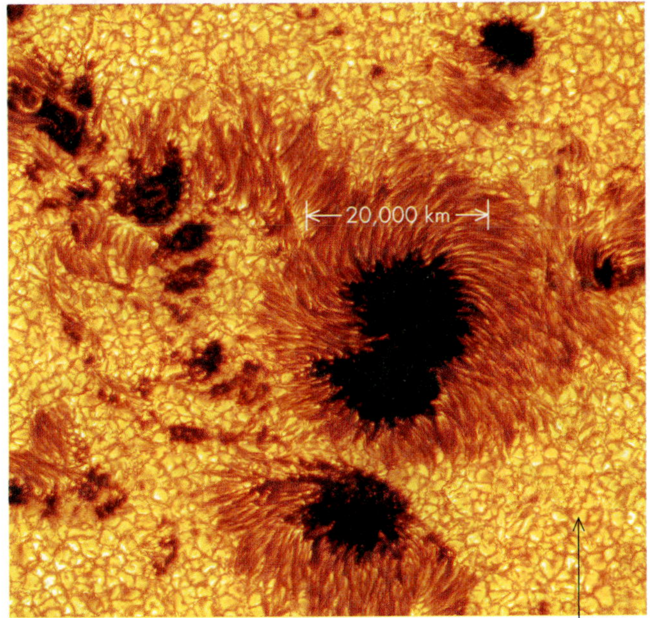

FIGURE 12.15
Visible-light photograph of a large group of sunspots. The darker areas are cooler gas, but they are still bright—they appear dark in the image just by contrast to the surrounding hotter regions of the photosphere.

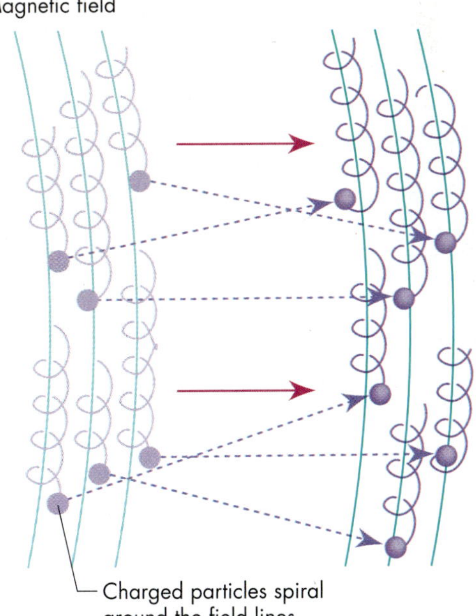

FIGURE 12.16
Charged particles are tied to magnetic field lines—spiraling back and forth along the lines. If forces move the gas containing the particles, it can drag the field with it, and vice versa.

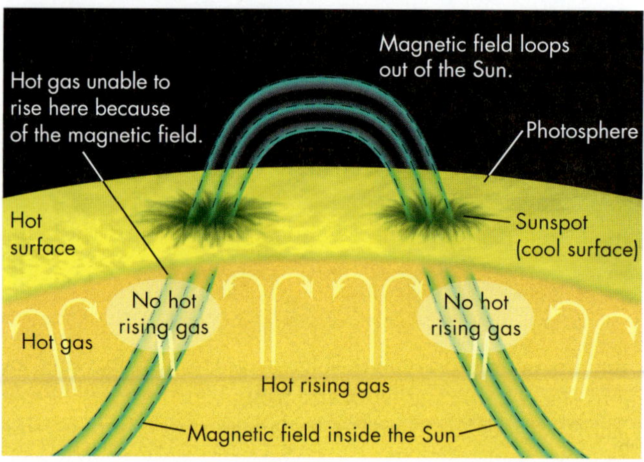

FIGURE 12.17
Sketch of how magnetic fields cause sunspots.

in the convection zone. With heat rising more slowly where the magnetic field is strong, the surface cools and becomes darker, making a sunspot, as shown in figure 12.17.

Prominences and Flares

Prominences and solar flares are magnetic disturbances in the low-density, virtually transparent hot gases above the Sun's visible surface. Prominences are huge plumes of glowing gas that jut from the lower chromosphere into the corona. Figure 12.18A

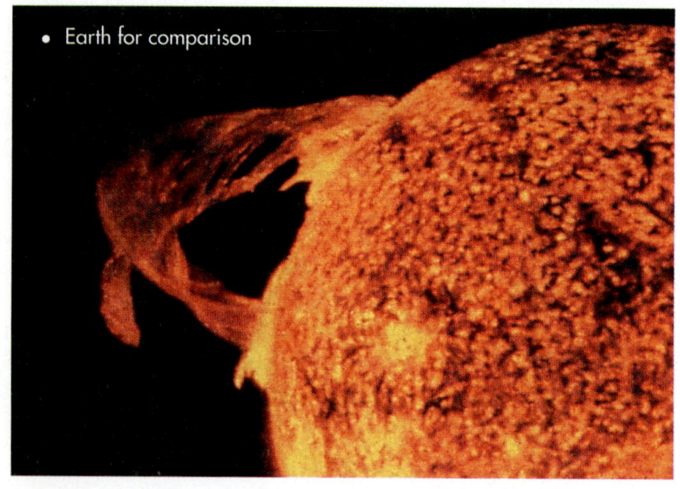

A

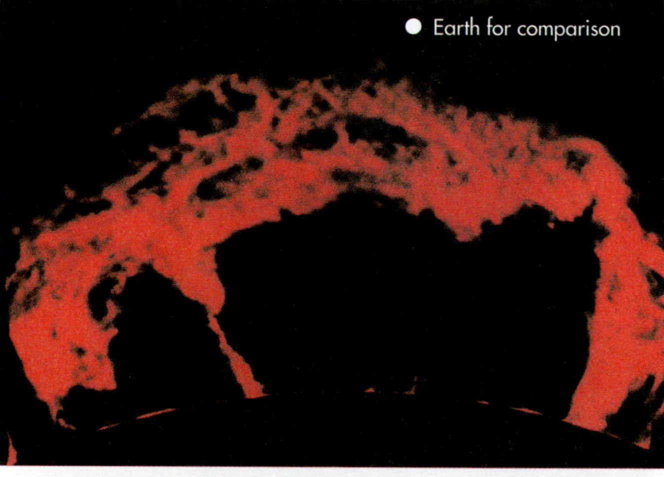

B

Cooling gas trapped
in magnetic field
Magnetic field
Sunspot pair

C

FIGURE 12.18
Solar prominences. (A) shows an ultraviolet image made from the *SOHO* satellite. (B) Photograph of a prominence in a spectral line of hydrogen. (C) Sketch illustrating how magnetic fields support a prominence.

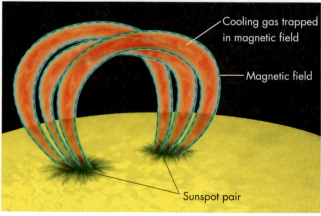

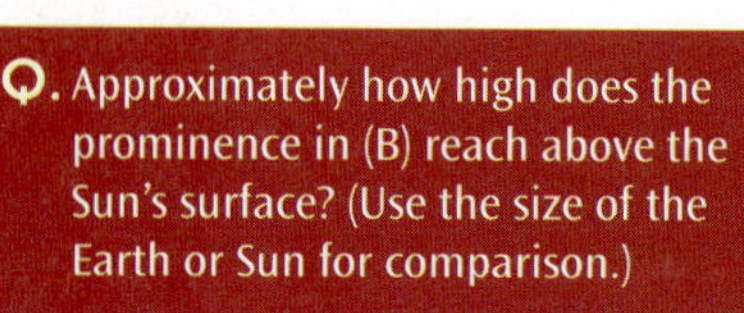

and B show some of their variety. You can get some sense of their immensity from the white dot, which shows the size of the Earth.

Prominences form where the Sun's magnetic field reduces heat flow to a region. They are *cooler* than the gas around them, which means, according to the perfect gas law, that the pressure inside is less than outside. Thus, the hot external gas "bottles up" the cooler gas of the prominence (fig. 12.18C). Under favorable conditions, this cooling gas, trapped in its magnetic prison, may glow for weeks.

Time-lapse movies show that gas streams through prominences, sometimes rising into the corona, sometimes raining down onto the photosphere. The flow is channeled by and supported by the magnetic field, which often arcs between sunspots. Thus, prominences also are related to sunspots.

ANIMATION
Prominences on the Sun's limb

Sunspots also give birth to **solar flares,** brief but bright eruptions of hot gas in the chromosphere (fig. 12.19). Over a few minutes or hours, gas near a sunspot may dramatically brighten. Such eruptions, though violent, are so localized they hardly affect the total light output of the Sun at all. Generally, you need a telescope to see their visible light, though they can increase the Sun's radio and X-ray emission by factors of a thousand in a few seconds.

Flares are poorly understood, but magnetic fields probably play an important role. One theory suggests that the field near a spot gets twisted by gas motions, a bit like winding up a rubber-band-powered toy. But such twisting can go only so far before the rubber band breaks. So, too, the magnetic field can be twisted only so far before it suddenly readjusts, whipping the gas in its vicinity into a new configuration. The sudden motion heats the gas, and it expands explosively. Some gas escapes from the Sun and shoots across the inner Solar System to stream down on the Earth. Such a burst created the spectacular auroral displays seen in March 1989 and illustrated in figure 12.20.

An aurora is a beautiful sight, but the stream of charged particles that causes them can have more serious consequences. Communications satellites have been disabled by some major flares, and even electrical power grids can be destabilized as the magnetic field carried by the stream of gas causes electric currents to surge in electrical transmission wires. The 1989 solar outburst caused blackouts in several locations across North America.

Astronomers today think that solar flares are only part of the cause of auroral displays here on Earth. A more important trigger appears to be *coronal mass ejections.* These ejections are enormous bubbles of hot gas and trapped magnetic fields that burst from the corona out into space. Our most spectacular auroral events occur when one of these ejections strikes Earth.

FIGURE 12.19
A solar flare—the bright spot on the right side of the image. This picture was taken through a special ultraviolet filter on the *SOHO* satellite.

FIGURE 12.20
Photograph of the great aurora of March 1989.

Q. Aurora was the goddess of dawn in Roman mythology. Why might that name have been chosen for this phenomenon?

EXTENDING *our reach*

DETECTING MAGNETIC FIELDS: THE ZEEMAN EFFECT

Astronomers can detect magnetic fields in sunspots and other astronomical bodies by the **Zeeman effect,** a physical process in which the magnetic field splits some of the spectrum lines of the gas into two, three, or more components. The splitting occurs because the magnetic field alters the atom's electron orbits, which in turn alter the wavelength of its emitted light.

Box figure 12.1A shows the Zeeman effect splitting spectrum lines in a sunspot. The line is single outside the spot but triple within, and by mapping the splitting across the Sun's face, astronomers can map the Sun's magnetic field, creating a magnetogram, as seen in box figure 12.1B. The colors in a magnetogram show the strength and polarity of the magnetic field (recall that magnetic fields have a north or south polarity indicating their direction). Notice that the field is strong around spots and weak elsewhere.

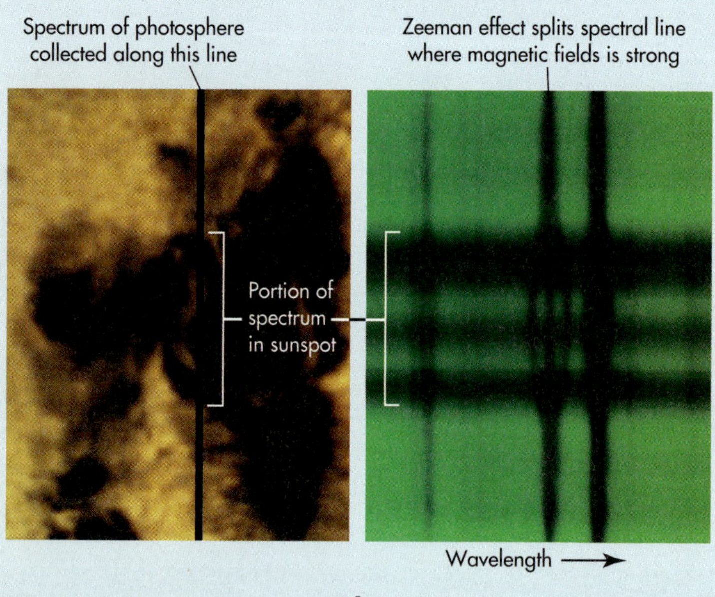

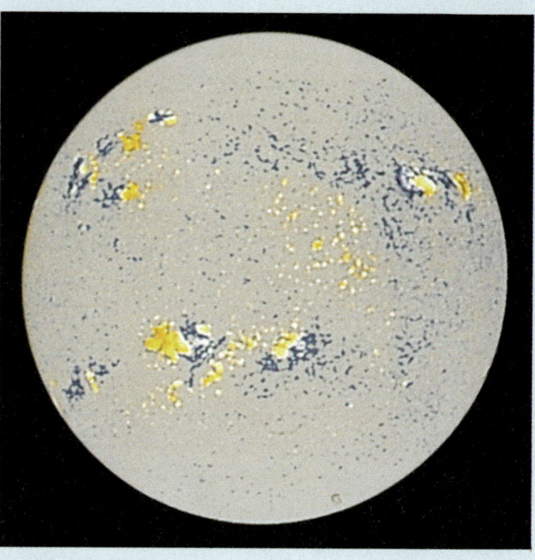

BOX FIGURE 12.1
(A) Spectrum of the Zeeman effect in a sunspot. Notice that the line is split over the spot where the magnetic field is strong but that the line is unsplit outside the spot where the field is weak or absent. (B) Magnetogram of the Sun. Yellow indicates regions with north polarity, and blue indicates regions with south polarity. Notice that the polarity pattern of spot pairs is reversed between the top and bottom hemispheres of the Sun. That is, in the upper hemisphere, blue tends to be on the left and yellow on the right, and the opposite In the lower hemisphere.

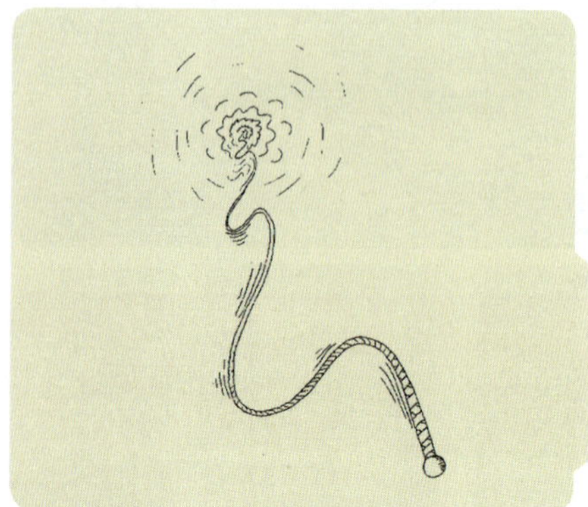

Heating of the Chromosphere and Corona

Although the Sun's magnetic field cools sunspots and prominences, it heats the chromosphere and corona. Other stars also have chromospheres and coronas, and so it is additionally important to understand the Sun's. To begin, we need first to recall that the temperature of a gas is a measure of how fast its particles are moving. Anything that speeds atoms up increases their temperature.

An analogy may help you understand how magnetic waves can heat a gas. When you crack a whip, a slow motion of its handle travels as a wave along the whip. As the whip tapers, the wave's energy of motion is given to an ever smaller piece of material. With the same amount of energy and less mass to move, the tip accelerates and eventually breaks the sound barrier. The whip's "crack" is a tiny sonic boom.

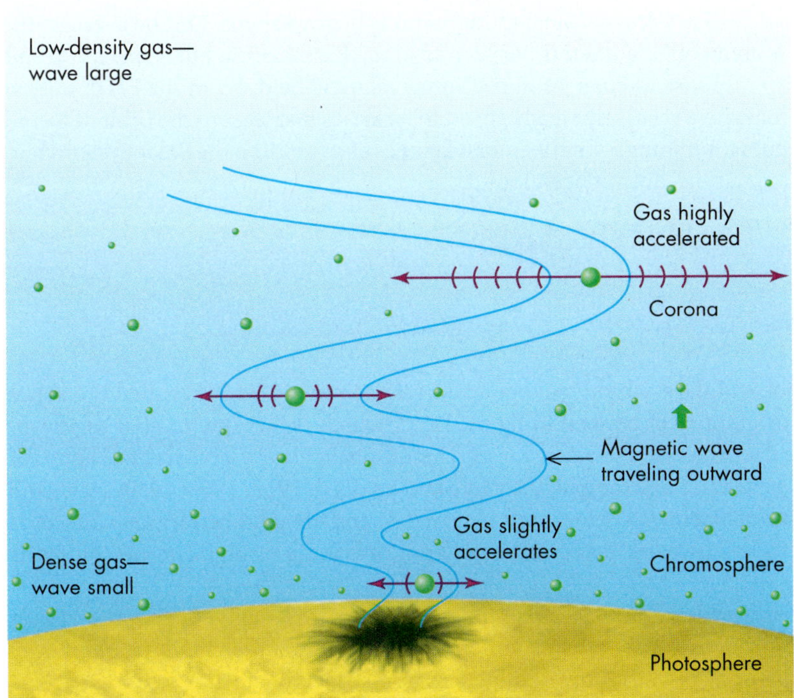

FIGURE 12.21
Diagram illustrating how magnetic waves (blue) heat the Sun's upper atmosphere. As the waves move outward through the Sun's atmosphere, they grow larger, imparting ever more energy to the gas (green dots) through which they move, accelerating and thereby heating it.

A similar speedup occurs in the Sun's atmosphere when magnetic waves formed in the photosphere move into the corona along the Sun's field lines (fig. 12.21). As the atmospheric gas thins, the wave energy is imparted to an ever smaller number of atoms, making them move faster in random directions. But "faster" in this case means hotter. Thus, the upper atmosphere heats up as the waves travel into it.

What generates the magnetic waves? They probably start in the convection zone where rising bubbles of gas shake the magnetic field and create magnetic waves just as shaking the loose end of a rope makes it wiggle. Thus, the high temperature of the chromosphere and corona is another example of the importance of the Sun's magnetic field and its convection zone. This theory is supported by observations of stars other than the Sun with active convection zones, which also have active chromospheres and coronas.

The Solar Wind

In addition to the mass it loses in the outbursts of flares, the Sun undergoes a steady, less dramatic loss of mass. The corona's high temperature gives its atoms enough energy to escape the Sun's gravity. As these atoms stream into space, they form the **solar wind,** a tenuous flow of mainly hydrogen and helium that sweeps across the Solar System. The amount of material lost from the Sun is small: less than 1 ten-trillionth of its mass each year. Nevertheless, the solar wind creates comet tails, as we saw in chapter 10, and auroras, as we saw in chapter 6.

The solar wind arises because, unlike the rest of the Sun, the corona is *not* in hydrostatic equilibrium. Recall that the temperature in the Sun's atmosphere increases with altitude, making the corona much hotter than the photosphere. The corona's high temperature, according to the perfect gas law, creates a pressure within it larger than we might otherwise expect for its distance above the photosphere. The pressure is in fact sufficient to overcome the Sun's gravitational force on gas in its upper atmosphere. As a result, it pushes that material outward into space. The expanding gas has a very low density (only a few hundred atoms in a cubic centimeter—the volume of a thimble*).

*For comparison, a thimbleful of the air we breathe contains about 10^{19} molecules!

The gas atoms begin their outward motion slowly but accelerate with increasing distance as the Sun's gravitational attraction on them weakens. On the average, the wind speed is about 500 kilometers per second at the Earth's orbit, but it speeds up and slows down in response to changes in the Sun's magnetic field. From the Earth outward, the wind coasts at a relatively steady speed that carries it at least to the orbit of Neptune. At some point, it impinges on the interstellar gas surrounding the Solar System, but where that happens has not yet been found.

12.5 The Solar Cycle

Sunspot and flare activity change from year to year in what is called the **solar cycle.** This variability can be seen in figure 12.22, which shows the number of sunspots detected over the last 140 years or so. The numbers clearly rise and fall approximately every 11 years. For example, the cycle had peaks in 1958, 1969, 1980, 1990, and 2001.

Samuel Heinrich Schwabe, a pharmacist and amateur astronomer, discovered the solar cycle. His work prevented him from observing at night; so he studied the Sun during the day.

Flares and prominences also follow the solar cycle, and climate patterns on Earth may, too. For this reason, astronomers have sought to understand not only the cause of the solar cycle but also how it influences terrestrial climate.

Cause of the Solar Cycle

As the Sun rotates, gas near its equator circles the Sun faster than gas near its poles; that is, it spins differentially, a property common in gaseous bodies (recall from chapter 10 that Jupiter and the other giant planets rotate differentially). The Sun's differential rotation is such that its equator rotates in about 25 days and its poles in about 35 days. Thus, a set of points arranged from pole to pole in a straight line would move over the course of time into a curve, as shown in figure 12.23.

Differential rotation should similarly distort the Sun's magnetic field, "winding up" the field below the Sun's surface.* Astronomers think such winding of the Sun's

*Spots and cycles of magnetic activity occur on other stars as well.

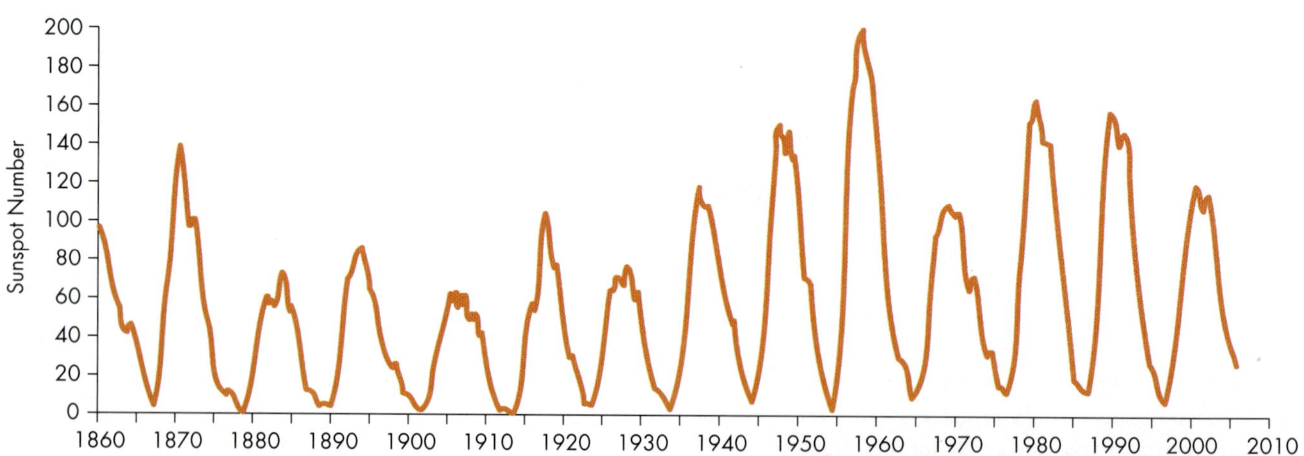

FIGURE 12.22
Plot of sunspot numbers showing solar cycle.

Q. Is the interval between the peaks always exactly 11 years? By how much does it change?

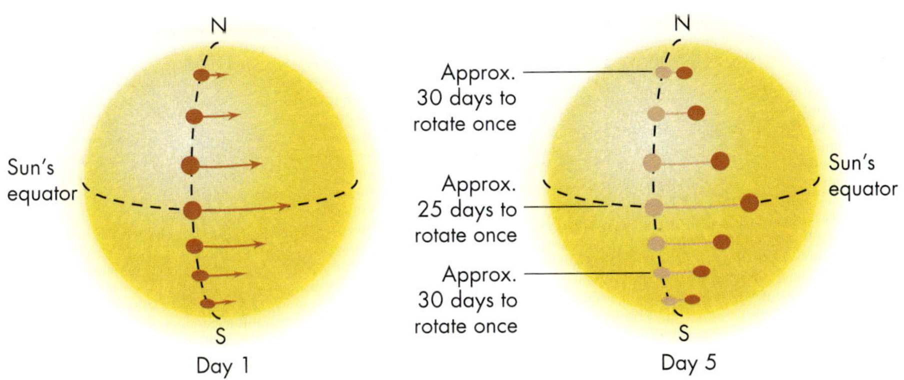

FIGURE 12.23
Sketch showing solar differential rotation. Points near the Sun's equator rotate faster than points near the poles.

magnetic field may cause the solar cycle, though the exact mechanism is still not well understood. According to one hypothesis, the Sun's rotation wraps the solar magnetic field into coils below the surface, making the field stronger and increasing solar activity: spots, prominences, and flares. The wrapping occurs because the Sun's magnetic field is "frozen" into the gas, as discussed in section 12.4. Thus, if the gas moves, so does the field, and vice versa.

Because the field and gas are tightly connected, differential rotation causes gas at the equator, which is moving faster than the gas at the poles, to drag the magnetic field with it so that a field, initially straight north to south, is wound into two subsurface loops, as shown in figure 12.24A. As the loops are wound tighter, they develop kinks, as when you twist a rubber band too tight. The cycle ends when the field twists too "tightly" and collapses, and the process repeats.

Sunspots form when kinks in the magnetic field rise to the Sun's surface and break through the photosphere (fig. 12.24B). Here, the field slows the outward flow of heat, making the surface cooler and darker than in surrounding areas and thereby creating sunspots, as we discussed in section 12.4. Each kink breaks the photosphere in *two* places—one where it leaves and one where it enters. We therefore expect that spots will occur in pairs or paired groupings, as sketched in figure 12.24C.

FIGURE 12.24
Sketch showing (A) the possible winding up of the subsurface magnetic field; (B) fields penetrating the Sun's surface; and (C) formation of a spot pair.

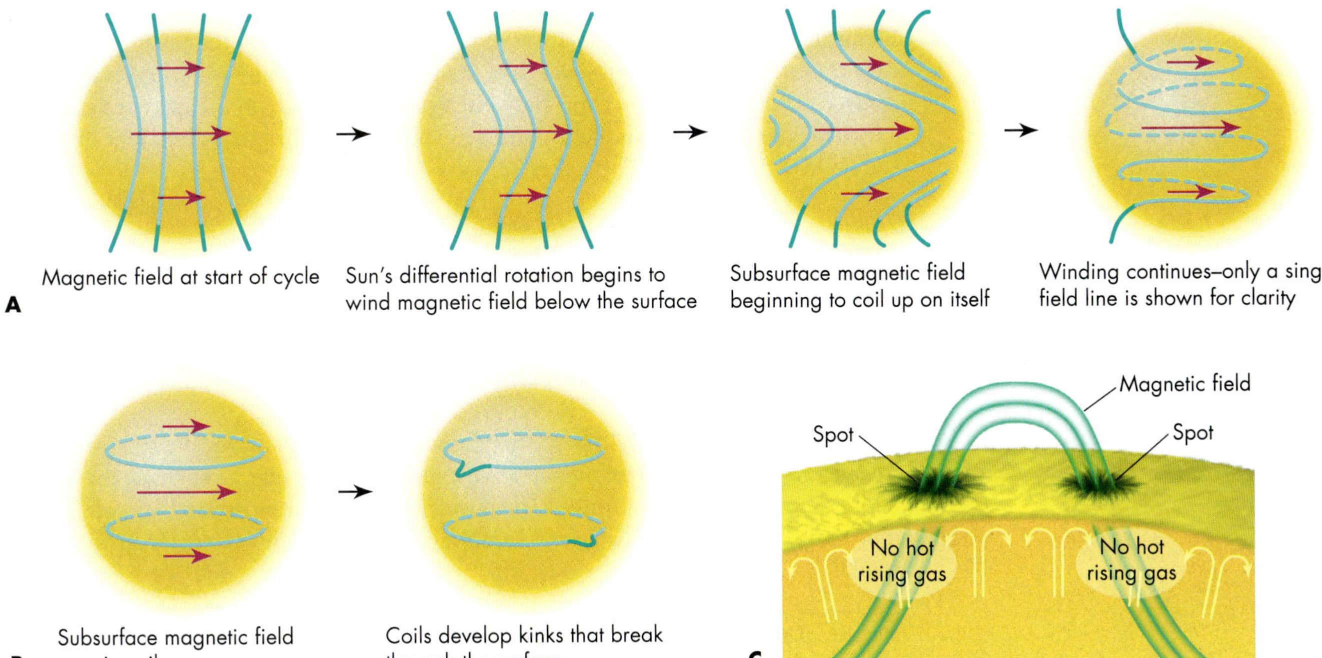

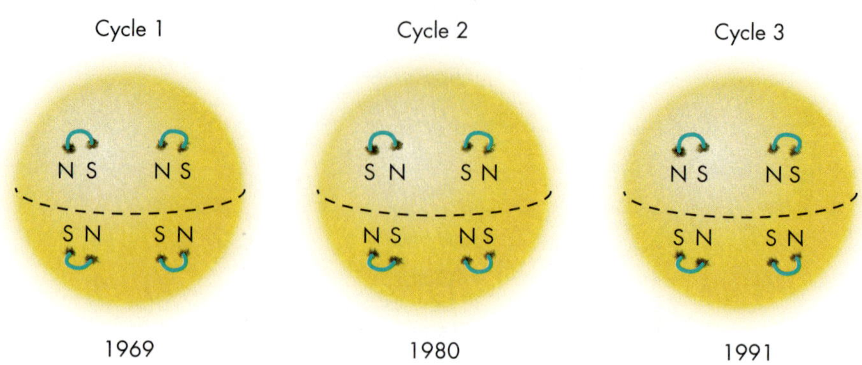

FIGURE 12.25
Diagram illustrating the approximate 22-year periodicity of solar magnetic activity.

In such pairs, one grouping has a north polarity and the other a south polarity.* That is, in one the field emerges from the surface, while in the other it descends so that the field direction is reversed. The false-color map of the Sun's field in Box figure 12.1B shows this effect clearly. Yellow areas have a north polarity and dark blue areas a south polarity. Notice that the pattern of the polarity differs between the two hemispheres of the Sun. In one, the yellow areas are on the right, while in the other they are on the left. This reversal arises because the subsurface field is coiled in opposite directions in the two hemispheres, additional evidence that sunspots and the solar cycle are caused by winding of the Sun's magnetic field.

ANIMATION
Wrapping of the Sun's subsurface magnetic field

Changes in the Solar Cycle

The solar cycle is not always 11 years: it may be as short as 7 or as long as 16 years. Moreover, if we consider the polarity of the spot groups, the cycle averages 22 years, rather than 11, because the polarity of the Sun's field reverses at the end of each 11-year cycle. It therefore takes *two* 11-year cycles for the field to return to its original configuration.

Figure 12.25 illustrates this effect. In the first frame, we see spot pairs as the cycle begins. A right-hand spot in the top hemisphere (technically, the "leading spot" because it leads in the direction the Sun rotates) has a south polarity, while a left-hand spot (technically called a "trailing spot") has a north polarity.

You can also see an additional feature of the cycle: all leading spots in one hemisphere have the same polarity. In the other hemisphere, the leading spots have the opposite polarity. Eleven years later, the polarity of the pairs will be reversed, as you can see by looking at the second frame in figure 12.25. Only in the next cycle, another 11 years later, will the spot fields have returned to their original directions. Thus, the full cycle of magnetic activity takes 22 years on the average.

Links Between the Solar Cycle and Terrestrial Climate

Climatologists find this 22-year period interesting because of a cycle in which droughts occur approximately every 22 years in the midwestern United States and Canada. Does the Sun's cycle affect the Earth's climate cycle? If so, how? One possibility is as follows.

The Sun's magnetic field heats the corona. The corona drives the solar wind. The solar wind alters the Earth's upper atmosphere; in particular, it changes the way the temperature varies with altitude. This in turn alters the atmosphere's circulation and may shift the jet stream to a new location. The jet stream steers storms and hence rainfall.

*Polarity of magnetic fields was described briefly in chapter 6, when we discussed the Earth's magnetic field.

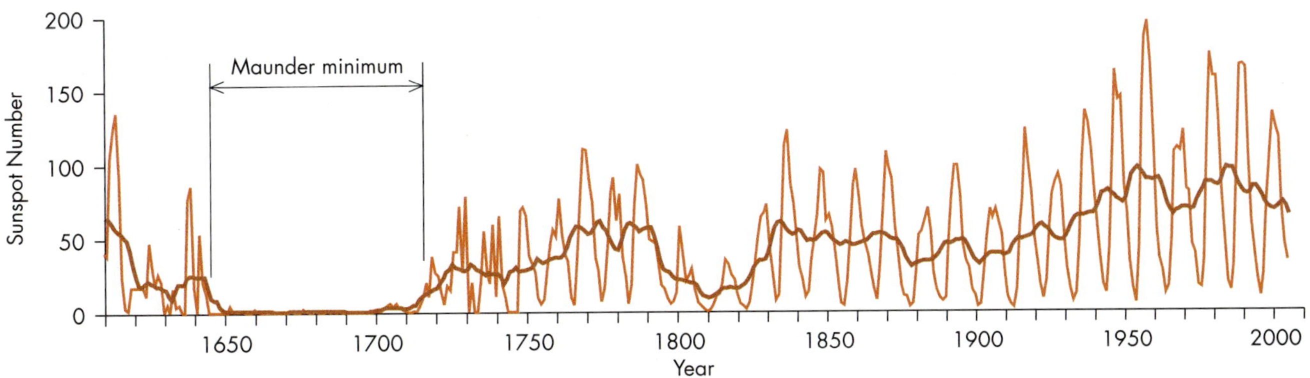

FIGURE 12.26
Plot illustrating that the number of sunspots changes with time, showing the Maunder minimum and the solar cycle.

Although this hypothesis cannot be verified yet, many scientists think that solar activity affects our climate. The evidence to support this hypothesis is based in part on the work of E. W. Maunder, a British astronomer who studied sunspots. Maunder noted in 1893 that, according to old solar records, very few sunspots were seen between 1645 and 1715 (fig. 12.26). He concluded that the solar cycle turned off during that period. The period is now called the **Maunder minimum** in honor of his discovery.

The Maunder minimum coincides with an approximately 70-year spell of abnormally cold winters in Europe and North America. Glaciers in the Alps advanced; rivers froze early and remained frozen late; the North Sea froze. The cold was so abnormal that meteorologists call the epoch part of the "little ice age." If only one such episode were known, we might dismiss the sunspot-climate connection as a coincidence, but three other cold periods have also occurred during times of low solar activity. This strengthens our belief that somehow the Sun's magnetic activity affects our climate.

Although scientists are still unsure about what creates the link between solar magnetic activity and our climate, they no longer doubt that such a link exists. Figure 12.27 shows how the ocean temperature (expressed as deviations from the normal average) changed from 1860 to 2000. The figure also shows how the number of sunspots changed over the same time span. Notice that from 1860 to 1980, when the

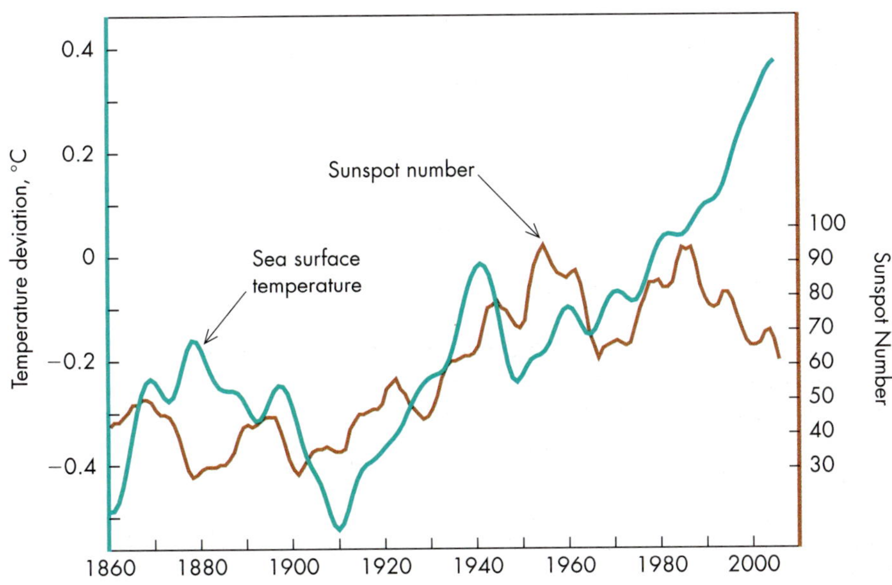

FIGURE 12.27
Curves showing the change in ocean temperatures on Earth and the change in sunspot numbers over more than a century. Notice that the curves were approximately in step until the 1980s. Over the last two decades the sea surface temperature has continued to rise while sunspot counts have declined. Astronomers deduce that solar activity affects our climate, but other factors must explain the recent temperature rise.

number of spots is high, the ocean is warmer than average and when the number of spots is low, the ocean is colder than average. But from 1990 on, the curves no longer are in step, which most scientists interpret as strong indication that increasing levels of greenhouse gases in Earth's atmosphere are to blame, not the Sun.

SUMMARY

The Sun is composed mostly of hydrogen and helium atoms. It is held together by gravity and supported against gravitational collapse by the pressure created by its high internal temperature. This temperature is high enough in the core that hydrogen can be converted into helium by nuclear fusion, thereby supplying the Sun with energy. That energy then flows to the surface by radiation and convection.

Above the Sun's visible surface (the photosphere) lie the chromosphere and corona—low-density, hot atmospheric layers. They are heated by magnetic waves traveling outward from the photosphere. In the corona, the temperature reaches a million degrees and drives the corona gases into space, creating the solar wind.

The Sun's magnetic field changes in structure and strength over time. As the subsurface field strengthens, sunspots form where the field breaks through the surface. Near spots, prominences and flares may occur. Spot numbers rise and fall with a roughly 11-year period. For reasons that are not yet well understood, this variation in solar magnetic activity also creates climatic variation on the Earth.

QUESTIONS FOR REVIEW

1. (12.1) How big is the Sun compared to the Earth?
2. (12.1) How can we measure the Sun's size, mass, and temperature?
3. (12.1) What is the Sun made of? How do we know this?
4. (12.1) What holds the Sun together?
5. (12.1–12.2) Why doesn't the Sun collapse?
6. (12.1) Why must the interior of the Sun be so hot?
7. (12.1) How does energy get to the Sun's surface from its core?
8. (12.1) What visible evidence do we have that the Sun has a convection zone?
9. (12.1) What are the photosphere, chromosphere, and corona? Which of these layers is hottest? How do we know this?
10. (12.2) How is solar energy generated? In what form(s) does it leave the core?
11. (12.3) What is meant by solar neutrinos? Why do astronomers say there is a discrepancy? Why is the discrepancy of interest?
12. (12.3) What is solar seismology? What does it tell us about the Sun?
13. (12.4) What is meant by solar activity?
14. (12.4) What role does magnetic activity play in solar activity?
15. (12.4) Why do sunspots appear dark?
16. (12.4) How do a prominence and a flare differ?
17. (12.4) How do we know there are magnetic fields in the Sun?
18. (12.5) What is the solar cycle?
19. (12.5) What is the period between maximum sunspot numbers? How does this differ from the period of the full solar cycle?
20. (12.5) What is the Maunder minimum? Why is it of interest?

THOUGHT QUESTIONS

1. (12.1) The Sun's corona has a temperature of 1 million K. Why does it not incinerate us?
2. (12.1–12.2) Why is hydroelectric power generation an indirect form of solar power?
3. (12.1–12.2) Why is sailing an indirect form of solar power?
4. (12.1–12.2) Why is it impossible that the Sun might be powered by a chemical process such as the combustion of hydrogen and oxygen to form water?
5. (12.1–12.2) The Sun clearly has a hot interior, and must contain the same kinds of materials as the rest of the Solar System. Why doesn't differentiation produce an iron core at the center as it did in other solar system bodies?
6. (12.1–12.2) What kinds of physics do you think would need to be included in the kind of model used to determine the interior structure of the Sun?
7. (12.1/12.4) Explain why Wien's law can't be used to determine the temperature of a prominence or the corona.
8. (12.2–12.3) The density in the ideal gas law is based on the number of atoms or molecules per volume, not the mass per volume. Suppose two identical containers are at the same temperature, but one contains hydrogen gas (H_2) and

the other contains deuterium gas (D_2). If the pressure is the same in the two containers, which container weighs more? If the two containers weigh the same, what is the relative pressure in the two containers?

9. (12.4) Because sunspots are a little cooler than the average temperature of the photosphere, they prevent some energy from being released from the part of the surface they occupy, but this energy is usually released from hotter and brighter than average areas nearby. What would happen to the Sun if this did not happen?

PROBLEMS

1. (12.1) Use the observation that the angular diameter of the Sun is 1/2 degree and that its distance is 1.5×10^8 kilometers to determine the Sun's diameter.
2. (12.1) Suppose you were an astronomy student on Jupiter. Use the orbital data for Jupiter (distance from Sun = 5.2 AU; period = 11.8 years) to measure the Sun's mass using the modified form of Kepler's third law.
3. (12.1) Show that the Sun's surface gravity is about 30 times Earth's.
4. (12.1) Calculate the escape velocity from the Sun. Compare this to the speed of rising material in the granulation given in section 12.1. If the escape velocity was less, what would happen at the center of the granulation cells?
5. (12.1) In this problem, you will calculate an estimate of the temperature of the Sun's core. You can do it either step by step or by writing out all the algebra to obtain a final result. You will need the following ideas: The pressure, P, in a gas is given by $P = \text{constant} \times \rho T$. If we measure length in meters, mass in kilograms, and T in Kelvin, then the constant has the value of about 8300 $m^2\ sec^{-2}\ K^{-1}$. The density of a body, ρ, is its mass per volume. The volume of a sphere is $4\pi R^3/3$. The pressure force from the interior is $P \times A$, where A is the area over which the pressure acts. You can take that as the Sun's cross section, πR^2, so the pressure force is $\pi R^2 P$. Finally, you need to invoke hydrostatic equilibrium: pressure forces must balance gravitational forces. Approximate the gravitational force holding the Sun together by assuming it is split into two equal halves and apply Newton's law of gravity to calculate the force between the halves. Assume they are separated by 1 solar radius.
6. (12.1) Calculate the Sun's density in grams per cubic centimeter. The Sun's mass is approximately 2×10^{33} grams. Its radius is approximately 7×10^{10} centimeters. How does the density you find compare with the density of Jupiter?
7. (12.2) The Sun's total energy output is 4×10^{26} Watts, and 1 watt is 1 Joule/second. Use the energy yield from the proton–proton chain to determine how many proton–proton chain fusion cycles must be happening each second in the solar core.
8. (12.2) Estimate the lifetime of the Sun. Assume that 10% of the Sun's total mass is processed through the proton–proton chain (there is only so much hydrogen available to the core), and that for each p–p chain reaction 6.6793×10^{-27} kg of mass is processed and 4.3×10^{-12} J of energy is liberated as per the values at the end of section 12.2. Assuming the Sun's power output given in table 1 is constant, how many years will it shine?

TEST YOURSELF

1. (12.1) The diameter of the Sun is about how large compared with the Earth's?
   (a) Twice as big
   (b) One-half as big
   (c) 10 times as big
   (d) 100 times as big
   (e) 10,000 times as big
2. (12.1) The Sun is supported against the crushing force of its own gravity by
   (a) magnetic forces.
   (b) its rapid rotation.
   (c) the force exerted by escaping neutrinos.
   (d) gas pressure.
   (e) the antigravity of its positrons.
3. (12.2) The Sun produces its energy from
   (a) fusion of neutrinos into helium.
   (b) fusion of positrons into hydrogen.
   (c) disintegration of helium into hydrogen.
   (d) fusion of hydrogen into helium.
   (e) electric currents generated in its core.
4. (12.2) According to the ideal gas law, if the temperature of a gas is made 4 times higher, which of the following is a possible result? (More than one answer may be correct.)
   (a) Its pressure increases by 4 times and its density remains the same.
   (b) Its density increases by 4 times and its pressure remains the same.
   (c) Its pressure and density both double.
   (d) Its pressure increases by 4 times while its density decreases by 4 times.
   (e) Its pressure and density both decrease by 2 times.
5. (12.3) During the daytime, about a trillion solar neutrinos per second pass through you. At night, the number is
   (a) zero.
   (b) about the same.
   (c) about half as much.
   (d) much, much smaller.
6. (12.3) The primary method astronomers use to measure oscillations on the surface of the Sun is by
   (a) comparing telescopic images.
   (b) magnetograms from measuring Zeeman splitting of spectral lines.
   (c) measuring the Doppler shift of absorption lines from the surface.
   (d) X-ray and ultraviolet imaging by satellites.
   (e) sonic detection.

7. (12.4) Sunspots are dark because
   (a) they are cool relative to the gas around them.
   (b) they contain 10 times as much iron as surrounding regions.
   (c) nuclear reactions occur in them more slowly than in the surrounding gas.
   (d) clouds in the cool corona block our view of the hot photosphere.
   (e) the gas within them is too hot to emit any light.
8. (12.5) About how many years elapse between times of maximum solar activity?
   (a) 3 (b) 5 (c) 11 (d) 33 (e) 105

KEY TERMS

chromosphere, 315
convection zone, 314
corona, 315
coronal hole, 316
cosmic rays, 321
granulation, 314
hydrostatic equilibrium, 316
Maunder minimum, 331
neutrino, 319
nuclear force, or strong force, 318
nuclear fusion, 318
perfect gas law, or ideal gas law, 317
photosphere, 313
pressure, 316
prominences, 324
proton–proton chain, 319
radiative zone, 314
solar cycle, 328
solar flare, 325
solar wind, 327
spicule, 315
sunspot, 323
Zeeman effect, 326

FURTHER EXPLORATIONS

Foukal, Peter V. "The Variable Sun." *Scientific American* 262 (February 1990): 34.

Hathaway, David H. "Journey to the Heart of the Sun." *Astronomy* 23 (January 1995): 38.

Holman, Gordon D. "The Mysterious Origins of Solar Flares." *Scientific American* 294 (April 2006): 38.

Kennedy, James R. "GONG: Probing the Sun's Hidden Heart." *Sky and Telescope* 92 (October 1996): 20.

Kippenhahn, Rudolph. *Discovering the Secrets of the Sun.* New York: Wiley, 1994.

Lang, Kenneth R. *Sun, Earth, and Sky.* New York: Springer-Verlag, 1995.

———. "Unsolved Mysteries of the Sun [Parts 1 and 2]." *Sky and Telescope* 92 (August 1996): 38; no. 92 (September 1996): 24.

McDonald, Arthur B., Joshua R. Klein, and David L. Wark. "Solving the Solar Neutrino Problem." *Scientific American* 288 (April 2003): 40.

Mims, Forrest M. "Sunspots and How to Observe Them Safely." *Scientific American* 262 (June 1990): 130.

Nesme-Ribes, Elizabeth, Sallie L. Baliunas, and Dmitry Sokoloff. "The Stellar Dynamo." *Scientific American* 275 (August 1996): 45.

Odenwald, Sten, and James L. Green. "Bracing for a Solar Storm." *Scientific American* 299 (August 2008): 80.

Schrijver, Carolus J. "The Science Behind the Sun's Corona." *Sky and Telescope* 111 (April 2006): 28.

Website

Visit the *Explorations* website at **http://www.mhhe.com/arny** for additional online resources on these topics.

FIGURE QUESTION ANSWERS

WHAT IS THIS? (chapter opening): This image shows Venus (the round dark circle on the right) passing directly between us and the Sun on the June 8, 2004, transit of Venus. You can also see red columns of hot gas in the Sun's chromosphere and corona projecting beyond the edge of the Sun at right.

FIGURE 12.4: Approx. 1000 km (600 miles). This is about the size of Texas or Egypt.

FIGURE 12.18: The prominence in (B) is ~200,000 km high.

FIGURE 12.20: The aurora's glow and sometimes rosy tints resemble the pale colors of dawn.

FIGURE 12.22: The intervals between the peaks are not always equal. They range from about 9 or 10 years to 12 or 13 years in the data shown here.

PROJECTS

1. **Measure the diameter of the Sun:** Take a piece of thin, dark cardboard and put a small hole in it. Hold it in sunlight about 1 meter (about 3 feet) from a piece of white paper so that a small image of the Sun appears on the paper (see Project fig. 12.1A).

   Carefully measure the distance (*d*) between the cardboard and the piece of paper and the size of the Sun's image (*s*) on the paper. On a separate piece of paper, draw two straight lines that cross with a small angle between them (see Project fig. 12.1B). Draw two small circles between the lines as shown in Project figure 12.1B. Convince yourself that if *D* is the distance to the Sun (1 AU) and *S* is its diameter, then $S/D = s/d$, where *s* is the size of the Sun's image on the paper and *d* is the distance from the paper to the cardboard. Look up the value of *D* in table 12.1. Then solve for *S*, the Sun's diameter. Does your value agree with the value you calculate from the value of the Sun's radius in table 12.1?
2. **Observing sunspots:** Mount a pair of binoculars on a tripod or some other fixed object. You can use duct tape if you cannot find a binocular clamp and tripod. Hold a piece of white paper about 60 centimeters (roughly 2 feet) from the eyepieces of the binoculars and adjust them so that the Sun's image falls on the paper (Project fig. 12.2). You may need to adjust the focus wheel of the binoculars to make the Sun's image sharp and clear on the paper. DO NOT LOOK AT THE SUN THROUGH THE BINOCULARS. THE SUN'S LIGHT CAN QUICKLY AND PERMANENTLY BLIND YOU IF YOU DO. Also, shade the binoculars or telescope from time to time so the eyepiece does not become too hot. Can you see any sunspots on the Sun's image on the paper? (They will appear as small dark blotches.) Draw a circle on the paper to match the size of the Sun's image.

   Mark the date on the paper and sketch the location of the spots that you see on the circle. Repeat your observations a few days later. Are the spots in the same location? If they have shifted their position, can you think of a reason why?
3. **Playing photon:** Try the following "game" to get a sense of why it takes energy so long to leak from the center of the Sun to its surface. Recall that in the Sun, photons are constantly absorbed and re-emitted, but in random directions. A given photon thus is as likely to be re-emitted back toward the Sun's core as toward it surface. This is much like what happens if you try to climb a flight of stairs using the flip of a coin to decide whether you go up a step or down.

   Stand at the bottom of a short flight of stairs—a flight of four or five steps is about right. If you can't find such a short flight, then put a book on the fourth step of a longer flight and pretend it's the top step. Flip a coin and go up a step if you get "heads" and down a step (or stay where you are if you are already at the bottom) if you get "tails." Record how many flips you need to reach the top step. Repeat half a dozen or so times. How many flips, on average, does it take you to reach the top step? What factors determine how long it takes you to climb the flight?
4. **Sunspot numbers:** Some solar observatories publish the daily full disk image of the Sun on their websites, and have archives available for up to several decades (you may need to find the "data archive" or the part of their website designed for astronomers to find more than the most recent image). Find an archive and make a record of the number of sunspots for a year during a solar maximum and a solar minimum period, or over several years. Since the Sun's rotation period is about 27 days, it should be sufficient to look at images every few weeks to identify new spots. Do any spots last more than one solar rotation? How different is the solar surface during solar minimum and maximum?

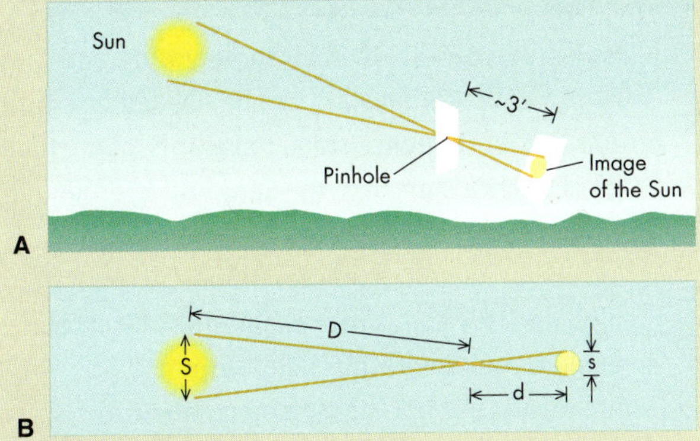

PROJECT FIGURE 12.1
A simple way to measure the Sun's diameter.

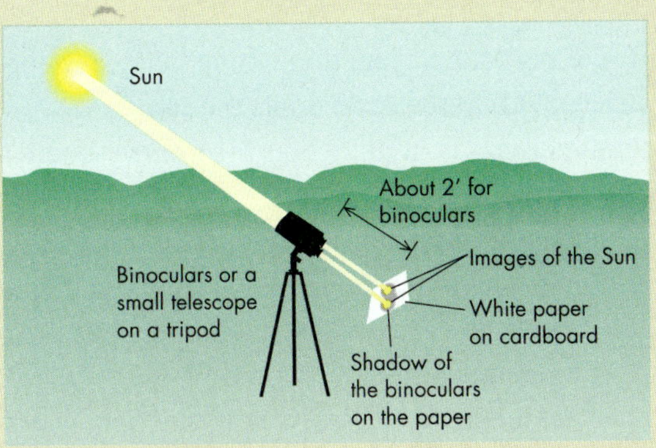

PROJECT FIGURE 12.2
Observing sunspots with a pair of binoculars.

The "Jewel Box," a cluster of stars in the southern constellation Crux.

KEY CONCEPTS

- Distances of stars can be found from tiny shifts they appear to make (parallax) as the Earth orbits the Sun.
- Starlight diminishes in brightness as the inverse square of the distance to the star.
- Astronomers can learn much about stars, such as their temperature and composition, from the spectra of their light.
- The amount of light a star emits is determined by its radius and its surface temperature.
- Many stars are members of binary star systems, in which two stars orbit one another; their orbits give information about their masses.
- Classification of stars by their temperature and intrinsic brightness shows that stars can be divided into three main groups: main-sequence stars, red giants, and white dwarfs.
- Main-sequence stars, like the Sun, fuse hydrogen to helium in their cores and obey the mass–luminosity law:
    - More-massive stars are more luminous.
- Red giants have cool surfaces and must be extremely large to explain their enormous light output.
- White dwarfs are hot, dense objects about the size of Earth.
- Some stars pulsate, their size and brightness changing regularly, with larger, more-luminous stars generally pulsating more slowly.

13 Measuring the Properties of Stars

CONCEPTS AND SKILLS TO REVIEW

- Relation between temperature of a hot object and the color of the light it emits (Wien's law) (4.3)
- Doppler shift (4.6)
- Modified form of Kepler's third law (3.6)

It is hard to believe that the stars we see in the night sky as tiny glints of light are in reality huge, dazzling balls of gas and that many are vastly larger and brighter than our Sun. They look dim to us only because they are so far away—several light-years (trillions of miles) to even the nearest. Such remoteness creates tremendous difficulties for astronomers trying to understand the nature of stars. We cannot physically travel to the stars, but in this chapter we will see how astronomers overcome the distance barrier that separates us from stars and how they learn many of the physical properties of these distant objects. How far away are they? How big? What are they made of?

The answers to these questions show us that most stars are remarkably like the Sun. For example, like the Sun, they are composed mostly of hydrogen and helium and most have fairly similar masses. A small percentage, however, are more than 30 times the Sun's mass (30 $M_{\odot}$) and are much hotter than the Sun and blue in color. Others are much less massive than the Sun, only one-tenth its mass, and are cool, red, and dim. Moreover, even stars similar to the Sun in composition and mass may differ enormously from it in their radii and density. For example, some giant stars have a radius hundreds of times larger than the Sun's—so big that were the Sun their size, it would extend beyond the Earth's orbit. On the other hand, some stars are white dwarfs, with as much material as the Sun packed into a volume the size of the Earth.

Astronomers can learn all these properties of stars by using physical laws and theories to interpret measurements made from the Earth. For example, theories of light yield the surface temperature, distance, and motion of a star; theories of atoms yield the composition of a star; and a modified form of Kepler's third law yields the mass of a star. In using such laws, astronomers may sometimes employ more than one method to determine a desired property of a star. For example, a star's temperature may be measured from either its color or its spectrum. Such alternative methods serve as checks on the correctness of the procedures astronomers use to determine the properties of stars.

Q: WHAT IS THIS? See end of chapter for the answer

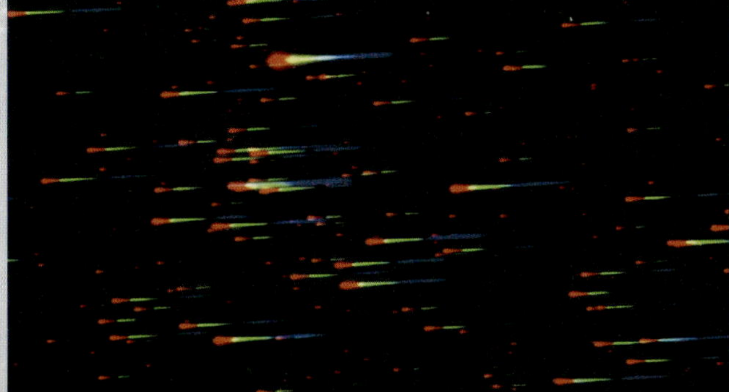

13.1 Measuring a Star's Distance

One of the most difficult problems astronomers face is measuring the distances to stars and galaxies. Yet knowing the distances to such bodies is vital if we are to understand their size and structure. For example, without knowledge of a star's distance, astronomers find it difficult to learn many of the star's other properties, such as its mass, radius, or energy output.

Using parallax to find distance may seem unfamiliar, but parallax creates our stereovision, the ability to see things three-dimensionally. When we look at something, each eye sends a slightly different image to the brain, which then processes the pictures to determine the object's distance. You can demonstrate the importance of the two images by covering one eye and trying to pick up a pencil quickly. You will probably not succeed in grasping it on the first attempt.

Measuring Distance by Triangulation and Parallax

Astronomers have several methods for measuring a star's distance, but for nearby stars the fundamental technique is **triangulation,** the same method we described for finding the distance to the Moon. In triangulation, we construct a triangle in which one side is the distance we seek but cannot measure directly and another side is a distance we can measure—a baseline, as shown in figure 13.1A. For example, to measure the distance across a deep gorge, we construct an imaginary triangle with one side spanning the gorge and another side at right angles to it and running along the edge we are on, as shown in figure 13.1B. By measuring the length of the side along the gorge edge and the angle *A*, we can determine the distance across the gorge either by a trigonometric calculation or from a scale drawing of the triangle.

Astronomers use a method of triangulation called **parallax** to measure the distance to stars. Parallax is a change in an object's apparent position caused by a change in the observer's position. An example—easy to demonstrate in your room—is to hold your hand motionless at arm's length and shift your head from side to side. Your hand seems to move against the background even though in reality it is your head that has changed position, not your hand.

ANIMATION
Parallax

This simple demonstration illustrates how parallax gives a clue to an object's distance. If you hold your hand at different distances from your face, you will notice that the apparent shift in your hand's position—its parallax—is larger if it is close to your face than if it is at arm's length. That is, *nearby objects exhibit more parallax than more remote ones,* for a given motion of the observer, a result true for your hand and for stars.

To observe stellar parallax, astronomers take advantage of the Earth's motion around the Sun, as shown in figure 13.2A. They observe a star and carefully measure its position

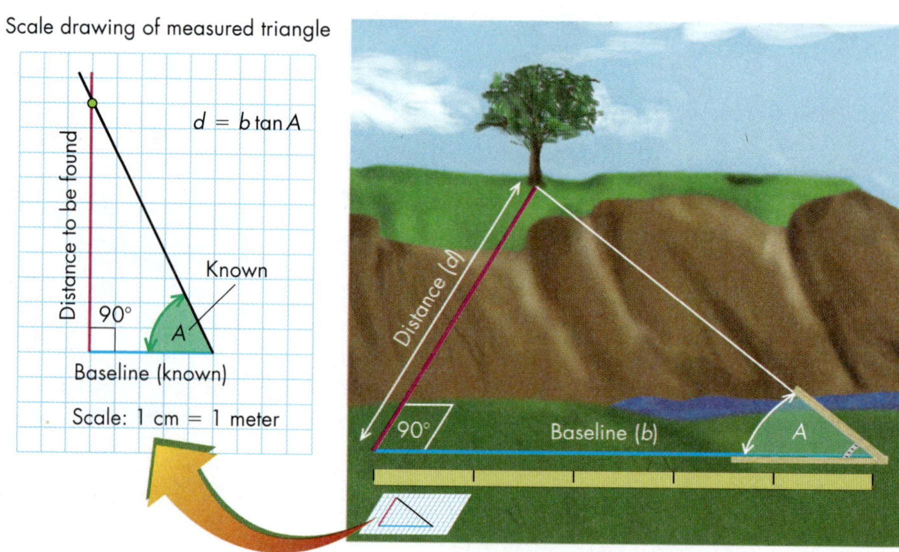

FIGURE 13.1
Sketch illustrating the principle of triangulation.

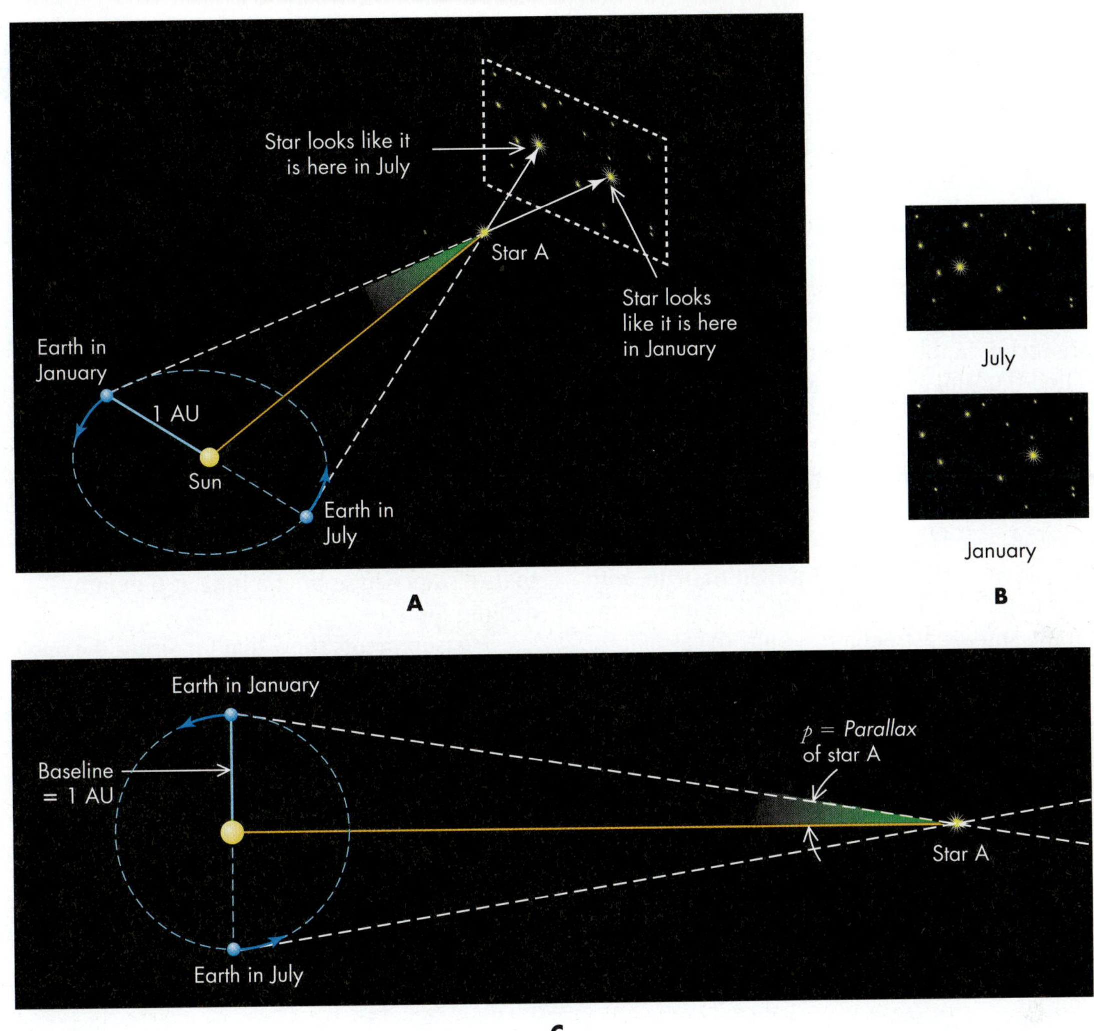

FIGURE 13.2
(A) Triangulation to measure a star's distance. The radius of the Earth's orbit is the baseline. (B) As the Earth moves around the Sun, the star's position changes as seen against background stars. (C) Parallax is defined as one-half the angle by which the star's position shifts. Sizes of bodies and their separation are exaggerated for clarity.

against background stars. They then wait 6 months until the Earth has moved to the other side of its orbit, a known distance of 2 AU (about 300 million kilometers), and make a second measurement. As figure 13.2B shows, the star will have a slightly different position compared to the background of stars as seen from the two points. The amount by which the star's apparent position changes depends on its distance from us. The change is larger for nearby stars than for remote stars, but for all stars it is extremely small—so small that it is measured not in degrees but in fractions of a degree called "arc seconds."*

For convenience, astronomers define a star's parallax, p, not by the angle its position appears to shift, but by half that angle (see fig. 13.2C). With that definition for parallax, the star's distance, d, is simply $1/p$ if we measure p in arc seconds and d, not in kilometers or light-years, but in a new unit called "parsecs" (abbreviated pc). That is,

$$d_{pc} = \frac{1}{p_{\text{arc seconds}}}$$

d_{pc} = distance measured in parsecs
$p_{\text{arc seconds}}$ = parallax measured in arc seconds

With this choice of units, one **parsec** equals 3.26 light-years (3.09×10^{13} kilometers). The word *parsec* comes from a combination of *par*allax and "arc *sec*ond."

*One arc second is $\frac{1}{3600}$ of a degree because an arc second is $\frac{1}{60}$ of an arc minute, which is $\frac{1}{60}$ of a degree.

EXTENDING *our reach*

MEASURING THE DISTANCE TO SIRIUS

We can measure the distance to Sirius (or any nearby star) using a geometric construction similar to the one we used in chapter 1 to relate an object's true diameter to its angular diameter and distance. We begin by measuring the star's apparent shift in the sky resulting from our planet's motion around the Sun, as illustrated in figure 13.2A. Let p be half that angle, as shown in figure 13.2C and Box figure 13.1, below. To determine the star's distance from the parallax angle, p, we construct the triangle ACS. The line AC is the radius of the Earth's orbit, with point A chosen so that an imaginary line from Sirius (S) to the Sun, C, will be perpendicular to AC. This makes ACS a right triangle, with angle C equal to 90° and the side AC—the distance from the Earth to the Sun—1 astronomical unit. To calculate Sirius's distance from the Sun, d, we draw a circle centered on Sirius and passing through the Earth and the Sun. We call the radius of the circle d, to stand for Sirius's distance from the Sun. We next form the following proportion: AC, the radius of the Earth's orbit, is to the circumference of the circle as p is to the total number of degrees around the circle, which we know is 360. That is,

$$\frac{AC}{2\pi d} = \frac{p}{360}.$$

If we measure d and AC in astronomical units, we can set AC = 1. Then, solving for d, we obtain

$$d = \frac{360}{2\pi p}\ \text{AU}$$

This equation takes on a much simpler form if we express p in arc seconds rather than degrees and d in parsecs (as defined above) rather than in AU. With these new units, the factors 360 and 2π disappear, leaving $d = 1/p$, as we claimed earlier. (In the above, we assume that Sirius itself has not changed its location significantly compared to its distance in so brief a time as 6 months—a very good assumption.)

Measurements show that for Sirius, the angle p is 0.377 arc seconds. Thus, its distance is $1/0.377 = 2.65$ parsecs. To express this in light-years, we multiply by 3.26 (the number of light-years in a parsec) to get Sirius's distance as about 8.6 light-years.

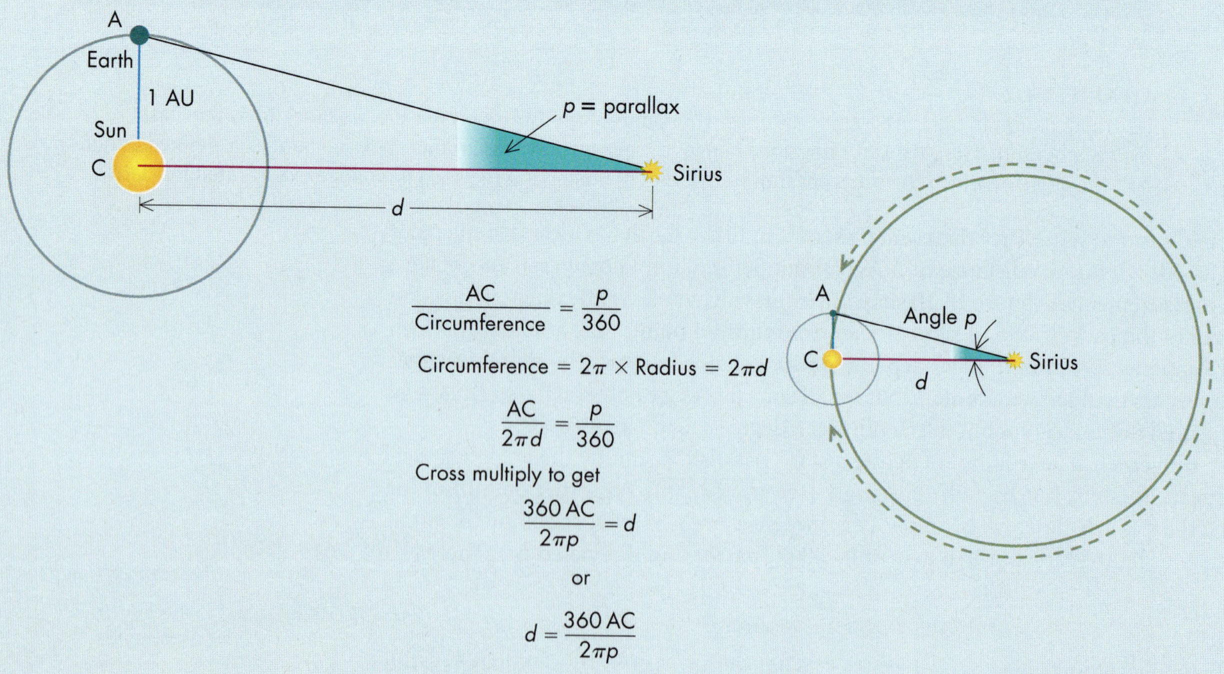

BOX FIGURE 13.1
How to determine the relation between a star's distance and its parallax.

To determine a star's distance, we measure its parallax, p, and use $d_{pc} = 1/p_{arc\ seconds}$. For example, suppose that from the shift in position of a nearby star we find that its parallax is 0.25 arc seconds. Its distance is then $d = \frac{1}{0.25} = 4$ parsecs. Similarly, a star whose parallax is 0.1 arc second is $\frac{1}{0.1} = 10$ parsecs from the Sun. From this technique, astronomers have discovered that at present the nearest star is Proxima Centauri (See "Looking Up: Centaurus and Crux, the Southern Cross" in the front matter), which lies 1.3 parsecs (4.3 light-years) from the Sun.* This spacing (about 1 parsec) is typical for stars near the Sun.

Keep in mind that the change in angle we are talking about here is very small. A shift of 1 second of arc is equivalent to looking at one edge of a U.S. penny 4 km (2.5 miles) away and then looking at its opposite edge.

Although the parallax–distance relation is mathematically a very simple formula, measuring a star's parallax to use in the formula is very difficult because the angle by which the star shifts is extremely small. It was not until the 1830s that the first parallax was measured by the German astronomer Friedrich Bessel at Königsberg Observatory (now in Kaliningrad). Even now, the method fails for most stars more than about 100 parsecs away, because the Earth's atmosphere blurs the tiny angle of their shift, making it almost unmeasurable. Astronomers can avoid such blurring effects by observing from above the atmosphere, and an orbiting satellite, *Hipparcos*, has done just that: making parallax measurements from space. With its data, astronomers are able to accurately measure distances to stars as far away as 250 parsecs. For more remote stars, they must use a different method, one based on how bright objects look to us.

Measuring Distance by the Standard-Candles Method

If you look at an object of known brightness, you can estimate its distance based on how bright it appears. For example, if you look at two 100-watt light bulbs, one close and one far away, you can tell fairly accurately how much farther away the dim one is. In fact, if you drive at night, your life depends on making such distance estimates when you see traffic lights or oncoming cars.

Astronomers call such distance measurements the **method of standard candles** and use a similar but more refined version of it to find the distance to stars and galaxies. But to use this method to measure an object's distance, astronomers must first determine the object's true brightness. So until we can make such brightness determinations, we must set this method aside, but we will return to it at the end of this chapter.

13.2 Measuring the Properties of Stars from Their Light

If we were studying flowers or butterflies, we would want to know something about their appearance, size, shape, colors, and structure. So, too, astronomers want to know the sizes, colors, and structure of stars. Such knowledge not only helps us better understand the nature of stars but also is vital in unraveling their life story. Determining a star's physical properties is not easy, however, because we cannot directly probe the star. But by analyzing a star's light, astronomers can deduce many of its properties, such as its temperature, composition, radius, mass, and motions.

Temperature

Stars are extremely hot by earthly standards. The surface temperature of even cool stars is far above the temperatures at which most substances vaporize, and so using a physical probe to take a star's temperature would not succeed, even if we had the technology to send the probe to the star. So, if astronomers want to know how hot a

*Proxima Centauri has two companions only slightly more distant. The bigger companion, Alpha Centauri A, is a bright star visible in the southern sky.

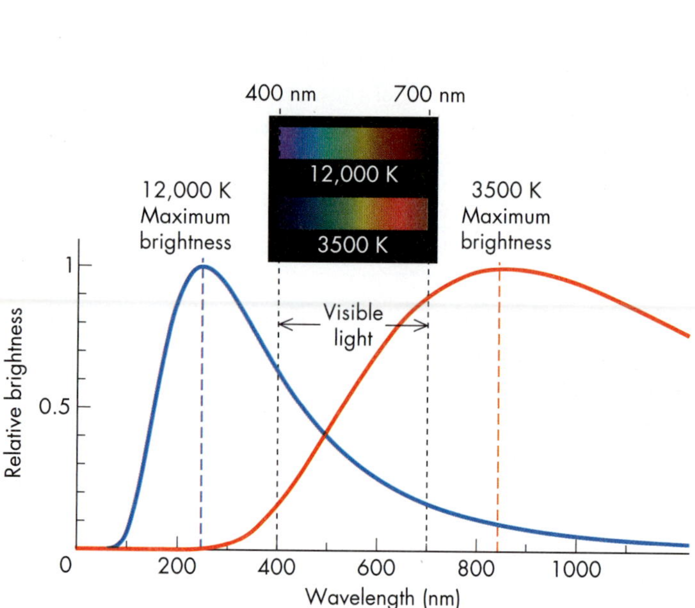

FIGURE 13.3
From the photograph at right you can see that Betelgeuse is red compared to Rigel. A plot of each star's relative brightness at different wavelengths shows that they peak at different wavelengths. Wien's law lets us determine their surface temperature from the wavelength at which their light output reaches a maximum.

star is, they must once more rely on indirect methods. Yet the method used is familiar. You use it yourself in judging the temperature of an electric stove burner.

An object's temperature can often be deduced from the color of its emitted light. As we saw in chapter 4, hotter objects emit more blue light than red. Thus, hot objects tend to glow blue and cooler ones red. You can see such color differences if you look carefully at stars in the night sky. Some, such as Rigel in Orion, have a blue tint. Others, such as Betelgeuse, are obviously reddish (fig. 13.3). Thus, even our naked eye can tell us that stars differ in temperature.

We can use color in a more precise way to measure a star's temperature with **Wien's law,** which we discussed in chapter 4. It states that an object's temperature, T, in Kelvin is given by the following:

T = Temperature of star in Kelvin
λ_{max} = Strongest emitted wavelength of starlight in nanometers

$$T = \frac{2.9 \times 10^6}{\lambda_{max}}$$

where λ_{max} is the wavelength in nanometers (nm) at which it radiates most strongly.* Thus, the longer the wavelength of the maximum emitted energy, the lower the temperature of the radiating object.

For an illustration of how we can use Wien's law to measure a star's temperature, suppose we pick Rigel. To determine the wavelength at which Rigel radiates most strongly, we need to measure its brightness at many different wavelengths, as illustrated in figure 13.3. We find that the strongest emission is at 240 nanometers, in the ultraviolet part of the spectrum. That is, λ_{max} = 240 nanometers. Inserting this value in Wien's law, we find

$$T = \frac{2.9 \times 10^6}{240} \approx 12{,}000 \text{ K}$$

*The subscript *max* on λ is to remind us that it is the wavelength at which the star's emitted energy is at a *maximum*.

For another example, we pick the red star Betelgeuse, which radiates most strongly at about 830 nanometers. Its temperature is therefore about

$$\frac{2.9 \times 10^6}{830} = \frac{2.9 \times 10^6}{8.3 \times 10^2} \approx 3.5 \times 10^3 = 3500 \text{ K}$$

Wien's law is just one way to determine a star's temperature from its light, as we will see in section 13.3, and there are several properties we can deduce from a star's light. For example, in some cases we can also measure the star's radius, but before we discuss how, we need to discuss briefly another general property of stars and other hot objects—the amount of energy they radiate.

Luminosity

Astronomers call the amount of energy an object radiates each second its **luminosity** (abbreviated as L). An everyday example of luminosity is the wattage of a light bulb: a typical table lamp has a luminosity of 100 watts, whereas a bulb for an outdoor parking lot light may have a luminosity of 1500 watts. Stars, of course, are enormously more luminous. For example, the Sun has a luminosity of about 4×10^{26} watts, which it obtains by "burning" its hydrogen into helium. Thus, a star's luminosity measures how fast it consumes its fuel, a vital quantity for determining its lifetime. Knowing a star's luminosity is also important because from it astronomers can measure a star's radius and distance. But to understand how such measurements are made, we must first discuss a relation between how bright an object appears, how bright it really is (its luminosity), and its distance—a relation known as the *inverse-square law.*

The Inverse-Square Law and Measuring a Star's Luminosity

The **inverse-square law** relates an object's luminosity to its distance and its apparent brightness—that is, how bright it looks to us. We all know that a light looks brighter when we are close to it than when we are farther from it. The inverse-square law puts that everyday experience into a mathematical form, describing how light energy spreads out from a source such as a star. As the light travels outward, it moves in straight lines, spreading its energy uniformly in all directions, as shown in figure 13.4A. Near the source, the light will have spread only a little, and so more enters an observer's eye, making it look brighter than if the observer were far away. That is, the farther away from us a source of light is, the less of its light enters our eyes. As a result, more-distant objects look dimmer (fig. 13.4B).

A

B

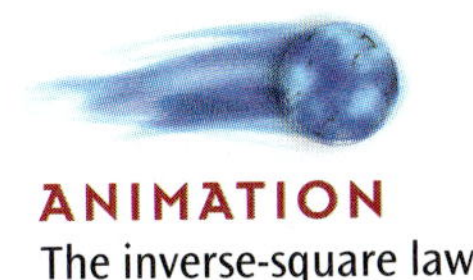

ANIMATION
The inverse-square law

FIGURE 13.4
Decrease of brightness with distance. (A) Light spreads out from a point source in all directions, so the rays become less concentrated. (B) You can see the inverse-square law at work in this picture of street lights.

Q. How far away would light have to travel in the figure so there was just one "photon" (dot) in each square?

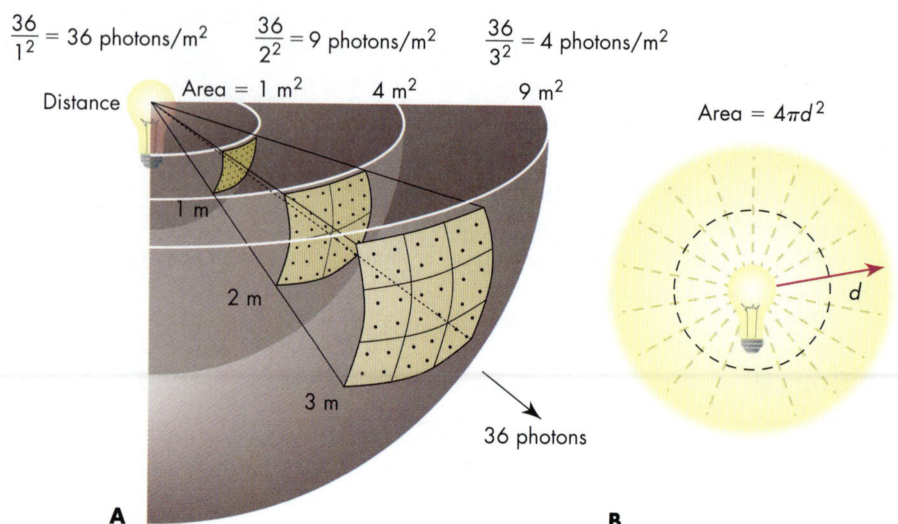

FIGURE 13.5
The inverse-square law. (A) As photons move out from a source, they are spread over a progressively larger area as the distance from the source increases. Thus, a *given* area intercepts fewer photons the farther it is from the source. (B) The area over which light a distance *d* from the source is spread is $4\pi d^2$.

This decrease in brightness with increasing distance can also be understood if you think of light as photons. Photons leaving a light source such as a star spread evenly in all directions. Now imagine a series of progressively larger spheres drawn around the source. If nothing absorbs the light, the same number of photons pass through each sphere. But because more-distant spheres are larger, the number of photons passing through any given *area* on any one sphere grows smaller as the spheres become more distant and larger, as shown in figure 13.5A. Similarly, as you move away from a light source, the photons reaching you are spread more widely—therefore fewer enter your eyes and the source appears dimmer. Thus, a more distant source appears dimmer because more of its light has spread along lines that never reach you.

We can use this argument to show the following: at a distance d from a light source, the source's luminosity, L, has spread over a sphere whose radius is d, as shown in figure 13.5B. (We use d to stand for radius here to emphasize that we are finding a distance.) Any sphere has a surface area given by $4\pi d^2$, and so the brightness we observe, B, is just

$$B = \frac{L}{4\pi d^2}$$

This relationship is called the inverse-square law because distance appears in the denominator as a square.

The inverse-square law is one of the most useful mathematical tools available to astronomers for measuring not only a star's luminosity but also its distance. To find a star's luminosity, astronomers measure its distance, d, by parallax, as described in section 13.1. Next, with a photometer, a device similar to the electric exposure meter in a camera, they measure how bright the star appears from Earth, B. Finally, with B and d known, they calculate the star's luminosity, L, using the inverse-square law. Such measurements reveal that the average star has a luminosity similar to that of our Sun. However, astronomers also find some stars that are millions of times more luminous than the Sun and other stars that are thousands of times less luminous than the Sun. Much of this range in luminosity comes from the great range in stellar radii. That is, some stars are vastly larger than others, as we will now show.

Radius

Common sense tells us that if we have two objects of the same temperature but of different sizes, the larger one will emit more energy than the smaller one. For example, three glowing charcoal briquettes in a barbecue emit more energy than just one briquette at the same temperature. Similarly,

> ***if two stars have the same temperature but one is more luminous than the other, the more-luminous star must have a larger surface area, therefore a larger radius than the dimmer star.***

Thus, if we know a star's temperature, we can infer its size from the amount of energy it radiates. To calculate the star's radius, however, we need a mathematical relation between luminosity, temperature, and radius—a relation known as the Stefan-Boltzmann law.

The Stefan-Boltzmann Law

Imagine watching an electric stove burner heat up. When the burner is on low and is relatively cool, it glows dimly red and gives off only a slight amount of heat. When the burner is on high and is very hot, it glows bright yellow-orange and gives off far more heat. Thus, you can both see and feel that raising a body's temperature increases the amount of radiation it emits per second—that is, its luminosity. This familiar situation is an example of a law deduced in the late 1800s by two German scientists, Josef Stefan and Ludwig Boltzmann, who showed that the luminosity of a hot object depends on its temperature.

The Stefan-Boltzmann law, as their discovery is now called, states that an object of temperature T radiates an amount of energy each second equal to σT^4 per square meter, as shown in figure 13.6A. The quantity σ is called the "Stefan-Boltzmann constant," and its value is 5.67×10^{-8} watts m^{-2} K^{-4}. The Stefan-Boltzmann law affords a mathematical explanation of what we noticed for the electric stove burner. That is, a hotter burner has a larger T, and therefore, according to the law, the burner radiates

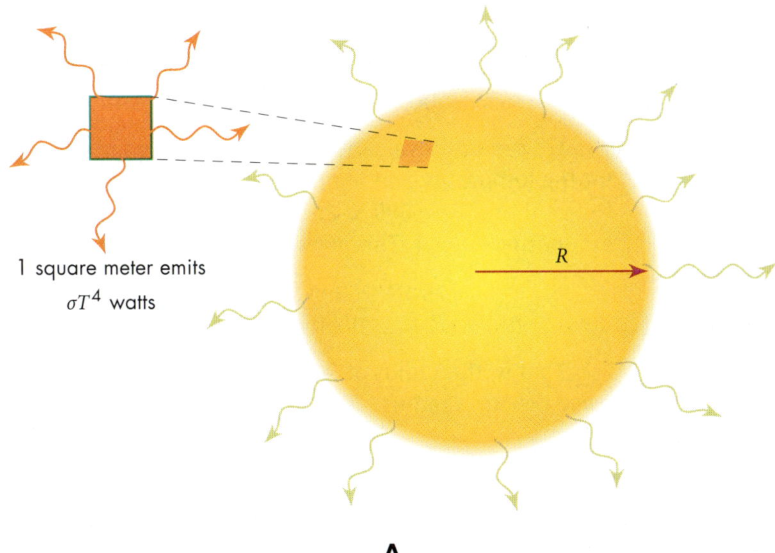

Total energy radiated per second by the star is its

Luminosity $= L$

$L =$ Energy emitted by 1 square meter $\times$ Number of square meters of its surface

$= \sigma T^4 \times$ Star's surface area

For a spherical star of radius R, the surface area is $4\pi R^2$

Thus, $L = \sigma T^4 \times 4\pi R^2$

or

$$R = \frac{\sqrt{L}}{\sqrt{4\pi\sigma} \times T^2}$$

B

$$L = 4\pi R^2 \sigma T^4$$

FIGURE 13.6
The Stefan-Boltzmann law can be used to find a star's radius. (A) Each part of the star's surface radiates σT^4.
(B) Multiplying σT^4 by the star's surface area ($4\pi R^2$) gives its total power output—its luminosity, $L = 4\pi R^2 \sigma T^4$. To find the star's radius, we solve this equation to get $R = \sqrt{L/(4\pi\sigma T^4)}$. Finally, we can use this equation and measured values of the star's L and T to determine R.

more strongly. Similarly, the Stefan-Boltzmann law is the basis for dimmer switches on lights: changing the amount of electricity going to the lamp changes its temperature and thereby its brightness. However, the Stefan-Boltzmann law does not apply to all hot objects. For example, it does not accurately describe the radiation from hot, low-density gas, such as that in fluorescent light bulbs or interstellar clouds.

We can apply the Stefan-Boltzmann law to determine a star's luminosity as follows. According to the law, if a star has a temperature T, each square meter of its surface radiates an amount of energy per second given by σT^4. We can find the total energy the star radiates per second—its luminosity, L—by multiplying the energy radiated from 1 square meter (σT^4) by the number of square meters of its surface area (fig. 13.5B). If we assume that the star is a sphere, its area is $4\pi R^2$, where R is its radius. Its luminosity, L, is therefore

$$L = 4\pi R^2 \sigma T^4$$

That is, a star's luminosity equals its surface area times σT^4. This relation between L, R, and T may at first appear complex, but its content is very simple: increasing either the temperature or the radius of a star makes it more luminous. Making T larger makes each square meter of the star brighter. Making R larger increases the number of square meters.

In the above expression, L is measured in watts, but the "wattage" of stars is so enormous that it is more convenient to use the Sun's luminosity as a standard unit. Likewise, it is easier to use the Sun's radius as a standard size unit rather than meters or even

EXTENDING *our reach*

MEASURING THE RADIUS OF THE STAR SIRIUS

From its color (the wavelength at which it radiates most strongly), we found that the temperature (T_s) of Sirius is about 10,000 K. From the amount of energy we receive from Sirius and its distance, which can be found by parallax, we can find that its luminosity (L_s) is about 25 $L_\odot$. We solve for the radius of Sirius (R_s) as follows. First write down the Stefan-Boltzmann law for Sirius to get $L_s = 4\pi R_s^2 \sigma T_s^4$. Then write down the same relation for the Sun: $L_\odot = 4\pi R_\odot^2 \sigma T_\odot^4$.

Next, we divide the expression for Sirius by the expression for the Sun to get

$$\frac{L_s}{L_\odot} = \frac{4\pi R_s^2 \times T_s^4}{4\pi R_\odot^2 \times T_\odot^4}$$

We can simplify this expression by canceling the identical 4π and σ factors to get

$$\frac{L_s}{L_\odot} = \frac{R_s^2 T_s^4}{R_\odot^2 T_\odot^4}$$

We next collect the R's and T's as separate factors, giving us

$$L_s/L_\odot = (R_s/R_\odot)^2 (T_s/T_\odot)^4$$

We now solve this expression for $(R_s/R_\odot)^2$ to get

$$(R_s/R_\odot)^2 = (L_s/L_\odot)(T_\odot/T_s)^4$$

Finally, we take the square root of both sides to get

$$R_s/R_\odot = (L_s/L_\odot)^{1/2} (T_\odot/T_s)^2$$

We can now evaluate $R_s/R_\odot$ by inserting the values for the luminosity and temperature of Sirius and the Sun. Notice that $T_\odot = 5800$ K, and by definition of our units for luminosity, $L_\odot = 1$. This gives us

$$\frac{R_s}{R_\odot} = \left(\frac{25}{1}\right)^{1/2} \left(\frac{5800\text{ K}}{10{,}000\text{ K}}\right)^2 = 5(0.58)^2 = 1.7$$

That is, the radius of Sirius* is 1.7 $R_\odot$, a little less than twice the Sun's radius.

*In classical times, the Persians, Greeks, and Romans sometimes called Sirius the Dog Star. It is the brightest star in the constellation Canis Major, the Big Dog, and was associated by many early people with misfortune and fevers. The ancient Greeks and Romans also blamed Sirius for the extreme heat of July and August. At that time of year, it rose at about the same time as the Sun and they believed it added its brilliance to the Sun's, making the season extra warm. The heat we receive from Sirius is negligible, of course, but like the ancient Greeks and Romans, we still refer to the hot days of late summer as the "dog days."

kilometers. If we need to convert to watts or meters, we simply remember that one solar luminosity, $L_\odot$, is 4×10^{26} watts and one solar radius, $R_\odot$, is 7×10^5 kilometers.

Because a star's luminosity depends on its radius and temperature, if we know its luminosity and temperature, we can find its radius, R. The method works because if we have a mathematical relation between three quantities (in this case L, T, and R) we can find any one of these given the other two. Thus, if we know a star's luminosity and its temperature, we can use the Stefan-Boltzmann law to solve for its radius. Extending Our Reach: "Measuring the Radius of the Star Sirius" explains how.

Astronomers can also measure a star's radius from its angular size, the technique we used in chapter 12 to measure the radius of the Sun. Unfortunately, the angular size of all stars except the Sun is extremely tiny because they are so far from the Earth, and so even in powerful telescopes, stars generally look like a smeary spot of light. That smearing is caused by the blurring effects of our atmosphere and by a physical limitation of telescopes called "diffraction," which we discussed briefly in chapter 5.

Diffraction limits a telescope's ability to measure tiny angular size, hopelessly blurring the light by an amount that depends on the diameter of the telescope's lens or mirror. Although such diffraction effects are less severe on larger-diameter telescopes, to measure the angular size of most stars, a telescope with truly immense diameter is needed. For example, to measure the angular size of a star like the Sun if it were 50 light-years away would require a telescope 300 meters in diameter, about three times the size of a football field. To avoid the need for such enormous (and expensive) telescopes, astronomers have therefore devised an alternative way to measure the angular size of stars by using not one huge telescope but two (or more) smaller ones separated by a large distance (for example, several hundred meters). Such a device is called an "interferometer" (discussed more fully in chapter 5), and its ability to measure angular sizes is equivalent to that of a single telescope whose diameter is equal to the distance that separates the two smaller ones. (Of course, a single enormous telescope would gather more light and enable astronomers to measure the sizes of very faint stars, but it may never be possible to build such a huge instrument.) A computer then combines the information from the two telescopes to give a crude picture of the star. Such interferometric observations are still hampered by blurring caused by our atmosphere. Those effects can be partially offset by a technique called "speckle interferometry" in which a high-speed camera takes many very short exposures of the star's light. A computer then combines the separate short exposures into a "deblurred" image, such as that of Betelgeuse, illustrated in figure 13.7. With interferometers, astronomers can measure the radius of several dozen nearby stars and a hundred or so more-distant giant stars.

The Stefan-Boltzmann law and interferometer observations show that stars differ enormously in radius. Although most stars have approximately the same radius as the Sun, some are hundreds of times larger, and astronomers call them **giants.** Smaller stars (our Sun included) are called **dwarfs.**

In using the Stefan-Boltzmann law to measure a star's radius, we have seen that L can be measured either in watts or in solar units for L. Astronomers sometimes use a different set of units, however, to measure stellar brightness.

Q. Use this image of Betelgeuse to measure its diameter. You will need the formula developed in the box in chapter 2, section 2.1, describing how to determine the diameter of a body from its angular size and distance. Betelgeuse is about 6×10^{15} kilometers away (200 parsecs).

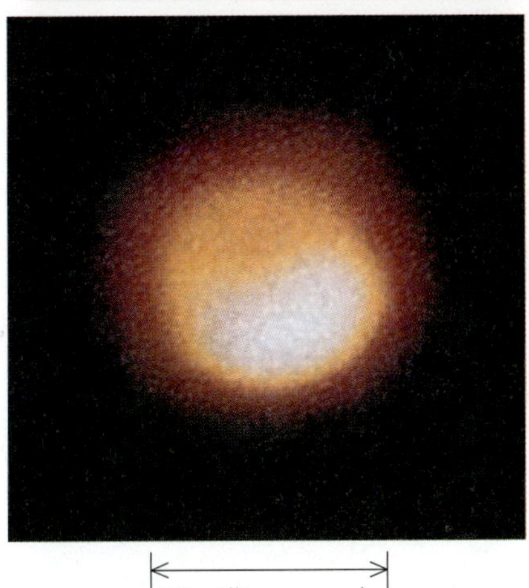

FIGURE 13.7
Image of the star Betelgeuse made by interferometry. The lack of detail in this image reflects the current difficulty in observing the disk of any star but our Sun. One milliarcsecond (mas) is approximately 2.8×10^{-7} degrees.

The Magnitude System

About 150 B.C., the ancient Greek astronomer Hipparchus measured the apparent brightness of stars in the night sky using units he called **magnitudes.** He designated the stars that looked brightest as magnitude 1 and the dimmest ones (the ones he could just barely see) as magnitude 6. For example, Betelgeuse, a bright red star in the constellation Orion, is magnitude 1, while the somewhat dimmer stars in the Big Dipper's handle are approximately magnitude 2. Astronomers still use this scheme to measure the brightness of astronomical objects, but they now use the term *apparent magnitude* to emphasize that they are measuring how bright a star *looks* to an observer. A star's apparent magnitude depends on its luminosity and its distance; it may look dim either because it has a small luminosity—it does not emit much energy—or because it is very far away.

Astronomers use the magnitude system for many purposes (for example, to indicate the brightness of stars on star charts), but it has several confusing properties. First, the scale is "backward" in the sense that bright stars have small magnitudes, while dim stars have large magnitudes. Moreover, modern measurements show that Hipparchus underestimated the magnitudes of the brightest stars, and so the magnitudes now assigned them are negative numbers.

Second, magnitude *differences* correspond to brightness *ratios*. That is, if we measure the brightness of a first-magnitude star and a sixth-magnitude star, the former is 100 times brighter than the latter. Thus, a *difference* of 5 magnitudes corresponds to a brightness *ratio* of 100; so, when we say a star is 5 magnitudes brighter than another, we mean it is a *factor* of 100 brighter. Each magnitude difference corresponds to a factor of 2.512… (the fifth root of 100) in brightness. Thus, a first-magnitude star is 2.512 times brighter than a second-magnitude star and is 2.512×2.512, or 6.310, times brighter than a third-magnitude star. Table 13.1 lists the ratios that correspond to various differences in magnitude. For example, let us compare the apparent brightness of the planet Venus with the star Aldebaran. At its brightest, Venus has an apparent magnitude of −4.2; Aldebaran's apparent magnitude is 0.8, and so the difference in their magnitudes is $0.8 - (-4.2) = 5.0$. Therefore we see from table 13.1 that Venus is 100 times brighter to our eye than Aldebaran.

Third, astronomers often use a quantity called "absolute magnitude" to measure a star's luminosity. Recall that a star's apparent magnitude is how bright it *looks* to an observer. But the apparent magnitude of a star depends on the distance from which we observe it. The same star will have one apparent magnitude if it is near to us but a different apparent magnitude if it is far from us. Astronomers therefore find it useful to have a way to describe a star's brightness that does not depend on the star's distance. One way to do that is to imagine how bright a star would look if we were to observe it from some *standard* distance. Astronomers have chosen 10 parsecs as that standard distance, and they call the apparent magnitude of a star seen from 10 pc its *absolute magnitude*. Because the absolute magnitude does not depend on distance, it is a measure of a star's true brightness. In other words, absolute magnitude is a measure of a star's luminosity. Table 13.2 illustrates how absolute magnitude is related to a star's luminosity.

TABLE 13.1 RELATING MAGNITUDES TO BRIGHTNESS RATIOS

Magnitude Difference	Ratio of Brightness
1	2.512:1
2	$2.512^2 = 6.31$:1
3	$2.512^3 = 15.85$:1
4	$2.512^4 = 39.8$:1
5	$2.512^5 = 100$:1
⋮	⋮
10	$2.512^{10} = 10^4$:1

TABLE 13.2 RELATING ABSOLUTE MAGNITUDE TO LUMINOSITY

Absolute Magnitude	Approximate Luminosity in Solar Units
−5	10,000
0	100
5	1
10	0.01

13.3 Spectra of Stars

A star's spectrum depicts the energy it emits at each wavelength and is perhaps the single most important thing we can know about the star. From the spectrum we can find the star's composition, temperature, luminosity, velocity in space, rotation speed, and some other properties as well. For example, under some circumstances we may also deduce the star's mass and radius.

INTERACTIVE
Stellar spectroscopy

Figure 13.8 shows the spectra of three stars. You can easily see differences between them. The middle spectrum is from a star similar to the Sun; the top spectrum is from a star hotter than the Sun; and the bottom spectrum is from a star cooler than the Sun. The hot star's spectrum has only a few dark lines, but they are strong and their spacing follows a regular pattern. The cool star's spectrum, however, shows a welter of lines with no apparent regularity. Understanding such differences and what they tell us about a star is one of the goals of studying stellar spectra.

Measuring a Star's Composition

As light moves from a star's core through the gas in its surface layers, atoms there absorb the radiation at some wavelengths, creating the dark absorption lines in the star's spectrum that we see in figure 13.8. Each type of atom—hydrogen, helium, calcium, and so on—absorbs at a unique set of wavelengths. For example, hydrogen absorbs at 656, 486, and 434 nanometers, in the red, blue, and violet part of the spectrum, respectively. Gaseous calcium, on the other hand, absorbs strongly at 393.3 and 396.8 nanometers, producing a strong double line in the violet. Because each atom absorbs a unique combination of wavelengths of light, each has a unique set of absorption lines. From such absorption lines, we can determine what a star is made of.

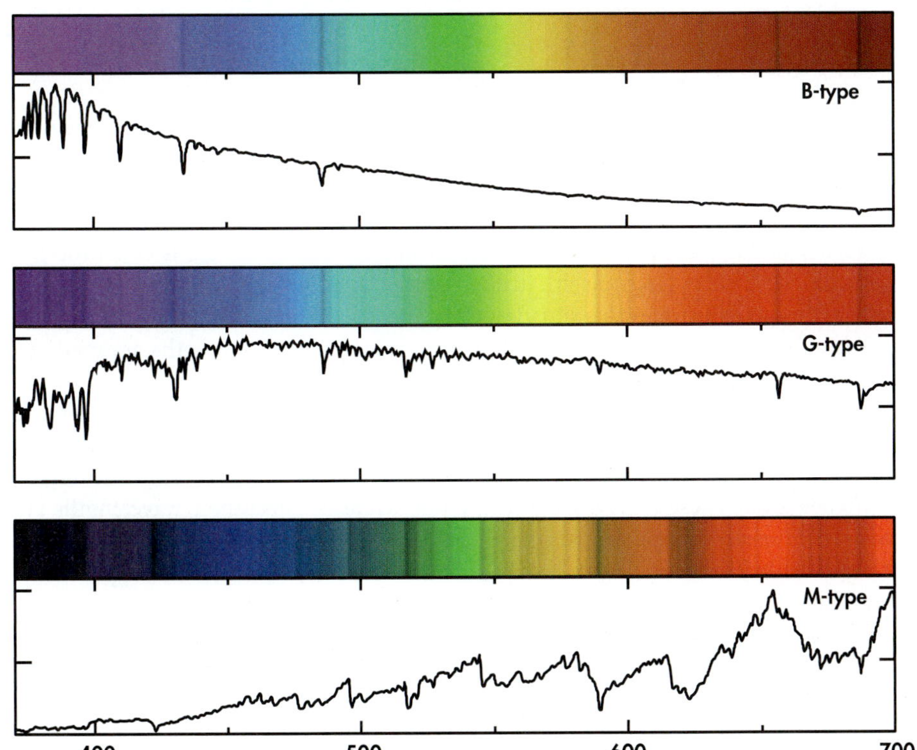

FIGURE 13.8
Spectra of three stars. The middle spectrum is similar to the Sun's. The one above it is hotter than the Sun. The one below is cooler than the Sun. A plot of the intensity of light at each wavelength is shown in the graph below each spectrum. Dark lines show up as dips in the graph.

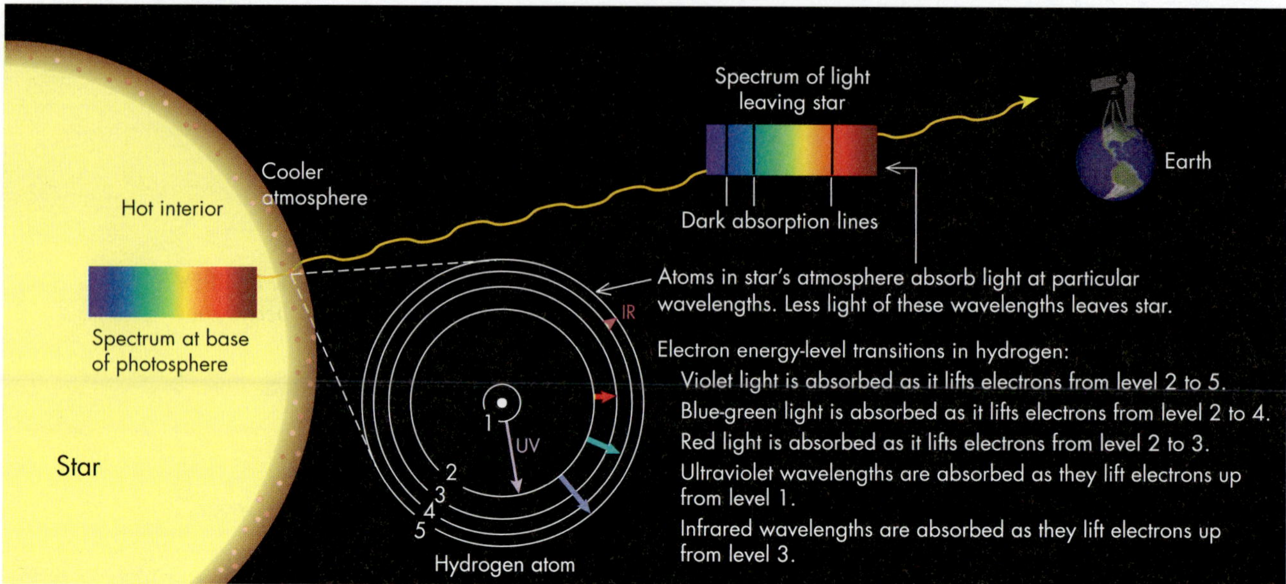

FIGURE 13.9
Formation of stellar absorption lines. Atoms in the cooler atmospheric gas absorb radiation at wavelengths corresponding to jumps between electron orbits.

To measure a star's composition, we make a spectrum of its light and then compare the absorption lines we see with tables that list the lines made by each atom. When we find a match between an absorption line in the star and a line in the table, we can deduce that the element listed exists in the star. To find the quantity of each atom in the star—each element's abundance—we use the darkness of the absorption line, as discussed in chapter 4. Such determinations of the identity and quantity of elements present are difficult, however. Even though an element may be present in the star, temperature effects might make the element's atoms unable to absorb light and make a spectral line.

How Temperature Affects a Star's Spectrum

To see why temperature affects a star's spectrum, recall that light is absorbed when its energy matches the energy difference between two electron orbits, as we discussed in chapter 4. For an atom to absorb light, its electrons must be in the proper orbit, or, more technically, energy level. An atom may be abundant in a star's atmosphere and create only very weak lines at a particular wavelength simply because the gas is either too hot or too cold, so that its electrons are in the "wrong" level to absorb light at that wavelength. Hydrogen illustrates this situation dramatically, as shown in figure 13.9.

The absorption lines of hydrogen that we see at visible wavelengths are made by electrons orbiting in its second level. These lines are sometimes called the "Balmer lines"* to distinguish them from other hydrogen lines in the ultraviolet and infrared wavelengths. Balmer lines occur at wavelengths where light has exactly the amount of energy needed to lift an electron from hydrogen's second energy level to the third level, or a higher level. These lines are especially important because their wavelengths are in the visible spectrum and are therefore easily observable. If the hydrogen atoms in a star have no electrons orbiting in level 2, no Balmer absorption lines will appear even though hydrogen may be the most abundant element in the star.

Hydrogen Balmer lines appear weakly or not at all if a star is either very cold or very hot. In a cool star, most hydrogen atoms have their electrons in level 1 and so they cannot absorb Balmer line radiation, which can be absorbed only if the electron is in

*Balmer lines are named for Johann Balmer, the scientist who first studied their pattern.

level 2. In a hot star the atoms move faster, and when they collide, electrons may be excited ("knocked") into higher orbits: the hotter the star, the higher the orbit. Moreover, in very hot stars the radiation is so energetic that the photons can knock electrons out of the atom, in which case the atom is said to be ionized. As a result of this excitation, proportionally more of the electrons in a very hot star will be in level 3 or higher. Only if the hydrogen has a temperature between about 8000 K and 15,000 K will enough atoms have their electrons in level 2 to make strong hydrogen Balmer lines. If we are therefore to deduce correctly the abundance of elements in a star, we must correct for such temperature effects. With these corrections, we discover that virtually all stars are composed mainly of hydrogen: their composition is similar to that of our Sun—about 71% H, 27% He, and 2% a mix of the remaining elements. But despite their uniform composition, stars exhibit a wide range in the appearance of their spectra, as astronomers noted when they began to study stellar spectra.

Classification of Stellar Spectra

Stellar spectroscopy, the study and classification of spectra, was born early in the nineteenth century when the German scientist Joseph Fraunhofer discovered dark lines in the spectrum of the Sun. He later observed similar lines in the spectra of stars and noted that different stars had different patterns of lines. In 1866, Pietro Angelo Secchi, an Italian priest and scientist, noticed that the line patterns depended on the star's color. He assigned stars to four color types—white, yellow, red, and deep red—and considered these types as evidence for stellar evolution—hot stars cooling from white heat to deep red as they aged.

Technological improvements have allowed astronomers to refine Secchi's system further. The first such improvement came in 1872, when Henry Draper,* a physician and amateur astronomer, recorded spectra on photographs. On Draper's death, his widow endowed a project at Harvard to create a compilation of stellar spectra, known as the Henry Draper Catalog.

FIGURE 13.10
Annie Jump Cannon.

The use of photography in compiling the Draper Catalog made it possible to obtain many more spectra and with better details than could be observed visually. In fact, details were so good that E. C. Pickering, the astronomer in charge of the project, began to use letters to subdivide the spectral types, assigning the letters *A* to *D* to Secchi's white type, *E* to *L* to the yellow type, and *M* and *N* to the red types.

About 1901, Annie Jump Cannon (fig. 13.10), the astronomer who was actually doing the classification for the Draper Catalog, discovered that the types fell in a more orderly sequence of appearance if rearranged by temperature. She therefore reordered Secchi's types to obtain the sequence *O*, *B*, *A*, *F*, *G*, *K*, and *M*, with *O* stars being the hottest and *M* stars being the coolest. Her work is the basis for the stellar **spectral types** we use today.

Stellar spectral types are based on the appearance of the spectrum. For example, *A*-type stars show extremely strong hydrogen lines, while *B*-type stars show helium and weak hydrogen lines. This scheme, however, despite its wide adoption, lacked a physical basis in that it employed the appearance of the spectra rather than the physical properties of the stars that produced them. For example, astronomers at that time did not know what made the spectra of *A* and *F* stars differ. That understanding came in the 1920s from the work of the American astronomer Cecilia Payne (later Payne-Gaposhkin) (fig. 13.11), who explained why the strength of the hydrogen lines depended on the star's temperature, as we described above. Her discovery was based on work by the Indian astronomer M. Saha, who showed how to calculate the level in which an atom's electrons were most likely to be found. Saha showed that the levels occupied by electrons—a quantity crucial for interpreting the strength of spectral lines—depend on the star's temperature and density. With Saha's equation, Payne then showed mathematically the

FIGURE 13.11
Cecilia Payne.

*Draper's father was a professor of chemistry and physiology, and the first person we know of to photograph an astronomical object—the Moon.

SCIENCE *at work*

NEW SPECTRAL TYPES

Since the establishment of the spectrum types *O* through *M*, astronomers, aided by new technology, have made further refinements in classifying stellar spectra. For example, much classification is now done automatically by computers that scan a star's spectrum and match it against standard spectra stored in memory. Other improvements come from more sensitive infrared detectors that allow astronomers to detect dim stars far cooler than any previously known. Some of these stars are so cool that not only do molecules form in their atmospheres but solid dust particles condense there as well. With atmospheres so different from those of ordinary stars, these extremely cool objects have spectra that are also radically different. Therefore, astronomers have devised new spectrum types—*L* and *T*—to describe these cool, dim objects. For example, *L* stars show strong molecular lines of iron hydride and chromium hydride, while *T* stars show strong absorption lines of methane.*

*Can you think up a mnemonic for the spectral types that includes these new types?

To distinguish finer gradations in temperature, astronomers subdivide each type by adding a numerical suffix—for example, *B*1, *B*2, . . . *B*9—with the smaller numbers indicating higher temperatures. With this system, the temperature of a *B*1 star is about 20,000 K, whereas that of a *B*5 star is about 13,500 K. Similarly, our Sun, rather than being just a *G* star, is a *G*2 star.

correctness and reasonableness of Cannon's order for the spectral types. She thereby demonstrated that a star's spectral type is determined mainly by its temperature.*

Payne's theory unfortunately left astronomers with Cannon's odd nonalphabetical progression for the spectral types, which—from hot to cold—ran *O*, *B*, *A*, *F*, *G*, *K*, and *M*. So much effort, however, had been invested in classifying stars using this system that it was easier to keep the types as assigned with their odd order than to reclassify them (Cannon, in her life, classified some quarter million stars). As a help to remember the peculiar order, astronomers have devised mnemonics. One of the first is "Oh, be a fine girl/guy. Kiss me." Another version is "Only brilliant, artistic females generate killer mnemonics." Choose whichever appeals most, or make up your own version, to help you learn this important sequence.

Definition of the Spectral Types

A star's spectral type is determined by the lines in its spectrum. Figure 13.12 illustrates the seven main types—*O* through *M*—and the differences in the line patterns are easy to see, even without knowing the identity of the lines. However, knowing which element makes which line can increase our understanding of the different line patterns. As mentioned earlier, almost all stars are made of the same mix of elements, but differing conditions, especially their surface temperature, result in differing spectra. For example, *O* stars have weak absorption lines of hydrogen but strong absorption lines of helium, the second most abundant element, because they are so hot. At an *O* star's high temperature, hydrogen atoms collide so violently and are excited so much by the star's intense radiation that the electrons are stripped from most of the hydrogen, ionizing it. With its electron missing, a hydrogen atom cannot absorb light. Because most of the *O* star's hydrogen is ionized, such stars have extremely weak hydrogen absorption lines. Helium atoms are more tightly bound, however, and most retain at least one of their electrons, allowing them to absorb light. Thus, *O* stars have absorption lines of helium but only weak lines of hydrogen.

Stars of spectral type *A* have very strong hydrogen lines. Their temperature is just right to put lots of electrons into orbit 2 of hydrogen, which makes for strong Balmer lines. Balmer lines also appear in *F* stars but are weaker. *F* stars are distinguished by

*Payne also made clear that stars are composed mainly of hydrogen, as mentioned earlier.

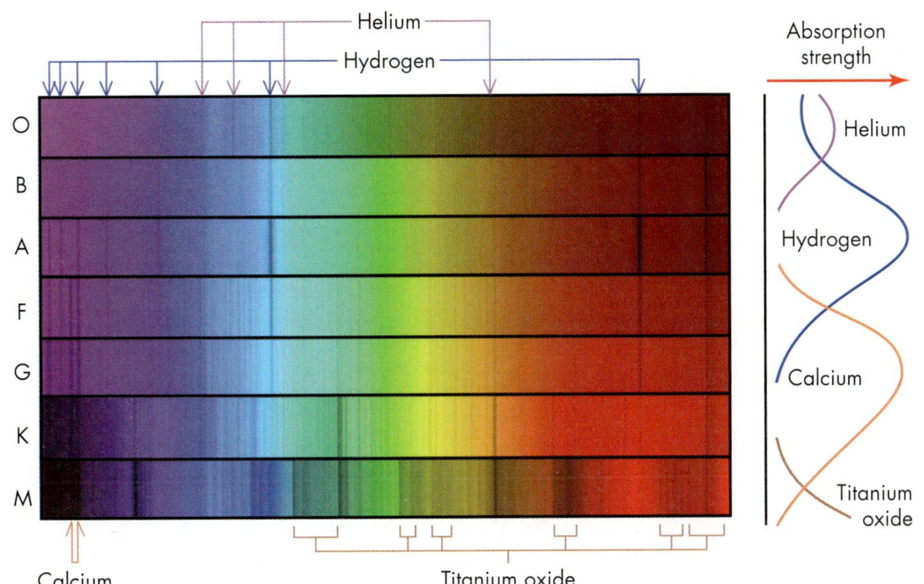

FIGURE 13.12
The stellar spectral types. Notice how the hydrogen line at right is weak in the top spectrum and grows stronger down to *A* stars, but then essentially disappears in *K* and *M* stars. The separate curves to the right of the spectra show how the strength of the hydrogen absorption lines and those of other chemical elements change with the spectral type.

the multitude of lines from metals such as calcium and iron, elements that also appear strongly in *G* and *K* stars. Such elements are present in hotter stars but are ionized and generally create only very weak spectral lines under those conditions. In cooler stars, however, metal lines, particularly ionized calcium, are moderately strong, and hydrogen lines become weak.

In the very cool *K* and *M* stars, hydrogen is almost invisible because its electrons are mostly in level 1 and therefore cannot make Balmer lines. These stars have such cool atmospheres that molecules form, and as we saw in chapter 4, they produce very complex spectra. As a result, *K* and *M* stars have numerous lines from such substances as the carbon compound cyanogen (CN) and the carbon radical methylidyne (CH), as well as gaseous titanium oxide (TiO). The small graph at the right-hand side of figure 13.12 shows how the strength of the absorption lines of hydrogen, calcium, and so on change with spectrum type. The changing strengths of these lines and others are summarized in table 13.3.

We have seen that *O* stars are hot and *M* stars are cool, but what, in fact, are their temperatures? Application of Wien's law and theoretical calculations based on Saha's

TABLE 13.3 SUMMARY OF SPECTRAL TYPES

Spectral Type	Temperature Range (K)	Features	Representative Star
O	Hotter than 30,000	Ionized helium, weak hydrogen	
B	10,000–30,000	Neutral helium, hydrogen stronger	Rigel
A	7500–10,000	Hydrogen very strong	Sirius
F	6000–7500	Hydrogen weaker, metals (especially ionized Ca) moderate	Canopus
G	5000–6000	Ionized Ca strong, hydrogen even weaker	The Sun
K	3500–5000	Metals strong, CH and CN molecules appearing	Aldebaran
M	2000–3500	Molecules strong, especially TiO and water	Betelgeuse
L	1300–2000	TiO disappears. Strong lines of metal hydrides, water, and reactive metals such as potassium and cesium	
T	~900–1300	Strong lines of water and methane	

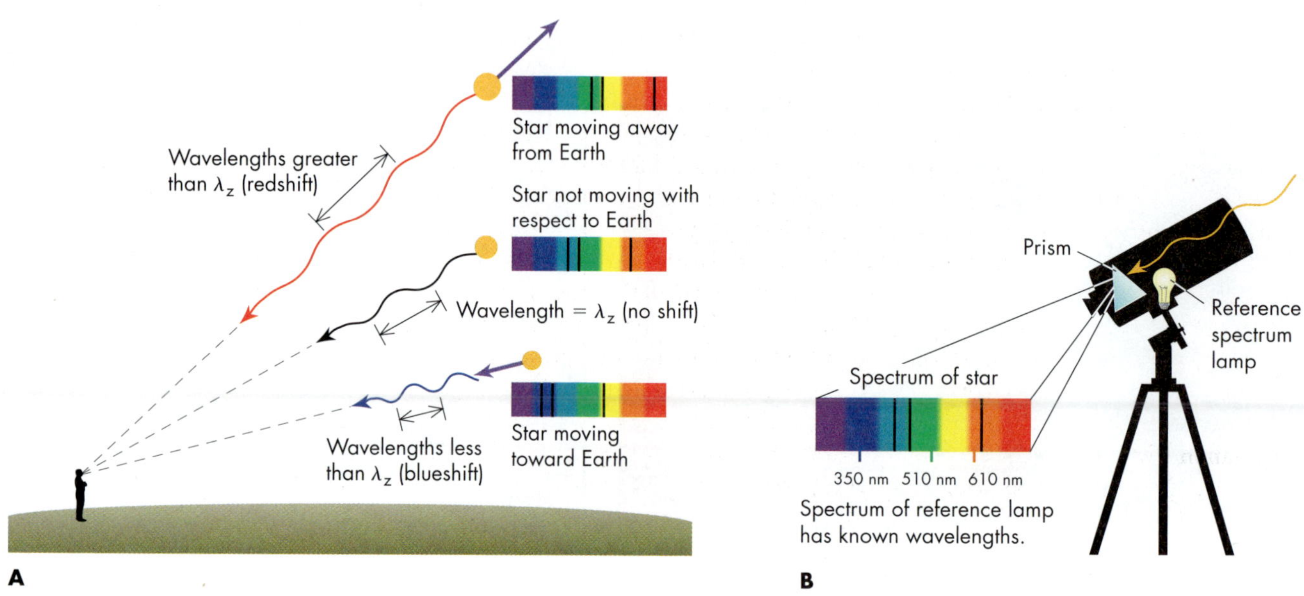

FIGURE 13.13
Measuring a star's radial velocity from its Doppler shift. (A) Spectrum lines from a star moving away from Earth are shifted to longer wavelengths (a redshift). Spectrum lines from a star approaching Earth are shifted to shorter wavelengths (a blueshift). (B) A standard lamp attached inside the telescope serves as a comparison spectrum to allow the wavelength shift of the star's light to be found.

law of how electrons are distributed in atomic orbits show that temperatures range from more than 30,000 K for *O* stars to less than 3500 K for *M* stars, with *A* stars being 7500 to 10,000 K and *G* stars, such as our Sun, being between 5000 and 6000 K.

Because a star's spectral type is set by its temperature, its type also indicates its color. We know that hot objects are blue and cool objects are red (recall Wien's law), and so we find that *O* and *B* stars (hot types) are blue, while *K* and *M* stars (cool types) are red.

Measuring a Star's Motion

All stars move through space and these motions turn out to be important clues to a star's history and are vital in determining star masses and the structure of galaxies.

We find a star's motion from its spectrum using the Doppler shift, which, as we saw in chapter 4, is the alteration of the wavelength of light from a moving source. If a source moves toward us, its wavelengths are shortened, while if it moves away, they are lengthened, as illustrated in figure 13.13A. The amount of the wavelength shift depends on the source's speed along the line of sight, what astronomers call **radial velocity.** We find the shift by subtracting the wavelength the star would absorb (or emit) if it were not moving relative to the Earth, λ_0, from the new wavelength, λ. We will call that difference $\Delta\lambda$. With these definitions, the Doppler shift law states that

$$\frac{\Delta\lambda}{\lambda_0} = \frac{V}{c}$$

$\Delta\lambda$ = Change in wavelength due to motion of light source
λ_0 = Wavelength of unmoving (0-velocity) light source
V = Radial velocity of light source*
c = Speed of light (300,000 km/sec)

*We assume here that $V << c$.

where V is the source's radial velocity, and c is the speed of light. That is, the change in wavelength created by motion of the star (or observer) divided by the wavelength in the absence of motion is just the radial velocity of the star divided by the speed of light. Thus, if we can measure $\Delta\lambda$, we can solve for V. The quantity $\Delta\lambda$ may turn out to be either positive or negative, depending on whether wavelengths are increased or decreased. If the wavelengths are increased, the distance between the source and observer is increasing, meaning that from the observer's point of view the source is moving away. If the wavelengths are decreased, the distance between the source and observer is decreasing, and so from the observer's point of view the source is approaching. Thus, from the observer's point of view, a positive $\Delta\lambda$ implies the source is receding, while a negative $\Delta\lambda$ implies it is approaching.

Astronomers measure the Doppler shift of a star by recording its spectrum directly beside the spectrum of a standard, nonmoving source of light attached to the telescope, as shown in figure 13.13B. They then measure the wavelength shift by comparing the spectrum of the star to the standard. From that shift, they calculate the star's radial velocity with the Doppler shift formula. Notice that a star will have a radial velocity only if its motion has some component toward or away from us. Stars moving across our line of sight, maintaining a constant distance from Earth, have no Doppler shift.

We illustrate how to measure a star's motion with the following example. Suppose we observe a star and see a strong absorption line in the red part of the spectrum with a wavelength of 656.380 nanometers. From laboratory measurements on nonmoving sources, we know that hydrogen has a strong red spectral line at a wavelength of 656.280 nanometers. Assuming these are the same lines, we infer that the wavelength is shifted by a $\Delta\lambda$ of 0.100 nanometers. The Doppler shift law tells us that

$$\frac{V}{c} = \frac{\Delta\lambda}{\lambda_0} = \frac{0.100}{656.28}$$

To find the radial velocity, V, we multiply both sides by c and obtain

$$V = \frac{0.100c}{656.28} = 45.7 \text{ kilometers per second}$$

Because $\Delta\lambda$ is positive, the star is moving *away* from us.

Such measurements reveal that all stars are moving and that those near the Sun share approximately its direction and speed of revolution (about 200 kilometers per second) around the center of our Galaxy, the Milky Way. Superimposed on this orbital motion, however, are small random motions of about 20 kilometers per second. Stars therefore move rather like cars on a freeway—some move a little faster and overtake slower-moving cars, which gradually fall behind. Stars are so far away, however, that their motion is imperceptible to the eye, much as a distant airplane seems to hang motionless in the sky. In fact, stellar motions across the sky are difficult to detect even in pictures of the sky taken years apart.

13.4 Binary Stars

Many stars have a special motion: they orbit around each other, held together as stellar companions by their mutual gravitational attraction, as illustrated in figure 13.14. Astronomers call such stellar pairs **binary stars** and value them greatly because they offer one of the few ways to measure stellar masses. To understand how, recall that the gravitational force between two bodies depends on their masses. The gravitational force, in turn, determines the stars' orbital motion. If that motion can be measured, we can work backward to find the mass.

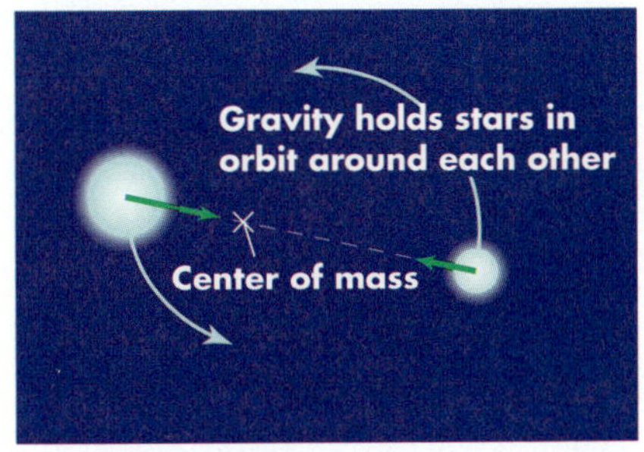

FIGURE 13.14
Two stars orbiting in a binary system, held together as a pair by their mutual gravitational attraction.

INTERACTIVE
Binary stars

A star's mass, as we will discover in chapter 14, critically controls its existence, determining both how long it lasts and how its structure changes as it ages. Thus, if we are to understand the nature of stars, we must know their mass. Fortunately, this is not difficult to determine for stars belonging to a binary system, and such stars are relatively common. Among the hottest stars (*O* and *B* type), roughly 80% have orbiting companions. For Sun-like stars, a little over half have companions. But for the coolest stars (*M* type, in particular), the majority lack companions. Because such cool, dim stars vastly outnumber more luminous stars, single stars are the rule, although binary stars predominate among the brighter stars. We have described binaries as star pairs, but in many cases more than two stars are involved. Some stars are triples, others quadruples, and in at least one case a six-member system is known.

Most binary stars are only a few astronomical units apart. A few, however, are so close that they "touch," orbiting in contact with each other in a common envelope of gas. These

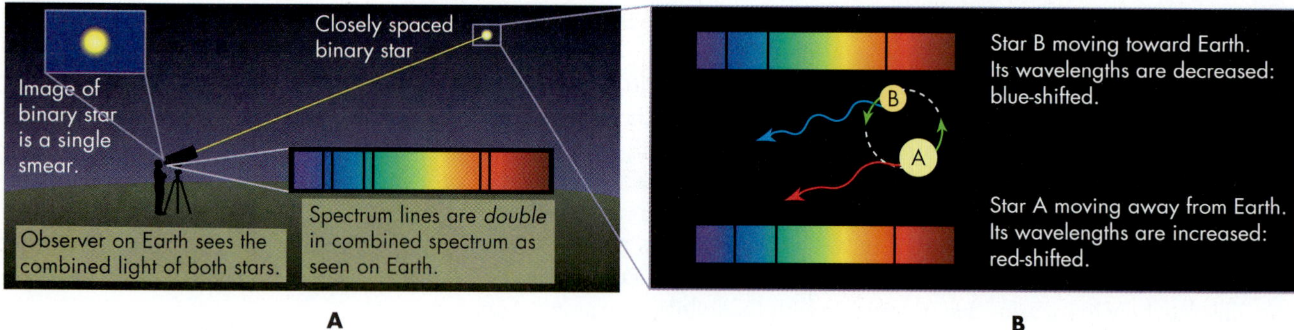

FIGURE 13.15
Spectroscopic binary star. (A) The two stars are generally too close to be separated by even powerful telescopes. (B) Their orbital motion creates a different Doppler shift for the light from each star. Thus, the spectrum of the stellar pair contains two sets of lines, one from each star.

contact binaries and other close pairs probably formed at the same time from a common parent gas cloud; that is, rather than becoming a single star surrounded by planets, as our Sun did, the cloud formed a close star pair. Widely spaced binaries, on the other hand, perhaps formed when one star captured another that happened to pass nearby, but astronomers still have many unanswered questions about how binary stars form.

A few binary stars are easy to see with even a small telescope. Mizar, the middle star in the handle of the Big Dipper, is a good example. When you look at this star with the naked eye, you will first see a dim star, Alcor, which is not a true binary companion but simply a star lying in the same direction as Mizar. With a telescope, however, Mizar can be seen to be two very close stars: a true binary. You can see where Alcor and Mizar are in the sky in "Looking Up: Ursa Major" in the front matter.

ANIMATION
Stages in the evolution of a low-mass binary

Visual and Spectroscopic Binaries

For some stars (such as Mizar) we can directly see orbital motion of one star about the other* by comparing images made several years apart. Such star pairs are called **visual binaries** because we can see two separate stars and their individual motion.

Some binary stars may be so close together, however, that their light blends into a single blob that defies separation with even the most powerful telescopes. In such cases, the orbital motion cannot be seen directly, but it may nevertheless be inferred from their combined spectra. Figure 13.15A shows a typical **spectroscopic binary,** as such stars are called. As the stars orbit each other about their shared center of mass, each star alternately moves toward and then away from Earth. This motion creates a Doppler shift, and so the spectrum of the star pair shows lines that shift—figure 13.15B—first to the red as the star swings away from us and then toward the blue as it approaches. From a series of spectra, astronomers can measure the orbital speed from the Doppler shift of the star, and by observing a full cycle of the motion, they can find the orbital period. From the orbital period and speed, they can find the size of the orbit, and with the size of the orbit and the period, they can find the stars' masses using a modified form of Kepler's third law, as we will now show.

ANIMATION
Orbital motion is reflected in the shifting spectral lines of a spectroscopic binary.

Measuring Stellar Masses with Binary Stars

In the early seventeenth century, Kepler showed that the time required for a planet to orbit the Sun is related to its distance from the Sun. If P is the orbital period in years

*More technically, the two stars orbit a common center of mass located on the imaginary line joining them. The exact location of the center of mass depends on the star masses—if the masses are the same, the center of mass lies exactly halfway between them.

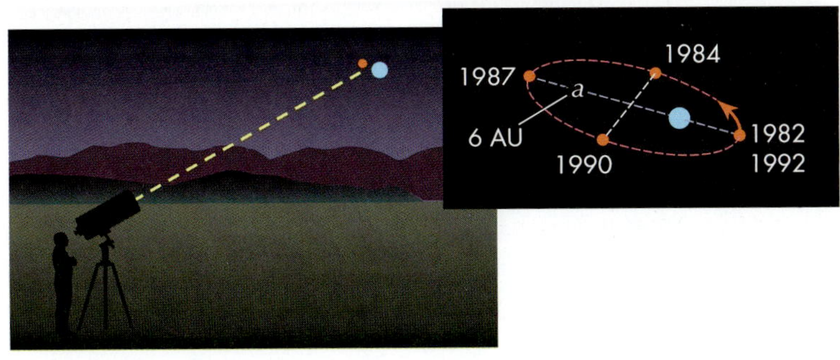

FIGURE 13.16
Measuring the combined mass of two stars in a binary system using the modified form of Kepler's third law.

and a is the semimajor axis (half the long dimension) of the planet's orbit in AU, then $P^2 = a^3$, a relation called Kepler's third law, as we discussed in chapter 2.

Newton discovered that Kepler's third law in a generalized form applies to any two bodies in orbit around each other. If their masses are m and M and they follow an elliptical path of semimajor axis, a, relative to each other with an orbital period, P, then

$$(m + M)P^2 = a^3$$

where P is expressed in years, a in astronomical units, and m and M in solar masses. This relation is our basic tool for measuring stellar masses, as we will now describe.

To find the mass of the stars in a visual binary, astronomers first plot their orbital motion, as depicted in figure 13.16. It may take many years to observe the entire orbit, but eventually the time required for the stars to complete an orbit, P, can be determined. From the plot of the orbit and with knowledge of the star's distance from the Sun, astronomers next measure the semimajor axis, a, of the orbit of one star about the other. Suppose $P = 10.0$ years and a turns out to be 6.00 AU. We can then find their combined mass, $m + M$, by solving the modified form of Kepler's law, and we get (after dividing both sides by P^2)

$$m + M = \frac{a^3}{P^2}$$

Inserting the measured values for P and a, we see that

$$m + M = \frac{6.00^3}{10.0^2}$$

$$= \frac{216}{100}$$

$$= 2.16\ M_\odot$$

That is, the combined mass of the stars is 2.16 times the Sun's mass.

Additional observations of the stars' orbits allow us to find their individual masses, but for simplicity we will omit the details here. From analyzing many such star pairs, astronomers have discovered that star masses fall within a fairly narrow range from about 30 to 0.1 $M_\odot$. We will discover in the next chapter why the range is so narrow, but once again, we find our Sun is about midway between the extremes.

In the above discussion, we assumed that we can actually see the two stars as separate objects, that is, that they are a visual binary. If the stars are spectroscopic binaries and cannot be distinguished as separate stars, their period and separation can be determined from a series of spectra.

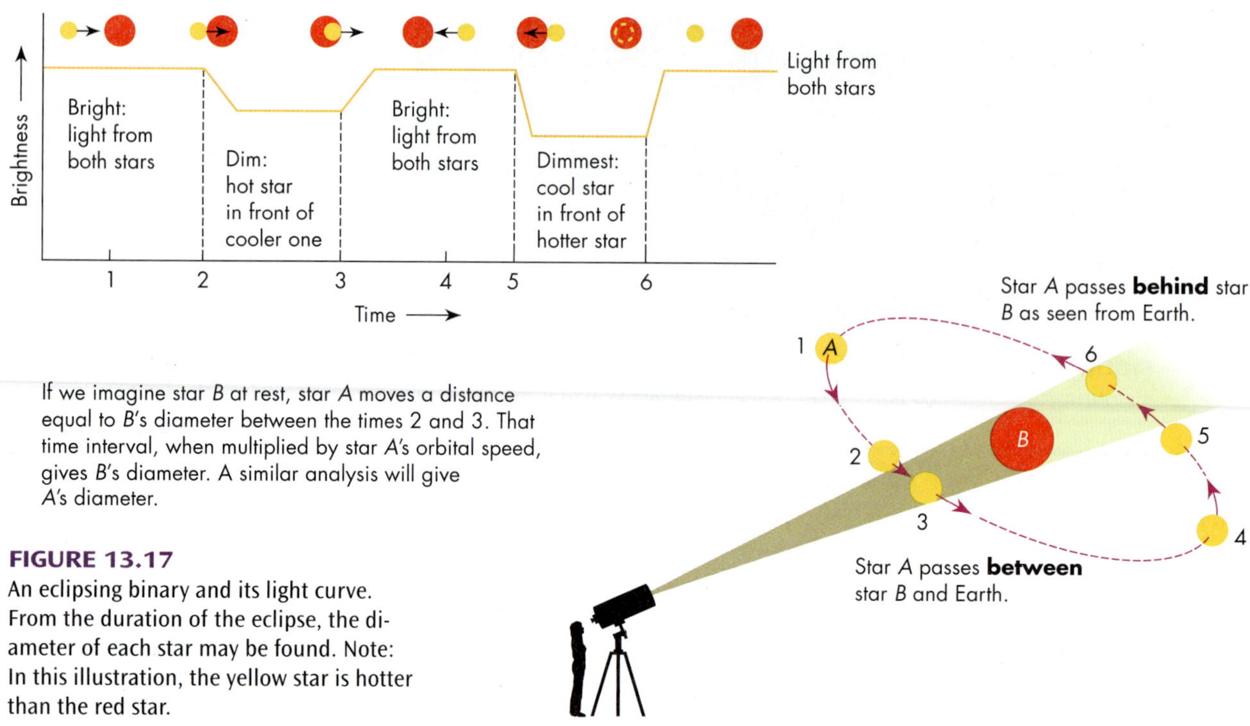

FIGURE 13.17
An eclipsing binary and its light curve. From the duration of the eclipse, the diameter of each star may be found. Note: In this illustration, the yellow star is hotter than the red star.

Eclipsing Binary Stars

On rare occasions, the orbit of a binary star will be almost exactly edge-on as seen from the Earth. Then, as the stars orbit, one will eclipse the other as it passes between its companion and the Earth. Such systems are called **eclipsing binary** stars, and if we watch such a system, its light will periodically dim. During most of the orbit, we see the combined light of both stars, but at the times of eclipse, the brightness of the system decreases as one star covers the other, producing a cycle of variation in light intensity called a "light curve." Figure 13.17A is a graph of such change in brightness over time.

ANIMATION
An eclipsing binary and its light curve

Eclipsing binary stars are useful to astronomers because the duration of the eclipses depends on the stars' diameter. Figure 13.17B shows why. The eclipse begins as the edge of one star first lines up with the edge of the other, as shown in the figure. The eclipse ends when the opposite edges of each star line up. During the eclipse, the covering star has moved relative to the other a distance equal to its diameter plus that of the star it has covered. From such observations and knowledge of the orbital velocities of the eclipsing stars, astronomers can calculate the diameter of each star.

Light curves of eclipsing stars can also give information about the stars' shape and the distribution of brightness across their disks. For example, "star spots"—analogous to sunspots—can be detected on some stars by this method. If, in addition, astronomers can obtain spectra of the eclipsing system, the stars' masses may be found by use of the method we described above for spectroscopic binaries. Thus, astronomers can obtain especially detailed information about eclipsing binary stars.

13.5 Summary of Stellar Properties

Measuring the properties of stars is not easy and is possible for only a tiny fraction of the stars known. Nevertheless, using the methods we have just described, astronomers can determine a star's distance, temperature, composition, radius, mass, and radial

TABLE 13.4 METHODS FOR DETERMINING STELLAR QUANTITIES

Quantity	Method
Distance	1. Parallax (triangulation)—for nearby stars (distance less than 250 pc)
	2. Standard-candle method for more distant stars
Temperature	1. Wien's law (color-temperature relation)
	2. Spectral type (*O* hot; *M* cool)
Luminosity	1. Measure star's apparent brightness and distance and then calculate with inverse square law.
	2. Luminosity class of spectrum (to be discussed later)
Composition	Spectral lines observed in star
Radius	1. Stefan-Boltzmann law (measure L and T, solve for R)
	2. Interferometer (gives angular size of star; from distance and angular size, calculate radius)
	3. Eclipsing binary light curve (duration of eclipse phases)
Mass	Modified form of Kepler's third law applied to binary stars
Radial velocity	Doppler shift of spectrum lines

velocity. Some of the methods used are summarized in table 13.4. We find from such measurements that all stars have nearly the same composition of about 71% hydrogen and 27% helium, with a trace of the heavier elements. Most have surface temperatures between about 30,000 and 3000 K and masses between about 30 and 0.1 $M_\odot$. We will see in the next chapter that the need to balance their internal gravity is what gives stars their properties, but before we do, we will discuss how astronomers illustrate diagrammatically many of the features we have described above.

13.6 The H-R Diagram

By the early 1900s, astronomers had learned to *measure* stellar temperatures, masses, radii, and motions but had *understood* little of how stars worked. What made an *A* star different from a *K* star? Why were some *M* stars more luminous and others much less luminous? What supplied the power to make stars shine? We have now reached a similar point in our study of stars. We know that there are many varieties of stars. But what creates that variety and what does it mean? A fundamental step in the discovery of the nature of stars remains: the Hertzsprung-Russell diagram.

In 1912, the Danish astronomer Ejnar Hertzsprung and the American astronomer Henry Norris Russell working independently found that if stars are plotted on a diagram according to their luminosity and their temperature (or, equivalently, spectral type), then most of them lie along a smooth curve. The nearly simultaneous discovery of this relation by astronomers working on opposite sides of the Atlantic Ocean is an interesting example of how conditions can be "ripe" for scientific advances.

A typical Hertzsprung-Russell diagram, now generally called an **H-R diagram** for short, is shown in figure 13.18. Stars fall in the diagram along a diagonal line, with hot luminous stars at the upper left and cool dim stars at the lower right. The Sun lies almost in the middle. The H-R diagram does *not* depict the position of stars at some location in space. It shows merely a correlation between stellar properties, much as a height-weight table does for people.

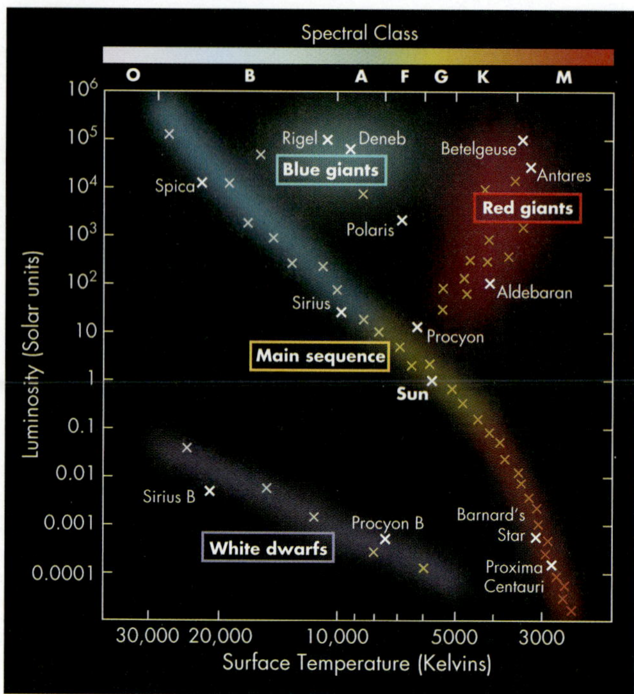

FIGURE 13.18
The H-R diagram. The long diagonal line from upper left to lower right is the main sequence. Giant stars lie above the main sequence. White dwarf stars lie below it. Notice that bright stars are at the top of the diagram and dim stars at the bottom. Also notice that hot (blue) stars are on the left and cool (red) stars are on the right.

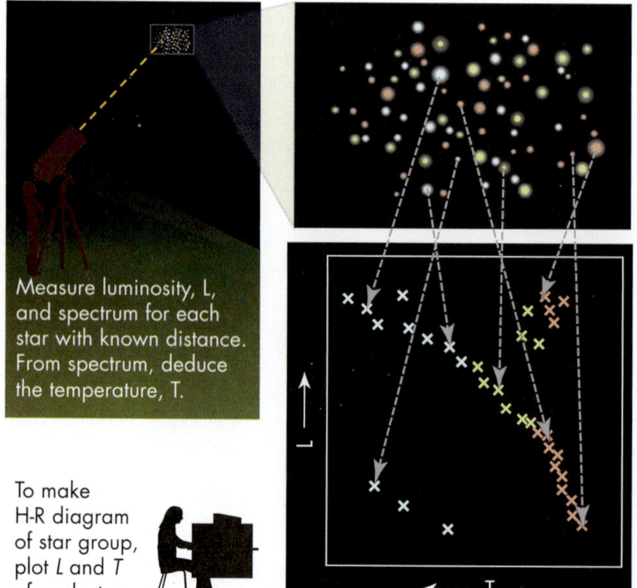

FIGURE 13.19
Constructing an H-R diagram from the spectra and luminosities of stars in a cluster.

Constructing the H-R Diagram

To construct an H-R diagram of a group of stars, astronomers plot each star according to its temperature (or spectral type) and luminosity, as shown in figure 13.19. By tradition, they put bright stars at the top and dim stars at the bottom—hot stars on the left and cool stars on the right. Equivalently, blue stars lie to the left and red stars to the right. Notice that temperature therefore increases to the left, rather than to the right, as is conventional in graphs. The approximately straight line along which the majority of stars lies is called the **main sequence.**

When a group of stars is plotted on an H-R diagram, generally about 90% will lie along the main sequence, but some will lie off it. Of these, a few will be in the upper right, where stars are cool but very luminous, and some will be below the main sequence to the lower center, where stars are hot but dim. What makes these stars different?

Analyzing the H-R Diagram

The Stefan-Boltzmann law, described in section 13.2, gives us the answer. A star's luminosity depends on its surface area and its temperature. If two stars have the same temperature, any difference in luminosity must reflect a difference in area. But a star's area depends on its radius—a large area means a large radius. Stars that lie in the upper right of the H-R diagram have the same temperature as those on the lower main sequence, but because they are more luminous, they must be larger. In fact they must be immensely larger, because some of these bright cool stars emit thousands of times more energy than main-sequence stars of the same temperature, implying that their surface areas are thousands of times larger. Astronomers therefore call these bright, cool stars "giants." The region in the H-R diagram populated by giants is sometimes called the "giant branch." Because many of the stars are cool and therefore red (recall that color is related to temperature), these huge cool stars are called **red giants.** For example, the bright star Aldebaran in the constellation Taurus is a red giant. Its temperature is about 4000 Kelvin and its radius is about 30 times larger than the Sun's.

A similar analysis shows that the stars lying below the main sequence must have very tiny radii if they are both hot and dim. In fact, the radius of a typical hot, dim star is roughly 100 times smaller than the Sun's, making it about the same size as the Earth. Because these stars are so hot, they glow with a white heat and are therefore called **white dwarfs.** Sirius B, a dim companion of the bright star Sirius, is a white dwarf. Its radius is about 0.008 the Sun's—a little smaller than the Earth's radius—and its surface temperature is about 27,000 Kelvin.

We can illustrate such extremes in size by drawing lines in the H-R diagram that represent a star's radius. That is, we use the Stefan-Boltzmann law, which relates the star's luminosity L, radius R, and temperature T. From this law, $L = 4\pi R^2/T^4$, we plot the values of L that result from a fixed choice of R as we vary T. Any stars lying along the plotted line must have equal radii. Figure 13.20

shows that such lines run diagonally from upper left to lower right. Notice that as we move to the upper right, we cross lines of progressively larger radius, implying that stars there are very large. That is, not only are they giants in luminosity, as we mentioned earlier, but they are also giants in radius. Comparing now the radii of stars across the H-R diagram, we see three main types: main-sequence stars, red giants, and white dwarfs.

Q. On the basis of these lines, what is the approximate radius of Betelgeuse? Does that agree with the number you found in figure 13.7?

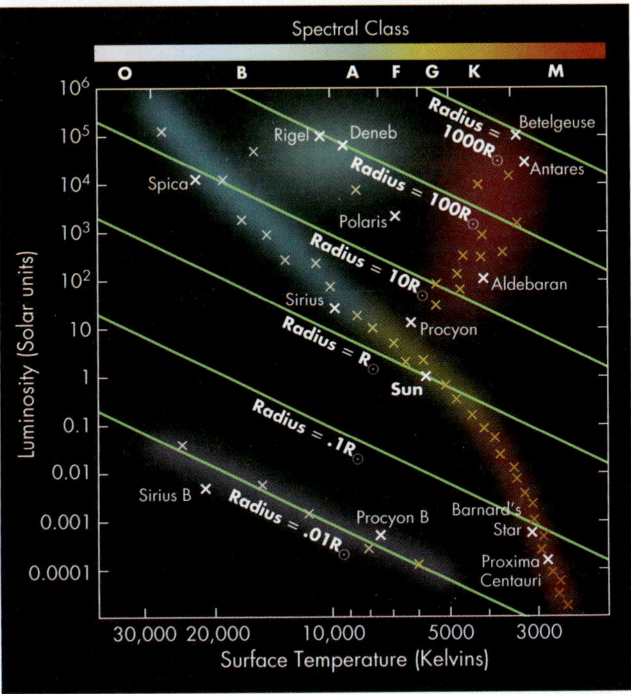

FIGURE 13.20
Lines showing where in the H-R diagram stars of a given radius will lie.

Giants and Dwarfs

Giants, white dwarfs, and main-sequence stars differ in more than just diameter. They also differ dramatically in density.* Recall that density is a body's mass divided by its volume. For a given mass, a larger body will therefore have a lower density, and so a giant star is much less dense than a main-sequence star if their masses are similar. For example, the density of the Sun, a main-sequence star, is about 1 gram per cubic centimeter, while the density of a typical giant star is about 10^{-6} gram per cubic centimeter—1 million times less.

Why do stars have such disparate densities? Russell suggested about 1918 that as a star evolves, its density changes. He believed that stars began their lives as diffuse, low-density red giants. Gravity then drew the stars' atoms toward one another, compressing their matter and making stars smaller, denser, and hotter, so that they turned into hot, blue, main-sequence stars. Russell, in fact, believed that gravity compressed these gaseous bodies so much that by the time they were on the main sequence, they had become liquid. Because liquids are difficult to compress, he believed that once a star reached the main sequence, it could no longer shrink. Unable to be compressed further, the star would have no source of heat and would therefore simply cool off and gradually evolve down the main sequence, growing dimmer and dimmer.

As we will discuss in the next chapter, today we know that stars evolve in the opposite direction. They begin life as main-sequence stars and turn into giants as they age. But although Russell was wrong about how stars evolve, his ideas led to many other important discoveries about stars, such as the mass–luminosity relation.

INTERACTIVE
Stellar evolution and the H-R diagram

The Mass–Luminosity Relation

In 1924, the English astrophysicist A. S. Eddington discovered that the luminosity of a main-sequence star is determined by its mass. That is, main-sequence stars obey a **mass–luminosity relation** such that the larger a star's mass, the larger its luminosity is, as illustrated in figure 13.21. A consequence of this relation is that stars near the top of the main sequence, brighter stars, are more massive than stars lower down.

Eddington made his discovery by measuring the masses of stars in binary systems and determining their luminosity from their distance and apparent brightness. Moreover, he discovered that the mass–luminosity relation could be expressed by a simple formula. If $\mathscr{M}$ and $\mathscr{L}$ are given in solar units, then for many stars

$$\mathscr{L} = \mathscr{M}^3$$

$\mathscr{L}$ = Luminosity of stars in solar units
$\mathscr{M}$ = Mass of stars in solar units

approximately. If we apply the relation to the Sun, its mass is 1 and its luminosity is 1 in these units, and the relation is clearly obeyed because $1 = 1^3$. If we consider

*Technically, we should use the term *average density* here and in the following.

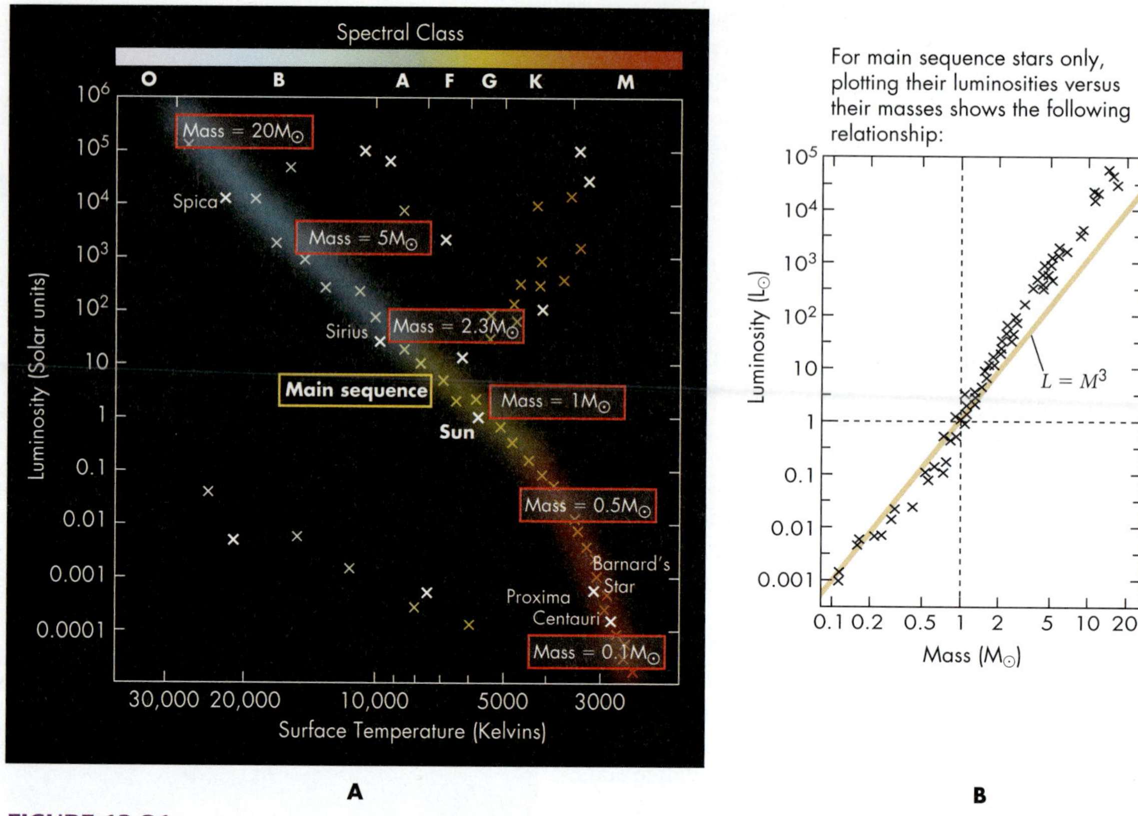

FIGURE 13.21

(A) The mass–luminosity relation shows that along the main sequence, more-massive stars are more luminous. The law does not work for red giants or white dwarfs. (B) On the H-R diagram, high-mass stars lie higher on the main sequence than low-mass stars.

instead a more massive star, for example, a 2-$M_{\odot}$ star, $\mathscr{L}$ (according to the relation) should be

$$2^3\ (=8)\ L_{\odot}$$

Eddington went on to show from the physics of stellar structure that just such a mass–luminosity relationship is required if a star is to avoid collapsing under its own gravitational forces and if its energy flows through its interior by radiation. In chapter 14, we will see how this relation arises and its importance in determining how long a star can live.

Q. All three of these stars have the same spectral type. Comparing these spectra to the ones in figure 13.12, what type is it?

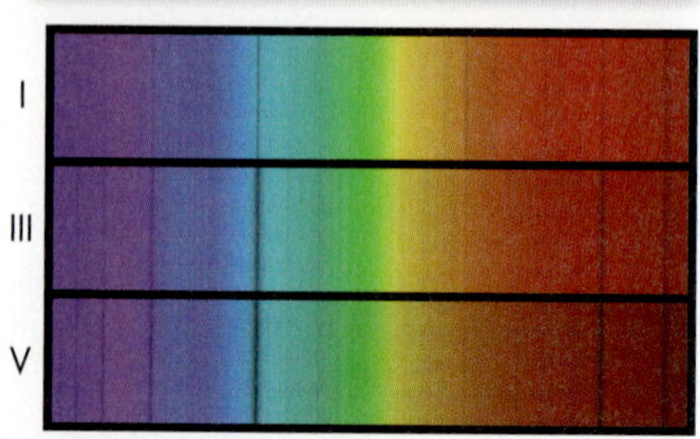

FIGURE 13.22

Spectral lines are narrow in giant (luminous) stars (I and III) and wide in main-sequence (dim) stars (V).

Luminosity Classes

Because astronomers use a star's luminosity to find its distance, radius, and life span, they have sought other ways to measure it. In the late 1800s, Antonia Maury, an early stellar spectroscopist at Harvard, noticed that absorption lines were extremely narrow in some stars compared with other stars of the same temperature. In the early 1900s, Hertzsprung, too, noticed that narrowness and, even before developing the H-R diagram, recognized that luminous stars had narrower lines than less luminous stars, as illustrated in figure 13.22.

The width of the absorption lines in a star's spectrum turns out to depend on the star's density: the lines are wide in high-density stars and narrow in low-density ones. The density of a star's gas is related in turn to its luminosity, because a large-diameter star has—other things being equal—a large surface area, causing it to emit more light, and a large volume, giving it a lower density. A small-diameter star, on the other hand, generally has a high density because its gas is compressed into a small volume. Such a star also tends to be less luminous because its surface area is small and thus emits less light.

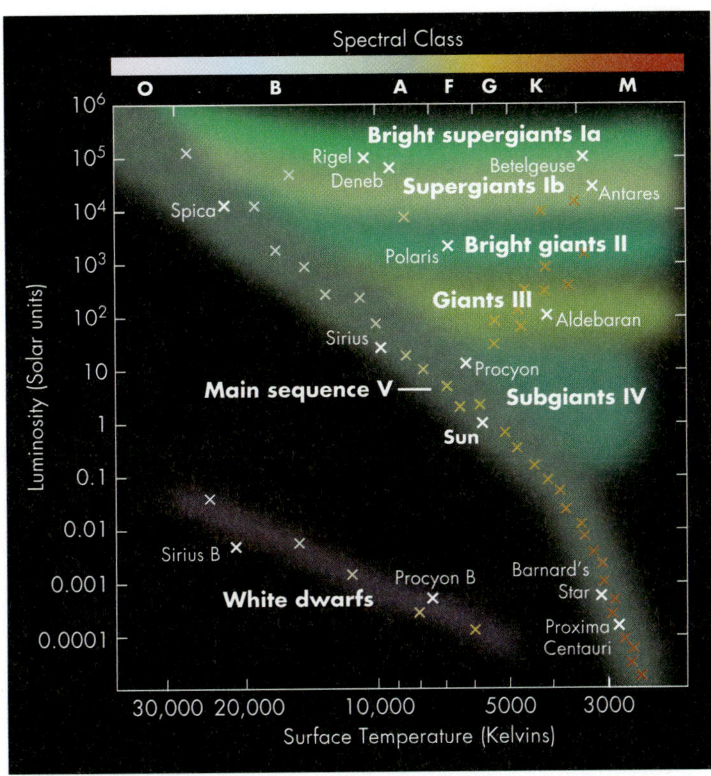

FIGURE 13.23
Stellar luminosity classes.

TABLE 13.5 STELLAR LUMINOSITY CLASSES

Class	Description	Example
Ia	Bright supergiants	Betelgeuse, Rigel
Ib	Supergiants	Antares (brightest star in Scorpius)
II	Bright giants	Polaris (the North Star)
III	Ordinary giants	Aldebaran (brightest star in Taurus)
IV	Subgiants	Procyon (brightest star in Canis Minor)
V	Main sequence	The Sun, Sirius

Using this relationship between spectral line width and luminosity, astronomers divide stars into five luminosity classes (fig. 13.23), denoted by the Roman numerals I to V. Class V stars are the dimmest, and class I stars are the brightest. In fact, class I stars are split into two classes, Ia and Ib. Table 13.5 shows the correspondence between class and luminosity, and although the scheme is not very precise, it allows astronomers to get an indication of a star's luminosity from its spectrum.

A star's luminosity class is often added to its spectral type to give a more complete description of its light. For example, our Sun is a *G*2V star, while the blue giant Rigel is a *B*8Ia star. The luminosity class is especially useful for indicating the difference in luminosity between main-sequence and giant stars of the same spectral type, such as the low-luminosity, nearby, main-sequence star 40 Eridani (*K*1V) and the high-luminosity giant star Arcturus (*K*1III).

Summary of the H-R Diagram

The H-R diagram offers a simple, pictorial way to summarize stellar properties. Most stars lie along the main sequence, with hotter stars being more luminous. Of these, the hottest are blue, and the coolest are red. Yellow stars, such as the Sun, have an intermediate temperature.

Along the main sequence, a star's mass determines its position, as described by the mass–luminosity relation. Massive stars lie near the top of the main sequence; they are hotter and more luminous than low-mass stars. Thus, from top to bottom along the main sequence, the star masses decrease but nevertheless lie in a relatively narrow range, from about 30 to 0.1 $M_{\odot}$.

Our study of the H-R diagram shows us that most stars fall nicely into one of three regions in the diagram: main-sequence stars, giants, or white dwarfs. But the H-R diagram also clearly shows that some stars fall into yet a fourth region—that of variable stars.

Q. Approximately how long is the Mira-type pulsation period?

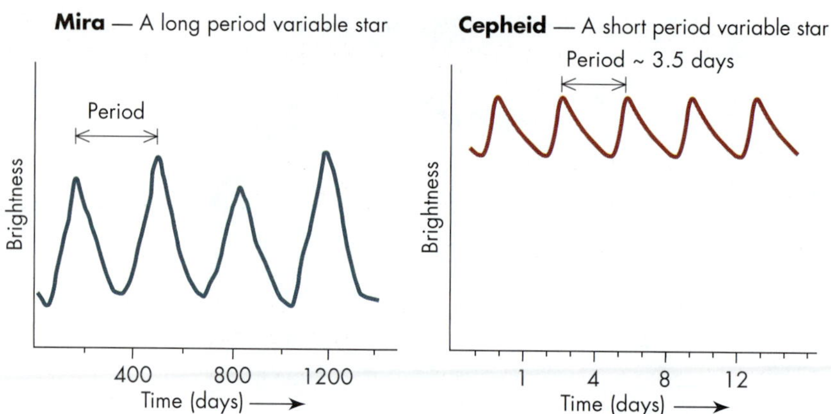

FIGURE 13.24
Two variable star light curves. Left: Sketch of the light variation of a long-period Mira-type variable. Right: Light variation of a Cepheid variable.

13.7 Variable Stars

Not all stars have a constant luminosity. Stars that change in brightness are called **variable stars.** Probably all stars vary slightly in brightness—even the Sun does in the course of its magnetic activity cycle—but relatively few change enough to be noticeable to the naked eye. Among those that do vary, however, are several varieties of stars that are very valuable to astronomers as distance indicators, as we will discover in chapter 14. Especially valuable are the pulsating variables—stars whose radius changes, rhythmically swelling and shrinking. As the radius changes, the star's luminosity must also change (recall that a star's luminosity depends on its radius and temperature). The luminosity change is complicated, however, because as the star pulsates, its temperature also changes, sometimes by as much as a factor of about 2. The result of both changes is that, as the star pulsates, its luminosity changes by a factor of 2 or 3. Although such large variations in luminosity are typical for many pulsating variables, ordinary stars also pulsate but much less dramatically. For them, the pulsations are more like the vibrations of a ringing bell. For example, our Sun's surface vibrates up and down in a complex pattern, as shown in figure 12.14.

The variation in luminosity and temperature is periodic for the majority of pulsating stars; that is, they repeat over some definite time interval called the **period,** as shown in figure 13.24. Astronomers call such a depiction of a star's brightness change its "light curve," and they use the shape of the curve and the time for the star to complete a cycle of brightness—the period—to classify variable stars. On the basis of such light curves, astronomers identify more than a dozen types of regularly pulsating variable stars. For example, some stars pulsate with a period of about half a day, while others take more than a year to complete a pulsation. If variable stars are plotted on the H-R diagram according to their average brightness, most lie in a narrow region called the "instability strip," illustrated in figure 13.25. Generally, the more luminous the star, the slower it pulsates. In chapter 14, we will learn why stars in this region tend to pulsate.

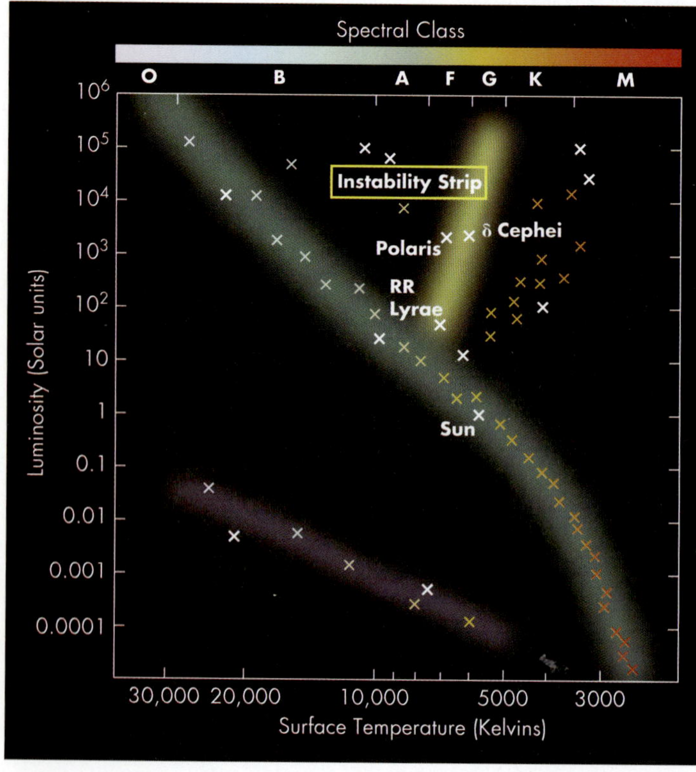

FIGURE 13.25
The instability strip in the H-R diagram. Stars that fall in this band generally pulsate.

Although many variable stars pulsate regularly, some variable stars change their brightness erratically. Such stars are called "irregular variables," and many very young and very old stars are of this category.

13.8 Finding a Star's Distance by the Method of Standard Candles

In our discussion at the start of this chapter, we claimed that astronomers can determine a star's distance from knowledge of its luminosity. As an example of the method, we mentioned using the brightness of an oncoming car's headlights to estimate its distance. We saw that this method works, however, only if we know the luminosity of the light source with some assurance. That is, the luminosity of the light source is some "standard" value. For that reason, astronomers sometimes call this distance-finding scheme the *method of standard candles,* and it is perhaps their single most powerful tool for finding distances.

To measure a star's distance, an astronomer first measures its apparent brightness, *B*, with a photometer, as mentioned in section 13.2. Next, the star's luminosity, *L*, must be determined by any of the number of ways astronomers have to deduce it. For example, they may use the star's spectrum, its position in the H-R diagram, or, for a pulsating variable star, its period (as we will discuss in chapter 14) to reveal the star's luminosity. With *B* and *L* known, the astronomers can then calculate the distance to the star, *d*, using the inverse-square law. Unfortunately, all of the above methods to determine the luminosity have uncertainties, and so the distance calculated is uncertain, too.

In using the inverse-square law, however, it is generally much easier to find a star's distance by comparing it to the distance of a known star. For example, suppose we have two stars whose luminosity we know to be the same because they have identical spectra. Suppose one star, the nearer one, is close enough that we can find its distance by parallax to be 10 parsecs. Furthermore, suppose the star whose distance we seek looks 25 times dimmer. We find its distance by writing out the inverse-square law for both stars:

$$B_{near} = \frac{L_{near}}{4\pi d_{near}^2}$$

$$B_{far} = \frac{L_{far}}{4\pi d_{far}^2}$$

L = Luminosity of star
B = Brightness of star
d = Distance to star

Next, we divide the bottom expression into the top one to obtain after cancellation

$$\frac{B_{near}}{B_{far}} = \frac{4\pi L_{near} d_{far}^2}{4\pi L_{far} d_{near}^2} = \frac{L_{near} d_{far}^2}{L_{far} d_{near}^2}$$

If the two stars have the same luminosity, we can also cancel the *L*s, leaving us with

$$\frac{B_{near}}{B_{far}} = \left(\frac{d_{far}}{d_{near}}\right)^2$$

That is, for two stars of equal luminosity, the brightness ratio is the inverse of the distance ratio squared. We can therefore find the distance ratio by taking the square root of both sides to get $d_{far}/d_{near} = B_{near}B_{far}$. For the problem we are solving, the brightness ratio is 25. Its square root is 5. Therefore, the farther star is 5 times more distant than the nearer one. Because we know that the nearer star is 10 parsecs away, the farther star is 50 parsecs from us.

This method may seem to be rather roundabout, but astronomers have few other options for finding star distances. Moreover, with a little practice, such calculations can be done quickly. **To find a star's distance compared to another star of the same luminosity, find the apparent brightness ratio and take its square root.** This is then the distance ratio. If the stars have different luminosities, the calculation is slightly more difficult. In that case, we must use the full equation above.

SUMMARY

Astronomers use many techniques to measure the properties of a star (fig. 13.26). Parallax—triangulation of the star from opposite sides of the Earth's orbit—gives a distance for nearby stars, with the nearest being about 1.3 parsecs (about 4 light-years) away from the Sun. The parallax method fails beyond about 250 parsecs because the angle becomes too small to measure. Astronomers must therefore use the inverse-square law, comparing apparent brightness and luminosity to find the distance of more remote stars.

Stars have a wide range of surface temperature as measured by their color. According to Wien's law, hot stars will be blue and cool stars red. Once the temperature and luminosity of a star are known, we can calculate its radius with the Stefan-Boltzmann law.

We can find a star's composition by looking at the lines in its spectrum. The spectrum can also be used to assign stars to the spectral types: *O*, *B*, *A*, *F*, *G*, *K*, *M*, where *O* stars are the hottest and *M* stars are the coolest.

We can find a star's mass if it is in a binary system—two (or more) stars bound together gravitationally. For such stars, the orbital period and size, when combined in the modified form of Kepler's third law, give the masses of the orbiting stars.

Once a star's luminosity and temperature are known, the star can be plotted on an H-R diagram. In this plot, most stars lie along a diagonal line called the main sequence, which runs from hot, blue, luminous stars to cool, red, dim ones. Main-sequence stars obey a mass–luminosity relation such that high-mass stars are more luminous than low-mass stars.

A small percentage of stars fall above the main sequence, being very luminous but cool. Their high luminosity implies they must have a large diameter, and so they are called red giants. Similarly, a few stars fall below the main sequence and are hot but dim. Their low luminosity implies a very small diameter, comparable to that of the Earth. These hot, small stars are called white dwarfs.

Some stars vary in brightness and are called variable stars. Many of these variables lie in a narrow region of the H-R diagram called the "instability strip."

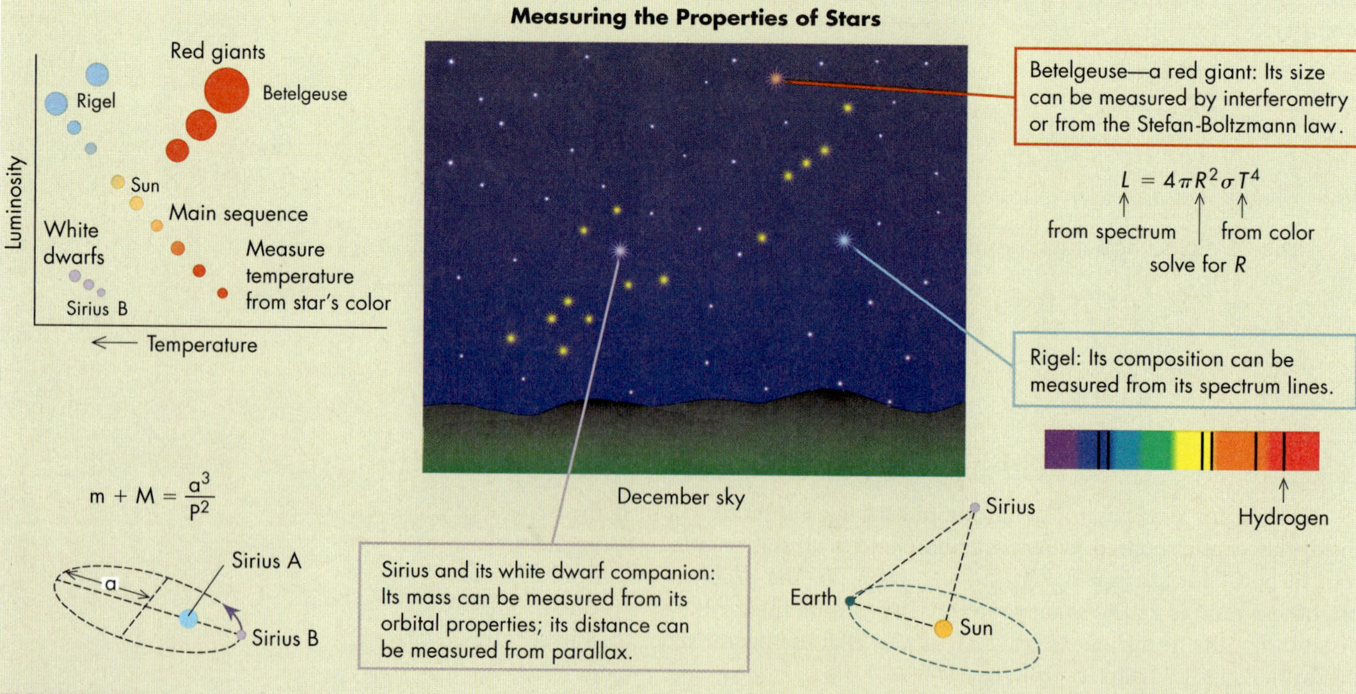

FIGURE 13.26
Summary of how astronomers measure the distance, temperature, mass, composition, and radius of stars.

QUESTIONS FOR REVIEW

1. (13.1) Describe the method of distance measurement by triangulation.
2. (13.1) How do astronomers triangulate a star's distance?
3. (13.1) How is the parsec defined? How big is a parsec compared with a light-year?
4. (13.2) How do astronomers measure a star's temperature?
5. (13.2) What do astronomers mean by an inverse-square law?
6. (13.2) What is luminosity? What two characteristics of a star determine its luminosity?
7. (13.2) What does a star's magnitude measure? Which is brighter, a star of magnitude 1 or a star of magnitude 3?
8. (13.3) What are the stellar spectral types? Which are hot and which cool?
9. (13.3) Why do stars have dark lines in their spectra?
10. (13.3) What is different about the spectra of different types of stars?
11. (13.4) What is a binary star? How are they useful to astronomers?
12. (13.4) How do visual and spectroscopic binaries differ?
13. (13.4) What is an eclipsing binary? What can be learned from eclipsing binaries?
14. (13.6) What is the H-R diagram? What are its axes?
15. (13.6) What is the main sequence?
16. (13.6) How do we know that giant stars are big and dwarf stars are small?
17. (13.6) How does mass vary along the main sequence?
18. (13.6) What is the mass–luminosity relation?
19. (13.7) What is a variable star? What is meant by the period of a variable star?
20. (13.7) Where in the H-R diagram are variable stars found?
21. (13.8) What is a standard candle and how is it used?

THOUGHT QUESTIONS

1. (13.1) Using the method of parallax, what observational result would disprove that Proxima Centauri is the closest star to Earth?
2. (13.1) Would it be easier to measure a star's parallax from Pluto? Why?
3. (13.2) Astronomers can directly measure the diameters of a few nearby, large stars using interferometry (see figure 13.7). We might wonder why interferometry is necessary. Looking at many of the images in this book, the stars do not look like points. Why can't you use a simple photograph of Betelgeuse (like fig. 13.3) to determine its size?
4. (13.3) The spectra of a B-type star and an M-type star feature different absorption lines from different elements and molecules (see fig. 13.12). Is this a result of a difference in the composition or the temperature of the star's surface? Explain.
5. (13.4) When we observe spectroscopic binaries, we often do not know how the plane of the orbit is oriented to our line of sight. How will the Doppler shift change if the same orbit is oriented edge-on, face-on, or somewhere in between?
6. (13.4) Redraw the brightness vs. time plot for the stars in figure 13.17 as it would appear if the smaller star were a red M type star and the larger star were a yellow giant.
7. (13.1/13.4) How could you tell the difference between a visual binary with a one-year orbit and a star moving as a result of parallax?
8. (13.6) If instead of plotting luminosity vs. temperature to make an H-R diagram, we made a plot of stars' distances vs. temperatures, what sort of pattern would we find?
9. (13.2/13.6) Suppose you are placing a star on an H-R diagram. Using a telescope you note a red star of apparent magnitude 3. What else would you need to know?
10. (13.2/13.8) Explain the difference between luminosity and brightness. In what way is this similar to the difference between force and pressure?

PROBLEMS

1. (13.1) Sirius has a parallax of 0.377 arc seconds. How far away is it?
2. (13.1) The parallax of Proxima Centauri is about 0.763 arc seconds. How far away is it?
3. (13.1) The parallax of the red giant Betelgeuse is just barely measurable and has a value of about 0.005 arc seconds. What is its distance? Suppose the measurement is in error by + or −0.003 arc seconds. What limits can you set on its distance?
4. (13.1) Estimate the distance from where you live to a neighboring building or object using triangulation (as shown in fig. 13.1 for a tree). Mark the location of the 90° angle with a stick. Pace off the distance along the baseline or measure it with a meterstick or tape measure. Now from the other end of the baseline, measure the angle A between the stick and the building with a protractor. Make a scale drawing (1/4 inch on the drawing to 1 foot in real life works well) to determine the distance.
5. (13.2) The star Rigel radiates most strongly at about 200 nm. What is its temperature? How does this compare to the Sun?
6. (13.2) The bright southern star Alpha Centauri radiates most strongly at about 500 nm. What is its temperature? How does this compare to the Sun's?
7. (13.2) Arcturus is about half as hot as the Sun but is about 100 times more luminous. What is its radius compared to the Sun's?
8. (13.2) A stellar companion of Sirius has a temperature of about 27,000 K and a luminosity of about 10^{-2} $L_{\odot}$. What is its radius compared to the Sun's? What is its radius compared to the Earth's?
9. (13.3) If a star has very faint Balmer lines, what are the next steps involved in determining its spectral type?

10. (13.3) A line in a star's spectrum lies at 400.0 nanometers. In the laboratory, that same line lies at 400.2 nanometers. How fast is the star moving along the line of sight? Is it moving toward or away from us?
11. (13.4) Two stars in a binary system have an orbital period, *P*, of 5 years and an orbital separation, *a*, of 10 AU. What is their combined mass?
12. (13.4) Two stars in a binary system have an orbital period, *P*, of 2 years and an orbital separation, *a*, of 4 AU. What is their combined mass?
13. (13.4) Two stars in a binary system are determined from their position on the H-R diagram and the mass–luminosity relation to have a combined mass of 8 $M_\odot$. Their orbital period, *P*, is 1 year. What is their orbital separation, *a*?
14. (13.6) Estimate the luminosity and temperature of both Rigel and Barnard's Star using one of the H-R diagrams in the chapter. Calculate the radii of both stars compared to a solar radius using the Stefan-Boltzman equation, and show the values agree with the radii in figure 13.20.
15. (13.6) If a main-sequence star has a luminosity of 5000 $L_\odot$, what is its mass in relation to the Sun's?
16. (13.6) Use the data in table 13.6 to plot an H-R diagram. Which star is a red giant? Which is a white dwarf ? Note: Plotting will be much easier if you plot the logarithm of the luminosity; that is, express it in powers of 10 and use the power. For example, if the luminosity is 100, plot it as 2 for 10^2. You can use a calculator to find the power: enter the luminosity in solar units and hit the "log" key. If the luminosity is 300,000, the answer you get should be 5.477. . . .
17. (13.2/13.8) The two stars τ Ceti and α Centauri B have nearly the same luminosity (and absolute magnitudes) but very different apparent magnitudes: τ Ceti is 3.5, and α Centauri B is 1.34. (A) Determine the ratio of brightness of α Centauri B to τ Ceti. (B) Use the method from section 13.8 to determine how much farther away τ Ceti is than α Centauri B.

TABLE 13.6 DATA FOR PLOTTING AN H-R DIAGRAM*

Star Name	Temperature	Spectrum	Luminosity (Solar Units)
Sun	6000	*G2*	1
Sirius	10,000	*A1*	25
Rigel	12,000	*B8*	100,000
Betelgeuse	3500	*M2*	100,000
40 Eridani B	15,000	*DA*†	0.01
Barnard's Star B	3000	*M5*	0.001
Spica	20,000	*B1*	2000

*These data have been rounded off to make plotting easier.

†Given the low luminosity of this star, what do you think the *D* in its spectral type stands for?

TEST YOURSELF

1. (13.1) A star has a parallax of 0.04 arc seconds. What is its distance?
   (a) 4 light-years (b) 4 parsecs (c) 40 parsecs (d) 25 parsecs (e) 250 parsecs
2. (13.2) A star radiates most strongly at 400 nanometers. What is its surface temperature?
   (a) 400 K (b) 4000 K (c) 40,000 K (d) 75,000 K (e) 7500 K
3. (13.2) A star that is cool and very luminous must have
   (a) a very large radius.
   (b) a very small radius.
   (c) a very small mass.
   (d) a very great distance.
   (e) a very low velocity.
4. (13.3) Which of the following stars is hottest?
   (a) An M star (b) An F star (c) A G star (d) A B star (e) An O star
5. (13.4) A binary system has a period of about 100 years and an average separation of 30 AU. Its combined mass is about ______ solar masses.
   (a) 27 (b) 4 (c) 9 (d) 3 (e) 1000
6. (13.6) In a sample of nearby stars, about what percentage will lie on the main sequence?
   (a) 99% (b) 90% (c) 50% (d) 35% (e) 9%
7. (13.6) In what part of the H-R diagram do white dwarfs lie?
   (a) Upper left
   (b) Lower center
   (c) Upper right
   (d) Lower right
   (e) Just above the Sun on the main sequence
8. (13.2/13.6) Which statement about stellar luminosities is FALSE?
   (a) A small drop in temperature has a much larger impact on the star's brightness than does a small change in size.
   (b) Most naked eye stars are more luminous than our own Sun.
   (c) Red stars with higher luminosity than our Sun must be much larger than our Sun.
   (d) Luminosities have a smaller range than masses.
9. (13.7) The fundamental cause of a variable star's change in luminosity is that the star's ______ is changing.
   (a) mass
   (b) temperature
   (c) radius
   (d) spectral type
   (e) period
10. (13.8) If two stars have the same luminosity, but one appears only one-quarter as bright as the other, you can conclude that the dimmer star is ______ times farther away.
   (a) 2 (b) 4 (c) 8 (d) 16 (e) 64

KEY TERMS

binary star, 355
dwarf, 347
eclipsing binary, 358
giant, 347
H-R diagram, 359
inverse-square law, 343
luminosity, 343
magnitude, 347
main sequence, 360
mass–luminosity relation, 361
method of standard candles, 341
parallax, 338
parsec, 339
period, 364
radial velocity, 354
red giant, 360
spectral type, 351
spectroscopic binary, 356
triangulation, 338
variable star, 364
visual binary, 356
white dwarf, 360
Wien's law, 342

FURTHER EXPLORATIONS

Berger, David H., Jason P. Aufdenberg, and Nils H. Turner. "Resolving the Faces of Stars." *Sky and Telescope* 113 (February 2007): 40.

Cannizzo, John K., and Ronald H. Kaitchuck. "Accretion Disks in Interacting Binary Stars." *Scientific American* 266 (January 1992): 92.

Davis, John. "Measuring the Stars." *Sky and Telescope* 82 (October 1991): 361.

Hearnshaw, John B. "Origin of the Stellar Magnitude Scale." *Sky and Telescope* 84 (November 1992): 494.

Kaler, James B. *Stars and Their Spectra: An Introduction to the Stellar Spectral Sequence.* New York: Cambridge University Press, 1989. (Much of this material is also covered in a series of articles that appeared in *Sky and Telescope* 71 [February 1986]: 129 and continued at 3-month intervals, culminating with "Journeys on the H-R Diagram," *Sky and Telescope* 75 [May 1988]: 482.)

______. "Stars in the Cellar: Classes Lost and Found." *Sky and Telescope* 100 (September 2000): 38.

Kidwell, P. A. "Three Women of American Astronomy." *American Scientist* 78 (May/June 1990): 244.

MacRobert, Alan M. "The Spectral Types of Stars." *Sky and Telescope* 92 (October 1996): 48.

______. "The Stellar Magnitude System." *Sky and Telescope* 91 (January 1996): 42.

Steffey, Philip C. "The Truth about Star Colors." *Sky and Telescope* 84 (September 1992): 266.

Terrell, Dirk. "Demon Variables (the Algol Family)." *Astronomy* 20 (October 1992): 34.

Website

Visit the *Explorations* website at **http://www.mhhe.com/arny** for additional online resources on these topics.

Q FIGURE QUESTION ANSWERS

WHAT IS THIS? (chapter opening): This is a photograph of the Hyades star cluster made through a thin prism. The prism spreads light from each star into a spectrum. The brightest spectrum toward the top is that of Aldebaran, a red giant.

FIGURE 13.5: The 36 dots would have to be spread out over 36 square meters. This would happen at a distance of 6 meters.

FIGURE 13.7: An angle of 1 arc second spans 1 AU at 1 parsec (by definition of the parsec). If Betelgeuse is 200 parsecs away, 1 arc second will span 200 AU. (An object's linear size is proportional to its angular size times its distance). An angle of 50 mas (milliarcseconds) is $50/1000 = 1/20$th of an arc second. Thus, Betelgeuse's diameter is $1/20$ of $200 \text{ AU} = 10 \text{ AU}$.

FIGURE 13.20: Betelgeuse lies just a little below the line for a radius of 1000 $R_{\odot}$, or about 7×10^8 km. We found in figure 13.7 that Betelgeuse's *diameter* is about $10 \text{ AU} = 1.5 \times 10^9$ km. Its radius will be half that number, or about 7.5×10^8 km, pretty close to what we find from reading it off the graph.

FIGURE 13.22: The spectra shown here are of spectral type A.

FIGURE 13.24: Measure the distance between the peaks and compare with the scale on the diagram. The time between peaks is about 11 months (340 days, or so).

PROJECT

Stellar Types: With the help of the star charts, find five of the following stars: Aldebaran, Altair, Antares, Betelgeuse, Capella, Castor, Deneb, Pollux, Procyon, Regulus, Rigel, Spica, Vega. What color does each star look to you? (Try not to let its twinkling affect the color you decide on.) Do your colors match those that you infer from the star's spectrum type as listed in appendix table 8? Which of these stars are giants? How many times bigger a radius does each have compared to the Sun's? Using the information in the appendix, about how long ago did the light from each of the stars start on its journey to Earth? Where were you when the light started its trip? Were you even born yet?

Hubble Space Telescope image of the Eagle Nebula, a region where stars are forming from interstellar dust and gas clouds.

KEY CONCEPTS

- Stars form from interstellar clouds of gas and dust.
- Once a star has formed, nuclear reactions in its core supply energy that replace the energy lost as starlight.
- For most of a star's life, its energy is supplied by the fusion of hydrogen atoms into helium atoms.
- Massive stars consume their fuel much more quickly and therefore have shorter lives.
- When a star uses up the hydrogen in its core, the core shrinks and heats and the outer layers of the star expand and cool. This makes the star turn into a red giant.
- The more massive a star is, the higher temperatures it can compress its core to, allowing it to fuse heavier elements.
- Stars like the Sun drive off much of their atmosphere in the red giant phase; their core becomes a cooling white dwarf.
- Fusion of iron and heavier elements does not release energy, so a star massive enough to form an iron core cannot support its weight and collapses.
- The collapse of star releases so much energy that the star explodes.

14 Stellar Evolution

CONCEPTS AND SKILLS TO REVIEW

- Structure of atoms (4.3)
- Relation between the temperature of a hot object and the color of the light it emits (Wien's law) (4.3)

To us, stars appear permanent and unchanging, but like many other things, they are born, grow old, and die. Stars are not alive, of course, they are immense balls of gas undergoing gradual changes in structure and appearance. Astronomers refer to these changes as stellar evolution, which is driven by the physical laws that govern a star's structure.

For example, gravity holds a star together, and the pressure of the star's gas supports it against gravity's inward pull. A star generates its supporting pressure from heat energy created by gravity compressing the material in its core. That heat energy steadily escapes from the star as the starlight we see, and it must be replenished or the star will collapse. A star replenishes its energy by nuclear reactions in its core. In the early stages of its evolution, the nuclear reactions in its core fuse hydrogen into helium. But eventually the star consumes the hydrogen in its core and changes its structure, swelling into a red giant. Stars like the Sun consume their core's hydrogen in about 10 billion years, but more-massive stars use up their hydrogen much faster. They therefore evolve much faster than the Sun and burn out in only a few million years. To understand why massive stars evolve faster, recall that according to the mass–luminosity relation, massive stars are far more luminous than stars like the Sun. Thus, to supply their greater luminosity, massive stars must burn their fuel faster than less-massive stars.

When a star runs out of fuel, its evolution is nearly finished. A star like the Sun dies slowly and relatively quietly and becomes a white dwarf. A more massive star explodes and leaves a neutron star or black hole as a remnant. Such stellar explosions, though marking the death of one star, may trigger the birth of another. Moreover, the explosion of a dying star blasts into space elements vital to human life, such as carbon and iron, elements that eventually become incorporated into new stars, planets, and, ultimately, people. We therefore owe our very existence to stellar evolution. The goal of this chapter is to illustrate how such evolution occurs.

Q: WHAT IS THIS?

Hint: The right and left pictures were taken several months apart. See end of chapter for the answer.

14.1 Overview of Stellar Evolution

Stars change so slowly that most stars we see today would have been visible to the dinosaurs. We observe very little changing in our lifetimes, or indeed in the entire history of astronomical observations. Astronomers can, however, deduce from observations some of the basic features of how a star evolves. For example, the existence of main-sequence stars, red giants, and white dwarfs suggests to astronomers a picture of how stars age, much the way a snapshot of a baby with its parents and grandparents suggests to a viewer a picture of human aging.

Astronomers have been able to piece together this evidence to deduce how stars are born, how long they live, and the spectacular ways they die. We can predict how much longer the Sun will live, and how the nature of our own Solar System depended on generations of stars before the Sun. In this first section of the chapter we will explore some of these general ideas about stellar evolution and summarize the lives of stars. We then will examine each phase in more detail in subsequent sections.

The Importance of Gravity

Gravity drives stellar evolution. From the moment of a star's birth, when an interstellar gas cloud collapses to become a star, to the moment of its death, gravity provides the inexorable force that crushes the star toward its final collapse. What begins as a gentle tug grows into a crushing force that heats up a star's interior to temperatures hotter than any furnace on Earth. The steady contraction caused by gravity pauses only if a star can generate internal pressures sufficient to counterbalance gravity's unrelenting pull.

Despite the titanic forces involved, the evolutionary changes at work in stars require millions to billions of years. While we cannot directly observe most of the changes that occur in a star's lifetime, astronomers have a powerful tool for examining the lives of stars. We can "watch" a star's aging with computer calculations that solve the equations that govern the star's structure. Stellar models, as such calculations are called, allow us to trace a star's life, revealing billions of years of a star's history. Figure 14.1 outlines a few of the most important physical processes governing a star's structure and evolution that go into such a calculation.

Gravity squeezes a star until it drives the gas temperature up so high that nuclear fusion occurs. No other source of energy is powerful enough to bring the star's gravitational contraction to a standstill. The energy from this fusion replenishes the heat that is constantly flowing out of the star, establishing a balance with gravity for as long as the nuclear fuel lasts. When the fuel is exhausted, a star then must readjust and enter a new phase of its lifetime.

All stars need fuel to keep their cores hot, and all begin their lives by fusing hydrogen into helium. When they use up their hydrogen fuel, their structure must change. For example, when a star runs out of hydrogen fuel in its core, gravity's force is no longer counterbalanced, and the core of the star resumes its contraction, heating up to even higher temperatures until it finds a new fuel to burn. The outer parts of the star respond to this change by expanding, and the star begins its red giant phase.

With these ideas in mind, let us follow the life story of a star from its birth to its death. In telling that story, we will discover that its ending depends critically on how much material the star contains—its mass. A star's mass determines not only how strong its gravity is, but how much fuel it has. Thus, stars that begin their lives with the same mass as the Sun live out lives very similar to the Sun's.

There are any number of small differences between stars as we move up to incrementally larger masses, but there is a major change in the fates of stars that weigh more than about eight times the Sun's mass. These high-mass stars have such a powerful gravitational pull that they undergo a cataclysmic collapse after they have exhausted all

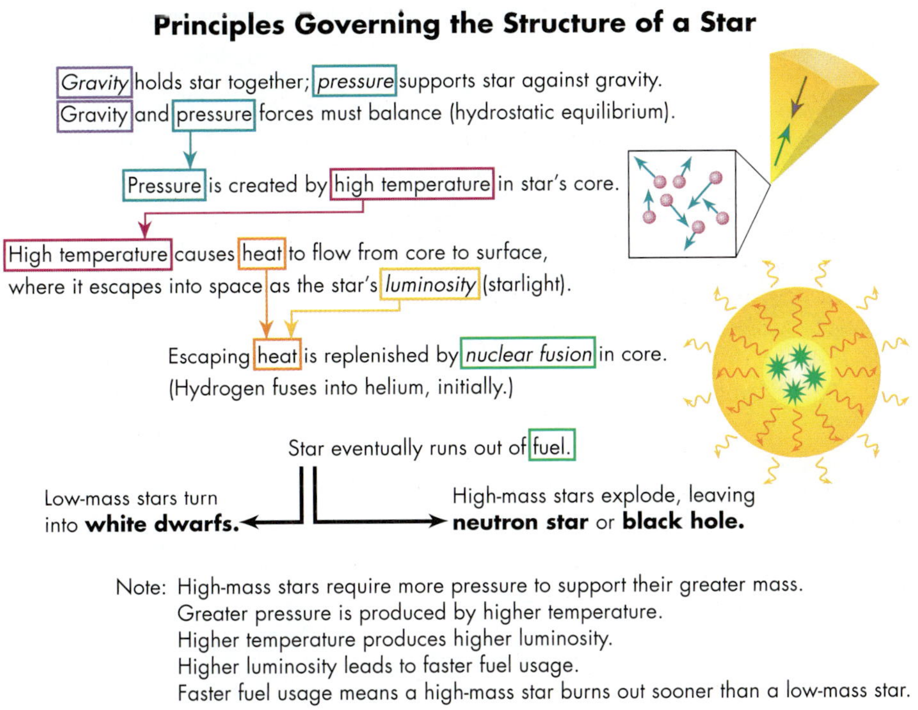

Principles Governing the Structure of a Star

Gravity holds star together; *pressure* supports star against gravity.
Gravity and pressure forces must balance (hydrostatic equilibrium).

Pressure is created by high temperature in star's core.

High temperature causes heat to flow from core to surface, where it escapes into space as the star's *luminosity* (starlight).

Escaping heat is replenished by *nuclear fusion* in core. (Hydrogen fuses into helium, initially.)

Star eventually runs out of fuel.

Low-mass stars turn into **white dwarfs.**

High-mass stars explode, leaving **neutron star** or **black hole.**

Note: High-mass stars require more pressure to support their greater mass.
Greater pressure is produced by higher temperature.
Higher temperature produces higher luminosity.
Higher luminosity leads to faster fuel usage.
Faster fuel usage means a high-mass star burns out sooner than a low-mass star.

FIGURE 14.1
Outline of the processes that govern the structure of stars.

possible nuclear fuels. We will therefore divide stars into two groups—low-mass stars, such as the Sun, and high-mass stars. We will briefly describe the stages that each of these kinds of stars passes through, beginning with one such low-mass star, our Sun.

The Life Story of the Sun—A Low-Mass Star

The Sun began its life as an **interstellar cloud,** a tenuous, cold, dark mass of gas drifting around the Milky Way Galaxy. The cloud—perhaps because of a collision with a neighboring cloud or the explosion of a nearby star—began to collapse. As it shrank under the influence of its increasing gravity, it grew smaller and hotter. Within a few million years—an instant, to the Sun—hydrogen began to fuse into helium in its core. The energy released by the fusion sustained the pressure inside the Sun, stopping its collapse and leaving it looking much as it does today: a small, yellow star. If we plotted it on the H-R diagram, it would lie on the main sequence.

The Sun will remain a main-sequence star until it consumes about 90% of the hydrogen in its core, a process that will take a total of 10 billion years, about half of which time the Sun has already used up. When the hydrogen in the Sun's core is nearly spent, its core will shrink and grow hotter. The rising temperature in the core will make the remaining hydrogen there burn faster, generating more energy. As that energy flows outward through the Sun's outer layers, it will lift them and they will cool as they expand farther from the source of heat. As the surface of this bloated Sun cools, it will turn red, and its light will reveal to our distant descendants a Sun swollen into a red giant. The Earth will have become uninhabitable, however, the intense luminosity of the Sun having boiled away our oceans and perhaps even melted Earth's crust.

ANIMATION
An H-R diagram showing the evolutionary track of a 1-$M_{\odot}$ star

The Sun will shine as a red giant for perhaps another billion years but will then shrink and grow hotter, transforming—when its core finally becomes hot enough to fuse helium—into a yellow giant. During this stage, it will pulsate as if taking slow, deep breaths, but these are dying gasps. Eventually, the Sun will consume most of the helium in its core and will once again change into a red giant, but a giant even larger and more luminous than before. That high luminosity is its death knell: radiation streaming outward through its atmosphere will drive its gas into space, stripping the Sun to its bare

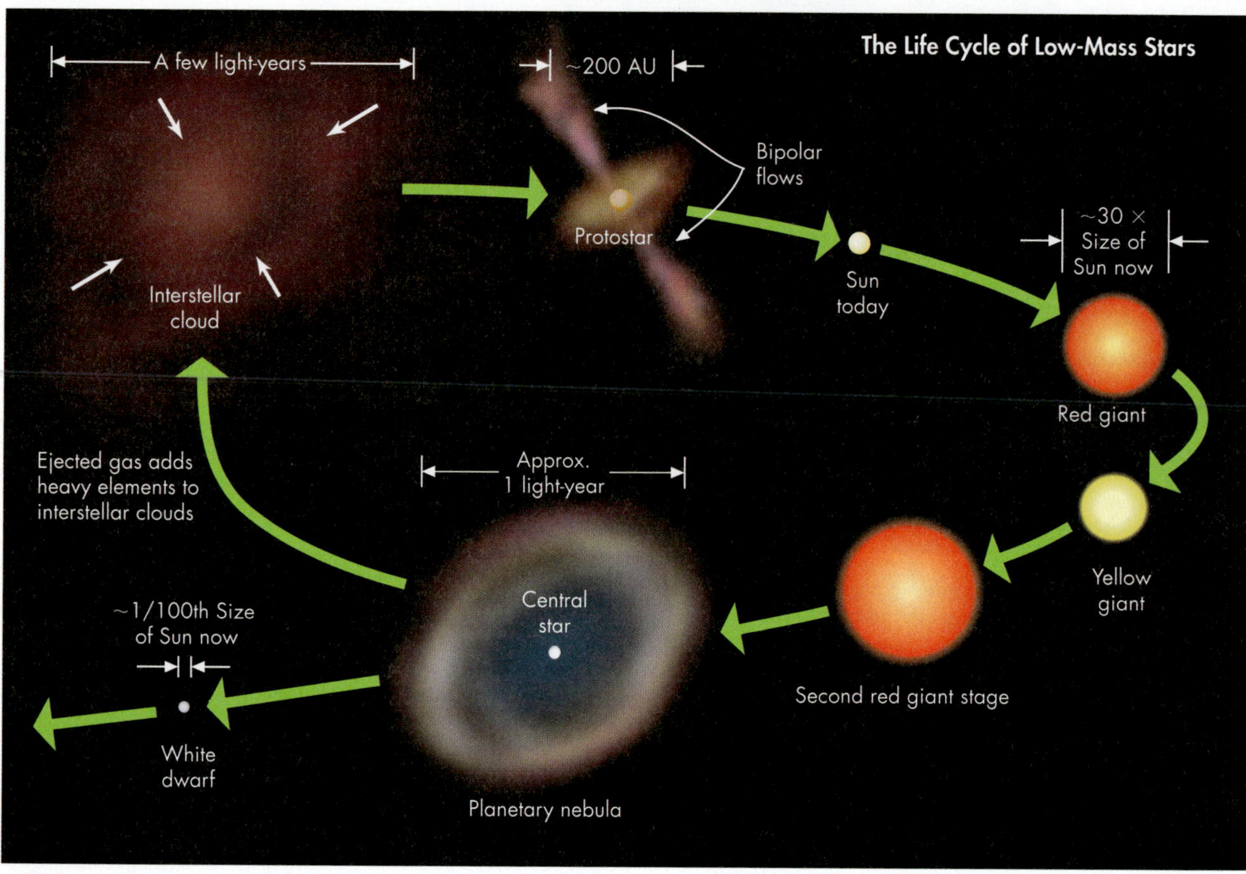

FIGURE 14.2
Evolution of a low-mass star such as the Sun. Stars with masses below 8 $M_\odot$ share a similar fate.

core. The ejected gas will form a shell that will gradually disperse. The tiny core, fiercely hot but with no further energy supply, will cool and dwindle to a white dwarf. These changes, summarized in figure 14.2, are the fate of the Sun.

The Life Story of a High-Mass Star

A very different fate awaits stars significantly more massive than the Sun: they explode when they run out of fuel. Astronomers are uncertain precisely what mass is needed to make a star explode but believe it is about 8 $M_\odot$. For that reason, we take 8 $M_\odot$ as the dividing line between high-mass and low-mass stars.

We chart the history of a high-mass star in figure 14.3, where we see that its early life is similar to the Sun's: it originates from the collapse of an interstellar cloud but from a larger and more massive clump than the one that formed the Sun. The clump's greater mass "squeezes" it more, thereby heating it more than the young Sun was heated. Thus, when the clump becomes a main-sequence star, it is much hotter, bluer, and more luminous than the Sun. Its greater luminosity implies that it burns its fuel faster, and so rather than taking 10 billion years to expend its hydrogen, it consumes its fuel in 100 million years or less, aging far more rapidly than the Sun. As it runs out of hydrogen, a massive star behaves much like a low-mass star, swelling steadily in size and growing cooler. However, a massive star passes through a pulsating yellow-giant stage before it turns into a red giant.

The greater mass of such a star now becomes a primary influence on its evolution, creating an intense gravitational compression of its core. As the core is compressed, its temperature rises, and it burns fuel ever more furiously, thereby maintaining the pressure that supports the star. The higher temperature also permits the star to fuse

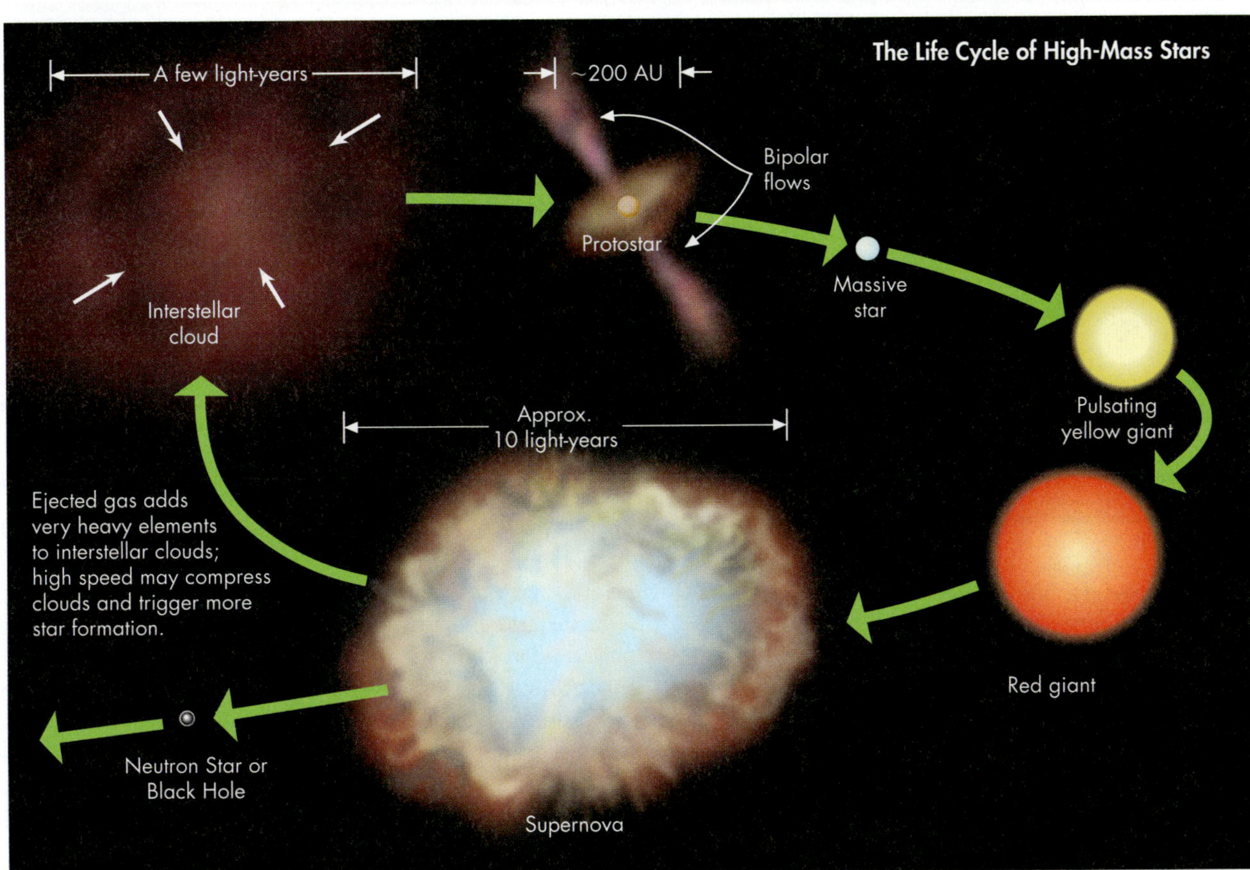

FIGURE 14.3
Evolution of a high-mass star.

progressively heavier elements to supply its energy—helium is fused into carbon, carbon into oxygen, and, ultimately, silicon into iron. Iron, however, does *not* release energy when it is fused, and thus a star with a massive iron core cannot support itself against gravity's inward pull. Lacking support, its core collapses in less than a second, triggering a cataclysmic explosion that showers into space the heavy elements it has made. Such a star does not leave a white dwarf; rather, its core becomes either a tiny, incredibly compressed ball of neutrons (a neutron star) or an even denser body—a black hole, gravity's ultimate triumph.

Stellar Recycling

Stars form from the matter in interstellar clouds, but they also return much of that matter—in some cases as much as 90% of their original mass—to interstellar space. The great energy they release during their lifetimes drives a steady flow of matter during their lifetimes, and as they die, they may blast huge quantities out at high speeds. Much of this material is the original gas that the star formed from, but added to this are heavy elements formed by fusion during the star's lifetime. This material may even trigger more stars to form as the gas collides with and compresses interstellar clouds.

This gas–star–gas cycle has steadily increased the amount of heavy elements in interstellar clouds. In the cloud from which the Sun formed, about 2% of its mass was composed of elements heavier than hydrogen and helium, created by several previous generations of stars. Indeed, the terrestrial planets would not exist without that material. This process has been going on for more than ten billion years and has produced more than 100 billion stars in our own Galaxy. We next examine each stage of this process in more detail, beginning with the formation of stars from interstellar clouds.

14.2 Star Formation

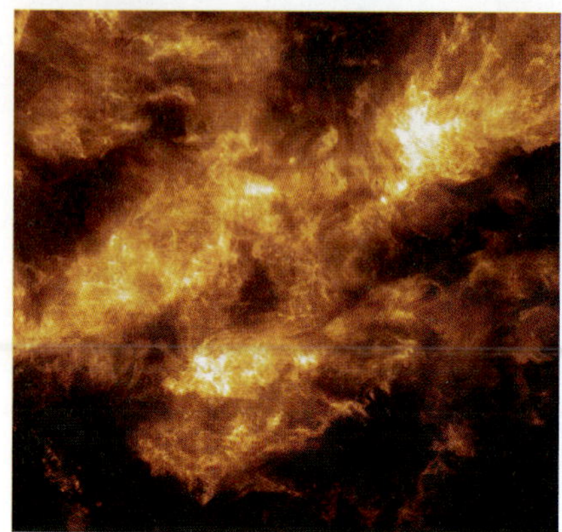

FIGURE 14.4
False-color radio map of a star-forming interstellar cloud in Taurus. The image shows a region about 60 light-years across.

Interstellar Gas Clouds

Stars form from interstellar matter—huge clouds of gas and dust that orbit with the stars inside our Galaxy. The gas is mostly hydrogen (71%) and helium (27%); the dust is composed of solid, microscopic particles made primarily of silicates, carbon, and iron compounds. Interstellar gas is generally cold, perhaps only 10 K (hundreds of degrees below zero on the Fahrenheit scale). At such low temperatures, atoms and molecules in a gas move too slowly to generate much pressure. As a result, the pressure in such cold gas may not be large enough to support the cloud against its own gravity; if the cloud is disturbed, it may then collapse. The collapse may be triggered by collision with a neighboring cloud, the explosion of a nearby star, or other processes.

Maps made with radio telescopes show that gas clouds are not uniform: they contain smaller clumps of gas where the density is higher than the average (fig. 14.4). When a cloud with such clumps collapses, each clump is also compressed, growing denser. Thus, a single cloud may break into many smaller pieces, each of which may form a star. The result is that stars generally form in groups, not in isolation, and all the stars within a group form at approximately the same time.

Astronomers picture the transformation from gas cloud to star as proceeding in several distinct stages. In the first stage, a dense clump within a cloud begins to collapse, its gas drawn inward by gravity, which compresses and heats it. In the second stage, any rotation of the gas clump makes it flatten into a disk, as we described in chapter 8 on the origin of the Solar System. In about a million years, the disk forms a small, hot, dense core at its center called a **protostar,** which marks the third stage. These stages are illustrated in figure 14.5.

Protostars

A protostar is hotter than the gas from which it condensed—perhaps 1500 K—but it is still much cooler than ordinary stars. Its relatively low temperature makes a protostar

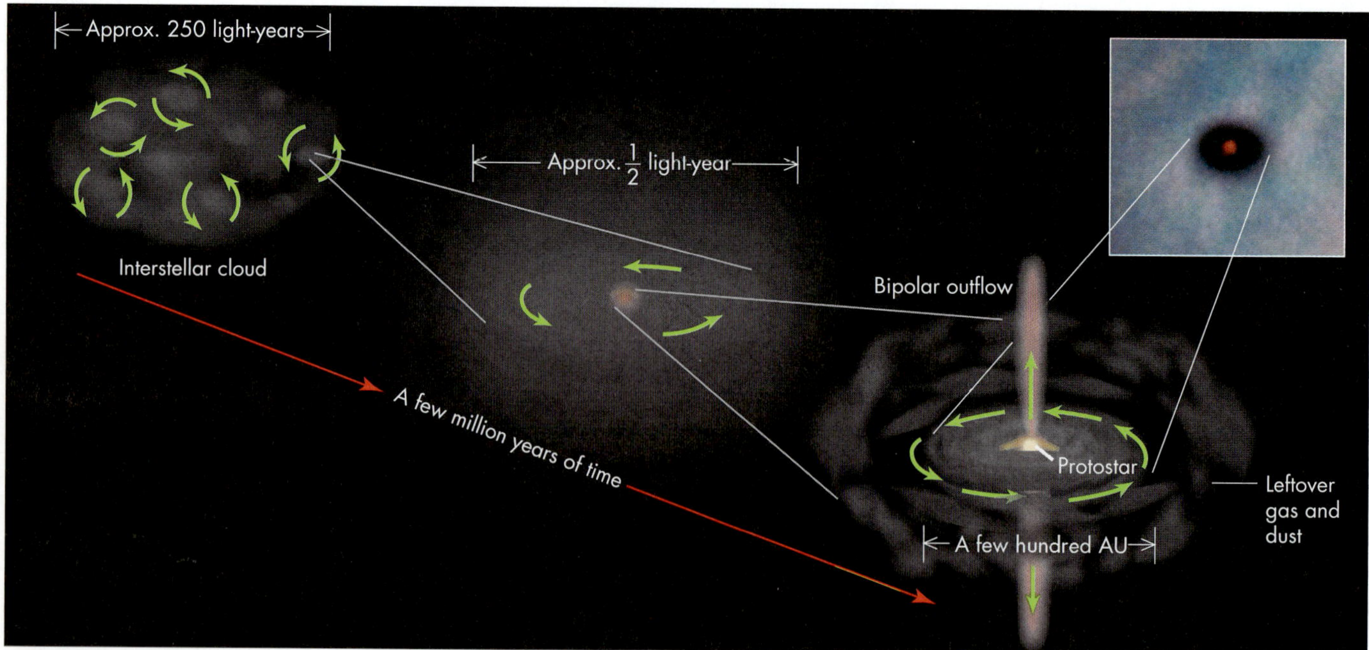

FIGURE 14.5
An artist's sketches depicting the birth of a star. (Insert, far right) A Hubble Space Telescope picture of the disk around a star forming in Orion.

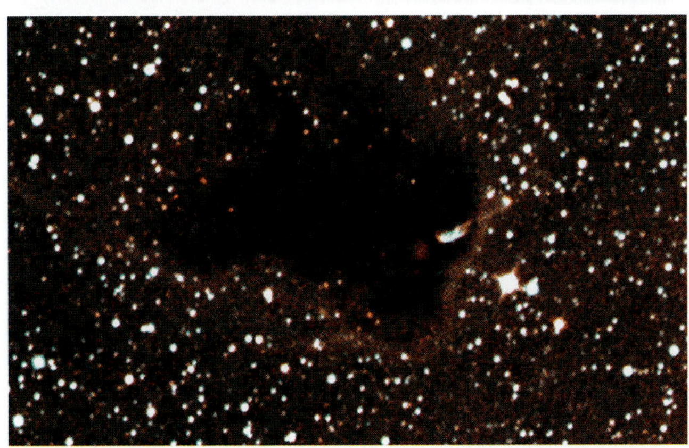

FIGURE 14.6
(Left) Visible-light image of a dense cloud in which star formation is occurring. The cloud is seen in silhouette against distant background stars. (Right) A false-color infrared image of the same region made with the Spitzer Space Telescope showing a bright protostar. Note also the outflows from the protostar. The images cover a region 3 × 2 light-years across.

"shine" mainly at infrared wavelengths rather than at visible ones. Thus, to observe a protostar, astronomers use telescopes that detect infrared wavelengths. Visible and infrared images of an interstellar cloud are shown in figure 14.6. A protostar shows up as the brightest object in the infrared image. It is virtually undetectable at visible wavelengths because it is cool and because the dust around it absorbs what little visible light it emits.

Protostars do not remain cool for long. As their gravity continues to draw surrounding material inward, they grow hotter from the gravitational energy released by the infalling material. Gravitational energy is released whenever something falls. For example, when a cinder block falls onto a box of tennis balls, the impact scatters the balls in all directions, giving them kinetic energy—energy of motion. In a gas, energy of motion appears as heat, raising the temperature; as material falls onto a protostar, its gas atoms speed up and it grows hotter.

When the temperature in the protostar's core reaches about 1 million Kelvin, some nuclear reactions begin there. These reactions fuse the trace amounts of a hydrogen isotope known as **deuterium** (^{2}H) into helium and release energy. Deuterium, sometimes called "heavy hydrogen," is the isotope of hydrogen that has one neutron in addition to a proton in the nucleus. It is about 100,000 times less abundant than the common form of hydrogen, which has no neutrons. The energy supplied by deuterium burning replenishes the energy that is constantly radiating away from the star, so the star stops contracting while this fuel is consumed. Surrounding material, especially from the disk, continues to fall onto the star for millions of years. The infalling matter releases gravitational energy, which supplies additional heat to the protostar. In fact, for many young stars, the gravitational energy of infalling matter greatly exceeds that released by nuclear reactions in the core. Once the deuterium is used up, the protostar resumes its contraction. A protostar passes through these stages in a time that is brief by astronomical standards: a few million years for a star like the Sun, and even less time for a more massive star.

Bipolar Flows from Young Stars

For all types of protostars, infall creates violent changes in brightness and, more strangely, ultimately creates a strong *outflow* of gas. The outflow of gas from protostars does not take place in all directions. Rather, it generally is "focused" into narrow jets of gas. Figure 14.6 shows a typical jet, but what powers the jet and what makes it so narrow are

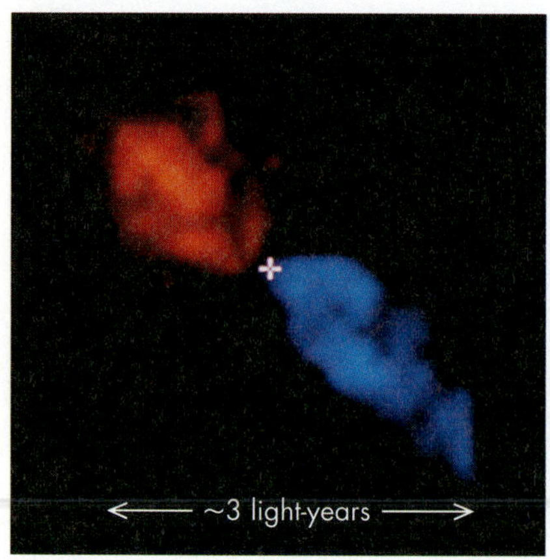

FIGURE 14.7
Bipolar flow of gas ejected from a young star. This is a false-color picture of the radio emissions from the gas. The gas receding from us is shown in red, and the gas approaching us is shown in blue. The cross marks the location of the young star.

not yet fully understood. Computer simulations suggest that the jet's power somehow comes from energy released as matter falls from the inner parts of the disk onto the star. The focusing of the jet probably results from magnetic fields in the star and disk.

Jets from young stars play an important role in "cleaning up" after a star has formed. They may also affect the formation of other stars in their vicinity. To better understand these outflows, let us look in more detail at figure 14.6. You can see there long, thin jets of gas squirting out from a dark cloud in which a young star lies. The jets have blown "bubbles" of hot gas whose edges are visible in green in the false-color infrared image. The jets of gas from the star spread when it hits surrounding gas, and the impact heats both the jet and the gas around it, making them glow, as shown in the picture. The jet also pushes gas outward from around the protostar to create **bipolar flows.** Bipolar flows (which get their name because there are two jets of gas ejected parallel to the star's polar axes) can be seen in figure 14.7, which shows a radio map of the area around a protostar. Here, a computer has been used to color red the gas moving away from us (that is, red-shifted material) and to color blue the gas moving toward us.

Bipolar flows clear away gas and dust from around protostars and thereby allow astronomers to see them directly with visible light. However, even powerful bipolar flows leave much of the surrounding material behind. Thus, many young stars are partially immersed in interstellar matter, as illustrated in figure 14.8A. Such stars often vary erratically in brightness and have gas streaming away from their surfaces, as can be detected from the Doppler shift of their spectral lines. This flow occurs in the star's upper atmosphere and is distinct from the bipolar flow we described above. It is more similar to the solar wind but far more intense. Young stars that exhibit variable light and outflowing gas are called **T Tauri stars,** and some of their variability is probably caused by magnetic activity much like that observed on our Sun. For example, T Tauri stars show evidence of giant "starspots" and magnetic flares thousands of times more intense than the activity we observe on our Sun. Astronomers cannot see the spots directly with optical telescopes, but by combining data on variations of the star's brightness obtained with X-ray, radio, and infrared telescopes, they can deduce the presence of spots on these peculiar young stars.

Q. What creates the red glow of the gas near these stars?

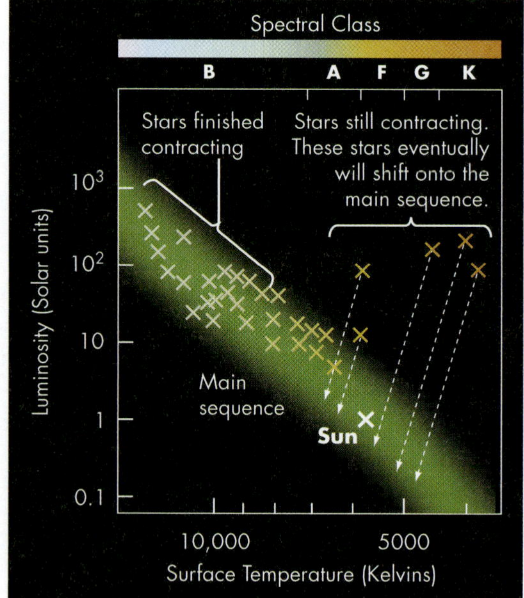

FIGURE 14.8
(A) The young star cluster NGC 2264 and (B) its H-R diagram. Notice the large amounts of leftover gas heated by the brilliant young stars.

Figure 14.8B shows the H-R diagram for the young star group shown in figure 14.8A. The T Tauri stars lie a little above the lower main sequence in the H-R diagram because they are still shrinking and are thus larger and brighter than main-sequence stars. The position of protostars in the H-R diagram for this and other young clusters shows that lower-mass stars take longer to contract than high-mass stars, which rapidly reach the main sequence.

If you look closely in the central regions of the star cluster in figure 14.8A, you can see small, dark blobs that may be protostars that have not yet become hot enough to emit visible light. These tiny, dark globs are called Bok globules in honor of Bart Bok, the astronomer who first studied them in detail. Globules exhibit a diversity of sizes, with diameters ranging from about 0.1 to 2 light-years and with masses ranging from a few up to about 200 times the Sun's mass.

Stellar Mass Limits

We saw in chapter 13 that the mass of most stars is between about 0.1 and 30 $M_\odot$. These limits are not absolute—higher- and lower-mass stars exist—but the theory of star formation described above can help us understand why very high- or very low-mass stars are uncommon.

Stars much smaller than about 0.1 $M_\odot$ are rarely seen because their mass is too small to compress them enough to begin steady nuclear burning. Such low-mass stars are very dim, and are called "brown dwarfs" because of the dim red light they emit. These objects, midway between stars and planets in mass, are massive enough to reach a temperature to fuse deuterium, but they are too low in mass to sustain hydrogen fusion. They might be very numerous, but their dimness makes them extremely hard to detect. Searches with sensitive detectors have begun only in recent years to detect these tiny, dim, failed stars.

Very high-mass stars are rare for a different reason. As their great gravity compresses them, they rapidly become extremely hot and luminous, emitting such intense radiation that they heat the gas around them, raising its pressure and preventing additional material from falling onto them. Thus, their high temperature limits the amount of material they can accumulate. Furthermore, once formed, high-mass stars are so luminous (recall the mass–luminosity relation) that their radiation strips gas from their outer layers, driving it into space. Thus, a 200-$M_\odot$ star would quickly "shine" itself apart.

We have followed the stages of a protostar's early development. Now we will proceed to the next stage of its life.

14.3 Main-Sequence Stars

A star stops being a protostar and becomes a main-sequence star when its core temperature rises above about 5 million Kelvin. At this stage of its life, the star's interior structure is like the Sun's: it contains a core, where normal hydrogen fuses into helium, and an envelope of gas around the core that transports energy to the star's surface. The properties of the core and envelope, however, depend greatly on the star's mass.

Why a Star's Mass Determines Its Core Temperature

We saw in chapter 13 that high-mass stars are both hotter and more luminous than low-mass stars. This relation between mass, temperature, and luminosity results from the fact that the gravity and pressure inside a star must be in balance, a condition we called "hydrostatic equilibrium." To see how this relation arises, we need to recall that

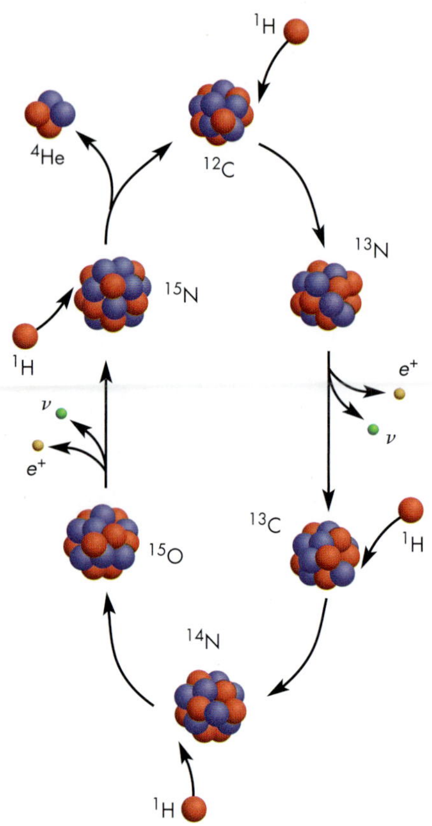

FIGURE 14.9
The CNO cycle. Carbon, nitrogen, and oxygen act as catalysts to help hydrogen fuse into helium.

a star forms by gravity drawing interstellar material together. As that material is pulled inward on itself, it is compressed, and that compression heats the material. The amount of heating is set by the forming star's mass, such that the larger the star's mass, the greater its heating.* Thus, high-mass stars end up hotter (and thereby more luminous) than low-mass stars.

Structure of High-Mass and Low-Mass Stars

The high temperature at the core of a massive star allows it to fuse its hydrogen into helium by a different mechanism from that of the Sun. In the Sun, hydrogen is converted to helium by the proton–proton chain: two protons fuse to form deuterium, to which a third proton is added to make ^3He. The ^3He then fuses to form ^4He. In a massive star, the conversion of hydrogen to helium takes place by means of the CNO cycle, whereby carbon, nitrogen, and oxygen atoms already present in the star's core act as catalysts to aid the reaction (fig. 14.9). The CNO cycle, however, requires a very high temperature to operate and thus can only generate energy at the very center of the star. Photons are unable to carry this huge energy flow from so small a volume, and so the gas in the core begins to rise in irregular clumps and carries the heat outward by convection.

As we saw in chapter 12, convective motions occur in approximately the outer third of the Sun just below its surface, where the radiant energy flow is impeded by the strong absorption of the cool subsurface gas. In a high-mass star, however, the convection currents are confined to the core. They do not extend close enough to the star's surface to allow hydrogen-rich gas from there to mix into its core and replenish the core's depleted fuel. Neither high-mass stars nor low-mass stars such as the Sun can use the fuel in their outer layers. Only very low-mass stars—less than about 0.4 $M_\odot$—mix completely. Thus, a star can be starved for fuel even though its outer layers are fuel-rich. This strongly affects a star's lifetime.

Main-Sequence Lifetime of a Star

The time a star stays on the main sequence is called its **main-sequence lifetime.** A star's main-sequence lifetime depends on its mass and luminosity. To see why, we can use a simple analogy. Suppose we want to know how long a car can run on a tank of gasoline. That time clearly depends on how much fuel it has and how rapidly the fuel is consumed. For example, if the car has a 16-gallon tank of gasoline and burns fuel at the rate of 2 gallons per hour, it can cruise for 8 hours ($\frac{16}{2}$) before running out of gas. When the same formula is applied to a star, its mass determines how much fuel it has, and its luminosity determines how rapidly the fuel is burned. Thus, to calculate the star's main-sequence lifetime, we divide the amount of fuel by the rate at which it is consumed.

The rate at which a star burns its fuel depends on its luminosity: more-luminous stars burn their fuel faster. We can show, using the information in chapter 12, that the Sun consumes its fuel at the rate of about 2×10^{19} kilograms of hydrogen per year. For a star whose luminosity is $\mathscr{L}$ solar units, the rate of fuel consumption in kilograms per year is therefore $2 \times 10^{19}\ \mathscr{L}$. Using the car analogy, the lifetime of the star, t, is then its mass divided by $2 \times 10^{19}\ \mathscr{L}$, or

$$t = \frac{\text{Mass (kg)}}{2 \times 10^{19}\ \mathscr{L}}\ \text{years.}$$

A star cannot use all its hydrogen for fuel, however. Only its core is hot enough for nuclear reactions to occur, and the core is generally only about one-tenth of its total mass. Therefore, the value of the mass we use above should be about 0.1 times the star's mass, and so for a star whose mass is M, the mass of available fuel is $0.1M$. For the Sun, whose mass is 2×10^{30} kilograms, the mass of available fuel is

*Technically, the larger mass releases a larger amount of gravitational energy.

$2 \times 10^{30} \times 0.1 = 2 \times 10^{29}$ kilograms. For a star whose mass is $\mathcal{M}$ times the Sun's, the mass of available fuel is therefore $2 \times 10^{29}\ \mathcal{M}$ kilograms. Using this expression for the amount of fuel available in our expression for a star's lifetime, t, we obtain

$$t = \frac{2 \times 10^{29}\ \mathcal{M}}{2 \times 10^{19}\ \mathcal{L}} \text{ years}$$

$$= 10^{10}\frac{\mathcal{M}}{\mathcal{L}} \text{ years}$$

where $\mathcal{M}$ is in solar masses and $\mathcal{L}$ is in solar luminosities.

For the Sun ($\mathcal{M} = 1$ and $\mathcal{L} = 1$), this is 10^{10} (or 10 billion) years. For a star such as Sirius, whose mass is about 2 $M_\odot$ and whose luminosity is about 20 $L_\odot$, its main sequence lifetime is $10^{10}(\frac{2}{20}) = 10^9$ years. For a star whose mass is 30 $M_\odot$ and whose luminosity is 10^5 $L_\odot$, the main-sequence lifetime is a mere 3×10^6 years. Notice that *massive stars have much shorter lifetimes* than the Sun, despite having more fuel. This is because they burn their fuel so much faster to supply their greater luminosity. They are like gas-guzzler cars that despite having 20-gallon fuel tanks, run out of fuel sooner than economy-size cars with only 10-gallon tanks.

The shortness of massive stars' lives implies that those we see must be relatively young. Because massive stars are blue, we can conclude that, in general, *blue stars have formed recently.** Their youth is also indicated by the fact that they are frequently found in the presence of the gas left over from their formation.

14.4 Giant Stars

Leaving the Main Sequence

When a main-sequence star has consumed most of the hydrogen in its core, its structure begins to change. With too little fuel available, energy cannot be generated fast enough. That in turn allows the weight of the star's outer layers to compress the core, which heats it and makes it denser, as shown in figure 14.10. As a result, the hydrogen

ANIMATION
The structure of a star when hydrogen burning in the shell begins

*Strictly speaking, we are considering only main-sequence stars here.

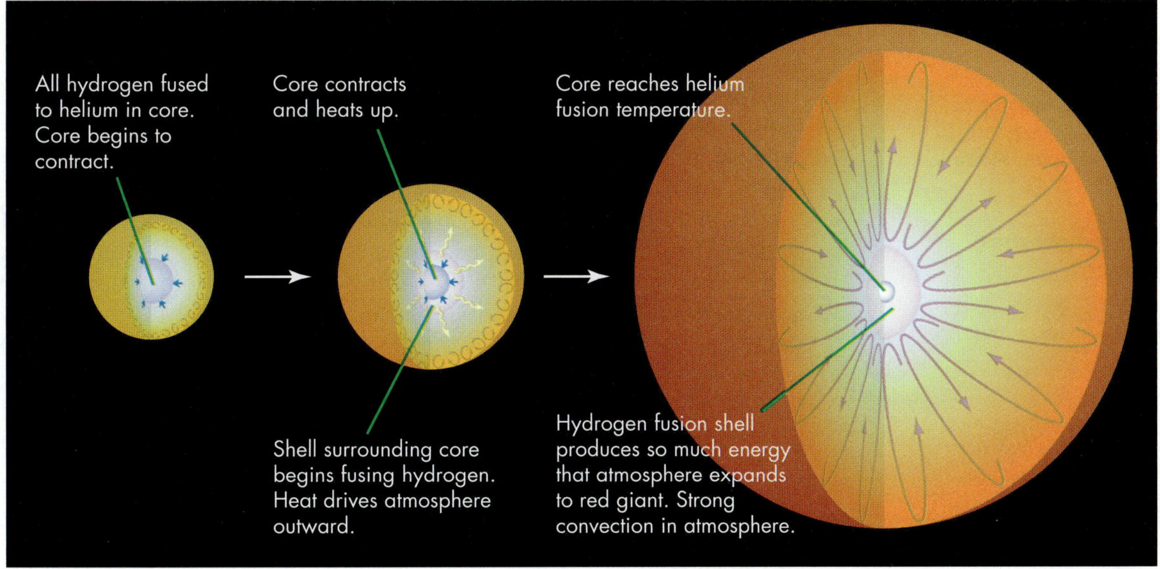
All hydrogen fused to helium in core. Core begins to contract.
Core contracts and heats up.
Core reaches helium fusion temperature.
Shell surrounding core begins fusing hydrogen. Heat drives atmosphere outward.
Hydrogen fusion shell produces so much energy that atmosphere expands to red giant. Strong convection in atmosphere.

FIGURE 14.10
The end of main-sequence fusion and the start of the red giant phase.

just *outside* the core now reaches a high enough temperature that it begins to fuse, creating what is called a **shell source,** as illustrated in the middle part of figure 14.10.

Heat from the shell source and contracting core raises the pressure around the core. That stronger pressure pushes the surrounding gas outward, making the star expand and its radius grow by a factor of anywhere from five to several hundred, depending on the star's mass. But this expansion cools the outer layers, and so the star grows red (recall that cooler bodies radiate more strongly at long [red] wavelengths). Because of its size and color, we call such a star a **red giant.** Thus, a main-sequence star becomes a red giant when it uses up the hydrogen in its core.

A giant star has a very different structure from that of a main-sequence star. As its name implies, its radius is much larger: hundreds of times larger than the Sun's, in some cases. But this enormous size is deceptive. Although most of the star's *volume* is taken up by its very low-density, tenuous atmosphere, most of its *mass* lies in a tiny, hot, compressed core not much larger than the Earth. The core, or its surrounding shell source, supplies the star's energy. The huge envelope is relatively cool and the gas in it readily absorbs the photons flowing from the core toward the star's surface. This slows the photons, and so convection rather than radiation carries energy to the star's surface. Some of that energy continues to come from burning hydrogen, but many stars switch to a new energy supply: helium burning.

Nuclear Fuels Heavier Than Hydrogen

Helium burning begins very differently in high-mass and low-mass stars and creates significant differences in their structures. To see why, we need to look in more detail at how nuclear fusion occurs and why it requires a very high temperature.

Two nuclei fuse when they are brought close enough together for the strong nuclear force to bind them to one another. That force, which is also what holds the protons and neutrons together in a nucleus, operates only over very short distances. If nuclei are more than a few diameters apart, the nuclear force is too weak to bond them. In fact, they will be repelled by the electrical force between the similarly charged protons, as illustrated in figure 14.11. That electric repulsion grows as the number of protons increases. It is weak for hydrogen, with its single proton, but stronger for heavier elements, with more protons. It is enormously more difficult, therefore, for two carbon atoms, with their six protons, to fuse than for two hydrogen atoms to do so.

Nuclei can overcome this electrical repulsion and fuse if they collide at enormously high speed. That speed is achieved in hot gases: the hotter a gas, the faster its nuclei move, and so the more easily it can fuse heavy nuclei. Hydrogen can fuse at about 5 million Kelvin, but helium, with its two protons, must be heated to about 100 million Kelvin. Carbon, with its six protons, must be still hotter to fuse.

If a star's core is as hot as 100 million Kelvin, helium nuclei there will fuse into carbon. The nuclear reaction combines 3 ^{4}He into a ^{12}C. Helium nuclei are sometimes called "alpha particles," and so this reaction is sometimes called the **triple alpha process,** or simply "helium burning." The reaction releases energy but only about one-tenth the amount per kilogram of fuel that hydrogen burning yields. Nevertheless, once a star has consumed all its core hydrogen, it must use its helium or collapse. How does a star make the switch from hydrogen to helium fusion? It must heat its core to 100 million K, and any star more massive than about 0.5 $M_{\odot}$ can do this by compressing its core.

A high-mass star needs to compress its core only a little before helium begins to fuse, because its core is already extremely hot. Thus, helium burning begins easily for high-mass stars. A low-mass star such as the Sun, however, must compress its core enormously to make it hot enough for helium fusion. The compression of such a star packs its gas atoms so closely that they no longer behave like an ordinary gas—in technical terms, the gas becomes a **degenerate gas.**

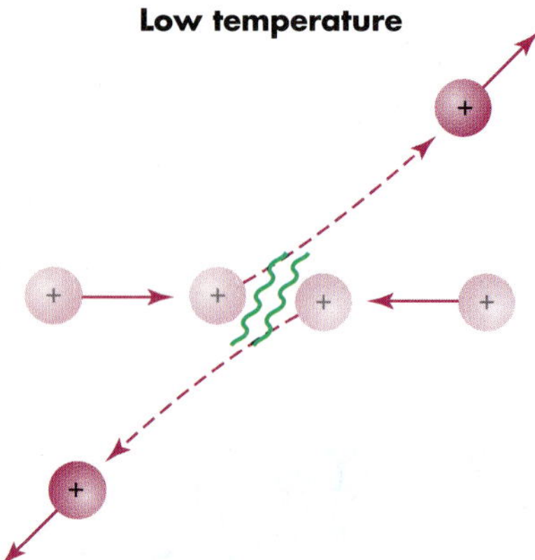

Nuclei move slowly and electric force repels them and pushes them apart. **No** fusion.

High temperature

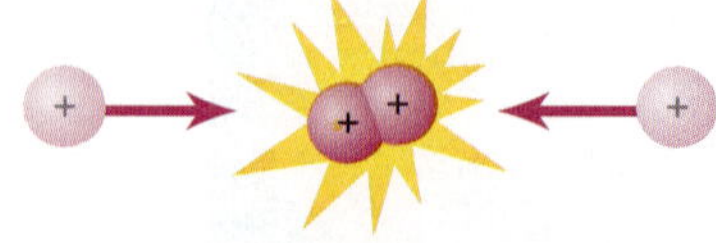

Nuclei move faster and electrical force repelling them is overwhelmed. They collide and **fuse.**

FIGURE 14.11
How a high temperature acts to overcome the electrical repulsion of the nuclei and brings them close enough for the nuclear force to fuse them.

Degeneracy in Low-Mass Stars

In a degenerate gas, the matter is so densely packed that the electrons interact not as ordinary charged particles but according to laws of subatomic physics. One consequence of these laws is that no more than two electrons of the same energy can be packed into the same submicroscopic volume.* This packing limit makes a degenerate gas "stiff," so it behaves almost like a solid. As a result, a degenerate gas does not expand on being heated any more than a heated brick swells up. This lack of response to heating strongly affects nuclear burning in a degenerate gas.

When a nuclear fuel begins to fuse in a normal gas, the energy released heats the gas, raising its pressure so that the gas expands. The expansion cools the gas and reduces the rate of nuclear burning, acting like a thermostat that switches off the furnace when the house gets too hot. The reason cooling slows nuclear burning is that only at high temperatures do nuclei collide violently enough to fuse. Thus, as a gas cools, the rate at which its atoms fuse decreases, so that less energy is released. In normal stars, this mechanism keeps them from collapsing or exploding. In degenerate matter, this thermostating mechanism fails.

For a low-mass star with a degenerate core, when a fuel ignites, the energy released does not raise the pressure because the pressure does not depend on the temperature. Thus, the gas does not expand and cool. Instead, it simply gets hotter and releases more energy, which makes it hotter still. Under these circumstances, the release of energy accelerates explosively in what is called the **helium flash.** In only a few minutes, the star's energy production increases several thousand times, yet this outburst is totally hidden from our view by the star's outer layers, just as a firecracker set off under a mattress creates little visible disturbance. However, the energy released heats the core enough to change it back to a normal gas, ending its degeneracy. With its degeneracy gone, the gas can now expand. That expansion in turn makes the star's outer layers adjust—shrinking, compressing, and heating. The star's reheated surface changes color from red to yellow, and the star becomes a yellow giant. The helium flash therefore ends the red-giant stage of low-mass stars.

Stars more than about twice as massive as the Sun do not have a helium flash. They are hot enough initially that their cores ignite helium with less compression, and so they do not become degenerate at this stage of their lives. They continue to expand and change gradually from yellow to red giants, but for both high-mass and low-mass stars, the onset of helium burning may trigger instabilities in the star and make it pulsate.

14.5 Yellow Giants and Pulsating Stars

When stars are plotted on the H-R diagram, some lie between the hot, luminous, blue, main-sequence stars and the red giants. Such stars are called "yellow giants." The most luminous yellow giants are aging high-mass stars. Less luminous yellow giants are old low-mass stars that have completed their first red-giant stage and are burning helium in their cores. Regardless of their mass, many yellow giants have the unusual property of swelling and shrinking rhythmically: they pulsate. As they change size, their luminosity changes, and so astronomers refer to them as "variables," or "pulsating stars."

Two particularly important types of pulsating stars are the RR Lyrae (pronounced *lie-ree*) and Cepheid (pronounced *sef-ee-id*) variables. RR Lyrae stars have a mass comparable to the Sun's and are yellow-white giants with about 40 times the Sun's luminosity. The time it takes them to complete a pulsation cycle from bright to dim and back to bright is called their "period," as illustrated in figure 14.12. RR Lyrae variables have periods of about half a day and are named for RR Lyrae, a star in the constellation Lyra, the harp, which was the first star of this type to be identified.

Cepheid variables are yellow supergiants that are more massive than the Sun and range in luminosity from several hundred times the Sun's luminosity to several tens of

*Technically, the size of the volume is given by the laws of quantum physics.

thousands times its luminosity. They are named for the star Delta Cephei, and their periods can be as short as about 1 day or as long as about 70 days.

Why Do Stars Pulsate?

Giant stars pulsate because their atmospheres trap some of their radiated energy. This heats their outer layers, raising the pressure and making the layers expand. The expanded gas cools, and the pressure drops, so gravity pulls the layers downward and recompresses them. The recompressed gas begins once more to absorb energy, leading to a new expansion. These stars continue alternately to trap and release the energy, and so they continue to swell and shrink, as shown in figure 14.13.

A covered pan of water boiling on a stove behaves similarly. The lid will trap the steam so that pressure inside rises. Eventually, the pressure becomes strong enough to tip the lid, and steam escapes. The pressure decreases, and the lid falls back. It again traps the steam, the pressure again builds up, and the cycle is repeated.

A similar process occurs in pulsating stars, with the role of steam played by the star's radiation and the role of the lid played by the star's atmosphere. For a star to trap radiation this way, its atmosphere must have special absorbing properties—technically called "opacity"—that occur only if its surface temperature and radius fall in a narrow range. That range, called the **instability strip,** is shown in the H-R diagram illustrated in figure 14.14A. A star that assumes these characteristics as it evolves will begin to pulsate and will continue to pulsate until its temperature or radius changes enough to remove it from the instability strip. For example, when some low-mass stars cross the region of the instability strip, they become RR Lyrae variable stars. When some high-mass stars cross, they, being more luminous, lie above the RR Lyrae variables in the instability strip and instead become Cepheid variables. The amount of time a given star spends in the instability strip depends on its mass. Massive stars such as the highly luminous Cepheids evolve across the strip in less than 1 million years, but some cross the region several times as their interior structures alter. Low-mass stars such as the RR Lyrae variables

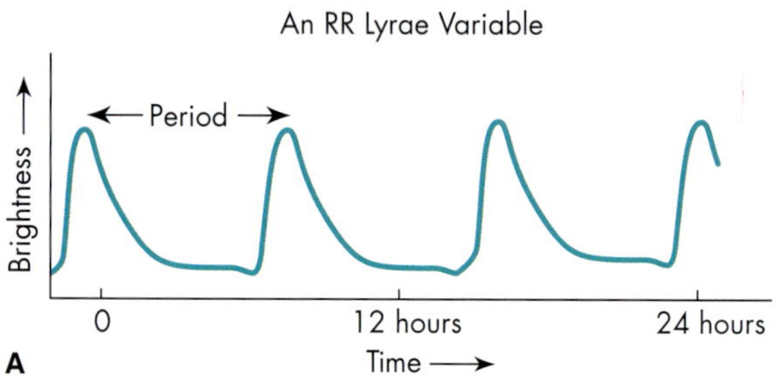

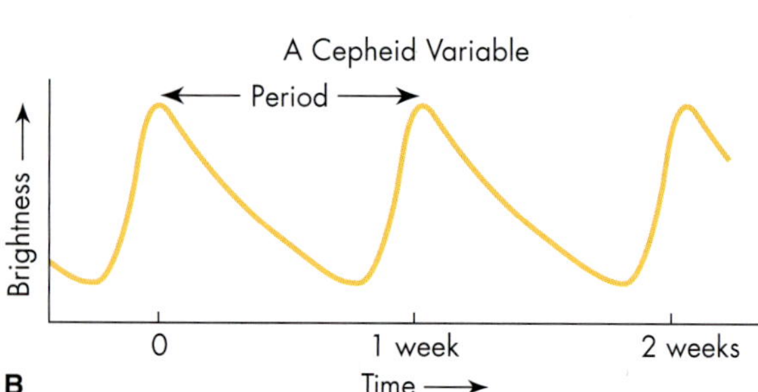

FIGURE 14.12
Schematic light curves of two pulsating variable stars: (A) an RR Lyrae star; (B) a Cepheid variable.

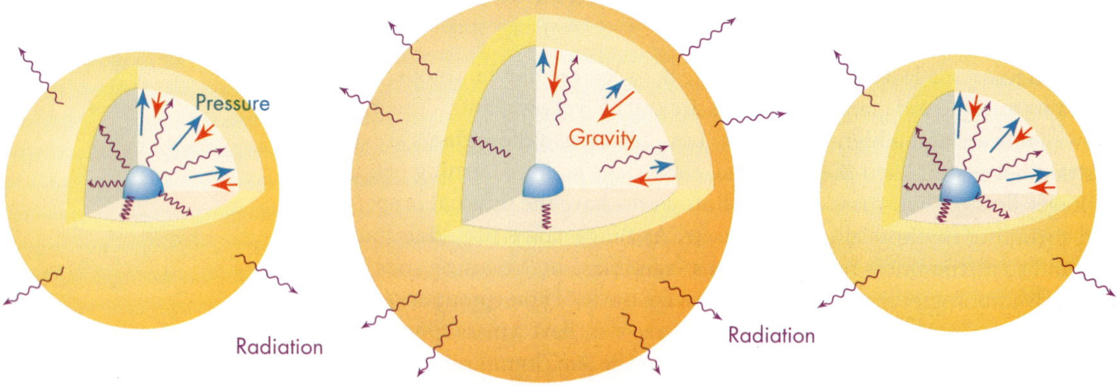

Radiation (purple wiggles) partially trapped. Pressure (blue arrows) increases, overcomes star's gravity (red arrows), and makes it "inflate" and expand.

Expansion allows trapped radiation to escape. Star cools and pressure decreases. Gravity now compresses star back to original size.

Cycle begins again.

FIGURE 14.13
Schematic view of a pulsating star.

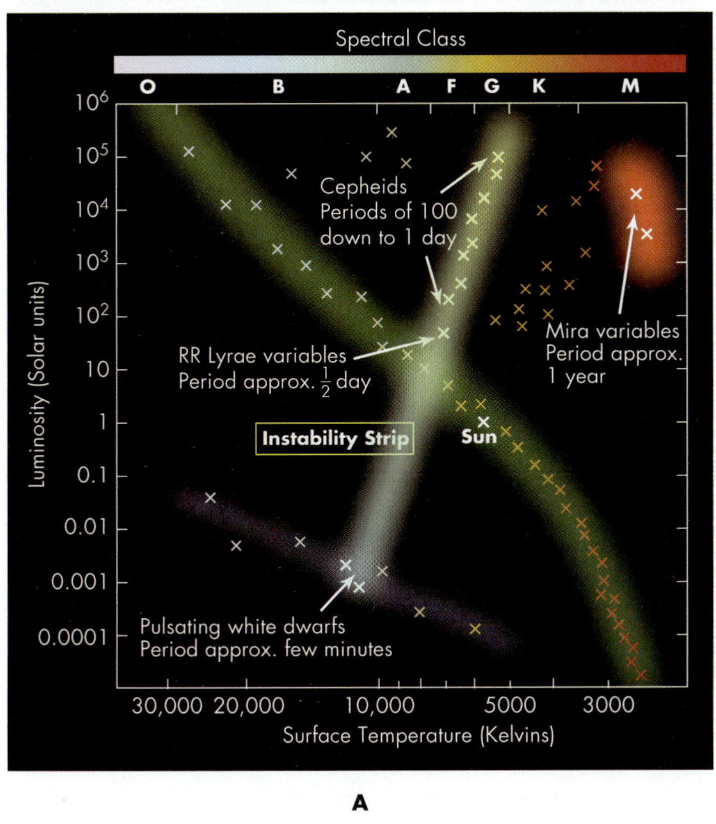

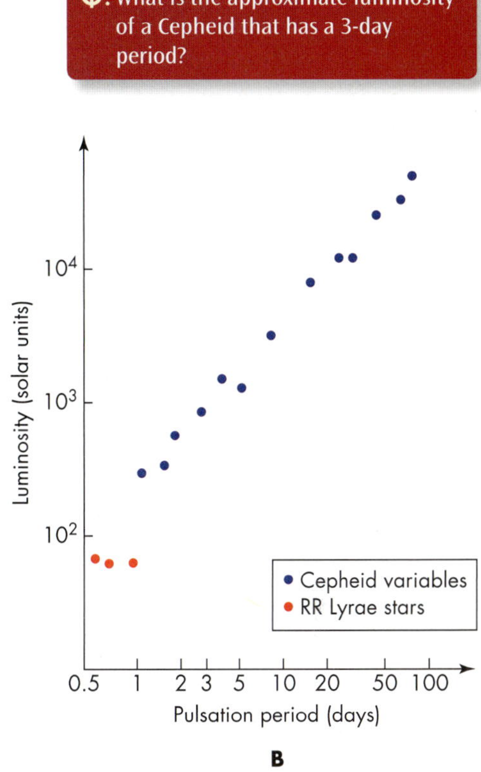

FIGURE 14.14
Properties of variable stars. (A) The instability strip in the H-R diagram. Stars in this narrow region pulsate. The locations of a few other types of variable stars are also indicated. (B) The period–luminosity law. More-luminous stars tend to pulsate more slowly.

spend more time in the strip, perhaps a few million years, but cross it less often. In either case, stars pulsate for only a brief portion of their lives.

Astronomers have identified many other types of pulsating variables besides Cepheids and RR Lyrae stars. For example, ZZ Ceti stars, a kind of pulsating white dwarf with periods as short as a few minutes, lie along a rough extension of the instability strip across the main sequence and downward toward the lower left of the H-R diagram. Another type, Mira variables, have pulsation periods of about 1 year, lie in the upper right of the H-R diagram. The Sun will probably become a Mira variable near the end of its lifetime, as we will discuss in the next section.

The Period–Luminosity Law

Many pulsating variable stars obey a law that relates their luminosity to their period—the time it takes them to complete a pulsation. Observations (and theoretical calculations) show that the more slowly a star pulsates, the more luminous it is (fig. 14.14B). This so-called **period–luminosity law** arises because, other things being equal, more-luminous stars have a larger radius than less-luminous ones. This follows simply from the fact that a bigger star has more surface, so it can emit more light (see section 13.2). Why, though, does the larger radius make the larger star pulsate more slowly? Again, the answer is simple. Because an object's gravity weakens with increasing distance, other things being equal, a large-radius star has a weaker surface gravity than a small-radius star. Hence, gravity pulls more feebly inward for a large-radius star than for a small-radius one, so its pulsation takes longer. As a result, big (and therefore bright) stars pulsate more slowly than small (and therefore dim) stars.

The period–luminosity law gives astronomers a powerful tool for measuring distances. By measuring the star's period, astronomers can find its luminosity. From its luminosity and apparent brightness, they find the star's distance, as discussed in chapter 13. We will see in chapters 16 and 17 some specific applications of this method to measuring the distance to galaxies.

14.6 Death of Stars Like the Sun

Once the Sun becomes a yellow giant, it will be nearing the end of its life, and as death approaches, its evolution will speed up. The Sun will spend about 10 billion years consuming its hydrogen and becoming a red giant. It will spend only about 100 million years (1% of its main-sequence lifetime) consuming its helium. The evolution will be faster because helium yields less energy than hydrogen when it is burned, and the Sun will be more luminous, so it will burn it faster. During the time it consumes the helium in its core, the star will lie between the main sequence and the giant region in the H-R diagram, a part of the diagram known as the horizontal branch.

Ejection of a Low-Mass Star's Outer Layers

As the core burns its helium into carbon, its radius will once again shrink, compressing and heating it. That compression is unable to make the core hot enough to burn carbon, but it does make it hot enough to increase significantly the rate at which the star burns its fuel. The more rapid fuel consumption makes the star more luminous, inflating it to an even larger radius than before, making it a bright red giant. As the star swells in size, its outer layers cool to about 2500 K, becoming a Mira variable.

The atmosphere is so cool that carbon and silicon atoms condense and form "flakes," just as water in our atmosphere forms snowflakes when it cools. These carbon and "rock" flakes (technically, what are called grains) do not fall into the star, however. Rather, they are pushed *outward* by the flood of photons pouring from the star's luminous core, just as dust in a comet is blown into a tail by the Sun's radiation. The rising grains drag gas with them and drive it into space, forming a huge shell around the star (fig. 14.15).* The carbon grains (as well as other solid particles that condense in the star's cool atmosphere) survive their trip out of the star and drift through space. Some have been found in meteoritic material within our Solar System, "fossils" from a bygone star.

Planetary Nebulas

The gas shell expands, thins, and becomes transparent, allowing us to see through to the star's furiously hot core. Because it is so hot, the core's radiation is rich in ultraviolet light, which heats and ionizes the shell around it, making it glow. Astronomers have observed many such glowing shells around dying stars (fig. 14.16) and call them **planetary nebulas.** This term is unfortunate because planetary nebulas have nothing

*Each pulsation of the Mira variable probably sends an outflow of gas and dust, and the loss of opacity from the dust plays a role in the pulsation process, similar to the covered-pan analogy in section 14.5.

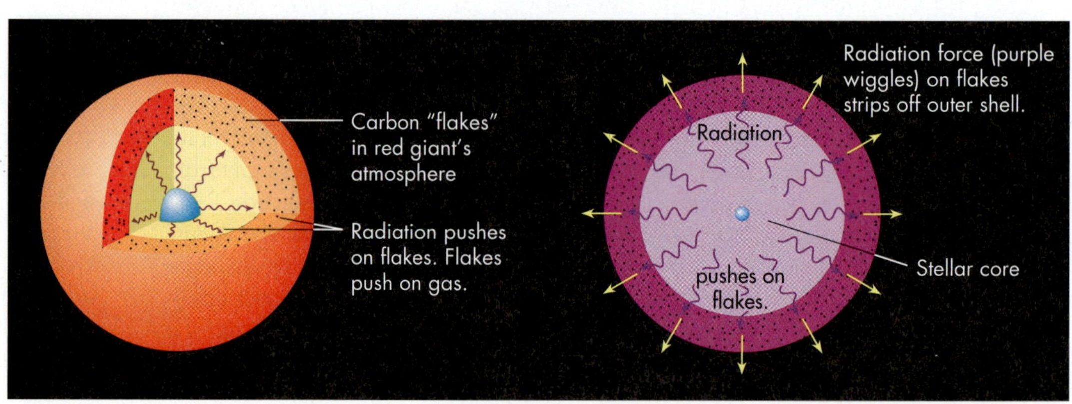
Carbon "flakes" in red giant's atmosphere
Radiation pushes on flakes. Flakes push on gas.
Radiation force (purple wiggles) on flakes strips off outer shell.
Radiation
pushes on flakes.
Stellar core

FIGURE 14.15 Artist's conception of how radiation pressure in a luminous red giant pushes on carbon flakes in its atmosphere and drives off the gas. Magnetic forces probably play a role too. The ejected gas forms a shell around the star—a planetary nebula.

SCIENCE *at work*

PLANETARY NEBULAS

For many years, astronomers thought that the typical planetary nebula was ejected uniformly in all directions so that it formed a huge "bubble" around the star's core, as you might infer from figure 14.16A. Astronomers knew of oddly shaped planetary nebulas but thought them to be the exception.

This view has changed markedly over the last decade, especially now that more detailed pictures (such as fig. 14.16C, made with the Hubble Space Telescope) are available. Many such pictures show that the shell is *not* spherical and that the gas has been ejected mainly along two directions, presumably the star's rotation axis. Such ejection creates not a bubble but two oppositely directed cones, as sketched in figure 14.16E. Moreover, many planetary nebulas have extremely narrow jets of hot gas that are directed more or less along the cones, as you can see in figure 14.16D. Why this happens is unclear, but it has forced astronomers to rethink their hypotheses about how planetary nebula form.

For example, some astronomers think that magnetic fields create the narrow jets and that the fields also create a pressure that helps expel the gas. Such magnetic effects occur on our Sun and help expel gas from it, although on a much smaller scale. Another hypothesis is that a ring of dust and gas in orbit around the star's equator blocks the ejection in those directions. Such a ring might form if a companion star's gravity draws some of the ejected material into a disk around both stars.

Whether such disks or rings give shape to all planetary nebulas is still unclear, but some of those that look circular merely have a rotation axis that happens to point toward us so we see their cones end-on, not from the side. Thus, we see how better data may require new ideas for its interpretation.

A

B

C

D

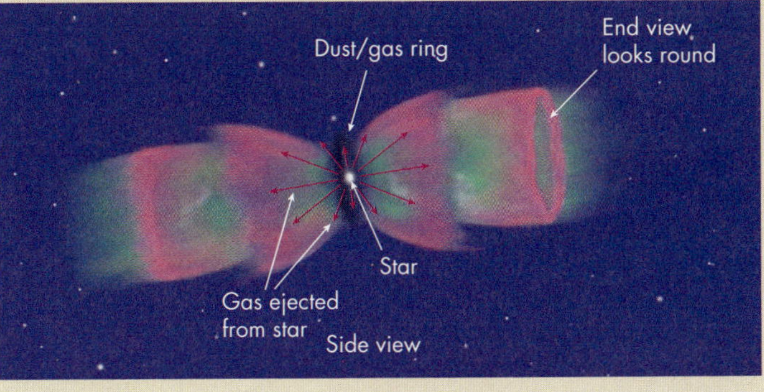

E

FIGURE 14.16
Pictures of several planetary nebulas. (A) Helix Nebula. (B) Ring Nebula. (C) Hourglass Nebula. (D) Butterfly Nebula. Notice the central star in each. Other stars that look as if they are inside the shell are foreground or background stars. (E) Sketch of how such a curious object as (D) might form.

Q. What will eventually happen to the central star? What will eventually happen to the nebula itself?

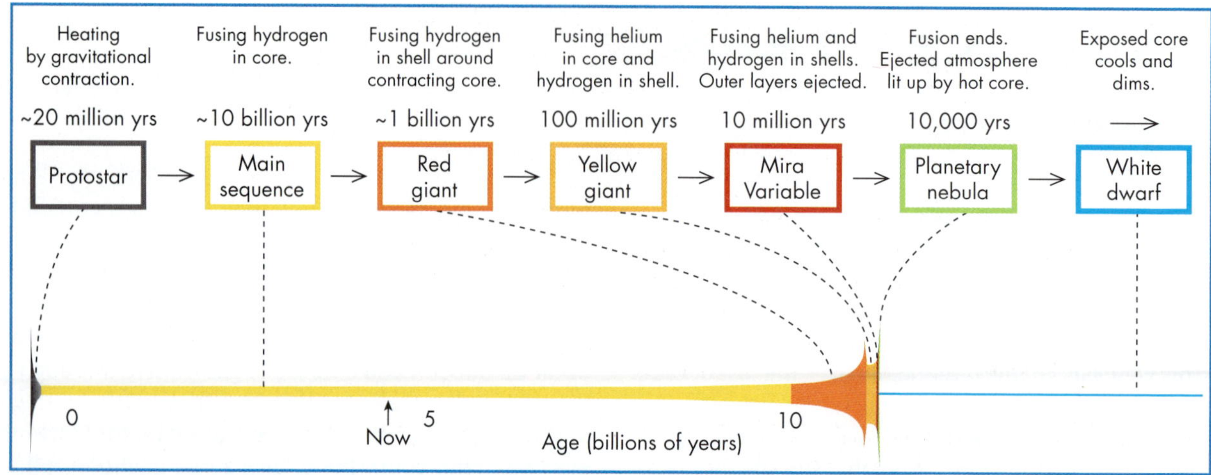

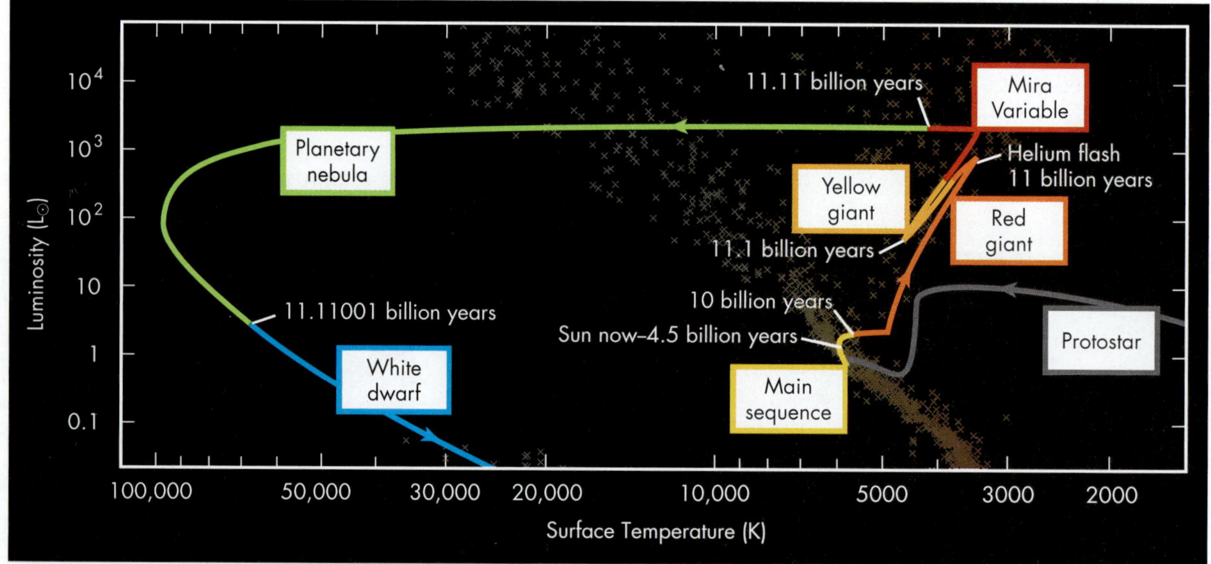

FIGURE 14.17
The evolution of a low-mass star such as the Sun plotted in an H-R diagram. The time line along the top illustrates the relative times spent in each stage and gives a general idea of the changing size of the star (not to scale). The fuel powering the star at each stage is described above the time line.

to do with planets. The usage survives from times when observers had only poor telescopes through which planetary nebulas looked like small disks, similar to planets.

A planetary nebula shell contains about one-fourth of a solar mass of glowing gas but may have as much as several solar masses of cooler, nonluminous gas around it. Shells are typically about one-fourth of a light-year in diameter and expand at about 20 kilometers per second. We can see this expansion by comparing photographs taken several decades apart. The shells eventually grow so big and diffuse that they mingle with the interstellar medium. Thus, the cycle of a star's life is complete: gas to star and back to gas. But the core of the star remains behind as a tiny, glowing ball that astronomers call a "white dwarf." Figure 14.17 shows these changes in an H-R diagram, affording a shorthand view of the life of a star like the Sun.

The Fates of Other Low-Mass Stars

Calculations indicate that all stars that begin their lives with less than 8 $M_{\odot}$ end up as white dwarfs, but their life stories may differ a little from the Sun's. The exact evolution depends on how rapidly the star loses mass in its red giant phase, and this is difficult to predict precisely. However, it appears that low-mass stars that begin with more than a few solar masses may even grow hot enough in their cores to fuse carbon into heavier elements toward the end of their lives. On the other hand, stars that have less than half the Sun's mass may never ignite helium, and their mass loss may be too little and too slow to form a planetary nebula.

The total lifetimes of low-mass stars vary greatly, from several *trillion* years for the lowest-mass stars to under 100 million years at the upper end of the mass range. The white dwarfs they leave behind all have masses less than about 1.4 $M_{\odot}$, and although they initially glow white hot from the leftover heat from fusion, they gradually darken and fade away over billions of years. A very different fate, however, awaits a more massive star.

14.7 Old Age of Massive Stars

Stars that begin their lives with more than 8 solar masses do not become planetary nebulas or white dwarfs. Their great mass compresses and heats their cores enough to ignite carbon and allows them to keep burning when their helium is gone. A variety of other nuclear reactions may also occur that create heavy elements and at the same time supply energy for the star. Astronomers call the formation of heavy elements by such nuclear burning processes **nucleosynthesis.** Nucleosynthesis in stars was discussed in detail by E. Margaret Burbidge, Geoffrey R. Burbidge, William A. Fowler, and Fred Hoyle, who proposed in 1957 that *all* the chemical elements in our Universe heavier than helium were made in this way. Their idea is one of the triumphs of stellar evolution theory, and an enormous amount of theory and observation support it.

Formation of Heavy Elements: Nucleosynthesis

Heavy elements form when two or more light nuclei combine and fuse into a single heavier one. The new nucleus must have a number of protons+neutrons equal to the sum of the numbers found in its parent nuclei.* For example, suppose ^{12}C and ^{16}O fuse. Carbon has 6 protons and 6 neutrons, and oxygen has 8 protons and 8 neutrons. Their fusion makes a nucleus with $6 + 8 = 14$ protons and $6 + 8 = 14$ neutrons (that is, an atom with 14 protons and 28 particles total in the nucleus). Reference to a table of the elements shows that such an atom is silicon, ^{28}Si. Table 14.1 shows some of the reactions that make heavy elements in high-mass stars. The table also lists the approximate temperatures needed for the reactions and their duration in a 25 $M_{\odot}$ star.

In massive stars, these reactions provide enough energy to support the stellar mass against gravity's unceasing pull. As each fuel—hydrogen, helium, carbon—is exhausted, the star's core contracts and heats by compression. The higher temperature then allows the star to burn still heavier elements, the "ashes" of one set of nuclear reactions becoming the fuel for the next set. Oxygen, neon, magnesium, and, eventually, silicon are formed, but because progressively higher temperatures are needed for each new burning process, each new fuel is confined to a smaller and hotter region around the star's core. Thus, the star develops a layered structure, as illustrated in figure 14.18. The star's surface, where the temperature is too low for nuclear reactions to occur, remains hydrogen, but beneath the surface lies a series of nested shells, each made of a heavier element than the one surrounding it. Helium surrounds a shell of carbon, which in turn surrounds a shell of neon, and so on to the core. By the time the star is

---

*Usually the total number of protons and neutrons each remain the same separately, but in some reactions, such as the proton–proton chain, a proton may turn into a neutron or a neutron into a proton.

TABLE 14.1 SEQUENCE OF MAJOR FUSION REACTIONS IN MASSIVE STARS

Fusion Reaction	Minimum Temperature	Duration in 25-$M_\odot$ Star
$4\,^{1}H \rightarrow {}^{4}He$	5,000,000 K	7,000,000 years
$3\,^{4}He \rightarrow {}^{12}C$	100,000,000 K	700,000 years
$^{12}C + {}^{4}He \rightarrow {}^{16}O$	200,000,000 K	
$^{12}C + {}^{12}C \rightarrow {}^{20}Ne + {}^{4}He$	600,000,000 K	300 years
$^{20}Ne + {}^{20}Ne \rightarrow {}^{24}Mg + {}^{16}O$	1,500,000,000 K	8 months
$^{16}O + {}^{16}O \rightarrow {}^{28}Si + {}^{4}He$	2,000,000,000 K	3 months
$^{28}Si + {}^{28}Si \rightarrow {}^{56}Fe$	2,500,000,000 K	1 day

KEY: H, hydrogen; He, helium; C, carbon; O, oxygen; Ne, neon; Mg, magnesium; Si, silicon; Fe, iron.

burning silicon into iron, its core has shrunk to a diameter smaller than the Earth's, and the core's temperature is about 2.5 billion K. A very massive star (20 $M_\odot$ or more) may take less than 10 million years from its birth to form an iron core. This brief lifetime results because (1) the star burns its fuel rapidly to offset the energy lost because of its high luminosity, and (2) the structure of atomic nuclei is such that fusion of elements heavier than hydrogen yields much less energy than the fusion of a similar amount of hydrogen, and so more fuel must be burned to supply the same amount of energy.

Some recent computer simulations of the collapse of a massive star show that its core may be set violently vibrating. These vibrations then heat the gas around the core and thereby help power the star's explosion.

Core Collapse of Massive Stars

The formation of an iron core signals the end of a massive star's life. Iron cannot burn and release energy: the iron nucleus turns out to be the most tightly bound of all nuclei. As a result, attempting to fuse additional protons or neutrons to it *weakens* the bonds and *absorbs* energy, rather than releasing it. Thus, nuclear fusion stops with iron, and a star with an iron core is out of fuel.

Fuel exhaustion causes a star's core to shrink and heat. But in high-mass stars, the shrinkage presses the iron nuclei so tightly together that a new reaction can occur: protons and electrons may themselves "merge," neutralizing their charge and becoming

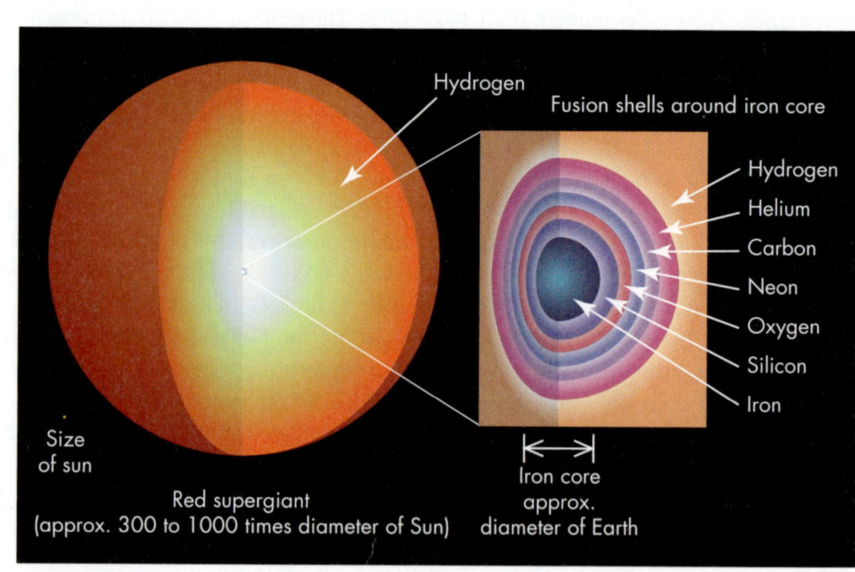

FIGURE 14.18
The layered structure of a massive star as it burns progressively heavier fuels in its core. (These inner zones are enormously magnified for clarity.)

neutrons. The shrinking core is thus transformed from a sphere of iron into a sphere of neutrons, with catastrophic results for the star. Most of the pressure that supported the core was supplied by electrons, but they have been absorbed by protons. Thus, the star's core pressure suddenly drops. Nothing remains to support the star, and so its interior begins to collapse. However, because the matter is so dense, its force of gravity is immense, and it crushes the core. In less than a second, the core drops from an iron ball the size of the Earth to a ball of neutrons about 10 kilometers (about 6 miles) in radius. The star's outer layers, with nothing to support them, plummet inward like a tall building whose first floor is suddenly blown away. The plunging outer layers of the star strike the neutron core, crushing it still more, while the impact heats the infalling gas to billions of degrees. The pressure surges and lifts the outer layers away from the star in a titanic explosion—a **supernova.**

FIGURE 14.19
A supernova explosion in a distant galaxy observed in 2002 shines brightly compared to the billions of stars in the galaxy.

Supernova Explosions

The explosion of a massive star* mixes the elements synthesized by nuclear burning during its evolution with the star's outer layers and blasts them into space. This incandescent spray expands away from the star's collapsed core at more than 10,000 kilometers per second. Depending on the star's mass, 10 or so solar masses of matter may be flung outward, ultimately to mix with surrounding material and, in time, to form new generations of stars. Interstellar gas is thereby enriched in heavy elements, the atoms needed to build the rock of planets and the bones of living creatures. Indeed, the supernova outburst itself generates even more heavy elements. The explosion creates free neutrons that rapidly combine with atoms in the star to build up heavy and rare elements such as gold, platinum, and uranium.

For the star, the supernova explosion is a quick and glorious death. In a few minutes, it releases more energy than it has generated by nuclear burning during its entire existence. In a matter of hours, the dying star brightens to several billion times the luminosity of the Sun, radiating energy in its death throes at a greater rate than all the other stars in the galaxy combined. Figure 14.19 is a picture of a galaxy with a supernova: the exploding star gleams like a beacon.

A supernova emits more than just visible light: most of the energy of its blast is carried by a burst of neutrinos, the same tiny, highly penetrating particles generated in the Sun as it converts hydrogen into helium. Just as the Sun's neutrinos escape freely into space, so the supernova's neutrinos escape as well. A pulse of such neutrinos was detected in February 1987 when a supernova—SN 1987A—blew up in the Large Magellanic Cloud. Despite its enormous distance, more than a trillion neutrinos from that explosion passed through each person on Earth.

Supernova Remnants

Gas ejected by the supernova blast plows through the surrounding interstellar space, sweeping up and compressing other gas that it encounters. The huge, glowing cloud of stellar debris—a **supernova remnant**—steadily expands. At the end of one year, it is 300 billion kilometers (about 0.03 light-years) across. At the end of one century, it is several light-years in diameter, but its expansion slows as it runs into more surrounding gas. Figure 14.20 shows pictures of several supernova remnants of different ages. Notice how ragged the remnants in figures 14.20 look compared with the smoothly ejected bubbles of planetary nebulas—evidence that massive stars die more violently than low-mass stars.

One such violent outburst was seen about 1000 years ago by astronomers in China and elsewhere in the Far East. On a date that according to our calendar was July 4,

*We will see in chapter 15 that low-mass stars can also explode as supernovas if they are members of a binary star system (see section 15.1).

FIGURE 14.20
A few supernova remnants: (A) Color photograph of the Vela supernova remnant. (B) Color picture of the Crab Nebula. (C) X-ray image of the Cassiopeia supernova remnant. In this image, the red on the left is radiation from hot iron atoms and the bright green is radiation from silicon and sulfur atoms.

1054, their records describe a "guest star" that appeared in the evening sky and that was visible even in broad daylight for several weeks. Today, with even a small telescope, you can still see the glowing gases ejected from that dying star. Figure 14.20B is a picture of this perhaps most famous of all supernova remnants, now known as the **Crab Nebula.** In chapter 15, we will learn that the star's core survived the blast and lies in the expanding debris.

A supernova explosion marks the death of a massive star, and we chart its complete life in the H-R diagram shown in figure 14.21. We can see that its path through

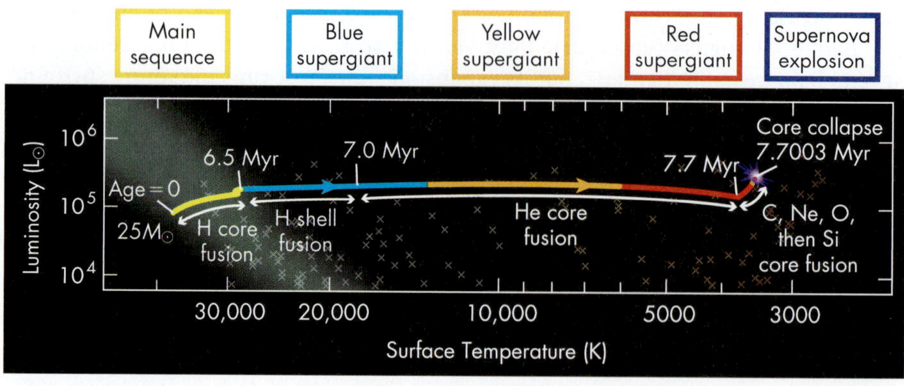

FIGURE 14.21
H-R diagram of a massive star's evolution.

FIGURE 14.22
Photograph of a portion of the Large Magellanic Cloud (a small, nearby galaxy) before Supernova 1987A exploded, and a second photograph taken during the event.

time is far simpler than that of a low-mass star. The massive star begins its life on the upper main sequence as a blue star. As its core hydrogen is consumed, it leaves the main sequence, swelling and cooling to become a yellow supergiant while it begins to burn helium in its core. When the helium there is consumed, the star switches to other fuels, all the while contracting and heating its core and expanding and cooling its outer layers to become a red supergiant. Overall, it keeps its luminosity approximately constant as it ages, but some massive stars, depending on how many heavy elements are incorporated in them at their birth, may return partway to the main sequence as they switch to new fuels. The resultant heating may turn them into blue supergiants before they explode as supernovas. Such was the behavior of Supernova 1987A, whose explosion is documented in figure 14.22.

Supernova remnants eventually slow down, cool, and mingle with the interstellar clouds around them. But the remnant's gas is rich in heavy elements that were formed in the core of the star that blew up as a supernova. Thus, when the remnant mixes with an interstellar cloud, the cloud is enriched in heavy elements. Eventually the enriched cloud may collapse and turn into a new generation of stars. These new stars will, therefore, contain more heavy elements than earlier generations of stars. In this way, the amount of heavy elements in stars increases from generation to generation.

What remains when the supernova remnant dissipates? For stars that began their lives with less than about 20 $M_{\odot}$, the core of the star—a few solar masses—survives as a ball of neutrons (a neutron star). Astronomers think that more-massive stars may end up as even more compressed bodies—black holes—as we will discover in chapter 15.

14.8 History of Stellar Evolution Theories

The theories we have discussed above about how stars work and evolve are relatively new. For example, Aristotle wrote more than 2000 years ago that stars are heated by their passage through the heavens. He never considered the possibility that they might evolve. In the eighteenth century, the German philosopher Immanuel Kant described stars slightly more scientifically, envisioning the Sun as a fiery sphere, formed when gases gravitated to the center of the solar nebula.

Early Stellar Models

The English physicist Lord Kelvin and the German physicist Hermann von Helmholtz were the first scientists to use the equations of basic physics to arrive at a mathematical description of what we would today call "model stars." In the 1850s and 1860s, they were searching for a mechanism that could explain the Sun's power supply, as discussed in chapter 12. Kelvin and Helmholtz independently solved the mathematical equations for a gas sphere in which pressure is balanced by gravity and showed that a star's core is compressed by the weight of the overlying gas and is much denser than the outer parts of the star. They further showed that as gravity compresses a star, that compression converts gravitational energy into heat, raising the star's temperature. However, the amount of energy available from gravitational energy is far too small to power it for billions of years.

EXTENDING our reach

MEASURING THE AGE OF A STAR CLUSTER

Astronomers can deduce the age of a star cluster from the shape of its H-R diagram. To understand why its H-R diagram reveals the cluster's age, recall that in a star cluster, all the stars form at approximately the same time. Moreover, at their birth, all the stars are burning hydrogen in their core. Because the main sequence is determined by the location of stars burning hydrogen in their core, all the stars in a newly formed cluster lie on or near the main sequence. Not all the cluster's stars burn their fuel at the same rate, however. To maintain their high luminosity, massive stars burn their fuel more rapidly than low-mass stars do. With their hydrogen used up, high-mass stars leave the main sequence and turn into red giants. The low-mass stars, on the other hand, still have hydrogen to burn, and so they remain on the main sequence longer.

Because the high-mass stars in an old cluster turn into red giants (and thus lie off the main sequence), a line in the H-R diagram connecting the position of the cluster's stars bends away to the right, as shown in Box figure 14.1. The point where that line bends away from the main sequence is called the **turnoff point**. At the turnoff point, a star is not quite old enough to have used up the hydrogen in its core. That is, its age is just a tiny bit less than its main-sequence lifetime. But all stars in the cluster are the same age, namely, the age of the cluster. To get the cluster's age, we therefore measure the age of the star at the turnoff point by calculating its main-sequence lifetime from its mass and luminosity, as described earlier. The answer we get is the age of the star cluster.

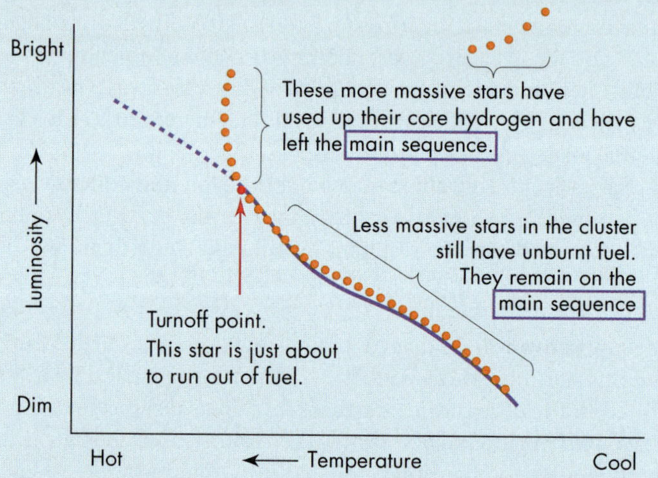

BOX FIGURE 14.1
The turnoff point in the H-R diagram of a star cluster indicates its age.

Box figure 14.2 shows a series of H-R diagrams for clusters ranging in age from very young to very old (a few million years for the youngest to more than 10 billion years for the oldest). Notice that old clusters have few, if any, stars on the upper part of the main sequence. Young clusters, on the other hand, do have stars on the upper main sequence.

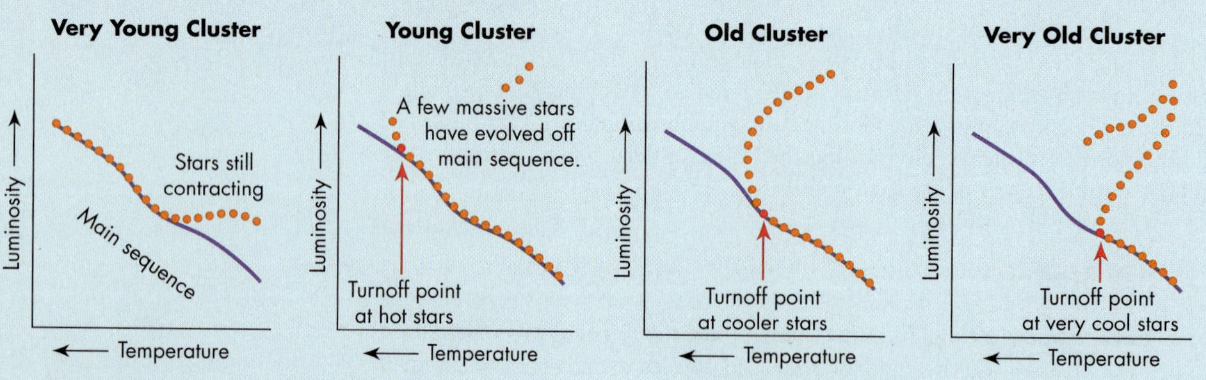

BOX FIGURE 14.2
Schematic H-R diagrams of clusters of different ages, illustrating the turnoff point and how it shifts down and to the right for older clusters.

Astronomers had to await discoveries about the nature of atoms in the early years of the twentieth century before they could solve the mystery of stellar energy. The British astrophysicist Sir Arthur Stanley Eddington, who discovered the mass–luminosity relation, was one of the first scientists to recognize the immense power available in the form of nuclear energy, but at the time too little was known about how atoms interacted to make models of stars powered by hydrogen burning. Moreover, Eddington discovered that to make a model star, more was required than simply balancing pressure and gravitational forces. He demonstrated that a correct model must also account for the outward leakage of energy from the star's core to its surface, a process that made solving the equations much more difficult.

The equations that govern a star's structure express a complicated set of relations between the star's pressure, density, temperature, gravitational force, energy generation, composition, and luminosity. The equations must balance at each point in the star and yield answers that agree with observations. Solving such complex equations is extremely difficult, but it became easier in the 1940s when astronomers began using mechanical calculators. These machines—incredibly primitive compared with a modern computer—were operated by turning a hand crank. Nevertheless, they were faster than pencil-and-paper solutions, and the models so produced revealed more clearly the physical processes in stars. Several months might be needed to make the calculations for a half-dozen models that showed how a main-sequence star could change into a red giant. Then, in the 1950s, when astronomers began to use electronic computers to calculate model stars, only a few minutes were needed to solve the equations of a stellar model. By combining the results of dozens of such models, each of which represented a star with successively more of its precious fuel consumed, astronomers could see how stars evolved. Such work continues to this day, with astronomers still trying to understand the late stages of a star's life, when it is losing mass and burning elements heavier than helium in its core.

Testing Stellar Evolution Theory

Although we now understand that main-sequence stars turn into red giants, as recently as the 1940s some astronomers thought the reverse. What evidence supports our current theories of stellar evolution? The best demonstration that the modern theory is correct comes from comparing the H-R diagrams of real groups of stars with theoretically determined H-R diagrams.

We learned in section 14.2 that stars probably form in groups—star clusters. The stars in a cluster are bound together by their mutual gravitational attraction, although each star moves along its own orbit within the cluster. All the stars within a given cluster form at approximately the same time and are therefore nearly the same age. If we look at the stars in a cluster shortly after its birth, when they have all completed their contraction stage from protostars, they will all lie on the main sequence. If we look at some later time, some of the stars will have consumed their core hydrogen and evolved off the main sequence. To check our theory of stellar evolution, we calculate evolutionary tracks, such as those in figures 14.17 and 14.21, for every star on the main sequence to show where each star will be at, say, 10 million years, 100 million years, and so on. The resulting curves show us what the H-R diagram of the entire cluster should look like 10 million years, 100 million years, and so on, after its birth. When we compare such curves with H-R diagrams of actual star clusters, the match is excellent. If our theory of stellar evolution were wrong, the shapes would be unlikely to agree so well. In addition, the models help astronomers understand why so many kinds of stars exist. For example, the models not only offer a natural explanation of main-sequence, red giant, and white dwarf stars but also help us see how such totally different objects, such as pulsating variables, planetary nebulas, and supernovas, fit into the scheme of stellar evolution. This success in interpreting such stellar diversity is evidence that the theory is essentially correct. Moreover, that success offers astronomers a way to measure the age of a star cluster.

SUMMARY

A star forms from interstellar gas drawn together by gravity, which compresses and heats the gas to form a protostar. Further heating causes the core of the protostar to fuse hydrogen into helium. The energy released keeps the core hot and thereby maintains an outward pressure that stops the protostar's collapse and changes it into a main-sequence star. When the hydrogen is exhausted in a star's core, the core shrinks and heats, making the star swell into a red giant.

A low-mass star such as the Sun continues to burn hydrogen in a shell during its first red giant stage but eventually compresses the helium in its core enough to ignite the gas in a helium flash. The energy released expands the core, and the outer layers shrink, turning the star into a yellow giant. As the star burns helium, its core again shrinks, and it once more becomes a red giant. It is so luminous now that its radiation drives off its outer layers to form a planetary nebula shell. The core remains as a small, dense star: a white dwarf.

Because of the tremendous heating from gravitational compression of their cores, high-mass stars more easily ignite fuels heavier than hydrogen and undergo less dramatic changes initially than low-mass stars do. They build in their cores layers of heavy elements: carbon, neon, silicon, and, eventually, iron. The iron core cannot burn and will collapse, triggering a supernova explosion. The heavy elements produced during the star's life and in the supernova that destroys it are blasted into space, enriching the interstellar medium. The core remains as a ball of neutrons—a neutron star—or may collapse entirely and become a black hole.

The lifetime of a star is dependent on its mass. High-mass stars burn their fuel quickly to supply the energy they need to support their great weight. These stars are blue because they are so hot. Thus, blue stars burn out quickly and must be young.

Both high- and low-mass stars pass through a stage late in their lives when they pulsate as yellow giants. Low-mass stars become RR Lyrae variables, whereas high-mass stars become Cepheids. Table 14.2 summarizes these stages.

TABLE 14.2 EVOLUTION OF LOW-MASS AND HIGH-MASS STARS

Low-Mass Star (Mass Less Than 8 $M_{\odot}$)	High-Mass Star (Mass More Than 8 $M_{\odot}$)
Protostar (gravity supplies energy)	Protostar (gravity supplies energy)
Main sequence (hydrogen burns in core) Core hydrogen is consumed.	Main sequence (hydrogen burns in core) Core hydrogen is consumed.
Red giant (hydrogen burns in shell) Core contracts, heats, and ignites helium (helium flash).	Yellow supergiant (helium burns in core) Star pulsates.
Yellow giant (star pulsates) Red giant (helium burns in shell) Outer layers ejected to form:	Red supergiant (heavier elements burn in a series of shells in the star's core) Buildup of iron core and its collapse.
Planetary nebula Core cools to become white dwarf.	Supernova explosion when core collapses; remnant may be a neutron star or black hole.

QUESTIONS FOR REVIEW

1. (14.1) What processes and forces determine the structure of stars?
2. (14.1) Through what stages will the Sun evolve? Through what stages will a high-mass star evolve?
3. (14.2) What heats a protostar? How can we observe protostars? Why are they surrounded by dust and gas?
4. (14.2) What is a bipolar flow?
5. (14.2) What is a T Tauri star?
6. (14.2) What is a Bok globule?
7. (14.2–14.3) What determines when a star becomes a main-sequence star?
8. (14.3) How long do stars stay on the main sequence?
9. (14.4) What makes a star move off the main sequence?
10. (14.4–14.5) Where do main-sequence stars end up as they evolve?
11. (14.4) Why is it easier for a high-mass star than for a low-mass star to burn helium?
12. (14.4) Why do high- and low-mass stars evolve differently as they become red giants?

13. (14.5) What is meant by a pulsating star? Why do stars pulsate?
14. (14.4/14.6) What happens to a solar-mass star when it starts to burn helium in its core? What does it turn into?
15. (14.6) What is a planetary nebula?
16. (14.6) What is one explanation for how a low-mass star expels its outer layers to make a planetary nebula?
17. (14.6) What is left when a planetary nebula dissipates?
18. (14.7) What makes a high-mass star's core collapse?
19. (14.7) Why do neutrons form in a massive star's iron core?
20. (14.7) What is a supernova explosion?
21. (14.7) What kind of subatomic particles have been observed when a supernova explodes?
22. (14.8) How are clusters of stars used to test theories of stellar evolution?

THOUGHT QUESTIONS

1. (13.6–13.7/14.1) Stars of all colors are visible to the naked eye, from Betelgeuse's red to Sirius's white to Spica's blue-white. If late in their lives, stars did not evolve into giants but just became dimmer and dimmer, what would you expect for the relative numbers of blue, white, yellow and red stars visible in the night sky? Would any kind of yellow stars be visible to the naked eye?
2. (14.2) Inflate a balloon and carefully measure its size. Put it in the freezer for a few hours. Does it look the same when it is cold? How does this relate to how stars form in cold regions of space?
3. (14.2) Describe how a protostar is believed to form.
4. (14.2–14.4) Take a plastic bottle and put a little soapy water in it. Run your finger across the mouth to make a soap film. Now, without breaking the film, run hot water over the bottle. What happens to the soap film? How does this relate to what happens to a star when it is heated?
5. (14.3–14.5) Suppose that stars could mix unburned fuel from their outer layers into their cores. Would that alter the way they evolve? Can you suggest some possible differences?
6. (14.6) Hold a clear cylindrical water glass below a bright light so that it casts a shadow on a piece of paper. If you hold the glass at different angles, you can create different shadow patterns. Can you produce shapes similar to the planetary nebula images in figure 14.16? Is there another shape that would create patterns more similar to some images?
7. (14.7) Hold a small rubber ball on top of a basketball, and drop them together toward the floor. What happens to the small ball? Does that help you understand what happens to the outer layers of a supernova as they collapse on the core?
8. (14.7) Thinking about why massive stars start fusing heavy elements, and examing the periodic table in the back of the book, explain why there should be more neon, magnesium, sulfur, argon, and calcium in a supernova remnant than fluorine, sodium, phosphorus, and chlorine.
9. (14.7) Explain why gold and platinum are much scarcer than iron and silicon (found in most rocks), thinking about how massive stars live and die.
10. (14.7/14.8) In some very old, dense star clusters, a few blue stars, known as "blue stragglers," are seen. Why are blue stars unexpected in such clusters? If stars that stick together mix their material thoroughly when they collide, how might you explain such blue stars?

PROBLEMS

1. (14.1) Sketch an H-R diagram of what you expect the evolutionary paths of a 2-$M_\odot$ and a 15-$M_\odot$ star to look like.
2. (14.2) The density of a cold interstellar cloud prior to collapsing might be about 1.67×10^{-20} kg/m^3. How many cubic meters of this cloud would contain enough mass to build the Sun? What is the ratio of this volume to the volume of the Sun? What is the volume of cloud required if written in cubic light-years?
3. (14.2) The jets from a T Tauri star are measured to have speeds of about 300 km/sec. At this speed, how long would it take the ejected material to travel 2 light-years (similar to the jets shown in figure 14.7)?
4. (14.3) Calculate the main-sequence lifetime of the Sun by first determining the rate at which it burns hydrogen (refer to chapter 12) and then dividing that rate into its core mass. If your answer is in seconds, convert to years, given that there are about 3.2×10^7 seconds in a year.
5. (14.4/14.6) Calculate the escape velocity from a red giant's atmosphere (use the formula for escape velocity from chapter 3). Assume that the star's mass is 1 $M_\odot$ and its radius is 100 $R_\odot$. How does this compare with the speed at which a planetary nebula shell is ejected?
6. (14.6) Calculate the average density of a red giant star in grams per cubic centimeter. Take the star's mass to be 2×10^{33} grams and its radius to be 3×10^{12} centimeters. How does this density compare with that of the air we breathe (about 10^{-3} gm/cm^3)?
7. (14.6) How long will it take a planetary nebula shell moving at 20 km/sec to expand to a radius of one-fourth of a light-year?
8. (14.7) Calculate the Doppler shift of X-rays emitted from shocked silicon in a supernova remnant if the ejected material is moving at 5000km/sec and the X-ray line is at 0.688 nanometers.
9. (14.8) Figure 14.23 shows the H-R diagram of a star cluster. How old is it? (The dashed line is the position of the stars. The solid line is the main sequence.)

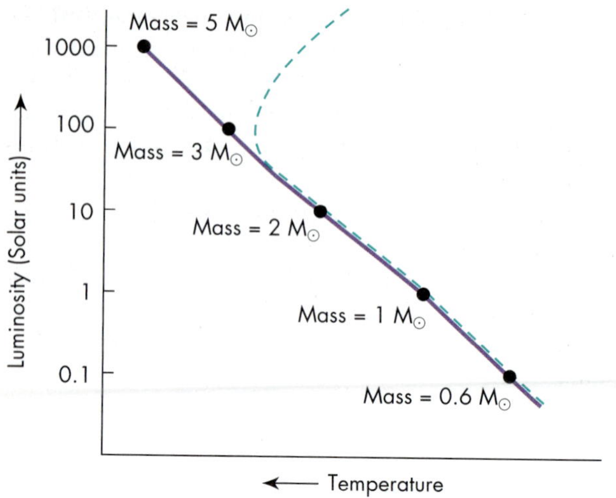

FIGURE 14.23
H-R diagram of a star cluster.

TEST YOURSELF

1. (14.1) Which of the following sequences correctly describes the evolution of the Sun from young to old?
   (a) White dwarf, red giant, main-sequence, protostar
   (b) Red giant, main-sequence, white dwarf, protostar
   (c) Protostar, red giant, main-sequence, white dwarf
   (d) Protostar, main-sequence, white dwarf, red giant
   (e) Protostar, main-sequence, red giant, white dwarf
2. (14.1) ________ and ________ drive what happens in every phase of a star's life.
   (a) Gravity, composition
   (b) Gravity, mass of the star
   (c) Fusion of hydrogen, gravity
   (d) Angular momentum, composition
   (e) Radiation pressure, angular momentum
3. (14.2) All of the following are terms related to star formation except
   (a) T Tauri stars.
   (b) Bok globules.
   (c) interstellar cloud.
   (d) bipolar outflow.
   (e) planetary nebula.
4. (14.3) A star whose mass is 5 times larger than the Sun's and whose luminosity is 100 times larger than the Sun's has a main-sequence lifetime about
   (a) 5 times longer than the Sun's.
   (b) 500 times longer than the Sun's.
   (c) 5 times shorter than the Sun's.
   (d) 100 times shorter than the Sun's.
   (e) 20 times shorter than the Sun's.
5. (14.4) The helium flash occurs in stars with a degenerate core because
   (a) degenerate gas expands faster than ideal gases.
   (b) stars with degenerate cores are larger.
   (c) a degenerate gas can get very hot over a large range very quickly.
   (d) stars with degenerate cores have much lower densities in the core.
6. (14.5) Variable stars are useful because (choose all that apply)
   (a) they can be used as standard candles.
   (b) they provide ways to test stellar structure theories.
   (c) their luminosity can be determined from the period.
   (d) many variable stars have large luminosities.
7. (14.6) A planetary nebula is
   (a) another term for the disk of gas around a young star.
   (b) the cloud from which protostars form.
   (c) a shell of gas ejected from a star late in its life.
   (d) what is left when a white dwarf star explodes as a supernova.
   (e) the remnants of the explosion created by the collapse of the iron core in a massive star.
8. (14.6) As a star like the Sun evolves into a red giant, its core
   (a) expands and cools.
   (b) contracts and heats.
   (c) expands and heats.
   (d) turns into iron.
   (e) turns into uranium.
9. (14.6–14.7) Stars like the Sun probably do not form iron cores during their evolution because
   (a) all the iron is ejected when they become planetary nebulas.
   (b) their cores never get hot enough for them to make iron by nucleosynthesis.
   (c) the iron they make by nucleosynthesis is all fused into uranium.
   (d) their strong magnetic fields keep their iron in their atmospheres.
   (e) None of the above.
10. (14.7) A supernova (select all that are correct)
    (a) creates elements heavier than iron.
    (b) is the final state of all stars.
    (c) disperses metals into the interstellar medium.
    (d) can outshine the rest of a galaxy.
    (e) can trigger star formation in a nearby cloud.
11. (14.8) An H-R diagram of a cluster created from data that included the spectral types of the stars could be used to determine (select all that are correct)
    (a) the age of the cluster.
    (b) the angular momentum of the stars in the cluster.
    (c) the gravitational force in the cluster.
    (d) the distance of the cluster.
    (e) None of the answers are correct.

KEY TERMS

bipolar flows, 378
Crab Nebula, 392
degenerate gas, 382
deuterium, 377
helium flash, 383
instability strip, 384
interstellar cloud, 373
main-sequence lifetime, 380
nucleosynthesis, 389
period–luminosity law, 385
planetary nebulas, 386
protostar, 376
red giant, 382
shell source, 382

FURTHER EXPLORATIONS

Balick, Bruce. "How the Sun will die." *Astronomy* 36 (December 2008): 38.

______, and Adam Frank. "The Extraordinary Deaths of Ordinary Stars." *Scientific American* 291 (July 2004): 51.

Bethe, Hans A., and Gerald Brown. "How a Supernova Explodes." *Scientific American* 252 (May 1985): 60.

de Grasse Tyson, Neil. "Forged in the Stars." *Natural History* 105 (August 1996): 72.

Frank, Adam. "Blowing Cosmic Bubbles." *Astronomy* 25 (February 1997): 36.

Herbst, William, and George C. Assousa. "Supernovas and Star Formation." *Scientific American* 241 (August 1979): 138.

Hillebrandt, Wolfgang, Hans-Thomas Janka, and Ewald Muller. "How to Blow Up a Star." *Scientific American* 295 (October 2006): 42.

Iben, Icko, Jr., and Alexander Tutukov. "The Lives of Stars: From Birth to Death and Beyond (Part I)." *Sky and Telescope* 94 (December 1997): 36.

Jastrow, Robert. *Red Giants and White Dwarfs.* New York: Norton, 1990.

Kawaler, Steven D., and Donald E. Winget. "White Dwarfs: Fossil Stars." *Sky and Telescope* 74 (August 1987): 132.

Kirshner, Robert P. "The Earth's Elements." *Scientific American* 271 (October 1994): 58. (How the elements formed.)

Kwok, Sun. "Butterflies and Crabs of the Southern Sky." *Sky and Telescope* 103 (January 2002): 48.

______. *Cosmic Butterflies: The Colorful Mysteries of Planetary Nebulae.* New York: Cambridge University Press, 2001.

______. "What Is the Real Shape of the Ring Nebula?" *Sky and Telescope* 100 (July 2000): 32.

Ray, Thomas P. "Fountains of Youth: Early Days in the Life of a Star." *Scientific American* 283 (August 2000): 42.

Shubinski, Raymond. "All about the Veil Nebula." *Astronomy* 36 (June 2008): 56.

Soderberg, Alicia. "X Rays Mark the Spot: A Newborn Supernova." *Sky and Telescope* 116 (November 2008): 26.

Stahler, Steve W. "The Early Life of Stars." *Scientific American* 265 (July 1991): 48.

Stephens, Sally. "Needles in the Cosmic Haystack." *Astronomy* 23 (September 1995): 50. (About brown dwarfs.)

Tayler, R. "The Birth of Elements." *New Scientist* 124 (December 16, 1989): 25.

Woosley, Stan, and Tom Weaver. "The Great Supernova of 1987." *Scientific American* 261 (August 1989): 32.

Website

Visit the *Explorations* website at **http://www.mhhe.com/arny** for additional online resources on these topics.

FIGURE QUESTION ANSWERS

WHAT IS THIS? (chapter opening): These two pictures show the red supergiant Mira, a long-period variable star, at its brightest and at its dimmest. The star's brightness changes with a period of about 332 days, as you can easily see by watching it once each month over a year. In diameter it is one of the largest known stars. Its size varies as it pulsates, but it is typically about 10 AU in diameter.

FIGURE 14.8: The red glow is caused by hydrogen emission lines.

FIGURE 14.14: About 1000 times the Sun's luminosity.

FIGURE 14.16: The central star will eventually cool off and become a white dwarf. The nebula will expand, becoming ever larger, and eventually will mingle with the surrounding interstellar gas.

PROJECT

Understanding the H-R Diagram: The solid line that illustrates the evolutionary path of a star like the Sun in figure 14.17 is a bit deceptive. It gives the impression that we are as likely to see stars like the Sun in their protostar phase as in their main-sequence phase. A star moves through some phases rapidly and others much more slowly. If instead of a line you put down one point for every 10 million years, you would get a better idea of where you would be likely to see stars like the Sun in the H-R diagram. Draw your own version of figure 14.17 with a very light line tracing the Sun's path. Now, darker and in a different color, try to place points along the line using the following numbers to estimate how many dots to put down over each section of the line corresponding to each part of the Sun's evolutionary path: protostar phase ~20 million years (Myr); main-sequence phase ~10,000 Myr; first red giant phase ~1000 Myr; yellow giant phase ~100 Myr; second red giant phase ~10 Myr; planetary nebula phase 0.01 Myr; white dwarf phase ~3000 Myr. (The star remains a white dwarf beyond this time, but fades from visibility.) Does any part of your graph resemble a solid line?

Composite X-ray, infrared, and optical image of Tycho's supernova remnant.

KEY CONCEPTS

- When stars use up all their available nuclear fuel, they are crushed to high density by gravity.
- Low-mass stars like the Sun become white dwarfs.
- White dwarfs are about the size of the Earth but have a mass about that of the Sun.
- High-mass stars (more than about eight times the Sun's mass) may explode as supernovas.
- The core of a high-mass star may survive as either a neutron star or a black hole.
- White dwarfs may also explode in a supernova if a companion star sheds enough mass onto them.
- Neutron stars have a mass slightly more than that of the Sun but are a mere 10 kilometers or so in diameter.
- A black hole forms if the core of a massive star becomes so dense that the escape velocity at the surface of the core reaches the speed of light.
- Black holes emit no electromagnetic radiation but may be detected by observing gas that is heated as it collides outside the hole.

15 Stellar Remnants: White Dwarfs, Neutron Stars, and Black Holes

CONCEPTS AND SKILLS TO REVIEW

- Density (6.1)
- Escape velocity (3.8)

During its life, every star supports itself against gravity by burning nuclear fuel in its core. When its fuel is spent, the star collapses. This fate awaits the Sun 5 billion years from now. Gravity will crush it into a white dwarf—a star about 100 times smaller than the present Sun and roughly the size of the Earth. More-massive stars surrender sooner and are more dramatically squeezed to even smaller dimensions, becoming either neutron stars or black holes.

Compact stars, as these three kinds of stellar remnants are known, are the end points of stellar evolution. Because nearly all stars must eventually reach this stage, the Galaxy is littered with their shriveled bodies. But unlike stars at earlier stages of evolution, no nuclear fuel makes them shine. They shine—if at all—with heat inherited from their previous state. Their matter, too, is unusual. In crushing compact stars to their tiny dimensions, gravity squeezes them into exotic materials. For example, a piece of white dwarf material the size of an ice cube would weigh about 16 tons. Matter in a neutron star is so compressed that electrons have merged with protons, making the star resemble a giant atomic nucleus. So unmercifully has gravity squeezed the most massive of compact stars that they have collapsed completely, their immense gravity warping the space around them so that no light escapes, making them black holes in space.

Compact, however, does not mean inconspicuous: some of these crushed stars radiate intensely despite having no fuel of their own. Some compact stars "parasitize" a companion star, capturing matter from it that may burn explosively and allow the crushed star to "rise from the dead" as a nova—a "new" star. And because of a compact star's intense gravity, any material that falls on it releases such immense amounts of gravitational energy that this dead star may be thousands of times brighter than a "live" star like the Sun. But this brightness is bought at a dreadful price. The star's mass may increase to the point that gravity causes it to collapse even further or, for some stars, to explode as a type Ia supernova, blasting the star into atoms.

See end of chapter for the answer.

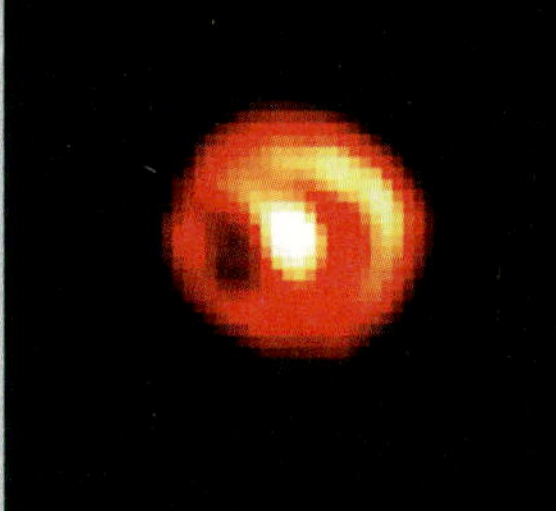

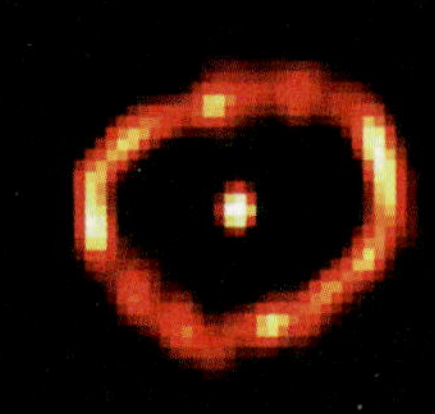

15.1 White Dwarfs

General Properties, Origin, and Fate

White dwarfs are hot, compact stars whose mass is comparable to the Sun's but whose diameter is about the same as the Earth's (about 1% that of the Sun). Their tiny size gives them so small a surface area that they are very dim despite being hot. Even the brightest white dwarfs have only about 1% the brightness of the Sun. Their pale light is not supplied by burning fuel, for they have exhausted all fuel supplies. Instead it comes from residual heat that leaks from the star's interior. Because they shine by residual heat and are therefore steadily cooling, white dwarfs show a wide range of surface temperatures. Some are as hot as 25,000 K; others are as cool as about 4500 K. A typical surface temperature might be about 10,000 K.

A white dwarf forms when a low-mass star such as the Sun nears the end of its life. When the star swells into a red giant, its intense radiation strips off its outer layers to form a planetary nebula shell. This process exposes the star's core, which becomes the white dwarf, as illustrated in figure 15.1. The composition of the core (and thus the white dwarf) is mainly carbon and oxygen, the end products of the parent star's nuclear burning. The white dwarf's surface is composed of a thin layer of hydrogen and helium, but its core has too little mass and therefore insufficient gravity to contract and heat itself to the ignition temperature of carbon. So, unable to burn further, the dying core simply cools and grows dim.

Although many white dwarfs form by a roughly solar-mass star ejecting a planetary nebula shell,* in theory, some could form from very low-mass stars (less than about 0.5 $M_{\odot}$) without such ejection. These stars are relatively cool throughout, and so their gas, like the relatively cool matter just below the Sun's surface, must carry energy

*"Roughly solar mass" in this case means up to about 8 $M_{\odot}$. Both theory and observation suggest stars as massive as 8 $M_{\odot}$ become white dwarfs. They lose most of their mass through a powerful stellar wind.

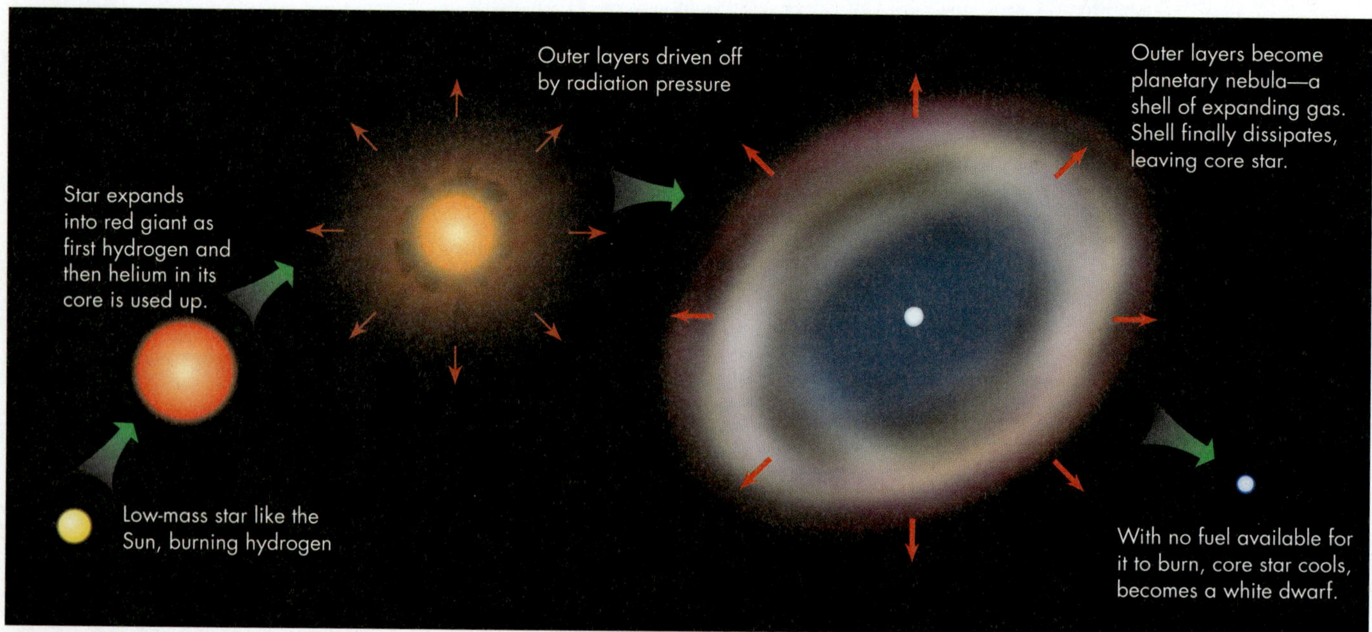
Outer layers driven off by radiation pressure
Outer layers become planetary nebula—a shell of expanding gas. Shell finally dissipates, leaving core star.
Star expands into red giant as first hydrogen and then helium in its core is used up.
Low-mass star like the Sun, burning hydrogen
With no fuel available for it to burn, core star cools, becomes a white dwarf.

FIGURE 15.1
Origin of a white dwarf. White dwarfs form when a low-mass star such as the Sun reaches the end of its life. Having expanded into a red giant for the second time (see fig. 14.2), the star loses its outer atmosphere and becomes a planetary nebula. Its core is exposed, and eventually it turns into a white dwarf, a star with no source of nuclear fuel. (Note: not all stages in a star's evolution are shown.)

by convection. The convection currents in these small stars extend throughout their spheres, mixing their gas and preventing a core from forming. Moreover, their weak gravity cannot compress the spent fuel to make them red giants. Because low-mass stars evolve so slowly, however, they require longer than the age of the Universe to reach this point in their evolution, and therefore none of these very low-mass stars has turned into white dwarfs yet.

What happens to a white dwarf as time passes? Although it is born very hot (recall that it was the core of a star), it has no fuel to burn to keep it hot. Like a dying ember, therefore, it cools rapidly (by stellar standards). Astronomers calculate that it takes only about 10 million years (roughly analogous to a month in a human lifetime) for the temperature of a white dwarf to drop to 20,000 K, and the whole age of the Universe for it to drop to 4000 K. In trillions of years it would be so dark as to be undetectable. Astronomers call such dead, dark stars **black dwarfs.** Cooling white dwarfs are probably very abundant in our Galaxy. How abundant is still controversial.

Structure of White Dwarfs

Because white dwarfs are very compact and have no fuel supply, their structure differs significantly from that of ordinary stars. Although they are in hydrostatic equilibrium, with pressure balancing gravity, their pressure arises from a peculiar interaction among the electrons that limits the number that can occupy a given volume. This gives white dwarfs an odd property: increasing their mass makes them shrink! Even more crucial, however, white dwarfs must have a mass below a critical limit, or they collapse. To see why, we must consider conditions in a white dwarf's interior.

White dwarfs are very dense, having formed from the shrunken cores of their parent stars. If we divide the mass of a white dwarf by its volume, we find that its density is about 10^5 grams per cubic centimeter (roughly 1 ton per cubic inch). This high density packs the star's atoms so closely that they are separated by less than the normal radius of an electron orbit. If electrons could remain attached to nuclei under these conditions, their orbits would overlap those of neighboring atoms. But electrons can easily escape from their nuclei at such high density and move freely between nuclei, allowing the electrons to interact in a peculiar way.

Degeneracy and the Chandrasekhar Limit

A law of subatomic physics called the **exclusion principle** limits the number of electrons that can be put into a given volume. This, in turn, limits how closely electrons can be squeezed, thereby creating a pressure that, to distinguish it from that of an ordinary gas, is called **degeneracy pressure.** Degeneracy pressure depends only on the gas density, not on its temperature. In an ordinary gas, pressure depends on temperature and density. When such a gas is compressed, both the heat and the higher density contribute to raising its pressure. When a degenerate gas is compressed, it heats up, but the heat does not raise its pressure as it does for an ordinary gas. This makes degenerate gas less "springy" than ordinary gas.

If mass is added to a white dwarf, its radius *shrinks.* The extra mass increases the star's gravity, compressing the star. Its pressure rises, but because the star is degenerate, the rise is small and is unable to prevent the star from being squashed into a smaller volume. Squashing the star decreases the distance between particles and makes its gravity larger still, squashing the star even more. In fact, gravity may ultimately overwhelm the star's pressure forces and crush the star completely. Thus, too much mass makes a white dwarf collapse, as was demonstrated by theoretical calculations made in 1931 by the Indian astrophysicist Subrahmanyan Chandrasekhar (pronounced *chan-dra-say-car*). Chandrasekhar won the Nobel Prize for Physics in 1983 for these

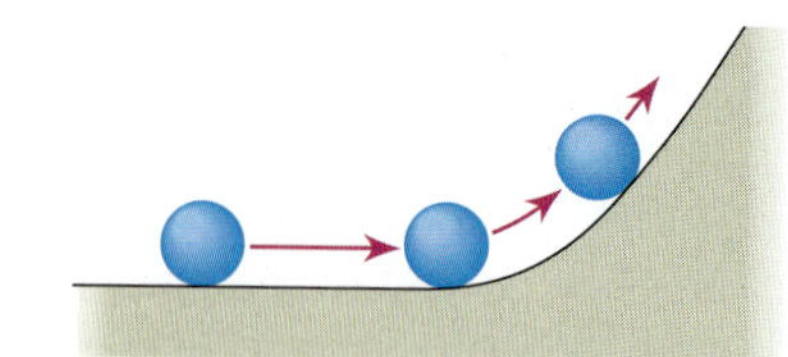

A Ball rolling up a hill loses energy and slows down.

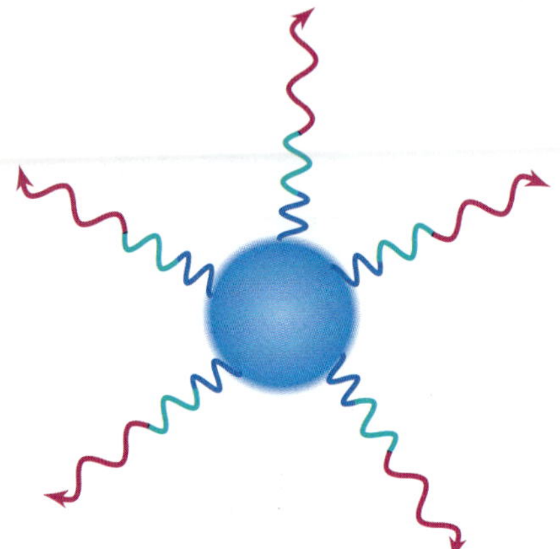

B Light moving away from a star loses energy

FIGURE 15.2
The gravitational redshift. (A) Just as a ball rolling up a hill loses energy, light loses energy as it moves away from a star. (B) Light does not slow down, however. Instead, because its energy, E, determines its wavelength, λ, according to the relation $E = \text{Constant}/\lambda$, as the light's energy decreases, its wavelength increases. Longer wavelengths correspond to redder colors, and so we refer to the lengthening of the light's wavelength caused by the star's gravity as the gravitational redshift.

calculations and related work. The limiting mass of a white dwarf is now called the **Chandrasekhar limit** in his honor. For a typical white dwarf, the limiting mass is about 1.4 $M_{\odot}$, but this depends somewhat on the white dwarf's composition.

Measurements of white dwarf masses after the completion of Chandrasekhar's work showed that no white dwarfs exceeded this limit. For those in binary star systems, the masses can be measured using the modified form of Kepler's third law. For isolated white dwarfs, calculations are based on the shift of their spectral lines caused by their immense gravity.

When light escapes from a body, it must work against the force of gravity. The escaping light acts like a ball rolled up a slope: as the ball moves higher, it slows down as it loses energy. Light cannot slow down—it must always travel at the constant speed, c—but it can lose energy, and it does. Light's energy, as you have learned, determines its wavelength: lower energy corresponds to longer wavelengths. Longer wavelengths, in turn, correspond to redder colors; thus, as light "climbs" away from a body, the body's gravity stretches the wavelengths and creates a **gravitational redshift,** as shown in figure 15.2. The amount of redshift depends on the star's mass and radius. The white dwarf's mass can be estimated from the redshift if the dwarf's radius is known* (from its luminosity and temperature, for example). The greater the redshift, the greater is the mass that causes it.

Spectral lines also permit astronomers to measure the magnetic fields of white dwarfs. We discussed in chapter 12 how the Sun's magnetic field can be measured from the splitting of its spectral lines caused by the Zeeman effect. Similar observations show that some white dwarfs have magnetic fields 10,000 times stronger than sunspots—10^8 times the strength of the Earth's magnetic fields. These immense field strengths occur because as the white dwarf shrinks, it draws its magnetic field lines closer together, amplifying the field.

White Dwarfs in Binary Systems: Novas and Supernovas of Type Ia

Isolated white dwarfs cool off and disappear from sight, but in binary systems they may heat up and explode. If a white dwarf has a nearby companion, gas expelled from the companion may fall onto the dwarf, as illustrated in figure 15.3. Coming from the companion's outer layers, such gas is generally rich in hydrogen, and it may briefly replenish the white dwarf's fuel supply. The new fuel forms a layer on the dwarf's surface, where

*To use this method successfully, astronomers need to know the original wavelengths of the spectrum lines. These can be found from the Doppler-shift law if the white dwarf's velocity is known from the motion of a companion star and if the kind of atoms creating the lines can be identified.

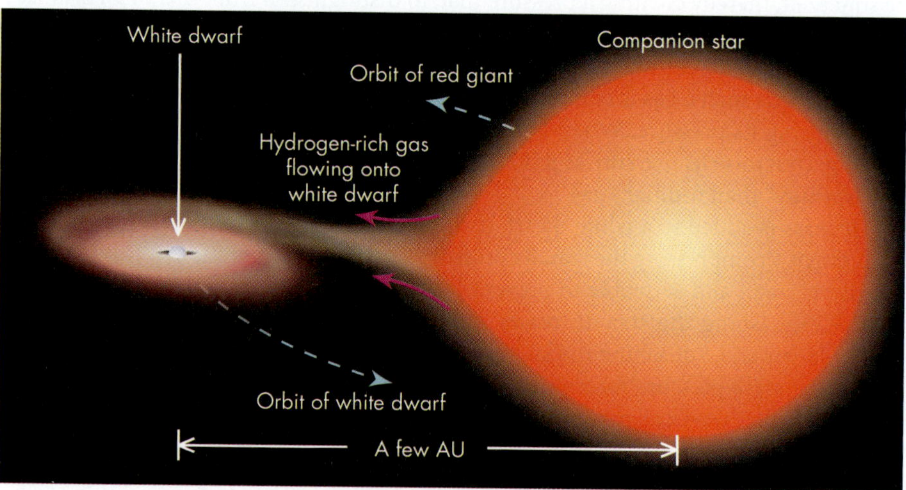

FIGURE 15.3
A white dwarf accreting (pulling in) mass from a binary star companion, which in this case is shown as a red giant. The typical separation of the stars is a few AU.

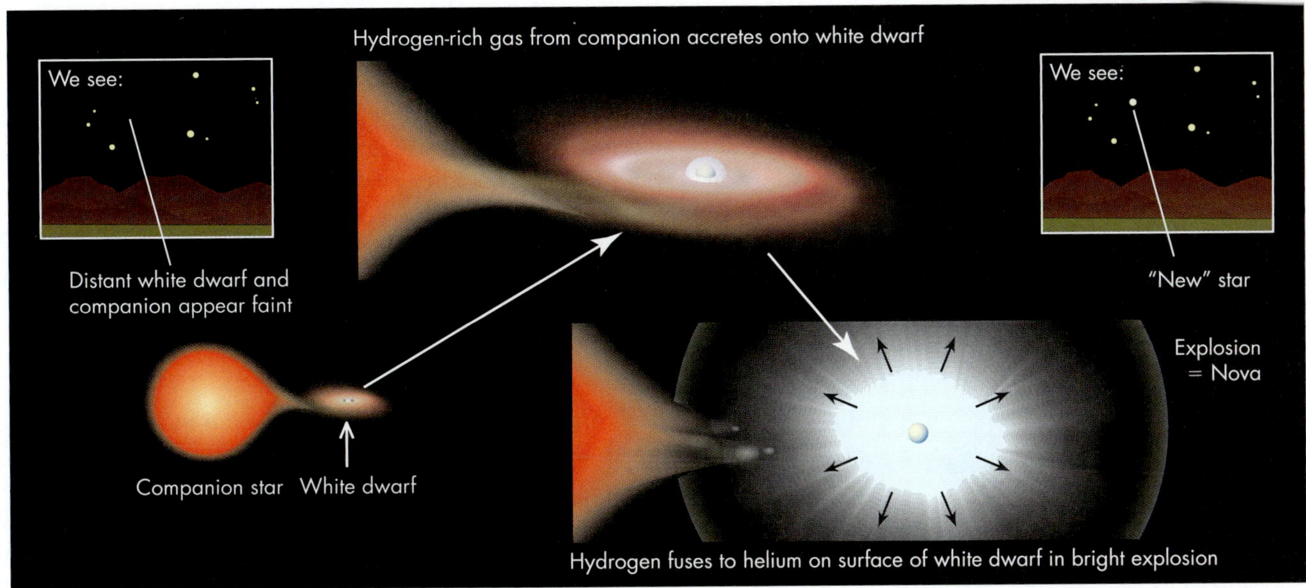
Hydrogen-rich gas from companion accretes onto white dwarf
We see:
Distant white dwarf and companion appear faint
Companion star
White dwarf
We see:
"New" star
Explosion = Nova
Hydrogen fuses to helium on surface of white dwarf in bright explosion

FIGURE 15.4
A white dwarf exploding as a nova. The hydrogen that falls onto the surface of a white dwarf from its companion may suddenly fuse into helium, creating an explosion that makes the star brighten.

gravity compresses and heats it. The gas layer eventually reaches the ignition temperature for hydrogen, but as we saw in chapter 14, nuclear burning in a degenerate gas can be explosive. The detonating hydrogen is blasted into space and forms an expanding shell of hot gas, as figure 15.4 shows. This shell radiates far more energy than the white dwarf itself. Sometimes these stellar explosions are visible to the naked eye. When early astronomers saw such an event, they called it a *nova stella,* from the Latin for "new star," because the explosion would make a bright point of light appear in the sky where no star was previously visible. Today, we shorten *nova stella* to **nova.**

A nova may erupt repeatedly if it does not accumulate too much mass. If enough gas is added to push it over the Chandrasekhar limit, however, the white dwarf will collapse. Carbon and oxygen in the collapsing star will be compressed and heated and, although initially too cold to burn, quickly reach their ignition temperature and begin nuclear fusion. Carbon and oxygen fuse to form silicon (^{12}C + ^{16}O becomes ^{28}Si), which in turn fuses into nickel (2 ^{28}Si becomes ^{56}Ni). The energy released in this burning is enough to blow the entire star apart, resulting in a **type Ia supernova** (fig. 15.5). The nickel, sprayed into space by the explosion, is highly radioactive and rapidly decays into cobalt (^{56}Co), which finally decays to iron (^{56}Fe), each decay adding additional energy to the outburst.

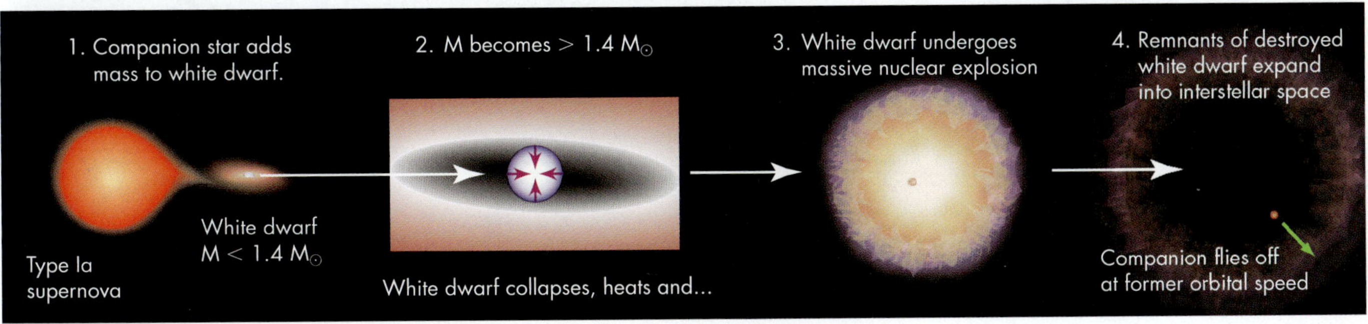
1. Companion star adds mass to white dwarf.
Type Ia supernova
White dwarf $M < 1.4\ M_\odot$
2. M becomes > 1.4 $M_\odot$
White dwarf collapses, heats and...
3. White dwarf undergoes massive nuclear explosion
4. Remnants of destroyed white dwarf expand into interstellar space
Companion flies off at former orbital speed

FIGURE 15.5
A white dwarf exploding as a type Ia supernova. If gas from a companion accumulates on the white dwarf and raises the mass above the Chandrasekhar limit (about 1.4 $M_\odot$), the white dwarf collapses. Heat from the resulting compression causes the white dwarf to explode.

A type Ia supernova leaves no remnant star, except for the white dwarf's binary companion. This star sails off at its former orbital speed because the gravitational pull from its companion has disappeared. All that is left of the white dwarf is a cloud of rapidly expanding gas rich in the leftover carbon and oxygen, plus the silicon, iron, and other elements made during the explosive nuclear burning during the supernova. Much of the iron in your blood was probably made when a white dwarf met its doom.

15.2 Neutron Stars

General Properties and Origin

INTERACTIVE
Neutron stars

Just a year after Chandrasekhar published his calculations about the collapse of white dwarfs, physicists in England studying subatomic particles discovered the neutron. In 1934 this led astronomers Walter Baade (pronounced *Bah-deh*), at Mount Wilson Observatory, and Fritz Zwicky, at the California Institute of Technology, to predict that a collapse could lead to the formation of a **neutron star**—an incredibly dense object made of neutrons, like a giant atomic nucleus. They were also the first to suggest that this collapse would lead to a massive explosion and coined the term *supernova*.

As described in chapter 14, this type of supernova occurs when a star more massive than about 8 $M_{\odot}$ forms a large iron core. Just as Chandrasekhar found for white dwarfs, an iron core cannot support its own weight when its mass grows larger than about 1.4 $M_{\odot}$ (or somewhat higher, since high temperatures are supplying part of the pressure against gravity). Because fusing iron does not release energy when it fuses, when the core collapses there is not a sudden ignition of nuclear burning as in a type Ia supernova. This is instead called a **type II supernova,** or core-collapse supernova,* as illustrated in figure 15.6.

Astronomers readily accepted Baade and Zwicky's idea that a massive star dies as a supernova—such violent explosions had been observed—but they paid little attention to their suggestion that a neutron star might be born in the blast. Such stars would be tiny compared even with white dwarfs and were thought to be unobservable. Nevertheless, several physicists calculated the properties of neutron stars and found that their radii should be about 10 kilometers (about 6 miles) and their masses should typically be between one and several times that of the Sun. The calculations also showed that, like white dwarfs, neutron stars have a maximum possible mass, which for them is about 2 to 3 solar masses.

---

*Before the exact mechanisms were understood, astronomers classified several different kinds of supernovas based on their spectra and pattern of brightening. Less common types known as Ib and Ic also come from massive stars whose iron cores collapse and only superficially resemble type Ia. Some astronomers therefore prefer the names nuclear detonation (type Ia) and core-collapse (types II, Ib, Ic) supernova.

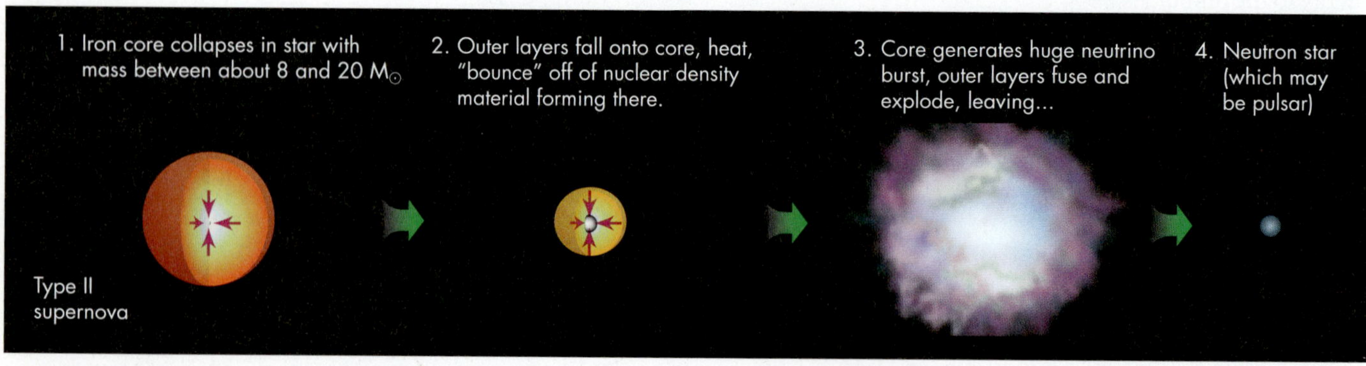
1. Iron core collapses in star with mass between about 8 and 20 $M_{\odot}$
2. Outer layers fall onto core, heat, "bounce" off of nuclear density material forming there.
3. Core generates huge neutrino burst, outer layers fuse and explode, leaving...
4. Neutron star (which may be pulsar)
Type II supernova

FIGURE 15.6
When a star that begins its life with more than about 8 $M_{\odot}$, it builds up an iron core that collapses to a neutron star. Surrounding materials are blasted outward by a type II supernova.

Pulsars and the Discovery of Neutron Stars

Despite this theoretical model of neutron stars, there was no evidence that they existed, and so Baade and Zwicky's concept of tiny, collapsed stars lay dormant for over 30 years. Then in 1967 a group of English astronomers led by Anthony Hewish was observing fluctuating radio signals from distant, peculiar galaxies. Jocelyn Bell, a graduate student working with the group, noticed an odd radio signal with a rapid and astonishingly precise pulse rate of one burst every 1.33 seconds. The precision of the pulses led some astronomers to wonder whimsically if they had perhaps stumbled onto signals from another civilization, and informally the signal was known as LGM-1, for "little green men #1." Over the next few months, Hewish's group found several more pulsating radio sources, which came to be called **pulsars** for their rapid and precisely spaced bursts of radiation (fig. 15.7).

The discovery of other pulsars convinced astronomers that they were observing a natural phenomenon, and so their explanations focused on peculiar varieties of pulsating stars. Ordinary pulsating stars, such as Cepheid variables, obey a period–density relation such that each star's pulsation period depends inversely on the square root of its density. That law applies in a generalized form to all pulsating stars. Thus, the extreme shortness of pulsar periods indicates that they must be extremely dense objects, and the first pulsars discovered had periods that were consistent with their being white dwarfs. But soon other pulsars were discovered whose periods were much shorter, implying densities far larger than those of white dwarfs, and so astronomers turned back to Baade and Zwicky's ideas about neutron stars.

Neutron stars, according to their theory, not only would be extremely dense (and thus have very short periods) but also would be associated with supernova explosions. Thus, when a very short-period pulsar was found in the Crab Nebula supernova remnant, it seemed to confirm Baade and Zwicky's model. But calculations of the pulsation period of neutron stars indicated that they should pulsate much faster than any of the pulsars so far observed. The solution to this puzzle came from the independent work of Franco Pacini and Thomas Gold, astrophysicists at Cornell University.

Even before the discovery of pulsars, Pacini had been studying the link between neutron stars and supernova remnants that had been proposed by Baade and Zwicky some 40 years earlier. He found that a neutron star remaining in the debris of a supernova explosion could supply large amounts of energy to the gas in its vicinity if it were rapidly spinning and if it had a very strong magnetic field. Rapid rotation and a strong magnetic field were also the keys to Gold's completely unrelated work, which was an attempt to explain, not the energizing of supernova debris, but the rapidity of pulsar pulses. Gold suggested that a pulsar is a rapidly *spinning*, not pulsing, neutron star. Furthermore, its magnetic field causes it to emit radiation in two narrow beams so that, as the star spins, its beams sweep across space, as illustrated in figure 15.8. We see a burst of radiation when the beam points at Earth. Thus, a pulsar shines like a cosmic lighthouse whose beams swing around as its lamp rotates. But the model worried some astronomers because at least one of the pulsars then known—that in the Crab Nebula

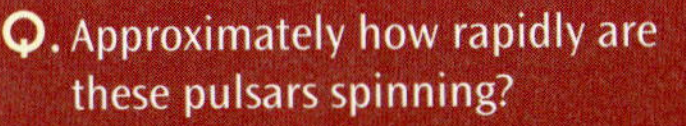

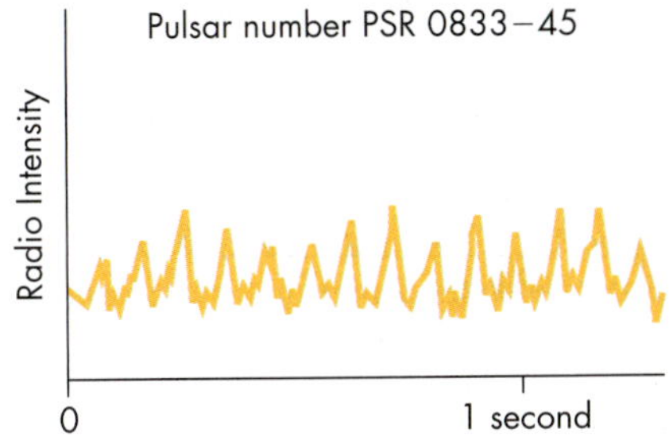

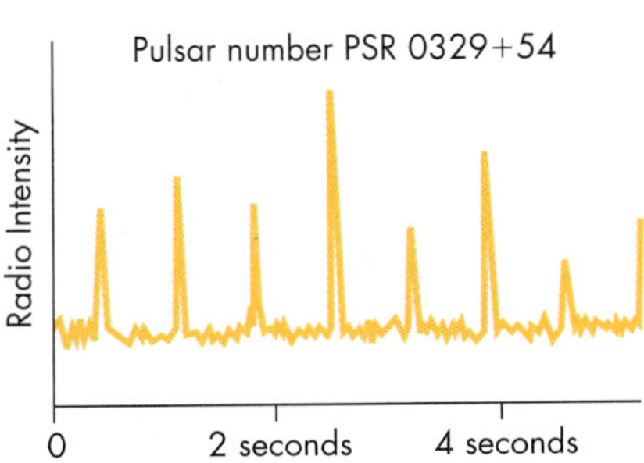

FIGURE 15.7
Pulsar signals recorded from a radio telescope.

ANIMATION
The oblique rotator model of a pulsar

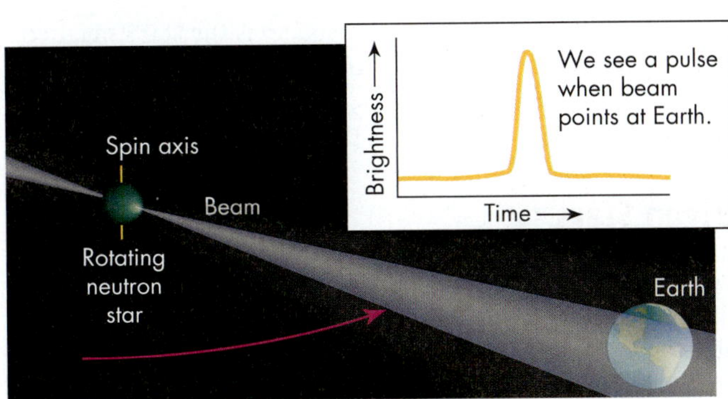

FIGURE 15.8
A pulsar's pulses are like the flashes of a lighthouse as its lamp rotates.

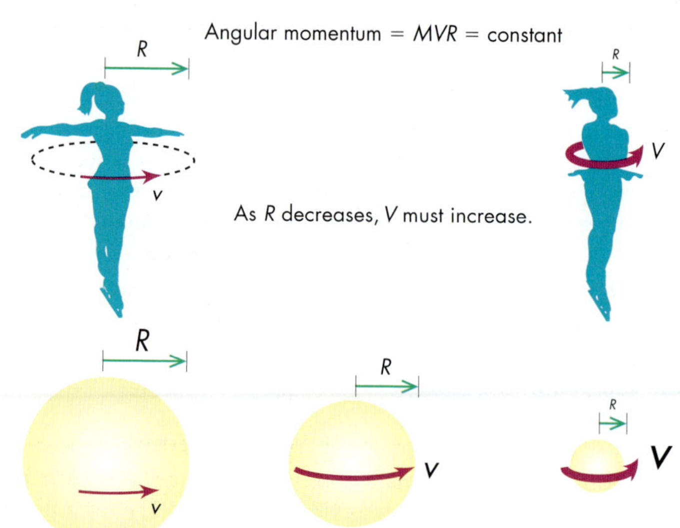

FIGURE 15.9
Conservation of angular momentum spins up a skater and a spinning star. Angular momentum is proportional to a body's mass, M, times its rotation speed, V, times its radius, R. Conservation of angular momentum requires that mathematically MVR = Constant. Thus, if a given mass changes its R, its V must also change. In particular, if R decreases, V must increase, as the ice skater demonstrates by spinning faster as she draws her arms in. Similarly, a rotating star spins faster if its radius shrinks.

supernova remnant—rotates 30 times per second. What could make a star spin that fast? For comparison, consider that the Sun spins once a month.

If pulsars are neutron stars, their rapid spin has a natural explanation. When a massive star collapses, forming a neutron star and exploding as a supernova, its core shrinks to a tiny radius—perhaps only 10 kilometers or so—and any rotation it has will be vastly increased. This rapid rotation results from the same effect that causes ice skaters to spin faster. As ice skaters draw their arms inward, an initially slow pirouette turns into a blur of rotation.

Any rotating body spins faster when it shrinks—a principle called the **conservation of angular momentum,** illustrated in figure 15.9. Angular momentum measures rotational inertia. For a spherical mass, the angular momentum is approximately given by the body's mass, M, times its equatorial rotation velocity, V, times its radius, R; or

M = Mass of an object
V = Rotation velocity at the object's equator
R = Radius of the object

$$\text{Angular momentum} = MVR$$

The principle of the conservation of angular momentum states that this product must be constant unless some force acts to brake or accelerate the spin. Therefore, when the mass, M, does not change, *if we make R smaller, V must increase* to keep the angular momentum constant. Thus, a spinning body that shrinks in size (decreasing its radius) must spin faster, and a simple calculation using this principle shows that if even a slowly spinning star such as the Sun were to shrink to the size of a neutron star, it would rotate several thousand times per second! Such a shrinkage is exactly what happens when the iron core of a massive star collapses. Thus, the rapid spin of neutron stars becomes entirely explainable.

Collapse to such a small radius has another effect: it squeezes the star's magnetic field into a denser state, amplifying the field's strength to about 1 trillion times that of the Earth's. Pulsars are thus extremely powerful magnets, and the combination of such a strong magnetic field and the rapid rotation generates the pulsar's radiation beams.

Emission from Neutron Stars

A varying magnetic field creates an electric field—a principle we use here on Earth to generate electricity by spinning magnets in dynamos. Similarly, the spin of a neutron star and its magnetic field generates powerful electric fields. These fields rip positively and negatively charged particles off the star's surface and accelerate them to nearly

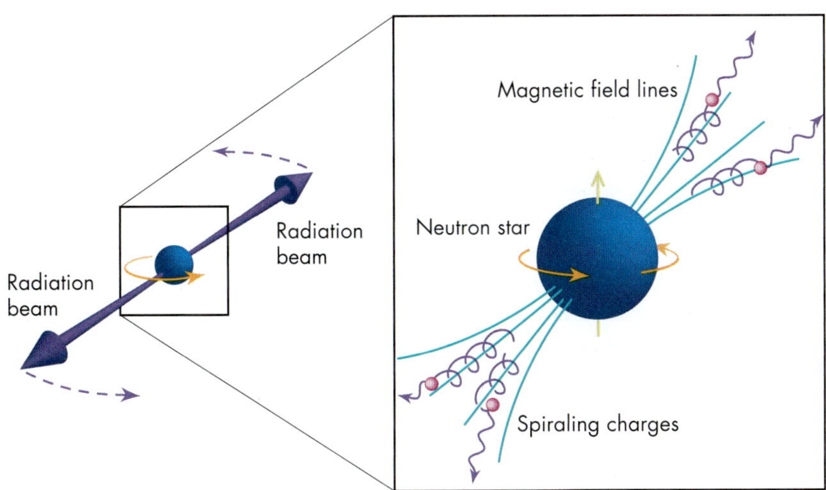

FIGURE 15.10
The strong magnetic field of a spinning pulsar generates a powerful electric field, which strips charges from its surface. The charges spiral in the star's magnetic field and emit radiation along their direction of motion. Because the charges stream in a narrow beam confined by the magnetic field, their radiation is also in a narrow beam at each magnetic pole of the star.

the speed of light. No metal wires confine the charges; rather, they are channeled by the pulsar's intense magnetic field. We saw in chapter 6 how the Earth's magnetic field directs solar particles toward the poles to create the aurora. The magnetic field of a spinning pulsar, in a like manner, generates two narrow beams of charged particles at the star's magnetic poles.

To explain how the pulsar's particle beams generate radiation, we invoke the physical law that accelerating charges radiate electromagnetic energy. A radio transmitter works according to this principle: electric currents are pulsed through the broadcast antenna, accelerating electrons in it; the accelerated electrons in turn produce the radio waves we detect. In a pulsar, too, the charges radiate as they accelerate, in this case along the magnetic field, spiraling as they go (fig. 15.10). The spiraling charges radiate electromagnetic energy over a range of wavelengths that depends on their energy. Higher-energy charges generate more energetic—and therefore shorter-wavelength—photons. The radiation is beamed because the charges are traveling along the field lines that, near the star's surface, form a tight cone. Because of this beaming, we see only those neutron stars whose sweeping beams point to Earth; those radiating in other directions are invisible to us. Thus, many more neutron stars must exist than the few hundred we have detected so far.

The emission created by the accelerating charges is called **nonthermal radiation** because its properties depend on the charges' acceleration and on the strength of the magnetic field, rather than on the temperature of a heated gas. It is also sometimes called **synchrotron radiation,** after a type of atomic particle accelerator in which such radiation was first observed on Earth. For most pulsars, the charges generate mostly low-energy radiation at radio wavelengths. However, in some very young pulsars the charges are so energetic that they produce radiation across the entire electromagnetic spectrum, including visible light and gamma rays. Figure 15.11 shows the very young pulsar in the Crab Nebula, which emits visible light flashes 30 times per second in addition to its radio waves. A high-speed television camera was used to capture the flashes caused by its rapid spin. Pulsars presumably emit some thermal radiation—ordinary visible light and X rays—from

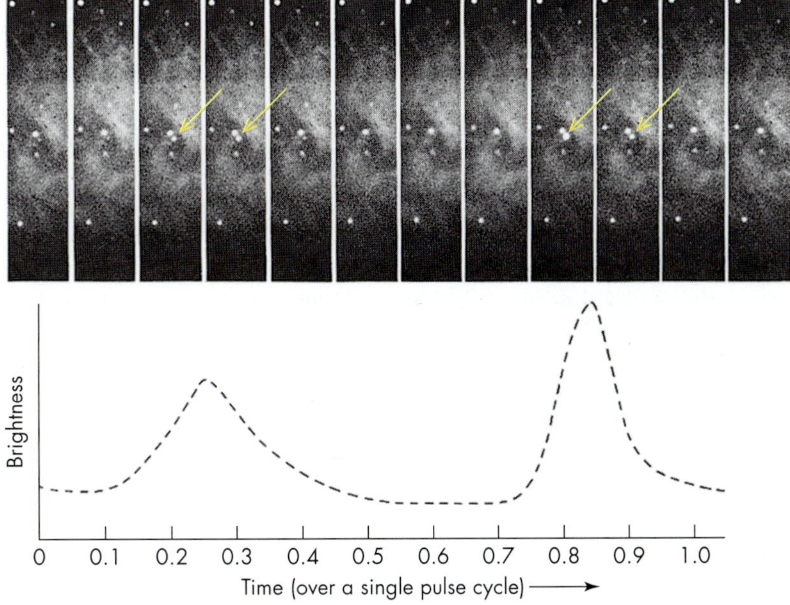

FIGURE 15.11
The rotation of the pulsar in the Crab Nebula makes the star appear to turn on and off 30 times per second as its beam of radiation sweeps across the Earth.

their hot surfaces, but because they are so tiny, that radiation is very dim and difficult to detect.

In everyday life, spinning objects slow down. Pulsars are no exception. As a pulsar spins, it drags its magnetic field through the particles that boil off its surface into the surrounding space. The magnetic field exerts a force on those particles and speeds them up. But as the particles speed up, they exert a reaction force back on the magnetic field that slows it and the pulsar down. This process is a little like what happens if you lower the beaters of a rapidly spinning electric mixer into a bowl of thick dough: the dough begins to spin, but the beater blades almost stop spinning. Astronomers measure this "spindown" of a pulsar by precisely timing the interval between pulses. Such measurements indicate that the time interval—the period of the spinning pulsar—is lengthening. The slower rotation also limits the energy of the radiation the pulsar emits. Young, rapidly spinning pulsars emit visible light and radio waves; old, slowly spinning pulsars generate only radio waves.

As the pulsar spins ever more slowly, its emission of electromagnetic radiation weakens, and ultimately the star becomes undetectable. Thus, there could be many neutron stars scattered throughout space that are invisible to us because they are no longer emitting pulses.

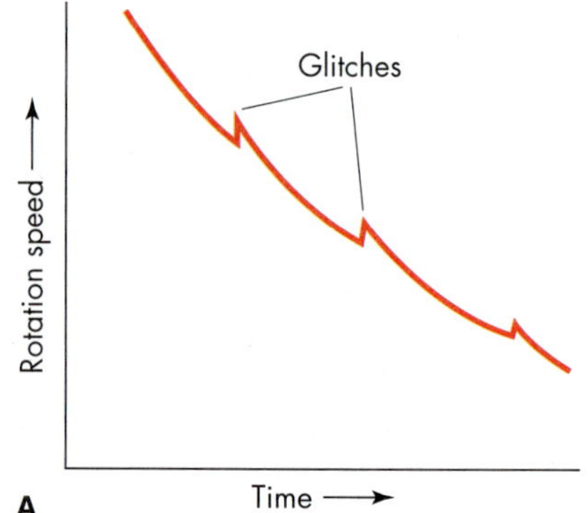

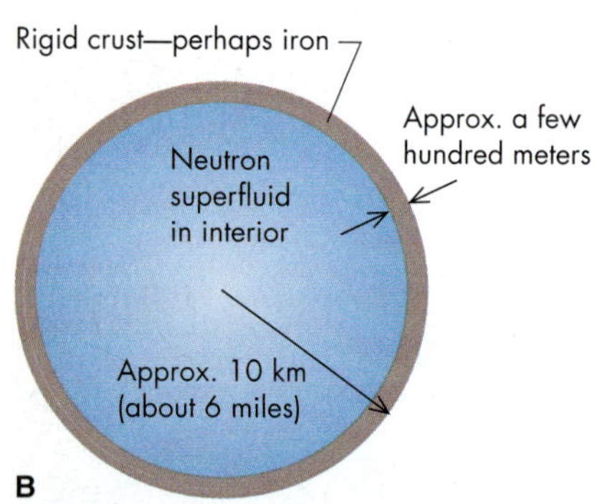

FIGURE 15.12
(A) A pulsar's rotation gradually slows down, but occasionally it will speed up briefly during a "glitch," which probably occurs when a rigid crust adjusts its shape. (B) Schematic structure of the interior of a neutron star.

Structure of Neutron Stars

Although in general pulsars are slowing down, some occasionally speed up suddenly. Such jumps in their rotation speed are called **glitches,** (fig. 15.12A) and they can tell us about the interior structure of the pulsar or neutron star. Astronomers study the structure of neutron stars with the same methods used to study ordinary stars; that is, they construct mathematical models of the neutron star. Such models are based on physical laws—for example, the requirement that, in a stable star, gravitational forces must balance pressure forces. From these models, astronomers can deduce the structure of the neutron star's interior—its density and pressure. Such calculations show that neutron stars contain at least three distinct regions: a thin, outer gaseous atmosphere about 1 millimeter thick; a thin, solid (perhaps iron) crust a few hundred meters thick, below the atmosphere; and the bulk of the star, a liquid "sea" of neutrons lying below the crust (fig. 15.12B). As the neutron star's spin slows, the bulge at the neutron star's equator decreases. Astronomers think that glitches occur when the rigid crust suddenly adjusts to the changing shape of the neutron star.

Neutron Stars in Binary Systems

Like other types of stars, some neutron stars belong to binary star systems. But such binary pairs are rare, for several reasons. First, the supernova explosion that creates the neutron star might expel so much mass that gravity is no longer strong enough to hold the pair of stars together—they rapidly escape from the debris of the explosion that spawned them. This helps explain why so few are seen in supernova remnants. Second, astronomers have calculated that the supernova blast or other processes "kick up" the neutron star to speeds of hundreds of kilometers per second, again making it difficult for its companion to hold on to it. In fact, a number of such speeding pulsars have been detected. Despite such difficulties, some neutron stars do remain bound to a companion, and to such odd couples we now turn.

X-Ray Binary Stars

Neutron stars in binary systems sometimes emit intense X-ray radiation. Such stars are called **X-ray binaries.** Astronomers have identified several types of X-ray binaries. In one type, the X-ray emission occurs in intense bursts at irregular time intervals, and so these stars are called X-ray bursters. According to one hypothesis, the X rays come from gas that falls onto the neutron star's surface, where the star's intense gravity compresses and heats it to millions of degrees. This heating triggers a thermonuclear explosion like those that cause nova outbursts. The exploding gas is so hot that most of its radiation is at X-ray wavelengths, in accordance with Wien's law (the hotter a gas, the shorter the wavelength at which it radiates). Such explosions on the surface of a neutron star emit thousands of times more energy than the Sun, an impressive achievement for a "dead" star.

In another type of X-ray binary, the X-ray emission comes in rapid, regular pulses every few seconds or less. The X-ray pulses from these stars are reminiscent of the radio pulses from ordinary pulsars, and so astronomers refer to such objects as "X-ray pulsars." An X-ray pulsar generates its emission from small "hot spots" near its magnetic poles. Gas from a companion star flows in along the field lines, much as gas particles from the Sun stream in toward Earth's magnetic poles, creating the aurora. As the gas falls toward the neutron star, the star's intense gravity accelerates the gas. As it crashes onto the neutron star at the magnetic poles (fig. 15.13), the gas is compressed and heated intensely, causing it to emit X rays. As the neutron star spins, the hot regions near each pole move into and then out of our field of view, creating the X-ray pulses that we observe.

Because pulsars generally slow down as they age, astronomers were initially puzzled when they discovered some extremely rapidly spinning—and therefore presumably very young—pulsars in several very old star clusters. Some of these seemingly young pulsars rotate approximately 1000 times per second, so that the pulses we receive from them are separated by little more than a thousandth of a second. Astronomers accordingly call them "millisecond pulsars." Millisecond pulsars are distinguished not only by their rapid rotation but also by having companion stars in most cases. In fact, astronomers think they probably all had companions at one time, for the reason outlined below.

If a neutron star has a companion that ejects mass, the neutron star may capture some of the mass by its gravitational attraction. However, the captured mass generally does not fall directly onto the neutron star. Instead, it temporarily goes into orbit around the neutron star, accumulating in a flat, spinning disk that astronomers call an **accretion disk.** The matter in the accretion disk eventually falls onto the neutron star, and when it does, the material carries some of its spinning motion—technically, angular momentum—with it. That spinning motion is imparted to the neutron star when the gas strikes its surface, making the neutron star spin faster. Thus, the usual slowdown of an aging pulsar may be reversed if it has a companion star. But what of those millisecond pulsars without companions? What made their rotations speed up? One possibility is that such pulsars originally had companions that have since disappeared. In some cases, the companion may have been lost if the pulsar passed too closely to a neighboring star. Gravitational forces from that neighbor may have pulled the pulsar's companion away. A more sinister possibility is that the pulsar destroyed its companion, a process that astronomers believe they may be witnessing in the case of the "black-widow pulsar." This peculiar pulsar has a low-mass companion star that is evaporating under the intense heating created by the pulsar. Given enough time, the companion star may totally evaporate, and future astronomers may see the millisecond pulsar without its original companion.

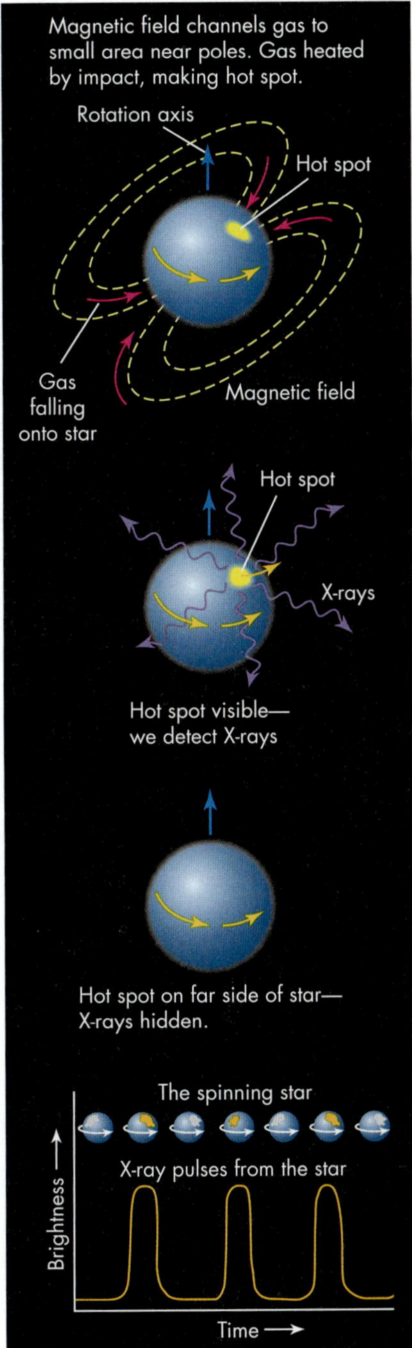

FIGURE 15.13
Gas falling onto a neutron star follows the magnetic field lines and makes a hot spot on the star's surface, creating X rays. As the star rotates, we observe X-ray pulses.

Gravitational Waves from Binary Neutron Stars

Astronomers have discovered a few neutron stars in tight orbits around each other. This unusual situation occurs when two massive stars that formed together both collapse without disrupting the binary system. The neutron stars probably spiralled close to each other during their red giant phases because of the frictional drag as they orbited within their companion's atmosphere. The intense gravity and rapid motion of these stars has

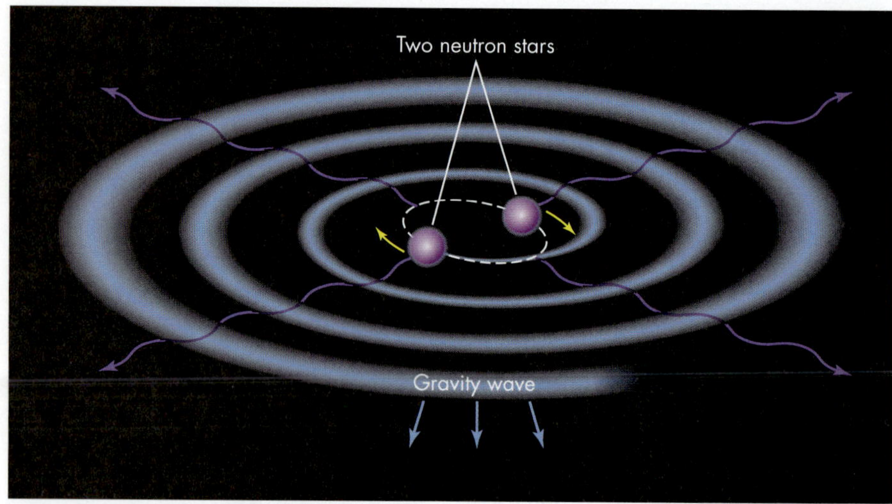

FIGURE 15.14
Gravitational waves generated by a rapidly orbiting pair of neutron stars (purple). Such waves are "ripples" in space and time—and, like ripples in water, they make objects bob (move up and down) slightly as they pass.

allowed astronomers to carry out a critical test of Albert Einstein's theory of gravity developed in 1916, known as general relativity (see essay 2). General relativity is based on the idea that gravity causes a **curvature of space.** According to general relativity, masses generate a curvature or distortion of space and time that causes objects to accelerate toward the masses.

Where gravity is weak, Einstein's theory matches the predictions of Newton's law of gravity, but intense gravitational fields produce some unexpected effects. The orbits of objects such as neutron stars can generate strong **gravitational waves.** Just as ripples spread away from a stone tossed into a pond, gravitational waves spread across space, sending waves of distortion in space and time as suggested in figure 15.14. Gravitational wave detectors are currently being built, but they are not yet sensitive enough to detect the phenomenon. Astronomical "detectors" offer a shortcut.

Early in the 1900s, Einstein predicted that rapidly orbiting stars should generate gravitational waves. These waves carry energy away from the orbiting stars, making them gradually move together. In 1974, Joseph Taylor and Russell Hulse, then at the University of Massachusetts, discovered a binary pulsar: two neutron stars in orbit around each other. From observations of the pulses, Taylor and Hulse deduced the orbital speed and separation of the neutron stars, and to their delight the speed and separation changed exactly in the manner predicted by Einstein's general theory of relativity. For this discovery, which showed that Einstein's theory of gravity was right and that gravitational waves do exist, Taylor and Hulse were awarded the 1993 Nobel Prize in Physics. These results provide strong support for the theory of general relativity, a theory that predicts even stranger objects—the existence of "rips" in the fabric of space and time called black holes.

15.3 Black Holes

When a star that was initially more massive than about 20 $M_\odot$ reaches the end of its life and collapses, it may create a compact star whose properties differ dramatically from those of white dwarfs or neutron stars. Its greater mass can compress its core so much that pressure is unable to support it, and it totally collapses to form what astronomers call a **black hole.**

Black holes have very strange properties, and to understand them, we need to review the concept of **escape velocity,** which we discussed in chapter 3. Escape velocity is the speed a mass needs to avoid being drawn back by another object's gravity. For a body of mass M and radius R, the escape velocity, V_{esc}, for an object moving away from the surface of that body is

V_{esc} = Velocity necessary to escape an object's surface
M = Object's mass
R = Object's radius

$$V_{esc} = \sqrt{\frac{2GM}{R}}$$

where G is the gravitational constant and has a value of 6.67×10^{-11} if V_{esc} is in meters per second, R is in meters, and M is in kilograms. For the Earth, the escape velocity is about 11 kilometers per second; for the Sun, it is about 600 kilometers per second. You can see from the formula that the escape velocity for a body of a given mass will be larger if the body has a small radius. The same mass squeezed to a smaller radius results in a larger surface gravity, making that body more difficult to escape. Thus, a white dwarf whose mass is the same as the Sun's has an escape velocity much larger than the Sun's because its radius is so small. In fact, because its radius is about 1% of the Sun's and because the escape velocity depends on the square root of the radius, its escape velocity is larger by a factor of $\sqrt{100}$ (=10) and equals about 6000 kilometers per second. An object leaving the surface of a white dwarf at a velocity of 6000 kilometers per second would therefore overcome the star's gravity and escape into space. The escape velocity for a neutron star, whose radius is about 10^5 times smaller than the Sun's but whose mass is comparable, is about 180,000 kilometers per second, more than half the speed of light. Thus, if a neutron star were a little smaller or a little more massive, its escape velocity would exceed the speed of light. Such an object would become a black hole.

As long ago as 1783, the English cleric John Michell discussed the possibility that there might exist objects with escape velocities exceeding the speed of light. A little more than a decade later, the French mathematician and physicist Pierre Simon Laplace entertained the same idea. Following their logic, we can calculate the radius at which a body of a given mass would become a black hole by equating its escape velocity to the speed of light. That is, $V_{esc} = c = \sqrt{2GM/R}$. Notice that the calculation involves only two quantities of the body itself: its mass and its radius.

The mathematically correct treatment of a black hole's structure using general relativity is much more difficult, but the German astrophysicist Karl Schwarzschild showed that relativity theory gives the same formula for a black hole as we derived above. Thus, we can solve for the **Schwarzschild radius,** R_S, at which a body of mass M becomes a black hole and find that

$$R_S = \frac{2GM}{c^2}$$

R_S = Schwarzschild radius—radius at which an object of a particular mass becomes a black hole
M = Object's mass
c = Speed of light

For a body of 1 solar mass, this turns out to be about 3 kilometers, or 1.9 miles. In other words, if the Sun could be compressed that small, it would become a black hole. Since the radius of a black hole is proportional to the mass, we find that the radius of any black hole is approximately $3\mathcal{M}$ kilometers, where $\mathcal{M}$ is the mass of the body in solar units.

The Nature of Space Around Black Holes

It was not until the advent of Einstein's general theory of relativity that astronomers obtained the mathematical and physical understanding to properly study black holes. As discussed in the previous section, relativity theory showed that gravity is related to the curvature of space and time. A black hole then is a place where the curvature of space and time become so extreme that motion away from the black hole becomes impossible.

An analogy may help illustrate how gravity and curvature of space are related. Imagine a waterbed on which you have placed a baseball (fig. 15.15). The baseball makes a small depression in the otherwise flat surface of the bed. If a marble is now placed near the baseball, it will roll along the curved surface into the depression. The bending of its environment made by the baseball therefore creates an "attraction" between the baseball and the marble. Now suppose we replace the baseball with a bowling ball. It will make a bigger depression and the marble will roll in farther and be moving faster as it hits the bottom. We therefore infer from the analogy that the strength of the

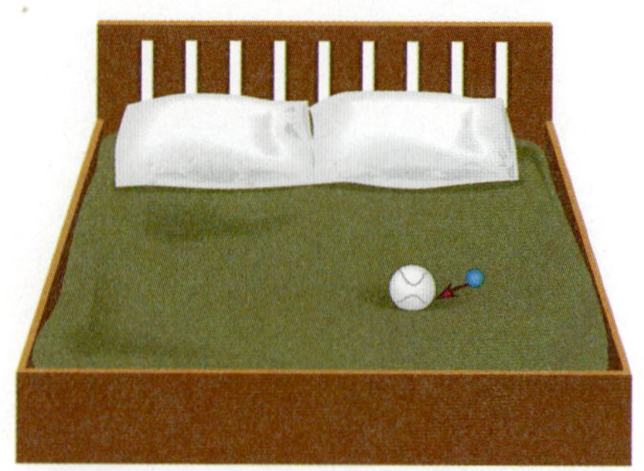

Baseball:
Marble rolls into depression

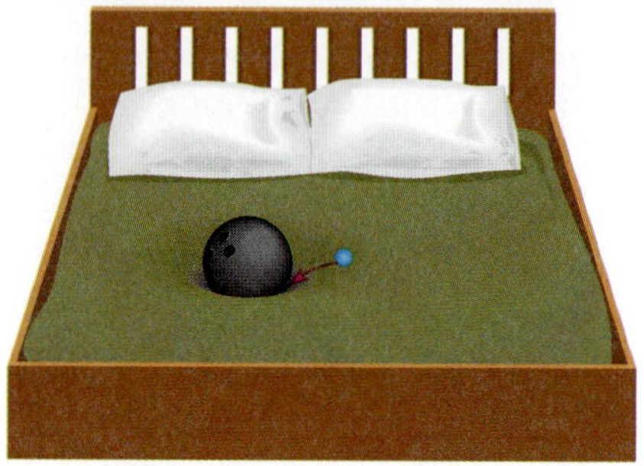

Bowling ball:
Marble rolls in faster

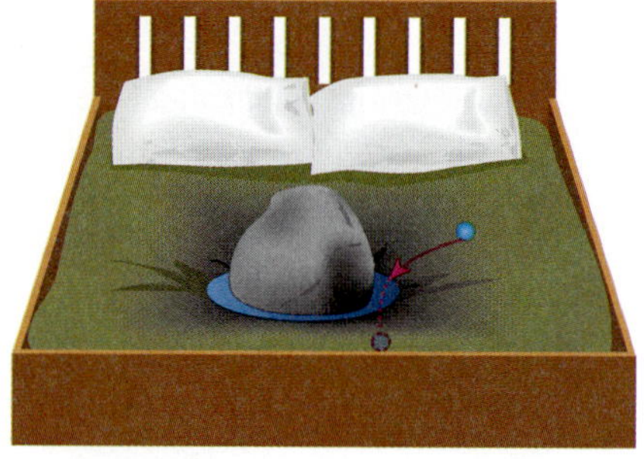

Big rock:
Marble disappears into hole

FIGURE 15.15
Objects on a waterbed make depressions analogous to the curvature of space created by a mass. According to the theory of general relativity, that curvature produces the effect of gravity. We can see the similarity by placing a marble at the edge of the depression and watching it roll inward, as if attracted to the body that makes the depression. Bigger bodies make bigger depressions, and so a marble rolls in faster. However, a very big body may tear the waterbed, creating an analog of a black hole.

attraction between the bodies depends on the amount by which the surface is curved. Gravity also behaves this way, according to the general theory of relativity. According to that theory, mass creates a curvature of space, and gravitational motion occurs as bodies move along the curvature.

We now extend our analogy to black holes by supposing that we remove the bowling ball and place a large boulder on the waterbed. This depresses the plastic even more and rips a hole in it. A marble placed on the torn surface will fall toward the boulder, drop into the hole, and disappear. Analogously, a black hole is a "rip" in space where the curvature has become so strong that the structure of space is disrupted.

All masses, including the Sun, curve space to some degree. Far from the Sun, space is "flat," nearly unaffected by the Sun's gravity. As we approach the Sun, however, the curvature increases. The path of light or radio waves is bent by gravity, so astronomers can *see* the curvature of space. In essay 2 we described how this curvature can also be thought of as an intrinsic flow of space toward massive objects. When traveling through space, then, the flow of space changes the direction that light travels, much as the direction of a boat is changed by currents in the water.*

The extreme curvature of a black hole corresponds to an inward flow of space that reaches the speed of light at the Schwarzschild radius, as illustrated in figure 15.16. Because nothing can travel through space faster than the speed of light, objects crossing this boundary cannot travel outward. Because nothing can go out, this boundary is sometimes called the **event horizon.** Just as the Earth's horizon blocks our view of what lies beyond it, so too we cannot see past the event horizon. All that happens within it is hidden forever from our view. No radiation of any sort, nor any material body—rocket, spacecraft, and so on—can break free of its gravity. Because we cannot observe the interior of a black hole even in principle, there is only a limited number of physical properties we can ascribe to it. For example, it is meaningless to ask what a black hole is made of. When it warps space inside its event horizon, we can no longer observe its interior—it could be made of neutrons, corn flakes, or gerbils. Only the amount of mass is important, not what composes it.

We can measure the mass of a black hole because the mass generates a gravitational field. In fact, at large distances, the gravitational field generated by a black hole is no different from that generated by any other body of the same mass. For example, if the Sun were suddenly to become a black hole with the same mass it has now, the Earth would continue to orbit it just as it does now: it would not be pulled in. Nor would Mercury. For something to fall into a black hole, it must approach the black hole to nearly its Schwarzschild radius (recall that this is about 3 kilometers for a solar-mass black hole). Because astronomical objects are generally so immensely far apart, the likelihood that any object will approach another to within a few kilometers is extremely small. In fact, at a given distance, a black hole exerts exactly the same gravitational force as *any* object of the same mass. That force is given (essentially) by Newton's law of gravity, just as it is for the Sun. So, we are no more likely to be pulled into a black hole than we are to be pulled into the Sun.

Black holes can make themselves felt beyond their Schwarzschild radius by more than just their gravity. They may possess electric charge and angular momentum. For example, a black hole may become charged if an excess of one kind of charge, say, positive, falls into it.

Angular momentum is much more likely to be a significant property of a black hole. Material nearly always has some angular momentum, and if it becomes a black hole, its angular momentum is retained. As a result, a spinning black hole will not be a

*This inward flow of space gives us another way of interpreting the gravitational redshift we discussed in section 15.1. Light leaving the surface of a white dwarf is redshifted by an amount that is the same as if the surface were moving away from us at the escape velocity of the white dwarf.

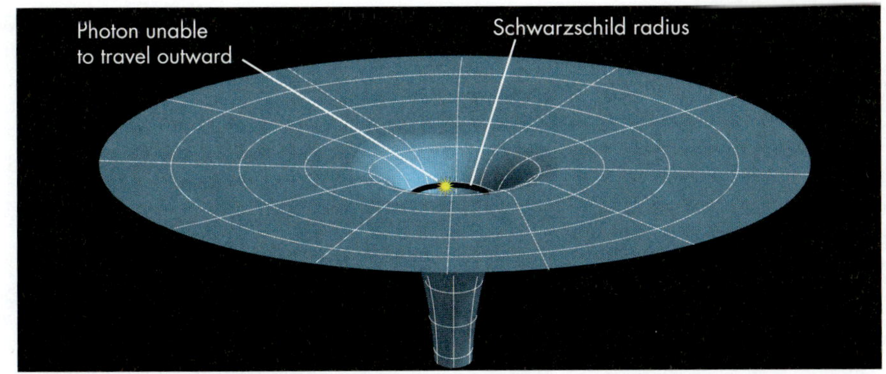
Photon unable to travel outward
Schwarzschild radius

FIGURE 15.16
Space "flows" toward masses like water down a drain. At the Schwarzschild radius, the inward flow reaches the speed of light, so nothing, not even light, can travel out from inside this radius.

sphere. The distance from its center to the event horizon is smaller along the rotation axis than it is perpendicular to it.

The Formation and Observation of Black Holes

Astronomers have excellent evidence that some supernova explosions form neutron stars, as we showed above. The step from a neutron star to a black hole does not require a breakdown of the degeneracy pressure between neutrons comparable to the way the degeneracy pressure fails in a white dwarf or a massive star's iron core. If we simply have more than about 3 $M_\odot$ of material of nuclear density, its Schwarzschild radius will exceed its size. We might imagine gently dropping matter onto a neutron star until its mass exceeded this limit. Light from the surface of the neutron star would become steadily more gravitationally redshifted, and then when the limit was exceeded, the star would disappear from sight.

In recent years, astronomers have attempted to model the collapse of very massive stars. A star that begins its life with more than about 20 $M_\odot$ will probably have a collapsing core that exceeds 3 $M_\odot$ at the end of its life. The models suggest some differences from a type II supernova for the death of these very massive stars, as illustrated in figure 15.17. The main difference is caused by the disappearance of matter straight into the black hole. In a type II supernova, much of the energy that blasts material outward comes from the impact of the collapsing material with the surface of the neutron star as well as the flood of neutrinos produced when protons and electrons combine to make neutrons. Without this energy source, the matter falls even farther in, crashing together to form a disk around the black hole, somewhat like the disks of material that collect around young stars (chapter 14). Like these young stars, extremely hot gas is ejected in jets perpendicular to the disk. These jets could explain intense bursts of gamma rays that are seen coming from distant galaxies (see the Extending Our Reach box in chapter 5) when the beams

FIGURE 15.17
Very massive stars may collapse directly into a black hole. When this happens, infalling matter may collapse to a thin disk and produce intense jets blasted out perpendicular to the disk.

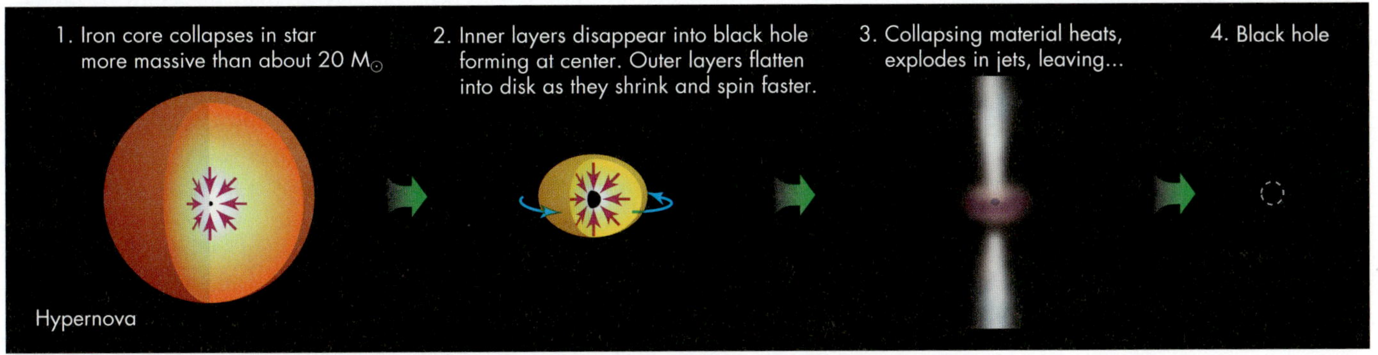
1. Iron core collapses in star more massive than about 20 $M_\odot$
2. Inner layers disappear into black hole forming at center. Outer layers flatten into disk as they shrink and spin faster.
3. Collapsing material heats, explodes in jets, leaving...
4. Black hole
Hypernova

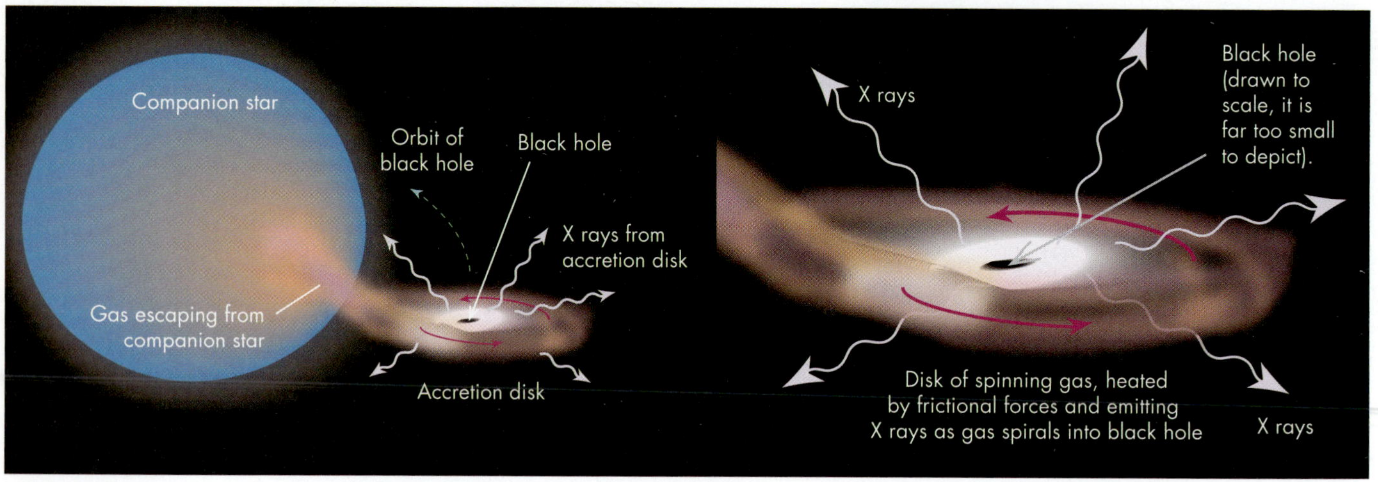

FIGURE 15.17
Black holes may reveal themselves by the X rays emitted by gas orbiting them in an accretion disk. Sizes and separations are not to scale.

happen to point toward Earth. Some astronomers have termed this type of collapse a "hypernova."

While these theoretical ideas are intriguing, it is a much more difficult matter to show by *observation* that black holes exist. An object that emits no light or other electromagnetic radiation is not easy to observe! But just as you may "see" the wind by its effect on leaves and dust, astronomers can see black holes by their effects on their surroundings.

Suppose a massive star in a binary system explodes, and its core mass is 5 $M_{\odot}$, rather than 2 $M_{\odot}$, so that the core collapses to become a black hole rather than a neutron star. Gas from the companion star may be drawn toward the hole by its gravity, as we know happens for neutron stars that are X-ray bursters. The infalling matter swirls around the black hole and forms an accretion disk whose inner edge lies just outside the hole's Schwarzschild radius, as depicted in figure 15.17. Here, where the disk orbits at nearly the speed of light, turbulence and friction heat the swirling gas to a furious 10 million K, making it emit X rays and gamma rays.

As the black hole orbits its neighbor, the X-ray-emitting gas may disappear from our view as it is eclipsed by the companion star. An X-ray telescope trained on such a star system will show a steady X-ray signal that disappears at each eclipse, as shown in figure 15.18. Such a signal might be the sign of a black hole, and astronomers have detected several X-ray sources that bear that signature. But how do we know an X-ray source is not just a neutron star?

ANIMATION
X rays from a black hole binary

Since X-ray sources are binary stars, we can measure their masses. If X-ray-emitting gas surrounds a body we cannot see but whose mass exceeds about 3 $M_{\odot}$, we can be reasonably confident that the invisible object is a black hole, because neutron stars cannot be that massive. For example, Cygnus X-1, the first X-ray source detected in the constellation Cygnus, the swan, consists of a *B* supergiant star and an invisible companion whose mass—based on an application of the modified form of Kepler's third law—is at least 15 $M_{\odot}$. Nothing we know of but a black hole can be so massive and yet invisible. An even better candidate is A0620−00, an X-ray-emitting object in the constellation Monoceros, the Unicorn. For A0620−00, the invisible star's mass is probably 16 $M_{\odot}$.

We have just described the evidence for the existence of black holes with masses just over ten times that of the Sun. Curiously, in the last decade astronomers have found convincing evidence that enormously more massive black holes—with masses millions or even billions of times that of the Sun—exist in the centers of many galaxies,

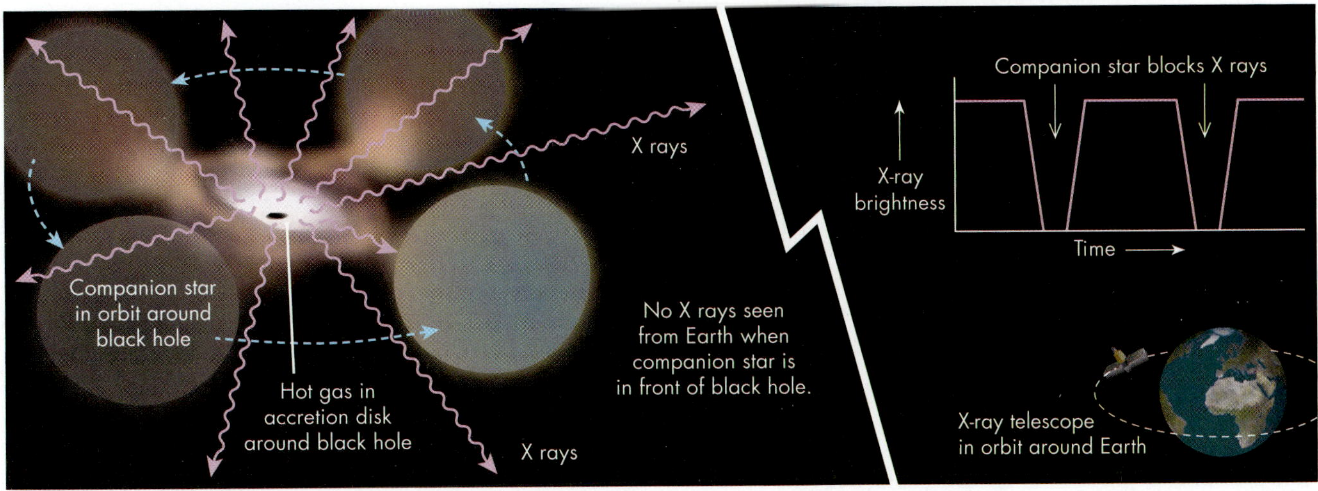

FIGURE 15.18
X-ray emission from gas around a black hole. The signal drops as the hole is eclipsed by its companion star. Sizes and separations are not to scale.

including our own. We will discuss the evidence for these supermassive black holes in chapters 16 and 17

Hawking Radiation

Up to this point, we have described black holes as intrinsically black. In 1974, however, the English mathematical physicist and cosmologist Stephen Hawking predicted that black holes can radiate! As so often happens in science, others were working on similar ideas. For example, in the previous year, the physicist Jacob Bekenstein, who was studying the relation between gravity and thermodynamics, noted that black holes can be assigned a temperature. But Hawking went further and showed that—astonishingly—a black hole emits blackbody radiation, such as we discussed in chapter 4. In fact, we can use Wien's law, which relates a body's temperature to the wavelength at which it radiates most strongly, to estimate the temperature of a black hole.

Hawking calculated that the wavelength of maximum emission is approximately 16 times the black hole's Schwarzschild radius (R_S), a result also found by Bekenstein. Wien's law states that an object's temperature is inversely proportional to the wavelength of the radiation it emits, and so the temperature of a black hole is equal to a constant divided by R_S. For a solar-mass black hole, this turns out to be about 6×10^{-8} K. This is very cold, but it is not absolute zero. Accordingly, a black hole, like any other body whose temperature is not absolute zero, emits energy in the form of electromagnetic waves, energy now known as **Hawking radiation.** However, for black holes of stellar mass, the amount of Hawking radiation emitted is far too small to be detected at this time. Nevertheless, the existence of such radiation means that black holes are not truly black.

Hawking radiation is created by quantum physical processes that allow energy to escape from black holes despite their intense gravity. Because the only source of energy available to a black hole is its mass, as a black hole "shines" (albeit very dimly), its mass must decrease. In other words, black holes must eventually "evaporate." However, the time it takes for a solar-mass black hole to disappear by "shining itself away" is *very* long: approximately 10^{67} years! This is immensely long—vastly longer than the age of the Universe—but the implications are important: even black holes evolve and "die."

SUMMARY

White dwarfs, neutron stars, and black holes are the remnants of dead stars. A white dwarf forms when a low-mass star expels its outer layers to form a planetary nebula shell and leaves its hot core exposed. The radius of a white dwarf is about the same as the radius of the Earth. Its matter is degenerate, and its mass must be less than the Chandrasekhar limit or it collapses. Having no fuel supply, a white dwarf gradually cools and dims. If a white dwarf is in a binary system, it may accrete mass from its neighbor and explode either as a nova or as a type Ia supernova.

A neutron star forms when a massive star's iron core collapses and triggers a supernova explosion. The collapse compresses the core's protons and electrons together to make neutrons, forming a ball of neutrons with a radius of a mere 10 kilometers that may contain up to about 2 to 3 solar masses. Conservation of angular momentum during its collapse accelerates its rotation rate to about 1000 times per second. This spin and the star's magnetic field generate beams of radiation that sweep across space, making the spinning neutron star a pulsar.

If the massive star's iron core contains more than about 3 $M_{\odot}$, a black hole, rather than a neutron star, is born. Gravity overwhelms the pressure forces of the core, crushing it and bending space around it so that light cannot escape. The black hole lies invisibly inside its Schwarzschild radius. If it has a companion star, however, we may be able to see X rays from an accretion disk as gas from the neighbor falls toward the hole.

QUESTIONS FOR REVIEW

1. (15.1) What are the approximate mass and radius of a white dwarf compared with those of the Sun?
2. (15.1) How does a white dwarf form?
3. (15.1) What keeps a white dwarf hot?
4. (15.1) Can a white dwarf have a mass of 10 solar masses? Why? What happens if a white dwarf increases in mass?
5. (15.1–15.22) What is meant by degeneracy pressure? How is it related to white dwarfs and neutron stars?
6. (15.1) Explain what makes a nova occur.
7. (15.2) What is a neutron star?
8. (15.2) What are the mass and radius of a typical neutron star compared with those of the Sun? Can a neutron star have a mass of 10 solar masses?
9. (15.2) How does a neutron star form?
10. (15.2) How do we observe neutron stars?
11. (15.2) What is a pulsar? Does it pulsate?
12. (15.2) Are all neutron stars pulsars? Are all pulsars neutron stars?
13. (15.2) What creates the beams of radiation seen in pulsars?
14. (15.2) What is nonthermal radiation?
15. (15.2) What happens when a gravitational wave moves? What does it affect? Compare this to how light waves move.
16. (15.3) What is a black hole? Are they truly "black"? What properties can they have?
17. (15.3) What is the Schwarzschild radius?
18. (15.3) Why might the distance to the event horizon of a black hole vary depending on which direction you measure from the center?
19. (15.3) Can astronomers see black holes? Explain.
20. (15.3) What is Hawking radiation?

THOUGHT QUESTIONS

1. (15.1) Some astronomers have suggested that cooled white dwarfs are made of diamond. Why might it be impractical to mine them?
2. (15.1) Is a white dwarf significantly different from the core of a low-mass main-sequence star? Thinking about how the speed of stellar evolution relates to the mass of the stars, and about what happens when novas occur, can you explain how you might have a binary system with an 8-solar-mass blue main-sequence star and a 1-solar-mass white dwarf?
3. (15.2) Is it surprising that a pulsar is not seen in every supernova remnant? Why?
4. (15.2) The period of a star is equal to its circumference divided by its rotational velocity. If a star collapses to $1/X$ of its previous radius, how many times shorter is the period?
5. (14.5/15.2) Explain how neutron star "pulses" are different from the "pulsing" of variable stars. Are neutron stars variable stars?
6. (15.1–15.3) White dwarfs, neutron stars, and black holes can be found in binary star systems with normal or giant companion stars. What happens in each of these systems?
7. (2.8/15.1–15.3) A 1-solar-mass black hole would be smaller than a 1-solar-mass neutron star, which in turn would be smaller than a 1-solar-mass white dwarf. If the Sun were replaced by each of these objects, what would be the impact on the Earth's orbit? Compare the conditions close to the surface or event horizon of these objects.
8. (15.3) How would you explain to an 8-year-old child that black holes may exist?
9. (15.3) Suppose you jumped into a black hole feet first. What would happen to you as your feet approached its

Schwarzschild radius? Hint: Think about tides on the Earth created by the Moon.

10. (15.3) What would be different if you were orbiting a 16-solar-mass black hole or a 16-solar-mass star at a distance of 5 AU?

PROBLEMS

1. (15.1) Calculate the density of a white dwarf star of 1 solar mass that has a radius of 10^4 kilometers.
2. (15.1–15.2) Calculate the escape velocity from a white dwarf and a neutron star. Assume that each is 1 solar mass. Let the white dwarf's radius be 10^4 kilometers and the neutron star's radius be 10 kilometers.
3. (15.2) A neutron star has a radius of about 10 km. What is the circumference of the equator of a neutron star? If you could stand on it, it would be a very visibly curved surface. Imagine walking along the neutron star's equator. How many kilometers along the circumference would correspond to moving 7° around the star? How far is 7° along the surface of the Earth (radius of 6400 km)? Seven degrees was the difference in angle Eratosthenes measured when he determined the circumference of the Earth in ancient times, a task that would be a little faster to do on a neutron star (if you could survive the high temperature and gravity!).
4. (15.2) The mass of a neutron is about 1.7×10^{-27} kg. Suppose the mass of a neutron star is about 3.4×10^{30} kg. How many neutrons does such a star contain?
5. (15.2) The volume of a neutron is about 10^{-45} cubic meters. Suppose you packed the number of neutrons you found for problem 4 (above) into a cube so that the neutrons touched edge to edge. How big would the volume of the cube be? How big across would the cube be? Hint: The volume of a cube with sides of length X is X^3. How does the cube's size compare to the size of a neutron star? What can you conclude about the spacing of neutrons in a neutron star?
6. (15.2) Very approximately, the core of a high-mass star about to collapse to form a neutron star in a type II supernova might be as large as 10^4 km and have a rotational period of around a half a day. If the core collapsed to a radius of 10 km, what would be the period of the resulting neutron star? (Use the answer to Thought Question 4). How does this compare to the periods of the pulsars discussed in the chapter?
7. (15.3) Calculate the Schwarzschild radius of the Sun.
8. (15.3) Calculate your Schwarzschild radius. How does that compare to the size of an atom? How does it compare to the size of a proton?
9. (4.2/13.2/15.3) Use Wien's law to determine the wavelength of light generated by 10 million K gas in an accretion disk. What type of photons have this wavelength?
10. (15.1–15.3) You observe a main sequence K0-type star that moves as if it is in a binary system, but no companion is visible. If the period of the system is 34 days and the semi-major axis is 0.5 AU, what is the mass of the system (remember to convert 234 days to years to use Kepler's law as discussed in section 13.4)? What is the mass of the companion (you can look up the mass of a K0 star in the Appendix, table 10)? What kind of compact star do you think the companion is? What other observational evidence would you look for to confirm this hypothesis?
11. (15.1–15.3) You observe a main sequence B5-type star that moves as if it is in a binary system, but no companion is visible. If the period of the system is 8.4 years and the semimajor axis is 8 AU, what is the mass of the system? What is the mass of the companion (you can look up the mass of a B5 star in the Appendix, table 10)? What kind of compact star do you think the companion is? What other observational evidence could you look for to confirm this hypothesis?
12. (15.1–15.3) You observe a main-sequence A0-type star that moves as if it is in a binary system, but no companion is visible. If the period of the system is 4 years and the semimajor axis is 4 AU, what is the mass of the system? What is the mass of the companion (you can look up the mass of an A0 star in the Appendix, table 10)? What kind of compact star do you think the companion is? What other observational evidence could you look for to confirm this hypothesis?

TEST YOURSELF

1. (15.1) If a quantity of hydrogen is added to a white dwarf, then: (select all that apply)
   (a) Its radius increases.
   (b) Its radius decreases.
   (c) Its density increases.
   (d) It may exceed the Chandrasekhar limit and collapse.
   (e) The hydrogen may explosively fuse to helium.
2. (15.1–15.3) For a white dwarf or a neutron star to shrink,
   (a) it must emit Hawking radiation.
   (b) some of the electrons or neutrons must gain higher energies.
   (c) material must be removed from the star through a burst event.
   (d) a black hole must develop inside the white dwarf or neutron star.
   (e) it must be losing material to a companion star.
3. (15.1–15.2) The spectrum of a supernova shows lots of iron and nickel, but no hydrogen. This is probably because
   (a) the explosion was of a white dwarf pushed over the Chandrasekhar limit.
   (b) the explosion was of an old star that had used up all its hydrogen.

(c) the hydrogen was fused into heavier elements in the explosion.
(d) the explosion was the result of core-collapse of a massive star.
(e) the shock of the explosion pushed all the hydrogen far away from the exploding stellar remnant.

4. (15.2) Which of the following has a radius (linear size) closest to that of a neutron star?
(a) The Sun
(b) The Earth
(c) A basketball
(d) A small city
(e) A gymnasium

5. (15.2) What causes the radio pulses of a pulsar?
(a) The star vibrates.
(b) As the star spins, beams of radio radiation from it sweep through space. If one of these beams points toward the Earth, we observe a pulse.
(c) The star undergoes nuclear explosions that generate radio emission.
(d) The star's dark orbiting companion periodically eclipses the radio waves emitted by the main star.
(e) A black hole near the star absorbs energy from it and re-emits it as radio pulses.

6. (15.2) In binary systems, accreting material that falls on a neutron star produces an X-ray burst while material that falls on a white dwarf produces a visible-light nova burst because
(a) the acceleration of gravity is stronger near the surface of a neutron star.
(b) neutron stars are hotter than white dwarfs.
(c) the accreting material comes from a hotter source in an X-ray binary.
(d) in a nova, the accreting matter is hydrogen, but in an X-ray burst it is iron.
(e) neutron stars rotate much faster than white dwarfs.

7. (15.3) What evidence leads astronomers to believe that they have detected black holes?
(a) They have seen tiny dark spots drift across the face of some distant stars.
(b) They have detected pulses of ultraviolet radiation coming from within black holes.
(c) They have seen X rays, perhaps from gas around a black hole, suddenly disappear as a companion star eclipses the hole.
(d) They have seen a star suddenly disappear as it was swallowed by a black hole.
(e) They have looked inside a black hole with X-ray telescopes.

8. (15.3) The Schwarzschild radius of a body is
(a) the distance from its center at which nuclear fusion ceases.
(b) the distance from its surface at which an orbiting companion will be broken apart.
(c) the maximum radius a white dwarf can have before it collapses.
(d) the maximum radius a neutron star can have before it collapses.
(e) the radius of a body at which its escape velocity equals the speed of light.

9. (15.1/15.3) Spectral lines emitted by material close to the event horizon of a black hole appear redshifted to us because
(a) blue light cannot escape from a black hole.
(b) the photons lose energy in the strong gravitational field.
(c) there is always a strong Doppler effect near the black hole.
(d) accreting material is heated and glows red.

10. (15.1–15.3) The Sun will ultimately
(a) suffer core collapse and explode in a type II supernova and become a neutron star.
(b) collapse without exploding and become a black hole.
(c) eject its outer layers and leave behind its core as a white dwarf.
(d) suffer core collapse and explode in a type II supernova and become a black hole.
(e) explode as a type Ia supernova.

KEY TERMS

accretion disk, 411
black dwarf, 403
black hole, 412
Chandrasekhar limit, 404
compact star, 401
conservation of angular momentum, 408
curvature of space, 412
degeneracy pressure, 403
escape velocity, 412
event horizon, 414
exclusion principle, 403
glitches, 410
gravitational redshift, 404
gravitational waves, 412
Hawking radiation, 417
neutron star, 406
nonthermal radiation, 409
nova, 405
pulsar, 407
Schwarzschild radius, 413
synchrotron radiation, 409
type Ia supernova, 405
type II supernova, 406
white dwarf, 402
X-ray binary, 410

FURTHER EXPLORATIONS

Bethe, Hans A., and Gerald Brown. "How a Supernova Explodes." *Scientific American* 252 (May 1985): 60.

Bernstein, Jeremy. "The Reluctant Father of Black Holes." *Scientific American* 274 (June 1996): 80. (About Einstein.)

Charles, Philip A., and R. Mark Wagner. "Black Holes in Binary Stars: Weighing the Evidence." *Sky and Telescope* 91 (May 1996): 38.

Drieger, Walter C. "The Care and Feeding of Black Holes." *Astronomy* 23 (May 1995: 19). (A satire.)

Greenstein, George. *Frozen Star.* New York: Freundlich Books, 1984.

Herbst, William, and George C. Assousa. "Supernovas and Star Formation." *Scientific American* 241 (August 1979): 138.

Hillebrandt, Wolfgang, Hans-Thomas Janka, and Ewald Muller. "How to Blow Up a Star." *Scientific American* 295 (October 2006): 42.

Jastrow, Robert. *Red Giants and White Dwarfs.* New York: W. W. Norton, 1990.

Kawaler, Steven D., and Donald E. Winget. "White Dwarfs: Fossil Stars." *Sky and Telescope* 74 (August 1987): 132.

Kirshner, Robert P. "The Earth's Elements." *Scientific American 271 (October 1994): 58.*

Lewin, Walter H. G. "The Sources of Celestial X-ray Bursts." *Scientific American* 244 (May 1981): 72.

Murdin, Paul, and Lesley Murdin. *Supernovae.* New York: Cambridge University Press, 1985.

Nadis, Steve. "On the hunt for magnetars." *Astronomy* 34 (November 2006): 32.

Piran, Tsvi. "Binary Neutron Stars." *Scientific American* 272 (May 1995): 52.

Ruthen, Russell. "Catching the Wave." *Scientific American* 266 (March 1992): 90.

Starrfield, Sumner, and Steven N. Shore. "The Birth and Death of Nova V1974 Cygni." *Scientific American* 272 (January 1995): 76.

Trimble, Virginia. "Neutron Stars and Black Holes in Binary Systems." *Contemporary Physics* 32 (March/April 1991): 103.

______. "White Dwarfs: The Once and Future Suns." *Sky and Telescope* 72 (October 1986): 348.

Weisberg, Joel M., Joseph H. Taylor, and Lee A. Fowler. "Gravitational Waves from Orbiting Pulsars." *Scientific American* 245 (October 1981): 74.

Williams, Robert E. "The Shells of Novas." *Scientific American* 244 (April 1981): 120.

Zimmerman, Robert. "The Life and Death of Super Suns." *Astronomy* 36 (July 2008): 34.

Website

Visit the *Explorations* website at **http://www.mhhe.com/arny** for additional online resources on these topics.

Q FIGURE QUESTION ANSWERS

WHAT IS THIS? (chapter opening): This image shows two pictures (made with the Hubble Space Telescope) of a nova, an exploding star. The pictures were taken 7½ months apart. You can clearly see that the shell of gas ejected by the nova has expanded

FIGURE 15.7: Count the peaks in a time interval of, say, 1 or 2 seconds. Then divide the number of peaks by the time interval. The left one spins 9 times per second. The right one spins 3 times in 2 seconds for a spin rate of 1.5 times per second. If pulses were visible from both poles, the number of peaks should be first divided by 2.

PROJECTS

Modeling Curved Space: You can get a bit of insight into distorted space by building a model like the waterbed example in the book. The model will be of a two-dimensional universe (why?). To make a stretchy two dimensional space-time, you will need about a square yard of a stretchy fabric. Attach this with clothespins, staples or other means across the top of a large, sturdy cardboard box at least two feet per side (tuck the flaps inside to make it sturdy). Be careful not to start any rips in the fabric. You could also mount the fabric on a large, old picture frame or on a hula hoop, but you'll need a way to hold up the frame or hoop sturdily. The flat fabric represents a two-dimensional universe with no mass—space is perfectly flat. Try rolling a marble across the fabric from one side to the other—does it travel more or less in a straight line? If we add an object with mass, space will stretch around the object. Create a circle with a piece of string, and compare the radius measured across it with what you get dividing the circumference by 2π. Now place an orange or apple on the fabric. Try rolling the marble across at different distances from the orange. How is its path affected? Try setting the marble in motion around the orange. Can you create elliptical orbits? Can you get the marble going in a circular orbit? Friction with the fabric will slow the marble. Pick up the orange and place the end of a fabric tape measure or a string under the center of the orange and run it straight out along the surface. Using another piece of string, measure the circumference of a circle about an inch or so from the orange (somewhere where the fabric appears stretched). If you divide this circumference by 2π, is the apparent radius of the circle the same as how far it is from the end of the tape measure to where your circle intersects it? Space has been stretched—made longer—in the direction toward the mass. Replace the orange with a heavier object (you can put something heavy in a bowl) and try the experiments again.

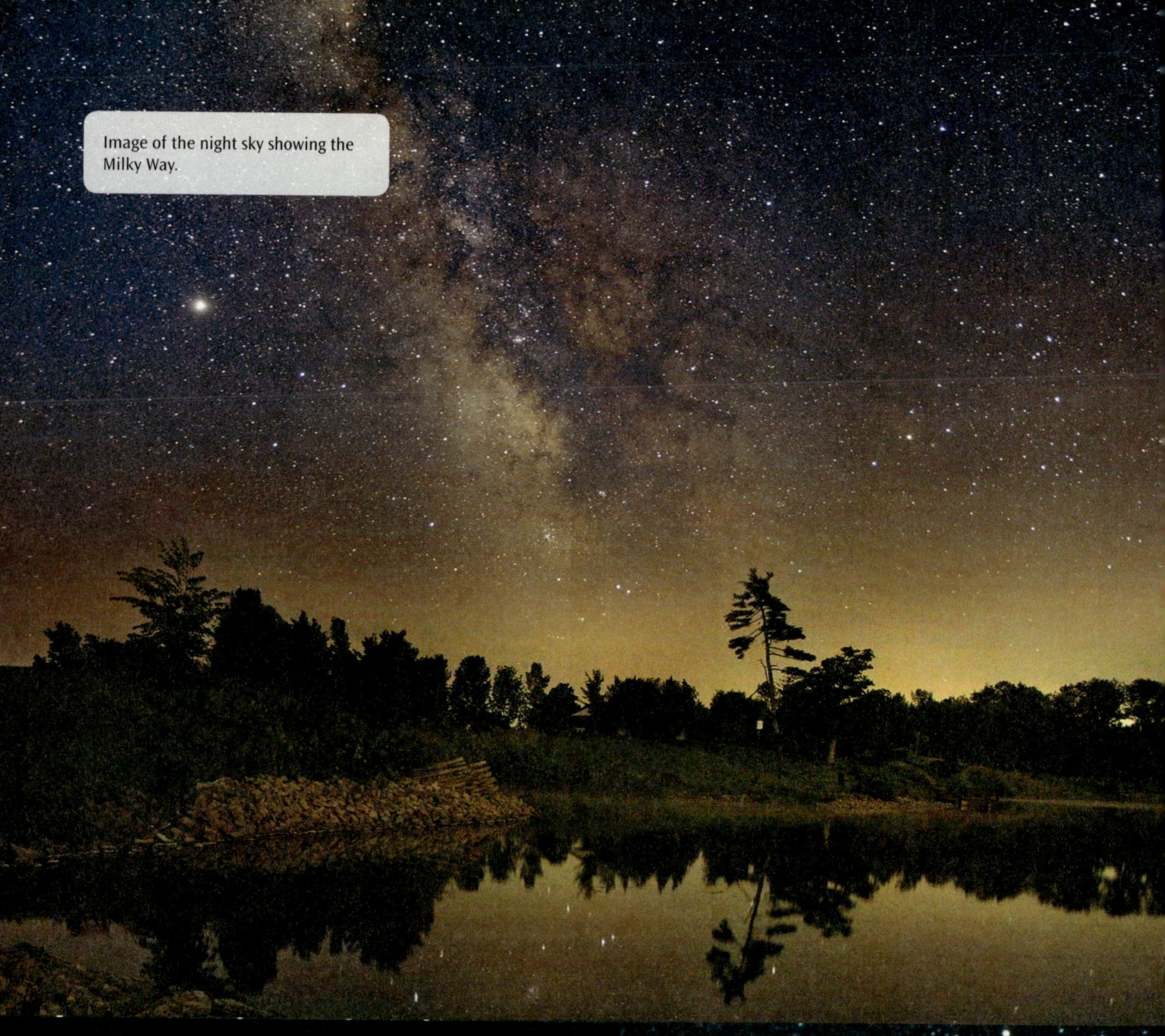

Image of the night sky showing the Milky Way.

KEY CONCEPTS

- We live in the Milky Way Galaxy.
- It contains about 100 billion stars (in addition to the Sun), which orbit within it.
- Two main populations of stars occupy the Milky Way:
  - Old red stars in the bulge and halo
  - Young, often blue, stars in the disk
- These populations probably formed in two separate stages as gas gravitationally collapsed to form our Galaxy.
- The Milky Way is about 100,000 light-years across.
- The Milky Way contains clouds of gas and dust in the immense spaces between the stars.
- Interstellar gas is most easily seen when it is heated by stars and emits light through spectral lines; interstellar dust blocks and reflects starlight.
- The Milky Way contains a large proportion of dark matter whose nature is unknown.
- Dark matter exerts a gravitational force on the stars and gas in our Galaxy, but it emits no as-yet detectable radiation.
- A black hole of more than a million times the Sun's mass lies at the center of the Milky Way.
- Mergers and interactions with other galaxies play an important role in the evolution of the Milky Way.

16 The Milky Way Galaxy

CONCEPTS AND SKILLS TO REVIEW

- Inverse-square law (13.2)
- Standard-candle method (13.1, 13.8)
- Newton's law of gravity (3.5)
- Modified form of Kepler's third law (3.7)
- Red light has a longer wavelength than blue (4.1)
- Red stars are cooler than blue stars (4.3, 13.2)

On a clear, moonless night, far from city lights, you can see a pale band of light spangled with stars stretching across the sky as shown in the chapter-opening image. The ancient Hindus thought this shimmering river of light in the heavens was the source of the sacred river Ganges. To the ancient Greeks, this dim celestial glow looked like milk spilled across the night sky, and so they called it the **Milky Way.** In the seventeenth century, Galileo showed that the Milky Way is millions of stars too dim to be seen as individual points of light. Now, in the twenty-first century, we know that these stars, along with our Sun, form a huge, slowly spinning disk—our Galaxy. The word *galaxy* itself comes from the ancient Greeks and their word for "milk"—*galactos*. Thus, "Milky Way" is both the name of the band of light across the night sky and the name of our Galaxy.

A view of the Milky Way on a clear, dark night is one of nature's finest spectacles. Superposed on the dim background glow are most of the bright stars and star clusters that we can see, which all belong to our Galaxy. Here and there, however, dark blotches interrupt the glowing backdrop of stars. Ancient Incan astronomers observing the Milky Way from their temple observatories in the Andes gave these dark areas names, just as peoples of the classical world named the star groups. Today, we know the dark regions are clouds of dust and gas that give birth to new stars.

Q: WHAT IS THIS? See end of chapter for the answer

Q. What do you think the whitish blobs along the horizon at the bottom are?

The first recorded use of "Milky Way" and "Galaxy" in English are found in Geoffrey Chaucer's 1380 poem "House of Fame":

See yonder, lo, the Galaxyë
Which men clepeth the Milky Wey,
For hit is whyt.

FIGURE 16.1
Wide-angle photo of a portion of the Milky Way taken from Mount Graham in Arizona. The dark ring around the edge of the picture is the horizon.

16.1 Discovering the Milky Way

ANIMATION
The Milky Way

Shape of the Milky Way

Our understanding of the Milky Way as a star system dates to the eighteenth century, when Thomas Wright, an English astronomer, and Immanuel Kant, the great German philosopher, independently suggested that the Milky Way is a flattened swarm of stars. They argued that if the Solar System were near the center of a spherical cloud of stars, we would see roughly the same number of stars in all directions. Instead, we see vastly more stars in the direction of the Milky Way than in directions perpendicular to it, as shown in the "fish-eye" photograph in figure 16.1. This is what we expect to see if the Milky Way is a disk rather than a ball of stars, as illustrated in figure 16.2.

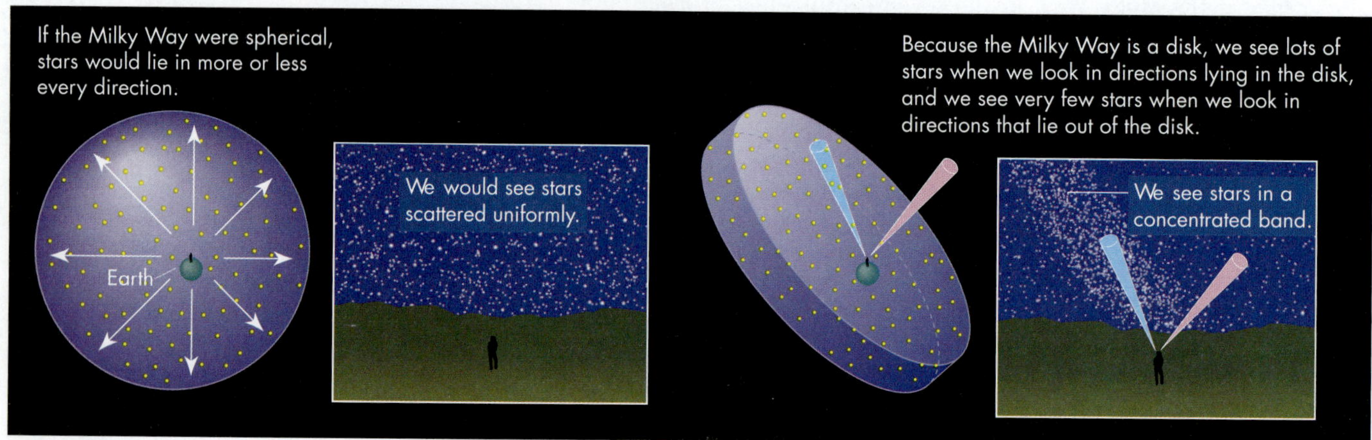

If the Milky Way were spherical, stars would lie in more or less every direction.
Earth
We would see stars scattered uniformly.
Because the Milky Way is a disk, we see lots of stars when we look in directions lying in the disk, and we see very few stars when we look in directions that lie out of the disk.
We see stars in a concentrated band.

FIGURE 16.2
If the Milky Way had a spherical shape, we would see about the same number of stars in every direction. However, because it is a disk, we see stars concentrated into a band around the sky.

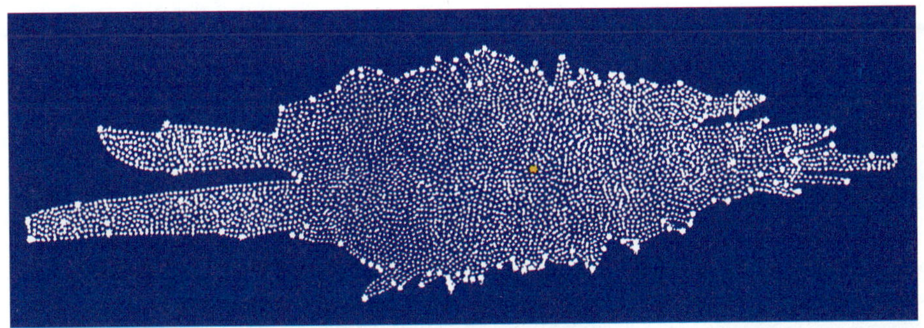

FIGURE 16.3
Herschel's 1784 sketch of the Milky Way. He based this cross-sectional view on the number of stars he could see in different directions. He correctly deduced that the Milky Way was wider than it was thick, but *incorrectly* concluded that the Sun (the orange dot) is near the Milky Way's center.

A simple analogy may help you better understand why we see the Milky Way as a band on the sky. When you are sitting in a classroom, if you look at the heads of other students throughout the room, they all lie roughly in a band that encircles you. You see no heads when you look up toward the ceiling or down toward the floor, but you see lots of heads lying in a plane roughly parallel to the floor of the classroom. So, too, it is with us and stars in the disk of the Milky Way.

These arguments were refined toward the end of the eighteenth century by the English astronomer Sir William Herschel, who counted the number of stars visible in different directions, and deduced that the disk was about five times wider than it was thick. Moreover, because he saw roughly the same number of stars in all directions around the disk, he concluded that the Sun must lie near its middle, as shown in figure 16.3. Based on the brightness of the dimmest stars he could see, Herschel estimated that the Milky was about 8000 light-years, or 2500 parsecs,* across. However, this was a very uncertain estimate because stellar parallaxes were not detected until the next century.

Size of the Milky Way

The size of the Milky Way was remeasured with much more precision in the early 1900s by the Dutch astronomer Jacobus C. Kapteyn (pronounced *cap-tine*) and the American astronomer Harlow Shapley. Like Herschel, Kapteyn concluded that the Milky Way is a disklike system, though about 18,000 parsecs in diameter, with the Sun located near the center.

Almost immediately after Kapteyn's publication of his model of the Milky Way, Shapley published a model in strong disagreement with it. Shapley argued that the Milky Way is larger than Kapteyn believed and that the Sun is not near the center but about two-thirds of the way out in the disk. Shapley drew his conclusions from a study of globular star clusters—dense groupings of up to a million stars, such as the one shown in figure 16.4. Because these clusters contain so many stars, they are very luminous and can be seen at large distances—across the Galaxy and beyond. Moreover, many of them lie above the disk, and so we have a clear view of them.

Shapley noticed that the globular clusters are not scattered uniformly across the whole sky but are concentrated in the same place that the stars are concentrated, that is, toward the constellation Sagittarius (see Looking Up #7 in the front matter for a

Kapteyn measured the size of the Milky Way by first finding distances to nearby stars from their parallax. He then estimated the distance to more remote stars from their motion, based on the following idea: As stars move in their orbits around the Galaxy, their positions in the sky change very slowly. Although such motion is not noticeable to the naked eye, it is detectable when pictures taken many years apart are compared. Kapteyn realized that nearby stars change their position faster than more distant stars, just as a ball tossed past your head moves across your field of view more rapidly than one thrown at the far end of a playing field. Using such motions for remote stars and comparing them with the motion of nearby stars whose distance he knew from parallax, he estimated the distance to the remote stars and thus the size of the Milky Way.

*In this and the remaining chapters, we will be dealing with huge objects at immense distances. Perhaps a quick review of some units and powers-of-ten notation is in order.

Astronomers measure large distances in either light-years (ly) or parsecs (pc). One parsec is 3.26 light-years and is about 3×10^{13} (30 trillion) kilometers. Recall that 10^{13} stands for 1 followed by 13 zeros.

Galaxy dimensions are so huge that even parsecs are inappropriate for measuring their size, and so astronomers often use kiloparsecs (kpc) for that purpose. One kiloparsec = 1000 parsecs. Later, we will find that an even larger unit, the megaparsec (Mpc) is a more appropriate unit for measuring the distances to galaxies.

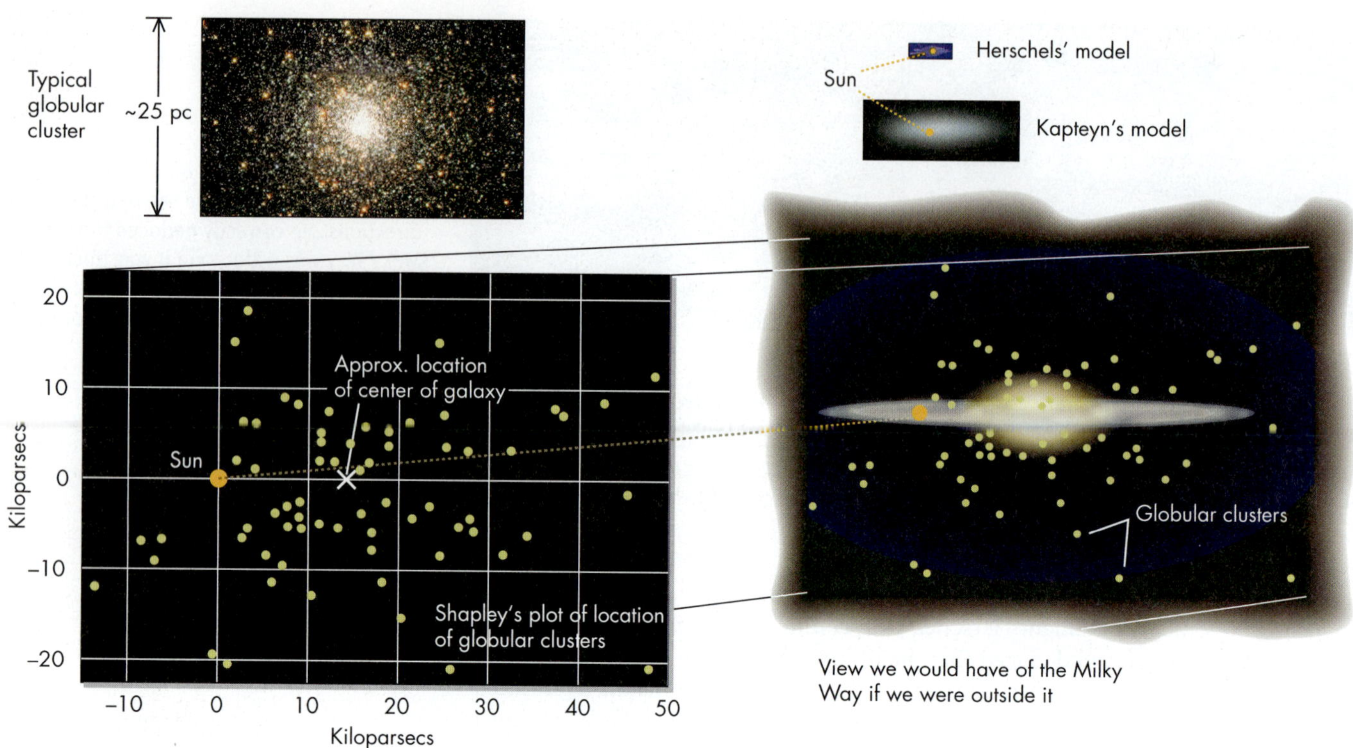

FIGURE 16.4
Schematic version of Shapley's plot of globular clusters from which he inferred the size of the Milky Way and the Sun's location in it. Notice that the clusters fill a roughly elliptical region and that the Sun is *not* at the center of their distribution. An image of a typical globular cluster is shown at top, and Kapteyn's and Herschel's models are shown to scale.

picture of this region). He hoped, therefore, that a detailed map of the distribution of the globular clusters would show him the shape of the Milky Way.

To make his map of the clusters, Shapley needed to know their distance from the Sun. He found the distances to the clusters by observing the variable stars in them. These stars obey a law that relates their luminosity to how long it takes them to brighten, dim, and brighten again (their period) that had just recently been discovered by Henrietta Leavitt. From the luminosity of the stars, their apparent brightness, and the inverse-square law, Shapley calculated the distances to the clusters they were in. Then, having found the distance to each cluster, Shapley plotted where they lay in the Milky Way. Figure 16.4 shows his results.

The clusters fill a roughly elliptical region about 100,000 parsecs (roughly 300,000 light-years) across. Moreover, the Sun lies *not* at the middle of the region, but about two-thirds of the way from its center. Shapley therefore concluded that our Galaxy is roughly 100,000 parsecs in diameter and that the Sun is nearer to its edge than to its center (fig. 16.4).

The great difference between Shapley's value for the diameter of the Milky Way and the Sun's location and Kapteyn's finding created a major controversy among astronomers. Like so many other controversies, both sides were partially right and partially wrong. For example, Kapteyn deduced a size for the Milky Way that was far too small because he assumed that space was transparent: he did not recognize the dimming effect of interstellar dust. On the other hand, Shapley was correct about the Sun not being at the center, but he greatly overestimated the diameter of the Milky Way, finding a value about three times larger than the one currently accepted. Part of the reason for his error was that Shapley also, like Kapteyn, did not recognize the dimming effect of interstellar dust. Thus, Shapley mistook the dimness of the globular clusters to mean that they were very far away. Moreover, he did not realize that he was dealing with two types of pulsating variables: RR Lyrae stars and Cepheids. Leavitt had determined the period–luminosity law for Cepheids, but Shapley was measuring the Galaxy's size with RR Lyrae stars. Because RR Lyrae stars are much dimmer than Cepheids, Shapley thought they were much farther away than they really are. Thus, he overestimated the size of the Milky Way. Kapteyn came closer to the presently accepted size of our Galaxy, but he was wrong about where the Sun lies in it.

Today, astronomers have revised and improved Shapley's model of the Milky Way. Much of that improvement comes from studies of other galaxies. For example, many external galaxies are flat disks with conspicuous spiral arms, as illustrated in figure 16.5. Such arms are extremely hard to detect in our own Galaxy because from our location in the Milky Way we cannot get an overview of its disk. Nevertheless, given that the Milky Way is a flat disk and that other disk galaxies have spiral arms, it is reasonable to deduce that ours does too. Thus, by combining observations of our own Galaxy with the general features known to occur in other galaxies, astronomers can assemble a detailed model for the Milky Way.

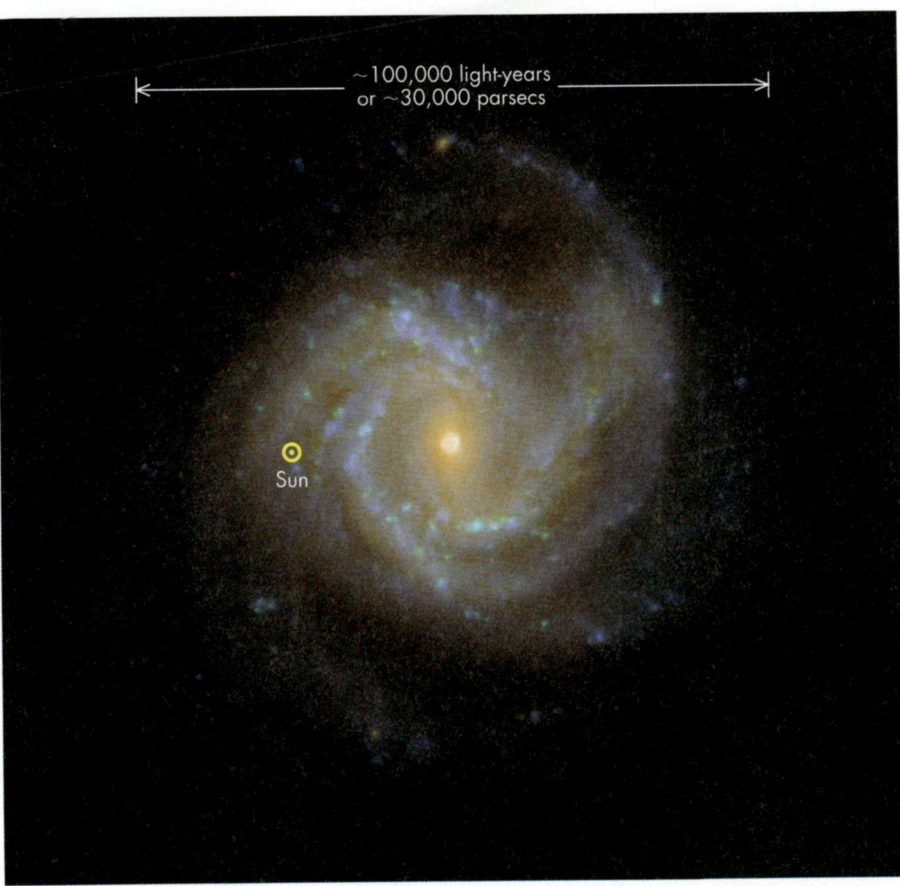

FIGURE 16.5
Image of the spiral galaxy NGC 4303 (also known as M61), thought to be quite similar to the Milky Way. A position corresponding to the Sun's distance from the center is also indicated.

16.2 Overview of the Milky Way

Content and Structure

Mingled with the stars of the Milky Way are huge clouds of gas and dust that amount to about 15% of the mass of stars. We can see some of these clouds by the visible light they emit, whereas others reveal themselves by blocking the light of background stars, as can be seen in figure 16.6. In fact, dust scattered throughout the Galaxy prevents

FIGURE 16.6
Clouds of gas and dust in the Milky Way. Some hot clouds glow pink with hydrogen emission, the dust in some cold clouds can be seen silhouetted against background stars.

Top view

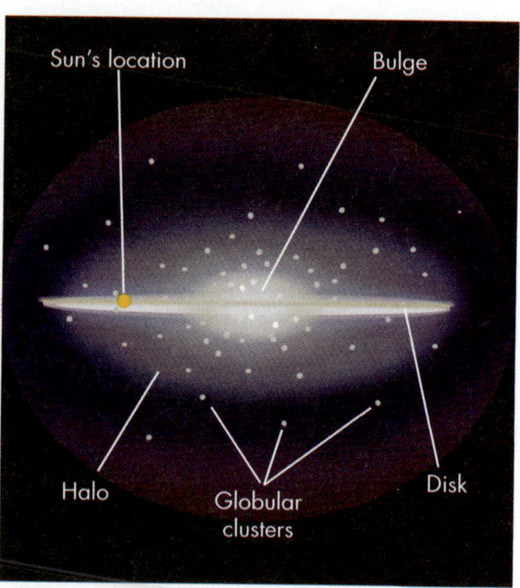

Side view

FIGURE 16.7
Artist's sketch of the Milky Way showing top and side views, illustrating the disk, bulge, and halo. Notice how thin the disk is and how the halo surrounds the disk much as a bun surrounds a hamburger. Because the Milky Way's stars gradually thin out and do not just stop at some distance, its size is only labeled approximately. The size of the dark matter halo in which the Milky Way is embedded is very uncertain, so it is shown only schematically. It is omitted from the top view for clarity.

us from seeing farther than about 3000 parsecs within the disk. Thus, we cannot see our **Galaxy's nucleus**—its core—at visible wavelengths. However, radio, infrared, and X-ray telescopes can "see" through the dust and reveal that the core of our Galaxy contains a dense swarm of stars and gas as well as a massive black hole.

The Milky Way's structure has three main parts, as illustrated in figure 16.7: a **disk** about 30 kiloparsecs (100,000 light-years) in diameter, a **bulge** at its center, and a **halo** that extends out to large distances. Most of the gas and dust clouds lie within the disk, along with numerous bright young stars that gather into **spiral arms** that wind outward from near the center. Our Solar System lies on the inner edge of a spiral arm about 8 kiloparsecs (roughly 26,000 light-years) from the center. The orbital plane of the planets in the Solar System is tilted by about 60° with respect to the Galaxy's disk. This tilt is the reason the band of the Milky Way on the sky is tipped with respect to the ecliptic.

Because the central bulge of our Galaxy is elongated, the Milky Way should be described as a "barred" spiral (see chapter 17).

The bulge at the center of the Milky Way is a dense concentration of stars primarily. The bulge is somewhat flattened and elongated, like the shape of a flattened football. The halo of the Galaxy contains the globular clusters, scattered stars, and hot gas, and is more spherically distributed. It extends from the center of the Galaxy outward, reaching out beyond where the disk can be detected. You may find the following analogy helpful in visualizing the scale of our Galaxy: if the Milky Way were the size of the Earth, the Solar System would be the size of a cookie.

The gravity of all the components holds our Galaxy together—but that gravity would draw everything inward and collapse the Milky Way, were it not for the orbital motion of the stars and gas. Astronomers have measured that motion for our Sun and neighboring stars: the Sun and its neighbors move around the center of the Milky Way at a speed of about 220 kilometers per second. That motion around the center is what originally gave the disk its flattened shape, much as the rotation of the presolar nebula made our Solar System flat. Moreover, just as planets near the Sun complete their orbits faster than planets farther out, the stars near the center of the Galaxy complete their orbits faster than stars at the edge of the Galaxy. Thus, the inner parts of the Galaxy spin faster than the outer parts, a phenomenon called "differential rotation." In section 16.5 we will see that this plays an important role in making the Galaxy's spiral arms.

The Sun's large orbital speed (roughly 10 times the speed of the Earth around the Sun) might make you think the Milky Way spins rapidly, but because our Galaxy is so huge, the Sun takes approximately 220 million years to complete one trip around it. Since the death of the dinosaurs about 65 million years ago, our Galaxy has traveled only about one-quarter of its orbit.

Within this swirling mass of stars, the Sun is nearly lost—like a speck of sand on a beach. But unlike the close packing of grains of sand, stars are widely separated. For example, near the Sun and throughout most of the disk, stars are typically several light-years apart: the Sun's nearest neighbor is 4.2 light-years away, a separation relative to its size like that of grains of sand 15 miles apart. Near the core of the Milky Way, stars are packed far more densely, with a separation roughly 1500 times less than that near the Sun, a separation like grains of sand at opposite ends of a football field. The density of stars, as measured by the number of stars per cubic light-year, is probably more than 10 million stars per cubic light-year near the core but only about 0.003 stars per cubic light-year near the Sun. Near the edge of the disk, stars are spread even more thinly, much as our atmosphere thins out as it merges into space. Thus, the Milky Way has no sharply defined outer edge.

Mass of the Milky Way and the Number of Stars

Astronomers can calculate the Milky Way's mass from the gravitational attraction needed to hold its stars in orbit. Such calculations show that the Milky Way must contain at least 10^{11} $M_{\odot}$ of matter. However, it must contain even more matter, because its gravitational effect on nearby galaxies implies a mass as large as 2×10^{12} $M_{\odot}$. This large mass is puzzling, because when astronomers add up the mass of the Milky Way's *visible* stars and gas, they find only about one-tenth the amount of mass needed to hold our Galaxy together. The other nine-tenths is undetectable except by its gravitational effect. Thus, astronomers conclude that the major part of our Galaxy is composed of some unseen material, which they therefore call **dark matter.**

Because dark matter emits no radiation of any sort, it is very difficult to study. Its composition is one of the major mysteries in astronomy today, and astronomers know few of its properties. They do know that dark matter is spread throughout our Galaxy within the halo. They also know that it does not interact with ordinary stars and gas except through its gravitational pull. Finally, as we will see in chapter 17, astronomers know that the Milky Way is not alone in having a massive dark-matter halo: other galaxies are also embedded in dark-matter haloes of their own.

Our uncertainty about the nature of dark matter and the value of our Galaxy's mass makes it difficult to determine the number of stars our Galaxy contains. However, if we assume that the average star in our Galaxy has a mass comparable to the Sun's (an assumption based on censusing stars, as we will see in section 16.3), then from the best estimates of the Galaxy's total mass, we deduce that it contains about 10^{11} (100 billion) stars, each moving along its own orbit, held in its path by the collective gravity of the other stars, gas, and dark matter.

Age of the Milky Way

The immensity of the Milky Way's size and number of stars is matched by the Milky Way's great age. Astronomers measure this age from the age of our Galaxy's oldest stars, using the techniques for measuring stellar ages that we discussed in chapter 14. Such measurements (and others) show that the oldest stars in our Galaxy have an age of approximately 13 billion years, an age we will take as that of the Milky Way itself.

Our Galaxy's stars were not all created at a single time in that distant past. As we discussed in chapter 14, stars are forming even today and astronomers can estimate the average number of stars born each year by dividing the total number of stars in the Galaxy by the Galaxy's age. This gives a birthrate of approximately 7 to 8 stars born in the Milky Way each year, although stars probably were born at a much higher rate when our Galaxy first formed. Given this rate and the amount of interstellar gas remaining in our Galaxy, we can be confident that our Galaxy will continue to shine for many billions of years into the future.

16.3 Stars of the Milky Way

Stellar Censuses

The Milky Way contains many types of stars: giants and dwarfs, hot and cold, young and old. Stellar censuses, which list all known stars of a given type in a given region of space, show that all the types described in chapters 13 to 15 occur in our Galaxy. By analyzing the types of stars and their relative numbers, astronomers can deduce some of the ancient history of our Galaxy, much as archeologists can learn about life in an ancient city from the kinds of buildings it contained. Such studies of the stellar contents of the Milky Way show that stars similar to or smaller than the Sun vastly outnumber more-massive stars.

The most numerous stars in our Galaxy turn out to be dim, cool, red dwarfs—stars that lie on the main sequence but whose mass is about 0.1 to 0.5 $M_\odot$. Astronomers do not understand yet what determines the distribution of stellar masses, though it probably depends on the conditions in the cloud where the stars form. The average mass for Milky Way stars is a little smaller than 1 $M_\odot$, though some are significantly more massive (up to 100 or so $M_\odot$), whereas others are much less massive (perhaps as small as 0.08 $M_\odot$). Stars much more massive than about 30 $M_\odot$ form only very rarely, as we discussed in chapter 14. Objects less massive than about 0.08 $M_\odot$ were once also believed to be rare, but astronomers now think they may be common and were simply missed in previous stellar surveys.

How could astronomers miss these objects if they are so abundant? Recall that low-mass stars tend to be cooler than the Sun. In fact, if a star is less massive than about 0.08 $M_\odot$, it is so cool that it is unable to fuse hydrogen into helium. Thus, these "brown dwarfs" (as we called them) are more like huge planets than tiny stars, and their extreme dimness and small size make them very difficult to detect. However, searching the sky at the wavelengths at which these objects are brightest greatly increases the chance of finding them. What wavelengths are best? Because they are so cool, brown dwarfs are brightest in the infrared, so it is with infrared surveys such as 2MASS (2 Micron All Sky Survey) that astronomers are now finding large numbers of these "failed" stars. In fact, astronomers now think brown dwarfs may outnumber ordinary stars, making them one of the most common kinds of astronomical objects in our Galaxy.

Selection effects can occur whenever one samples a population. For example, when the U.S. Census Bureau counts our population, it must make corrections for people who are homeless or who otherwise might be overlooked. Similarly, astronomers must guard against drawing false conclusions from data that are biased by selection effects.

Despite their great numbers, low-mass stars are hard to see because they are so intrinsically dim. Thus, when we look out at the night sky, the stars we see tend to be giants. Such stars are rare, but because we can see them even at great distance, we see greater numbers of them. Astronomers refer to this difference between what is seen and what is truly present as a **selection effect.**

Two Stellar Populations: Population I and Population II

Hidden in the great diversity of star types described above is an underlying simplicity astronomers first noted in the 1940s. At that time, the 100-inch reflector at Mount Wilson Observatory, located outside Los Angeles, California, was the largest telescope in the world and the best for observing galaxies. The glow from city lights, however, made it hard to see faint galaxies. But blackouts during World War II darkened the night sky, and Walter Baade, an astronomer at Mount Wilson, took advantage of the darkness to make a series of photographs of neighboring galaxies. He noticed that stars in these nearby galaxies were segregated by color. Red stars were concentrated in the bulges and halos of the galaxies, whereas blue stars were concentrated in their disks and especially in the spiral arms. To distinguish these groups, Baade called the blue stars in the disk **population I** and the red stars of the bulge and halo **population II** (Pop I and Pop II for short). As he pursued his studies of the stellar populations of other galaxies, he saw that Milky Way stars showed the same division.

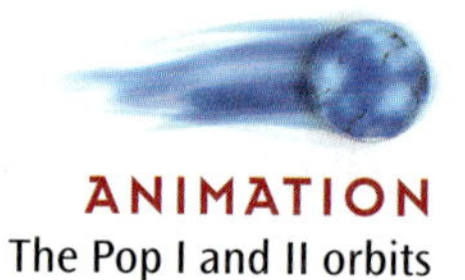

ANIMATION
The Pop I and II orbits

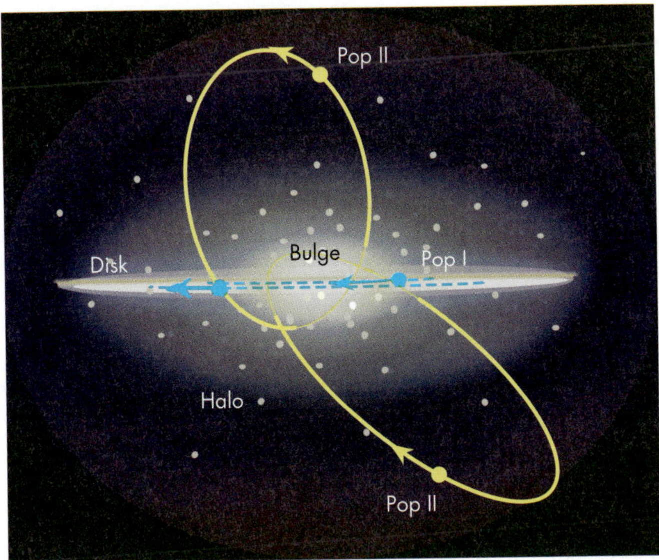

Side view

A

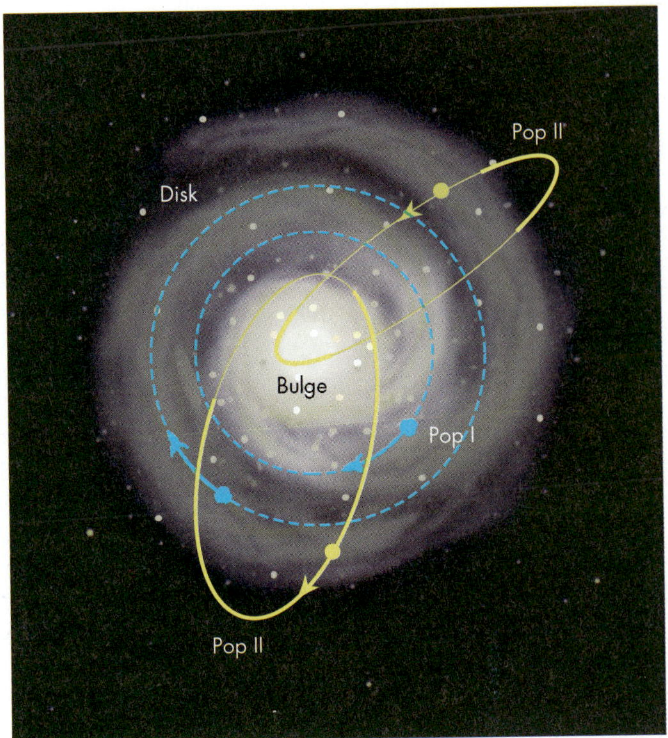

Top view

B

FIGURE 16.8
Stellar orbits in the Milky Way. Pop I stars orbit in the disk (blue lines). Pop II stars orbit in the halo (yellow lines). Notice the elongation and inclination of their orbits.

TABLE 16.1 PROPERTIES OF POPULATION I AND II STARS

	Pop I	Pop II
Age	Young to old (10^6 to a few times 10^9 years)	Old (about 10^{10} years)
Color	Blue (overall)	Red
Location	Disk and concentrated in arms	Halo and bulge
Orbit	Approximately circular in disk	Plunging through disk
Heavy-element content	High (similar to Sun's)	Low (1% to 0.1% of Sun's)

Population I and population II stars generally differ in five ways: location in Galaxy, motion, composition, color, and age. Population I stars lie in the disk and follow approximately circular orbits, as shown in blue in figure 16.8. Their composition is like the Sun's, mostly hydrogen and helium, but about 3% of their mass is heavy elements such as carbon and iron. They are generally young (10^6 to a few times 10^9 years old) compared to the Galaxy, and many of them are blue.

Population II stars, on the other hand, lie in the bulge and halo of the Galaxy and move along highly elliptical orbits that may be tilted strongly with respect to the disk, as shown in yellow in figure 16.8. They too are mostly hydrogen and helium, but only a few hundredths percent of their mass is composed of heavy elements, roughly a hundred times less than in stars like the Sun. Unlike population I stars, population II stars are very old (about 10^{10} years old) and generally red. Table 16.1 summarizes some of these properties.

In fact, some astronomers now prefer the simple labels of *disk, bulge,* and *halo* stars to describe the stellar populations of the Milky Way and other galaxies.

Division of all stars into two broad categories is an oversimplification. For example, the Sun fits in neither category precisely, but it is considered a Pop I star because of its relatively high heavy-element content and approximately circular orbit in the disk. Although not all stars fit within the precise division of these two populations,

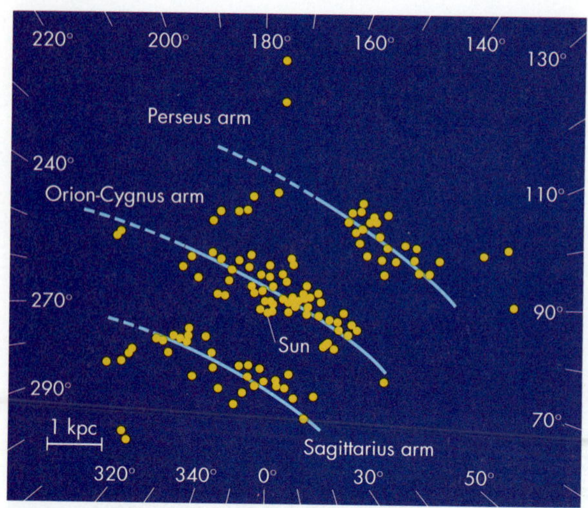

FIGURE 16.9
The positions of O and B stars gives some indication of the spiral arms near the Sun. The stars are plotted in the plane of the Milky Way, with 0° indicating the direction toward the Galaxy's center.

the division still provides a fairly good description of the majority of stars. Many astronomers think that the presence of these two different populations show that star formation has not occurred continuously in the Milky Way. Population II stars probably formed in a major burst at the time of the Galaxy's birth during its initial collapse, whereas population I stars formed later and continue forming even today. This hypothesis explains the differences between the two populations, as we will discuss in greater detail in section 16.8.

One of the first uses astronomers made of stellar population differences was to map the Milky Way's spiral arms. As Baade noted, the spiral arms of other galaxies are demarcated by their Pop I stars, especially the hot blue ones of spectral classes *O* and *B*. It is therefore logical to assume that the same is true for the Milky Way. By measuring the location of *O* and *B* stars near the Sun as shown in figure 16.9, astronomers can make a crude picture of the spiral arms near us, even though they cannot view the arms from a distance. This method is limited, however, because dust in space prevents us from seeing even the brilliant *O* and *B* stars if they are farther from the Sun than about 3000 parsecs (10,000 light-years). Nevertheless, such maps were the first evidence that we live in a spiral Galaxy.

Star Clusters

Some of the Milky Way's stars are bound together gravitationally into groups called **star clusters.** Within a cluster, each star moves along its own orbit about the center of mass of the cluster, held to it by the group's gravity. You can see several such clusters with your naked eye. The most obvious one is the Pleiades, or the Seven Sisters (shown in fig. 16.10), named for the daughters of the giant Atlas, who in Greek mythology carried the world on his shoulders. The Pleiades (pronounced *plee-a-dees*) are visible from late August through March as a tiny group of stars north of the V in the constellation of Taurus (see Looking Up #5 in the front matter). With binoculars, you can see that the Pleiades contain many more than the half-dozen stars visible to the naked eye. Likewise, by scanning the Milky Way with binoculars, you will see many additional clusters dimmer than the Pleiades.

The Milky Way has two main types of star clusters: open clusters and globular clusters. **Open clusters**—the Pleiades, for example—contain up to a few hundred members in a volume with a radius of typically 7 to 20 light-years. They are called "open" because their stars are scattered loosely, as you can see in figure 16.10. Sometimes they are also called "Galactic clusters" because most of them lie in the disk of the Galaxy.

Astronomers think that open clusters form when giant, cold, interstellar gas clouds move into the Galaxy's spiral arms. The clouds are compressed by the stronger gravitational field within the arm and collapse, breaking up into hundreds of stars whose mutual gravity binds them into the cluster. Once formed, an open cluster continues to orbit the Galaxy in the disk, but over hundreds of millions of years, stars gradually escape from it and so the cluster eventually dissolves. Assuming that open clusters are as abundant elsewhere in the Galaxy as within the regions visible from Earth, astronomers estimate that our Galaxy contains about 20,000 open clusters, though dust hides most of them from our view. Our own Sun probably once was a member of such a star group, its companion stars having long since scattered across space.

Within our Galaxy's spiral arms, extremely young stars sometimes occur in loose groups called **associations** that are a few hundred light-years across. Associations typically spread out from a single large open cluster near their center and may contain other, smaller star groupings. Moreover, the stars in associations are usually still mingled with the massive clouds of dust and gas from which they formed.

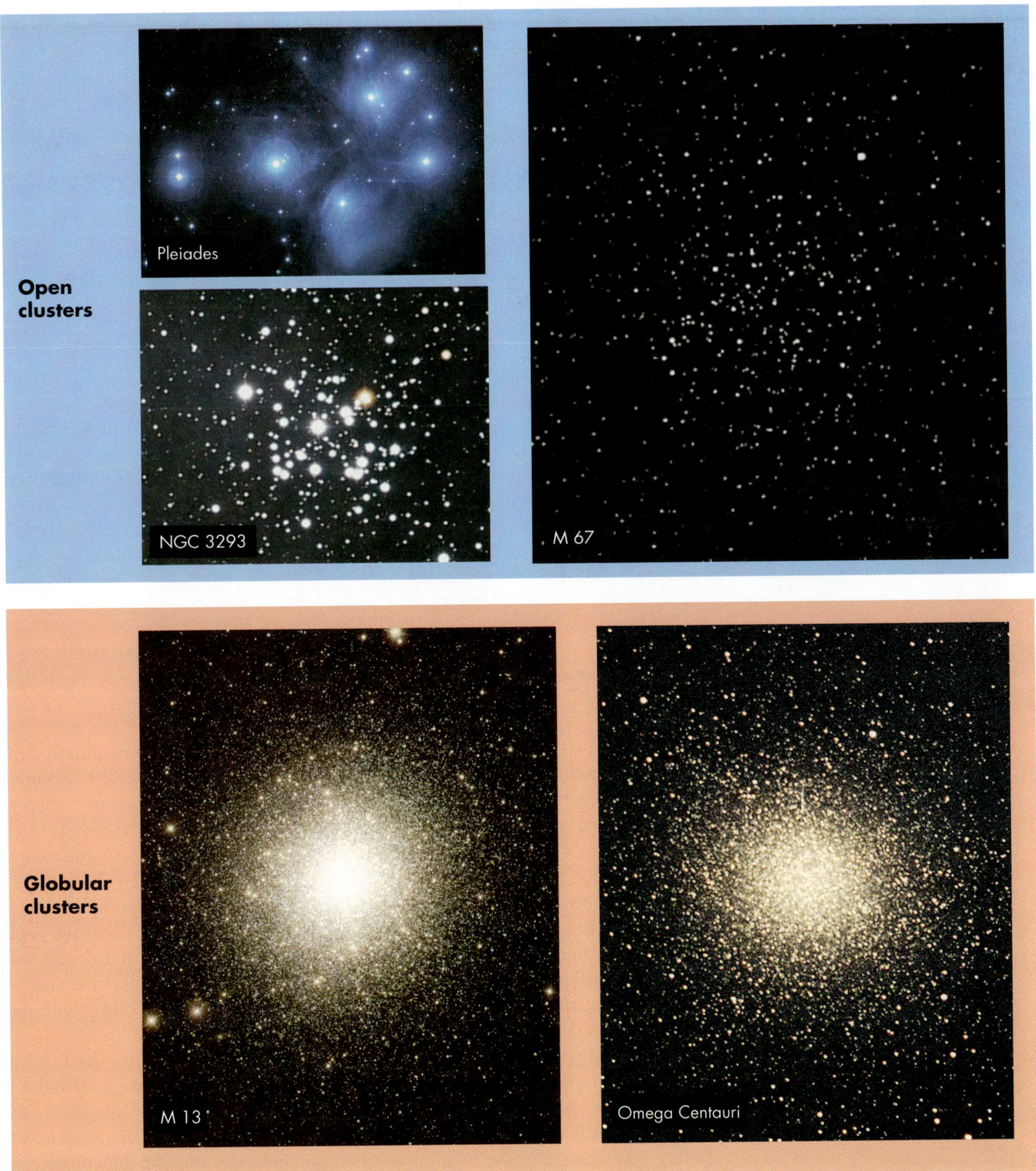

FIGURE 16.10
Star clusters: both open and globular types.

Globular clusters contain far more stars than found in open clusters, from a few hundred thousand to several million per cluster. We described them briefly in our discussion of Shapley's study of the shape of the Milky Way. They have larger radii than open clusters have—40 to 160 light-years—but the larger number of stars in them creates a stronger gravity than that found in open clusters. This stronger gravity pulls their

TABLE 16.2 PROPERTIES OF CLUSTERS AND ASSOCIATIONS

Type	Number of Stars	Radius*	Location
Open cluster	Tens to a few thousand	7 to 20 ly (2 to 6 pc)	Along spiral arms and in disk
Globular cluster	10^5 to 10^6	40 to 160 ly (12 to 50 pc)	Halo and bulge of galaxy
Associations†	5 to 70 O or B stars	65 to 325 ly (20 to 100 pc)	Spread along a spiral arm

*Because star clusters do not have sharp edges but instead gradually thin out, quoted dimensions differ substantially.
†Astronomers identify several other types of associations as well. For example, T associations are regions with above-average numbers of T Tauri stars.

stars into a denser ball, as figure 16.10 shows. Astronomers estimate that about 150 to 200 globular clusters orbit the Milky Way. Table 16.2 summarizes some properties of open clusters, globular clusters, and associations.

Open clusters and globular clusters reflect the properties of Pop I and II stars that we discussed earlier. Open clusters are generally Pop I, whereas globular clusters are always Pop II. This difference in their stellar population makes clusters especially useful to astronomers for studying the structure of the Milky Way. For example, globular clusters (among the oldest stars in the Galaxy, many of them roughly 12 billion years old) outline the halo. On the other hand, the younger open clusters (some a mere million years old) trace the Galaxy's arms.

16.4 Gas and Dust in the Milky Way

The space between stars is not empty: it contains **interstellar matter** composed of gas and dust particles. By terrestrial standards, this space is almost a perfect vacuum. On the average, each cubic centimeter contains only a few atoms of gas. For comparison, the air we breathe contains about 10^{19} atoms per cubic centimeter. The density of gas in interstellar space compared to air is like one marble in a box 5 miles on a side compared to the same box filled completely with marbles.

Despite interstellar matter's very low density, it plays several important roles in the Milky Way. First, it is the material from which new stars form, as we discussed in chapter 14. With no interstellar matter, star birth would stop—though star deaths would, for a while, replenish the material with gas and dust blown out of stars by stellar winds or explosions. Thus, interstellar matter is a kind of cosmic compost heap with matter from dying stars being recycled back into new stars.

A second reason interstellar matter is important is that the dust mixed in with it strongly limits what we can observe within the Milky Way. Because interstellar dust dims and blocks our view of distant parts of our Galaxy, it seriously affects our ability to measure distances to stars and even to detect stars in the more distant regions of the Milky Way. But before we turn to how astronomers overcome the difficulties created by interstellar dust, it will be helpful to look more closely at some general properties of interstellar matter.

Distribution and Composition of Interstellar Matter

Interstellar matter is scattered throughout the Milky Way, but gravity has pulled most of it into a thin layer in the disk, as can be seen by the dark band of dust in other galaxies (fig. 16.11). Within this layer, gravity and gas pressure have further clumped the dust and

Q. Look carefully above and below the disk of the Galaxy. Do you see some small, fuzzy, round spots? On the basis of their location, what do you think they are?

FIGURE 16.11
Interstellar matter, outlined by dark dust clouds, is concentrated in the plane of a galaxy, forming a dark band across it. This galaxy is sometimes called the "Sombrero galaxy." Technically, it is called either NGC 4594 or M104.

gas into clouds whose sizes range from a few light-years to a few hundred light-years. We can see such clouds directly (see fig. 16.6), but we can also detect them by their effect on light from distant stars.

As light from a background star passes through an interstellar cloud, the cloud's gas and dust imprint narrow absorption lines in the star's spectrum (fig. 16.12A). By analyzing these imprinted lines, astronomers can deduce the cloud's composition, temperature, density, and speed. Such observations show that the gas in interstellar clouds has a composition similar to that of the Sun and other stars, namely about 71% hydrogen and 27% helium, with trace amounts of heavier elements such as carbon and neon. Analysis of the light absorbed by the dust particles shows that they are made of silicates and carbon compounds, coated with a thin layer of ices made of molecules such as water, carbon monoxide, and methyl alcohol. These particles are extremely tiny, ranging from about a micrometer to a nanometer or so in diameter (roughly the size of a single large molecule).

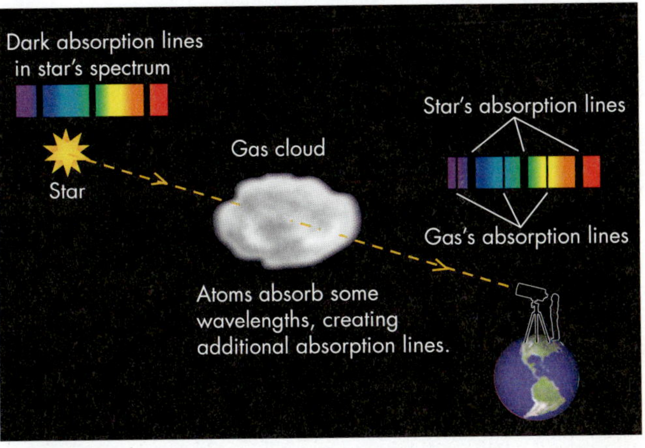

A

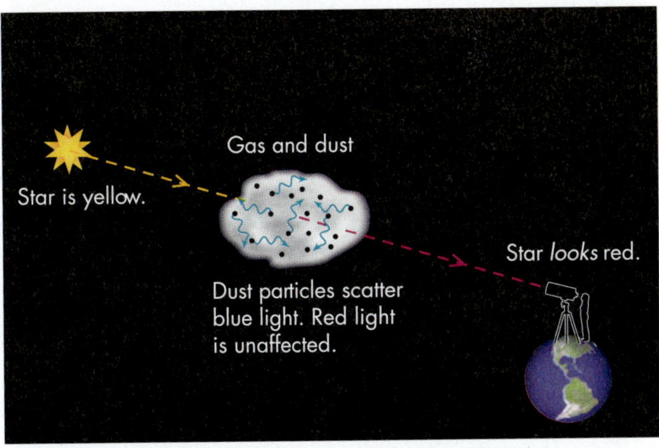

B

FIGURE 16.12
(A) Gas atoms in an interstellar cloud between us and a distant star absorb some of the star's light, adding their own absorption lines to those in the star's spectrum. (B) Dust in an interstellar cloud scatters the blue light from a distant star, removing it from the radiation reaching Earth and making the star look redder than it really is.

Interstellar Dust: Dimming and Reddening

We know there is dust in interstellar space because it dims the brightness and alters the color of the stars we see through it, much as haze in our atmosphere dims and reddens the setting Sun. Dimming and reddening result when light strikes a tiny particle such as a dust grain. The light is reflected from it in random directions in a process called **scattering.** A particle's ability to scatter light depends on its size compared to the light's wavelength. Light is scattered strongly if its wavelength is smaller than the particles, which for interstellar dust and molecules in our atmosphere means that short-wavelength blue and ultraviolet light is scattered extremely well. Scattering of sunlight creates the blue color of our sky. Scattering of starlight in space also creates visible effects.

Scattering dims a star's light because radiation that would have reached us in the absence of dust is sent off in random directions and is lost to our sight. Because red light scatters less than blue, more of the star's red light reaches us than its blue. As a result the star looks redder than it really is, a phenomenon called **reddening** (fig. 16.12B). The same effect occurs in our atmosphere and makes the setting Sun look red.

Q. How is this similar to the red appearance of the Sun when it sets?

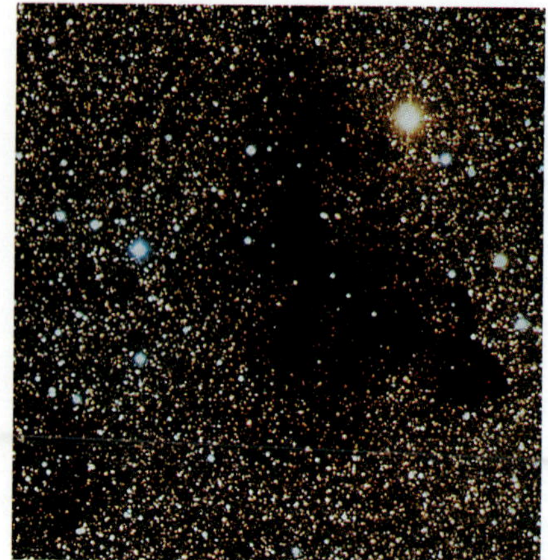

FIGURE 16.13
A dark nebula in the Milky Way. Dust blocks our view of background stars.

If a dust cloud is extremely thick or dense, it may dim the light of background stars to the point where they become invisible, an effect called "obscuration." This creates a dark region, seemingly empty of stars, which astronomers call a **dark nebula.** Figure 16.13 shows a region of the Milky Way in which a dark nebula is visible, silhouetted against a swarm of faint background stars and blotting out their light.

Only about 1% of the interstellar matter is dust, but this dust scatters visible light so well that it strongly limits our view of the Milky Way. We can see no more than a few thousand light-years away from the Sun into the disk because the dust is concentrated there. Only by looking above or below the disk can we see more-distant objects. Dust in the disk also hides galaxies that lie behind it, creating a band around the sky called the **zone of avoidance,** in which essentially no galaxies can be seen. Figure 16.14 shows a chart of over 34,000 galaxies illustrating this effect. The narrow horizontal region where no galaxies appear is the zone of avoidance and coincides with the plane of the Milky Way. We know today that galaxies do occur in this region because they can be seen with radio and infrared telescopes sensitive to radiation that can penetrate the dust.

FIGURE 16.14
Dust limits our view in the disk of the Milky Way creating the "zone of avoidance," where we see no galaxies. (A) A plot showing how galaxies are distributed on the sky. Notice how few lie along the central line (the Galactic equator). This blank region is the zone of avoidance. (B) A sketch illustrating why we see so few galaxies along the plane of the Milky Way. The gray blobs represent dust clouds that block our view toward directions that lie in the Milky Way's plane. The galaxies are in reality about the same size as the Milky Way and are, of course, much farther away than can properly be shown in the figure.

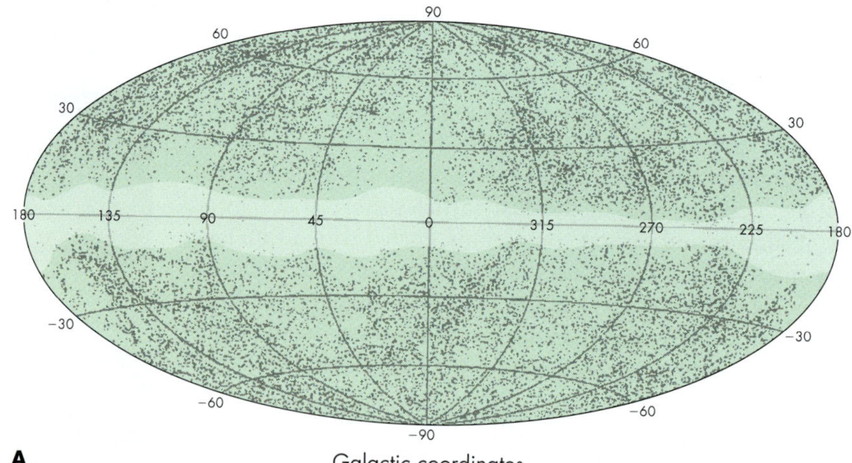

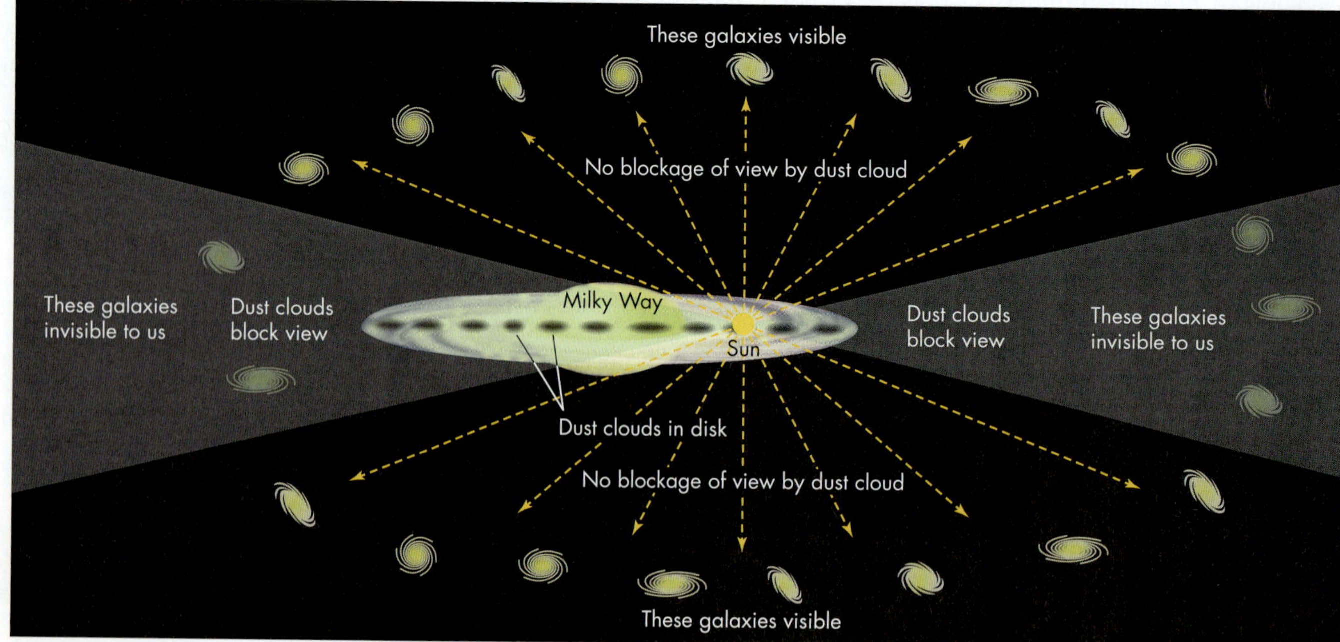

FIGURE 16.15
An all sky view of our Galaxy, as we see it from Earth. The image was made by plotting on a map of the whole sky the millions of stars observed in the infrared by the 2-Micron All Sky Survey (2MASS). It is therefore like a map of the Earth, flattened out so as to show all continents. The map clearly reveals the flatness of our Galaxy and the bulge of its central regions.

Infrared and radio waves pass more readily through dusty clouds than visible light does because their wavelengths are much larger than the dust particles. This allows infrared pictures to reveal the structure of the Milky Way far better than visible light pictures can, as illustrated in figure 16.15. The disk and central bulge of our Galaxy are easy to see in this wide-angle picture, which was made by plotting the positions of millions of stars observed as part of the 2MASS infrared survey of the sky.

Dust is not all bad. Most of the atoms that compose the Earth and our bodies came from such dust grains. Dust grains may also scatter enough light to become visible, thereby creating beautiful **reflection nebulas** (fig. 16.16). Scattered starlight makes the interstellar dust shine, just as scattered sunlight makes dust in the air visible. Reflection nebulas are not only lovely but additional proof that dust exists in space.

Interstellar Gas

Interstellar gas plays several crucial roles in our Galaxy. It is the material from which stars form, and it is the repository of matter blown off dying stars. Without interstellar

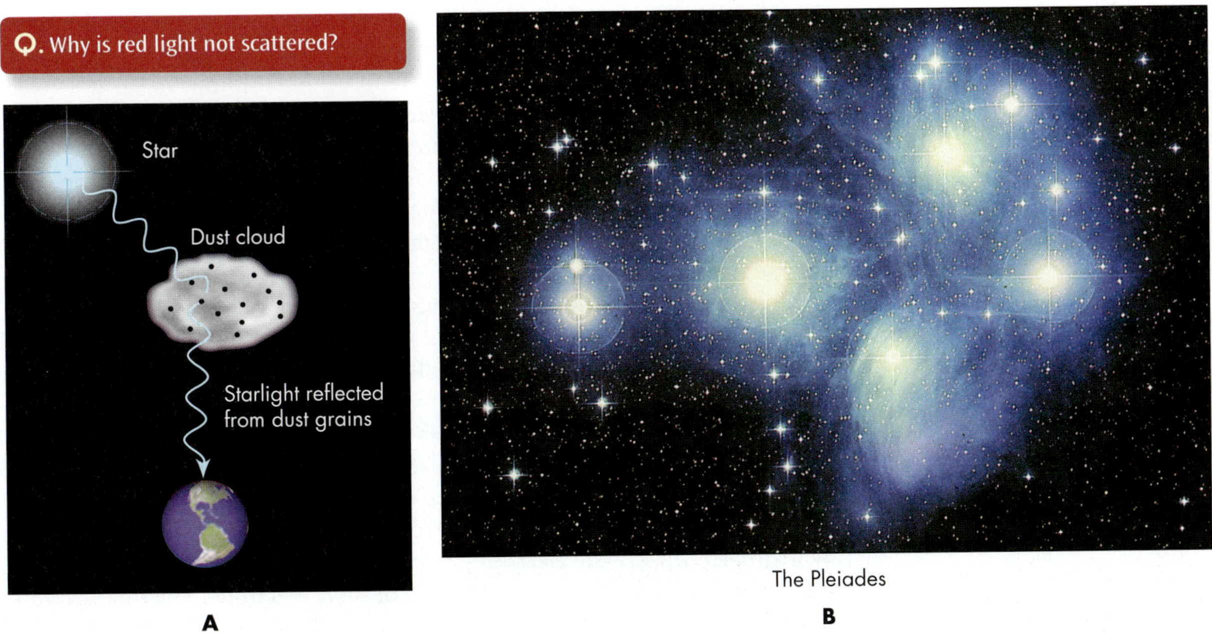

FIGURE 16.16
Reflection nebula. (A) Light from a star near a dust cloud is reflected from the dust making it visible. (B) The Pleiades star cluster. Note the bluish reflection nebula around many of the brighter stars.

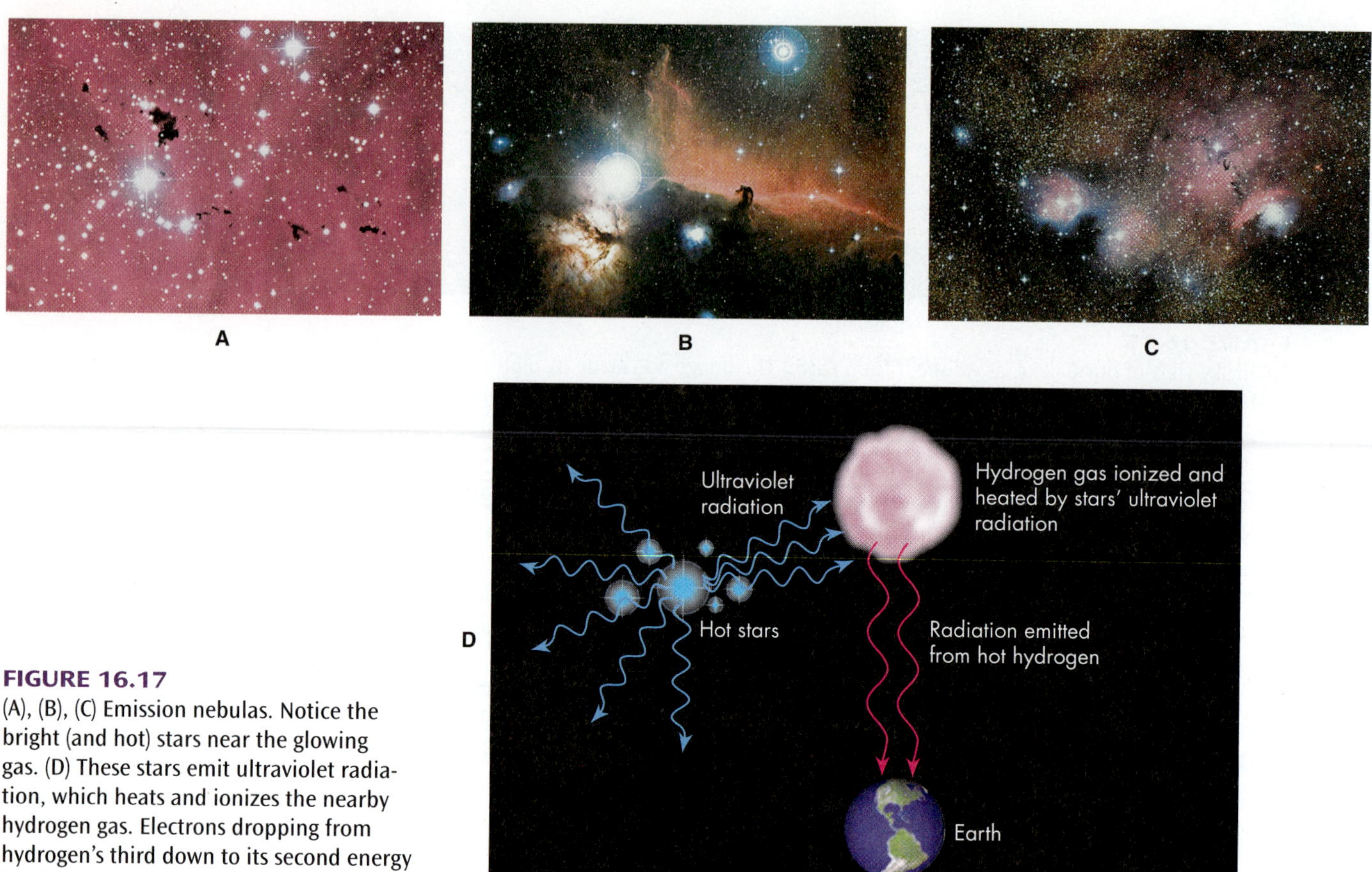

FIGURE 16.17
(A), (B), (C) Emission nebulas. Notice the bright (and hot) stars near the glowing gas. (D) These stars emit ultraviolet radiation, which heats and ionizes the nearby hydrogen gas. Electrons dropping from hydrogen's third down to its second energy level generate the red light.

matter to make new stars to replace those that die, the Milky Way would be far dimmer and filled only with stellar corpses and low-mass stars. Interstellar gas is also a help to astronomers seeking to map the Milky Way. Dust blocks our view of the Milky Way at visible wavelengths, making it difficult for us to observe its structure, but interstellar gas emits radiation that penetrates the dust and allows us to "see" regions of the Galaxy that would otherwise be invisible to us. Finally, interstellar gas creates some of the most spectacular structures in our Galaxy: huge glowing clouds heated by stars embedded in them, as shown in figure 16.17. Another example of such a gas cloud is the Great Nebula in Orion, which is easy to see with a small telescope or binoculars (see Looking Up #6 in the front matter). If you look at the middle star in Orion's sword you will see a pale, fuzzy glow created by luminous interstellar gas surrounding a dense star cluster.

Interstellar gas clouds that emit visible light are called **emission nebulas.** Emission nebulas give off their own light in visible and other wavelengths, unlike the dust in reflection nebulas, which simply reflects and scatters light from neighboring stars. However, a cloud must be hot to emit visible light, and so such clouds need a source of heat (generally a nearby star), as shown in figure 16.17D. Hot, blue stars are especially effective at heating because they emit large amounts of ultraviolet radiation whose short wavelength is much more energetic than that of visible light. Ultraviolet light whose wavelength is less than 91.2 nm is so energetic that when it is absorbed by hydrogen, it tears the electron free of the nucleus, ionizing the gas. In an ionized gas (one in which atoms are missing one or more electrons), the free electrons collide with atoms of oxygen, nitrogen, and other elements and excite them, making them emit light of their own. Eventually, the electrons recombine with the ionized hydrogen atoms and emit more radiation, especially the red light of hydrogen alpha, which gives these nebulas their lovely reddish color. Because hydrogen in these hot gas clouds is

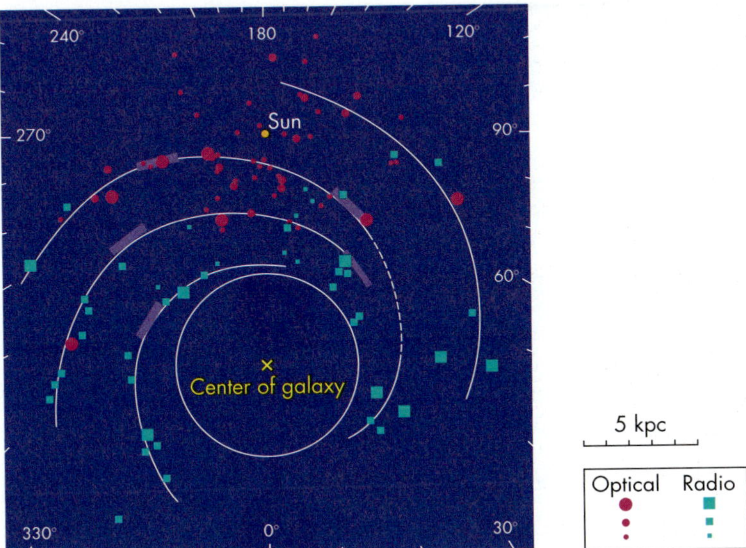

FIGURE 16.18
Map of the Milky Way made by combining radio observations (since radio can penetrate dust and thereby allow one to see all the way across the Milky Way's disk) and optical observations of the position of HII regions. Such regions outline the spiral arms of other galaxies. In our Galaxy, too, they form a spiral pattern.

ionized, they are sometimes called **HII regions,** the *H* denoting the hydrogen and the *II* indicating that it is ionized.

HII (pronounced *H-two*) regions have a temperature of approximately 10,000 K and are heated by the electrons freed by the ionizing ultraviolet radiation. The liberated electrons collide with gas atoms around them and speed them up, increasing their energy and thus their temperature. The freed electrons are themselves a powerful source of radio emission, emission that helps astronomers see HII regions that might otherwise be hidden from us by dust. Because these HII regions are generally located in spiral arms, radio maps of their location can reveal the spiral structure of our Galaxy. In fact, maps made by combining radio observations and optical observations of HII regions are the best evidence we have that the Milky Way has spiral arms (fig. 16.18).

Although we can easily see the glow from gas clouds that happen to be near hot stars, much interstellar gas is too cold to emit visible light, and we can detect it only with radio telescopes. Clouds more than a few tens of light-years from hot stars may be only a few Kelvin (more than 400° below zero on the Fahrenheit scale). Such cold material emits no visible light because it has too little energy to generate visible photons. However, it can emit radio waves of very low energy (recall that radio waves carry much less energy than visible light does), as described in Extending Our Reach: "Mapping the Milky Way with Radio Waves" (page 440).

16.5 Motion of Stars and Gas in the Milky Way

The motion of stars and gas gives important clues to the structure of the Milky Way. We saw earlier that in accordance with Newton's first law (the law of inertia), if no force acted on the stars, each would move in a straight line. But each star is acted on by a force: the gravitational force created collectively by all the other stars and gas. That force acts to draw each star inward toward the center of the Galaxy. The result is that each star follows its own orbit around the center of the Galaxy, just as each planet moves in orbit about the Sun.

Although all stars within the Milky Way move around its center, the paths followed by stars in the disk and halo are very different, as we discussed in section 16.3. We saw there that all the stars in the disk orbit in the same direction and in nearly the same plane, giving the disk its flat shape, just as the planets' motion in approximately the same plane gives the Solar System its flattened shape. On the other hand, stars in the halo and bulge follow orbits that are steeply inclined to the disk's plane (see fig. 16.8).

EXTENDING *our reach*

MAPPING THE MILKY WAY WITH RADIO WAVES

One of the most important sources of radio emission is cold hydrogen atoms,* which radiate at a wavelength of 21 centimeters. This radiation, called **21-centimeter radiation,** arises because subatomic particles such as protons and electrons possess a property, known as "spin," that makes them behave like tiny magnets. Magnets have north and south poles with the property that like poles repel and unlike poles attract. If an electron has the same alignment as the proton it is orbiting—both with north poles "up," for example—the electron will flip over, as shown in box figure 16.1. As it flips, the changed magnetic interaction between the electron and the nucleus makes the electron shift to a slightly smaller orbit with a slightly lower energy. However, energy must be conserved, and so when the electron shifts to the new orbit, it emits the excess energy as electromagnetic radiation. We discussed in chapter 3 that this process can produce visible light. In cold hydrogen, however, the flip of the electron liberates too little energy to make visible radiation and can create only radio emission at a wavelength of 21 centimeters.

The 21-centimeter radiation has proved extremely valuable for studying the Milky Way and other galaxies. Hydrogen is abundant in space, and so the signal of the 21-centimeter radiation is strong. From the strength of the signal, astronomers can deduce the amount of hydrogen in an area. If the gas is moving, the wavelength will be Doppler-shifted, from which the gas's speed along our line of sight can be found. Radio observations thereby allow astronomers to map not only where the gas is concentrated in the Galaxy but also how it is moving. Although details in such 21-centimeter maps are uncertain because of assumptions that must be made about how the gas moves, they nevertheless are extremely useful because the radiation is not absorbed by interstellar dust, thereby allowing astronomers to "see" the entire Milky Way in its radiation. Such maps show that the gas is confined to a thin disk and its distribution is suggestive of a spiral pattern, confirming the picture deduced by other means.

The 21-centimeter radiation is not a perfect tool, however. It is emitted only by hydrogen if its electron is attached. If the hydrogen is heated so that its electron is torn free—that is, the hydrogen is ionized—it produces no 21-centimeter waves. Likewise, at extremely low temperatures (15 K or so), hydrogen atoms bond to one another to make hydrogen molecules and, in its molecular form, hydrogen cannot emit 21-centimeter radiation. However, in the cold, dense regions where hydrogen molecules form, other molecules that do emit radio waves form as well.

More than 100 interstellar molecules have been identified so far, including common compounds such as

*Cold hydrogen atoms with their electrons attached are called "HI" (pronounced H-one) to distinguish them from ionized hydrogen atoms, HII.

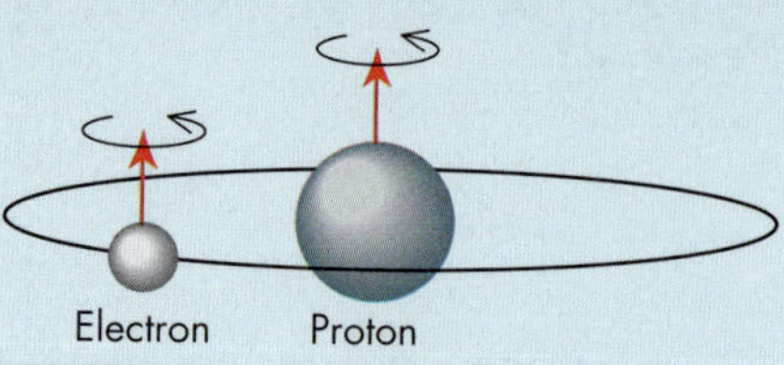

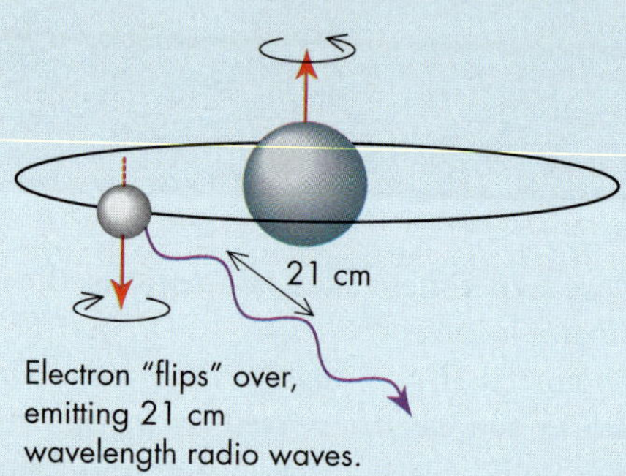

21 cm
Electron "flips" over,
emitting 21 cm
wavelength radio waves.

BOX FIGURE 16.1
Radio radiation at 21 centimeters is emitted by cold hydrogen as the electrons "flip."

TABLE 16.3	SOME INTERSTELLAR MOLECULES
H_2	Molecular hydrogen
OH	Hydroxyl radical
CN	Cyanogen radical
CO	Carbon monoxide
H_2O	Water
HCN	Hydrogen cyanide
NH_3	Ammonia
HCOH	Formaldehyde
HCCH	Acetylene
HCOOH	Formic acid
CH_3OH	Methyl alcohol (methanol)
CH_3CH_2OH	Ethyl alcohol (ethanol)

carbon monoxide (CO), formaldehyde (HCHO), and ethyl alcohol (CH_3CH_2OH), and also more exotic substances. Table 16.3 lists a few of the many interstellar molecules now known. Collisions among neighboring molecules make them spin and give them energy, as shown in box figure 16.2, but they quickly radiate that energy away and reduce their spin by emitting radio waves. Each molecule has a distinctive set of wavelengths at which it can emit its rotational energy, just as each atom has a distinctive set of wavelengths (its spectral lines) at which it can radiate energy from electron jumps. For example, carbon monoxide emits at wavelengths of 2.6 and 1.3 millimeters, and ethyl alcohol emits at 3.5 millimeters. Both these molecules emit at many other wavelengths as well.

The strength of the molecular radio emission from a gas cloud depends on the number of molecules it contains, giving astronomers information on its density. If the gas is moving, the wavelength of the emission will be slightly altered by the Doppler shift, displacing and broadening the spectrum lines. From such shifts and broadening, astronomers can deduce the speed of the cloud and its rotation. With that information, they can then map out where in the Milky Way the molecular gas is located and thereby obtain additional information about the Galaxy's structure.

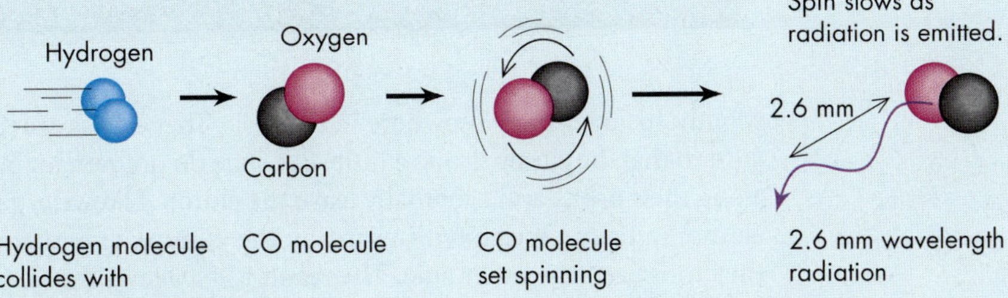

BOX FIGURE 16.2
Emission of radio waves by spinning molecules in cold interstellar clouds. Collisions set the molecules spinning. When their rotation speed changes, they emit radio waves. The wavelength emitted depends on the kind of molecule and by how much its rotation speed changes.

Motion within the disk also creates the Milky Way's spiral arms. We know that gravity holds a galaxy together, and so we might infer that gravity is also what makes a spiral arm. Gravity does in fact help hold a spiral arm together, but a spiral arm is not a fixed mass of material. Rather, the stars and gas in an arm slowly drift out and are replaced by new stars and gas, according to the density-wave theory of spiral structure. In the **density-wave model,** waves of stars and gas sweep around the Galactic disk. The wave is not primarily an up-and-down motion like an ocean wave; instead, it is a place where the density of stars and gas is large compared to the surroundings. This denser region is the wave crest, and we see it as a spiral arm. Spiral waves are not unusual in rotating matter. Batter stirred in a bowl with an electric mixer forms spiral waves, but to better understand how arms form in the Milky Way, consider the following analogy.

As cars move along a freeway, most move at nearly the same speed and keep approximately the same separation. If one car moves slightly slower than the others, however, traffic will begin to bunch up behind it. Cars can pass the slower vehicle but must change lanes to do so. The result is a clump of cars behind the more slowly moving one. But that clump is not composed permanently of the same cars. Cars join the clump from behind, pass the slower moving car, and then leave the clump at its front.

A similar phenomenon happens to stars orbiting in the Milky Way. As stars circle the Galaxy, a few closely spaced stars at a given location will make the local gravity

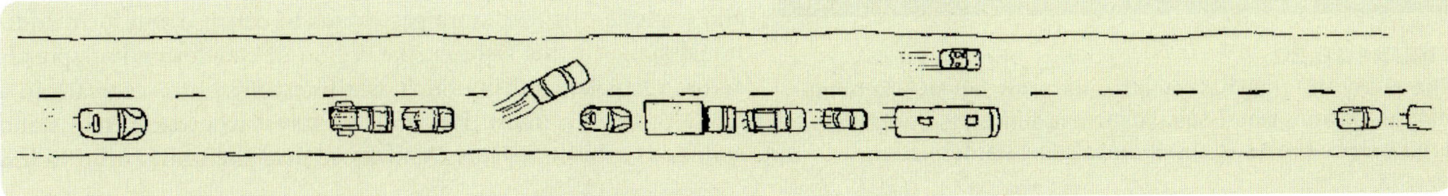

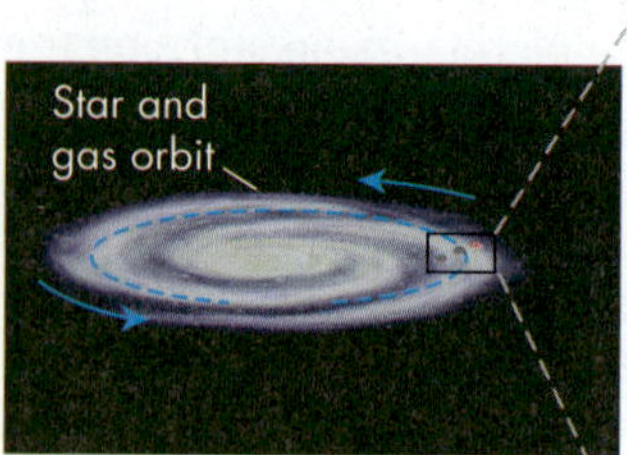
Star and
gas orbit

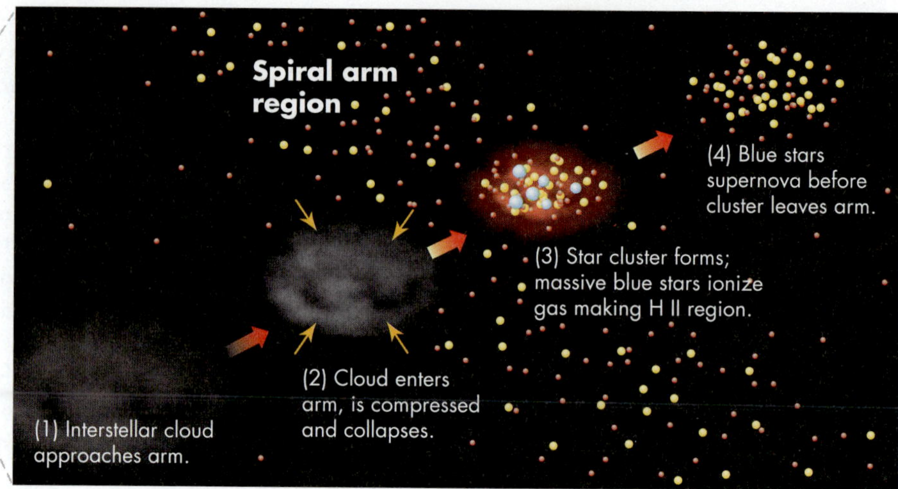
Spiral arm
region
(4) Blue stars
supernova before
cluster leaves arm.
(3) Star cluster forms;
massive blue stars ionize
gas making H II region.
(2) Cloud enters
arm, is compressed
and collapses.
(1) Interstellar cloud
approaches arm.

FIGURE 16.19
Sketch of a density wave in a galaxy, showing the progression from dust and gas to stars across the arm.

slightly stronger than elsewhere in the disk. That excess gravity draws stars into the region so that the clump grows a little. But stars do not remain in the clump. They continue along their orbits and eventually leave the clump. Likewise, gas clouds drift through the clump, initially accelerated inward by the clump's gravity and then slightly slowed as they move out the other side. The result is a wave in the Galaxy's disk where stars and gas accumulate before moving through. The accumulation of gas compresses clouds and makes them collide with one another, which in turn makes some of the clouds collapse and form stars. This is the first stage of the star formation process described in chapter 14. The brilliant young *O* and *B* stars so born illuminate the arms and make them stand out against the more ordinary stars of the disk. In fact, most of a galaxy's stars form in its spiral arms, and as we will discover in chapter 17, galaxies without arms generally contain extremely few young stars.

Although the density-wave theory has difficulty explaining the longevity of spiral arms, many observations support the theory. For example, by combining radio and visual pictures of a galaxy, astronomers can see a progression through a spiral arm as gas enters on one side, is compressed, and forms stars, taking a few million years to sweep through the arm. On one side of the arm, you can see gas and young stars; on the other side, older, more evolved stars, as indicated in figure 16.19. Doppler-shift studies of the gas motion also clearly show the gas moving into the arm on one side and leaving on the other.

Although most spiral galaxies probably make their arms this way, another theory has been proposed to explain the very ragged-appearing arms in some galaxies, such as the bottom one shown in figure 16.20. This second theory is called **self-propagating star formation.**

The idea of this second theory is that star formation starts at random points in the disk of a galaxy when gas clouds collapse and turn into stars. As the stars heat the gas around them and explode as supernovas, they generate a disturbance that makes the surrounding gas clouds collapse and turn into stars. The original stars burn out, but the new stars—formed as the old ones evolve and explode—trigger more gas clouds to collapse and form additional stars. In this fashion, the region of star formation spreads across the galaxy's disk, much as a forest fire burns outward in a circle through a forest. But unlike a forest where the trees stand still, stars orbit in a galaxy, and those near the center orbit in less

FIGURE 16.20
The galaxy M81 (top) has smooth spiral arms, but M63 (bottom) has ragged arms whose spiral pattern is difficult to trace. M63's arms may be the result of self-propagating star formation.

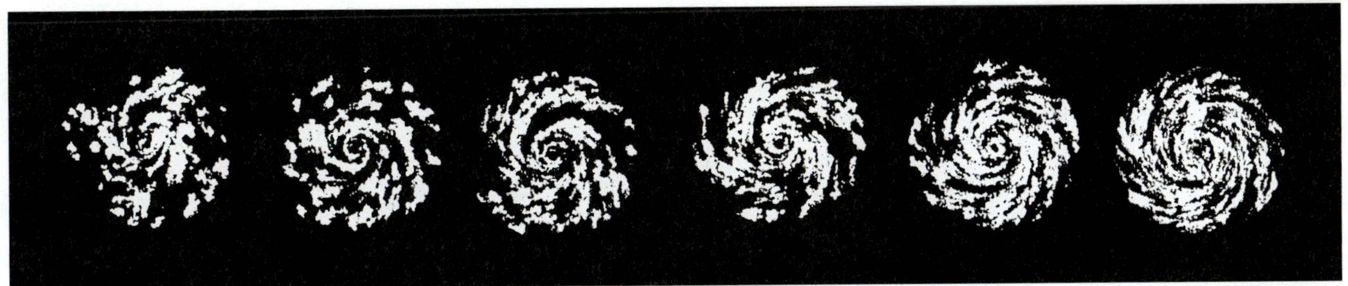

FIGURE 16.21
Computer model of self-propagating star formation.

time than those farther out. Thus, the zone in which star formation occurs is drawn out into a spiral by the difference in rotation rate between the inner and outer parts of the disk. The result is spiral structure, as illustrated in figure 16.21.

Astronomers are not certain which process—density-wave or self-propagating star formation—is most important in forming spiral arms. There is probably a continuum, with both processes contributing different amounts in different galaxies.

16.6 Measuring the Milky Way

We described earlier in this chapter how astronomers measure the basic properties of the Milky Way, such as its radius and mass. In this section, we look in greater detail at those methods.

Diameter of the Milky Way

Astronomers have several ways to measure the Milky Way's diameter, but all such methods depend on first knowing how far from the center the Sun is. In section 16.1 we saw how Shapley used globular clusters to map out the structure of the Galaxy and thereby find its size and where the Sun lies in it. This method remains one of the best techniques for measuring our Galaxy's scale. Astronomers measure the distance and direction from the Sun to a large number of globular clusters or some type of star that can be seen for great distances and then plot their positions, as shown in figure 16.22. They next locate the point around which the clusters seem to be most symmetrically located and assume that this point is the center of the Galaxy. Finally, they measure the distance from that point to the Sun, as shown in the figure. This method is somewhat uncertain and gives distance estimates of about 6 to 10 kiloparsecs to the center.

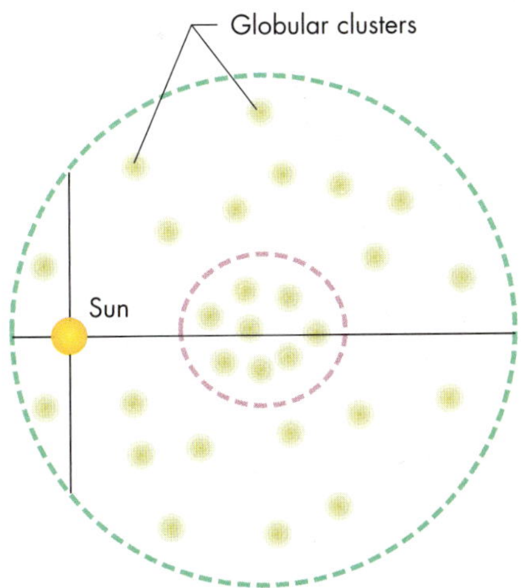

FIGURE 16.22
Finding the distance to the center of the Milky Way using globular clusters. Make a scale plot of where the clusters lie with respect to the Sun. Outline the region filled by the clusters and note where they are most concentrated. The center of their concentration is presumably the center of the Galaxy. Measure the distance from the center to the Sun and from the known scale of the plot determine Sun–center distance.

Some newer methods give a more precise distance estimate by using radio observations of cool giant stars. Some such stars emit powerful beams of energy at radio wavelengths analogous to the light emitted by lasers. These are called "maser sources." The word *maser* is an acronym for *m*icrowave-*a*mplification by *s*timulated *e*mission of *r*adiation. The radiation comes from molecules in an expanding shell of gas around the star. Maser sources are especially common in the inner bulge of the Milky Way because of the many red giants there, and with the masers, astronomers can measure the distance from the Sun to the Milky Way's center.

The maser sources expand away from the star they surround. Some move along our line of sight and therefore show a Doppler shift, which gives their velocity. Others move perpendicular to our line of sight, and we can detect that sideways motion by comparing observations a few years apart. To obtain their distance, astronomers assume that, on the average, maser sources moving along the line of sight have the same speed as sources moving across the line of sight. From the change of position and

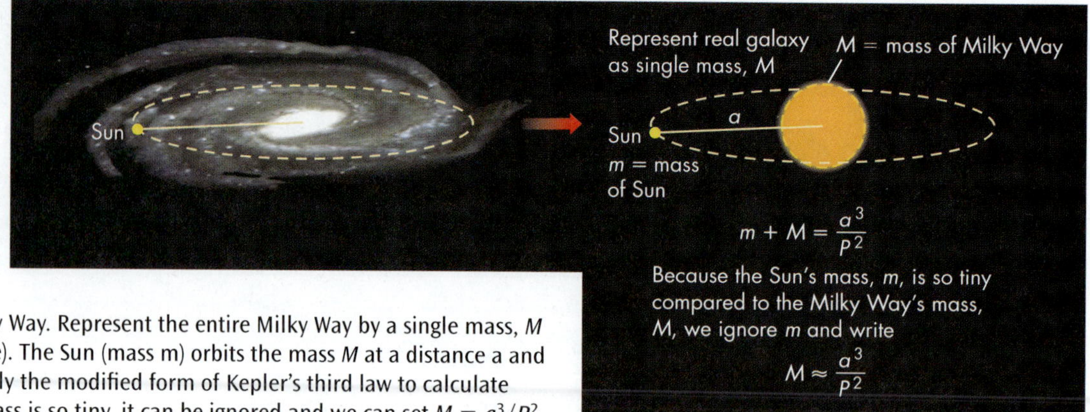

FIGURE 16.23
Finding the mass of the Milky Way. Represent the entire Milky Way by a single mass, M (the orange blob in the figure). The Sun (mass m) orbits the mass M at a distance a and with an orbital period P. Apply the modified form of Kepler's third law to calculate $M + m$. Because the Sun's mass is so tiny, it can be ignored and we can set $M = a^3/P^2$.

speed, astronomers can deduce the stars' distances and therefore the distance to the center of the Milky Way (the point about which the stars orbit).

Recently, astronomers have also been able to detect stars in orbit near the center of the Galaxy by observing at infrared wavelengths. The estimates of the distances for these stars and the maser sources give a distance of about 8 kiloparsecs (approximately 26,000 light-years). Once the Sun's distance from the center is known, the overall size of the Milky Way follows when the distance to objects lying near the Galaxy's edge is measured and the two distances are added. From such measures, astronomers estimate that the Milky Way has an overall diameter spanning 40 kiloparsecs or more, but much of the vast outer part beyond the Sun's orbit is extremely sparsely populated.

Mass of the Milky Way

Astronomers can deduce the Milky Way's mass from its gravitational attraction on matter in and near it. The procedure they use is very similar to the one described in chapter 13 to determine the mass of stars. You may recall that if one star is in orbit around another, we can calculate their combined mass using the modified form of Kepler's third law. Similarly, astronomers can find the mass of the Milky Way by treating it as a scaled-up version of a binary star in which one star, the Sun, for example, orbits under the gravitational influence of the entire Galaxy (fig. 16.23). The details are given in Extending Our Reach: "Measuring the Mass of the Milky Way."

More-refined methods build on this same technique but account for the mass of the entire Galaxy and use the rotation speed of stars at a variety of distances (the so-called **rotation curve**). The rotation curve, shown in figure 16.24, shows how fast the Galaxy spins at each distance from its core. That speed must be exactly sufficient to offset the gravity of the Milky Way that pulls orbiting stars in toward the center. The

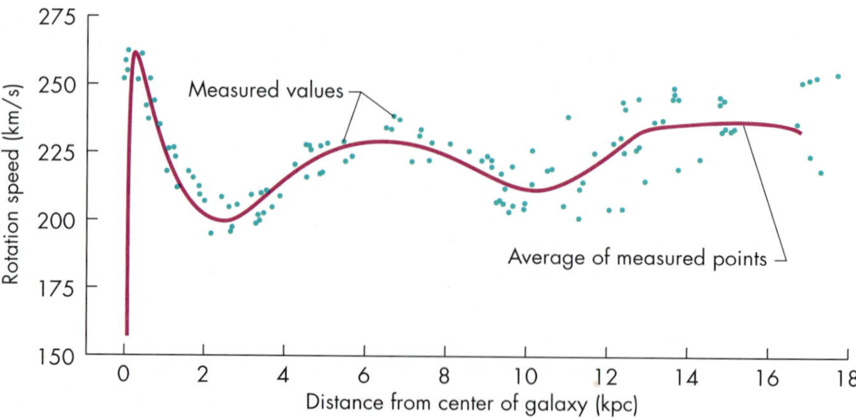

FIGURE 16.24
Schematic rotation curve of the Milky Way.

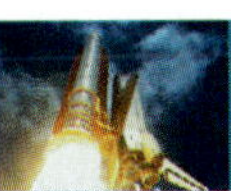

EXTENDING *our reach*

MEASURING THE MASS OF THE MILKY WAY

In our calculation, we will make use of a result from physics that, if the Milky Way were spherical, only that part of it inside the Sun's orbit would affect the Sun's motion. Moreover, the gravity of that inner part acts as if all the mass interior to the Sun's orbit were packed into a point at the orbit center. The Milky Way is not spherical, but detailed calculations using its true shape differ only slightly from what we will find with our simple model. A more serious difficulty, however, is that our method of calculation omits all the mass in the Milky Way that lies outside the Sun's orbit. That is, we can assert the Milky Way contains *at least* the amount of mass given by our calculation.

We proceed as follows. We represent the mass of the Milky Way inside the Sun's orbit as a single mass, M, and assume the Sun (with mass m) moves in a circular orbit of radius a about M, as shown in figure 16.23. We now use the modified form of Kepler's third law described in chapter 13 for measuring the masses of binary stars. The modified form of Kepler's third law states that

$$m + M = a^3/P^2$$

m = Mass of smaller object
M = Mass of larger object
a = Distance between objects
P = Years to complete one orbit

where m and M are the masses of the orbiting bodies in solar masses, P is their orbital period in years, and a is their separation in astronomical units. The orbital period of the Sun around the Milky Way is about 220 million years (we will see below how this number is determined). Its orbital radius is about 8 kiloparsecs, or about 1.65×10^9 AU. If we insert these values into the equation, we find that the mass of the Sun (m) plus that of the Milky Way (M) is

$$\begin{aligned} m + M &= (1.65 \times 10^9)^3/(2.2 \times 10^8)^2 \\ &= 1.65^3 \times 10^{27}/(2.2^2 \times 10^{16}) \\ &= 9.3 \times 10^{10}\ M_\odot \approx 10^{11}\ M_\odot \end{aligned}$$

Because m, the mass of the Sun, is so tiny compared with the mass of the Galaxy, we can ignore the Sun's mass and conclude that the Milky Way's mass is at least $10^{11}\ M_\odot$.

gravity in turn depends on the Milky Way's mass, and so from the rotation curve and Newton's laws of motion and gravity, astronomers can calculate the mass of our Galaxy. The first task, therefore, is to measure the rotation curve.

To determine the rotation curve, we must remember that we are observing from the Earth, which is traveling with the Sun in the Sun's orbit around the center of the Galaxy. Astronomers thus have the same problem that the police in a moving patrol cruiser have in measuring the speed of a traffic violator. That is, to find the speed of the star around the Milky Way, they must measure how fast it moves with respect to the Sun and then add the Sun's motion about the Galaxy. Astronomers have several methods for making this measurement. For example, we can measure the Doppler shifts of globular clusters. These Doppler shifts reflect not only their orbital motions but the Sun's as well. Because these clusters fill an approximately spherical volume around the Galaxy, they, as a group, presumably revolve only slowly compared with the flat disk. Thus, the Sun's average speed with respect to the globular clusters is approximately its own rotation speed around the Milky Way.

From many such measurements, astronomers conclude that the Sun is orbiting the Galaxy with a velocity of about 220 kilometers per second (about 130 miles per second!). From the Sun's orbital speed, we can calculate its orbital period, P, by dividing the speed, V, into the circumference of its orbit. That is,

$$P = 2\pi R/V$$

P = Sun's orbital period
R = Radius of Sun's orbit
V = Sun's orbital speed

where R is the radius of the Sun's orbit, which we showed earlier is about 8 kiloparsecs. Inserting the values of R and V determined above (in appropriate units), we find that P is about 2.2×10^8 years, the number we used earlier to find the Milky Way's mass.

Unfortunately, the measurements we have described above are difficult to do accurately, so the speed of the Sun is uncertain. Because that influences the mass that we calculate for the Milky Way, our Galaxy's mass is also uncertain. A far more serious problem is accounting for the large mass of our Galaxy that such calculations imply. That is, the calculated mass greatly exceeds the mass of detectable stars and gas. As we saw in section 16.2, this discrepancy has led astronomers to hypothesize that our Galaxy is embedded in a huge, massive halo of dark matter. The dark matter exerts a gravitational force on the detectable stars and gas, but it emits no as-yet observed radiation of its own. Moreover, observations of its gravitational effects suggest that this immense halo has a radius of at least 100 kiloparsecs, so that it dwarfs the visible Milky Way. What can this dark matter be made of? Although some of it may be cool, dim stars, the nature of the rest remains unknown and is one of the greatest mysteries in astronomy today.

16.7 The Galactic Center

The asterisk in Sagittarius A* is part of its name, said as "Sagittarius A star."

The Sun lies at the center of the Solar System. What lies at the center of the Milky Way Galaxy? Even though the core of our Galaxy is difficult to observe because interstellar dust clouds totally block the visible light emitted by objects there, radio, infrared, X-ray, and even gamma-ray telescopes allow astronomers to see through the dust and reveal a dense swarm of stars and gas swirling around the center of our Galaxy (fig. 16.25). With such observations, astronomers have detected an intense source of radio waves known as Sagittarius A*, which lies at the very core of our Galaxy. Measurements of this source show that it is less than 10 AU in diameter (about the size of Jupiter's orbit) and that hot gas and stars are in rapid orbit around it. Over the past few years, astronomers have tracked the motion of a number of those stars—some of which are moving at hundreds of kilometers per second (fig. 16.26). From these motions, and using the modified form of Kepler's third law, astronomers calculate that the stars are orbiting an object containing about 5 million solar masses. Only such an

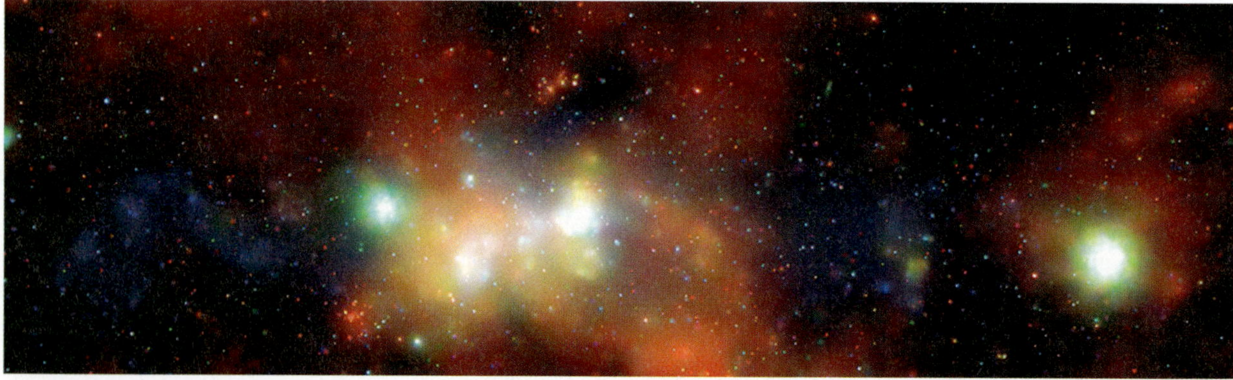

FIGURE 16.25
The central region of the Milky Way as observed at X-ray wavelengths by the orbiting X-ray telescope *Chandra*. The supermassive black hole at the Milky Way's core lies inside the bright white patch near the center of the image, which shows an area 900 light-years across. In this false-color image, the colors represent X rays of different energies. The small bright spots are white dwarfs, neutron stars, and probably gas surrounding a few black holes. The diffuse glow is from extremely hot gas surrounding our Galaxy's core.

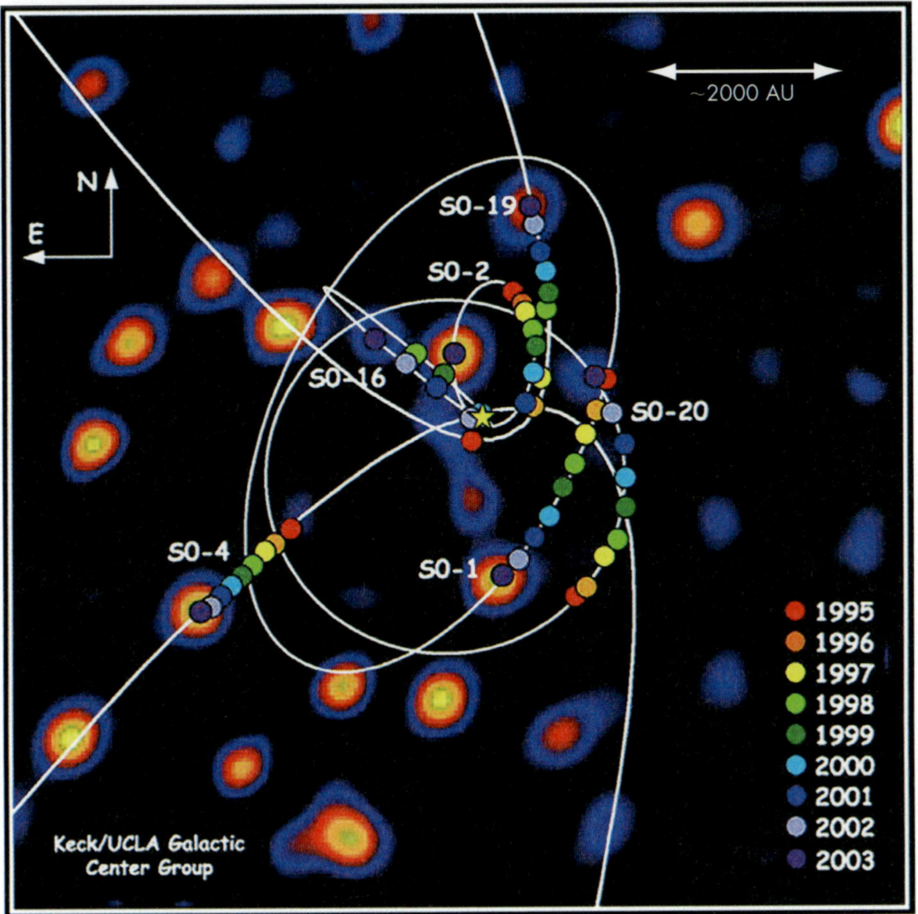

FIGURE 16.26
A plot of the position of several stars as they orbit the massive black hole at the core of the Milky Way. The stars are colored to show their positions in successive years. (Yellow "star" marks center of Milky Way.)

enormous mass could keep such high-speed stars from escaping from our Galaxy's core. Yet no object is visible there with any kind of telescope. Thus, the almost inescapable conclusion is that a massive black hole lies at the center of the Milky Way and holds those stars in its grip. In fact, these observations are perhaps the best evidence yet that black holes exist.

How might such a huge black hole form? We saw in chapter 14 that when a massive star reaches the end of its life, it may explode as a supernova, leaving a black hole as a remnant. Such black holes contain only 10 or so solar masses, but a black hole that formed near the core of the Milky Way during our Galaxy's youth had opportunities for growth. At that long-ago time, gas naturally tended to fall inward toward the core, where any black hole lurking there might capture it. As such a black hole swallowed material, its mass and hence its gravity would increase, allowing it to capture material ever more easily. In addition, gas lost from stars by explosion, stellar wind, or tidal force stripping from neighbors in that crowded environment might also be drawn into the black hole. With so many potential sources of matter, once a black hole formed in the center of the Milky Way, it would not stop growing until it ran out of available matter. In that way, a small black hole might have grown to $10^6\ M_\odot$ in less than 1 billion years.

The Milky Way is not alone in having a black hole at its center. Other galaxies also contain massive black holes, as we will discuss in chapter 17. For now we will simply say that, according to some hypotheses, such black holes may be a nearly inevitable byproduct of galaxy formation.

16.8 Evolution of the Milky Way

One of the major unsolved problems in astronomy is how galaxies form. For many years astronomers thought that the process was simply a large-scale version of star formation; that is, instead of a small gas cloud collapsing under its own gravity to form a single star, a huge cloud collapses to form billions of stars. Although this hypothesis can explain many properties of the Milky Way, it fails to explain some of its other features. Thus, astronomers have modified their hypothesis by suggesting that the formation of moderate to large galaxies (such as the Milky Way) requires several steps. In the first step, as proposed before, a gas cloud collapses and forms a midsize galaxy. But now, in a second step, the midsize galaxy grows by merging with other similar or smaller galaxies.

In trying to explain the origin of the Milky Way, astronomers draw on a variety of observations. We saw, in discussing the origin of the Solar System, that its disklike shape, the orbital regularities and common age of the planets, and the existence of two families of planets all were important clues to how it formed. Similar clues guide astronomers in studying the formation of the Milky Way. For example, our Galaxy has a flattened shape and contains two main groups of stars—populations I and II—that differ greatly in their properties. In particular, Pop II stars are old, contain few heavy elements, and follow orbits that are steeply inclined to the disk. Pop I stars, on the other hand are younger, contain more heavy elements, and follow roughly circular orbits that lie in the disk. Using such information, British astronomer Donald Lynden-Bell and American astronomers Olin J. Eggen and Allan R. Sandage proposed in 1962 a two-stage collapse model to explain the birth of the Milky Way, as described below.

Birth of Population I and II Stars

Our Galaxy probably began as a vast, slowly rotating gas cloud, perhaps a million light-years in diameter and containing about 10^{11} $M_{\odot}$ of gas. The cloud was composed of almost pure hydrogen and helium because no stars yet existed to make the heavier elements. As gravity began shrinking this immense cloud, clumps of gas within it grew more dense and gradually turned into stars (fig. 16.27A). These stars, composed of essentially pure hydrogen and helium, may have been exceptionally massive compared to stars forming today. Because of their greater mass, they evolved quickly and blew up

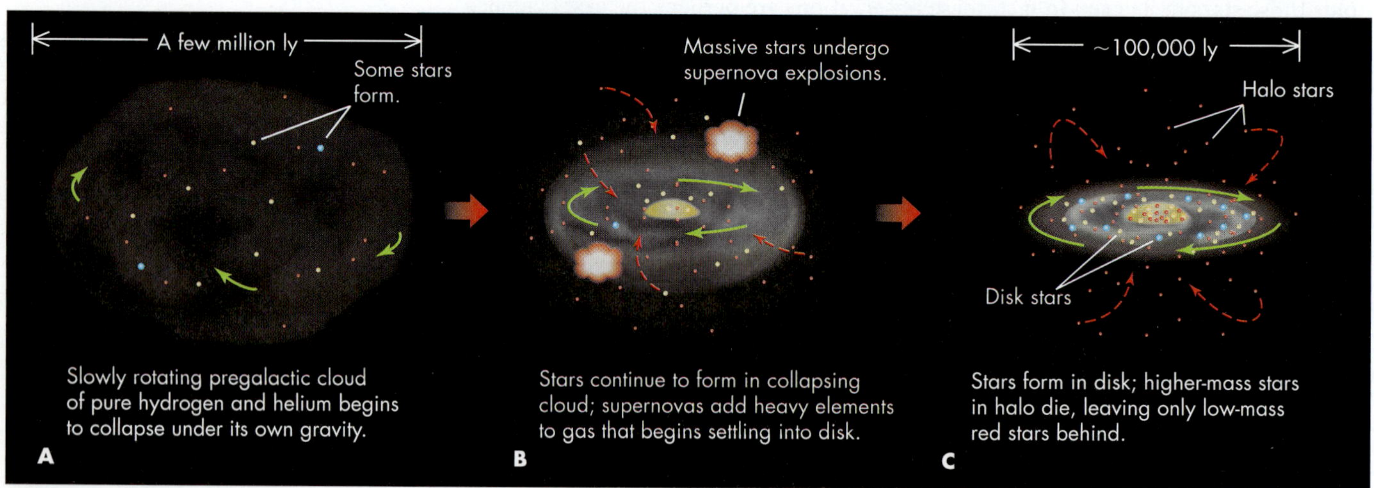

FIGURE 16.27
(A) Birth of the Milky Way as a gas cloud collapses. A first generation of stars (population II) forms. (B) Collapse continues with more population II stars forming and some exploding as supernovas. Heavy elements created by the stars exploding as supernovas are mixed into the gas. Gas not used in making stars settles into a rotating disk. (C) Population I stars form in the disk.

as supernovas, which added the first heavy elements to the gas cloud even before it had finished collapsing.

It may seem odd that stars could form and die before the cloud's collapse was complete. But collapse of such an immense cloud takes hundreds of millions of years, whereas massive stars evolve in only a few tens of millions of years—10 or so times faster.

As the collapse continued, more stars formed, now with at least some heavy elements. Because these stars formed from infalling gas, they follow elongated "plunging" orbits—becoming the population II stars that we see today.

Star formation during this initial stage was not efficient enough to consume all the collapsing gas, and some gas survived to form a rotating disk, as sketched in figure 16.27B. That disk gas, now rich in heavy elements from the death of Pop II stars, then began to form the first population I stars. These stars followed the motion of the gas they formed from. Pop I stars, therefore, contain more heavy elements than Pop II stars and move along approximately circular orbits lying in the disk, as illustrated in figure 16.27C.

Computer simulations support this collapse model for the birth of the Milky Way. For example, figure 16.28 shows how an unstructured region of gas, of the appropriate size and rotation speed, collapses. You can see that it forms a thin disk with spiral arms, surrounded by a halo. Such simulations also show that the unseen dark matter mentioned earlier plays an important role in making galaxies, because computer models that include dark matter much more closely resemble real galaxies than models that do not include dark matter.

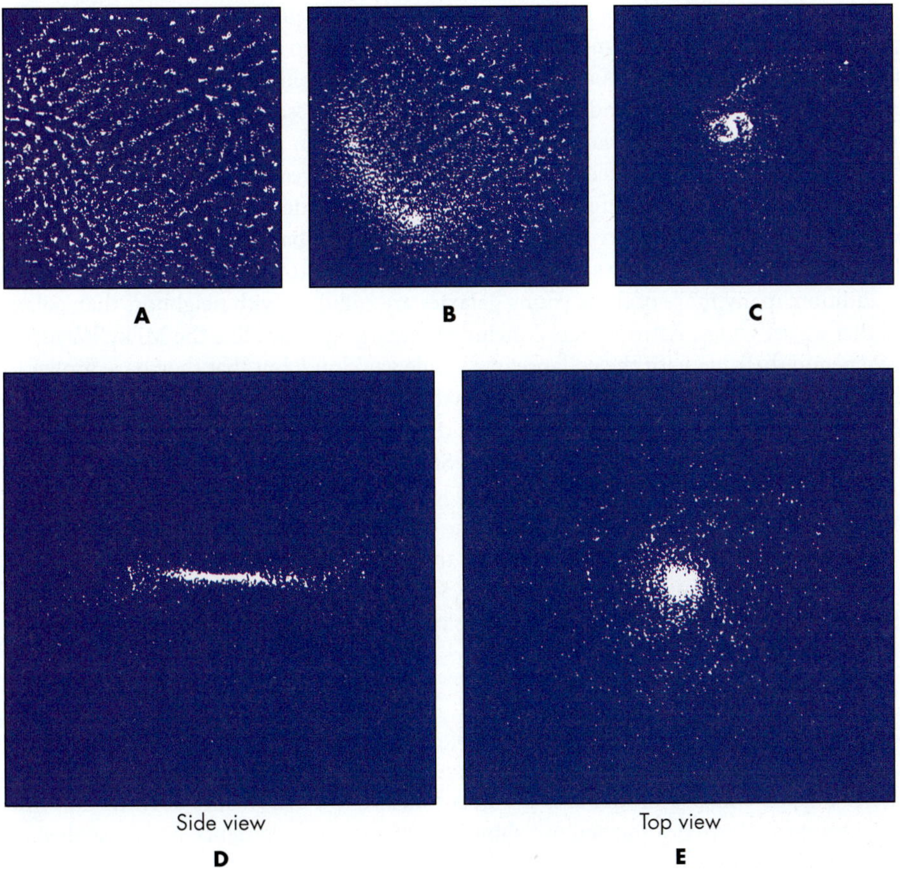

FIGURE 16.28
Computer simulation of the birth of a galaxy similar to the Milky Way. The top three frames (A), (B), and (C) show the initial state of the gas and the development of a clump that becomes the galaxy, illustrated in the bottom two frames (D) and (E). Notice the thin disk and spiral arms. Mergers are not part of this simulation.

FIGURE 16.29
Two colliding galaxies known as The Mice or more formally as NGC4676. The gravitational interaction between galaxies pulls out long streams of stars.

Evolution by Mergers

Although the collapse hypothesis explains many of the features of our Galaxy, it fails to explain some details of stellar populations. In the collapse model, population II stars should have all formed during the relatively brief period of the Galaxy's initial collapse. But observations of Pop II stars show that they formed over a much longer time span than the model predicts. There is also evidence that Pop I stars did not form steadily, but show some periods of rapid formation and others where the Galaxy was relatively quiescent. Astronomers now believe an additional process plays a major role in shaping galaxies: collisions with other galaxies.

As galaxies move through space, they occasionally collide with one another. These collisions may cause bursts of star formation when interstellar gas is compressed by the gravitational disturbance. Sometimes these collisions lead to a merger of the two galaxies into a single larger system. Astronomers today can see collisions occurring in a number of systems (fig. 16.29), and with the Hubble Space Telescope they can see that such collisions and mergers were far more common in the past. How can the Hubble Space Telescope see into the past? Recall that light takes 5 billion years to reach us from a galaxy 5 billion light-years away. Thus, when we see distant galaxies, we observe them as they were long ago, when they were young. With this "window" back in time, astronomers see that many young galaxies are smaller than galaxies we see today, such as the Milky Way. In addition, many more of these young galaxies are colliding with neighbors than galaxies that we see today. Astronomers conclude that large galaxies like the Milky Way may have begun their life with the collapse of a single gas cloud, but that they have grown to their present size by swallowing smaller galaxies.

This hypothesis is supported by recent discoveries of a number of dim, dwarf galaxies orbiting very close to the Milky Way. Several of these are so close that the Milky Way's gravity is tearing them apart,* leaving long faint streams of stars within the Milky Way's halo. The streams are stars that appear to have been ripped out of small galaxies as they were falling into the Milky Way, and these stars will eventually be widely distributed in the Milky Way's halo. For example, figure 16.30 illustrates one such stream of stars detected in an infrared survey of the sky. Thus, galaxy evolution is a dog-eat-dog event.

Population III

The two-stage collapse model also predicts that the low-mass stars born in that first generation that have not burned out should have no heavy elements, but no such stars have ever been observed. We should expect such a composition if our Galaxy formed from a cloud in which no stars had yet made heavier elements. Up to now, all stars found in the Milky Way contain at least a small percentage of heavy elements (typically

*This is similar to how gravitational tidal forces can pull apart satellites orbiting within the Roche limit. See chapter 10.

FIGURE 16.30
Diagram illustrating a stream of stars pulled out of a dwarf galaxy orbiting the Milky Way. Based on estimated distances to the stars in the stream, the figure shows what the stream (shown in red) might look like from outside our Galaxy (shown as a blue spiral).

at least 1/1000 the amount seen in the Sun), although one star has recently been observed with a mere 5×10^{-6} times the heavy element abundance of the Sun.

Astronomers have devised different explanations for why they have so far seen no **population III stars,** as these hypothetical ancient pure hydrogen and helium stars are called. First, conditions in the early Galaxy may have allowed only short-lived massive stars to form. For example, the gas may have been too hot or turbulent to make small stars that would have lived long enough for us to see them today. Alternatively, because massive stars can form and die before very low-mass stars even complete their contraction, the surface layers of Pop III stars may have been contaminated by gas ejected during the deaths of more-massive neighboring stars. Thus, it is possible that there exist Pop III stars that are masquerading as Pop II stars because they contain heavy elements that were not part of their original makeup.

The Future of the Milky Way

We know that stars change dramatically as they age. Will the Milky Way change as well? Astronomers think the answer is yes. Some major mergers may lie in the distant future. Two relatively large satellite galaxies of the Milky Way—the Large and Small Magellanic Clouds (see Looking Up #9 in the front matter)—are spiraling into our Galaxy and will eventually merge with it. From the speed of our motion toward the even larger galaxy M31 in Andromeda (see Looking Up #3), it appears that the Milky Way will itself be "eaten" in about 5 billion years.

Meanwhile, as our Galaxy ages, it steadily uses up its interstellar gas to make stars. Although on reaching the end of their lives, stars return some of their material to space as planetary nebula shells, supernova remnants, and stellar winds, eventually that gas will also end up in stellar remnants such as white dwarfs, neutron stars, and black holes, leaving the Milky Way (or our future even larger merged galaxy) no gas from which to make new stars. From time to time new gas may be captured when other galaxies merge, but eventually that gas will also be used up. Tens of billion years from now our Galaxy will contain only dead stars and dim, low-mass red dwarf stars. Gradually even these dwarfs will burn out, and our Galaxy will fade, slowly spinning in space, a dark disk of stellar cinders.

SUMMARY

Our Galaxy, the Milky Way, consists of a disk about 100,000 light-years across embedded in a halo that extends out several times farther. Together, the disk and halo contain about 10^{11} stars. Near the Sun, stars are a few light-years apart. Because the disk rotates, it is flat, and it contains star groups arranged in a pattern of spiral arms. Near its center, the Galaxy is thicker (several kiloparsecs), forming the Galactic bulge, which we can see toward the constellation Sagittarius. The Sun lies about two-thirds of the way out in the disk. From Earth, we see the Milky Way's disk as a band of dim light stretching across the night sky.

The Milky Way's stars form two main groups—population I and population II—which differ in many ways. Population I stars are found mainly in the disk. They are relatively young, tend to be blue, move in circular orbits, and are rich in heavy elements. Population II stars are found mainly in the halo and bulge. They are old, tend to be red, move in elongated and tilted orbits, and are poor in heavy elements.

About 15% of the Galaxy's mass is gas and dust clouds, which orbit with the stars and dim their light. The dust makes it difficult for astronomers to see beyond 10,000 light-years in the disk. The gas, mostly hydrogen and helium, plays a crucial role in the evolution of the Milky Way because it is the source of new stars. Radio emission from the gas, both cold hydrogen emitting 21-centimeter radiation and interstellar molecules, also helps us map the Galaxy and see through the dust that obscures our view of its structure.

At the very center of the Milky Way lies the Galactic nucleus, a dense collection of stars and gas. Astronomers think a massive black hole, about $10^6\ M_{\odot}$, lies at the center because stars orbiting there move so rapidly.

Astronomers can measure the diameter of the Milky Way and the Sun's location in it by plotting the position of globular star clusters. They measure its mass by applying the modified form of Kepler's third law to the motion of the Sun and other stars, but such measures give values of the mass that disagree with the value arrived at when stars are simply counted and their masses are added. Astronomers, therefore, infer that the Milky Way is embedded in a halo of dark matter whose nature is unknown.

The existence of two different stellar populations, I and II, leads astronomers to believe that the Milky Way formed by the collapse of a vast gas cloud about 14 billion years ago. The first stars to form became the population II stars. Some gas did not turn into stars during collapse but instead collected into a rotating disk. From supernova explosions of population II stars, heavy elements became part of the disk of gas. Stars eventually formed from this material to become the population I stars such as our Sun. Subsequently the Milky Way has captured smaller galaxies and grown larger. Eventually the Milky Way will convert all its available gas to stars, and as the surviving stars die, our Galaxy will grow dark.

QUESTIONS FOR REVIEW

1. (16.1) What does the Milky Way look like in the night sky?
2. (16.1) How do we know our Galaxy is a flat disk?
3. (16.1) How did Shapley deduce the Milky Way's size and the Sun's position in it?
4. (16.1–16.2) Draw a sketch of the Milky Way Galaxy and label its major components. Where is the Sun in the Milky Way?
5. (16.2) Roughly how big in diameter is our Galaxy, and how much mass does it contain?
6. (16.3) What are some differences between Pop I and Pop II stars?
7. (16.3) Summarize the differences between open and globular clusters. Are both still forming today in the Milky Way?
8. (16.4) How do we know interstellar matter exists? What different kinds and forms are there?
9. (16.4) How do we know that some interstellar matter is dust? What is the other main category of interstellar matter?
10. (16.4) How does interstellar dust affect our observations of stars, the Milky Way, and other galaxies? What is the zone of avoidance?
11. (16.4) What is the difference between an emission and a reflection nebula?
12. (16.4) What are some ways that we know cold clouds of gas exist in space?
13. (16.4–16.5) What evidence makes astronomers conclude the Milky Way has spiral arms?
14. (16.5) How is star formation related to spiral arms?
15. (16.6) How can we determine the Milky Way's diameter?
16. (16.6) How can we determine the Milky Way's mass?
17. (16.6) What is meant by dark matter? Why do astronomers conclude the Milky Way may contain such as-yet unobserved material?
18. (16.7) What is the evidence for a black hole at the center of the Milky Way? How does it compare to the black holes discussed in chapter 15?
19. (16.8) Describe one model for the origin of the Milky Way. How does this model explain the difference between Pop I and Pop II stars?
20. (16.8) How do mergers affect the evolution of the Milky Way?
21. (16.8) What is meant by Pop III stars?

THOUGHT QUESTIONS

1. (16.1–16.7) What would the Milky Way look like in the night sky to an observer on a planet at the very edge of the Galaxy? What would it look like to an observer in the Galactic bulge?
2. (16.3) Suppose the Sun were a population II star in the halo of the Milky Way. In what ways would that make studying the Milky Way easier?
3. (16.3) Knowing whether a selection effect is affecting your data is very important in astronomy. Imagine you were an alien visitor who only had an hour to study how long humans are babies, children, young adults, middle-aged, and elderly. Describe how your results would be affected by selection effects if you visited a shopping mall for an hour on a Wednesday afternoon in October, if you visited an elementary school, if you visited the mall at 9 P.M., or if you visited the mall on the weekend. What if instead you were studying human clothing, and you landed in Chicago or Florida in December?
4. (16.3) Understanding possible selection effects is very important in polling to make sure results are accurate. If you were conducting a telephone poll about two presidential candidates, one of whom was more likely to appeal to older people and one more likely to appeal to younger people, what effects might you have to think about if you (1) could only call people with land lines (no cell phones), (2) could only call people with a cell phone, (3) could call both groups?
5. (16.5) If you were explaining the density-wave theory to another person, what additional examples can you think of besides the slow car described in section 16.5 where a density-wave causes clumps of people, cars, or something else?
6. (16.5) Cut four circles of cardboard whose diameters are 2, 4, 6, and 8 inches. Fasten them together through their centers with a brass paper fastener. Draw a circle about 2.5 inches in radius, centered near the edge of the 4-inch disk that marks all the disks. Now turn the circles in a way that reflects differential rotation. Use what you see to explain how self-propagating star formation might make spiral arms.
7. (16.5–16.6) Are spiral arms permanent structures? Explain, using the idea of the density-wave model in the book. What would happen to a spiral arm if the same stars stayed in it? (Consider how long it would take a star on the end of an arm to orbit once around compared to a star halfway in on the arm, if both have a velocity of about 225 km/sec).
8. (16.6) How could we measure the mass of the Milky Way using a distant orbiting globular cluster?
9. (15.2–15.3/16.7) How might you argue that there is a black hole at the center of the Milky Way, instead of, for example, a region filled with many neutron stars? What observations would help you rule out such a case?
10. (16.2/16.4/16.8) What would the night sky look like if there were no dust or gas in the Milky Way, but the Galaxy were otherwise the same as it is today? What will the night sky look like when the Milky Way is very old and has used up all its star-forming materials?

PROBLEMS

1. (16.1–16.2) Make a scale diagram of the Galaxy looking down from above. Let 1 inch = 20,000 light-years, and draw a 5-inch diameter circle to represent the disk of the Galaxy (see fig. 16.7). The Sun is about 26,000 light-years out from the center, so place a small dot 1.3 inches out from the center. The most distant stars potentially visible to the naked eye (a 100,000 solar luminosity star that appears at an apparent magnitude of 6) would be no more than about 15,000 light-years from the Earth, most bright naked-eye stars are less than 5000 light-years away. Draw circles around the Sun at distances corresponding to 5000 and 150,000 light-years. Are any of the stars visible in the night sky from outside the Milky Way Galaxy?
2. (16.1–16.2) Consider again the models in the previous question. If you were looking up or down through the disk of our Galaxy, about how far would the farthest star be that you could see with your naked eye (assume the Sun is midway through the disk)? Can we see to the edge of the disk? If another galaxy were at least as far away as half the width of the Milky Way, could you see any individual stars in it with the naked eye?
3. (16.6) Given that the Sun moves in a circular orbit of radius 8 kiloparsecs around the center of the Milky Way and that its orbital speed is 220 kilometers per second, work out how long it takes the Sun to complete one orbit of the Galaxy. (*Hint:* Remember to convert kiloparsecs to kilometers and convert your final answer to years, not seconds.)
4. (16.6) Gas is observed to be orbiting in a circle at 200 km/sec about a dark central mass. If the gas's orbit has a radius of 1 parsec, how much mass must be present to hold it in orbit?
5. (15.3/16.6) Suppose the mass in the previous problem were packed into a sphere whose radius is 20 AU. Would it be a black hole?
6. (16.6) According to figure 16.24, the Milky Way's rotation speed is about 230 km/sec at a distance of about 16 kpc from its center. Use the method discussed in the box on page 445 to find the mass of the Milky Way implied by these values. *Hint:* These values correspond to a distance of about 3.3×10^9 AU and an orbital period of about 4.3×10^8 years.
7. (16.6–16.7) How long is the period for a star orbiting the black hole at the center of our Galaxy that is about 2800 AU from the black hole at the farthest part of its orbit, but that gets as close as 1200 AU at the nearest? Take the mass of the black hole to be 5×10^6 solar masses. Compare your answer to the path of star S0-20 in Figure 16.26.
8. (16.7) Calculate the Schwarzschild radius of the black hole at the center of the Galaxy using 5×10^6 solar masses for its mass. How many AU is the Schwarzschild radius?
9. (16.2/16.7) The 5×10^6 solar mass black hole at the center of the Galaxy may have gained its first million solar masses in the first billion years of the Milky Way's existence. How much mass must it have absorbed on average each year since then to reach 5 million solar masses today?

TEST YOURSELF

1. (16.1) One way astronomers deduce that the Milky Way is a disk is that they
   (a) see stars arranged in a circular region around the north celestial pole.
   (b) see far more stars along the band of the Milky Way than in other directions.
   (c) see a large dark circle silhouetted against the Milky Way in the Southern Hemisphere.
   (d) see the same number of stars in all directions in the sky.
   (e) None of the above
2. (16.2) Select all that are correct. Most of the stars in the Milky Way
   (a) are not visible to the naked eye.
   (b) have plunging orbits around the Galactic center.
   (c) have the same or less mass than the Sun.
   (d) are surrounded by cold gas clouds.
   (e) burn out before they leave a spiral arm.
3. (16.3) A young blue star moving along a circular orbit in the disk is a
   (a) Pop I (b) Pop II star (c) Pop III star.
4. (16.3) Globular clusters have mainly ______ stars.
   (a) Pop I (b) Pop II (c) Pop III
5. (16.4) Astronomers know that interstellar matter exists because (select all that apply)
   (a) they can see it in dark clouds and clouds that absorb light.
   (b) the matter creates narrow absorption lines in the spectra of some stars.
   (c) X-rays cannot penetrate the clouds toward the center of the Galaxy.
   (d) they can detect radio waves coming from atoms and molecules in the cold gas.
   (e) spacecraft have sampled clouds near Orion.
6. (16.5) Astronomers think the Milky Way has spiral arms because (select all that are correct)
   (a) they can see them unwinding along the celestial equator.
   (b) radio maps show that gas clouds are distributed in the disk with a spiral pattern.
   (c) young star clusters, HII regions, and associations outline spiral arms.
   (d) globular clusters outline spiral arms.
   (e) other galaxies that also have gas and dust have spiral arms.
7. (16.6) The modified form of Kepler's third law allows astronomers to determine the Milky Way's
   (a) mass.
   (b) age.
   (c) composition.
   (d) shape.
   (e) number of spiral arms.
8. (16.7) Astronomers think there is a massive black hole at the center of the Galaxy because (select all that are correct)
   (a) without it, there does not seem to be enough mass to hold the Galaxy together.
   (b) the center of the Galaxy is dark in visible light images.
   (c) other galaxies have massive black holes at their centers.
   (d) they have plotted the orbits of stars and calculated a large, dark mass is present there.
   (e) an enormous supernova explosion was observed in the direction of the center.
9. (16.8) Faint streams of gas in the Milky Way's halo are interpreted as evidence that
   (a) parts of the Milky Way are still collapsing.
   (b) the Milky Way is losing gas and will eventually have none.
   (c) supernova explosions are pushing gas out of the disk.
   (d) the Milky Way is stripping material from smaller dwarf galaxies.
   (e) the Milky Way does not have enough mass to hold all the gas and dust in the disk.

KEY TERMS

association, 432
bulge, 428
dark matter, 429
dark nebula, 436
density-wave model, 441
disk, 428
emission nebula, 438
globular cluster, 433
halo, 428
HII region, 439
interstellar matter, 434
Milky Way, 423
nucleus of galaxy, 428
open cluster, 432
population I stars, 430
population II stars, 430
population III stars, 451
reddening, 435
reflection nebula, 437
rotation curve, 444
scattering, 435
selection effect, 430
self-propagating star formation, 442
spiral arm, 428
star clusters, 432
21-centimeter radiation, 440
zone of avoidance, 436

FURTHER EXPLORATIONS

Basri, Gibor. "The Discovery of Brown Dwarfs." *Scientific American* 282 (April 2000): 76.

Berendzen, Richard, Richard Hart, and Daniel Seeley. *Man Discovers the Galaxies.* New York: Science History Publications, 1976.

Binney, James. "The Evolution of Our Galaxy." *Sky and Telescope* 89 (March 1995): 20.

Bromm, Volker. "Out of the Dark Ages: The First Stars." *Sky and Telescope* 111 (May 2006): 30.

Combes, Francoise. "Ripples in a Galactic Pond." *Scientific American* 293 (October 2005): 42.

Dorminey, Bruce. "Is Our Galaxy Running out of Gas?" *Astronomy* 37 (January 2009): 38.

Hoskin, Michael. "William Herschel and the Making of Modern Astronomy." *Scientific American* 254 (February 1986): 106.

Jayawardhana, Ray. "How the Milky Way devours Its Neighbors." *Astronomy* 36 (March 2008): 34.

Loeb, Abraham, and T. J. Cox. "Our Galaxy's Date with Destruction." *Astronomy* 36 (June 2008): 28.

Reynolds, Ronald J. "The Gas Between the Stars." *Scientific American* 286 (January 2002): 34.

Scoville, Nick, and Judith S. Young. "Molecular Clouds, Star Formation, and Galactic Structure." *Scientific American* 250 (April 1984): 42.

Trimble, Virginia, and Samantha Parker. "Meet the Milky Way." *Sky and Telescope* 89 (January 1995): 26.

Wakker, Bart P., and Philipp Richter. "Our Growing, Breathing Galaxy." *Scientific American* 290 (January 2004): 38.

Website

Visit the *Explorations* website at **http://www.mhhe.com/arny** for additional online resources on these topics.

Q FIGURE QUESTION ANSWERS

WHAT IS THIS? (chapter opening): This infrared image shows the inner 2 light-years of the core of the Milky Way. The exact center of the Galaxy is indicated by the two small arrows. Observations of the motion of a star near this spot are strong evidence that a black hole lies there.

FIGURE 16.1: The glows are distant city lights.

FIGURE 16.11: They are globular clusters.

FIGURE 16.12: The Sun looks red when it sets because molecules and dust in our atmosphere interact more strongly with the blue light than with the red light and scatter the blue light. As a result, less blue light reaches us, so the Sun looks redder. Interstellar dust does the same thing to starlight—it preferentially removes the blue light, leaving the star looking redder than it really is.

FIGURE 16.16: Red light has a longer wavelength than blue and so it does not interact as strongly with the dust. It is therefore scattered less.

PROJECTS

1. **Observe the Milky Way:** Find a spot where the sky is dark and you can see most of the sky. Sketch the path of the Milky Way across the sky, as best you can see it. Do you notice any particularly bright stars along the Milky Way or elsewhere? Can you see any dark regions along the Milky Way where the number of stars looks small? Mark them on your sketch. Then compare your sketch to one of the star maps inside the covers of this book.
2. **Reddening:** You can get a feeling for how reddening and scattering work if you have a clear glass container, a flashlight, and some powdered milk. It works best if one direction along the container is longer than the others (a rectangular pan, or a tall glass). Fill the container with water and turn the lights down in the room. If you are using a glass, shine the flashlight up from the bottom and look down into the top. Record the color of the light. Now, look at the glass from the side—is much light scattered out the side? Put in a small amount of powdered milk—you can keep adding it later for a stronger effect. Now, put the flashlight underneath and look down into the glass. What color is the light? What color is the light if you look at the side of the glass? If you add more milk, what happens? Which side is like looking at a dark nebula, and which is like looking at a reflection nebula? The powdered milk grains are actually fairly close in size to interstellar dust grains.

A cluster of galaxies that contains a diverse assortment of galaxy types.

KEY CONCEPTS

- Galaxies have a variety of shapes, but three types dominate: spirals, ellipticals, and irregulars.
- Ellipticals have the least amount of gas and dust; irregulars have the most.
- Deep observations of distant galaxies allow us to see what galaxies looked like when they were young.
- Galaxies appear to have grown in size over the history of the Universe through mergers and cannibalism.
- Galaxies may change type through interactions with other galaxies or intergalactic gas.
- All but the nearest galaxies are receding from us and show a redshift in their light.
- The farther a galaxy is from us, the faster it recedes (Hubble law).
- Galaxies are mostly made of dark matter, which may be a kind of elementary particle that does not interact with the "normal" matter we are made of.
- Many large galaxies have an active galactic nucleus that has a very high luminosity and may emit jets of hot, high-speed gas.
- The power source for an active galaxy is probably a supermassive black hole.
- Galaxies are often found in groups and larger clusters.

CONCEPTS AND SKILLS TO REVIEW

- Inverse-square law (13.2)
- Standard-candle method (13.1, 13.8)
- Red light has a longer wavelength than blue (4.1)
- Newton's law of gravity (3.5)
- Modified form of Kepler's third law (3.7)

Beyond the edge of the Milky Way and filling the depths of space are billions of other star systems. These remote, immense star clouds are called galaxies, named after our own Galaxy, and their properties are the subject of this chapter.

While some galaxies resemble the Milky Way, many others differ in shape, content, and size. Some are "egg-shaped" and contain mostly old stars, while others are irregular and contain predominantly young stars. Some are rich in interstellar matter, while others have hardly any at all. The largest galaxies contain millions of times more stars than the smallest galaxies. Yet other galaxies have a tremendously powerful energy source at their core that emits as much energy as the entire Milky Way but from a region less than a few light-years in diameter. Many of these so-called active galactic nuclei eject gas in narrow jets at nearly the speed of light.

Galaxies are not spread uniformly throughout space. Just as stars often lie in clusters, so too galaxies often lie in galaxy clusters, and galaxy clusters themselves lie in clusters of galaxy clusters. These immense collections of galaxies are not stationary. They are separating from one another, caught up in the expansion of the Universe, like snowflakes carried on the wind.

Astronomers can see all this diversity in galaxies, but as yet they only partially understand how it arises. Unlike stars, whose structure and evolution are well understood, galaxies remain mysterious in many ways. Astronomers do not yet agree about why some galaxies are disks while others are egg-shaped. Although this lack of certainty can be frustrating, it offers us a chance to see how scientists work, and, in particular, how astronomers begin with basic observations and go on to construct hypotheses to explain what they observe. For example, many galaxies appear to be undergoing collisions with their neighbors. Could past interactions of this kind account for the different galaxy shapes?

Q: WHAT IS THIS? See end of chapter for the answer.

17.1 Discovering Galaxies

A **galaxy** is an immense swarm of hundreds of millions to hundreds of billions of stars and vast clouds of interstellar gas. Each star moves along its own orbit, held within the galaxy by the combined gravitational force of the rest of the matter within the galaxy. Each galaxy (apart from those undergoing collisions) is therefore an independent and isolated star system.

Early Observations of Galaxies

All galaxies (except ours) are extremely distant from Earth, and few are visible without a telescope. Of those similar in size to the Milky Way, the nearest is about 2.5 million light-years away—M31, the Great Galaxy in Andromeda (fig. 17.1). You can see M31 on clear, dark nights in autumn and winter in the Northern Hemisphere. It looks like a pale elliptical smudge on the sky (see Looking Up #3 in the front matter). Early observers such as the tenth-century Persian astronomer Al-Sufi noted this object, and it appeared on early star charts as the "Little Cloud." In the Southern Hemisphere, you can easily see with the naked eye the Large and Small Magellanic Clouds, two satellite galaxies of the Milky Way (fig. 17.2 and Looking Up #9). Today, astronomers have studied millions of galaxies besides the Magellanic Clouds and M31, and they estimate that the visible Universe contains tens of billions more.

The study of other galaxies began in the eighteenth century, when the French astronomer Charles Messier (pronounced *mess-yay*) accidentally discovered many galaxies during his searches for new comets. He noticed a large number of faint, diffuse patches of light, and to avoid confusing them with comets, he assigned them numbers and made a catalog of their positions. Although many of Messier's objects have since been identified as star clusters or glowing gas clouds in the Milky Way, several dozen are galaxies. These and the other objects in Messier's catalog are still known by their Messier, or M number, such as M31, mentioned above.

FIGURE 17.1
Photograph of M31, the Andromeda galaxy, the nearest spiral galaxy to us.

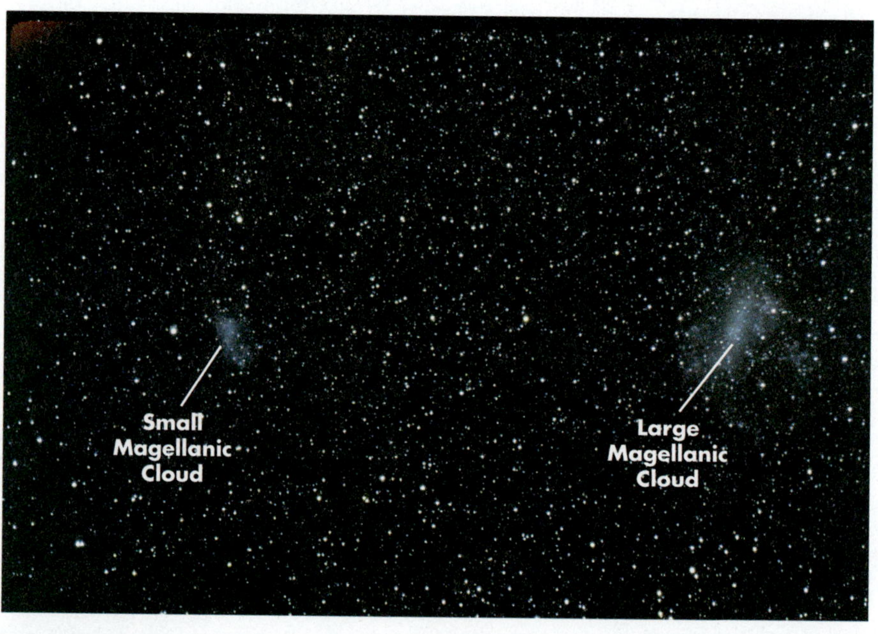

FIGURE 17.2
Photograph of the Large and Small Magellanic Clouds.

In the nineteenth century, other astronomers, such as Sir William Herschel, began systematically to map the heavens, finding and cataloging numerous astronomical objects, including galaxies. Herschel's work, revised and added to by his son and John Dreyer, is now known as the New General Catalog (or NGC, for short), and it gives names to many galaxies, such as NGC 1275, a peculiar, active galaxy. But many galaxies appear in more than one catalog and so bear several names. For example, Messier's M82 is the same galaxy as NGC 3034.

We know today that galaxies are immense star systems like our Milky Way, but this idea was not widely accepted by astronomers until about 1920. Before that time, astronomical opinion was strongly divided as to whether galaxies were huge star systems like the Milky Way or merely small satellites of our own Galaxy. This controversy was aired in detail during a debate organized by the National Academy of Sciences, in which two American astronomers, Harlow Shapley and Heber Curtis, took opposing sides. Shapley argued that the Milky Way is huge and that other galaxies (at that time called nebulas) are merely small, nearby companions. Curtis, on the other hand, argued that the Milky Way is smaller than Shapley claimed, and that the nebulas are star systems like our own and are immensely distant from it.

Historians differ on who "won" the debate. Although Shapley was more nearly correct about our Galaxy's size, Curtis was right that the nebulas are distant systems like the Milky Way. In any case, once the distance and true nature of galaxies were understood, astronomers could more successfully study galaxies' properties.

Types of Galaxies

Even early nineteenth-century observers noticed that not all galaxies look the same. However, it was Edwin Hubble, an American astronomer working first as a graduate student at the University of Chicago and later at Mount Wilson in California in the 1920s, who demonstrated that galaxies could conveniently be divided on the basis of their shape into three main types: spirals, ellipticals, and irregulars. **Spiral galaxies** have two or more arms winding out from the center. We saw in chapter 16 that the Milky Way is of this type, and figure 17.3 shows photographs of several others seen with their disks oriented at different angles to our line of sight.

A

B

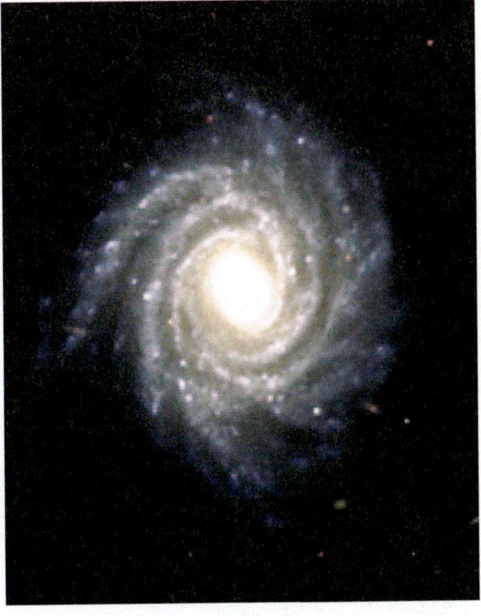

C

FIGURE 17.3
Pictures of typical spiral galaxies. (A) NGC 4565; (B) NGC 4414; and (C) NGC 1288. These examples show spiral galaxies with their disks oriented from nearly edge-on (A) to nearly face-on (C).

FIGURE 17.4
Pictures of a few elliptical galaxies: NGC 205, M49, and M87.

The second type shows no signs of spiral structure. These galaxies have a smooth and featureless appearance and a generally elliptical shape, as can be seen in figure 17.4. Accordingly, astronomers call them **elliptical galaxies.** These galaxies have no disk or arms—the stars are distributed around the center in all three dimensions, some nearly spherical and others more squashed or football-shaped.

Galaxies of the third major type show neither arms nor a smooth uniform appearance. In fact, they generally have stars and gas clouds scattered in random patches. For this reason, they are called **irregular galaxies** (fig. 17.5).

Spiral, elliptical, and irregular galaxies are sometimes denoted by *S*, *E*, and *Irr*, respectively. However, in addition to these three main types, astronomers recognize two additional galaxy types closely related to spiral systems. The first of these are **barred spiral galaxies,** which have arms that emerge from the ends of an elongated central region, or bar, rather than from the core of the galaxy, as shown in figure 17.6. It is this bar that gives them their name, and they are denoted *SB* galaxies to distinguish them from normal spirals. The second of these that Hubble identified is a kind of "spiral without a spiral." These are galaxies with disks with no evidence of arms (fig. 17.7), which he called *S0* (pronounced "S"-zero) galaxies.

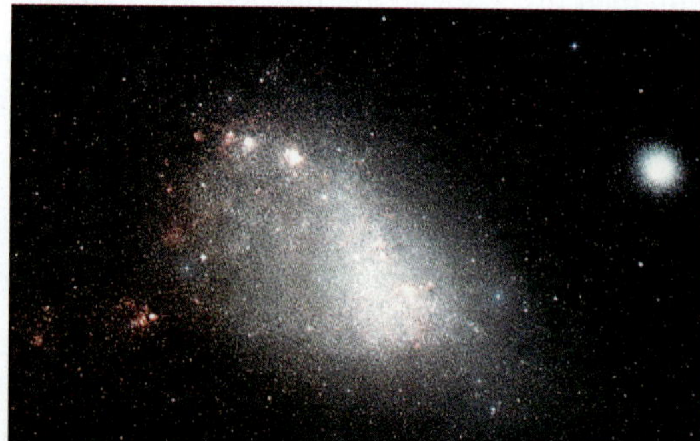

FIGURE 17.5
Photographs of two irregular galaxies, NGC 6822 and the Small Magellanic Cloud.

FIGURE 17.6
Barred spirals: NGC 1365 and NGC 5236 (M83). The "bar" refers to the elongated shape of the galaxy's central bulge.

FIGURE 17.7
Two S0 galaxies, NGC 3384 and NGC 4552. These galaxies look similar to ellipticals, but some of the stars are distributed in a disk, as occurs in spiral galaxies.

Hubble refined his classification system one step further, noticing that both spiral and elliptical galaxies can be subdivided into additional classes that, when properly ordered, show a smooth transition from spherical E type of galaxies to flatter E galaxies, to spirals, and to irregulars. This sequence of types was illustrated by Hubble in the "tuning fork" diagram, illustrated in figure 17.8.

Astronomers continue to use this diagram today as a useful way to organize the galaxy types and subtypes that Hubble described; however, the distinction into barred and unbarred spirals is now regarded as less significant. Computer models of galaxies

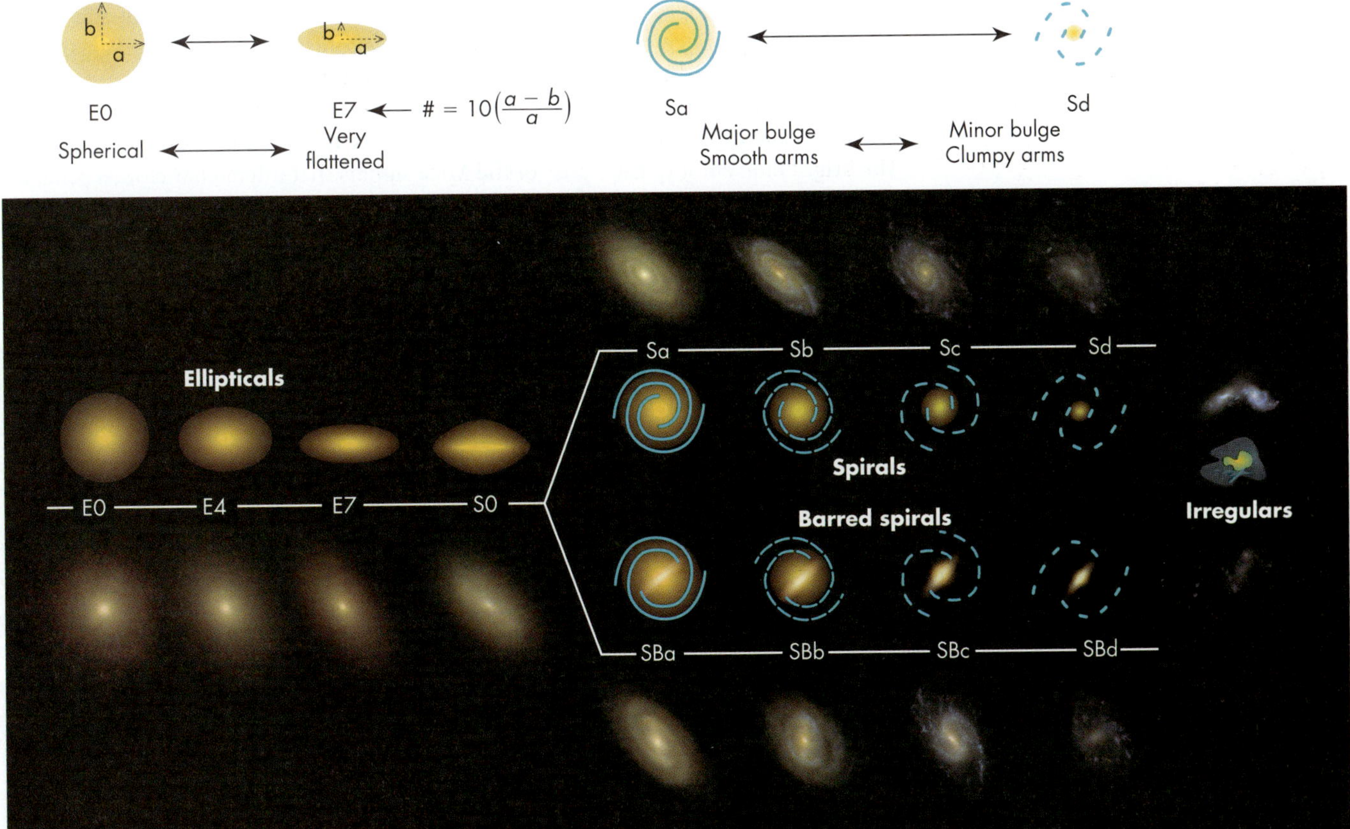

FIGURE 17.8
The Hubble tuning-fork diagram. Elliptical galaxies are subdivided by how "flattened" they look. Spirals are subdivided according to how large their bulges are and how tightly wound their arms are. (Images are drawn from the Sloan Digital Sky Survey, with foreground stars removed to make the galaxies' structures clearer.)

show that most disk-shaped systems will form a bar if they undergo a gravitational disturbance, perhaps as the result of a close encounter with a neighboring galaxy. Even our own Milky Way has a weak bar. The bar forms because a gravitational disturbance alters the orbits of the stars, causing them to bunch up in an elliptical pattern that we see as the bar. Some models also suggest that over hundreds of millions of years, the orbits spread out again and the bar dissolves. Thus, bars may appear and disappear, and so an ordinary spiral galaxy may change into a barred spiral and then back again.

Since Hubble's time, it has also been recognized that the irregular class is composed of two very different types. The majority are smaller galaxies, such as the Magellanic Clouds. They are often called dwarf galaxies, and they lack a coherent structure. Examples of these are shown in figure 17.5. However, some large spirals and ellipticals can also become classified as irregulars when they undergo interactions with neighboring galaxies that distort their shapes.

When galaxies collide,* the gravitational force of the galaxies alters the orbits of their stars and thereby changes the shape of each galaxy. Figure 17.9 shows a computer simulation of such a close passage, and you can see the stars torn from each galaxy by their mutual gravitational force and flung outward in long, luminous arcs. The fifth picture in the sequence shows an image of two real galaxies with just such plumes of stars.

Although interactions of this sort are disruptive to galaxies, they rarely harm the individual stars. Because stars are so far apart within a galaxy, most stars move harmlessly past each other. Thus, galaxy collisions are much like tossing two handfuls of sand together: most of the grains simply pass by one another without hitting. Clouds of gas and dust in the galaxies are not so lucky. Being much larger than stars, clouds do collide and their impact may compress them enough to trigger a burst of star formation within them. Such "starburst galaxies," as these collisionally stimulated galaxies are called, are among the most luminous galaxies known.

ANIMATION
Galaxies with tails

A burst of star formation caused by a collision is illustrated in figure 17.10A, two galaxies (called the "Mice") that have been severely distorted by each other's gravity. The bright blue color of the "tails" of the Mice signals that interstellar clouds collapsed to form stars even as they were being flung away from the galaxies.

*Astronomers use the word *collision* to describe any close interaction between galaxies, even if the galaxies do not directly hit each other.

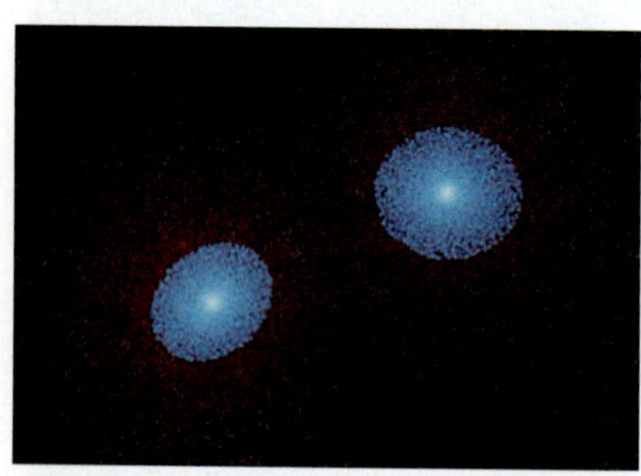

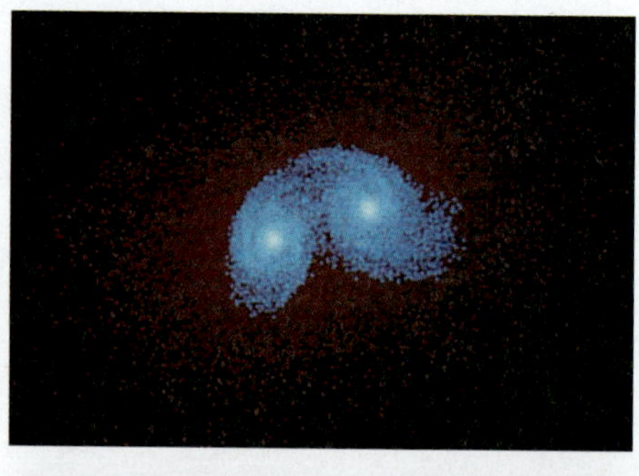

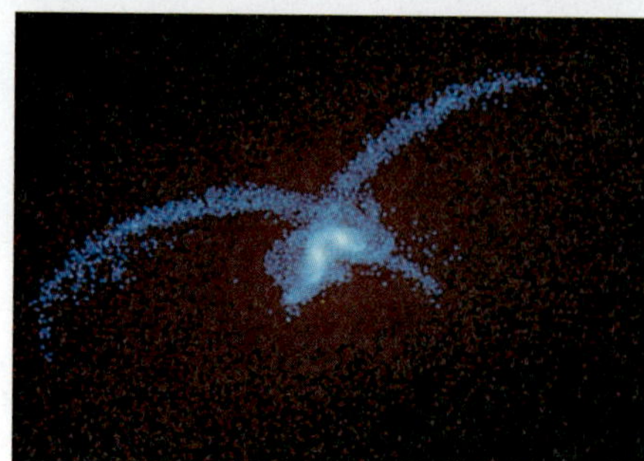

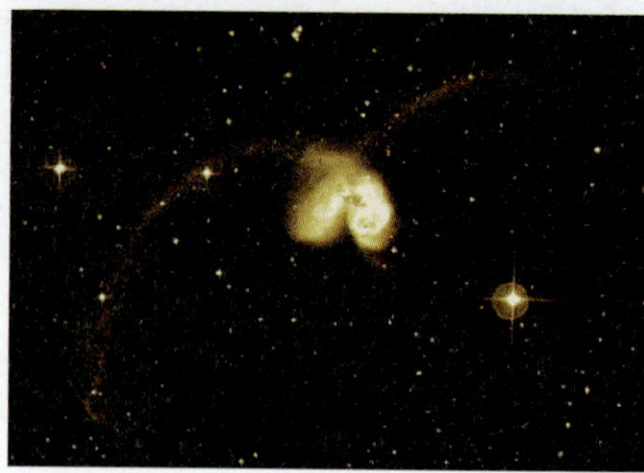

FIGURE 17.9
Computer simulation of two galaxies colliding and photograph of system called the "Antennae" undergoing just such a collision.

A

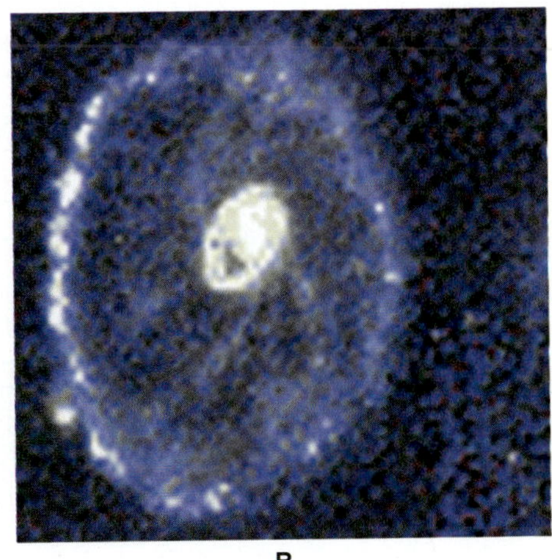

B

FIGURE 17.10
(A) The "Mice." This is a pair of galaxies that collided, causing star formation in the interstellar gas even as the material was flung out in long streams of stars. (B) The "Cartwheel" galaxy. The white central "blob" is the bulge of this ring-shaped galaxy. The galaxy that passed through this galaxy is far off to the right of this image.

Figure 17.10B shows what can happen when one galaxy plows directly into another. The head-on collision created what astronomers call a ring galaxy. The colliding galaxy sent a compression wave traveling out through the disk, causing interstellar clouds to collapse and form stars. This is somewhat like the compression we discussed in the arms of the Milky Way, but more extreme. We might imagine that over time, after the brightest young stars have died and the galaxy's rotation has smeared out the stars formed here, that the galaxy will again show a spiral pattern.

While the odd shapes of colliding galaxies might cause them to be temporarily labeled as "irregulars," they will probably settle back into spirals or ellipticals after the collisions. This is a possible way for galaxies to change shape over time, which we will examine in more detail a little later, but first we want to catalog other fundamental distinctions between galaxy types.

Differences in the Stellar and Gas Content of Galaxies

Since Hubble's time, astronomers have discovered that spiral, elliptical, and irregular galaxies differ not only in their shape but also in the type of stars they contain. For example, spiral galaxies contain a mix of young and old stars (Pop I and II, see chapter 16), but elliptical galaxies contain mostly old (Pop II) stars. This difference in the kinds of stars composing S and E galaxies is understandable in terms of their different gas content. To make young (Pop I) stars, a galaxy must contain dense clouds of gas and dust. Spiral galaxies like the Milky Way have about 15% of the mass of their disk in the form of such interstellar clouds, and so they can easily make the young stars of their spiral arms. Radio images of hydrogen emission (fig. 17.11) show the presence of gas in spirals and irregulars. On the other hand, ellipticals contain much less interstellar matter. In fact, astronomers used to think that the E type of galaxies contained no interstellar gas or dust at all. However, more recent observations made with X-ray telescopes show that many E galaxies contain very low-density but very hot (10^7 K) gas. Such hot gas does not readily form stars, and so it is no surprise that elliptical galaxies rarely contain young, blue stars like the ones we see in the disks of spiral systems.

Although young blue stars are rare in elliptical systems, such stars are common in most irregular galaxies. Many of these Irr systems also contain large amounts of

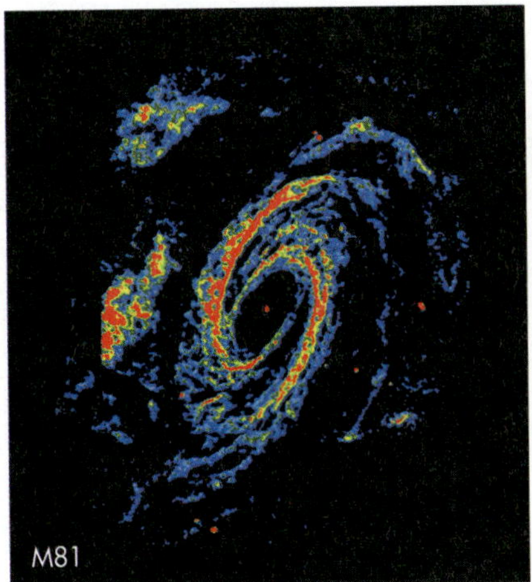

FIGURE 17.11
False-color map of 21-centimeter radio emission from hydrogen in the spiral galaxy M81. Hydrogen is also visible in a faint irregular galaxy to its left. Note the concentration of hydrogen in M81's spiral arms, and the lack of it in the bulge.

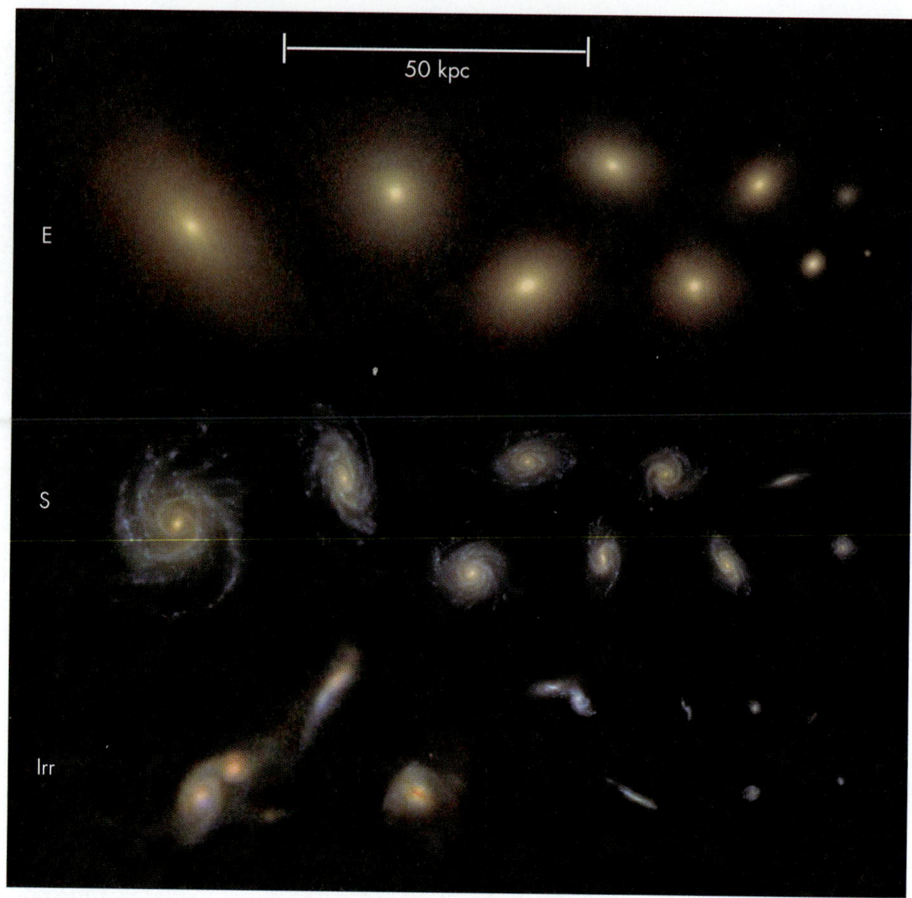

FIGURE 17.12
Images of galaxies shown with their correct relative sizes. The spirals are all in Hubble's Sc class (see fig. 17.8). The irregulars are a mix of interacting galaxies and dwarf galaxies. Blue colors indicate current star formation. (Galaxy images were drawn from the Sloan Digital Sky Survey. Foreground stars have been digitally removed to make the galaxy structure clearer.)

interstellar matter. In fact, some have more mass in gas than in stars. Accordingly, astronomers think that these galaxies, though possibly just as old as other types, have simply not yet used up much of their gas in making stars.

Apart from their different star and interstellar matter content, S, E, and Irr galaxies have few other differences to generalize about. For example, if we look at the mass and radius of galaxies, we do not find that all ellipticals are huge and all spirals are small. Instead, we discover that elliptical galaxies range enormously in size. Some are not much larger than globular clusters and contain only 10^7 solar masses of material; others are monster galaxies, 10 to 100 times the mass of the Milky Way. This spread is shown in figure 17.12, which shows sets of ellipticals, spirals, and irregulars all reproduced to their correct relative sizes.

What fraction of galaxies makes up each type? Even that question has no simple answer. If we make a census of the galaxies near the Milky Way, we find that most galaxies are dim dwarf E and Irr systems, sparsely populated with stars. These dim galaxies are undetectable at greater distances, where we can see only the more luminous galaxies. If we make a census of those luminous objects, we find that the majority of galaxies are spirals, but the percentage depends on whether we look in crowded or in more sparsely populated regions. Where galaxies are crowded together, only about 10% of the members are spirals, but in less crowded regions that proportion is about 80%.

The Evolution of Galaxies: Collisions and Mergers

The diversity of galaxy form and content requires an explanation, and a major area of astronomical research today is exploring why galaxies come in such distinct and different forms. In recent years, with the aid of the Hubble Space Telescope, astronomers have begun to develop a picture of how galaxies form and evolve. That instrument's ability to see fine details in distant, ancient galaxies has allowed astronomers to observe galaxies at various ages all the way back to when they were less than a billion years old.

The most distant galaxies are so faint that they require hundreds of hours to detect. Because it would take millions of years to study the whole sky at this depth, astronomers selected two tiny "blank" regions on the sky to concentrate their efforts. They produced "deep fields" by pointing the Hubble Space Telescope at these regions for hundreds of hours, allowing them to examine galaxies all the way back to near the beginning of time. A portion of one of these deep fields is shown in figure 17.13. This region is tiny—if you hold a dime at arm's length, the eye of Theodore Roosevelt could completely cover this area—yet it contains thousands of galaxies. The galaxies appear crowded together, but they are actually very spread out—distributed from nearby to almost 13 billion light-years' distance.

FIGURE 17.13
The Hubble Space Telescope observed two "deep fields," small regions of the sky, for more than 100 hours to detect extremely distant galaxies. Nearly every dot in this image is a galaxy. The galaxies are not close together in space, but are spread out over more than 10 billion light-years along our line of sight.

Many of the most distant and therefore youngest galaxies are much smaller than systems like the Milky Way. Some examples of distant galaxies detected with the Hubble Space Telescope are shown in figure 17.14. Not only are these young galaxies small, they are more numerous than ordinary galaxies today. Thus, these young galaxies must somehow disappear. How can a galaxy disappear? From what is currently known, astronomers tend to think that they collide and merge into bigger systems. This would account both for the formation of large systems like the Milky Way and for the smaller number of galaxies we see today.

The importance of interactions, or what some astronomers call "harassment," is strikingly confirmed in the images of distant systems made with the Hubble Space Telescope (fig. 17.14). Remote galaxies show signs of disturbance by neighboring galaxies—the odd shapes, bent arms, and twisted disks caused by collisions—much more frequently than nearby galaxies.

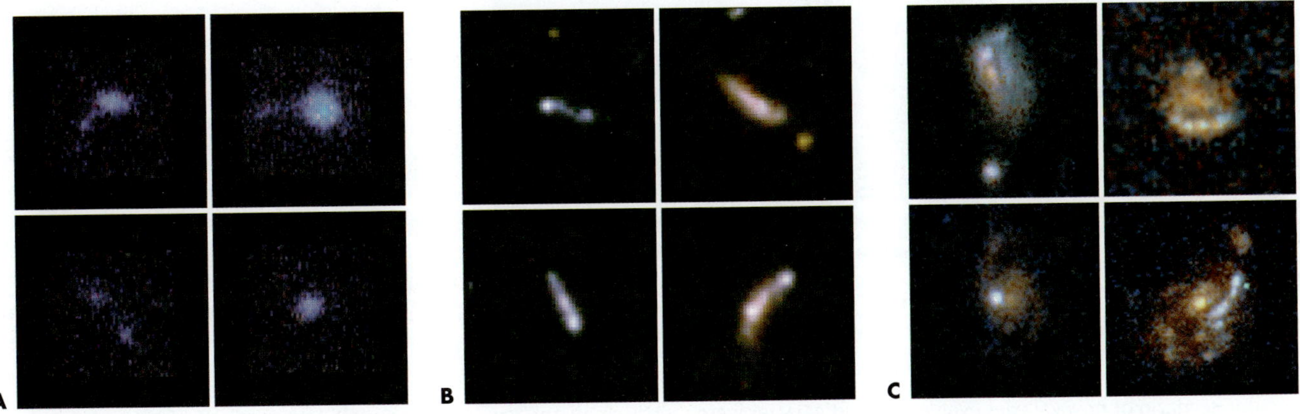

FIGURE 17.14
Hubble Space Telescope images of distant galaxies. Each small square shows an area approximately 25 kiloparsecs on a side, about the size of the Milky Way's disk today. The panels show galaxies (A) less than 1 billion years old, (B) a few billion years old, and (C) about 6 billion years old—half the current age of the Milky Way. (A) and (B) are from the Hubble Ultra Deep Field, (C) is from the Medium Deep Survey.

FIGURE 17.15
Streams of stars around the galaxy NGC 5907 are the remnants of a small galaxy that has been pulled apart by the larger galaxy's gravity.

We saw in chapter 16 that our own Galaxy, the Milky Way, shows signs of past mergers. For example, astronomers can see streams of stars in our Galaxy's halo, which appear to be debris from small galaxies that have fallen into the Milky Way and been torn apart by its gravity. A past history of mergers also explains a puzzling feature of stars in the bulge of the Milky Way. Astronomers at one time thought that the stars in our Galaxy's bulge must have formed in one single, massive event at the Galaxy's birth. However, stars in the bulge have a wide range of ages, suggesting that the bulge grew gradually over time. Such gradual growth is easy to explain by mergers. Each time a large galaxy (such as the Milky Way) swallows a smaller companion, some of the gas from the companion sinks to the center of the larger system, where it forms new stars. Successive mergers lead to successive generations of stars, so that the bulge stars have the wide age range that is observed.

Streams of stars from small galaxies that have been torn apart are seen around other galaxies as well, as shown in figure 17.15. Astronomers call this process of a large galaxy capturing and absorbing a small galaxy **galactic cannibalism.** Our neighboring galaxy, M31, also appears to have cannibalized a small galaxy. Astronomers have detected a small bright clump of stars near the center of M31 that may be the remains of a galaxy it absorbed.

Cannibalism is one of the main mechanisms by which galaxies grow, but what makes one series of mergers result in a spiral galaxy, while another becomes an elliptical galaxy? Astronomers have proposed a model illustrated in figure 17.16. Most galaxies start out as small, gas-rich systems, which actively form stars and are therefore blue in color. Subsequently, when these galaxies collide and grow into larger systems, they undergo bursts of star formation, and perhaps consume much of their gas. Computer models suggest that if enough stars form at once, the supernova explosions of many massive stars may drive gas out of the galaxy. If the collision/merger leaves little gas remaining in the system, an elliptical galaxy results. Without gas to form new stars, the elliptical galaxy grows redder with time. If the collision/merger leaves moderate to large amounts of gas in the system, a large spiral results. The spiral continues to form stars and remains relatively blue in color. Such interactions potentially allow some ellipticals to turn back into spirals. This can happen if an elliptical captures a gas-rich system. The gas then becomes the disk of a "new" spiral, while the elliptical becomes its bulge.

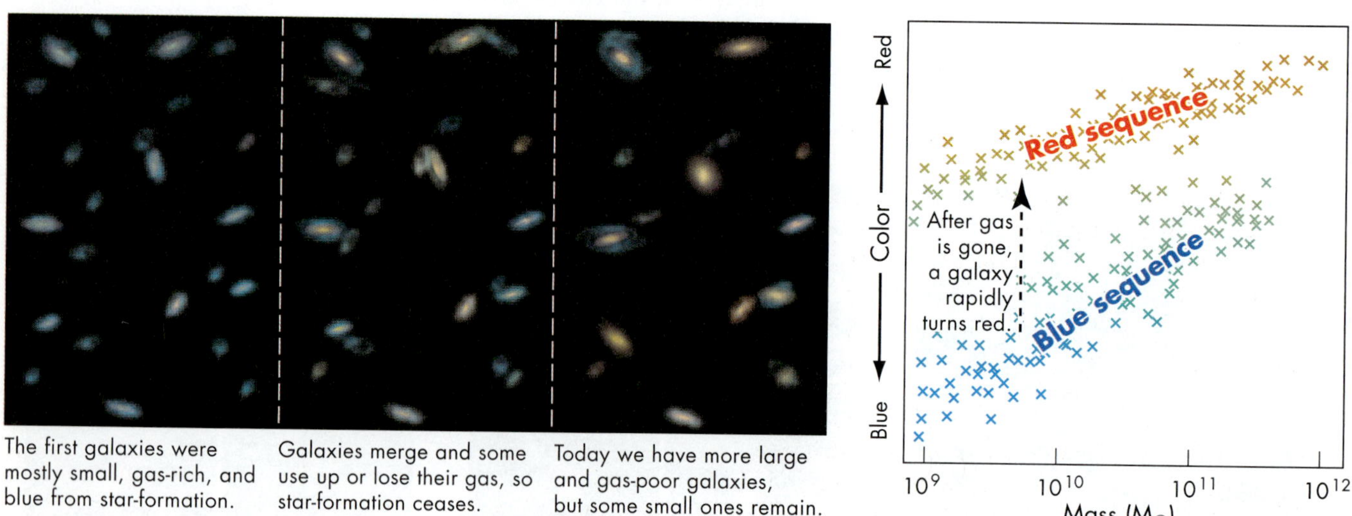

The first galaxies were mostly small, gas-rich, and blue from star-formation.
Galaxies merge and some use up or lose their gas, so star-formation ceases.
Today we have more large and gas-poor galaxies, but some small ones remain.
Red sequence
Blue sequence
After gas is gone, a galaxy rapidly turns red.
Blue
Color
Red
10^9
10^{10}
10^{11}
10^{12}
Mass ($M_\odot$)

FIGURE 17.16
A possible scenario of how galaxies evolve. The first galaxies were mostly small and gas-rich (left panel). Larger galaxies formed through mergers, as illustrated by the next two panels. Some of these galaxies consumed their gas or lost it during collisions, and ceased forming stars, so they grew redder in color. Others merged to form large spiral galaxies that continue to form stars and remain blue in color. When we plot the colors and magnitudes of galaxies today (right panel), galaxies are distributed along a "blue sequence," which are probably galaxies that grew through mergers without losing their gas. After losing their gas, they shift to the "red sequence."

A

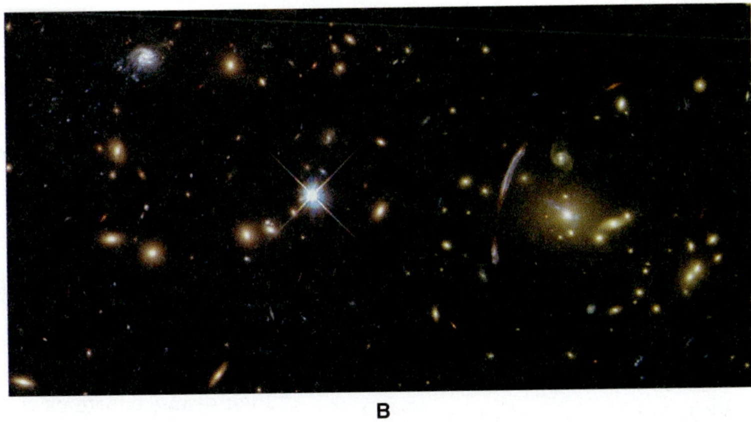
B

FIGURE 17.17
Galaxies changing character. (A) A merger in which two spirals are colliding and forming an elliptical galaxy. (B) As the spiral galaxy near the middle of this image orbits within its galaxy cluster, gas is being stripped from the galaxy. The blue blobs below and to the left of the galaxy are regions where stars are forming in the stripped gas. The cluster, Abell 2667, is a little over 3 billion light-years distant.

This hypothesis of the origin of spiral and elliptical systems may explain why galaxies exhibit a division into red and blue "sequences" when their colors are plotted against their masses as shown in the graph in figure 17.16. Galaxies grow more massive through mergers, using up more gas each time, so the more massive galaxies have fewer young blue stars. And once the gas is exhausted, the galaxies shift to the red sequence once the short-lived blue stars die. Through repeated cannibalism, a galaxy may grow into a giant, which perhaps explains why abnormally large elliptical galaxies are found at the centers of groups and clusters of galaxies (see section 17.6). Cannibalism may also account for much of the activity observed in some galaxies (see section 17.4).

Astronomers have identified some galaxies that appear to be undergoing the conversion from spirals to ellipticals. For example, the galaxies in figure 17.17A are colliding. Both galaxies show evidence of clouds of dust and gas like spirals, but the collision has disrupted any clear disklike pattern of circular rotation in the larger galaxy. Stars have been sent into new orbits oriented in various directions, and the resulting distribution of stars resembles an elliptical galaxy.

S0 galaxies are particularly interesting because they appear to be transitional between ellipticals with almost no gas and spirals with substantial gas content, yet they still have stars orbiting in a flat disk. According to one hypothesis, S0 galaxies are spirals whose gas has been stripped by their motion through the hot, tenuous gas in intergalactic space. The impact of such gas from outside might "sweep" gas and dust out of the moving spiral, much like a leaf-blower cleans debris from a sidewalk. Such stripped gas is visible down and to the left of the blue spiral galaxy shown in figure 17.17B. (The blue blobs in the stripped gas are regions of forming stars.) Alternatively, the original gas in a disk galaxy might all have turned into stars early in the galaxy's youth, perhaps because of an interaction with a neighboring galaxy that compressed the gas and made it collapse into stars in one large burst. In either case, without gas the disk galaxy can make no new stars and hence it shows no spiral structure.

In this discussion of the factors that might shape a galaxy, we have said nothing about how spiral arms form. In chapter 16, we discussed some hypotheses about how the Milky Way may have formed its spiral arms. For example, we saw that spiral arms might form as a density wave travels around a galaxy's disk, compressing gas clouds and triggering star formation in a spiral pattern. Astronomers assume that this mechanism operates in other galaxies and forms their arms, too. We also discussed in chapter 16 a different model in which spiral arms form as the galaxy's rotation stretches regions of star formation into elongated arcs. This mechanism probably also forms arms in other galaxies.

So far in our discussion of galaxy types and their possible evolution, we have ignored the methods by which the size and mass of a galaxy are found. An even more important point that we have overlooked is how to find the distance to a galaxy. Without knowing how far away a galaxy is, we can learn little about its properties. Thus, before we look further at galaxies, we need to discuss how their properties are measured.

17.2 Measuring the Properties of Galaxies

Our ability to study galaxies is hindered by their great distance from us. In fact, the extent of their distance from us was for centuries a major controversy among astronomers. For example, in 1755 the German philosopher Immanuel Kant, whose views on the origin of the Solar System were so ahead of their time, suggested that galaxies were remote star systems—Island Universes—similar to the Milky Way. On the other hand, as we saw in section 17.1, the American astronomer Harlow Shapley argued as recently as the 1920s that galaxies were merely small satellites of the Milky Way. The reason for such disagreement comes from the difficulty of *measuring* how far away from us galaxies are.

Galaxy Distances

Galaxies are separated from each other by enormous distances. Even the Large Magellanic Cloud, a small companion of the Milky Way, is more than 150,000 light-years distant. Astronomers cannot use parallax to measure such vast distances because the angle by which the galaxy's position changes as we move around the Sun is too tiny to be detected. Astronomers must therefore use other ways to find distances, such as the method of standard candles.

We discussed the method of standard candles in detail in chapter 13. We saw there that if we know the luminosity of a light source, we can use the inverse-square law to deduce its distance from how bright it looks to us. This was the technique we used in chapter 16 to find the size of the Milky Way, and we can apply it here to find the distance to galaxies.

Astronomers use many different astronomical objects as standard candles, but one of the most reliable is the class of luminous, pulsating variable stars known as the "Cepheids" (described in chapter 14). Cepheids have a luminosity about a million times the luminosity of the Sun, and because of their brilliance, they are easy to see in nearby galaxies. Furthermore, because their brightness changes as they pulsate, they are easy to distinguish. Far more important, however, is that the time it takes these variable stars to go from bright to dim to bright again (their period) is related to their luminosity. (Recall our discussion of the period–luminosity relation in chapter 14.) This then allows astronomers to determine a Cepheid's luminosity by simply measuring how long it takes the star to change its luminosity over one cycle. Once the luminosity of a variable star is known, astronomers can find its distance from a simple measurement of how bright the star appears as seen from Earth and an application of the inverse-square law. (See the Extending Our Reach box on the next page.)

Cepheids are luminous enough that astronomers can use them to measure the distance to nearby galaxies, but despite their luminosity, Cepheids are not observable in very distant galaxies. Therefore, astronomers use other standard candles such as supergiant stars, planetary nebulas, or supernovas when determining the distance to such remote systems. Such objects are not ideal standard candles, however, because their luminosity may not be accurately known. Or they may be rare, and therefore not present, in the galaxy whose distance one wants to measure. Astronomers therefore have sought other ways to measure the distance to galaxies, as we will discuss in the next section.

EXTENDING *our reach*

MEASURING THE DISTANCE OF A GALAXY USING CEPHEID VARIABLES

Suppose we detect a Cepheid in a galaxy and find that the Cepheid appears 100 million (10^8) times dimmer than a Cepheid of the same pulsation period in the Milky Way. Because the stars have the same period, we assume that they have the same luminosity. Let us denote the brightness and distance of the Cepheid in the Milky Way as B_* and d_*, respectively. Similarly, we will denote the brightness and distance of the Cepheid in the distant galaxy by B_g and d_g. According to the inverse-square law, if two objects have the same luminosity, their apparent brightnesses B_* and B_g, are related to their respective distances, d_* and d_g, by

$$\frac{B_*}{B_g} = \left(\frac{d_g}{d_*}\right)^2$$

B_* = Brightness of Cepheid in Milky Way
B_g = Brightness of Cepheid in distant galaxy
d_* = Distance to Cepheid in Milky Way
d_g = Distance to Cepheid in distant galaxy

To find their distances, we take the square root of both sides, getting

$$\sqrt{\frac{B_*}{B_g}} = \frac{d_g}{d_*}$$

In other words, the square root of the ratio of their brightness equals the ratio of their distances. Because the distant Cepheid in our problem looks 10^8 times dimmer from Earth than the Cepheid in the Milky Way, $B_*/B_g = 10^8$. Thus, according to the inverse-square law, the ratio of the distance of the Cepheid in the galaxy to the distance of the Cepheid in the Milky Way must be

$$\frac{d_g}{d_*} = \sqrt{\frac{B_*}{B_g}} = \sqrt{10^8} = 10^4$$

Therefore, $d_g = 10^4 d_*$. That is, the Cepheid in the remote galaxy is 10^4 times farther away than the one in our galaxy.

Suppose we know that the Cepheid in our own galaxy is 1000 light-years away. Then the Cepheid in the other galaxy must be 10^4 times 1000 light-years away. The distant Cepheid, and hence the galaxy in which it is located, is therefore 10^7 light-years distant (Box fig. 17.1).

$$\frac{B_*}{B_g} = \left(\frac{d_g}{d_*}\right)^2 \quad \text{or} \quad \frac{d_g}{d_*} = \sqrt{\frac{B_*}{B_g}}$$

BOX FIGURE 17.1
Finding the distance to a galaxy using Cepheid variables as standard candles. B_* and B_g are the apparent brightness of the Cepheid in the Milky Way and in the Cepheid distant galaxy, respectively. Their distances are d_* and d_g, respectively. From a measurement of the brightness ratio of the Cepheids, astronomers can calculate the ratio of their distances and, knowing the distance of the Cepheid in the Milky Way, they can therefore find the distance to the other galaxy.

The Redshift and the Hubble Law

We saw in earlier chapters that we can learn much about a star from its spectrum. Likewise, we can learn much about a galaxy from its spectrum.* One of the most important such discoveries was made in the early 1900s by the American astronomer Vesto Slipher. Slipher found that of the galaxies he observed, most of their spectral lines were strongly

*We need to keep in mind, however, that a galaxy's spectrum is the spectrum of all of its stars, gas clouds, and other contents added together.

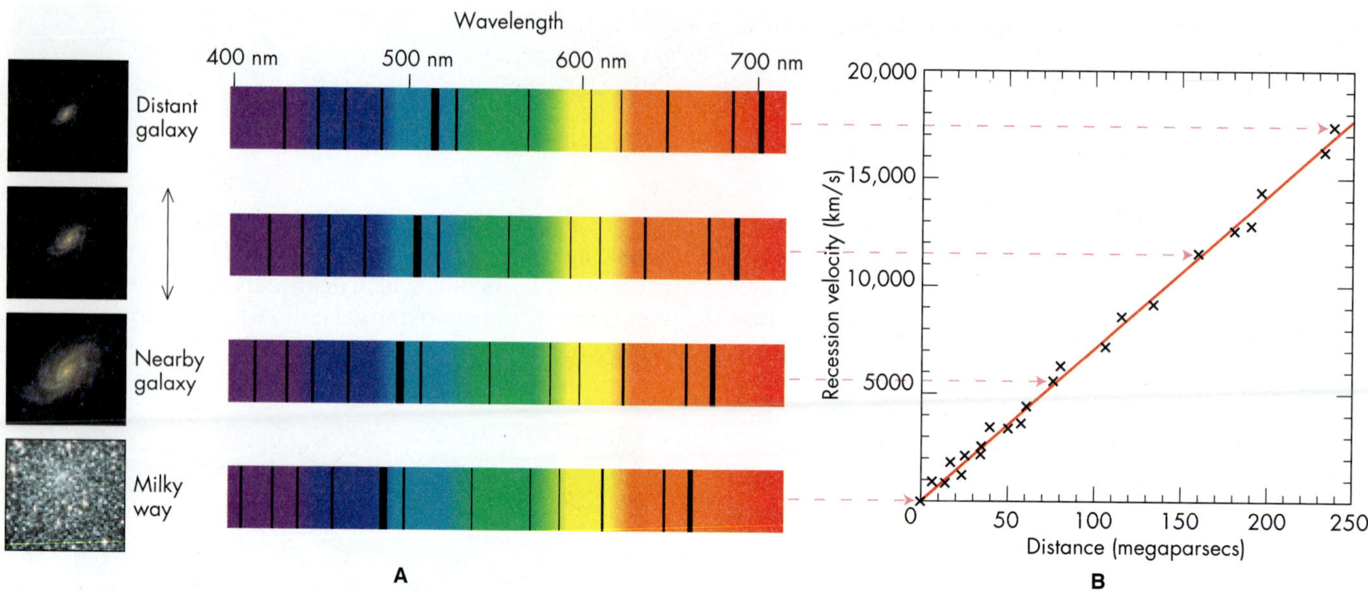

FIGURE 17.18
Redshift and recession velocity. (A) The more distant a galaxy, the larger its redshift. The spectrum of the total light of the Milky Way (which has zero redshift) is shown for comparison. (B) The Hubble law: Galaxies farther away recede faster.

shifted toward longer wavelengths—the red end of the spectrum. This so-called **redshift** is seen in nearly all galaxies. (The exceptions are a few nearby galaxies, whose motion is influenced by the gravitational attraction of the Milky Way and its neighbors.)

Galaxy redshifts are *not* simply Doppler shifts. Rather, as astronomers know today, they result from the expansion of the Universe, as we will discuss in chapter 18. In Slipher's time, however, the cause of galaxy redshifts was a major puzzle. This puzzle became even more mysterious when Edwin Hubble and others noticed that the redshift was larger for dimmer (and therefore presumably more distant) galaxies. This implied that the speed with which a galaxy moves away from us—its **recession velocity**—increases with distance, as figure 17.18 shows. That is, *the faster a galaxy recedes, the farther away it is.* This, then, becomes a powerful tool for measuring galaxy distances. In 1929, Edwin Hubble discovered that a simple formula related the recession velocity, *V*, and the distance, *d*. He found that

$$V = \mathrm{H}d$$

where H is a constant. Because of his discovery, this relation is called the **Hubble law,** and H is called the **Hubble constant.**

The value of the Hubble constant depends on the units used to measure *V* and *d*. Astronomers generally measure *V* in kilometers per second and *d* in **megaparsecs,** where 1 megaparsec (abbreviated Mpc) is a million parsecs (= 3.26 million light-years). In these units, H is about 70 kilometers per second per megaparsec. For many decades astronomers disagreed about the value of H. The reason for this controversy was that, to determine H, one had to know *both* the distance and the velocity of at least a few galaxies in order to calibrate the law. H was then simply the average value of each calibration galaxy's velocity divided by its distance. However, as we have seen, it is difficult to accurately measure the distance to even a nearby galaxy. A new technique for determining H has essentially resolved this problem—establishing that H = 70 kilometers per second per megaparsec. This new method uses information about irregularities in the brightness of radiation emitted by the young Universe, which we will discuss in chapter 18. For now, we will simply say that the method appears to give the most precise value yet for H.

Once a value for H is known, we can turn the method around and use the Hubble law to find the distance to a galaxy. The method is as follows: Take a spectrum of the galaxy whose distance you want to know. From the spectrum, we measure the shift of the spectral lines. From the Doppler-shift formula, we calculate the recession velocity. Finally, we use that velocity in the Hubble law to find the distance, as we show below.

We begin by writing out the Hubble law

$$V = \mathrm{H}d$$

V = Recession velocity of galaxy
D = Distance of galaxy
H = Hubble constant

To find d, we divide both sides by H, obtaining

$$d = V/\mathrm{H}$$

Finally, we insert the measured value of V and our choice of H and solve for the distance. For example, suppose we find from a galaxy's spectrum that its recession velocity is 56,000 kilometers per second. We find its distance by inserting this value for V in the expression above and then divide by H.

$$D = (56{,}000\ \mathrm{km/sec})/\mathrm{H}$$

For H = 70 kilometers per second per megaparsec, the value of d is then

$$d = \frac{56{,}000\ \mathrm{km/sec}}{70\ \mathrm{km/sec/Mpc}} = 800\mathrm{Mpc}$$

Thus, the galaxy is 800 megaparsecs away from us.

In applying this method, care must be used about the units. When we divide V in kilometers per second by H in kilometers per second per megaparsec, the answer we get will be in megaparsecs. Thus, the 800 found in this example is not in parsecs or light-years but in megaparsecs. Moreover, the Hubble law applies only to galaxies that are reasonably far away. The velocity of nearby galaxies is strongly affected by the gravitational attraction of their neighbors, and so the method fails when applied to such systems. (Of course, it also does not apply to stars or other objects in the Milky Way.)

With the distance to a galaxy known, we can now measure its radius and mass, as we will now show.

Measuring the Diameter of a Galaxy

Astronomers measure the diameter of a galaxy using the same method we discussed in chapter 2 for measuring the diameter of the Moon or the Sun. There, we showed that an object's true diameter, D, is given by

$$D = \frac{2\pi A d}{360}$$

where A is its angular size on the sky in degrees and d is its distance (fig. 17.19). The units of d are the same as for D; for example, in measuring galaxy sizes, we might express both in light-years or both in parsecs.

To illustrate the method, we will measure the diameter of M31, a close neighbor of the Milky Way. M31 is a spiral galaxy, so when we speak of its diameter, we are referring to the diameter of its disk. Even that is somewhat ambiguous, because galaxies do not have sharp edges—the stars just thin out gradually—and it is hard to say exactly where the galaxy ends and intergalactic space begins.

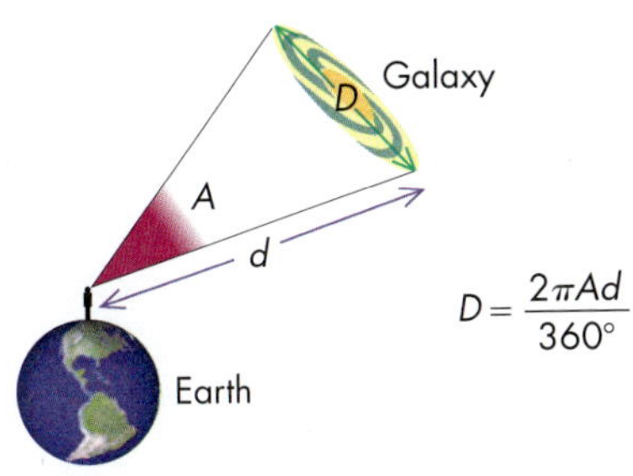

FIGURE 17.19
Measuring the diameter of a galaxy from its angular diameter.

EXTENDING *our reach*

OTHER WAYS TO MEASURE A GALAXY'S DISTANCE

The Hubble law is a powerful tool for finding a galaxy's distance, but in some cases it cannot be applied. As we said earlier, the Hubble law does not work for nearby galaxies because their speed is determined by their gravitational interaction with one another rather than by the expansion of the Universe. The Hubble law is also difficult to apply to very remote galaxies. The light from such distant systems has taken billions of years to reach us, and so we see those galaxies as they were billions of years ago. At that distant time, the Universe's rate of expansion was different from its rate today, and thus the value of the Hubble constant was not the same as it is now. With these difficulties in mind, astronomers have sought other ways to find a galaxy's distance.

One method is similar to the simple observation that you have made if you look closely at a picture printed in a newspaper or comic book. Holding such a picture near your eyes, you can see that it consists of many tiny separate dots. As you move the picture farther from your eyes, the dots disappear and merge together to form the picture. Similarly, in a galaxy near us, we can easily see the separate spots of its more luminous stars. However, in more distant galaxies, the stars blend together into a uniform blur of light. Careful measurement of the graininess of a galaxy's picture can therefore give an indication of its distance.

Another method widely used for finding galaxy distances is based on the standard-candle method but uses the galaxies themselves as the standard candle. This method relies on the observation that the rotation velocity of a spiral galaxy is related to its luminosity. In particular, the more luminous a galaxy, the more rapidly it rotates. We might anticipate this result because if a galaxy rotates very rapidly, it must be very massive to hold itself together gravitationally. Moreover, a massive galaxy contains more stars and thus emits more light than a small galaxy. Thus, the faster a galaxy rotates, the more luminous it is.

Astronomers can find a galaxy's rotation speed by obtaining a spectrum of the galaxy at optical or radio wavelengths. The spectral lines will shift to shorter wavelengths on the side moving toward us, and to longer wavelengths on the other side, by an amount that tells us the rotation speed according to the Doppler formula. From the speed of rotation, astronomers can estimate the galaxy's luminosity, and the distance can then be calculated from the luminosity and apparent brightness using the inverse-square law. This procedure is known as the "Tully-Fisher method," after the two astronomers who devised it.

Keep in mind that most methods for finding the distance to a galaxy have *many* underlying assumptions. For example, many methods require knowing the distance to remote stars or other objects in our own galaxy, and those distances depend in turn on the distance to nearby stars. Astronomers refer to such multistep distance determinations as the distance ladder, and they must account for the inevitable errors and uncertainties at each step of the ladder. As a result, the final answer to how far away a given galaxy is can be quite uncertain.

The angular diameter, A, of M31 is about 5° and its distance, d, is about 750,000 parsecs. Using the expression for D above and inserting these quantities, we find that the diameter, D, of M31 is

A = Angular size of object
D = Diameter of object
d = Distance to object

$$D = \frac{2\pi A d}{360}, \quad \text{or}$$

$$D = \frac{2\pi \times 5 \times 750{,}000 \text{ pc}}{360}$$

$$\approx 65{,}000 \text{ pc}$$

Measuring the Mass of a Galaxy

We can measure the mass of a galaxy the same way we measure the mass of a star or planet: by an application of the modified form of Kepler's third law. As you may recall, this method uses the principle that the orbital motion of one object around another is set by their mutual gravity. Their gravity is, in turn, set by their combined mass. Thus, from the orbital motion we can find the mass.

We went through the details of determining the mass of our own Galaxy, the Milky Way, in chapter 16. There we explained that we begin by measuring the orbital speed of a star far out in the disk of the Milky Way. With the star's speed determined, and knowing the radius of its orbit, we then found its orbital period (see section 16.6 for details). Finally, we inserted the orbital period and radius in the modified form of Kepler's third law to give us the Milky Way's mass.

To measure the mass of another galaxy, we follow the same procedure. We use the Doppler shift of the galaxy's spectrum lines to find the orbital speed of a star* near the galaxy's rim. Then, from the size of the star's orbit and the modified form of Kepler's third law, we calculate the galaxy's mass.

The above method is an excellent and accurate way to measure a galaxy's mass, yet a puzzling discrepancy arises when astronomers compare that mass with the mass of the visible material in the galaxy. Almost without exception, the mass calculated from the modified form of Kepler's law is *much larger* than the mass of stars and interstellar matter detectable by optical, infrared, and radio telescopes. Thus, galaxies seem to contain matter that cannot be seen, what astronomers therefore call "dark matter."

17.3 Dark Matter

Dark matter is the material that astronomers believe accounts for the discrepancy between the mass of a galaxy as found from the modified form of Kepler's third law and the mass observable in the form of stars and gas. The amount of dark matter required to resolve this discrepancy is very large, amounting in typical galaxies to about 10 times their visible mass. That is, when we look at a galaxy, we may in some cases be seeing only one-tenth of its matter.

The strongest evidence that dark matter exists comes from galaxy-rotation curves. We discussed such curves in chapter 16 when we described the rotation of the Milky Way, but let us review them here briefly. A galaxy's rotation curve is a plot of the orbital velocity of the stars and gas moving around it at each distance from its center (curve A in fig. 17.20). The rotation curve is useful for studying dark matter because the orbital velocity depends on the mass, visible or not, that lies between the star and the galaxy's center. We can understand how that dependence arises by using the formula for orbital velocity that we developed in chapter 2 or by simply thinking about what keeps a star in orbit.

ANIMATION
Dark matter

According to Newton's first law, if an object does not move in a straight line (as when, for example, a star moves in a circular orbit around a galaxy), a force must be acting on it. We know that for a star orbiting in a galaxy, that force is gravitational attraction. We also know that the strength of the gravitational force depends on the mass of the attracting body (in this case, the galaxy). All other things being equal, a greater mass therefore implies a greater force. It turns out, however, that to a good approximation, *only the material between the star's orbit and the galaxy's center contributes* to the gravitational force. Stars near the center of a galaxy will therefore have only a small force acting on them because there is little mass between them and the center. Because the force acting on them is small, such stars cannot be orbiting very rapidly, or they would have flown out of the galaxy. On the other hand, stars orbiting far from the center should also feel a small gravitational force acting on them from the galaxy. They feel the effect of the galaxy's full mass, but the gravitational force on them is nevertheless weak because they are so far from the main mass (recall that the force of gravity grows rapidly weaker at greater distances from a mass). These outermost stars must also be

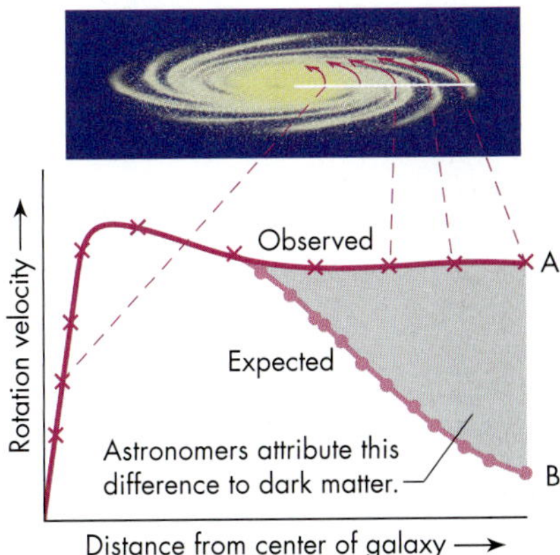

FIGURE 17.20
A schematic galaxy-rotation curve. The line with dots represents the curve expected if the galaxy's mass comes only from its luminous stars and gas. The line with crosses represents the observed curve, implying that the galaxy contains dark matter.

*In using this method, astronomers generally get, not a spectrum of a single star, but rather the spectra of groups of stars or gas clouds moving together around the galaxy. Moreover, the spectra do not need to be at visible wavelengths. Many of the most accurate observations of galaxy masses come from radio telescope observations of the Doppler shift.

SCIENCE *at work*

DARK MATTER OR MAYBE NOT

Although the great majority of astronomers think that the existence of dark matter is the best way to explain the observed high rotational speed of disk galaxies and the high speed of galaxies within clusters of galaxies, not all astronomers agree. A few have proposed that Newton's laws of motion need a correction for situations where the acceleration is very small, such as occurs with motions on the immense scale of galaxies.

One such proposal is referred to as the MOND hypothesis, for Modified Newtonian Dynamics. Proponents of the MOND hypothesis have had success in using the theory to explain motions within galaxies (including the Milky Way) and within galaxy clusters. Moreover, they claim that MOND models fit the observed motions as well as dark-matter models without requiring the existence of dark matter. On the other hand, observations using the theory of gravitational lenses to map the distribution of mass in galaxy clusters strongly favor the dark-matter hypothesis, and MOND does not appear to be consistent with general relativity.

For the time being, MOND provides an interesting example of how science works. Alternative ideas are proposed and explored by a subset of scientists. And perhaps if the prevailing theory reaches an impasse, scientists will find that a competing idea, such as MOND, offers a better explanation of what we observe.

orbiting slowly if they are not to fly out of the galaxy. Thus, both stars near the center and stars far from the center should have low orbital velocities. Accordingly, the galaxy's rotation curve should have a peak, as indicated in curve B of figure 17.20.

Detailed mathematical models bear out this analysis and show that a galaxy's rotation curve should start at small velocities near the core. The curve should then rise to higher velocities at middle distances and finally drop to lower velocities again far from the center. However, when astronomers *measure* the rotation curves of galaxies, they essentially *never* find this behavior. The rotation curves do rise near the center, but they then flatten out and virtually never drop off to low velocities again. What can explain such a major discrepancy between our hypothesis and observation? If our laws of physics are correct, only one thing can explain such behavior: galaxies must contain large amounts of unseen mass in their outer parts. As we saw in chapter 15, astronomers find evidence of such unseen mass in the Milky Way and call it dark matter. The dark matter exerts a gravitational force on stars in the outer parts of a galaxy, holding them in orbit despite their large velocity. The large gravitational force needed to bind such stars in a galaxy implies a huge mass of dark matter (many times the galaxy's luminous mass). Thus, we deduce that each galaxy is embedded in its own dark-matter halo, vastly more massive and larger than the "visible" galaxy.

What can this dark matter be? Astronomers are puzzled. They can rule out ordinary dim stars because such objects would emit at least some infrared radiation that would be detectable. They can rule out cold gas because that would be detectable by radio telescopes. They can rule out hot gas because that would be detectable by optical, radio, or X-ray telescopes. Astronomers cannot rule out tiny planetesimal-sized bodies, extremely low-mass cool stars, dead white dwarfs or neutron stars, or black holes, but it is difficult to develop a plausible model in which such objects would be common enough to explain the amount of dark matter indicated by the observations. Astronomers now suspect that some kind of subatomic particle predicted by theory but not yet observed in the laboratory, such as photinos or WIMPS (weakly interacting massive particles), makes up the dark matter. These particles are good candidates because they do not interact with normal matter except through their gravitational pull. The nature of the dark matter remains one of the central mysteries in astronomy today. Moreover, as we will see in section 17.6 and in chapter 18 as well, dark matter is not confined to galaxies. It also permeates galaxy clusters and perhaps the Universe itself.

17.4 Active Galaxies

Active galaxies are galaxies whose centers (nuclei) emit abnormally large amounts of energy from a tiny region in their core. Approximately 10% of galaxies are found to be active, but astronomers suspect that all large galaxies may go through active phases during their lives, especially when they are young and still actively merging. Not only is the emitted radiation intense, but it also usually fluctuates in intensity. The centers of these active systems are sometimes called AGNs (active galactic nuclei).

Almost a century ago, astronomers began identifying galaxies that are intensely bright in their centers. Later they identified other galaxies that exhibit intense radio emission and other activity outside their main body. These were initially thought to be separate classes of activity, but in recent years astronomers have been able to link all of these different forms of activity using a single unified model. We will begin by describing some of the unusual features of these galaxies, and examine how the evidence points toward an extraordinary source for all this activity: supermassive black holes at the heart of almost all large galaxies.

The Discovery of Nuclear Activity

Even before astronomers had determined what galaxies were, they noticed that some had peculiar features at their centers. For example, in the early 1900s, astronomers noticed that the bright elliptical galaxy M87 (fig. 17.21) has a "jet" of gas coming out of its nucleus. By measuring the shift in position of the individual blobs of gas in the jet, astronomers in recent years have shown that the gas in this jet is being ejected from the nucleus at nearly the speed of light.

In the 1940s, the American astronomer Carl Seyfert (pronounced see-firt) identified a class of unusual spiral galaxies. These spirals looked normal except that their nuclei were remarkably luminous. An example of such a **Seyfert galaxy** is shown in figure 17.22.

FIGURE 17.21
Hubble Space Telescope image of the center of the elliptical galaxy M87. A jet of hot gas can be seen coming from the bright nuclear source at the center of the galaxy.

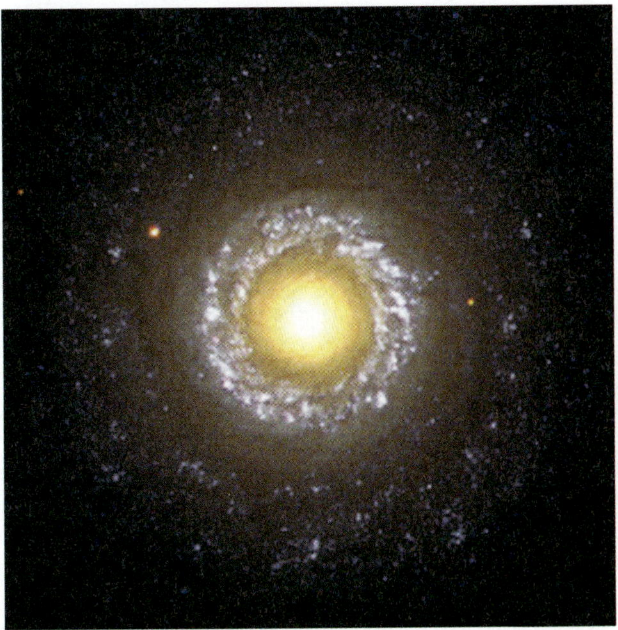

FIGURE 17.22
The Seyfert galaxy NGC 7742, imaged with the Hubble Space Telescope. The nucleus of this face-on spiral galaxy is exceptionally luminous.

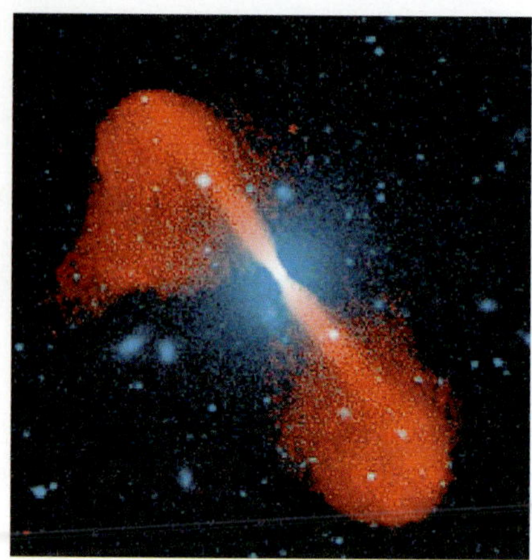

FIGURE 17.23
The radio galaxy NGC 5532. A visible-light photograph is shown in blue, and radio emission is shown in red. Note the large radio lobes on either side outside of the galaxy.

The luminosity of the nucleus of a typical Seyfert galaxy is immense, amounting to the entire radiation output of the Milky Way, but coming from a region far less than a light-year across. The nuclear regions of Seyfert galaxies also contain gas clouds moving at speeds of about 10,000 kilometers per second or more. In some instances, the nucleus is also ejecting gas in jets, as in M87. The intense radiation from a Seyfert galaxy's core can also fluctuate rapidly in intensity, sometimes changing appreciably in a few minutes.

By the 1960s, astronomers began to identify another kind of active galaxy, which they labeled **radio galaxies.** As their name suggests, these galaxies emit large amounts of energy in the radio part of the spectrum, typically millions of times more than the radio energy emitted by a normal galaxy. Most are elliptical galaxies, and the radio emission comes primarily from regions outside the galaxy on either side of it, as you can see in figure 17.23. This figure shows superimposed visible and radio wavelength images of the galaxy, with red representing the radio emission.

The radio-emitting regions visible outside the radio galaxy in figure 17.23 are called "radio lobes." These lobes, often located symmetrically on opposite sides of the galaxy, may span hundreds to millions of light-years. Although most of the radio emission comes from the lobes, some comes from within the visible body of the galaxy, from narrow jets and a tiny nuclear region much less than a light-year across.

What causes this intense radio emission? Astronomers can tell from its spectrum that the emission is synchrotron radiation, created by electrons traveling at nearly the speed of light and spiraling around magnetic field lines. We discussed this mechanism in chapter 15, where we described radio emission from pulsars. Pulsars generate high-speed electrons at their surface, but radio galaxies generate them in their core. The electrons are then shot out of the core in narrow **jets** (such as the ones shown in figs. 17.21 and 17.23) that generally point at the radio lobes. Streaming away from the end of the jet, the electrons eventually collide with the gas around the galaxy. There, they spread out to form the lobes, all the while emitting synchrotron radiation.

Quasars

Q. What evidence in this picture suggests this might be an active galaxy?

Quasars get their name from a contraction of the words *quasi-stellar radio source*, where *quasi-stellar* means "almost starlike." This name refers to their appearance in photographs like figure 17.24, in which they look like dim stars. However, some quasars are among the brightest radio sources in the sky despite their dimness at visual wavelengths.

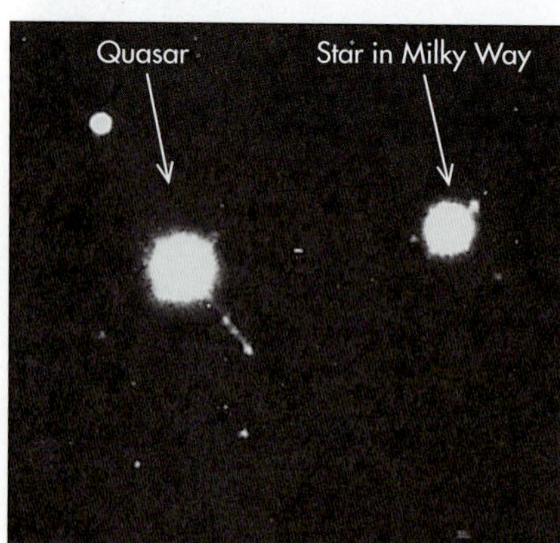

FIGURE 17.24
Photograph of the quasar 3C273. This is the brightest quasar in the sky, but it looks like little more than a dim star. ("3C" stands for *Third Cambridge Catalog of Radio Sources.*)

Quasars were initially thought to be some kind of exotic star in our Galaxy, but today they are known to be extremely luminous, extremely distant, active galaxies. We deduce their huge distance from the huge redshift of their spectra (recall that a galaxy's distance can be found from its redshift using the Hubble law). On the basis of their redshifts, the most distant ones are more than 10 billion light-years from Earth. To be visible at such an immense distance, an object must be immensely luminous, and a typical quasar turns out to be about 1000 times more luminous than the Milky Way.

Despite their huge power output, quasars, like Seyfert galaxies, fluctuate in brightness. Slow changes occur over months (fig. 17.25), but short-term flickerings occur in times as brief as hours. Rapid changes of this sort give clues to the size of the emitting object, as discussed in the Extending Our Reach box on the following page. However, astronomers need not rely solely on light variations to determine the size of an active galaxy's core. Observations with radio interferometer telescopes show that the active cores are extremely small (smaller than a fraction of a light-year, in some cases).

These findings tell us that some quasars have the power output of trillions of stars packed into a volume smaller than the Solar System. This is such an extraordinary claim that astronomers searched for alternative possibilities for quasars for a number of years. For example, the extraordinary luminosities derived for quasars are based on their huge distances, so some astronomers worked on the hypothesis that quasars' large redshifts were perhaps caused by gravitational redshifts instead (see chapter 15).

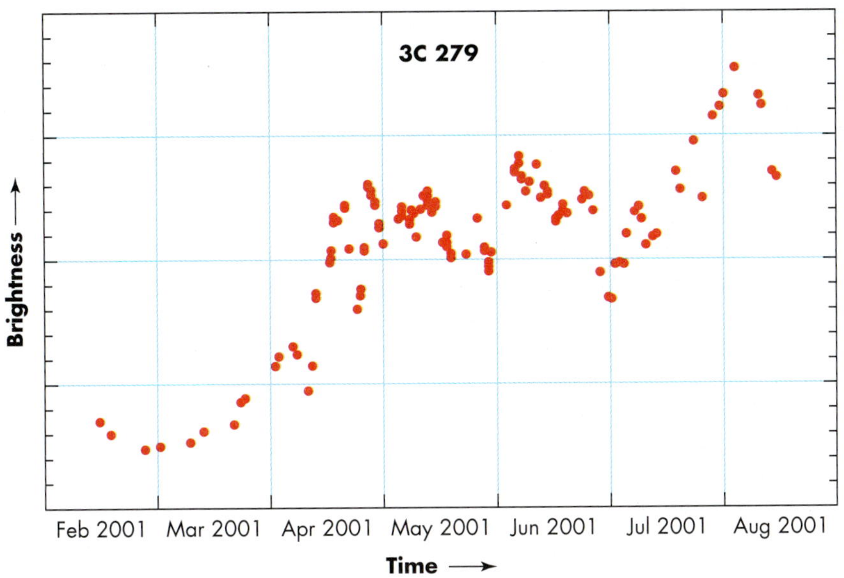

FIGURE 17.25
Plot of the light variation of a quasar.

FIGURE 17.26
PG 0052+251 as viewed by the Hubble Space Telescope. The image shows the galaxy surrounding the quasar, demonstrating that it is an active galactic nucleus. ("PG" stands for *Palomar Green* survey.)

If quasars were nearby, the implied luminosity would be smaller, but this model led to a host of other problems and self-contradictions.

That quasars actually *are* distant and almost unbelievably luminous was ultimately settled by images such as the one in figure 17.26, which clearly reveal that quasars lie in the centers of distant galaxies. The host galaxy of a quasar is difficult to see except from space because of the intense brightness of the quasar and the small angular size of the galaxy at its huge distance. It is now clear that quasars are the most powerful extreme of a broad range of active galaxies. Astronomers also now recognize that some other kinds of peculiar objects are related to quasars. For example, BL Lacertae objects (BL Lac, for short) were originally believed to be peculiar variable stars within the Milky Way. They are now known to be distant radio galaxies in which a jet of ejected matter is pointing almost straight at the Solar System.

Today, astronomers have cataloged hundreds of thousands of quasars and active galaxies. Apparently, something goes on at the very center of galaxies that can generate enormous power. What could possibly produce such vast amounts of energy within such a small volume?

A Unified Model of Active Galaxies

Because the different classes of active galaxies share many features in common, astronomers have sought a model for nuclear activity that could explain this huge power output. At the same time, a successful model must also explain the differences between the types of active galaxies. For example, why are some powerful radio emitters and others not? Why are the most powerful, the quasars, seen at large distances but not nearby?

Any model for active galaxies must explain how such a small central region can emit so much energy over such a broad range of wavelengths. For example, some quasars are powerful sources of both radio waves and gamma rays. Astronomers therefore hypothesize that the core must contain something very unusual. No ordinary single star could be so luminous. No ordinary group of stars that might produce so much radiation could be packed into so small a region. One kind of object that is very small but that can also emit intense radiation is the accretion disk around a black hole. We saw in chapter 15 that such accretion disks can emit intense X-ray

EXTENDING our reach

ESTIMATING THE DIAMETER OF ASTRONOMICAL OBJECTS BY USING THEIR LIGHT VARIABILITY

Changes in brightness give an important clue to the size of a quasar's or active galaxy's core. Suppose we have a ball of glowing material that we can turn on and off. Suppose further that when we turn it on, it begins to glow everywhere at the same time, and when we turn it off, it stops glowing everywhere at the same time. If we now look at this ball as it begins to glow, light from its nearest part, A, will reach us before light from its farthest part, B (Box fig. 17.2). An exactly similar effect occurs with sound produced from a lightning stroke. The lightning stroke lasts only fractions of a second, but the sound it generates lasts many seconds. The reason the sound is drawn out for such a long time is that sound travels relatively slowly (only about 1000 feet/second). Thus, if the stroke is 5000 feet long (a typical size for a bolt of lightning), sound from the near part reaches us 5 seconds before sound from the far part does. So, too, even though the ball begins to glow everywhere on its surface at exactly the same moment, the light we see does not turn on instantly. Instead, it gradually increases in brightness over a time interval equal to the time required for light to travel across the ball. That is, if the ball has a diameter of 1 light-second (the distance light travels in 1 second = 3×10^5 kilometers), the light reaching an observer does not reach full brilliance in less than 1 second. Likewise, if it is a light-month across (about 10^{12} kilometers) and it is turned off instantly, it does not dim away to darkness in less than 1 month.

From this, we can conclude that if a quasar varies in brightness in a time as short as a few days (and assuming simultaneous illumination and dimming), then the emitting region can be no more than a few light-days across. A core a few light-days across is tiny compared to the size of a typical galaxy (100,000 light-years across). In fact, such a core is about 10 million times smaller than a galaxy, or about 200 AU in diameter, about 5 times the size of Pluto's orbit around the Sun.

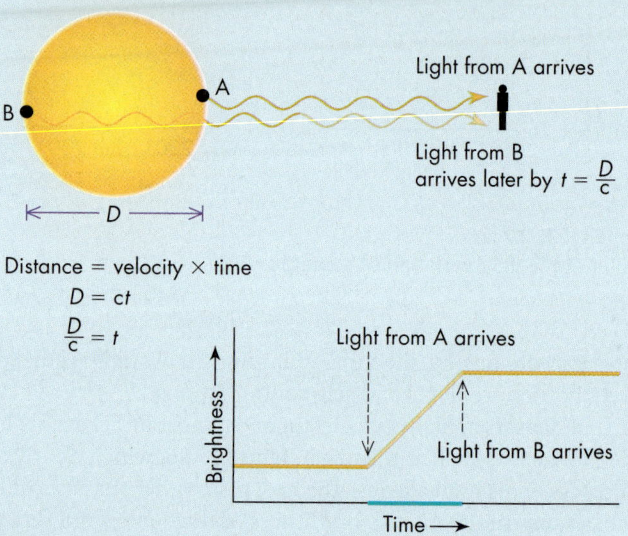

BOX FIGURE 17.2
Sketch illustrating that when a light source turns on, we see the light from its near side before we see the light from its far side, and so we do not see its light turn on instantly. For the objects that we encounter in everyday life, the time delay is so short we do not notice it. However, for astronomical-size bodies, the effect can create delays ranging from days to months, from which we can deduce that the emitting body is light-days or light-months across.

radiation. Could the center of an active galaxy contain an immense black hole? Could the radiation we observe from the core be hot gas in an accretion disk around such a black hole? This is precisely the model for active galaxies most widely favored by astronomers today. According to this model, an active galaxy has in its core a **supermassive black hole** that contains anywhere from millions to up to several billions of solar masses of matter. Around the black hole an accretion disk as large as the entire Solar System swirls furiously under the intense gravitational pull of the black hole. Orbiting under the black hole's fatal attraction, the gas in the accretion disk is heated to incandescence by frictional forces as it collides with other gas in the disk at a significant fraction of the speed of light.

Figure 17.27, an image made with the Hubble Space Telescope, supports this model. The image shows just such a disk in the central regions of the radio galaxy NGC 4261. Gas falling into the disk releases gravitational energy, which heats the material to temperatures as hot as several million K. Although most of the mass ultimately falls

FIGURE 17.27
The radio galaxy NGC 4261. The image on the left shows this elliptical galaxy and its radio jets in a composite optical and radio picture. The jets extend outward about 50,000 light-years. The image on the right, made with the Hubble Space Telescope, shows the central 1000 light-years, in which a dark, dusty accretion disk surrounds the central black hole.

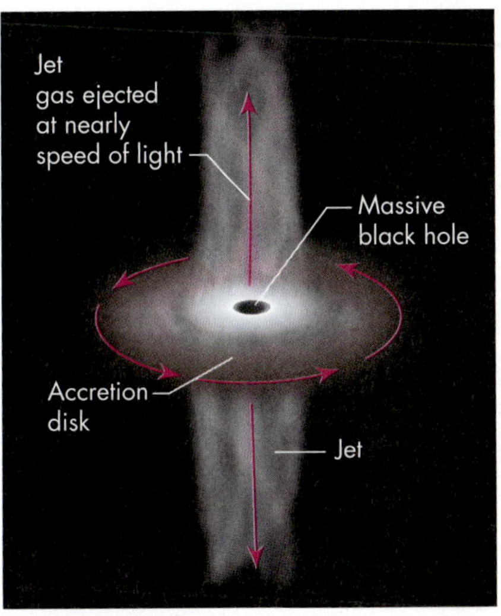

FIGURE 17.28
Sketch of jets formed by an accretion disk. Spiralling magnetic fields may help confine the jets.

into the black hole, some material from the accretion disk "boils off" at enormously high temperatures and escapes. This outflow is channeled into two oppositely directed jets by the black hole's magnetic field, as illustrated in figure 17.28. Computer models show that the intense magnetic fields become tightly twisted as the black hole spins. Magnetic fields (as you may recall in the discussion of pulsars in chapter 15) trap charged particles that exist in very hot gas. The spinning and twisting of the magnetic field can, therefore, capture hot gas from the accretion disk and pump it outward along the field. This model may help explain the narrowness of the jets and their kinks and turns. Moreover, it provides a more powerful supply of energy for the jet—the rotational energy of the spinning black hole.

SCIENCE *at work*

SUPERLUMINAL JETS

The jets observed in some quasars and other active galaxies exhibit a very curious property: they contain clumps that appear to move faster than the speed of light. Astronomers refer to such seeming faster-than-light motion as superluminal motion. According to Einstein's theory of relativity, no object can travel faster than light, but under certain circumstances it is possible for an object to *appear* to move faster than light.

This appearance of superluminal motion can occur when an object is moving almost straight toward us at close to the speed of light. Because the object is traveling toward us at nearly the same speed as the light it emits, we observe its motion in a compressed way. For example, light emitted by a clump a thousand years ago and light emitted now may arrive at Earth just a year apart if the object has moved 999 light-years closer to Earth in the meantime. From Earth, if the motion of the clump causes it to shift by more than a light-year relative to our line of sight during those thousand years, it will appear to us that it is moving faster than the speed of light.

While superluminal motion is a kind of optical illusion, it tells something important about the objects that power active galaxies: they must be able to accelerate matter to near the speed of light. This is something that the intense gravity of black holes is capable of doing.

How might such a massive black hole form? We saw in chapter 16 that astronomers have strong evidence that our own galaxy has a black hole at its center. That black hole has a mass of a few million solar masses. This is much smaller than the black holes thought to be in some active galaxies, but it is still enormously larger than the black holes produced when the most massive stars die. Astronomers hypothesize that growing such massive black holes proceeds in several steps. First, a massive star in the central region of a galaxy reaches the end of its life, explodes as a supernova, and forms a black hole perhaps as small as 5 $M_{\odot}$. Next, this small black hole grows as interstellar matter falls into it. Recall that this matter is most concentrated at the centers of galaxies, so a black hole there has greatest opportunity to grow by this means—especially when the galaxy is young and less of its gas is converted into stars. Because the Schwarzschild radius of a black hole is only a few kilometers for each solar mass it contains (chapter 15), the likelihood of an encounter close enough to capture a star is extremely unlikely. On the other hand, gas in the accretion disk loses energy through friction, so it spirals into the black hole.

As gas falls in, the hole's mass and radius increase, making it easier for the hole to attract and swallow yet more material. Eventually the hole becomes large enough to swallow entire stars, at which point it grows rapidly to a million or more solar masses. Some galaxies may have had more mass at their cores to begin with, or they may have gained mass by collisions with small, neighboring galaxies. In some galaxies this extra mass thereby allowed the galaxy's central black hole to grow far larger than the black hole in the Milky Way.

Astronomers today have evidence that many (perhaps even all) large galaxies have a massive black hole at their core. They come to this conclusion from the observation that stars and gas at the core of many galaxies are orbiting the core at extremely high speed. To prevent such rapidly moving material from flying out of the galaxy's core requires the huge masses of millions or even billions of times the Sun's mass. Yet this huge mass seems to emit no light or other radiation, and astronomers know of no other objects that are both so massive and dark. Thus, by elimination, they are led to think that many galaxies have a massive black hole at their core.

Another recent discovery has led astronomers to think that the growth of massive black holes must be an integral part of the growth of galaxies. For galaxies in which the mass of the central black hole has been measured, that mass closely correlates with the mass of the galaxy's central stellar bulge (fig. 17.29). If growth of supermassive black holes were a chance event, this correlation would be unlikely. Instead, the central black hole must grow at the same rate as the bulge grows, from the initial collapse of gas when a galaxy first forms its bulge to the mergers and cannibalism that cause the bulge to grow. It may also indicate that central black holes merge when galaxies merge.

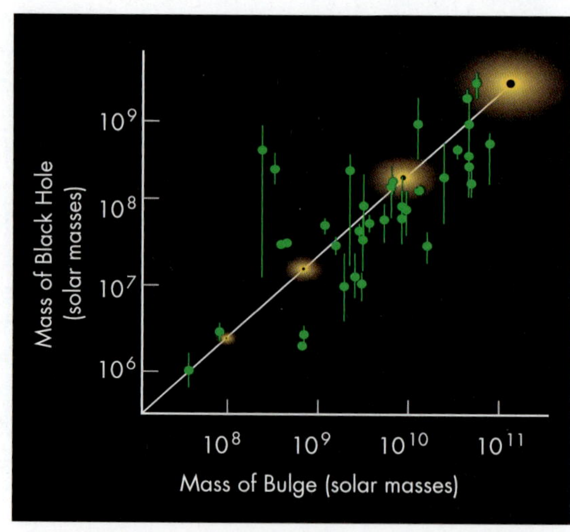

FIGURE 17.29
The correlation between central black-hole mass and the masses of the galaxies' central bulge. The green dots indicate the masses of central black holes and bulges in more than 30 galaxies studied with the Hubble Space Telescope. The vertical green bars indicate the uncertainty in the measurements. The mass of central black holes grows with galaxy mass, as illustrated by the background diagrams of galaxy bulges.

If black holes are so common in galaxies, why are none of the galaxies close to us as active as quasars? The youth of quasars may be the key. Because quasars are so distant, we see them the way they were in the distant past, not the way they are "now." For example, when we look at a quasar that is 10 billion light-years away, we are seeing light that has taken 10 billion years to reach us. Thus, we are seeing the quasar as it was 10 billion years ago. The fact that most quasars are billions of light-years away implies that the phenomenon causing them occurred billions of years ago, when the Universe was younger. Galaxies whose cores were once active quasars probably surround us, but their activity has long since died away. They still harbor huge black holes, but they have used up the gas that powered their activity.

Left to itself, every active galaxy's accretion disk would gradually drain into the central black hole and activity would cease. But the accretion disk may be replenished with matter from at least two sources. Sometimes stars within the galaxy may approach too closely to the black hole and be torn apart by its tidal forces. Their gaseous debris then falls into the accretion disk as a new source of material. Alternatively, from time to time, collisions with neighboring galaxies may add new material as the active galaxy cannibalizes smaller neighbors. Each such event gives the accretion disk a new burst of energy, causing the active galaxy to revive and increase in brightness.

17.5 Probing Intergalactic Space

Quasar Absorption Lines

Although interesting in their own right, quasars are also useful to astronomers as probes of intergalactic space. Because quasars are so distant, their light passes through other galaxies and intergalactic gas clouds on its way to Earth. Even if the intervening object is too dim to see, its matter may imprint its signature on the quasar's light, allowing us to detect these otherwise invisible objects.

For example, many quasars show absorption lines in their spectra that have very different Doppler shifts from those of the emission lines in the quasars themselves (fig. 17.30). These absorption lines are created by cool gas clouds lying between us and the quasar, much as similar lines in stars are created by cool interstellar clouds, as we discussed in chapter 16. The cool gas responsible for the absorption lines may be within galaxies, or it may lie in clouds that have not yet formed into galaxies. Matter within a galaxy cluster between us and a quasar may also affect the quasar's light in an even more remarkable way, creating a gravitational lens.

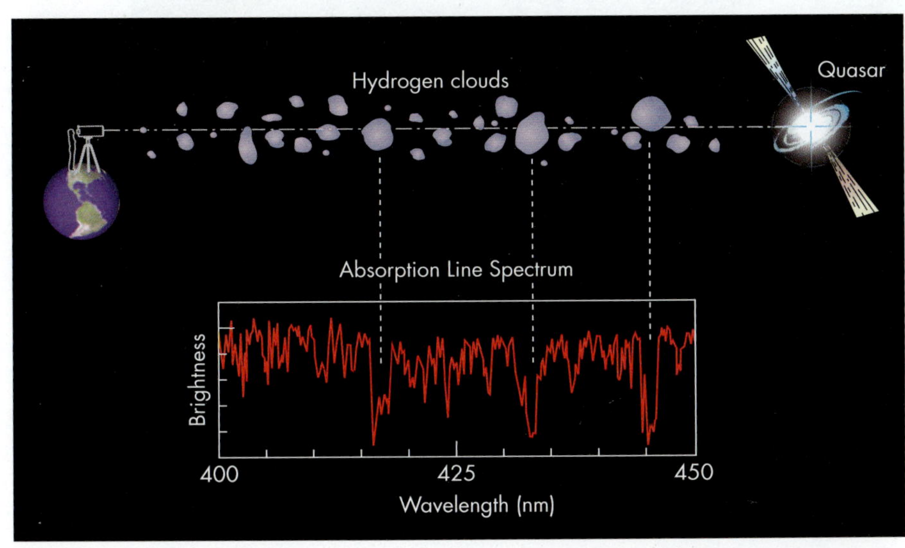

FIGURE 17.30
The light from a distant quasar produces a uniform light source against which we can see hundreds of absorption lines produced by hydrogen clouds along the line of sight to the quasar.

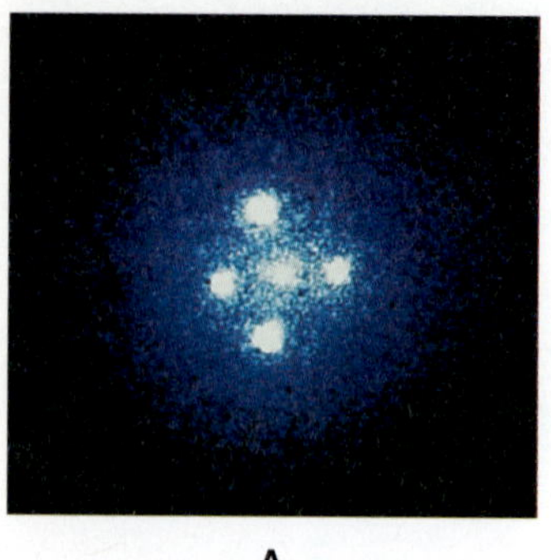

A

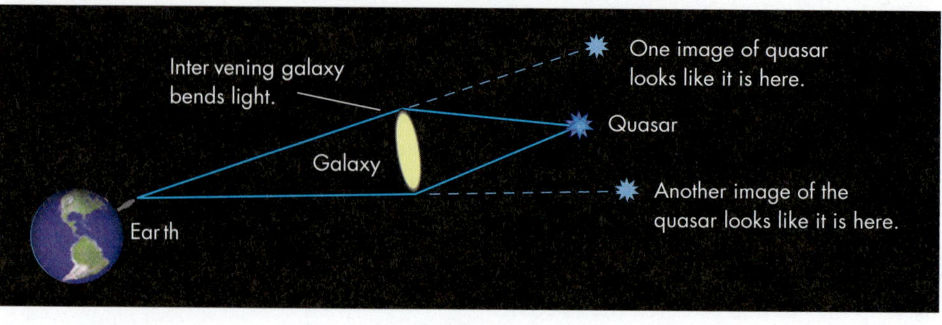

B

FIGURE 17.31
(A) An Einstein cross, a complex image of a quasar created by a gravitational lens. (B) Sketch of how a gravitational lens forms an image.

Gravitational Lenses

A number of quasars seem to have nearby companion quasars with essentially identical spectra but slightly different brightnesses and shapes. The existence of two so nearly identical objects so close together is unlikely, and astronomers now think that the "companions" are actually images of a single quasar created by a gravitational lens (fig. 17.31A).

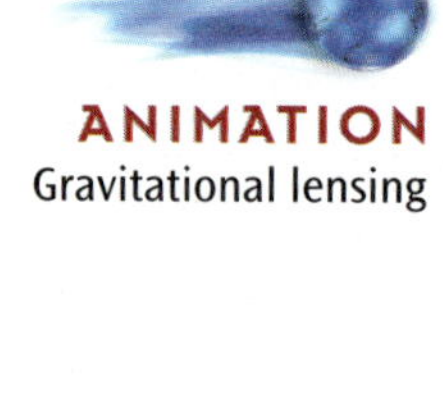

ANIMATION
Gravitational lensing

An ordinary lens forms an image because light bends as it passes through the lens's curved glass. A **gravitational lens** forms an image because light bends as it passes through the curved space around a massive object such as a galaxy. The galaxy's gravitational force bends the space around it so that light rays that would otherwise travel off in other directions and never reach the Earth are bent so that they do, as depicted in figure 17.31B. The existence of gravitational lenses thus supports Einstein's theory that gravity can bend space, as we described in essay 2. We saw in chapter 7 that even planets can bend light. Moreover, we will learn in the next chapter that such bending may occur on such an enormous scale that it affects the shape of the Universe.

Gravitational lenses may create forms other than multiple images of quasars. For example, figure 17.32 shows arcs created when light from a distant galaxy was bent

FIGURE 17.32
The arcs in this picture of the galaxy cluster Abell 2218 are the image of a distant galaxy distorted by the gravitational lens effect created by the galaxy cluster.

into elongated shapes as it passed through a galaxy cluster. Because the shape of these arcs depends on the mass of material within the cluster, astronomers can use the arcs to measure the cluster's mass. The masses found show that galaxy clusters contain large amounts of dark matter—about ten times more than all the stars and gas seen.

17.6 Galaxy Clusters

We have mentioned several times earlier in this chapter that galaxies are often found in groupings called **galaxy clusters.** A galaxy cluster is a group of galaxies held together by their mutual gravity. Each galaxy moves along its own orbit, just as within a galaxy each star moves on its own orbit. Galaxy clusters are huge. A typical cluster is more than several million light-years across and may contain anywhere from a handful to a few thousand member galaxies (fig. 17.32).

The Local Group

Our own Galaxy, the Milky Way, belongs to a very small cluster or "group" called the **Local Group.** The Local Group contains over forty members and is sketched in figure 17.33. Its three largest members are spiral galaxies: the Milky Way, M31, and M33.

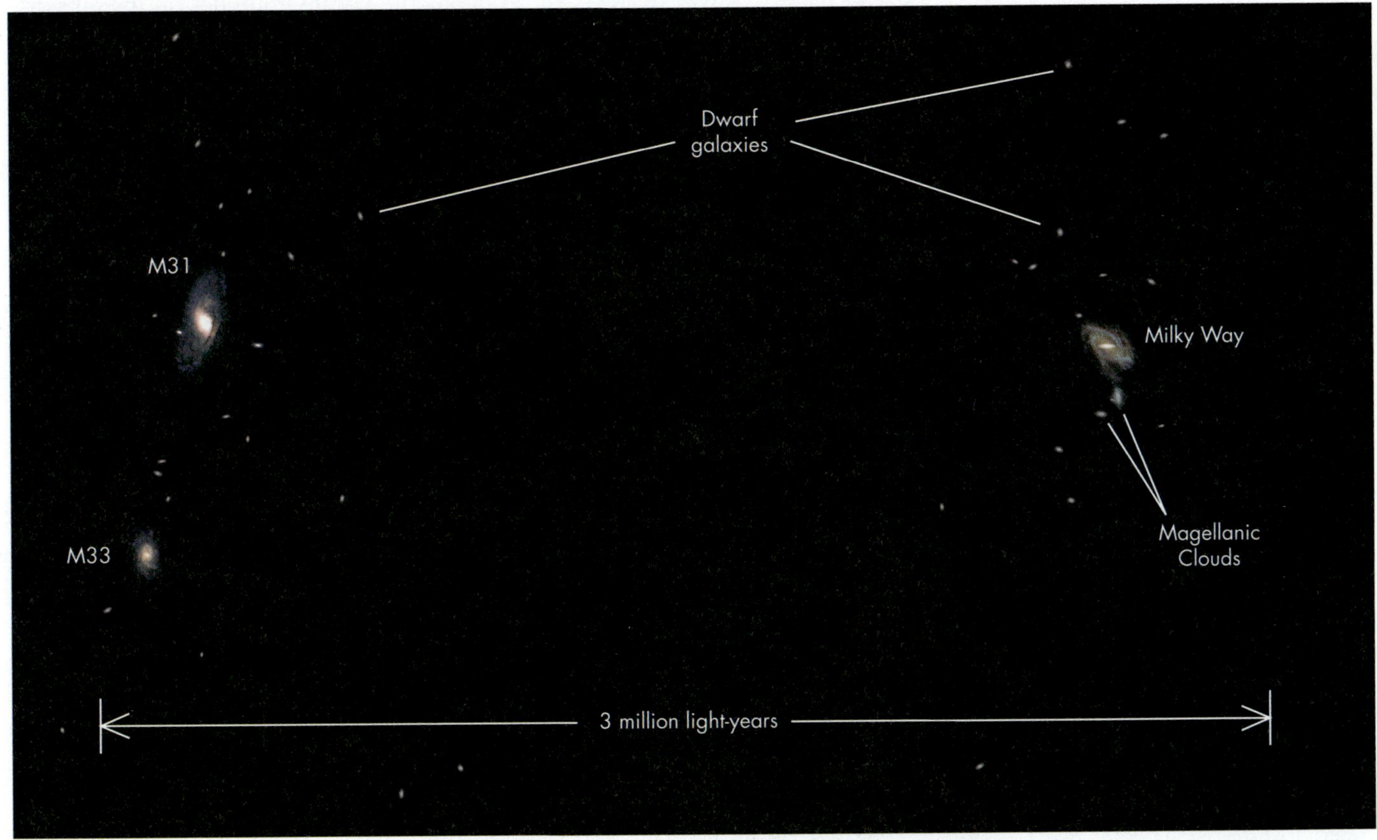

FIGURE 17.33
Images of the galaxies in the local group show what the Local Group might look like from outside. The Local Group's largest member is M31. The Milky Way is intermediate in size between it and M33. The Milky Way is here represented by an image of galaxy M67, because it is thought to be similar in appearance. Two small satellites of M31 are visible in this picture, as are as the Magellanic Clouds. The locations of several dozen other dwarf galaxies in the group are also illustrated.

FIGURE 17.34
The dwarf galaxy Leo I, a member of the Local Group of galaxies.

The smaller members include the satellite galaxies orbiting M31 and the Milky Way, the most famous of which are the Large and Small Magellanic Clouds, which we have referred to many times before.

The Magellanic Clouds (fig. 17.2) are about one-tenth the size of the Milky Way. They appear to be dwarf irregular galaxies, although the large cloud has a faint barlike structure. Both galaxies are approaching the Milky Way along very elliptical orbits and are being pulled into a distorted shape by the Milky Way's gravity.

The Local Group is particularly interesting to astronomers, not only because it is our home galaxy cluster, but also because it allows us to study the "demographics" of galaxies. That is, we can see the true population of galaxies in this region of space "close" to home. For example, most of the galaxies in the Local Group are faint, small "dwarf" galaxies. Figure 17.34 shows a typical dwarf galaxy. It is a dim, small group of stars with little or no interstellar gas. Such galaxies are too faint to be visible except in clusters close by. Thus, without the Local Group to let us see the real abundance of these small systems, we might greatly underestimate their true numbers.

The Relationship of Cluster Size and Galaxy Type

"Poor" clusters, or groups, like the Local Group, contain only a few large galaxies. Large clusters of galaxies may contain hundreds to thousands of large galaxies of the size of the Milky Way or M31. For example, the Local Group lies on the outskirts of the Virgo Cluster, a moderately large cluster in the direction of the constellation Virgo (fig. 17.35).

Not only do clusters and groups differ in their number of galaxies, but they also tend to contain different types of galaxies. The more galaxies a cluster contains, the more elliptical and S0 galaxies it has. Moreover, the few spiral galaxies that they contain tend to be found in the outer parts of the cluster. On the other hand, groups tend to have a high proportion of spiral and irregular galaxies. Spirals are rare in the inner regions of large clusters, probably because of galaxy collisions. In the core of a cluster, the galaxies are close together, and so collisions between them are frequent. As we saw earlier, collisions can change spiral galaxies into ellipticals. Thus, although clusters may have once contained many spirals, today they contain few, and those are the lucky ones that escaped running into a neighbor. The many collisions in the crowded inner part of these clusters also create giant elliptical galaxies by "cannibalism," as we described in section 17.1.

Q. What types of galaxies can you identify in this photograph?

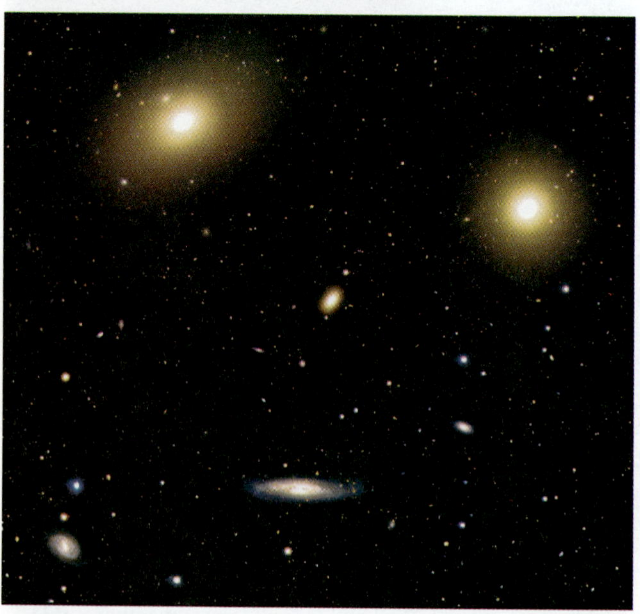

FIGURE 17.35
The Virgo galaxy cluster. Many of the small dots in the image are not stars, but small galaxies within this cluster.

Observations with X-ray telescopes show that many large clusters contain large amounts (10^{12} to 10^{14} $M_{\odot}$) of extremely hot gas that emits X rays (fig. 17.36). In fact, some clusters contain ten times more hot gas than they do stars in their member galaxies. Some of this gas may be left over from the cluster's formation. Some may have been ejected from the galaxies within the clusters. For example, when a supernova explodes, it blasts material to such a high velocity that the material can escape a galaxy's gravity and be driven into intergalactic space.

Gas also escapes from galaxies when they collide, as they move in the crowded cluster environment. Such collisions affect the stars hardly at all. Because they are so small compared to the distance between them, nearly all of stars in colliding galaxies simply shoot harmlessly by one another. The gas, on the other hand, is spread widely in clouds throughout the galaxies, so it cannot escape collisions. Thus, in a collision, the gas in one galaxy is driven at high speed into the gas of the other, compressing the gas and heating it to millions of degrees. In addition, as the stars of one galaxy move past the stars of the other, much of the gas may remain behind. Such gas, stripped from galaxies by collisions, generally contains heavy elements synthesized in stars within the galaxies. It then mixes into the otherwise pure hydrogen and helium gas from which the galaxies formed. But the essential point is that, regardless of its source, clusters contain huge amounts of hot gas.

In addition to X-ray-emitting gas, galaxy clusters also contain large amounts of dark matter. In fact, some of the first evidence for dark matter came from observations of the motion of galaxies within clusters. Just as stars within galaxies seem to orbit too rapidly for the gravitational force that can be attributed to their luminous mass, so too galaxies in clusters orbit too rapidly for their luminous mass. Additional evidence for dark matter in galaxy clusters comes from the hot gas they contain. To hold such hot gas within the cluster and prevent its expansion, the cluster must exert a strong inward gravitational pull on the gas. The gravity needed to confine the gas depends on the cluster's total mass, and, once again, the calculated mass greatly exceeds the mass directly observable—a signature of dark matter. Combining the evidence from these very different methods leads astronomers to deduce that the amount of dark matter in some galaxy clusters is so immense that it amounts to perhaps 100 times the amount of mass contained in stars.

Figure 17.37 shows how the dark matter is distributed in a galaxy cluster that is forming from the merger of two smaller clusters. Although the dark matter is, of course, invisible, its location can be mapped by its gravitational lensing effect on the light of background galaxies and is shown in blue. In this collision, the gas from the two clusters has collided and been compressed. The compression has heated the gas to millions of degrees, so that it emits X rays (shown here in red). The separate masses of dark matter in the two clusters did not affect each other as they collided. Rather, they simply passed harmlessly by one another. This figure also shows an example of how clusters form, according to current astronomical thinking. That is, large, rich clusters grow by the collision and merging of smaller

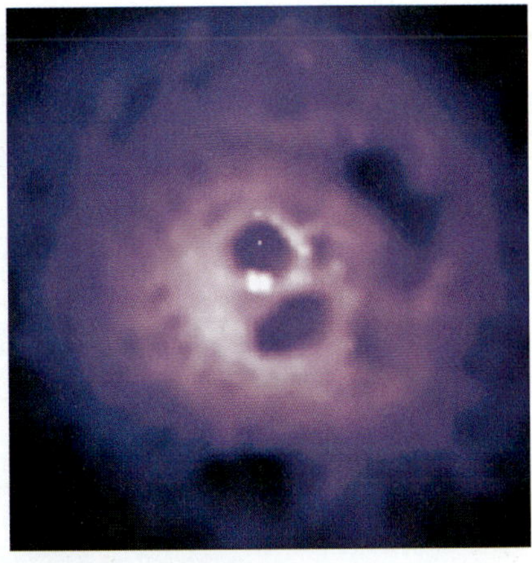

FIGURE 17.36
A false-color X-ray image of the Perseus galaxy cluster. The purple glow is hot gas in the cluster. The picture shows a region about 600,000 light-years across. The dark spots near the center are "holes" in the gas blown out by jets from the active galaxy at the cluster's core.

FIGURE 17.37
A map of the location of dark matter (illustrated in blue) after the collision of two galaxy clusters. The image was made by superposing three images: a picture made at optical wavelengths (showing the galaxies in the two clusters); an image made at X-ray wavelengths (showing in red the hot gas in each cluster); and a map of the dark matter in the two clusters (the blue blobs) as deduced from the gravitational lensing of the background galaxies. In the collision, the dark-matter blobs have passed through each other because the dark matter interacts with itself and other matter only gravitationally. The gas in the two clusters, on the other hand, collides and has clumped up in a bullet-like form, giving it its name: *The Bullet Cluster.*

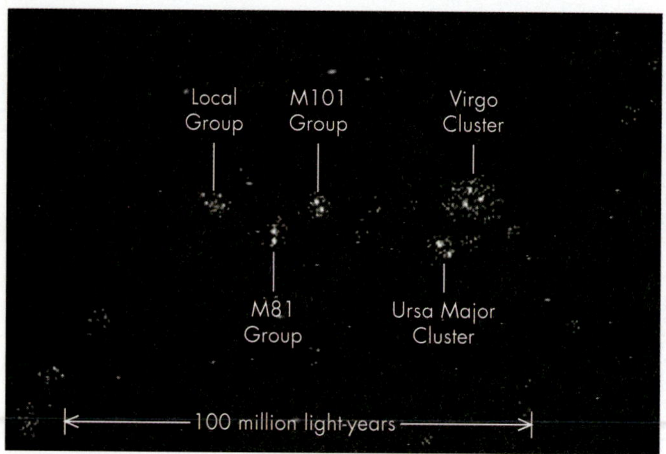

FIGURE 17.38
Artist's view of the Local Supercluster.

clusters. Further evidence for this hypothesis of how clusters form comes, once again, from our ability to "see back in time" by looking out into space. The galaxies in nearby clusters are generally fairly smoothly spread within the clusters. On the other hand, in some remote clusters the galaxies have a clumpy distribution, as if they have recently joined it and have not had time to spread themselves smoothly.

From such evidence, astronomers think that galaxy clusters form by galaxies attracting one another into small groups. Over time, these small groups attract other galaxies and even other entire clusters to make yet larger clusters, eventually growing into the major clusters we see today.

Superclusters

The great mass of galaxy clusters creates gravitational interactions between them and draws them together into larger structures called "superclusters." A **supercluster** contains a half dozen to several dozen galaxy clusters spread throughout a region of space tens to hundreds of millions of light-years across. That is, a supercluster is a cluster of galaxy clusters. For example, the Local Group is part of the Local Supercluster, which contains the Virgo galaxy cluster and more than a dozen others spanning about 150 million light-years (about 50 megaparsecs). It is sketched in figure 17.38.

Superclusters have irregular shapes and are themselves often part of yet larger groupings, but astronomers find it difficult to study the structure of such immense systems. To map the location of galaxies accurately, astronomers need to know not only how far away they are but also their distances relative to each other. Unfortunately, determining the distance to very remote galaxies is extremely hard. However, one feature is clear: many clusters appear to form chains or shells surrounding regions nearly empty of galaxies, as you can see in figure 17.39. These empty spaces are called *voids*, and although they are not totally empty, the few galaxies they contain are small and dim. Whether dark matter exists within voids is not yet known. At the vast distances involved, even the largest telescopes now available allow us to see only the brightest galaxies. Thus, clusters of superclusters and voids mark the end of the structure of the Universe that we can currently see. But although we may have difficulty *seeing* distant large-scale structures, our Galaxy and others can "feel" their presence by their gravitational attraction. That attraction pulls on the Local Group and many nearby galaxies, altering their velocity from what is predicted by the Hubble law. From such observations, astronomers have discovered the "Great Attractor." The Great Attractor is probably an assemblage of superclusters about 160 million light-years (50 megaparsecs) away from us and about 260 million light-years (80 megaparsecs) in diameter. It may contain about 3×10^{16} $M_{\odot}$, the equivalent of 10,000 or more galaxies like the Milky Way.

The existence of such immense concentrations of galaxies and their effect on their neighbors is also revealed in computer simulations of the large-scale structure of the Universe. In such simulations, astronomers write out the equations that govern the motion of matter (basically Newton's laws) and then solve the equations assuming that the material is expanding in a way that mimics the expansion of the Universe. They then watch to see what happens as the gravitational force of the material acts within the simulated volume. Figure 17.40 shows a series of images from such a simulation. You can see that matter that initially had a relatively smooth distribution is drawn by gravity into a network of filaments. Astronomers identify the place where filaments meet as locations where galaxies form. By adjusting the

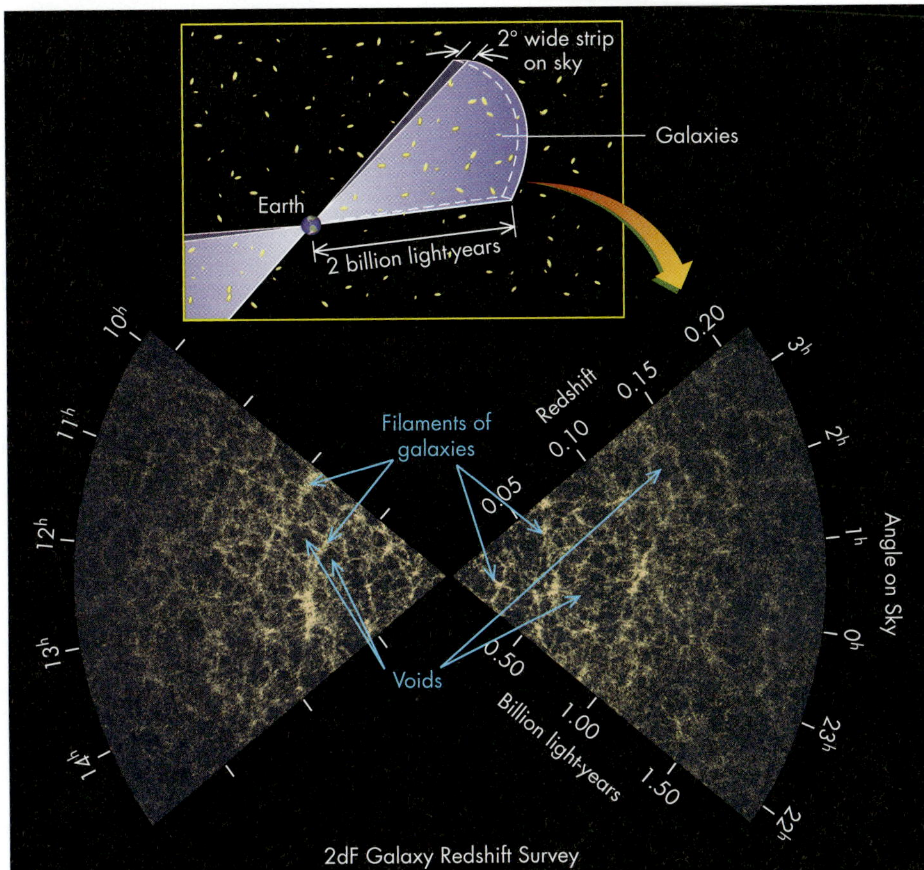

FIGURE 17.39
A map of how galaxies are distributed in space. The shape of the region shown in the figure is like two slices of pizza, tip-to-tip, as illustrated in the upper sketch. Each dot in the lower figure represents one of some 100,000 galaxies that lie in the wedge-shaped regions that extend out to about 2 billion light-years from Earth. Note the nearly empty regions (voids), the stringy structures, and the overall cobwebby appearance. The distances to the galaxies were measured from their redshifts. The angles refer to a set of coordinates on the sky.

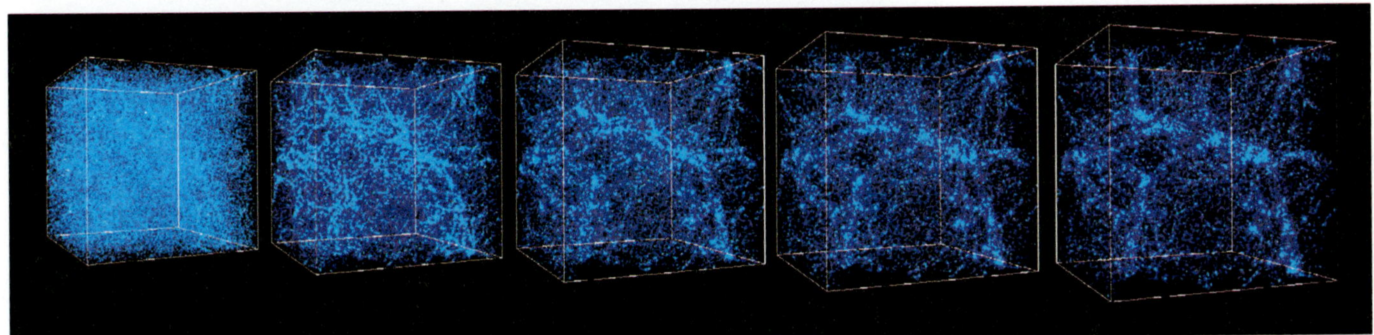

FIGURE 17.40
A computer simulation of the growth of structure as the Universe expands. You can see how an initially smooth distribution of matter becomes stringy, much like the features in figure 17.37. Each box spans about 140 million light-years. The first box on the left represents a time when the Universe was roughly one-tenth its present age. The last box on the right shows that same piece of the model of the Universe as it is today.

amount of ordinary and dark matter and the initial lumpiness, astronomers then “experiment” to find what mix best creates structures similar to what they actually observe. The outcome of many such simulations by many different astronomers confirms that without dark matter, the large-scale distribution of galaxies fails to match what we observe—which is yet additional evidence of the existence of that mysterious matter.

But at least one larger structure remains yet for us to explore: the Universe itself.

SUMMARY

Astronomers classify galaxies into three main types: spirals, ellipticals, and irregulars. Spirals are disklike, with arms of young stars and gas winding out from the nucleus. They contain a mixture of population I and II stars, and about 15% of their mass is in the form of interstellar gas and dust. Elliptical galaxies are spherical or egg-shaped and contain hardly any population I stars because they have essentially no dense interstellar matter from which new stars could form. Irregular galaxies lack a symmetry to their shape, but many are extremely rich in interstellar matter and thus contain many young, population I stars.

Galaxies form from the collapse of primordial gas clouds. Astronomers suspect that most galaxies are born as gas-rich irregular systems. Subsequently, mergers of these galaxies may produce spirals with gas orbiting in a circular disk, or may consume the gas and "scramble" the orbits of their stars and turn the merged galaxy into an elliptical.

Almost all galaxies are moving away from the Milky Way and each other as part of a universal expansion. The motion away from us causes galaxies' light to be redshifted. Hubble's law relates the speed of recession to the distance by a simple constant (Hubble's constant). Astronomers frequently use Hubble's law to estimate a galaxy's distance based on its observed redshift.

Studies of the motions of stars in galaxies reveal that most of the mass of a galaxy is made up of dark matter, not stars or gas. Astronomers do not yet know the composition of this dark matter, but it may be some exotic kind of matter that does not interact with normal matter.

Many galaxies have extremely small but very energetic and luminous nuclei. These include radio and Seyfert galaxies as well as quasars. The amount of energy generated may dwarf the light output from the galaxy's stars and, for a quasar, outshine the Milky Way by a factor of a thousand. According to current models, the activity results from matter falling in toward a supermassive black hole at the galactic center. These black holes have masses that are as large as several billion solar masses packed into a region not much larger than the Solar System. An accretion disk forms around the hole as matter falls in, and hot gas boils off the disk, forming jets and clouds of radio-emitting gas. Studies suggest that the black holes grow in size as galaxies grow and merge. Large galaxies today may have supermassive black holes at their centers that are quiet because no matter is falling into them.

Galaxies are usually grouped into clusters a few million light-years across. Small clusters or "groups" tend to contain spiral galaxies. The Milky Way belongs to a small cluster called the Local Group, which contains over forty member galaxies, mostly small "dwarf" galaxies. Large clusters contain thousands of galaxies, which are primarily ellipticals. Clusters themselves appear to be clustered into superclusters that are many tens of millions of light-years across and are arranged into long "filaments" surrounding voids.

QUESTIONS FOR REVIEW

1. (17.1) What are the three main types of galaxies? How do the basic galaxy types differ in shape, stellar content, and interstellar matter?
2. (17.1) Why is it not surprising that elliptical galaxies contain mostly Pop II stars?
3. (17.1) Why are galaxy collisions of interest?
4. (17.1) What happens to the stars when two galaxies collide?
5. (17.2) Explain several ways astronomers measure the distance to nearby galaxies. Why is it important to know the distance?
6. (17.2) What is the Hubble law? Why does the Hubble law *not* apply to nearby galaxies?
7. (17.2) How do astronomers measure the mass of a galaxy?
8. (17.3) What is meant by dark matter? What observations support its existence?
9. (17.4) What are the three main types of active galaxies? List the main characteristics of each.
10. (17.4) Make a sketch of a radio galaxy illustrating its structure and where its radio emission comes from. Why do astronomers call these objects "radio" galaxies?
11. (17.4) What process generates the radio emission in radio galaxies?
12. (17.4) Why do some astronomers think that quasars are very distant?
13. (17.4) How is it known that active galaxies have small core regions?
14. (17.4) What mechanism has been suggested as powering active galaxies?
15. (17.4) How might a large black hole form in a galaxy's core?
16. (17.5–17.6) What is a gravitational lens? What can observations of lenses tell us about clusters?
17. (17.6) What are some differences between clusters that contain large and small numbers of galaxies?
18. (17.6) What is the Local Group, and why is it a group and not a cluster?
19. (17.6) What is the Local Supercluster? How many clusters might be in a supercluster and how big is one?
20. (17.6) What is a void?
21. (17.6) What does the large-scale structure of the Universe look like? How do astronomers study it?

THOUGHT QUESTIONS

1. (17.1) Hubble proposed that his tuning-fork diagram might represent evolutionary changes with E galaxies gradually changing to S galaxies. What evidence based on the amount of interstellar matter and kind of stars in E and S galaxies indicates that this is very unlikely?
2. (17.1) Take a look at an egg from different angles. Do you think the classification of elliptical galaxies is always accurate? Examining figure 17.8, could an E0 be mistaken for an E7, or an E7 for an E0? What does this imply about the actual number of each type compared to what is observed?
3. (16.3/17.1/17.6) Explain how selection effects might make it difficult to obtain an accurate census of galaxy types. What might happen at near and far distances? What about if you picked an area in the middle of a supercluster, or an area in a void? (You don't know what's there until you start looking!)
4. (17.2) The value of the Hubble constant was only recently determined to high accuracy. If H were equal to 100 km/(sec Mpc) instead of H = 70 km/sec/Mpc, as some astronomers thought it would be, how many times larger or smaller would the distance to a distant galaxy be? How much larger or smaller would the diameter of the galaxy be?
5. (17.2/17.4) If you wanted to look at your mother when she was a baby, how far out in space would you need to be? (Assume you have a sufficiently powerful telescope and pretend that you can travel instantly to any distance.) What if you wanted to look in on astronomers in ancient Egypt around 3000 B.C.? How does this relate to astronomers studying young galaxies?
6. (14.2/17.4) Are the jets that come out of active galaxies similar to bipolar outflows? Explain your answer.

PROBLEMS

1. (17.2) A Cepheid variable in a nearby galaxy looks 10^6 times fainter than an identical Cepheid in the Milky Way. The Cepheid in the Milky Way is 1000 parsecs away. How far away is the nearby galaxy?
2. (17.2) A galaxy has a recession velocity of 13,000 kilometers per second. What is its distance in megaparsecs?
3. (4.6/17.2) A quasar's spectrum shows a shift in spectral lines ($\Delta\lambda/\lambda$) of 0.158. Use the Doppler-shift formula to determine what recession velocity this shift indicates, and then use the Hubble law to determine the distance to the quasar in Mpc if H = 70 km/sec/Mpc. Be sure to calculate the recession velocity in kilometers per second.
4. (17.2) A galaxy has an angular diameter of 0.1°. Its distance is 10 megaparsecs. What is its true diameter?
5. (17.2) What would be the angular diameter of the Milky Way as seen by an alien astronomer in the Virgo Cluster at a distance of 18 Mpc?
6. (17.2) A galaxy has a rotation speed of 200 kilometers per second at a distance from its center of 10 kiloparsecs. What is its mass within this distance?
7. (17.4) A radio galaxy shows lobes that extend 1.5 megaparsecs away from the galaxy on each side. If the accelerated electrons in the lobe travel at nearly the speed of light, how long will it take them to travel to the edge of the lobe?
8. (17.4) A nearby radio galaxy 50 megaparsecs from the Earth has a 1 billion solar mass black hole at its center. If the diameter of this black hole is twice its Schwarzschild radius, calculate the angular size of the black hole as seen from Earth. Convert the angular size from degrees to arc seconds (3600 arc seconds per degree). Can we directly image the black hole as a dark spot in the accretion disk? Compare your result with 1×10^{-5} arc seconds, which is about the best resolution currently possible with the radio astronomy technique of very long baseline interferometry.
9. (17.4) What is the minimum size of a quasar that shows variations in brightness over a period of 36 hours? Express your answer in kilometers.
10. (17.6) Estimate from figure 17.33 the ratio of the diameter of the local group to the diameter of the Milky Way (100,000 light-years across). Compare this with the ratio of the diameter of the Milky Way to the diameter of the Solar System (including the Oort cloud, the Solar System's diameter is about 200,000 AU).

TEST YOURSELF

1. (17.1) A large galaxy contains mostly old (Pop II) stars spread smoothly throughout its volume, but it has little dust or gas. What type galaxy is it most likely to be? Select all that are possible.
   (a) Irr (b) S (c) SB (d) E
2. (17.1) Which of the following would be a probable result of a collision between two galaxies?
   (a) Complete destruction of all the stars
   (b) Formation of a giant spiral galaxy from two elliptical ones
   (c) Formation of a single open star cluster
   (d) Formation of a ring galaxy
   (e) All of the above
3. (17.2) A galaxy's spectrum has a redshift of 28,000 kilometers per second. If the Hubble constant is 70 kilometers per second per megaparsec, how far away from Earth is the galaxy?
   (a) 6 megaparsecs
   (b) 1960 megaparsecs

(c) 400 megaparsecs
(d) 4 megaparsecs
(e) 1.960 megaparsecs

4. (17.3) Astronomers believe that dark matter exists because
   (a) they can detect it with radio telescopes.
   (b) the outer parts of galaxies rotate faster than expected on the basis of the material visible in them.
   (c) the galaxies in clusters move faster than expected on the basis of the material visible in them.
   (d) it is the only way to explain the black holes in active galaxies.
   (e) Both (b) and (c) are correct.
5. (17.3) Most of the mass of a galaxy is contained in
   (a) the massive O and B stars in the galaxy.
   (b) the cold interstellar gas.
   (c) the central black hole of the galaxy.
   (d) the dark matter of the galaxy.
   (e) the disk of the galaxy.
6. (17.3) The ratio of dark matter to visible matter in a typical elliptical or spiral galaxy is about
   (a) 100:1
   (b) 10:1
   (c) 1:1
   (d) 1:10
   (e) 1:100
7. (17.4) A spiral galaxy has a small bright central region, and its spectrum shows that it contains hot, rapidly moving gas. It is most likely a ________ galaxy.
   (a) barred spiral
   (b) Seyfert
   (c) radio
   (d) BL Lac
   (e) quasar
8. (17.6) The Local Group (select all that are correct)
   (a) contains about 40 member galaxies.
   (b) is one of the smaller galaxy clusters.
   (c) is the galaxy cluster to which the Milky Way belongs.
   (d) is one of the larger galaxy clusters.
   (e) contains mostly medium-size elliptical galaxies.
9. (17.6) The largest clusters probably contain many elliptical galaxies because
   (a) gas is stripped out of galaxies during collisions.
   (b) the dark matter in them preferentially attracts elliptical galaxies over spirals.
   (c) they are so old, all the galaxies have used up all their gas.
   (d) they were created at the same time in the history of the Universe as all the globular clusters.
10. (17.6) Galaxy clusters may contain _____ as much mass in hot gas as in stars in their member galaxies.
   (a) 100
   (b) 10
   (c) 1
   (d) 1/10
   (e) 1/100

KEY TERMS

active galaxy, 475
barred spiral galaxy, 460
dark matter, 473
elliptical galaxy, 460
galactic cannibalism, 466
galaxy, 458
galaxy cluster, 483
gravitational lens, 482
Hubble constant, 470
Hubble law, 470
irregular galaxy, 460
jets, 476
Local Group, 483
megaparsec, 470
quasar, 476
radio galaxy, 476
recession velocity, 470
redshift, 470
Seyfert galaxy, 475
spiral galaxy, 459
supercluster, 486
supermassive black hole, 478

FURTHER EXPLORATIONS

Barnes, Joshua, Lars Hernquist, and Francois Schweizer. "Colliding Galaxies." *Scientific American* 265 (August 1991): 40.

Bartusiak, Marcia. *Through a Universe Darkly: A Cosmic Tale of Ancient Ethers, Dark Matter, and the Fate of the Universe.* New York: HarperCollins, 1993.

Beck, Sara C. "Dwarf Galaxies and Starbursts." *Scientific American* 282 (June 2000): 66.

Bothun, Gregory D. "Beyond the Hubble Sequence." *Sky and Telescope* 99 (May 2000): 36.

———. "The Ghostliest Galaxies." *Scientific American* 276 (February 1997): 56.

Disney, Michael. "A New Look at Quasars." *Scientific American* 278 (June 1998): 52.

Dressler, Alan Michael. *Voyage to the Great Attractor: Exploring Intergalactic Space.* New York: Knopf/Random House, 1994.

Ellis, Richard S. "Quest for the First Galaxies." *Astronomy* 36 (March 2008): 22. See also "What Happens When Galaxies Collide?" by Steve Nadis and "Galaxies on Fire" by Daniel Pendick in the same issue.

Ferris, Timothy. *Galaxies.* San Francisco: Sierra Club Books, 1982.

Ford, Holland, and Zlatan I. Tsvetanov. "Massive Black Holes in the Hearts of Galaxies." *Sky and Telescope* 91 (June 1996): 28.

Henry, J. Patrick, Ulrich G. Briel, and Hans Böhringer. "The Evolution of Galaxy Clusters." *Scientific American* 279 (December 1998): 52.

Hodge, Paul. "Our New! Improved! Cluster of Galaxies [The Local Group]." *Astronomy* 22 (February 1994): 26.

Kauffmann, Guinevere, and Frank Van den Bosch. "The Life Cycle of Galaxies." *Scientific American* 286 (June 2002): 46.

Kraan-Kortewey, Renée C., and Ofer Lahav. "Galaxies Behind the Milky Way." *Scientific American* 279 (October 1998): 50.

Macchetto, F. Duccio, and Mark Dickinson. "Galaxies in the Young Universe." *Scientific American* 276 (May 1997): 92.

Martinez-Delgado, David, and R. Jay Gabany. "Galaxy Archeology with Amateurs." *Sky and Telescope* 117 (January 2009): 92.

Naeye, Robert. "Ghosts in the Cosmic Machine: The Answer to the Dark Matter Mystery May Lie in the Murky Realm of Ghostly Elementary Particles." *Astronomy* 24 (October 1996): 48.

Panek, Richard. "The Quest for Dark Matter." *Sky and Telescope* 116 (August 2008): 30.

West, Michael. "Galaxy Clusters: Urbanization of the Cosmos." *Sky and Telescope* 93 (January 1997): 30. See also related articles in the same issue.

Website

Visit the *Explorations* website at **http://www.mhhe.com/arny** for additional online resources on these topics.

FIGURE QUESTION ANSWERS

WHAT IS THIS? (chapter opening): This is not a comet, but a radio picture of gas being stripped out of the Large and Small Magellanic Clouds by the gravitational (tidal) pull of the Milky Way. The Magellanic Clouds are the two bright blobs on the left of the image. The main arc of hydrogen gas is about 300,000 ly long, three times the diameter of the Milky Way.

FIGURE 17.24 There is a jet pointing out of the blob labeled "Quasar" at an angle of about "5 o'clock."

FIGURE 17.35 A spiral (Sb) galaxy lies near the bottom center, and a barred spiral (SBa) at bottom left. An elliptical (E1) galaxy is near the upper right corner, an elliptical (E3) or S0 near the upper left corner and a small elliptical (E5) near the center of the image.

PROJECTS

1. **The Andromeda Galaxy:** If the season allows, use a pair of binoculars or small telescope to find M31, the great galaxy in Andromeda (see Looking Up #3 in the front matter). What shape does it appear to have? Use the information in the text to estimate how long it took the light you are seeing to reach you. Note: M31 is visible from most of North America, Europe, and Asia in the evening from August through February. The nebula lies about midway along a line from the eastern edge of the Great Square in Pegasus to Cassiopeia. (Check a detailed star chart for its exact location.)

2. **Images of Galaxies:** Go to the Hubble Space Telescope's public website (http://hubblesite.org) gallery and find images of (a) a few different spiral, elliptical, and irregular galaxies; (b) examples of galaxy mergers and "galactic cannibalism"; (c) the central galaxy in a radio galaxy or quasar or both (you may even find an image with the radio image superimposed), and a Seyfert galaxy; and (d) a galaxy cluster and a galaxy group (you may even find an image with an X-ray map of gas superimposed on a large cluster).

The Hubble Ultra Deep Field. This image shows a tiny portion of the constellation Fornax, containing about 10,000 galaxies. Some of the galaxies are so far away that we are seeing them the way they appeared when the Universe was less than one billion years old.

KEY CONCEPTS

- The Universe is expanding.
- From the speed of its expansion, we can estimate its age.
- The Universe began in a hot, dense state.
- We detect the remnant radiation of its beginnings as the cosmic microwave background.
- Observations of small variations in the brightness of the cosmic microwave background allow astronomers to determine the exact age of the Universe and its shape.
- The Universe is 13.7 billion years old.
- Overall, the Universe has no curvature ("flat") and will expand forever.
- Measurement of the speed of recession of the most distant galaxies shows that the Universe's expansion is speeding up.
- Dark energy may be the cause of that acceleration.
- The Universe is composed primarily of dark energy and dark matter.
- Shortly after its formation, the Universe went through a very brief, very fast expansion stage known as inflation.

18 Cosmology

CONCEPTS AND SKILLS TO REVIEW

- Hubble law (17.2)
- Density (6.1)
- Newton's law of gravity (3.5)
- Red light has a longer wavelength than blue (4.1)
- Forces (Preview, p. 8)

Cosmology is the study of the structure and evolution of the Universe as a whole. Cosmologists seek answers to questions such as: How big is the Universe? What shape is it? How old is it? How did it form? What will happen to it in the future?

Given our insignificant size in the cosmos, such questions may seem futile or even arrogant, but most cultures have tried to answer them. Many of those answers have become part of humanity's religious heritage. For example, Hindus believe that the Universe is the work of several gods who give the Universe form from nothingness and then destroy it in a cyclical pattern. Jews, Christians, and Muslims share a belief that there is a single god who began the Universe in a single act of creation. This is expressed in the King James version of the Bible, in the book of Genesis (1:1–3) in the following way:

> In the beginning God created the heaven and the earth; and the earth was without form and void; and darkness was upon the face of the deep and the spirit of God moved upon the face of the waters. And God said, "Let there be light," and there was light. And God saw the light, that it was good.

These creation stories give us powerful poetry and imagery of the birth of the Universe, but they offer few specifics of what our Universe was like at its birth.

In this final chapter, we will learn how contemporary scientists view the creation and evolution of our Universe, beginning with the observational discoveries that the Universe is expanding and that it is filled with a very low-energy background radiation. This expansion and radiation imply that the Universe was born 13.7 billion years ago in a hot, dense, violent burst of matter and energy called the **Big Bang.** Within the last few decades, cosmologists have sought to extend our knowledge of the Universe to the earliest moments of the Big Bang and have discovered that the entire Universe we see today may have expanded, during an instant known as the "inflationary stage," out of a volume smaller than a proton.

Q: WHAT IS THIS? See chapter summary for the answer.

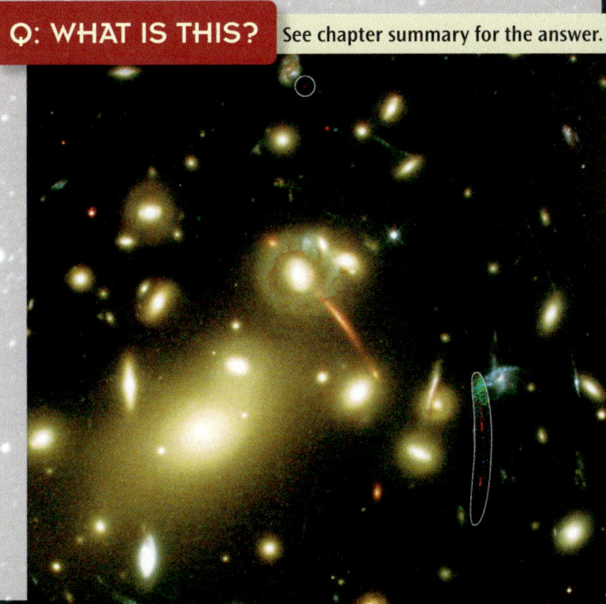

18.1 Observations of the Universe

In the early years of the twentieth century, most astronomers thought that the Universe consists of the Milky Way Galaxy and a number of small companions. Immanuel Kant, the German natural philosopher, suggested in the eighteenth century that the Milky Way is merely one of a multitude of galaxies, but his views were not widely accepted until the 1920s, when astronomers showed that other galaxies are millions of light-years from the Milky Way and comparable to it in size. About the same time, astronomers such as Hubble showed that galaxies are moving away from one another—in short, the Universe is expanding.

Distribution of Galaxies

Pictures of the night sky show that distant galaxies are scattered across it approximately uniformly, as shown in figure 18.1A. This is in sharp contrast to how stars are distributed in our Milky Way: its stars are concentrated in a narrow band from which we deduce that our Galaxy is a disklike system. The smooth distribution of galaxies leads astronomers to conclude that the Universe is filled approximately uniformly with galaxies, in much the way that raisins are distributed uniformly through a fruitcake.* Thus our first deduction is that the Universe is approximately the same in all directions, as portrayed in figure 18.1B. In addition to this uniformity in space, the Universe also exhibits a uniformity of motion.

Motion of Galaxies

In chapter 17, we learned that the spectral lines of nearly all galaxies are redshifted, and their shift to a longer wavelength implies that the galaxies are moving away from us. Moreover, the redshift increases with distance, d, in such a way that the recession velocity, V, of a galaxy obeys the Hubble law:

$$V = \mathrm{H}d$$

where H is the Hubble constant.

*We ignore here the clumping of galaxies into groups and clusters.

A

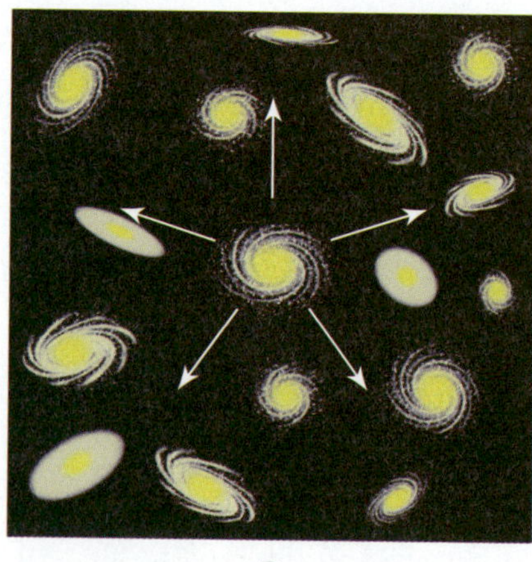

B

FIGURE 18.1
(A) An all-sky plot of the positions of over a million galaxies detected by 2MASS, an infrared survey of the sky. Each white dot represents a galaxy, or in clusters and superclusters of galaxies, they merge to form brighter areas. The blue-shaded portion of the diagram shows where the plane of the Milky Way crosses through this map. (B) The approximate constant number of galaxies in all directions implies that the Universe is uniform.

The motion of galaxies away from us is caused by the expansion of the Universe. This expansion is the result of space itself expanding. It is *not* like a bomb burst flinging fragments in all directions from a single central location. Instead, it is like the motion of buttons glued on a balloon when the balloon is inflated (fig. 18.2). Or, returning to the fruitcake analogy, it is like a giant, raisin-filled fruitcake expanding as it bakes. However, unlike a fruitcake, the Universe has no center to its expansion. Every point moves away from every other point.

The motion of both the buttons on the balloon* and the raisins in the cake is caused by the motion of the medium in which they are set. So too for galaxies: space itself is expanding, and galaxies are carried apart by that expansion, not by any motion of their own. However, although space expands, the objects themselves—galaxies, stars, planets, and so forth—do not. Gravity and other forces maintain such material bodies at a constant size. On the other hand, light waves *are* altered by the expansion of space, and that expansion creates the redshift of the galaxies that we observe. As a light wave from a galaxy moves through space, it is stretched as the space expands, just as the coils in a spring are stretched if you pull on the ends. If you draw a squiggle on a balloon, you can see a similar effect—the "wavelength" of the squiggle increases as you blow up the balloon (fig. 18.2).

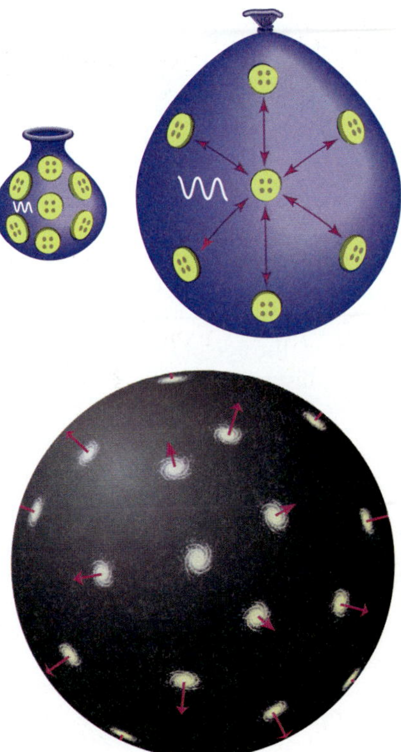

FIGURE 18.2
Representations of the expansion of the Universe by the surface of a balloon. The surface represents only two of the three dimensions of space. As it expands, the galaxies are carried farther apart.

Age of the Universe

The above model of the Universe as an expanding space filled uniformly with galaxies allows cosmologists to estimate its age and to calculate conditions near the time of its birth. Alexander Friedmann, a Russian scientist, made such calculations in the 1920s, but his work attracted little attention. Not until 1927, when Abbé Georges Lemaître, a Belgian cosmologist and priest, independently made similar calculations, did astronomers appreciate that conditions at the birth of the Universe could be deduced from what is observed today.

Lemaître argued that because galaxies are moving apart now, they must in the past have been closer together. He went on to propose that at one time in the distant past, all matter in the Universe was packed into a region approximately the size of the present Solar System. With so much mass in such a relatively small volume, the density of the Universe was enormous. In fact, at that remote time in the past, the density was so immense that the protons and neutrons that now compose the far-flung stars and galaxies were touching, just as today the protons and neutrons within a single atom touch. For that reason, Lemaître called this dense stage of the early Universe the Primeval Atom.

Lemaître argued that the Primeval Atom "decayed" much like an ordinary radioactive atom such as uranium decays. Moreover, he identified this decay with the birth of the Universe, and he calculated how long ago that occurred. Astronomers no longer accept Lemaître's theory of the birth of the Universe, but they do think that it was created from a highly compressed state in the distant past, and they measure its age from that time. To determine that age, we can make a simple calculation, as explained below, but an analogy may help us better understand this calculation.

Suppose a car leaves a starting point in a long race and drives steadily at 60 miles per hour. Suppose the driver looks at the mileage and notes that the car has traveled 120 miles. How long ago did the car leave the starting point? You can easily figure that out by dividing the distance traveled (120) by the speed (60), to get 2 hours. So, too, with measuring the age of the Universe. Astronomers pick some galaxy at a known distance from the Milky Way, d. If they know the galaxy's speed, V (from the Hubble law, for example), then they can determine that the two galaxies began moving apart at a time, $t = d/V$. We will take that time as the beginning of the Universe. A little

*You might wonder in the balloon analogy what lies inside of the balloon and what it is expanding into. Because we are describing space-time and because the balloon's surface represents the three dimensions of *space*, the balloon is expanding in *time*. Inside it lies the past. Outside it lies the future.

EXTENDING *our reach*

ESTIMATING THE AGE OF THE UNIVERSE

To calculate the age of the Universe, t, we proceed as follows: let two representative galaxies be a distance, d, apart and separating with a velocity V. For simplicity, we will assume that V has remained constant. Then in the time t, the age of the Universe, they will have separated by a distance given by

$$d = Vt$$

d = Distance between two galaxies
V = Velocity at which galaxies are separating
t = Age of the Universe

We can find how long ago they began moving apart by dividing both sides by V. This gives

$$t = \frac{d}{V}$$

Thus, in this simple model, the age of the Universe is d/V. The Hubble law, however, relates a galaxy's velocity to its distance by $V = d\mathrm{H}$. Therefore, substituting this expression for V into the expression for the age of the Universe, we find

$$t = \frac{d}{V} = \frac{d}{d\mathrm{H}} = \frac{1}{\mathrm{H}}$$

where H is the Hubble constant.

math (worked out in Extending Our Reach: "Estimating the Age of the Universe" and sketched in fig. 18.3) then shows that

$$t = 1/\mathrm{H}.$$

To evaluate t in years, we must know H, the Hubble constant. However, we cannot simply insert the value of H in the expression for t because H is expressed in units of kilometers/second/megaparsec. If we were to use H in these units, we would get an answer for the age of the Universe in equally odd units. We must therefore express H in units of 1/seconds by converting megaparsecs to kilometers so that the distance units in H cancel out. One megaparsec is 3.1×10^{19} km, so if the Hubble constant is 70 km/sec/Mpc, this gives

$$\mathrm{H} = 70\ \text{km/second}/(3.1 \times 10^{19}\ \text{km})$$
$$= \frac{70}{3.1} \times 10^{-19}\ \text{seconds}^{-1}$$
$$= 23 \times 10^{-19}\ \text{seconds}^{-1}$$
$$= 2.3 \times 10^{-18}\ \text{seconds}^{-1}$$

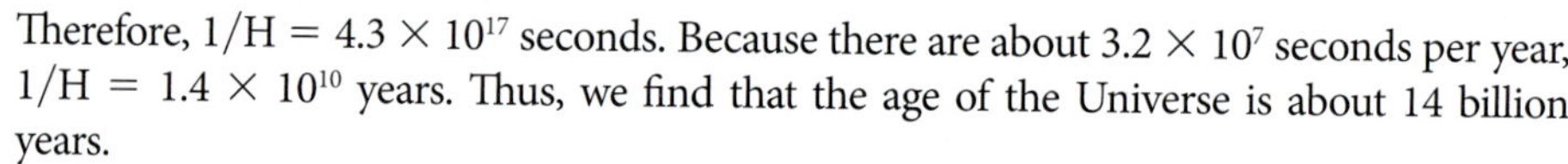

Therefore, $1/\mathrm{H} = 4.3 \times 10^{17}$ seconds. Because there are about 3.2×10^{7} seconds per year, $1/\mathrm{H} = 1.4 \times 10^{10}$ years. Thus, we find that the age of the Universe is about 14 billion years.

Astronomers find this value for the age of the Universe reassuring because it is much older than the Earth (4.6 billion years, as we saw in chapter 6) and is also older than the oldest known stars (about 13 billion years). Something would be seriously wrong with our model if the Universe turned out to be younger than its stars, a clearly impossible situation.

$d = Vt$

Therefore, $t = \frac{d}{V}$

But according to the Hubble law,
$V = d\mathrm{H}$

Therefore, $t = \frac{\not{d}}{\not{d}\mathrm{H}} = \frac{1}{\mathrm{H}}$

FIGURE 18.3
Estimating the age of the Universe from the recession of one galaxy from another galaxy.

Are We at the Center of the Universe?

The recession of distant galaxies from the Milky Way often leads to the mistaken belief that the Milky Way is at the center of the Universe. However, because space itself is expanding, no matter where you are in the Universe, galaxies will be moving away from you. A simple example illustrating this is shown in figure 18.4, which portrays galaxies distributed along a line. Each galaxy is separated from its nearest neighbor by 10 megaparsecs. Suppose the Milky Way is the third galaxy in the line and that

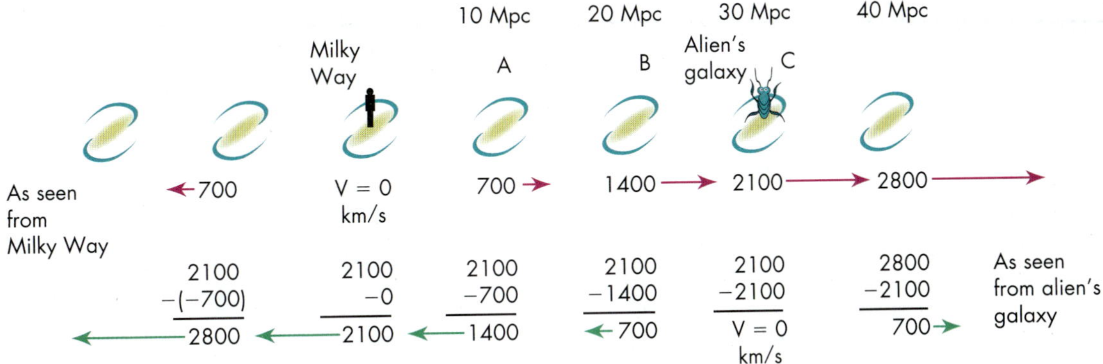

FIGURE 18.4
A line of galaxies illustrating that in a Universe that obeys a Hubble law of expansion ($V = Hd$), an observer on any galaxy sees the same law.

galaxies A, B, and C obey the Hubble law, with each receding at a velocity given by $V = 70d$. That is, A recedes from us at 700, B at 1400, and C at 2100 kilometers per second. Next, suppose we could communicate with an alien on galaxy C and ask what it sees. It would see us receding at 2100 kilometers per second, A receding at $2100 - 700 = 1400$ kilometers per second, and B receding at $2100 - 1400 = 700$ kilometers per second, as shown along the bottom portion of figure 18.4. In other words, the alien would see galaxies receding from it exactly the same way we see galaxies receding from us. Furthermore, the galaxies seen by the alien will obey the alien's version of the Hubble law.

A similar argument can be made for an observer on any of the galaxies. This means that because the expansion of the Universe causes the velocity to increase in proportion to the distance, *every observer, no matter where in the Universe he/she/it is, will see the Universe expanding the same way.* The Universe has no preferred "center" of expansion if it expands uniformly. In fact, observers in another galaxy will have the impression that they are at the center of the expansion. Astronomers describe this lack of a preferred location as the **cosmological principle.** It is a statement of cosmic modesty: there are no special positions. We are not at the center of the Universe, nor is anyone else. In fact, the notion of a "center" to the Universe is meaningless, as we will discuss further in section 18.4.

18.2 Looking Back Toward the Beginning of Time

Olbers' Paradox

The Universe's finite age can also be demonstrated by the simple experiment of stepping outside at night and noting that the sky is dark. This relation between the darkness of the night sky and the structure of the Universe is known as **Olbers' paradox,** after the German astronomer Heinrich Olbers, who in 1823 presented an argument that if the Universe extends forever and has existed forever, the night sky should be bright. The paradox arises from following reasonable premises (or so it seemed at that time) to a clearly nonsensical conclusion. The argument is as follows:

> Although the paradox bears Olbers' name, more than two centuries earlier Kepler pointed out that in an infinite Universe, the night sky should be bright, and he therefore concluded the Universe was finite in size.

Suppose you stand in a small grove of trees and look out between them to the surrounding landscape. If the grove is small, your line of sight will be blocked by trees in a few directions but will pass between the trees in most directions (fig. 18.5 A–C). If the grove is larger, more-distant trees in it will block your view in what were previously clear gaps. Then, no matter what direction you look, your line of sight is intercepted by a tree.

In a small grove of trees, only a few block your view.
Lots of space to view between trees.

A

In a larger forest, more distant trees block your view.
No open space visible between trees.

B

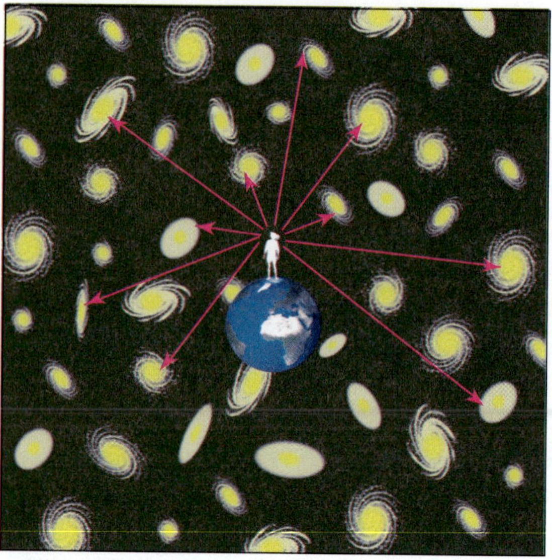

In a large enough universe, every line of sight eventually encounters a star within a galaxy

C

FIGURE 18.5

(A) An observer in a small grove of trees can see out through gaps between them to the surrounding countryside. (B) In a larger forest, distant trees block the gaps so that no matter where you look, your line of sight ends on a tree. (C) So, too, in a universe that extends forever and has existed forever, your line of sight will always end on a star. Therefore, the night sky should be bright.

Now suppose that, rather than looking between trees, you look out into space through the galaxies and stars that compose the Universe. If space extends sufficiently far and is populated with galaxies and stars such as we see around us in the Local Group, then no matter in what direction you look, your line of sight will ultimately intercept a star just as in a sufficiently large forest your line of sight ultimately hits a tree. Therefore, the sky should be covered with stars, each glowing brilliantly, with no dark spaces between them. Even though such distant stars are very faint to us, there are so many of them that their collective brightness should be very large. In other words, the night sky should not be dark. This argument must be fallacious because the most obvious observation one can make about the sky is that it is dark at night. That is the paradox: the night sky should be bright, but it is not. Where, therefore, is the error in our reasoning?

The paradox is avoided if there are no stars beyond some distance. That is, just as we can see out of the woods if there are only a few trees, so too we will see a dark night sky if there is only a limited number of stars. The paradox is also avoided if we account for the time it takes light from the stars to reach us. Light from a star 1 billion light-years away takes a billion years to get here. To see such light requires that the Universe has existed long enough for the light to arrive. That is, if stars have not been shining for billions of years, the night sky will be dark. A final way to avoid the paradox is that light from remote stars is redshifted by the expansion of the Universe. The redshift reduces the light's energy (recall that longer-wavelength radiation carries less energy) and thereby reduces the brightness of the most distant stars and galaxies. An expanding Universe can shift visible light to wavelengths we cannot see with our eyes. Therefore as we look out to the farthest distances we can see, we look back to the time when the first stars and galaxies formed, but to see what is there we must look at much longer wavelengths than we observe nearby stars and galaxies.

The Cosmic Horizon

The age of the Universe limits the distance we can see. Because light requires time to move from one point to another, we can see no farther than the distance light travels in a time span equal to the age of the Universe. If the Universe did not expand, calculating the distance that light travels in that time span would be simple. For example, if the Universe were not expanding and were 14 billion years old, we could see nothing farther than 14 billion light-years away (fig. 18.6). The Universe may well extend beyond that distance, but the maximum distance that light can travel in the Universe's age therefore defines a **cosmic horizon.** Astronomers call the space within the horizon the visible Universe.

Because the Universe is expanding, it turns out that there is no simple expression for the distance to the cosmic horizon, or even a distant galaxy, for that matter. A simple example will show the problem. Recall that when we see a galaxy, we see it the way it was at the time when its light left it, perhaps millions or even billions of years ago. But when it emitted its light, the galaxy was closer to us than it is now because the Universe was smaller. Furthermore, since the time when the light we now receive was emitted, the galaxy has moved yet farther from us, carried by the expansion of the Universe. Thus, which distance do we apply to the galaxy? Its distance when the light was emitted? Its distance when the light was received? Some other distance? To avoid such confusion (and for other reasons), astronomers often use the term "*light travel time distance*" when speaking of the distance to remote galaxies.

The light travel time distance measures how far the light from the galaxy (or other object) would have traveled if the Universe were not expanding. This distance is simply how far the galaxy's light has traveled in the time since it was emitted. With that definition of distance, we can then say that a galaxy, which we observe today and whose light was emitted 10 billion years ago, is 10 billion light-years away from us. Defining distance in this way allows us to be more precise when we say that the cosmic horizon

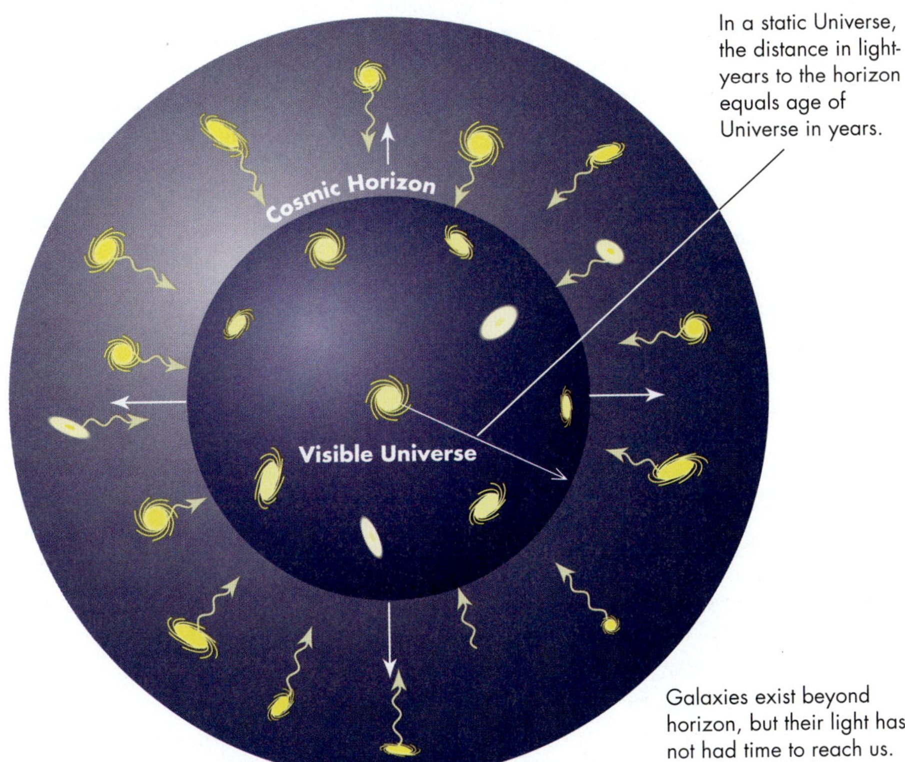
In a static Universe, the distance in light-years to the horizon equals age of Universe in years.
Cosmic Horizon
Visible Universe
Galaxies exist beyond horizon, but their light has not had time to reach us.

FIGURE 18.6
We cannot, even in theory, see beyond the cosmic horizon. Light takes more time than the age of the Universe to reach us from there. (The straight arrows indicate that the horizon moves outward as the Universe ages.)

is approximately 14 billion light-years away. Thus, we will take the size (radius) of the visible Universe to be approximately 14 billion light-years.

The Cosmic Microwave Background

Because the Universe is expanding, as we look farther away we are also seeing the Universe when it was younger and more dense. Pursuing this idea, George Gamow, a Russian astrophysicist who fled to the United States before World War II, realized that it must also have been extremely hot. His argument was based on the observation that compression heats a gas, and so the enormous compression of the early Universe suggests that it too was as hot and dense as the interior of a star when the Universe was young. Gamow's ideas were developed further by two of his collaborators, the American physicists Ralph Alpher and Robert Herman. In 1948 they predicted that after the young Universe had expanded for several hundred thousand years,* it would have cooled and expanded enough that the Universe became transparent. Astronomers often call this the **recombination era** because the ionized atoms would have combined with their electrons to make a transparent neutral gas. From this time onward, radiation was able to travel large distances and reach us today, though in altered form. To understand how that alteration occurred, we need to look at the properties of the radiation in greater detail.

According to Wien's law, a dense gas radiates most intensely at a wavelength determined by its temperature, as discussed in chapter 4. In particular, hot matter radiates most strongly at short wavelengths. The Universe became transparent when the temperature dropped to about 3000 K (similar to the surface of a cool star). At this temperature, the wavelength of maximum emission is about 1000 nanometers, as we found in chapter 13. As space expanded, it stretched the wavelength of radiation traveling through it, increasing its wavelength as illustrated in figure 18.7. (Recall the analogy used earlier of a squiggle on an expanding balloon.) Calculations show that as the Universe expanded to its present size, the wavelength of the radiation emitted at this early time was stretched by a factor of about 1000, so it is now about 1 millimeter (10^6 nanometers). This is in the microwave portion of the electromagnetic spectrum. Thus, Alpher and Herman showed that the Universe today should be filled with faint, microwave radiation created when the Universe was very young. Their prediction was confirmed in 1965, when Arno Penzias and Robert Wilson of the Bell Telephone Laboratories accidently discovered cosmic microwave radiation having just that property.

*Alpher and Herman first estimated this would take about 10 million years, but subsequent calculations have revised this downward to 380,000 years.

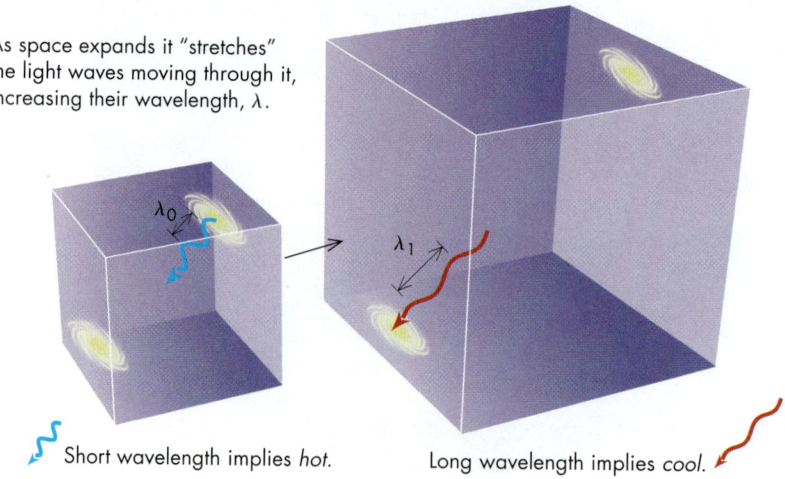
As space expands it "stretches" the light waves moving through it, increasing their wavelength, λ.
λ0
λ1
Short wavelength implies *hot.*
Long wavelength implies *cool.*

FIGURE 18.7
As space expands, it stretches radiation moving through it, making the wavelength longer. Because longer wavelengths are associated with cooler objects, the stretching of the radiation has the effect of cooling it.

Penzias and Wilson were trying to identify sources of background noise on telephone satellite links and noticed a weak radio signal with the unusual property that its strength was constant no matter in which direction of the sky they pointed their detector. They found that the signal came not from isolated objects such as stars or galaxies but rather from all space. Initially mystified by the signal, they soon discovered that scientists at Princeton University had repeated Alpher, Herman, and Gamow's calculations and were searching for the predicted radiation. Once aware of that work, Penzias and Wilson realized that the background interference they had found was radiation created in the young Universe. Their discovery of the **cosmic microwave background (CMB),** as it is now called, won them the Nobel Prize in Physics in 1978.

Figure 18.8 shows measurements of the intensity of the cosmic microwave background (CMB, for short) and demonstrates that it has a perfect blackbody spectrum, most intense at about 1 millimeter as theory predicts. From this wavelength and using Wien's law ($\lambda_{max} = 2.9 \times 10^6/T$, where λ_{max} is the wavelength in nm where the emission is strongest and T is the temperature in K), the temperature that describes the radiation turns out to be about 3 K. A more precise measurement of λ_{max} shows that the temperature that describes the CMB is 2.725 K, only a little warmer than absolute zero, the lowest temperature anything can have.

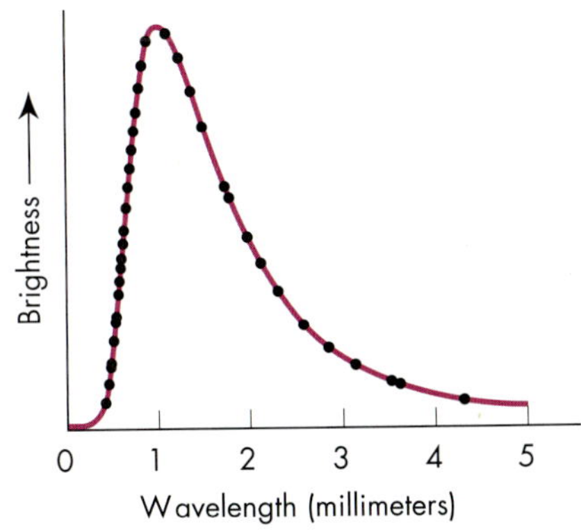

FIGURE 18.8
Spectrum of the cosmic microwave background. The shape of the spectrum matches perfectly a blackbody with a temperature of 2.7 K. Notice the wavelength at which it peaks. The red curve shows the theoretical curve, while black dots show data points schematically.

We see this radiation coming from every direction because as we look out in every direction we look out into the past, eventually seeing nearly to the cosmic horizon when the Universe was a few hundred thousand years old. Note that when the Universe was young this radiation would have appeared as hot as the surface of a star in every direction, and Olbers' prediction would have been true. However, the expansion of the Universe has stretched short wavelengths to long ones, so the CMB radiation looks like the emission from an extremely cold body today.

The discovery of this predicted radiation is one of the cornerstones of the **Big Bang** theory, that the Universe was born in a hot, dense state and expanded rapidly. No other theory has been successful at explaining the presence of this radiation. Another cornerstone of the Big Bang theory comes from even earlier times during the Universe's expansion. Although we cannot see light coming from this earlier time, the even hotter, denser conditions during the first few minutes after the Big Bang left another trace of the Universe's violent beginning.

Composition of the Oldest Stars

If we extrapolate to times before the CMB radiation was produced, the Universe must have been even hotter and more dense. Gamow realized that conditions in the young Universe must have been similar to those in the core of stars at early enough times, and so he predicted that heavy elements might have been created by nuclear fusion just as they are in stars. Gamow proposed that at its birth, the Universe was composed only of neutrons, which rapidly decayed into protons and electrons. Some of the protons then fused to form helium and heavier elements.

Gamow's hypothesis that most of the helium formed in the young Universe is supported by more-recent theory as well as by observations. According to modern theory, however, the young Universe consisted of protons, neutrons, and electrons, rather than just neutrons, as Gamow believed. Nevertheless, his basic idea is correct: in its youth, the Universe was hot and dense enough for nuclear reactions to fuse some of the protons and neutrons into helium. Moreover, current calculations show that the amount of helium and other light elements formed depends on how hot and dense the Universe was and how long those conditions lasted. The conditions at this early time can be related to the density of matter in the Universe today. If the Universe was lower in density, less ^{4}He would have been produced, because the particles would not have collided and fused as often. This also would leave higher amounts of some partial fusion products like deuterium (^{2}H) and ^{3}He. (See chapter 12, where the steps in the proton-proton chain are described.) By about 10 minutes after the expansion began, the Universe had

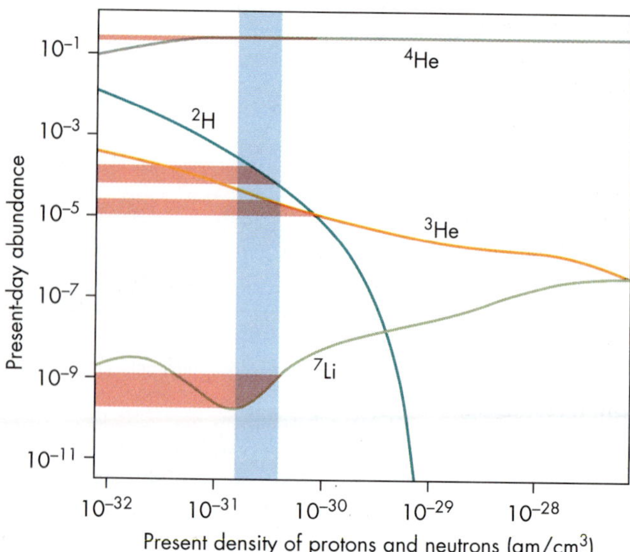

FIGURE 18.9
The amount of different elements expected to be produced by nuclear fusion in the young Universe. The curves show how the amount of each element varies depending on the density of matter in the Universe. The values along the bottom axis show the density of normal matter in the Universe today, while the vertical axis indicates the amount of each element that would be produced. The red horizontal bars indicate the best estimates of the amount of each element measured in the oldest stars and gas clouds. The vertical blue bar shows the range of densities consistent with these measurements.

cooled to about 1 billion K and spread apart so much that fusion stopped, leaving about 24% of the matter transformed into ^{4}He. The curves in figure 18.9 show theoretical predictions of how much of each element should form, depending on the Universe's density.

This prediction can be tested by measuring the amount of helium and other light elements in the oldest stars in the Milky Way and other galaxies. Those stars formed from the gas left over from the Big Bang, and therefore they have abundances of matter produced in the Big Bang. Although additional fusion has occurred deep in their cores during their long life, the composition of the outer layers of these ancient stars is essentially the same as produced by the Big Bang. What is that abundance? It turns out to be 24% helium, in excellent agreement with the Big Bang model. This is not, however, the abundance of helium in stars like the Sun, which formed well after the Milky Way's birth. These later-generation stars have formed from interstellar gas that has been enriched in helium from earlier generations of stars that have exploded or lost mass in other ways. Consequently, the Sun—born some 9 billion years after the Big Bang—contains about 27% helium. The amounts of other light elements and isotopes also agree well with predictions based on the Big Bang theory, as shown in figure 18.9. This provides further strong support that the theory is correct.

Deductions from Basic Observations of the Universe

The recession of galaxies and their uniform distribution leads us to conclude that the Universe was born in a cataclysmic event about 14 billion years ago. At that time, the Universe must have been much hotter and denser than now.

Confirmation that this hot, dense state existed comes from (1) the cosmic microwave background—the radiation created during that hot stage but cooled by the Universe's subsequent expansion—and (2) from the percentage of helium and other light elements found in the oldest stars and gas clouds. The high temperature and rapid expansion from a hot, dense state led the English astronomer Sir Fred Hoyle to jokingly refer to this theory of the birth of the Universe as the "Big Bang," a name that has stuck despite his sarcasm. Hoyle long championed an alternative "steady-state" model, proposed by him and two other British astronomers, Hermann Bondi and Thomas Gold, in the early 1950s. Their idea was that new matter forms continually out of nothing to create new galaxies in the empty, expanding space between older galaxies. This keeps the average density of the Universe constant so the Universe on the average looks the same at all times. This unchanging appearance (which gives the steady-state Universe its name) is unlike that of the Big Bang theory, where space becomes progressively emptier as the galaxies move apart. But many observations (especially the cosmic microwave background) now persuade almost all astronomers that the Big Bang theory for the Universe and its predictions match our observations of the Universe far better than do the predictions of the steady-state model.

18.3 Evolution of the Universe: Expansion Forever or Recollapse?

The Universe is currently expanding. But will it expand forever, or will it at some time stop expanding and contract? This question is fundamental to the nature of the Universe, for according to the Big Bang theory, all the hydrogen in the Universe was created in the Big Bang. Thus, if the Universe expands forever, its stars will one by one consume their hydrogen and burn out, leaving the Universe a black, empty space—a dismal prospect. However, recollapse of the Universe means that all the objects and atoms (and us) within it will be compressed into what some astronomers have called the "Big Crunch." The Big Crunch might lead to a rebirth of the Universe, perhaps leading to a new Big Bang. Such a universe might oscillate, expanding and collapsing repeatedly.

INTERACTIVE
Cosmology and the cosmos's fate

To determine which fate awaits the Universe, we need to understand what factors affect its expansion or recollapse. But before we look at details, a simple analogy may help. Suppose you throw a stone into the air. As the stone moves upward, Earth's gravitational force slows its rise and eventually stops it so the stone falls back to Earth. On the other hand, if you could throw the stone faster than the escape velocity, it would rise forever. Thus, the stone's behavior depends on the strength of gravity and the upward force given to it. So, too, with the Universe.

Near the time of its birth, some process began the Universe's expansion. Since that time, the gravitational force between its galaxies and other contents has generally acted to slow its expansion. If the gravitational force is weak or if the impulse of expansion is strong, we might expect the Universe to expand forever. On the other hand, if the gravitational force is strong or the initial impulse is weak, we might expect that the expansion will stop and the Universe will collapse.

Is there some way we can measure the relative strength of these effects for the Universe? The answer is yes. We can compare the gravitational energy working to hold the Universe together with the energy of its expansion. If gravity dominates, the expansion will stop. If the expansion energy dominates, the expansion will continue forever (fig. 18.10). Thus, we can learn (at least in theory) whether the Universe will expand forever or recollapse by seeing if the Universe has a gravity strong enough to stop its expansion.

We saw in earlier chapters that the gravitational force produced by an object depends on its mass—a result we can extend to the Universe itself. In particular, if astronomers know the Universe's mass, they can calculate the gravitational force it exerts to slow its expansion. However, to make such a calculation, it is easier to use the Universe's density (its mass divided by its volume) rather than its mass alone. What density do astronomers find? We will answer that question in the next section and also see there what effect this density has on the Universe's expansion.

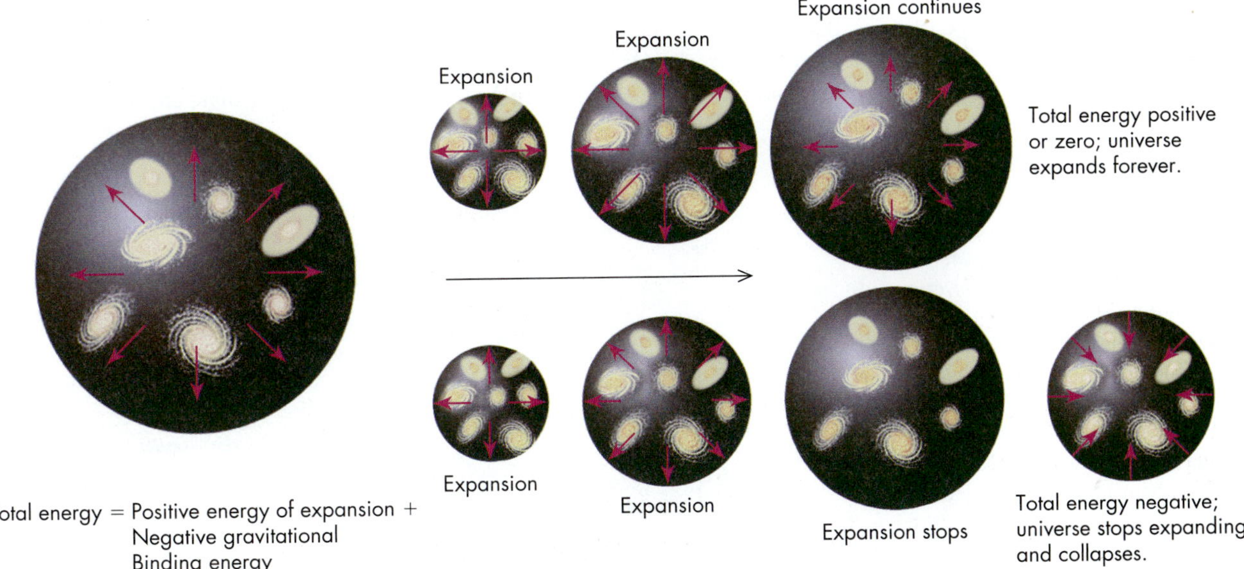
Expansion
Expansion
Expansion continues
Total energy positive or zero; universe expands forever.
Expansion
Expansion
Expansion stops
Total energy negative; universe stops expanding and collapses.
Total energy = Positive energy of expansion + Negative gravitational Binding energy

FIGURE 18.10
A sketch of how gravity and the energy of expansion determine the behavior of the Universe.

The Density of the Universe

To find the density of the Universe, astronomers choose a volume of space and add up the mass of all the material in it. In doing so, they must be sure to include the mass of dark matter as well as the visible matter—including dark matter is important because, as we have seen, it greatly outweighs the visible matter. They then divide the mass by the volume to obtain the density. When such a calculation is made, the density of the combined visible and dark matter in the Universe turns out to be about 3×10^{-30} gm/cm^3. This value is about ten times higher than the density of matter made of protons and neutrons based on the fusion of elements during the first few minutes of the Universe, which makes sense based on the percentage of dark matter found in galaxies and galaxy clusters (chapter 17).

To determine whether the Universe will expand forever or recollapse, astronomers compare the observed density with a theoretically calculated **critical density.** If the observed density is greater than the critical density, the Universe will recollapse; if it is less, the Universe will expand forever as discussed above. It turns out mathematically that the critical density is $\rho_c = 3H^2/(8\pi G)$, where H is the Hubble constant, and G is the gravitational constant. Notice that the critical density depends on the Universe's rate of expansion (the Hubble constant), because if the expansion rate is very large, the critical density must also be large in order to have enough gravity to stop the expansion.

Astronomers use a shorthand notation for describing the Universe's density: they take the ratio of the actual density of matter ρ_M to the critical density ρ_c, which they call Ω_M. This is the Greek letter Omega, and the subscript "M" denotes that this is the density of matter. Cosmologists originally predicted that if Ω_M $(= \rho_M/\rho_c)$ is greater than 1, then the Universe would eventually recollapse. This is illustrated in figure 18.11, where different curves show how the size of the Universe would change, depending on the value of Ω_M.

When we compare the observed and critical densities, what do we find? If we use the value of H determined by observation (70 kilometers per second per megaparsec), we find that the critical density is about 10^{-29} grams per cubic centimeter, the equivalent of six hydrogen atoms per cubic meter. Thus, the actual density of matter is approximately one-third the critical density, which implies that the Universe should expand forever.

Although astronomers were pleased that the question of whether the Universe would expand forever or recollapse appeared to have been settled, they were troubled by the value found for the Universe's density. According to the leading model for how the Big Bang began, the actual density should *exactly* equal the critical density, rather than be approximately one-third of that value, as was measured. Was this model, therefore, wrong? Or had some component of the Universe been overlooked? Perhaps, for example, there is even more dark matter than our estimate above suggests.

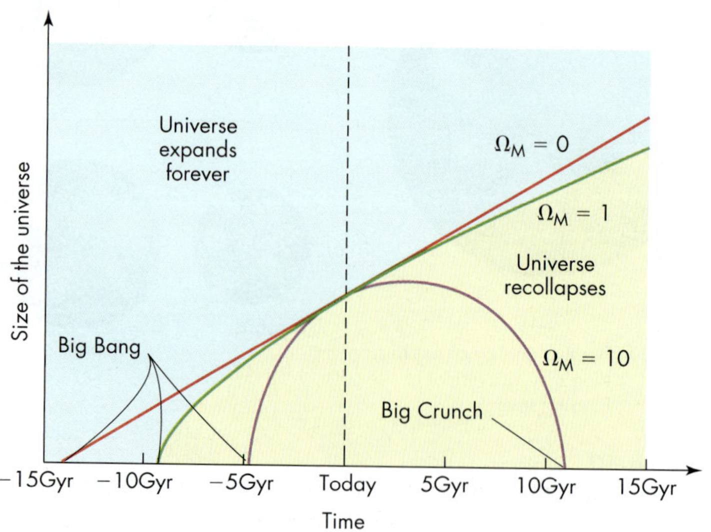

FIGURE 18.11
The size of the Universe in the past and future as calculated for different scenarios. Curves are shown for three cases, in which: there is no matter ($\Omega_M = 0$), so the Universe expands without deceleration; the pull of gravity and the expansion rate balance ($\Omega M = 1$), so the Universe constantly slows down but never recollapses; and when the pull of gravity greatly exceeds the energy of expansion ($\Omega_M = 10$), so the Universe recollapses in about 12 billion more years. All curves have the same slope today, which is determined by the Hubble constant. Curves lying in the blue region of the diagram have enough energy to expand forever, while those in the yellow region will eventually collapse back on themselves.

To confirm their models of expansion, astronomers sought to compare the present expansion speed of the Universe to its speed in the past. Note that each of the curves in figure 18.11 has a different history of sizes at different times. The curves all match the rate of expansion of the Universe today, which is determined by the Hubble constant, but each makes a different prediction of the distance to objects when the Universe was, for example, half its current size. If astronomers could determine the actual distances to extremely remote galaxies, they could compare those results to the different expansion curves. But how might such a comparison be made?

ANIMATION
The future of expansion

Dark Energy

Astronomers are fortunate in that they, in fact, have such a way to determine the distances to extremely remote galaxies. By observing Type Ia supernova blasts in remote galaxies, they can find both the distances and the redshifts of these systems. Recall that Type Ia supernovas are produced when a white dwarf collapses (chapter 15). This process occurs when a white dwarf accretes enough material to exceed a specific mass, and so the explosions are always very similar to each other. This makes them an excellent "standard candle."

At the same time, from the galaxy's redshift, we know how much bigger the Universe is today than it was at the time the light left the galaxy. This allows astronomers to predict how faint the supernova should appear, depending on its redshift for different cosmological models, as shown in figure 18.12. Note that the nearby galaxies (in the lower left-hand part of the diagram) lie in a part of the diagram where the curves are so close together, they might be consistent with $\Omega_M = 1$. However, the more remote galaxies (upper right) clearly do not follow that law. Rather, the distant galaxies lie *below* the line predicted for $\Omega_M = 0$—a Universe containing no matter at all! This shows that expansion in the past was slower than it is now. That is, the expansion of the Universe seems to have *sped up!* How is this possible? Gravity's pull of one galaxy on another should slow the expansion.

One possible explanation of this remarkable result comes from Einstein's early efforts to understand the structure of the Universe. In his general theory of relativity, Einstein proposed that matter and energy create the force of gravity and bend space to give the Universe its structure. From his theory, he devised and solved equations to describe this structure. His solution involved a constant whose value he had no way of knowing. However, if he made the constant zero, the Universe would either collapse under gravity's influence or expand forever.

At the time of Einstein's work, astronomers had not yet discovered that the Universe is expanding. In fact, it had not yet even been discovered that there were other galaxies than the Milky Way. Einstein therefore thought that the Universe must be at

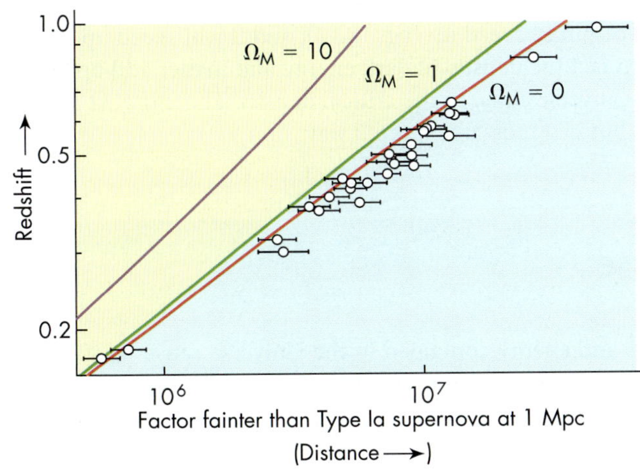

FIGURE 18.12
A graph of the redshift and brightness of Type Ia supernovas in distant galaxies. The brightness is labeled according to how many times fainter the supernovas appeared compared to how bright they would have been if they were in a galaxy 1 Mpc away. The curves show the predicted brightness if the Universe has expanded according to models of the Universe in which Ω_M = 0, 1, or 10 (as in fig. 18.11). The horizontal lines through each data point show the uncertainty for each supernova distance measurement. The graph is based on observations by S. Perlmutter and collaborators. Note that the distant galaxies tend to lie below any of the models—indicating that the light is dimmer than any model in which the Universe has undergone only deceleration.

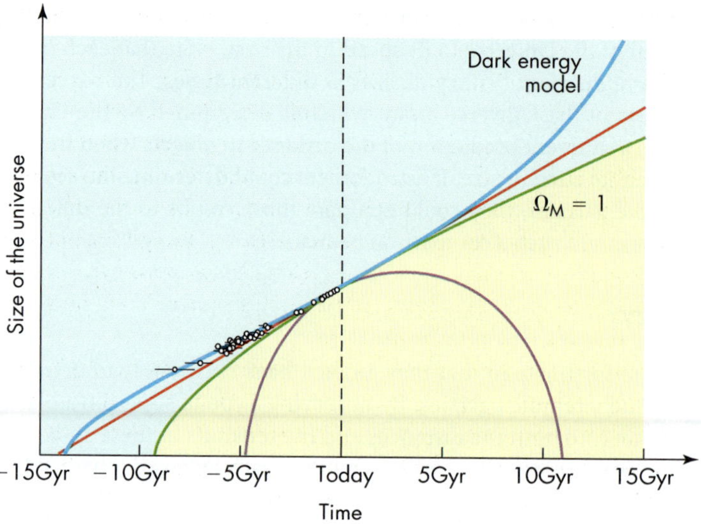

FIGURE 18.13
The data from Type Ia supernovas is shown in a graph of the expansion of the Universe. At times in the past, the data lie above curves based on expansion with no dark energy. The best-fit curve, in which the Universe contains nearly three times as much dark energy as matter, is shown in blue. A Universe with dark energy will accelerate more and more rapidly in the future.

rest overall. To achieve that, his equations needed a cosmic repulsive force to offset gravity's attraction. Making the constant different from zero had exactly this effect, and the term in his equations became known as the **cosmological constant.**

According to theory, the repulsion effect of the cosmological constant is far too weak to detect in the laboratory or even on the scale of the Solar System. Only over distances of millions of light-years does it become powerful enough to offset gravity and prevent the Universe's collapse. Within a decade of Einstein's introduction of the cosmological constant, astronomers discovered that the Universe was in fact expanding and so Einstein thought the constant was not needed and that he should have set it equal to zero all along. However, the discovery that the Universe's expansion is speeding up makes astronomers think once more that the cosmological constant might not be zero.

If the cosmological constant is not zero, it contributes an energy to the Universe. We cannot directly detect this energy, and so astronomers call it **dark energy,** in analogy to dark matter. Dark energy is quite different from other forms of energy because it is a property of space, not of objects. You cannot bottle dark energy or make an antigravity machine out of it. It acts only to make space expand faster and faster.

Although it is invisible, dark energy's effect on the Universe appears to be immense. In addition to creating the repulsive force that speeds up the recession of galaxies, dark energy contributes a colossal mass to the Universe. Using the relation $E = mc^2$ of relativity theory, astronomers find that dark energy contributes over two-thirds of the Universe's total mass—more than twice as much as that of all the other mass in the Universe, dark and observable, combined. Thus, the Universe appears to be *mostly* dark energy.

When the Universe was young, the density of matter exceeded the density of dark energy, but the matter grew more dilute as the expansion proceeded. Dark energy remained a constant of space, and now exceeds the density of matter, so the Universe's expansion has begun to speed up (fig. 18.13), and it will speed up exponentially faster in the future. In fact the density of dark energy and matter add up to make the overall density of the Universe equal to the critical density. Therefore $\Omega = 1$, as astronomers had predicted, but in a most unexpected way.

18.4 The Curvature of the Universe

All of the mass and energy contained in the Universe causes space to be "curved" according to Einstein's general theory of relativity (essay 2). We have previously described how space is curved by the gravity of isolated objects such as exoplanets, black holes, and galaxy clusters. For example, astronomers can measure the bending of the path of light when one object lies in front of another.

How can we picture the **curvature of the Universe?** Our home planet, Earth, offers a simple example. Athletic fields and parking lots appear flat and, if we ignore hills, the surface of the Earth certainly *looks* flat. We know, however, that if we could walk in a "straight" line along this "flat" surface, we would return to our starting point. We are therefore like an ant that accidently crawls into a garden hose. The unbent hose represents a flat, uncurved space, and so, if the ant crawls in a straight line, it will reach the hose's end and emerge into the garden again. However, suppose we pick up the hose with the ant still in it and bend the hose so that its ends meet. Now the ant could crawl forever and would never reach the hose's end.

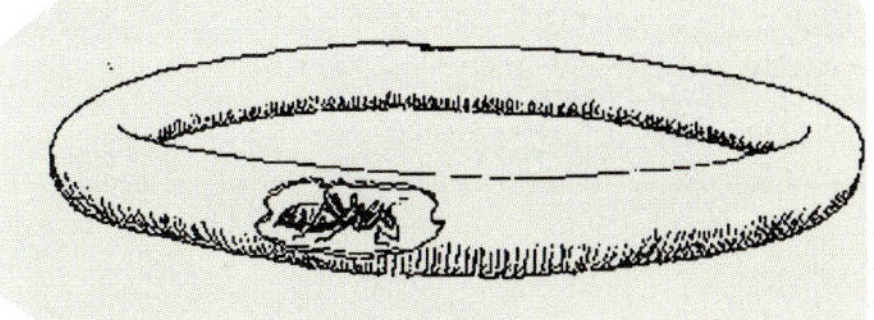

A similar situation exists for a Universe curved around on itself. Such a space looks as if it extends forever in all directions, but if we could travel in a spaceship along a "straight" line, we would eventually return to our starting position. Thus, we would have moved through space forever without coming to an end point. A universe with this property is said to have **positive curvature** and is "closed" because it has no "outside." We can infer that the Earth has a "positively curved surface" because if we travel in a straight line on it, we return to our starting point. Positively curved surfaces have other properties as well: lines that start out parallel eventually meet if they extend far enough, and the sum of the interior angles of a triangle drawn on such surfaces is greater than 180°, as illustrated in figure 18.14A.

The Universe might have other types of curvature, however. It might be a **flat Universe** (that is, have no curvature) or have **negative curvature.** Flat space is what most people picture when they think of space. In it, parallel lines do not meet, and a triangle's interior angles add up to exactly 180°, as illustrated in figure 18.14B. Negatively curved space is harder to visualize, but you can think of it as being bent into a saddle-like shape, as shown in figure 18.14C. Such curvature corresponds to bending space one way along one line and the opposite way along another line. In negatively curved space, the sum of the interior angles of a triangle is less than 180°.

Can we measure the curvature of space and determine whether it is positive, negative, or flat? Yes. After nearly half a century of attempts that gave only inconclusive results, astronomers at last have a method for measuring the curvature of the Universe that works with good accuracy. This method uses observations of the small temperature fluctuations of the cosmic background radiation, shown in figure 18.15. The size of these fluctuations gives information not only about the curvature of the Universe but also about the Universe's age and the amount of ordinary and dark matter.

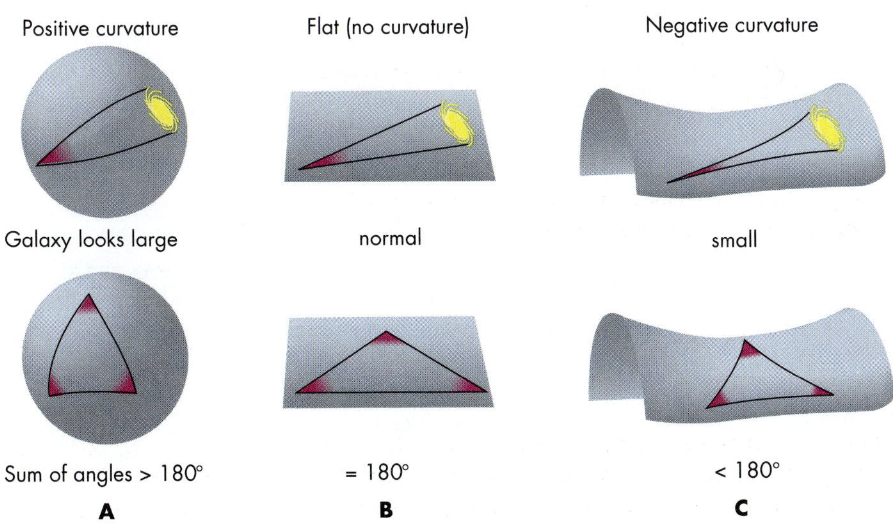

FIGURE 18.14
Figures illustrating how the shape of a surface determines the geometry of a triangle on it. (A) The interior angles of a triangle on the surface of the Earth add up to more than 180°. (B) The interior angles of a triangle on a flat surface add up to 180°. (C) The interior angles of a triangle on a saddle-shaped surface add up to less than 180°. The top row of figures illustrates that a distant object will therefore appear larger if the curvature is positive (A) and smaller if the curvature is negative (C).

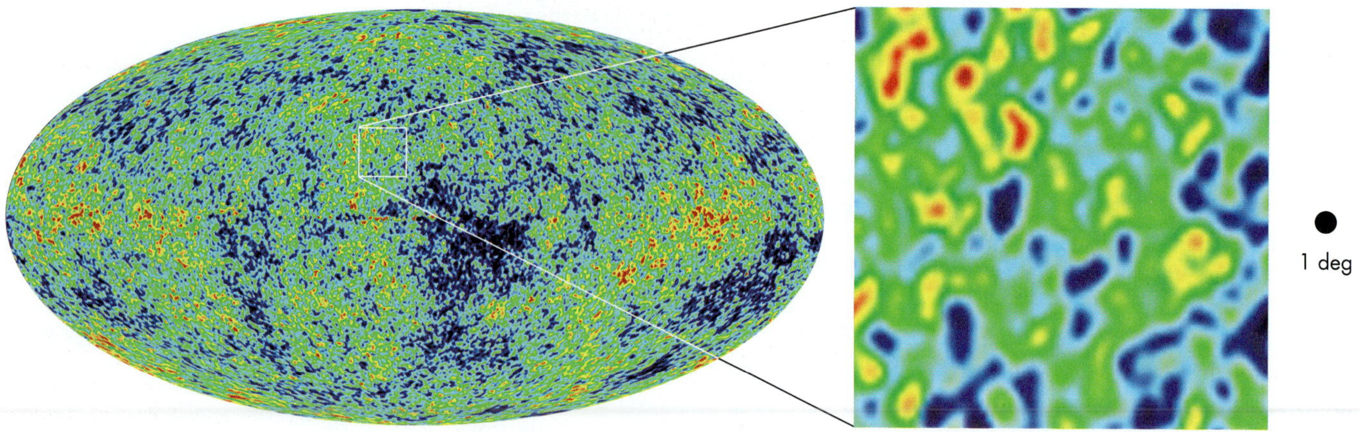

FIGURE 18.15
A map of the cosmic microwave background, made by the *Wilkinson Microwave Anisotropy Probe* satellite (WMAP). The map at left shows the brightness distribution of the cosmic microwave background across the entire sky, much as a map of the Earth depicts its features on a flat surface. A small region is blown up on the right to show the size of fluctuations in the CMB. Red regions are brighter (warmer), and blue regions are fainter (colder). They show irregularities in the temperature of the Universe at the time the Universe became transparent to radiation, about 380,000 years after the Big Bang. The satellite that gathered the data for the map was named for David T. Wilkinson, a pioneer in the study of the cosmic microwave background. The term *anisotropy* refers to the lack of uniformity from point to point across the sky.

To understand the method, it may help to think of a simple experiment. Imagine blowing across the mouth of an empty bottle. As you blow, you make the air in the bottle vibrate and generate sound waves. The pitch (tone) of sound waves is set by their wavelength, with shorter wavelengths having a higher pitch. You can demonstrate that effect by putting some water into the bottle and blowing across it again. The pitch of the sound will now be higher. This happens because the water gives the sound waves less space in the bottle and makes their wavelength shorter. Thus, the length (pitch) of the sound waves gives us a clue to the amount of air in the bottle.

In earlier chapters, we saw how scientists used waves to probe objects. For example, seismic waves allow us to probe the interior of the Earth, and waves in the Sun's atmosphere allow us to determine properties of the Sun's interior. We can use similar techniques to probe the Universe.

As the Universe expanded in earliest times, waves formed in its hot gas. Those waves took the form of fluctuations in its density and temperature. We can observe those waves today as variations in the brightness of the CMB radiation across the sky as seen in figure 18.15. Because the radiation comes from a time 380,000 years after the Big Bang, regions could have interacted with each other only if they were less than 380,000 light-years

SCIENCE *at work*

THE UNIVERSE'S FATE

For many years astronomers have used the terms *open* and *closed* to describe the shape of the Universe and whether its expansion will continue forever or not. The discovery that the cosmological constant (usually expressed by Λ, the Greek letter lambda) is probably not equal to zero greatly complicates the use of these two seemingly simple terms.

If $\Lambda = 0$, then an open Universe expands forever and a closed Universe recollapses. If $\Lambda \neq 0$, that simple linkage does not hold. *Open* and *closed* still describe the curvature of the Universe, but not its fate. An open Universe may recollapse and a closed Universe may expand forever. It is too bad that we lose that easy way to describe the Universe's behavior, but the Universe doesn't care about our convenience.

apart (the cosmic horizon). As a result, the most common size of the fluctuations is determined by the size of the cosmic horizon at that time. Combining this fact with the observation that these fluctuations are about 1° across allows us to determine the Universe's curvature. *WMAP*'s measurements show that the Universe is flat to very high precision, so its overall density is equal to the critical density.

More-detailed analyses of the brightness variations in the CMB allow astronomers to deduce how much ordinary and dark matter is present. This is because waves in dark matter, which is frictionless, behave differently than waves in normal matter, which collides and heats. In addition, the observations show that the age of the Universe is 13.7 billion years (to within a few percent). Finally, when combined with measurements such as the supernova results described in the previous section, astronomers can deduce the amount of dark energy, and the relative amount of various kinds of ordinary matter. Figure 18.16 shows the fraction of the Universe composed of each of these components, based on this type of analysis.

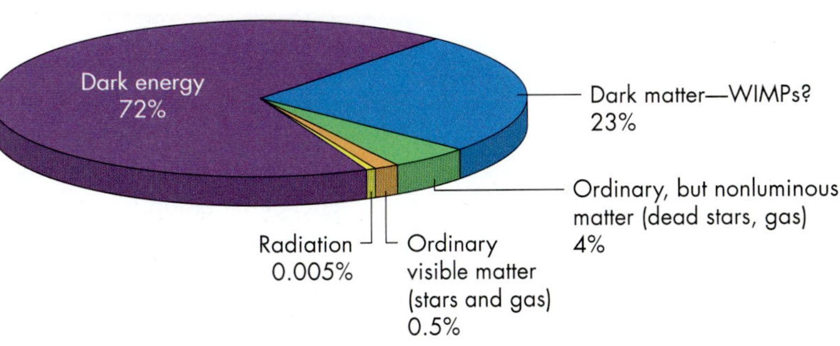

FIGURE 18.16
The makeup of the Universe as deduced from observations of the brightness variations in the cosmic microwave background and other data. The percentages have been rounded off, so they do not add up to exactly 100%.

18.5 The Origin of the Universe

Where did the Universe come from, and what was it like in its early stages? Cosmologists study that remote time by extrapolating the present conditions in the Universe backward in time. That is, they use the laws of physics and present conditions to infer what the Universe was like at the time of its birth. We have already described one important conclusion of such extrapolations: the early Universe was hot and dense (as confirmed by the cosmic background radiation). Figure 18.17 shows how its temperature has changed over the last approximately 14 billion years.

You can see from the figure that the young Universe was extremely hot—trillions of degrees in its earliest moments. That very high temperature implies that in its youth, the Universe was a mix of gas and radiation, much like the core of our Sun, although far hotter. Such high-temperature gases turn out to obey relatively simple laws. For example, the temperature and density depend in known ways on the Universe's size. This allows astronomers to deduce the Universe's past properties from its present properties, which is how figure 18.17 was worked out.

The young Universe's enormous temperature also implies that its matter and radiation mingled in a manner unlike their sharp distinction today. The mingling occurred because, in a high-temperature gas, radiation can be so energetic that it can be converted into matter. Thus, in the early Universe, matter and radiation behaved almost as a single entity.

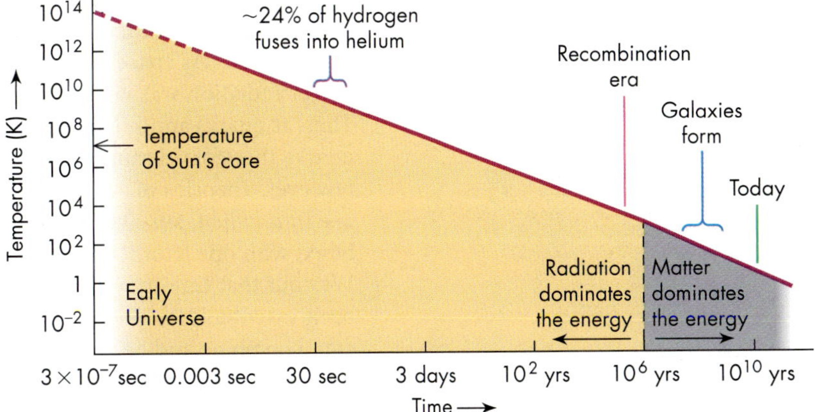

FIGURE 18.17
A plot of how the temperature of the Universe changes with time, according to the Big Bang theory. Note that the horizontal scale is chosen to bring out details of the young Universe.

The Early Universe: Radiation, Matter, and Antimatter

Einstein's equation $E = mc^2$ equates an amount of energy, E, to an amount of mass, m, times the speed of light, c, squared. It tells us that stars can obtain their energy from mass lost when, for example, hydrogen is converted to

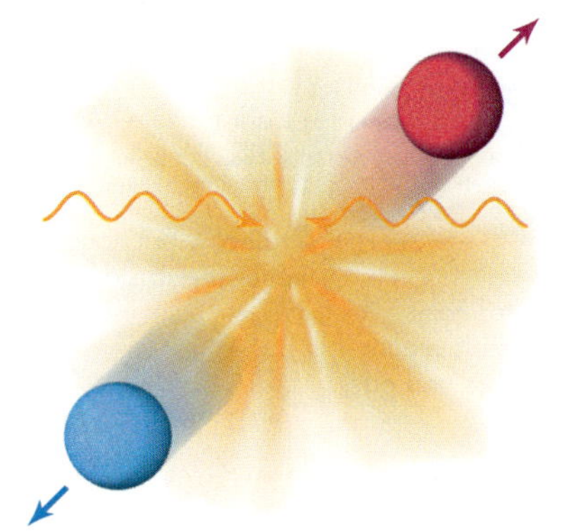

A Radiation creates particle and antiparticle.

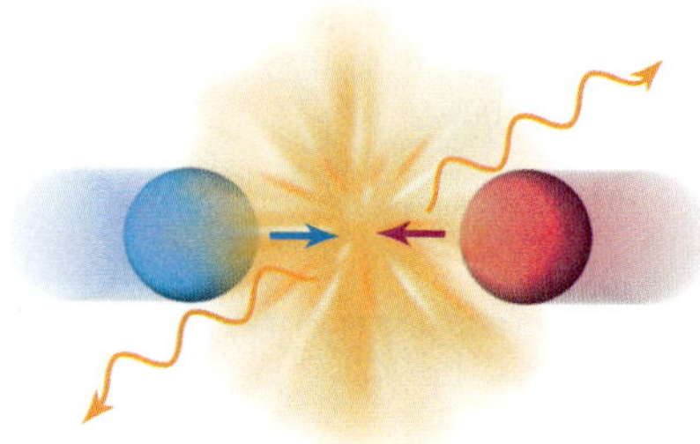

B Particle and antiparticle annihilate, creating radiation.

FIGURE 18.18
(A) The energy of electromagnetic radiation can be converted into mass with the formation of a particle and its antiparticle. (B) The collision of a particle and its antiparticle leads to their annihilation and conversion of their mass back into energy.

ANIMATION
How matter goes to energy and vce versa

helium, as discussed in chapters 12 and 14. In the early Universe, the implication of this energy–mass relation was reversed: a quantity of energy E may have turned into particles whose mass equaled E/c^2. Scientists know that matter can be created this way: in laboratory experiments they can observe high-energy radiation (gamma rays) forming particles. Given that energy can change into matter, it seems likely that high-energy radiation in the early Universe must have done the same. Such transformations of energy into mass, however, always create a *pair* of particles, as illustrated in figure 18.18A. Moreover, the particle pair has two special properties: the particles must have opposite charge (or no charge), and one of the pair must be made of ordinary matter, whereas the other must be made of what is called **antimatter.**

We briefly discussed antimatter in chapter 12, where we saw that the electron has an antimatter companion particle, the positron, with identical mass but opposite electric charge. Protons also have a corresponding antiparticle called the antiproton, and all other subatomic particles are similarly paired.

Antimatter has the important property that a particle and its antiparticle are annihilated on contact, leaving only high-energy radiation, as depicted in figure 18.18B. Thus, radiation (electromagnetic energy) can become matter, and matter can become radiation. In the early Universe, this shifting between mass and energy happened continually as particles formed and were annihilated.

Energetic radiation cannot turn into just any pair of particles: the pair's combined mass multiplied by c^2 must be less than the radiation's energy. For example, to become an electron-positron pair, the radiation needs an energy of about 1.6×10^{-13} joules, equivalent to a wavelength of about 0.0012 nanometers (a gamma ray). To become a proton-antiproton pair, the radiation needs roughly 1800 times more energy, making it a very high-energy gamma ray. Radiation with such large energies occurs only at very high temperatures. For example, to have enough energy to make an electron-positron pair, the temperature must be about 10^{10} Kelvin. To make a proton-antiproton pair, the temperature must be about 10^{13} Kelvin. Such high temperatures are not reached in ordinary stars, but they occurred in the Big Bang.

History of Matter and Radiation in the Early Universe

Astronomers estimate that approximately 1 microsecond after the Big Bang, the Universe had a temperature of about 10^{13} K, hot enough for its radiation to be converted into not only protons and antiprotons but even quarks and antiquarks, the particles that make up protons and neutrons. At that time, the Universe we now see was packed into a volume smaller than the Solar System. This volume, an inferno jammed with quarks and antiquarks created from the high-energy radiation, was expanding at a fantastic rate. As space expanded, this dazzlingly hot matter and radiation cooled. The cooler radiation was no longer energetic enough to create new quark-antiquark pairs. Thus, as quarks annihilated antiquarks, they were no longer re-created, and most of the quarks therefore disappeared from the Universe. This self-annihilation was not total, however. Theories of subatomic particles predict that a tiny excess of quarks, literally, *one in a billion,* survived with no antiquark to annihilate. These survivors then combined with one another to make the protons and neutrons of the Universe around us. Without that tiny imbalance in the annihilation of quarks, our Universe would have been essentially empty. Every particle would have had an antiparticle with which it could combine and be annihilated into radiation, leaving the Universe filled only with the cosmic background radiation and containing no matter at all.

The next landmark in the history of the early Universe occurred about 3 minutes after the Big Bang, when the expansion of space had cooled the Universe to a few hundred million degrees. Such conditions are similar to those in the core of a very hot star, and much as a star can fuse protons into helium, so did the Universe. Roughly one-quarter of the Universe's protons and neutrons fused into helium at that time. As we saw earlier, that helium, mixed with the more abundant hydrogen, permeates all space, explaining why the earliest generation of stars have the same proportion of hydrogen to helium in their outer layers.

For about the next 400,000 years, little else happened in the young Universe apart from further expansion and cooling. Eventually the temperature dropped to about 3000 K, and electrons and protons had low enough energy to bind together to form hydrogen atoms. This period (known as the recombination era because of the combination of electrons with nuclei) marks the time when the Universe's radiation and matter ceased to interact strongly. Before recombination, electrons and radiation collided violently enough to exchange energy and maintain the same temperature. However, when electrons became attached to nuclei and formed hydrogen and helium atoms, their interactions with radiation became much weaker because the atoms have no overall electric charge. Thus, before recombination the Universe behaved as a single blend of matter and radiation, but after recombination its matter and radiation acted as separate ingredients.

The changing behavior of the early Universe's matter and radiation is summarized in figure 18.19. The diagram shows how the temperature of the Universe has changed as the Universe has expanded and cooled and changed from quarks to atoms of relatively cool hydrogen and helium. One vital stage remains to link that ancient time to the present: the formation of galaxies.

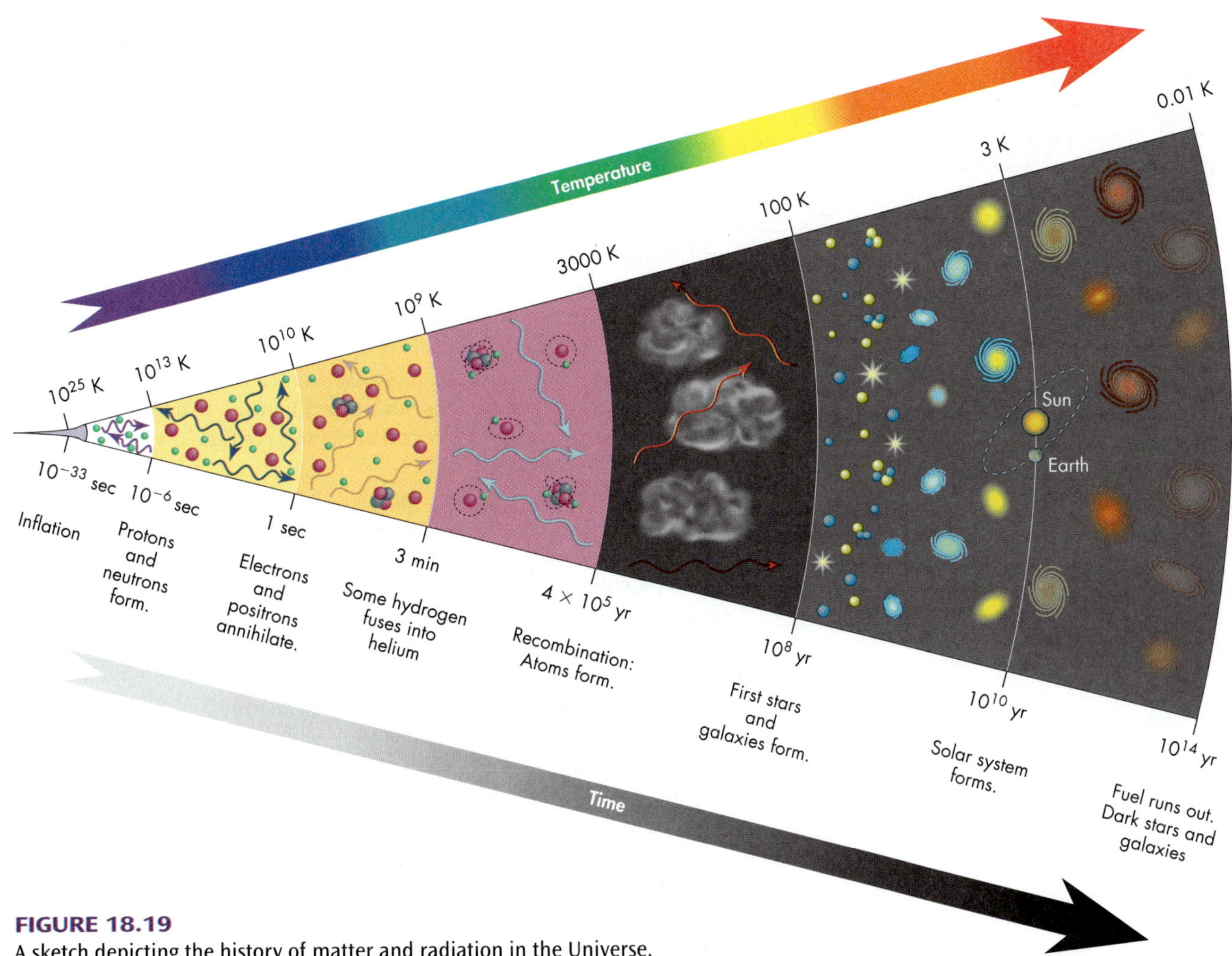

FIGURE 18.19
A sketch depicting the history of matter and radiation in the Universe.

The Formation of Galaxies

From the ages of the stars within galaxies, astronomers deduce that the Milky Way and other galaxies are about 13 billion years old. In fact, gravity began to pull together the regions of slightly higher density soon after protons and electrons recombined to make the first hydrogen atoms in the early Universe.

Astronomers think that galaxies formed by a process similar to the way stars form, but on a much larger scale. That is, gravity drew the gas created in the Big Bang into gigantic clouds, which eventually collapsed under gravitational attraction and formed stars, as we discussed in describing the birth of the Milky Way in chapter 16. But to observe this process occurring in the first galaxies is extremely difficult because it happened so long ago. That is, to look back 13 billion years in time, we need to look out into space 13 billion light-years. At such enormous distances even objects the size of galaxies appear very small and dim. Nevertheless, with the Hubble Space Telescope, astronomers have captured images of that remote time, as shown in the chapter-opening figure and discussed in chapter 17.

To understand galaxy formation, astronomers also rely heavily on computer simulations. Such simulations show how small irregularities in the Universe's matter grow into galaxies (see fig. 17.40 for example), and they nicely create structures similar to the observed distribution of galaxies and voids, as we discussed in chapter 17. The simulations also show that dark matter plays an important role in creating a gravity that is strong enough to allow galaxies to form quickly while the Universe is still relatively young. The simulations further suggest that the large-scale structure of the Universe—the voids and superclusters of galaxies that we see today—probably began from the tiny variations of density that arose at the very earliest moments of the Big Bang.

18.6 The Inflationary Universe

The classic Big Bang theory allows cosmologists to understand the Universe back to about 1 microsecond after its birth. But what happened earlier? We can extrapolate that the Universe was even hotter, denser, and more energetic at earlier times. Cosmologists hypothesize that in its infancy, the visible Universe was packed into a volume smaller than a proton and that its temperature was trillions of times hotter than trillions of degrees—about 10^{27} K. It was so hot, in fact, that the Universe was simply intensely high energy. Some theories imply that under such energetic conditions there can be a very large cosmological constant, creating a violent expansion, which cosmologists call **inflation.**

Q. By what factor has the Universe increased *since* inflation?

According to these theories, the inflation stage of the early Universe began about 10^{-35} seconds after its birth and lasted until about 10^{-32} seconds. During this time, space expanded explosively in size, its radius swelling by a factor of 10^{25}, as shown in figure 18.20. If the period at the end of this sentence expanded by a similar factor, it would end up the size of the Local Group of galaxies. During that expansion, the Universe cooled (relatively), and the forces within it evolved to their present form. The inflationary stage "launched" the Big Bang.

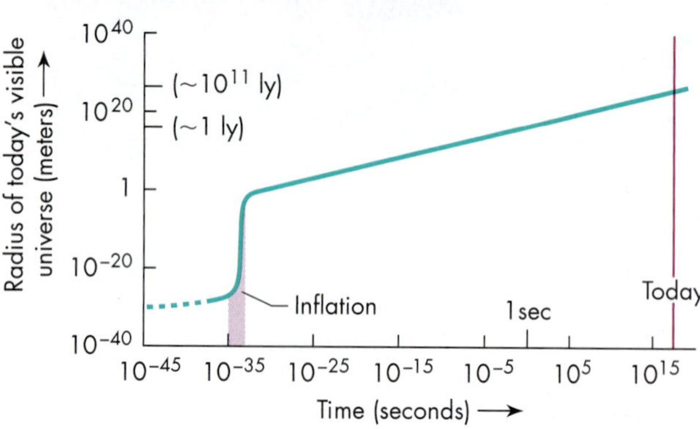

FIGURE 18.20
The size of the Universe through time. The size of today's visible Universe is shown through time. Note the sudden inflation in size starting at 10^{-35} seconds.

Inflationary models of the first moments of the Universe mark the frontier of our understanding of the cosmos and offer tentative explanations of some of its most mysterious features. For example, the models suggest that the Universe may have been created literally from nothing. Some models imply the existence of other universes forever separated from our own. Still others suggest that space has more dimensions than the four we are familiar with: three of space

EXTENDING *our imagination*

OTHER UNIVERSES?

The inflationary Universe has yet another curious property: it suggests that our Universe is just a tiny part of a vast "multiverse." Different regions within the multiverse began inflating at different times and at different rates. Such isolated regions might well be considered separate universes, independent of ours and completely unobservable by us. Furthermore, according to inflationary theories, some of these other universes may be expanding much faster or recollapsing.

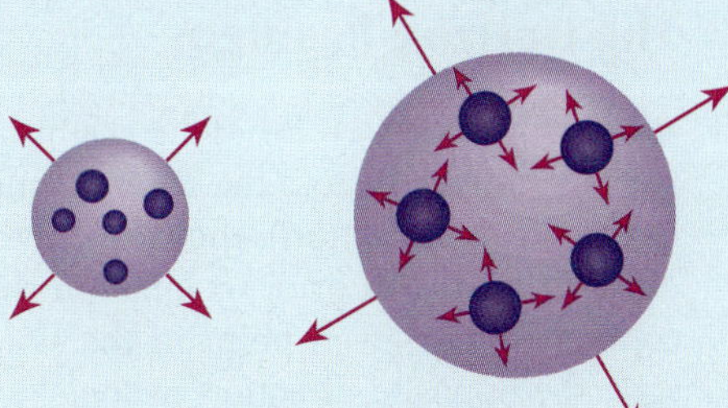

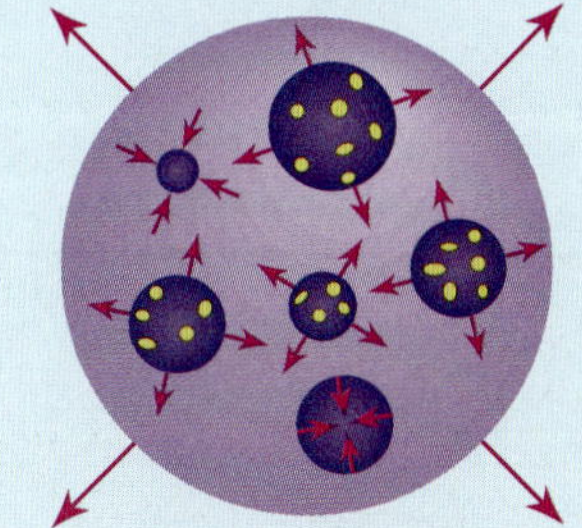

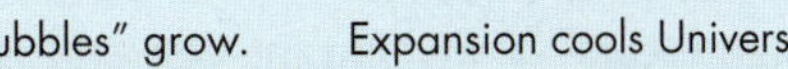
"Bubbles" grow.
Expansion cools Universe.

"Bubbles" become separate universes?

BOX FIGURE 18.1
A sketch depicting bubbles that might become separate universes forming in an inflating universe.

Inflation creates such sections of space much as bubbles form in a pot of boiling water. As water begins to boil, regions of the liquid suddenly change from liquid to gas (steam) and begin expanding. We see that region of steam as a bubble. Some bubbles expand, rise to the surface, and burst, whereas others collapse. So, too, in inflationary theories, entire universes may form and dissolve, entirely unknowable to us (Box fig. 18.1). It would be amazing indeed if all the wonderful intricacy and beauty of our Universe is merely a single bubble in an even vaster sea of space, a space that we can never see.

and one of time. In fact, according to some versions of these models, space has as many as 10 or 11 dimensions. Inflationary models, however, also deal with less exotic questions, such as why (at least as far as we can tell) the Universe is all ordinary matter with essentially no antimatter. Perhaps the most intriguing aspect of these theories is that they hypothesize links between the Universe at large, the properties of subatomic particles, and the fundamental forces of nature. Here we will deal with only two of the curious puzzles of our Universe that inflation neatly explains: why the Universe has just the critical density, making it "flat" and not curved, and why the Universe (especially the cosmic microwave background) is so uniform. Astronomers refer to the first of these puzzles as the "flatness problem," and they refer to the second one as the "horizon problem."

The Flatness Problem

We saw in section 18.4 that space is curved. Yet when astronomers measure the curvature of our Universe, they find that it is flat, implying that the density of the Universe equals the critical density that we referred to earlier. This is really quite surprising because astronomers knew of no reason why the Universe's curvature should be anywhere close to being flat. In fact, if the curvature had been even slightly different from flat, say $\Omega = 0.9999$, when the Universe was young, it would have rapidly evolved to a much smaller value today. In fact any value of Ω other than 1 is highly unstable, so the Universe must have begun with $\Omega = 1$.

Inflation explains in the following elegant way why this is so. Recall that during the time of inflation, the observable Universe expanded by a colossal factor—from smaller than a proton to a few centimeters in size. Such an enormous expansion of an object flattens any curvature that it has. To see why, imagine inflating a balloon. While the balloon is small, the curvature of its surface is very obvious. However, if the balloon could be inflated to the size of the Earth, its surface, for all practical intents, would be flat, just as the surface of the Earth is "flat." The rapid inflation of the early Universe does the same to space, taking an initially curved space and flattening it.

The Horizon Problem

We saw in section 18.2 that the brightness of the cosmic microwave background is extremely uniform across the sky. Although very small variations exist (as we discussed in the previous section), they are only about 1 part in 100,000. Such uniformity was difficult to explain prior to inflation theory. For example, if we look at the radiation coming from a given point in the sky and then look at radiation coming from exactly the opposite direction in the sky, we are seeing radiation that is coming from locations that are roughly the diameter of the observable Universe—about 28 billion light-years—apart. But the Universe is only about 14 billion years old, so no signal can possibly have traveled from one of these regions to the other in order to smooth out any variation. Using the terminology we used earlier, these regions lie outside each other's cosmic horizon. But the cosmic horizon is far smaller during the time when the Universe "inflates"—far smaller even than a proton when inflation begins. Because of this tiny size, energy moving at the speed of light has plenty of time to cross it many times, which ensures that each part has essentially the same physical conditions as every other part. That in turn makes the cosmic microwave background almost exactly the same in every direction.

Finally, inflation creates tiny irregularities in the density of the young Universe, which gravity will ultimately draw into clusters of galaxies. Thus, the first steps toward the formation of galaxies and their intervening voids may have been taken when the Universe was smaller than an atom.

SUMMARY

Because distant galaxies are receding from the Milky Way and each other, the Universe in its youth was much smaller, hotter, and denser than it is today. By extrapolating the present recessional motion of galaxies back in time, astronomers deduce that the Universe was created approximately 14 billion years ago in the Big Bang, a release of energy that initiated the expansion of space and the creation of matter. This age is consistent with that of other astronomical objects, in that it is greater than the age of the Earth and the age of the oldest stars.

Additional evidence for a hot, dense early Universe comes from the cosmic microwave background. This weak radio-wavelength radiation was originally emitted as visible light long before galaxies formed and when the Universe was a glowing fireball of matter and energy. As the Universe expanded and cooled, so did the radiation; now its temperature is only 3° above absolute zero.

Whether the Universe will expand forever or collapse depends on its density and rate of expansion. If the density is less than a critical value, not enough mass is present for gravity to stop the cosmic expansion, and the Universe will expand forever. Observations, in fact, imply that this is the case.

Astronomers have discovered that the expansion of the Universe is speeding up. To explain this acceleration, astronomers hypothesize that the Universe contains dark energy that causes the speedup. Dark energy appears to make up approximately 70% of the Universe, but its nature is almost completely unknown.

Analysis of tiny brightness variations in the cosmic microwave background allows astronomers to measure the shape of the Universe and its age to much higher accuracy than was previously possible. They deduce from these measurements that the Universe is flat and 13.7 billion years old—both estimates accurate to a few percent.

The flatness of the Universe and the extreme uniformity of the cosmic microwave background lead astronomers to deduce that almost immediately after its formation, the Universe underwent a very brief but very rapid increase in volume—a process called inflation.

QUESTIONS FOR REVIEW

1. (18.1) Why do astronomers think that the Universe is expanding?
2. (18.1) What is meant by the age of the Universe? How old is the Universe? How is its age found?
3. (18.2) What is Olbers' paradox?
4. (18.2) Is our cosmic horizon the same as that of beings in another galaxy? Why?
5. (18.2) What is meant by a cosmic horizon?
6. (18.2) What is the cosmic microwave background? What is its origin?
7. (18.2/18.5) Why do astronomers think that the early Universe was hot and dense?
8. (18.2/18.5) How was helium formed from hydrogen in the early Universe?
9. (18.2/18.5) What explanation do astronomers offer for why the relative abundance of hydrogen and helium is uniform?
10. (18.3) What determines whether the Universe will expand forever or recollapse? Explain what the critical density is.
11. (18.3) What role does the cosmological constant play in cosmology? Is this the same role Einstein originally created it for? What evidence makes astronomers think it may exist?
12. (18.3) What is dark energy? What does its presence imply for the future of the Universe?
13. (18.4) What do astronomers mean when they say that the Universe is flat? What are the other possibilities?
14. (18.4) What evidence do astronomers have that the Universe is flat?
15. (18.5) According to the Big Bang theory, what were the major eras in the history of the Universe? What was the Universe like in each of them? Approximately how long did each last?
16. (18.3/18.5) Why is Einstein's equation for the equivalence of mass and energy important when thinking about the early Universe?
17. (18.5) What happened at the recombination era? Why is this important today?
18. (18.6) What is meant by inflation in cosmology?
19. (18.5/18.6) What problems of the Big Bang theory are resolved by inflationary models?

THOUGHT QUESTIONS

1. (18.1–18.6) Does it bother you that the Universe may expand forever?
2. (18.1) M31, the great galaxy in Andromeda and a member of the Local Group of galaxies, is approaching us. Does this mean that the Hubble law is wrong? Why?
3. (18.1) Explain why the cosmological redshift is not caused in the same way as a Doppler shift (4.6) even though galaxies are being moved away from us as the Universe expands.
4. (18.2/18.5) Why do most stars, including our own Sun, contain more helium than the 24% formed during the Big Bang?
5. (18.3) Why are there points below the green line in figure 18.12 (instead of above it), if the Universe is expanding faster today than in the past?
6. (18.4) If you draw three lines that connect at their corners on a globe (like in fig. 18.14A), have you drawn one triangle, or two? Explain your answer. If you draw three lines that connect on a flat or saddle-shaped surface, have you drawn one or two triangles?
7. (18.5) If the Universe had not expanded since the recombination era, what would the night sky look like? What would the daytime sky look like?
8. (18.2/18.5) Discuss three different kinds of observational evidence for the Big Bang theory you could present if you were challenged by a skeptical child.
9. (18.3/18.6) The recession velocities of distant galaxies suggest that the Universe is expanding faster today than in the early Universe. How is this similar to the process of inflation in the early Universe?
10. (18.2/18.3/18.6) Dark energy accelerates the expansion of the Universe, but galaxies and solar systems hold themselves together while the distances between them increase. If the Universe continues to expand faster and faster, what will happen to our cosmic horizon? Will studying cosmology be easier or harder in the very distant future?

PROBLEMS

1. (18.1) Until recently, experimental results for the Hubble constant ranged from about 50 to 100 km/sec/Mpc. Calculate the age of the Universe for both of these extremes. Do you see any potential conflicts with other astronomical observations if H had been either of these numbers?
2. (18.2) The temperature of Universe at recombination was about 3000 K. If the ratio of the temperature of the background radiation at recombination to the temperature of the CMB radiation today is equal to the ratio of the wavelength of light today to the wavelength of the background radiation at recombination, how many times wider is the Universe today than it was at recombination? Remember that as the Universe expands it stretches the light.
3. (18.2/18.5) Using the same logic as the previous question, how many times wider is the Universe today than it was when the average temperature was 3×10^9 K, sometime between 1 and 3 minutes old? If you shrank the Milky Way's disk by that factor, would it fit inside the Solar System?
4. (18.2/18.5) If the change in the previous problem is the change in diameter, how many times larger is the volume of the Universe today than when the temperature was 3×10^9 K?

5. (13.3/18.2/18.5) The temperature of the Universe at recombination was about 3000 K. Use Wien's law to calculate the peak wavelength of the radiation at that temperature. Which spectral type of star has this surface temperature?
6. (18.6) When the Universe was extremely young, it increased in size "exponentially" until it was 10^{25} times larger than it began. How many times would the Universe have to double in size to grow this much larger than it started? (If you are unfamiliar with logarithms, you can estimate this by testing different powers of 2 on a calculator by trial and error.)

TEST YOURSELF

1. (18.1) From what evidence do astronomers deduce that the Universe is expanding?
   (a) They can see the disks of galaxies getting smaller over time.
   (b) They see a redshift in the spectral lines of distant galaxies.
   (c) They detect cosmic background X-ray radiation.
   (d) They can see distant galaxies dissolve, pulled apart by the expansion of space.
   (e) All of the above.
2. (18.1) If Hubble's constant were half of the value we think it is, the Universe would be ______ as we think it is.
   (a) 4 times as old (c) the same age (e) 1/4 as old
   (b) 2 times as old (d) 1/2 as old
3. (18.2) Why do astronomers think the Universe was much hotter at its birth than it is now? (Select all that apply)
   (a) They detect cosmic microwave background radiation.
   (b) Dark matter in galaxy clusters appears to be heated.
   (c) They detect intense X-ray emission from gas in galaxy clusters.
   (d) The amount of helium observed in all stars implies that the early Universe was hot enough to fuse some of its hydrogen into helium.
   (e) The Universe was originally positively curved and today appears flat.
4. (18.2) What is meant by the cosmic microwave background?
   (a) It is radiation from distant quasars.
   (b) It is radiation from hot gas in intergalactic space.
   (c) It is radiation from the first stars formed when the Universe was young.
   (d) It is radiation created in the early Universe.
   (e) It is the explanation of Olbers' paradox.
5. (18.3) What would have happened if the density of the Universe had been greater than the critical density and the Universe contained no dark energy?
   (a) Matter would never have formed.
   (b) Galaxies would not have formed.
   (c) Too many galaxies would have formed.
   (d) All the matter formed would have been neutrons.
   (e) The Universe would recollapse.
6. (18.3) Dark energy (select all that apply)
   (a) is represented in cosmological equations by the cosmological constant.
   (b) is the energy associated with dark matter.
   (c) acts against gravity.
   (d) causes space to expand ever faster.
   (e) is distributed evenly through space.
7. (18.4) Current measurements suggest the Universe is composed of about
   (a) 90% regular matter, 10% dark matter.
   (b) 23% dark matter, 4% dark energy, 72% regular matter.
   (c) 72% dark energy, 23% dark matter, 4% regular matter.
   (d) 33% dark matter, 33% regular matter, 33% dark energy.
   (e) 72% dark matter, 23% dark energy, 4% regular matter.
8. (18.5) For most of the time since the beginning of the Universe, conditions were
   (a) much hotter than today.
   (b) not very different from today.
   (c) similar to the inside of an atomic nucleus.
   (d) like the core of a star.
   (e) like the surface of a star.
9. (18.6) What is meant by inflation in the early Universe?
   (a) The force of gravity suddenly grew stronger in the distant past.
   (b) Protons expanded to the size of stars, which was how our Sun formed.
   (c) The Universe increased dramatically in size in an extremely brief period of time.
   (d) The number of galaxies that we see at large distances is much greater than the number we can see near us.
   (e) The diameter of distant galaxies is much greater than the diameter of galaxies near us.
10. (18.6) Which of the following are explained by a brief period of inflation? (select all that apply)
   (a) The high uniformity of the cosmic microwave background in multiple directions
   (b) The large number of black holes in the Universe
   (c) The fact that the Universe appears flat
   (d) The ratio of helium to hydrogen in very old stars
   (e) The existence of more matter than antimatter in the Universe

KEY TERMS

antimatter, 510
Big Bang, 493
cosmic horizon, 499
cosmic microwave background (CMB), 501
cosmological constant, 506
cosmological principle, 497
cosmology, 493
critical density, 504
curvature of the Universe, 507
dark energy, 506
flat Universe, 507
inflation, 512
negative curvature, 507
Olbers' paradox, 497
positive curvature, 507
recombination era, 500

FURTHER EXPLORATIONS

Aguirre, Anthony. "Where Did It All Come From?" *Sky and Telescope* 112, (December 2006): 36.

Arkani-Hamed, Nima, Savas Dimopoulos, and Georgi Dvali. "The Universe's Unseen Dimensions." *Scientific American* 283 (August 2000): 62.

"Brave New Cosmos." *Scientific American* 284 (January 2001): 37. [Five articles on recent cosmological developments.]

Carroll, Sean. "Dark Energy and the Preposterous Universe." *Sky and Telescope* 109 (March 2005): 32.

Conselice, Christopher J. "The Universe's Invisible Hand." *Scientific American* 296 (February 2007): 34.

Dvali, Georgi. "Out of the Darkness." *Scientific American* 290 (February 2004): 68.

Freedman, Wendy L., and Michael S. Turner. "Cosmology in the New Millennium." *Sky and Telescope* 106 (October 2003): 30.

Guth, Alan H. "The Inflationary Universe." *Scientific American* 250 (May 1984): 116.

Hinshaw, Gary, and Robert Naeye. "Decoding the Oldest Light in the Universe." *Sky and Telescope* 115 (May 2008): 18.

Hu, Wayne, and Martin White. "The Cosmic Symphony." *Scientific American* 290 (February 2004): 44.

Krauss, Lawrence M. *Quintessence: The Mystery of the Missing Mass in the Universe.* New York: Basic Books, 2000.

Kruesi, Liz. "Will Dark Energy Tear the Universe Apart?" *Astronomy* 37 (February 2009): 34.

Lineweaver, Charles H., and Tamara M. Davis, "Misconceptions About the Big Bang." *Scientific American* 292 (March 2005): 36.

Livio, Mario. *The Accelerating Universe.* New York: John Wiley, 2000.

Nadis, Steve. "Sizing Up Inflation." *Sky and Telescope* 110 (November 2005): 32.

______. "Searching for the Shape of the Universe." *Astronomy* 36 (April 2008): 28.

Naeye, Robert. "Delving into Extra Dimensions." *Sky and Telescope* 105 (June 2003): 38.

Osterbrock, Donald E., Joel A. Gwinn, and Ronald S. Brashear. "Edwin Hubble and the Expanding Universe." *Scientific American* 269 (July 1993): 84.

Panek, Richard. "Going Over the Dark Side." *Sky and Telescope* 117 (February 2009): 22.

Rees, Martin. *Just Six Numbers: The Deep Forces That Shape the Universe.* New York: Basic Books, 2000.

Riess, Adam G., and Michael S. Turner. "From Slowdown to Speedup." *Scientific American* 290 (February 2004): 62.

Rowan-Robinson, Michael. *Nine Numbers of the Cosmos.* Oxford, England: Oxford University Press, 1999.

Singh, Simon. *Big Bang: The Most Important Scientific Discovery of All Time and Why You Need to Know About It.* New York: Fourth Estate Press, 2005.

Strauss, Michael A. "Reading the Blueprint of Creation." *Scientific American* 290 (February 2004): 54.

Tegmark, Max. "Parallel Universes." *Scientific American* 288 (May 2003): 40.

Website

Visit the *Explorations* website at **http://www.mhhe.com/arny** for additional online resources on these topics.

Q FIGURE QUESTION ANSWERS

WHAT IS THIS? (chapter opening): This dim, thin red line in the white outline is an image of one of the most distant galaxies known. Another image of the galaxy is dimly visible in the small circle at the top of the picture. The galaxy itself lies behind a massive cluster of galaxies, and we see the distant galaxy's light distorted into the line shape by the gravitational force of the cluster that lies between it and us. The galaxy cluster is about 2 billion light-years away. The remote galaxy itself is about 13 billion light-years from us.

FIGURE 18.20 By a factor of about 10^{25}, similar to the factor during inflation.

PROJECT

Cosmic Beach. For this project you will need at least a couple of buckets of wet sand, a protractor, and a couple of straight sticks. You can easily replicate the shapes shown in figure 18.14 out of sand, and if you have a few friends, make them quite large. First, sketch out a triangle on the flat sand and measure the angles to verify they add up to 180°. Next, create a hemispherical dome of sand to represent half of a positively curved Universe. Try creating a couple of different triangles and measuring the angles. What is the largest "triangle" you can make? What do its angles add up to? Notice that if you set a crab or an insect walking around the equator of your hemisphere, it could eventually come back to where it started. Finally, mound up some sand to form a large saddle shape and repeat the experiments once more. This shape is hard to create using paper because it needs to curve downward "across" the saddle, and "upward" from the front to back, but it is easy to make a section of it out of the sand.

ESSAY 4

Life in the Universe

The evolution of the Universe from Big Bang to galaxies, stars, and planets has along the way created us. This evolutionary path from raw quarks to thinking beings who seek to understand how they fit into this grand scheme—what eighteenth-century scientists called "the great chain of being"—is one of many marvels of the Universe and is the theme of this section.

In earlier chapters, we described astronomical evolution and how the Earth and Sun fit into the cosmos. Now we turn to life on Earth and how it might have formed and developed. We will see that the existence of life on our own planet is suggestive of the possibility of its existence elsewhere, and so we will also speculate about whether we are alone, leading to the question, "Do other intelligent civilizations exist?" Astronomers are strongly divided about such possibilities. Some argue that such life is highly probable, whereas others argue that it is highly improbable. One goal of this section is to describe the evidence on both sides of this debate.

We will begin our discussion by examining the long history of life on Earth, looking at the factors thought to play a role in its development here. That history began quite soon after the Earth formed, as illustrated by the time line shown in figure E4.1.We will also discuss the likelihood of similar conditions elsewhere, discovering that planets suitable for life are probably numerous. In addition, we will describe how some astronomers have been motivated by such arguments to search for radio signals from other civilizations. Finally, we will speculate about whether our existence may even say something about the nature of the Universe itself.

LIFE ON EARTH

Life has existed on Earth for nearly the entire history of our planet. For example, fossil bacteria such as those illustrated in figure E4.2 occur in rocks perhaps as old as 3.5 billion years. The dates of these rocks have been measured by radioactivity in their associated rock, as described in chapter 6. For example, the ancient life forms "cyanobacteria" (cyan is a blue-green color) form colonies called "stromatolites," which are found in fossils more than 3 billion years old and look identical to the stromatolites that exist today in shallow, warm, salty bays along the coast of western Australia (fig. E4.3).

FIGURE E4.1
A time line illustrating the history of life on Earth.

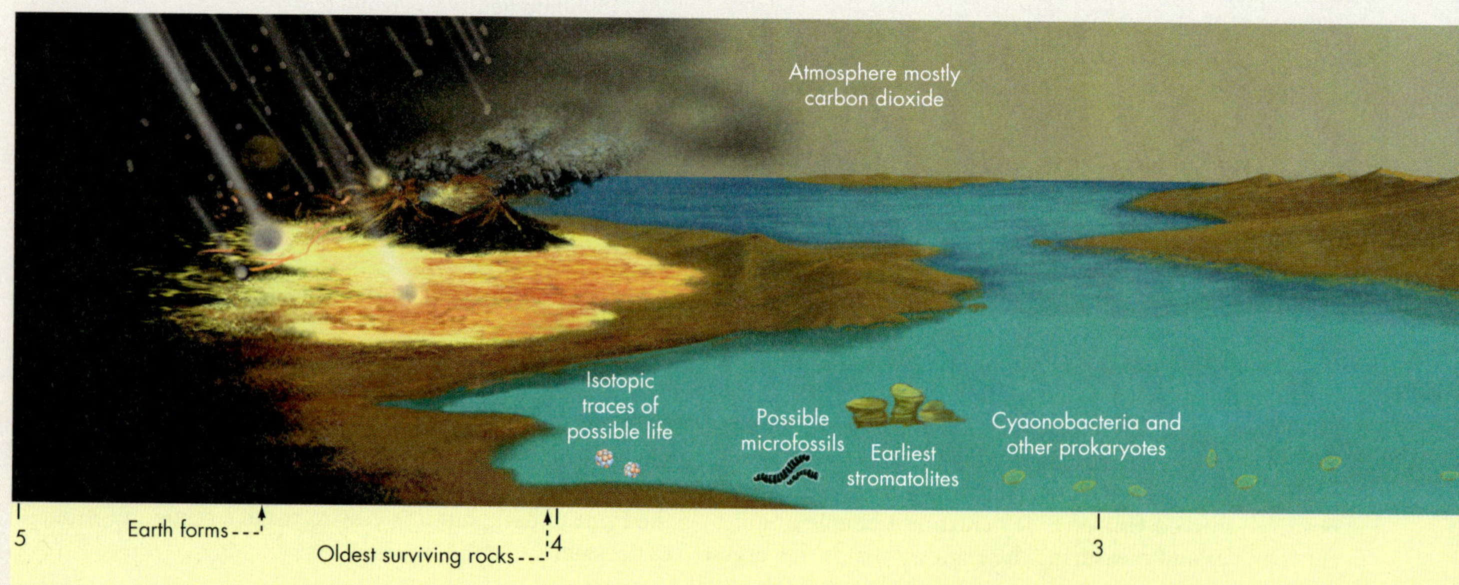

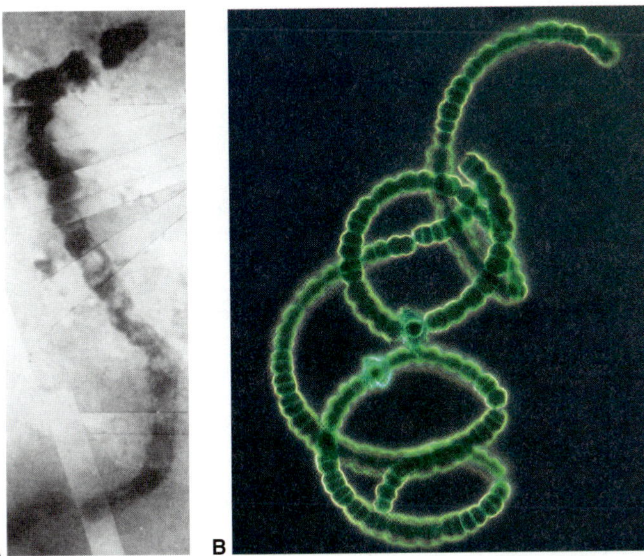

FIGURE E4.2
(A) Photograph of probable microfossil cyanobacteria in rock 3.5 billion years old. (B) Microscopic picture of modern cyanobacteria. Cyanobacteria are a kind of photosynthesizing bacteria. Not all scientists agree that these microscopic features are fossils. More widely accepted fossils are found in increasing numbers over the subsequent billion years.

FIGURE E4.3
Stromatolites in western Australia, are formed by colonies of a cyanobacteria. They trap sediments, forming a layered structure in this characteristic dome shape. The inset shows a 3.4-billion-year-old stromatolite fossil, sliced to show the layered growth pattern.

There are possible signs of even earlier life in the isotopes of carbon found in deposits as much as 3.8 billion years old. Living organisms are selective in choosing lighter isotopes of carbon, so deposits of carbon containing abnormally low amounts of the heavier isotopes may indicate that life was present. Given that Earth formed about 4.6 billion years ago and that its surface was probably molten rock for several hundred million years, it appears that life began quite quickly.

Scientists conclude from the fossil evidence that only extremely simple life forms such as bacteria and algae existed for roughly three-fourths of Earth's history. Then, approximately 600 million years ago, more-complex forms such as invertebrates developed, and about 500 million years ago, only a little more than one-tenth of the Earth's lifetime, shells and crustaceans appeared. Mammals and dinosaurs both appeared roughly 250 million years ago, but the latter were wiped out about 65 million years ago in the great Cretaceous extinction,

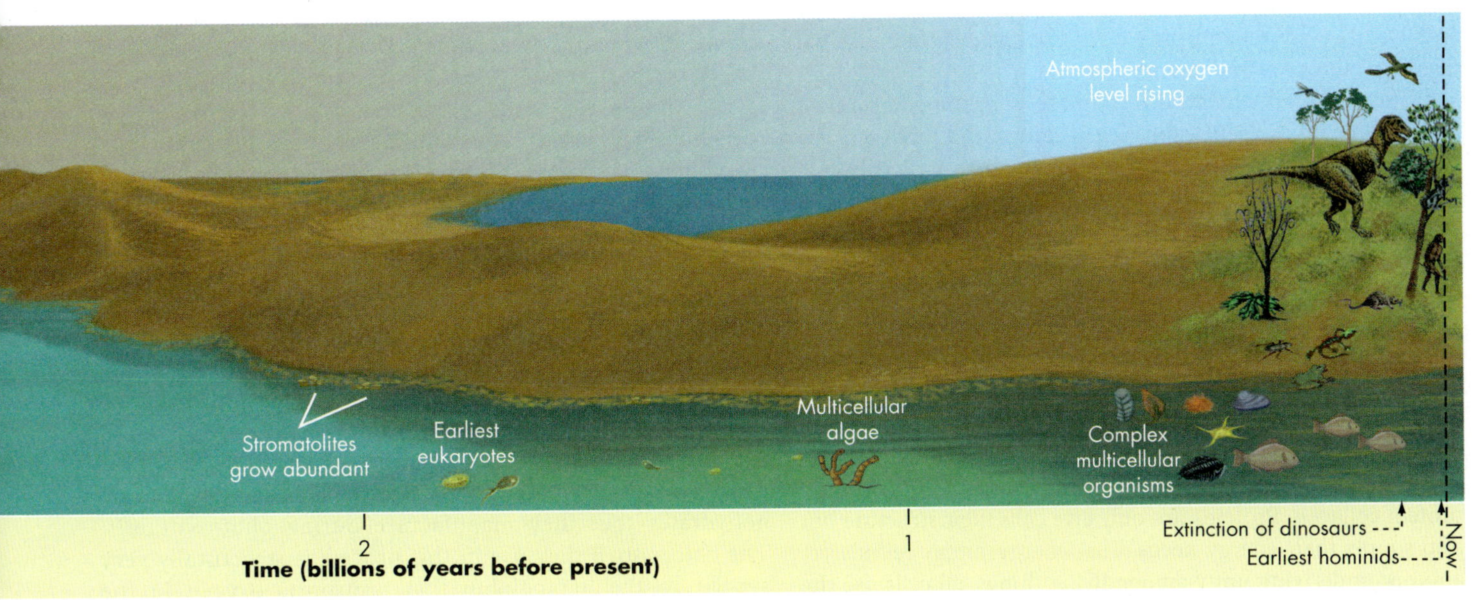

FIGURE E4.4
A few of Earth's many complex life forms.

probably caused by a totally chance event—an asteroidal collision, as described in chapter 11. Early hominids, our immediate ancestors, appeared roughly 5.5 million years ago, but our own species, *Homo sapiens,* evolved only about 500,000 years ago. Earth has therefore existed as a planet roughly 10,000 times longer than we have as a species. Thus, if all Earth's history were compressed into a year, humans would appear only in the last hour, and civilization only in the last few minutes.

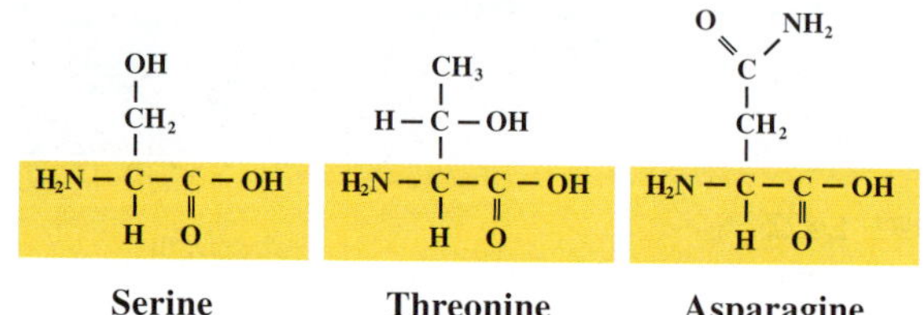

FIGURE E4.5
Diagrams illustrating the structure of several amino acids. H, C, N, and O represent atoms of hydrogen, carbon, nitrogen, and oxygen. The basic building block (shown in yellow) of amino acids is the same.

THE UNITY OF LIVING BEINGS

Life on Earth shows a dazzling variety of forms, from butterflies to birds, worms to whales, mushrooms to maples, as shown in figure E4.4. Despite that variety, all living beings on Earth have an underlying unity of structure, reproduction, and metabolism. All living organisms use the same kinds of atoms for their structure and function (hydrogen, oxygen, carbon, nitrogen, and iron, for example). These atoms are not only widely used, but they are also abundant throughout the Universe. Thus, life on Earth generally is very economical in its use of the same chemical substances in widely differing organisms. Moreover, life tends to draw on the materials with which our planet is most richly endowed: we are made primarily of the same substances that make up our ocean and atmosphere.

The chemical elements that compose living things are linked into long-chain molecules for their structure and function. The primary structural units of many of these chain molecules are about 20 amino acids. **Amino acids** are organic molecules containing hydrogen, carbon, nitrogen, and oxygen atoms arranged in a certain pattern (fig. E4.5). A few amino acids also contain one or more sulfur atoms.

Moving up to the next level of complexity, we find that all living things use amino acids as the structural units of more-complicated molecules called **proteins.** Protein molecules give living things their structure. For example, the scales of reptiles and fish are composed of the protein keratin, and the stiff cartilage of our own bodies and other animals is composed of the protein collagen. Proteins not only give cells their structure but also supply their energy needs. That energy supply reveals yet another underlying unity among living things: all cells use the same kind of molecule, adenosine triphosphate (ATP) to supply energy for action and growth.

This unity of structure and energy supply is echoed in the reproduction of life forms. All single-celled organisms divide to create new cells; all multicelled organisms produce egg cells, which divide to create new cells. Moreover, in all cases, "parents" pass on genetic information to their offspring by the same molecule, **DNA.**

DNA stands for deoxyribonucleic acid. It is a massive molecule consisting of two extremely long chains of smaller molecules wound around each other and linked by molecules called "base pairs." This gives DNA the appearance of a twisted ladder with the role of the connecting rungs played by the base pairs (fig. E4.6). When a cell reproduces, the DNA separates into two individual chains. Each chain then adds a new base pair at each point along its length to match the one previously present on the other chain. Thus, each strand makes a copy of the other, so that after division and replication, a new DNA molecule exists identical to the original one. Copying errors—mutations—sometimes occur. Most are harmful and lead to the organism's death. Nonlethal ones make evolution possible. That all terrestrial organisms use DNA for their reproduction gives life a truly remarkable unity.

Unity can also be seen in the way higher organisms function. For example, plants derive their food from carbohydrates that they manufacture during photosynthesis. The chlorophyll they use in this process is structurally very similar to the hemoglobin that transports oxygen in the

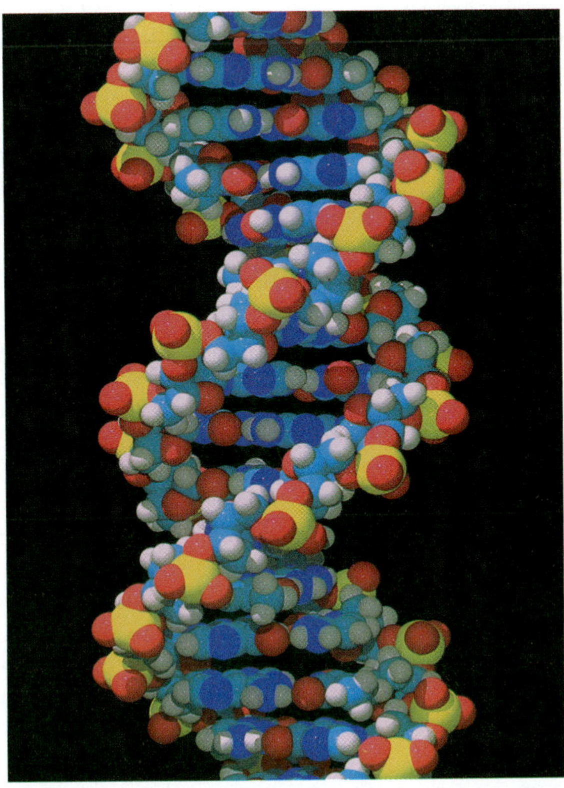

FIGURE E4.6
A diagram illustrating the structure of the DNA molecule. The blue horizontal structures in this model are the base pairs.

blood of animals. These two molecules differ primarily in that chlorophyll is built around a magnesium atom, whereas hemoglobin is built around an iron atom. Likewise, cellulose, the structural material of plants, has a molecular structure very similar to chitin, the structural material in insect bodies and crustacean shells.

Finally, living systems exhibit a unity in their functions, such as the chemical processes that move biologically important molecules into and out of cells.

DEDUCTIONS FROM THE UNITY OF LIFE AND THE TIME LINE

What conclusions can we draw from the chemical similarities of living beings and their ancient history, as deduced from the time line? First, the chemical similarity almost inescapably suggests that all life on Earth had a common origin. Second, the antiquity of life suggests that under suitable conditions, life develops rapidly from complex molecules. Thus, we might infer that wherever such conditions occur, so too will life.

We should treat these conclusions with great caution, however, for they are based on only a single case: life on Earth. For example, if the first time you hit a golf ball you make a lucky hole-in-one, it does not mean that you will do so again anytime soon. Thus, the fact that life exists on Earth may say next to nothing about its chances elsewhere. Supposing, however, we do choose to speculate, what can we say about the origin of life and its likelihood elsewhere?

THE ORIGIN OF LIFE

Most scientists today think that terrestrial life originated from chemical reactions among complex molecules present on the young Earth. This idea dates back to at least the time of Charles Darwin, but it was strongly bolstered in 1953 by an experiment performed by Stanley Miller—then a graduate student at the University of Chicago—and his professor, Harold Urey. Miller and Urey filled a sterile glass flask with water, hydrogen, methane, and ammonia—gases thought to be present in the Earth's early atmosphere (fig. E4.7). They then passed an electric spark through this mixture. The spark generated both visible and ultraviolet radiation, which triggered reactions in the gas-water mixture. At the end of a week, they analyzed the mixture, finding in it a variety of organic molecules as well as five of the amino acids used to make proteins. They therefore concluded that at least some of the ingredients necessary for life could have been generated spontaneously on the early Earth.

The success of Miller and Urey's experiment motivated other researchers to perform similar but more complex experiments. Subsequent experiments created conditions even more similar to the young Earth, and produced even more types of amino acids as well as the nucleic acids, which are the building blocks of DNA. Even the energy needs of living things can be produced in Miller-Urey type of experiments. For example, the Sri Lankan-American biochemist Cyril Ponnamperuma and

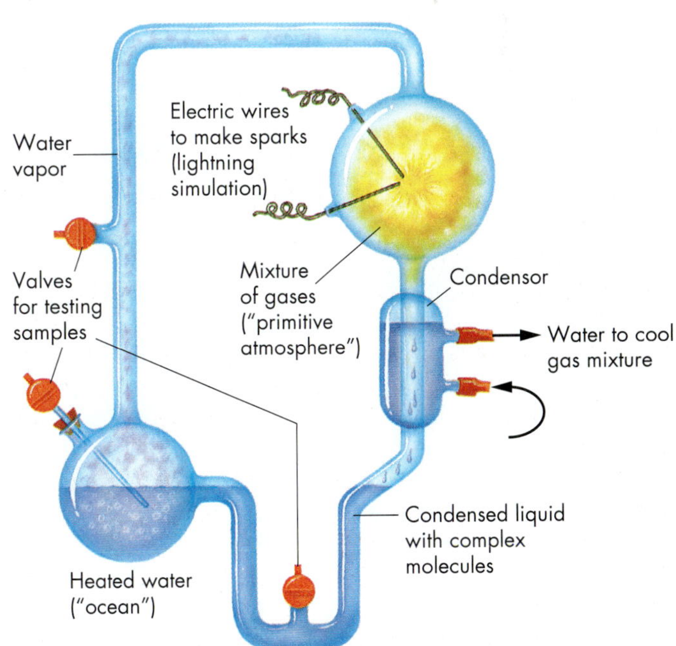

FIGURE E4.7
The Miller-Urey Experiment simulated conditions on the early Earth and showed that complex organic molecules could form then.

the U.S. astronomer Carl Sagan showed that ATP can be readily synthesized from compounds similar to those generated by Miller and Urey, demonstrating one additional step in creating living from nonliving material.

The U.S. chemist Sydney Fox demonstrated that another essential step in the process of forming life may occur by simple chemical processes. He took complex organic molecules such as those made by Miller and Urey and by repeatedly heating and cooling them in water was able to create short strands of proteins called "proteinoids." More interesting, however, Fox discovered that these proteinoids spontaneously formed small spheres reminiscent of cells. These small spheres have no power to reproduce, nor do they show any of the complex structure commonly found in living cells, but they do demonstrate that organic material can spontaneously build a structure that could wall off complex molecules from their environment, where they might take the next steps toward reproduction and life.

Why then do we not observe similar processes occurring today on Earth? One reason is that our atmosphere now contains oxygen, which quickly attacks organic molecules, breaking them down. Thus, the absence of oxygen earlier in the Earth's history—an absence shown by analysis of ancient rocks—was probably crucial for the development of life. Another reason was given by Charles Darwin in his speculations on life's origin: today's living things immediately consume such organic material.

The importance of low oxygen abundance to the origin of life has led some researchers to speculate that life may have originated deep in the ocean near sea-floor vents. These environments are rich in minerals and energy, and today they support a variety of life forms (fig. E4.8). In ancient days, these same environments might have been especially congenial to the assembly of complex molecules. Even now, bacteria and other life forms live in hot, oxygen-poor environments such as the boiling hot springs at places like Yellowstone National Park.

A

B

FIGURE E4.8
(A) Undersea volcanic vents expel hot sulfur-rich minerals that provide the nutrients for very primitive types of bacteria. (B) These bacteria in turn are at the bottom of the food chain for colonies of unusual undersea creatures at such depths that sunlight never reaches them.

Even without the success of the Miller-Urey experiment in creating organic material, scientists studying the origin of life on Earth are confident that such material would be present on the young Earth. One source of organic material is interstellar matter. We saw in chapter 16 that many interstellar clouds contain a rich mix of organic molecules, including many that are precursors to amino acids. Because the Solar System and Earth formed from such clouds, complex organic molecules must have been present on or near the Earth from the time of its birth. Even if these molecules were destroyed during the Earth's formation, fresh molecules may have been carried to its surface by planetesimals toward the end of their heavy bombardment of Earth's surface. In fact, carbonaceous chondritic meteors—probably surviving fragments of the ancient planetesimals—are rich in organic material, and some even contain amino acids, as we saw in chapter 11. Finally, some scientists have demonstrated experimentally that the compression and heating of gases resulting from planetesimal bombardment can be as fruitful as the electric discharges used by Miller and Urey in forming complex organic materials. Thus, astronomers are confident that organic molecules were common on the ancient Earth.

On the other hand, even if one assumes that organic matter existed on the early Earth (whether created by lightning, ultraviolet radiation, or planetesimal bombardment), many steps are needed to take it from amino-acid-filled droplets in a primordial ocean to living cells. For example, the structures created must be able to reproduce, a feature not yet seen in any origin-of-life experiments.

Cells replicate today using the double-chained molecule DNA. Some researchers have speculated, however, that the earliest life forms used the slightly simpler molecule, ribonucleic acid (RNA). In fact, biochemists have shown that if the proper chemicals are present (especially certain enzymes), RNA can replicate in the absence of living cells. Moreover, they have also discovered several other molecules far simpler in structure than RNA that can replicate, but so far only in a "mechanical" sense. That is, if such molecules are put in a solution containing the necessary raw materials, they will make copies of themselves. In fact, one particular mix of molecules not only replicates but also can "mutate" into a slightly better replicator. Even so, far more is needed to produce life as we know it. For example, simple life forms must be able to form cells that work cooperatively.

ORIGIN OF COMPLEX ORGANISMS

The original cellular life on Earth was probably far simpler than the cells we see today. For example, the earliest cells probably had no nucleus, a condition found now only in bacteria and certain algae. **Prokaryotes,** as these cells without nuclei are known, may accidentally have merged with other cells, a merger that benefited both cells by allowing them to grow and reproduce more rapidly. A cell good at storing energy, for instance, might have combined with a cell good at reproducing, thereby creating a more complex cell. From such simple beginnings, **eukaryotes** (cells with nuclei) probably evolved and, after them, multicelled organisms. Today, most cells have a complex membrane that holds them together and allows food to enter and waste products to leave. Moreover, most cells other than bacteria and some algae have a nucleus in which genetic information is stored, as well as mitochondria, which are tiny bodies within the cell that convert food into energy that the cell utilizes.

The greater complexity of multicelled organisms would in some cases allow them to exploit resources unavailable to simpler beings. For example, an organism that can move from nutrient-poor to nutrient-rich environments may be able to reproduce more prolifically than a stationary creature. Such reproductive advantages combined with changing environments have led to the incredible diversity of life that we see around us today. In fact, such reproductive advantage (natural selection) and random mutations are crucial to evolution and the development of higher life forms, as Charles Darwin explained in 1859 in his monumental work *On the Origin of Species by Means of Natural Selection.*

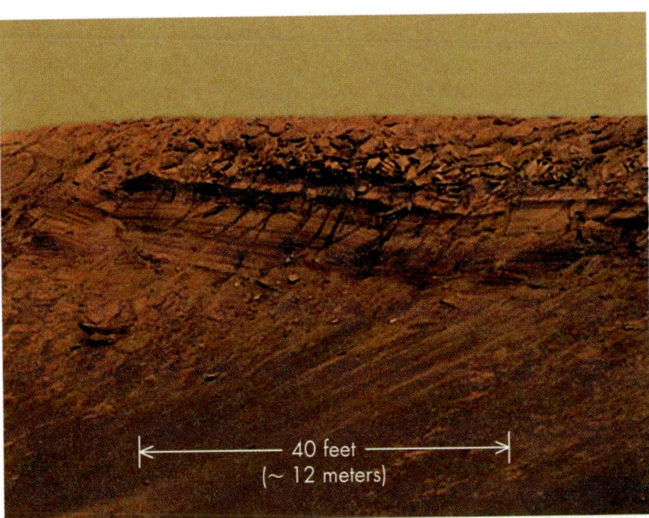

FIGURE E4.9
Cliffs on the inner wall of Endurance Crater, which was visited by *Opportunity,* one of the Mars rovers. The layers (strata) that you can see here suggest that this area on Mars once was covered with water. Because water is vital to life here on Earth, scientists are anxious to explore regions on other planets and moons where water is, or was once, present—in the hope of finding traces of present or past life on other worlds.

LIFE ELSEWHERE IN THE UNIVERSE

Humans have speculated for thousands of years about life elsewhere in the Universe. Around 300 B.C., the ancient Greek philosopher Epicurus wrote in his "Letter to Herodotus" that "there are infinite worlds both like and unlike this world of ours." Epicurus went on to add that "in all worlds there are living creatures and plants." Likewise, the Roman scholar Lucretius wrote, around 50 B.C., that "it is in the highest degree unlikely that this Earth and sky is the only one to have been created."

Such views were not universal, however. For example, Plato and Aristotle argued that the Earth was the only abode of life, a view that prevailed through most of the Middle Ages. Today, however, influenced by science fiction movies, television, and books, many people find no difficulty at all in believing that extraterrestrial life exists. In fact, supermarket tabloids make silly claims almost weekly about encounters with aliens. If we look at the facts, however, there is not a shred of evidence that life exists elsewhere. This does not mean there is no extraterrestrial life; it simply means we do not *know* if there is. Thus, with no evidence, what can we say meaningfully about life elsewhere in the Universe?

SEARCHING FOR LIFE ELSEWHERE

Is there any evidence for life elsewhere in the Universe? Not really, although in the summer of 1996 two groups of scientists suggested that traces of organic chemicals that they detected in meteorites from Mars may indicate that life once existed there (see chapter 9 for more details). The absence of hard evidence may seem discouraging, but as British astronomer Sir Martin Rees has said, "Absence of evidence is not the same as evidence of absence."

Even if the Universe were teeming with life, we have so far examined only a tiny portion of it: some half dozen spots on the Moon and on Mars (by robot space-landers), and some asteroidal fragments picked up on Earth. The Moon's lack of atmosphere and water makes it so inhospitable that astronomers did not expect to find life there. Mars seemed more promising because its environment, though harsh, is more like the Earth's than that of any other planet.

Laboratory experiments have demonstrated that some terrestrial organisms (bacteria and lichens) could in fact survive Martian conditions, though they did not reproduce or grow. We know that the Martian environment was less harsh in the past. For example, Mars once had flowing water on its surface, an indication that a warmer climate and denser atmosphere existed long ago (fig. E4.9). Nevertheless, the two *Viking* spacecraft that landed there in 1976 found no trace of life, nor have the *Mars Pathfinder,* the *Spirit* and *Opportunity* rovers, or the *Phoenix* lander, which landed on Mars between 1997 and 2008. Given the harsh conditions presently on Mars, it is possible that if there ever were microscopic organisms, they have retreated deeper underground where liquid water may remain.

Another intriguing possibility is that life may have formed in the oceans under the icy crust of moons like Europa, Ganymede, or Enceladus. Tidal and internal radioactive heating keep the interiors of these icy moons liquid (see chapter 10). Conditions there may be similar to the ocean on the young Earth, with underwater volcanic vents. Astronomers are considering how we might deliver a deep-sea vehicle to explore one of these moons' oceans. Just one of the challenges of such a mission will be to get through the kilometers of ice on the moon's surface. Perhaps the lander can heat the area beneath itself, melting the ice and gradually sinking down to the liquid interior.

PANSPERMIA

Most scientists think that life originated on Earth spontaneously, but is that necessarily true? The discovery of a meteorite from Mars with possible traces of life revived an older idea that life originated elsewhere and was later transported to Earth either accidentally or deliberately.

The idea that terrestrial life descended from organisms created elsewhere in the Universe is sometimes called **panspermia.** According to this idea, simple life forms (perhaps bacteria) from some other location drifted from their place of origin across space to Earth. Bacteria spawned on some distant planet, for instance, may have hitched a ride on rocks or soil that were knocked into space by a meteorite impact. Astronomers had initially assumed that most of the debris of a meteorite impact would be superheated and sterilized, but the meteorite from Mars containing organic molecules made them reexamine that assumption. Although the immediate impact area of a collision is intensely heated, the shock wave that travels out from the impact site can launch substantial material at higher than the escape velocity (chapter 3). Because such collisions were even more common in the early Solar System, it is even possible that we are all descended from microscopic Martians who arrived at the Earth when it was young.

It is not just Mars that might be a source for such material. Once in space, this material might travel almost anywhere. Although space is highly hostile to advanced life forms, simple organisms may be able to survive long passages through it if embedded in a rock. Eventually these rocks may have arrived in the Solar System and been captured by the Earth's gravity, initiating life on our planet.

Panspermia has a certain appeal, in that it suggests we may have distant cousins on remote planets elsewhere in our Galaxy. The theory was very popular in the 1920s, but critics point out that it does not really simplify the problem of the origin of life, it just shifts it elsewhere. It does, however, provide additional reasons for speculating that life may exist elsewhere.

ARE WE ALONE?

Do other life forms exist in the Universe? Is there somewhere, hundreds of light-years from Earth, a "student alien" reading a book discussing the possibility of life on planets such as Earth? Scientists do not know, and in fact they are strongly divided into two groups. One group of scientists (let us call them "many-worlders") believes that millions of planets with life exist in the Milky Way and that many of these have advanced civilizations. The other group (let us call them "loners") argues that we are the only intelligent life in the Galaxy. We will discuss below how these two radically different points of view arise.

ARGUMENTS FOR MANY WORLDS

Many-worlders argue as follows: Earth-like planets are common. Life has formed on Earth. Therefore, it would be surprising if life did not exist elsewhere. In fact, even if only a small fraction of such planets have life, many advanced civilizations probably exist. Such arguments may actually vastly *underestimate* the number of habitable planets, because life forms elsewhere may be able to live in environments intolerable to terrestrial organisms. To see how such numbers are deduced, let us begin by estimating the number of Earth-like planets in the Milky Way.

We start by asking how many Sun-like stars are in the Milky Way. We limit ourselves to such stars because luminous blue ones burn their fuel so rapidly that they will burn out and explode as supernovas before life has much chance to evolve, and dim red stars are so cool that planets must be very close to them for conditions to be warm enough for a terrestrial type of life. Given those restrictions on star types, we turn to censuses of stars, which show that stars like the Sun make up about 10% of our Galaxy's stars, for a total of about 10^{10} such bodies.

Of these 10^{10} stars, we next ask, How many of them have Earth-like planets? Our own system has two planets out of a total of eight that might harbor life as we know it: Earth and Mars. Unfortunately, although astronomers have detected many planets around other stars, Earth-size planets are still very hard to detect and so the probability of their occurrence is still unknown. If we are conservative, therefore, and say that other systems might not be quite so lucky as we are in the Solar System, we might choose the probability of a life-sustaining planet around a solar type of star to be 1 in 10, or 10%. Taking 10% of the 10^{10} Sun-like stars leaves us with 10^9 habitable planets.

How many of these actually have life on them? That will depend on how easy it is for life to form, a probability that the Miller-Urey experiments seem to indicate is quite high. The difficulty in assessing that probability is that we have only the single case of life here as our basis. If, however, we use Earth as the pattern, the rapidity with which life developed on Earth and the Miller-Urey experiment's success in making molecular precursors to life indicate that the chance of life starting is fairly high, say, 1 in 100 rather than 1 in a million. Given that probability, we end up with $10^9 \times 0.01 = 10^7$ = 10 million worlds with life in our Galaxy.

We now go to the next step and ask, On how many of these life-bearing planets do technical civilizations form? It is extremely difficult to assess the probability that higher life will evolve. Moreover, even if life succeeds in starting, perhaps it will be annihilated by asteroidal impact or, if it develops a high technology and is careless or thoughtless, life may destroy itself by pollution, nuclear war, or some other disaster. On the other hand, a disaster for one species provides an opportunity for another species to evolve—just as we might never have evolved if the dinosaurs had not been wiped out by a major impact. Some have speculated that major evolutionary changes may occur until a species arises that is capable of technology that can prevent cataclysmic events like an asteroid impact. Suppose we make a conservative guess and say that if life forms, it has a 1 in 1000 chance of developing a technical civilization. Thus, we find that the number of advanced civilizations in our Galaxy is $10^7 \times 0.001 = 10{,}000$. When we further consider that there are billions of other galaxies, it seems very likely that we are not alone.

The above argument was first presented by the U.S. astronomer Frank Drake, who pointed out that one can estimate the number of civilizations by multiplying together the probability of each condition necessary for it, a result now called the **Drake equation.** Unfortunately, the accuracy of the probabilities is extremely uncertain, but they do help us to focus the argument. It is interesting that even using fairly conservative estimates, the equation suggests that *many* civilizations other than ours exist in our Galaxy. Assuming that they exist, we might then ask, How close is the nearest one? We can estimate the distance as follows.

Suppose the Milky Way contains N civilizations scattered randomly throughout its disk, as illustrated in figure E4.10. We can calculate their average separation, d, as follows. Imagine drawing a sphere of radius $d/2$ around each civilized world. Then N such spheres must cover the disk of the Milky Way. That is,

$$N\pi(d/2)^2 = \pi R^2$$

where we assume that the Milky Way's disk has radius R. We can now cancel π to obtain

$$Nd^2/4 = R^2$$

Moving the N and 4 to the right-hand side of the equation gives

$$d^2 = 4R^2/N$$

Taking the square root of both sides gives

$$d = 2R/\sqrt{N}$$

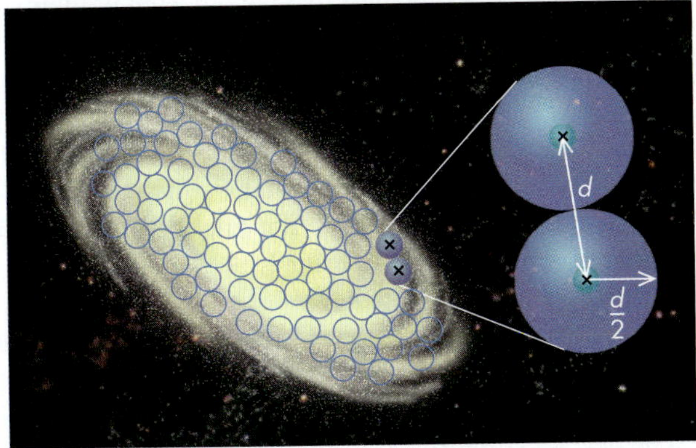

FIGURE E4.10
A sketch illustrating how to estimate the distance between civilizations in the Milky Way.

To evaluate d and thereby find how far away our neighbors are, we set R equal to the Milky Way's radius (taken to be 50,000 light-years) and $N = 10{,}000$, the value we deduced for the number of technical civilizations. These choices then give a value for d of 1000 light-years. Thus, even if such a large number of civilizations exist in our Galaxy, they are all extremely far from us. Such immense separations also make it very unlikely that we will be able to exchange information with other civilizations.

ARGUMENTS THAT WE ARE ALONE

Loners suggest that we are the only advanced life in the Milky Way. They point out that any of the probabilities going into the Drake equation could be much smaller than we have guessed. Perhaps so small that $N = 1$. Us.

For example, the British astrophysicist Fred Hoyle compared the likelihood of life arising spontaneously to the likelihood of a box containing several hundred tons of aluminum being shaken and by chance assembling itself into a 747 jumbo jet. While this comparison may seem silly, it illustrates the divergence of opinion on the likelihood of life forming. Perhaps Earth-like planets are far rarer than we have estimated. Perhaps evolution on other planets produces life forms that are so totally unlike those of Earth that we could not recognize them as being alive. And if life forms in oceans on worlds with an ice crust over the ocean, is there any reason to think that there is even a remote chance of a civilization arising? Furthermore, even if life forms similar to us do evolve elsewhere, the time period during which we could communicate with them might be so narrow that the opportunity passes. For example, if a civilization had existed on Mars even as recently as 100 years ago, we could not have communicated with them by radio, and if in the meantime they destroyed themselves somehow, the chance would have been lost forever.

The loners also base their argument on the rapidity with which our species has spread across Earth and the potential for its rapid spread across the Milky Way. This argument was first advanced by the Italian physicist Enrico Fermi, and is often called the **Fermi paradox.** Fermi argued that given the increasing rate of our technological advancement, it seems likely we will develop the ability for interstellar travel within a thousand years. And even if it takes us a thousand years to travel to a nearby star, hopping from star to star, it will take only a few million years to colonize our entire Galaxy. While this is long in human terms, it is extremely short compared to the long history of our Galaxy. The paradox Fermi pointed out is that if other civilizations exist, they should have already arrived at Earth long ago, and evidence of them would be plentiful. Thus, loners argue, we are probably alone in the Milky Way.

However, this argument, like that of the many-worlders for a multiplicity of civilizations, rests on several unproved assumptions. For example, it assumes that (1) most civilizations are driven to colonize, (2) they seek rather than avoid contact, and (3) they have mastered interstellar flight and communication.

FIGURE E4.11
When completed, the Allen Telescope Array, being constructed in California, will be able to simultaneously carry out astronomical surveys while it conducts a deep, targeted search for signals from other civilizations. The array will be made up of hundreds of small radio telescopes, using interferometry to simultaneously examine large areas of the sky over a wide part of the radio spectrum.

RADIO SEARCHES

Given the failure to find life near to us, and the difficulty of directly sampling distant locales, a few astronomers have searched for radio signals from other civilizations. Such signals might be either broadcast to us deliberately or communications directed elsewhere that we might simply "overhear." Astronomers began "listening" for extraterrestrial radio signals in the 1960s with Project Ozma, which monitored radio emission from several nearby star systems, hoping to detect emission from other than natural processes. More-recent projects, called **SETI** (*S*earch for *E*xtra-*T*errestrial *I*ntelligence), use receivers that automatically scan billions of radio wavelengths to search for signals from other civilizations (fig. E4.11).*

Searching for signals is not easy, because even if they exist, they are probably very weak and buried in cosmic static. Thus, astronomers use considerable care when choosing the wavelengths to monitor. For example, at very long wavelengths, interstellar gas in the Galaxy emits strongly and will overwhelm all but the most powerful signals. On the other hand, at very short wavelengths, molecules in our atmosphere block the signals. It turns out that the optimum wavelengths are near 21 centimeters, the wavelength emitted by neutral hydrogen in the cold gas clouds of our Galaxy. This is also near the wavelength of OH (hydroxyl) molecules, which emit strongly at 18 centimeters. These two wavelengths are very distinctive, and any other intelligent civilization would probably have discovered their significance and have equipment that could receive or broadcast them. Thus, several searches have been made for signals in this wavelength region, which is sometimes called the "waterhole" because it is bracketed on one side by emission from H and on the other by emission from OH. However, Ozma and all subsequent searches have proved fruitless so far.

One possible drawback to such searches is that other civilizations might not be transmitting signals. Like us, they may be listening. However, if other civilizations are like ours, they inadvertently broadcast a wide range of signals. For example, television stations and defense radar transmit extremely powerful signals that would be detectable many light-years from Earth. Moreover, on at least one occasion, a ceremony in 1974 celebrating the renovation of the huge Arecibo radio telescope in Puerto Rico, astronomers deliberately broadcast a signal toward the globular cluster M13. This cluster is more than 20,000 light-years away from us, so the signal has only begun its journey.

LIFE AND THE TRANSFORMATION OF PLANETS

The many difficulties associated with radio searches and the limited reach of our space probes have led scientists to consider other ways to search for signs of extraterrestrial life. One interesting method involves looking for unexpected gases in the atmosphere of a planet, gases that might be produced by living things.

In the late 1960s, James Lovelock, a British chemist, noted that several of the gases making up the Earth's atmosphere would disappear if Earth had no living beings to replenish them. For example, oxygen is very reactive and readily combines with surface rock. Lovelock went on to argue that the presence of such gases in a planet's atmosphere is an indication of life there. Thus, rather than searching for signals or sending robot spacecraft to explore a planet's surface, astronomers might find evidence of life by analyzing the light from a planet's atmosphere.

Astronomers have succeeded in detecting gases in the atmospheres of some exoplanets that pass in front of their stars, so that the gases in the atmosphere produce absorption lines in the star's spectrum. Indeed, oxygen was found in one gas giant, although it appears to be from the breakdown of water vapor as gas boils off this hot planet. Astronomers have also spectroscopically detected methane from Mars—this is another gas that might be associated with life, but it might also be produced by volcanic activity in Mars's interior.

While these results are inconclusive so far, these ideas have led to the realization that life may transform a planet in even more profound ways. In 1974, Lovelock and the U.S. microbiologist Lynn Margulis suggested that life creates a single "larger entity" with a planet, a "symbiosis" of life and planet, which they called the **Gaia hypothesis.** They chose this name from the Greek goddess of the Earth, Gaia (pronounced *Guy-uh* in this context but *Jee-uh* elsewhere).

*Such searches for life elsewhere require huge amounts of computer time. The data in one project is e-mailed from the central search telescope to a network of millions of volunteers who allow their home computers to process the information when they are not using their machines.

According to the Gaia hypothesis, life does not merely respond to its environment but actually alters its planet's atmosphere and temperature to make it more hospitable. As we discussed in chapter 6, plant life has almost certainly done just that on Earth. For example, by photosynthesis, plants have created an oxygen-rich atmosphere on Earth that shields them from dangerous ultraviolet radiation. Similarly, plants may alter a planet's temperature through the greenhouse effect. If the planet is too cold, plant metabolism is slowed, and less CO_2 is converted to oxygen. The greater CO_2 abundance warms the planet and enhances plant growth. Conversely, if the planet gets too warm, plant growth, according to the hypothesis, runs amok and produces lots of oxygen, reducing the CO_2 abundance and therefore also reducing the greenhouse warming.

Lovelock and Margulis have also proposed that many other facets of the environment, such as humidity, salinity of the oceans, sea level, and even plate tectonics, may be controlled by living organisms so as to optimize their reproduction. Such an intimate linkage between the living and nonliving is a very appealing idea. It almost makes the planet "alive." But is the Gaia hypothesis correct? Some aspects certainly are, but many scientists are skeptical of its more extreme claims (for example, that life can control plate tectonics). Nevertheless, living beings do exert a remarkable control over their local environment, and many unsuspected links between life and the physical world surely remain to be found. One such possible connection has its roots in cosmology, a connection called the "anthropic principle."

THE ANTHROPIC PRINCIPLE

In 1961, Robert Dicke, a physicist at Princeton University, wrote a paper concerning some remarkable cosmological coincidences. Dicke noted, for example, that the age of the Universe is very close to the lifetime of stars like the Sun and went on to argue that this "coincidence" is in fact not remarkable at all. Rather, it is a *necessity* if cosmologists are to exist to note it. Dicke pointed out that life requires elements such as carbon, silicon, and iron, which are made in massive stars. Thus, for life to form, enough time must pass for massive stars to evolve and make the heavy elements and then eject them into space with a supernova explosion. Moreover, additional time is needed for the ejected material to be incorporated back into interstellar clouds and form new stars. Only with this second generation of stars does life become likely. Dicke therefore concluded that intelligent life is impossible in a Universe only a few thousand years old. He likewise argued that life requires stars, and so it can exist only in a Universe young enough that stars have not all burnt out. Thus, the very existence of an intelligent observer in the Universe requires that the age of the Universe fall within certain limits.

In 1974 the English physicist Brandon Carter took Dicke's idea further and proposed what he called the **anthropic principle.** According to the anthropic principle, "what we can expect to observe must be restricted by the conditions necessary for our presence as observers." Since Carter's work, many scientists have shown how "fine-tuned" the Universe must be for life to exist. Thus, no matter how unlikely some aspects of the Universe may appear to us, if those aspects are necessary for life, then that is what we will observe. For example, we might argue how truly marvelous it is that conditions on Earth are just right for life. According to the anthropic principle, the rude reply is, "Of course conditions are just right for life. If they were not, there would be no life here to do the marveling."

SUMMARY

Life has probably existed on Earth for at least 3.8 billion years, a large portion of our planet's existence. That long duration and the underlying similarities of all life forms in terms of their structure, energy supply, and methods of reproduction indicate that life had a common origin relatively soon after Earth's formation. Experiments such as Miller and Urey's show that the precursors to life can be easily created under natural conditions. Thus, many astronomers believe that life has formed in numerous other places in the Universe. All searches for life, however, have found nothing, but given the vast distances between stars, that negative result is not surprising.

Once life develops on a planet, it may subtly alter its environment to its own ends, a proposition known as the "Gaia hypothesis." Such alterations may be detectable more readily than signals from other civilizations and may therefore be signatures of extraterrestrial life.

QUESTIONS FOR REVIEW

1. What is the earliest evidence for life on Earth?
2. How old are the earliest life forms thought to be?
3. What features of life are suggestive of a common origin?
4. How might the building blocks of life have come to exist on the early Earth?
5. What is meant by panspermia?
6. What is the Drake equation?
7. What are the main arguments for whether life is common or uncommon in our Galaxy?
8. What is the Fermi paradox?
9. Why is 21 centimeters believed to be a good wavelength region to search for signals from other civilizations?
10. What is meant by the Gaia hypothesis?
11. What is meant by the anthropic principle?

THOUGHT QUESTIONS

1. Would you be upset if it turned out that we were the only intelligent life in the Universe? Why or why not?
2. If a long time ago the Sun formed along with many other stars from a single large collapsing cloud, how might that

affect the likelihood of finding life on planets around nearby stars?
3. Do you think the "harder" step in the development of intelligent life is the initial development of life itself (self-replicating, very simple creatures) or the development of an intelligent civilization? Try to justify your argument with an appeal to—or argument about—the time these steps took on Earth.
4. During the Apollo program, returning astronauts were briefly quarantined to ensure that any germs or other hazards they might have carried back from the Moon were not let loose on Earth. The danger may in fact be the opposite situation. How important do you think it is to ensure that spacecraft visiting Mars and the icy moons of Jupiter and Saturn have been sterilized? How might such planetary protection measures affect the search for primitive life on other planets in our Solar System?
5. Suppose astronomers did detect signals from another civilization. What effect do you think it would have on our society? What effect would it have on you?
6. Do you think it is a good idea to send out signals in the hope that aliens might detect them? Suppose, for example, they were more technologically advanced than us and could do to us what we have done to less developed societies on Earth.
7. Based on what you know about the composition of the Milky Way, are there are any places in it you think we would be more or less likely to find a civilization? If you were sending messages, to which parts of the Galaxy would you try sending them?
8. What message would you send to an alien civilization?

PROBLEMS

1. Look up a couple of the nearest Sun-like stars in the Appendix Table 10. How long would it take a meteoroid to reach one of them traveling at 10 kilometers per second? If bacteria were harbored inside such meteorites and survived the trip, how long would it take before some meteorites traveled about 50,000 light-years—effectively covering the whole Galaxy?
2. Use a method similar to the Drake equation to estimate the number of people getting a haircut at some given moment. That is, multiply the probabilities of each such event times the number of people in, for example, the United States.
3. Using the equation discussed in the section, how many current civilizations would you need in the Milky Way to expect one to be within 30 light-years of our own (permitting one to ask one or two questions and get one or two answers over a lifetime of conversation by radio transmissions)?

TEST YOURSELF

1. Evidence that life on Earth is very ancient comes from
   (a) fossil algae in rocks more than 3 billion years old.
   (b) fossils buried thousands of feet below the surface.
   (c) fossil algae in rocks a few million years old.
   (d) the discovery of silicon-based life forms in ancient rocks.
   (e) the discovery of silicon-based life forms in a Miller-Urey type of experiment.
2. The Miller-Urey experiment demonstrated that
   (a) very simple bacteria can be created from the chemicals present in the atmosphere of the ancient Earth.
   (b) Earth's early atmosphere was too harsh for life to have formed until recently.
   (c) conditions on the early Earth were suitable for the creation of many of the complex organic molecules found in living things.
   (d) if simple life forms landed on the early Earth, they could have survived.
   (e) early life forms were silicon-based.
3. The Drake equation attempts to determine
   (a) the conditions under which life originated on Earth.
   (b) the optimum wavelength for communicating with extraterrestrials.
   (c) the age of life on Earth.
   (d) the number of other technically advanced civilizations.
   (e) the lifetime of our own civilization.
4. Those who argue we are the only civilization in the Milky Way might support their argument by claiming that (select all that apply)
   (a) other civilizations might keep to themselves out of fear of being colonized.
   (b) otherwise we should have been colonized by another more advanced civilization.
   (c) primitive life may have been delivered to Earth in meteorites.
   (d) otherwise we should be inundated by radio signals from other civilizations.
   (e) just because life started early on Earth doesn't mean it is easy to start life—we might be an exception.
5. A planet on which life forms create an atmosphere that makes the planet more hospitable for life is an example of
   (a) the Drake equation.
   (b) the Gaia hypothesis.
   (c) SETI.
   (d) the anthropic principle.
   (e) the Miller-Urey experiment.

6. According to the anthropic principle,
   (a) we are the only civilization in the Universe.
   (b) there are 10^{12} civilizations more advanced than us in the Milky Way.
   (c) human beings are the highest form of life.
   (d) the Universe could not differ radically from its present properties, or we would not be here to observe it.
   (e) the greenhouse effect is beneficial to us.

KEY TERMS

amino acids, 520
anthropic principle, 527
DNA, 520
Drake equation, 525
eukaryotes, 523
Fermi paradox, 525
Gaia hypothesis, 526
panspermia, 524
prokaryotes, 522
proteins, 520
SETI, 526

FURTHER EXPLORATIONS

Crowe, M. J. *The Extraterrestrial Life Debate: 1750–1900.* New York: Cambridge University Press, 1986.

Diamond, Jared. "Alone in a Crowded Universe." *Natural History* 99 (June 1990): 30.

Dick, S. J. *Plurality of Worlds.* New York: Cambridge University Press, 1982.

______. *The Biological Universe.* New York: Cambridge University Press. 1996.

Dorminey, Bruce. "Did Molecules from Space Seed Life in the Cosmos?" *Astronomy* 36 (April 2008): 50.

Drake, Frank, and Dava Sobel. *Is Anyone Out There? The Scientific Search for Extraterrestrial Intelligence.* New York: Delacorte Press, 1992.

Falk, Dan. "A Universe Fine-Tuned for Life? The Anthropic Principle's Surprising Resurgence." *Sky and Telescope* 107 (March 2004): 42.

Fox, Douglas. "Did Life Evolve in Ice?" *Discover* (February 2008): 52.

Gould, Stephen J. "The Evolution of Life on the Earth." *Scientific American* 271 (October 1994): 84.

______. "War of the Worldviews." *Natural History* 105 (December 1996): 22.

Impey, Chris. "How Life Could Thrive on Hostile Worlds." *Astronomy* 36 (December 2008): 54.

Knoll, A. H. "End of the Proterozoic Eon." *Scientific American* 265 (October 1991): 64.

LePage, Andrew J. "Habitable Moons." *Sky and Telescope* 96 (December 1998): 50.

LePage, Andrew J., and Alan M. MacRobert. "SETI Searches Today." *Sky and Telescope* 96 (December 1998): 44.

Lovelock, James E. *The Ages of Gaia: A Biography of Our Living Earth.* New York: Norton, 1988.

______. *Gaia, A New Look at Life on Earth.* New York: Oxford University Press, 1979.

Lovelock, James E., and Lynn Margulis. "The Gaia Hypothesis." *Tellus* 26 (1974): 1.

Nadis, Steve. "Does Life Really Need Water?" *Astronomy* 34 (November 2006): 38.

Orgel, Leslie E. "The Origin of Life on the Earth." *Scientific American* 271 (October 1994): 76.

Schilling, Govert. "The Chance of Finding Aliens: Reevaluating the Drake Equation." *Sky and Telescope* 96 (December 1998): 36.

Shklovski, I. S., and Carl Sagan. *Intelligent Life in the Universe.* New York: Dell Publishing Co., 1966.

Shostak, Seth. "Listening for Life." *Astronomy* 20 (October 1992): 26.

Soffen, Gerald A. "Life in the Solar System." In *The New Solar System,* edited by J. Kelly Beatty, Carolyn Collins Petersen, and Andrew Chaikin. 4th ed. Cambridge, England: Sky Publishing Co. and Cambridge University Press, 1999.

Ward, Peter D., and Donald Brownlee. *Rare Earth: Why Complex Life Is Uncommon in the Universe.* New York: Copernicus, 2000.

Weinberg, Steven. "Life in the Universe." *Scientific American* 271 (October 1994): 44.

Website

Visit the *Explorations* website at **http://www.mhhe.com/arny** for additional online resources on these topics.

ANSWERS TO TEST YOURSELF

Preview	1–c; 2–c; 3–b; 4–c; 5–e; 6–a,c,d,e
Chapter 1	1–d; 2–c; 3–c; 4–a,c,d; 5–d; 6–b; 7–c; 8–d; 9–c; 10–c; 11–b
Chapter 2	1–d; 2–b; 3–b; 4–d; 5–a; 6–e; 7–c
Essay 1 (Backyard)	1–c,d; 2–b; 3–a; 4–d; 5–b,d
Chapter 3	1–a,b,c,d; 2–d; 3–c; 4–e; 5–d; 6–a; 7–c; 8–b; 9–a
Chapter 4	1–e; 2–b; 3–c; 4–e; 5–b; 6–d; 7–c; 8–a
Essay 2 (Relativity)	1–e; 2–c; 3–b; 4–d,a,b,c
Chapter 5	1–d; 2–c; 3–a; 4–d; 5–b,e; 6–a; 7–d; 8–b
Chapter 6	1–e; 2–d; 3–c; 4–d; 5–a,d,e; 6–a; 7–c; 8–b; 9–a
Essay 3 (Keeping Time)	1–c; 2–e; 3–a; 4–e; 5–b
Chapter 7	1–d; 2–d; 3–a; 4–c; 5–d; 6–a
Chapter 8	1–c; 2–c; 3–e; 4–d; 5–a; 6–d
Chapter 9	1–d; 2–a; 3–c; 4–d; 5–b; 6–d; 7–b; 8–b
Chapter 10	1–c; 2–d; 3–a; 4–d; 5–e; 6–b; 7–a; 8–b,d
Chapter 11	1–d; 2–b; 3–a,b,c,d; 4–d; 5–e; 6–e; 7–d; 8–a; 9–c; 10–a,d,e
Chapter 12	1–d; 2–d; 3–d; 4–a; 5–b; 6–c; 7–a; 8–c
Chapter 13	1–d; 2–e; 3–a; 4–e; 5–d; 6–b; 7–b; 8–d; 9–c; 10–a
Chapter 14	1–e; 2–b; 3–e; 4–e; 5–c; 6–a,b,c,d; 7–c; 8–b; 9–b; 10–a,c,d,e; 11–a,d
Chapter 15	1–b,c,d,e; 2–b; 3–a; 4–d; 5–b; 6–a; 7–c; 8–e; 9–b; 10–c
Chapter 16	1–b; 2–a,c; 3–a; 4–b; 5–a,b,d; 6–b,c,e; 7–a; 8–c,d; 9–d
Chapter 17	1–d; 2–d; 3–c; 4–e; 5–d; 6–b; 7–b; 8–a,b,c; 9–a; 10–b
Chapter 18	1–b; 2–b; 3–a,d; 4–d; 5–e; 6–a,c,d,e; 7–c; 8–b; 9–c; 10–a,c
Essay 4 (Life in the Universe)	1–a; 2–c; 3–d; 4–b,d,e; 5–b; 6–d

APPENDIX

Scientific Notation

Scientific or powers-of-ten notation is a shorthand method for expressing and working with very large or very small numbers. With this method, we express numbers as a few digits times ten to a power or exponent. The power indicates the number of times that ten is multiplied by itself. For example, $100 = 10 \times 10 = 10^2$. Similarly $1{,}000{,}000 = 10 \times 10 \times 10 \times 10 \times 10 \times 10 = 10^6$. Note that we do not always need to write out $10 \times \ldots\ldots$ Instead, we can simply count the zeros. Thus, 10,000 is 1 followed by 4 zeros, so it is 10^4.

We can write 1 and 10 itself in powers-of-ten notation as well: $1 = 10^0$ and $10 = 10^1$.

To write a number like 300, we break it into two parts, $3 \times 100 = 3 \times 10^2$. Similarly, we can write $352 = 3.52 \times 100 = 3.52 \times 10^2$.

We can also write very small numbers (numbers less than 1) using powers of ten. For example, $0.01 = 1/100 = 1/10^2$. We can make this even more concise, however, by writing $1/10^2$ as 10^{-2}. Similarly, $0.0001 = 10^{-4}$. Note that for numbers less than 1, the power is 1 more than the number of zeros.

We can write a number like 0.00052 as $5.2 \times 0.0001 = 5.2 \times 10^{-4}$.

Suppose we want to multiply numbers expressed in powers of ten. The rule is simple: we add the powers. Thus, $10^3 \times 10^2 = 10^{3+2} = 10^5$. Similarly $2 \times 10^8 \times 3 \times 10^7 = 2 \times 3 \times 10^8 \times 10^7 = 6 \times 10^{15}$. In general, $10^a \times 10^b = 10^{a+b}$.

Division works similarly, except that we subtract the exponents. Thus, $10^5/10^3 = 10^{5-3} = 10^2$. In general, $10^a/10^b = 10^{a-b}$.

The last operations we need to consider are raising a power-of-ten number to a power and taking a root. In raising a number to a power, we multiply the powers. Thus, $(10^3)^4 = 10^{3\times4} = 10^{12}$. Care must be used if we have a number like $(4 \times 10^2)^3$. Both the 4 and the 10^2 are raised to a power, so the result is $4^3 \times (10^2)^3 = 64 \times 10^{2\times3} = 64 \times 10^6$.

Taking a root is equivalent to raising a number to a fractional power. Thus, the square root of a number is the number to the $\frac{1}{2}$ power. The cube root is the number to the $\frac{1}{3}$ power, and so forth. For example, $\sqrt{100} = (100)^{1/2} = (10^2)^{1/2} = 10^1 = 10$.

Some Useful Formulas

$$\text{Distance} = \text{velocity} \times \text{time } (d = Vt)$$

$$\text{Circumference of circle} = 2\pi R$$

$$\text{Area of a circle} = \pi R^2$$

$$\text{Surface area of a sphere} = 4\pi R^2$$

$$\text{Volume of a sphere} = \frac{4\pi R^3}{3}$$

Solving Distance, Velocity, Time (d, V, t) Problems

A large number of problems in this book (and in science in general) involve the motion of something. In such problems, we often know two of the three quantities (d, V, t) and want to know the third. For example, we have something moving at a speed V and want to know how far it will travel in a time t. Or we know that something travels with a speed V and want to find out how long it takes for the object to travel a distance d. We all can solve such problems in our heads if the motion involves automobiles. For example, if it is 150 miles to the city and we travel at 50 miles per hour, how long does it take to get there? Or how far can we drive in 2 hours if we are traveling at 45 miles per hour? Because we all do such problems routinely, you might find it easier to think of astronomical problems in terms of cars. Regardless of your approach, the method of solution is simple.

Begin by making a simple sketch of what is happening. Draw some arrows to indicate the motion. Label the quantities known and put question marks beside the things you want to find. Then write out the basic relation $d = Vt$. If you want to find d and know V and t, just multiply them for the answer. If you want to find the time and are given V and d, solve for t by dividing both sides by V to get $t = d/V$. If you want the velocity, divide d by t. In some problems, the motion may be in a circle of radius R. In that case, the distance traveled will be related to the circumference of the circle, $2\pi R$. For such cases, you may need to use the expression $d = 2\pi R$.

In most problems, you will find that it is extremely helpful to write the units of the quantities in the equation. For example, suppose you are asked how long it takes to travel 1500 km at a velocity of 30 kilometers per hour. First write out $d = Vt$. Then rewrite it as

$$d/V = t$$

Next insert the quantities so that

$$t = \frac{1500 \text{ km}}{30 \text{ km/hr}} = 50 \text{ hr}$$

Note that the units of km cancel out and leave us with units of hours, as the problem requires.

TABLE A.1 PHYSICAL AND ASTRONOMICAL CONSTANTS

Physical Constants	
Velocity of light (c)	$= 2.99792458 \times 10^8$ m/sec $\approx 3 \times 10^5$ km/sec
Gravitational constant (G)	$= 6.67259 \times 10^{-11}$ $m^3 \cdot kg^{-1} \cdot sec^{-2}$ (or newton$\cdot m^2 \cdot kg^{-2}$)
Planck's constant (h)	$= 6.62608 \times 10^{-34}$ joule·sec
Mass of hydrogen atom (M_H)	$= 1.6735 \times 10^{-27}$ kg
Mass of electron (M_e)	$= 9.1094 \times 10^{-31}$ kg
Stefan-Boltzmann constant (σ)	$= 5.6705 \times 10^{-8}$ $watts \cdot m^{-2} \cdot deg^{-4}$
Constant in Wien's law ($T \times \lambda_{max}$)	$= 2.8978 \times 10^6$ K·nm
Astronomical Constants	
Astronomical unit (AU)	$= 1.495978706 \times 10^{11}$ m $\approx 1.5 \times 10^8$ km
Light-year (ly)	$= 9.4605 \times 10^{15}$ m $= 9.4605 \times 10^{12}$ km $= 6.324 \times 10^4$ AU
Parsec (pc) = 3.26 ly	$= 3.085678 \times 10^{16}$ m $= 3.085678 \times 10^{13}$ km $=$ 206,265 AU
Sidereal year	$= 365.256366$ days $= 3.155815 \times 10^7$ sec
Mass of Earth	$= 5.974 \times 10^{24}$ kg
Mass of Sun ($M_\odot$)	$= 1.989 \times 10^{30}$ kg
Equatorial radius of Earth	$= 6378.0$ km
Radius of Sun ($R_\odot$)	$= 6.96 \times 10^8$ m $= 6.96 \times 10^5$ km
Luminosity of Sun ($L_\odot$)	$= 3.83 \times 10^{26}$ watts

TABLE A.2 CONVERSION BETWEEN AMERICAN AND METRIC UNITS*

Length			
1 km	= 1 kilometer	= 1000 meters	= 0.6214 mile
1 m	= 1 meter	= 1.094 yards	= 39.37 inches
1 cm	= 1 centimeter	= 0.01 meter	= 0.3937 inch
1 μm	= 1 micrometer	= 10^{-6} meter	= 10^{-4} cm = 3.93×10^{-5} inch
1 nm	= 1 nanometer	= 10^{-9} meter	= 10^{-7} cm
1 mile	= 1.6093 km		
1 inch	= 2.5400 cm		
Mass			
1 metric ton	= 10^6 grams	= 1000 kg	= 2.2046×10^3 lb
1 kg	= 1000 grams	= 2.2046 lb	
1 g	= 1 gram	= 0.0022046 lb	= 0.0353 oz
1 lb	= 0.4536 kg		
1 oz	= 28.3495 g		

* You will find it easier to use the metric system if you remember the following prefixes that denote very small to very large quantities:

nano	=	10^{-9}	=	1 billionth	kilo	=	10^3	=	1 thousand
micro	=	10^{-6}	=	1 millionth	mega	=	10^6	=	1 million
milli	=	10^{-3}	=	1 thousandth	giga	=	10^9	=	1 billion

TABLE A.3 PHYSICAL PROPERTIES OF THE PLANETS*

Name	Radius (Eq) (Earth units)	Radius (Eq) (km)	Mass (Earth units)	Mass (kg)	Average Density (g/cm^3)
Mercury	0.383	2440	0.055	3.30×10^{23}	5.43
Venus	0.949	6052	0.815	4.87×10^{24}	5.24
Earth	1.00	6378	1.00	5.97×10^{24}	5.52
Mars	0.532	3396	0.107	6.42×10^{23}	3.94
Jupiter	11.21	71,492	317.9	1.90×10^{27}	1.33
Saturn	9.45	60,268	95.16	5.68×10^{26}	0.69
Uranus	4.01	25,559	14.54	8.68×10^{25}	1.27
Neptune	3.88	24,764	17.15	1.02×10^{26}	1.64
Pluto	0.188	1160	0.0021	1.31×10^{22}	2.04

* Pluto, now officially a dwarf planet, is retained here for historical reasons.

TABLE A.4 ORBITAL PROPERTIES OF THE PLANETS*

Name	Distance from Sun† (AU)	Distance from Sun† (10^6 km)	Period Years	Period Days	Inclination of Orbit‡	Eccentricity of Orbit§
Mercury	0.387	57.9	0.2409	(87.97)	7.00	0.206
Venus	0.723	108.2	0.6152	(224.70)	3.39	0.007
Earth	1.00	149.6	1.0	(365.26)	0.00	0.017
Mars	1.524	227.9	1.8809	(686.98)	1.85	0.093
Jupiter	5.204	778.6	11.8622	(4332.59)	1.31	0.048
Saturn	9.582	1,433.5	29.4577	(10,759.22)	2.49	0.056
Uranus	19.20	2,872.5	84.011	(30,685.4)	0.77	0.046
Neptune	30.05	4,495.1	164.79	(60,189)	1.77	0.010
Pluto	39.48	5,906	247.68	(90,465)	17.15	0.248

* Pluto, now officially a dwarf planet, is retained here for historical reasons. †Semimajor axis of the orbit. ‡With respect to the ecliptic. §Eccentricity ranges from 0 for a circular orbit to a maximum of 1 for the most extremely elongated orbit.

TABLE A.5 SATELLITES OF THE SOLAR SYSTEM*

Planet Satellite	Radius† (km)	Distance from Planet (10^3 km)	Orbital Period (days)	Mass (10^{20} kg)	Density (g/cm^3)
Earth					
Moon	1738	384.4	27.322	734.9	3.34
Mars					
Phobos	$13 \times 11 \times 9$	9.38	0.319	1.3×10^{-4}	2.2
Deimos	$8 \times 6 \times 5$	23.5	1.263	1.8×10^{-5}	1.7
Jupiter‡					
Metis	20	127.96	0.295	9×10^{-4}	—
Adrastea	$13 \times 10 \times 8$	128.98	0.298	1×10^{-4}	—
Amalthea	$135 \times 82 \times 75$	181.3	0.498	8×10^{-2}	—
Thebe	$58 \times 49 \times 42$	221.9	0.6745	1.4×10^{-3}	—
Io	1821	421.6	1.769	893.3	3.57
Europa	1565	670.9	3.551	479.7	2.97
Ganymede	2634	1070.0	7.155	1482	1.94
Callisto	2403	1883.0	16.689	1076	1.86
Leda	~8	11,094	238.7	4×10^{-4}	—
Himalia	92.5	11,480	250.6	8×10^{-2}	—

Planet Satellite	Radius† (km)	Distance from Planet (10^3 km)	Orbital Period (days)	Mass (10^{20} kg)	Density (g/cm³)
Lysithea	~18	11,720	259.2	6×10^{-4}	—
Elara	~38	11,737	259.7	6×10^{-3}	—
Ananke	~15	20,200	631R	4×10^{-4}	—
Carme	~20	22,600	692R	9×10^{-4}	—
Pasiphae	~25	23,500	735R	1.6×10^{-3}	—
Sinope	~18	23,700	758R	6×10^{-4}	—
Saturn‡					
Pan	10	133.58	0.574	4.2×10^{-6}	—
Atlas	23 × 19 × 10	137.64	0.602	1.6×10^{-4}	—
Prometheus	74 × 50 × 34	139.35	0.613	1.4×10^{-3}	0.27
Pandora	55 × 44 × 31	141.7	0.629	1.3×10^{-3}	0.42
Epimetheus	69 × 53 × 53	151.42	0.694	5.6×10^{-3}	0.63
Janus	99 × 95 × 76	151.47	0.695	2.0×10^{-2}	0.65
Mimas	210 × 197 × 193	185.52	0.942	0.370	1.12
Enceladus	256 × 247 × 244	238.02	1.370	0.65	1.00
Tethys	536 × 528 × 526	294.66	1.888	6.17	0.98
Calypso	15 × 8 × 8	294.66	1.888	4×10^{-5}	—
Telesto	15 × 13 × 8	294.67	1.888	6×10^{-5}	—
Dione	559	377.4	2.737	10.8	1.49
Helene	18 × 16 × 15	377.4	2.737	1.6×10^{-4}	—
Rhea	764	527.04	4.518	23.1	1.24
Titan	2575	1221.85	15.945	1345.5	1.88
Hyperion	185 × 140 × 112	1481.1	21.277	0.28	—
Iapetus	720	3561.3	79.331	15.9	1.0
Phoebe	115 × 110 × 105	12,952	550.4R	0.1	—
Uranus‡					
Cordelia	13	49.75	0.336	1.7×10^{-4}	—
Ophelia	16	53.77	0.377	2.6×10^{-4}	—
Bianca	22	59.16	0.435	7×10^{-4}	—
Cressida	33	61.77	0.465	2.6×10^{-3}	—
Desdemona	29	62.65	0.476	1.7×10^{-3}	—
Juliet	42	64.63	0.494	4.3×10^{-3}	—
Portia	55	66.1	0.515	1×10^{-2}	—

(continued)

TABLE A.5 SATELLITES OF THE SOLAR SYSTEM* (*Continued*)

Planet Satellite	Radius† (km)	Distance from Planet (10^3 km)	Orbital Period (days)	Mass (10^{20} kg)	Density (g/cm^3)
Rosalind	29	69.93	0.560	1.5×10^{-3}	—
Belinda	33	75.25	0.624	2.5×10^{-3}	—
Puck	77	86.00	0.764	5×10^{-3}	—
Miranda	240 × 234 × 233	129.8	1.413	0.66	1.26
Ariel	581 × 578 × 578	191.2	2.520	13.5	1.65
Umbriel	584.7	266.0	4.144	11.7	1.44
Titania	788.9	435.8	8.706	35.2	1.59
Oberon	761.4	582.6	13.463	30.1	1.50
Caliban	~36	7231	579.7R	$\sim 2.5 \times 10^{-3}$	—
Sycorax	~75	12,179	1288.3R	$\sim 2.3 \times 10^{-2}$	—
Prospero	~25	16,256	1978.3R	$\sim 8.5 \times 10^{-4}$	—
Setebos	~24	17,418	2225.2R	$\sim 7.5 \times 10^{-4}$	—
Stephano	~10	20,901	2805.5R	$\sim 5.4 \times 10^{-5}$	—
Neptune‡					
Naiad	29	48.2	0.296	1.4×10^{-3}	—
Thalassa	40	50.0	0.312	4×10^{-3}	—
Despina	74	52.5	0.333	2.1×10^{-3}	—
Galatea	79	62.0	0.396	3.1×10^{-2}	—
Larissa	104 × ? × 89	73.6	0.554	6×10^{-2}	—
Proteus	218 × 208 × 201	117.6	1.121	0.6	—
Triton	1352.6	354.59	5.875R	214	2.0
Nereid	170	5588.6	360.125	0.31	—
Pluto (Dwarf Planet)					
Charon	603	17.53	6.38718	15.2	1.83
Nix	44	48.71	24.9	1×10^{-2}	—
Hydra	36	64.75	38	4×10^{-3}	—

*Note: Authorities differ substantially on many of the values above. "R" means orbit is retrograde.

†a × b × c values for the radius are the approximate lengths of the axes for irregular moons.

‡Astronomers have recently found many new moons orbiting each of the outer planets. All are small and many are as yet unnamed. Their orbital properties and sizes are uncertain. They may be captured asteroids. A relatively up-to-date list for each planet is available at the website solarsystem.nasa.gov/planets. Click on the planet of interest and then click on the tab labeled "moons."

TABLE A.6 PROPERTIES OF SOME OF THE SOLAR SYSTEM'S DWARF PLANETS

Name	Distance from Sun (semi-major axis of orbit in AU)	Orbital Period (years)	Inclination of Orbit to Ecliptic (degrees)	Eccentricity of Orbit	Diameter (km)	Mass (kg)
Ceres	2.77	4.599	10.6	0.08	975×909	9.5×10^{20}
Pluto	39.48	248.1	17.1	0.25	2300	1.3×10^{22}
Haumea	43.3	285	28.2	0.19	1200	4.2×10^{21}
Makemake	45.8	310	29.0	0.16	1600	4×10^{21}
Eris	67.7	557	44.2	0.44	2400	1.7×10^{22}
Sedna*	532	12,260	11.9	0.86	~1500	$\sim 4 \times 10^{21}$
Orcus*	39.42	247.5	20.6	0.23	~1300	7×10^{20}
Quaoar*	43.4	286	8.0	0.03	~1300	$\sim 2 \times 10^{21}$

*These objects are not yet officially designated as dwarf planets, but they (and at least a dozen others) appear to meet the criteria.

TABLE A.7 METEOR SHOWERS

Shower	Dates	Hourly Rate	Radiant R.A.*	Radiant Dec.*	Associated Comet
Quadrantids	Jan. 2–4	30	$15^h 24^m$	50°	2003 EH1
Lyrids	Apr. 20–22	8	$18^h 4^m$	33°	Thatcher
η Aquarids	May 2–7	10	$22^h 24^m$	0°	Halley
δ Aquarids	July 26–31	15	$22^h 36^m$	−10°	?
Perseids	Aug. 10–14	40	$3^h 4^m$	58°	Swift-Tuttle
Orionids	Oct. 18–23	15	$6^h 20^m$	15°	Halley
Taurids	Nov. 1–7	8	$3^h 40^m$	17°	Enke
Leonids	Nov. 14–19	6	$10^h 12^m$	22°	Tempel-Tuttle
Geminids	Dec. 10–13	50	$7^h 28^m$	32°	Phaethon

*R.A. = right ascension; Dec. = declination.

TABLE A.8 THE BRIGHTEST STARS*

Star	Name	Apparent Visual Magnitude	Spectral Type	Luminosity Class	Distance (ly)	Luminosity ($L_\odot$)
α CMa A	Sirius	−1.44	A1	V	8.6	26
α Car	Canopus	−0.74	F0	Ib/II	310	14,000
α Cen	Rigel Kentaurus	−0.01	G2	V	4.3	2.3
α Boo	Arcturus	−0.05	K2	III	37	220
α Lyr	Vega	0.03	A0	V	25	59
α Aur	Capella	0.08	G8	III	43	91
β Ori	Rigel	0.10	B8	Ia	860	50,000
α CMi	Procyon	0.40	F5	IV/V	11.5	7
α Ori	Betelgeuse	0.45	M2	Iab	500	80,000
α Eri	Achernar	0.45	B3	V	140	3000
β Cen AB	Hadar	0.61	B1	III	390	74,000
α Aql	Altair	0.77	A7	V	16.7	11
α Tau A	Aldebaran	0.87	K5	III	65	400
α Cru	Acrux	0.78	B0.5	IV	320	40,000
α Vir	Spica	0.98	B1	III-IV	250	15,000
α Sco A	Antares	1.02	M1	Ib	550	60,000
α PsA	Fomalhaut	1.17	A3	V	25	17
β Gem	Pollux	1.16	K0	III	34	46
α Cyg	Deneb	1.26	A2	Ia	1400	50,000
β Crux	Mimosa	1.25	B0.5	IV	350	30,000

*Luminosity and distance are uncertain for many of these stars, especially distant giants.

TABLE A.9 THE NEAREST STARS

Name	Distance (ly)	Spectral Type	Absolute Visual Magnitude	Visual Luminosity ($L_\odot$)	Apparent Visual Magnitude
Sun		G2	4.83	1.0	−26.8
Proxima Centauri	4.24	M5.5V	15.5	0.00005	11.09
α Cen A	4.36	G2V	4.4	1.6	0.01
B	4.35	K0V	5.7	0.5	1.34
Barnard's Star	5.96	M4V	13.2	0.0004	9.53
Wolf 359	7.78	M6V	16.6	0.00002	13.4
Lalande 21185	8.29	M2V	10.4	0.005	7.47
Sirius A	8.58	A1V	1.5	26.0	−1.43
B	8.58	DA2 (white dwarf)	11.3	0.003	8.44
Luyten 726-8 A	8.73	M5.5V	15.3	0.00006	12.41
B	8.73	M6V	16.1	0.00003	13.2
Ross 154	9.68	M3.5V	13.1	0.0005	10.43
Ross 248	10.3	M5.5V	14.8	0.0001	12.29
ϵ Eridani	10.5	K2V	6.2	0.3	3.73
Lacaille 9352	10.7	M1.5V	9.75	0.01	7.34
Ross 128 (FI Virginis)	10.9	M4V	13.5	0.0003	11.13
Luyten 789-6 ABC (EZ Aquarii)	11.3	M5V	14.7	0.0001	12.33
Procyon A	11.4	F5IV–V	2.7	7.0	0.38
B	11.4	DA (white dwarf)	13.0	0.0005	10.7
61 Cygni A	11.4	K5V	7.5	0.08	5.22
B	11.4	K7V	8.3	0.04	6.03
BD + 59° 1915 A (Struve 2398)	11.5	M3V	11.2	0.003	8.90
B	11.5	M3.5V	12.0	0.001	9.68
Groombridge 34 A (GX Andromedae)	11.6	M1.5V	10.4	0.006	8.08
B (GQ Andromedae)	11.6	M3.5V	13.4	0.0004	11.07
ϵ Indi	11.8	K5Ve	7.0	0.13	4.68
GJ 1111 (DX Cancri)	11.8	M6.5V	17.0	0.00001	14.79
τ Ceti	11.9	G8Vp	5.8	0.4	3.50
GJ 1061	12.0	M5.5V	15.2	0.00007	13.03

Data based mainly on RECONS (Research Consortium on Nearby Stars) website.

Note: Many of the stars in the above table are known by several names. In a few instances, such alternate names are in parentheses. BD stands for Bonner Durchmusterung catalog and GJ for Gliese-Jahreiss catalog; *p* indicates a peculiar spectrum; *e* indicates that the star shows emission lines in its spectrum.

TABLE A.10 PROPERTIES OF MAIN-SEQUENCE STARS*

Spectral Type	Luminosity ($L_\odot$)	Temperature (K)	Mass ($M_\odot$)	Radius ($R_\odot$)
O5	790,000	44,500	60	12.0
B0	52,000	30,000	17.5	7.4
B5	830	15,400	5.9	3.9
A0	54	9,520	2.9	2.4
A5	14	8,200	2.0	1.7
F0	6.5	7,200	1.6	1.5
F5	3.2	6,440	1.3	1.3
G0	1.5	6,030	1.05	1.1
G5	0.79	5,770	0.92	0.92
K0	0.42	5,250	0.79	0.85
K5	0.15	4,350	0.67	0.72
M0	0.08	3,850	0.51	0.6
M5	0.01	3,240	0.21	0.27
M8	0.001	2,640	0.06	0.1

*Authorities differ substantially on many of the above values, especially at the upper and lower mass values. Note also that the values are generally not consistent with the Stefan-Boltzmann law.

The data in the preceding tables come from many sources, including

Lang, Kenneth R. *Astrophysical Data: Planets and Stars.* New York: Springer-Verlag, 1992. NOTE: Be sure to get a sheet of corrected errors, however.

Landolt-Bornstein: *Zahlenwerte und Funktionen aus Naturwissenschaften und Technik/Numerical Data and Functional Relationships in Science and Technology.* Neue Serie/New Series. Group VII: Astronomy, Astrophysics, and Space Research, vols. 1–2, Astronomy and Astrophysics. Berlin, New York: Springer-Verlag, 1965–1982.

Donald K. Yeomans of JPL for satellite dimensions and masses.

TABLE A.11 KNOWN AND SUSPECTED MEMBERS OF THE LOCAL GROUP OF GALAXIES

Name of Galaxy	Right Ascension (hours and minutes)	Declination (degrees and minutes)	Galaxy Type	Distance (Mpc)	Diameter (kpc)	Visual Apparent Magnitude	Approximate Luminosity (millions of solar luminosities)
WLM	0 02	−15 28	Irr	1.0	3	11	50
IC10	0 20	+59 18	Irr	0.7	1	11	300
Cetus	0 26	−11 02	E	0.8	—	14	1
NGC 147	0 33	+48 30	E5	0.7	2	10	100
Andromeda III	0 35	+36 30	E5	0.8	1	15	1
NGC 185	0 39	+48 20	E3	0.7	2	9	150
NGC 205	0 40	+41 41	E5	0.8	5	8	300
Andromeda VIII	0 42	+40 37	E	0.8	14	9	150
M32	0 43	+40 52	E2	0.8	2	8	300
M31	0 43	+41 16	Sb	0.8	40	3	25,000
Andromeda I	0 46	+38 02	E3	0.8	0.5	14	4
SMC	0 53	−72 50	Irr	0.06	5	2	600
Sculptor	1 00	−33 42	E3	0.09	1	9	1
Pisces	1 04	+21 53	Irr	0.8	0.5	18	1
IC 1613	1 05	+02 07	Irr	0.7	3	9	100
Andromeda V	1 10	+47 38	E	0.8	—	15	1
Andromeda II	1 16	+33 25	E2	0.7	0.7	13	4
M33	1 34	+30 40	Sc	0.8	16	6	3,000
Phoenix	1 51	−44 27	Irr	0.4	0.6	12	1
Fornax	2 40	−34 27	E3	0.14	0.7	9	15
UGCA 92	4 32	+63 36	Irr	1.44	0.6	14	5
LMC	5 23	−69 45	Irr	0.05	0.6	1	2,000
Carina	6 42	−50 58	E4	0.10	0.7	16	1
Canis Major	7 35	−28 00	Irr	0.08	220	—	—
Leo T	9 35	+17 03	Irr	0.420	0.34	16	0.04
Leo A	9 59	+30 45	Irr	0.7	1	12	3

(continued)

TABLE A.11 KNOWN AND SUSPECTED MEMBERS OF THE LOCAL GROUP OF GALAXIES (*Continued*)

Name of Galaxy	Right Ascension (hours and minutes)	Declination (degrees and minutes)	Galaxy Type	Distance (Mpc)	Diameter (kpc)	Visual Apparent Magnitude	Approximate Luminosity (millions of solar luminosities)
Sextans B	10 00	+05 20	Irr	1.41	1.6	12	30
NGC 3109	10 03	−26 09	Irr	1.38	5	10	100
Antlia	10 04	−27 20	E3	1.41	0.6	15	2
Leo I	10 08	+12 18	E3	0.2	0.7	10	4
Sextans A	10 11	−04 43	Irr	1.60	1.8	12	35
Sextans	10 13	−01 37	E	0.09	1	12	1
Leo II	11 13	+22 09	E0	0.2	0.7	12	1
DDO 155	12 59	+14 13	Irr	2.4	0.4	14	8
Ursa Minor	15 09	+67 13	E5	0.06	0.5	11	1
Draco	17 20	+57 55	E3	0.08	0.8	10	1
Milky Way	17 46	−29 00	SBc	0.01	40	—	20,000
Sagittarius	18 55	−30 29	Irr	0.03	—	10	30
SagDIG	19 30	−17 41	Irr	1.2	1.5	15	5
NGC 6822	19 45	−14 48	Irr	0.5	2	9	200
Aquarius	20 47	−12 51	Irr	1.0	0.6	14	2
Tucanae	22 42	−64 25	E4	0.9	0.5	15	1
UGCA 438	23 26	−32 33	Irr	1.4	0.5	14	5
Cassiopeia	23 27	+50 42	E	0.7	0.5	13	5
Pegasus	23 29	+14 45	Irr	0.8	1	13	7
Pegasus II	23 52	+24 35	E	0.8	0.9	14	3

TABLE A.12 THE BRIGHTEST GALAXIES BEYOND THE LOCAL GROUP*

Name of Galaxy	Right Ascension (hours and minutes)	Declination (degrees and minutes)	Galaxy Type	Distance (Mpc)	Diameter (kpc)	Visual Apparent Magnitude	Approximate Luminosity (millions of solar luminosities)
NGC0253	00 48	−25 17	Sc	2.6	21	7.3	17,000
NGC0300	00 55	−37 41	Scd	2.2	13	8.2	2,700
NGC1068 (M77)	02 43	−00 01	Sb	13.8	28	8.7	60,000
NGC1291	03 17	−41 06	S0-a	9.2	27	8.7	25,000
IC 342	03 47	+68 06	SBc	2.5	15	8.5	51,000
NGC2403	07 37	+65 36	Sc	3.4	23	8.4	6,800
NGC3031 (M81)	09 56	+69 04	Sab	3.7	23	6.9	33,000
NGC3034 (M82)	09 56	+69 41	Irr	3.7	10	8.3	15,000
NGC4258 (M106)	12 19	+47 18	Sb	7.4	41	8.4	34,000
NGC4472 (M49)	12 30	+08 00	E	16.3	43	8.3	110,000
NGC4486 (M87)	12 31	+12 24	E	15.6	32	8.6	81,000
NGC4594 (M104)	12 40	−11 37	Sa	9.2	24	8.1	58,000
NGC4736 (M94)	12 51	+41 07	Sab	5.2	11	8.1	14,000
NGC4826 (M64)	12 57	+21 41	Sab	7.4	17	8.4	30,000
NGC4945	13 05	−49 28	SBc	4.6	29	8.6	37,000
NGC5055 (M63)	13 16	+42 02	Sbc	7.7	21	8.6	25,000
NGC5128 (Cen A)	13 25	−43 01	S0	3.7	32	6.7	32,000
NGC5194 (M51)	13 30	+47 12	Sbc	8.0	26	8.3	33,000
NGC5236 (M83)	13 37	−29 52	Sc	4.6	15	7.3	28,000
NGC5457 (M101)	14 03	+54 21	Sc	7.4	47	7.8	34,000
NGC6744	19 10	−63 51	Sb	10.7	65	8.4	62,000

*Galaxies in this table were selected from the HyperLeda database (http://leda.univ-lyon1.fr) based on their total visual apparent magnitude.

GLOSSARY

A

absorption the process in which light or other electromagnetic radiation gives up its energy to an atom or molecule. For example, ozone in our atmosphere absorbs ultraviolet radiation.

absorption-line spectrum a spectrum showing dark lines at some narrow color regions (wavelengths). The lines are formed by atoms absorbing light, which lifts their electrons to higher orbits.

acceleration a change in an object's velocity (either its speed or its direction).

accretion the addition of matter to a body. Examples are gas falling onto a star and asteroids colliding and sticking together.

accretion disk a nearly flat disk of gas or other material held in orbit around a body by its gravity.

achondrite a meteorite lacking chondrules, associated with larger bodies whose gravity and internal heating has caused them to differentiate.

active galaxy a galaxy whose central region emits abnormally large amounts of electromagnetic radiation from a small volume. Examples are radio galaxies, Seyfert galaxies, and quasars.

adaptive optics a technique for adjusting a telescope's mirror or other optical parts to compensate for atmospheric distortions, such as seeing, thereby giving a sharper image.

æther a substance at one time proposed to be the substance that made it possible for light waves to travel through otherwise empty space.

AGN active galactic nucleus. The core of an active galaxy.

alpha particle a helium nucleus: two protons plus (usually) two neutrons.

altitude an object's angular distance above the horizon.

amino acid a carbon-based molecule used by living organisms to build protein molecules.

angstrom unit a unit of length used in describing wavelengths of radiation and the sizes of atoms and molecules. One angstrom $\nabla 10^{-10}$ meters.

angular momentum a measure of an object's tendency to keep rotating and to maintain its orientation. Mathematically, it depends on the object's mass, M, radius, R, and rotational velocity, V, and is proportional to MVR.

angular size a measure of how large an object *looks* to you. It is defined as the angle between lines drawn from the observer to opposite sides of an object. For example, the angular diameter of the Moon is about $\frac{1}{2}^\circ$.

annular eclipse an eclipse in which the body in front does not completely cover the other. In an annular eclipse of the Sun, a bright ring of the Sun's disk remains visible around the black disk of the Moon. We therefore see a ring (annulus) of light around the Moon.

anthropic principle the principle that the properties we observe the Universe to possess are limited to those that make our existence possible.

antimatter a type of matter that, if brought into contact with ordinary matter, annihilates it, leaving nothing but energy. The positron is the antimatter analog of the electron. The antiproton is the antimatter analog of the proton. Antimatter is observed in cosmic rays and can be created from energy in the laboratory.

aphelion the point in an orbit where a body is farthest from the Sun.

association a loose grouping of young stars and interstellar matter.

asterism an easily identified grouping of stars, often part of a larger constellation. For example, the Big Dipper.

asteroid a small, generally rocky, solid body orbiting the Sun and ranging in diameter from a few meters to hundreds of kilometers.

asteroid belt a region between the orbits of Mars and Jupiter in which most of the Solar System's asteroids are located.

astronomical unit (AU) a distance unit based on the average distance of the Earth from the Sun.

atmospheric window a wavelength band in which our atmosphere absorbs little radiation. For example, on Earth the visible window ranges from about 300 to 700 nanometers, allowing the light we can see with our eyes to pass through the atmosphere.

atom a submicroscopic particle consisting of a nucleus and orbiting electrons. The smallest unit of a chemical element.

aurora the light emitted by atoms and molecules in the upper atmosphere. This light is a result of magnetic disturbances caused by the solar wind. Often called the northern or southern lights.

autumnal equinox the autumn equinox in the Northern Hemisphere. Fall begins on the autumnal equinox, which is on or near September 22.

averted vision looking slightly to one side of a dim object so that you see it slightly away from the center of your field of view. This allows you to see a faint object better, although at a sacrifice of sharpness.

azimuth a coordinate for locating objects on the sky. Azimuth is the angle measured eastward from due north to the point on the horizon below the object.

B

barred spiral galaxy a galaxy in which the spiral arms wind out from the ends of a central bar rather than from the nucleus.

Big Bang the event that, according to many astronomical theories, created the Universe. It occurred about 13.7 billion years ago and generated the expanding motion that we observe today.

binary stars two or more stars in orbit around each other, held together by their mutual gravity.

bipolar flow narrow columns of high-speed gas ejected by a protostar in two opposite directions.

blackbody an object that is an ideal radiator when hot and a perfect absorber when cool. It absorbs all radiation that falls upon it, reflecting no light; hence, it appears black. Stars are approximately blackbodies. The radiation emitted by blackbodies obeys Wien's law and the Stefan-Boltzmann law.

black dwarf a collapsed star that has cooled to the point where it emits little or no visible radiation.

black hole an object whose gravitational attraction is so strong that its escape velocity equals the speed of light, preventing light or any radiation or material body from leaving its "surface."

BL Lac object a type of active galaxy named for the peculiar galaxy BL Lac. These objects generally are strong radio sources, and their visible light varies rapidly and erratically.

blueshift a shift in the wavelength of electromagnetic radiation to a shorter wavelength. For visible light, this implies a shift toward the blue end of the spectrum. The shift can be caused by the motion of a source of radiation toward the observer or by the motion of an observer toward the source. For example, the spectrum lines of a star moving toward the Earth exhibit a blueshift. *See also* **Doppler shift.**

Bode's rule a numerical expression for the approximate distances of most of the planets from the Sun.

Bok globule small, dark, interstellar cloud, often approximately spherical. Many globules are the early stages of protostars.

brown dwarf a star that has a mass too low for it to begin nuclear fusion.

bulge the dense, central region of a spiral galaxy.

C

carbonaceous chondrite a type of meteorite containing many tiny spheres (chondrules) of rocky or metallic material stuck together by carbon-rich material.

CCD charge-coupled device. An electronic device that records the intensity of light falling on it. CCDs have replaced film in most astronomical applications.

celestial equator an imaginary line on the celestial sphere lying exactly above the Earth's equator. It divides the celestial sphere into northern and southern hemispheres.

celestial pole an imaginary point on the celestial sphere directly above the Earth's North or South Pole.

celestial sphere an imaginary sphere surrounding the Earth representing the sky. Ancient astronomers pictured celestial objects as attached to it.

Cepheid a class of yellow-giant pulsating stars. Their pulsation periods range from about 1 day to about 70 days. Cepheids can be used to determine distances. *See also* **standard candle.**

Chandrasekhar limit the maximum mass of a white dwarf above which it collapses. Approximately 1.4 solar masses. Named for the astronomer who first calculated that such a limit exists.

chondrite a meteorite containing small spherical grains called chondrules.

chondrule a small spherical grain embedded in a meteorite.

chromosphere the lower part of the Sun's outer atmosphere that lies directly above the Sun's visible surface (photosphere).

cluster a group of objects (stars, galaxies, and so forth) held together by their mutual gravitational attraction.

CNO cycle/process a reaction involving carbon, nitrogen, and oxygen (C, N, and O) that fuses hydrogen into helium and releases energy. The process begins with a hydrogen nucleus fusing with a carbon nucleus. Subsequent steps involve nitrogen and oxygen. The carbon, nitrogen, and oxygen act as catalysts and are released at the end of the process to start the cycle again. The CNO cycle is the dominant process for generating energy in main-sequence stars that are hotter and more massive than the Sun.

coma the gaseous atmosphere surrounding the head of a comet.

comet a small body in orbit around the Sun, consisting of a tiny, icy core and a tail of gas and dust. The tail forms only when the comet is near the Sun.

compact stars very dense stellar remnants or "dead" stars whose radii are much smaller than the Sun's. These stars include white dwarfs, neutron stars, and black holes.

condensation conversion of free gas atoms or molecules into a liquid or solid. A snowflake forms in our atmosphere when water vapor condenses into ice.

conjunction the appearance of two astronomical objects in approximately the same direction on the sky. For example, if Mars and Jupiter happen to appear near each other on the sky, they are said to be in conjunction. *Superior conjunction* refers to a planet that is approximately in line with the Sun but on the far side of the Sun from the Earth. *Inferior conjunction* refers to a planet that lies approximately between the Sun and the Earth.

conservation of angular momentum a principle of physics stating that the angular momentum of a rotating body remains constant unless forces act to speed it up or slow it down. Mathematically, conservation of angular momentum states that MVR is a constant, where M is the mass of a body moving with a velocity V in a circle of radius R. One extremely important consequence of this principle is that if a rotating body shrinks, its rotational velocity must increase.

conservation of energy a principle of physics stating that energy is never created or destroyed, although it may change its form. For example, energy of motion may change into energy of heat.

constellation a grouping of stars on the night sky. Astronomers divide the sky into 88 constellations.

continuous spectrum a spectrum with neither dark absorption nor bright emission lines. The intensity of the radiation in such a spectrum changes smoothly from one wavelength to the next.

convection the rising and sinking motions in a liquid or gas that carry heat upward through the material. Convection is easily seen in a pan of heated soup on a stove.

convection zone the region immediately below the Sun's visible surface in which its heat is carried by convection.

Coriolis effect a deflection of a moving object caused by its motion across the surface of a rotating body. The Coriolis effect makes storms on Earth spin, generates large-scale wind systems, and creates cloud belts on many of the planets.

corona the outer, hottest part of the Sun's atmosphere.

coronal hole a low-density region in the Sun's corona. The solar wind may originate in these regions.

cosmic horizon the maximum distance one can see out into the Universe at a given time. The horizon lies at a distance in light-years approximately equal to the age of the Universe in years.

cosmic microwave background (CMB) radiation from the young Universe that began traveling through space after hot gas from the Big Bang expanded and cooled enough to become transparent. The radiation is visible in all directions and appears to have a temperature of only 2.73 K because of the redshift caused by the expansion of the Universe.

cosmic rays extremely energetic particles (protons, electrons, and so forth) traveling at nearly the speed of light. Some rays are emitted by the Sun, but most come from more-distant sources, perhaps exploding supernovas.

cosmological constant a term in the equations that Einstein developed to describe the expansion of the Universe. The cosmological constant has the effect of a repulsive "force" opposing gravity. See also "dark energy."

cosmological principle the hypothesis that, on average, the Universe looks the same to every observer, no matter where he or she is located in it.

cosmology the study of the structure and evolution of the Universe.

crater a circular pit, generally with a raised rim and sometimes with a central peak. Crater diameters on the Moon range from centimeters to several hundred kilometers. Most craters on bodies such as the Moon are formed by the impact of solid bodies, such as asteroids.

critical density the minimum overall density of the Universe needed to provide enough gravity to make it eventually stop expanding and collapse. If the density is less than or equal to the critical density, the Universe will expand forever.

crust the rigid surface of a planet, moon, or other solid body.

curvature of space the bending of space by a mass, as described according to Einstein's general theory of relativity. Black holes bend the space around them, curving it so that the region within the black hole is cut off from the rest of the Universe. The Universe too may be curved in such a way as to make its volume finite.

curvature of the Universe the overall curvature of space in the Universe produced by a combination of all its matter and energy. The Universe appears to be very close to the critical density, at which there is no overall curvature, a condition astronomers describe as a "flat" Universe.

D

dark adaption the process by which the eye changes to become more sensitive to dim light.

dark energy a form of energy (or a cosmological constant) detected by its effect on the expansion of the Universe. It causes the expansion to speed up. The nature and properties of dark energy are unknown.

dark matter matter that emits no detectable radiation but whose presence can be deduced by its gravitational attraction on other bodies.

dark nebula a dense cloud of dust and gas in interstellar space that blocks the light from background stars.

daughter atoms the atoms produced by the decay of a radioactive element. For example, uranium decays into lead. These lead atoms are daughter atoms.

daylight saving time the time kept during summer months by setting the clock ahead one hour. This gives more hours of daylight after the workday.

declination one part of a coordinate system for locating objects in the sky north or south of the celestial equator. Declination is analogous to latitude on the Earth's surface.

degeneracy pressure the pressure created in a dense gas by the interaction of its electrons. Degeneracy pressure does not depend on temperature.

degenerate gas an extremely dense gas in which the electrons and nuclei are tightly packed. The pressure of a degenerate gas does not depend on its temperature.

density the mass of a body or region divided by its volume.

density-wave model a theory to account for the spiral arms of galaxies. According to the theory, waves of higher density traveling through the disk of a galaxy pull stars and interstellar gas into a spiral pattern.

deuterium a form of hydrogen in which the nucleus contains a neutron in addition to a proton. Sometimes called "heavy hydrogen."

differential gravitational force the difference between the gravitational forces exerted on an object at two different points. The effect of this force is to stretch the object. Such forces create tides and, if strong enough, may break up an astronomical object. *See also* **Roche limit.**

differentiation the separation of previously mixed materials inside a planet or other object. This is the same separation that occurs when a dense material, such as iron, settles to the planet's core while low-density material floats to the surface.

diffraction a bending of the path of light (or other electromagnetic waves) as it passes through an opening or around an obstacle. Diffraction limits the ability to distinguish fine details in images.

disk the flat, round portion of a galaxy. The Sun lies in the disk of the Milky Way.

dispersion the spreading of light or other electromagnetic radiation into a spectrum. A rainbow is an example of dispersion of light caused by raindrops.

DNA deoxyribonucleic acid. The complex molecule that encodes genetic information in all organisms here on Earth.

Doppler shift the change in the observed wavelength of radiation caused by the motion of the emitting body or the observer. The shift is an increase in the wavelength if the source and observer move apart and a decrease in the wavelength if the source and observer approach. *See also* **redshift** and **blueshift.**

Drake equation a formula, named after the astronomer who proposed it, used to estimate the number of civilizations that might be present in the Milky Way that are capable of communicating with us.

dwarf a small, dim star.

dwarf planet an object that orbits the Sun and is massive enough that its gravity compresses it into an approximately spherical shape. However, it has not swept its orbital region clear of other objects of comparable mass.

dynamo a physical process for generating magnetic fields in astronomical bodies. In many cases, the process involves the generation of electric currents from an interaction between rotation and convection.

E

eclipse the blockage of light from one astronomical body caused by the passage of another between it and the observer. The shadow of one astronomical body falling on another. For example, the passage of the Moon between the Earth and Sun can block the Sun's light and cause a solar eclipse.

eclipse seasons the times of year, separated by about 6 months, when eclipses are possible. At any given eclipse season, both a solar eclipse and lunar eclipse generally occur.

eclipsing binary a binary star pair in which one star periodically passes in front of the other, totally or partially blocking the background star from view as seen from Earth.

ecliptic the path that the Sun appears to make around the celestial sphere as the Earth moves along its orbit. The path gets its name because eclipses can occur only when the Moon crosses the ecliptic.

electric charge the electrical property of objects that causes them to attract or repel one another. A charge may be either positive or negative.

electric force the force generated by electric charges. It is attractive between unlike charges (≶) but repulsive between like charges (≶≶ or).

electromagnetic force the force arising between electrically charged particles or between charges and magnetic fields. Forces between magnets are a special case of this force. This force holds electrons to the nucleus of atoms, makes moving charges spiral around magnetic field lines, and deflects a compass needle.

electromagnetic radiation a general term for any kind of electromagnetic wave.

electromagnetic spectrum the assemblage of all wavelengths of electromagnetic radiation. The spectrum includes the following wavelengths, from long to short: radio, microwave, infrared, visible light, ultraviolet, X rays, and gamma rays.

electromagnetic wave a wave consisting of alternating electric and magnetic energy. Ordinary visible light is an electromagnetic wave, and the wavelength determines the light's color.

electron a low-mass, negatively charged subatomic particle. Electrons orbit the atomic nucleus but may at times be torn free. *See also* **ionization.**

element a fundamental substance, such as hydrogen, carbon, or oxygen, that cannot be broken down into a simpler chemical substance. Approximately 100 elements occur in nature.

ellipse a geometric figure related to a circle but flattened into an oval shape.

elliptical galaxy a galaxy in which the stars smoothly fill an ellipsoidal volume. Abbreviated E galaxy. The stars in such systems are generally old (Pop II).

emission the production of light, or more generally, electromagnetic radiation by an atom or other object.

emission-line spectrum a spectrum consisting of bright lines at certain wavelengths separated by dark regions in which there is no light.

emission nebula a hot gas cloud in interstellar space that emits light.

energy a measure of the ability of a system to do work or cause motion.

energy level any of the numerous orbitals that an electron can occupy in an atom or molecule, roughly corresponding to an electron orbit.

epicycle a fictitious, small, and circular orbit superimposed on another circular orbit and proposed by early astronomers to explain the retrograde motion of the planets.

equator the imaginary line that divides the Earth (or other body) symmetrically into its northern and southern hemispheres. The equator is perpendicular to a body's rotation axis.

equinox the time of year when the Sun appears to cross the celestial equator. At this time, the number of hours of daylight and night are approximately equal. The vernal and autumnal equinoxes mark the beginning of the spring and fall seasons.

escape velocity the speed an object needs to move away from another body in order not to be pulled back by its gravitational attraction. Mathematically, the escape velocity, V_{esc}, is defined as $\sqrt{2GM/R}$ where M is the body's mass, R is its radius, and G is the gravitational constant.

eukaryotes cells with nuclei. Most cells in current terrestrial organisms have nuclei and are thus eukaryotes.

Evening Star the planet Venus seen low in the western sky after sunset. (Sometimes used for other bright planets.)

event horizon the location of the "boundary" of a black hole. An outside observer cannot see in past the event horizon.

excited the condition in which the electrons of an atom are not in their lowest energy level (orbit).

exclusion principle the condition that no more than two electrons may occupy the same energy state in an atom. This limitation leads to degeneracy pressure.

exoplanet a planet not orbiting our Sun.

F

false-color picture/photograph a depiction of an astronomical object in which the colors are not the object's real colors. Instead, they are colors arbitrarily chosen to represent other properties of the body, such as the intensity of radiation, that we cannot see.

Fermi paradox an argument that we are probably the only technological civilization that has ever arisen in the Milky Way. Fermi suggested that if any such civilization had existed it would have covered the whole Galaxy in a time much shorter than the evolution of humans.

fission the splitting of an atomic nucleus into two or more smaller nuclei.

flare an outburst of energy on the Sun. *See also* **solar flare.**

flat universe a universe that extends forever with no curvature. Its total energy is zero.

fluorescence the conversion of ultraviolet light (or other short-wavelength radiation) into visible light.

focus (1) one of two points within an ellipse used to generate the elliptical shape. Planets orbit along ellipses with the Sun at one focus of the ellipse. (2) a point in an optical system in which light rays are brought together. The location where an image forms in such systems.

frequency the number of times per second that a wave vibrates.

fundamental forces the four basic forces of nature: gravitation, electromagnetism, the weak force, and the strong force. Electromagnetism and the weak force are now recognized to both be part of an "electroweak" force. According to some modern theories, all of the forces are different forms of a single, more fundamental, unified force.

G

Gaia hypothesis the hypothesis that life does not merely respond to its environment but actually alters its planet's atmosphere and temperature to make the planet more hospitable. For example, by photosynthesis, plants have created an oxygen-rich atmosphere on Earth, which shields the plants from dangerous ultraviolet radiation.

galactic cannibalism the capture and merging of one galaxy into another.

galaxy a massive system of stars held together by their mutual gravity. Typical galaxies have a mass between about 10^7 and 10^{13} solar masses. Our Galaxy is the Milky Way.

galaxy cluster a group of galaxies held together by their mutual gravitational attraction. The Milky Way belongs to the Local Group galaxy cluster.

Galilean relativity a method for determining the relative speeds of motion seen by observers moving with respect to each other. This method works satisfactorily for motions at low speed, but it fails when the speed becomes an appreciable fraction of the speed of light.

Galilean satellites the four moons of Jupiter discovered by Galileo: Io, Europa, Ganymede, and Callisto.

general relativity Einstein's theory of gravity. The theory describes how mass and energy "curve" space and time so that objects move along the paths described by the gravitational "force."

geocentric models models of the Solar System centered on the Earth. Many of the earliest attempts to describe the Solar System were geocentric in that they supposed that the planets moved around the Earth rather than around the Sun.

giant a star of large radius and large luminosity.

glitches abrupt changes in the pulsation period of a pulsar, perhaps as the result of adjustments of its crust.

global warming a phenomenon in which the Earth's surface temperature has been observed to increase significantly over the last century. Most scientists attribute the change to increasing levels of gases that cause the greenhouse effect, released by human activities such as burning fossil fuels and deforestation.

globular cluster a dense grouping of old stars, containing generally about 10^5 to 10^6 members. They are often found in the halos of galaxies.

globule *see* **Bok globule.**

granulation texture seen in the Sun's photosphere. Granulation is created by clumps of hot gas that rise to the Sun's surface.

grating a piece of material that creates a spectrum by reflecting light from, or passing it through, many very fine and closely spaced parallel lines.

gravitational lens an object that bends space (and thereby the light passing through the space) by its gravitational attraction and focuses the light to create an image of a more distant object. *See also* **curvature of space.**

gravitational lensing the bending of light from a distant object to form an image, usually strongly distorted by the gravitation of a mass between the distant object and the observer. See *gravitational lens.*

gravitational redshift the shift in wavelength of electromagnetic radiation (light) created by a body's gravitational field as the radiation moves away from the body. Only extremely dense objects, such as white dwarfs, produce a significant redshift of their radiation.

gravitational waves a wavelike bending of space generated by the acceleration of massive bodies.

gravity the force of attraction that is between two bodies and is generated by their masses.

greatest elongation the position of an inner planet (Mercury or Venus) when it lies farthest from the Sun on the sky. Mercury and Venus are particularly easy to see when they are at greatest elongation. Objects may be at greatest eastern or western elongation according to whether they lie east or west of the Sun.

greenhouse effect the trapping of heat by a planet's atmosphere, making the planet warmer than would otherwise be expected. Generally the greenhouse effect operates if visible sunlight passes freely through a planet's atmosphere but the infrared radiation produced by the warm surface cannot escape readily into space, because of gases such as carbon dioxide or water vapor.

Gregorian calendar the calendar devised at the request of Pope Gregory XIII and essentially the civil calendar used throughout the world today. It omits the leap year for century years not divisible evenly by 400.

H

HII region a region of ionized hydrogen. HII regions generally have a pink/red glow and often surround luminous, hot, young stars.

halo the approximately spherical region surrounding spiral galaxies that contains mainly old stars, such as the globular clusters. The halo also contains large amounts of dark matter.

Hawking radiation radiation that black holes are hypothesized to emit as a result of quantum effects. This radiation leads to the extremely slow evaporation of black holes.

heliocentric models models of the Solar System centered on the Sun. Compare to **geocentric models.**

helium flash the beginning of helium fusion in a low-mass star. The fusion begins explosively and causes a major readjustment of the star's structure.

highlands the old, heavily cratered regions on the Moon.

horizon the line separating the sky from the ground. *See also* **cosmic horizon.**

H-R diagram a graph on which stars are located according to their temperature and luminosity. Most stars on such a plot lie along a diagonal line, called the main sequence, which runs from cool, dim stars in the lower right, to hot, luminous stars in the upper left.

Hubble constant the multiplying constant H in the Hubble law, $V \nabla Hd$. The reciprocal of the Hubble constant (in appropriate units) is approximately the age of the Universe.

Hubble law a relation between a galaxy's distance, *d*, and its recession velocity, V, which states that more distant galaxies recede faster than nearby ones. Mathematically, $V \nabla Hd$, where H is the Hubble constant.

hydrogen burning nuclear fusion of hydrogen into helium. It is not "burning" like ordinary fire but is instead the transformation of one kind of atom into another accompanied by the release of energy.

hydrostatic equilibrium the condition in which pressure and gravitational forces in a star or planet are in balance. Without such balance, bodies will either collapse or expand.

hypothesis an explanation proposed to account for some set of observations or facts.

I

ideal gas law See **perfect gas law.**

inclination the tilt angle of an astronomical object's spin or its orbit.

inertia the tendency of an object at rest to remain at rest and of a body in motion to continue in motion in a straight line at a constant speed. *See also* **mass.**

inferior conjunction *see* **conjunction.**

inferior planet a planet whose orbit lies between the Earth's orbit and the Sun. Mercury and Venus are inferior planets.

inflation the rapid expansion of the early Universe by an enormous factor.

infrared a wavelength of electromagnetic radiation longer than visible light but shorter than radio waves. We cannot see these wavelengths with our eyes, but we can feel many of them as heat. The infrared wavelength region runs from about 700 nm to 1 mm.

inner core the innermost part of a planet, also called the *solid core.* The Earth's inner core is a mixture of solid iron and nickel.

inner planet a planet orbiting in the inner part of the Solar System. Sometimes taken to mean Mercury, Venus, Earth, and Mars.

instability strip a region in the H-R diagram in which stars pulsate.

interferometer a device consisting of two or more telescopes connected together to work as a single instrument. Used to obtain a high resolving power, the ability to see small-scale features. Interferometer telescopes have been contructed operating at radio, infrared, and visible wavelengths.

international date line an imaginary line from the Earth's North to South Pole, running approximately down the middle of the Pacific Ocean. It marks the location on Earth at which the date changes.

interstellar cloud a cloud of gas and dust in between the stars. Such clouds may be many light-years in diameter.

interstellar grains microscopic solid dust particles in interstellar space. These grains absorb starlight, making distant stars appear dimmer and redder than they truly are.

interstellar matter matter in the form of gas or dust in the space between stars.

inverse-square law (1) any law in which some property varies inversely as the square of the distance, *d*. Mathematically, as 1 d^2. (2) the law stating that the apparent brightness of a body decreases inversely as the square of its distance.

ionization the removal of one or more electrons from an atom, leaving the atom with a positive electric charge. Under some circumstances an extra electron may be attached to an atom, in which case the atom is described as negatively ionized.

ionized a condition in which the number of an atom's electrons does not equal the number of its protons. Typically, this means the atom is missing one or more electrons.

irregular galaxy a galaxy lacking a symmetric structure.

J

jets narrow streams of gas ejected from any of several types of astronomical objects. Jets are seen near protostars and in many active galaxies.

jet stream a narrow stream of high-speed wind that blows in the atmosphere of a planet. Such winds occur on Earth and many other planets.

joule a unit of energy. Expending one joule per second equals one watt of power.

Jovian planet one of the giant, gaseous planets: Jupiter, Saturn, Uranus, and Neptune, or most exoplanets. The name *Jovian* was chosen because the structure of Jupiter (or Jove) is representative of the others.

Julian calendar a 12-month calendar devised under the direction of Julius Caesar. It includes a leap year every four years.

K

Kepler's three laws laws that describe the motion of planets around the Sun. The first law states that planets move in elliptical orbits with the Sun off-center at a focus of the ellipse. The second law states that a line joining the planet and the Sun sweeps out equal areas in equal times. The third law relates a planet's orbital period, *P*, to the semimajor axis of its elliptical orbit, *a*. Mathematically, the law states that $P^2 \nabla a^3$, if *P* is measured in years and *a* in astronomical units.

Kirkwood gaps regions in the asteroid belt with fewer than the average number of asteroids. The gaps result from the gravitational force of Jupiter removing asteroids from orbits that have orbits with periods that are a simple fraction of Jupiter's period.

Kuiper belt a region from which some comets come. The region appears to extend from the orbit of Neptune, past Pluto, out to approximately 55 AU.

L

law of gravity a description of the gravitational force exerted by one body on another. The gravitational force is proportional to the product of the bodies' masses and the inverse square of their separation. If the masses are *M* and *m* and their separation is *d*, the force between them, *F*, is $F \nabla GMm\ d^2$, where G is a physical constant.

light electromagnetic energy.

light-gathering power a measure of the ability of a telescope (or other device) to collect light. It is generally proportional to the *area* of the telescope's mirror or lens. For example, a telescope with a 6-inch diameter lens has a four times larger light-gathering power than a telescope with a 3-inch diameter lens.

light-year a unit of distance equal to the distance light travels in one year. A light-year is roughly 10^{13} km, or about 6 trillion miles.

liquid core the molten interior of a planet, also called the *outer core.*

Local Group the small cluster or group of several dozen galaxies to which the Milky Way belongs.

Lorentz factor a term that designates how much an object's time, space, and mass are altered as a result of its motion. The factor is very close to 1 except at speeds approaching the speed of light.

luminosity the amount of energy radiated per second by a body. For example, the wattage of a lightbulb defines its luminosity. Stellar luminosity is usually measured in units of the Sun's luminosity (approximately $4 \quad 10^{26}$ watts).

lunar eclipse the passage of the Earth between the Sun and the Moon so that the Earth's shadow falls on the Moon.

M

Magellanic Clouds two small companion galaxies of the Milky Way.

magnetic field a representation of the means by which magnetic forces are transmitted from one body to another. A compass needle points along the direction of the Earth's magnetic field.

magnetic lines of force fictitious lines used to visualize the orientation and strength of a magnetic field.

magnitude a unit for measuring stellar brightness. The smaller the magnitude, the brighter the star.

main sequence the region in the H-R diagram in which most stars, including the Sun, are located. The main sequence runs diagonally across the H-R diagram from cool, dim stars to hot, luminous ones. Stars on the main sequence fuse hydrogen into helium in their cores. *See also* **H-R diagram.**

main-sequence lifetime the time a star remains a main-sequence star, fusing hydrogen into helium in its core.

mantle the solid, outer part of a planet. This part is immediately below the crust and may flow very slowly due to heat rising from the core.

mare a vast, smooth, dark, and congealed lava flow filling a basin on the Moon and on some planets. Many maria have roughly circular shapes.

maria plural of *mare.*

maser an intense radio source created when excited gas amplifies some background radiation. *Maser* stands for *m*icrowave *a*mplification by *s*timulated *e*mission of *r*adiation.

mass a measure of the amount of material an object contains. A quantity measuring a body's inertia.

mass–luminosity relation a relation between the mass and luminosity of main-sequence stars. Higher-mass stars have higher luminosity.

Maunder minimum the time period, from about A.D. 1600 to 1740, during which the Sun was relatively inactive. Few sunspots were observed during this period.

mean solar day the standard 24-hour day. The mean solar day is based on the average day length over a year. (The time interval from solar noon to solar noon varies slightly throughout the year.)

megaparsec a distance unit equal to 1 million parsecs and abbreviated Mpc.

metal astronomically, any chemical element more massive than helium. Thus, carbon, oxygen, iron, and so forth are called metals.

meteor the bright trail of light created by small solid particles entering the Earth's atmosphere and burning up. A "shooting star."

meteor shower an event in which many meteors occur in a short space of time, all from the same general direction in the sky. The most famous shower is the Perseids in mid-August.

meteorite the solid remains of a meteor that falls to the Earth.

meteoroid the technical name for the small, solid bodies moving within the Solar System.

When a meteoroid enters our atmosphere and heats up, the trail of luminous gas it leaves is called a meteor. When the body lands on the ground, it is called a meteorite. ("A meteoroid is in the void. A meteor above you soars. A meteorite is in your sight.")

method of standard candles *See* standard candle.

Milky Way Galaxy the Galaxy to which the Sun belongs. Seen from Earth, the Galaxy is a pale, milky-white band in the night sky.

Miller-Urey experiment an experimental attempt to simulate the conditions under which life might have developed on Earth. Miller and Urey discovered that amino acids and other complex organic compounds could form from the gases that are thought to have been present in the Earth's early atmosphere, if the gases are subjected to an electric spark or ultraviolet radiation.

millisecond pulsar a pulsar whose rotation period is about a millisecond.

minute of arc a measure of angle equal to one-sixtieth of a degree.

model a theoretical representation of some object or system.

molecule two or more atoms bonded into a single particle, such as water, H_2O (two hydrogen atoms bonded to one oxygen) or carbon dioxide, CO_2 (one carbon atom bonded to two oxygen atoms).

Moon illusion the illusion that the Moon appears larger when near the horizon than when seen high in the sky.

Morning Star the planet Venus seen in the eastern sky before dawn. (Sometimes applied to other bright planets.)

N

nanometer a unit of length equal to 1 billionth of a meter (10^{-9} meters) and abbreviated nm. Wavelengths of visible light are several hundred nanometers. The diameter of a hydrogen atom is roughly 0.1 nm.

neap tide the abnormally small tide occurring when the Sun's and Moon's gravitational effects on the ocean partially offset each other.

near-Earth object an asteroid with an orbit that crosses Earth's orbit or comes close to it.

nebula a cloud in interstellar space.

negative curvature a form of curved space sometimes described as being "open" in that it has no boundary. Negative curvature is analogous to a saddle shape.

neutrinos tiny neutral particles with little or no mass and immense penetrating power. These particles are produced in great numbers by the Sun and other stars as they fuse hydrogen into helium, and also by supernova explosions.

neutron a subatomic particle of nearly the same mass as the proton but with no electric charge. Neutrons and protons compose the nuclei of atoms.

neutron star a very dense, compact star composed primarily of neutrons.

Newton's first law of motion the law that a body continues in a state of rest or uniform motion in a straight line unless made to change that state by forces acting on it. *See also* **inertia.**

Newton's second law of motion $F \nabla ma$. In words, the amount of acceleration, a, that a force, F, produces depends on the mass, m, of the object being accelerated.

Newton's third law of motion the law that when two bodies interact, they exert equal and opposite forces on each other.

nonthermal radiation radiation emitted by charged particles moving at high speed in a magnetic field. The radio emission from pulsars and radio galaxies is nonthermal emission. More generally, nonthermal means "not due to high temperature."

north celestial pole the point on the celestial sphere directly above the Earth's North Pole. Objects on the sky appear to circle around this point.

North Star any star that happens to lie very close to the north celestial pole. Polaris has been the North Star for about 1000 years, and it will continue as such for about another 1000 years, at which time a star in Cepheus will be nearer the north celestial pole.

nova a process in which a surface layer of hydrogen builds up on a white dwarf and then fuses explosively into helium making the white dwarf visible for a few weeks. Nova explosions do not destroy the white dwarf and may be recurrent.

nuclear fusion the binding of two light nuclei to form a heavier nucleus, with some nuclear mass converted to energy. For example, the fusion of hydrogen into helium. This process supplies the energy of most stars and is commonly called "burning" by astronomers.

nucleosynthesis the formation of elements, generally by the fusion of lighter elements into heavier ones. For example, the formation of carbon by the fusion of three helium nuclei.

nucleus the core of an atom around which the electrons orbit. The nucleus has a positive electric charge and constitutes most of an atom's mass.

nucleus of a comet the core, typically a few kilometers across, of frozen gases and dust that make up the solid part of a comet.

nucleus of a galaxy the central region of a galaxy, often containing a supermassive black hole.

O

Olbers' paradox an argument that in an infinite Universe, the sky should be bright at night, filled with the light from many distant stars and galaxies.

Oort cloud a vast region in which comet nuclei orbit. This cloud lies far beyond the orbit of Neptune.

opacity the blockage of light or other electromagnetic radiation by matter.

open cluster a loose cluster of stars, generally containing a few hundred members.

opposition the configuration of a planet when it is opposite the Sun in the sky. If a planet is in opposition, it rises when the Sun sets and sets when the Sun rises.

orbit the path in space followed by a celestial body.

orbital a description of an electron's possible location in an atom as it "orbits" the nucleus. At these tiny scales, the wave nature of matter only allows a description of position in terms of probabilities.

outer core the molten interior of a planet; also called the *liquid core.*

outer planet a planet whose orbit lies in the outer part of the Solar System. Jupiter, Saturn, Uranus, and Neptune are outer planets.

ozone a form of oxygen consisting of three oxygen atoms bonded together. Its chemical symbol is O_3. The ozone in our atmosphere absorbs, and thus shields us from, the Sun's harmful ultraviolet radiation.

P

panspermia a theory that life originated elsewhere than on Earth and came here across interstellar space either accidently or deliberately.

parallax the shift in an object's position caused by the observer's motion. A method for finding distances based on that shift.

parsec a unit of distance equal to about 3.26 light-years (3.09 10^{13} km), defined as the distance at which an observer sees the maximum angle between the Sun and the Earth to be one arc second.

perfect gas law a law relating the pressure, density, and temperature of a gas. It states that the pressure is proportional to the density times the temperature. This is also called the *ideal gas law.*

perihelion the point in an orbit closest to the Sun.

period the time required for a repetitive process to repeat. For example, orbital period is the time it takes a planet or star to complete an orbit. Pulsation period is the time it takes a star to expand and then contract back to its original radius.

period–luminosity law a law stating that the longer the period of a pulsating variable star, the more luminous it is.

phases the changing illumination of the Moon or other body that causes its apparent shape to change. The following is the cycle of lunar phases: new, crescent, first quarter, gibbous, full, gibbous, third quarter, crescent, new.

photo dissociation the breaking apart of a molecule by intense radiation.

photon a particle of visible light or other electromagnetic radiation.

photosphere the visible surface of the Sun. When we look at the Sun in the sky, we are seeing its photosphere.

planet a body in orbit around a star that is large enough to have taken on a round shape, and which has cleared the path of its orbit of all bodies of comparable size.

planetary nebula a shell of gas ejected by a low-mass star late in its evolutionary lifetime. Planetary nebulas typically appear as a glowing gas ring around a central star.

planetesimal one of the numerous small, solid bodies that, when accreted together, form a planet.

plate tectonics the idea that the crust of the Earth (or some other planet) is divided into large regions (plates) that move very slowly over the planet's surface. Interaction between plates at their boundaries creates mountains and activity such as earthquakes.

polarity the property of a magnet that causes it to have a north and south pole.

population (Pop) I the younger stars, some of which are blue, that populate a galaxy's disk, especially its spiral arms.

population (Pop) II the older, redder stars that populate a galaxy's halo and bulge.

population (Pop) III a hypothetical stellar population consisting of the first stars that formed in a galaxy, composed of only hydrogen and helium.

positive curvature bending of space leading to a finite volume. A space that is "closed." A universe with positive curvature is analogous to a spherical shape.

positron a subatomic antimatter particle with the same mass as the electron but a positive electric charge. An electron's antiparticle.

precession the slow change in direction of the pole (rotation axis) of a spinning body or of the orientation of an orbit.

pressure the force exerted by a substance such as a gas on an area divided by that area. That is, pressure area ∇ force.

principle of equivalence an idea developed by Einstein to understand that gravity is equivalent to an accelerating frame of reference.

prokaryotes cells without nuclei. The first life forms on Earth were prokaryotes.

prominence a cloud of hot gas in the Sun's outer atmosphere. This cloud is often shaped like an arch, supported by the Sun's magnetic field.

protein any of many complex organic molecules composed of a chain of amino acids. Proteins serve many functions in cells, including structure and metabolism.

proton a positively charged subatomic particle. One of the constituents of the nucleus of an atom along with neutrons.

proton–proton chain the nuclear fusion process that converts hydrogen into helium in stars like the Sun and thereby generates their energy. This is the dominant energy-generation mechanism in cool, low-mass stars.

protostar a star still in its formation stage, heated by gravitational contraction.

pulsar a spinning neutron star that emits beams of radiation that happen to sweep across the Earth each time the star spins. We observe the radiation as regularly spaced pulses.

pulsate to expand and contract regularly. For example, pulsating variable stars swell and shrink in a predictable, regular fashion.

Q

quantized the property of a system that allows it to have only discrete values.

quasar a peculiar galaxy characterized by a large redshift, high luminosity, and an extremely small, active core. Quasars are among the most luminous and most distant objects known to astronomers.

R

radial velocity the velocity of a body along the line of sight. That is, the part of its motion directly toward or away from the observer.

radiant the point in the sky from which meteors in showers appear to come. *See also* **meteor showers.**

radiation pressure the force exerted by radiation on matter.

radiative zone the region inside a star where its energy is carried outward by radiation (that is, photons).

radioactive decay the breakdown of an atomic nucleus by the emission of subatomic particles.

radioactive element an element that undergoes radioactive decay and breaks down into a lighter element.

radio galaxy a galaxy, usually an elliptical, that emits abnormally large amounts of radio energy from huge lobes of hot gas ejected from the galaxy.

rays long, narrow, light-colored markings on the Moon or other bodies that radiate from young craters. Rays are debris "splashed" out of the crater by the impact that formed it.

recession velocity the velocity of an external galaxy (or other object) away from our Galaxy due to the expansion of the Universe.

recombination era a period of time about 400,000 years after the Big Bang when matter cooled enough for electrons to combine with nuclei to form neutral atoms.

reddening the alteration in a star's color as seen from Earth as the star's light passes through an intervening interstellar dust cloud. The dust preferentially scatters the blue light from the beam, leaving the remaining light redder.

red giant a cool, luminous star whose radius is much larger than the Sun's.

redshift a shift in the wavelength of electromagnetic radiation to a longer wavelength. For visible light, this implies a shift toward the red end of the spectrum. The shift can be caused by a source of radiation moving away from the observer or by the observer moving away from the source. For example, if a star is moving away from Earth, its spectrum lines exhibit a redshift. *See also* **Doppler shift.**

reflection nebula an interstellar cloud in which the dust particles reflect starlight, making the cloud visible.

reflector a telescope that uses a mirror to collect and focus light.

refraction the bending of light when it passes through one substance and enters another.

refractor a telescope that uses a lens to collect and focus light.

regolith the surface rubble of broken rock on the Moon or other solid body.

resolving power the ability of a telescope or instrument to discern fine details. Larger-diameter telescopes have greater (that is, better) resolving power.

resonance a condition in which the repetitive motion of one body interacts with the repetitive motion of another so as to reinforce the motion. Sliding back and forth in a bathtub to make a big splash is an example.

rest frame a system of coordinates that appear to be at rest with respect to the observer.

retrograde motion the drift of a planet westward against the background stars. Normally planets shift eastward because of their orbital motion. The planet does not actually reverse its motion. The change in its direction is caused by the change in the position from which we view the planet as the Earth overtakes and passes it.

rifting the breaking apart of a continental plate.

right ascension a coordinate for locating objects on the sky, analogous to longitude on the Earth's surface. Measured in hours and minutes of time.

rilles narrow canyons on the Moon or other body.

Roche limit the distance from an astronomical body at which its gravitational force can pull apart another astronomical body.

rotation axis an imaginary line through the center of a body about which the body spins.

rotation curve a plot of the rotation velocity of the stars or gas in a galaxy at different distances from its center.

RR Lyrae stars a type of white, giant, pulsating variable stars with a period of about one day or less. They are named for their prototype star, RR Lyrae.

runaway greenhouse effect greenhouse effect in which heat trapping grows as temperature rises, so the planet's atmosphere becomes extremely hot, as has occurred on Venus.

S

satellite a body orbiting a planet.

scattering the random redirection of a light wave or photon as it interacts with atoms or dust particles.

Schwarzschild radius the radius of a black hole. The distance from the center of a black hole to its event horizon.

scientific method the process of observing a phenomenon, proposing a hypothesis on the basis of the observations, and then testing the hypothesis.

scientific notation a shorthand way to write numbers using ten to a power. For example, $1{,}000{,}000 \nabla 10^6$ and $2000 \nabla 2 \quad 10^3$. Also called "powers-of-ten notation."

scintillation the twinkling of stars caused by the atmosphere. See **seeing**.

seeing a measure of the steadiness of the atmosphere during astronomical observations. Under conditions of bad seeing, fine details are difficult to see. Bad seeing results from atmospheric irregularities moving between the telescope and the object being observed.

seismic waves waves generated in the Earth's interior by earthquakes. Similar waves occur in other bodies. Two of the more important varieties are S and P waves. The former can travel only through solid material; the latter can travel through either solid or liquid material.

selection effect an unintentional selection process that omits some set of the objects being studied and leads to invalid conclusions about the objects.

self-propagating star formation a model that explains spiral arms as arising from stars triggering the birth of other stars around them. The resulting pattern is then drawn out into a spiral by the galaxy's rotation.

semimajor axis half the long dimension of an ellipse.

SETI *S*earch for *E*xtra*t*errestrial *I*ntelligence. Some such searches involve automatic "listening" to millions of radio frequencies for signals that might be from other civilizations.

Seyfert galaxy variety of active galaxy with a small, abnormally bright nucleus containing hot gas. Named for the astronomer who first drew attention to these objects.

shell source a region in a star where the nuclear energy generation occurs around the core rather than in it.

shepherding satellite one or more satellites that by their gravitational attractions prevent particles in a planet's rings from spreading out and dispersing. Saturn's F-ring is held together by shepherding satellites.

short-period comet a comet whose orbital period is shorter than 200 years. For example, Halley's comet has a period of 76 years.

sidereal day the length of time from the rising of a star until it next rises. The length of the Earth's sidereal day is 23 hours 56 minutes.

sidereal period the time it takes a body to turn once on its rotation axis or to revolve once around a central body, as measured with respect to the stars.

sidereal time a system of time measurement based on the motion of stars across the sky rather than the Sun.

silicates material composed of silicon and oxygen, and generally containing other substances as well. Most ordinary rocks are silicates. For example, quartz is silicon dioxide.

solar cycle the cyclic change in solar activity, such as sunspots and solar flares, rising and declining about every 11 years.

solar day the time interval from one sunrise to the next sunrise or from one noon to the next noon. That time interval is not always exactly 24 hours but varies throughout the year. For that reason, we use the mean solar day (which, by definition, is 24 hours) to keep time.

solar eclipse the passage of the Moon between the Earth and the Sun so that our view of the Sun is partially or totally blocked. *See also* **total eclipse.**

solar flare a sudden increase in brightness of a small region on the Sun. This flare is caused by a magnetic disturbance.

solar nebula the rotating disk of gas and dust from which the Sun and planets formed.

solar nebula theory the theory that the Solar System formed from a rotating cloud of gas and dust, the solar nebula.

Solar System the Sun, planets, their moons, and other bodies that orbit the Sun.

solar wind the outflow of low-density, hot gas from the Sun's upper atmosphere. It is partially this wind that creates the tail of a comet, by blowing gas away from the comet's immediate surroundings.

solid core the inner iron-nickel core of the Earth or another planet. Despite its high temperature, the core is solid because it is under great pressure. Also called the *inner core.*

solstice (winter and summer) the beginning of winter and summer. Astronomically the solstice occurs when the Sun is at its greatest distance north (June) or south (December) of the celestial equator.

south celestial pole the imaginary point on the celestial sphere directly over the Earth's South Pole.

special relativity a theory developed by Einstein to explain why the speed of light is always the same, regardless of the motion of the source of the light or the observer. The theory shows (and experiment has confirmed) that an object's mass, length, rate of time passage, and other quantities change depending on the speed of the observer relative to the object.

spectral type an indicator of a star's temperature. A star's spectral type is based on the appearance of its spectrum lines. The fundamental types are, from hot to cool: *O, B, A, F, G, K,* and *M.*

spectrograph a device for making a spectrum.

spectroscopic binary a type of binary star for which the spectrum lines exhibit a changing Doppler shift as a result of the orbital motion of one star around the other.

spectroscopy the study and analysis of spectra.

spectrum electromagnetic radiation (for example, visible light) spread into its component wavelengths or colors. The rainbow is a spectrum produced naturally by water droplets in our atmosphere.

spicule a hot, thin column of gas in the Sun's chromosphere.

spiral arm a long, narrow region containing young stars and interstellar matter that winds outward in the disk of spiral galaxies.

spiral galaxy a galaxy with a disk in which its bright stars form a spiral pattern.

spring tide the abnormally large tides that occur at new and full moon.

standard candle a type of star or other astronomical body in which the luminosity has a known value, allowing its distance to be determined by measuring its apparent brightness and applying the inverse-square law: Cepheid variable stars, supernovas, and so forth.

standard time a uniform time kept within a given region so that all clocks there agree.

star a massive, gaseous body held together by gravity and generally emitting light. Normal stars generate energy by nuclear reactions in their interiors.

star cluster a group of stars, numbering from hundreds to millions, held together by their mutual gravity.

Stefan-Boltzmann law the law that the amount of energy radiated from 1 square meter in 1 second by a blackbody of temperature T is proportional to T^4.

stratosphere a layer of Earth's atmosphere extending from about 12 to 50 km above the surface. A layer of ozone is found there. I

strong force the force that holds protons and neutrons together in the atomic nucleus. Sometimes called the *nuclear force.*

subatomic particles particles making up an atom, such as electrons, neutrons, and protons, or other particles of similar small size.

subduction the sinking of one crustal plate where it encounters another.

sunspot a dark, cooler region on the Sun's visible surface created by intense magnetic fields.

supercluster a cluster of galaxy clusters. One of the largest structures in the Universe.

supergiant a very large-diameter and luminous star, typically at least 10,000 times the Sun's luminosity.

superior conjunction *see* **conjunction.**

superior planet a planet orbiting farther from the Sun than the Earth. Mars, Jupiter, Saturn, Uranus, and Neptune are superior planets.

supermassive black hole a huge black hole, containing millions to billions of solar masses, at the center of most large galaxies.

supernova an explosion that occurs at the end of some stars' evolution. Astronomers identify two main kinds of supernovas: Type Ia and II. Type Ia occurs in a binary system in which one star is a white dwarf. The explosion is triggered when mass from a companion star falls onto the white dwarf, raising its mass above the Chandrasekhar limit and causing the star to collapse. Collapse heats the white dwarf so that its carbon and oxygen fuse explosively, destroying the star and leaving no remnant. Type II probably occurs when a massive star's iron core collapses. A Type II supernova leaves either a neutron star or a black hole, depending on the mass of the collapsing core.

supernova remnant the debris ejected from a star when it explodes as a supernova. Typically, this material is hot gas, expanding away from the explosion at thousands of kilometers or more per second.

surface gravity the acceleration caused by gravity at the surface of a planet or other body.

synchronous rotation the condition that a body's rotation period is the same as its orbital period. The Moon rotates synchronously as it orbits the Earth.

synchrotron radiation a form of nonthermal radiation emitted by charged particles spiraling at nearly the speed of light in a magnetic field. Pulsars and radio galaxies emit synchrotron radiation. The radiation gets its name because it was first seen in synchrotrons, a type of atomic accelerator.

synodic period the time between repeated configurations of a planet or moon. For example, the time between oppositions of a planet or between full moons.

T

21-centimeter radiation a spectrum line at radio wavelengths produced by non-ionized (neutral) hydrogen.

tail the plume of gas and dust from a comet. The plume is produced by the solar wind and radiation pressure acting on the comet. The tail points away from the Sun and gets longer as the comet approaches perihelion.

terrestrial planet a rocky planet similar to Earth in size and structure. The terrestrial planets are Mercury, Venus, Earth, and Mars.

tidal braking the slowing of one body's rotation as a result of gravitational forces exerted on it by another body.

tidal bulge a bulge on one body created by the gravitational attraction on it by another. Two tidal bulges form, one on the side near the attracting body and one on the opposite side.

tides the rise and fall of the Earth's oceans created by the gravitational attraction of the Moon. Tides also occur in the solid crust of a body and its atmosphere.

time dilation the slowing of time due to the effects of special and general relativity.

time zone one of 24 divisions of the globe every 15 degrees of longitude. In each zone, a single standard time is kept. Most zones have irregular boundaries.

total eclipse an eclipse in which the eclipsing body totally covers the other body. Only at a total solar eclipse can we see the Sun's corona.

transit the passage of a planet directly between the observer and the Sun. At a transit, we see the planet as a dark spot against the Sun's bright disk. From Earth, only Mercury and Venus can transit the Sun.

trans-Neptunian objects (TNOs) numerous small, icy objects orbiting in the outer part of the Solar System beyond the orbit of Neptune.

triangulation a method for measuring distances. This method is based on constructing a triangle, one side of which is the distance to be determined. That side is then calculated by measuring another side (the base line) and the two angles at either end of the base line.

triple alpha process the fusion of three helium nuclei (alpha particles) into a carbon nucleus. This process is sometimes called helium burning, and it occurs in many old stars.

troposphere the lowest layer of the Earth's atmosphere extending up to about 12 km, where convection leads to clouds and precipitation.

T Tauri star a type of extremely young star that varies erratically in its light output.

tuning-fork diagram a diagram devised by Hubble to classify the various forms of spiral, elliptical, and irregular galaxies. The diagram is named for its shape.

turnoff point the location on the main sequence where a star's evolution causes it to move away from the main sequence toward the red giant region. The location of the turnoff point can be used to deduce the age of a star cluster.

twin paradox a supposed paradox in special relativity arising from the difference in time measurement of two observers moving relative to each other. It is usually expressed in terms of a twin who travels to a distant star and returns to Earth to find that the other twin has aged much more than the first twin. The paradox is explained because the motion of the twins is not symmetric. The twin traveling away from Earth experiences accelerations that the twin remaining on Earth does not experience.

Type Ia supernova *see* supernova

U

ultraviolet a portion of the electromagnetic spectrum with wavelengths shorter than those of visible light but longer than those of X rays. By convention, the ultraviolet region extends from about 10 nm to 400 nm.

Universal time the time kept at Greenwich, England. Universal time is the same as Greenwich mean time. Most local times (in the continental United States, these are Eastern, Central, Mountain, and Pacific) differ from it by a whole number of hours.

Universe the largest astronomical structure we know of. The Universe contains all matter and radiation and encompasses all space.

V

Van Allen radiation belts doughnut-shaped regions surrounding the Earth containing charged particles trapped by the Earth's magnetic field.

variable star a star whose luminosity changes over time.

vernal equinox the spring equinox in the Northern Hemisphere. Spring begins on the vernal equinox, which is on about March 20.

Virgo Supercluster the cluster of galaxy clusters in which the Milky Way is located. The Local Group is one of its members.

visible spectrum the part of the electromagnetic spectrum that we can see with our eyes. It consists of the familiar colors violet, blue, green, yellow, orange, and red, extending from about 400 nm to 700 nm.

visual binary star a pair of stars held together by their mutual gravity and in orbit about each other, and that can be seen with a telescope as separate objects.

visual double star two stars that appear to lie very close together on the sky but in reality are at greatly different distances.

W

wavelength the distance between wave crests. It determines the color of visible light and is usually denoted by the Greek letter λ.

wave–particle duality the theory that electromagnetic radiation may be treated as either a particle (photon) or an electromagnetic wave.

weak force the force responsible for radioactive decay of atoms. Now known to be linked to the electromagnetic force at high energies and therefore called the electroweak force.

white dwarf a dense star whose radius is approximately the same as the Earth's but whose mass is comparable with the Sun's. White dwarfs burn no nuclear fuel and shine by residual heat. They are the end stage of stellar evolution for stars like the Sun.

white light visible light exhibiting no color of its own but composed of a mix of all colors. Sunlight and many artificial light sources are "white."

Wien's law a relation between a body's temperature and the wavelength at which it emits radiation most intensely. Hotter bodies radiate more intensely at shorter wavelengths. Mathematically, the law states that $\lambda_{max} = 3 \times 10^6/T$, where λ_{max} is the wavelength of maximum emission in nanometers and T is the body's temperature on the Kelvin scale.

X

X rays a portion of the electromagnetic spectrum with wavelengths of about 0.01 nm to 10 nm.

X-ray binary a binary star system in which one of the stars, or the gas associated with a star, emits X rays intensely. Such systems generally contain a collapsed object such as a neutron star or a black hole. *See also* **X-ray burster.**

Y

year the time it takes the Earth to complete its orbit around the Sun; that is, the period of the Earth's orbit.

Z

Zeeman effect the splitting of a single spectrum line into two or three lines by a magnetic field. A method for detecting magnetic fields in objects from their spectra.

zenith the point on the celestial sphere that lies directly overhead at your location.

zodiac a band running around the celestial sphere in which the planets move.

zone of avoidance a band running around the sky in which few galaxies are visible. It coincides with the Milky Way and is caused by dust that is within our Galaxy. This dust blocks the light from distant galaxies.

CREDITS

PHOTOGRAPHS

Preview

Preview 1: NASA; P.2A: © Vol. 74 PhotoDisc/Getty RF; P.2B: Dr. F.A. Ringwald; P.3(all): Courtesy of NASA/JPL; P.4(Sun): Courtesy of SOHO, ESA & NASA; (Jupiter): Courtesy of NASA/JPL/University of Arizona; (Earth): Courtesy of NASA; P.7A: Courtesy of 2MASS/UMass/IPAC-Caltech/NASA/NSF; P.7B: NASA/JPL – Caltech; P.8A: Courtesy of David Malin, Anglo-Australian Telescope Board; P.8B: Courtesy of Anglo-Australian Telescope Board; P.11(Milky Way): NASA/JPL – Caltech; (Earth): NASA.

Chapter 1

Opener: © English Heritage Photo Library; p. 15: © CNES/Jean-Pierre Haignere, 1999; 1.2a(left): Roger Ressmeyer/Corbis, digitally enhanced by Jon Alpert; (right): © Roger Ressmeyer/Corbis; 1.2b(left): Roger Ressmeyer/Corbis, digitally enhanced by Jon Alpert; (right): Roger Ressmeyer/Corbis, digitally enhanced by Jon Alpert; 1.11b(both): Steve Schneider; 1.12a: © English Heritage Photo Library; 1.12b: Courtesy Mag. Iván Ghezzi; 1.15(right): Steve Schneider; 1.16(left): John Walker; 1.18(left): 2002 South African solar eclipse composite photo by Wendy Carlos and Jonathan Kern Copyright 2002 Serendip LLC. All Rights Reserved; 1.18(right): Courtesy of Richard Wainscoat.

Chapter 2

Opener: The Bridgeman Art Library/Getty Images; p. 39: Laurent Laveder; 2.1a: Copyright 2009 Anthony Ayiomamitis; 2.8: Courtesy Tunc Tezel; 2.11: © Art Resource/Erich Lessing; 2.14: © Corbis – Bettmann; 2.15, 2.18: © Art Resource/Erich Lessing; 2.20: Courtesy of Bausch and Lomb Optical Co.

Essay 1

E1.4: Courtesy of Bausch and Lomb Optical Co.; E1.6A-B: © Roger Ressmeyer/Corbis; E1.10: Courtesy of Dominique Dierick.

Chapter 3

Opener: NASA/Rick Wetherington/Tim Powers; p. 73 & 3.9: Courtesy of NASA.

Chapter 4

Opener: Terje O. Nordvik; p. 89: Courtesy of NASA; 4.3(left): Photo © Ewing Galloway, Inc./American Stock; 4.5(Pulsar): Courtesy of NASA/CXC/ISAO; (Sun): SOHO (ESA & NASA); (Stars, Cloud): Courtesy of NOAO and Anglo-Australian Observatory, photograph by David Malin; (Cosmic microwave): Courtesy of NASA/WMAP Science Team; (Galaxy): NRAO/AUI/NSF; 4.8A-B: © The McGraw-Hill Companies, Inc./Joe Franek, photographer; 4.18: © Courtesy of Mees Solar Observatory, University of Hawaii; Bx Fig 4.1A: Courtesy of Richard Wainscoat; Bx Fig 4.1B: Courtesy of NRAO; Bx Fig 4.1C: Courtesy of NASA/CXC/SAO.

Essay 2

E2.1: Everett Collection, Inc.; E2.5: © Hulton-Deutsch/Corbis.

Chapter 5

Opener: W.M. Keck Observatory Library/Archives; p. 125: Courtesy of MSSS, JPL, NASA; 5.3: Photo courtesy of Yerkes Observatory; Box Fig. 5.1: Tom Arny; Box Fig. 5.3C: Photo courtesy of Patrick Watson; 5.5: Large Binocular Telescope Observatory; 5.7(left): Ángel L. Aldai; (right): Photograph courtesy of Gara Mora, IAC; 5.8a: © E.R. Degginger; 5.8b: R. Thompson, (University of Arizona) and NASA; 5.9: Courtesy of Steve Criswell; 5.10A-B: Courtesy of Andrea Ghez, UCLA; 5.11: California Institute of Technology; 5.12a: Courtesy of NRAO/AUI; 5.12B: Courtesy of NASA/CXC/SAO/Rutgers/J. Hughes; 5.14(top & bottom left): Courtesy of NASA/JPL; 5.14(top & bottom right): Courtesy of NASA, ESA, and Max Planck Institute for Extraterrestrial Physics, Germany; 5.16(both): Courtesy of NASA and The Hubble Heritage Team; 5.17(left): Courtesy of USAF; 5.17(right): Canada France Hawaii Telescope; 5.18A: Courtesy of Kitt Peak National Observatory; 5.18B: Courtesy Astronomical Society of the Pacific; 5.18C: W.T. Sullivan, © 1993; 5.19: ESO.

Chapter 6

Opener: NASA; p. 149: NASA/GSFC/METI/ERS-DAC/JAROS and U.S./Japan ASTER Science Team; 6.1A – 6.2B: NASA; 6.5: © Taxi/Getty Images; 6.7: NASA; 6.13A: Copyright by Marie Tharp 1977/2003. Reproduced by permission of Marie Tharp Oceanographic Cartographer, One Washington Ave., South Nyack, New York 10960; 6.14(all): Plate Tectonic and Paleogeographic maps and animations by C.R. Scotese, © 2009, PALEOMAP Project (www.scotese.com); 6.15(right): Tom Arny; 6.17A: Courtesy Paul Schneider; 6.17B: NASA; p. 165: Stephen E. Schneider; 6.19: © Corbis/Punchstock: 6.21, 6.22A: SVS/TOMS/NASA; 6.25: Courtesy of NOAA; 6.26: Courtesy NASA/JPL; p. 173: Courtesy of Martin Hackworth.

Chapter 7

Opener: Dr. Walter Koprolin; p. 187: NASA; p. 188: NASA/JPL/USGS; 7.2A-B: UC Regents; UCO/Lick Observatory image; 7.3A-B: Courtesy of NASA; 7.4B: © John Gillmoure/CORBIS; 7.5A, 7.7: Courtesy of NASA; 7.9: Astrogeology Team, U.S. Geological Survey, Flagstaff, Arizona; 7.16B(both): Tom Arny.

Chapter 8

Opener: NASA/JPL-Caltech/T. Pyle (SSC); p. 205: Mark McCaughren (Max-Planck-Institute for Astronomy), C. Robert O'Dell (Rice University), and NASA; 8.1: Courtesy SOHO/ESA/NASA; 8.4: Courtesy of Alan Dyer; 8.6(sun): Courtesy SOHO/ESA/NASA; (planets): Courtesy of NASA/JPL; 8.10: Courtesy of Anglo-Australian Telescope Board, photo by David Malin; 8.12A: Courtesy of Space Telescope Science Institute; 8.12B: NASA, ESA, D. Golimowski (Johns Hopkins University), D. Ardila (IPAC), J. Krist (JPL), M. Clampin (GSFC), H. Ford (JHU), and G. Illingworth (UCO/Lick) and the ACS Science Team; 8.13: © Comstock Images/PictureQuest; 8.15 (Mercury): Courtesy NASA's Planetary Geology and Geophysics Program/Arizona State University; (Mars, Manigougan Crater on Earth, Dione, Tethys): Courtesy of NASA; 8.16: NASA, ESA, P. Kalas, J. Graham, E. Chiang, E. Kite (University of California, Berkeley), M. Clampin (NASA Goddard Space Flight Center), M. Fitzgerald (Lawrence Livermore National Laboratory), and K. Stapelfeldt and J. Krist (NASA Jet Propulsion Laboratory).

Chapter 9

Page 228 (Mercury): NASA/John Hopkins University Applied Physics Laboratory/Carnegie Institution of Washington; (Venus): Courtesy of NASA/JPL/USGS; (Earth): NASA; (Mars): NASA/JPL/MSSS; (E-J): Courtesy of NASA/JPL/USGS; (swirling clouds): NASA Johnson Space Center-Earth Sciences and Image Analysis (NASA-JSC-ES&IA); (L-P): Courtesy of NASA/JPL/USGS; p. 229: Courtesy of NASA/JPL/Malin Space Science Systems; Fig 9.1(Mercury): NASA/Johns Hopkins University Applied Physics Laboratory/Carnegie Institution of Washington; (Moon): NASA; 9.2, 9.3A: Courtesy of NASA; 9.3B,C: NASA/Johns Hopkins University Applied Physics Laboratory/Carnegie Institution of Washington; 9.4, 9.8: Courtesy of NASA; 9.9: Courtesy of Alan Dyer; 9.10, 9.11(both): Ted Stryk; 9.12(both): NASA/JPL/USGS; 9.13(all): Courtesy of NASA/JPL; 9.15: NASA, J. Bell (Cornell U.) and M. Wolff (SSI); 9.16: NASA/JPL; 9.17: NASA/JPL/Arizona State University; 9.18B: Courtesy of A.S. McEwen, USGS; 9.19A: Courtesy of NASA/JPL; 9.19B: Courtesy of A.S. McEwen, USGS; 9.19C: NASA/JPL-Caltech/University of Arizona; 9.20, 9.21A: Courtesy of NASA; 9.21B: NASA/Calvin J. Hamilton; 9.22: Courtesy of NASA/JPL; 9.23: Courtesy of Mars Exploration Rover Mission, JPL, NASA; 9.24: Courtesy of NASA/JPL/CORNELL/ARC; 9.25: Courtesy of Malin Space Science Systems/NASA; 9.26: Courtesy of NASA/JPL/Malin Space

Science Systems; 9.27A-B: Courtesy of NASA/JPL; 9.28A-B: Courtesy of NASA; 9.29: Courtesy of NASA/JPL; 9.30: Courtesy of NASA.

Chapter 10

Opener: NASA Jet Propulsion Laboratory (NASA-JPL); p. 259: NASA/JPL/Space Science Institute; 10.1, 10.3C: Courtesy of NASA; 10.4: Courtesy of NASA/JPL; 10.5A: Courtesy of J. Clarke and G. Ballester, University of Michigan; J. Trauger and R. Evans, Jet Propulsion Laboratory and NASA; 10.5B: Courtesy of NASA/JPL; 10.6 (Saturn): NASA/JPL/Space Science Institute; (Jupiter): NASA/JPL; (Uranus): NASA, ESA, and M. Showalter (SETI Institute); (Neptune): NASA/JPL; 10.7–10.9C: Courtesy of NASA/JPL; 10.10: Courtesy of Cassini Imaging Team/SSI/JPL/ESA/NASA; 10.12: Courtesy of NASA/JPL; 10.13: Courtesy of NASA/JPL/Space Science Institute; 10.14: Courtesy of NASA/JPL; 10.15: Courtesy of NASA; 10.16: Courtesy of STScI/HST; 10.17: Courtesy of NASA/JPL/Univ. of Arizona; 10.18A–D: NASA/JPL; 10.19(all): NASA/JPL/Space Science Institute; 10.20, 10.24(both)-10.27: Courtesy of NASA/JPL; 10.28: UC Regents; UCO/Lick Observatory image; 10.30: Courtesy of STScI; 10.32(Triton, Orcus, Sedna, Haumea, Quasar, Makemake, Eris): NASA Jet Propulsion Laboratory (NASA-JPL); (Pluto, Charon): Images courtesy of Marc W. Buie/Lowell Observatory; (Earth): © Stocktrek/Getty Images RF; (Moon): © Digital Vision RF/Punchstock.

Chapter 11

Opener: Robert McNaught/Photo Researchers, Inc.; p. 287: Courtesy of Juan Carlos Casado and Isabel Graboleda; 11.1: Courtesy of Ronald A. Oriti, Santa Rosa Junior College, Santa Rosa, Calif.; 11.3A-B: Stephen E. Schneider; 11.5A: Courtesy of NASA/JPL; 11.5B: Courtesy of NASA, ESA, J. Parker (Southwest Research Institute), P. Thomas (Cornell University), L. McFadden (University of Maryland, College Park), and M. Mutchler and Z. Levay (STScI); 11.6A: Courtesy of NASA/JPL; 11.6B: ISAS/JAXA; 11.10A: Copyright 1986, Max Planck Institute for Astronomy; courtesy of Harold Reitsema, Ball Aerospace; 11.10B: NASA Jet Propulsion Laboratory (NASA-JPL); 11.11(all): NASA/JPL-Caltech/UMD; 11.15: Courtesy of Mike Skrutskie, University of Virginia; 11.19: Courtesy of David Roddy Meteor Crater, Northern Arizona, USA; (inset): Photo by Tom Arny; 11.20: Courtesy of Landsat/EOS; 11.21: NASA Jet Propulsion Laboratory (NASA-JPL); 11.22(right): A. Hildebrand, M. Pilkington, and M. Connors.

Chapter 12

Opener: M. Aschwanden et al. (LMSAL), TRACE, NASA; p. 311: Courtesy of Stefan Seip/www.astromeeting.de; 12.4: Hinode JAXA/NASA/PPARC; 12.5: Courtesy of Jacques Guertin, Ph.D.; 12.6: Courtesy of NOAO/AURA/NSF; 12.7: 1988 eclipse image courtesy High Altitude Observatory (HAO), University Corporation for Atmospheric Research (UCAR), Boulder, Colorado. UCAR is sponsored by the National Science Foundation; 12.8: SOHO-EIT; 12.13: Courtesy of Ernest Orlando, Lawrence Berkeley National Laboratory; 12.14: Courtesy of MDI/SOHO, ESA, NSF; 12.15: Courtesy of Royal Swedish Academy of Sciences; 12.18A-B: Courtesy of NOAO/AURA/NSF; 12.19: Courtesy of SOHO-EIT Consortium, ESA, NASA; 12.20: Courtesy of Eugene Lauria; Box Fig. 12.1(all): Courtesy of NOAO/AURA/NSF.

Chapter 13

Opener: Johannes Schedler/Panther Observatory; p. 337: Courtesy of Dean Ketelsen; 13.3: Courtesy of Harvard College Observatory; 13.7: Courtesy of Mullard Radio Observatory; 13.8: NOAO/AURA/NSF; 13.10: Courtesy of Harvard College Observatory; 13.11: Courtesy of Katherine Haramundanis; 13.12, 13.22: NOAO/AURA/NSF.

Chapter 14

Opener: Courtesy of NASA, HST, J. Hester & P. Scowen (ASU); p. 371: © Akira Fujii/DMI; 14.4: Five College Radio Astronomy Observatory (FCRAO); 14.5(inset): © European Southern Observatory; 14.6(left): NASA/JPL-Caltech/A. Noriega-Crespo (SSC/Caltech), Digital Sky Survey; (right): NASA/JPL-Caltech/R. Hurt (SSC/Caltech); 14.7: Courtesy of Ronald Snell, FCRAO; 14.8A: Courtesy of Anglo-Australian Observatory; photograph by David Malin; 14.16A: Courtesy of Hubble Heritage Team (AURA, STScI/NASA); 14.16B: Courtesy of Anglo-Australian Observatory; photograph by David Malin; 14.16C: Raghvendra Sahai and John Trauger (JPL), the WFPC2 Science Team, and NASA; 14.16D: Bruce Balick, University of Washington; Vincent Icke, Leiden University, Netherlands; Garrelt Mellema, Stockholm University/NASA; 14.19: Pablo Candia; 14.20A: Courtesy of Anglo-Australian Observatory; photograph by David Malin; 14.20B: Courtesy of Richard Wainscoat; 14.20C: Courtesy of NASA/CXC/SAO/Rutgers/J. Hughes; 14.22(both): Courtesy of Anglo-Australian Observatory, photograph by David Malin.

Chapter 15

Opener: X-ray: NASA/CXC/SAO, Infrared: NASA/JPL-Caltech; Optical: MPIA, Calar Alto, O. Krause et al.; p. 401: Courtesy of Paresce, R. Jedrezejewski (STScI); 15.11: Photos courtesy of NOAO.

Chapter 16

Opener: Kerry-Ann Lecky Hepburn (Weather and Sky Photography); p. 423: Courtesy of Rainer Schodel (MPE) et al., NAOS-CONICA, ESO; 16.1: Courtesy of Steward Observatory and NOAO; 16.4(cluster): Courtesy of Hubble Heritage Team (AURA/STScI/NASA); 16.5: Courtesy Sloan Digital Sky Survey; 16.6: Courtesy of Anglo-Australian Observatory; photograph by David Malin; 16.10(Pleiades): NASA, ESA and AURA/Caltech; (NGC 3293, M67): Courtesy of Anglo-Australian Observatory; photograph by David Malin; (M 13): Robert Lupton and the Sloan Digital Sky Survey Consortium; (Omega Centauri): Max Kilmister; 16.11: Courtesy of NASA and The Hubble Heritage Team (STScI/AURA); 16.13: Courtesy of Anglo-Australian Observatory; photograph by David Malin; 16.15: Atlas Image courtesy of 2MASA/UMass/IPAC-Caltech/NASA/NSF; 16.16B-16.17C: Courtesy of Anglo-Australian Observatory; photograph by David Malin; 16.20(top): Stefan Seip/Adam Block/NOAO/AURA/NSF; (bottom): Bruce Hugo and Leslie Gaul/Adam Block/NOAO/AURA/NSF; 16.25: Courtesy of NASA, UMass, D. Wan et al; 16.28(all), 16.29: Courtesy of Neal Katz, University of Massachusetts, and James E. Gunn, Princeton University.

Chapter 17

Opener: NASA, ESA, and The Hubble Heritage Team (STScI/AURA); p. 457: Mary E. Putman (University of Michigan), Lister Staveley-Smith (Australia Telescope National Facility), Kenneth C. Freeman (Australian National University), and Brad K. Gibson and David G. Barnes (both Swinburne University of Technology). The Parkes telescope is operated by CSIRO Australia Telescope National Facility; 17.1: Courtesy of George Greaney; 17.2: Courtesy of William C. Keel; 17.3A: Courtesy of Bruce Hugo and Leslie Gaul, Adam Block, NOAO, AURA, ASF; 17.3B: Courtesy of Hubble Heritage Team, AURA/STScI/NASA; 17.3C: © European Southern Observatory; 17.4(NGC 205): California Institute of Technology; (M49): Courtesy of NOAO; (M87)-17.6: Courtesy of Anglo-Australian Observatory, photographs by David Malin; 17.7-17.9(all): Courtesy Sloan Digital Sky Survey; 17.10A: Courtesy of NASA; 17.10B: Courtesy of STScI and NOAO; 17.11: Courtesy of NRAO/AUI; 17.12: Courtesy Sloan Digital Sky Survey; 17.13: Roff/Lowell Observatory; 17.14A-B: Courtesy of NASA and The Hubble Heritage Team; 17.14C: Richard Griffiths (JHU), STScI, NASA; 17.15: R Jay Gabany (Blackbird Observatory)-collaboration; D.Martínez Delgado (IAC, MPIA), J.Peñarrubia (U.Victoria) I. Trujillo (IAC) S.Majewski (U.Virginia), M.Pohlen (Cardiff); 17.17A: Martin Pugh; 17.17B: NASA, ESA, and J. P. Kneib (Laboratoire d' Astrophysique de Marseille); 17.21: NASA and The Hubble Heritage Team (STScI/AURA); 17.22: AURA/STScI/NASA; 17.23: Images courtesy of NRAO/AUI; 17.24: Courtesy of NOAO; 17.26: Courtesy of John Bahcall, Institute for Advanced Study; 17.27(left): Courtesy of W. Jaffe, Leiden Observatory, and H. Ford, Johns Hopkins University, Space Telescope Science Institute and NASA; (right): Courtesy of NRAO and California Institute of Technology; 17.31A: Courtesy of NASA; 17.32: Courtesy of STScI.

Chapter 18

Opener & p. 493: Courtesy of S. Beckwith and the HUDF Working Group (STScI), HST, ESA, NASA; 18.1A: Two Micron All Sky Survey, a joint project of the University of MA and the Infrared Processing and Analysis Center/CA Institute of Technology, funded by NASA/National Science Foundation/T.H. Jarrett, J. Carpenter, & R. Hurt. Perha; 18.15: Courtesy of NASA/WMAP Science Team.

Essay 4

E4.2A: Reproduced with permission from Dr. J. William Schopf, 1992–1993, CSEOL/UCLA; E4.2B: Dwight Kuhn; E4.3: © Paul Hoffman; (inset): Courtesy of Donald R. Lowe; E4.4(whale): © Brandon Cole/Visuals Unlimited; (butterfly): © Corbis RF; (giraffe): © DV/Getty RF; E4.6: Courtesy of Ken Eward/BioGrafx; E4.8(both): © Dudley Foster/ Woods Hole Oceanographic Institution; E4.9: Courtesy of NASA/JPL/Cornell; E4.11: Seth Shostak.

LOOKING UP FEATURES

Northern Circumpolar Constellations

Background (both): © Akira Fujii/DMI; M52: Hartmut Frommelt; M81 and M82: © Robert Gendler; M101: Adam Block/NOAO/AURA/NSF.

Ursa Major

Background: © Akria Fujii/DMI; M97: Gary White and Verlenne Monroe/Adam Block/NOAO/ AURA/NSF.

M31 & Perseus

All photos: © Akira Fujii/DMI.

Summer Triangle

Background: © Akira Fujii/DMI; M57: Courtesy of H. Bond et al., Hubble Heritage Team (STScI/ AURA), NASA; M27: © IAC/RGO/Malin; Albireo: Courtesy of Randy Brewer.

Taurus

Background: © Akira Fujii/DMI; M1: Courtesy of R. Wainscoat; M45: Courtesy of Anglo-Australian Observatory, photographs by David Malin; Hyades; © Akira Fujii/DMI.

Orion

Background: © Akira Fujii/DMI; Horsehead Nebula: Courtesy of Anglo-Australian Observatory, photograph by David Malin; Close-up of Horsehead Nebula: Courtesy NOAO/AURA/NSF; Betelgeuse: Courtesy A. Dupree (CFA), NASA, ESA; Pink Orion Nebula: Courtesy of Carol B. Ivers; Orion Nebula with Dust and Gas: Courtesy Gary Bernstein (U. Pennsylvania), Copyright U. Michigan, Lucent; Protoplanetary Disk; Courtesy of A.M. Lagrange, D. Mouillet, and J.L. Beuzit, Grenoble Observatory/CNRS/ESO.

Sagittarius

Background: Courtesy of Till Credner, AlltheSky.com; M16(Close-up): Courtesy of NASA, HST, J. Hester & P. Scowen (ASU); M16 with Gas Cloud: Courtesy of Bill Schoening/NOAO/AURA/NSF; M20: Courtesy of Jason Ware; M22: Courtesy of N.A. Sharp, REU program/NOAO/AURA/NSF.

Centaurus and Crux, The Southern Cross

Background: Anglo-Australian Observatory/ David Malin Images; Centaurus A: Courtesy Peter Ward, 2004; Omega Centauri: Al Kelly; Eta Carinae: J. Hester/Arizona State University, NASA; The Jewel Box: Research School of Astronomy and Astrophysics, the Australian National University.

Southern Circumpolar Constellations

Background: Christopher J. Picking; Hourglass Nebula: Raghvendra Sahai and John Trauger (JPL), the WFPC2 science team, and NASA; Thumbprint Nebula: Courtesy of STScI; Tarantula Nebula: WFI/2.2-m/ESO.

LINE ART

Chapter 4

Table 4.4: Reprinted from *Gerchim Cosmochim Acta* 53 (1989): 1987, by Anders and Grevesse. Copyright 1989, with permission from Elsevier; 4.17: Courtesy of Stephen M. Larson, University of Arizona; 4.18A: Courtesy of Doug McGonagle, FCRAO; Figure 4.18B: Courtesy of P.F. Winkler, Middlebury College.

Chapter 7

7.15A: Courtesy of Robin Canup, Southwest Research Institute.

Chapter 8

8.19: Courtesy of G. Marcy at University of California at Berkeley, from SFSU.

Chapter 9

9.8: Based on a computer simulation, courtesy of W. Benz, University of Arizona; W. Slattery, Los Alamos; A.G.W. Cameron, CFA.

Chapter 11

11.4: Courtesy of E.L.G. Bowell, Lowell Observatory; 11.9: Courtesy of E.L.G. Bowell, Lowell Observatory.

Chapter 15

15.6: Courtesy of M.I. Large, University of Sydney; R.N. Manchester, Australia Telescope, CSIRO; and Joseph H. Taylor, Princeton University.

Chapter 16

16.14A: Courtesy of NOAO; 16.18: Courtesy of Y.M. and Y.P. Georgelin; 16.21: This sequence was generated by P. Seiden (deceased) and H. Gerola, *Ap. J.* 233, 56, 1979. Reproduced by permission; 16.24: Based on diagram by Dan P. Clemens, *Ap J.* 295, 422.

Chapter 17

17.16B: Data courtesy of Robert P. Kirshner, CFA; 17.24: Courtesy of Tom Balonek and Jeyhan Kartaltepe, Colgate University.

Chapter 18

18.7: Based on curve supplied by NASA/Goddard Spaceflight Center: COBE Science Working Group.

INDEX

Note: Page references with *n* (e.g., 108*n*) refer to footnotes.

A

B

D

I

N

O

P

T

U

V

W

X

Y

Z

SPRING NIGHT SKY

WHEN TO USE THIS STAR MAP

Early March:	2 a.m.
Late March:	1 a.m.
Early April:	midnight
Late April:	11 p.m.
Early May:	10 p.m.
Late May:	Dusk

This star chart is most accurate if used within an hour or so of the times listed and is plotted for observers located between 30° and 50° north latitude. All times are standard time; if daylight-saving time is in effect, add one hour.

To use this chart, hold it in front of you and rotate it so that the yellow label corresponding to the direction you are facing is positioned at the bottom, right-side up. The stars in the sky should match those depicted on the chart. The center of the chart is the zenith, the point in the sky directly overhead.

SUMMER NIGHT SKY

This star chart is most accurate if used within an hour or so of the times listed and is plotted for observers located between 30° and 50° north latitude. All times are standard time; if daylight-saving time is in effect, add one hour.

To use this chart, hold it in front of you and rotate it so that the yellow label corresponding to the direction you are facing is positioned at the bottom, right-side up. The stars in the sky should match those depicted on the chart. Ignore all the parts of the map above horizons you are not facing.

WHEN TO USE THIS STAR MAP

Early June:	1 a.m.
Late June:	midnight
Early July:	11 p.m.
Late July:	10 p.m.
Early August:	9 p.m.
Late August:	Dusk

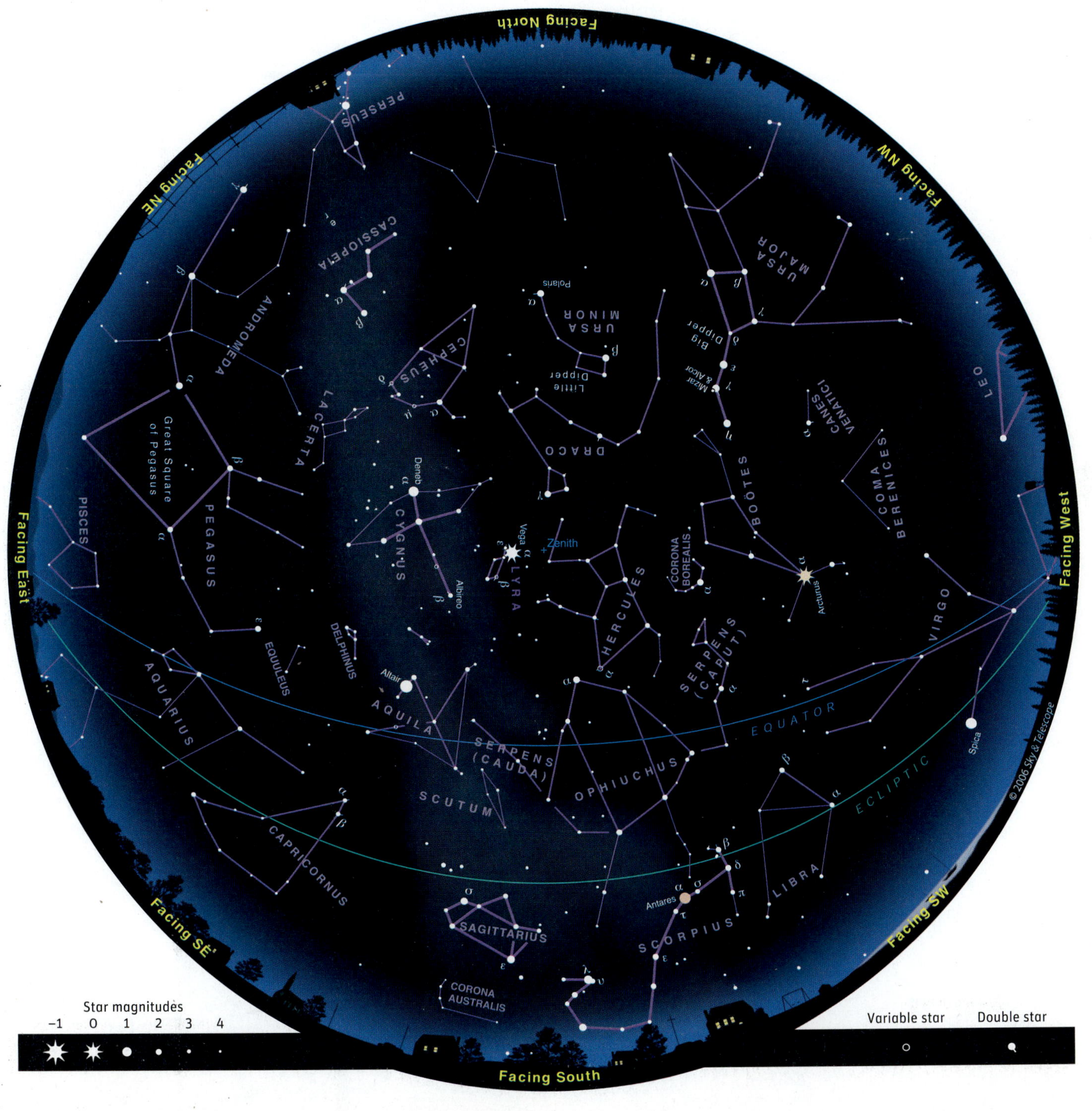

AUTUMN NIGHT SKY

WHEN TO USE THIS STAR MAP

Early September:	midnight
Late September:	11 p.m.
Early October:	10 p.m.
Late October:	9 p.m.
Early November:	8 p.m.
Late November:	7 p.m.

This star chart is most accurate if used within an hour or so of the times listed and is plotted for observers located between 30° and 50° north latitude. All times are standard time: if daylight-saving time is in effect, add one hour.

To use this chart, hold it in front of you and rotate it so that the yellow label corresponding to the direction you are facing is positioned at the bottom, right-side up. The stars in the sky should match those depicted on the chart. The farther up from the map's edge they appear, the higher they'll be shining in your sky.

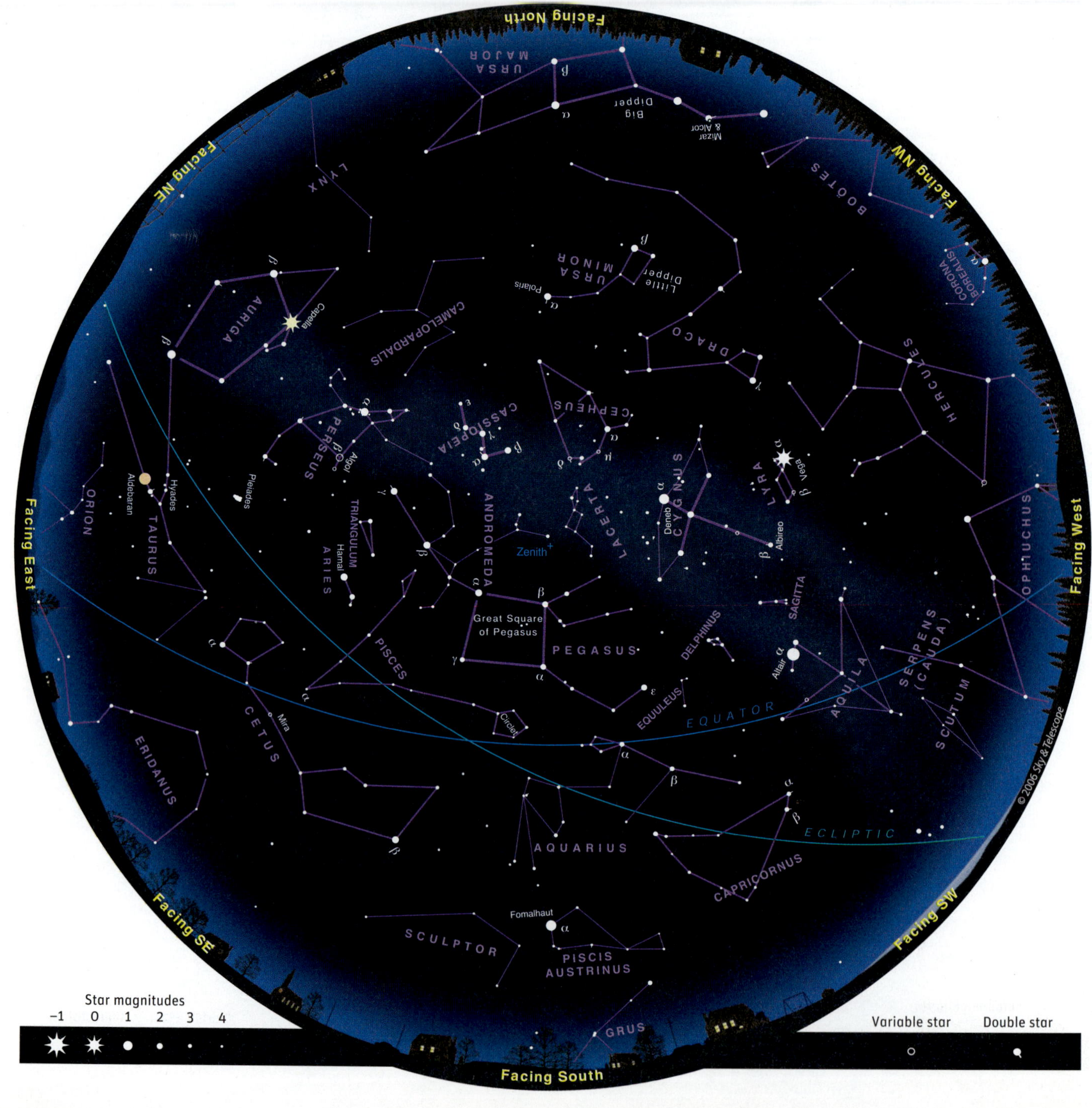

WINTER NIGHT SKY

This star chart is most accurate if used within an hour or so of the times listed and is plotted for observers located between 30° and 50° north latitude. All times are standard time.

To use this chart, hold it in front of you and rotate it so that the yellow label corresponding to the direction you are facing is positioned at the bottom, right-side up. The stars in the sky should match those depicted on the chart. Remember that star patterns will look much larger in the sky than they do here on paper.

WHEN TO USE THIS STAR MAP

Early December:	midnight
Late December:	11 p.m.
Early January:	10 p.m.
Late January:	9 p.m.
Early February:	8 p.m.
Late February:	Dusk

Using the Foldout Star Charts➤

These star charts cover the entire sky and can be used to identify stars and constellations from any location at any time. They were produced at Stephen F. Austin State University by Professor Dan Bruton and his students, with additions by the publisher. The large rectangular chart covers a wide region around the celestial equator, and the two circular charts are centered on the northern and southern celestial poles. The charts are labeled with Right Ascension and Declination coordinates, which are like longitude and latitude on the celestial sphere (see Essay 1). Stars visible to the naked eye have sizes on the charts that depend on their brightness (or apparent magnitude—see Chapter 13). Lines connecting the stars help to identify the pattern and primary stars in each constellation. The pale blue region shows the approximate location of the Milky Way.

For viewers in the northern hemisphere, to observe the northern half of the sky, use the **Northern Region** chart. If you are looking at the sky around 8 P.M., you should locate the closest date along the outer perimeter of the northern region chart. Rotate the chart so that the correct date is at the top. For each hour earlier/later that you are outside, you should rotate the map so that a Right Ascension an hour earlier/later is at the top. For example, if it is 8 P.M., on September 18, you would rotate the map so that "Sep 20" (Right Ascension of 20^h) is at the top. On the same date at 9 P.M., you would rotate the map so that a Right Ascension of 21^h is at the top. Held with this orientation, the chart should match what you see over the northern sky. The stars that are due north will match what you see at the center of the chart. Stars at the bottom of the chart will be below the horizon and stars near the top of the chart will be overhead.

Looking southward from the northern hemisphere, use the **Equatorial Region** chart. Find the date along the top of the chart, and adjust to an earlier or later Right Ascension according to how much earlier or later than 8 P.M. you are observing the sky as explained for the Northern Region chart. Stars located at the Right Ascension you have identified will lie straight up from the point on the horizon that is due south of you, along the "meridian" (see Essay 3). For observers north of latitude 30°, some stars along the bottom of the chart will never rise above the horizon.

The curving "sine wave" line running through the middle of the Equatorial Region chart is the ecliptic (Chapter 1), which marks the path of the Sun among the stars. The dates along this line indicate where the Sun is located throughout the year. The Moon and planets also remain close to this line as they move around the celestial sphere.

The star charts can be used from the southern hemisphere using the same instructions but swapping the words "north" and "south" wherever they occur. Looking northward from the southern hemisphere, you will also need to hold the Equatorial Region chart upside down.

Using the Moon and Planet Finder Tables

Along the bottom of the foldout chart, information is provided to help you find where the Moon and five bright planets will be located each month. The Moon will be in its new phase and will lie close to the ecliptic at the Sun's position on the date given in the table. It then shifts about 13° to the east each subsequent day. If the date of the new moon is in a black circle, there will be a total solar eclipse on that date. The dates of total lunar eclipses of the *full moon* are also provided, listed in red circles.

The locations of Venus, Mars, Jupiter, and Saturn each month are indicated by the abbreviations for the constellations they are in. The abbreviation is listed in green when the planet is primarily in the morning sky and listed in blue for the evening sky. The listing is in black when the planet is in opposition to the Sun, indicating that it rises at sunset and sets at sunrise. The word *Sun* is listed when the planet is nearly in conjunction and therefore difficult to see. For Venus, the month when it is at its greatest elongation from the Sun is marked by an asterisk.

Mercury is always fairly close to the Sun and can be seen only shortly after sunset or shortly before sunrise. The date of its greatest elongation is given in green when the planet is in the morning sky and in blue when it is in the evening sky. The best opportunity for seeing Mercury is generally within about one week of this date.